# NEW TECHNOLOGIES FOR COMPUTER CONTROL 2001
## (NTCC 2001)

*A Proceedings volume from the IFAC Conference,*
*Hong Kong, China, 19 - 22 November 2001*



Edited by

H.B. VERBRUGGEN
*Faculty of Information Technology,*
*Delft University of Technology,*
*Delft, The Netherlands*

C.W. CHAN
*Department of Mechanical Engineering,*
*The University of Hong Kong,*
*Hong Kong, China*

and

R.A. VINGERHOEDS
*ENIT-LGP, Tarbes, France*

Published for the

INTERNATIONAL FEDERATION OF AUTOMATIC CONTROL

by

PERGAMON
An Imprint of Elsevier Science

ELSEVIER SCIENCE Ltd
The Boulevard, Langford Lane
Kidlington, Oxford OX5 1GB, UK

Elsevier Science Internet Homepage
http://www.elsevier.com

Consult the Elsevier Homepage for full catalogue information on all books, journals and electronic products and services.

IFAC Publications Internet Homepage
http://www.elsevier.com/locate/ifac

Consult the IFAC Publications Homepage for full details on the preparation of IFAC meeting papers, published/forthcoming IFAC books, and information about the IFAC Journals and affiliated journals.

---



First edition 2002

**Library of Congress Cataloging in Publication Data**

A catalogue record for this book is available from the Library of Congress

**British Library Cataloguing in Publication Data**

A catalogue record for this book is available from the British Library

ISBN 0-08-043700 1
ISSN 1474-6670



*Printed in Great Britain*

---

**To Contact the Publisher**

Elsevier Science welcomes enquiries concerning publishing proposals: books, journal special issues, conference proceedings, etc. All formats and media can be considered. Should you have a publishing proposal you wish to discuss, please contact, without obligation, the publisher responsible for Elsevier's industrial and control engineering publishing programme:

Dr Martin Ruck
Publishing Editor
Elsevier Science Ltd
The Boulevard, Langford Lane
Kidlington, Oxford
OX5 1GB, UK

Phone: +44 1865 843230
Fax: +44 1865 843920
E.mail: m.ruck@elsevier.com

General enquiries, including placing orders, should be directed to Elsevier's Regional Sales Offices – please access the Elsevier homepage for full contact details (homepage details at the top of this page).

# IFAC CONFERENCE ON NEW TECHNOLOGIES FOR COMPUTER CONTROL 2001

*Sponsored by*
International Federation of Automatic Control (IFAC)
Technical Committees on:
- Distributed Computer Control Systems
- Real-Time Software Engineering
- Artificial Intelligent in Real-Time Control
- Algorithms and Architectures for Real-Time Control
- Safety of Computer Control Systems

The University of Hong Kong
City University of Hong Kong

*Organised by*
Control, Automation and Instrumentation Division
The Hong Kong Institution of Engineers

*International Programme Committee* (IPC)
Verbruggen, H.B. (NL) (Chair)
Chan, C.W. (HK) (Co-Chair)

Albertos, P. (ES)
Araki, M. (JP)
Arzen, K.E. (SE)
Boullart, L. (BE)
Crespo, A. (ES)
van Cauwenberghe (BE)
Fleming, P.J. (UK)
Garbot, B. (FR)
Halang, W. (DE)
Harris, C.J. (UK)
Hellendoorn, H. (NL)
Irwin, G.W. (UK)
Isermann, R. (DE)
Kutti, S. (AU)
Kwon, W.H. (KR)
Lewis, F.L. (US)
Lei Guo (CN)
Lee, T.H. (SG)
Man Lin (CA)
MacLeod, I.M. (ZA)
Man, K.F. (HK)
Motus, L. (EE)
Masten, M. (US)
Ng, T.S. (HK)
Yoh-Han Pao (US)
Park, J. (KR)
Pereira, C.E. (BR)
de la Puente, J. (ES)
Ming Rao (CA)
Ruano, A. (PT)
Suski, G. (US)
Sa da Costa, J. (PT)
Strmcnik, S. (SI)
Vingerhoeds, R.A. (FR)
Wong, K.P. (AU)
Zalewski, J. (US)

*National Organizing Committee* (NOC)
Leung, A.Y.T. (Chairman)
Cheung, K.C. (Secretary)

Chan, C.W.
Fung, E.H.K.
Law, S.S.
Lock, L.
Man, K.F.
Sze, W.S.

# Foreword

The advent of computers has gradually led to a second industrial revolution. We can perform many tasks now with high reliability and accuracy that we have never dreamt possible before. Our ability to perform increasingly powerful tasks continues to grow. In the area of control, we benefit from the development of new technologies, such as networking of computers, ever-increasing computing power, mechatronics and intelligent sensors for improving the reliability and safety of control systems. Implementing sophisticated control algorithms and structures are within easy reach. It is the right moment at the beginning of the new millennium to review the status of the current development and to work out a road map for computer control in the 21$^{st}$ century, which is the main goal of this multi-track Conference.

The initial idea to organize the Conference was put forward five years ago, and was finally taken up during the 1999 IFAC World Congress held in Beijing by the Co-ordinate Committee on Computer Control together with its five Technical Committees (TC): TC on Distributed Computer Control Systems, TC on Real-Time Software Engineering, TC on Artificial Intelligent in Real-Time Control, TC on Algorithms and Architectures for Real-Time Control, and TC on Safety of Computer Control Systems. As these Technical Committees cover important and essential aspects of computer control, there were in-depth discussions in each of these technical topics and more importantly provided ample opportunity for cross-fertilization.

The International Program Committee (IPC) received 140 regular papers, which included papers for three Special Sessions. The submitted papers covered a broad range of topics including: parallel and distributed computing, software architecture and hardware for distributed computer control, real-time operating systems, real-time communication architectures, intelligent control, evolutionary control techniques, safety aspects of control systems. It is interesting to note that the authors of the submitted papers came from 31 countries. Out of a total of 140 submitted papers, 104 of them are accepted for presentation at the Conference. However, only 93 camera-ready papers were received, and there were 11 "no-show" papers. The IPC was quite impressed with the quality of the accepted papers, grouped in 22 technical sessions. Six plenary papers were delivered by world-renowned experts in Computer Control, and were presented in the first three days of the conference. The plenary papers covered a range of important topics in computer control including: standards for real-time computer systems, delivery of time-sensitive information, evolutionary computing for control engineers, transparent neuro-fuzzy modeling and an interesting application area relating to the computer control of combustion engines. The Conference concluded with a plenary panel discussion on the Road Map for Computer Control organized by Henk Verbruggen and chaired by George Irwin.

We would like to take this opportunity to thank all IPC members, all 5 Technical Committee Chairmen, and all reviewers for their valuable contributions, which were essential for the success of the conference. We would also like to thank the National Organizing Committee for organizing an excellent conference.

C. W. Chan and H. B. Verbruggen
IPC Chairmen
NTCC-2001

# CONTENTS

## *PLENARY PAPERS*



## *DISTRIBUTED AND NETWORKED SYSTEMS*



## *FAULT DETECTION AND DIAGNOSIS*



## *ADVANCED MODELLING AND CONTROL DESIGN*

## *SAFETY CRITICAL REAL TIME SYSTEMS*



## *ALGORITHMS*



## *INTELLIGENT CONTROL WITH INDUSTRIAL APPLICATIONS (SPECIAL SESSION)*

## *APPLICATIONS I*



## *PERFORMANCE OF CONTROL SOFTWARE*



## *INDUSTRIAL APPLICATIONS*



## *REAL TIME SOFTWARE ENGINEERING*



## *SCHEDULING AND COMMUNICATIONS*

## *NEURAL NETWORK APPLICATIONS IN NONLINEAR MODELING AND CONTROL OF BIOCHEMICAL PROCESSES (SPECIAL SESSION)*



## *REAL-TIME SYSTEMS*



## *MECHANICAL APPLICATIONS*



## *ADVANCED MONITORING AND DIAGNOSTICS*



## *FUZZY CONTROL*

## *ROBOTICS*



## *FUZZY AND NEURAL*



## *APPLICATIONS II*



## *HIGH INTEGRITY PROGRAMMABLE ELECTRONIC SYSTEMS*



## *SAFETY OF COMPUTER CONTROL SYSTEMS (SPECIAL SESSION)*

## *NOVEL CONSTRUCTION AND LEARNING TECHNIQUES*

IFAC Publications
www.elsevier.com/locate/ifac

# DESIGN OF COMPUTER CONTROLLED COMBUSTION ENGINES

**Rolf Isermann and Norbert Müller**

*Darmstadt University of Technology*
*Institute of Automatic Control*
*Laboratory of Control Systems and Process Automation*
*Landgraf-Georg-Str. 4, D-64283 Darmstadt, Germany*
*Tel: +49/6151-162114, Fax: +49/6151-293445*
{RIsermann,NMueller}@iat.tu-darmstadt.de

Abstract: Globalization and growing new markets, as well as increasing emission and fuel consumption requirements force the car manufacturers and their suppliers to develop new engine control strategies in shorter time periods. This can mainly be reached by development tools and an integrated hardware and software environment enabling rapid implementation and testing of advanced engine control algorithms.
The structure of a Rapid Control Prototyping (RCP) system is explained, which allows fast measurement signal evaluation, and rapid prototyping of advanced engine control algorithms. A hardware-in-the-loop simulator for Diesel engine control design is illustrated, simulation results for a 40 tons truck are presented. Providing efficient engine models for the proposed development tools, a dynamic local linear neural network approach is explained and applied for modelling the $NO_X$ emission characteristics of a 1.9 liter direct injection Diesel engine. Furthermore the application of a RCP system is exemplified by the application of combustion pressure based closed-loop ignition timing control for a SI engine. Experimental results are shown for a 1.0 liter SI engine on a dynamic engine test stand. *Copyright © 2001 IFAC*

Keywords: computer-aided control system design, computer-aided testing, engine control, identification, engine modelling, learning systems, feedforward control

## 1. INTRODUCTION

The last two decades in the automotive industry have seen an ever increasing usage of electronics. In the middle 1970s car manufacturers introduced microprocessor-based engine control systems to meet state regulations and customer demands of high fuel economy, low emissions, best possible engine performance and ride comfort. Especially the American regulations (clean air act 1963, on-board diagnosis (OBD I 1988, OBD II 1994)) and the corresponding European regulations EURO 1 (1992), EURO 2 (1996), EURO 3 (2000) had a considerable effect on the development of new engine control methods. New sensors with electrical output and new actuators had to be developed. Also auxiliary units like electro-mechanical controlled carburetors, low pressure injection systems for SI engines, and high pressure injection systems for Diesel engines have shown a development from pure mechanical to electro-mechanical devices with electronic control. New actuators were added like for exhaust gas recirculation (EGR), camshaft positioning and variable geometry turbochargers (VTG). Today's combustion engines are completely microcomputer controlled with many *actuators* (e.g. electrical, electro-mechanical, electro-hydraulic or electro-pneumatic actuators, influencing spark timing, fuel-injector pulse widths, exhaust gas recirculation valves), several *measured output variables* (e.g. pressures, temperatures, engine rotational speed, air mass flow, camshaft position, oxygen-concentration of the exhaust gas), taking into account different *operating phases* (e.g. start-up, warming-up, idling, normal operation, overrun, shut down). The microprocessor-based control has grown up to a rather complicated control unit with 50-120 look-up tables, relating about 15 measured inputs and about 30 manipulated variables as outputs. Because many output variables like torque and emission concentrations are mostly not available as measurements (too costly or short life time) a majority of control functions is feedforward.

In the future increasing computational capabilities using floating point processors will allow advanced estimation techniques for non-measurable quantities like engine torque or exhaust gas properties and precise feedforward and feedback control over large ranges and with small tolerances. New electronically controlled actuators and new sensors entail additional control functions for new engine technologies (VTG turbo chargers, swirl control, dynamic manifold pressure, variable valve timing (VVT) of inlet valves, combustion pressure based engine control).

The following chapter gives an overview on engine control structures of state-of-the-art SI and Diesel engines. Modern development and testing tools applied for engine control algorithms are outlined in chapter 3. Chapter 4 and 5 explain more detailed some basics and the structure of Rapid Control Prototyping (RCP) and Hardware-in-the-Loop simulation, respectively. Chapter 6 is devoted to modelling techniques using local linear neural networks, since efficient process modelling is a fundamental prerequisite for all of the proposed development tools. The last chapter explains a combustion pressure based ignition control system as a sophisticated application example for a RCP system.

## 2. CONTROL STRUCTURE FOR INTERNAL COMBUSTION ENGINES: STATE OF THE ART

Modern IC engines obtain increasingly more actuating elements. SI engines have - except the classical inputs like amount of injection $m_{inj}$, ignition angle $\theta_{ign}$, injection angle $\theta_{inj}$ - additionally controlled air/fuel ratio, EGR and VVT, see **Fig. 1**. The location of sensors and actuators are shown in **Fig. 2**.

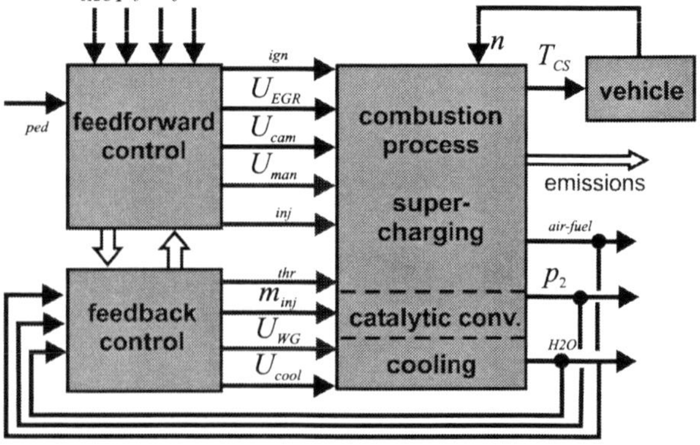


Fig. 1. Simplified control structure of a SI engine

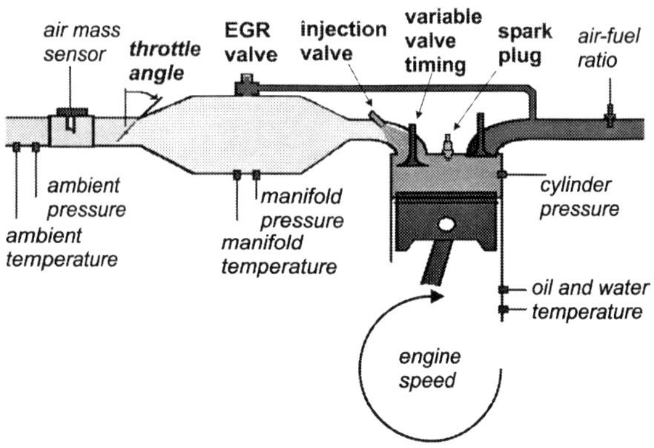


Fig. 2. Location of *sensors* and **actuators** of a SI engine (all of them except cylinder pressure sensors are state-of-the-art in current engine control units)

Diesel engines were until 1987 only manipulated by injection mass and injection angle. Now they have additional manipulated inputs like EGR, variable geometry turbochargers, common rail pressure and modulated injection, see **Fig. 3**.

The engine control system has therefore to be designed for 5-10 main manipulated variables and 5-8 main output variables, leading to a complex nonlinear multiple-input

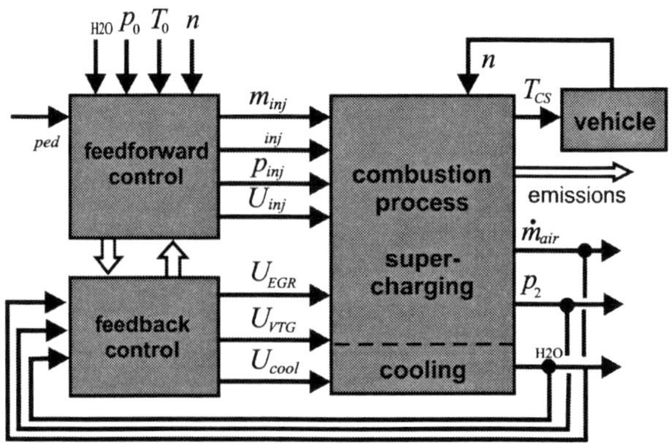


Fig. 3. Simplified control structure of a Diesel engine with turbocharger

multiple-output system. Because some of the design requirements contradict each other suitable compromises have to be made. Since the majority of control functions are realized by feedforward structures, precise models are required.

*Feedback control* is used in the case of SI engines for example for the λ-control (keeping a stoichiometric relative air/fuel ratio $\lambda=1$ for best conversion efficiency of the catalytic converter), for the electronic throttle control and for the ignition angle in case of knock. Diesel engines with turbochargers possess a charging pressure control with waste gate or variable vanes and speed overrun break away. Both engines have idling speed control and coolant water temperature control. Additionally, there are several auxiliary closed loop controls like fuel pressure control and oil pressure control. All control functions have to be defined and tested for all possible operating phases.

Most *feedforward control* functions are implemented as grid-based two-dimensional (i.e. two input signals) look-up tables or as one-dimensional characteristics. This is because of the strongly nonlinear static and dynamic behavior of the IC engines and the direct programming and fast processing in microprocessors with fixed point arithmetics. Some of the functions are based on physical models with correction factors, but many control functions can only be obtained by systematic experiments on dynamometers and with vehicles.

## 3. DEVELOPMENT TOOLS FOR ENGINE CONTROL SYSTEMS

Without appropriate tools for the design, implementation, and testing of new engine control algorithms, the tight schedules in the development process of engine control systems cannot be met anymore. Especially in the auto-motive industry, the concurrent design of engine, body and electronics requires efficient development methods. In this context the simulation increasingly plays an important role in all steps of the development process, (Hanselmann, 1996). **Figure 4** shows the employment of simulation techniques for the design and testing of engine control unit (ECU) functions. As in many cases basic ECU functions are already existing, this case is considered. Depending on the application and the development stage, the following simulation methods can be used, **Fig. 5**:

(1) **Software-in-the-Loop simulation (SiL)**: In the first step of the function development some basic analysis has to be done to develop the control structure and actuator configuration. A sufficient method is the software-in-the-loop simulation, where a software version of the control function is tested with the simulated process.

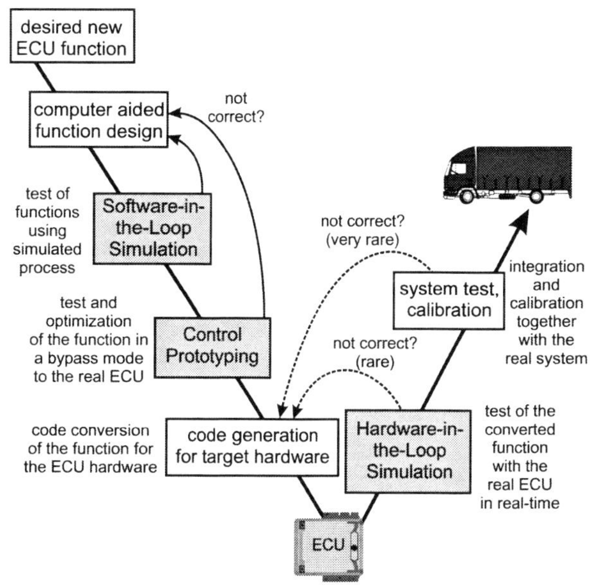


Fig. 4. Design and simulation steps for ECU function development of internal combustion engines (Schaffnit, 2001)

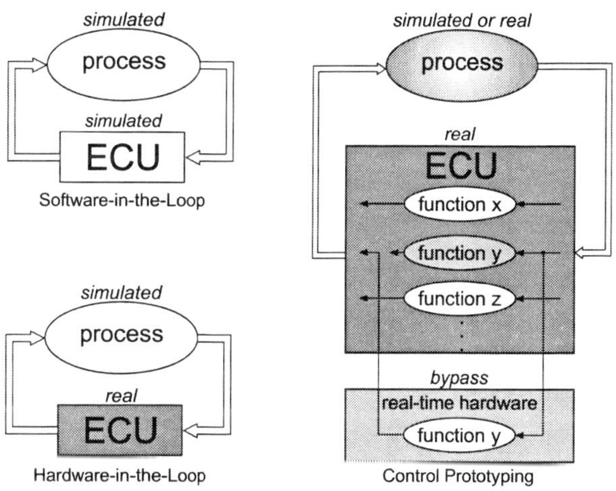


Fig. 5. Simulation methods for the development of electronic control unit (ECU) functions

(2) **Rapid Control Prototyping (RCP)**: The Control Prototyping is used to test the new control function together with the real already existing control unit and the simulated or the real process. In contrast to the software-in-the-loop simulation, only a selected subset of the ECU functions is realized as a special software implementation. Only these interesting parts are calculated on a real-time computer system in a bypass mode to the real ECU, **Fig. 5**.

(3) **Hardware-in-the-Loop simulation (HiL)**: After the target code generation of the control functions, the Hardware-in-the-Loop simulation is employed for testing the implemented ECU functionality. In

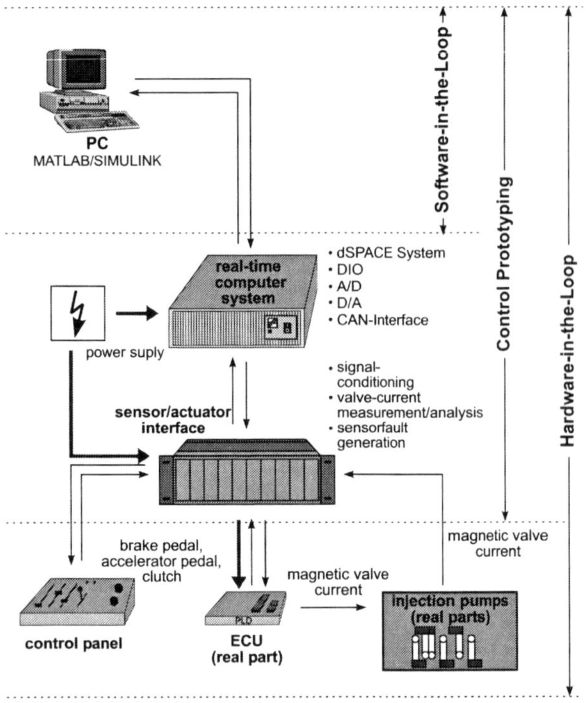


Fig. 6. Simulation system family for Software-in-the-Loop, Control Prototyping and Hardware-in-the-Loop

this configuration the real ECU hardware operates together with the simulated process in real-time, (Sinsel, 2000). To couple the real-time computer system with the engine control unit, it is necessary to generate the required sensor signals (e.g. pulses of the crankshaft and camshaft inductive speed sensors, temperatures and pressures) and to process the actuator signals, (Isermann *et al.*, 1998).

**Figure 6** shows a simulation family system for all three simulation catagories, SiL, RCP and HiL, (Schaffnit and Isermann, 2000).

## 4. RAPID CONTROL PROTOTYPING AND EXPERIMENT CONTROL

In order to allow *development and testing* of new control functions *on-line in conjunction with the real engine* in a comfortable manner, powerful real-time computer systems are required. RCP systems allow the implementation and testing of new algorithms together with the already existing ECU. Thus programming and modifications of the production ECU are avoided.

The structure of a RCP system is explained in the following, it is capable for fast measurement signal evaluation and advanced engine control algorithms, as it is required for instance in case of combustion pressure based engine control. It consists of two subsystems, the RCP and an

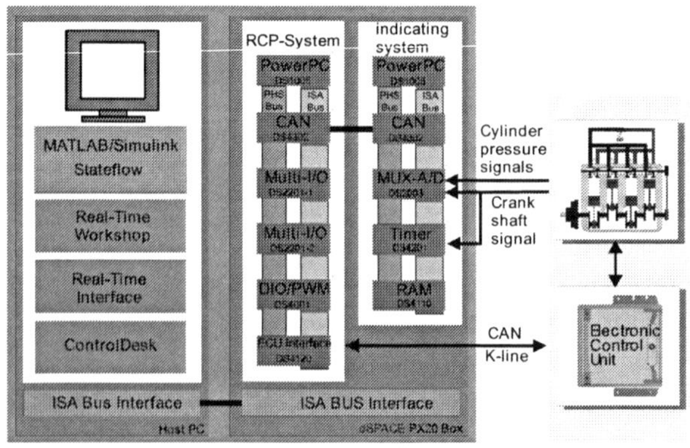


Fig. 7. Rapid control prototyping system configuration

indication system, see **Fig. 7**. Both systems operate in parallel to the production car's ECU, and are based on PowerPCs, type Motorola MPC 750, 480 MHz, which are designed for real-time use and are programmed in high level language.

The indicating system allows to evaluate cylinder pressure signals in real time at a resolution of 1° crankshaft angle (CA) for four cylinders. Thus it operates in a crank synchronous manner. Thermodynamic and signal-based cylinder pressure features, like mean indicated pressure, crank angle of the center of combustion, and location of peak pressure, are calculated by the indicating system and are transmitted to the RCP system.

The RCP system of the company dSPACE allows a very fast and easy implementation and testing of new control concepts on real-time hardware. It computes the control and optimization algorithms and operates in a time-synchronous manner, at a sampling rate of 1 kHz. The user is enabled to code newly developed algorithms from block diagrams (e.g. MATLAB/Simulink) and download the code by means of an automatic code generation software to the real-time hardware with a mouse click. A complete design iteration can thus be accomplished within a few minutes, (Hanselmann, 1996).

The RCP system uses the production car sensors, additional test stand sensors and the output signals and messages of the ECU. By evaluating the production car sensor information, the standard-ECU control settings can be investigated and then be modified, also independent own control strategies can be implemented. The actuator signals, which are calculated in real-time, are then sent to the actuators or the ECU by means of a CAN bus or by PWM-signals.

The described real-time hardware system enables very fast and easy implementation and testing of complex algorithms including extensive data preprocessing for the cylinder pressure, even under the hard real-time conditions of combustion engines.

Fig. 8. Asynchronous motor coupled to the combustion engine

**Fig. 8** shows the dynamometer, consisting of an 160 kW asynchronous motor coupled to the combustion engine. The depicted engine was used for testing the control algorithms described in chapter 7.

## 5. HARDWARE-IN-THE-LOOP SIMULATION FOR TRUCK DIESEL ENGINES

After the evaluation of new control functions by using RCP systems, the program code is implemented on the real ECU hardware and tested with the real ECU in real-time, see Fig. 4. **Figure 9** shows the setup of a HiL simulation test bench for testing new control functions *of the real engine control unit* of a truck Diesel engines together *with the simulated engine*. It may be subdivided in the following parts (Isermann *et al.*, 1999):

- real-time computer system including I/O-modules,
- periphery, consisting of the sensor and actuator interface,
- PLD control unit including real actuator components, stand-alone or in combination with the real FMR,
- PC with graphical user interface,
- control panel.

### 5.1 *The HiL-Simulator*

The real-time computer system for this simulator is based on a dSPACE-system equipped with digital signal processors and a DEC Alpha processor. This system has the advantage of high computing power which makes a parallel calculation unnecessary. It offers also the possibility to realize all the models in MATLAB/Simulink to use all the benefits of a graphical simulation environment. Special I/O-modules (Digital-I/O-module, D/A

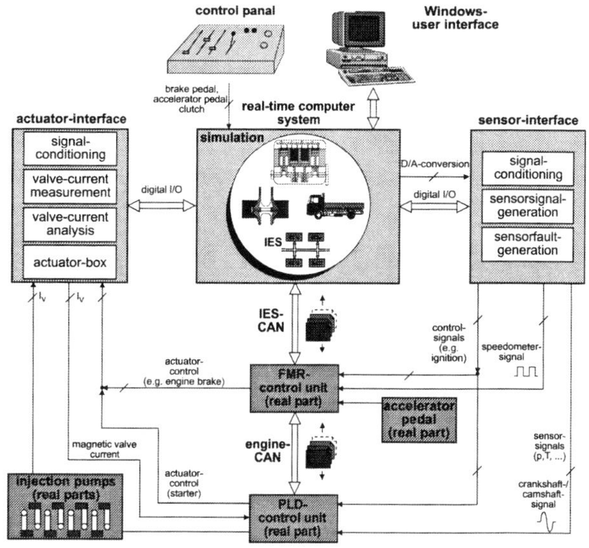


Fig. 9. Hardware-in-the-Loop test bench with simulated engine and vehicle and with real engine and vehicle and with real engine control unit and real injection actuators.
FMR: vehicle management system
PLD: pump line nozzle

converter, A/D converter and CAN-interface) are used for the communication with the periphery. The coupling of the simulator and the control unit is implemented with special periphery which can be subdivided into a sensor and an actuator interface.

The *sensor interface* generates the necessary sensor signals like temperatures and pressures. The pulses of the camshaft and crankshaft inductive speed sensors are generated with a board specifically designed for high speed signal generation. For that purpose a lookup-table with the pulse-signals versus the crankshaft angle is stored off-line in memory. During the simulation, the signals are periodically read out, synchronous to the simulated engine speed. This realization guarantees a high flexibility in forming the pulses and adapting different gear wheels. The sensor interface also contains a relay electronic to simulate sensor faults like interruptions and short circuits.

The *actuator interface* mainly consists of the injection pumps (pump-line-nozzle injection system) which are integrated in the simulation test bench as real components, because the combination of the PLD control unit and the injection pumps represent a mechatronic unit which is difficult to model. A special electronic device measures the magnetic valve currents to reconstruct the real valve opening time and to determine the pulse width and the beginning of injection. These quantities are transferred to the real-time computer system for engine simulation. By this way the real behavior of the injection pumps is included.

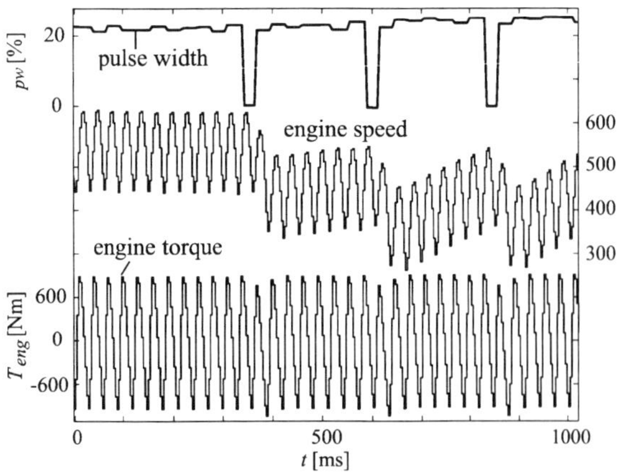


Fig. 10. HiL-simulation of a single injection pump valve cut off

The simulator test bench was set up with the objective of testing the PLD control unit stand alone or in combination with the FMR control unit. In the first case, the necessary FMR functions are simulated by the computer system. The data transfer is done via the engine CAN-bus. In the second case the PLD and the FMR control units are connected directly via the engine CAN-bus. The computer system emulates other IES units in this operation mode by transmitting the data via the IES-CAN-bus to the FMR. For an efficient use of the simulator, a comfortable experimental environment is needed.

The simulator operation is performed with a windows user interface on the host-PC which copies the functionality of a real truck-cockpit. All relevant simulation quantities can be visualized on-line or be recorded for off-line analysis. To ensure reproducible results a driver simulation is implemented which can automatically follow a given speed cycle by operating the accelerate pedal, brake, clutch and gear. As alternative, an interactive "driving" of the simulator can be performed manually with a control panel where the most important cockpit functionalities are realized.

### 5.2 *Simulation Results*

In the following three HiL simulation examples for an 8-cylinder truck engine (420 kW) are represented in order to document the applicability and the performance of the simulator. **Figure 10** demonstrates the effect of switching off a single injection pump valve. The gearbox is in neutral position and at the beginning the engine runs with idle speed. The cyclic decrease of the engine torque and the engine speed after the fuel shut off can directly be seen. The control unit gradually compensates for the missing torque of one cylinder by increasing the pulse width in order to keep the desired idle speed.

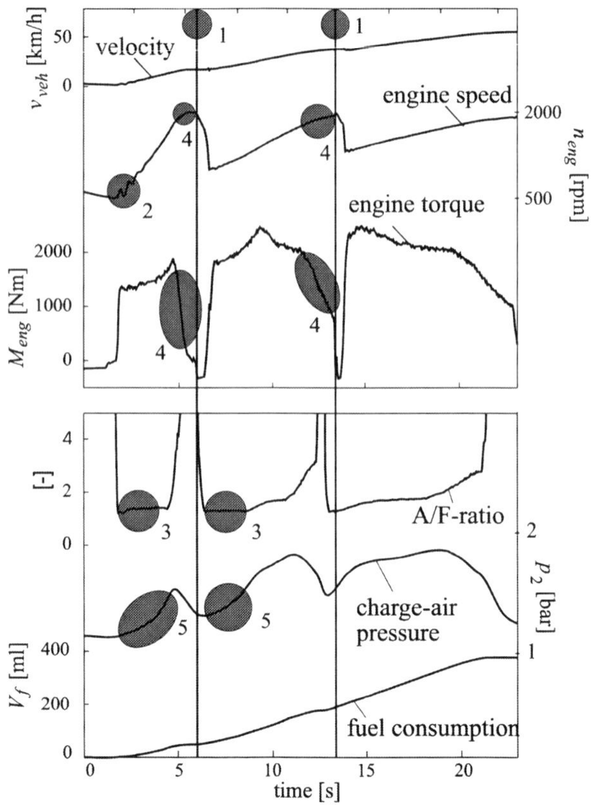


Fig. 11. HiL-simulation of a full power acceleration of a 40 tons truck

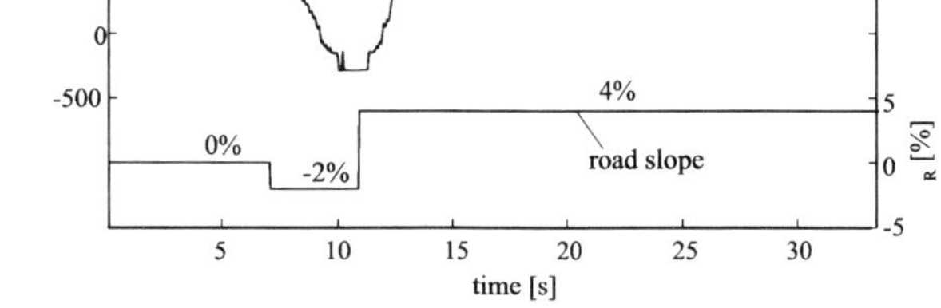


Fig. 12. HiL-simulation, evaluating the cruise control function of a 40 tons truck

A full power acceleration of a 40 tons truck including two gear shifts (1) is depicted in **Fig. 11**. The following effects can be observed:

- oscillations in the drive chain (2)
- maximum speed limit regulation (4)
- limitation of soot (3)
- lagged reaction of the turbo charging pressure (5)

**Figure 12** shows the simulation results evaluating the cruise control function. The 40 tons truck runs at a constant speed of $v_{veh} = 60\,\text{km/h}$. After a road slope of $\alpha_R = -2\,\%$ the engine control unit reduces engine torque $M_{eng}$ until all injection pumps are switched off (at 10 s). After the change of the road slope to $+4\,\%$ engine torque is increased in order to compensate for the deviation between desired vehicle speed and simulated vehicle speed.

The various simulation results illustrate the performance of the HiL simulator. It allows repeatable testing of engine control units and control algorithms under various conditions. Thus expensive and probably dangerous engine experiments and driving maneuvers are avoided. The behaviour of the ECU during sensor or actuator faults can be easily studied.

Several of theses HiL systems were developed for DaimlerChrysler AG, Stuttgart, Germany.

## 6. NONLINEAR IDENTIFICATION OF EXHAUST GAS COMPONENTS

Mathematical models of engines are of basic importance not only for HiL simulators as shown in the preceding section, but also for feedforward and feedback control design of engine control units. The approach for modelling the $NO_X$ emissions of a 1.9 liter direct injection Diesel engine is explained in the following.

In order to minimize computational requirements, mathematical models have to describe the behaviour of a system as compact as possible. Based on physical relations, *theoretical modeling* can be applied to simulate e.g. the mechanics of pistons and the crankshaft, leading to so called "white-box" models. However, in case of thermodynamic or chemical processes, extravagant requirements in calculating power rule out physical modelling for control design or real-time usage. Theoretical modelling of exhaust gas concentrations for instance depends on complex thermodynamic and chemical equations, as well as on boundary conditions like swirl, tumble, quenching, and local temperatures (Heywood, 1988).

This motivates the application of "black-box" or *experimental models*, which model the input-output behavior of a system using universal approximators, capable for multi-dimensional modelling. In contrast to look-up tables, which are only suitable for one- or two-dimensional problems, *neural networks* (Isermann *et al.*, 1997) have been successfully applied for high order problems, allowing to consider all relevant influencing variables. Especially neural networks based on the local linear model approach (LOLIMOT) showed to be highly efficient and applicable to engine modelling (Nelles, 1999). Depending on the complexity of the engine, more than 5-7 inputs might be necessary to consider the most relevant variables concerning the exhaust gas formation. Among these are e.g. the engine speed, load, injection angle and -pressure, EGR and the position of the guide blades of turbochargers with VTG. The application of *dynamic* neural nets allows to model the *dynamic behavior* of exhaust gas emissions. Furthermore static emission maps can be calculated by means of the dynamic neural network (Hafner *et al.*, 2000).

In the following section the structure and training procedure of LOLIMOT, a local linear model approach, is explained more detailed.

### 6.1 *Fast Local Linear Neural Networks*

LOcal LInear MOdel Tree (LOLIMOT) is an extended radial basis function network (Nelles, 1999), which is extended by replacing the output layer weights with a linear function of the network inputs. Thus, each neuron represents a local linear model with its corresponding validity function. Furthermore, the radial basis function network is normalized, that is the sum of all validity functions for a specific input combination sums up to one. The Gaussian validity functions determine the regions of the input space where each neuron is active. The input space of the net is divided into $M$ hyper-rectangles, each represented by a linear function.

The output $\hat{y}$ of a LOLIMOT network with $n$ inputs $x_1, \ldots, x_n$ is calculated by summing up the contributions of all $M$ local linear models

$$\hat{y} = \sum_{i=1}^{M} (w_{i0} + w_{i1}x_1 + \ldots + w_{in}x_n) \cdot \Phi_i(\underline{x}) \quad (1)$$

where $w_{ij}$ are the parameters of the $i$-th linear regression model and $x_i$ is the model input. The validity functions $\Phi_i$ are typically chosen as normalized Gaussian weighting functions.

$$\Phi_i(\underline{x}) = \frac{\mu_i}{\sum\limits_{j=1}^{M} \mu_j} \quad (2)$$

with

$$\mu_i = \exp\left(-\frac{1}{2}\frac{(x_1 - c_{i1})^2}{\sigma_{i1}^2} - \ldots - \frac{1}{2}\frac{(x_n - c_{in})^2}{\sigma_{in}^2}\right)$$

with center coordinates $\underline{c}_{ij}$ and dimension individual standard deviations $\underline{\sigma}_{ij}$. LOLIMOT is trained as follows. In an outer loop the network structure is optimized. It is determined by the number of neurons and the partitioning of the input space, which is defined by the centers and standard deviations of the validity functions.

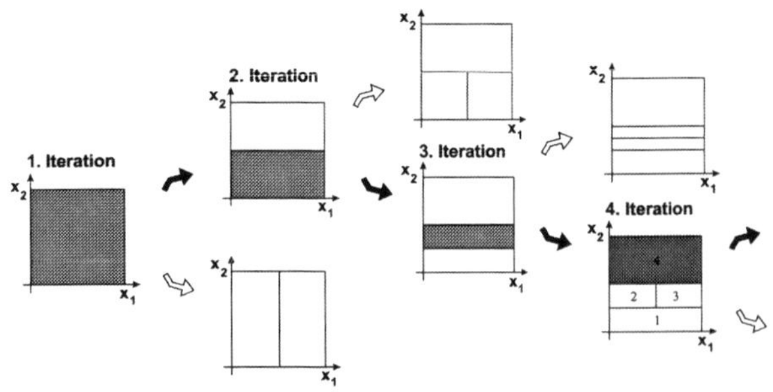

Fig. 13. First four iterations of LOLIMOT training algorithm

An inner loop estimates the parameters and possibly the structure of the local linear models. The network structure is optimized by a tree construction algorithm that determines the centers and standard deviations of the validity functions, see **Fig. 13**.

The LOLIMOT algorithm partitions the input space in hyper-rectangles. In the center of each hyper-rectangle, the validity function of the corresponding linear model is placed. The standard deviations are chosen proportional to the size of the hyper-rectangle. This makes the size of the validity region of a local linear model proportional to its hyper-rectangle extension. Thus, a model may be valid over a wide operating range of one input variable but only in a small area of another one. At each iteration of the outer loop, one additional neuron is placed in a new hyper-rectangle, which is derived from the local linear model with the worst local error measure

$$J_{local} = \sum_{j=1}^{N} \Phi_i(\underline{x}(j)) \cdot (y(j) - \hat{y}(j))^2.$$

In other words, the local error is the sum of the squared errors weighted with the corresponding validity function $\Phi_i$ over all data samples $N$. The new hyper-rectangle is then chosen by testing the possible cuts in all dimensions and taking the one with the highest performance improvement.

In an inner loop the parameters of the local linear models are estimated by a local weighted least-squares technique. The prediction errors are weighted with the corresponding validity function. Each local linear model is estimated separately, that is the overlap between neighbored local models is neglected. This approach is very fast and robust. It is especially important to note that due to the local estimation approach the computational demand increases only linearly with the number of local models.

In addition to approximating stationary relations of nonlinear processes, LOLIMOT is also capable of simulating the dynamic behavior of processes. In order to model multivariate, nonlinear, dynamic processes, the

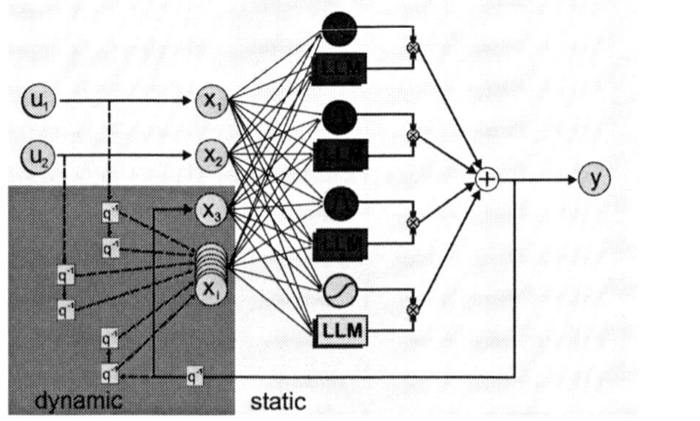


Fig. 14. LOLIMOT net with external dynamics, LLM: Local Linear Models The filters are chosen as simple time delays $q^{-1}$

following time-delay neural network approach can be taken.

$$\hat{y}(t) = f(\underline{x}(t))$$
$$\underline{x}(t) = [u_1(t-1), ..., u_1(t-m), ..., u_p(t-1)..., u_p(t-m), \hat{y}(t-1), ..., \hat{y}(t-m)]^T$$

where $m$ is the dynamic order of the model, $u_1(t), \ldots, u_p(t)$ are the $p$ model inputs and $\hat{y}(t)$ is the model output at time $t$, see **Fig. 14**.

Some of the main advantages of the LOLIMOT approach is its fast training time of some 10..30 s, compared to many minutes to hours with other neural networks, its applicability to adaptive problems (Fischer *et al.*, 1998) and the interpretability of the net structure and parameters in a physical sense.

### 6.2 *Modelling the $NO_X$-Emissions of a Diesel Engine*

Since *experimental* modelling is based on *measurement data*, engine experiments are essential. When recording the training data for *dynamic* models, one should keep in mind, that the measured data has to contain the whole range of amplitudes and dynamics of the process in order to get satisfying results. APRBS signals (amplitude modulated pseudo random binary signal) are often suitable as process inputs because they excite the process within a wide range of amplitudes and frequencies. In this example, step responses of differing amplitudes have proven to be suitable for the training of the $NO_X$-models.

**Figure 15** shows for the training data set the dynamically measured $NO_X$ emissions and the three input signals, fuel mass $m_{inj}$, engine speed $n_{eng}$ and crankshaft angle of injection $\theta_{inj}$ (EGR and VTG were disabled during this test).

The neural network approach based on LOLIMOT networks consists of dynamic local linear models, each of

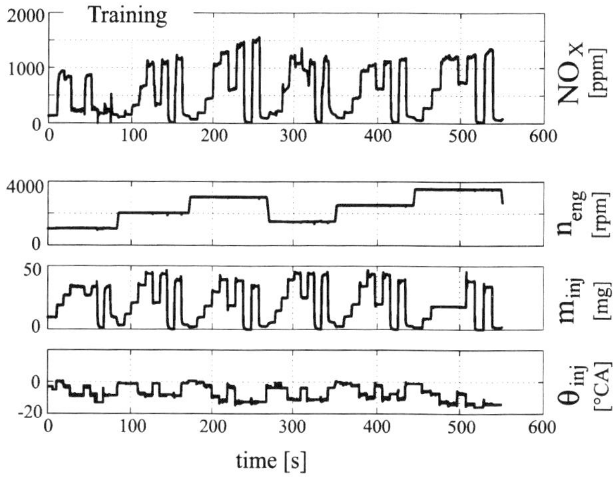


Fig. 15. Training data set for a dynamic NOx model

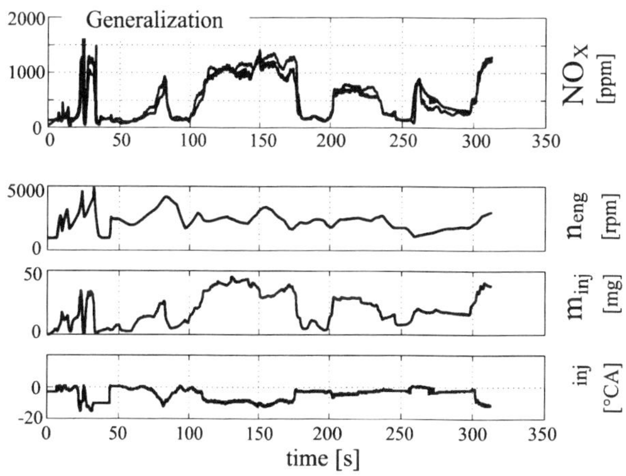


Fig. 16. Generalization of the NOx model trained with the data in Fig. 15

them being valid in a specific operating domain. The structure of each local linear model was selected as

$$\hat{NO}_X(k) = f_{LOLIMOT}(m_{inj}(k-1), \\ n_{eng}(k-1), \theta_{inj}(k-1), \hat{NO}_X(k-1)\,, \tag{3}$$

giving a simple linear model of first order. After building the neural model, the model can be applied to a new (unknown) data set and simulates the $NO_X$ according to the respective input data (on-line in the vehicle, if necessary).

**Figure 16** illustrates the good generalization performance of a first order dynamic net with 15 neurons for the $NO_X$ emissions. Despite the excitation of the input signals in Figs. 15 and 16 were of quite different quality, the simulated $NO_X$ is able to follow the measured values with an error of just a few percent, which was quite satisfying.

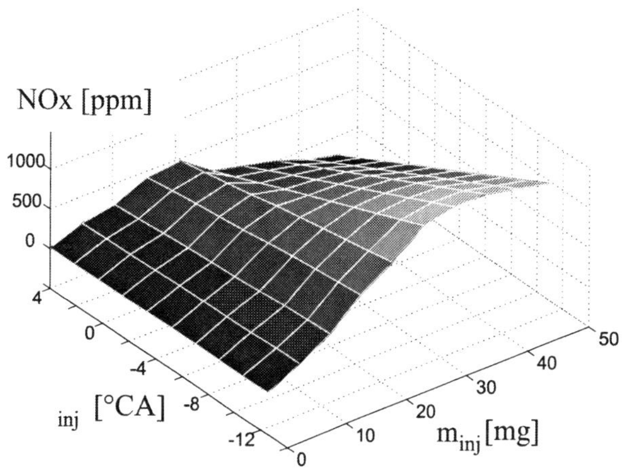


Fig. 17. NOx-Map, calculated from dynamic neural net ($n_{eng} = 2500$ 1/min)

Another important feature of these dynamic neural models is the possibility of deriving stationary maps from dynamic models. The data in **Fig. 15** led to a dynamic model according to (3). This model was then used to calculate the stationary two-dimensional look-up table shown in **Fig. 17**. Thus, it is possible to get a good visual overview on the engine's characteristics by maps derived from dynamic measurements. Even more importantly, this feature allows a fast dynamic measurement of the engine's characteristics and a calculation of the static engine maps by means of the dynamic neural model. Consequently, there is no need to measure the whole look-up table using measurement points on equidistant grids, which helps to save expensive measurement time.

The obtained dynamic model for the $NO_X$ emissions can be used for off-line or on-line optimization of exhaust gas emissions, and has been implemented on a RCP system as explained in chapter 4. In the same way nonlinear models are obtained for torque, fuel-consumption, CO, HC and soot, (Hafner *et al.*, 2000).

## 7. ADAPTIVE FEEDFORWARD CONTROL OF IGNITION ANGLE WITH RAPID CONTROL PROTOTYPING FOR AN SI ENGINE

Cylinder pressure signals contain valuable information for closed loop engine control. For using this information low cost combustion pressure sensors with high long-term stability have been developed (Anastasia and Pestana, 1987; Herden and Küsell, 1994) and are starting to be installed into production engines (Inoue *et al.*, 1993). Real-time cylinder pressure evaluation is, however, a demanding task and requires powerful computational resources. Sampling of cylinder pressure signals is usually performed in a crankshaft synchronous manner, a typical resolution for engine research is 1° crank angle (CA), which results in sampling rates up to 40 kHz, depending on engine rotational speed. Nevertheless, increasing computational performance as well during the control design stage using RCP systems, as for future engine control units entail an increasing interest in benefiting from combustion pressure information.

A variety of engine control functions could be improved or implemented using cylinder pressure sensors. One of them is the implementation of a closed-loop ignition control system, as it is explained in the following.

The objective of *ignition control* is to achieve optimum engine efficiency for each combustion event. Generally factors that influence the optimal ignition angle are engine specifications like configuration of the combustion chamber, operating conditions like engine speed, load, temperature and EGR flow rate, as well as ambient conditions such as air temperature, air pressure and humidity in the atmosphere.

Standard ignition control systems are based on feedforward control and therefore rely heavily upon calibration of look-up tables. The database values are initially calibrated from an analysis of a nominal engine under fixed environmental conditions. However, changing environmental conditions, aging effects, and manufacturing tolerances usually change an engine's characteristics and lead to a deteriorating performance. This motivates the development of closed-loop control systems.

Combustion pressure sensors allows to optimize the point of ignition of each cylinder separately. The variable which is to be controlled is calculated from the "mass fraction burned" (MFB) signal, which can be derived from cylinder pressure evaluations (Isermann and Müller, 2001). Measurements and theoretical analysis reveal, that optimal ignition, i.e. maximum torque from each combustion event, can be obtained if 50 % of fuel mass have been burned until a crank angle of 8° after top dead center (TDC) (Bargende, 1995). This crank angle is also referred to as the "center of combustion". The proposed approach calculates the crankshaft angle of 50 % MFB $\theta_{50\%\,MFB}$ for each combustion cycle and controls it to $\theta_{50\%\,MFB} = 8°$ CA after TDC.

### 7.1 *Control Structure*

Applying closed-loop ignition control using standard controllers (e.g. of PI type) cannot provide acceptable control performance under fast changing operating conditions, since dead times and noisy signals prohibit the tuning of controller gains with high control effort. The dead times are due to the fact, that after ignition of the air-fuel mixture, the combustion cannot be influenced any further. Therefore the ignition angle can only be computed for the next cycle, based on measurements

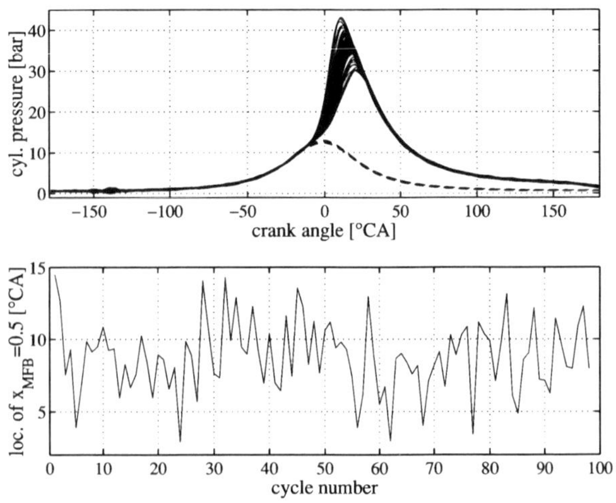


Fig. 18. Upper diagram: measured cylinder pressure signals and reconstructed polytrope; lower diagram: calculated crank angle locations of 50% MFB. $n_{mot} = 3000$ rpm, 50% load

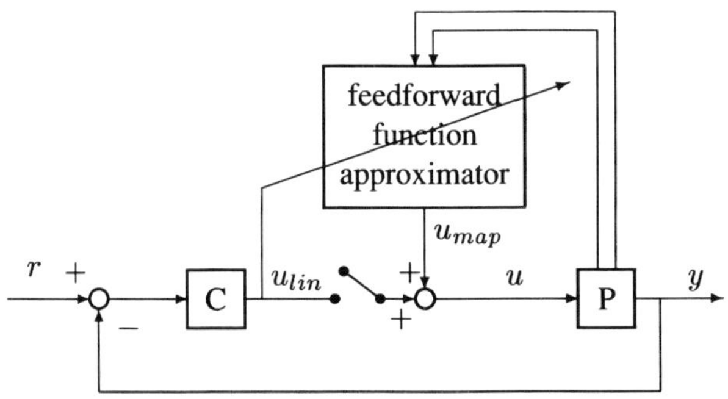


Fig. 19. General structure of a LFFC (Learning Feedforward Controller). The controller $C$ is basically used for the adaptation of the feedforward map

from the present engine cycle, and a dead time of one cycle is inherent. Moreover as there exist significant cycle fluctuations even under steady operating conditions, the calculated cylinder pressure features of several consecutive engine cycles have to be averaged (e.g. a moving average over 10 cycles is used). However, because the system error usually differs from one engine operating condition to another, using controllers with small control gains, it takes a certain amount of time after engine transients, to "integrate" to a new ignition angle. The stochastic nature of the combustion events can be seen in **Fig. 18**. The upper diagram shows the measured cylinder pressure signals of the compression and the power strokes of 100 consecutive cycles of an arbitrary cylinder. The dotted lines represent the reconstructed polytrope, which is the component of the cylinder pressure which is due to the piston motion (also called "towed" or "motored" pressure) . The lower diagram depicts the corresponding crank angle locations of 50% MFB, which are to be controlled to 8° CA.

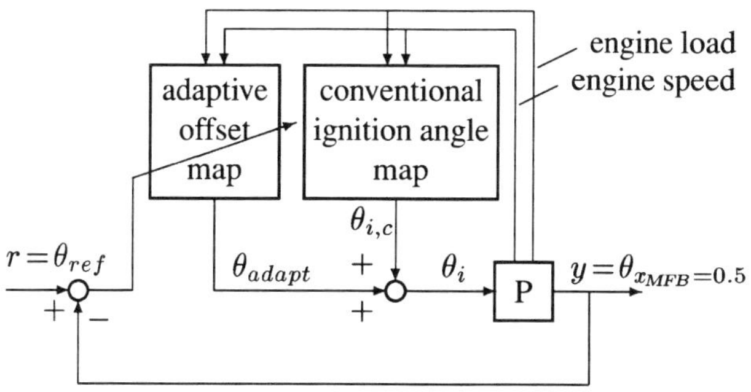


Fig. 20. Structure of the adaptive feedforward ignition control system

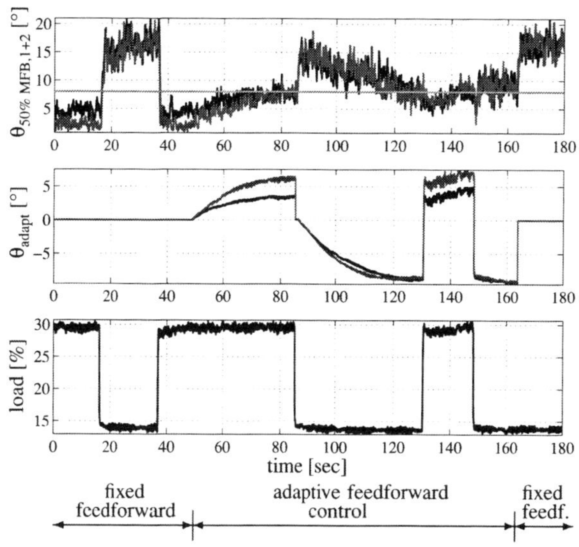


Fig. 21. Control performance of the ignition control system during three equivalent load changes. The adaptive look-up table is initialized with zero, after 50 s the adaptive feedforward control is activated.

This motivates the use of Learning Feedforward Control (LFFC), whose general structure is shown in **Fig. 19**. The linear controller $C$ is used to compensate random disturbances and to provide a reference signal for the learning feed-forward controller. It does not need to have a high performance and can be designed in such a way that it provides a robust stability. A function approximator works in parallel to the linear controller, it can be represented by a neural network or, in the simplest case, by a two-dimensional look-up table. Since the learning function approximator acts instead of a controller's integral term, the linear controller is preferably a simple proportional gain. It can also be deactivated for noise suppression.

For the ignition control system the function approximator is divided into the conventional, fixed ignition look-up table and into the adaptive offset map, see **Fig. 20**. The adaptive look-up table is trained on-line using NLMS or RLS training algorithms (Isermann and Müller, 2001). The learning process is performed during normal operation of the engine, without determining special excitation signals. The learning system is just collecting and processing input-output samples at the individual operating points. The operating condition is determined by the engine load (a normalized value calculated from the intake manifold pressure signal) and the engine's rotational speed. The conventional look-up table determines an approximate value of the ignition angle $\theta_{i,c}$ and is valid for all cylinders. For each cylinder the location of 50 % MFB is calculated and compared to the reference location of 8° CA. Correction angles $\theta_{adapt}$ are calculated and memorized by the adaptive offset look-up table of each cylinder, at the corresponding operating condition. In order to reduce noise in the controller output signal $\theta_i$, no linear proportional controller $C$ is used in parallel to the function approximator, see **Fig. 19**.

### 7.2 *Measurement Results*

In **Fig. 21** at a constant engine speed of 2500 rpm a certain load change sequence is repeated for three times, see the third diagram. The first diagram shows the crank angle locations of 50% MFB for two cylinders, which are to be controlled to 8° CA after TDC. The second diagram depicts the correction values $\theta_{adapt}$ for the ignition angles of the corresponding cylinders. For the first 50 s only the conventional feedforward control is active. The considerable deviations of the crank angle locations of 50% MFB from the optimal value for best engine efficiency at 8° CA after TDC is visible, see the upper diagram. Between 50 and 168 s the adaptive feedforward controller is active. Since the adaptive input-output map of each cylinder had been initialized to zero, it first has to be adapted for both operating conditions. The control performance for the already adapted feedforward control is shown between 130 and 160 s. Then it is switched back to the conventional, fixed feedforward control.

In spite of the considerable, stochastic nature of the combustion events, the adaptive feedforward control allows to maintain the mean crank angle location of 50% MFB around its optimal value at about 8° CA after TDC.

## 8. CONCLUSIONS

The demands for improved engine emissions, performance and efficiency, as well as the tight schedules in the development process, require efficient methods and tools for the design, implementation and calibration of engine control units.

Rapid Control Prototyping systems allow a fast and easy implementation and testing of new engine control algorithms. Graphical user interfaces and automatic code generating methodologies allow the computation of control functions, in bypass to the engine control unit, on a real-time hardware.

Hardware-in-the-Loop simulators allow testing of engine control units and control algorithms under various conditions and in a repeatable manner. Thus expensive and probably dangerous engine experiments are avoided and valuable time on dynamometers can be saved. The behaviour of the engine control unit during sensor faults can be studied.

Efficient engine models, as required for all proposed development tools, can be obtained by using dynamic local linear neural networks. The basic structure of the network type LOLIMOT is explained and applied for modeling and online simulations of the dynamic $NO_X$ emissions of a direct injection Diesel engine.

To exemplify the potential of Rapid Control Prototyping systems, a combustion pressure based ignition control system was presented. An indicating system allows the online evaluation of cylinder pressure signals in real-time, in a crank synchronous manner. Learning Feed-forward control is implemented on a time synchronous RCP, allowing to optimize the ignition angle of each cylinder separately. The proposed control system is capable to compensate for manufacturing tolerances, fuel quality variations and long-term effects such as aging or wear of the engine.

## 9. REFERENCES


Anastasia, C. and G.W. Pestana (1987). A cylinder pressure sensor for closed loop engine control. In 'SAE Technical Paper Series'. number 870288.

Bargende, M. (1995). 'Most optimal location of 50 % mass fraction burned and automatic knock detection components for automatic optimization of SI-engine calibrations'. *MTZ Worldwide*.

Fischer, M., O. Nelles and R. Isermann (1998). 'Adaptive predictive control of a heat exchanger based on a fuzzy model'. *Control Engineering Practice* **6**(2), 259–269.

Hafner, M., M. Schüler and R. Isermann (2000). 'Model-based optimization of ic engines by means of fast neural networks - part 2: Static and dynamic optimization of fuel consumption versus emissions'. *MTZ Worldwide*.

Hanselmann, H. (1996). Automotive control: From concept to experiment to product. In 'IEEE International Symposium on Computer-Aided Control System Design'.

Herden, W. and M. Küsell (1994). A new combustion pressure sensor for advanced engine management. In 'SAE Technical Paper Series'. number 940379.

Heywood, J. B. (1988). *Internal Combustion Engine Fundamentals*. McGraw-Hill.

Inoue, T., S. Matsushita, K. Nakanishi and H. Okano (1993). Toyota lean combustion system - the third generation system. In 'SAE Technical Paper Series'. number 930873.

Isermann, R. and N. Müller (2001). Nonlinear identification and adaptive control of combustion engines. In 'IFAC-Workshop on Adaptation and Learning in Control and Signalprocessing'. Como, Italy.

Isermann, R., J. Schaffnit and S. Sinsel (1999). 'Hardware-in-the-loop simulation for the design and testing of engine control systems'. *Control Engineering Practice* **7**, 643–653.

Isermann, R., S. Ernst and O. Nelles (1997). Identification with dynamic neural networks - architectures, comparisons, applications -. In 'IFAC Symposium on System Identification'. Fukuoka, Japan.

Isermann, R., S. Sinsel and J. Schaffnit (1998). Modeling and real-time simulation of diesel engines for control design. In 'SAE Technical Paper Series'. number 980706.

Nelles, O. (1999). Nonlinear System Identification with Local Linear Neuro-Fuzzy Models. PhD thesis. Darmstadt University of Technology. Darmstadt, Germany.

Schaffnit, J. (2001). Simulation und Control Prototyping zur Entwicklung von Steuergerätefunktionen für aufgeladene Nutzfahrzeug-Dieselmotoren. PhD thesis. Darmstadt University of Technology. Darmstadt, Germany, to appear in 2001.

Schaffnit, J. and R. Isermann (2000). Hardware-in-the-loop simulation and rapid prototyping for the development of control functions for diesel engines. In '1st IFAC Conference on Mechatronic Systems, MECHATRONICS2000'. Darmstadt, Germany.

Sinsel, S. (2000). Echtzeitsimulation von Nutzfahrzeug-Dieselmotoren mit Turbolader zur Entwicklung von Motormanagementsystemen. PhD thesis. Darmstadt University of Technology. Berlin.

IFAC Publications
www.elsevier.com/locate/ifac

# CORBA AND CONTROL SYSTEMS RESEARCH

**Ricardo Sanz**

*Ricardo.Sanz@etsii.upm.es*
*Autonomous Systems Laboratory*
*Universidad Politécnica de Madrid, Spain*

Abstract: This paper describes the important role that CORBA and Real-time CORBA can play in advanced control systems development and in particular as a foundation for research in advanced control systems with higher degrees of autonomy and intelligence. *Copyright © 2001 IFAC*

Keywords: Integration, distributed control, interoperation, standardization, real-time object oriented systems, constructability, complexity.

## 1. INTRODUCTION

In the way to full autonomy, requirements for control systems do get tough continuously as happens in all other engineering activities for different reasons. For those of us involved in the use of *computers for control*, this toughness has a special meaning, because if forces us to have in mind several technological aspects that are traditionally considered collateral to our field of automatic control.

An example of the problems we're facing inside the controls community is the practical marginality of real-time computing research in the minds of hard-nosed control scientists.

Many control scientists and engineers neglect real-time systems as a *just implementational issue* ignoring that some of the theoretical problems posed by closing control loops using computing systems are in many cases well beyond our present reach.

In this paper we will not focus on deep hard real-time computing issues nor theoretical aspects of it. Will take the wider perspective of human research and development of very complex control systems; a journey trough freedom and imagination in the land of *building the controllers of the future.*

In this paper we will address the extremely important role that real-time distributed object technologies are going to play in advanced control systems development and in particular as a foundation for research in control systems that include higher degrees of intelligence.

## 2. COMPLEX CONTROL SYSTEMS ENGINEERING

Engineers in the field of complex controllers (Åström *et al.*, 2000) confront the double challenge of *a)* designing an extremely complex application and *b)* putting it to practice. The design challenge uses theoretical and experiential tools. The implementation challenge is mostly confronted using just experience. Except marginally minor systems, it is not possible at all to formalize and automate the development of controllers.

But both challenges cannot be considered in isolation from a practical point of view, because, in real conditions, the second step follows necessarily the first one (implementation technology reverse-influences design).

This means that control designs must take into account the implementation technology to be used –as is the case in any other type of engineering process. While this idea should seem obvious for everybody, it is not as widely understood and fostered as it should. In many cases, software system failure to fulfill requirements is due not to bad designs or to bad implementation technology, but from a bad coupling of both aspects of the life-cycle (Rombach *et al.*, 1992). Full life-cycle approaches are necessary to attack the problem of complexity not just at the level of companies but also in applied research.

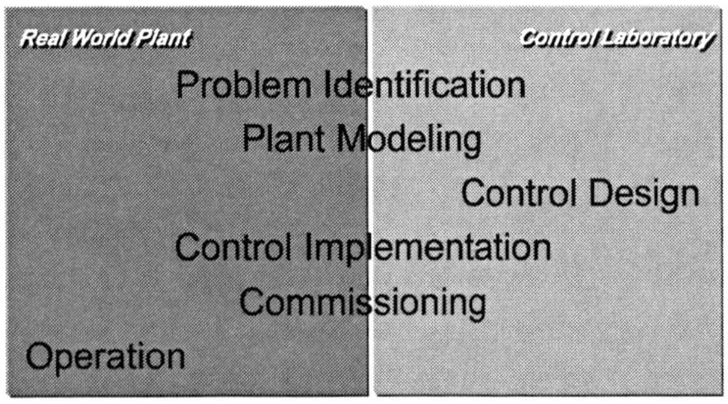


Fig. 1. From conception to operation. There is a strong need of full life-cycle approaches to complex control systems construction.

In some sense, complex control systems engineering can be considered a subset of software engineering. Software-based control systems can range from a small SCADA to C3PO control software. In an immediate effort we can make a quick and dirty classification of software systems based on factors that induce systems complexity:

- Conventional: the shoe shop database.
- Real-time: meeting deadlines.
- Embedded: run within limited resources.
- Fault Tolerant: good behavior under faults.
- Distributed: run on several interacting computers.
- Intelligent: solving ill-posed problems.
- Large: millions of lines of code.
- Integrated: interoperate with alien systems.
- Heterogeneous: run on heterogeneous platforms.

In any case, the implementation activity depends critically on the availability of proper programming languages, proper real-time computing platforms and proper development tools for the effective implementation of the controllers. In the fields where complexity is bigger (for example distributed control, intelligent control or fault-tolerant control), the need for such resources is critical.

But resources are not enough. Complex real-time systems cannot be built using methods suited for small, embedded applications. There is a strong need for *modularity* to support modular construction and testing. There is a strong need for *quality* to build truly dependable systems. There is a strong need of *reusability* to be cost-effective. In summary, there is a strong need for *development methodologies and supporting tools* to approach the problem from a global perspective.

In this paper we will focus on the approach to systems construction using distributed real-time objects. In particular we will address a specific technology under continuous development that will prove extremely useful in our field. This technology is referred-to as CORBA (Common Object Request Broker Architecture) and is being fostered by the Object Management Group (OMG, 2000).

The main goal of OMG is to create a standardized object-oriented architectural framework for distributed applications. Objectives include the reusability, portability, and interoperability of object-oriented software components in heterogeneous environments (hardware, operating systems, programming languages). To achieve these objectives, the OMG adopts interface and protocol specifications, based on commercially available object technology, that together define an Object Management Architecture (OMG, 1998*a*).

## 3. COMPLEX CONTROL SYSTEMS RESEARCH

Research in control systems pursues the development of new theories, technologies and methods of achieving better dynamical system performance. This explodes in an extremely wide and heterogeneous spectrum of activities that range from the development of mathematical theories to the details of real-time scheduler implementation. In all but simple cases, controller engineering confronts difficult problems.

In many cases the difficulty of building a proper control system relies in the lack of good solutions for difficult mathematical problems. But in many other cases the problems are more related with the whole life-cycle approach mentioned before, specially when we reach complexity levels far beyond single human reach.

There are several approaches to overcome complexity, and one of the most effective ones is divide and conquer. The methodology of building control systems using modular pieces is gaining credit as more and more systems are built this way.

This modularity enables independent construction of functionally designed components, reuse, modular testing, incremental construction, etc. But to be truly effective, modular components should be built to be flexibly integrated. This means that the design compromises in the external interfaces of the components should be reduced to a minimum. Examples of these design compromises are the use of specific programming languages, interaction modes, fixed ontologies, *etc.*

Having flexibly integrable modules simplify engineering processes, but also, and more important for us, enables fast prototyping of experimental systems. For example, replacing a conventional controller of a CST reactor by an experimental fuzzy controller keeping in place subsystems for sensing and actuation. Another example, replacing a locally running planner for an autonomous robot by a planner remotely running on a Beowulf cluster.

This integrational flexibility combined with the component oriented model of development and the minimization of design compromises are the signature of CORBA.

## 4. WHAT IS CORBA?

CORBA is a standardized architecture for heterogeneous distributed object systems. The basic CORBA paradigm is that of a request for services from a remote object.

The services that an object provides are given by its interface. Interfaces are defined in OMG's Interface Definition Language (IDL). Distributed objects are identified by object references, which are typed by IDL interfaces.

Figure 2 graphically depicts a request to a remote server. A client holds an object reference to a distributed server object. The Object Request Broker (ORB) delivers the request to the server object and returns any results to the client object.

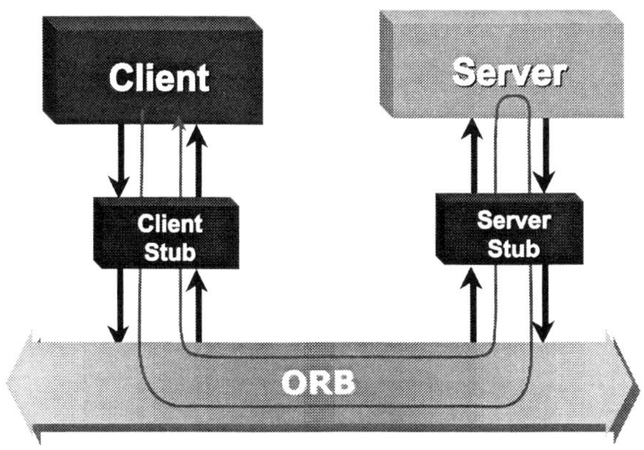


Fig. 2. Request forwarding done by the Object Request Broker.

The CORBA acronym refers to the base specification (OMG, 2000) but is also used to refer to the complete collection of specifications related with the Object Management Architecture (OMA, See (OMG, 1998*a*)): brokering, common services, facilities, language mappings, *etc.* The basic structure is depicted in Figure 3.

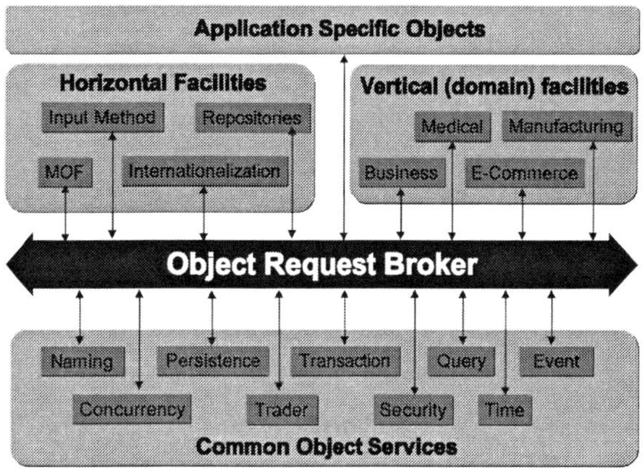


Fig. 3. OMA Overview. The main parts are shown but only some of the services are detailed.

The core technology provided by the OMA can be grouped in:

**Object Request Broker (ORB):** is the runtime integration vehicle that forwards requests and responses.

**Interface Definition Language (IDL):** is the interface definition mechanism for implementation independence.

**Language Mappings:** are the mappings from IDL to several programming languages for the implementation of client an servers.

**Repositories:** are stores that provide run-time information about interfaces and implementations.

**Interoperability:** between CORBA systems and with external entities (like Microsoft COM).

**Predefined Services:** interfaces providing a specific functionality for a domain.

OMG technology and standardization process is public, open and in continuous evolution.

## 5. CORBA IN CONVENTIONAL CONTROL SYSTEMS

CORBA will obviously be of extreme help in the implementation and deployment phase of control systems as any other type of software technology. But the core concepts we can find in it establish a sound, modular design guideline that enable the incremental construction of complex control systems. This is specially true after the incorporation of specifications for real-time, embedded and fault tolerant systems (OMG, 1999*b*; OMG, 1998*b*; OMG, 1999*a*).

Present-day distributed control systems (Honeywell, Siemens, ABB, etc) are built as modular controllers deployed over a collection -sometimes vendor specific- of computing systems. However, present trends in industrial systems, lead to more open architectures where plant-specific functions could be added.

Most companies are using today distributed object concepts for the implementation of their systems but only at higher layers because they base their system on COM technology (OPC) and it does not offer good performance for tight scheduling. New CORBA technology is being used today to push distributed object concepts to the very low levels of intelligent sensors and basic controllers (Complete Vertical Integration).

Embedded and real-time CORBA will help on this effort providing a platform for evolutionary, incremental development of highly integrated, intelligent, distributed control systems: from sensor-embedded processors to high level cognitive architectures. All them can be developed and deployed using this technology with a minimal coding effort, letting us focus on more critical activities in our research and development.

Sample applications can be found in all control system domains where distribution or heterogeneous object interaction is needed. Figures 4,5,6 and 7 show some examples of varying complexity.

Figure 4 is taken from recent IST project that uses a CORBA compliant broker to implement IEC 61850 models (TC-57, 2000) for substation automation systems (control and communication in electric utilities). This application provides interoperability in multivendor equipment, dynamic management and enhanced systems configuration. CORBA helps bridge the gap between real-time and non real-time worlds in distributed control systems (see (Sanz *et al.*, 2001*b*)).

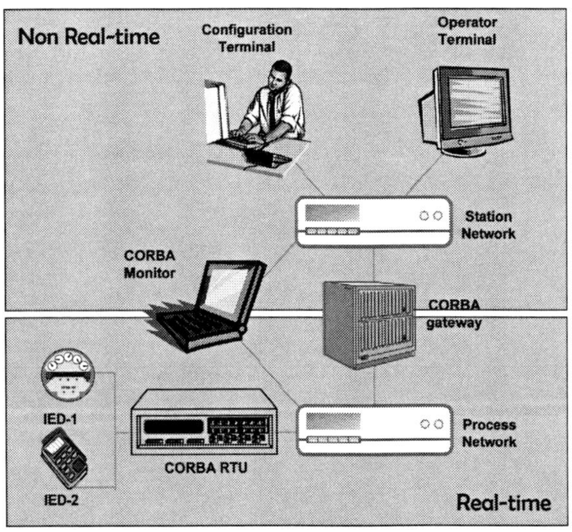


Fig. 4. CORBA in substation automation systems do provide interoperability across the real-time frontier (IED stands for *Intelligent Electronic Device*)

Figure 5 shows the use of CORBA in distributed information sharing in airborne military applications. This corresponds to an experimental implementation done by Boeing (see (Sharp, 2001)).

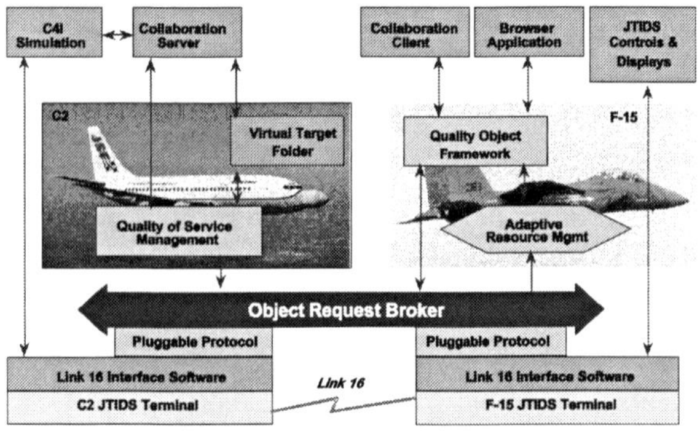


Fig. 5. Example CORBA application in the Joint Tactical Information Distribution System.

Figure 6 summarily describes the process of reengineering a cooperative architecture for autonomous robots (see (Sanz *et al.*, 2001*c*)). The cumbersome and unexpandable message passing policy has been substituted by alternative methods based on conventional requests and event channels. This is just a demonstration application of the reduced effort needed to enhance, from a software engineering point of view, an specific control architecture.

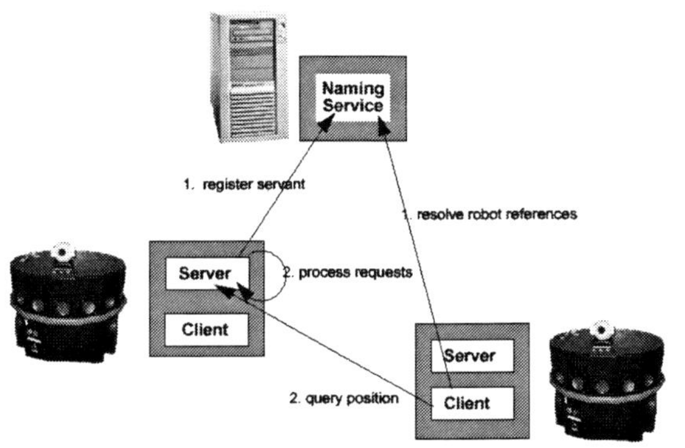


Fig. 6. Cooperative robot architectures can be easily reengineered using CORBA to gain openness, flexibility and control.

Figure 7 contains a high level view of a application for strategic decision making in the cement industry. This application is described in detail in another paper in this conference.

This application incorporates as CORBA objects neural network models of cement kilns, advanced user interfaces, alarm management expert systems and complex full plant mathematical models. The application was deployed on a network of heterogeneous systems (Digital UNIX, Windows NT, Linux, etc) in a cement plant.

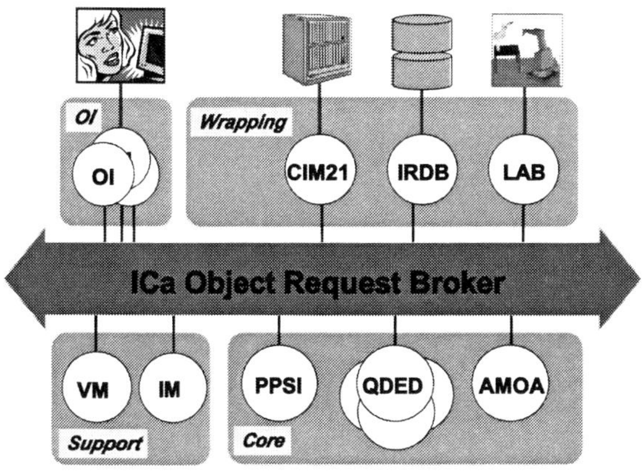


Fig. 7. The PikMac system is built as a collection of active CORBA objects that provide specific pieces of functionality: core systems, legacy application wrapping, operator interface and system support.

## 6. CONTROL SYSTEMS ENGINEERING IN AN IMPASSE

My hope is not only that object oriented middleware will help us in better building control systems, but that the development model it proposes can be scaled up at the research laboratory to the levels of complexity needed for extremely advanced control systems.

For many people, the discipline of control systems is a little bit at an impasse: no clear roads, not enough funding, *etc.* Lack of visibility is claimed as the one of the main reasons for lack of funding and/or interest by people outside the area (specially politicians). But from my personal perspective there are two intertwined problems here.

The first problem is the visibility problem. The huge ignorance of the external world about the critical issues we are dealing with. Controllers are mostly hidden entities and their importance is somewhat dismissed except for those few that know enough about them or get much from them. When controllers work properly, the success is assigned to the computer that supports the controller, and the controller does not get any credit.

But the second problem is perhaps harder because the origin is in our own perspective. We have been misled to think that there exists a distinct, separate, neat discipline of control systems. There isn't. What journals like *Automatica* or *Transactions on Automatic Control* publish is no more no less than just mathematics and physics. Our leading journals are just bits and pieces of the mathematics sub-discipline of dynamical systems.

We must understand that the scientific part of our discipline is just physics and mathematics and that what is most different from them is *what we do with it*: engineering of control systems.

I can see this impasse manifest in the field where I am more knowledgeable: distributed intelligent control. I have not seen any major advance in the last years, only small steps mostly centered on applications in specific domains.

We should abandon that perspective about this CC that reads *"Computer Control belongs to the Support Technologies which are of eminent importance for implementing control algorithms"*.

The question here is that it is not possible to have control systems without implementations. Without implementations all you have are toys and theories. This means that control engineering is not just systems mathematics as medicine is not just physiology or architecture is not just brick physics.

Control systems engineering goes from conception to retirement and henceforth all phases are equally critical and not just "supportive". In fact, for the hands-on control engineer, what is "supporting theory" is what is what is published in theoretical journal (because real control engineering is what he does).

This *computing-is-just-supporting-technology-for-control* vision must be eradicated even inside this CC. Controller implementation is not "just implementation" but a real complex, hard-to-solve and highly underestimated problem.

The reason why *Automatica* mostly publishes mathematical analysis and design papers is not because they are more profound and critical for the business. The reason in that there are lots of people that cannot confront -with sound mathematical methods- life cycle phases that are not so amenable to simple mathematical modeling.

This means that more theory (i.e. more work) is needed just where many people think it is not necessary of just collateral. Perhaps control engineering is just mathematical modeling and design. I don't think so.

We need to learn more about all aspects of controllers to have the proper background to understand all the factors that affect controller operation. Perhaps -in the future- control engineers will make specifications and the rest will go straight (like in some fields of mechanical design).

But it is not the case in control engineering. Not yet. Our theory must go beyond ODEs, hybrid, fuzzy, etc. to encompass all aspects of *information processing for proper action.* This is what *control* actually is. And computers are not marginal, supporting entities in this endeavor.

Let's stop complaining and look at what can we offer. Look at what are the major scientific breakthroughs that are at our reach.

## 7. RESEARCH TRENDS IN CONTROL SYSTEMS

Others can describe better than myself what are the research trends in the community. In fact a milestone report is being prepared to be distributed among us in the next months. What I will try to do instead, is to propose a new, bold and extremely risky research line.

The first step is to recognize the fact that controllers are not isolated entities. The first step is to recognize that our field is not *control* but *control* + *computing* + *communications* (or to put it simply $Co^3$).

We have seen that CORBA is a potential foundation for controller modularity, openness and extensibility. In some sense, these are the properties that enables controller evolution to do quantum leaps in complexity (Albus, 1981). Hierarchies, metalevel organizations, hot replacement and component evolution are based on these properties (Wills *et al.*, 2001).

If we are open and a little bit courageous, we can become the leaders of the coming revolution on intelligent systems engineering. The revolution of human and machine consciousness (Crick, 1994). Because we have the knowledge, tools and methods to understand complex systems dynamics (Sanz *et al.*, 2001*a*). Because based on this, we can design and *build* proper controllers for any endeavor. Even controllers –minds– for human like robots.

The ongoing discussions about the nature of intelligent behavior are more or less as sterile as they were one hundred years ago, because there exist a misunderstanding on the nature of intelligent behavior. Research in intelligent control is passing a dessert of lack of ideas of how to achieve the higher levels of functionality needed by intelligent systems. The main problem we confront here is a problem of scaling present intelligent control technology to build large scale artificial intelligences (Holland and Goodman, 2001).

Embedded and real-time CORBA will help on this effort providing a platform for evolutionary, incremental development of highly integrated intelligent control systems: from sensor-embedded processors to high level cognitive architectures will be developed and deployed using this technology. We can use CORBA as a foundation for *true intelligence.*

The automatic control sciences can break one of the frontiers of present science: the explanation of mind. Take for example schizophrenia. Tell psychiatrists what will happen if in a complex, distributed, highly connected control system someone runs twice a top-level intelligent controller in the same host computer. They will easily identify the symptoms: Less powerful high level thinking, interference between non-coherent actions, data flows coming from nowhere, *etc.* They will easily identify schizophrenia.

Complex controllers are just staring to be like us. An they will grow thanks to the possibilities offered by modern software engineering methodologies like CORBA.

The last step for us is to recognize that our field is not *control* or $Co^3$ but *control* + *computing* + *communications* + *cognition*+ *consciousness*, *i.e.* $Co^5$.

## REFERENCES


Albus, James S. (1981). *Brains, Behavior and Robotics.* BYTE Books - McGraw-Hill. Peterborough, N.H.

Åström, Karl, Isidori, Alberto, Albertos, Pedro, Blanke, Mogens, Schaufelberger, Walter and Sanz, Ricardo, Eds.) (2000). *Control of Complex Systems.* Springer. Berlin.

Crick, Francis (1994). *The Astonishing Hypothesis.* Simon and Schuster. New York.

Heck, B.S., L.M. Wills and G.J. Vachtsevanos (2001). Software enabled control: background and motivation. In: *Proceedings of the American Control Conference.* Vol. 5. pp. 3433–3438.

Holland, Owen and Rodney Goodman (2001). Consciousness and adaptive model-based predictive controllers for mobile robots. In: *Proceedings of the Toward a Science of Consciousness Conference, TSC'2001.* Skövde, Sweden.

OMG (1998*a*). A discussion of the object management architecture. Technical report. Object Management Group. Falls Church, USA.

OMG (1998*b*). Minimum CORBA specification. Technical Report orbos/98-08-04. Object Management Group. Falls Church, USA.

OMG (1999*a*). Fault tolerant CORBA joint revised submission. Technical Report orbos/99-10-05. Object Management Group. Falls Church, USA.

OMG (1999*b*). Real-time CORBA adopted specification. Technical Report orbos/99-06-02. Object Management Group. Falls Church, USA.

OMG (2000). Common object request broker architecture and specification. release 2.4. Technical report. Object Management Group. Falls Church, USA.

Rombach, H. Dieter, Bradford T. Ulery and Jon D. Valett (1992). Toward full life cycle control: Adding maintenance measurement to the SEL. *The Journal of Systems and Software* **18**(2), 125–138.

Sanz, Ricardo, Fernando Matía and Eugenio A. Puente (1999). The ICa approach to intelligent autonomous systems. In: *Advances in Autonomous Intelligent Systems* (Spyros Tzafestas, Ed.). Chap. 4, pp. 71–92. Microprocessor-Based and Intelligent Systems Engineering. Kluwer Academic Publishers. Dordretch, NL.

Sanz, Ricardo, Ignacio López and Juan Escasany (2001*a*). Cybernetic consciousness. In: *Proceedings of the Toward a Science of Consciousness Conference, TSC'2001*. Skövde, Sweden.

Sanz, Ricardo, José A. Clavijo, Miguel Segarra, Angel de Antonio and Mariano Alonso (2001*b*). Corba-based substation automation systems. In: *IEEE International Conference on Control Applications, CCA'2001*. Mexico D.F.

Sanz, Ricardo, Mariano Alonso, Ignacio López and Carlos A. Garc´ýa (2001*c*). Enhancing control architectures using corba. In: *IEEE International Symposium on Intelligent Control, ISIC'2001*. Mexico D.F.

Sharp, David (2001). Object-oriented real-time computing for reusable avionics software. In: *Proceedings of the 4th IEEE International Symposium on Object-Oriented Real-Time Distributed Computing, ISORC'2001*. Hong-Kong.

TC-57 (2000). Communication networks and systems in substations part 1: Basic principles. Draft Standard 61850-1. IEC.

Wills, L.M., S. Kannan, S. Sander, M. Guler, B. Heck, J.V.R. Prasad, D. Schrage and G.J. Vachtsevanos (2001). An open platform for reconfigurable control. *IEEE Control Systems Magazine*.

Wohlever, Steven, Victor Fay-Wolfe, Richard Freedman and John Maurer (1999). Building adaptable real-time command and control systems using corba. In: *Proceedings of the 4th IEEE International Workshop on Object-Oriented Real-Time Dependable Systems, WORDS'1999*. Santa Barbara, USA.

IFAC Publications
www.elsevier.com/locate/ifac

# EVOLUTIONARY COMPUTING: OPPORTUNITIES FOR CONTROL

**P.J. Fleming and R.C. Purshouse**

*Department of Automatic Control and Systems Engineering*
*University of Sheffield, UK*
*email: P.Fleming@sheffield.ac.uk*

Abstract: Control systems engineers regularly use optimisation in design and on-line applications. However, increasingly, due to the nature of the search space or because of the decision variable data types, for example, these problems cannot be solved satisfactorily using traditional approaches. Evolutionary computing (EC) offers certain advantages over traditional approaches for many of these problems, and those features that confer particular benefits for control applications are described here. As a result of these new opportunities, evolutionary algorithms have been applied to controller design and model identification, for both parameter and structural optimisation and for handling multiple competing performance objectives. Examples, including real-time and industrial uses of EC are described, together with other applications involving fault diagnosis, system analysis, robotics, sensor-actuator placement and optimisation for fuzzy and neural control. The rapid developments in computer technology will permit further realisation of the potential of EC and future perspectives are discussed. *Copyright © 2001 IFAC*

**Keywords:** control system design; genetic algorithms; identification; multiobjective optimisation; on-line control.

## 1. INTRODUCTION

The history of the application of optimisation to control often suggests that its use has been compromised by the need to formulate problems suitable for traditional numerical optimisation approaches. Real-life problems may often be difficult to formalise mathematically and may be poorly understood. Such problems typically require a search over multimodal landscapes, the consideration of multiple and non-commensurate performance criteria, and/or control variables that are a mix of continuous and discrete data types. Other assumptions that are necessary for the application of classical optimisation methods, involving system properties such as time-variance and noise, may also be inappropriate. As a result control engineers have considered alternative methods in order to obtain a satisfactory solution. Evolutionary computing (EC) has proved to be a popular alternative.

EC is a robust search and optimisation methodology that is able to cope with ill-behaved problem domains, exhibiting attributes such as multimodality, discontinuity, time-variance, randomness, and noise. Compared with classical approaches, it permits an exceptional level of flexibility with regard to performance assessment and design specification. However, EC is not without its own challenges. Much care must be taken to produce an effective search algorithm and the methodology suffers from the image of computational inefficiency. Application to real-time and safety-critical systems is highly limited at present.

We describe how the EC methodology has been applied to control systems engineering. The suitability of the scheme is discussed, and methods for incorporating typical control problem characteristics are outlined in Section 3, following a brisk overview of evolutionary computing in Section 2. A broad range of applications is described in Section 4, which explores the use of EC for parameter optimisation, structure optimisation and multiobjective optimisation in design. We attempt to provide an account of published work that exploits the special nature of EAs in applications, rather than those papers that describe relatively straightforward use as a traditional optimiser. Key application areas considered are controller design, model identification, robust stability analysis, and fault diagnosis. Robotics, manufacturing and systems applications are also described. On-line applications, although relatively few in number because of the difficulties associated with using EC in real-time, are included. The paper closes with perspectives on the future of evolutionary computing, paying particular attention to issues of concern to the control engineer.

## 2. AN OVERVIEW OF EVOLUTIONARY ALGORITHMS

### *2.1 Introduction*

Evolutionary algorithms (EAs) are global, parallel, search and optimisation methods, based on Darwin's principles of natural selection. The genetic algorithm (Holland, 1975) is perhaps the most popular of these algorithms that include evolutionary programming (EP) and evolution strategies (ES) (Fogel *et al*, 1966; Schwefel, 1965; Rechenberg, 1973). EAs work with a population of potential solutions to a problem. Each individual within the population represents a particular solution to the problem, generally expressed in some form of genetic code. The population is evolved, over generations, to produce better solutions to the problem.

Most evolutionary algorithms (EAs) can be described thus (Michalewicz and Fogel, 2000):

$$x[t+1] = v(s(x[t])), \quad (1)$$

where $x[t]$ is the population under a representation $x$ at time $t$, $v$ is (are) the variation operator(s) and $s$ is the selection operator. Eq.(1) describes the probabilistic course of evolution over each generation. The initial population, $x[0]$, must be determined (usually randomly).

### *2.2 Genetic Algorithms*

A schematic of a simple *genetic algorithm* is shown in Fig.1. Each individual within the population represents a particular solution to the problem, generally expressed in some form of genetic code. The population is evolved, over generations, to produce better solutions to the problem. Each individual within the population is assigned a fitness value, which expresses how good the solution is at solving the problem. The selection operator, *s,* probabilistically determines how successful the individual will be at propagating its genes (its code) to subsequent generations.

Evolution is performed using a set of stochastic genetic operators, *v,* which manipulate the genetic code. Most evolutionary algorithms include operators that produce new individuals based on those selected, and determine the composition of the population at the subsequent generation. *Recombination* and *mutation* are two well-known operators.

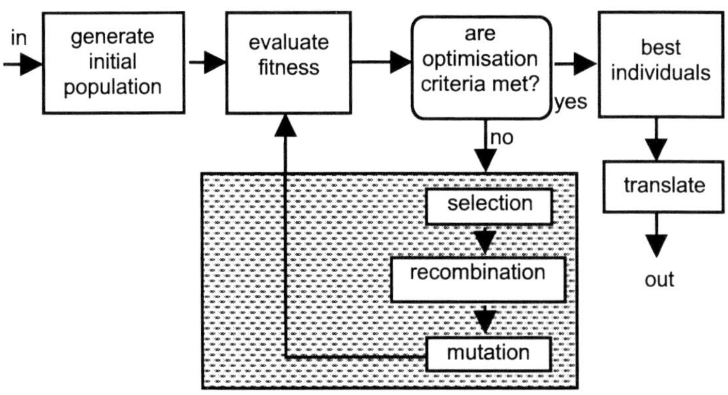


Fig.1 A simple genetic algorithm

The recombination operator involves the exchange of genetic material between chromosomes (parents), in order to create new chromosomes (offspring). The mutation operator makes small, random, changes to a chromosome. Usually considered as a background operator, its role is often regarded as providing a guarantee that the probability of searching any given string will never be zero (thereby enhancing multimodal search) and providing an opportunity to recover good genetic material that may be lost through the action of selection and crossover. Thus, a new population, $x[t+1]$, is constructed.

GAs, and EAs in general, are robust tools, able to cope with discontinuities and noise in the problem landscape. They have proved useful at tackling problems that cannot be solved using conventional means. Inclusion of domain-specific heuristics is not a pre-requisite, although it may improve the performance of an EA.

### *2.3 Representation, Evaluation, and Operators*

The specific details of each EA will vary from application to application. The choice of representation, evaluation, and genetic operators must be made with the particular application in mind. A short summary of the key issues and developments made thus far is presented below. Refer to Michalewicz and Fogel (2000) for a comprehensive discussion.

*Representation* – The EA works on an encoding of the decision variables. The classic choice is to encode each variable as a binary string, and then concatenate each variable to form a single solution. However, any representation is permitted for which suitable variation operators can be devised. Direct floating-point representations are becoming increasingly popular (Michalewicz, 1996).

*Evaluation* – The EA requires a scalar value representing the performance of each solution. This may be an absolute quantity or may be relative to other solutions. Techniques for the conversion of raw performance to an EA fitness value have received much attention in the literature. Ranking of solutions is a popular approach, especially for multiobjective problems (see later).

*Selection* – Individual solutions must be selected for application of the genetic operators and for subsequent reinsertion into the population if a generational gap is used. Commonly used selection methods are:- tournaments between two individuals (which achieves good parallelism), and stochastic universal sampling (which is unbiased), (Baker, 1987). A trade-off has been shown to exist between these two aims.

*Variation* – Recombination tends to direct the search to superior areas of the search space, whilst mutation acts to explore new areas of the search space and to ensure that genetic material cannot be irretrievably lost. Choice of genetic operators must be made together with choice of representation. Choices of values for the parameters of the operators, such as mutation rate and mutation size, are critical to the success of the algorithm. (Guidance is available in the literature.) Also, parameters may vary adaptively over the course of a run in order to improve the search results.

*Population* – EAs evolve a population of solutions, usually comprising a static number of between 20 and 100 individuals, over a number of generations. The population is initialised with random values, possibly seeded with some known good solutions. At each subsequent generation, the population will comprise the results of genetic manipulation, some remnants of the past population, and possibly a few randomly generated individuals. An EA is defined as *elitist* if it deliberately reintroduces past high-performance individuals.

### *2.4 Multiobjective Optimisation*

Real-world problems usually involve the simultaneous consideration of multiple performance criteria. These objectives are often non-commensurable and are frequently in conflict with one another. *Trade-offs* exist between some objectives, where advancement in one objective will cause deterioration in another. It is very rare for problems to have a single solution; rather a family of *non-dominated* solutions will exist. These *Pareto-optimal* (PO) solutions are those for which no other solution can be found which improves on a particular objective without detriment to one or more other objectives. The concept of Pareto optimality in the two-objective case is illustrated in Fig. 2.

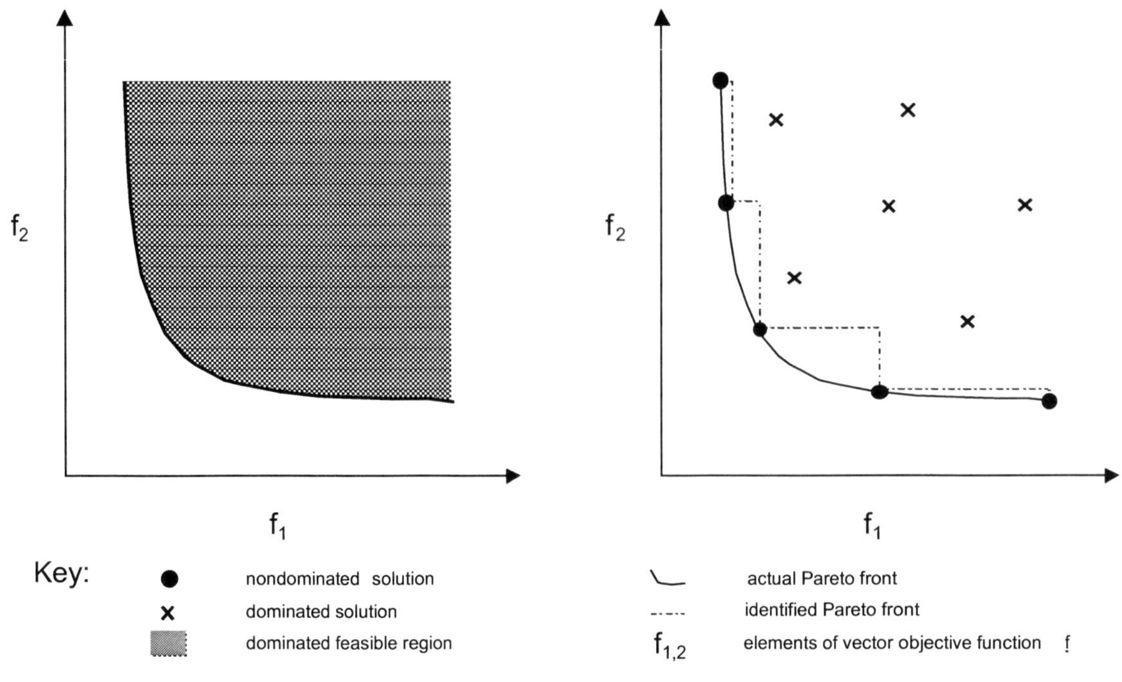


Fig. 2: Pareto optimality

Evolutionary algorithms are well suited for multiobjective optimisation. Due to their population-based nature, they are capable of supporting several different solutions simultaneously. One of the first approaches to utilise the concept of Pareto optimality was the multiobjective genetic algorithm (MOGA) of Fonseca and Fleming (1993) that tends to be the favoured approach for control engineers. Several excellent surveys of multiobjective evolutionary algorithm (MOEA) activity can be found (e.g. Van Veldhuizen and Lamont, 2000).

In the past, multiobjective problems have been cast as, effectively, single objective problems by constructing a utility function describing the relative importance of each objective. For example, in linear quadratic regulator design, the competing objectives of error and control size have been combined as a weighted-sum of quadratic measures. The cost function is defined prior to the optimisation procedure. It requires in-depth information concerning the various trade-offs and valuation of each individual. This data is not commonly fully available in practice. Furthermore, the weighted-sum approach is unable to identify non-convex parts of the trade-off surface, potentially missing important areas for compromise. In contrast, the GA selection operator can be used to identify degrees of Pareto optimality, thus enabling objectives to be handled individually. Hence, the requirement for a forced combination of objectives and the need for *a priori* information are both avoided. Indeed, this kind of search can help identify the existence and nature of specific trade-offs.

## 3. EC IN CONTROL: SPECIAL FEATURES

### *3.1 Suitability*

The evolutionary approach has proved particularly successful in problems that are difficult to formalise mathematically, and which are therefore not conducive to analysis. This includes systems that are highly non-linear, that are stochastic, and that are poorly understood. Problems involving these classes of process tend to be difficult to solve satisfactorily using conventional methods. The EA's lack of reliance on domain-specific heuristics makes it attractive for application in this area. Very little *a priori* information is required, but this can be incorporated if so desired.

A control engineering problem can contain a mixture of decision variable formats. This can prove significantly problematic for conventional optimisers that require variables of a single mathematical type or scientific unit. Since the EA operates on an encoding of the parameter set, diverse types of variable can be represented (and subsequently manipulated) within a single solution. For example, the decision vector {temperature, 18.3°, 1+j, 2π } does not pose an intrinsic problem to the EA.

The EA is a robust search and optimisation method that is well able to cope with ill-behaved cost landscapes, exhibiting such properties as multimodality, discontinuity, time-variance, randomness, and noise. Each of these properties can cause severe difficulties to traditional computational search methods, in addition to the lack of amenity to an analytical solution. Furthermore, an EA search is *directed* and, hence, represents potentially much greater efficiency than a totally random or enumerative search.

*3.2 Unsuitability*

For problems that are well understood, that are approximately linear, and for which trusted solutions exist, the EA is unlikely to produce competitive results. If a problem can be solved analytically with an acceptable level of assumptions then that approach is probably best. If such a solution cannot be found, and other problem-specific techniques are at an embryo stage, then the use of an EA could prove to be highly profitable and worthwhile.

Mission-critical and safety-critical applications do not appear, initially, to be favourable towards EA usage due to the stochastic nature of the evolutionary algorithm. No guarantee is provided that the results will be of sufficient quality for use on-line. There is also the matter of how individuals will be evaluated if no process model is available (as may well be the case). Some supporting theory exists for genetic algorithms, but is unlikely to prove sufficient to win the approval of standards and certification agencies. Much care would clearly be needed for critical systems.

Real-time performance is of particular interest to the engineer. However, EAs can be very computationally intensive, often requiring massively parallel implementations in order to produce results within an acceptable timeframe. Hence, on-line application to real-time control is largely infeasible at present. Processes with long time constants represent the most feasible application.

*3.3 Representation: Strings and Trees*

*Strings* - Each solution to a problem consists of a number of design parameters (decision variables). These are encoded as a chromosome, which can then be manipulated by genetic operators. Thus, a mapping is required between the actual decision variables (*phenotypes*) and their genetic equivalent (*genotypes*). Choice of representation is very flexible; indeed any is acceptable so long as suitable genetic operators can be developed to support the representation. The classic approach is to encode the set of parameters in a concatenated binary string, as shown by the simple example in Fig. 3.

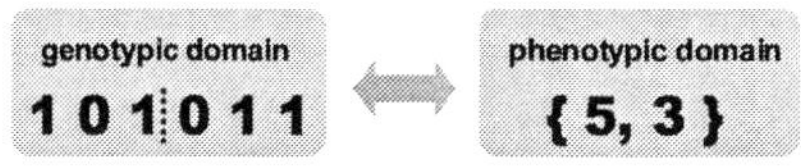


Fig. 3: Mapping between GA chromosome and decision variables

Both discrete and continuous variables can be catered for. A binary, or *n*-ary, coding can be used to represent variables to an arbitrary resolution. It should be noted, however, that the size of the search space increases exponentially with chromosome length. *Gray* encoding is often preferred to standard binary since it maintains the closeness of different solutions in both genotypic and phenotypic space. In recent years, floating-point representations have become popular alternatives to *n*-ary codings (Michalewicz, 1996). This development has been motivated in the main by users' comfort with one-gene-one-variable correspondence, but float-encoding has several benefits: it is faster to manipulate, it has been shown empirically to be more consistent from run to run, it permits much higher precision, and it is intuitively closer to the problem space. This final point is particularly beneficial since it enables easier incorporation of domain-specific knowledge.

*Trees* - The representation of decision variables need not be restricted to a linear string format. The evolutionary algorithm, known as *genetic programming* (GP), facilitates the representation and manipulation of variable length structures in a tree-based representation (Koza, 1992). These trees are composed of functions, ***G***, and terminals, ***T***, appropriate to the problem domain, providing a dynamic and variable representation. Fig. 2 shows an example of a hierarchical tree that represents the following expression

$$c + y(t-2) + y(t-1)^2 + y(t-1)y(t-2) \\ + y(t-1)u(t-1) + y(t-2)u(t-1)$$

The internal nodes of the tree structure are elements from the function set (operations), and leaf nodes are the input data from the terminal set. In this example ***T*** = {*c, y(t-1), y(t-2), u(t-1), u(t-2)* } and ***G*** = {+, -, *, /}, and it will be noted that not all are used for this representation.

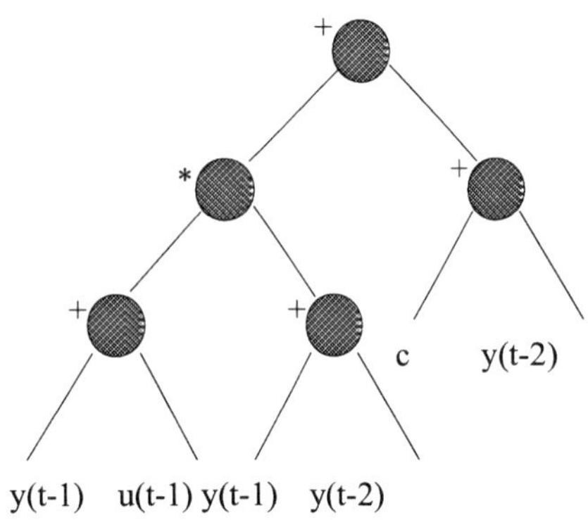


Fig. 3. Hierarchical tree encoding.

Like the standard GA, the two main operators are recombination and mutation. The recombination operation produces a pair of trees that inherit characteristics from both parent trees by selecting a random node in each of the structures of the parents and exchanging the associated sub-expressions. Because of the dynamic representation used in genetic programming, the parents are typically of different size, shape and content, and the offspring are also generally different. Mutation is performed by randomly selecting a node that can be an internal or terminal node, and replacing the associated sub-expression with a randomly generated sub-tree.

While tree representations have been used to evolve polynomial expressions, for example, for use in system identification, the same structure can be used to evolve block diagram representations.

*3.4 Cost Functions*

Many engineering problems consist of multiple objectives. For the purposes of the EA, these must be combined to form a single value. The (flawed) weighted-sum approach has proved popular in the literature, since it is amenable to a solution by conventional EA methods, but Pareto-based techniques are likely to surpass this in the future.

Traditionally, the root-mean-square (RMS) of the error between the desired output and the actual output has been taken as the cost of a particular solution. This measure is amenable to conventional search techniques. No such restrictions apply to the EA, so this limiting approach is no longer necessary.

Many engineering problems have the attribute of multimodality. This can cause premature convergence for conventional search techniques. EAs are not immune to this problem, although the parallel nature of the search process is a significant benefit.

## 4. EC APPLICATIONS IN CONTROL

This section cites certain landmark or novel uses of EC in control with brief notes on their relevance. The reader is invited to pursue those references of interest.

*4.1 Controller Design*

*4.1.1 Parameter optimisation*

Here EAs are used to select controller parameters in a variety of settings, where the controller structure has been prespecified.

*PID controller tuning*

*Oliveira et al (1991)* : early use of GAs.

*Wang and Kwok (1992)*: stressed the benefit of flexibility with regard to cost function and, in particular, alluded to the concept of Pareto-optimality (providing the potential to simultaneously address multiple objectives, such as transient performance and disturbance rejection).

*Porter and Jones (1992)*: simple, generic, method of tuning digital PID controllers.

*Vlachos et al (1999)*: tuning of decentralised PI (Proportional-Integral) controllers for multivariable processes.

*MBPC*

*Onnen et al (1997)*: determination of an optimal control sequence in model-based predictive control (non-linear systems with input constraints). Applied to a simulated batch-fed fermenter, with favourable results reported (compared to the traditional *branch-and-bound* method) for long control horizons.

*Direct design - H-infinity, multiple performance criteria, QFT*

*Chen and Cheng (1998)*: a structure specified H-infinity controller. A GA was used to search for good solutions within the admissible domain of controller parameters (obtained via the Routh-Hurwitz stability criterion).

*Fonseca and Fleming (1998)*: multiobjective optimisation of the low-pressure spool speed governor of a *Rolls-Royce Pegasus* gas turbine engine. Multiple design objectives were defined, in both the time- and frequency-domains.

*Tan et al (2001)*: multiobjective optimisation approach used in the automation of the Quantitative Feedback Theory approach for multivariable control system design.

*Indirect design - LQR, H-infinity*

A further approach to controller design using GAs is to apply the methodology *indirectly*. In such a scheme, the GA manipulates input parameters to an established controller design process, which in turn produces the final controller. The linear quadratic Gaussian (LQG) method (*Mei and Goodall, 2000*) and the H-infinity control scheme (*Dakev et al, 1997*) have both been utilised in this manner, the latter employing the use of multiple objectives.

*4.1.2 Controller structure*

Genetic programming approaches have been used to select controller structures, where only elements of the structure are pre-specified.

*Koza et al (2000)*: duplicated existing patents (for PI and PID controllers) and rediscovered old ones (a controller making use of the second derivative of the error between the reference signal and the output signal).

*Chipperfield and Fleming (1996):* used a MOGA to select controller structure and suitable parameters for a multivariable control system for a gas turbine engine.

*4.1.3 Fuzzy-neural control*

Evolutionary algorithms have been used in attempts to optimise various aspects of intelligent controllers. In fuzzy control, an EA can be used to generate the fuzzy rulebase, and to tune the associated membership function parameters. In neural control, an EA can function as an alternative choice to learning the weight values. EAs have also been mooted as capable of optimising the topology of a neural network.

*Linkens and Nyongesa (1996):* comprehensive appraisal of both approaches.

*Ichikawa and Sawa (1992):* neural network (NN) as a direct replacement for a conventional controller. The weights were obtained using a GA approach.

*Sette et al. (1998)*: combine neural networks and genetic algorithms to optimise a cotton spinning process, in terms of both the tenacity and elongation of the yarn.

Work in the area of GAs applied to fuzzy control is broadly split into two categories:

(a) tuning of the membership functions, and

(b) elicitation of the rulebase in addition to tuning.

Practitioners of approach (a) tend to argue that the form of the rules is likely to be known *a priori*, and that most uncertainty lies in the development of the associated membership functions. Use of a static rulebase also reduces the necessary level of computational complexity, which may be a further reason for the popularity of this approach.

*Kim et al (1995)*: used an asynchronous GA to optimise membership functions in order to facilitate rapid prototyping of fuzzy controllers.

*Matsuura et al (1996)*: used a GA to obtain optimal control of sensory evaluation of the saké mashing process. The GA learned rules for a *fuzzy inference* mechanism.

*Linkens and Nyongesa (1995)*: on-line and off-line application of GAs to fuzzy control.

*4.2 System Identification*

System identification can be decomposed into two, inter-related, problems:

- *Selection* of a suitable model structure, and subsequent
- *Estimation* of model parameters.

Well-developed techniques exist for parameter estimation of linear models and linear-in-the-parameters non-linear models. Techniques for the selection of structure and for non-linear-in-the-parameters estimation are still the subject of ongoing research.

The application of EAs to black-box and grey-box model identification has received considerable interest.

*Kristinsson and Dumont (1992)*: seminal paper describes the application of GAs to the system identification of both continuous- and discrete-

time systems. The technique employed can be used in on-line as well as off-line applications.
One of the key problems in system identification is the choice of the input, output, and delay terms that are to contribute to the model. EAs provide a simple method for searching the structure space for terms that make the most significant contributions to process output.
*Fonseca et al (1993)*: subset selection to select nonlinear model terms for NARMAX (Nonlinear AutoRegressive Moving Average eXogenous) models (Leontaritis and Billings, 1985).
*Luh and Wu (1999)*: migration-based GAs to identify NARX (Nonlinear AutoRegressive eXogenous) models.

GP has established itself as a popular technique for evolutionary system identification.
*Gray et al (1998)*: non-linear model structure identification using GP. They considered two representations: block diagrams (using SIMULINK) and equations (differential and integro-differential).
*Marenbach et al (1997)*: general methodology for structure identification using GP with a block diagram library. Typical blocks included time delays, switches, loops, and domain-specific elements. It was applied to a biotechnological fed-batch fermentation process.
*Rodriguez-Vázquez et al (1997)*: used *multiobjective genetic programming* to identify a NARMAX model for a gas turbine engine. Seven objectives were optimised simultaneously: the number of model terms, model degree, model lag, residual variance, long-term prediction error, the auto-correlation function of the residuals, and the cross-correlation between the input and the residuals.
*Choi et al (2000)*: used an *evolution strategies* algorithm to identify the parameters of static (Karnopp) and dynamic (LuGre) friction models. The model structures were predefined and the results were used in a friction compensation control system, which utilised a sliding-mode controller.
*Billings and Mao (1998)*: non-linear rational model identification. Both structure and parameter information was encoded in each individual to facilitate simultaneous optimisation of both elements.

### *4.3 Fault Diagnosis*

The tasks of a fault diagnosis system can be split into three areas, namely:

- *Detection* of the presence of a fault,
- *Isolation* of the fault, and
- *Identification* or classification of the fault,

commonly referred to as *FDI* (fault detection and isolation).
A popular approach to FDI involves the generation and analysis of *residuals*. These are signals that reflect inconsistencies between nominal and faulty operation.

*Patton et al (1997)*: formulated model-based FDI as a multiobjective optimisation problem, in which the task was to maximise the effect of faults on the residual, whilst minimising the effect of uncertainty. The approach was applied to the detection of sensor faults in a flight control system.
*Kowalczuk et al (1999)*: robust observers for fault detection using a Pareto-based approach with various performance criteria (including previously unused frequency-domain information).
*Marcu et al (1997)*: applied EAs to FDI methods that are not based on the concept of residuals. They formulated the model-based diagnosis of process faults as a feature selection and classifier design problem. A multiobjective evolutionary algorithm was used for both off-line learning of regions corresponding to *known* fault and fault-free conditions (using component shapes), and for on-line identification of process coefficients (the so-called *symptom vector*).
*Miller et al (1993)*: applied a GA to the generalised task of determining the correct problem (fault) from a collection of problems given a set of symptoms that indicate that a problem exists. In this approach, the *relative likelihood* of a diagnosis given observable manifestations was considered. The method relied upon the availability of *a priori* probabilities that a particular disorder caused a particular symptom.

### *4.4 Reliable Systems*

*Painton and Campbell (1995):* developed a technique for improving overall system reliability by considering component-level choices. For each possible component, a failure rate distribution and a cost were defined. A GA was used to search the component-choice landscape for the most reliable system, in terms of mean time-before-failure (MTBF), within a pre-defined cost ceiling.
*Coit and Smith (1996)*: considered multiple component choices, but also introduced redundancy by using a GA to find optimal design configurations for series – parallel systems.

*Ren and Bechta Dugan (1998)*: search for optimal design configurations using a *fault tree* approach. The methodology has been applied to a cardiac-assist system used to treat heart failure.

*4.5 System analysis*

*Marrison and Stengel (1997)*: used GAs in combination with Monte Carlo evaluations to evolve efficient robust control system designs. Stengel and co-workers have refined this technique in later work.
*Fadali et al 1999*: evolutionary search for system poles that lie outside the unit circle in connection with robust stability analysis of discrete-time systems.
*Thompson et al (1999)*: application of MOGA to the multidisciplinary task of control systems architecture selection.

*4.6 Robotics*

Evolutionary algorithms have been utilised in robotics for both path planning and the design of behavioural controllers.
*Rana and Zalzala (1997)*: examined collision-free path planning of robot arms.
*Erkmen et al (2000)*: used GAs to improve the *dexterity* of robot manipulators. Dexterity can be defined as the capability to manipulate objects in crowded, changing, and partially known environments and the ability to recover from failing grasps through a re-grasping procedure.
*Dorigo and Columbetti (1994)*: used reinforcement learning to *shape* a robot to perform predefined target behaviour. Learning *classifier systems* performed the behavioural control (a classifier is a cognitive system capable of recognising objects and events in its environment and of making appropriate responses). A GA was used as a rule discovery algorithm to generate the classifiers.

*4.7 Control-related Combinatorial Problems*

Combinatorial problems are those in which an optimum set of choices must be made from a, usually significantly large, pool of potential choices. Many control problems are combinatorial, or have an element thereof, and can be generally viewed as a set of decisions, each with its own set of possible choices. Hence, a solution to a problem can be defined as a set of decision-choice pairs. The effects of the choices on the output may be non-separable, and dependencies may exist between certain decision-choice pairs. Various combinatorial problems can be identified in the earlier discussions, including selection of model structure, selection of controller structure, and system component selection. A further example is that of sensor and actuator placement.

*Krishnakumar et al (1994)*: simultaneous optimisation of actuator-sensor placement and feedback controller gains as a means of providing optimal structural control.
*Dimopoulos and Zalzala (2000)*: broad review of evolutionary algorithms in manufacturing optimisation.

*4.8 On-Line Applications*

On-line applications present a particular challenge to the EA. Successful applications in this field have been somewhat limited to date. The benefits of an EA for on-line control systems engineering applications are the same as those discussed for off-line applications. However, an on-line EA approach must be used with particular caution. There are several considerations to be made. It is important that an appropriate control signal is provided at each sample instant. If unconstrained, the actions of the 'best' current individual of the EA may inflict severe consequences on the process. This is unacceptable in most applications, especially in the case of a safety- or mission-critical system.

Given that it may not be possible to apply the values represented by the best individual in an EA population to the system, it is clear that evaluation of the complete, evolving, population cannot be performed on the actual process. The population may be evaluated using a process model, assuming that such a model exists, or performance may be *inferred* from system response to actual input signals. Inference may also be used as a mechanism for reducing processing requirements by making a number of full evaluations and then computing estimates for the remainder of the population based on these results.
There are three broad approaches to the use of GAs for on-line control (*Linkens and Nyongesa, 1995*):

- Utilise a process *model*.
- Utilise the process *directly*.
- Permit *restricted tuning* of an existing controller.

*Lennon and Passino (1999)*: applied GAs on-line in a so-called *genetic model reference adaptive control* system. The particular application was 'base-braking'. This involves ensuring that a motor vehicle's brakes perform consistently as the driver commands.

*Chang and Sim (1997)*: GAs as on-line optimisers for automatic train operation systems, running on mass rapid transit networks with automatic train protection.

Few applications have been reported on actual (rather than simulated) real-time use of GAs for control.

*Ahmad et al (1997)*: a GA has been used to tune the parameters of a PI controller in real-time for the on-line regulation of temperature in a heating system.

*Moriyama and Shimizu (1996)*: utilised a real-time GA to determine the optimal temperature profile for an ethanol fermentation process. The scheme was implemented on a dual processor system, with sensor – actuator processing and optimisation being handled in parallel.

GAs have also been utilised for the on-line tuning of controllers, *prior* to actual on-line usage of the system.

*Schroder et al (2001)*: tuned a controller of a canned, electrical pump running on active magnetic bearings, dramatically improving upon existing controller designs.

## 5. FUTURE PERSPECTIVES

The future use of evolutionary algorithms in control is tied fundamentally to advances in the general field. Due to the largely generic nature of the methodology, developments that arise in control applications in one domain are likely to prove beneficial to other domains.

The continued progress in computer technology will permit further realisation of the EA methodology's potential. The main frustration with any EA is the large amount of computation required to formulate a good solution. Hence, the increased availability of low-cost, high-performance computing, coupled with advancements in parallel architectures, will undoubtedly be of great benefit. In addition to increasing the effectiveness of the technique for off-line applications, computing developments should also make new on-line applications, such as evolutionary adaptive control, feasible for the first time. Using approximate models, such as neural networks (Duarte *et al,* 2000) and response surface models (Stewart and Fleming, 2001), as replacements for part of the search process is proving effective in realising performance speedup.

A second key area for future development is the field of multiobjective optimisation and decision support. The MOEA embodies an exciting opportunity to realise the full potential of the EA concept. The development of population-based, high-performance, robust, search techniques is just one facet of this research area; of crucial importance is the incorporation of, potentially multiple, decision-makers into the search process. Preference articulation and decision-space (and criterion-space) visualisation are both issues of ongoing research.

The combination of an EA search and optimisation method with complementary techniques to create hybrid algorithms is seen as vital in order to create a fully effective tool. For example, the use of a GA with a local hill-climber provides the benefit of a robust, global, search with an efficient fine-tuning mechanism. Hybrids involving such algorithms as Bayesian belief networks, intelligent agents, and ant foraging are expected to emerge in the literature. Neural network, fuzzy logic, and simulated annealing hybrids can already be found. Refer to Michalewicz and Fogel (2000) for further details.

The emergence of a standard EA toolbox has been predicted throughout the 1990s. However, despite the availability of packages developed by several research laboratories, a standard set of tools has not been adopted. Furthermore, no major design tool manufacturer has integrated an EA tool into their software package. This is perhaps largely due to the required customisation as mentioned above. Commercial off-the-shelf (COTS) EA packages are expected to become more widely available in the future. These may be entirely software based, or, perhaps, hardware will form part of the package. A possible development is the emergence of consultancies that offer EA methods as a part of a decision-support service.

In summary, significant future EA applications are likely focus on two main areas: off-line, multicriterion, design-aid tools and robust, on-line, search and improvement algorithms.

## ACKNOWLEDGEMENT

Robin Purshouse wishes to acknowledge UK EPSRC for research studentship support.

## REFERENCES


Ahmad, M., Zhang, L., and Readle, J. C., 1997, On-line genetic algorithm tuning of a PI controller for a heating system, *Proc Int Conf on Genetic Algorithms in Engineering Systems: Innovations and Applications*, pp510-515.

Baker, J. E., 1987, Reducing bias and inefficiency in the selection algorithm, *Proc 2nd International Conference on Genetic Algorithms*, pp14-21.

Billings, S. A. and Mao, K. Z., 1998, Structure detection for nonlinear rational models using genetic algorithms, *Int J Systems Science*, **29**, No. 3, pp223-231.

Chang, C. S. and Sim, S. S., 1997, Optimising train movements through coast control using genetic algorithms, *IEE Proc – Electric power applications*, **144**, No. 1.

Chen, B. S. and Cheng, Y. M., 1998, A Structure-Specified H-Infinity Optimal Control Design for Practical Applications: A Genetic Approach, *IEEE Trans on Control Systems Technology*, **6**, pp707-718, 1998.

Chipperfield, A. and Fleming, P., 1996, Multiobjective Gas Turbine Engine Controller Design Using Genetic Algorithms, *IEEE Trans on Industrial Electronics*, **43**, pp1-5.

Choi, S. H., Lee, C. O., and Cho, H. S., 2000, Friction compensation control of an electropneumatic servovalve by using an evolutionary algorithm, *Proc Inst Mech Eng*, **214**, Part I, pp173-184.

Coit, D. W. and Smith, A. E., 1996, Reliability Optimization of Series – Parallel Systems Using a Genetic Algorithm, *IEEE Trans on Reliability*, **45**, pp254-260.

Dakev, N. V., Whidborne, J. F., Chipperfield, A. J., and Fleming, P. J., 1997, Evolutionary H-infinity design of an electromagnetic suspension control system for a maglev vehicle, *Proc Inst Mech Eng*, **211**, Part I, pp345-355.

Dimopoulos, C. and Zalzala, A. M. S., 2000, Recent Developments in Evolutionary Computation for Manufacturing Optimization: Problems, Solutions, and Comparisons, *IEEE Trans on Evolutionary Computation*, **4**, pp93-113.

Dorigo, M. and Columbetti, M., 1994, Robot shaping: developing autonomous agents through learning, *Artificial Intelligence*, **71**, pp321-370.

Duarte N.M., Ruano A.E., Fonseca C.M. and Fleming P.J., 2000, Accelerating multi-objective control system design using a neuro-genetic approach, *Proc 8$^{th}$ IFAC Symposium on Computer Aided Control Systems Design (CACSD 2000)*, pp87-92..

Erkmen, I., Erkmen, A. M., and Günver, H., 2000, Robot Hand Preshaping and Regrasping Using Genetic Algorithms, *Int J Robotics Research*, **19**, N, pp857-874.

Fogel, L.J., Owens, A.J. and Walsh, M.J., 1966, *Artificial Intelligence Through Simulated Evolution*, New York: John Wiley.

Fonseca, C. M. and Fleming, P. J., 1998, Multiobjective Optimization and Multiple Constraint Handling with Evolutionary Algorithms – Part I: A Unified Formulation & Part II: Application Example, *IEEE Trans on Systems, Man, and Cybernetics – Part A*, **28**, pp26-37 & pp38-47.

Fonseca, C. M., Mendes, E. M., Fleming, P. J., and Billings, S. A., 1993, Non-linear model term selection with genetic algorithms, *IEE/IEEE Workshop on Natural Algorithms in Signal Processing* , **2**, pp27/1-27/8.

Gray, G. J., Murray-Smith, D. J., Li, Y., Sharman, K. C., Weinbrunner, T., 1998, Nonlinear model structure identification using genetic programming, *Control Engineering Practice*, **6**, pp1341-1352.

Holland, J. H., 1975, *Adaptation in Natural and Artificial Systems*, Ann Arbor: The University of Michigan Press.

Ichikawa, Y. and Sawa, T., 1992, Neural Network Application for Direct Feedback Controllers, *IEEE Trans on Neural Networks*, **3**, pp224-231.

Kowalczuk, Z., Suchomski, P., and Bialaszewski, T., 1999, Evolutionary Multi-Objective Pareto Optimisation of Diagnostic State Observers, *Int J Appl Maths and Comp Sci*, **9**, pp689-709.

Koza, J. R., 1992, *Genetic Programming – On the Programming of Computers by Means of Natural Selection*, The MIT Press.

Koza, J. R., Keane, M. A., Yu, J., Bennett III, F. H., Mydlowec, W., 2000, Automatic Creation of Human-Competitive Programs and Controllers by Means of Genetic Programming, *Genetic Programming and Evolvable Machines*, **1**, pp121-164.

KrishnaKumar, K., Swaminathan, R., and Montgomery, L., 1994, Multiple Optimal Solutions for Structural Control Using Genetic Algorithms with Niching, *J Guidance & Control*, **17**, pp1374-1377.

Kristinsson, K. and Dumont, G. A., 1992, System Identification and Control Using Genetic Algorithms, *IEEE Trans Systems, Man, and Cybernetics*, **22**, pp1033-1046.

Lennon, W. K. and Passino, K. M., 1999, Intelligent Control for Brake Systems, *IEEE Trans on Control Systems Technology*, **7**, pp188-202.

Leontaritis, I. J. and Billings, S. A., 1985, Input Output Parametric Models for Non-Linear Systems, Part 1: Deterministic Non-Linear Systems & Part 2: Stochastic Non-Linear Systems, *Int J Control*, **41**, pp303-328 & pp329-344.

Linkens, D. A. and Nyongesa, H. O., 1995, Genetic algorithms for fuzzy control – Part 1: Offline system development and application & Part 2: Online system development and application, *IEE Proc – Control Theory and Apps*, **142**, pp161-176 & pp177-185.

Linkens, D. A. and Nyongesa, H. O., 1996, Learning systems in intelligent control: an appraisal of fuzzy, neural and genetic control applications, *IEE Proc – Control Theory and Apps*, **143**, pp367-386.

Luh, G. C. and Wu, C. Y., 1999, Non-linear system identification using genetic algorithms, *Proc Inst Mech Eng*, **213**, Part I, pp105-118.

Marcu, T., 1998, A multiobjective evolutionary approach to pattern recognition for robust diagnosis of process faults, *Proc IFAC Conf SAFEPROCESS '97*, pp1183-1188.

Marenbach, P., Bettenhausen, K. D., Freyer, S., Nieken, U., and Rettenmaier, H., 1997, Data-driven structured modelling of a biotechnological fed-batch fermentation by mean of genetic programming, *Proc Inst Mech Eng*, **211**, Part I, pp325-332.

Marrison, C. I. and Stengel, R. F., 1997, Robust Control System Design Using Random Search and Genetic Algorithms, *IEEE Trans on Aut Control*, **42**, pp835-839.

Matsuura, K., Shiba, H., Hirotsune, M., and Nunokawa, Y., 1996, Optimal control of sensory evaluation of the sake mashing process, *J Process Control*, **6**, pp323-326.

Mei, T. X. and Goodall, R. M., 2000, LQG and GA solutions for active steering of railway vehicles, *IEE Proceedings – Control Theory and Apps*, **147**, pp111-117.

Michalewicz, Z., 1996, *Genetic Algorithms + Data Structures = Evolution Programs*, 3rd Ed., Springer-Verlag, Berlin.

Michalewicz, Z. and Fogel, D. B., 2000, *How to Solve It: Modern Heuristics*, Springer-Verlag, Berlin.

Miller, J. A., Potter, W. D., Gandham, R. V., and Lapena, C. N., An Evaluation of Local Improvement Operators for Genetic Algorithms, *IEEE Trans on Systems, Man, and Cybernetics*, **23**, pp1340-1351.

Moriyama, H. and Shimizu, K., 1996, On-line Optimisation of Culture Temperature for Ethanol Fermentation Using a Genetic Algorithm, *J Chem Technology & Biotechnology*, **66**, pp217-222.

Oliveira, P., Sequeira, J., and Sentieiro, J., 1991, Selection of Controller Parameters using Genetic Algorithms, *Engineering Systems with Intelligence. Concepts, Tools, and Applications*, pp431-438.

Onnen, C., Babuška, R., Kaymak, U., Sousa, J. M., Verbruggen, H. B., and Isermann, R., 1997, Genetic algorithms for optimization in predictive control, *Control Engineering Practice*, **5**, pp1363-1372.

Painton, L. and Campbell, J., 1995, Genetic Algorithms in Optimization of System Reliability, *IEEE Trans on Reliability*, **44**, pp172-178.

Patton, R. J., Chen, J., and Liu, G. P., 1997, Robust fault detection of dynamical systems via genetic algorithms, *Proc Inst Mech Eng*, **211**, Part I, pp357-364.

Porter, B. and Jones, A. H., 1992, Genetic tuning of Digital PID Controllers, *Electronics Letters*, **28**, pp843-844.

Rana, A. S. and Zalzala, A. M. S., 1997, Collision-free motion planning of multiarm robots using evolutionary algorithms, *Proc Inst Mech Eng*, **211**, Part I, pp373-384.

Rechenberg, I., 1973, *Evolutionsstrategie: Optimierung Technischer Systeme nach Prinzipien der Biologischen Evolution*. Sttutgart: Frommann-Holzboog.

Ren, Y. and Bechta Dugan, J., 1998, Design of Reliable Systems Using Static and Dynamic Fault Trees, *IEEE Trans on Reliability*, **47**, pp234-244.

Rodríguez-Vázguez, K., Fonseca, C. M., and Fleming P. J., 1997, Multiobjective Genetic Programming: A Nonlinear System Identification Application, *Late Breaking Papers, Genetic Programming Conference*, pp207-212.

Sette, S., Boullart, L., and Van Langenhove, L., 1998, Using genetic algorithms to design a control strategy of an industrial process, *Control Engineering Practice*, **6**, pp523-527.

Schroder, P., Green, B., Grum, N., and Fleming, P. J., 2001, On-line evolution of robust control systems: an industrial active magnetic bearing application, *Control Engineering Practice*, **9**, pp37-49.

Schwefel, H.P., 1965, *Kybernetische Evolution als Strategie der Experimentellen Forschung in der Strömungstechnik.* Dipl Thesis, Tech Univ Berlin.

Spears, W. M., De Jong, K. A., Bäck, T., Fogel, D. B., and de Garis, H., 1993, An Overview of Evolutionary Computation, Machine Learning: ECML-93, *Lecture Notes in Artificial Intelligence*, No. 667, pp442-459.

Stewart, P.G. and Fleming. P.J., 2001, The response surface methodology for real-time distributed simulation, *Proc IFAC Conference on New Technologies for Computer Control,* ibid.

Tan K.C., Lee T.H. and Khor, E.F., 2001, Automating the design of multivariable QFT control system design via evolutionary computation, *Proc Inst Mech Eng*, Part I, (in press).

Thompson, H. A., Chipperfield, A. J., Fleming, P. J., and Legge, C., 1999, Distributed aero-engine control systems architecture selection using multi-objective optimisation, *Control Engineering Practice*, **7**, pp655-664.

Van Veldhuizen, D. A. and Lamont, G. B., 2000, Multiobjective Evolutionary Algorithms: Analyzing the State-of-the-Art, *Evolutionary Computation*, **8**, pp125-147.

Vlachos, C., Williams, D., and Gomm, J. B., 1999, Genetic approach to decentralised PI controller tuning for multivariable processes, *IEE Proc – Control Theory and Apps*, **146**, pp58-64.

Wang, P. and Kwok, D. P., 1992, Autotuning of Classical PID Controllers Using an Advanced Genetic Algorithm, *Proc Int Conf on Industrial Electronics, Control, Instrumentation and Automation (IECON 92)*, **3**, pp1224-1229.

IFAC Publications
www.elsevier.com/locate/ifac

# TEMPORALLY CONSISTENT DELIVERY OF TIME-SENSITIVE INFORMATION: SOLUTIONS AND CHALLENGES

**Krithi Ramamritham**

Department of Computer Science and Engineering
Indian Institute of Technology Bombay
Mumbai, India 400076
krithi@cse.iitb.ernet.in

Abstract: The web and the internet have become the dominant infrastructures for information access and dissemination. Even time varying information such as currency and stock prices, real-time data needed for tele medicine, etc., are being delivered via the Web. Clearly, the Quality of Service needs for delivering such time-critical data are much more stringent than for static Web pages. This paper examines the nature of the dynamics on the Web, studies the suitability of the current infrastructure for disseminating time varying information, and discusses fresh approaches tailored for dynamic data. A challenging and fruitful research direction is to examine the use of a control theoretic framework to address the problem of tuning the algorithms needed for adaptive data delivery. *Copyright © 2001 IFAC*

Keywords: dynamic data, data dissemination, consistency, continuous queries, quality of service.

## I. INTRODUCTION

An ever increasing fraction of the data delivered using today's internet resources is time-varying (i.e., changes frequently). Examples of such data include sports information, news, and financial information such as stock prices. Tele medicine is another application for which the web is used. An important issue in the dissemination of time varying data in such applications, i is the maintenance of temporal coherency. The coherency requirements on a data item depends on the nature of the item and user tolerances. To illustrate, a user may be willing to receive sports and news information that may be out-of-sync by a few minutes with respect to the server, but may desire stronger coherency requirements on the status of patients' critical parameters.

The World Wide Web itself can be viewed as a global virtual database. When time varying data is stored in this database, it is not only necessary to keep this virtual database consistent with the state of the real world that it is supposed to capture, but also essential to transmit the contents of this virtual database to the consumers of the data so that the discrepancies between what the consumers see and what the real world's state is, is within acceptable tolerance. In other words, the temporal quality of the data viewed by consumers should be acceptable. We call this requirement *temporal consistency maintenance*

On the Web, the sources of data are autonomous with no regulations on the way information is represented. The data that we have to deal with is of a temporal nature, characterized by having a) rapid and b) unpredictable changes. Given a choice between the methods of obtaining data, the system must be capable of choosing the right method depending on the prevailing conditions, the major factors in the consideration being server capabilities, user tolerance and the network bandwidth available. At the same time, the user must not be made to request the same data again and again, but should be able to run continual queries on this dynamic data.

Given these challenges, we have been working towards a system with the following goals:

- It should be capable of dealing with Dynamic Data which, as we mentioned before changes rapidly and unpredictably.
- It should be able to retrieve data from multiple sources.
- It should provide high fidelity to the users. We define fidelity as the percentage of time for which user's requirements are met.
- It should be scalable with respect to the number of users, network capacity, computations, state space.
- It should be tolerant to faults at the data sources.
- It should forward only relevant information to the users.

In this paper, we discuss the progress made till now in meeting these requirements and also discuss the challenges that remain.

We start with the considerations involved in the dissemination of individual data items to understand the scope of the problem and the possible solutions. Then we examine the problem when several data items, e.g., several stock prices, need to be viewed together, introducing the need to address the issue of mutual consistency. Finally, we extend the problem domain to consider the issues arising from the need to execute arbitrary continuous queries, i.e., long-running queries, over the virtual database. We end the paper with a discussion of the challenges that remain, including a formal, perhaps control-theoretic, framework to dynamically determine the tunable parameters and to make adaptive decisions.

## II. DISSEMINATING INDIVIDUAL DYNAMIC DATA

To maintain coherency of the cached data, data as seen by a user must be periodically refreshed with the copy at the server. We assume that a temporal coherency requirement *tcr* is attached to each cached item of interest. The value of *tcr* denotes the maximum permissible deviation of the cached value from the value at the server and thus constitutes the user-specified tolerance. Let $S(t)$ and $U(t)$ denote the value of the data item at the server and the user, respectively. Then, to maintain temporal coherency (*tc*) we should have

$$|U(t) - S(t)| \leq c.$$

In addition to specifying the coherency requirement $c$, users can also specify their fidelity requirement $f$ for each data item so that an algorithm that is capable of handling users' fidelity requirements (as well as *tcrs*) can adapt to users' fidelity needs. In the following sections we explain how one can use user specified constraints to offer high fidelity with efficient use of available resources.

### *A. Client Pull vs. Server Push*

The *Pull* approach is the most traditional approach for maintaining the temporal coherency of dynamic data. In this model, each data item is assigned a certain *time-to-refresh (TTR)* when the data object is brought to a data store at the user's end, or cache. Since the arrival of the data item and until time equal to *TTR* elapses, all the requests for the data object are satisfied from the cache without looking up the values in the data sources. Thus in this approach, the user is responsible for obtaining the data from the server, by issuing a GET request to the server and the server just delivers the required data.

Many current applications do **WebCasting [11].** Users periodically poll the server for this data. This approach is equivalent to setting the *TTR* value of a cached item *statically*. However, this can lead to a very high network overheads in case the polling period is too low or may cause the user to miss some changes of interest if the polling period is too high. Clearly, this technique is useful only if rate of change of data is constant or relatively low (such as news). If the rate of change itself is varying, then this technique of assigning frequency apriori is not suitable (as in the case of stock quotes). But still, currently this is the most popular data delivery technique as it can be purely web-based (because of HTTP) and does not need any special resources (like push capability in servers or modification of HTTP).

Since dynamic data changes independently and unpredictably, we cannot use standard prediction and forecasting algorithms for predicting the next *TTR* value to be assigned to the data object. We have developed a novel method for assigning *TTR* values in [9]. Given a user's coherency requirement, this technique allows a proxy to adaptively vary the TTR value based on the rate of change of the data item. The TTR decreases dynamically when a data item starts changing rapidly and increases when a hot data item becomes cold. To achieve this objective, the *Adaptive TTR* approach takes into account (a) static bounds so that TTR values are not set too high or too low, (b) the most rapid changes that have occurred so far and (c) most recent changes to the polled data.

In the *Server* Push method, the server is responsible for delivering the relevant data to the user. The server does not behave in *request-response* mode, where the server delivers some data only when there is an explicit request for it. But instead, server pushes the data into channel without any explicit request. As before, the server can push the data either periodically or aperiodically.

In *periodic push*, the server is not aware of the exact coherency needs of a particular client, but only of the general demand for data items. So, based on the general demand distribution of data items, the server creates a schedule for the periodic dissemination of data items. A data item with higher demand will be disseminated with higher frequency and vice-versa. All data items get divided into frequency bands, where data items belonging to one frequency band have similar demands. Once the push schedule is created using these frequency bands, it is not changed. The server then repeats this schedule periodically. The Broadcast Disks [1] approach is one such approach where the frequency bands are termed as broadcast disks. This approach also provides for client feedback. An interesting property of this approach is that it treats the channel like a medium and tries to decide on the "format" in which the channel should hold the data.

It is clear that since the server is not aware of the specific needs of the users, the push schedule may not be adequate for users who desire high fidelity and specific results. So, the approach may lead to low fidelity and/or wastage of bandwidth. But, the approach is useful for data which is generic and changes not very rapidly (e.g. news, digests, entertainment).

This basic idea has been extended to adaptively disseminate data in asymmetric communication environments in [6] in such a way as to take into account the temporal properties of the data as well as users' deadline requirements. More generally, in the *aperiodic push* approach, a user registers with a server, identifying the data of interest and the associated *tcr*, i.e., the value *c*. Whenever the value of the data changes, the server uses the *tcr* value *c* to determine if the new value should be pushed to the user; only those changes that are of interest to the user (based on the *tcr*) are actually pushed.To achieve this objective, the server needs to maintain state information consisting of a list of users interested in each data item, the *tcr* of each user and the last update sent to each user.

The key advantage of this approach is that it can meet stringent coherency requirements---since the server is aware of every change, it can precisely determine which changes to push and when. A limitation of push-based servers is that the amount of state that needs to be maintained can be large, especially for popular data items. A server can optimize the state space overhead by combining requests from all proxies with identical *tcr*s into a single request; all proxies are notified if the change to the data item *D* exceeds a specified *tcr*. Even with such optimizations, the state space overhead can be excessive, which in turn limits the scalability of the server. A further limitation of the approach is that it is not resilient to failures. The state information is lost if the server fails and requires the proxy to detect the failure and re-register its *tcr* for the data item.

Thus, the Push and Pull approaches have complementary properties with respect to fidelity, network utilization, scalability and resiliency [4]:

- A pull-based approach does not offer high fidelity when the data changes rapidly or when the coherency requirements are stringent (i.e., small values of *c*). Moreover, the pull-based approach imposes a large communication overhead (in terms of the number of messages exchanged) when the number of clients is large. But it may suffice for requests which have large values of *c*.
- A push-based algorithm can offer high fidelity for rapidly changing data and/or stringent coherency requirements. However, it incurs a significant computational and state-space overhead resulting from a large number of open push connections and their serving processes/threads. Moreover, the approach is less resilient to failures due to its stateful nature.

These properties indicate that a push-based approach is suitable when a client expects its coherency requirements to be satisfied with high fidelity, or when the communication overheads are a bottleneck. A pull-based approach is better suited to less frequently changing data or for less stringent coherency requirements, and when resilience to failures is important. The complementary properties of the two approaches indicate the need for having an approach which combines the advantages of both while not suffering from any of their disadvantages.

### B. Adaptive Combinations of Push and Pull

As is clear from the previous discussion, neither push nor pull alone is sufficient for efficient dissemination of dynamic data. Since these two techniques have complementary properties, a few attempts have been made in the past to combine these two canonical techniques. *Adaptive leases* [5] and *Volume leases* [7] are two examples. The former is used for maintaining strong cache consistency in the World Wide Web while the latter is used for caches holding a large number of data items. None of these is meant for highly dynamic data. Nor do they take user requirements into account. In this section we describe the Adaptive Leases approach as well as two algorithms, PaP and PoP, that we have developed for better scalability and coherency for dynamic data.

*Leases are like contracts given to a lease holder over some property* . Whenever some client requests server for a certain document, server returns that document along with a lease. In other words, a server takes the responsibility of informing the client about any changes during the lease period. Once a lease expires, a client must contact the server and renew the lease. Client can use the cached copy while it has a valid lease over the data item. During valid lease period, client remains in push mode and is switched back to pull mode after the lease expires. Thus the client is alternatively served in push and pull modes.

Clearly, pure leases are not very useful for dynamic data. It is very important to choose a good lease period. For a very high value, client remains in push mode for most of the time and scalability problem may arise. On the other hand, for small values the lease renewal cost may prove very high. *Adaptive Leases* try to dynamically adjust the lease duration. The decision is based on many parameters like popularity of the data item, server state space available and network bandwidth available[5].

In the PaP [4]approach, the proxy operates in pull mode using some TTR algorithm, while the server is in push mode and knows the constraint. Using this constraint and user access patterns, the server tries to predict when a client is going to poll next. If it determines that within this predicted time the client is likely to miss a change of interest, it pushes that change to the client. For predicting the client connection times, the server may run the TTR algorithm in parallel with client or use some simpler approximation of it. Given the network delays, a server waits for the client to pull within a small window around the predicted time. So, once the first few changes are intimidated to the client (by pull or push), the rest of the successive changes can be pulled by the client since the client would have appropriately lowered its TTR as a result of the few pushes.

In the ideal case, the fidelity offered will be 100%, but due to synchronization problems and other factors, it will be slightly less. But, it will always be much greater than pull. Because of the pull component the resiliency of the system will be high. And due to the push component, communication overheads will also be low. PaP also provides for fine tuning of its behavior. It has a few parameters which swing it towards more push or more pull, and thus its performance in terms of fidelity and resiliency can be controlled[4].

Another possibility is to divide incoming clients at the server into either push or pull clients and dynamically switch them to one or the other mode[4]. If resources are plentiful, every client is given a push connection irrespective of its fidelity requirements. This ensures that the best fidelity is offered. As more and more clients start requesting the service, resource contention may arise at the server leading to scalability problems. Few clients are then shifted to pull mode. Thus valuable resources are freed and system scales properly. Contrary to this, when resources again become available few high priority clients are switched back to push mode thus ensuring high fidelity.

The most important issue is how to assign priorities to different clients. Few of the possible parameters are the access frequency of each client, temporal coherency requirement, fidelity requirement and network bandwidth available. Clearly no single criterion suffices but collectively they have the potential to offer high average fidelity still keeping the system scalable[4].

In summary, *PaP* achieves resiliency and temporal coherency while PoP achieves scalability by classifying users as per their capabilities and requirements. Now if we replace the push component of *PoP* with *PaP*, then we have achieved all the major objectives, namely resiliency, scalability and coherency in a single algorithm. This new algorithm is *PoPoPaP* i.e. Push or Pull or *PaP*.

## III. From Accessing Individual Data Items to Continuously Querying Data from a Domain

Till now we were dealing with systems tailored to specific instances of queries, such as "Tell me when IBM stocks go up by $c". We will now consider a system that gives users the freedom to ask any query on the database. Continual Queries (CWs) allow users to obtain new results from databases without having to issue the same query again and again. CQs are especially useful in an environment like the Web, where the amount of frequently changing information is extremely large.

A continual query (CQ) is defined as a triple $(A,T,R)$ (**A**ction, **T**rigger, **T**ermination), where $A$ is a normal query written in some traditional query language (like SQL), $T$ is the trigger condition of the CQ and $R$ is the termination condition. A CQ is "installed" when it is first specified. The initial execution of $A$ takes place on the existing data, when a CQ is installed. The result of this query is returned to the user as the first result of the CQ. The subsequent executions of $A$ depend on values of $T$ and $R$. The subsequent execution of $A$ takes place when the trigger condition $T$ becomes true, provided that the termination condition $R$ is false.

Developing the infrastructure for executing continuous queries on Web data needs several subproblems to be addressed: Consider a continuous query where a user wants to be informed when the net worth of his/her stock portfolio changes by more than a specified threshold. To this end, a user might execute the following query:

show v1=(100 * MSFT + 200 * INTL + 300 * IBM)
for stocks when change(v1) > $500 till end-of-the-month;

Executing this query implies obtaining the current prices of MSFT, INTL and IBM from (their respective) web sources

- using one of the approaches mentioned earlier (Pull, PaP, PoP and PoPoPaP, etc.),

- in such a way that the data is mutually consistent, and
- such that users view of stock portfolio is the same as the actual real-world state of the portfolio.

Formally, the following equation should be satisfied to achieve fidelity

$$if \sum_{all\ stocks} |S(t)xN(t) - S(t')|xN(t) \geq threshold$$

then

$$\sum_{all\ stocks} |C(t)xN(t) - S(t')xN(t)| \geq threshold$$

Where

- *S(t)* is the current stock price at the server
- *S(t')* is the initial stock price at the server
- *C(t)* is the current stock price at the client
- *N(t)* is the number of shares of stock in the portfolio
- *threshold* specifies the coherency requirement for the stock portfolio.

It may happen that due to loss of coherency, the user may perceive the wrong state for his/her portfolio. There are two classes of misperceptions: (i) *false positive*, when a user thinks that the threshold has been exceeded, whereas it is well within the threshold, and (ii) *false negative*, when the portfolio changes have actually exceeded the threshold but the user is unaware of this. In some sense, the latter could be more harmful than the former because in the case of a *false positive* the proxy can always verify the value of the portfolio by polling all the stocks. But no such solution exists in case of false negatives. So it is important to avoid or at least minimize false negatives. We now discuss approaches to achieving these diverse requirements.

As we have seen earlier, the HTTP protocol used in the Web is inherently pull based. Users need to frequently pull the data from the server based on the dynamics of the data and the coherency requirement of the user. The proxy can use the if-modified-since HTTP requests to obtain any changes in the data value at the server. Using push, the number of false positives and false negatives can be made small, barring network and computational delays. But, very few servers provide push support and this service often comes at a price, namely space and computation overheads at the server. Also, the servers have to be re-engineered to deal with portfolio queries. A final reason for using pull based technique is that the push based approach will require cooperation between the servers of the individual data items, and this may not be feasible. For these reasons, our goal is to work within the standard HTTP protocol and hence use pulls intelligently to maintain the coherence of portfolios within user allowed thresholds values.

To achieve this, we must impose a coherency requirement on each of the data items constituting a portfolio, depending on (a) the overall contribution of the data item to the portfolio and (b) the coherency requirement of the portfolio itself. A very naive way to assign *cr*s to data items would be to assign a small (i.e., very stringent) coherency to a data item that makes the greatest contribution to the portfolio (and hence such a data item will be polled more often). But the problem with this approach is that it does not consider the dynamics of the data items. Depending on the way the values of the various data items are changing the *cr*s of each of the data items can be manipulated, thereby saving a lot of network overhead. For example, consider a portfolio consisting of 100 stocks of company A and 400 stocks of company B. Now if the company B stock price hasn't changed for a long time and the company A stock sees a lot of trading and hence its price fluctuates a lot, then the *cr* of company A should be reduced and that of company B should be increased. This will help in reducing the false negatives and it will also help in saving a lot of unnecessary pulls of the company B stock. Hence there is a need to judiciously assign the *cr* to each of the data items depending on the dynamics of the data. We have developed an algorithm [2] that continuously changes the *cr* associated with each of the data item constituting the portfolio based on the dynamics of the data, as well as the relative performance of the data item with respect to the other items in the portfolio. If false positives and false negatives are not to occur then the proxy must present a logically consistent view of the data items cached [10]. Our algorithm achieves this while aiming to reduce the network overhead – by pulling very rarely when there is very little chance of the threshold being exceeded. The algorithm uses the *Adaptive TTR Algorithm* [9] to compute the time after which a data item should be polled so that the coherency associated with that data item is not violated. How to achieve this is the crux of [2], wherein we provide experimental evidence for the efficiency of the approach. The algorithm requires some amount of history to be kept, which can be easily done. A useful feature of our technique is that it has several adjustable parameters which can be used to tune the algorithm to get the desired fidelity and network overhead characteristics.

If a client communicates the *cr* associated with a data item to the source of that data item, we could even make the server push the changes to the client according to the specified *cr*. We are currently developing an algorithm that adjusts the *cr*s in a fashion similar to that of pull based approaches, but in this case, the adjusted *cr* is communicated to the server of the data. It is even possible to envisage a non-symmetric window of temporal coherency as opposed to the symmetric window implied by the *cr* based constraint.

In summary, we achieve our objective by considering the stocks in a portfolio as a semantic unit, in the sense that it reduces the polling frequency of all the stock of the portfolio if the stock prices are changing such that there is very little chance of the portfolio being satisfied. Whereas developed in the context of financial information such as stock prices, our approach can also be used to answer similar queries on any kind of dynamic data (e.g., traffic conditions, sensor data).

## IV. Towards the Continuous Querying of Arbitrary Time-varying Data

Finally, in this section, we discuss a proposal for a full fledged system to handle continuous queries over a virtual database containing data from arbitrary data sources. For example, a patient status monitoring system, unlike a stock portfolio monitoring system, will have to cater to diverse data such as blood pressure, body temperature, etc. each measured in its own unit and not combinable in the way a portfolio was built from individual stock prices wherein all contributors were measured using the same unit. Since such a system must "think" on the fly and adapt its decision making, we have named it DeepThought, which can be configured as an appendage to a proxy on the Web.

The main features (and design goals) of the DeepThought system are:

- **Tailored**. The system is specifically designed for dynamic data. All operations are *memory based* and the system tries to minimize the time taken for processing at the proxy. Also, the delivery of data from the server to the proxy will be governed according to the query conditions with minimum possible delays. The DeepThought query language is also specifically tailored for dynamic data.
- **Distributed**. To ensure scalability and availability on the Internet scale.
- **Flexible**. Since there are various algorithms for data delivery between the server and proxy, the system chooses the best algorithm for a particular query/connection combination. DeepThought can use any of Pull, PoP, PaP or PoPoPaP for data retrieval.

Figure 1 shows the complete structure of the DeepThought system. The main interface of the DeepThought system to the user side is via the CQ Server Interface (CQSI), which may be a separate machine or a router machine. The user initially connects with the CQSI and asks for service. If the CQSI determines that the user can be serviced (depending on load at the proxy), then it responds with the query interface, otherwise a negative acknowledgement is sent and the user may attempt again later. The user, after receiving the query interface, enters the query (either manually or using some tool) and this query is sent to CQSI as a text string. We are not including the syntax description of DTQL, the DeepThought Query language for want of space, but it is a language specifically tailored for dynamic data. The following actions are taken after this,

- When a CQ is issued, the CQ Parser (CQP) breaks it up into A, T and R,
- The CQ Installer installs the CQ
- A thread/process called CQ Thread Monitor (CQTM) is started for the further processing of the CQ. This is preferably done on another machine in order to achieve *load sharing*. The CQTM does the job of evaluating the action part (A) of the CQ and coordinate the various tasks between components.
- the CQSI returns the CQ identifier to client as well as the chosen CQTM machine identity. The client now connects to the chosen CQTM and downloads an applet to obtain data from the machine.

The Source Monitor (SM) is the component which actually brings the data from the data source. There is only one SM for each data item in the whole system. This saves a lot of effort and bandwidth. SM uses an existing algorithm module to deliver data to TH. The algorithm modules may be for algorithms like Pull, Push, PaP, PoP [9],[4] etc. The algorithm module is pluggable, and SM provides a uniform interface for registry of the module, negotiation of the parameters and data delivery. The SM chooses the best algorithm depending on various parameters like server capabilities, volume of data through the connection, load on the SM etc. So we can use the most suitable algorithm for data retrieval. The SM is a separate process which may reside on a different machine (thus improving the bandwidth utilization).

Once the CQ has been parsed and installed, the actual execution of the query starts. The trigger, T, is installed by a separate entity called the Trigger Handler, following which:

- The CQTM passes the trigger part T to TH, together with some meta information and the list of items to be monitored.
- The TH installs T and gets it ready for repetitive execution.
- For each of the data items TH tries to determine if there is already a Source Monitor for a data item. If there is, then it is used or a new SM must be started. The SM pushes a new data value for a data item to the TH, whenever it is available.
- When new data value arrives, the TH evaluates the trigger T and the termination condition R, and returns the resulting data to the CQTM which can then take action A on it.

When the SM uses Pull-based algorithms, TH conveys polling intervals to the SM for each data item. For Push-based algorithms, this is not required. This must be done by TH since the final goal of the trigger is known only to TH through the trigger

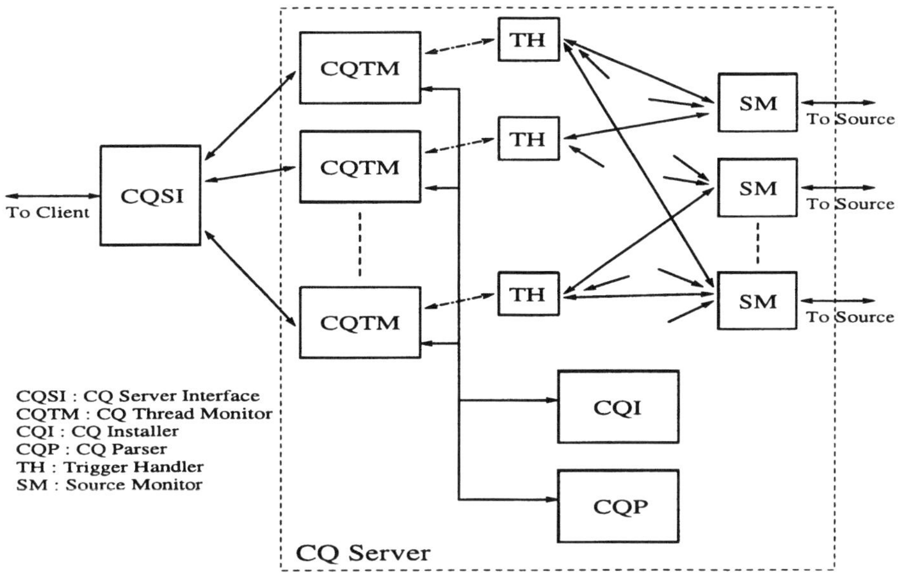


Fig. 1. The DeepThought System Design

condition T. Depending on T, TH decides polling frequencies for each of the data items in T. The process of determining polling frequencies will be a variation of the one developed for portfolio queries. We are in the process of developing such a system and experience with its functioning will be reported in a future article.

## V. Challenges in Handling Dynamic Data on the Web

Pull and push approaches are typically used to retrieve data from the Web. Due to the specific nature of dynamic data, none of these approaches is a perfect fit: While pull suffers from lack of fidelity, push suffers from lack of resiliency and scalability. Our work demonstrates the need for various combinations of pull and the push approaches. Most notably, we show how fidelity, resiliency and scalability of the system improves by using PaP and PoP. This science of combination of push and pull can be put to innovative use in continual queries made on dynamic data. Our continual query system is probably the first one to deal specifically with dynamic data.

But there are many challenges in catering to the many complex types of queries that are likely to arise. For example, the queries that have been seen so far only consider positive threshold levels. In the general case, there could be portfolio queries wherein the proxy could be asked to monitor the difference between two data items. Such queries can be called *differential queries*. Examples of such kind of queries could be "Tell me when the difference between the price of Cisco stock and Intel stock becomes greater than $ 20" [10] or "Tell me when the blood pressure of a patient goes up by 10 units and his pulse crosses 100". Another type of query is what could be called a *gradient query*, whose result is based on the trend in data values. The problem in using the standard methods for differential queries is that constraints cannot be assigned directly to the constituent stocks. One way to handle such queries is to assign individual thresholds to each of the stocks. Further, the slope of the price change of the stocks also needs to be monitored and based on the direction of this change relative to the other changes, the thresholds need to be adjusted. With portfolios constituting of many stocks, the problem can be quite computationally expensive and efficient mechanisms need to be devised to solve this.

An astute reader would have noted the feedback driven nature of many of the decisions made by algorithms such as PoP, PaP, as well as their continual query counterparts. One of the challenges is the adjustment of the parameters used by these algorithms in order to achieve the feedback. The parameters should themselves adapt taking into account the current input and system dynamics. Currently we are investigating the usefulness of utilizing control theoretic principles to address the tuning of the parameters that drive such feedback. The talk will illustrate the results of this line of research.

There are several topics which deserve further attention in the context of time sensitive data dissemination:

- In a proxy/cache based environment, a proxy acts as a client to the provider of its data and as a server to its clients downstream. Each such server-client pair needs to decide the dissemination approach to use between them. Making these decisions in a decentralized fashion while globally achieving temporal coherency with low overheads and high resiliency is an interesting problem.
- We need to develop algorithms for disseminating time sensitive content through a content distribution network (CDN). Currently deployed CDNs are

mostly concerned with storing static content, such as images, close to the potential clients of such data. It is perhaps even more important to use such an infrastructure for time varying data because of the potential for CDNs (and the underlying application level multicasting) to reduce communication and computation overheads.

- Support for disseminating time sensitive data in mobile and broadcast environments is another challenge. Given the wireless bandwidth restrictions in such environments, it is important to cater to user needs without explicit requests from users. This calls for push based approaches, but the challenge lies in obtaining user request profiles on a dynamic basis. An extension of the approach suggested in [?] might prove useful: a fraction of the bandwidth is allocated to each mode, periodic broadcast and on-demand broadcast. (This can be viewed as the broadcast analog of PoP.) The fraction allocated to each mode is varied with time depending on the access frequency distribution of data. To achieve this, the server needs some feedback from the clients about their access patterns.

**Acknowledgements**

This work was supported by Tata Consultancy Services through their funding for the Center for Intelligent Internet Research at IIT Bombay. Additional support was provided by the National Science Foundation Grant CCR-0098060. I am grateful to Mukund Agarwal, Manish Bhide, and Faizaan Marolia for their comments on an earlier version of this paper and to Prashant Shenoy for his collaboration on several aspects of the work reported here.

**REFERENCES**

[1] S. Acharya, M.J. Franklin and S.B. Zdonik: Balancing Push and Pull for Data Broadcast, *Proceedings of the ACM SIGMOD Conference, May 1997.*

[2] M. Bhide, K. Ramamritham and P.~Shenoy, Efficiently Maintaining Stock Portfoilis up-to-date on the Web, (to appear in) *IEEE Research Issues in Data Engineering (RIDE'02) workshop, March 2001.*

[3] J. Chen, D. Dewitt, F. Tian and Y. Wang, NiagraCQ: A Scalable Continuous Query System for Internet Databases}. *Proceedings of the 2000 ACM SIGMOD International Conference on Management of Data, May 16-18, 2000, Dallas, Texas, USA.*

[4] P. Deolasee, A. Katkar, A. Panchbudhe, K. Ramamritham and P. Shenoy, Adaptive Push-Pull: Dissemination of Dynamic Web Data, 10th International World Wide Web Conference , Hong Kong, May 2001.

[5] V. Duvvuri, P. Shenoy and R. Tewari, *Adaptive Leases: A Strong Consistency Mechanism for the World Wide Web. InfoCom March 2000.*

[6] J. Fernandez and K. Ramamritham, Adaptive Dissemination of Data in Asymmetric Communication Environments, ACM Mobile Networks Applications Journal (to appear).

[7] Krishnamurthy and C. Wills: Piggyback Server Invalidation for Proxy Cache Coherency. Proc. World Wide Web Conference, April 1998.

[8] L. Liu, C. Pu and W. Tang, Continual Queries for Internet Scale Event-Driven Information Delivery}, *IEEE Trans. on Knowledge and Data Engg., July/August 1999.*

[9] R. Srinivasan, C. Liang and K. Ramamritham, Maintaining Temporal Coherency of Virtual Data Warehouses, *The 19th IEEE Real-Time Systems Symposium (RTSS98), Madrid, Spain, December 2-4, 1998.*

[10] B. Urgaonkar, A. Ninan, M. Raunak, P. Shenoy and K. Ramamritham, Maintaining Mutual Consistency for Cached Web Objects *International Conference on Distributed Computing Systems , April 2001.*

[11] M.J. Franklin and S.B. Zdonik: "Data In Your Face": Push Technology in Perspective. SIGMOD Conference, Seattle, Washington, May-June 1998.

IFAC Publications
www.elsevier.com/locate/ifac

# DATA-DRIVEN CONSTRUCTION OF TRANSPARENT NEURO-FUZZY SYSTEMS

**Robert Babuška and Henk Verbruggen**

*Delft University of Technology, Faculty ITS, Systems and Control Engineering Group, Mekelweg 4, P.O. Box 5031, 2600 GA Delft The Netherlands, tel: +31 15 27 85117, fax: +31 15 27 86679
e-mail: r.babuska@its.tudelft.nl*

Abstract: In recent years, we have witnessed a strong emphasis on high performance and precision of fuzzy and neuro-fuzzy systems. Many publications are focused on data driven approaches, i.e., the construction of such systems from data and applying them in areas like data mining, pattern recognition, prediction or control. In these applications, neuro-fuzzy system inevitably must be compared with other inductive methods, like neural networks, machine learning or statistical techniques. The most prominent feature that distinguishes neuro-fuzzy systems from many other techniques is their transparency and interpretability, which as assumed to be automatically "inherited" from linguistic fuzzy models. In the current literature, however, surprisingly little attention has been devoted to the study of the interplay between interpretability and precision. These objectives are to a certain degree conflicting and attention must be paid to both of them. This paper addresses these issues in the context of parameter estimation in neuro-fuzzy systems identification. *Copyright © 2001 IFAC*

Keywords: neuro-fuzzy systems, nonlinear identification, ANFIS network, Takagi–Sugeno fuzzy system.

## 1. INTRODUCTION

Neuro-fuzzy modeling has been recognized as a powerful tool that can facilitate the effective development of models by combining information from different sources, such as empirical laws, heuristics and data. Neuro-fuzzy models describe systems by means of fuzzy *if-then* rules in the form '*If $x$ is small then $y$ is large*.' At the computational level, these rules are represented as a network, to which learning algorithms known from the area of artificial neural networks can be applied. Thanks to this structure, neuro-fuzzy models are to a certain degree transparent to interpretation and analysis, i.e., can be better used to explain solutions to users than completely black-box models such as neural networks.

Little attention has been so far devoted to the study of the interplay between accuracy of the model and its interpretability in terms of relevant if-then rules (Setnes *et al.*, 1998). In this paper, an overview of fuzzy systems and neural networks is first given in Section 2. Neuro-fuzzy systems are then addressed in Section 3. Various methods for parameter estimation are given in Section 4 and two numerical examples can be found in Section 5: approximation of a static function and dynamic modeling of the pH process. Section 6 concludes the paper.

## 2. FUZZY SYSTEMS AND NEURAL NETWORKS

Both neural networks and fuzzy systems are motivated by imitating human reasoning processes. In fuzzy systems, relationships are represented explicitly in the form of if–then rules. In neural networks, the relations are not explicitly given, but are 'coded' in the network and its parameters. In contrast to knowledge-based techniques, no explicit knowledge is needed for the application of neural nets.

Neuro–fuzzy systems combine the semantic transparency of rule-based fuzzy systems with the learning capability of neural networks. This section gives the background on nonlinear input–output modeling, fuzzy systems and neural nets, which is essential for understanding the rest of this article.

## 2.1 *Nonlinear System Identification*

A wide class of nonlinear dynamic systems with an input $u$ and an output $y$ can be described in discrete time by the NARX (nonlinear autoregressive with exogenous input) input–output model:

$$y(k+1) = f(\mathbf{x}(k)) \quad \text{with} \quad \mathbf{x}(k) = \qquad (1)$$
$$[y(k) \ \dots \ y(k-n_y+1) \ u(k) \ \dots \ u(k-n_u+1)]^T$$

where $y(k+1)$ denotes the output predicted at the future time instant $k+1$ and $\mathbf{x}(k)$ is the *regressor* vector, consisting of a finite number of past inputs and outputs. The dynamic order of the system is represented by the number of lags $n_u$ and $n_y$. Although for simplicity stated with a scalar input and output, the NARX model can also be used for multivariable systems. In that case, however, the number of regressors usually becomes large and one may prefer the nonlinear state-space description:

$$\begin{aligned} \boldsymbol{\xi}(k+1) &= g(\boldsymbol{\xi}(k), \mathbf{u}(k)) \\ \mathbf{y}(k) &= h(\boldsymbol{\xi}(k)) \end{aligned} \qquad (2)$$

The problem of nonlinear system identification is to infer the unknown function $f$ in (1) or the functions $g$ and $h$ in (2) from some sampled data sequences $\{(u(k), y(k)) \mid k = 1, 2, \dots, N\}$.

In black-box modeling, these functions are approximated by some general function approximators such as neural networks, neuro-fuzzy systems, splines, interpolated look-up tables, etc. If the aim of modeling is only to obtain an accurate predictor for $y$, there is not much difference between these models, as they all can approximate smooth nonlinear systems arbitrarily well. Often, however, besides accurate predictions, one wants to have a model that can be used to learn something about the underlying system and analyze its properties. From this point of view, fuzzy and neuro-fuzzy systems are more transparent than most other black-box techniques.

## 2.2 *Fuzzy Models*

A mathematical model which in some way uses fuzzy sets is called a *fuzzy model*. In system identification, rule-based fuzzy models are usually applied. In these models, the relationships between variables are represented by means of if–then rules with imprecise (ambiguous) predicates, such as:

If *heating is high* then *temperature will increase fast.*

This rule defines in a rather qualitative way the relationship between the heating and the temperature in a room, for instance. To make such a model operational, the meaning of the terms 'high' and 'fast' must be defined more precisely. This is done by using fuzzy sets, i.e., sets where the membership is changing gradually rather than in an abrupt way. Fuzzy sets are defined through their membership functions (denoted by $\mu$) which map the elements of the considered universe to the unit interval $[0, 1]$. The extreme values 0 and 1 denote complete membership and non-membership, respectively, while a degree between 0 and 1 means partial membership in the fuzzy set. Depending on the structure of the if–then rules, two main types of fuzzy models can be distinguished: the Mamdani (or linguistic) model and the Takagi–Sugeno model.

### 2.2.1. *Mamdani Model.*

In this model, the antecedent (if-part of the rule) and the consequent (then-part of the rule) are fuzzy propositions:

$$\mathcal{R}_i\text{: If } x \text{ is } A_i \text{ then } y \text{ is } B_i, \quad i = 1, 2, \dots, K. \qquad (3)$$

Here $A_i$ and $B_i$ are the antecedent and consequent linguistic terms (such as 'small', 'large', etc.), represented by fuzzy sets, and $K$ is the number of rules in the model. The linguistic fuzzy model is useful for representing qualitative knowledge, illustrated in the following example.

*Example 1.* Consider a qualitative description of the relationship between the oxygen supply to a gas burner ($x$) and its heating power ($y$):

$\mathcal{R}_1$: If $O_2$ flow rate is *Low* then power is *Low*.
$\mathcal{R}_2$: If $O_2$ flow rate is *OK* then power is *High*.
$\mathcal{R}_3$: If $O_2$ flow rate is *High* then power is *Low*.

The meaning of the linguistic terms {*Low, OK, High*} and {*Low, High*} is defined by membership functions such as the ones depicted in Fig. 1. Membership functions can be defined by the model developer based on prior knowledge or by using data (in this example, the membership functions and their domains are selected quite arbitrarily).

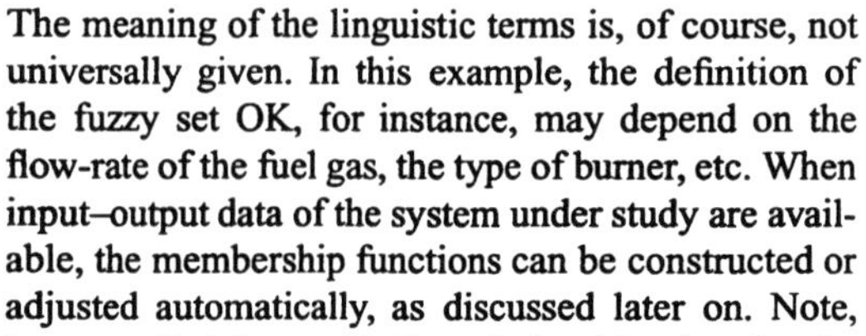


Fig. 1. Membership functions for the Mamdani model.

The meaning of the linguistic terms is, of course, not universally given. In this example, the definition of the fuzzy set OK, for instance, may depend on the flow-rate of the fuel gas, the type of burner, etc. When input–output data of the system under study are available, the membership functions can be constructed or adjusted automatically, as discussed later on. Note, however, that the qualitative relationship given by the rules is usually expected to be valid for a range of conditions. □

2.2.2. *Takagi–Sugeno Model.* The Mamdani model is typically used in knowledge-based (expert) systems. In data-driven identification, the model due to Takagi and Sugeno has become popular. In this model, the antecedent is defined in the same way as above, while the consequent is an affine linear function of the input variables:

$$\mathcal{R}_i\text{: If } \mathbf{x} \text{ is } A_i \text{ then } y_i = \mathbf{a}_i^T\mathbf{x} + b_i, \quad i = 1, \ldots, K, \tag{4}$$

where $\mathbf{a}_i$ is the consequent parameter vector and $b_i$ is a scalar offset. This model combines a linguistic description with standard functional regression: the antecedents describe fuzzy regions in the input space in which the consequent functions are valid. The output $y$ is computed by taking the weighted average of the individual rules' contributions:

$$y = \frac{\sum_{i=1}^{K} \beta_i(\mathbf{x}) y_i}{\sum_{i=1}^{K} \beta_i(\mathbf{x})} = \frac{\sum_{i=1}^{K} \beta_i(\mathbf{x})(\mathbf{a}_i^T\mathbf{x} + b_i)}{\sum_{i=1}^{K} \beta_i(\mathbf{x})} \tag{5}$$

where $\beta_i(\mathbf{x})$ is the *degree of fulfillment* of the $i$th rule. For the rule (4), $\beta_i(\mathbf{x}) = \mu_{A_i}(\mathbf{x})$, but it can also be a more complicated expression, as shown later on. The antecedent fuzzy sets are usually defined to describe distinct, partly overlapping regions in the input space. The parameters $\mathbf{a}_i$ are then (approximate) local linear models of the considered nonlinear system. The TS model can thus be regarded as a smooth piece-wise linear approximation of a nonlinear function or a parameter-scheduling model. Note that the antecedent and consequent variables may be different.

*Example 2.* Consider a static characteristic of an actuator with a dead-zone and a non-symmetrical response for positive and negative inputs. Such a system can conveniently be represented by a TS model with three rules each covering a subset of the operating domain that can be approximated by a local linear model, see Fig. 2. The corresponding rules and the input-output equation are:

$$\begin{aligned}
\mathcal{R}_1 &: \text{If } u \text{ is } \textit{Negative} \text{ then } y_1 = a_1 u + b_1 \\
\mathcal{R}_2 &: \text{If } u \text{ is } \textit{Zero} \text{ then } y_2 = a_2 u + b_2 \\
\mathcal{R}_3 &: \text{If } u \text{ is } \textit{Positive} \text{ then } y_3 = a_3 u + b_3
\end{aligned}$$

$$y = \frac{\mu_{Neg}(u) y_1 + \mu_{Zero}(u) y_2 + \mu_{Pos}(u) y_3}{\mu_{Neg}(u) + \mu_{Zero}(u) + \mu_{Pos}(u)}$$

□

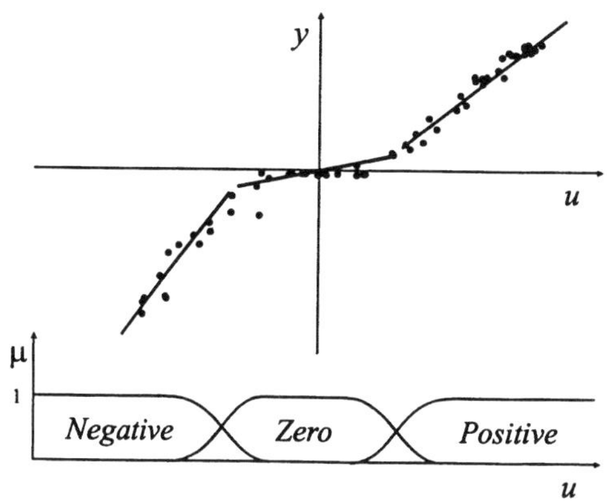


Fig. 2. A Takagi–Sugeno fuzzy model as a piece-wise linear approximation of a nonlinear system.

As the consequent parameters are first-order polynomials in the input variables, model (4) is in the literature also called the *first-order TS model.* This is in order to distinguish it from the zero-order TS model whose consequents are simply constants (zero-order polynomials):

$$\mathcal{R}_i\text{: If } \mathbf{x} \text{ is } A_i \text{ then } y_i = b_i, \qquad i = 1, 2, \ldots, K\,. \tag{6}$$

For this model, the input–output equation (5) reduces to:

$$y = \frac{\sum_{i=1}^{K} \beta_i(\mathbf{x}) b_i}{\sum_{i=1}^{K} \beta_i(\mathbf{x})} \tag{7}$$

The above model can also be obtained as a special case of the Mamdani system (3) in which the consequent fuzzy sets degenerate to singletons (real numbers):

$$\mu_{B_i}(y) = \begin{cases} 1, & \text{if } y = b_i, \\ 0, & \text{otherwise}. \end{cases} \tag{8}$$

2.2.3. *Fuzzy Logic Operators.* In fuzzy systems with multiple inputs, the antecedent proposition is usually represented as a combination of terms with univariate membership functions, by using logic operators 'and' (conjunction), 'or' (disjunction) and 'not' (complement). In fuzzy set theory, several families of operators have been introduced for these logical connectives. Table 1 shows the two most common ones.

Table 1. Commonly used functions for fuzzy logic operators.

| | $A$ and $B$ | $A$ or $B$ | not $A$ |
|---|---|---|---|
| Zadeh | $\min(\mu_A, \mu_B)$ | $\max(\mu_A, \mu_B)$ | $1 - \mu_A$ |
| probabilistic | $\mu_A \cdot \mu_B$ | $\mu_A + \mu_B - \mu_A \cdot \mu_B$ | $1 - \mu_A$ |

As an example, consider the commonly used *conjunctive form* of the antecedent, which is given by:

$$\begin{aligned}
\mathcal{R}_i &: \text{If } x_1 \text{ is } A_{i1} \text{ and } \ldots \text{ and } x_p \text{ is } A_{ip} \\
&\quad \text{then } y_i = \mathbf{a}_i^T\mathbf{x} + b_i
\end{aligned} \tag{9}$$

with the degree of fulfillment

$$\begin{aligned}
&\beta_i(\mathbf{x}) = \min\left(\mu_{A_{i1}}(x_1), \ldots, \mu_{A_{ip}}(x_p)\right) \\
&\quad \text{or} \\
&\beta_i(\mathbf{x}) = \mu_{A_{i1}}(x_1) \cdot \mu_{A_{i2}}(x_2) \cdots \mu_{A_{ip}}(x_p)
\end{aligned} \tag{10}$$

for the minimum and product conjunction operators, respectively. The complete set of rules (9) divide the input domain into a lattice of overlapping axis-parallel hyperboxes. Each of these hyperboxes is a Cartesian product intersection of the corresponding univariate fuzzy sets.

2.2.4. *Dynamic Fuzzy Models.* In the modeling of dynamic systems, fuzzy models are used to parameterize of the nonlinear functions $f$ in (1) or $g$ and $h$ in (2). Consider, for instance, the TS NARX model:

$$\mathcal{R}_i: \text{ If } \mathbf{x}(k) \text{ is } A_i \text{ then}$$
$$y_i(k+1) = \sum_{j=1}^{n_y} a_{ij} y(k-j+1) \quad (11)$$
$$+ \sum_{j=1}^{n_u} b_{ij} u(k-j+1) + c_i$$

where the antecedent regressor $\mathbf{x}(k)$ is generally given by (1), but it may of course contain only some of the past inputs and outputs or even other variables than $u$ and $y$. Similarly, state-space models can be represented in the TS framework by:

$$\mathcal{R}_i: \text{ If } \boldsymbol{\xi}(k) \text{ is } A_i \text{ and } \mathbf{u}(k) \text{ is } B_i \text{ then} \quad (12)$$
$$\begin{cases} \boldsymbol{\xi}_i(k+1) = \boldsymbol{\Phi}_i \boldsymbol{\xi}(k) + \boldsymbol{\Gamma}_i \mathbf{u}(k) + \mathbf{a}_i \\ \mathbf{y}_i(k) = \mathbf{C}_i \boldsymbol{\xi}(k) + \mathbf{c}_i \end{cases}$$

An advantage of the state-space modeling approach is that the structure of the model can easily be related to the physical structure of the real system, and, consequently, the model parameters are physically relevant. This is not necessarily the case with input-output models. In addition, the dimension of the regression problem in state-space modeling is often smaller than with input–output models.

## 2.3 *Artificial Neural Networks*

Artificial neural nets (ANNs), originally inspired by the functionality of biological neural networks can learn complex functional relations by generalizing from a limited amount of training data. Neural nets can thus serve as black-box models of nonlinear, multivariable static and dynamic systems and can be trained by using input–output data observed on the system. The most common ANNs consist of several layers of simple processing elements called neurons, interconnections among them and weights assigned to these interconnections. The information relevant to the input–output mapping of the net is stored in the weights.

2.3.1. *Multi-Layer Neural Network.* A feedforward multi-layer neural network (MNN) has one input layer, one output layer and an number of hidden layers between them. For illustration purposes, consider a MNN with one hidden layer (Fig. 3).

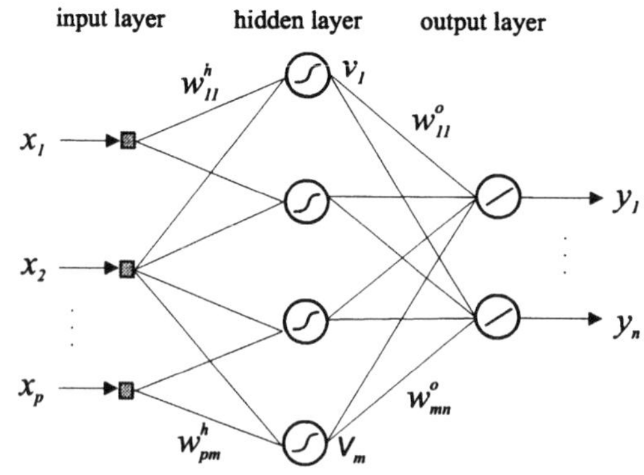


Fig. 3. A feedforward with one hidden layer.

The input-layer neurons do not perform any computations, they merely distribute the inputs $x_i$ to the weights $w^h_{ij}$ of the hidden layer. In the neurons of the hidden layer, first the weighted sum of the inputs is computed

$$z_j = \sum_{i=1}^{p} w^h_{ij} x_i = (\mathbf{w}^h_j)^T \mathbf{x}, \quad j = 1, \ldots, m. \quad (13)$$

It is then passed through a nonlinear *activation* function, such as the tangent hyperbolic:

$$v_j = \frac{1 - \exp(-2z_j)}{1 + \exp(-2z_j)}, \quad j = 1, 2, \ldots, m. \quad (14)$$

Other typical activation functions are the threshold function (hard limiter) and the sigmoidal function. The neurons in the output layer are linear, i.e., only compute the weighted sum of their inputs:

$$y_l = \sum_{j=1}^{h} w^o_{jl} v_j = (\mathbf{w}^o_l)^T \mathbf{x}, \quad l = 1, \ldots, n. \quad (15)$$

Training is the adaptation of weights in a multi-layer network such that the error between the desired output and the network output is minimized. Two steps are distinguished in this procedure:

(1) *Feedforward computation.* From the network inputs $x_i$, the outputs of the first hidden layer are first computed. Then using these values as inputs to the second hidden layer, the outputs of this layer are computed, etc. Finally, the output of the network is obtained.
(2) *Weight adaptation.* The output of the network is compared to the desired output. The difference of these two values, the error, is then used to adjust the weights first in the output layer, then in the layer before, etc., in order to decrease the error (gradient-descent optimization). This backward computation is called error backpropagation (Rumelhart *et al.*, 1986).

A network with one hidden layer is sufficient for most approximation tasks. More layers can give a better fit, but the training takes longer. Choosing the right number of neurons in the hidden layer is essential for a good result. Too few neurons give a poor fit, while too many neurons result in overtraining of the net (poor generalization to unseen data). A compromise is usually sought by trial and error methods.

2.3.2. *Dynamic Neural Networks.* A dynamic network can be realized by using a static feedforward network combined with an external feedback connection. The output of the network is fed back to its input through delay operators $z^{-1}$. This is in fact a realization of the NARX model (1). Figure 4a shows an example of a first-order system $y(k+1) = f_{\rm nn}(y(k), u(k))$.

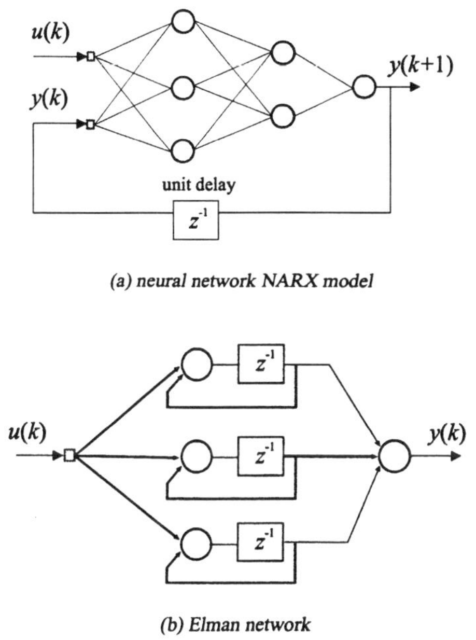


Fig. 4. Neural-network models of dynamic systems.

Another possibility is to use recurrent networks in which neurons are arranged in one or more layers and feedback is introduced either internally in the neurons, to other neurons in the same layer, or to neurons in preceding layers. Examples of these networks are the Elman network (Fig. 4b) or the Hopfield network.

2.3.3. *Error Backpropagation.* Consider for simplicity a MNN with one output. A set of $N$ input-output data pairs $\{(\mathbf{x}_k, y_k^*) \mid k = 1, 2, \ldots, N\}$ is available. We represent this set as a matrix $\mathbf{X} \in \mathbb{R}^{N \times p}$, having the input vectors $\mathbf{x}_k$ in its rows, and a column vector $\mathbf{y}^* \in \mathbb{R}^N$, containing the desired outputs $y_k^*$:

$$\mathbf{X} = [\mathbf{x}_1, \ldots, \mathbf{x}_N]^T, \quad \mathbf{y}^* = [y_1^*, \ldots, y_N^*]^T. \quad (16)$$

The difference between the desired output $y^*$ and the output of the network $y$ is called the error. This error is used to adjust the weights in the net via the minimization of the following cost function:

$$J = \frac{1}{2} \sum_{k=1}^{N} e_k^2 \quad \text{with} \quad e_k = y_k^* - y_k\,.$$

Note that the network's output $y$ is nonlinear in the weights $\mathbf{w}$ (for notational convenience, all the weights are lumped in a single vector $\mathbf{w}$). The training of a MNN is thus a *nonlinear optimization problem* to which various methods can be applied:

- Error backpropagation (first-order gradient).
- Newton, Levenberg-Marquardt methods (second-order gradient).
- Genetic algorithms and many others techniques

First-order gradient methods are based on the following general update rule for the weights:

$$\mathbf{w}(n+1) = \mathbf{w}(n) - \alpha(n) \nabla J(\mathbf{w}(n)), \quad (17)$$

where $\mathbf{w}(n)$ is the weight vector at iteration $n$, $\alpha(n)$ is a (variable) learning rate (a parameter) and $\nabla J(\mathbf{w})$ is the Jacobian of the network

$$\nabla J(\mathbf{w}) = \left[ \frac{\partial J(\mathbf{w})}{\partial w_1}, \frac{\partial J(\mathbf{w})}{\partial w_2}, \ldots, \frac{\partial J(\mathbf{w})}{\partial w_M} \right]^T . (18)$$

The nonlinear optimization problem is thus solved by using the first term of its Taylor series expansion (the gradient). Second-order gradient methods make use of the second term as well:

$$J(\mathbf{w}) \approx J(\mathbf{w}_0) + \nabla J(\mathbf{w}_0)^T (\mathbf{w} - \mathbf{w}_0) + \frac{1}{2} (\mathbf{w} - \mathbf{w}_0)^T \mathbf{H}(\mathbf{w}_0)(\mathbf{w} - \mathbf{w}_0),$$

where $\mathbf{H}(\mathbf{w}_0)$ is the Hessian. Second order methods are usually more effective than first-order ones.

2.3.4. *Radial Basis Function Network.* The radial basis function (RBF) network is a two-layer network with an architecture depicted in Fig. 5. This network is represented by the following function:

$$y = f(\mathbf{x}) = \sum_{i=1}^{m} w_i \phi_i(\mathbf{x}) \quad (19)$$

where the usual choice for the basis functions $\phi_i(\mathbf{x})$ is the Gaussian function:

$$\phi_i(\mathbf{x}) = \exp\left( -\frac{\|\mathbf{x} - \mathbf{c}_i\|^2}{2\sigma_i^2} \right)$$

Note that adjustable weights are only present in the output layer. The connections from the input layer to the hidden layer are fixed to unit weights. The free parameters of RBF nets are the output weights $w_i$ and the parameters of the basis functions (centers $\mathbf{c}_i$ and radii $\sigma_i$). Since the network's output (19) is linear in the weights $w_i$, these weights can be estimated by least-squares methods. For each data point $\mathbf{x}_k$, first the outputs of the neurons are computed:

$$v_{ki} = \phi_i(\mathbf{x}_k)$$

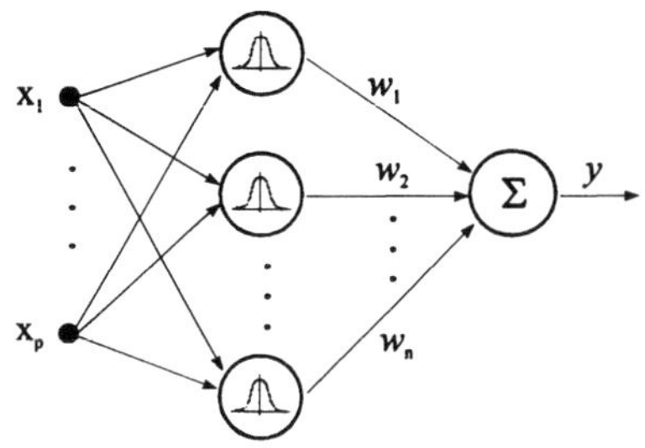


Fig. 5. Radial basis function network.

and put in the matrix $\mathbf{V} = [v_{ki}]$. Introducing the weight vector $\mathbf{w} = [w_1, w_2, \ldots, w_m]$, we can write the following matrix equation for the whole data set:

$$\mathbf{y}^* = \mathbf{V}\mathbf{w}\,.$$

The least-square estimate of the weights $\mathbf{w}$ that minimize the network error $\mathbf{e} = \mathbf{y}^* - \mathbf{y}$ is:

$$\mathbf{w} = \left[\mathbf{V}^T\mathbf{V}\right]^{-1}\mathbf{V}^T\mathbf{y}^*\,. \tag{20}$$

The adaptation of the RBF parameters $\mathbf{c}_i$ and $\sigma_i$ is a nonlinear optimization problem that can be solved by the gradient-descent method (17).

## 3. NEURO-FUZZY MODELING

At the computational level, a fuzzy system can be seen as a layered structure (network), similar to artificial neural networks of the RBF type (Jang and Sun, 1993). In order to optimize parameters in a fuzzy system, gradient-descent training algorithms known from the area of neural networks can be employed. Hence, this approach is usually referred to as neuro-fuzzy modeling (Brown and Harris, 1994; Jang, 1993; Jang *et al.*, 1997).

Consider first a simple example of a zero-order TS fuzzy model with the following two rules:

If $x_1$ is $A_{11}$ and $x_2$ is $A_{21}$ then $y = b_1$
If $x_1$ is $A_{12}$ and $x_2$ is $A_{22}$ then $y = b_2$

Figure 6 shows a network representation of these two rules. The nodes in the first layer compute the membership degree of the inputs in the antecedent fuzzy sets. The product nodes Π in the second layer represent the antecedent connective (here the 'and' operator). The normalization node N and the summation node Σ realize the fuzzy-mean operator (5). This system is called ANFIS – Adaptive Neuro-Fuzzy Inference System (Jang, 1993).

Typically, smooth antecedent membership functions are used, such as the Gaussian functions:

$$\mu_{A_{ij}}(x_j; c_{ij}, \sigma_{ij}) = \exp\left(-\frac{(x_j - c_{ij})^2}{2\sigma_{ij}^2}\right). \tag{21}$$

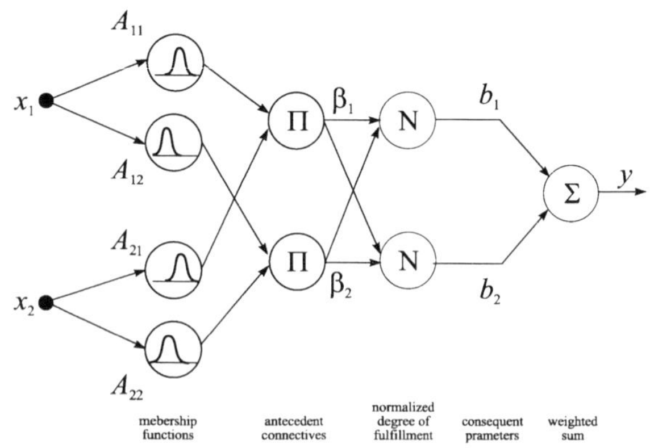


Fig. 6. An example of a zero-order TS fuzzy model with two rules represented as a neuro-fuzzy network.

The input–output equation of a general zero-order TS model with the conjunctive form antecedent is:

$$y = \sum_{i=1}^{K} \gamma_i(\mathbf{x}) b_i \tag{22}$$

$$\text{with} \quad \gamma_i(\mathbf{x}) = \frac{\prod_{j=1}^{p} \exp\left(-\frac{(x_j - c_{ij})^2}{2\sigma_{ij}^2}\right)}{\sum_{i=1}^{K}\prod_{j=1}^{p} \exp\left(-\frac{(x_j - c_{ij})^2}{2\sigma_{ij}^2}\right)}$$

The first-order TS fuzzy model can be represented in a similar fashion. Consider again the example with two rules:

If $x_1$ is $A_{11}$ and $x_2$ is $A_{21}$ then
$y_1 = a_{11}x_1 + a_{12}x_2 + b_1$
If $x_1$ is $A_{12}$ and $x_2$ is $A_{22}$ then
$y_2 = a_{21}x_1 + a_{22}x_2 + b_2$

for which the corresponding network is given in Figure 7.

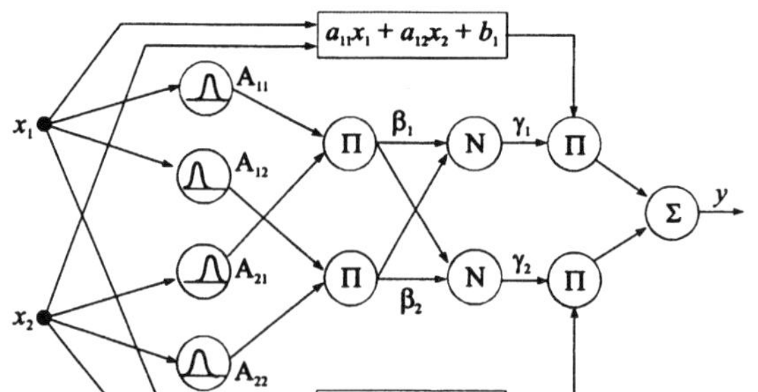


Fig. 7. An example of a first-order TS fuzzy model with two rules represented as a neuro-fuzzy network called ANFIS.

The input–output equation of this first-order TS model is:

$$y = \sum_{i=1}^{K} \gamma_i(\mathbf{x}) \left(\mathbf{a}_i^T \mathbf{x} + b_i\right) \qquad (23)$$

with $\gamma_i(\mathbf{x})$ given by (23).

### 3.1 *Constructing Neuro-Fuzzy Networks*

Both prior knowledge and process data can be used to construct neuro-fuzzy systems. Prior knowledge can be of a rather approximate nature (qualitative, heuristics). Two main approaches to the integration of knowledge and data can be distinguished:

(1) Expert knowledge is formulated as a collection of if–then rules. In this way, an initial model is created. The parameters of this model (the membership functions, consequent parameters) are then fine-tuned by using process data.
(2) Fuzzy rules (including the associated parameters) are constructed from scratch by using numerical data. In this case, the advantage of using a neuro-fuzzy model is the possibility to interpret the obtained result (which is not possible with truly black-box structures like neural networks). An expert can confront the information stored in the rule base with his own knowledge, can modify the rules, or supply additional ones to extend the validity of the model, etc.

The above techniques can, of course, be combined, depending on the problem at hand.

### 3.2 *Structure and Parameters*

The two basic steps in system identification are *structure identification* and *parameter estimation*. The choice of the model's structure is very important, as it determines the flexibility of the model in the approximation of (unknown) systems. A model with a rich structure can approximate more complicated functions, but, at the same time, will have worse generalization properties. Good generalization means that a model fitted to one data set will also perform well on another data set from the same process. In neuro-fuzzy models, the structure selection process involves the following main choices:

- *Selection of input variables.* This involves not only the physical inputs $\mathbf{u}$ but also the dynamic regressors, defined by the input and output lags, $n_y$ and $n_u$ respectively. Prior knowledge, insight in the process behavior and the purpose of the modeling exercise are the typical sources of information for the choice of an initial set of possible inputs. Automatic data-driven selection can then be used to compare different structures in terms of some specified performance criteria.
- *Number and type of membership functions, number of rules.* These two structural parameters are mutually related (for more membership functions more rules must be defined) and determine the level of detail, called the granularity, of the model. The purpose of modeling and the amount of available information (knowledge and data) will determine this choice. Automated, methods can be used to add or remove membership functions and rules.

### 3.3 *Gradient-Based Learning*

It is quite straightforward to derive the gradient-descent learning rule for the $b_i$ $c_{ij}$ and $\sigma_{ij}$ parameters. The procedure is identical to the derivation of the backpropagation formulas for neural nets. Consider the zero-order ANFIS model, given by rules (6). For the consequent parameters $b_i$, we have the Jacobian (18):

$$\frac{\partial J}{\partial b_i} = \frac{\partial J}{\partial e} \cdot \frac{\partial e}{\partial b_i} = -\gamma_i e, \quad i = 1, \ldots K, \qquad (24)$$

and the update law:

$$b_i(n+1) = b_i(n) + \alpha(n)\gamma_i e \,. \qquad (25)$$

For the centers and spreads of the Gaussian membership functions (21) we apply the chain rule for differentiation and after some algebra, the following update formulas are obtained:

$$c_{ij}(n+1) = c_{ij}(n) + 2\alpha(n)\gamma_i e \frac{x_j - c_{ij}}{\sigma_{ij}^2} [b_i - y] \,.$$

and

$$\sigma_{ij}(n+1) = \sigma_{ij}(n) + 2\alpha(n)\gamma_i e \frac{(x_j - c_{ij})^2}{\sigma_{ij}^3} [b_i - y] \,.$$

The parameter-update equations for the first-order ANFIS model can be derived in a similar fashion.

## 4. INTERPRETABILITY ISSUES

We have already noticed that the output-layer parameters in RBF networks can be estimated by linear least-squares (LS) techniques (20). As LS methods are more effective than the gradient-based update rule (25), hybrid methods are often applied that combine one-shot least-squares estimation of the consequent parameters with iterative gradient-based optimization of the membership functions (Mathworks, 2001).

In terms of error minimization, the choice of a particular least-squares estimation method is not crucial. If, however, the consequent parameters are to be interpreted as local models, for instance, great care must be taken in the choice of the estimation method. The problem is that the ANFIS models, especially the first-order one, tend to be over-parameterized for most approximation problems. This may lead to numerical problems, over-fitting and meaningless parameter estimates. The following example demonstrates this problem.

*Example 3.* Assume we wish to approximate a second-order polynomial $y = f_s(u) = 3u^2 - 5u + 6$ by a first-order ANFIS model. First we choose two points $t_1$ and $t_2$ and define initial triangular membership functions for $t_1 \le u < t_2$:

$$\mu_{A_1}(u) = \frac{u - t_1}{t_2 - t_1}, \quad \mu_{A_2} = 1 - \mu_{A_1} \tag{26}$$

The model consists of two rules:

$$\mathcal{R}_i\text{: If } u \text{ is } A_i \text{ then } y_i = a_i u + b_i, \qquad i = 1, 2$$

By substituting the membership functions (26) into (5), the output of the TS model is obtained (after some elementary algebra):

$$\begin{aligned} y = & \frac{a_1 - a_2}{t_2 - t_1} u^2 \\ & + \frac{t_2(a_1 - a_2) + a_2(t_2 - t_1) + (b_1 - b_2)}{t_2 - t_1} u \\ & + \frac{t_2(b_1 - b_2) + b_2(t_2 - t_1)}{t_2 - t_1} \end{aligned}$$

As this is a second order polynomial in $u$, our model can perfectly represent the given nonlinear system. However, it has four free parameters ($a_1$, $a_2$, $b_1$ and $b_2$) while three are sufficient to fit the polynomial – it is thus over-parameterized. This is a very simple example, but the essence of the over-parameterization problem remains the same when approximating complex unknown systems. □

### 4.1 *Consequent Parameter Estimation*

To circumvent over-parameterization, the basic least-squares criterion can be combined with additional criteria for local fit, or with constraints on the parameter values. In the following, several techniques are discussed.

4.1.1. *Global Least-Squares Estimation.* The global least-squares estimation method yields parameters that minimize the following prediction error criterion:

$$\boldsymbol{\theta} = \arg\min \sum_{k=1}^{N} \left( y_k^* - \sum_{i=1}^{K} \gamma_i(\mathbf{x}_k) \left[\mathbf{x}_k^T 1\right] \boldsymbol{\theta}_i \right)^2 .$$

where $\boldsymbol{\theta}^T = [\boldsymbol{\theta}_1^T \ldots \boldsymbol{\theta}_K^T]$ is the concatenation of all the individual rules' parameter vectors. For the data matrices (16), this criterion can be rewritten in a matrix form:

$$\boldsymbol{\theta} = \arg\min \; (\mathbf{y}^* - \boldsymbol{\Lambda}\boldsymbol{\theta})^T (\mathbf{y}^* - \boldsymbol{\Lambda}\boldsymbol{\theta}) \tag{27}$$

with $\boldsymbol{\Lambda} = [\boldsymbol{\Gamma}_1 \boldsymbol{\varphi} \ldots \boldsymbol{\Gamma}_K \boldsymbol{\varphi}]$ where $\boldsymbol{\varphi} = [\mathbf{X}\ \mathbf{1}]$ and $\boldsymbol{\Gamma}_i = \text{diag}\,(\gamma_i(\mathbf{x}_1) \ldots \gamma_i(\mathbf{x}_N))$, i.e., a diagonal matrix having $\gamma_i(\mathbf{x}_k)$ as its $k$th diagonal element. The optimal solution of (27) is then directly obtained by using matrix pseudo-inverse:

$$\boldsymbol{\theta} = \left(\boldsymbol{\Lambda}^T \boldsymbol{\Lambda}\right)^{-1} \boldsymbol{\Lambda}^T \mathbf{y}^* . \tag{28}$$

4.1.2. *Local Least-Squares Estimation.* While the global solution gives the minimal prediction error, it may bias the estimates of the consequents as parameters of local models. If locally relevant model parameters are required, a weighted least-squares approach applied per rule should be used. This is done by minimizing a set of $K$ weighted local LS criteria:

$$\begin{aligned} \boldsymbol{\theta}_i = \arg\min \; (\mathbf{y}^* - \boldsymbol{\varphi}\boldsymbol{\theta}_i)^T \boldsymbol{\Gamma}_i (\mathbf{y}^* - \boldsymbol{\varphi}\boldsymbol{\theta}_i), \\ i = 1, 2, \ldots, K \end{aligned} \tag{29}$$

for which the solutions are

$$\boldsymbol{\theta}_i = \left(\boldsymbol{\varphi}^T \boldsymbol{\Gamma}_i \boldsymbol{\varphi}\right)^{-1} \boldsymbol{\varphi}^T \boldsymbol{\Gamma}_i \mathbf{y}^*, \quad i = 1, 2, \ldots, K .$$

In this case, the consequent parameters of the individual rules are estimated independently of each other, and therefore the result is not influenced by the interactions of the rules. At the same time, however, a larger prediction error is obtained than with global least squares.

*Example 4.* The application of local and global estimation to the TS model from Example 3 results in the consequent models given in Fig. 8. Note that the consequents estimated by local least squares describe properly the local behavior of the function, but do not give a good fit. For global least squares, the opposite holds – a perfect fit is obtained, but the consequents are not relevant for the local behavior of the system. □

When interpreting ANFIS models obtained from data, one has to be aware of the trade-offs between local and global estimation. Constrained and multicriteria optimization can also be applied to restrict the freedom in the parameters.

4.1.3. *Constrained Estimation.* Knowledge about the dynamic system such as its stability, minimal or maximal static gain, or its settling time can be translated into convex constraints on the consequent parameters (see Fig. 9). By using input-output data, optimal parameter values are then found by means of quadratic programming, instead of least squares. There are two types of constraints, global and local. Local constraints represent detail knowledge pertaining to each specific rule, while global constraints apply to the entire model and should thus refer to some global system properties such as the overall stability. To see this, realize that the affine TS model (4) can be regarded as one quasi-linear system

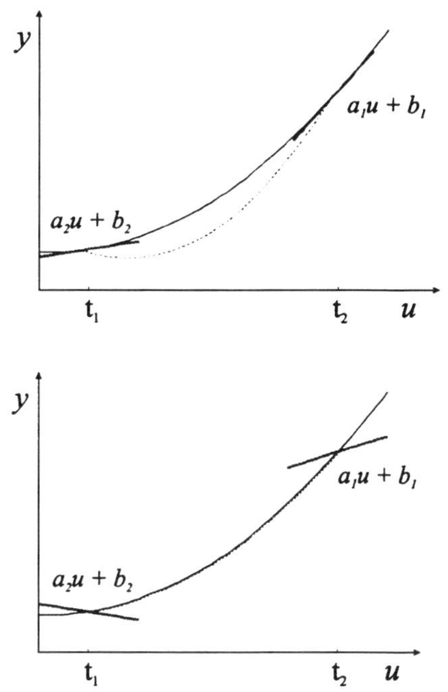


Fig. 8. Result of local (top) and global (bottom) estimation of the consequent parameters. The dashed line is the output of the model.

$$y = \left( \sum_{i=1}^{K} \gamma_i(\mathbf{x}) \mathbf{a}_i^T \right) \mathbf{x} + \sum_{i=1}^{K} \gamma_i(\mathbf{x}) b_i \qquad (30)$$
$$= \mathbf{a}^T(\mathbf{x})\mathbf{x} + b(\mathbf{x}) .$$

with input-dependent 'parameters' $\mathbf{a}(\mathbf{x})$, $b(\mathbf{x})$ which are convex linear combinations of the individual consequent parameters $\mathbf{a}_i$ and $b_i$, i.e.:

$$\mathbf{a}(\mathbf{x}) = \sum_{i=1}^{K} \gamma_i(\mathbf{x}) \mathbf{a}_i, \quad b(\mathbf{x}) = \sum_{i=1}^{K} \gamma_i(\mathbf{x}) b_i . \quad (31)$$

This property allows us to define global convex constraints for the entire model. Besides, it also facilitates the analysis of TS models in the framework of polytopic systems (linear differential inclusions). Methods have been developed to design controllers with desired closed loop characteristics and to analyze their stability (Tanaka *et al.*, 1998).

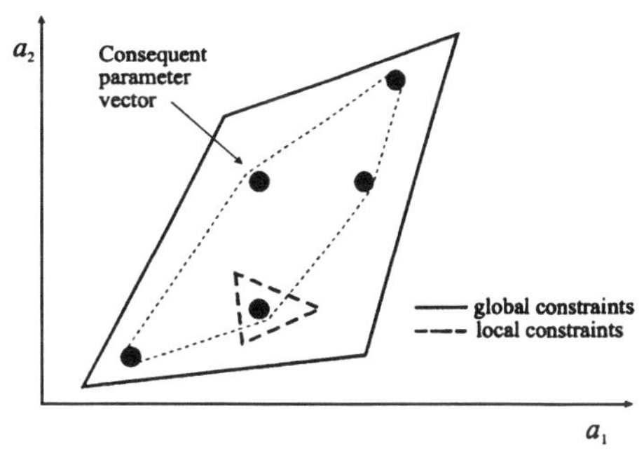


Fig. 9. Convex constraints on the consequent parameters.

4.1.4. *Multi-Objective Optimization.* Another possibility is to regularize the estimation by penalizing undesired local behavior of the model. This can dramatically improve the robustness of the construction algorithm, eventually leading to more relevant (interpretable) parameter estimates. One way is to minimize the weighted sum of the global and local identification criteria (27) and (30):

$$\boldsymbol{\theta} = \arg\min \Big\{ (\mathbf{y}^* - \boldsymbol{\Lambda}\boldsymbol{\theta})^T (\mathbf{y}^* - \boldsymbol{\Lambda}\boldsymbol{\theta}) + \sum_{i=1}^{K} \delta_i (\mathbf{y}^* - \varphi\boldsymbol{\theta}_i)^T \boldsymbol{\Gamma}_i (\mathbf{y}^* - \varphi\boldsymbol{\theta}_i) \Big\} .$$

The weighting parameters $\delta_i \geq 0$ parameterize the set of Pareto-optimal solutions of the underlying multi-objective optimization problem and thus determine the tradeoff between the possibly conflicting objectives of global model accuracy and local interpretability of the parameters.

### 4.2 *Initialization of Antecedent Membership Functions*

For a successful application of gradient-descent learning to the membership function parameters, good initialization is important. Several initialization methods are briefly reviewed in this section.

4.2.1. *Template-Based Membership Functions.* With this method, the domains of the antecedent variables are *a priori* partitioned by a number of membership functions. These are usually evenly spaced and shaped. The rule base is then established to cover all the combinations of the antecedent terms. A severe drawback of this approach is that the number of rules in the model grows exponentially. Furthermore, if no knowledge is available as to which variables cause the nonlinearity of the system, all the antecedent variables are usually partitioned uniformly. However, the complexity of the system's behavior is typically not uniform. Some operating regions can be well approximated by a local linear model, while other regions require a rather fine partitioning. In order to obtain an efficient representation with as few rules as possible, the membership functions must be placed such that they capture the non-uniform behavior of the system.

4.2.2. *Discrete Search Methods.* Iterative tree-search algorithms can be applied to decompose the antecedent space into hyper-rectangles by axis-orthogonal splits. In each iteration, the region with the worst local error measure is divided into two halves (or other portions). Splits in all dimensions of the input are tested and the one with the highest performance improvement is chosen. This successive partitioning stops when a specified error goal is met or when the desired number of rules is reached. The first four steps of such an algorithm are illustrated in Fig. 10. An advantage of this approach is its effectiveness for high-dimensional data and the transparency of the obtained partition. A drawback is that the tree building procedure is sub-optimal (greedy) and hence the number of rules obtained can be quite large (Nelles *et al.*, 2000).

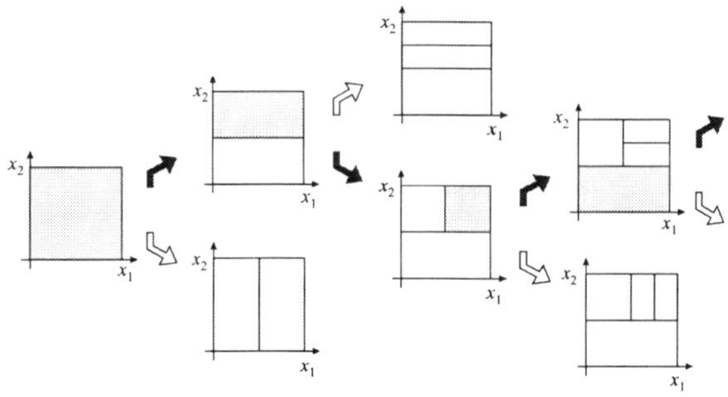

Fig. 10. Antecedent space decomposition by a heuristic search algorithm. The dark areas represent rules with the worst local fit in the given step.

4.2.3. *Fuzzy Clustering.* Construction methods based on fuzzy clustering originate from data analysis and pattern recognition, where the concept of fuzzy membership is used to represent the degree to which a given data object is similar to some prototypical object. The degree of similarity can be calculated by using a suitable distance measure. Based on the similarity, data vectors are clustered such that the data within a cluster are as similar as possible, and data from different clusters are as dissimilar as possible.

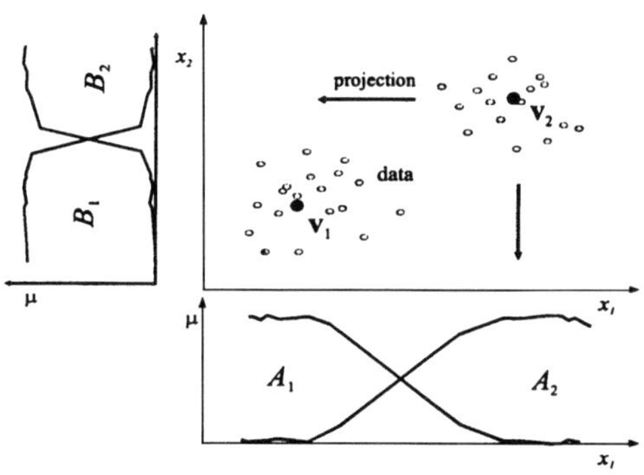


Fig. 11. Identification of membership functions through fuzzy clustering.

Figure 11 gives an example of two clusters in $\mathbb{R}^2$ with prototypes $\mathbf{v}_1$ and $\mathbf{v}_2$. The partitioning of the data is expressed in the *fuzzy partition matrix* $\mathbf{U} = [\mu_{ij}]$ whose elements are the membership degrees of the data vectors $\mathbf{x}_k$ in the fuzzy clusters with prototypes $\mathbf{v}_j$. The antecedent membership functions are then extracted by projecting the clusters onto the individual variables. For the initialization of first-order ANFIS models, the prototypes can be defined as linear subspaces or the clusters are ellipsoids with adaptively determined shape. The number of clusters in the data can either be determined *a priori* or sought automatically by using cluster validity measures and merging techniques (Babuška, 1998).

## 5. SIMULATION EXAMPLES

In this section, two simulation examples are given to illustrate several important issues related to the training of neuro-fuzzy systems. The first example is a simple fitting problem of a univariate static function. It demonstrates the typical construction procedure of a neuro-fuzzy model. Numerical results show that an improvement in performance is achieved at the expense of obtaining if-then rules that are not completely relevant as local descriptions of the system.

The second example, the modeling of a nonlinear dynamic system, illustrates that the performance of a neuro-fuzzy model does not necessarily improve after training. This is due to overfitting which in the case of dynamic systems can easily occur when the data only sparsely cover the domains.

### 5.1 *Static Function*

Let us approximate a univariate function $y = \sin(u)$ by the ANFIS model with linear consequent functions. We choose the number of rules to be five and construct an initial model by clustering the data $U \times Y$, using a methodology based on the Gustafson-Kessel algorithm (Babuška, 1998). The following rules are obtained:

$$
\begin{aligned}
&\text{If } u \text{ is } A_1 \text{ then } y = 5.721u + 0.030\\
&\text{If } u \text{ is } A_2 \text{ then } y = 0.035u + 0.904\\
&\text{If } u \text{ is } A_3 \text{ then } y = -5.302u + 2.380\\
&\text{If } u \text{ is } A_4 \text{ then } y = 0.734u - 1.413\\
&\text{If } u \text{ is } A_5 \text{ then } y = 6.283u - 5.623
\end{aligned}
$$

The fit of the function with this initial model is shown in Fig. 12a. The membership functions and the corresponding local models are given in Fig. 12b. The membership functions are denoted from left to right by $A_1$ through $A_5$.

Note that this initial model can easily be interpreted in terms of the local behavior (the rule consequents) and it is already reasonably accurate (the root mean squared error is RMS= 0.0258). However, by using the ANFIS method, the model parameters can be fine-tuned and the approximation accuracy can be significantly improved. A model obtained after 100 learning epochs of hybrid learning using the `anfis` function of the MATLAB Fuzzy Logic Toolbox (Mathworks, 2001) is described by the following fuzzy rules:

$$
\begin{aligned}
&\text{If } u \text{ is } A'_1 \text{ then } y = 5.275u + 0.065\\
&\text{If } u \text{ is } A'_2 \text{ then } y = 0.442u + 0.899\\
&\text{If } u \text{ is } A'_3 \text{ then } y = -3.206u + 1.405\\
&\text{If } u \text{ is } A'_4 \text{ then } y = 0.977u - 1.693\\
&\text{If } u \text{ is } A'_5 \text{ then } y = 5.062u - 4.388
\end{aligned}
$$

The performance has improved to that degree that no approximation error is visible (not shown). The root mean squared error is now RMS= 0.0011, which is about 23 times better than the initial model. The membership functions have only been slightly modified, but the change in the local models is more apparent.

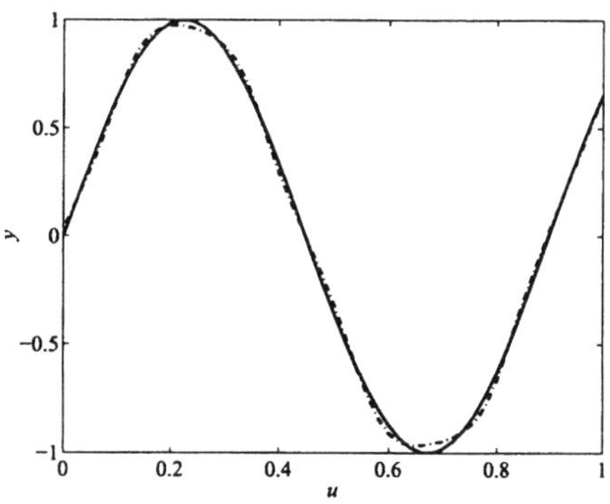


*(a) The data (solid) and the model (dashed-dotted).*

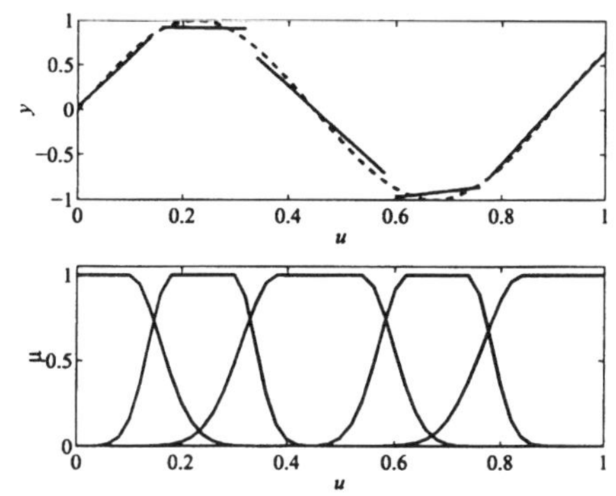


*(b) Local models (top) and membership functions (bottom).*

Fig. 12. Approximation by the initial TS fuzzy model. The root mean squared error is RMS= 0.0258.

After learning, the local models are much further from the true local description of the function. To quantify this, we can compute the difference between the consequent parameters of the fuzzy models, denoted by $\boldsymbol{\theta}$, and the 'true' local estimates $\boldsymbol{\theta}_0$, computed by least squares for the data in $\text{core}(A_i)$. For the initial fuzzy model, we have $\|\boldsymbol{\theta}-\boldsymbol{\theta}_0\| = 1.81$, while for the ANFIS model we have $\|\boldsymbol{\theta} - \boldsymbol{\theta}_0\| = 5.30$. The rules of the fine-tuned neuro-fuzzy model are thus less accurate in describing the system locally. This contradiction between local and global approximation accuracy is inherent to TS fuzzy systems with linear consequents (Abonyi and Babuška, 2000) and thus also to the ANFIS network. Great care must be exercised when one attempts to interpret rules in trained neuro-fuzzy models.

### 5.2 *pH Neutralization Process*

A neutralization tank with three influent streams (acid, buffer and base) and one effluent stream is considered. The identification and validation data sets are obtained by simulating the model by Hall and Seborg (1989) for random changes of the influent base stream flow rate $Q$. The influent buffer stream and the influent acid stream are kept constant. The output is the pH in the tank. The identification data set contains $N = 499$ samples with the sampling time of 15 s. It was obtained from (Johansen, 1994). The process is approximated as a first-order discrete-time NARX model:

$$\text{pH}(k+1) = f\big(\text{pH}(k), Q(k)\big),$$

where $k$ denotes the sampling instant, and $f$ is an unknown relationship approximated by a neuro-fuzzy model. Based on prior knowledge about the process, it was decided to include only $Q(k)$ in the antecedent (it is known that the main source of nonlinearity is the titration curve, which is the steady-state characteristic relating $Q$ to pH). The number of membership functions (and thus also rules) was set to three. The initial membership functions were evenly spread over the domain, as shown in the left panel of Fig. 13.

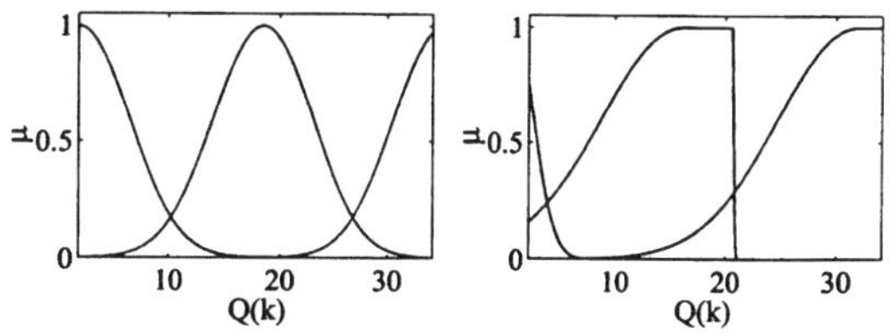


Fig. 13. Membership functions before (left) and after training (right). The membership functions are denoted from left to right by 'Low', 'Medium' and 'High'.

The initial rule base, with the consequent estimated by weighted local least squares (30), is given by:

If $Q(k)$ is *Low* then
$\quad \text{pH}(k+1) = 0.83\text{pH}(k) + 0.09Q(k) + 0.03$
If $Q(k)$ is *Medium* then
$\quad \text{pH}(k+1) = 0.83\text{pH}(k) + 0.09Q(k) + 0.10$
If $Q(k)$ is *High* then
$\quad \text{pH}(k+1) = 0.46\text{pH}(k) + 0.02Q(k) + 5.09$

After 1000 epochs of hybrid learning using the ANFIS function of the MATLAB Fuzzy Logic Toolbox (Mathworks, 2001), the following rule base has been obtained:

If $Q(k)$ is *Low*$'$ then
$\quad \text{pH}(k+1) = 0.37\text{pH}(k) - 0.05Q(k) + 2.14$
If $Q(k)$ is *Medium*$'$ then
$\quad \text{pH}(k+1) = 0.91\text{pH}(k) + 0.06Q(k) - 0.23$
If $Q(k)$ is *High*$'$ then
$\quad \text{pH}(k+1) = 0.40\text{pH}(k) + 0.03Q(k) + 5.63$

Note that the consequent model in the first rule has a negative coefficient for $Q(k)$. As this is a physically impossible value, not interpretation can be given to these parameters and this trained model has become a complete black-box. Also notice in Fig. 13 that the membership functions were adjusted in a very peculiar way by the gradient-descent optimization method.

Table 2 shows that while the numerical performance in terms of the root-mean-square error improved for the training data, it has become worse for the validation data. This is a typical example of overtraining. This

Table 2. Comparison of RMS before and after training.

| | before training | after training |
|---|---|---|
| training data set | 0.90 | 0.82 |
| validation data set | 0.81 | 0.89 |

can also be observed in Fig. 14 where the predictions generated by the model are less accurate after training than before. Clearly, this kind of behavior is difficult to predict for a new problem at hand. The importance proper model validation can thus hardly be overemphasized.

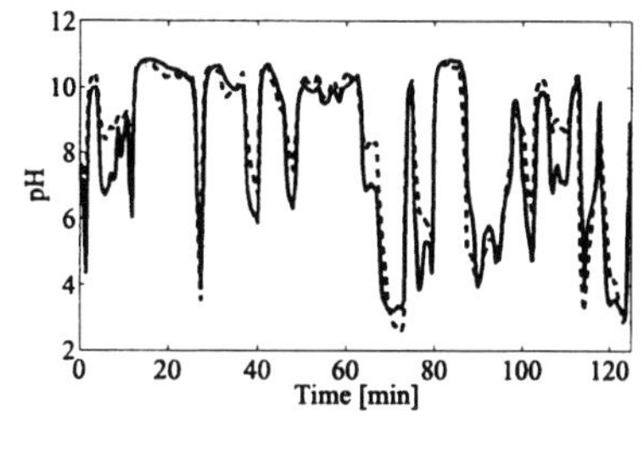


*(a) initial model*

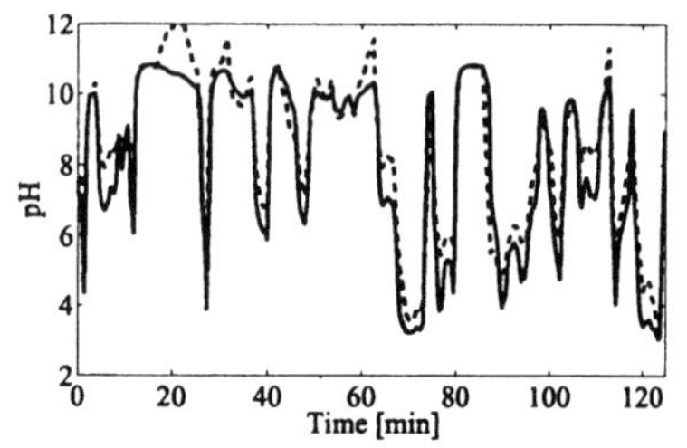


*(b) after training*

Fig. 14. Performance of the initial and trained model on the validation data set (solid – data, dashed – model).

## 6. CONCLUDING REMARKS

Neuro-fuzzy modeling is a flexible framework in which different paradigms can be combined, providing, on the one hand, a transparent interface with the designer and, on the other hand, a tool for accurate nonlinear modeling and control. The rule-based character of neuro-fuzzy models allows for the analysis and interpretation of the result. Conventional methods for numerical validation can be complemented by human expertise, which often involves heuristic knowledge and intuition.

A drawback of neuro-fuzzy modeling is that the current techniques for constructing and tuning fuzzy models are rather complex, and their use requires specific skills and knowledge. In this sense, it will probably never become a 'one-button', fully automated identification technique. Neuro-fuzzy modeling should rather be seen as an interactive method, facilitating the active participation of the user in a computer-assisted modeling session. This holds, to a certain degree, also for other, more established methods. Modeling of complex systems will always remain an interactive approach.

## 7. REFERENCES


Abonyi, J. and R. Babuška (2000). Local and global identification and interpretation of parameters in Takagi–Sugeno fuzzy models. In: *Proceedings 9th IEEE International Conference on Fuzzy Systems*. San Antonio, USA. pp. 835–840.

Babuška, R. (1998). *Fuzzy Modeling for Control.* Kluwer Academic Publishers. Boston, USA.

Brown, M. and C. Harris (1994). *Neurofuzzy Adaptive Modelling and Control.* Prentice Hall. New York.

Hall, R.C. and D.E. Seborg (1989). Modelling and self-tuning control of a multivariable pH neutralization process. Part I: Modelling and multiloop control. In: *Proceedings American Control Conference.* Vol. 2. Pittsburgh, U.S.A.. pp. 1822–1827.

Jang, J.-S.R. (1993). ANFIS: Adaptive-network-based fuzzy inference systems. *IEEE Transactions on Systems, Man & Cybernetics* **23**(3), 665–685.

Jang, J.-S.R. and C.-T. Sun (1993). Functional equivalence between radial basis function networks and fuzzy inference systems. *IEEE Transactions on Neural Networks* **4**(1), 156–159.

Jang, J.-S.R., C.-T. Sun and E. Mizutani (1997). *Neuro-Fuzzy and Soft Computing; a Computational Approach to Learning and Machine Intelligence.* Prentice-Hall. Upper Saddle River.

Johansen, T.A. (1994). Operating Regime Based Process Modelling and Identification. PhD dissertation. The Norwegian Institute of Technology – University of Trondheim. Trondheim, Norway.

Mathworks, The (2001). Fuzzy logic toolbox for use with MATLAB. User's guide, version 2. The Mathworks, Inc.. Natick, MA, USA.

Nelles, O., A. Fink, R. Babuška and M. Setnes (2000). Comparison of two construction algorithms for Takagi–Sugeno fuzzy models. *International Journal of Applied Mathematics and Computer Science* **10**(4), 835–855.

Rumelhart, D.E., G.E. Hinton and R.J. Williams (1986). Learning internal representations by error propagation. In: *Parallel Distributed Processing* (D.E. Rumelhart and J.L. McClelland, Eds.). MIT Press. Cambridge, MA.

Setnes, M., R. Babuška and H.B. Verbruggen (1998). Rule-based modeling: Precision and transparency. *IEEE Transactions on Systems, Man and Cybernetics, Part C: Applications and Reviews* **28**(1), 165–169.

Tanaka, K., T. Ikeda and H.O. Wang (1998). Fuzzy regulators and fuzzy observers: relaxed stability conditions and LMI-based designs. *IEEE Transactions on Fuzzy Systems* **6**(2), 250–265.

IFAC
Publications
www.elsevier.com/locate/ifac

# SAFETY OF COMPUTER CONTROL SYSTEMS: CHALLENGES AND RESULTS IN SOFTWARE DEVELOPMENT

**Janusz Zalewski* Wolfgang Ehrenberger** Francesca Saglietti*** Andrew Kornecki******

* *University of Central Florida*
*Orlando, FL 32816-2450, USA*
jza@ece.engr.ucf.edu
** *University of Applied Science*
*36039 Fulda, Germany*
wolfgang.ehrenberger@informatik.fh-fulda.de
*** *Institute for Safety Technology, Forschungsgelände*
*85748 Garching / Munich, Germany*
SAF@grs.de
**** *Embry-Riddle Aeronautical University*
*Daytona Beach, FL 32114, USA*
korn@db.erau.edu

Abstract: This paper reviews some results in improving software safety in computer control systems. The discussion covers various aspects of the software development process, as opposed to the product features. Software diversity, off-the-shelf software, rigorous and formal software development are discussed. *Copyright © 2001 IFAC*

Keywords: Software safety, software diversity, off-the-shelf software, rigorous development, formal verification, UML, software for computer control.

## 1. INTRODUCTION

The key aspect of safety in computer control systems is to minimize the risk of harm, which can lead to loss of life, limbs, or large financial losses due the failure of hardware or software of a computer control system embedded in a larger application. Safety aspects are important in all modern computer applications, where computer control is embedded in larger systems, for example: all means of transportation, such as cars, railways, airplanes, ships; large plants, such as nuclear power plants, chemical plants, etc.; as well as smaller devices, such as medical electronic devices.

New challenges show up every day, because computer-controlled systems are being increasingly applied in other areas related to human safety. In addition to traditional applications concerning safety listed above, several new areas have emerged, related to safety of computer control, including: telecommunication systems, banking systems, fire protection systems on oil/gas platforms, and others. Assuring safety becomes more and more important, because of an increasing concern about safety of computers and computer-controlled systems in a modern society as a whole.

The technologies of providing safety assurance in computer control systems are often based on the architecture of feedback control, but they rarely use, if at all, results from continuous systems. Due to the discrete nature of the safety problem, methods of discrete mathematics are usually applied.

In this view, verification of hardware and software, with respect to safety, is a key challenge. Methods currently offered are so complex that they are manageable only for simple systems, while the complexity of safety-related applications is usually high and continues to increase dramatically with the progress of computing technologies.

In this paper, we present selected approaches to increase safety of computer control systems via the appropriate software design process. We focus on the development process to improve software safety, as opposed to engineering the design structure of the controller itself.

## 2. SOFTWARE SAFETY: PRODUCT VERSUS PROCESS

Safety is usually defined as a 'negative' property that asserts simply that nothing bad happens (Rushby 1994). Safety is addressed through a hazard analysis process, which is normally not conducted in non-critical software development. Attention of the developers must be focused on preventing hazards that are conditions that lead to a mishap, rather than preventing mishaps directly.

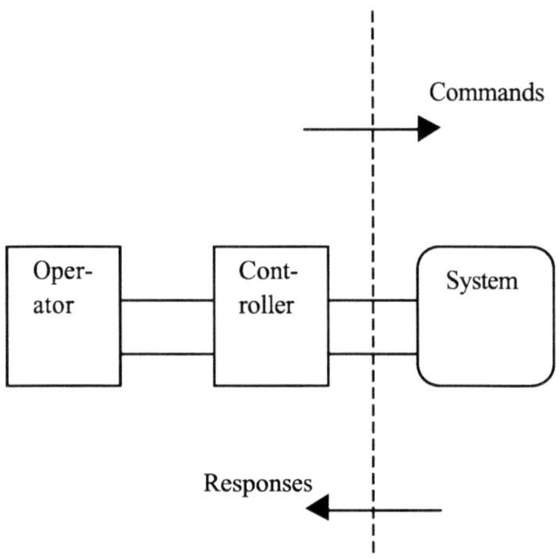


Fig. 1. Context for Discussing Safe Software.

In terms of a control system (Fig. 1), hazards are caused by various failures of the controller. An 'omission failure' is caused by a failure of the controller to react correctly to a given system state. The controller providing incorrect commands or inputs to a system causes a 'commission failure'. The controller responding correctly but outside of the required timing constraints causes a 'timing failure'.

There have been many technical solutions proposed to ensure software safety (Leveson 1996, Hilburn and Zalewski 1996). Some of the authors' own research (Anderson et al. 1999, Sahraoui et al. 2000) included the development of a software safety shell, that depends on a guard to ensure a safe behavior of the system (Fig. 2). The guard is a low-level construct that detects the danger and then forces the system to go to a safe state instead of hazardous one, if at all possible. This approach was applied in a traffic-light control case study to detect the conditions that, if unchecked, would lead to a failure.

All theoretical methods, which rely on engineering the proper design of the controller's structure to detect hazardous states, however, ultimately need to go through the software development process. This means that just designing the controller is necessary but not sufficient for ensuring safety, and during software development additional care has to be taken to retain or even improve safety properties embedded in the design.

One traditional method of developing software for safety is to use diversity, that is, have independent teams apply different principles to develop the software product to the same specifications. Even this simple approach, however, has its flaws, which we discuss in Section 3. A further problem arises when re-using software components, typically commercial off-the-shelf (COTS) software, which nowadays is more and more often applied in safety-related systems. New principles of dealing with COTS in safety-related systems have to be developed, which we discuss in Section 4.

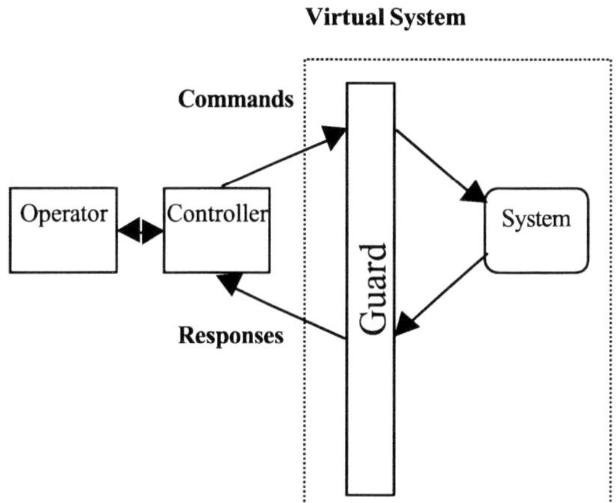


Fig. 2. System View of a Safety Guard.

Applying diversity during development and using systematic methods for dealing with COTS software might increase the confidence in software correctness, but in general are not sufficient to guarantee full product safety. In order to further improve software safety and lower the risk associated with its use, approaches relying on more formal mathematical techniques are required. One step in this direction is to support the design phase in the development process by formal design verification, which makes the entire process more rigorous.

We discuss it in Section 5 in relation to Petri nets. However, purely formal techniques are not sufficient to capture completely the reality to be modeled and analyzed; a very promising approach to the development of safety-related software consists of combining formal methods with established engineering techniques throughout the entire software life cycle. We discuss this issue in Section 6, for verifying timing requirements with UML and extended timed graphs.

The rest of this article presents an overview on the techniques mentioned above; in the following four chapters each of the co-authors summarizes the results and insight gained in his / her own area of research.

## 3. DIVERSITY IN THE SOFTWARE PROCESS

### *3.1 Common Failures in Diverse Variants*

Since a long time it has been of interest whether independently developed variants of software could contribute to the increase of safety. This concept has been particularly attractive to reduce the licensing costs of safety-related programs. The basic idea has been as follows:

- Step 1. Produce independently two variants of software to solve the same task.
- Step 2. Demonstrate that each single variant has achieved a certain reliability, for example, that its failure probability per demand $p_1 \approx p_2$ is below a certain limit, say $10^{-4}$.
- Step 3. Conclude for the probability of common failure per demand of both variants

$$p_{12} \approx p_1 * p_2 \approx p_1^2 < 10^{-8} \qquad (1)$$

This way of thinking may lead to commercially attractive solutions under certain circumstances. Doubling the effort - we have to produce two variants - would result in the *square of the relevant reliability figure*. In other words, *linear effort increase during production could provide exponential reliability gain*. The gain would improve safety properties, at the expense of a loss of availability properties. Under such assumptions diversity looks to be economically attractive.

Of course, during producing two variants one would take any precaution to avoid common faults among variants. The collection of diversity techniques proposed in (Saglietti et al. 1992) could be taken to select a proper set of techniques.

There were, however, severe objections against conclusion (1). In a theoretical work (Eckhardt and Lee 1985) it was shown that (1) is to be augmented by a measure of the varying "difficulty" throughout the input data space, namely the variance of the probability of committing programming errors over the input domain of the two diverse program variants:

$$p_{12} \approx p_1^2 + \mathrm{Var}(q) \qquad (2)$$

where q represents the probability of an arbitrary variant failing on input x. The variance is to be taken over all inputs x. An accurate estimate of q would require examination of a large number of software variants for any practical project, which is obviously not viable. Var(q) may exceed $p_1^2$ significantly.

Further investigation leads to the result derived in (Ehrenberger and Saglietti 1993). If the input domain of a diverse software system is processed by K disjoint input channels, and if the software of the individual channels fails independently, the probability of common failures is increased by a factor close to K:

$$p_{12KChannels} \approx p_{12} * K \approx p_1^2 * K \qquad (3)$$

### *3.2 Estimation of the Variance*

Littlewood and Miller (1987) developed equation (2) further and a more general result has been derived, including the particular case of *dissimilarity being enforced* during the development of the diverse variants:

$$p_{12} = p_1^2 + \mathrm{Cov}(q_1, q_2) \qquad (4)$$

where $q_1(x)$ and $q_2(x)$ represent average probabilities of failure on input x of two arbitrary variants developed by using different methodologies (forced diversity). Formula (4) degenerates to (2), if no dissimilarity was enforced. In principle, $\mathrm{Cov}(q_1,q_2)$ can become negative. Also here no estimate of the size of the covariance is known for an arbitrary concrete system.

Further details on theoretical considerations are presented in (Ehrenberger 2001b). As they do not provide a formal proof, experimental evidence is needed for support. As mentioned above, such evidence is difficult to get from software development, because programming a large number of variants would be too expensive. However, there may be a way out.

If we believe that *programming is a human activity similar to writing exams,* we may consider the results of student exams in place of the results of programs. Table 1 shows the envisaged analogies. The table of the exam results is given in (Ehrenberger 2001a), where 8 questions had to be answered, including some on deriving algorithms. A total of 13 exam papers were considered worth being used in the evaluation, because the number of faulty answers was reasonably small. They formed 78 pairs.

Table 1 Analogies between programming faults and faults in exams.

| Aspect | Solution of exam | Program |
|---|---|---|
| Requirements from | question sheet from professor | program specification |
| Result | exam paper | developed code |
| Form of requirement | answer question x | react on input x |
| Faults | faults in solving the question | program faults |
| Form of failure | human failure | human failure |
| Single task | single question, single answer | single program part |
| Single fault | fault in answer x | fail. on input x |
| Characteristic: probability of fault per question or program part | probability of fault per question $\frac{\text{number of incorrect answers}}{\text{total number of questions}}$ each answer, which does not score fully, is considered faulty | probability of failure per program part: $\frac{\text{number of failing program parts}}{\text{total number of program parts}}$ each program part, which has a fault is faulty |
| Characteristic: probability of fault for an answer or an input portion | fault characteristic of an answer per exam: $\frac{\text{number of exam papers with faults in this answer}}{\text{total number of exam papers}}$ | fault characteristic per input portion: $\frac{\text{number of variants with faults in this input area}}{\text{total number of programs}}$ |
| Variance to be calculated | over all answers, over the probability of the faulty answer | over the individual input domain portions, over the probability of faulty treatment of an input |

Any correct answer was qualified with 0, any incorrect one with 1. For each question the number of failed answers was counted. The $q_i$ of question i was calculated as:

$$\frac{\text{number of exam papers with faults in answer i}}{\text{number of all exam papers}}$$

and the variance of the difficulty of the questions, with N representing the number of questions:

Var(question difficulty) =

$$\frac{1}{N}\sum_{i=1}^{N}\left(q_i - \frac{1}{N}\sum_{i=1}^{N} q_i\right)^2 \qquad (5)$$

On the other hand, the probabilities of common faults of two students (i,j) can be taken from the related matrix as $p_{ij}$ equal to the following:

$$\frac{\text{number of answers with faults in both exam papers}}{\text{number of all questions}}$$

The probabilities of individual faults per solution $p_i$ and $p_j$ could also be evaluated:

$p_{mean} = 0.365$; $(p_{mean})^2 = 0.133$

Var(question difficulty) = $0.023 < (p_{mean})^2$.

$p_{ij} < p_i * p_j$ for 28 pairs

$p_{ij} = p_i * p_j$ for13 pairs

$p_{ij} > p_i * p_j$ for 37 pairs

$p_{ij} \leq p_i * p_j$ for 41 pairs

$p_{ij} \leq 2 * p_i * p_j$ for 70 pairs

These observations together with the theoretical considerations suggest the following estimate for the common failure probability of a two-fold diverse system:

$$p_{12} \approx p_1^2 + \mathrm{Var}(q) \approx 2 * p_1^2 \qquad (6)$$

Regarding the exams' results one has to remember that no type of forced diversity had been applied. As forced diversity will be the rule for any real project with diverse software, failure independence is likely to be higher than in the exams.

### *3.3 Results*

If one accepts the *analogy of the fault producing processes in students' exams and programming*, one may consider data from university exams as a basis for studying software diversity. The investigation of the answers of a student exam shows that in general the intuitive result (1) is too optimistic. A correction by the factor 2 is appropriate. This correction is more conservative than result (4).

If the program reduces information during its execution by a selection mechanism, the number of channels involved in that reduction has to be considered as a further source of common failure of independent variants. In many cases of industrial software application the influence of the information reduction will be more substantial than the influence of the varying difficulty of the input domain. Under ideal circumstances and bringing together results (6) and (3), we may conclude that:

$$p_{12,KChannels} \approx p_{12} * K \approx 2 * p_1^2 * K \qquad (7)$$

Under ideal circumstances the probability of common failure of two diverse software variants may be estimated as two times the square of the value of the failure probability of the singular variant times the number of diversity channels that entail information reduction.
The results of the experiment on diverse software development can be summarized as follows:

- Errors are not made independently of the questions to be answered.
- The value of the variance of the difficulty over the input domain is close to the value of common failure probability.
- Under ideal circumstances the probability of common failure equals the square of the failure probability of one variant multiplied by the doubles number of channels.

## 4. OFF-THE-SHELF SOFTWARE

For evident economical reasons, COTS software components are increasingly (re)used in different application areas, including safety-critical ones. Due to their origin, the available information on the underlying development process of these components is often only fragmentary.

A very limited knowledge on the component production, as well as differences between past and future usage (such as, different reliability demands or different usage profiles) create a serious challenge for the software engineering community facing the problem of assessing the suitability of COTS components for new development projects.

The strategy suggested to approach this difficult problem consists of five successive decision phases, as illustrated in Figure 3:

- Phase 1. Identification of safety demands at system level;
- Phase 2. Analysis of role of COTS software within the system (safety relevance & sensitivity);
- Phase 3. Qualitative (subjective) assessment of the software process and product quality;
- Phase 4. Quantitative (objective) assessment of past (testing &) operating experience;
- Phase 5. Validation of component interfaces within the integrated system.

Considerations on phases 2, 3 and 4 were presented in (Saglietti 2001, Saglietti 2000b and Saglietti 2000a), respectively. Each phase is discussed below in terms of key factors to be considered and necessary actions to be taken during development.

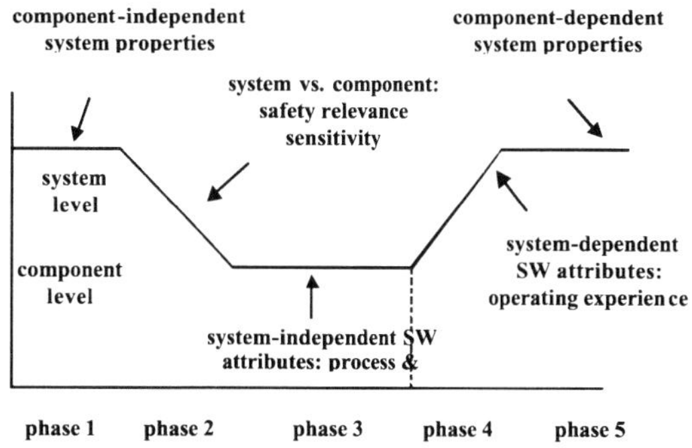


Fig. 3. Phased Approach to Component Analysis.

Phase 1: Identification of safety and reliability demands at the system level.

Key factors: risk analysis (black-box, component-independent).

- Risk analysis: identification of occurrence of failure-initiating events and loss caused by them; comparison of hazards involved in automation with those inherent to the application as such, i.e., in an uncontrolled mode; identification of critical events to be controlled; classification of event criticality if not under control.

- Analysis of results: determination of minimum quantitative target for reliability demands to be demonstrated, in terms of minimum probability of operational survival, i.e. of correct or acceptable performance under application-specific operational demand profile.

Phase 2: Analysis of the role of COTS components within the system (Saglietti 2001).

Key factors: COTS safety relevance, sensitivity (objective, quantitative, system dependent), safety criticality and sensitivity.

- Safety Criticality: identification of potentially critical failure propagation through COTS modules, evaluation of individual responsibility of COTS components in terms of safety-relevance to distribute verification and validation effort accordingly, determination of bottlenecks for system safety, definition of fault tolerant architectural properties accordingly.

- Sensitivity: determination of impact of component reliability on system reliability to know whether modest reliability figures at component level are acceptable for systems with higher reliability demands, in order to derive to which extent a post-qualification of pre-developed software allows to expect an increase of system reliability.

Phase 3: Qualitative assessment of COTS at component level (Saglietti, 2000b).

Key factors: COTS process and product (subjective, qualitative, system-independent), assessment of non-operational evidence.

Qualitative judgment on non-operational aspects from all life-cycle phases preceding operation, should include the following aspects:

- Development process (structured, semi-formal, formal methods)
- Safety culture (awareness of consequences, ethical and financial)
- Documentation (accuracy and consistency of reports and dependencies)
- Resources (human labor and mechanical aid)
- Informal checks (manual analysis of code and documents)
- Automated static analysis (syntactic and semantic checks by automatic tools)
- Non-operational tests (execution for non-representative inputs).

Phase 4: Quantitative assessment of COTS components within the system (Saglietti 2000a).

Key factors: COTS behavior during testing and operation (objective, quantitative, system-dependent), assessment of operational evidence.

- Quantitative judgment of operational aspects: in spite of deterministic nature of logical faults, a probabilistic approach is justified by input randomness, representing physical variables subject to unpredictability of state transitions in technical processes under control.

- Estimation of software reliability by statistical testing on the basis of significant amount of operational evidence, correct execution of a large number of independent, operationally representative scenarios; based on sampling theory, estimation of upper bound of failure probability at a given confidence level.

Phase 5: Validation at integrated system level.

Key factors: interface analysis (white box, component dependent, context dependent).

Inconsistencies between physics and logical specification have to be analyzed, including:

- Violated global properties, such as in Airbus incident in Warsaw.

- Violated local properties at interfaces, such as:

(a) Mars Climate Observer, where force reference system has not been specified by NASA; JPL controllers expected Newton (British system), Lockheed Martin Astronautics provided metric units.
(b) Berlin Fire Brigade – Millennium: different operating systems in different system parts, inconsistent communication of date format.
(c) Ariane 5 Explosion, where conversion routine originally developed for flight trajectory of Ariane 4 used insufficient range.
(d) Blood Databank, where program designer developed the database manager for a single computer application without considering networked applications; could not anticipate that simultaneous access to a record by two users would lead to a hazard.

## 5. RIGOROUS DEVELOPMENT PROCESS

The fundamental objective of achieving software safety is to guarantee that the software does not cause or contribute to a system reaching a hazardous state. This may be supported by the rigor and conformance to well defined engineering standards.

### *5.1 Safety Analysis and Verification*

Mojdehrakhsh et al. (1994) describe a process for safety improvement. The process must assure that the software safety analysis is closely integrated with the system safety analysis and the software safety is explicitly verified.

The following activities for software safety program are geared toward achieving the stated objective, with emphasis on analysis first and verification next. The analysis phase involves conducting system safety analysis to:

- Identify the key inputs into the software requirements specification, such as hazardous commands, limits, timing constraints, sequence of events, voting logic, failure tolerance, etc.
- Create and identify the specific software safety requirements in the body of the conventional software specification.
- Identify which software design components are safety critical.

The initial analyses and subsequent system and software safety analyses identify when software is a potential cause of a hazard or will be used to support the control of a hazard. Using specific software design and implementation techniques and processes is crucial to reduce potential hazards introduced by software. The associated verification phase involves the following steps:

- Apply specific verification and validation techniques to ensure appropriate implementation of the software safety requirements.
- Create test plans and procedures to satisfy the intent of the software safety verification requirements.
- Introduce any necessary corrective actions resulting from the software safety verification.

The process is fundamental to the identification of the safety-critical functions and the causal factors, including the factors that may be software-induced or controlled. The identification shall include also creation of an appropriate taxonomy of hazards (commission, omission, timing error).

Some of the techniques used for analysis and verification include (Bishop, 1990): Fault Tree Analysis (FTA), Event Tree Analysis (ETA), Hazard Operability Analysis (HAZOP), Failure Mode Effect Analysis (FMEA), Failure Mode Effect and Criticality Analysis (FMECA), Common Mode Failure Analysis (CMF), Cause Consequence Diagrams (CCD), and Petri Nets (PN).

Most of these methods allow for rigorous treatment of the development process. The one we found particularly suitable for studying safety properties is Petri nets (Saglietti 1998, Kornecki at al. 1998). We have applied it to the analysis of the TCAS software specification.

*5.2 Petri Nets in Design Verification*

TCAS is a Traffic Collision Avoidance System, introduced to reduce the risk of mid-air collisions between aircraft. The TCAS equipment is a small electronic device, consisting of a computer and software, a directional antenna, a transponder and cockpit display and controls (Fig. 4 and 5).

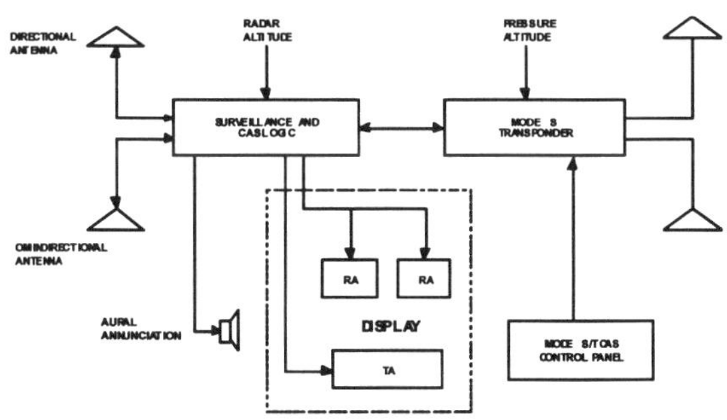


Fig. 4. TCAS Architecture.

TCAS continuously monitors aircraft within 10 nautical miles to identify the potential threat. If an aircraft is determined to be a threat, a Traffic Advisory is issued showing the available information about the potential threat, including the call sign, relative position, altitude difference, and altitude vector.

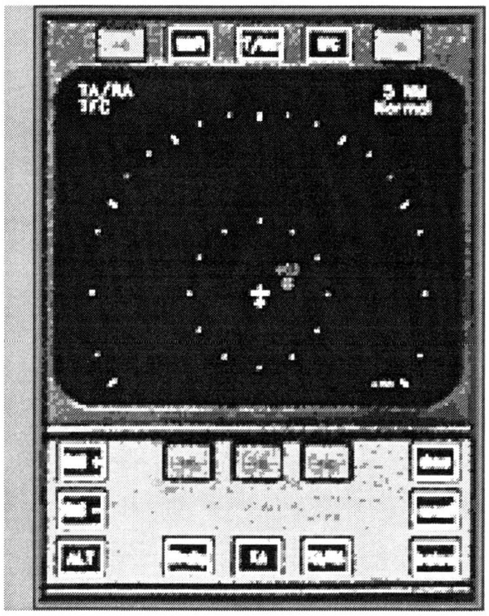

Fig. 5. TCAS Cockpit Display

If an aircraft persists to be a threat, the TCAS evaluates whether collision might occur within the next 23 seconds. In such case, a Resolution Advisory is issued, which requires the protected aircraft to execute one of the range vertical maneuvers.

Based on the TCAS specification, a Petri net model was built and verified using a public domain tool Cabernet (Kornecki et al. 1998). The objective was to check formally, whether certain states, having impact on safety, can be achieved as a result of inconsistencies in the specification.

For example, state changes to the Emergency State, described under respective conditions, as below, converted to predicates:

IntruderStatus = Continuing &
Threat = Establishes &
RA = VerticalSpeedLimit &
Intruder is TCAS-equipped &
Own Mode S ID is higher &
Threat has the sane sense as Own &
!(RA display deferred)

were analyzed using the reachability graphs. In effect, several properties were verified with respect to safety aspects. Verification of a sample property of the original TCAS specification, that has been found "not-satisfied", is shown in Fig. 6.

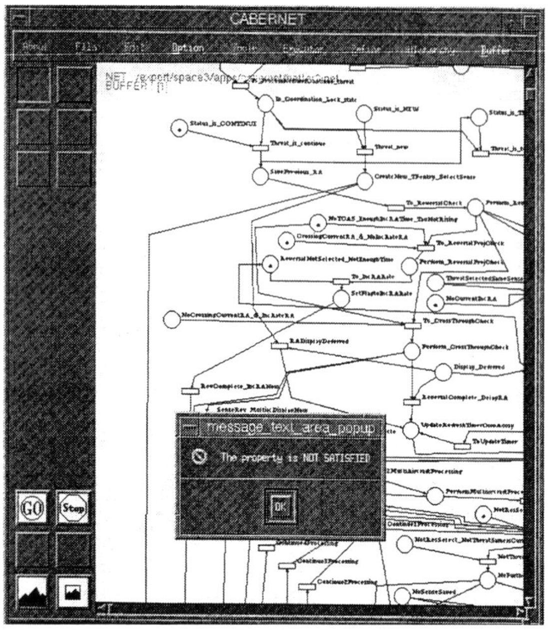


Fig. 6. Sample Output from Cabernet.

The automatic verification was very successful, since it allowed to identify several flaws in the TCAS specification. The results showed, however, that a significant knowledge of Petri nets (timed environment relationship nets, in this case) is needed to build a meaningful TCAS model. In order to enhance usability, it is useful to support formalization by means of established engineering practice. One such practice emerging recently, with supporting software tools, is the use of a Unified Modeling Language (UML).

## 6. VERIFYING TIMING REQUIREMENTS USING UML

This section presents a method of enhancing UML by dealing with timing requirements via statecharts. We propose to combine traditional engineering tools, such as those relying on the UML notation, with formal methods tools, such as model checkers. Combining both types of tools into a single integrated system via a common user interface, to express and verify timing properties, leads to an architecture of a verification system (Al-Daraiseh et al. 2001) as in Fig. 7.

Two tools, a typical design tool based on UML, and a typical model checker based on a formal method, are interfaced to a user via a GUI. For expressing timing requirements, UML statecharts are converted to Extended Timed Graphs (XTGs), which in turn are converted to textual representation and analyzed by a model checker. A new software tool has been developed to ease the interfacing of models built in XTG formal language to other tools.

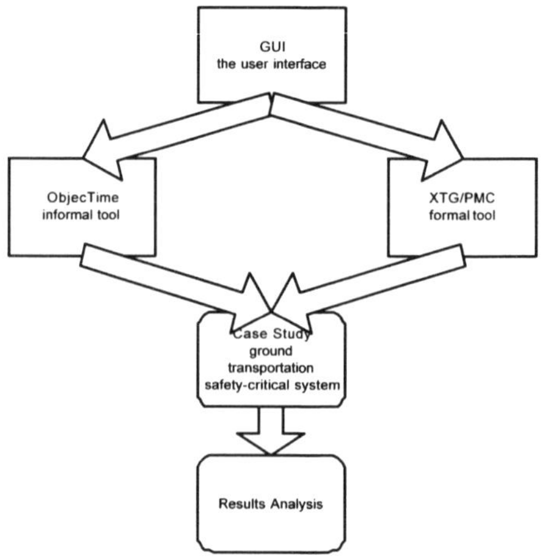


Fig. 7. Architecture of the Integrated System.

### *6.1 Integrated Methodology*

Both statecharts and XTG's are visual tools to model the behavior of software. Since UML has no means to express real-time properties, we propose a method to convert the UML statecharts to XTGs, so that statechart models can be model checked using the PMC model checker (van Katwijk et al. 2000). Converting a UML model expressed in statecharts, to XTG graphs, requires thorough understanding of their mutual correspondence.

XTG is a new engineering notation for describing real-time systems based on timed automata (Ammerlaan et al. 1998). It provides a simple representation for high-level specification languages and is a suitable notation for those languages that allow extensive modeling of data, having a maximal progress semantics, and modeling interprocess communication by value passing through data channels.

A UML statechart consists of states and transitions. A state is an ontological condition that persists for a significant period of time, is distinguishable from other such conditions, and is disjoint from them. A distinguishable state means that it differs from other states in the events it accepts, the transitions it takes as a result of accepting those events, or the actions it performs. A transition is a response to an event that causes a change in state.

A simple statechart diagram is shown in Fig 8. In the IDLE state nothing is happening and the timer is not counting down, waiting for the start command "StartCmd". The StartCmd carries a single value, which is the starting time the timer shall count down.

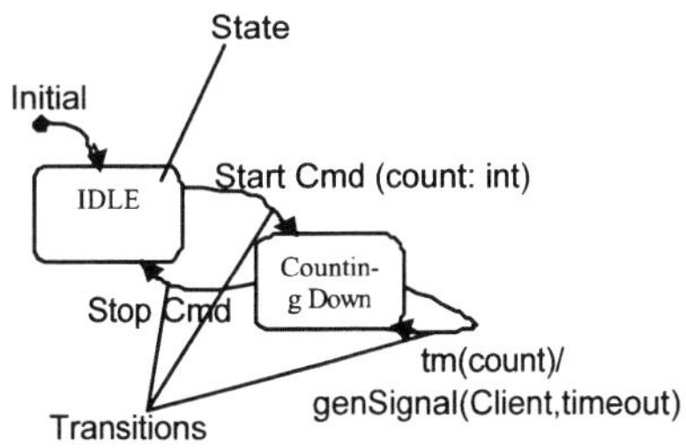


Fig. 8. Simple Statechart Representing a Counter.

When this event occurs the system transfers to the Counting Down state and the timer starts counting down again. There is also a timeout signal "tm". When the timeout occurs the "tm" transition is taken and the action from its action list will be executed. The actions are shown after the slash "/". In this case the action is to send a signal to a client object. Actions in UML are usually short but they run until completion before the transition completes. After that, the counting state is re-entered, and the counting starts again until the Stop Cmd event occurs. Then the system transfers to IDLE waiting for the Start.

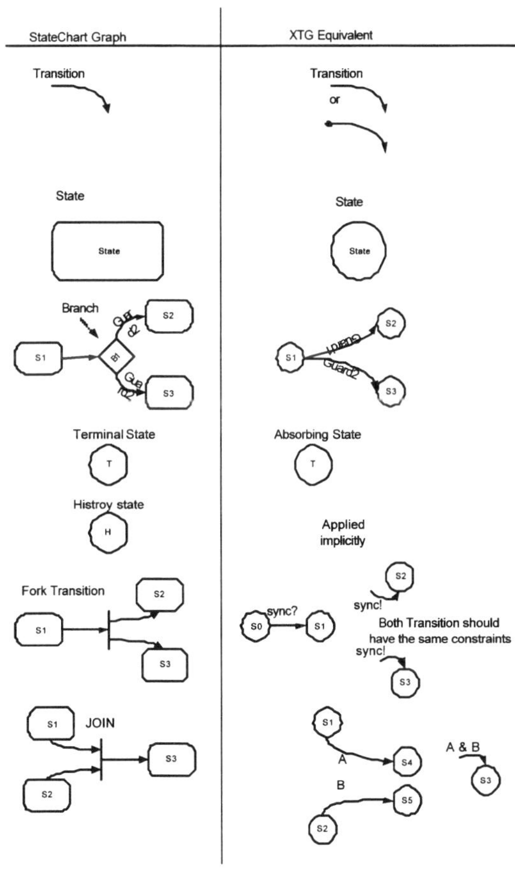


Fig. 9. Statechart to XTG Graphs Conversion.

To convert statecharts to XTG graphs, we developed an equivalence table (partially presented in Fig. 9), that contains the statechart symbols and the XTG equivalent symbols. As shown in Fig. 9 the state symbols are equivalent but their shapes are different. Transitions in both cases are the same. According to the table, the equivalent XTG graph for the Counter example is shown in Fig. 10.

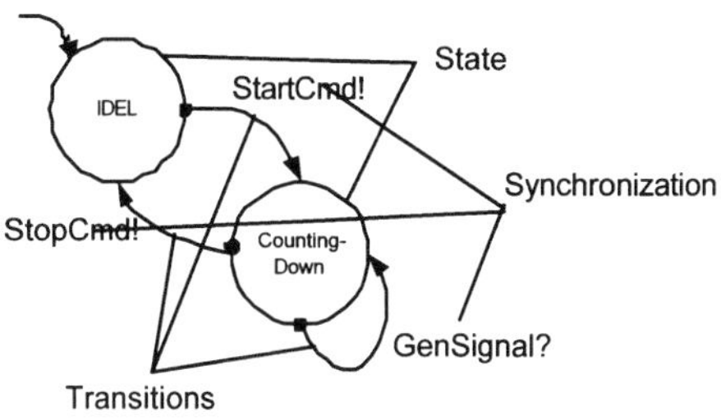


Fig. 10. XTG Graph Equivalent to Fig. 8.

## *6.2 Railroad Crossing Case Study*

To check the usefulness of the concept of model conversion and the applicability of the tool developed for this purpose, we applied the entire procedure to the railroad crossing system. The system consists of three parts: a set of sensors detecting the passing trains, a gate controller, and a gate. The sensors process can be in one of three states namely: far from the crossing, T0, near the crossing, T1, and in the crossing, T2. A function called g(t) is defined to represent the gate formally, where output of this function $g(t) \in [0,90]$ means that at 0 the gate is closed and at 90 the gate is open.

<u>UML Design.</u> The behavior of the system is described with UML in Fig. 11. The safety and utility properties are expressed in temporal logic (z represents clock) as follows:

1. AG(train@in => gate@closed)
2. AG(gate@closed => (z=0).
   AF (gate@open and z <=5))

The timing properties are missing on the diagram, since they cannot be expressed in UML.

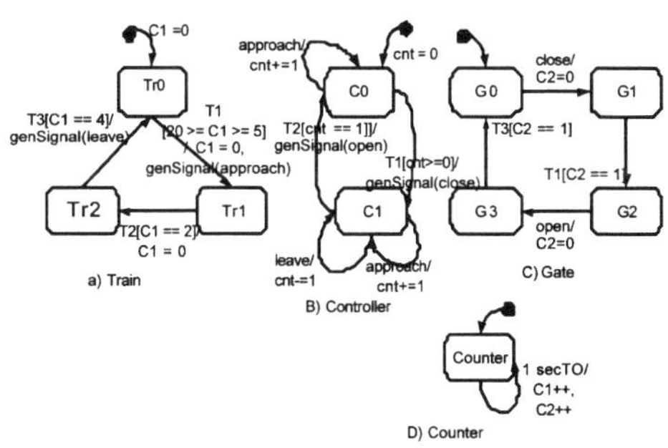


Fig. 11. UML Model of System Behavior.

As shown in this figure, the UML representation has to include four processes instead of three, as suggested previously, since UML doesn't have the clock variable. The sensors process can be in one of three states:

- Far from the crossing, Tr0
- Near the crossing, Tr1
- In the crossing, Tr2.

The railroad crossing example was implemented in ObjecTime tool as shown in Fig. 12, using four actors: TrainActor, GateActor, ControllerActor and CounterActor. ObjecTime behavioral diagram is similar to UML statecharts. The behavior of the GateActor is simulated internally. The principle of operation is the same as described in the UML design.

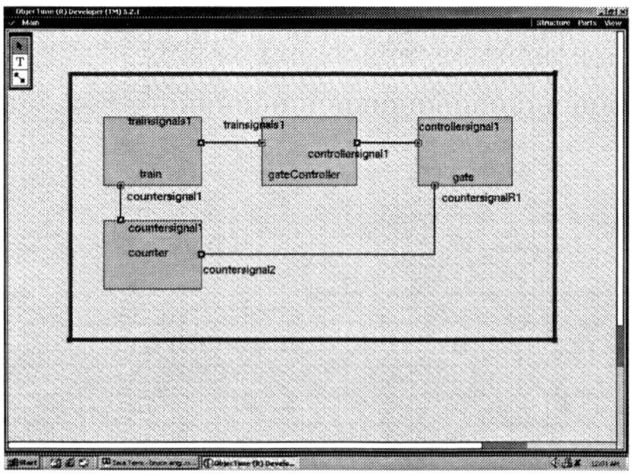


Fig. 12. Railroad Crossing ObjecTime Model.

With ObjecTime model simulation, we could only check sequencing and a limited timing behavior via animation, but there was no proof of the safety constraints since the tool does not have this capability.

_XTG Design._ Next the UML description is translated into XTG's. The gate has four possible states. Initially it is open (G0). When receiving a *close* signal, it transfers to the closing state (G1). One time unit is allowed to close the gate, enforced by [C2==1] clock constraint.

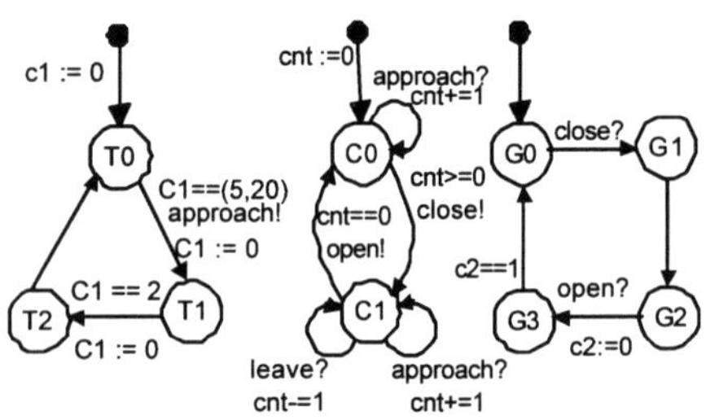


Fig. 13. Three XTG Processes Representing the System: Sensors, Controller and Gate.

This description is sufficient to use the XTGConverter tool for producing an XTG code, form the XTG graphs, and prepare for automatic model checking. Results for railroad crossing are presented in (van Katwijk et al. 2000, and Al-Daraiseh et al. 2001). A portion of the generated XTG code for the sensors process is shown in Fig.14.

```
-- this is the RailRoadCrossing System
system RailRoadCrossing
. . .
processes
        Sensors         Train1 ;
        Gate            Gate1 ;
        GateCont        GateCont1;
composition
        Train1 | Gate1 | GateCont1

graph Sensors
-- Local variable assignment added here
state
-- clocks and variables here
clock C1:= 0;
ports
--all the ports are added here
        out approach
        out leave
--initial state comes after init keyword
init
        T0
locations
. . .
T0
{
when C1 == (5,20)
sync approach!
do  C1 := 0;
goto    T1
}
```

Fig. 14. XTG Code for Model Checking.

Finally, after running the generated code through the PMC model checker, the verification results are obtained. The properties to be verified in the railroad crossing system are:

```
1. AG(train@in => gate@closed)
2. AG(gate@closed => (z=0).
   AF(gate@open and z <=5))
```

This means that, at any time, if there are trains in the crossing the gate should be closed, and at any time if the gate is closed, it should open in the future.

## 7. CONCLUSION

Modern safety related systems can be so complex and so software dependent, that dealing with the software production process becomes indispensable to achieve the required level of trustworthiness in software.

Usual verification methods, such as simulation and informal prototyping, are dealing only with the products, not with the processes, and may work well for checking individual, safety related or timing properties, but not for the verification of the correctness of development processes. On the other hand, the use of diverse software for safety critical systems is process-related, basing on the assumption of dissimilar development methodologies. Also the evaluation of commercial off-the-shelf software to be re-used in a safety-critical context requires a careful consideration of the original development process.

On the whole, the approaches described in this article are complementary with respect to their suitability to address different problems at different levels of formalism. An adequate combination is considered to be promising. One way to proceed is to incorporate formal techniques into engineering practice and support it by a combined use of automatic tools, so that formal verification of safety properties may become more automated and better accessible to engineers and software developers

## REFERENCES


Al-Daraiseh A., J. Zalewski, H. Toetenel (2001), Software Verification in Ground Transportation Systems, *Proc. SCI2001, 5th World Multiconference on Systemics, Cybernetics and Informatics*, Orlando, Fla., July 22-25

Ammerlaan M., R.L. Spelberg, H. Toetenel (1998), XTG – An Engineering Approach to Modelling and Analysis of Real-Time Systems, *Proc. 10th Euromicro Workshop on Real-Time Systems,* IEEE Computer Society Press, pp. 88-97

Anderson E., J. van Katwijk, J. Zalewski (1999), New Method of Improving Software Safety in Mission Critical Real-Time Systems, *Proc. 17th Int'l System Safety Conference,* Orlando, FL, August 16-21, System Safety Society, Unionville, Va., pp. 587-596,

Bishop P., ed. (1990), *Dependability of Critical Computer Systems: Guidelines. Techniques Directory*. Elsevier Applied Science, London

Eckhardt Q.E. Jr., L. D. Lee (1985), A Theoretical Basis for the Analysis of Multiversion Software Subject to Coincident Errors, *IEEE Trans. on Software Engineering,* Vol. SE-11, No. 12, pp. 1511-1517

Ehrenberger W. (2001a), Software Diversity: Some Considerations on Failure Dependency, *Proc. SCI2001, 5th World Multiconference on Systemics, Cybernetics and Informatics*, Orlando, Fla., July 22-25

Ehrenberger W. (2001b), Software-Verifikation, Hanser Verlag, Munich

Ehrenberger W., F. Saglietti (1993), Architecture and Safety Qualification of Large Software Systems, *Proc. ESREL'93, European Safety and Reliability Conference,* Munich, Germany, May 12-14, Elsevier, Amsterdam, pp. 985-999

Hilburn T., J. Zalewski (1996), Real-Time Safety Critical Systems: An Overview, *Proc. 2nd IFAC Workshop on Safety and Reliability in Emerging Control Technologies,* Elsevier, Oxford, pp. 127-138

van Katwijk J., H. Toetenel, A.E.K. Sahraoui, E. Anderson, J. Zalewski (2000), Specification and Verification of a Safety Shell with Statecharts and Extended Timed Graphs, *Proc. SAFECOMP 2000, 19th Int'l Conf. On Computer Safety, Reliability and Security,* Springer-Verlag, Berlin, pp. 37-52

Kornecki A., B. Nasah, J. Zalewski (1998), TCAS Safety Analysis Using Timed Environment-Relation Petri Nets, *Proc. ISSRE'98, Int'l Symposium on Software Reliability Engineering,* Paderborn, Germany, November 4-7

Leveson N. (1996), *Safeware: System Safety and Computers*, Addison-Wesley, Reading, Mass.

Littlewood B., D. R. Miller (1987), A Conceptual Model of Multi-Version Software, *Proc. FTCS-17, Int'l Symp. On Fault-Tolerant Computing,* IEEE Computer Society Press, pp. 170-175

Mojdehrakhsh R., W.T. Tsai, S. Kirani, L. Elliott (1994), Retrofitting Software Safety in an Implantable Medical Device, *IEEE Software,* Vol. 11, No. 1, pp. 41-50

Rushby J. (1994), Critical System Properties: Survey and Taxonomy, *Reliability Engineering and System Safety*, Vol. 43, pp. 189-219

Saglietti F., W. Ehrenberger, M. Kersken (1992), *Software Diversität für Steuerungen mit Sicherheitsverantwortung,* Report BAU-Forschungsbericht FB 664, Bundesanstalt für Arbeitsschutz, Dortmund

Saglietti F. (1998), Integration of Logical and Physical Properties of Embedded Systems by Use of Timed Petri Nets, *Proc. SAFECOMP'98, 17th Int'l Conf. On Computer Safety, Reliability and Security,* Springer-Verlag, Berlin, pp. 319-328

Saglietti F. (2000a), Evaluation of Pre-developed Software for Use in Safety-Critical Systems, *Proc. EUROMICRO'2000, 26th Euromicro Conference on Software Process and Product Improvement,* IEEE Computer Society Press, Vol. 2, pp. 193-199

Saglietti F. (2000b), Statistical Significance of Expert Judgement for Ultrahigh Software Reliability Demands, *Proc. 5th Int'l Conference on Probabilistic Safety Assessment and Management,* Osaka, Japan, Nov. 27-Dec. 1, Universal Academy Press, Tokyo

Saglietti F. (2001), Criticality and Sensitivity Analysis for Off-the-Shelf Components in Safety-Relevant Systems, *Proc. SCI2001, 5th World Multiconference on Systemics, Cybernetics and Informatics,* Orlando, Fla., July 22-25

Sahraoui A.E.K., E. Anderson, J. van Katwijk, J. Zalewski (2000), Formal Specification of a Safety Shell in Real-Time Control Practice, *Proc. WRTP'2000, 25th IFAC Workshop on Real-Time Programming,* Elsevier, Oxford, pp. 117-123

Copyright © IFAC New Technologies for Computer Control,
Hong Kong, 2001

IFAC Publications
www.elsevier.com/locate/ifac

# DESIGN AND PERFORMANCE EVALUATION OF DCS USING DCOM, CORBA AND TSPACE

**Weonjoon Kang, Hongseong Park, *Hogil Lee**

Dept. of Electrical and Computer Eng. Kangwon Univ. chuncheon, Kangwon, 200-701
*Korea Institute of Industrial Technology, chonan, Chungnam, 330-825
E-mail: tailhook@control.kangwon.ac.kr, hspark@kangwon.ac.kr, *leehg@kitech.re.kr

Abstract: This paper designs three types of DCS based on typical object oriented middleware of DCOM, CORBA and Java VM and extends and applies DCS to the device level based on Ethernet. It is very important to guarantee the real-time features of the system when object oriented middleware is used for DCS. Therefore, this paper analyzes the performance of DCS based on each object-oriented middleware and presents some advanced service modules to reduce network delays. *Copyright © 2001 IFAC*

Keywords: distributed control system, object-oriented, DCOM, CORBA, Java, TSpace

## 1. INTRODUCTION

As the software techniques based on object-oriented are growing frequently, many researches are focused to adopt and integrate those techniques into the industrial environments such as factories, power plants and power supply systems etc, because the techniques is purposed for constructing the distributed system (Redlich et al, 1998). Specially, integrating the object oriented techniques into a distributed control system which is consist of a plenty of devices distributed on the network provides many benefits to the users and the system designers, because the devices, such as sensors, actuators, and controllers, are simply able to be represented as an object. In addition, since the off-the-shelf object-oriented middleware like DCOM, CORBA and Java VM supports communication protocols that are transparent to operating systems, the users can design and manage whole system easily without expensive costs. But, when we decide to use middleware to implement a DCS, we often meet some considerations as follows: Firstly, the performance of the system should not be lowered after integration. Adding or deleting some objects to and from the system should be possible with keeping the system stability. Secondly, the real-time features of the system should be guaranteed. Since DCOM, CORBA and Java VM have comparatively long time delays in communication between remote objects, most researches in Object oriented DCS region has distributed the devices such as controllers, sensors, actuators in either a local station or a bounded real-time network such as fieldbus.(Mock et al, 2000; Shoop et al). Figure 1 (a) shows the generic object oriented DCS.

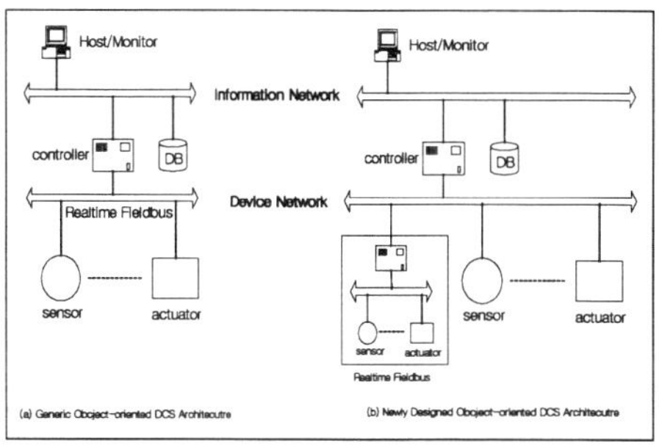


Fig. 1. Generic and New DCS Architectures

But, in this paper, we design three types of distributed control systems and each of them is implemented based on DCOM, CORBA and TSpace implemented on Java VM, and the system performance for each DCS is analyzed in view of network delay. As shown in Figure 1 (b), all field devices are distributed in a device network. In our implemented systems, a control loop consists of a controller, a sensor and an actuator and the roundtrip time for a control loop is measured to compare the performance for each middleware. We adopt the designed systems to Ball and Beam position control system and showed the results. And also, we design and adopt the common object type for the field devices and additionally, two advanced modules to reduce transmission delay are suggested. The organization of this paper is as follows: Section 2 describes middleware principles and the transmission time comparisons. Section 3 presents Object design and Section 4 explains the implementation. Finally, Section 5 states the conclusions.

## 2. THE OFF-THE-SHELF MIDDLEWARES

The middleware, while it is laid on operating system, support many inter-object services such as inter-process communication, inter-object linking, object access security, event and transaction management. Figure 2 shows the general middleware structure. Object1 doesn't need to know where Object2 is located because Broker executes all interactions with Object2. Therefore Object1 always should rely on Broker.

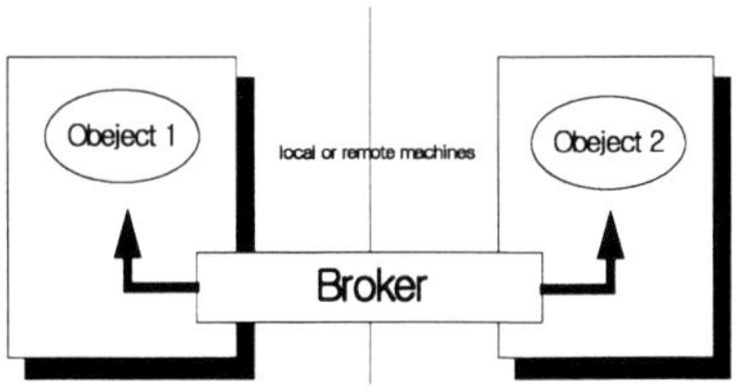


Fig. 2.General Middleware Structure

But Broker generally has many tasks to create and communicate between two Objects and this makes serious time delay not less than Fieldbus used in the industrial environments. So, we should considerate about the time features of the off-the-shelf middleware because DCOM, CORBA and Java VM have designed not for the physical devices level but for integrating and linking of the software components level. Due to the time delay, the off-the-shelf middleware trend to be applied in soft-real time area, for example a monitoring system.

### *2.1 DCOM*

As shown in Figure 3, the client object is linked with the server object through COM Library. That is, the Client has information of a server object and requests to COM to access server object. Then COM searches the location of server object in the system registry and creates the server object; return the pointer of server object to the client object.

If the server object is located in the remote system, COM sends the request to the remote COM through RPC (Remote Procedure Call) channel. However, the client object doesn't need to know where the server object is. Because the proxy, which represents the server object in the client system, and the stub, which represents the client object, roles as virtual counterpart object in local system.

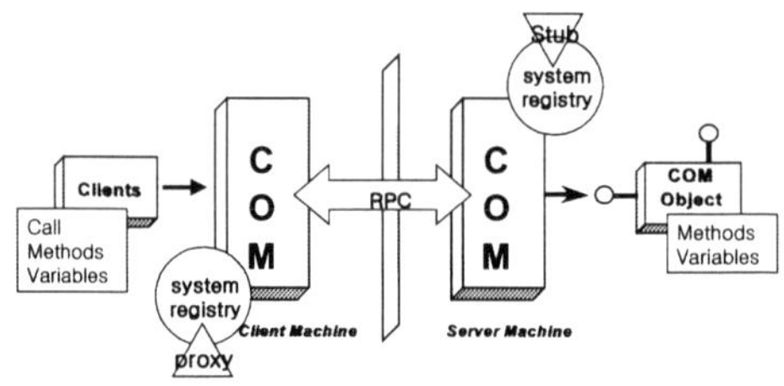


Fig. 3. DCOM Object Model

### *2.2 CORBA*

CORBA is similar to DCOM because ORB(Object Request Broker) works same as COM in Figure 3 and interfaces are used to access other objects. CORBA has an interface repository and an implementation repository to store object information. Figure 4 shows the CORBA Object Model.

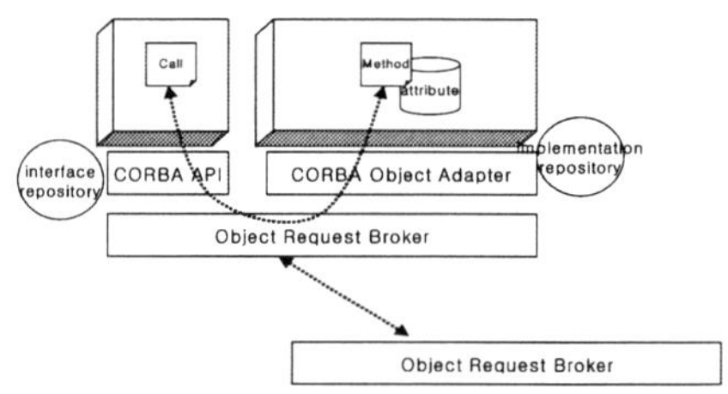


Fig. 4. CORBA Object Model

### *2.3 TSpace*

TSpace, developed by IBM, is lightweight global data storage and expected to be used as the network middleware. Since TSpace is implemented as java classes and has small size, it can be ported to the small java-embedded device. TSpace Model is shown in Figure 5. When client 1 wants to send a tuple *aaa1* to Client 2, Client 1 makes a tuple space in TSpace Server and writes *aaa1* to the TSpace Server. Then Client 2 requests to read *aaa1* from the TSpace Server and the Server answers. Actually, TSpace is not like DCOM and CORBA but its simple model is good to share data between clients and to implement the system with very low costs.

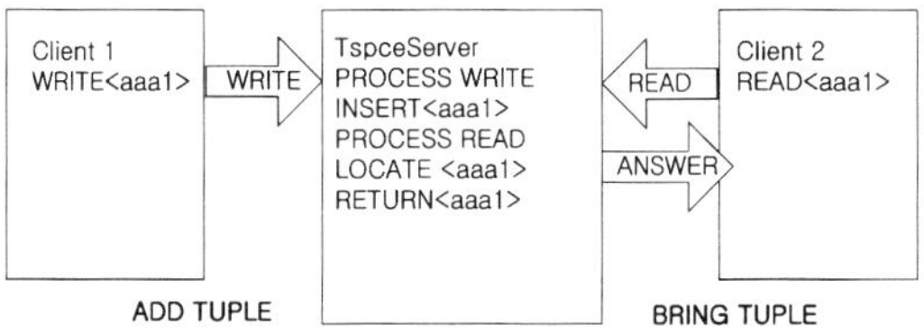


Fig. 5. TSpace Model

## 3. OBJECT DEFINITIONS AND DESIGN

The fundamental objects for the control system are mainly I/O objects such as sensor or actuator and control objects. In addition, new objects were introduced in this paper such as gateway objects for linking with real-time fieldbus, event objects for event notification and broadcast objects for multi transmission.

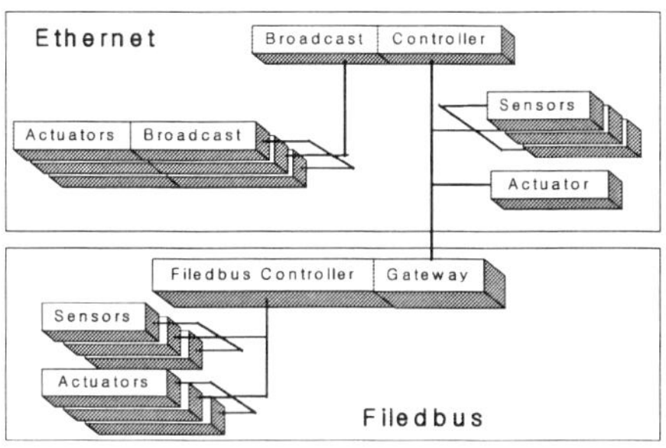


Fig. 6. Simple Object-oriented distributed system model

Control objects calculate control algorithm with the data got from sensor objects, which represent input devices, and transmit the result to actuator objects, which represent output devices. Since DCOM and CORBA doesn't support simultaneously data transmission to multi-nodes, broadcast object was introduced to broadcast data. This is expected to save time to transmit from a control object to multi actuator objects. Figure 6 shows a simple object-oriented distributed system model.

### *3.1 Definitions of Object Attributions*

An object is basically composed of variables and methods. And each attributes is composed of Out_variables and Out_methods, which are exposed by an interface to other objects, and In_variables and In_methods, which are used only within an object. This is shown in Figure 7. As shown in Figure 6, an object can be accessed via an interface that exposes Out_methods and Out_variables. In_methods and In_variables can be used to implement internal object functions. In case of I/O objects, device driver APIs provided by the manufacturer are contained in In_methods. Table 1 is the attributes for the basic object. All objects inherit these attributes and each

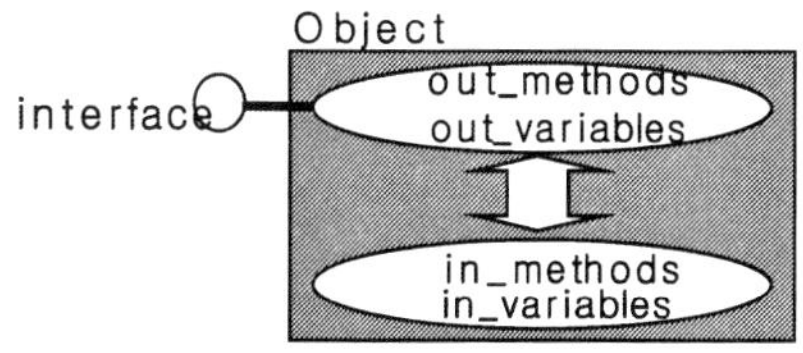


Fig. 7.Simple Object Structure

object has its additional attributes.

Table 1 The attributes for the basic object

| Type | Attributes | Meanings |
|---|---|---|
| Variables | IID | Interface ID |
| | Name | Object Name |
| | CMD | Controller command parameters |
| Methods | Get() | data output from the object |
| | Put() | data input into the object |

### *3.2 I/O Object*

I/O objects are sensor objects and actuator objects. These objects should have methods to make controller access to their data. Therefore, I/O objects inherit Get and Put attributes from the basic object, but the two attributes only are used as Out_methods. As shown in Figure 8, I/O objects have several In_methods to implement interactions with physical devices. These interactions can be done through the device driver's APIs. Table 2 shows the attributes for I/O object.

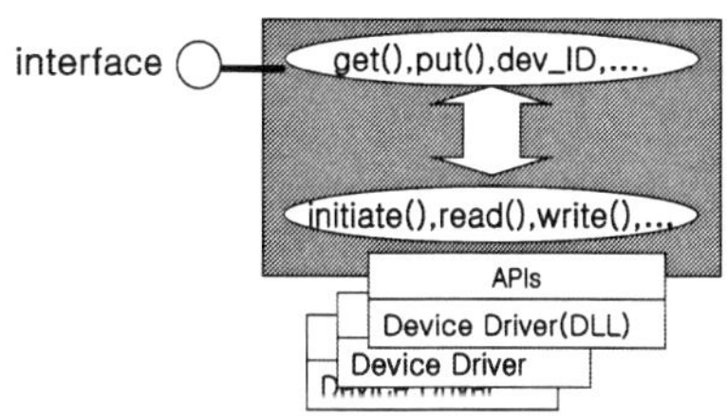


Fig. 8. I/O Object Structure

Table 2 The attributes fot the I/O object

| Type | Attributes | Meanings |
|---|---|---|
| Out_Variables | IID | Interface ID |
| | Name | Object Name |
| | CMD | Controller command parameters |
| | Dev_ID | Physical device ID |
| Out_Methods | Get() | data output from I/O object |
| | Put () | data input into the I/O object |
| In_Methods | Initiate() | Device Initialize |
| | Read() | data output from the device |
| | Write() | data input into the device |
| | Diag() | device self-diagnostic |

### *3.3 Control Object*

The control object, which is placed in the controller, creates all related object instances such as sensor objects, actuator objects, gateway objects and do exchange data with them. Also, the control objects roles a data distributor from sensor object to actuator object.

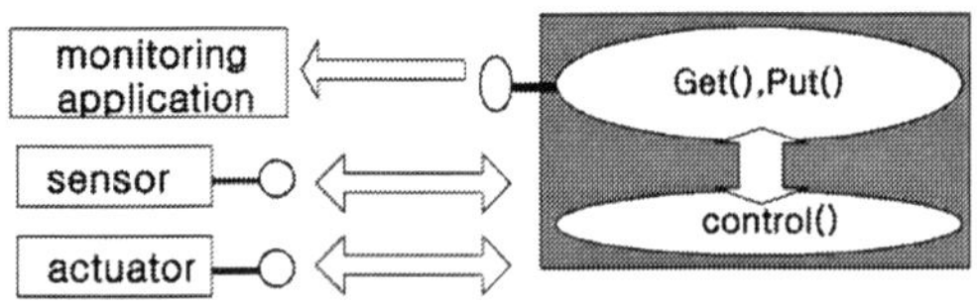


Fig. 9. Control Object Structure

Before transmitting to actuator object, the control object calculates control algorithm with data from sensor object. Another role of control object is to provide monitoring data to the user application when it requires. Figure 9 shows the structure of control object. The control object has various attributes as shown in Table 3.

Table 3 The attributes for the control object

| Type | Attributes | Meanings |
|---|---|---|
| Out_Variables | IID | Interface ID |
| | Name | Object Name |
| | IOVAL | monitoring data |
| Out_Methods | Get() | data output to monitoring application |
| | Put () | data input from monitoring application |
| In_Methods | control() | control algorithm |

*3.4 Gateway object*

Since DCOM and CORBA and Tspace just work on Ethernet and TCP/IP, Gateway object was introduced to link with the objects on non-Ethernet network, such as real-time fieldbus. Gateway object is similar to I/O object. In other words, as shown in Fig. 10, Fieldbus driver's APIs are contained in In_method, The APIs should provided by the fieldbus manufacturer.

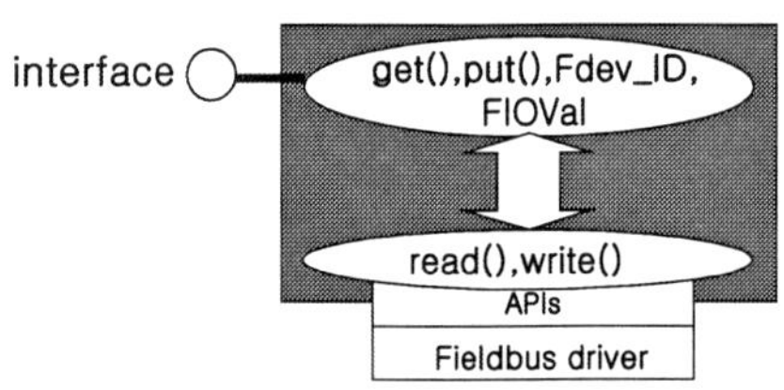


Fig.10. Gateway object structure

The attributes for the gateway object are presented in Table 4.

Table 4 The attributes for the Gateway Object

| Type | Attributes | Meanings |
|---|---|---|
| Out_Variables | IID | Interface ID |
| | Name | Object Name |
| | Fdev_ID | device ID on Filedbus |
| | FIOVal | I/O value to/from Filedbus |
| Out_Methods | Get() | data output from I/O object |
| | Put () | data input into the I/O object |
| In_Methods | Read() | data output from Field device object |
| | Write() | data input to Field device object |

*3.5 Broadcast service*

Since DCOM and CORBA communicate with remote object in Peer-to-Peer, it is not possible to transmit data to multi-nodes at the same time. That is, when a control object sends a value to various actuator objects, some time gap necessary to happen between the first actuator and the last one, because DCOM and CORBA transmit the value several times to each actuator object. As a result, increasing unnecessary network overheads will introduces time gap and data inconsistency. To prevent this time gap, a broadcast module was introduced in this paper. A broadcast object is a simple send and receive daemon built on UDP and included in a control object and an actuator object that want to broadcast data. Therefore broadcast module doesn't inherit basic object attributes. As shown in Figure 6 and Figure 11, a control object may send a value either using broadcast module or not.

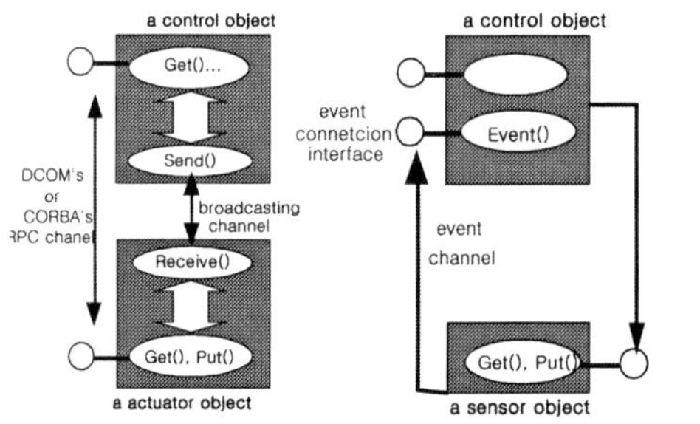


(a) broadcast service module (b) event service module

Fig. 11. The broadcast service module and the event Service Model

A control object just has a broadcast send methods and an actuator object just has a broadcast receive method.

*3.6 Event service*

A control object mainly gets data from sensor objects and put them into actuator objects periodically. But, when a sensor object catches some urgent data, like events or alarms that happens non-periodically, it is necessary to notify the event to the control object as soon as possible. Fortunately, DCOM and CORBA provide the event notification service. To implement event service, an interface is added in control object side. This interface is used by sensor object to notify event and exposes event execution methods. Figure 11 shows a control object and a sensor object with event service.

## 4. IMPLEMENTATIN AND PERFORMANCE ANAYSIS

Implementation above designed objects is done in two ways, one for DCOM and CORBA and the other for TSpace. Because TSpace supports some limited object-related service. The implemented system consists of a control object, a sensor object and an actuator object. The goal of the implemented system is the performance comparison with time features of each DCOM, CORBA and TSpace, when it is applied in a simple distributed control system. The test-bed is the Ball position control system, as shown in Figure 12.

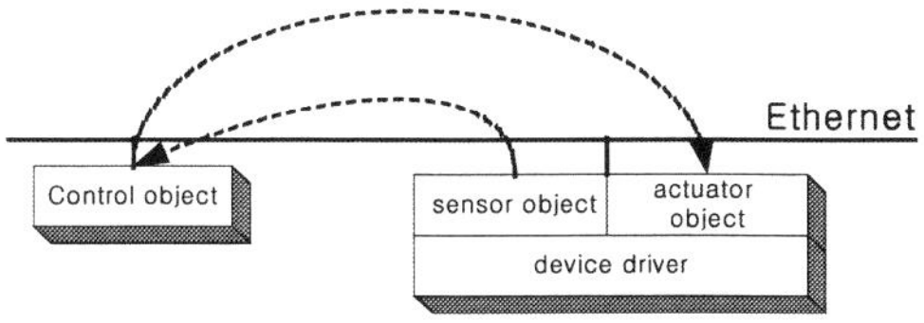


Fig. 12. Ball and Beam position control system

TSpace based test system is different as shown Figure 13. But each system, based on DCOM, CORBA and TSpace, has same object number and attributes and placed in same system.

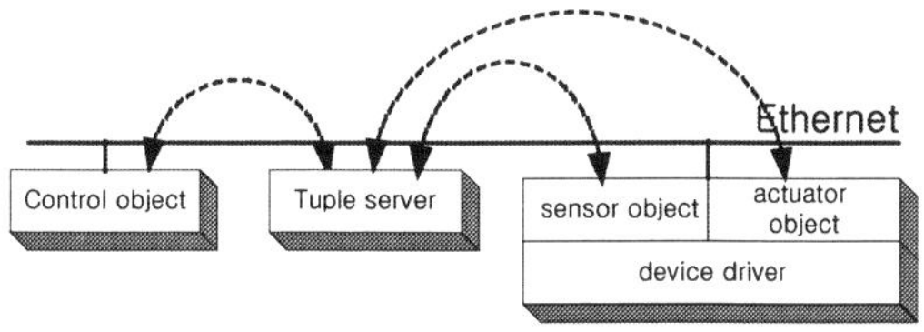


Fig. 13.The control system for TSpace

Figure 14 shows the time of a round trip between control object and sensor object, actuator object. The start time was stamped when a control object polls a value from a sensor object and the end time was stamped after sending the controlled value to the actuator object. The minimum roundtrip time to control ball position is 80ms. But TSpace has a great time delay than CORBA and DCOM because there is no synchronization in the tuple server that notifies data incoming.

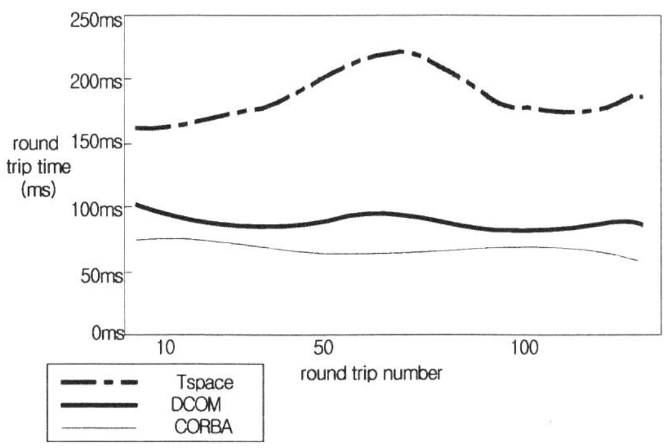


Fig. 14. Time features of a simple distributed control system

Therefore TSpace is not suitable for the control system. DCOM showed unstable control state because its roundtrip time is near by minimum time value. However, CORBA roundtrip time always under the minimum roundtrip time and showed stable ball position control state. As mentioned in Section 3, to reduce communication time delays, we suggested two advanced modules, broadcasting and event. Follows are the measurement results of applied module into DCOM. As shown in Figure 15 (a), when broadcasting module applied, although node number are increased, all node could receive data simultaneously. Figure 15 (b) illustrates the time performance when event module applied. By using event module, the network resident time is reduced by half.

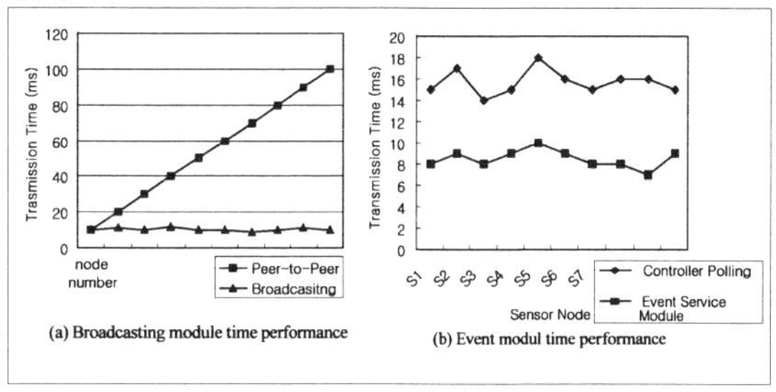


Fig. 15. Advanced module performance

## 5. CONCLUSIONS

Applying the off-the-shelf middleware, DCOM, CORBA and TSpace, to the distributed control system have some considerations. The most is their transmission time delays due to the middleware's internal structure. In this paper, three simple distributed control systems, based on DCOM, CORBA and TSpace, were designed and implemented and the system performances were compared and evaluated. TSpace showed a great time delay between clients and TSpace server and DCOM is not suitable for DCS for the highly time delay. Comparatively, CORBA showed small time delay than DCOM and suitable for our simple DCS. But, by using advanced modules suggested in this paper, DCOM can be used in DCS.

## REFERENCES

M. Mock et al, Monitoring Distributed Real-Time Activities in DCOM. *ISORC2000*, pp.26-33, March, 2000

R. Schoop et al, Agent-Oriented Material Flow Control System Based on DCOM. *ISORC2000*, pp.342-345, Mar, 2000

J. P. Redlich, et al, Distributed Object Technology for Networking, IEEE Comm. Maga. pp.100-111, Oct, 1998

E. Chang et al, A Large Scale Distributed Object Architecture – CORBA&COM for Real Time Systems. ISORC98, pp.338-341

S. D. Urban, L. Fu, J. Shah, The Implementation and Evaluation of the Use of CORBA in an Engineering design Application, *Software-Practice and Experience*, vol. 29, pp.1313-1338, 1999

G. Bollella, T. J. Lehman, Real Time T-Spaces, *IECON'99*, pp.837-842, Nov, 1999

IFAC Publications
www.elsevier.com/locate/ifac

# A REAL-TIME NETWORK MANAGER FOR DISTRIBUTED IMPRECISE COMPUTATION

**Hidenori Kobayashi** *,1 **Nobuyuki Yamasaki** *,2
**Yuichiro Anzai** *,3

* *Faculty of Science and Technology,*
*Keio University,*
*3-14-1, Hiyoshi, Kohoku-ku,*
*Yokohama, Kanagawa, 223-8522, JAPAN*

Abstract: This paper describes a real-time network manager which is suitable to client-server models in distributed real-time systems, where the server tasks are based on the imprecise computation model. An imprecise task is a task that is capable of trading off the quality of the result to the amount of the execution time. However, the task lacks the ability to determine the appropriate amount of work. The proposed real-time network manager determines, on behalf of the server task, the adequate amount of execution time as well as the adequate transmission time by holding the intermediate results in an EDF manner. *Copyright © 2001 IFAC*

Keywords: Real-time communication, Real-time operating systems, Distributed models, Communication environments, Multimedia

## 1. INTRODUCTION

Significant advances in recent processors and network technologies have made it possible to construct real-time systems over a network. In these distributed real-time systems, tasks communicating over the network are often in the client-server model. This model is very useful in systems where the resources are physically distributed. An example of such systems is a video-on-demand system where the client and the server are located on different nodes. Another example is a robot which consists of several nodes where sensors and actuators are attached to different nodes. In the latter example, the model is usually called the sensor-actuator model.

One of the important characteristics of the distributed real-time systems is that the end-to-end timing constraints often must be met over the network. There are several crucial and distinct problems in guaranteeing the end-to-end timing constraints between a client and a server. They are related to the network, the operating system, and the application software.

Firstly, the greatest source of unpredictability is the network. The network traffic generally varies from time to time, so does the network latency. An exclusive line or a centralized management scheme is generally necessary to cope with this problem.

Secondly, the main issue in operating systems is the scheduling algorithm. Besides selecting an appropriate scheduling algorithm, the protocol processing of messages must be integrated to it in order to guarantee end-to-end timing constraints.

Lastly, two problems are concerned with the execution time of a server task. One of them is that the execution time of the server needed to produce the result of the same quality is not always the same. For example, execution time of a multiple target tracking task varies according to

[1] kobahide@ny.ics.keio.ac.jp
[2] yamasaki@ics.keio.ac.jp
[3] anzai@ics.keio.ac.jp

the input data and the number of the tracks the task was holding at the time. In addition, different clients may request different timing constraints and different quality of results for the same operation. To cope with such situation, a server task must be able to dynamically trade off the quality of the result for its execution time. The other problem is the estimation of the optimal amount of work, which should be completed before the result is actually transmitted back to the client, provided that the server is allowed to terminate at an arbitrary point. Since a task does not have any information on the available execution time nor network latency to the destination node, it is impossible for a server task to determine its appropriate execution length.

One solution to all the above problems is to statically schedule all the communications and the executions of tasks on every node together, which certainly is an unrealistic approach in practice. In contrast, the main purpose of this paper is to provide a real-time guarantee for dynamically established client-server relations. It describes a network manager architecture that, by imprecise computation, solves both the operating system and the application design problem stated above. The network manager also handles unexpected network congestion to some extent by means of imprecise computation and some features of the communication hardware, *Responsive Link*, described in (Yamasaki, 1997).

## 2. IMPRECISE COMPUTATION

Imprecise computation was introduced in (Lin *et al.*, 1987) as a technique to trade off the quality of the result for the amount of execution time. A task based on the imprecise computation model consists of two parts, the mandatory and the optional part. The mandatory part merely produces an acceptable result, while the optional part enhances the quality of the previous result. In other words, in real-time systems, the mandatory part must always complete in time, while the optional part need not execute when an overload occurs or when the timing constraint was too strict.

A number of researches on imprecise computation were carried out from various aspects. Some of them regard imprecise computation as a means to achieve graceful degradation in quality of services. For example, (Feng and Liu, 1996) applies the imprecise computation technique to real-time communication services in an ATM network. The scheme uses the cell loss priority bit in the header to distinguish mandatory packets from optional packets. Some of other works focus on controlling the executions of imprecise tasks so that the performance of the whole system would be maximized. This includes vast numbers of scheduling algorithm. Examples are found in (Feng and Liu, 1997), (Kim *et al.*, 1998), and (Baruah and Hickey, 1998). Others are related to the implementation of imprecise computations. In (Hull *et al.*, 1996), Imprecise Computation Environment (ICE), an environment for developing imprecise systems is described. The ICE helps the application programmer by automatically generating support code for imprecise tasks.

## 3. DISTRIBUTED CLIENT-SERVER MODEL

In this section, characteristics of the client-server model, especially in distributed real-time systems, are pointed out. As previously stated in section 1, the required execution time of the server is not constant at all. Accordingly, it is reasonable to assume that the server task is based on the imprecise computation model. Under this supposition, there are two issues concerning the distributed client-server model.

One of them is related to the amount of the execution time given to the server task. Since the server task is based on the imprecise computation model, longer execution of its optional part usually leads to higher quality in the result. However, longer execution possibly results in higher deadline miss ratio of the client task. In order to find out the balancing point of this trade off, both information on the logical structure of the imprecise task and the available execution time is necessary. The reason that the logical structure of the task is needed is that even an imprecise task would not produce any result before the completion of its mandatory part. In addition, the result produced by an actual imprecise task is often a staircase function of allocated time. This means that the optimal point to terminate the task is at where the results of higher quality are produced. As a matter of course, a task itself is capable of determining the optimal point that it should terminate if the length of available execution time is given. However, in a multiprogrammed environment, a task has no knowledge on the amount of execution time available to it. Note that making the task ask the scheduler about its available execution time is not an efficient method, as the available execution time is subject to dynamic task arrivals and deletions. Thus, it introduces too much overhead and no guarantee. On the other hand, the operating system conversely has no knowledge of the logical structure of the task. Therefore, neither the task nor the operating system can set the termination point of the imprecise task appropriately. This problem will be solved shortly in section 4

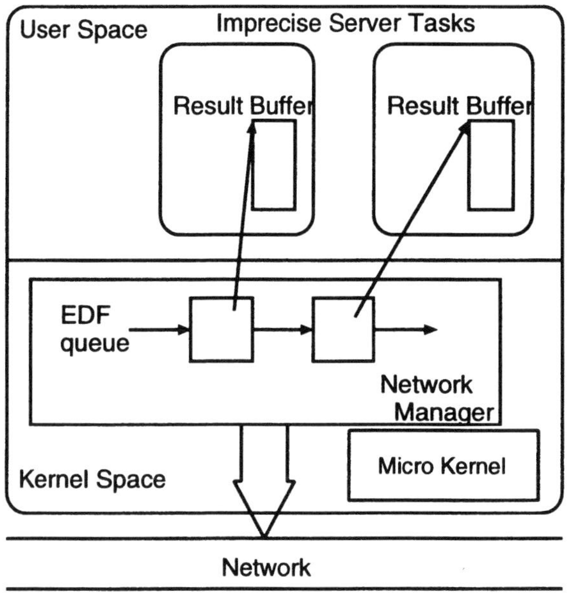


Fig. 1. Overall architecture

The other issue is the variation in the network latency, which also is a great source of a deadline miss. In a network, where communications happen dynamically, it is impossible to provide a full guarantee on the network latency, unless some kind of centralized management scheme exists, which often leads to very low utilization of the network. In this paper, this problem is first solved by using exclusive lines. As this approach is neither efficient nor always achievable, a method for optimistic control of network latency is presented in section 5.

## 4. HARDWARE INDEPENDENT ISSUES

### 4.1 *The Real-Time Network Manager*

The real-time network manager lies between the application tasks and the network, whose overall architecture is shown in figure 1. Its aim is to transmit in time the best possible result produced by an imprecise task server without actually figuring out the balancing point of the previously stated trade off in execution time. This is mainly achieved through interactions between the tasks and the real-time network manager.

A typical case of such interaction is illustrated in figure 2. At time $t_{REQ}$, the client task makes a request through the local and the remote real-time network manager. The message contains the server identifier, the request identifier, and the latest time the result should arrive, $t_{DUE}$. The request arrives at $t_{ST}$, and at the same time the server task starts to execute a corresponding service routine. The server task completes its mandatory part at time $t_{MEND}$. At this time, a pointer to the result, the size, and the deadline of the message, $t_{DUE}$, is handed to the network manager

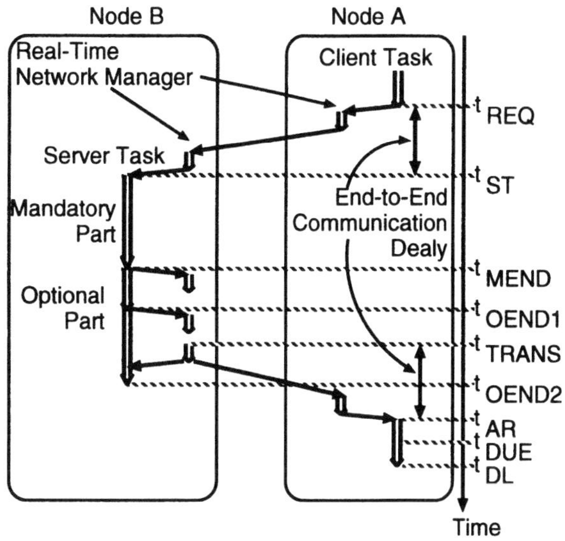


Fig. 2. Interactions between tasks and the real-time network manager

with the transmission request. In contrast to the case where a task is responsible for deciding when to transmit the data, the network manager is responsible for it, while the server task now has to take care not to overwrite the previous result contained in the buffer. The network manager holds the requests in a queue in the ascending order of the latest transmission time. For the sake of simplicity, it is assumed here that no other tasks put requests to the network manager. In this case, the network manager immediately sets itself to sleep until $t_{TRANS}$. Meanwhile, the server task continues to execute its optional part and produces a higher quality result at $t_{OEND1}$. This time the server task tries to update the information held within the network manager. The network manager recalculates the latest transmission time and again sets itself to sleep. As shown in the figure 2, the server task would have produced a result of even higher quality if it were allowed to execute till $t_{OEND2}$. Suppose this more precise result can not be delivered to the client node in time. Then, the result is of no use to the client and the server task had just executed in vain. This situation can be avoided by the network manager. It wakes up at $t_{TRANS}$ and transmits the result. And then, it signals the server task to stop its execution.

In summary, the proposed real-time network manager guarantees the best possible result be delivered to the client in time, by reserving necessary transmission time for intermediate results. As a matter of fact, the server task had no idea on the shortage of its execution time until it was signaled to terminate its execution.

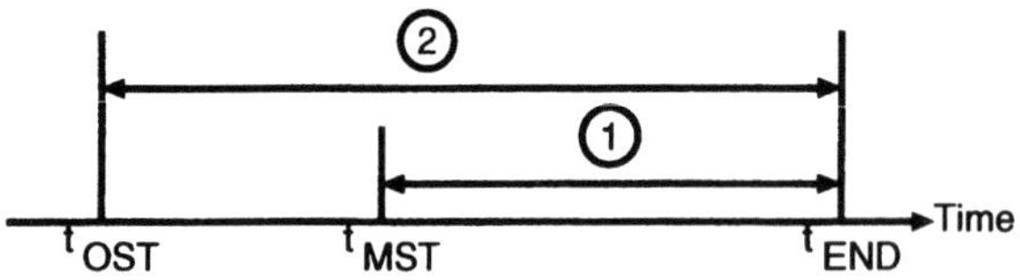


Fig. 3. Reserved transmission time

### 4.2 *Transmission Queue Management*

On processing the request, the network manager treats two types of data differently. In this Paper, the data that must be transmitted timely is called the *mandatory data* or the *mandatory result*. On the other hand, the data that does not have to be transmitted if it can not meet its timing requirements is called the *optional data* or the *optional result*. For example, a result of a mandatory part is a typical kind of the mandatory data, where the entire data must be transmitted to the opponent timely in order to make sense. In comparison, a result of an optional part could be categorized as both cases. If the result of the optional part was something that should be added to the previous result of the mandatory part, the added data is an optional result. If the data were to replace the previous result, the new complete data will also be categorized as a mandatory result.

The proposed real-time network manager fully utilizes this flexible nature of results to resolve contentions between transmission requests. First, the network manager calculates the latest time which the transmission of the result should end at by subtracting the network latency and the reception protocol processing overhead from the latest acceptable arrival time of the result. This is labeled as $t_{END}$ on the time axis in figure 3. It then calculates the latest time the transmission of the whole data and the mandatory part should begin, $t_{OST}$ and $t_{MST}$ respectively, by subtracting the transmission protocol processing overheads of the corresponding size from $t_{END}$ . From these calculated values, the manager checks whether any conflict between requests exists. Passed requests are held in the ascending order of $t_{END}$. The only check is made on whether the necessary time to transmit the mandatory result, ① in figure 3, overlaps that of the neighboring requests. The overlap in transmission time of the whole data, ② in figure 3, is not checked at all here. One reason for this is that even checking both overlaps can not always provide full guarantee that the whole data is transmitted on time. If the network latency varies unexpectedly and suppose it was larger at the start of the transmission, it may well be appropriate to discard the optional part in order to meet the deadline. In conclusion, checking both overlaps does not pay.

If a new transmission request was rejected, the task has no guarantee in transmitting the result.

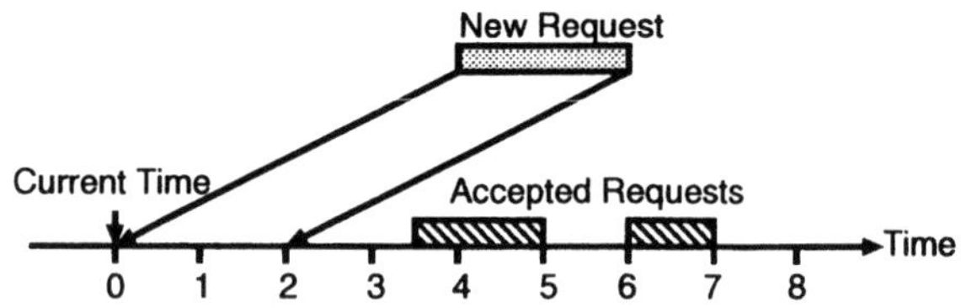


Fig. 4. Transmission of a denied request

However, it does not mean that there is no time to transmit the data. That is, the server task still has a chance to transmit the result in time. This is illustrated in figure 4. Boxes on the time axis show previously accepted requests. Suppose a newly arrived request had $t_{MST}$ at 4 and $t_{END}$ at 6. This request conflicts with one of the accepted requests, hence can not be accepted by the network manager. However, transmission of the data is possible at time zero. This sort of transmission is not always possible, so the server task should stop its execution immediately on the rejection of its first request and try to transmit the result as early as possible. If it was an update request, the server task just terminates. The conflict between this direct transmission and the reserved transmission by the real-time network manager is avoided by scheduling the network manager with the highest priority.

### 4.3 *Imprecise Transmission*

The network manager wakes up at $t_{OST}$ to transmit the result. At this moment, it recalculates $t_{OST}$ and $t_{MST}$ using newer values of corresponding network latency. These recalculated values are used to check whether the reserved amount is still adequate or not. If the current time of the system was earlier than $t_{OST}$, it transmits the whole result. Otherwise, it transmits only the mandatory data. Since the network manager neither checks the existence of any optional data nor validates its timing requirements, this is an optimistic algorithm.

## 5. HARDWARE DEPENDENT ISSUES

This section describes issues related to the actual implementation of the proposed scheme. Furthermore, a method for mitigating the effect of unexpected variations in network latency is also described.

### 5.1 *Responsive Processor and Responsive Link*

The proposed real-time network manager is implemented in a distributed system, which consists of *Responsive Processors* connected via maximum of four *Responsive Links*. Details of *Responsive*

*Processor* and *Responsive Link* are described in (Yamasaki, 1997). One *Responsive Link* consists of one *Event Link* and one *Data Link*, whose packet size is 16 bytes and 64 bytes respectively. The network manager uses the *Event Link* to transmit requests and the *Data Link* to transmit results. A packet has two priority bits in its header field. This is used for two purposes in the switch. One is the arbitration and the other is the routing of the packet. The priority may be changed when the packet crosses the switch, depending on the routing table. This paper takes these advantages to rank the messages. The last feature is that it uses DMA to transmit and receive data. Although this allows faster transmission, it introduces great unpredictability to the CPU memory references.

### 5.2 *Calculation of Communication Delay*

The end-to-end communication delay between two tasks is calculated as a summation of software protocol processing overhead in the sender node, network latency, the software protocol processing overhead in the receiver node. Note that the delay caused by the local task scheduling is not included. This is because a node in distributed systems never holds the latest information of remote node schedules. In a system that uses *Responsive Link*, protocol processing overheads are calculated by a function of the number of the packet. These overheads will remain the same regardless of the network congestion. On the other hand, network latency can be calculated by a function of the hop count, provided that no contention occurs. Note that the data size does not affect the network latency. This is because, as soon as the first packet comes in, the network manager starts execution so that the reception of the subsequent packet overlaps the protocol processing of the preceding packets. However, if any contention exists, throwing probing packets may be the only way.

The proposed real-time network manager calculates the latency from the hop count if it was the first transmission to the destination node. The network manager receives a feedback from the destination node if any message did not arrive in time. It then holds that value in a table. This value will be used on next transmission to the node and will be updated on another deadline miss.

### 5.3 *Priority Assignment to Packets*

In a system where exclusive lines are not available, all the communication must be controlled by a centralized management scheme. However, in systems where this approach is not feasible, there must be at least some way to discriminate real-time messages from non real-time messages. Since the *Responsive Link* switch is capable of handling four levels of priority, the real-time network manager assigns four levels of priorities to each message in the following manner. The lowest priority is assigned to non real-time messages, the second lowest to soft real-time messages, the second highest to the optional result, and the highest to the mandatory result.

## 6. EVALUATION

An experiment was performed to show that the proposed real-time network manager actually is capable of realizing graceful degradation in the quality of the result produced by an imprecise server task.

The experiment system consists of two *Responsive Processor* nodes connected by a *Responsive Link*. Throughout the evaluation, the SPARClite was set to 100 MHz and the *Responsive Link* 43 Mbaud. Under this condition, preliminary experiments were made to formalize the communication delay of the *Data Links*. Based on the result, it was possible to calculate the sender's overhead($S$), network latency($N$), and the receiver's overhead($R$) by the following equations, where $P$ is the number of the packet and $H$ is the hop count. All the values are in micro seconds.

$$S = 1.46 * P + 17.05 \tag{1}$$

$$N = 1.13 * H + 18.23 \tag{2}$$

$$R = 8.88 * P + 31.30 \tag{3}$$

In addition to these values, an end-to-end deadline guarantee requires that the time between the packet reception and the actual wakeup time of the destination thread be also taken into account. Here, it is assumed that it takes as much as 1 tick long, which is 100 $\mu s$.

The experiment was performed in the following manner. On one of the nodes runs a client task and the network manager. On the other node runs a server task, the network manager, and some periodic tasks. The server task is based on the imprecise computation model. The worst case execution time of the server task is set to 10 $ms$, where the mandatory part occupies 2.5 $ms$. The rest of the server task belongs to the optional part. The optional part consists of five parts, each of which requires 1.5 $ms$ to enhance the quality of the former result. This represents the staircase function in producing imprecise results. The quality of the result produced by an imprecise task is measured in terms of value. The value is zero at the point where only the mandatory part is completed and will be one if all of the optional part is completed. The periodic tasks on the server

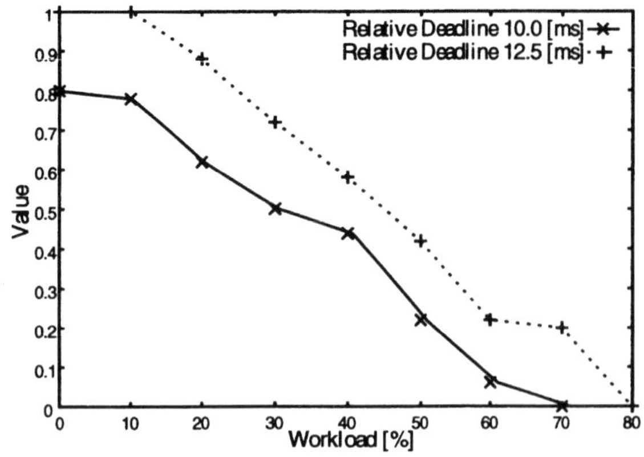


Fig. 5. Graceful degradation in the quality of the result

node are used to estimate the effect of dynamic arrival and deletion of tasks with higher priorities. This is accomplished by scheduling both nodes according to the fixed priority algorithm, where priorities are assigned in the descending order to the network manager, the client task, the periodic tasks, and then the server task.

The result is shown in Figure 5. The horizontal axis represents the CPU workload of the server node and the vertical axis represents the value of the result obtained by the client. The client requests the server in two different ways with respect to the relative deadline. The solid line shows the case where the relative deadline is set to 10 *ms* and the dotted line 12.5 *ms*. Since the periodic tasks have higher priorities, the available execution time of the server task will decrease in response to increases in the total amount of their execution time. If the relative deadline is 10 *ms*, the value obtained by the client never reaches 1, even there is no workload at all. This is because the requested deadline is too tight. Although the relative deadline just covers the worst execution time of the server task, it will not be sufficient when the communication delay and other software overheads are taken into account. At the point where the workload is 70 %, the sum of the CPU utilization of the periodic workload and the mandatory part of the server task approaches 1 if the relative deadline is 10 *ms*. On the other hand, when the relative deadline is 12.5 *ms*, it actually sums up to 1 if the workload is 80 %. In either cases, the server task merely has enough time to complete the mandatory part. When the workload of periodic tasks gets higher than these points, the server task will not have enough time to complete even the mandatory part. Hence, all the transmission request from the server task will be rejected by the network manager.

As for the transmission requests which were accepted by the network manager, the average time it took for the client task to receive the result were 9.68 *ms* and 12.2 *ms* respectively. This result confirms that the result is transmitted on adequate time. As a matter of fact, deadline misses were scarcely observed during the experiment.

## 7. CONCLUSION

In distributed real-time systems, it is crucially important that the end-to-end timing constraints be met. However, in a dynamic real-time system, it is a difficult problem to determine an appropriate amount of execution time for a server task. The proposed real-time network manager solves this problem by means of imprecise computation. The experiment showed that the proposed scheme is capable of determining the adequate amount of execution time as well as the adequate transmission time.

Future works are planned on the management of the network latency. This includes developments of routing algorithms and the management of virtual circuit that are suited to *Responsive Link*.

## 8. REFERENCES


Baruah, Sanjoy K. and Mary Ellen Hickey (1998). Competitive On-Line Scheduling of Imprecise Computations. *IEEE Transactions on Computers* **47**(9), 1027–1032.

Feng, W. and J. W.-S. Liu (1996). Performance of a Congestion Control Scheme on an ATM Switch. In: *Proceedings of the International Conference on Networks.*

Feng, W. and J. W.-S. Liu (1997). Algorithms for Scheduling Real-Time Tasks with Input Error and End-to-End Deadlines. *IEEE Transactions on Software Engineering* **23**(2), 93–106.

Hull, David, W. Feng and J. W.-S. Liu (1996). Operating System Support for Imprecise Computation. In: *Proceedings of the AAAI Fall symposium on Flexible Computation.*

Kim, Jai-Hoon, Kihyun Song, Kyunghee Choi, Gihyun Jung and SeunHun Jung (1998). Performance Evaluation of On-Line Scheduling for Imprecise Computation. In: *Proceedings of Fifth International Conference on Real-Time Computing Systems and Applications.* pp. 217–222.

Lin, K., S. Natarajan and J. W.-S. Liu (1987). Imprecise Results: Utilizing Partial Computations in Real-Time Systems. In: *Proceedings of the IEEE 8th Real-Time Systems Symposium.* pp. 210–217.

Yamasaki, N. (1997). A Functionally Distributed Responsive Micro Controller for Distributed Real-Time Processing. In: *Proceedings of IEEE/RSJ International Conference on Intelligent Robots and Systems.* Vol. 2. pp. 793–798.

IFAC Publications
www.elsevier.com/locate/ifac

# Distributed Client/Server Network Control Using Web-Based Technologies

**Jiang B. Liu and Igor V. Motov**

*Bradley University*
*Peoria, IL 61625, U.S.A.*

ABSTRACT: Distributed computer systems control modeling is a well-studied topic in network control and management, yet the problem of finding optimal approaches that are not too complex to implement on the real-time distributed systems remain unsolved. We have observed that many computer networks are naturally clustered. Client/Server Web-computing can be easily implemented on the clustered computer network model. Our approach is to use such natural clustering to develop and implement the control schemes using Web-based technologies in clustered networks. *Copyright © 2001 IFAC*

Keywords: Distributed Control, Clustered Network Model

## 1. INTRODUCTION

Distributed control over the network using the Internet protocols has many advantages. Network infrastructure that supports the IP-based applications has grown exponentially in the last two decades. Optical network growth rate has exceeded the Silicon and Storage growth rates by a large margin in the last few years. The broadband networks plus the rapid rises of the eCommerce and web services are driving the control applications towards IP-based real-time client/server computing platform. It has an enormous economical advantage to develop a suitable network service model to provide such a platform. In this paper, we will present one such a distributed client/server control model based on the clustered network structure to meet such a challenge.

Distributed computer networks have many attractive features, including the capability of sharing processing of jobs in the event of unbalanced resources within the system. There are many resources sharing models developed in the last decade to tackle the optimal load scheduling with respect to a variety of system structures (Bonomi and Kumar, 1990). However, finding the global optimal solution of a distributed system resource sharing has been shown to be NP-complete in general (Price and Krishnaprasad, 1984). The problem of finding an optimal control scheme that is not too complex to solve or implement on the real-time distributed computer networks remains unsolved.

In a lightly loaded distributed computer system (DCS), resource sharing is simple, since clients can choose any server and route that best serves their needs. However, when the number of the tasks increases, the problem becomes more complex. A competition of resources results in significantly nonlinear behaviors of the system. It creates the nonproportional job loads, where some resources in the system are overused and others are underused, that decreases the overall system performance and client's satisfaction. In worst cases it can cause deadlocks and overall system congestion.

In this research we have developed a solution of the resource sharing problem by decomposing it according to the clustering structures of the network. Our approach is motivated from the observation that most computer networks are naturally clustered (Figure 1). The imbalance jobs usually occur between the clusters rather than inside the clusters. The computers inside a building or an organization are usually connected on a local area network (LAN). The LANs are in turn connected on a metropolitan area network (MAN) or a wide area network (WAN).

Broadband-based network has the logical over physical topology that naturally supports the clustered network control. The clustered control model has the advantages of multi-leveled control strategy, such as the Inter-cluster and Intra-cluster controls, these are well suited for today's Web-based network structures. Each cluster, whether it is

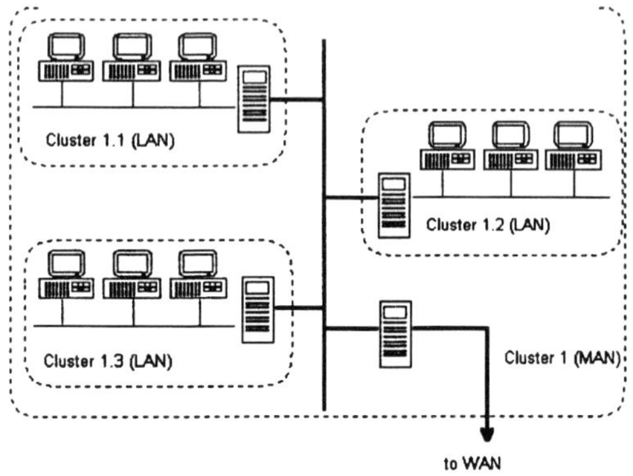


Fig. 1. Naturual Clustering of the network

application-based or topology-based, has a control server that processes all the jobs submitted from the clients within the cluster. The server will either process the job locally or forward it to other cluster, based on the control policy specified. HTML form-based clients are used to submit the simple jobs with few parameters and less interactions to the web server. For more complex jobs, Java applet clients are used. The Web server directs the job process request to the control server that is implemented by a set of Java servlets and Java ServerPages (JSPs) (Asbury and Weiner, 1999). The job process resources can also be fetched from a cluster database via the JDBC or a set of Enterprise JavaBeans (EJBs). The control server also communicates with other clusters using either Java RMI or CORBA.

Our goal is to develop network scheduling policies for allocating computational resources among computers of a network, which could then be easily implemented in a real-time, general-purpose, naturally clustered DCS. The theoretical analysis and simulation results of our resource sharing model and algorithm showed that our approach is valid and efficient as a practical resource sharing control scheme on the clustered computer networks. The web-based client/server controls using Java technologies are robust, efficient, and easy to implement on different computing platforms (different hardware, different operating systems) on the network.

## 2. Analytical Model of a clustered network

During last three decades, many models for distributed computer networks were developed. Janusz Filipiak (Filipiak, 1988) listed five distinguished approaches, grouped according to the model types and their usages.

The most detailed models are the simulation models. They are often used for analysis of congestion controls and routings. However, it is difficult and time consuming to apply such models for a large system, especially when the reliable statistics have to be collected. The most accurate models are the analytic description of a queueing system that can be obtained by using probability distribution of all events. Less accurate models, but more simpler to implement, are the models that specify only the averages and variances of the distribution. Diffusion models are the examples of such models. Dynamic flow models utilize time-dependent average quantities. The static flow models describe systems by means of long-term averages only.

Our analytic model built on the clustering structure is based mainly on the diffusion models. It has the advantages of being both accurate and quick to implement.

The performances of the models also depend on the scheduling policies used. A DCS using the central scheme contains a single job dispatcher that receives all the jobs for the system, makes job allocation decisions, and sends the jobs to the processors. The DCS using the distributed scheme has no central control units; all computers (processors) in the system are completely independent. This DCS requires a more complex solution than the DCS with central control. The independence of the computers within a system results in much difficulty in sharing free resources, especially when the significant cooperation is required. A number of solutions for this job scheduling problem have been developed in previous years. Most of them were based on the decomposition. A good summary of them can be found in (Larsen, McEntire, and O'Reilly, 1982). Some original approaches such as the application of Bayesian decision theory, using preferred lists of four-cube systems, or using contract net protocol have also appeared.

Usually, DCS with the centralized control is considered to be simpler and faster, but decentralized control makes systems more reliable and efficient.

The control scheme in our model is an alternative to the above control schemes. It uses a clustering approach for network resource scheduling. It can be considered as an implementation of the conceptual level principles of a "domule" developed in (Tenney and Sandel, 1981) and (Eager, Lazowska, and Zahorjan, 1986). The domule is a combination of a decision agent and its subsystem model. Our DCS consists a decision agent and a number of clusters. The decision agent and the clusters are connected to the network. Each cluster consists of a set of computers and a job scheduler. A job scheduler in a cluster should decide whether to perform the job inside the cluster or send it to another cluster. The job scheduler performs *a nondeterministic routing.* The parameters of this routing are calculated by a decision agent. The decision agent regularly collects information about the performance of clusters, finds optimal routing parameters for each cluster, and sends the parameters to the clusters.

The computer network is assumed to consist of $n$ clusters and can be considered as *loosely coupled.* It means that the communication time between processors is an important system parameter. The clusters can be identified either by the geographic characteristics or by the operational structures of the network. Figure 2 shows the logical structure of the cluster. The physical structure of clusters can be different but it will not affect our algorithm as long as the cluster has a job scheduler with the nondeterministic routing.

We assume that the job stream generated in each cluster follows a Poisson distribution and all the service requirements are exponentially distributed. The service time distribution adapted in this research is based on Eager, et al., paper (Eager, Lazowska, and Zahorjan, 1986).

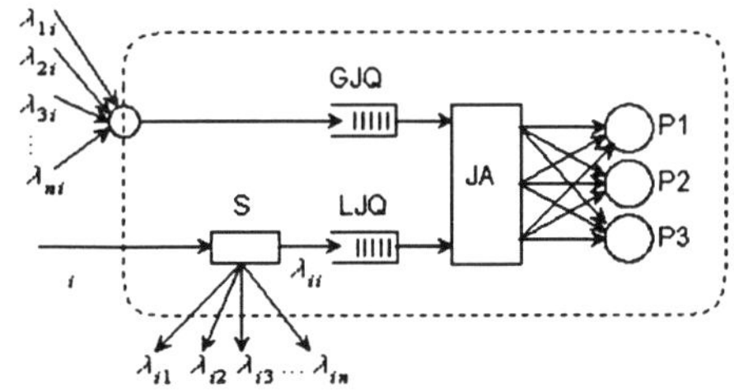


The cluster $i$ receives a locally generated stream of jobs at rate $g_i$. The job dispatcher **S** performs *the nondeterministic routing.* An arriving job is routed to cluster $j$ with probability $p_{ij}$ where the routing decision is based on the outcome of an independent trial. So, a part of this stream with rate $l_{ii}=g_i\, p_{ii}$ leaves in the cluster $i$ by placing it in a local job queue (**LJQ**). The rest are dispatched to other clusters with rates $l_{i1}$, $l_{i2}$, ..., $l_{in}$, where

(1) $$\lambda_{ij} = \gamma_i p_{ij}$$

The cluster $i$ also receives jobs from other clusters with rates $l_{i1}$, $l_{i2}$, ..., $l_{in}$. All jobs from other clusters are placed into a global job queue **GJQ** and cannot be sent to another cluster. The job arbitrator **JA** of the cluster $i$ loads jobs from queues into computers of the cluster by a dispatch algorithm. This algorithm can be nondeterministic as well as deterministic.

Since the mean parameters of our network change very slowly, it is possible to compute a mean system time (service time plus time of waiting in queue) for any local job generated in the cluster of our network. Any local job that is generated in the cluster $i$ will be computed in the cluster $k$ with probability

(2) $$p_{ik} = \frac{\lambda_{ik}}{\sum_{j=1}^{n} \lambda_{jk}}$$

It means that mean system time for any job that comes to cluster $i$ will be equal to

(3) $$M_i = b_i + \sum_{j=0, j\neq i}^{n} (B_j + t_{ij}) p_{ij}$$

where $b_i$ is the mean system time for all jobs in local job queue of cluster $i$;

$B_j$ is the mean system time for all jobs in global job queue of cluster $j$;

$t_{ij}$ is the mean time of sending job from cluster $i$ to cluster $j$;

$p_{ij}$ is the probability of sending job from cluster $i$ to cluster $j$;

In general, $b_i$, $B_j$, and $t_{ij}$ are functions from the cluster's state. Behavior of these functions is determined by a number of factors such as cluster structure, job arbitrator policies, transmission delays between the clusters, and network architecture and topology. It can be computed by theoretical analysis of the cluster structure or built from a simulation model of a cluster. The function $b_i$, $B_j$, and $t_{ij}$ for different cluster structures are detailed in Section 5. Note that they usually depend only on current load of a cluster or a channel:

(4) $$b_i = b_i\left(\lambda_{ii}, \sum_{k=1}^{n} \lambda_{ki}\right);$$

(5) $$B_j = B_j\left(\lambda_{jj}, \sum_{k=1}^{n} \lambda_{kj}\right)$$

(6) $$t_{ij} = t_{ij}(L);\ L = \{\lambda_{ij}\}$$

In order to fully utilize system resources it is advantageous to dispatch the jobs among the clusters in such a way that the mean system time of any job in the system is minimized. This process is commonly called *load balancing.*

The mean system time of any job in system is

(7) $$T = \sum_{i=1}^{n} \frac{\gamma_i}{\sum_{k=1}^{n} \gamma_k} \left( b_i p_{ii} + \sum_{j=0, j\neq i}^{n} (B_j + t_{ij}) p_{ij} \right)$$

Now the optimal job scheduling problem can be formulated as follows.

Find the matrix $\boldsymbol{P} = \{p_{ij}\}$ such that the following objective function will be minimized.

$$\sum_{i=1}^{n} \gamma_i (b_i(\gamma_i p_{ii}, \sum_{k=1}^{n} \gamma_k p_{ki}) + \sum_{j=0, j\neq i}^{n} (B_j(\gamma_j p_{jj}, \sum_{k=1}^{n} \gamma_k p_{kj}) + t_{ij}(\gamma, \boldsymbol{P})) p_{ij}) \rightarrow \min$$

(7*)

subject to:

$$\forall i,j : 1 \leq i \leq n \wedge 1 \leq j \leq n : 0 \leq p_{ij} \leq 1,$$

$$\forall i : 1 \leq i \leq n : \sum_{j=1}^{n} p_{ij} = 1$$

$$\gamma = \begin{pmatrix} \gamma_1 \\ \gamma_2 \\ \ldots \\ \gamma_n \end{pmatrix}$$

The constant $\sum_{k=1}^{n} \gamma_k$ can be removed from the function (7) because it does not affect minimization.

## 3. LOAD BALANCING ALGORITHM

The rate of an input job stream $g_i$=$g_i(t)$ can be assumed to be constant throughout the interval. The measurements are averaged over an interval of time D$t$ that usually is much longer than the transmission delay and service time. These measurements are subject to periodic adjustments in a dynamic environment. The scheduling is also conducted periodically.

Let $\gamma_i(t_k)$ denotes the mean input rate for the period of time from $t_k$ to $t_k$+D$t$. At the moment $t_{k+1}$ we need to make a routing decision which will affect the routing at the next period $t_{k+1}$+D $t$. In order to make a proper decision, it is necessary to know $\gamma_i(t_{k+1})$. The future rate is assumed to be unknown, but it is possible to forecast it by analysis of the previous history. Different types of extrapolation can be used here, including the simplest case when supposed value of the rate is $\tilde{\gamma}_i(t_{k+1}) = \gamma_i(t_k)$, which we will use in this project.

As shown in Figure 3, the decision agent **D** periodically collects information about the state of the clusters **C1**, **C2**, and **C3**. Ideally, this information will contain only a vector of mean input stream rates $\gamma_i(t_k)$ for the period of time from $t_k$ to $t_k$+D$t$. But for real DCS it is necessary to include the information about current server characteristics as a feedback. After collecting this information, a decision agent can find a solution for (7) and send an optimal scheduling policy for the next period of time $t_{k+1}$+D$t$ in the form of a vector of the loading parameters

$$\mathbf{P}_i = \{p_{ij}\}_{j=1}^{n} \tag{8}$$

back to clusters. The problem (7*) is a problem of multidimensional minimization. The behavior of the objective function depends on the behavior of the function of a mean system time for clusters. If all these functions are concave then the objective function is also concave as a sum of concave functions. In this case, problem (7*) has a single minimum, and can be solved by any one of multidimensional minimization techniques. Fortunately, the functions (6) for common queue models such as M/M/1 or M/M/n are concave. However, if at least one of the functions (6) is convex or concave-convex, the function (7*) can also be concave-convex. This is quite possible if clusters are using non-FIFO queuing discipline. In this case, a desired global minimum can be hidden among many, poorer, local extrema. A conventional searching technique for finding optimal solutions is to use gradient descent, which finds a direction for the next iteration from the gradient of the objective function. For complicated problems such as that, the gradient descent techniques often get stuck at a local minimum where the objective function has surrounding barriers. In this research we were using Downhill Simplex Method (Press, Flannery, Teukolsky, and Vetterling, 1992) for the first case and it's simulated annealing modification for the second (Lee, 1991).

To increase the speed of implementation for both methods, we also could modify the problem (7*). It is possible to decrease the number of dimensions in the original problem in more than two times. First of all, by using substitution (9), we can remove all variables $p_{ii}$ from the function (7).

$$p_{ii} = 1 - \sum_{k=1}^{n} p_{ik} \tag{9}$$

The network transmission delay is always positive. A solution cannot be optimal if two clusters are both sending jobs to each other, i.e., if for any $i \neq j$.

$$p_{ij}p_{ji} \neq 0 \tag{10}$$

In order to enforce the optimal solution and reduce number of variables, we can introduce a new set of variables:

$$p'_{ij} = \begin{cases} p_{ij} & \text{if } p_{ij} \neq 0 \\ -p_{ji} & \text{otherwise} \end{cases}, 1 \le i \le n, i < j \le n \tag{11}$$

Now our problem has only *(n-1)n/2* variables instead of $n^2$ variables as it has before. To remove the constraints from the problem, we should transform the objective function (7*). A new objective function is a function from $\mathbf{P}' \in \boldsymbol{R}^{n(n-1)/2}$ to $\boldsymbol{R}$. The function argument $\mathbf{P}'$ can be build from $\mathbf{P}$ by removing all elements of a diagonal and transformation

$$P' = \begin{bmatrix} p'_{1,2} & p'_{1,3} & p'_{1,4} & \cdots & p'_{1,n} \\ & p'_{2,3} & p'_{2,4} & \cdots & p'_{2,n} \\ & & p'_{3,4} & \cdots & p'_{3,n} \\ & & & & \vdots \\ & & & & p'_{(n-1),n} \end{bmatrix} \tag{12}$$

$$b'_i(\mathrm{P}') = b_i(\gamma_i(1 - \sum_{j=1}^{n} p_{ij}), \sum_{j=1}^{n} \gamma_j p_{ji}) \tag{13}$$

$$B'_i(\mathrm{P}') = B_i(\gamma_i(1 - \sum_{j=1}^{n} p_{ij}), \sum_{j=1}^{n} \gamma_j p_{ji}) \tag{14}$$

$$T'_{ij}(\mathrm{P}') = T_{ij}(\gamma, \mathrm{P}) \tag{15}$$

By using substitutions (12)-(15) and considering the constraints, a new objective function can be written:

$$f(\mathrm{P}') = \begin{cases} \sum_{i=1}^{n} \gamma_i(b'_i(\mathrm{P}') + \sum_{j=0, j\neq i}^{n} (B'_j(\mathrm{P}') + T_{ij}(\mathrm{P}'))p_{ij}), \\ \text{if } (\forall (i,j): 1 \le i \le n \wedge i < j \le n : 0 \le p'_{ij} \le 1) \wedge \\ \quad \forall i \sum_{j=1}^{n} p_{ij} \le 1 \\ \infty, \text{otherwise} \end{cases} \tag{16}$$

where all encounters of $p_{ij}$ should be substituted according to (11). The function (16) can be minimized by the downhill simplex method or it's simulated annealing modification.

## 4. Cluster Structures and Load Balancing

As was mentioned earlier, $b_i$, $B_j$, and $t_{ij}$ are functions of the cluster states. This section will describe several examples of these functions for different cluster structures. In this research, we considered three general cluster structures: a single processor with common queue, n processors with common queue, and a single processor with priority queueing. All the clusters are using a single network model that can be represented as a M/M/1. With this assumption transmission delay is

$$T'_{ij}(P') = \frac{1}{C - \sum_{k=1}^{n} \gamma_k \sum_{\substack{m=1 \\ m \neq k}}^{n} p_{km}}$$

(17)

where C is a network capacity.

### 4.1 Single Processor Cluster with Common Queue

A single processor cluster with a common queue can be simulated by a M/M/1 queue. This cluster does not make any differentiation between local and global jobs. Thus the local job queue and the global job queue of the cluster can be combined into a single job queue. This model can be used to represent a single-processor server with public accesses.
According to the queueing theory, a mean system time for any job in this cluster is

$$B'_i(P') = b'_i(P') = \frac{1}{\mu_i - (\gamma_i(1 - \sum_{j=1}^{n} p_{ij}) + \sum_{j=1}^{n} \gamma_j p_{ji})}$$

(18)

where $m_i$ is an average service rate of the cluster $I$.

### 4.2 m-Processors Cluster with Common Qucuc

Another cluster structure can be modeled as an M/M/m with common queue. This model can be used as a worst-case scenery of the deterministic routing (Asbury and Weiner, 1999) for a cluster of m computers or as a model of a m-processors server with public accesses. All local and global jobs are placed into a single queue. Whenever a server is free, the first job in a queue is assigned to that server.
According to (Kleinrock, 1975), if $l$ is an arrival rate, $1/m$ is an average service time, $m$ is a number of servers, and $r = 1\ /m\ m$ then an equilibrium probability of finding k jobs in the system is

$$p_k = \begin{cases} p_0 \frac{(m\rho)^k}{k!} & k \leq m \\ p_0 \frac{(\rho)^k m^m}{m!} & k \geq m \end{cases}$$

(19)

where

$$p_0 = \left[ \sum_{k=0}^{m-1} \frac{(m\rho)^k}{k!} + \frac{(m\rho)^m}{m!(1-\rho)} \right]^{-1}$$

(20)

The mean number of jobs in the system is

$$\bar{N} = \sum_{k=1}^{\infty} k p_k$$

(21)

$$\bar{N} = p_0 \left[ \sum_{k=1}^{m-1} \frac{(m\rho)^k}{k!} k + \frac{m^m}{m!} \sum_{k=m}^{\infty} \rho^k k \right]$$

(22)

$$\bar{N} = p_0 \left[ \sum_{k=1}^{m-1} \left( \frac{(m\rho)^k}{k!} k \right) + \sum_{k=1}^{m-1} \left( \frac{\rho}{(1-\rho)^2} - \rho^k k \right) \frac{m^m}{m!} \right]$$

(23)

$$\bar{N} = \rho m + \frac{\rho}{(1-\rho)\left( (1-\rho) m! \sum_{k=0}^{m-1} \frac{(m\rho)^{k-m}}{k!} + 1 \right)}$$

(24)

Using Little's result and (22) we can now find the system time:

$$T = \frac{\rho}{\lambda} m + \frac{\frac{\rho}{\lambda}}{(1-\rho)\left( (1-\rho) m! \sum_{k=0}^{m-1} \frac{(m\rho)^{k-m}}{k!} + 1 \right)}$$

(25)

after substitution $r = 1\ /m\ m$, (23) becomes:

$$T = \frac{1}{\mu} + \frac{\frac{1}{\mu}}{\left(1 - \frac{\lambda}{\mu m}\right)\left( (1 - \frac{\lambda}{\mu m}) m! \sum_{k=0}^{m-1} \frac{\left(\frac{\lambda}{\mu}\right)^{k-m}}{k!} + 1 \right)}$$

(26)

In this model a mean system time for any job in this cluster is

$$B'_i(P') = b'_i(P') = \frac{1}{\mu_i} +$$

$$+ \frac{\frac{1}{\mu_i}}{\left(1 - \frac{\eta_i}{\mu_i m_i}\right)\left( (1 - \frac{\eta_i}{\mu_i m_i}) m_i! \sum_{k=0}^{m-1} \frac{\left(\frac{\eta_i}{\mu_i}\right)^{k-m}}{k!} + 1 \right)}$$

where

$$\eta_i = \gamma_i (1 - \sum_{j=1}^{n} p_{ij}) + \sum_{j=1}^{n} \gamma_j p_{ji}$$

(27)

### 4.3 Cluster with Head-of-the-Line Priorities Queue

A cluster model with head-of-the-line priorities is another example of a cluster structure. It most naturally represents the behavior of a user workstation in a DCS. This model has both local and global job queues and a single processor. Jobs in the local job queue have highest priority. Jobs in the global job queue have lowest priority. Jobs from the global job queue can be performed only if the local job queue does not contain any jobs. We also assume that our system has a resume type of a preemptive queueing discipline. A preemption makes global jobs "invisible" for a local user. This model is a preemptive head-of-the-line (HOL) system with two "priority" groups according to (Kleinrock, 1975). The mean system time for a job with priority p in the HOL system with preemption is

(28) $$T_p = \frac{\bar{x}_p(1-\sigma_p) + \sum_{i=p}^{P} \lambda_i \bar{x}_i^2/2}{(1-\sigma_p)(1-\sigma_{p+1})}$$

where

(29) $$\sigma_p = \sum_{i=p}^{P} \rho_i$$

We assume that all jobs have the same mean service time:

(30) $$\bar{x}_p = 1/\mu$$

So, in this model mean time in the system for any job in the cluster is

(31) $$b_i'(\mathrm{P}') = \frac{1}{\mu_i - \gamma_i(1 - \sum_{j=1}^{n} p_{ij})}$$

$$B_i'(\mathrm{P}') = \frac{\mu - \dfrac{\gamma_i(1 - \sum_{j=1}^{n} p_{ij}) + \sum_{j=1}^{n} \gamma_j p_{ji}}{2}}{(\mu - (\gamma_i(1 - \sum_{j=1}^{n} p_{ij}) + \sum_{j=1}^{n} \gamma_j p_{ji}))(\mu - \gamma_i(1 - \sum_{j=1}^{n} p_{ij}))}$$

(32)

## 5. Client/Server Network Control

We can implement the clustered control model with current web-based Java technologies. All jobs will be submitted from a client GUI. Simple jobs with few parameters and less interactions to the web server can use the HTML form-based clients. For more complex jobs, Java applet clients are used. The Web server directs the job process request to the control server that is implemented by a set of Java servlets for control policies and Java ServerPages (JSPs) for dynamic interactions. The job process resources can also be fetched from a cluster database via the JDBC or a set of Enterprise JavaBeans (EJBs). The control server communicates with other clusters using either Java RMI or CORBA. Our experience indicated that the web-based client/server controls using Java technologies are robust, efficient, and easy to implement on different computing platforms (different hardware, different operating systems) on the network.

## 6. Conclusion

In this research we have demonstrated how the conceptual level principles of a "domule" for developing a DCS can be applied to the resource sharing problem.

We have developed a distributed network control model that can be used in the clustered network. This model can be implemented for a wide range of cluster structures and suitable for the popular client/server computing using web-based technologies. This model was tested on the simulation model that we have developed early (Liu and Motov, 1996).


## Acknowledgements

This research was supported by the National Science Foundation under contract No. CCR-9504137.


## References


F. Bonomi and A. Kumar (1990), "Adaptive Optimal Load Balancing in a Nonhomogeneous Multiserver System with a Central Job Scheduler," IEEE Trans. Comput., vol. 39, no. 10, pp. 1232-1250.

C. C. Price and S. Krishnaprasad (1984), "Software allocation models for distributed systems," in Proc. 5th Int.' Conf. on Distributed Computing, pp. 40-47.

Stephen Asbury and Scott R. Weiner (1999), Developing Java Enterprise Applications, John Wiley & Sons.

J. Filipiak (1988), "Modeling and control of dynamic flows in communication networks," Sprinfield-Verlag.

R. E. Larsen, P. L. McEntire, and J. G. O'Reilly (1982), "Tutorial: Distributed Control." Silver Spring, MD: IEEE Computer Society Press.

R. R. Tenney and N. R. Sandel, Jr. (1981), "Strategies for Distributed Decision making," IEEE Trans. Syst., Man, Cybern., vol. SMC-11, pp. 527-538.

D. Eager, Y. Lazowska, and J. Zahorjan (1986), "Adaptive load sharing in homogeneous Distributed Systems," IEEE Trans. Software Eng., vol. SE-12, no. 5, pp. 662-675.

W. H. Press, B. P. Flannery, S. A. Teukolsky, W. T. Vetterling (1992), "Numerical Recipes in C: the Art of Scientific Computing," Cambridge University Press.

B. W. Lee (1991) "Hardware Annealing in Analog VLSI Neurocomputing," Kluwer Academic Publishers, Boston/Dordrecht/London.

L. Kleinrock (1975), "Queueing Systems", John Wiley & Sons.

J. B. Liu, I. V. Motov (1996), "Computer-aided Education in Network Simulation," Proceedings of ISCA 9th International Conference on Computer Applications in Industry and Engineering.

IFAC Publications
www.elsevier.com/locate/ifac

# SOAP MESSAGE STRUCTURE FOR WEB-BASED DISTRIBUTED CONTROL SYSTEMS

**Changho Lee * Hongseok Kim ** Daewon Kim *** Jaehyun Park ******

* *Dept. of Automation Engineering, Inha University, Inchon, Korea, Email:chlee@resl.inha.ac.kr*
** *Korea Institute of Industrial Technology, Cheonan, Korea, Email:hskim@chaos.kitech.re.kr*
*** *MyongJi University, YongIn, Korea, Email:dwkim@mju.ac.kr*
**** *Dept. of Automation Engineering, Inha University, Inchon, Korea, Email:jhyun@inha.ac.kr*

Abstract: Emerging IT technologies, specially Internet communication and web-based applications are adopted to the modern distributed control systems. This paper defines a web-based application domain for a distributed control system connected by Internet. XML(eXtensible Markup Language) is used for representing a control system and control devices. These IT technologies makes a distributed control system more flexible and scalable than existing distributed systems with the standard RMI protocols such as CORBA or DCOM. 

Keywords: Open architecture, distributed control system, SOAP, XML, real-time control.

## 1. INTRODUCTION

Modern Internet and information technology(IT) enables control systems to utilize the distributed environment connected by high speed networks, such as field-buses, Ethernet, ATM, and wireless networks. With adopting these information technologies in WAN(Wide Area Network), the geometric range for a distributed control system becomes wider and wider, and in turn, a new software framework is required for this new environment(Lee and Park, 2000).

A software for the Internet-based distributed control systems is usually designed using the standard software components that operate on the top of the industry standard middlewares such as CORBA (Common Object Request Broker Architecture) from OMG (Object Management Group) (OMG, 1996; Orfali *et al.*, 1998) and COM (Component Object Model) from Microsoft(Microsoft, 1996; Caron, 2000). With using these kinds of middlewares that support the network-based software components, a distributed control system becomes more flexible and manageable.

However, because these middlewares are developed primarily for the general Internet-based applications, they have inherently several problems to be directly used for the control systems(Grimes, 1997; Daniel and Vallee, 1999).

- Since two major middleware standards, DCOM and CORBA, are not compatible with each other, it is hard to maintain the interoperability between them(Chang *et al.*, 2000).
- Since the network messages for these middlewares hardly pass through the security barrier like a firewall, they are not likely used in the distributed control systems across the WAN boundaries.

* The work reported in this paper is partially supported by Korea G7 project(grant number G7/5-3-1-4).

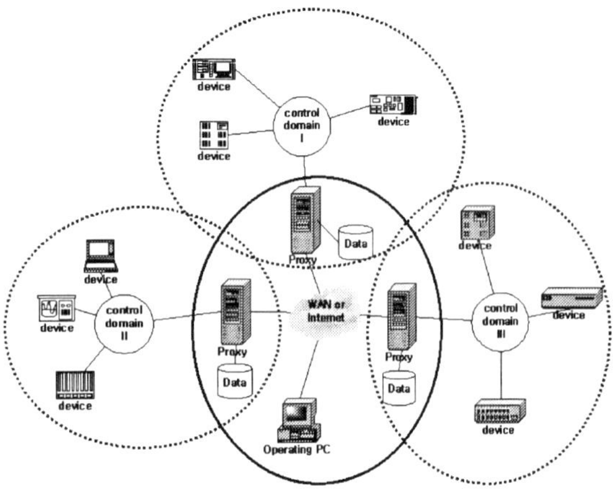


Fig. 1. Global Control Domain

- Since these middlewares are primarily developed for the general-purpose operating systems such as Microsoft Windows or Unix, it is difficult to apply them to the embedded control systems directly(Chen *et al.*, 1999; Chisholm, 1997).

To overcome these difficulties in using CORBA or DCOM, a new protocol, SOAP (Simple Object Access Protocol), is proposed as a RMI(Remote Method Invocation) for the Web-based applications. Since SOAP uses HTTP(Hyper Text Transfer Protocol), it is suitable for designing a control systems in WAN environment. However, the current version of SOAP standard defines only a simple protocol format for a remote method invocation, a SOAP message structure should be defined in detail to use SOAP for a distributed control system. This paper defines the SOAP message structure for a distributed control system for the remote monitoring and control.

This paper is organized as follows. Section 2 introduces the background technologies that are needed to design the control system. Section 3 proposes a control domain and SOAP message structure. Section 4 demonstrates the implementation example.

## 2. BACKGROUND TECHNOLOGY

### 2.1 *eXtensible Markup Language(XML)*

XML, a standard protocol defined by W3 organization, is a meta-language to describe other languages(Bray *et al.*, 2000). Since the underlying philosophy of XML aims flexibility and portability, it is possible to design a platform-independent RMI using XML. XML can be considered as two kinds of objects: a document and data. XML as a document, is used to make a document and define tags and technologies. XML as a data, is considered as a transfer syntax as well as data types(Box, 2000*a*). In addition, hierarchical data structure of XML is very useful to convey the internal data structure between the distributed control systems. Recently, to utilize XML data more efficiently, XML Information Set(XML InfoSet) is proposed. With XML InfoSet, network traffic can be reduced for the real-time communication.

### 2.2 *Simple Object Access Protocol*

SOAP (Simple Object Access Protocol) is a HTTP-based RMI(Remote Method Invocation) protocol and has a message structure defined by XML standard(Sturm, 2000). Since only a web server module is necessary to use SOAP, ranges of embedded control systems are able to maintain interoperability with other platforms using SOAP. The packet format of SOAP is similar to that of existing ORPC(Object Remote Procedure Call) that is used in DCOM or CORBA IIOP/GIOP. This similarity makes it easy to convert most of the existing objects and methods written in DCOM and CORBA to SOAP-based objects and methods. For the object references and method requests, SOAP uses URL(Uniform Resource Locator) and URI(Uniform Resource Identifier)(Box, 2000*b*). Even SOAP has the same function as ORPC, SOAP has following advantages over other RMI methods: First, since SOAP is a plain text, it is easy to bind various protocols. Second, while IIOP(Internet Inter-ORB Protocol) of CORBA and Java RMI require very bulky infrastructure for a remote object reference or a garbage collection, SOAP doesn't require it.

## 3. SOAP MESSAGE FOR CONTROL DOMAIN

This section defines the SOAP message format and XML document for the application software for a distributed control system.

A large distributed control system, defined as a control domain, consists of several sub-domains that include multiple devices as shown in Figure 1. Inter-networking between control domains is gated by a *proxy server*. The proxy server between control sub-domains gives several merits. First, the proxy server can moderate the network traffic, that is very important issue for the real-time operation of control systems. If a real-time control system is open to the public Internet, it experiences high network load as well as attacking from outside. For the real-time operation, the network between control devices should be isolated from the public Internet. This isolation is easily achieved by using the proxy server, with which each control devices exchange their data with each other without being interrupted from the outside traffic. Second, the proxy server can

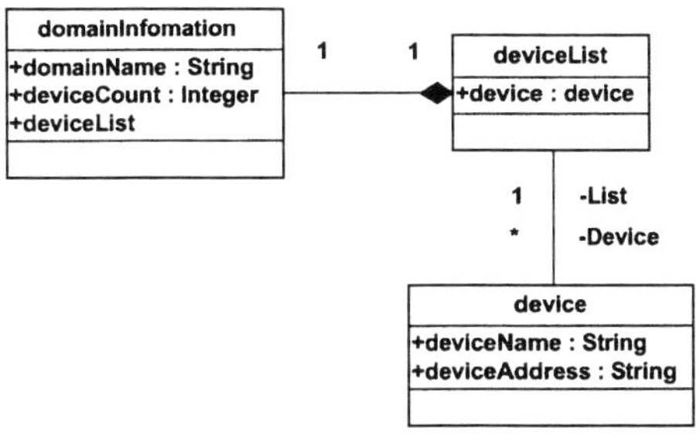


Fig. 2. Control Domain DDL

maintain a high level of security. Real-time control system can be efficiently protected from the active attacks such as DOS(Denial of Service). Last but important point is to compensate the inherit stateless operation of SOAP. Without any state information, maintaining the control information efficiently is almost impossible and large network load is usually injected. But with a proxy server, state-oriented communication can be efficiently managed even SOAP itself doesn't support it(Microsoft, 2000).

### 3.1 *Control Domain Abstraction*

To use SOAP for a control domain, a control domain itself as well as the services provided by the control devices should be well defined in XML format. These definitions are called *Domain Description Language (DDL)* and *Service Description Language (SDL)* respectively, and they reside in the proxy server that represents a particular control domain.

Figure 2 shows the proposed XML schema of DDL for a control domain. A DDL consists of three major elements: `domainName`, `deviceCount`, and `deviceList`. `domainName` represents the name of current control domain. `deviceCount` means the number of devices registered in a control domain. The devices within a control domain is described in the `deviceList` that has two elements: `deviceName` and `deviceAddress`. With `deviceName` element, user can fetch its SDL (deviceName.sdl) from a proxy server, and, in turn, get services and parameters. A `deviceAddress` element is used to inform the proxy server of the physical address of a device. In Internet, `deviceAddress` is represented by IP Address or hostname.

### 3.2 *SOAP Message for Control Domain*

This section proposes a SOAP message format for a control domain whose abstraction is defined in the previous section. Since current SOAP standard defines only the minimal set of SOAP

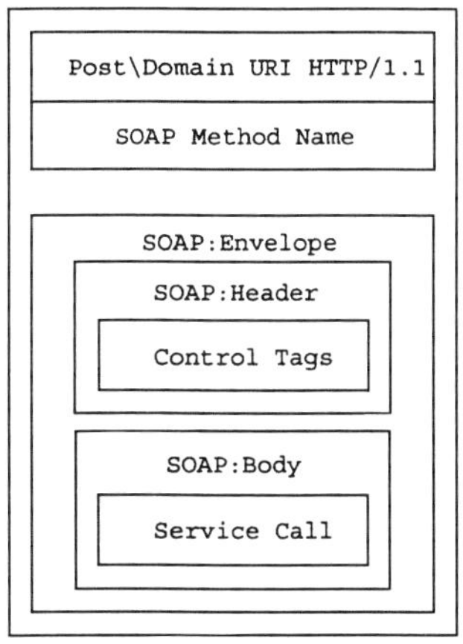


Fig. 3. SOAP Message Format

```
<!ELEMENT SOAP-ENV:Envelope (SOAP-ENV:Header)>
<!ATTLIST SOAP-ENV:Envelope     xmlns:xsi   CDATA #REQUIRED
          xmlns:SOAP-ENV  CDATA #REQUIRED >

<!ELEMENT SOAP-ENV:Header      (trans:transfer , const:constraint )>
<!ATTLIST SOAP-ENV:Header   xsi:type CDATA #REQUIRED >
<!ELEMENT trans:transfer    (trans:to , trans:from )>
<!ATTLIST trans:transfer    xmlns:trans CDATA #REQUIRED
         xmlns:agr   CDATA #REQUIRED
         SOAP-ENV:mustUnderstand CDATA #REQUIRED>
<!ELEMENT trans:to (trans:address )>
<!ELEMENT trans:address EMPTY>
<!ATTLIST trans:address     xsi:type    CDATA #IMPLIED >
<!ELEMENT trans:from    (trans:address )>
<!ELEMENT const:constraint  (const:timestamp ,  const:timeout  )>
<!ATTLIST const:properties  xmlns:const CDATA #REQUIRED
             SOAP-ENV:mustUnderstand CDATA #REQUIRED>
<!ELEMENT const:timestamp       (#PCDATA )>
<!ELEMENT const:timeout     (#PCDATA )>
```

Fig. 4. SOAP Control Tags

header, the detail structure of header and body for a specific domain should be defined. Figure 3 shows the SOAP message structure for the control domain proposed in this paper.

#### 3.2.1. *Control Tags*

Figure 4 shows the internal structure of SOAP header used in this paper. As described above, the control domain defined in this paper uses a proxy server, all information related to message transfer should be registered in the proxy server. This information includes address information and timing constraint required for the real-time control. To provide information of control devices for the proxy server, *control tag* is defined in SOAP header, that includes two tags.

`transfer tag`
In the control domain proposed in this paper, since control devices are isolated from the external domains as described above, user should provide all of the information to drive a device in order that a proxy server could interface with control devices instead of clients. For this purpose, a mandatory element, `transfer tag`, is defined. `transfer tag` includes the location information of SOAP message source and destination.

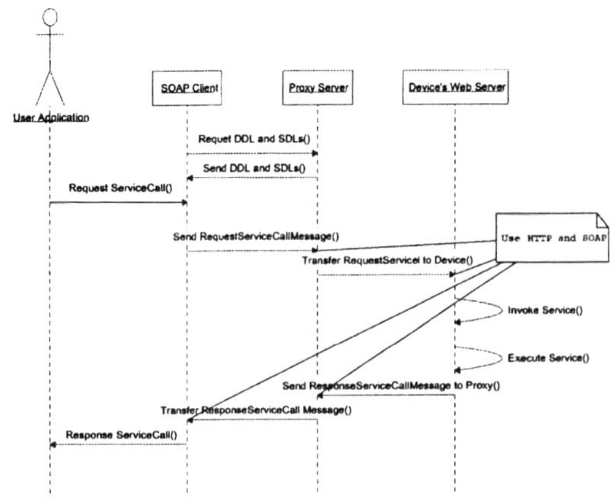


Fig. 5. SOAP-Compliant Client Processing SOAP

```
<?xml version="1.0" ?> <serviceDescription
    name="SoapService"
    xmlns="urn:schemas-xmlsoap-org:sdl.2000-01-25"
    xmlns:dt="http://www.w3.org/1999/XMLSchema"
    xmlns:ss ="#SimpleStuff">
    <import namespace="#SimpleStuff"
        location ="#SimpleStuff"/>
    <soap xmlns="urn:schemas-xmlsoap-org:soap-sdl-2000-01-25">
        <service>
            <requestResponse name="Method">
                <request ref="ss:MethodReq" />
                <response ref="ss:MethodRes" />
            </requestResponse>
        </service>
    </soap>
    <ss:schema  id="SimpleStuff"
            targetNamespace="http://server-url/services.xml"
            xmlns:dt="http://www.w3.org/1999/XMLSchema"
                xmlns="http://www.w3.org/1999/XMLSchema">
        <element name="MethodReq"/>
            <type>
                <element name="Input" type="dt:int" />
            </type>
        <element name="MethodRes">
            <type>
                <element name="Output" type="dt:int" />
            </type>
        </element>
    </ss:schema>
</serviceDescription>
```

Fig. 6. Service Description Language Example

`constraint tag`
A constraint tag contains the timing information to decide whether a tranferred message meets the real-time constraints or not. Sender records the timestamp in a `timestamp` element, and timeout value in a `timeout` element. If the timeout value is expired, a message is considered as an invalid message.

3.2.2. *Service Call* For the security reason, a user is required to be authorized for accessing the control system. A proxy server can authorize user to access the control domain using an industry standard authentication mechanism. After authentication process, users can obtain the domain information written in DDL and SDL from the proxy server. With the information obtained, the service request and the response messages are exchanged as shown in Figure 5. The internal structure of the request and response message written in SDL and the message example are shown in Figure 6 and Figure 7, respectively.

```
<!-- Request Message -->
POST HTTP /SomeProxy HTTP/1.1
Host:www.someproxy.com
Content-Type:text/xml
Content-Length:nnnn

SOAPMethodName:Some-Namespace-URI#MethodName
<SOAP:Envelope xmlns:SOAP="urn:schemas-xmlsoap-org:soap.v1" >
    <SOAP:Header>
        <to>Device1 Address</to>
        <from>User Address</from>
    </SOAP:Header>
    <SOAP:Body>
        <m:Method xmlns:m="Some-Namespace-URI">
            <Input>Value</Input>
        </m:Method>
    </SOAP:Body>
</SOAP:Envelope>

<!-- Response Message -->
HTTP/1.1 200 OK
Content-Type:text/xml
Content-Length:nnnn

SOAPMethodName:Some-Namespace-URI#MethodNameResponse
<SOAP:Envelope xmlns:SOAP="urn:schemas-xmlsoap-org:soap.v1" >
    <SOAP:Header>
        <to>User Address</to>
        <from>Device1 Address</from>
    </SOAP:Header>
    <SOAP:Body>
        <m:MethodRes xmlns:m="Some-Namespace-URI">
            <Output>Value</Ouput>
        </m:MethodRes>
    </SOAP:Body>
</SOAP:Envelope>
```

Fig. 7. SOAP Service Request and Response

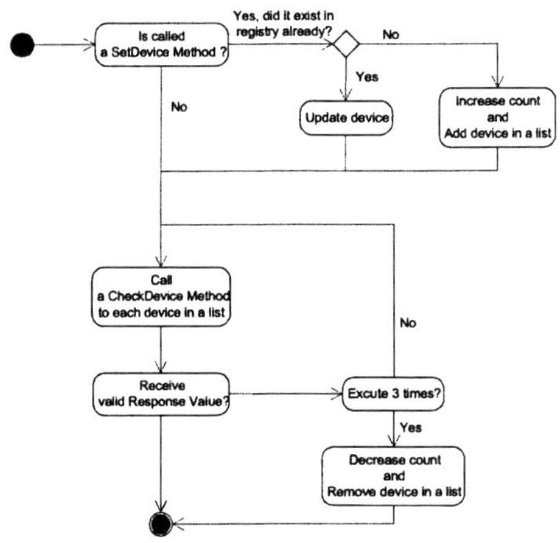


Fig. 8. Registration and Management Process State Diagram at a Proxy Server

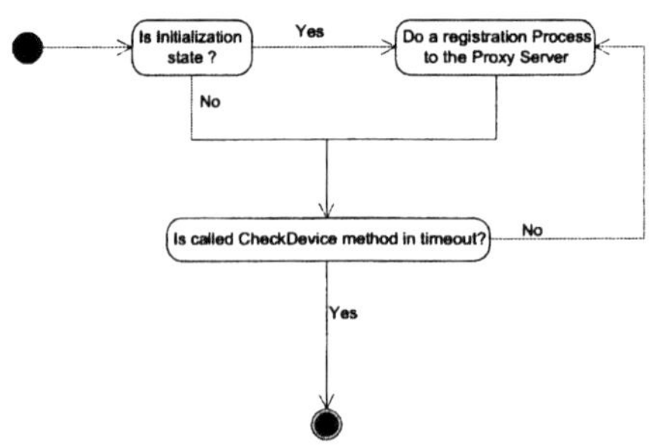


Fig. 9. Registration Process State Diagram at a Control Device

### 3.3 *Registration and Management of Control Devices*

Since DDL locates in the proxy server, each device must register its services to the proxy server. A control device can register its IP address and SDL

```
ProtoType : boolean SetRegister(string, string, string);
            int     CheckDevice(int);

<ss:schema  id="RegisterStuff"
        targetNamespace="http://server-url/services.xml"
        xmlns:dt="http://www.w3.org/1999/XMLSchema"
            xmlns="http://www.w3.org/1999/XMLSchema">
    <element name="SetRegisterReq"/>
        <type>
            <element name="Addr" type="dt:string" />
            <element name="DomainName" type="dt:string" />
            <element name="SDL" type="dt:string" />
        </type>
    <element name="SetRegisterRes">
        <type>
            <element name="RetureCode" type="dt:boolean" />
        </type>
    </element>
</ss:schema>

<ss1:schema  id="HeartbeatStuff"
        targetNamespace="http://server-url/services.xml"
        xmlns:dt="http://www.w3.org/1999/XMLSchema"
            xmlns="http://www.w3.org/1999/XMLSchema">
    <element name="CheckDeviceReq"/>
        <type>
            <element name="ConnectionReference" type="dt:int" />
        </type>
    <element name="CheckDeviceRes">
        <type>
            <element name="ReferenceValue" type="dt:int" />
        </type>
    </element>
</ss1:schema>
```

Fig. 10. SDL of Registration and Management Methods

to the proxy server through a web service defined in Figure 10. As shown in Figure 8, when the proxy server receives a registration message from a device, it determines whether a device is already registered in the `deviceList` or not. If a device is not registered in the list, the proxy server adds a new device into the list. Otherwise, it just updates the information of the device in the list.

After the registration process completes, the proxy server uses the `CheckDevice` method to check the device is removed from a domain. If the device returns failure code or doesn't respond to the `CheckDevice` message, the proxy server assumes that it is not in a domain and removes it from the list. On the contrary, if there is no `CheckDevice` message from the proxy server during a certain period, a control device requests a registration process as shown in Figure 9.

### 3.4 *User Interface*

One of the most interesting point using the SOAP protocol is that users can communicate with devices using a SOAP-compatible application or a web browser. It means any kind of web-browser such as Microsoft Internet Explorer or Netscape browser as well as a dedicated SOAP-compatible application program can be used for control and/or monitoring the control devices. A web server supplies a style sheet document so that user can supervise system's data easily through web browser.

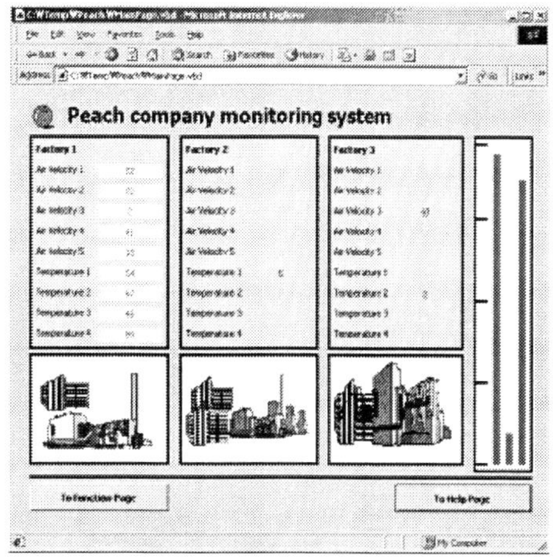


Fig. 11. Implemented Device's Methods

## 4. IMPLEMENTATION

Using the SOAP message proposed in this paper, a demonstration system is implemented. Target application domain is a distributed controller for a semiconductor manufacturing machine, Samsung's commercial chip mounter (Samsung CP-40). To demonstrate the O/S-independent characteristics of the protocol, the devices in the domain are designed on two different operating systems: Microsoft Windows and embedded Linux.

SEMI (Semiconductor Equipment and Material International Inc.) introduced the SECS (SEMI Equipment Communication Standard) protocol to reduce the cost and to improve communication ability between hosts and equipments. This protocol consists of SECS-I for the message communication and SECS-II for the message format. SECS-I uses point-to-point protocol such as RS-232. Recently, HSMS (High-Speed SECS Service) based on TCP/IP is used widely. GEM(Generic Model for Communications and Control of SEMI Equipment) is the protocol that is used between semiconductor equipment and host computer on the top of SECS-I, II and HSMS. In this paper, the proposed SOAP message is used to integrate GEM-based semiconductor manufacturing machines.

Implemented system consists of a proxy server and several devices. The proxy server is implemented on a Microsoft Windows 2000 platform and IIS5.0. It stores and supplies the device's information in DDL and exchange various SOAP messages. As an authentication protocol, the *Kerberos* authentication protocol is used, that is also independent of platforms. Various target control devices are implemented with Microsoft Windows family(2000, CE) and embedded Linux. Figure 11 shows a demonstration system.

## 5. CONCLUSIONS

Emerging IT technology, specially Internet and web-based communication, is likely to be adopted to distributed control systems recently. This paper defines web-based control domain connected by Internet, and proposes a SOAP message structure for it. Control domains defined in this paper are connected with each other via a proxy server. Under the control of proxy server, the real-time operation and the security of control domain can be maintained systematically. For openness, a control domain and control devices are abstracted in XML, and SOAP is used for RMI(remote method invocation). With adopting these IT technologies, a distributed control system becomes more flexible and scalable than existing distributed systems even they use a standard RMI method such as CORBA or DCOM. At last, the proposed protocols can be easily implemented on the control devices regardless of operating systems, total costs and developing time can be reduced. One drawback of the proposed protocol is high overhead caused by raw XML data and HTTP protocols. These problems can be, however, reduced with XML Infoset mechanism and other methods.

## 6. REFERENCES


Box, Don (2000*a*). *Essential XML beyond Markup*. Addison Wesley.

Box, Don (2000*b*). HTTP + XML=SOAP. *MSDN Magazine* **15**(3), 67–81.

Bray, Tim, Jean Paoli, C. M. Sperberg-McQueen and Eve Maler (2000). Extensible markup language (xml) 1.0 (second edition). Technical report. W3C.

Caron, Rob (2000). Web services and the soap for visual studio 6.0. *MSDN Magazine* **15**(8), 62–73.

Chang, E., D Annal and F Grunta (2000). A large scale distributed object architecture - corba and com for real time systems. *Third IEEE International Symposium* pp. 338–341.

Chen, Deji, Aloysius Mok and Mark Nixon (1999). Real-time support in com. In: *Proceedings of the 32nd Hawaii International Conference on Systems Sciences.* Hawaii, USA.

Chisholm, Al (1997). OPC Overview. Technical report. OPC Foundation.

Daniel, Jerome and Bruno Traversonand Vincent Vallee (1999). Active com : An inter-working framework for corba and dcom. *Proceedings of the International Symposium, Distributed Objects and Applications.*

Grimes, Rechard (1997). *Professional DCOM Programming.* Wrox.

Lee, Wongoo and Jaehyun Park (2000). A design of infrastructure for control/monitoring system in the distributed computing environment. In: *Procceedings of 15th Korea Automation Control Conference.* Yongin, Korea.

Microsoft (1996). Dcom technology overview. Technical report. Microsoft Corporation.

Microsoft (2000). Universal plug and play device architecture. Technical report. Microsoft Corporation.

OMG (1996). Corba services:common object services specification. Technical report. OMG,Inc.

Orfali, Robert, Dan Harkey and Jeri Edwards (1998). *Instant CORBA.* John Wiley & Sons.

Sturm, Jake (2000). *Developing XML Solutions.* Microsoft Press.

IFAC Publications
www.elsevier.com/locate/ifac

# SCHEDULING ANALYSIS FOR NETWORKED CONTROL SYSTEMS

Dong Sung Kim, Young Sam Lee, Wook Hyun Kwon

*ERC-ACI, School of Electrical and Computer Engr.*
*Seoul National University, Seoul, 151-742, Korea*
*Email:dskim@cisl.snu.ac.kr*

**Abstract:** This paper proposes a new method to obtain a maximum allowable delay bound and a scheduling analysis of networked control systems. The proposed method is formulated in terms of linear matrix inequalities (LMI) and can give a much less conservative delay bound than the existing methods. A scheduling analysis is presented based on the delay bound obtained through the proposed method. *Copyright © 2001 IFAC*
**Keywords:** Scheduling Analysis, Networked Control Systems, Maximum Allowable Delay Bounds

## 1 INTRODUCTION

In distributed control systems, a feedback control loop is closed through a network. Distributed control systems with networks are called networked control systems (NCSs). In an NCS, various delays with variable length occur due to sharing a common network medium, which are called network-induced delays. Network-induced delays can vary widely according to the transmission time of messages and the overhead time. Hence it is necessary to present some methods to make these network induced delays bounded and smaller, which are called network scheduling methods for the NCS.

In feedback control systems, it is important that sampled data should be transmitted within a sampling period and that stability of control systems should be guaranteed. While a shorter sampling period is preferable in most control systems, for some purposes it can be lengthened up to a certain bound within which stability of the system is guaranteed in spite of the performance degradation. This certain bound is called a maximum allowable delay bound (MADB). Therefore, it is necessary to find the MADB for stability of the NCS, and then to find an appropriate network scheduling method that limits the network-induced delay to less than the MADB. A network scheduling method is required to reduce network-induced delays within the MADBs, while guaranteeing real-time transmission of sporadic and periodic data, and to minimize network utilization for non-real time message.

An MADB has been obtained from stability conditions of control systems. There have been some results on the stability of NCSs [1]-[3], but these were concerned with obtaining stability conditions of the NCS with a given delay. There have been also some results on the MADB for stability in non-networked control systems [4][6]. In these papers, the MADB is obtained using the Ricatti equation approach, which yields conservative delay bounds. Less conservative results on the MADB in non-networked control systems are reported in [7][8]. However, these results still remains to be improved [9]. The MADB thus obtained can be extended as a maximum bound of a sampling period in the NCS. That is, the sampling periods determined by the proposed sampling period decision algorithm can be set to values less than the MADB.

A network scheduling method considering an MADB with three types of data based on a multi-input and multi-output (MIMO) system was proposed [11]. In this paper, the estimation of MADB using the Ricatti equation is too conservative, which means the estimated MADB is too small and the network scheduling method discussed in this paper is somewhat heuristic. In [12]-[14], calculation methods of MADBs and stability analysis of NCSs were presented. However, these results were conservative to be of practical use and still remains to be improved.

In this paper, a new method to obtain the MADB guaranteeing a stability of the NCS is proposed in terms of linear matrix inequalities (LMI). The proposed method gives a much less conservative delay bound than the existing methods. The paper includes a scheduling analysis considering three types of data based on a MIMO system. The network scheduling method is based on the result in [11].

This paper is organized as follows. In the following section, an NCS model is described. In Section 3, an MADB for the stability of the NCS is derived by LMI formulation. In Section 5, simulations are given to show that the method is useful. Finally, the conclusions are presented in Section 6.

## 2 Modelling of Networked Control Systems

The NCS can be described as in Fig. 1. The NCS consists of multiple control loops. A control loop is composed of a plant, a controller, sensors, and actuators. The sensors, actuators, and controller share a common communication medium with those of other control loops. A control loop can affect other control loops by transferring sporadic and aperiodic data to other control loops. To guarantee the stability of NCS, every control loop has to be analyzed. A plant in a

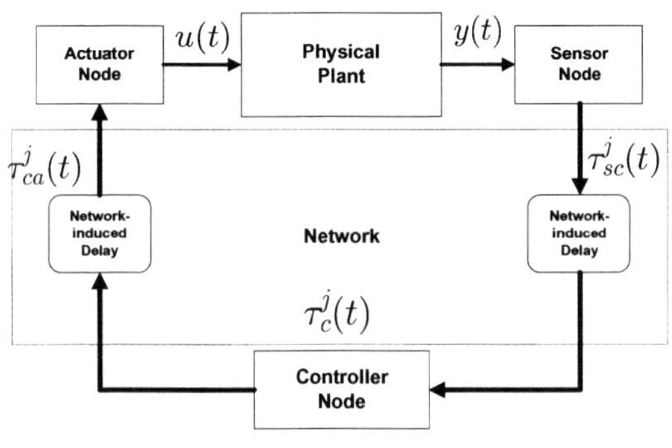


Figure 1: Diagram of a networked control system

single control loop $j$ can be described in the following state space form:

$$\begin{aligned} \dot{x}_p^j(t) &= A_p^j x_p^j(t) + B_p^j u_p^j(t) \\ y_p^j(t) &= C_p^j x_p^j(t), \end{aligned} \quad (1)$$

where $u_p^j(t) \in R^{l_A^j}$, $y_p^j(t) \in R^{l_S^j}$, $x_p^j(t) \in R^{l_{NP}^j}$. $l_A^j$, $l_S^j$, and $l_{NP}^j$ are respectively the numbers of the actuators, the numbers of the sensors, and the dimension of the plant system, in the control loop $j$. $A_p^j$, $B_p^j$, and $C_p^j$ are matrices of appropriate dimensions.

A controller in the control loop $j$ can be described by

$$\begin{aligned} \dot{x}_c^j(t) &= A_c^j x_c^j(t) + B_c^j u_c^j(t) \\ y_c^j(t) &= C_c^j x_c^j(t-\tau_c^j) + D_c^j u_c^j(t-\tau_c^j), \end{aligned} \quad (2)$$

where $u_c^j(t) \in R^{l_S^j}$, $y_c^j(t) \in R^{l_A^j}$, $x_c^j(t) \in R^{l_{NC}^j}$ and $l_{NC}^j$ is the dimension of the control system in the control loop $j$. $A_c^j$, $B_c^j$, and $C_c^j$ are matrices of appropriate dimensions. $0 \le \tau_c^j \le \bar{\tau}_c^j$, $\tau_c^j$ is the computation delay in the controller $j$, and $\bar{\tau}_c^j$ is the maximum computation delay in the controller $j$. The communication delays in the control loop $j$ are modelled as

$$\begin{aligned} u_c^j(t) &= y_p^j(t-\tau_{sc}^j) \\ u_p^j(t) &= y_c^j(t-\tau_{ca}^j), \end{aligned} \quad (3)$$

where $0 \le \tau_{sc}^j \le \bar{\tau}_{sc}^j$, $0 \le \tau_{ca}^j \le \bar{\tau}_{ca}^j$, $\tau_{sc}^j$ and $\bar{\tau}_{sc}^j$ are, respectively, the communication delay and the maximum communication delay from the sensors to the controller, and $\tau_{ca}^j$ and $\bar{\tau}_{ca}^j$ are, respectively, the communication delay and the maximum communication delay from the controller to the actuators.

Using Equations (1), (2), and (3), a control system in the control loop $j$ can be described as

$$\begin{aligned} \dot{x}^j(t) &= \begin{bmatrix} A_p^j & 0 \\ 0 & A_c^j \end{bmatrix} x^j(t) + \begin{bmatrix} 0 & 0 \\ B_c^j C_p^j & 0 \end{bmatrix} x^j(t-\tau_{sc}^j) \\ &+ \begin{bmatrix} B_p^j D_c^j C_p^j & 0 \\ 0 & 0 \end{bmatrix} x^j(t-\tau_{sc}^j-\tau_{ca}^j-\tau_c^j) \\ &+ \begin{bmatrix} 0 & B_p^j C_c^j \\ 0 & 0 \end{bmatrix} x^j(t-\tau_{ca}^j-\tau_c^j), \end{aligned} \quad (4)$$

where $x^j(t) = \begin{bmatrix} x_p^{j\,T}(t) & x_c^{j\,T}(t) \end{bmatrix}^T$. Then the above equation can be rewritten as

$$\begin{aligned} \dot{x}^j(t) = \; & A^j x^j(t) + A_1^j x^j(t-\tau_1^j) + A_2^j x^j(t-\tau_2^j) \\ & + A_3^j x^j(t-\tau_3^j) \end{aligned} \quad (5)$$

where

$$A^j = \begin{bmatrix} A_p^j & 0 \\ 0 & A_c^j \end{bmatrix}, A_1^j = \begin{bmatrix} 0 & 0 \\ B_c^j C_p^j & 0 \end{bmatrix},$$
$$A_2^j = \begin{bmatrix} B_p^j D_c^j C_p^j & 0 \\ 0 & 0 \end{bmatrix}, A_3^j = \begin{bmatrix} 0 & B_p^j C_c^j \\ 0 & 0 \end{bmatrix},$$
$$0 \le \tau_1^j = \tau_{sc}^j \le \bar{\tau}_{sc}^j = \bar{\tau}_1^j$$
$$0 \le \tau_2^j = \tau_{sc}^j + \tau_{ca}^j + \tau_c^j \le \bar{\tau}_{sc}^j + \bar{\tau}_{ca}^j + \bar{\tau}_c^j = \bar{\tau}_2^j$$
$$0 \le \tau_3^j = \tau_{ca}^j + \tau_c^j \le \bar{\tau}_{ca}^j + \bar{\tau}_c^j = \bar{\tau}_3^j.$$

Each control loop in the NCS can be described as in Equation (5) using three types of delay.

## 3 AN MADB FOR STABILITY IN A SINGLE CONTROL LOOP

To generalize results to the multiple state-delayed case, consider the following system.

$$\begin{aligned} \dot{x}(t) &= Ax(t) + \sum_{i=1}^{N} A_i x(t-\tau_i), \\ x(t) &= \phi(t), \quad t \in [-\bar{\tau}, 0] \end{aligned} \tag{6}$$

where $x(t) \in \mathbf{R}^n$ is the state, $\tau_i > 0$ is the delay of the system, $\phi(\cdot)$ is the initial condition, $A$, $A_i$ are real constant matrices with appropriate dimensions, and $\bar{\tau}$ is upper bound of $\tau_i$.

In obtaining the results of this paper, the following upper bound for the inner product of two vectors plays an important role:

$$-2a^T b \le \inf_{X,Y,Z} \begin{bmatrix} a \\ b \end{bmatrix}^T \begin{bmatrix} X & Y-I \\ Y^T - I & Z \end{bmatrix} \begin{bmatrix} a \\ b \end{bmatrix} \tag{7}$$

where $\begin{bmatrix} X & Y \\ Y^T & Z \end{bmatrix} \ge 0$ and $I$ denotes an identity matrix with an appropriate dimension. Extending the idea of Equation (7), the following lemma is derived.

**LEMMA 1.** *[9] Assume that $a(\cdot) \in \mathbf{R}^{n_a}$, $b(\cdot) \in \mathbf{R}^{n_b}$, and $\mathcal{N}(\cdot) \in \mathbf{R}^{n_a \times n_b}$ are defined on the interval $\Omega$. Then, for any matrices $X \in \mathbf{R}^{n_a \times n_a}$, $Y \in \mathbf{R}^{n_a \times n_b}$, and $Z \in \mathbf{R}^{n_b \times n_b}$, the following holds:*

$$-2\int_\Omega a^T(\alpha)\mathcal{N} b(\alpha) d\alpha \le \tag{8}$$

$$\int_\Omega \begin{bmatrix} a(\alpha) \\ b(\alpha) \end{bmatrix}^T \begin{bmatrix} X & Y-\mathcal{N} \\ Y^T - \mathcal{N}^T & Z \end{bmatrix} \begin{bmatrix} a(\alpha) \\ b(\alpha) \end{bmatrix} d\alpha \tag{9}$$

*where*

$$\begin{bmatrix} X & Y \\ Y^T & Z \end{bmatrix} \ge 0. \tag{10}$$

**THEOREM 1.** *If there exist $P > 0$, $Q_i > 0$, $X_i$, $Y_i$ and $Z_i$, $i = 1, \cdots, N$, such that*

$$\begin{bmatrix} \mathcal{P}_{11} & \mathcal{A}^T \mathcal{Z} \\ \mathcal{Z}\mathcal{A} & -\Gamma \end{bmatrix} < 0, \quad \begin{bmatrix} X_i & Y_i \\ Y_i^T & Z_i \end{bmatrix} \ge 0, \; i = 1, \cdots, N, \tag{11}$$

*where*

$$\mathcal{P}_{11} \triangleq \begin{bmatrix} \begin{matrix}(A^T P + PA + \\ \sum_{i=1}^N \{Y_i + Y_i^T + \bar{\tau} X_i + Q_i\})\end{matrix} & PA_1 - \mathcal{Y} \\ A_1^T P - \mathcal{Y} & -\mathcal{Q} \end{bmatrix},$$

$$\begin{aligned} \mathcal{A} &\triangleq [A \;\; A_1 \;\; \cdots \;\; A_N], \\ \mathcal{A}_1 &\triangleq [A_1 \;\; \cdots \;\; A_N], \\ \mathcal{Y} &\triangleq [Y_1 \;\; \cdots \;\; Y_N], \\ \mathcal{Z} &\triangleq \bar{\tau}[Z_1 \;\; \cdots \;\; Z_N], \\ \mathcal{Q} &\triangleq \mathrm{diag}\{Q_1, \cdots, Q_N\}, \\ \Gamma &\triangleq \bar{\tau}\mathrm{diag}\{Z_1, \cdots, Z_N\}, \end{aligned}$$

*then the system (6) is asymptotically stable for any time-delay $\tau_i$ satisfying $0 \le \tau_i \le \bar{\tau}_i$, $i = 1, \cdots, N$.*

This theorem extends the results in [9], which deals with a MADB for a single state-delayed systems. The system equation (5) can be represented by Equation (6) with $N = 3$ and $\bar{\tau}$ can be interpreted as $\max\{\bar{\tau}_1, \bar{\tau}_2, \bar{\tau}_3\}$.

To generalize results to the discrete case, consider the following system. In discrete system, the single delay is only considered.

$$\begin{aligned} x(k+1) &= Ax(k) + A_1 x(k-\tau), \\ x(k) &= \phi(k), \quad k \in [-\bar{\tau}, 0] \end{aligned} \tag{12}$$

where $\tau_i > 0$ is the delay of the system, $\phi(\cdot)$ is the initial condition, $A$, $A_1$ are real constant matrices with appropriate dimensions, and $\bar{\tau}$ is upper bound of $\tau_i$.

**THEOREM 2.** *If there exist $P > 0$, $Q > 0$, $X$, $Y$ and $Z$, such that*

$$\begin{bmatrix} \mathcal{P}_{11} & -Y & A^T P & \bar{\tau}(A-I)^T Z \\ -Y^T & -Q & A_1{}^T P & \bar{\tau} A_1{}^T Z \\ PA & PA_1 & -P & 0 \\ \bar{\tau} Z(A-I) & \bar{\tau} Z A_1 & 0 & -\bar{\tau} Z \end{bmatrix} < 0, \tag{13}$$

$$\begin{bmatrix} X & Y \\ Y^T & Z \end{bmatrix} \ge 0 \tag{14}$$

*where*

$$\begin{aligned} \mathcal{P}_{11} &\triangleq [-P + \bar{\tau} X + Y + Y^T + Q], \\ \mathcal{Z} &\triangleq \bar{\tau}[Z_1 \;\; \cdots \;\; Z_N], \\ \mathcal{Q} &\triangleq \mathrm{diag}\{Q_1, \cdots, Q_N\}, \\ \Gamma &\triangleq \bar{\tau}\mathrm{diag}\{Z_1, \cdots, Z_N\}, \end{aligned}$$

*then the system (12) is asymptotically stable for any time-delay $\tau$ satisfying $0 \le \tau \le \bar{\tau}$.*

Theorem 2 can be proved via the similar procedure in Appendix A. For the proof of Theorem 2, the conditions of $\Delta V(k) = V(k+1) - V(k) < 0$, $\Delta V(k) = \Delta V_1(k) + \Delta V_2(k) + \Delta V_3(k) < 0$ can be obtained by Lemma 1.

The delay bound of each control loop is used as a major parameter in the determination of the sampling period and allocation of bandwidth.

## 4 SIMULATIONS

### 4.1 Analysis of an MADB in a Control Loop

For an analysis of an MADB in a control loop, the following state space model in [12][13] is considered.

$$\begin{aligned} \dot{x} &= \begin{bmatrix} 0 & 1 \\ 0 & -0.1 \end{bmatrix} x + \begin{bmatrix} 0 \\ 0.1 \end{bmatrix} u \\ y &= [1 \;\; 0]x \end{aligned} \tag{15}$$

Table 1: MADB ($ms$) by model (15)

| Theorem 1 | By [13] | *By* [11] | *By* [13] | By [12] |
|---|---|---|---|---|
| $26.3 \times 10^3$ | $1.7\times10^3$ | 53.8 | 17.5 | 0.45 |

By using the model (15), the MADB is obtained as 53.8 $ms$ from Lemma 1 in [11] and $26.3 \times 10^3$ $ms$ from Theorem 1. The MADBs were obtained from methods in [12], is 0.45 $ms$. In [13], the MADB was obtained as $1.7 \times 10^3$ $ms$ by stability region technique. The MADB is obtained as $26.3 \times 10^3$ $ms$ from Theorem 1. So the MADB of this example can be obtained as the smaller value than $T_s = 26.3 \times 10^3$ $ms$.

### 4.2 Examples of NCS model

As an example for verification of the proposed method, a plant with six DC motors is considered. Each motor has an armature position controller with two sensors and one actuator, which are linked via the network. This configuration of six motors can be assumed to be a part of a six-axis robot. If the armature inductance ($L_a$) and viscous frictional coefficient ($B_m$) are negligible, the motor dynamics can be modelled by

$$\begin{aligned} \dot{x}_p &= F_p x_p + G_p u_p \\ &= \begin{bmatrix} -K_i K_b / R_a J & 0 \\ 1 & 0 \end{bmatrix} x_p + \begin{bmatrix} K_i / R_a J \\ 0 \end{bmatrix} u_p \quad (16) \\ y_p &= x_p, \quad (17) \end{aligned}$$

where $x_p = [\omega\ \theta]^T$, $u_p$ is applied voltage ($V$), and $\omega$ and $\theta$ are respectively, the rotor angular velocity (rad/sec) and displacement (rad). $R_a$, $K_i$, $K_b$, and $J$ represent respectively, the armature resistance, torque constant, back emf constant, and inertia of rotor and load. If a constant gain ($K$) is used as a state feedback controller, the system (5) is changed to

$$\dot{x}_p(t) = F_p x_p(t) + G_p * K x_p(t - \tau) \quad (18)$$

as a single control loop in the NCS, where $\tau = \tilde{\tau}_c + \tilde{\tau}_{sc} + \tilde{\tau}_{ca}$.

For the simulations, the motor in each loop has nominal values such that $R_a$ ($\Omega$), $K_i$ = 10 (oz-in/A), $K_b$ = 0.075 (V/rad/sec), and $J$ = 0.006 (oz-in-$sec^2$). The tested motors in each loops have the same nominal values as the previous one except $R_a$. Other two motors have the value of $R_a = 10$ ($\Omega$), $R_a = 13$ ($\Omega$), $R_a = 14$ ($\Omega$), $R_a = 19$ ($\Omega$), $R_a = 21$ ($\Omega$) and $R_a = 25$ ($\Omega$), respectively. Using Lemma 1 in [11] and the given parameters of the motors, the MADBs are calculated as 1.4, 2.6, 3.1, 6.1, 7.5 and 10 $ms$ in Table 3. Using Therorem 1 in [3] and the given parameters of the motors, the MADBs are calculated as 0.16, 0.21, 0.22, 0.30, 0.33 and 0.39 $ms$. So the final MADB is obtained as 1.4 $ms$ by Lemma 1. Using Theorem 1, the MADBs are calculated as $1.812 \times 10^3$, $1.802 \times 10^3$, $1.799 \times 10^3$, $1.785 \times 10^3$, $1.780 \times 10^3$, and $1.771 \times 10^3$ $ms$. Therfore the final MADB is obtained as $1.771 \times 10^3$ $ms$ by Theorem 1.

Table 2: MADB ($ms$) in the each loop

| Control Loop | by [11] | by [3] | by Theorem 1 |
|---|---|---|---|
| $R_a(\Omega) = 10$ | 1.4 | 0.16 | $1.812\times10^3$ |
| $R_a(\Omega) = 13$ | 2.6 | 0.21 | $1.802\times10^3$ |
| $R_a(\Omega) = 14$ | 3.1 | 0.22 | $1.799\times10^3$ |
| $R_a(\Omega) = 19$ | 6.1 | 0.30 | $1.785\times10^3$ |
| $R_a(\Omega) = 21$ | 7.5 | 0.33 | $1.780\times10^3$ |
| $R_a(\Omega) = 25$ | 10 | 0.39 | $1.771\times10^3$ |

### 4.3 Application of a Scheduling Method

For the simple test of a scheduling method [11] using MADBs, one motor has nominal values such that $R_a$ = 14 ($\Omega$), $K_i$ = 10 (oz-in/A), $K_b$ = 0.075 (V/rad/sec), and $J$ = 0.006 (oz-in-$sec^2$). Other two motors have the same nominal values as the previous one except $R_a$. Other two motors have the value of $R_a = 19$ ($\Omega$) and $R_a = 21$ ($\Omega$), respectively. Using Lemma 1 in [11] and the given parameters of the motors, the MADBs are calculated as 3.1, 6.1, and 7.5 $ms$. Using Theorem 1, the MADBs are calculated as $1.799 \times 10^3$, $1.785 \times 10^3$ and $1.780 \times 10^3$ $ms$ respectively. From now, the MADBs of each loop can be set as 3, 6, and 7 $ms$ for convenience of calculations. It is assumed that data for the sensor and the actuator have four bytes.

Using the notation in [11], the followings are given:
$N_C^j = N_A^j = 1$ for $j = 1,2,3$,
$N_S^j = 2$ for $j = 1,2,3$, $\quad N^j = 3$ for $j = 1,2,3$,
$N^* = 1, P = 3, \quad N = \sum_{j=1}^{P}(N_C^j + N_S^j) + N^* = 10$,
$D_{S_i}^j = 0.1$ $ms$ for $i = 1,2,\ j = 1,2,3$,
$D_{C_1}^j = 0.1$ $ms$ for $j = 1,2,3$,
$T_D^1 = 3$ $ms$, $T_D^2 = 6$ $ms$, $T_D^3 = 7$ $ms$,
$N_S^M = 2$,
$U_N^m = 0.16$,
where $N^*$ is total number of extra nodes which do not participate in control loops in the NCS. $D_{S_i}^j$ or $D_{C_i}^j$ is the $i$-th sensor or controller delay in the $j$-th loop, respectively. Applying steps 1 to 6 of the sampling period decision algorithm to the example, $N_P^T$

is calculated as 12 and $N_E$ is 2, while $N_P^M$ is 14 in the polling control network. Hence 14 nodes can be scheduled in the polling control network, within a basic sampling period. From this calculation, therefore, it can be shown that all nodes cannot be scheduled using token control, but can be, using polling control. Then, following the repetition steps of the sampling period decision algorithm, the sampling period can be reduced in the case of the polling control network. In the polling control case, as the overhead time in each node is bounded by a constant value, calculation results in the last section are similar to the real values. A basic sampling period can be determined from 1.4 $ms$ to 3.0 $ms$. Hence, the basic sampling period can be reduced by about 1.4 $ms$ in this case. Results from calculations are matched to those of the allocations. If 1.4 $ms$ basic sampling period is selected, the bandwidth can be allocated as shown in Fig. 2. However

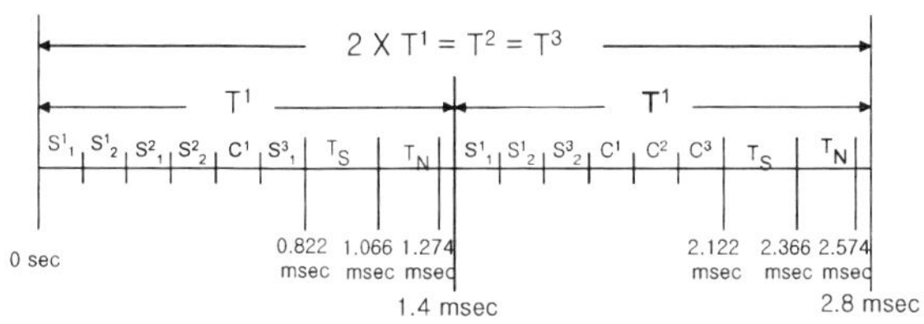


Figure 2: Bandwidth allocation result using polling control.

the above example is conservative to be practical use of NCSs. As an example, in building control systems, the NCSs with many nodes must be considered to a real time networking and scheduling within basic sampling time. In the case of $T_D^1 = 1.799 \times 10^3$ $ms$, $T_D^2 = 1.785 \times 10^3$ $ms$, $T_D^3 = 1.780 \times 10^3$ $ms$, more than 300 nodes and 2100 nodes can be scheduled in the token control and the polling control network respectively, within a basic sampling period. So a scheduling analysis of the large-scaled NCS is available. From this calculation, therefore, it can be shown that the large-scaled NCSs with many nodes be scheduled using token control and polling control.

## 5 Conclusions

In this paper, the MADBs are obtained for the stability of the NCS, and are used as the basic parameter for a scheduling method for the NCS can adjust the sampling period as small as possible, allocate the bandwidth of the network for three types of data, and exchange the transmission orders of data for sensors and actuators.

In an NCS, the presented method is useful, as it provides a solution to determine the sampling periods of each control loop and it can indicate whether the predetermined network protocol is possible for the given control system or not. An example is presented to show the usefulness of the proposed method for the NCS.

As future works, delays due to packet losses, buffering, and packetizing will be investigated for the analysis of NCSs.

## References


[1] Yoram, H. and Asok, R. 1988. Integrated Communication and Control Systems: Part I - Analysis, *Journal of Dynamic Systems, Measurement, and Control,* **110**, Dec., 367-373.

[2] Asok, R. and Yoram, H. 1988. Integrated Communication and Control Systems: Part II - Design Consideration, *Journal of Dynamic Systems, Measurement, and Control,* **110**, Dec., 374-381.

[3] Krtolica, R., Özgüner, Ü., Chan, H., Göktas, H., Winkelman, J. and Liubakka, M. 1994. Stability of Linear Feedback Systems with Random Communication Delays, *Int. J. Control,* **59**, (4), 925-953.

[4] Mori, T., Fukuma, N. and Kuwahara, M. 1981. Simple stability criteria for single and composite linear systems with time delays, *International Journal of Control,* **34**, (6), 1175-1184.

[5] Su, T-J and Huang, C-G. 1992. Robust Stability of Delay Dependence for Linear Uncertain Systems, *IEEE Tr. on AC,* **37**, (10), 1656-1659.

[6] Liu, P-L and Su, T-J. 1995. On a Sufficient Condition for Robust Stability of Delay Systems, *Proc. of ACC,* 328-329.

[7] Li, X and de Souza, C. E. 1997. Delay-dependent robust stability and stabilization of uncertain linear delay systems: a linear matrix inequality approach, *IEEE Trans. Automat. Contr.*, **42**, (8), 1144-1148

[8] Park, P.G. 1999. A Delay-dependent stability criterion for systems with uncertain time-invariant delays, *IEEE Trans. Automat. Contr.*, **44**, (4), 876-877.

[9] Moon, Y.S., Park, P.G., Kwon, W.H.and Lee, Y.S. 2001. Delay-dependent robust stabilization of uncertain state-delayed systems, to be appeared to *International Journal of Control.*

[10] Hong, S.H. 1995. Scheduling Algorithm of Data Sampling Times in the Integrated Communication and Control Systems, *IEEE Tr. on CST,* **3**, (2), 225-231.

[11] Park, H.S., Kim, Y.H., Kim, D.S and Kwon,W.H. 1999. A Scheduling Method for Network Based Control Systems, *IEEE Tr. on CST,* in Submission.

[12] Branicky, M.S., Phillips, S.R. and Wei Zhang. 2000. Stability of networked control systems: explicit analysis of delay, *Proceedings of the American Control Conference*, **4**, pp. 2352 -2357.

[13] Wei Zhang, Branicky, M.S. and Phillips, S.M. 2001. Stability of networked control systems, *IEEE Control Systems Magazine*, **21**, (1), pp. 84-99.

[14] Walsh, G.C. and Hong Ye. 2001. Scheduling of networked control systems, *IEEE Control Systems Magazine*, **21**, (1), pp.57 -65

## APPENDIX A

*Proof.* (Theorem 1) Choose a Lyapunov functional as follows:

$$V(x(t-\alpha), \alpha \in [0, \bar{\tau}]) = V_1 + V_2 + V_3 \qquad (19)$$

where

$$V_1 \triangleq x^T(t)Px(t), \qquad (20)$$

$$V_2 \triangleq \sum_{i=1}^{N}\left\{\int_{-\tau_i}^{0}\int_{t+\beta}^{t} \dot{x}^T(\alpha)Z_i\dot{x}(\alpha)d\alpha d\beta\right\}, \qquad (21)$$

$$V_3 \triangleq \sum_{i=1}^{N}\left\{\int_{t-\tau_i}^{t} x^T(\alpha)Qx(\alpha)d\alpha\right\}. \qquad (22)$$

Since x(t-$\tau_i$) $= x(t) - \int_{t-\tau_i}^{t}\dot{x}(\sigma)d\sigma$ $= x(t) - \int_{t-\tau_i}^{t}\left[Ax(\sigma) + \sum_{i=1}^{N}\{A_i x(\sigma-\tau_i)\}\right]d\sigma$, the system (6) can be written as

$$\dot{x}(t) = \left(A + \sum_{i=1}^{N} A_i\right)x(t) - \sum_{i=1}^{N}\left\{A_i\int_{t-\tau_i}^{t}\dot{x}(\alpha)d\alpha\right\}$$

and thus the derivative of $V_1$ satisfies the relation $\dot{V}_1 = 2x^T(t)\left[P\left(A + \sum_{i=1}^{N} A_i\right)\right]x(t)$ $-2\sum_{i=1}^{N}\left\{x^T(t)PA_i\int_{t-\tau_i}^{t}\dot{x}(\alpha)d\alpha\right\}$. Defining $a(\cdot)$, $b(\cdot)$, and $\mathcal{N}$ in Equation (1) as $a(\alpha) \triangleq x(t)$, $b(\alpha) \triangleq \dot{x}(\alpha)$, and $\mathcal{N} \triangleq PA_i$ for all $\alpha \in [t-\tau_i, t]$ and applying Lemma 1 will supply $\begin{bmatrix} X_i & Y_i \\ Y_i^T & Z_i \end{bmatrix} \geq 0, \quad i = 1, \cdots, N$, and

$$\begin{aligned}\dot{V}_1 \leq\ & 2x^T(t)\left[P\left(A + \sum_{i=1}^{N} A_i\right)\right]x(t) + \sum_{i=1}^{N}\tau_i x^T(t)X_i x(t) \\ & +2\sum_{i=1}^{N}\left\{x^T(t)(Y_i - PA_i)\int_{t-\tau_i}^{t}\dot{x}(\alpha)d\alpha\right\} \\ & +\sum_{i=1}^{N}\left\{\int_{t-\tau_i}^{t}\dot{x}^T(\alpha)Z_i\dot{x}(\alpha)d\alpha\right\} \\ \leq\ & x^T(t)\left[A^TP + PA + \sum_{i=1}^{N}\{\bar{\tau}X_i + Y_i + Y_i^T\}\right]x(t) \\ & +2\sum_{i=1}^{N}\{x^T(t)(PA_i - Y_i)x(t-\tau_i)\} \\ & +\sum_{i=1}^{N}\left\{\int_{t-\tau_i}^{t}\dot{x}^T(\alpha)Z_i\dot{x}(\alpha)d\alpha\right\}.\end{aligned}$$

Since $\dot{V}_2$ and $\dot{V}_3$ yield the relation

$$\begin{aligned}\dot{V}_2 =\ & \sum_{i=1}^{N}\left\{\tau_i\left[Ax(t) + \sum_{i=1}^{N}\{A_i x(t-\tau_i)\}\right]^T Z_i\right. \\ & \left.\sum_{i=1}^{N}\{A_i x(t-\tau_i)\}\right] - \sum_{i=1}^{N}\left\{\int_{t-\tau}^{t}\dot{x}^T(\alpha)Z\dot{x}(\alpha)d\alpha\right\},\end{aligned}$$

$$\dot{V}_3 = \sum_{i=1}^{N}\left\{x^T(t)Q_i x(t) - x^T(t-\tau_i)Q_i x(t-\tau_i)\right\},$$

we have the derivative of $V$ as follows:

$$\dot{V} = \dot{V}_1 + \dot{V}_2 + \dot{V}_3 \leq \bar{x}^T\begin{bmatrix} X_{11} & X_{12} \\ X_{12}^T & X_{22} \end{bmatrix}\bar{x},$$

where

$$\bar{x} \triangleq [x^T(t)\ \ x^T(t-\tau_1)\ \ \cdots\ \ x^T(t-\tau_N)]^T,$$

$$X_{11} \triangleq A^TP + PA + \sum_{i=1}^{N}\{Y_i + Y_i^T + \bar{\tau}X_i + Q_i + \bar{\tau}A^TZ_iA\}$$

$$X_{12} \triangleq \left[PA_1 - Y_1 + \sum_{i=1}^{3}\bar{\tau}A^TZ_iA_{1N} - Y_N + \sum_{i=1}^{3}\bar{\tau}A^TZ_iA_N\right.$$

$$X_{22} \triangleq -\text{diag}\{Q_1, \cdots, Q_N\} + \begin{bmatrix} \sum_{i=1}^{N}\bar{\tau}A_1^TZ_iA_1 \cdots \sum_{i=1}^{N}\bar{\tau}A_1^TZ_iA_N \\ \vdots \ddots \vdots \\ \sum_{i=1}^{N}\bar{\tau}A_N^TZ_iA_1 \cdots \sum_{i=1}^{N}\bar{\tau}A_N^TZ_iA_N \end{bmatrix}.$$

Therefore, if

$$\begin{bmatrix} X_{11} & X_{12} \\ X_{12}^T & X_{22} \end{bmatrix} < 0, \qquad (23)$$

the system (6) is asymptotically stable according to the Lyapunov-Krasovskii stability theorem. Equation (23) can be rewritten into $\begin{bmatrix} Y_{11} & Y_{12} \\ Y_{12}^T & Y_{22} \end{bmatrix} + \mathcal{A}^T\mathcal{Z}\Gamma^{-1}\mathcal{Z}\mathcal{A} < 0$, where

$$Y_{11} \triangleq A^TP + PA + \sum_{i=1}^{N}\{Y_i + Y_i^T + \bar{\tau}X_i + Q_i\} \qquad (24)$$

$$Y_{12} \triangleq [PA_1 - Y_1\ \ \cdots\ \ PA_N - Y_N], \qquad (25)$$

$$Y_{22} \triangleq -\text{diag}\{Q_1, \cdots, Q_N\}, \qquad (26)$$

By the Schur complement, Equation (26) is equivalent to the first inequality in Equation (11). This completes the proof. □ □

IFAC Publications
www.elsevier.com/locate/ifac

# MULTI-AGENT SYSTEMS FOR MODEL-BASED FAULT DIAGNOSIS

**X. Ren * S. M. Hargrave * H. A. Thompson ***
**P. J. Fleming *,[1]**

* *Department of Automatic Control and Systems Engineering, University of Sheffield, Sheffield, S1 3JD, UK*

Abstract: Model-based methods are commonly used for fault diagnosis. But for modern complex processes, the analytical model of the overall process is not usually available. In this paper, a multi-agent framework for distributed fault diagnosis and isolation is proposed. The distributed fault diagnosis and isolation is achieved by a group of cooperative intelligent agents. The structure of the multi-agent framework is described. The learning algorithms employed by agents for building the fuzzy analytical model of the local process is detailed. The implementation of this agent-based approach on aero gas turbine engine fault diagnosis is also discussed. 

Keywords: Intelligent agents, model-based fault diagnosis, fuzzy inference systems

## 1. INTRODUCTION

The increasing demand to make dynamic systems safer and more reliable has motivated the development of fault diagnosis and isolation (FDI) systems. A model-based method is a popular approach in FDI research. Its focus is on dynamic consistency (parity) relations. The basic idea is to generate the analytic symptoms from an input-output measurement. These generated analytic symptoms and heuristic symptoms are then analysed in the diagnostic stage for finding the type, size and location of the fault (Gertler, 1998; Isermann, 1997; Chen and Patton, 1999).

Modern industrial machines often include a large number of working units or sub-components. Traditional centralised model-based FDI approaches require an overall plant analytical model and the measurements of all the plant variables. This is particularly resource intensive requiring both bandwidth and computation time. Model inaccuracy and measurement uncertainty present significant challenges to achieve accurate fault diagnosis. Also, the identification of the root cause is difficult because the degradation of some components have influence on others. To overcome these shortfalls, distributed FDI and qualitative based methods are highly recommended (Henry and Clarke, 1993; Patton *et al.*, 2000).

Intelligent agent technology and multi-agent systems are receiving considerable attention within the industrial field. An agent is an autonomous system that is able to act without the direct intervention of human or other agents with the ability to control its own actions and internal states. An intelligent agent is an agent that has the capability to learn, to reason and to adapt to changes in its environment. Consequently, an intelligent agent is capable of flexible autonomous action. Key features of an intelligent agent include reactivity, proactivity and cooperativity (Wooldridge and Jennings, 1995).

Multi-agent systems (MAS) are concerned with coordinating problem solving behaviour amongst a collection of agents. It is convenient to define a multi-agent system as an agent society. Each agent in this society represents a specific set of

[1] Corresponding author. Email: p.fleming@sheffield.ac.uk

problem solving skills and experience. The intention of the MAS is to coordinate the skills, knowledge, plans and experience of different agents to achieve a common high level system goal.

Intelligent agent technology and multi-agent systems are receiving greater attention in the industrial field and so it is natural to apply these approaches to the problem of distributed fault diagnosis. Typical reasons for using a multi-agent approach in the industrial domain for control, diagnosis and repair include reducing high data loading, integration of legacy components, system flexibility and following the naturalness of the distributed systems (Nejdl and Werner, 1994). Agent-based applications in the field of control engineering can be found in (Velasco and Gonzalez, 1996; Parunak, 1999; Jennings and Wooldridge, 1998).

A MAS-based architecture for distributed fault diagnosis is proposed in this paper. In this general framework, several software agents cooperate with each other to monitor the local processes and draw their own diagnostic conclusion based on their qualitative and quantitative knowledge. A higher-level agent is employed to supervise local diagnostic agents and draw the final conclusion of the global fault diagnosis and isolation.

This paper is organised as follows: the second section presents an overview of the proposed multi-agent framework for distributed fault diagnosis and isolation. The type and role of agents in this framework are summarised. In the third section, the details of the learning algorithms employed by agents are presented. Local diagnostic agents use these proposed learning algorithms to generate a fuzzy model as the analytical model for model-based FDI. Initial experiments for gas turbine engine FDI are given in section 4 and finally, some concluding remarks are made.

## 2. MULTI-AGENT FRAMEWORK FOR FAULT DIAGNOSIS AND ISOLATION

A MAS-based architecture for distributed fault diagnosis is illustrated in figure 1. This is a two-level framework. At the lower level, local diagnostic agents focus on local process variable monitoring based on their knowledge about the process, local observations and information provided by other agents as requested. At a higher level, the supervisor agent coordinates the local diagnostic agents to generate global fault diagnosis and isolation.

The type and roles of agents in this framework are described in the following.

**Local fault diagnostic agents.** The main function of the local fault diagnostic agents is to monitor process state variables by using a model-based

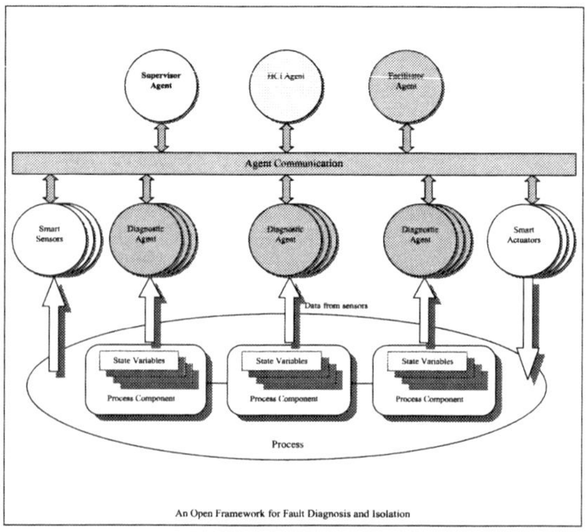


Figure 1. Open Agent Framework for FDI

method. Each agent has its own observation, reasoning mechanism and domain specific knowledge embedded. Based on its observations, a local fault diagnostic agent uses an analytical model embedded to predict the value of the state variable and then compare it with the real value of the process. A local fault diagnostic decision is made based on the evaluation of any discrepancy. A fuzzified message is then sent to trigger the upper level supervisor agent if a possible fault is detected. For doing local fault diagnostics, an analytical model of local process is needed for each agent. The algorithms for building the fuzzy inference systems are presented in section 3. Furthermore, a local fault diagnostic agent needs to communicate with other agents to acquire information such as the value of the relevant process variables or state of the other agents. Agents can also offer useful information to other agents for cooperation. All of the communication is performed using an agent communication language (ACL) such as KQML or FIPA.

**Supervisor agent for high-level coordination and decision making.** While each local diagnostic agent gives a partial conclusion on one process variable, all of these partial diagnostic results based on local knowledge are coordinated by a coordinator agent called the supervisor agent. The supervisor agent utilises process operating conditions and sequences of event messages, which contain partial detection information reported by the local diagnostic agents, to reason and make decisions based on the higher level abstract knowledge about the process. The higher level abstract knowledge can be presented as a qualitative model of the overall system. When the local diagnostic agents report their fault detection findings to supervisor agent, the supervisor agent uses its qualitative model combined with its own observation of the process state to validate these findings and draw the possible causality conclusions if faults exist. The possible qualitative modelling approaches for supervisor agents include production systems, causal models, Markov models, discrete event models and Bond graph models. Given a proper qualitative model, qualitative reasoning techniques will be used to generate an approximate solution to the overall FDI problem. Unlike the local fault diagnostic agents, the supervisor agent uses a symbolic model rather than an analytical model of the process. The concept of the symbolic model-based approach comes from the artificial intelligence field and is more suitable for the high-level reasoning and decision making task. By using this qualitative model and reasoning approach, the supervisor agent 'thinks and reasons' in a manner closer to a human being's thinking and reasoning.

**Other support agents.** For supporting the multi-agent coordination, other support agents such as the facilitator agent, agent manager and database manager are employed in this framework. The facilitator agent is used to direct the agent communication. The agent manager is used to manage the agent registration in the framework. The database manager agent maintains the global database that includes the real-time data of the process. A blackboard architecture is used to support the data updating. The human computer interface (HCI) agents are used to allow the human operator to assign tasks to agents and display the process diagnosis information.

By using this multi-agent framework, the complexity of large-scale system fault diagnosis and isolation is divided into simple sub-tasks, which can be solved in a parallel fashion. Since the local diagnostic agents only focus on one local process variable, the model-based fault diagnosis is simplified. By using soft-computing methods, model-based quantitative knowledge is combined with qualitative knowledge of the plant from human expertise. Agents can use this unified knowledge to deal with model inaccuracy and uncertainty in the fault diagnosis. Furthermore, the higher level supervisor agent uses the symbolic model of the system to deal with the fault isolation. This works like a human operator. This proposed multi-agent framework is flexible and extensible. New agents can be added by simply registering itself with the agent manager. Likewise, other elements such as smart sensors and smart actuators can also be included in this framework.

## 3. FUZZY INFERENCE SYSTEM FOR LOCAL FAULT DIAGNOSTIC AGENTS

As mentioned above, by using a model-based FDI approach the local diagnostic agents can detect process abnormality. Apart from more traditional use of non-linear and robust observers, parity space methods and hypothesis testing theory, soft-computing approaches like neural networks and fuzzy logic are now being investigated as powerful modelling and decision making tools (Patton *et al.*, 2000). In this section, the learning algorithms to generate the fuzzy analytical model are described. This learning procedure is used within the local specialised knowledge acquisition procedure. The local diagnostic agents employ the generated fuzzy model to predict the behaviour of its environment and to generate diagnosis reference by inference. In this fashion, the knowledge of the global process is actually distributed into a set of local agent knowledge bases.

### 3.1 *Fuzzy Inference Systems*

The fuzzy set theory formulated by Zadeh (Zadeh, 1965) has been considered an appropriate presentation method for systems which have imprecise or uncertain descriptions. The Mamdani-type fuzzy inference system is the most commonly used fuzzy inference system (Lee, 1990). The fuzzy rule of Mamdani fuzzy inference system is:

$$R^k\text{: if } x_1 \text{ is } A_1^i \text{ and } x_2 \text{ is } A_2^i, \ldots, \text{ and } x_m \text{ is } A_m^i \text{ then } y \text{ is } B^i$$

Where $R^k$ is the $k_{th}$ fuzzy rule $(k = 1, 2, \ldots, M)$. $x_j(j = 1, 2, \ldots, m)$ are the input fuzzy variables and $y$ is the output fuzzy variable. $A_j^i(j = 1, 2, \ldots, m; i = 1, 2, \ldots, n_j)$ and $B^i(i = 1, 2, \ldots, n_b)$ are fuzzy sets defined in the universe of discourse in which the fuzzy variables $x_j(j = 1, 2, \ldots, m)$ and y are defined.

The performance of the fuzzy inference system (FIS) depends on the input and output variables, fuzzy sets, proposition rules and the underlying control mechanisms that tie all these together into a cohesive reasoning environment. Once the input and output variables are defined and fuzzy inference method is chosen, the proposition rules and membership functions assigned with each fuzzy set are critical to a fuzzy inference system. Given a fuzzy inference system with $m$ input fuzzy variables and each fuzzy variable has $n_j$ fuzzy linguistic descriptions, the complete set of fuzzy rules contains $\prod_{j=1}^{m} n_j$ fuzzy rules. When the amount of input variables is large, the number of fuzzy if-then rules is usually too large to be included in a single FIS. Furthermore, the fuzzy rules will be computationally intensive to generate and use. For example, people face this problem when using the adaptive neuro fuzzy inference system (ANFIS) (Jang, 1993) to generate the FIS. The GA based rule selection method proposed in (Linkens and Nyongesa, 1995) is also incapable of handling such a large number of fuzzy if-then rules.

In this paper, Mamdani-type FIS is used for modelling the non-linear local process. By using the learning algorithms proposed in this paper, this analytical model of local process is generated with little prior process knowledge other than input-output data. A three-step procedure is used to build a compact fuzzy inference system for this modelling purpose. Firstly, locating the membership functions is achieved with a simple K-Mean clustering method to group the input-output data into a number of clusters. Each cluster is then covered by one membership function. Secondly, the well-known Wang's method for fuzzy rule generation is modified. Finally, the tentative fuzzy model is fine-tuned using genetic algorithms.

### 3.2 *Fuzzy Rule Generation*

Wang and Mendel define four steps for building a fuzzy rule base (Wang and Mendel, 1992). Firstly, fuzzy sets are defined to cover the input and output spaces. Secondly, one dominant rule is generated from one input-output pair. The generated rule has the format:

$$R^k\text{: if } x_1 \text{ is } A_1^i \text{ and } x_2 \text{ is } A_2^i, \ldots, \text{ and } x_m \text{ is } A_m^i \text{ then } y \text{ is } B^i \text{ with } W = W^k$$

where $x_j$ (j=1,2,...,m)is the $j_{th}$ input, and $y$ is the output. $A_j$ (j=1,2,...,m) are the fuzzy sets defined for $x_j$. $B$ are the fuzzy sets defined for $y$. $W^k$ is the weight of the generated rule $R^k$. For each input $x_j$, $A_j^*$ with the maximal degree of membership is chosen. Similarly, For output $y$, $B^*$ with the maximal degree of membership is chosen. On the third step, a weight is assigned to each rule. The weight of the rule is defined as the product of all the degrees of membership in the antecedent and conclusion part. Finally, the fuzzy rule base is created. When conflict rules appear, the winner is the rule with maximum weight. The rules with smaller weight are deleted.

The winner-takes-all rule selection method above is a shortcoming of Wang's method. When disturbances exist in the training data, one undesired data sample may generate a winner rule that masks all of the other possibly right rules. To overcome this, the following algorithm is used to select the winner rule and assign its weight. *The data samples that generate the same rule from second step above are clustered together. The cluster with dominant amount data samples is chosen as the rule generator. The rule generated by this winner data cluster is the winner rule, and the average weight of rules generated by this winner cluster is assigned as the weight of the final rule.*

Experiments have demonstrated that the proposed method is particularly suitable when the training data is rich. In this instance, the modified rule selection method has obvious advantage over Wang's method.

### 3.3 *Tuning the FIS by Using GAs*

As depicted above, the proposition rules and the membership functions for the linguistic properties are the most important parts of a fuzzy inference system. After the generation of the fuzzy rules by using the proposed algorithm, the location and shape of each membership function needs to be tuned to achieve better performance. Genetic algorithms are used in this tuning step. For using GAs, firstly the parameters that decide the location and shape of the membership function are encoded by means of chromosomes. Then, a set of chromosomes is generated as the initial generation. This initial generation is evolved under evolutionary operations to generate new generations. The following describes the details of this tuning procedure:

**Coding and decoding.**The parameters that need to be coded are the crucial parameters of the membership function. Binary Gray coding is used to encode these parameters to corresponding genes. A chromosome is a combination of all these genes. The purpose of decoding is to obtain the phenotype of each gene in a chromosome. Through this mapping, the value of the parameters in the real problem domain is obtained.

**Initialisation.** As mentioned before, the input-output data has been grouped by using K-Mean clustering algorithm and each cluster is covered by one membership function. The cluster centres are used as a reference for the centre of the membership function. The possible solutions are uniformly distributed around these cluster centres. Other parameters of the membership function are also uniformly distributed with their possible positions.

**Fitness evaluation.** Corresponding to the data modelling requirement, a normalised mean-square error is used as the objective function. For each chromosome in the new population, the learning algorithm for fuzzy rule generation in section 3.1 is applied and the fuzzy model with this new generated rule base and new membership functions is evaluated with test data. Since the objective is to identify the fuzzy model with the smallest difference to the real system, the fitness value is assigned to each chromosome according to the mean square error of the fuzzy model generated. Smaller mean square errors result in higher fitness value being assigned to the chromosome.

**Creation of offspring through selection, crossover and mutation.** After evaluation, a new population is evolved from the current generation. Selection operation is used to select the elitists from the current generation. A pre-set probability is used to keep the elitists with above average fitness value and replace below average members. Crossover is applied to the parents to hopefully generate children that keep the good features of their parents. A given crossover probability is used for crossover operation. While crossover exploits the current gene potentials, the mutation operation is used to find totally new solutions that parents do not have. Mutation is also useful to prevent the search being trapped in local optima.

**Termination condition.** The algorithm stops when the best fitness value is not improved significantly on several generations or a pre-set maximal number of generations is reached.

## 4. IMPLEMENTATION ON THE AERO GAS TURBINE ENGINE FAULT DIAGNOSIS

A modern aero gas turbine engine is a complex machine comprising thousands of components. Since the engine usually works in a harsh environment, performance deterioration is inevitable and can result in additional fuel consumption. Furthermore, component failure can have serious consequences. Thus there is a need to develop a reliable, efficient fault detecting and diagnostic system to enhance the safety and economical operation of the engine. As mentioned in the introduction, a multi-agent framework has been proposed for gas turbine engine FDI. In this multi-agent system, the low-level diagnostic agent only focuses on one local process variable, while at the upper-level a supervisor agent coordinates all agents' work to validate their fault reports and tries to find the possible cause and location of a fault.

The high pressure spool speed (NHP), low pressure spool speed (NLP), turbine blade temperature (TBT), effective fuel flow (WFE) are monitored for fault diagnosis. These variables, along with changes in the intake air temperature and pressure, altitude, and pilot's lever angle are all separately monitored by a group of local diagnostic agents. For example, an NHP agent is used for high pressure spool speed monitoring and a TBT agent for turbine blade temperature monitoring, etc.

A non-linear dynamic simulation of the engine using SIMULINK has been used to generate the training and test data for local process modelling. The training data set was generated according to a test profile shown in figure 2. This test profile covers a range of accelerations and decelerations, at different altitude and flight speeds. In the following example, a fuzzy model of the turbine blade temperature (TBT) diagnostic agent is shown. Analysis highlighted that this turbine blade temperature is most dependent upon the effective fuel flow (WFE), high pressure spool speed (NHP) and low pressure spool speed (NLP). Thus these three engine variables with the addition of altitude and rate of change of high pressure spool speed are used for creating the fuzzy model. Following the algorithms proposed in section 3, the analytical model for the TBT agent is generated. Figure 3 illustrates the actual TBT and the estimated TBT based on this fuzzy model.

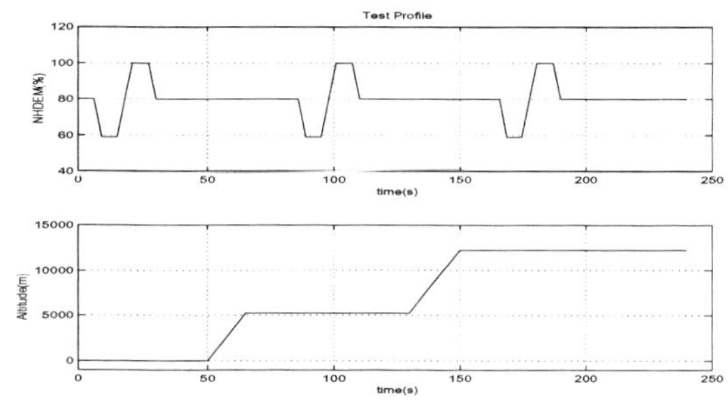


Figure 2. A Test Profile

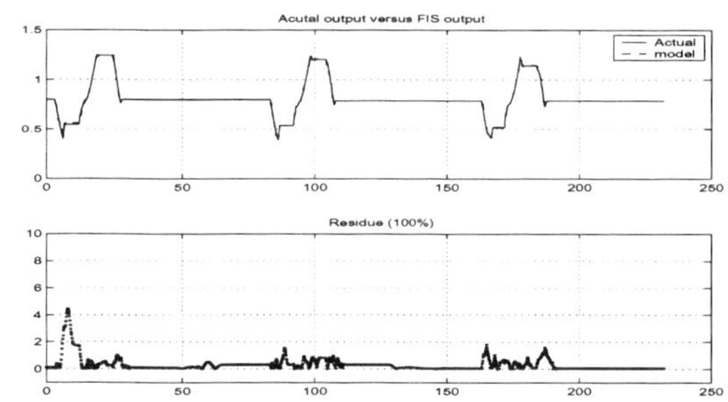


Figure 3. Actual and Model Output of TBT

A TBT sensor bias fault is injected for testing purposes and is illustrated in figure 4. Steady state detection is achieved with a threshold set at 3% above normal. During transient operation, detection with a threshold set at 8% above normal. This indicates that the supervisor agent needs to distinguish the process state and use different criteria within the diagnosis.

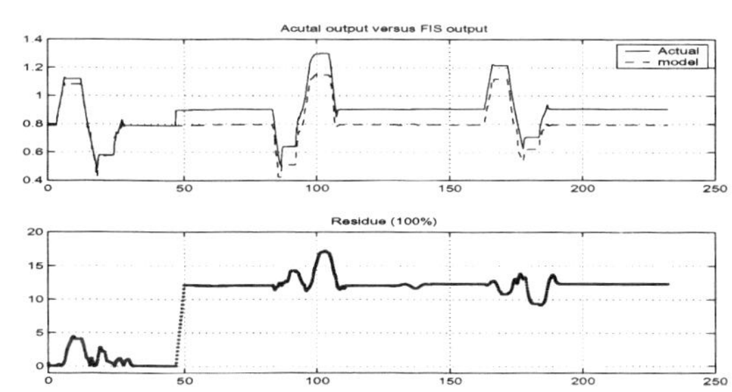


Figure 4. TBT Sensor Bias Fault Scenario

Following the same procedure as described for the TBT agent, more agents are employed to monitor other variables of the gas turbine engine. Figure 5 shows the actual and model output of NHP agent under the same sensor bias fault. Results show that this sensor fault has no effect on the NHP agent because this fault is compensated by the controller. Figure 6 shows the model output of the NLP agent. This agent does detect a fault and hence it will report a fault to supervisor agent. It should be noticed that the fault pattern is different to that of a NLP sensor fault. This difference will help the upper-level supervisor agent to find the location of fault.

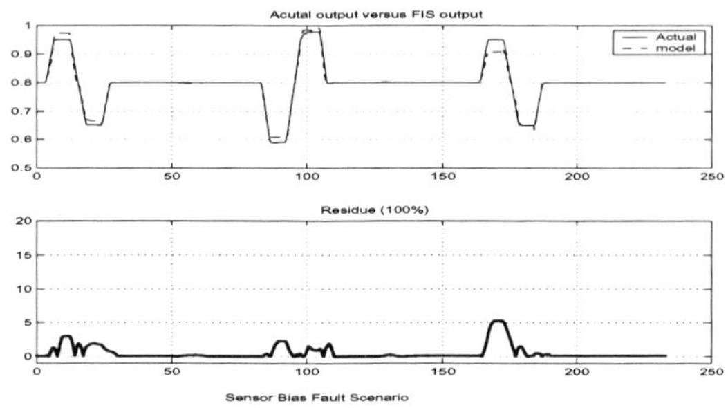


Figure 5. Actual and Model Output of NHP under TBT Sensor Bias Fault

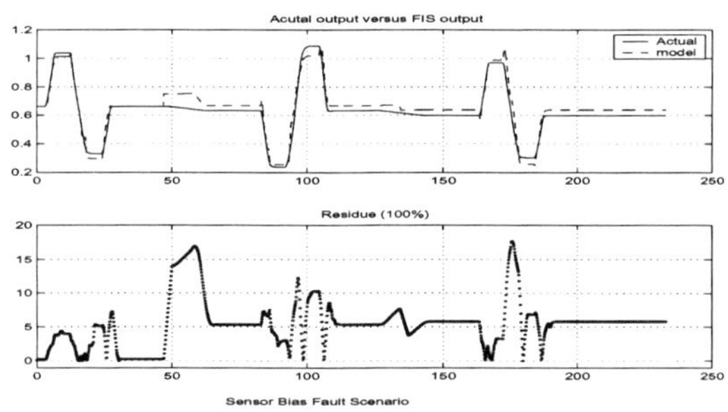


Figure 6. Actual and Model Output of NLP under TBT Sensor Bias Fault

## 5. CONCLUDING REMARKS

This paper has proposed a multi-agent framework for distributed fault diagnosis and isolation. This multi-agent framework can be used to facilitate the design and implementation of FDI for complex and large-scale systems. A learning procedure is proposed to build the local analytical model by using a fuzzy inference system. Preliminary experimental results carried out on the gas turbine engine have shown that both the agent-based architecture and the fuzzy modelling approach for building local process model are suitable for distributed FDI. Nevertheless, this work can be extended by investigating more efficient methods for both analytical and symbolic modelling. Further research will integrate high level coordination for fault isolation using qualitative reasoning techniques. For the gas turbine engine FDI, practical demonstrations of this research is ongoing.

## 6. REFERENCES

Chen, J. and R. J. Patton (1999). *Robust Model-based Fault Diagnosis for Dynamic Systems.* Kluwer Academic Publishers. USA.

Gertler, Janos (1998). *Fault Detection and Diagnosis in Engineering Systems.* Marcel Dekker.

Henry, M. and D. Clarke (1993). The self-validating sensor: rational, definitions and examples. *Control Engineering Practice* **1**(4), 585–610.

Isermann, R. (1997). Supervison, fault-detection and fault diagnosis methods - an introduction. *Control Engineering Practice* **5**(4), 639–652.

Jang, Jyh-Sing Roger (1993). Anfis: Adaptive-network-based fuzzy inference system. *IEEE Transactions on Systems, Man, and Cybernetics* **23**, 665–684.

Jennings, N. and M. Wooldridge (1998). *Agent Technology: Foundations, Applications and Markets.* Springer-Verlag. New York.

Lee, C. C. (1990). Fuzzy logic in control systems: Fuzzy logic controller (part 1). *IEEE Transactions on Systems, Man and Cybernetics* **20**(2), 404–435.

Linkens, D. A. and H. O. Nyongesa (1995). Genetic algorithms for fuzzy control part 1: Offline system development and application. *IEE proceeding on Control Theory and Application* **142**(3), 161–176.

Nejdl, Wolfgang and Martin Werner (1994). Distributed intelligent agents for control, diagnosis and repair. Technical report. Informatik V, RWTH Aachen. Aachen, Germany.

Parunak, H. Van Dyke (1999). *Industrial and Practical Applications of DAI.* Chap. 9, pp. 377–421. Multiagent systems ed.. The MIT Press. Cambridge, MA.

Patton, R. J., F. J. Uppal and C. J. Lopez-Toribio (2000). Soft computing approaches to fault diagnosis for dynamic systems: A survey. In: *4th IFAC Symposium on Fault Detection supervision and Safety for Technical Processes.* Budapest, Hungary. pp. 198–211.

Velasco, J. and J. Gonzalez (1996). Multiagent-based control systems: A hybrid apporach to distributed process control. *Control Engineering Practice* **4**(6), 839–845.

Wang, L. and J. Mendel (1992). Generating fuzzy rules by learning from examples. *IEEE Transactions on Systems, Man and Cybernetics* **22**(2), 1414–1427.

Wooldridge, M. and N. Jennings (1995). Intelligent agents: Theory and pratice. *Knowledge Engineering Review* **10**(2), 115–152.

Zadeh, L. A. (1965). Fuzzy sets. *Information and Control* **8**, 338–353.

IFAC Publications
www.elsevier.com/locate/ifac

# INTELLIGENT FAULT DIAGNOSIS BASED ON NEUROFUZZY NETWORKS FOR NONLINEAR DYNAMIC SYSTEMS

**Wang Y., Chan C.W. and Cheung K.C.**

*Department of Mechanical Engineering*
*The University of Hong Kong*
*Pokfulam Road, Hong Kong, China*
*Email: mechan@hkucc.hku.hk Fax: (852)28585415*

**Abstract:** In this paper, a new fault diagnosis technique based on B-spline neurofuzzy networks is proposed for nonlinear dynamical systems. The main reasons for using neurofuzzy networks here are that they can approximate nonlinear functions with arbitrary accuracy, and can be readily trained online. In the proposed scheme, the fault is approximated by the output of a neurofuzzy network, which is to be trained online, and the fault is isolated from the fuzzy rules extracted from the neurofuzzy network. However, as the number of fuzzy rules generated by neurofuzzy networks can be quite large, it is necessary to reduce these rules, which can be a difficult process. A simple approach proposed here is to delete fuzzy rules that have too small a confidence level. It is shown in the simulation example involving a nonlinear single-link robot arm that the proposed procedure can detect and isolate the fault. *Copyright © 2001 IFAC*

**Keywords:** fault diagnosis, neurofuzzy networks, nonlinear observer, online training

## 1. INTRODUCTION

Fault diagnosis for nonlinear systems has attracted a lot of attentions recently, since most practical systems are complex and nonlinear. A number of approaches have been proposed (see e.g., the survey papers, Frank (1990), and Isermann (1993), and the book (Chen and Patton, 1999)). In comparison with linear systems, fault diagnosis for nonlinear systems is much more difficult, and a general fault diagnosis scheme is difficult, if not impossible, to derive. Fault diagnosis for a class of nonlinear dynamical systems with the fault a nonlinear function of the state and the input of the system is considered in Chan, *et al.* (1999). It was proposed that the fault is estimated on-line using a neural network, and fault is diagnosed from its estimate. The result in Chan, *et al.* (1999) is extended to systems with model uncertainties in Wang, *et al.* (1999). Two groups of neural networks are used, one of which is to approximate the model uncertainties, and the other, to estimate the fault on-line.

The main emphasis of most existing fault diagnosis schemes is mainly on fault detection rather than fault isolation. In this paper, an intelligent fault isolation scheme using B-spline neurofuzzy networks is proposed based on the fault diagnosis schemes proposed in Chan, *et al.* (1999), and Wang, *et al.* (1999). As neurofuzzy networks incorporate the qualitative reasoning via fuzzy logic, and the quantitative learning from training data via neural networks, it combines intelligent qualitative and quantitative reasoning ($IQ^2$) within a single framework (Harris *et al.*, 1999). Zhang and Morris (1994) proposed a neurofuzzy network for on-line process fault diagnosis by inserting a fuzzification layer to a conventional feed-forward neural network. In the fault classifier proposed by Ayoubi (1995), a neurofuzzy network is used to model the system, and an initial knowledge base was extracted from a set of training data for fault diagnosis. Benkhedda and Patton (1996) used B-spline neural networks to generate residuals for fault diagnosis, and fuzzy rules are also used to interpret the results obtained from fault diagnosis. In this paper, the fault is estimated online first by a neurofuzzy network, and fuzzy rules are then extracted from the network to isolate the fault.

The paper is organized as follows. A brief review of the online fault estimation scheme proposed in Chan *et al.* (1999) is presented in Section 2. The procedure to extract fuzzy rules from B-spline neurofuzzy networks is described in Section 3, and is illustrated by a simulation example involving a nonlinear single-link robot arm is given in Section 4.

## 2. ON-LINE FAULT ESTIMATION

Consider the following nonlinear dynamic system,

$$\begin{aligned} x(k+1) &= Ax(k)+\alpha(y(k),u(k))+\varphi(k)+B\phi(x(k)),u(k)) \\ y(k) &= Cx(k) \end{aligned} \quad (1)$$

where $x \in R^n$, $y \in R^l$ and $x \in R^m$ are respectively the state, the output and the input of the system. Assume that the measurement of $u(k)$, $y(k)$ are available, but not that of $x(k)$. The system matrix $A \in R^{n\text{x}n}$, the output matrix $C \in R^{l\text{x}n}$ and the fault matrix $B \in R^{n\text{x}p}$ are assumed known. The known nonlinear term, $\alpha : R^l \text{ x } R^m \rightarrow R^n$, the unknown model uncertainties $\varphi : R^+ \rightarrow R^n$, and the unknown fault $\phi : R^n \text{ x } R^m\ R^+ \rightarrow R^p$ are assumed to be smooth nonlinear functions.

Though only the case that $\alpha$ is a function of $u(k)$ and $y(k)$ is considered in this paper, the method can be readily extended to the case that $\alpha$ is a function of $x(k)$ and $u(k)$. Note that $\phi(.)$ is a function of the input $u(k)$ and the state $x(k)$, the fault is much more difficult to estimate, as $x(k)$ is unknown. If there is no fault, $\phi$ in (1) becomes zero.

The fault diagnosis scheme proposed in (Chan, *et al*, 1999, Wang, *et al*, 1999) is extended here for discrete systems. Let

$$\phi(x,u) = [\phi_1(x,u) \quad \phi_2(x,u) \quad \cdots \quad \phi_p(x,u)]^T$$

and the estimate of $\phi(.)$ is denoted by

$$\hat{\phi}(\hat{x},u) = [\hat{\phi}_1(\hat{x},u) \quad \hat{\phi}_2(\hat{x},u) \quad \cdots \quad \hat{\phi}_p(\hat{x},u)]^T$$

where "^" denotes the estimate of $\phi(.)$. The nonlinear observer for system (1) is

$$\begin{aligned} \hat{x}(k+1) &= A\hat{x}(k) + \alpha(y(k),u(k)) + B(\hat{\phi}(k) - \delta(k)) \\ &\quad + K(y(k) - \hat{y}(k)) \\ \hat{y}(k) &= C\hat{x}(k) \end{aligned} \quad (2)$$

where $K \in R^{n\text{x}l}$ is the observer gain matrix, which is chosen such that $A - KC$ is stable. The robust vector $\delta(k)$ is for compensating the estimation error $\phi(x,u) - \hat{\phi}(\hat{x},u)$, which will be defined later. Let the nonlinear fault $\phi_i(x,u)$ be modeled by the neurofuzzy network,

$$\phi_i(k) = W_i^T \mu_i(x(k),u(k)) + \varepsilon_i(k) \quad (3)$$

where $\mu_i$ is the reception field function of the $i^{th}$ neurofuzzy network with appropriate dimension, and $W_i$ is the weight of the network. It is assumed that the optimal weights are located in a convex region, i.e., $\|W_i\|_F \le W_i^M$, where $W_i^M$ are pre-selected. As $W_i$ is unknown, its estimate, $\hat{W}_i$, is obtained from the training of the network. The estimate of the fault, $\hat{\phi}(\hat{x},u)$, is given by

$$\hat{\phi}_i(k) = \hat{W}_i^T(k)\mu_i(\hat{x}(k),u(k)) \quad (4)$$

From (1), (2), (3) and (4), the dynamical equation for the error is

$$\begin{aligned} \tilde{x}(k+1) &= A_c\tilde{x}(k) + \varphi(k) + B(\tilde{\phi}(k) + \delta(k)) \\ \tilde{y}(k) &= C\tilde{x}(k) \end{aligned} \quad (5)$$

where $\tilde{W}_i = W_i - \hat{W}_i$, $\tilde{x} = x - \hat{x}$, $\tilde{y} = y - \hat{y}$, and

$$\begin{aligned} \phi_i(k) - \hat{\phi}_i(k) &= W_i^T \mu_i(x(k),u(k)) - \hat{W}_i^T(k)\mu_i(\hat{x}(k),u(k)) + \varepsilon_1(k) \\ &= W_i^T \mu_i(x(k),u(k)) - W_i^T \mu_i(\hat{x}(k),u(k)) \\ &\quad + W_i^T \mu_i(\hat{x}(k),u(k)) - \hat{W}_i^T(k)\mu_i(\hat{x}(k),u(k)) + \varepsilon_1(k) \\ &= \tilde{W}_i^T \mu_i(\hat{x}(k),u(k)) + \varpi_i(k) + \varepsilon_1(k) \end{aligned}$$

and

$$\varpi_i(k) = W_i^T \mu_i(x(k),u(k)) - W_i^T \mu_i(\hat{x}(k),u(k)).$$

The online training of the network (4) is shown schematically in Fig. 1.

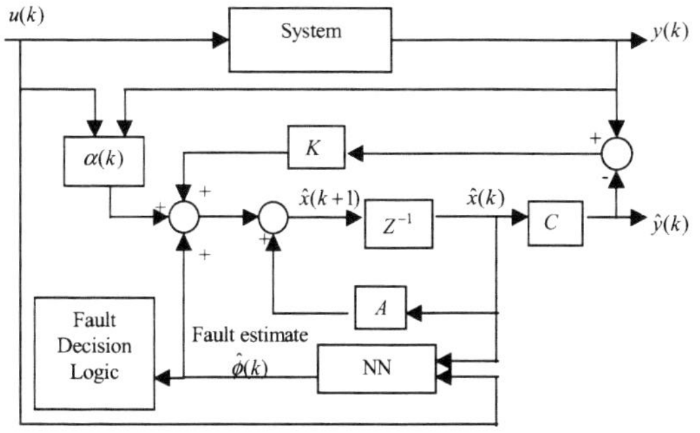


Fig. 1 Online training of the neurofuzzy network approximating the fault

An on-line training algorithm for $\hat{W}_i$ of the network (4) is given below (Wang, 2001).

$$\Delta\hat{W}_i(k)=\Gamma[\mu_i(\hat{x}(k),u(k))\bar{y}_i(k+1)-\theta\hat{W}_i(k)] \quad i=1,\cdots,p \quad (6)$$

$$\delta_i(k)=-\eta_i\cdot sign(\bar{y}_i(k+1)) \quad i=1,\cdots,p \quad (7)$$

where $sign(.)$ is the sign function, $\eta_i$ is the bounded value of $\bar{\omega}_i$, $\Gamma_i$, the size of the training step, is a matrix chosen to ensure the training of the network is sufficiently fast, $\bar{y}_i$ is the $i^{th}$ entry of $\bar{y}$. Note that $\bar{y}=M\tilde{y}$, where $M$ is chosen satisfying $B^TP = MC$, and $P$ is a positive definite matrix given by $(A - KC)^T P(A - KC) - P < 0$. It is shown in Wang (2001) that the training algorithm given by (6) and (7) is stable, and the estimated errors, though bounded from above, do not necessarily approach zero asymptotically. Further, the upper bound of the estimated errors depends on $K$. Consequently, the upper bound of the estimate errors can be reduced further by a suitable choice of $K$.

With certain conditions, fuzzy rules can be extracted from the B-spline neural networks. This is a useful feature in isolating faults (Harris, *et al.*, 1999), as discussed in the following section.

## 3. FUZZY RULES EXTRACTIONS WITH NEUROFUZZY NETWORK

As discussed in Section 2, the fault $\phi(k)$ is to be approximated by a B-spline neurofuzzy network (Brown and Harris, 1994), which is trained on-line. The estimate of the fault, $\hat{\phi}(k)$, is given by

$$\hat{\phi}(k)=\hat{W}^T(k)\mu(k) \quad (8)$$

Though the fault can be detected from $\hat{\phi}(k)$ given by (8), isolating the fault from $\hat{\phi}(k)$ is not straightforward. If *a-priori* knowledge of the system is available, then it is possible to isolate the fault, otherwise further assumptions may have to make. Under certain conditions, the qualitative description of the fault can be obtained from the fuzzy rules extracted from the B-spline neurofuzzy network. Consider, for example, the membership functions of the fuzzy sets of $\hat{\phi}(k)$ are chosen to be triangular B-spline functions, as shown in Fig. 2. Then the weight $w_i$ is given by (Brown and Harris, 1994),

$$w_i=\sum_{j=1}^{q} c_i^j \sigma_c^j \quad (9)$$

where $c_i^j$ is defined as the confidence of the $j^{th}$ fuzzy set, $\sigma_c^j$ is the center of the $j^{th}$ B-spline fuzzy set, and $q$ is the number of fuzzy sets. In Fig. 2, $q$ = 5, and $c_i=\begin{bmatrix}0 & 0 & c_i^3 & c_i^4 & 0\end{bmatrix}^T$. Note that $c_i^j$ can also be expressed in terms of $w_i$, as follows,

$$c_i^j=\mu_\phi(w_i) \quad (10)$$

where $\mu_\phi$ is the B-spline basis function. The mapping from the confidence level of the fuzzy sets to the weights and the inverse mapping from the weight to confidence level of the fuzzy sets are given respectively by (9) and (10). Hence for a given weight, the confidence of the fuzzy sets, $c_i$ can be evaluated from (10). For example, $\hat{\phi}(k)$ shown in Fig. 2 can be interpreted as follows:

(1) If $\hat{x}(k)$ is $B_x^j$ and $u(k)$ is $B_u^r$, then $\hat{\phi}(k)$ is $B_\phi^i$ (Rule confidence = $c_i^3$)

(2) If $\hat{x}(k)$ is $B_x^j$ and $u(k)$ is $B_u^r$, then $\hat{\phi}(k)$ is $B_\phi^{i+1}$ (Rule confidence = $c_i^4$)

As $\hat{\phi}(k)$ is an estimate of the fault, a qualitative knowledge of the fault can be obtained from these fuzzy rules. The isolation of fault using this approach depends on that $\hat{\phi}(k)$ is a sufficiently good approximation of the fault. This approach is used to isolate the fault in a nonlinear single-link robot arm, as presented in the next section.

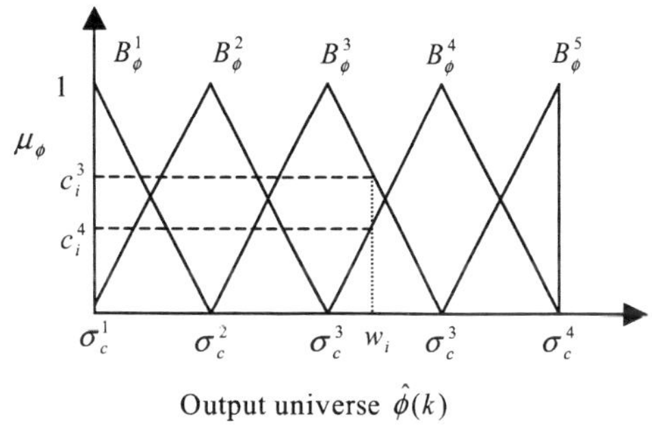


Fig. 2 Confidence of fuzzy sets for weight $w_i$

## 4. SIMULATION EXAMPLE

Consider a single-link robot arm given by the following equation (Marino and Tomei, 1995),

$$M\ddot{\beta}+\frac{1}{2}mgl\sin\beta=u\ ;\quad y=\beta \tag{11}$$

where $\beta$ is the angle, $u$, the input torque, $M$, the moments of inertia, $g$, the gravitation acceleration and $m$ and $l$ are respectively the mass and the length of the link. Let

$m$ = 1 kg, $l$ = 1m, $M$ = 0.5kg and $g$ = 9.8m/s$^2$ and the state be defined as, $x=[x_1\ \ x_2]^T=[\beta\ \ \dot{\beta}]^T$, then the state space mode of the system is

$$\begin{bmatrix}\dot{x}_1\\ \dot{x}_2\end{bmatrix}=\begin{bmatrix}0 & 1\\ 0 & 0\end{bmatrix}\begin{bmatrix}x_1\\ x_2\end{bmatrix}+\begin{bmatrix}0\\ 1\end{bmatrix}\frac{u-0.5mgl\sin x_1}{M}$$
$$y=\begin{bmatrix}1 & 0\end{bmatrix}\begin{bmatrix}x_1\\ x_2\end{bmatrix} \tag{12}$$

Assuming that the model uncertainties and the noise are bounded by $[0.01\ \ 0.02]^T$, and the input is given by $u$ = sin2$t$ + cos5$t$. Consider a fault occurring at $t$ = 5$s$ that reduces $M$ by 40%, which arises from a sudden loss of the load. Then $M$ is given by

$$M=\begin{cases}M_0 & t<5s\\ 0.6M_0 & t>5s\end{cases}$$

where $M_0$ denotes the value of $M$ under "healthy" condition. The discrete state-space equation of the robot is as follows,

$$\begin{bmatrix}\hat{x}_1(k+1)\\ \hat{x}_2(k+1)\end{bmatrix}=A\begin{bmatrix}\hat{x}_1(k)\\ \hat{x}_2(k)\end{bmatrix}+B\frac{u(k)-0.5mgl\sin x_1(k)}{M}+B\hat{\phi}(k)$$
$$+K(y(k)-\hat{y}(k)) \tag{13}$$
$$\hat{y}(k)=C\begin{bmatrix}x_1(k)\\ x_2(k)\end{bmatrix}$$

where $\hat{\phi}(k)$ is the on-line estimate of the fault obtained from the neurofuzzy network (8), and $A=\begin{bmatrix}1 & 0.005\\ 0 & 1\end{bmatrix}$, $B=\begin{bmatrix}0\\ 0.005\end{bmatrix}$, $C=\begin{bmatrix}1 & 0\end{bmatrix}$ and $K$ is chosen to be $\begin{bmatrix}1.5\\ 112\end{bmatrix}$.

As shown in Fig. 3, the estimate of the fault by the B-spline neurofuzzy network is reasonably accurate. Assuming the location of the fault is not known, and a qualitative knowledge of the fault is to be obtained from the fuzzy rules extracted from the neurofuzzy network, as discussed in the previous section. Suppose the fuzzy rules are extracted from the network at $t$ = 15$s$. A total of nine weights, $\{w_1, \ldots w_9\}$, are obtained as given below.

$$W=[0.1834\ \ 5.3359\ \ 2.1448\ \ -4.3694\ \ 0.6949$$
$$3.3201\ \ -1.1928\ \ -4.9298\ \ -0.6431]^T$$

Memberships of the fuzzy sets for $u(k)$ and $\hat{x}_1(k)$ are given by B-spline functions, as shown in Fig. 4. Fuzzy sets for the output $\hat{\phi}(k)$ are chosen to be "Negative (N)", "Positive (P)", "Big (B)" and "Small (S)" respectively. Fuzzy rules are extracted as follows. For example, for the 7th weight, $w_7=-1.1928$, and the confidence level of the fuzzy sets are $c_7^1=0.8012$ and $c_7^2=0.1988$ respectively. Considering the combination of the fuzzy rules, 18 rules are obtained, as presented below.

(1) ** If $\hat{x}_1(k)$ is NB and $u(k)$ is NB, then the fault is S (Rule confidence is 0.9694)

(2) * If $\hat{x}_1(k)$ is NB and $u(k)$ is NB, then the fault is PB (Rule confidence is 0.0306)

(3) * If $\hat{x}_1(k)$ is NB and $u(k)$ is S, then the fault is S (Rule confidence is 0.1107)

(4) If $\hat{x}_1(k)$ is NB and $u(k)$ is S, then the fault is PB (Rule confidence is 0.8893)

(5) ** If $\hat{x}_1(k)$ is NB and $u(k)$ is PB, then the fault is S (Rule confidence is 0.6425)

(6) * If $\hat{x}_1(k)$ is NB and $u(k)$ is PB, then the fault is PB (Rule confidence is 0.3575)

(7) * If $\hat{x}_1(k)$ is S and $u(k)$ is NB, then the fault is S (Rule confidence is 0.2718)

(8) If $\hat{x}_1(k)$ is S and $u(k)$ is NB, then the fault is NB (Rule confidence is 0.7282)

(9) ** If $\hat{x}_1(k)$ is S and $u(k)$ is S, then the fault is S (Rule confidence is 0.8842)

(10) * If $\hat{x}_1(k)$ is S and $u(k)$ is S, then the fault is PB (Rule confidence is 0.1158)

(11) ** If $\hat{x}_1(k)$ is S and $u(k)$ is PB, then the fault is S (Rule confidence is 0.4466)

(12) If $\hat{x}_1(k)$ is S and $u(k)$ is PB, then the fault is PB (Rule confidence is 0.5534)

(13) ** If $\hat{x}_1(k)$ is PB and $u(k)$ is NB, then the fault is S (Rule confidence is 0.8012)

(14) * If $\hat{x}_1(k)$ is PB and $u(k)$ is NB, then the fault is NB (Rule confidence is 0.1988)

(15) * If $\hat{x}_1(k)$ is PB and $u(k)$ is S, then the fault is S (Rule confidence is 0.1784)

(16) If $\hat{x}_1(k)$ is PB and $u(k)$ is S, then the fault is NB (Rule confidence is 0.8216)

(17) ** If $\hat{x}_1(k)$ is PB and $u(k)$ is PB, then the fault is S (Rule confidence is 0.8928)

(18) * If $\hat{x}_1(k)$ is PB and $u(k)$ is PB, then the fault is NB (Rule confidence is 0.1072)

Now the fuzzy rules are reduced by deleting rules that have a small confidence level, as marked by "*". For the ten rules remaining, the rules marked by "**" are also deleted, since only faults that are either positive big or negative big are considered. After deleting these rules, only four rules remain.

(i) If $\hat{x}_1(k)$ is NB and $u(k)$ is S, then the fault is PB (Rule confidence is 0.8893)

(ii) If $\hat{x}_1(k)$ is S and $u(k)$ is NB, then the fault is NB (Rule confidence is 0.7282)

(iii) If $\hat{x}_1(k)$ is S and $u(k)$ is PB, then the fault is PB (Rule confidence is 0.5534)

(iv) If $\hat{x}_1(k)$ is PB and $u(k)$ is S, then the fault is NB (Rule confidence is 0.8216)

From rules (ii) and (iii), it is concluded that when $u(k)$ is positive big (or negative big), the fault is also positive big (or negative big). From rules (i) and (iv), when $\hat{x}_1(k)$ is positive big (or negative big), the fault is also negative big (or positive big), giving a direct relationship between $u(k)$ and the fault, but an inverse relationship between $\hat{x}_1(k)$ and the fault. From (13), $\hat{x}_1(k)$ varies between $-\frac{\pi}{2}$ and $\frac{\pi}{2}$ for the given input $u$, implying that $\sin(x)$ is directly related to $x$. Consequently, the fault can only occur in $M$ or in $u$, as the chance of a fault in $m$ or $l$ is very small, and also that sensor fault is not being considered here. Consequently, it is concluded that the fault occurs in $M$, and it is a decrease in $M$.

## 5. CONCLUSIONS

In this paper, an intelligent fault diagnosis scheme based on B-spline neural networks is proposed for nonlinear dynamical systems. The proposed scheme involves estimating first the fault using neurofuzzy networks, from which fault is detected. The isolation of the fault is achieved by interpreting fuzzy rules extracted from the network. Although only a relatively simple nonlinear system involving a single-link robot arm is considered in the simulation example, it is demonstrated that the proposed procedure is able to detect and isolate the fault in the system. The proposed technique will form a basis for further development in fault isolation by extracting fuzzy rules from neurofuzzy networks.

## REFERENCES


Ayoubi, M. (1997). Neuro-fuzzy systems for diagnosis, *Fuzzy sets and systems*, **89**, pp. 289-307.

Benkhedda H. and Patton, R. J. (1996). Fault diagnosis using quantitative knowledge integration, *Proceedings of UKACC Int. Conf. on Control*, pp. 849-854.

Brown, M. and Harris, C. J. (1994). Neurofuzzy adaptive modelling and control, Prentice Hall.

Frank, P.M. (1990). Fault diagnosis in dynamical systems using analytical and knowledge based redundancy – a survey and new results, *Automatica*, **26**, 459-474.

Harris, C.J., Wu, Z.Q. and Gan, Q. (1999). Neurofuzzy state estimators and their applications, *IFAC Annual Reviews in Control*, **23**, pp. 149-158.

Isermann, R. (1993). Fault diagnosis of machines via parameter estimation and knowledge processing – tutorial paper, *Automatica*, **29(4)**, pp. 815-836.

Chan, C.W., Cheung, K.C., Wang, Y. and Chan, W.C. (1999). On-line fault detection and isolation of nonlinear Systems, *Proceedings of ACC*, San Diego, CA, pp. 3980-3984.

Chen, J. and Patton, R.J. (1999). Robust Model-based Fault Diagnosis for Dynamical Systems, *Kluwer Academic Publishers*, USA.

Wang, Y. (2001). On-line fault diagnosis of nonlinear dynamical systems using recurrent neural networks, Chapter, 5, Ph.D. thesis, *The University of Hong Kong*.

Wang, Y., Chan, C.W., Cheung, K.C. and Chan, W.C. (1999). Fault estimation for a class of nonlinear dynamical systems, *Proceedings of 38th CDC*, Phoenix, AZ, USA, pp. 3128-3129.

Wang Y., Cheung, K.C. and Chan, C.W. (2000). On-line fault estimation for discrete dynamical with unknown nonlinearities. *Proc. UKACC - CONTROL 2000*, Cambridge, UK.

Zhang, J. and Morris, A.J. (1994). On-line process fault diagnosis using fuzzy neural networks, *IEE Intelligent Sys. Engineering*, **3(1)**, pp. 37-47.

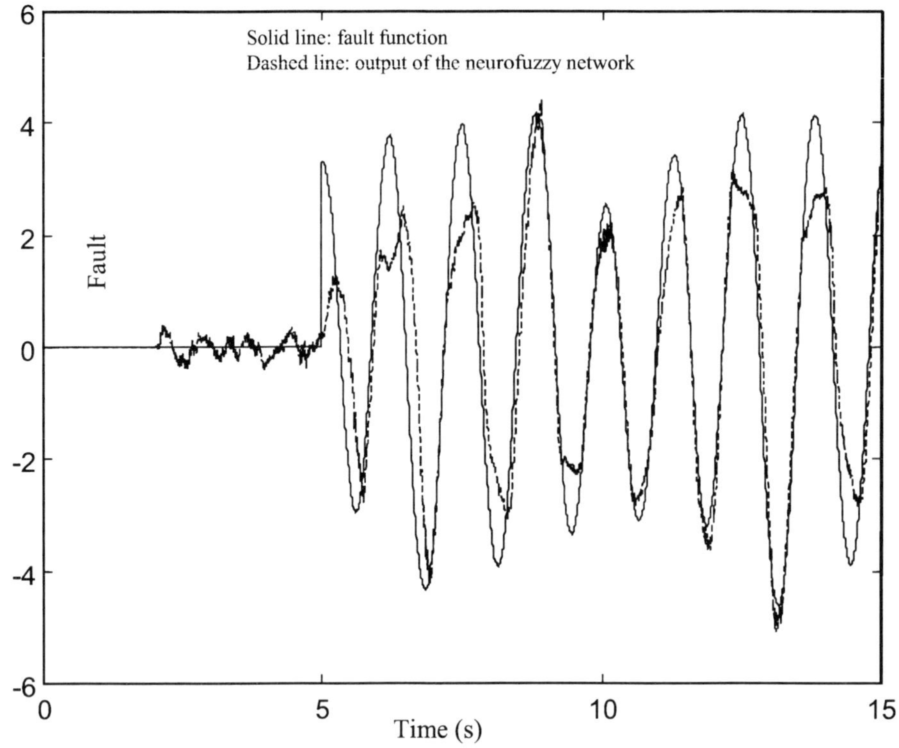


Fig. 3 Fault function and its on-line estimate by the B-spline neural network

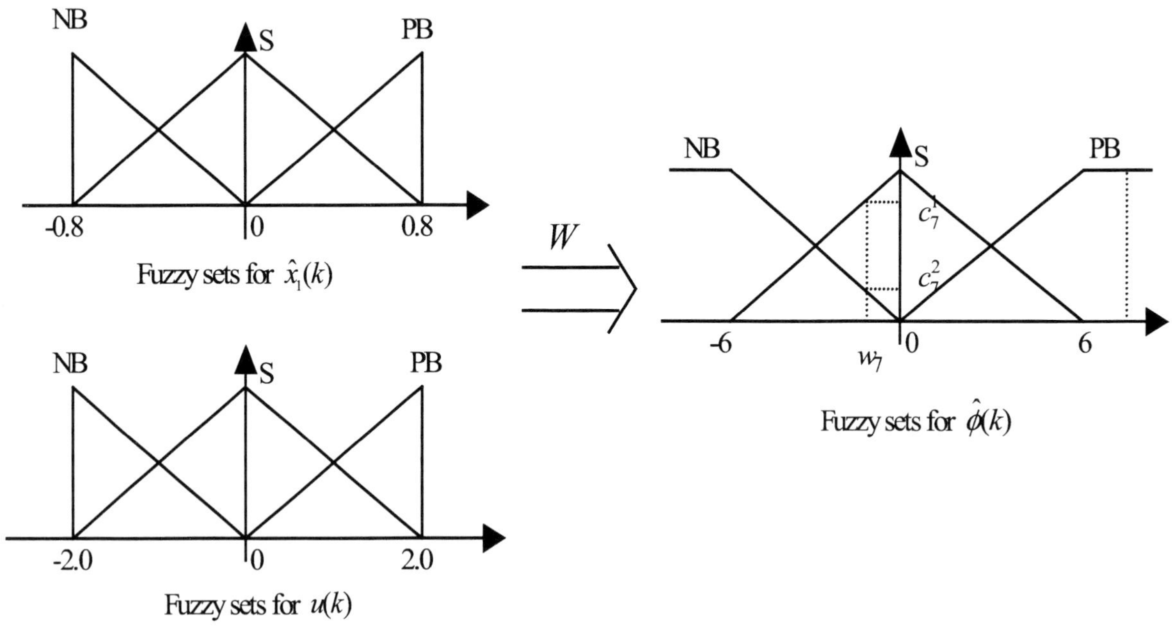


Fig. 4 Fuzzy interpretation for the fault diagnosis scheme

IFAC Publications
www.elsevier.com/locate/ifac

# LAYERED-INTEGRATED ANN DIAGNOSIS MODELS AND APPLICATIONS IN REMOTE DIAGNOSIS SYSTEMS

**Hong-sheng Li    Wen-wu Chen    You-xin Yuan    Zhao Zhao**

*Department of Automation, Wuhan University of Technology*
*Wuhan 430070, P. R. China*
*Fax: +86-27-87867763, Email: lihs8936@public.wh.hb.cn*

Abstract: ANN is a parallel information disposal network and is widely used in remote diagnosis systems. But the single-structure ANN is too massive. So the layered-structure ANN model is applied. However, it is inconvenient to implement Internet-based remote diagnosis systems. In order to solve it effectively, layered-Integrated ANN diagnosis models are presented in the paper. The construction of the models is composed of sub-network partition, confirming network structure, building expansion sample set, study and training. The performance of the models is analyzed from three aspects. It proves appropriate to apply the models to remote diagnosis systems through the tests and applications. *Copyright © 2001 IFAC*

Keywords: Neural Networks, Diagnosis, Diagnostic Inference, Remote Control, Monitored Control Systems, Intelligence.

## 1. INTRODUCTION

The network technologies, especially Internet/Intranet and its application technologies, have evolved very rapidly over the past few years, which means IRDS (Internet-based Remote Diagnosis System), large-area network diagnosis systems integrated next generation diagnosis technology, is coming soon (Robert. Itschner, *et al.*, 1998). In recent ten years, many countries have made great progress with theoretical approaches and application researches in the fault diagnosis field. Stanford University and MIT cooperated to start the Next Generation Remote Diagnostics Project on Oct.31, 1997. Wuhan University of Technology (China) also opened an elementary Internet-based Remote Diagnosis Test System on Nov. 1997 and consequent researches had been going on (Li Hongsheng, *et al.*, 2000). Fault diagnosis on complex devices, control systems or production processes could generally be transformed to the pattern-classification and pattern-recognition problems of working status in typical complicated systems (T.Kim, *et al.*, 1995). Pattern-classification could realize the mapping from pattern presentation space to explanation space (Sorsa, *et al.*, 1993). Many researches indicate that ANN as a parallel information disposal network composed of massive non-linear disposed unites, provides an effective tool for this mapping (J.A.Leonard, *et al.*, 1993).

Generally, the structure taken in constructing diagnosis model is single structure or layered sub-network structure. Each has unique features. But to complex diagnosis problems, the single-structure ANN is too massive, and the layered sub-network structure ANN is too complex, when implementing IRDS. In order to combine their advantages, an integrated ANN-based diagnosis model is introduced in the paper. By applying the diagnosis model to IRDS prototype, it proves that it is effective and appropriate to be applied to general RDE (Remote Diagnostic Engine).

## 2. NEW ANN DIAGNOSIS MODELS

Since the single-structure ANN and layered-structure ANN have their own disadvantages when constructing remote diagnosis model respectively, a new model— layered-integrated ANN diagnosis model is given.

### *2.1 Construction*

As the improvement of single-structure ANN diagnosis model, the layered-structure ANN diagnosis model in IRDS may be applied to two situations: *a*. Cooperation diagnosis mode. In this mode every diagnosis model is same to remote diagnosis servers in different areas. *b*. As remote diagnosis server. In this situation, the implementation would be inconvenient because the features of each sub-network are different. An integrated ANN diagnosis model as shown in Fig. 1 is introduced to improve the situation.

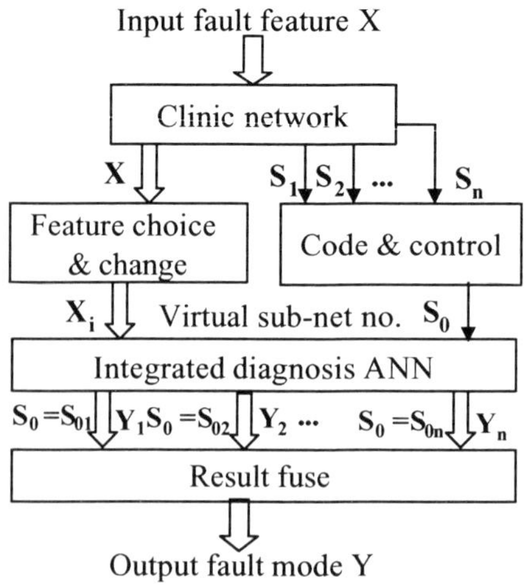


Fig.1 Integrated ANN diagnosis model

The model is composed of three-layer networks and the second layer is the primary part of construction. The first layer is a clinic network, whose input is fault feature vector X and output is expert (sub-network) choice vector S. The second layer integrates all sub-network diagnosis ANN. The third layer is a fuse network, which lastly integrate several sub-network outputs and get output fault mode vector Y. Its constructing approach is listed as below.

*Sub-network Partition*. According to fault disassembly, the second layer is disassembled into several sub-network i (i=1,2 ..n). That is, to disassemble output mode space Y ($Y \in R^M$, $M \in Z^+$) into several sub-spaces $Y_i$,

$$Y_i \in R^{L_i} (L_i \in Z^+; i=1,2,\cdots,n; \sum_{i=1}^{n} L_i = M); Y = \bigcup_{i=1}^{n} Y_i, Y_p \cap Y_q = \phi, p \neq q)$$

Suppose that, input feature space $X \in R^N$, $N \in Z^+$, is also accordingly graded into n sub-spaces $X_i$, and $X_i \subseteq X, i = 1,2,\cdots,n.$ The feature sample set D would be accordingly disassembled into subset $D_i$. If a sample is relevant to several sub-room $Y_i$, adding the sample into all relevant $D_i$.

*Confirming Network Structure*. Let $X_i = X$, and $L = \max L_i$ for simplification. Construct an integrated ANN J, whose input is defined as expansion fault feature $XJ=(X_0, X)$, $XJ \in R^{N+1}$, $X_0 \in [0,1]$, when $S=S_i$, $X_0=S_{0i}$, i=1,2 ..n; and $S_{0i} \in [0,1]$, $S_{0p} \neq S_{0q}$, $p \neq q$. And the output of the network J is defined as integrated fault mode $YJ \in R^L$. FF-ANN would be chosen, and its hidden layer would be decided by training results.

*Building Expansion Sample Set*. To every sub-network, the input is $X_J$, and the output is $Y_J$. Corresponding to subset $D_i$, let $X_J=(S_{0i}, X)$. If $L_i<L$, the latter $L-L_i$ dimensions in corresponding subset $Y_J$ would be vacant. To different $S_{0i}$, $Y_J$ would have a group of mode interpretation corresponding to the subset. Lastly integrate several expansion sample subsets would get the whole expansion sample set.

*Study and Training*. It's same to general single-structure ANN. Choose appropriate study arithmetic to train the integrated ANN in the expanded sample set, could get a uniform and smaller scale diagnosis model with all sub-network knowledges.

### *2.2 Implementation*

In the following part, the implementation of the integrated ANN mode is introduced.

The input of the first layer of the diagnosis inference model—the clinic network, is multi-parameter fault feature sample. With some arithmetic and tactics, one or several virtual sub-networks would be chosen to run as the clinic network.

The fault features of each sub-network would be transformed and organized by the dinic network, and sub-networks would be controlled to work. By that time, the computing entity of virtual sub-network would be produced.

According to computing circumstance, each computing entity of sub-network would draw the inference and compute in either serial or parallel way, then output each inference result.

The inference results of each sub-network would be disposed by the fuse network. In other words, in order to get better results it would combine the results of each sub-network in same fault models, and integrate the results of each sub-network in different fault modes. At last, the whole diagnosis results would be got.

### *2.3 Performance Analysis*

The performance of layered-integrated network will be analyzed from three aspects.

*Diagnosis Decision-Making Fields.* Compared with the decision-making fields of single-structure network and layered-structure network, the fields of layered-integrated network are composed of the fields of each sub–network in logic while integrative in physical. As its input feature space increases one dimension, some distance in superior dimension would separate sub-decision fields.

*Diagnosis Performance*. As ANN diagnosis decision-making is essentially an "analogy" thought, the distance between sub-decision-making fields would have some influence on inference results; the shorter the distance, the bigger the disadvantage. The distance for single-structure network is the smallest. Sub-network connection of integrated network is looser, and the distance is further than single-structure network. Increasing separate parameter and adjusting sub-network's array position would increase the isolation degree of sub-network for integrated network. To integrated network, the proper isolation would make the network have certain ability for diagnosing relevant fault or integrated fault after appropriate training.

*Network Size*. The size of integrated network is basically equivalent to that of one sub-network of layered-structure network, much smaller than that of single-structure network.

## 3. REMOTE DIAGNOSIS APPLICATIONS

### *3.1 Remote Diagnosis Model and Its Implementation*

The remote diagnosis model of IRDS prototype, using integrated ANN diagnosis model, is shown as Fig.2, taking the popular Browser/Server three-layered structure in Web-based application.

*Construction of The Model and System.* Firstly,

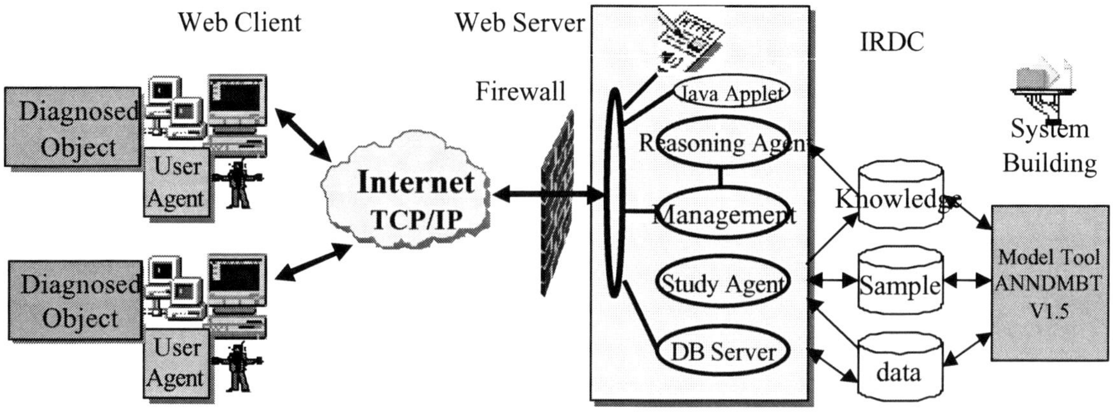


Fig.2 Approach of Remote Diagnosis System (IRDS1.0)

layered-integrated ANN diagnosis model is applied to this system. Windows NT 4.0 is taken as the system-computing platform of remote diagnosis server, and MS IIS (Internet Information Server) bound with Windows NT is taken as Web server. The demonstration system of IRDS is established with the prototype system using IIS, and Web database server is SQL server 6.5. IRDS client would apply standard universal JVM-enabled browser. The approaches of implementing GUI modules (Asking, Diagnosis, Sending) and function modules include CGI, DHTML, Jscript and Java Applet. VC++ is used to build the self-study Agent of ANN RDE.

*Asking for Diagnosis*. The client asking for diagnosis accesses the Website of IRDS through URL by Web way in Browser. After registering identification, it would download the HTML Homepage containing the asking GUI. According to the GUI' s requirement, choosing the type of diagnosed object, parameters and observed data, transmitting the information back to the Web server through network.

*Diagnosis Inference*. The Web server would organize the corresponding ANN diagnosis inference Applets, according to the asking information, and embed the Applets in the HTML Homepage as the diagnosis GUI (reading corresponding diagnosis knowledge document). Loading the diagnosis GUI to the client browser. The diagnosis Applet would be started to run in the client's JVM-enabled browser. According to the diagnosis GUI's requirement, interactively inputting fault feature data (applying IRDS's remote signal analysis and feature extracting function), starting the diagnosis inference engine and then getting diagnosis results with corresponding help interpreting information.

*Later Self-study of Integrated ANN RDE Agent*. The clients could add specific field fault samples and data to the corresponding database of diagnosed objects in IRDS through using the sending GUI by operating Web database. Self-study Agents of integrated ANN RDE run in IRDS server. The execution of self-study Agents has two tactics: one is to run periodically, through setting the time task in Windows server to implement; the other is to detect the database timely, when changing to start and run. The former is fit for the early stage of constructing IRDS, while having a large number of entering data; the latter is fit for the diagnosis models with perfect performance and full knowledge.

## 4. CONCLUSION

During the construction of the prototype system of IRDS (V1.0), the systems whose diagnosis models are based on integrated ANN RDE and rule-inference expert system have been respectively implemented. However, the performance of ANN RDE is better in Internet computing circumstance. The corresponding examples are presented in Fig.3~Fig.10.

The remote diagnosis approach by Web computing based on integrated ANN diagnosis model has following good features.

It is simple and efficient for Browser/Server-based structure and Web-based distributed application to implement the remote diagnosis services for massive network clients in different areas. It's easy to manage and upgrade. Developing complex user GUI is simplified by using standard browsers as the public uniform user supporting circumstance; across-platform application is simply realized and the cost of training, installation and maintenance is greatly reduced. It is easy to expand system functions and add new users to the diagnosis service network because of the open structure of IRDS.

The study and training will cost computing resource, while they could be separated from the inference simply and rapidly in integrated ANN diagnosis models. It could not only take advantage of applying computing resource by web, but also sustain large network flow of asking for diagnosis. In one word, integrated ANN diagnosis model is fit for Internet computing and remote diagnosis application circumstance.

Since integrated ANN model is used as RDE, sharing diagnosis knowledge becomes easy. Some proper approaches could be adopted to overcome the fault feature differences among congener diagnosed objects, and then diagnosis quality will be improved.

It is convenient to construct self-study Agents of integrated ANN RDE with the concept and system of Agent (H. Vogler, *et al.* 1997), which could well satisfy the requirements of RDE.

REFERENCES

H. Vogler, et al. (1997). Distributed Transaction Processing as a Reliability Concept for Mobile Agents. In *Proceeding of IEEE Computer Society Workshop on Future Trends of Distributed Computing Systems.* Los Alamitos CA USA, pp. 59–64.

J.A.Leonard, et al. (1993). Diagnosing Dynamic Faults Using Modular Neural Nets. *IEEE Expert*, pp. 44–53.

Li Hongsheng, et al. (2000). Integrated ANN Diagnostic Model-based Remote Diagnosis System on Internet. In Proceeding of IFIP The 16th WCC (World Computer Congress)-IIP2000, pp.232–237. Publishing House of Electronics Industry, Beijing, China.

R. Itschner, C. Pommerell, and M. Rutishauser (1998). GLASS: Remote Monitoring of Embedded Systems in Power Engineering, *IEEE Internet Computing*, **2**, No. 5, pp.46–52.

T.Kim, et al. (1995). Feed-forward Neural Networks for Fault diagnosis and Severity Assessment of a Screw Compressor. *Mechanical Systems and Signal Processing*, **9**, No. 5, pp.485–496.

T. Sorsa, et al. (1993). Application of Artificial Neural Networks in Process Fault Diagnosis. *Automatica*, **29**, No. 4, pp. 843–849.

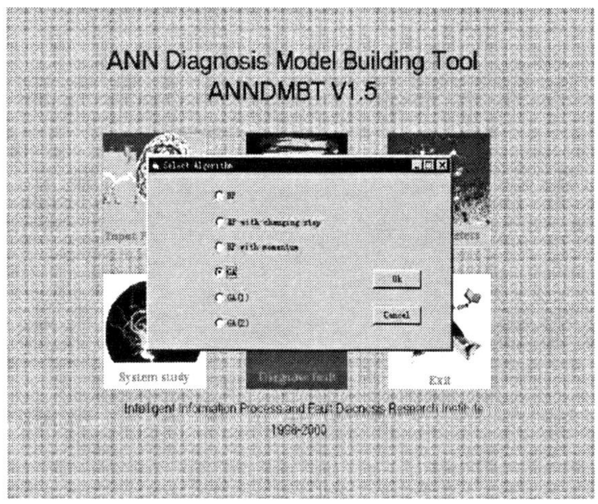


Fig.3 ANN Diagnosis Model Building Tool (1)

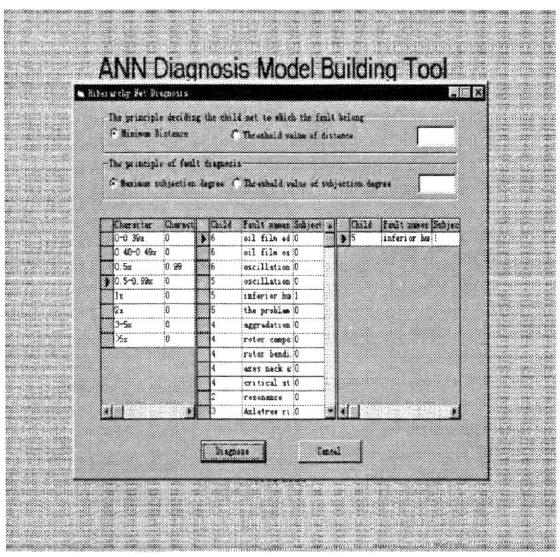


Fig.4 ANN Diagnosis Model Building Tool (2)

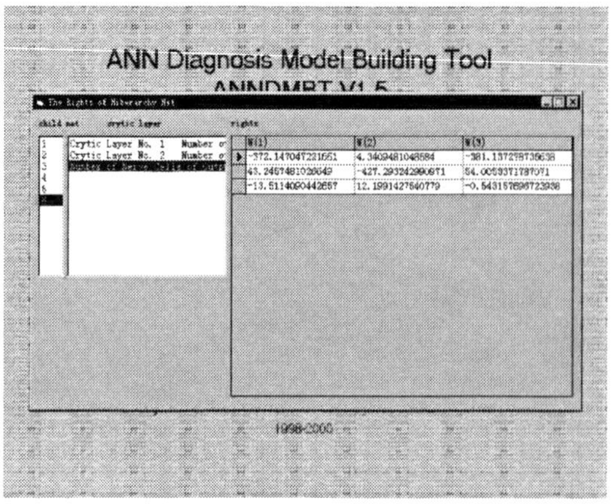


Fig.5 ANN Diagnosis Model Building Tool (3)

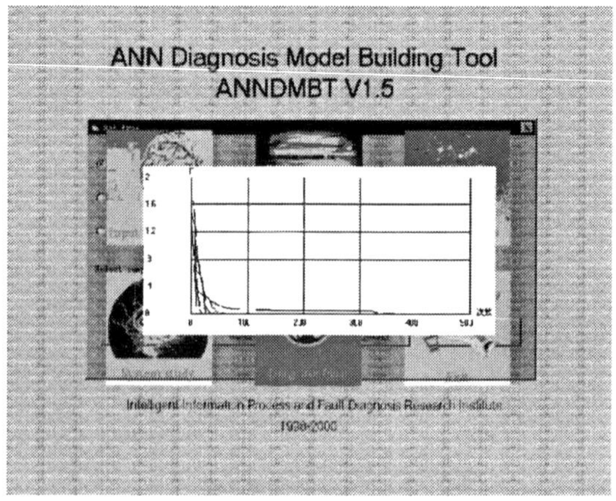


Fig.6 ANN Diagnosis Model Building Tool (4)

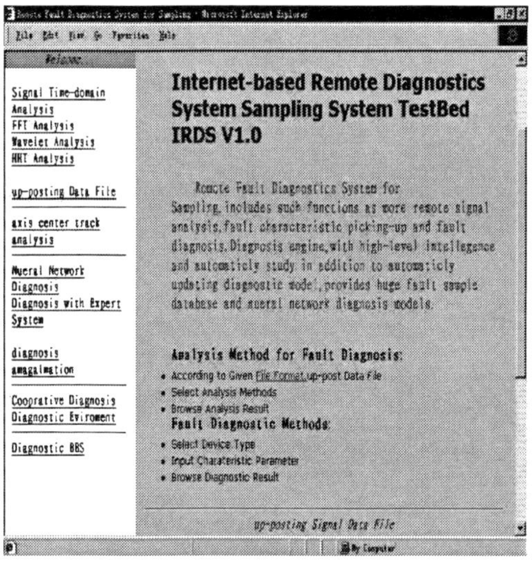


Fig.7 Remote Diagnosis System (1)

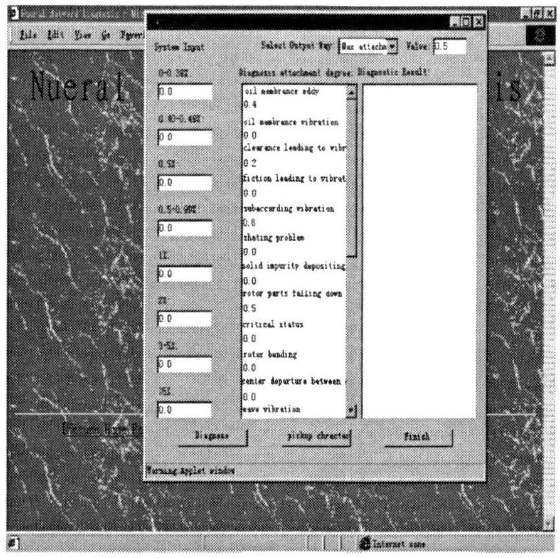

Fig.8 Remote Diagnosis System (2)

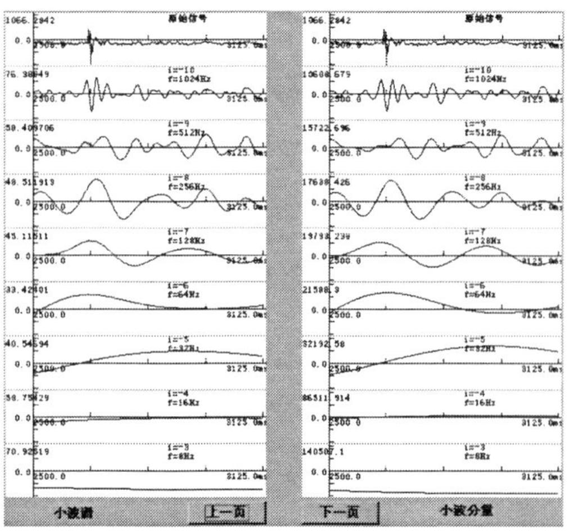

Fig.9 Remote Wavelet Signal Analysis Tool by Java Applet

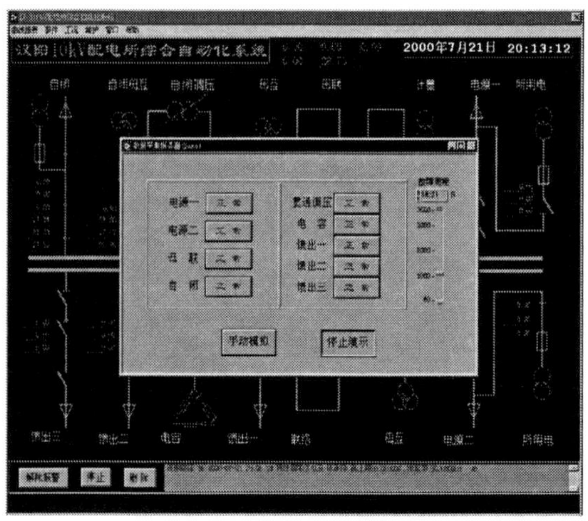


Fig.10 An Example of Remote Monitoring and Diagnosis System

IFAC Publications
www.elsevier.com/locate/ifac

# FAULT DETECTIONAND ISOLATION OF SYSTEMS WITH SLOWLY TIME-VARYING FAULTS USING THE OPTIMAL PARITY TEST

**H. Jin[1], C.W. Chan[2] and H.Y. Zhang[3]**

[1] *The Fourth Research Institute of The Second Artillery Corps. Beijing 100085, China, Email: hjin@ihw.com.cn*
[2] *Department of Mechanical Engineeirng, The University of Hong Kong Pokfulam Road, Hong Kong, Email: mechan@hku.hk*
[3] *Department of Automatic Control,Beijing University of Aeronautics and Astronautics, Beijing 100083, China, Email: buaa301@mimi.cnc.ac.cn*

**Abstract**: To improve the reliability in detecting and isolating actuator and sensor faults, a fault detection and isolation (FDI) technique based on the optimal parity test is derived in this paper. Fault isolation is achieved by designing optimal parity vectors such that they are sensitive only to specific actuator and sensor faults. The definition of the generalized input fault vector is presented, and the geometric invariability of collinear vectors is also discussed. An simulation examples is presented to illustrate the performance of the proposed FDI technique. *Copyright © 2001 IFAC*

**Keywords**: Fault detection and isolation, temporal redundancy, optimal parity test.

## 1. INTRODUCTION

A common approach in fault detection and isolation (FDI) is to generate the residuals first, from which faults are detected and are then isolated. A popular model-based FDI technique is the optimal parity test (OPT). It is proposed in [1,2] the residuals for FDI are generated from the "optimal robust parity equation", such that they are less sensitive to the modelling errors, but more sensitive to the system faults. In [3], the space of the parity vector is generalized by relaxing the constraints in the parity vectors in the generation of the residuals, and an optimal selection of the parity vector from the parity vector space is derived. However, fault detection is not always reliable, and fault isolate can still be difficult. To overcome this problem, the optimal parity test proposed in [4, 5], is more efficient in detecting and isolating faults by generating the residual, which are more sensitive to specific system faults and less sensitive to others factors. In this paper, the OPT is extended to detect and isolate slowly time-varying fault.

## 2. PRELIMINARIES

Consider the following linear discrete system,

$$\begin{cases} x(k+1) = Ax(k) + B_x u(k) + E_x w_x(k) + G_x \gamma_x(k) \\ y(k) = Hx(k) + B_y u(k) + E_y w_y(k) + G_y \gamma_y(k) \end{cases} \quad (1)$$

where $x(k) \in \Re^n$ is the state; $u(k) \in \Re^l$, the input; $y(k) \in \Re^m$, the measurement; $\gamma_x(k) \in \Re^{r_x}$, the actuator fault; $\gamma_y(k) \in \Re^{r_y}$, the sensor fault; $w_x(k) \in \Re^{t_x}$, the dynamic noise; $w_y(k) \in \Re^{t_y}$, the measurement noise; $A$, $B_x$, $H$, $B_y$, $E_x$, $E_y$, $G_x$, $G_y$, known constant matrices with appropriate dimensions. There are mainly two types of analytic redundancy, the direct and the temporal redundancy. In this paper, only fault detection based on the the parity relation using the temporal redundancy is considered.

### 2.1 Residual based on temporal redundancy

Scaling the residual obtained from (1) by $\nu$ gives

$$\rho = \nu'(Y(k) - A_1 U(k)) \quad (2)$$

where $Y(k) = \begin{pmatrix} y(k-s) \\ y(k-s+1) \\ \vdots \\ y(k) \end{pmatrix}, U(k) = \begin{pmatrix} u(k-s) \\ u(k-s+1) \\ \vdots \\ u(k) \end{pmatrix}$,

$$A_1 = \begin{pmatrix} B_y & 0 & 0 & 0 \\ HB_x & B_y & 0 & 0 \\ HAB_x & HB_x & 0 & 0 \\ \vdots & \vdots & \vdots & \vdots \\ HA^{s-1}B_x & HA^{s-2}B_x & HB_x & B_y \end{pmatrix},$$

$\nu \in R^{(s+1)m}$ is a column vector, and $s > 0$ is the order of the parity relation. System (1) can be expressed as [6],

$$\begin{aligned} Y(k) = H_A x(k-s) + A_1 U(k) + A_2 W_x(k) \\ + A_3 \Gamma_x(k) + A_4 W_y(k) + A_5 \Gamma_y(k) \end{aligned} \quad (3)$$

where $\Gamma_x(k) = \begin{pmatrix} \gamma_x(k-s) \\ \gamma_x(k-s+1) \\ \vdots \\ \gamma_x(k) \end{pmatrix}, \Gamma_y(k) = \begin{pmatrix} \gamma_y(k-s) \\ \gamma_y(k-s+1) \\ \vdots \\ \gamma_y(k) \end{pmatrix}$

$$W_x(k) = \begin{pmatrix} w_x(k-s) \\ w_x(k-s+1) \\ \vdots \\ w_x(k) \end{pmatrix}, W_y(k) = \begin{pmatrix} w_y(k-s) \\ w_y(k-s+1) \\ \vdots \\ w_y(k) \end{pmatrix}$$

$$H_A = \begin{pmatrix} H \\ HA \\ \vdots \\ HA^s \end{pmatrix},$$

$$A_2 = \begin{pmatrix} 0 & 0 & \cdots & 0 & 0 \\ HE_x & 0 & \cdots & 0 & 0 \\ HAE_x & HE_x & \cdots & 0 & 0 \\ \vdots & \vdots & \vdots & \vdots & \vdots \\ HA^{s-1}E_x & HA^{s-2}E_x & \cdots & HE_x & 0 \end{pmatrix}$$

$$A_3 = \begin{pmatrix} 0 & 0 & \cdots & 0 & 0 \\ HG_x & 0 & \cdots & 0 & 0 \\ HAG_x & HG_x & \cdots & 0 & 0 \\ \vdots & \vdots & \vdots & \vdots & \vdots \\ HA^{s-1}G_x & HA^{s-2}G_x & \cdots & HG_x & 0 \end{pmatrix}$$

$$A_4 = Block\ Diag(E_y \cdots E_y) \in \Re^{(s+1)x(s+1)}$$

$$A_5 = Block\ Diag(G_y \cdots G_y) \in \Re^{(s+1)x(s+1)}$$

$H_A$ is a constant matrix with a dimension of $s$, and $(C, H)$ is assumed to be observable, and $H$ has full row rank [7]. Let $w_x(k)$, and $w_y(k)$ be uncorrelated noise sequence having the following statistic,

$$E[w_x(k)] = 0, Cov[w_x(k)] = Q_k \geq 0$$

$$E[w_y(k)] = 0, Cov[w_y(k)] = R_k > 0$$

From (3),

$$Y_u(k) = Y(k) - A_1 U(k)$$

$$Y_u(k) = H_A x(k-s) + A_3 \Gamma_x(k) + A_5 \Gamma_y(k) + \varepsilon(k) \quad (4)$$

where $\varepsilon(k) = A_2 W_x(k) + A_4 W_y(k)$ is zero mean with a covariance matrix $\Xi(k)$,

$$\Xi(k) = A_2 \Xi_x(k) A_2' + A_4 \Xi_y(k) A_4'$$

$$\Xi_x(k) = Block\ Diag(Q_{k-s}, \cdots, Q_k)$$

$$\Xi_y(k) = Block\ Diag(R_{k-s}, \cdots, R_k)$$

Clearly, $\Xi(k)$ is a symmetric and positive definite. If $\gamma_x$ and $\gamma_y$ are constants, then (4) becomes

$$Y_u(k) = H_A x(k-s) + A_6 \gamma_x + A_7 \gamma_y + \varepsilon(k) \quad (5)$$

where

$$A_6 = \begin{pmatrix} 0 \\ \sum_{i=1}^{1} HA^{i-1}G_x \\ \sum_{i=1}^{2} HA^{i-1}G_x \\ \vdots \\ \sum_{i=1}^{s} HA^{i-1}G_x \end{pmatrix}, A_7 = \begin{pmatrix} G_y \\ G_y \\ G_y \\ \vdots \\ G_y \end{pmatrix}$$

### 2.2 Approximation by a general temporal redundant system

The slowly time-varying fault can be considered as a constant fault over a very short time interval, and the dynamic system can be approximated by the general redundant measurement equation,

$$M = H_A x + \Phi_x \gamma_x + \Phi_y \gamma_y + \Psi \varepsilon^* \quad (6)$$

where $M$ is obtained from the measurement data; $\Phi_x = (\phi_{x1}, \cdots, \phi_{xr_x})$ and $\Phi_y = (\phi_{y1}, \cdots, \phi_{yr_y})$ are respectively the input matrix of $\gamma_x$ and $\gamma_y$; $\varepsilon^*$ is a zero-mean error with a covariance matrix of $\sigma^2 I$, $I$, the identity matrix; and $\Psi$ is a nonsingular input matrix. The FDI of dynamic system (1) can now be investigated using the general redundant measurement system (6).

## 3. PROPERTIES OF GENERALIZED FAULT INPUT

### *3.1 Parity relations*

Let the space spanned by the column vectors of $H_A$ be the generalized measurement space, and the orthogonal space be the generalized parity space [8]. A vector from the generalized parity space is referred to as a parity vector. Let $V$ be full column rank and be constructed from a basis of the generalized parity space such that

$$V'H_A = 0 \tag{7}$$

A popular method to choose $V$ is that $V$ satisfies [9],

$$V'V = 1 \tag{8}$$

**Lemma 1 [9]** Let $P_V$ be the orthogonal projection matrix of the generalized parity space, then

$$P_V = V(V'V)^{-1}V' = I - H_A(H_A'H_A)^{-1}H_A' \tag{9}$$

is an idempotent matrix unrelated to $V$. When the constraint (8) is satisfied, then $P_V = V'V$.

### *3.2 Generalized faulty input vector (GFIV)*

From (6) and (7), the residual vector can be rewritten as

$$\xi = V'M = V'(\Phi_x\gamma_x + \Phi_y\gamma_y) + V'\Psi\varepsilon^* \tag{10}$$

where the mean and covariance matrix of $\xi$ are

$$\begin{cases} E[\xi] = V'(\Phi_x\gamma_x + \Phi_y\gamma_y) \\ Cov[\xi] = \sigma^2 V'\Psi\Psi'V \end{cases} \tag{11}$$

The generalized residual space is constructed from the generalized residual vector $\xi$. Since $V'\psi$ is full row rank, a non-singular matrix U exists such that

$$V'\Psi\Psi'V = UU' \tag{12}$$

Rewriting (10) gives,

$$\xi = \sum_{i=1}^{r_x} V'\phi_{xi}\gamma_{xi} + \sum_{i=1}^{r_y} V'\phi_{yi}\gamma_{yi} + U\varepsilon \tag{13}$$

where $\varepsilon$ is a zero-mean noise with a covariance matrix of $\sigma^2 I$. Let

$$\begin{cases} d_{xi} = V'\phi_{xi} \\ d_{yj} = V'\phi_{yj} \end{cases} \quad (i = 1,\cdots,r_x,\ j = 1,\cdots,r_y) \tag{14}$$

Then $d_{xi}$ can be interpreted as the generalized faulty input vector (*GFIV*) of the $i$th actuator fault $\gamma_{xi}$ and $d_{yi}$, the *GFIV* of the $j$th sensor fault $\gamma_{yi}$. For a chosen V, $d_{xi}$ and $d_{yi}$ can be pre-determined.

### *3.3 Geometric invariability of collinear GFIVs*

The geometric relationship between the *GFIV*s in the generalized residual space is presented in this section.

*(i) Collinear of two vectors*

**Definition 1**: The vectors $x$ and $y$ are said to be collinear if the following relationship holds

$x = cy$

where $c$ is a non-zero constant. The projection of a vector $x$ onto the generalized parity space is $P_V x$. The condition for the projection of vectors $x$ and $y$ onto the generalized parity space to be collinear is given below.

**Thereom 1** For the vectors $x$ and $y$, their projection onto the generalized parity space are collinear, if and only if,

$V'x = cV'y$

**Proof** If $P_V x = cP_V y$, then

$V(V'V)^{-1}V'x = c\ V(V'V)^{-1}V'y$

Rearranging gives

$$\begin{aligned} V'x &= V'V(V'V)^{-1}V'x = V'[c\ V(V'V)^{-1}V'y] \\ &= c\ V'V(V'V)^{-1}V'y = cV'y \end{aligned}$$

If $V'x = cV'y$, then

$P_V x = V(V'V)^{-1}V'x = V(V'V)^{-1}\ cV'y = cP_V y$

*(ii) Constraint (8) satisfied*

From (14) and Lemma 1,

$$\|d_{xi}\|^2 = \phi_{xi}'VV'\phi_{xi} = \phi_{xi}'P_V\phi_{xi} = \phi_{xi}'P_V'P_V\phi_{xi} = \|P_V\phi_{xi}\|^2$$

$$d_{xi}\cdot d_{yj} = d_{xi}'d_{yj} = \phi_{xi}'VV'\phi_{yj} = \phi_{xi}'P_V\phi_{yj}$$

Similarly,

$$\|d_{yi}\|^2 = \phi_{yj}'P_V\phi_{yj} = \|P_V\phi_{yj}\|^2$$

$$d_{xi}\cdot d_{xj} = d_{xi}'d_{xj} = \phi_{xi}'P_V\phi_{xj}$$

$$d_{yj}\cdot d_{yj} = d_{yi}'d_{yj} = \phi_{yj}'P_V\phi_{yj}$$

then

$$\begin{aligned} \cos < d_{xi}, d_{yj} > &= \frac{d_{xi}\cdot d_{yj}}{\|d_{xi}\|\cdot\|d_{yj}\|} = \frac{\phi_{xi}'P_V\phi_{yj}}{\|P_V\phi_{xi}\|\cdot\|P_V\phi_{yj}\|} \\ \cos < d_{xi}, d_{xj} > &= \frac{d_{xi}\cdot d_{xj}}{\|d_{xi}\|\cdot\|d_{xj}\|} = \frac{\phi_{xi}'P_V\phi_{xj}}{\|P_V\phi_{xi}\|\cdot\|P_V\phi_{xj}\|} \\ \cos < d_{yi}, d_{yj} > &= \frac{d_{yi}\cdot d_{yj}}{\|d_{yi}\|\cdot\|d_{yj}\|} = \frac{\phi_{yi}'P_V\phi_{yj}}{\|P_V\phi_{yi}\|\cdot\|P_V\phi_{yj}\|} \end{aligned} \tag{15}$$

Since $\phi_{xi}$ and $\phi_{yj}$ are determined for a given $s$ before generating the residual, and from Lemma 1, $P_V$ is also determined, hence the angles $<d_{xi}, d_{yj}>$, $<d_{xi}, d_{xj}>$, and $<d_{yi}, d_{yj}>$ are independent of $V$ for all $i$ and $j$. It implies that if $V$ satisfies (8), the geometric relation between any two *GFIV*s is determined in the generalized residual space independent of $V$. If $d_{xi}$ and $d_{yj}$

are collinear, i.e., $V'\phi_{xi} = cV'\phi_{yj}$, then from Theorem 1 $P_V\phi_{xi} = cP_V\phi_{yj}$, giving

$$\phi'_{xi}P_V\phi_{yj} = \phi'_{xi}P'_VP_V\phi_{yj} = c\left\|P_V\phi_{yj}\right\|^2$$

and cos $<d_{xi},d_{yj}>$ = 1 $sign(c)$.

*(iii) Constraint (8) not satisfied*

For any $V_0$ that satisfies (8), there exists a nonsingular matrix $C_0$ such that

$V_0 = VC_0$

The *GFIV*s based on $V_0$ are

$$d^*_{xi} = V'_0\phi_{xi} = C'_0d_{xi}$$

$$d^*_{yj} = V'_0\phi_{yj} = C'_0d_{yj}$$

From (15), $< d^*_{xi}, d^*_{xj} >$, $< d^*_{xi}, d^*_{yj} >$ and $< d^*_{yi}, d^*_{yj} >$ are determined for any given $V_0$, and hence which of the *GFIV*s are collinear can also be determined. Let $d^*_{xi}$ and $d^*_{yj}$ are collinear. Since $C_0$ is nonsingular, it is readily shown that the corresponding *GFIV*s, $d_{xi}$ and $d_{yj}$ are also collinear. Similarly, if $d_{xi}$ and $d_{yj}$ are collinear for some $V$, their corresponding *GFIV*s $d^*_{xi}$ and $d^*_{yj}$ are also collinear.

### *3.4 Isolation of system faults*

**Definition 2**: Two system faults are said to be isolable if their corresponding *GFIV*s are not collinear.

From Definition 1, two faults can only be isolated if their *GFIV*s are not collinear. Clearly, two faults with their respecive *GFIV*s having opposite signs cannot be isolated. When the angle between two *GFIV*s is small, it becomes difficult to isolate the corresponding system faults.

**Example 1** Consider system (1), where $E_x$, $E_y$, $G_x$ and $G_y$, and $Gy$ are identity matrices, $A = \begin{pmatrix} 1 & 0 \\ 1 & 1 \end{pmatrix}$, and $H = \begin{pmatrix} 2 & 1 \\ 0 & 1 \end{pmatrix}$. Four *GFIV*s are shown in Fig.1, and the angle between each pair of these four *GFIV*s for different $s$ are given in Table 1. From Table 1, the *GFIV*s, $d_{x2}$ and $d_{y2}$ are identical to each other, and the *GFIV*, $d_{y1}$ is collinear with both $d_{x2}$ and $d_{y2}$, but in the opposite direction. As the dimension of the generalized residual space in this example is two, only two faults can be isolated. The first one is the fault arising from the first actuator $\gamma_{x1}$, and the second one is caused by $\gamma_{x2}$, $\gamma_{y1}$ or $\gamma_{y2}$. As shown in Table 1, the directions of these two faults are approximately orthogonal to each other.

As dicussed earlier, the first actuator fault $\gamma_{x1}$ can be isolated from $\gamma_{x2}$, $\gamma_{y1}$ and $\gamma_{y2}$, but the latter three faults cannot be isolated from each other.

| s | $< d_{x1}, d_{x2} >$ | $< d_{x1}, d_{y1} >$ | $< d_{x1}, d_{y2} >$ |
|---|---|---|---|
| 1 | 65.9052 | 114.0948 | 65.9052 |
| 3 | 73.0755 | 106.9245 | 73.0755 |
| 5 | 78.4682 | 101.5318 | 78.4682 |
| 10 | 84.9838 | 95.0162 | 84.9838 |
| 20 | 88.3786 | 91.6214 | 88.3786 |

Table 1 Angles between generalized fault input

| s | $< d_{x2}, d_{y1} >$ | $< d_{x2}, d_{y2} >$ | $< d_{y1}, d_{y2} >$ |
|---|---|---|---|
| 1 | 180 | 0 | 180 |
| 3 | 180 | 0 | 180 |
| 5 | 180 | 0 | 180 |
| 10 | 180 | 0 | 180 |
| 20 | 180 | 0 | 180 |

Table 1 cont'd

## 4. THE OPT METHOD

From (6), the scaled residual is

$$\rho = v'M = \sum_{i=1}^{r_x} v'\phi_{xi}\gamma_{xi} + \sum_{i=1}^{r_y} v'\phi_{yi}\gamma_{yi} + v'\Psi\varepsilon \quad (16)$$

with a variance of $Var(\rho) = \sigma^2\left\|v'\Psi\right\|^2$, where $v$ is a parity vector obtained in some optimal sense, as discussed below.

### *4.1 Design of general parity vector*

In [1], the robust vector $v$ is designed such that the residual (15) is more sensitive to system faults and less sensitive to unknown disturbances or noise $\varepsilon$, i.e., $\left|v'\phi_{xi}\right|$ and $\left|v'\phi_{yj}\right|$ are as large as possible, whilst $\left|v'\Psi\right|$ is as small as possible. Two common choice of the cost functions for the design of $v$ are:

$$\text{(i)} \max_{v'H=0}\left(\sum_i (v'\phi_{xi})^2 + \sum_j (v'\phi_{yj})^2 - \left\|v'\Psi\right\|^2\right) \quad (17)$$

$$\text{(ii)} \max_{v'H=0} \frac{\sum_i (v'\phi_{xi}) + \sum_j (v'\phi_{yj})^2}{\left\|v'\Psi\right\|^2} \quad (18)$$

A solution of (17) is presented in [10], and a solution of (17) is given in [4]. In the following discussion, (16) is used.

*4.2 Design of optimal parity vector*

Maximizing (16) yields an optimal parity vector $v_x$ that is more sensitive to actuator fault $\chi_{xi}$,

$$\max_{v_x' H=0}\left( (v_x'\phi_{xi})^2 - \sum_{j\neq i}(v_x'\phi_{xj})^2 - \sum_{j}(v_x'\phi_{yj})^2 - \|v_x'\Psi\|^2 \right) \quad (19)$$

The corresponding scaled residual is

$$\rho_x = v_x' M \quad (20)$$

Similarly, the optimal parity vector $v_y$ that is more sensitive to sensor fault $\chi_{xi}$ is given by

$$\max_{v_y' H=0}\left( (v_y'\phi_{yi})^2 - \sum_{j\neq i}(v_y'\phi_{yj})^2 - \sum_{j}(v_y'\phi_{xj})^2 - \|v_y'\Psi\|^2 \right) \quad (21)$$

The corresponding scale residual is

$$\rho_y = v_y' M \quad (22)$$

When a fault is detected, it is possible to isolate it into an actuator fault or a sensor fault as described below. Two optimal parity vectors are designed, such that the first one is more sensitive to actuator faults, and the second one, more sensitive to sensor faults. More optimal parity vectors have to be designed, if it is necessary to isolate further which actuator is at fault. The main difference between the general parity test and the OPT method is that the latter is more suitable to isolate faults [5, 8].

*4.3 Fault detection and isolation*

The scaled residuals $\rho_{xi}$ and $\rho_{yj}$ for all $i$ and $j$ are computed first, then normalized as given below.

$$\rho_{xi}^0 = \rho_{xi} / \sqrt{\mathrm{var}(\rho_{xi})}$$

$$\rho_{yj}^0 = \rho_{yj} / \sqrt{\mathrm{var}(\rho_{yj})}$$

A fault is detected if $\rho_{opt} = \max_{i,j}(\rho_{xi}^0, \rho_{yj}^0)$ is greater than a given threshold, but the system operates normally otherwise. After a system fault is detected, it can be isolated into an actuator or a sensor fault by comparing the magnitude of all standardized residuals.

## 5. SIMULATION

The system given in example 1 is simulated for $Q = 2.5*10^{-5}I$, $R = 10^{-6}I$, and $u(k) = 0$. A fault with a magnitude of -0.5+0.005$k$, where $k$ is the number of samples is assumed to occur in either actuator 1 or 2. When the fault occurred in the first actuator, the residuals computed using the general parity test and the OPT methods are shown in Fig. 2, whilst those for the fault occurring in the second actuator are shown in Fig. 4. Comparing $|\rho_{opt}|$ generated by the OPT method (solid line) and $|\rho|$ generated by the general parity test (dashed line) shown in Figs. 2 and 4, the magnitude of the residuals generated by the OPT method are much greater than that generated by the general parity test, indicating that the OPT method is more effective in detecting the actuator faults.

For the fault in the first actuator, $|\rho_{x1}|$ is clearly greater than $|\rho_{x2}|$, $|\rho_{y1}|$ and $|\rho_{y2}|$, as shown in Fig. 3, since the *GFIV* $d_{x1}$ is not collinear with the other three *GFIV*s. Therefore, the fault in the first actuator can be isolated by the OPT method. Since the *GFIV*s $d_{x2}$, $d_{y1}$ and $d_{y2}$ are collinear, the fault in the second actuator cannot be isolated, as shown in Fig. 5, though a system fault is detected.

## 6. CONCLUSION

It is shown in this paper that faults with *GFIV*s that are not collinear in the generalized residual space can be detected and isolated by the OPT method. Since the *GFIV*s of the actuators and sensors in a time-invariant system can be computed *a-priori*, the ability to detect and isolate sensor and actuator faults can be pre-determinated. As denomstrated in the simulation example, the OPT method is much better than the general parity test in detecting and isolating slowly time-varying system faults, and that faults with *GFIV*s that are not collinear in the generalized residual space can be readily detected and isolated.

## REFERENCES


Lou, X.C. Willsky, A.S. and Verghese, G.C. (1986). Optimally Robust Redundancy Relations for Failure Detection in Uncertain Systems, *Automatica*, **22**, pp. 333-343.

Chow, E.Y. and Willsky, A.S. (1984). Analytical redundancy and the design of robust failure detection systems, *IEEE Trans. Auto. Contr.*, **AC-29**, pp. 603-614.

Zhang, H.Y. and Patton, R.J. (1993). Optimal Design of Robust Analytical Redundancy for Uncertain Systems, *IEEE Conf. Comp. Comm. Contr. and Power Eng.*, Beijing, China.

Jin, H. and Zhang, H.Y. (1996). Failure Detection Using Optimal Parity Vector Sensitive to Special Sensor Failure, *Proc. IEEE/SICE/RSJ Int. Conf. Multisensor Fusion and Integr. for Int. Sys.*, pp. 55-61.

Jin, H. and Zhang, H.Y. (1999). optimal parity vector sensitive to designated sensor fault, *IEEE Trans. on Aero. and Electronic Sys.*, **35**, pp. 1122-1128.

Jin, H. (1998). Study on accuracy and fault-tolerant performance of navigation systems, Ph.D. Thesis, Dept. of Auto. Contr., *Beijing Univ. of Aeronautics and Astronautics*.

Ding X., Guo, L. and Jeinsch, T. (1999). A characterization of parity space and its application to robust fault detectio, *IEEE Trans. on Auto. Contr.*, **AC- 44**, pp. 337-343.

Jin, H. and Zhang, H.Y. (1997). Configuration of redundant sensor system and its fault detection using parity vector metho, *IFAC Sym. Fault Det., Super. and Safety for Tech. Processes*, **2**, pp. 843-848.

Potter, J.E. and Suman, M.C. (1977). Thresholdless redundancy management with arrays of skewed instruments, *Agardograph-224, Contr. Sys.*, pp. 15-1~15-25.

Jin, H. and Zhang, H.Y. (1997). Fault detection of integrated navigation system based on measurement model, *Proc. 2nd Asian Contr. Conf.*, Seoul, pp. III-551-III-554.

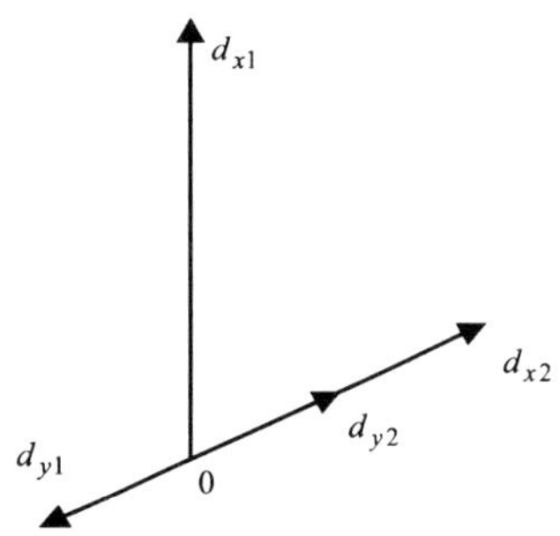


Fig.1 Four GFIVs in generalized parity space

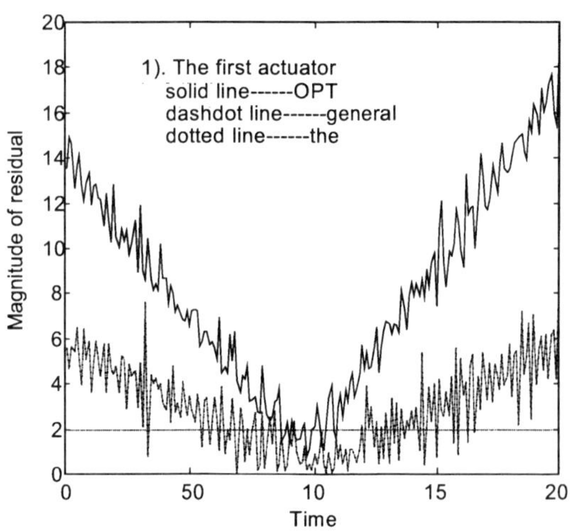


Fig. 2 Residuals generated by OPT and general parity test method for fault in first actuator

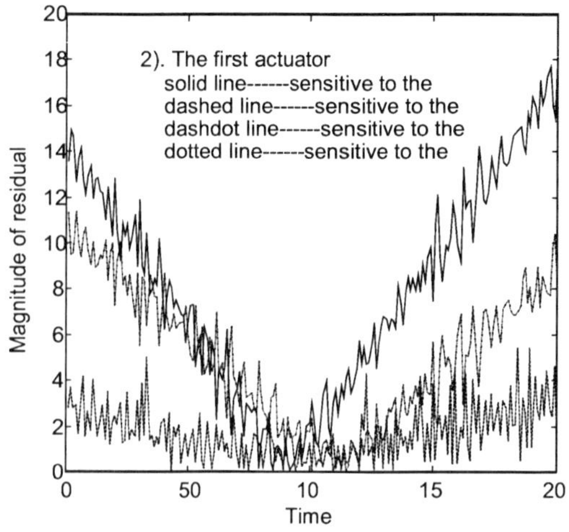


Fig. 3 Residuals for fault in first sensor

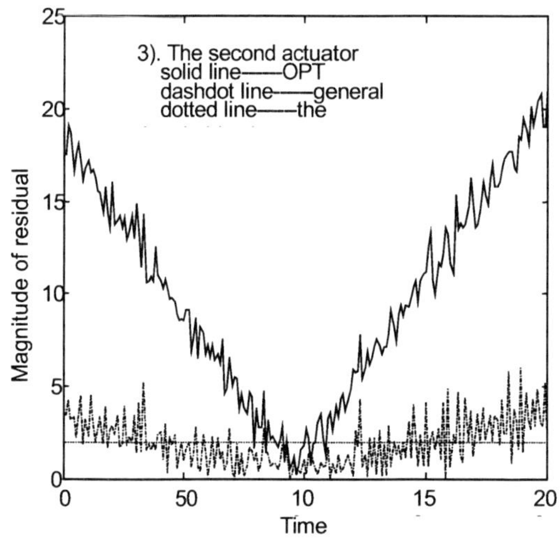


Fig. 4 Residuals generated by OPT and general parity test method for fault in seond actuator

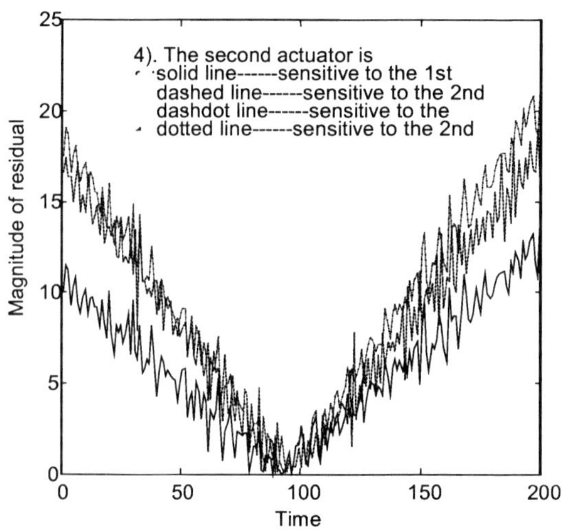


Fig. 5 Residuals for fault in second sensor

IFAC Publications
www.elsevier.com/locate/ifac

# FAULT DETECTION WITH SENSOR ERRORS

**Min You Hong Zhuan Qiu Hong Yue Zhang**

*Beijing University of Aeronautics and Astronautics, Beijing 100083, China*
*Email: hoyzhang@public3.bta.net.cn*

Abstract: A scheme for fault detection with sensor errors is discussed. Sensor output generally contains scale factor error and misalignment error as well as constant bias. In this paper the neural network is applied to compensate the residuals calculated by optimal parity vector method in order to remove the effect of sensor errors. And then compensated residuals are used to detect and isolate fault by analytic method and neural network. Thereby the accuracy of fault detection and isolation is improved.


Keywords: fault detection, fault isolation, neural network, sensor systems, sensitivity functions

## 1. INTRODUCTION

The reliability of inertial navigation system (INS) is most important as a reference system in integrated navigation system. Usually it can be increased by using the redundant and fault-tolerant techniques. Now there are many methods, for example the Generalized Likelihood Test (GLT) and Optimal Parity Vector (OPT), can be used to detect and isolate failure in redundant sensor system. Jin and Zhang (1999a, b) presented optimal parity vector without considering sensor errors. But in fact sensor error is affirmed to be existing which can restrict the ability of fault detection and isolation. In this paper neural network is applied to compensate the residuals calculated by OPT in order to remove the effect of sensor errors, and then compensated residuals are used to detect and isolate failure by analytic method and neural network. Thereby the accuracy of fault detection and isolation is improved.

## 2. OPTIMAL PARITY VECTOR (OPT)

Consider redundant measurement systems:

$$\mathbf{y} = \mathbf{Hx} + \mathbf{g}_i \gamma_i + \sum_{j \neq i} \mathbf{g}_j \gamma_j + \mathbf{E}\boldsymbol{\varepsilon} \tag{1}$$

Where $\mathbf{x} \in \Re^n$ is the state vector. $\mathbf{H} \in \Re^{m \times n}$ is the geometry matrix of sensor configuration. $\mathbf{y} \in \Re^m$ is the $m$ dimensional measurement vector. $\boldsymbol{\varepsilon}$ is measurement noise which is assumed to be Gaussian white noise with zero mean and covariance matrix $\Xi = \sigma^2 \mathbf{I}$ . $\gamma = (\gamma_1, \cdots, \gamma_m)'$ is the sensor failure vector. $\mathbf{G} = (\mathbf{g}_1, \cdots, \mathbf{g}_m)$ is failure input matrix and $\mathbf{g}_i$ is input vector of $i$ th sensor. $\mathbf{E}$ is noise input matrix.

A new performance criterion for robust failure detection is proposed (Jin and Zhang, 1999a). Due to this performance criterion, the parity equation is most sensitive to specified sensor failure and most insensitive to other sensor failures and system noise.

In the following a brief introduction of this method is given.

In order to design optimal parity vector sensitive to specified sensor, consider a performance criterion

$$\underset{\mathbf{v}'\mathbf{H}=0}{Sup}\left((\mathbf{v}'\mathbf{g}_i)^2 \Big/ [\left\|\mathbf{v}'\mathbf{E}\right\|^2 + \textstyle\sum_{j\neq i}(\mathbf{v}'\mathbf{g}_j)^2]\right) \tag{2}$$

Where numerator is sensitivity to the fault of $i$ th sensor, and denominator is sensitivity to noise and fault of all other sensors. In addition, constrained condition of parity: $\mathbf{v}'\mathbf{H}=0$ is still kept. Parity vector $\mathbf{v}$ can be expressed as a linear combination of all columns of $\mathbf{V}$

$$\mathbf{v} = \mathbf{V}\mathbf{c} \tag{3}$$

Where $\mathbf{V}$ is a matrix formed by a set of basis in parity space, $\mathbf{c}$ is a weight vector. Substituting (3) into (2), the extremum problem can be rewritten as

$$\underset{\mathbf{c}\neq 0}{Sup}(\mathbf{u}'\mathbf{c})^2 / \mathbf{c}'\mathbf{B}\mathbf{c} \tag{4}$$

Where $\mathbf{B} = \mathbf{V}'(\mathbf{E}\mathbf{E}' + \mathbf{G}\mathbf{G}' - \mathbf{g}_i\mathbf{g}_i')\mathbf{V}$ is a symmetrical matrix with dimension *m-n*, $\mathbf{u} = \mathbf{V}'\mathbf{g}_i \in \Re^{m-n}$. If $\mathbf{B}$ is positive matrix, then

1) When $\mathbf{c} = k \cdot \mathbf{B}^{-1}\mathbf{u}$, (4) achieves maximum.

2) Optimal parity vector is $\mathbf{v} = k \cdot \mathbf{V}\mathbf{B}^{-1}\mathbf{u}$. Where $k$ is an arbitrary real number not equal to zero. Suppose $\mathbf{v}_i$ is an optimal parity vector sensitive to $i$ th sensor fault, corresponding residual is:

$$\zeta_i = \mathbf{v}_i'\mathbf{y} \tag{5}$$

In order to determine detection threshold, assume that $\varepsilon$ is a normal distribution of multiple variables. Thereby residual obtained from (5) is a normal random variable as follows:

1) If $\gamma = 0$, $\zeta_i \sim N_1(0, \mathbf{v}_i'\mathbf{E}\Xi\mathbf{E}'\mathbf{v}_i)$

2) If $\gamma \neq 0$, $\zeta_i \sim N_1(\mathbf{v}_i'\mathbf{G}\gamma, \mathbf{v}_i'\mathbf{E}\Xi\mathbf{E}'\mathbf{v}_i)$

Let normalized residual be $\zeta_i^* \triangleq \dfrac{\zeta_i}{\sqrt{\mathbf{v}_i'\mathbf{E}\Xi\mathbf{E}'\mathbf{v}_i}}$,

1) If $\gamma = 0$, $\zeta_i^* \sim N_1(0,1)$

2) If $\gamma \neq 0$, $\zeta_i^* \sim N_1(\mathbf{v}_i'\mathbf{G}\gamma / \sqrt{\mathbf{v}_i'\mathbf{E}\Xi\mathbf{E}'\mathbf{v}_i}, 1)$

So, $m$ normalized residuals $\zeta_i^*$ $(i = 1, \cdots, m)$ sensitive to their own sensor fault are obtained. Then normalized residual with maximal absolute value can be determined. If $|\zeta_k^*| = \max_i |\zeta_i^*|$, it indicates that $k$ th normalized residual is maximum. For given false alarm rate $\alpha$, detection threshold $T_D$ is determined by quantity $\alpha/2$ for normal distribution. If $|\zeta_k^*| \leq T_D$, there is no failure occurred. If $|\zeta_k^*| > T_D$, fault occurs in the $k$ th sensor and can be isolated simultaneously.

## 3. APPLICATION OF NEURAL NETWORK TO COMPENSATE RESIDUAL

In practice the true value for the geometry matrix $\mathbf{H}$ can not be obtained due to the mounting and scale factor errors of inertial measurement units (Qin and Zhang, 1998). The parity vector $\mathbf{v}$ designed as above will be affected by the state and this will decrease the performance of failure detection. When considering sensor errors, measurement systems without fault is:

$$\mathbf{y} = (\mathbf{I} + \mathbf{H}_{se})[(\mathbf{H}_n + \mathbf{H}_{me})\mathbf{x} + \mathbf{b}_f + \varepsilon_1] \tag{6}$$

Where $\mathbf{H}_n \in \Re^{m\times n}$ is nominal measurement matrix, $\mathbf{H}_{se} \in \Re^{m\times m}$ is scale factor error matrix, $\mathbf{H}_{me} \in \Re^{m\times n}$ is misalignment matrix, $\mathbf{b}_f$ is sensor bias, and $\varepsilon_1$ is Gaussian white noise. Let

$$\begin{cases} \mathbf{b} = (\mathbf{I} + \mathbf{H}_{se})\mathbf{b}_f \\ \varepsilon = (\mathbf{I} + \mathbf{H}_{se})\varepsilon_1 \\ \mathbf{H}_e = \mathbf{H}_{me} + \mathbf{H}_{se}\mathbf{H}_n + \mathbf{H}_{se}\mathbf{H}_{me} \end{cases} \tag{7}$$

Then equation (6) is rewritten as:

$$\mathbf{y} = (\mathbf{H}_n + \mathbf{H}_e)\mathbf{x} + \mathbf{b} + \varepsilon \tag{8}$$

Due to not knowing $\mathbf{H}_e$, optimal parity vector can only be designed by nominal measurement matrix $\mathbf{H}_n$. Suppose $\mathbf{v}_i$ is an optimal parity vector sensitive to $i$ th sensor fault, corresponding residual is:

$$\zeta_i = \mathbf{v}_i'\mathbf{H}_e\mathbf{x} + \mathbf{v}_i'\mathbf{b} + \mathbf{v}_i'\varepsilon \quad \text{(Fault free)} \tag{9}$$

$$\zeta_i = \mathbf{v}_i'\mathbf{H}_e\mathbf{x} + \mathbf{v}_i'\mathbf{b} + \mathbf{v}_i'\mathbf{G}\gamma + \mathbf{v}_i'\varepsilon \quad \text{(With fault)} \tag{10}$$

From above formula, the residual $\zeta_i (i = 1,...,m)$ is not only sensitive to its own sensor fault but also to the state of the system and bias error. If the aircraft takes a big maneuver, it is possible that two former terms in equation (9) become so big as to exceed detection threshold and result in false alarm. Also in equation (10) it is possible that two former terms are bigger than the fault term and result in false alarm. To remove the effect of system state and constant bias on the residual, neural network is applied to compensate residual to enhance the ability of failure detection as follows.

Because neural networks have good ability of nonlinear mapping, the multilayer perceptron is used to compensate the residual. Taking the residuals without fault as the samples, multilayer perceptron is trained off-line. The compensated residuals are the difference between the output of multilayer perceptron trained and the residuals calculated by optimal parity vector. It is shown in Fig.1 where NN is multilayer perceptron and H is real measurement matrix. Compensated residuals are sensitive to their own sensor faults and not affected by system state and bias error. After that statistical test technique as shown in section II and Kohonen networks are applied to detect and isolation failure respectively. The former one must have the knowledge of statistical property of noise but the latter one does not. Kohonen networks has a good capability of pattern classification and is an effective technique in fault classification.

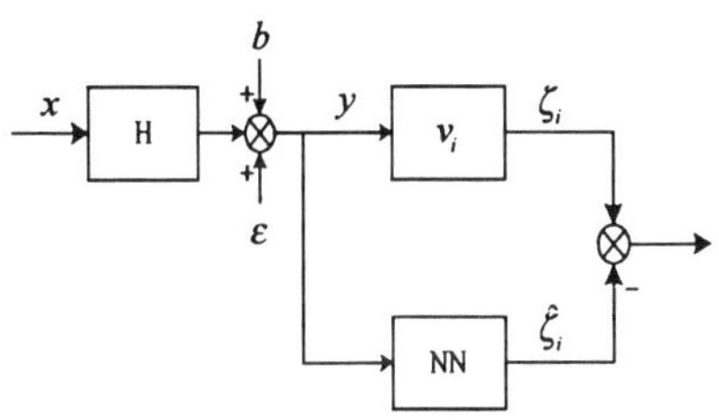


Fig.1: Compensating residuals by neural network

## 4. SIMULATION EXAMPLE

The measurement matrix **H** of a redundant sensor system consisting of six sensors mounted on dodecahedron is as follows

$$\mathbf{H}=\begin{bmatrix}\sin\theta & -\sin\theta & \cos\theta & \cos\theta & 0 & 0\\ 0 & 0 & \sin\theta & -\sin\theta & \cos\theta & \cos\theta\\ \cos\theta & \cos\theta & 0 & 0 & \sin\theta & -\sin\theta\end{bmatrix}^T$$

Where $\theta = 31.72^\circ$ is the exact mounting angle. Suppose misalignment angle is $20''$, the value of scale factor is $10^{-4}$, bias error is $0.1^\circ/h$, the covariance of measurement noise is $0.01^\circ/h$. Simulation runs 2000 seconds. The aircraft trajectory is consisted of sinusoidal maneuver along three orthogonal axes with angular rate $0.5^\circ/s$. At $t = 500$, swift sinusoidal maneuver with angular rate $2^\circ/s$ is added on $z$ axis. Suppose the first sensor develops a fault from $t = 1000$. The magnitude of fault is $0.3^\circ/h$.

Uncompensated residual is shown in fig.2 (a). Compensated residual by multilayer perceptron is shown in fig.2 (b). Where $z(1)$ and $z(2)$ represent the residual sensitive to the 1st sensor fault and the 2nd sensor fault respectively. Apparently, system state affects the uncompensated residual seriously. Compensated residual is not affected by system state and an obvious jump occurs only after there is a fault. Maximum normal residual is shown in Fig.3. It indicates that the maneuver can easily make the uncompensated maximal normalized residual exceed detection threshold resulting in a false alarm, and compensated residual is not changed along with system state but is sensitive to the fault. Let false alarm $\alpha$ be 0.01, so detection threshold $T_D$ is 2.33 from normal distribution table. The simulation result shows that the first sensor has developed a fault at the time $t = 1000$. Statistical test has good ability of detecting fault as above. In the following, Kohonen networks (Kohonen, 1988) in Fig.4 is introduced which does not need the statistical property of noise.

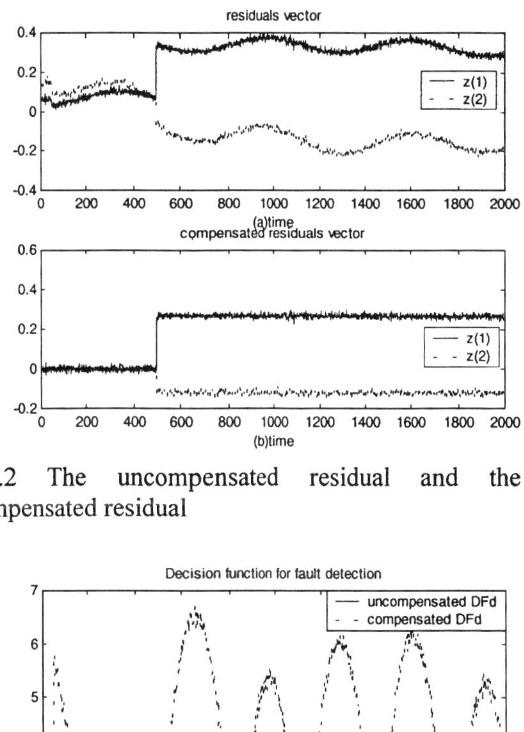


Fig.2 The uncompensated residual and the compensated residual

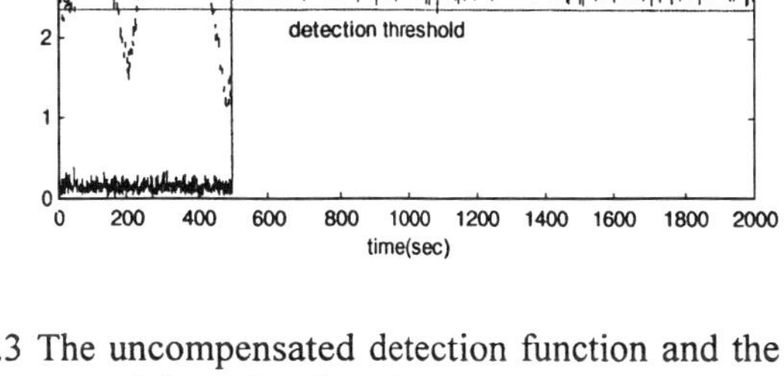


Fig.3 The uncompensated detection function and the compensated detection function

For six sensors of redundant system, seven output neurons are required in Kohonen networks (Jin, 1998). The six neurons are chosen such that each one of them represents the fault occurring at each sensor, and the seventh one is for the normal operation of the system. As shown in Fig.4, the input of Kohonen networks (KN) is the residual $\zeta^*$ which has been compensated by the multilayer perceptron. The weighting vector is $\mathbf{w}_i = (w_{i1}, w_{i2}, \cdots w_{i6})'$. Where $w_{ij}$ is the weighting coefficient linking the $j$th input $\zeta_j^*$ and the $i$th output $z_i, z_i = \mathbf{w}_i \zeta$. The competition rule in the KN is to determine the winning output neuron by comparing the distance between the weighting vector and the input before updating the corresponding weighting vector,. The Euclidean distance is defined as the input intensity of the $i$th output neuron of the KN:

$$\Pi_i = \left\| \zeta^* - \mathbf{w}_i \right\|^2, i = 1,2,\cdots 7 \qquad (11)$$

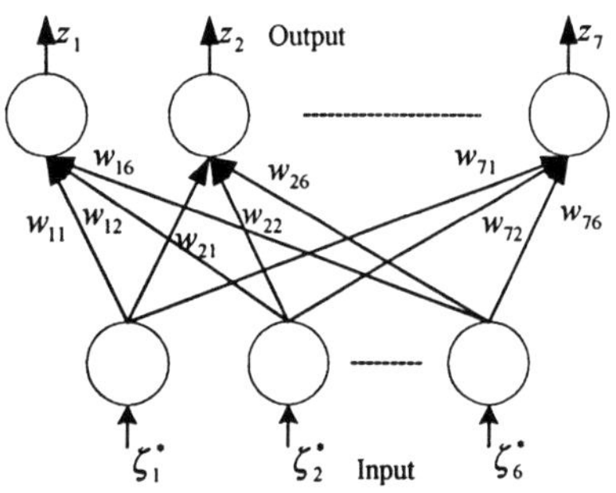


Fig.4 Construction of kohonen networks

The output neuron with the smallest Euclidean distance is the winner and its output is set to one, whilst all outputs of other neurons are set to zero. Let the $i$th output neuron be the winner, the weighting vector $w_i$ is adjusted by the following learning algorithm

$$\begin{aligned} \mathbf{w}_i^{new} &= (1-\eta)\mathbf{w}_i^{old} + \eta \zeta \\ \mathbf{w}_j^{new} &= \mathbf{w}_j^{old} \qquad j = 1,\ldots 7, j \neq i \end{aligned} \qquad (12)$$

Where $\eta$ is the learning rate. Suppose the first sensor develops a fault between $t = 1000$ and $t = 1400$, whilst the other sensors are operating normally. The magnitude of fault is $3^\circ / h$. The input intensities of all seven neurons are shown in Fig.5.For $t$ between $1000$ and $1400$, the input intensity of the first neuron is the smallest one indicating that a fault has developed in first sensor. At all other time periods, the input intensity of the seventh neuron is the smallest one indicating that the system is operating normally. Further more, from the competition rule of KN, the output of the first neuron is 1 for $t$ between $1000$ and $4000$, whilst the output of the seventh neuron is 1 in all other time periods. The outputs of other neurons are zero all the time. Consequently, it can deduce that the first sensor has developed a fault during the time period from $1000$ to $4000$, and the system operates normally in the all other time periods.

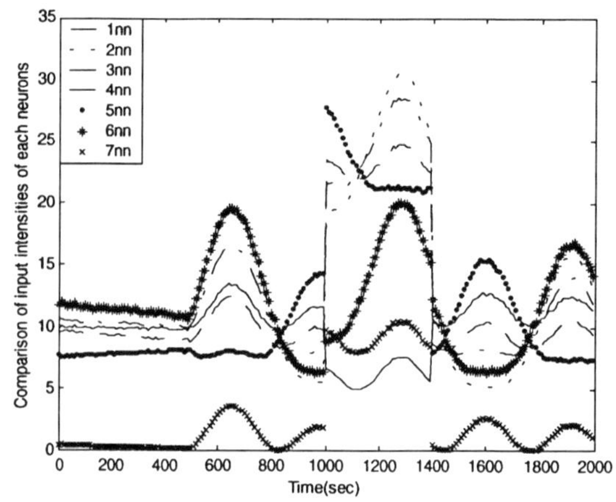


Fig.5 Comparison with the input intensities of all neurons

## 5. CONCLUSIONS

The residual generated using optimal parity vector method with sensor errors is not only sensitive to specified sensor fault but also to the state of the system and bias error. To solve this problem a approach applying neural network to compensating the residual is proposed. Compensated residual is sensitive to specified sensor fault and without the effect of sensor errors. Then statistic test and Kohonen networks are used to detect the fault. The effectiveness of two techniques in fault detection is illustrated by an example of a redundant sensor system consisting of six sensors.

## REFERENCES


Jin H. and Zhang H.Y.(1999a). Comparison of two techniques used to FDI of redundant system. *Proceeding of the first Conference on technique process and security in China,* Beijing.

Jin H. and Zhang H.Y.(1999b). Optimal parity vector sensitive to designated sensor fault, *IEEE Trans.on Aerospace and Electronic Systems,* Vol.35, No.4, pp1122-1128.

Qing Y.Y. and Zhang H.Y.(1998). Principle of Kalman filter and integrated navigation, Northwestern Polytechnical University Publishing House.

Hecht-Nielsen R.(1990).Neurocomputing. Macmillan College Publishing Company, pp101-112.

Jin H.(1998). Study on accuracy and fault-tolerant performance of navigation system. *Doctor dissertation,* Beijing University of Aeronautics and Astronautics.

Kohonen T.(1988). Self-organization and associative memory, 2nd ED., Springerverlag Berlin.

IFAC Publications
www.elsevier.com/locate/ifac

# MULTIOBJECTIVE OPTIMISATION APPROACH TO ROBUST CONTROLLER DESIGN

**Ian Griffin BEng MSc[1]**
**Peter Schroder BEng PhD[2]**
**Andrew Chipperfield BSc PhD MAIAA AMIEE[1]**
**Peter Fleming BSc PhD CEng FIEE FInstMC MBCS[1]**

*1. The University of Sheffield, Department of Automatic Control and Systems Engineering, UK*
*2. McLaren International, Woking, Surrey, UK*

Abstract: A control system design procedure based on the optimisation of multiple objectives is used to address the control design requirements of a simulated gasification plant. The non-linear gasifier is represented by three linear models relating to three separate operating points. The $H_\infty$ LSDP is used to guarantee the stability of the linear controller at it's design point while the multiobjective (MO) search method is used in order to optimise the robustness of the controller in terms of both stability and performance. A linear controller emerges that provides closed loop stability at all three operating points and is capable of meeting all the performance requirements except for those of the gas pressure at the 0% load operating point. *Copyright © 2001 IFAC*

Keywords: Genetic algorithms, H-infinity control, Multiobjective optimisation, Multivariable control, Robust control.

## 1. INTRODUCTION

Modern awareness of environmental issues has led to the desire for low pollution power generation techniques. One such technique, studied by ALSTOM, involves the combustion of pulverised coal using an Integrated Gasification Combined Cycle (IGCC) power plant. The operation of this Pilot Integrated Plant (PIP) is based on the Air Blown Gasification Cycle (ABGC). Limestone, which is required in order to minimise sulphur emission, is added to the pulverised coal and this mixture is then fluidised in a stream of air and steam and conveyed into the gasification plant (gasifier). A low calorific value fuel gas is produced by the reaction between the air and steam and the volatile constituents of the coal. The limestone, ash and residual carbon are extracted as bed material from the base of the gasifier or elutriated from the product gas which is then cleaned and used to power a gas turbine.

The specified objective of the work summarised in this paper was to perform a controller design procedure for the gasifier. The nature of the input and output constraints categorise the gasifier as a critical system (Zakian, 1989; Whidbourne and Lui, 1993). The loop-shaping design procedure (LSDP) (McFarlane and Glover, 1990) used here is suitable for multiple input multiple output (MIMO) systems. Weighting functions are employed to shape the open-loop transfer function followed by $H_\infty$ optimisation of a normalised coprime factorisation of the nominal plant description. This procedure results in the state space realisation of a linear controller.

The gasifier is a non-linear, multivariable system. The non-linear operating envelope is represented by three linear state space models which represent the open loop plant at 100%, 50% & 0% load operating points. The control system design procedure was performed using the 100% load linear model as this is the operating condition at which the gasifier spends the majority of its operational life. In order to ensure that the resulting linear controller was closed loop stable and met performance requirements at all three operating points, a multiobjective search method was employed to select suitable weighting functions. Hence, the robustness required for the linear controller to be applicable to a non-linear operating range was achieved through multiobjective optimisation.

## 2. $H_\infty$ LOOP-SHAPING DESIGN PROCEDURE

The normalised left coprime factorisation (NLCF) of a plant $G$ is given by $G = M^{-1}N$. A perturbed plant model $G_p$ is then given by,

$$G_p = (M + \Delta M)^{-1} (N + \Delta N) \tag{1}$$

To maximise this class of perturbed models such that the configuration shown in Fig. 1 is stable, controllers $K_s$ that stabilise the nominal closed-loop system and minimise $\gamma$ must be found where

$$\gamma = \left\| \begin{bmatrix} K_s \\ I \end{bmatrix} (I - GK_s)^{-1} M^{-1} \right\|_\infty \tag{2}$$

This is the problem of robust stabilisation of normalised coprime factor plant descriptions (Glover and McFarlane, 1989). From the small gain theorem (Skogestad and Postlethwaite, 1996), the closed-loop plant will remain stable if,

$$\left\| \begin{bmatrix} \Delta N \\ \Delta M \end{bmatrix} \right\|_\infty < \gamma^{-1} \tag{3}$$

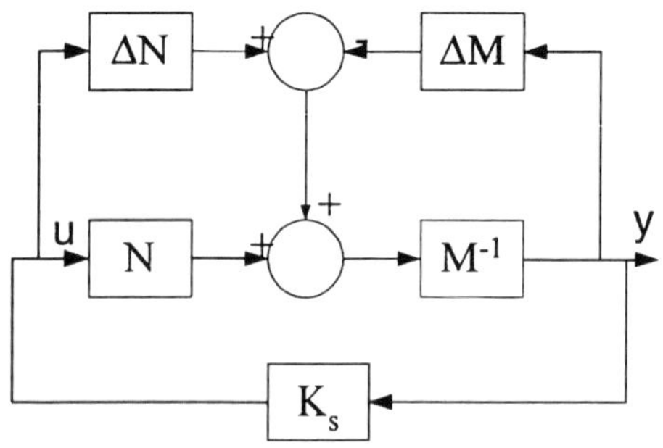


Fig. 1 Robust Stabilisation with respect to coprime factor uncertainty.

The lowest possible value of $\gamma$ and hence the highest achievable stability margin is given by
$\gamma_{min} = (1 + \rho(ZX))^{½}$, where $Z$ and $X$ are the solutions to the following algebraic Ricatti equations,

$$(A - BS^{-1}D^TC)Z + Z(A - BS^{-1}D^TC)^T - ZC^TR^{-1}CZ + BS^{-1}B^T = 0 \tag{4}$$

$$(A - BS^{-1}D^TC)^TX + X(A - BS^{-1}D^TC) - XBS^{-1}B^TX + C^TR^{-1}C = 0 \tag{5}$$

where $A$, $B$, $C$, and $D$ are the matrices of the state space representation of G and,

$$R = I + DD^T, \qquad S = I + D^TD. \tag{6}$$

By solving these equations the state space controller, $K$, can be generated explicitly (Skogestad and Postlethwaite, 1996). This controller gives no guarantee of the system's performance, only that it is stable with reasonable robustness.

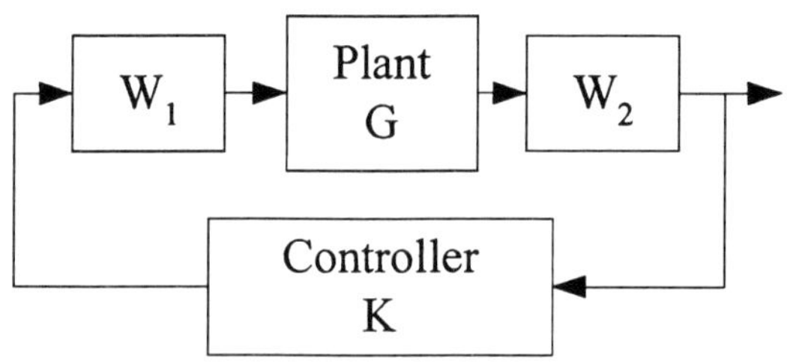


Fig. 2 Loop Shaping Controller Structure

It is therefore necessary to shape the system's response with both pre- and post-plant weighting function matrices $W_1$ and $W_2$, as shown in Fig. 2. This will ensure that the closed-loop performance meets the specifications. It is throught adjusting these weighting functions that the dsigner influences the procedure.

## 3. PROBLEM DESCRIPTION

The gasifier is a highly non-linear multivariable component. The plant has four manipulated inputs and four measured outputs.

Manipulated inputs:

- WCHR – char extraction flow (kg/s).
- WAIR – air mass flow (kg/s).
- WCOL – coal flow (kg/s).
- WSTM – steam mass flow (kg/s).

Measured outputs:

- CVGAS – calorific value of fuel gas (J/kg).
- MASS – Bed mass (kg)
- PGAS – Pressure of fuel gas ($N/m^2$)
- TGAS – Temperature of fuel gas ($^0K$)

Limestone is a dependent input, being introduced at a ratio of 10:1 coal:limestone. Disturbances to the plant take the form of changes in pressure (PSINK) upstream of the turbine that the gasifier is powering. These are due to adjustments to the gas turbine fuel valve. The system is required to maintain the fluctuations of the measured outputs within certain limits during prescribed disturbance conditions, namely step and sinusoidal changes in the value of PSINK. Actuator limitations are represented by saturation and rate limits applied to the manipulated inputs.

The gasifier is simulated in the MATLAB/Simulink environment using three linear models. These represent the plant at the 100%, 50% and 0% load operating points. The gasifier is run at 100% load for the vast majority of its operating life. Therefore, the $H_\infty$ controller is to be designed for the 100% load operating point but with sufficient robustness to meet the performance requirements at the other two operating points.

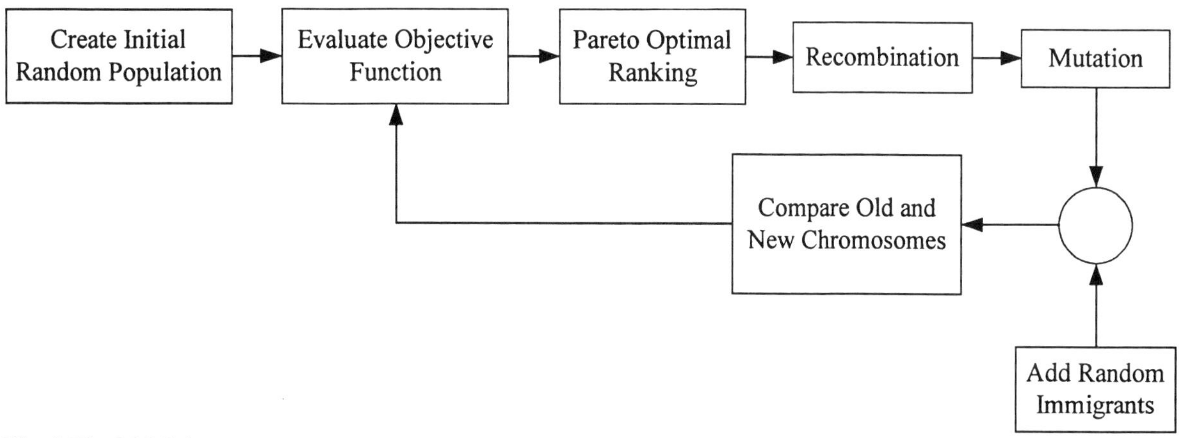


Fig. 3 The MOGA

## 4. MULTIOBJECTIVE GA STRUCTURE

A multiobjective genetic algorithm (MOGA) is used to find a set of optimal weighting functions. It is implemented using the GA Toolbox for MATLAB (Chipperfield, *et al.*, 1994), developed in house, with additional extensions to accommodate multiobjective ranking, sharing and mating restrictions (Fonseca and Fleming, 1993). The salient features of this MOGA are shown in Fig. 3. In order for the MOGA to rank the prospective controllers, an objective function is required to evaluate the controller's performance against a number of objectives. The objective function first has to construct the $H_\infty$ controller by solving the algebraic Ricatti equations (4) & (5). This was done using a state space linear model of the gasifier at the 100% load operating point and the weighting functions formed from the individual under evaluation. The weighting function structures used were those of a diagonal matrix of first order lags for $W_1$ and a diagonal matrix of gains for $W_2$. The first order lag structure of $W_1$ was considered necessary to break any algebraic loop which may appear in simulation due to the non-zero *D* matrices in the linear models. The terms in $W_2$ were specified as stateless in order to minimise the order of the resultant controllers. The chromosome is structured as a binary string in which the binary representations of the decision variables are right-concatenated. The controller's performance was then evaluated by running closed loop simulations using the linear models representing the 100%, 50% and 0% load operating points of the gasifier. Each simulation involved subjecting the closed-loop system to a step and a sine wave disturbance in turn.

As the optimisation philosophy of MOGA is to minimise objective function values, the linear plant models were not off-set, relative values about the operating point being preferred to absolute input/output values. This allowed the peak deviation of each output from its operating point value to be assessed by taking the maximum absolute value of each output vector.

Table 1. The Objectives

| Objective No. | Objective Description |
|---|---|
| 1 | Peak fluctuation of CVGAS from 100% operating point |
| 2 | Peak fluctuation of MASS from 100% operating point |
| 3 | Peak fluctuation of PGAS from 100% operating point |
| 4 | Peak fluctuation of TGAS from 100% operating point |
| 5 | Peak fluctuation of CVGAS from 50% operating point |
| 6 | Peak fluctuation of MASS from 50% operating point |
| 7 | Peak fluctuation of PGAS from 50% operating point |
| 8 | Peak fluctuation of TGAS from 50% operating point |
| 9 | Peak fluctuation of CVGAS from 0% operating point |
| 10 | Peak fluctuation of MASS from 0% operating point |
| 11 | Peak fluctuation of PGAS from 0% operating point |
| 12 | Peak fluctuation of TGAS from 0% operating point |
| 13 | Maximum continuous eigenvalue of closed loop system |
| 14 | $H_\infty$ robustness measure $\gamma$ |

This was done for each disturbance condition and each candidate controller in turn. Input constraints, representing actuator limitations, were adhered to by placing saturation and rate-limit blocks on the input lines of the closed loop Simulink system representation. These blocks contained relative values appropriate to the operating point under investigation.

Closed- loop stability at all three operating points was guaranteed by calculating the maximum closed-loop continuous eigenvalue and discarding any individual in the population which did not result in this value being less than zero across all the three operating points. One further objective attempted to minimise the $H_\infty$ norm, $\gamma$, in order to maximise the robustness of the closed loop control system. The objectives are shown in Table 1 above.

## 5. RESULTS

Figure 4 shows a typical trade off graph for the gasifier. The $x$-axis shows the design objectives from Table 1 and the $y$-axis shows the objective domain performance of the controllers.

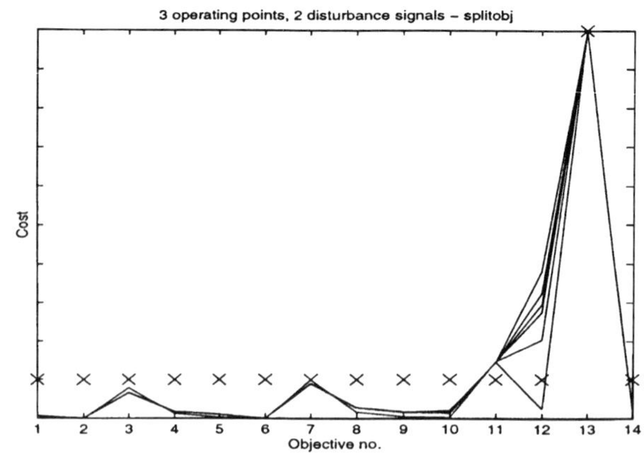

Figure 4. Trade off graph

With the exception of objective 13, the displayed ranges of each objective are normalised to leave the 'x' marks representing the optimisation goals one tenth of the way up the cost axis. Trade-offs between adjacent objectives result in the crossing of lines whereas concurrent lines represent non-competing objectives. Here, the goals relating to maximum output fluctuation (objectives 1-12) are set to the maximum allowed control errors for the associated output. These targets are specified as constraints in order to guarantee that controllers represented on the trade off graph satisfy the output constraints over the run-time of the simulation. From fig. 4 it can be seen that all the controllers represented here offer excellent control over peak bed mass fluctuation (objectives. 2, 6 & 10). Therefore the bed mass fluctuation, as an objective, does not compete heavily with any other objective.

Figures 5 & 6 show output time responses for the closed loop system with a selected controller at the 100% and 0% load operating points respectively. The performance requirements for the gasifier which these time responses are required to conform to are as follows (figures relate to the relative deviation of the output value from the operating point during disturbance conditions).

Performance Requirements:

$$-10000 < CVGAS\ (J/kg) < 10000$$
$$-500 < MASS\ (kg) < 500$$
$$-10000 < PGAS\ (N/m^2) < 10000$$
$$-1 < TGAS\ (^0K) < 1$$

## 6. DISCUSSION OF RESULTS

Results show a robustly stable controller design capable of exerting effective control and meeting all constraints over all the outputs at the 100% and 50% load operating points. For the case of the 100% load operating point, the step disturbance results are shown in Fig. 5. These show all four outputs remaining within their maximum deviation limits but with varying degrees of steady state error. The fuel gas calorific value and bed mass are controlled very effectively with deviation of less than 2% of the values allowed by the performance specifications. Control of the gas temperature is also achieved reasonably tightly with a maximum deviation of around 15%. However, extended run-time simulation beyond 300 seconds show that the responses are not returning to the set point. Whilst control of the gas pressure is not achieved as tightly as for the other three outputs, the maximum deviation from the operating point is well within the performance specifications. Again, however, the steady state error persists for extended simulations. The response at the 100% operating point to a sine wave disturbance shows similar characteristics to those of the step disturbance shown in Fig. 5. The tightest control was achieved over the calorific value and bed mass. Again, whilst remaining within the constraints, the gas pressure proves the most challenging control problem.

At the 0% load operating point, the shapes of the responses are similar to the previous operating points (see Fig. 6). Tightest control was again achieved for the calorific value and bed mass and for the step disturbance, steady-state errors were present on all the outputs which did not return to the set point during extended simulations. However, for the 0% operating point, it did not prove possible to keep the peak fluctuation of the gas pressure within the maximum deviation limits. For the step responses shown in Fig. 6 there is a constraint breach which could not be contained. However, in each case the breach is less than 50% of the size of the limits specified in the performance requirements. The settled value of the step response was within the specified limits. The same characteristics were present in the sinusoidal disturbance responses.

## 7. ADVANTAGES AND LIMITATIONS OF LSDP

The evolutionary algorithm approach to controller design employed here is advantageous in that it results in the designer having a choice of controllers rather than one specific design. The final selection between varying performance characteristics can be made in the knowledge that all comply with the restrictions imposed on the system. The trade off graph informs the designer of the implications that his choice will have on all the explicit design objectives. The approach greatly aided the selection of suitable weighting functions, searching a far greater space than would be readily achieved by conventional techniques. A particularly strong feature of the GA approach is the ability to use a mixture if discrete and continuous parameters in the problem formulation. This leads to an intuitive and natural representation for the problem being considered and simplifies the process of efficiently searching a large space.

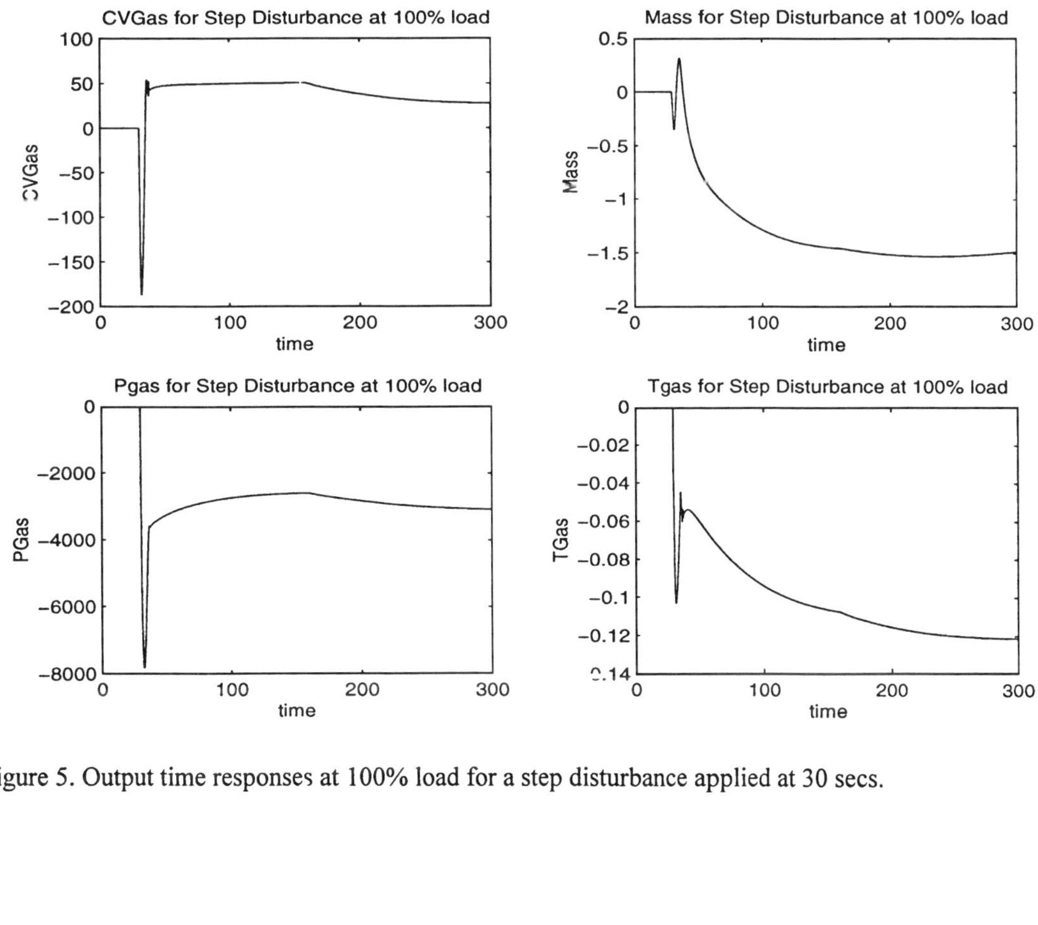


Figure 5. Output time responses at 100% load for a step disturbance applied at 30 secs.

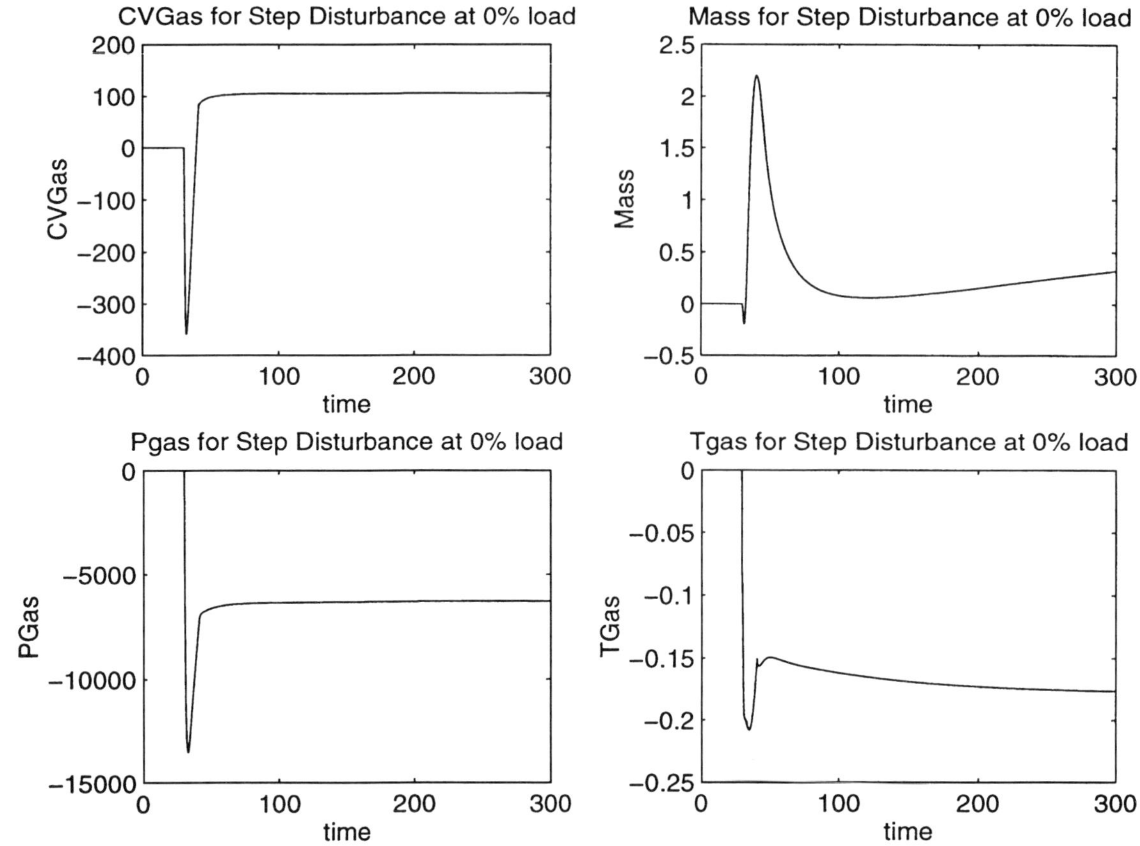


Fig. 6 Output time responses at 0% load for a step disturbance applied at 30 secs.

One limitation of this approach is that it produces solutions to a very specific formulation of the design problem. Whilst the response of the chosen controller may be optimal in terms of the objectives applied in the objective functions, it may not be optimal to similar objectives which the plant could reasonably be expected to face.

## 8. CONCLUDING REMARKS

An evolutionary algorithm approach has been proposed for the $H_\infty$ controller design for a coal burning gasification plant. The controller design procedure was applied to a primary linear operating point and robustified through optimisation. In this way, a linear controller has been fitted to a non-linear operating envelope. It should be noted that the mixed optimisation approach allows other design parameters to be included in the problem formulation. For example the magnitude of the real part of the largest eigenvalue and the robustness metric, $\gamma$, were declared as explicit design criteria.

## REFERENCES

Chipperfield, A. J. and Fleming, P. J. and Pohlheim, H. P. *A Genetic Algorithm Toolbox for MATLAB,* Proc. Int. Conf. Systems Engineering, Coventry, UK, pp. 200-207, Sept., 1994.

Fonseca, C. M. and Fleming, P. J. An Overview of Evolutionary Algorithms in Multiobjective Optimisation. *Evolutionary Computation,* **Vol. 1**, *No. 1, pp 25-49, 1993.*

Glover, K. and McFarlane, D. C. Robust stabilisation of normalised coprime factor plant descriptions with $H_\infty$ bounded uncertainties. *IEEE Trans. Autom. Control,* 1989, **34**(8), 821-830.

McFarlane, D. C. and Glover, K. *Robust Controller Design Using Normalised Coprime Factor Plant Descriptions,* **Vol. 138**, Lecture Notes on Control and Information Science, 1990 (Springer-Verlag, Berlin).

Skogestad, S. and Postlethwaite, I. Multivariable Feedback: Control, Analysis and Design. *John Wiley & Sons Ltd, Chichester, England,* 1996.

Whidborne, J. F. and Lui, G. P. *Critical Control Systems: Theory, Design and Applications,* 1993 (Research Studies Press, Taunton).

Zakian, V. Critical systems and tolerable inputs. *Int. J. Control, 1989,* **49**(4), 1285-1289.

IFAC Publications
www.elsevier.com/locate/ifac

# BLENDING LOCAL MODELS WITHIN A VELOCITY-BASED MULTIPLE MODEL NETWORK

**Séamus C. McLoone, George W. Irwin & Seán F. McLoone**

*Intelligent Systems and Control,*
*School of Electrical and Electronic Engineering,*
*Queen's University Belfast, Belfast, BT9 5AH, N. Ireland.*
*e-mail: seamus.mcloone@eeng.may.ie; g.irwin@ee.qub.ac.uk, s.mcloone@ee.qub.ac.uk*

Abstract: Velocity-based local model (LM) networks overcome many of the disadvantages of the conventional LM network approach. In the former, a set of *velocity-based linear* local models are interpolated using appropriate weighting functions whereas the latter employs *affine* local models. In contrast to the conventional network, the global dynamics of the velocity-based nonlinear model comprise a simple linear combination of the local model dynamics. This paper examines in detail the blending of local .nodels which is a key issue in the velocity-based approach. Various methods of blending the sub-models are studied and it is highlighted that only one of these is suitable. *Copyright © 2001 IFAC*

Keywords: Nonlinear, identification, multiple models, networks, interpolation.

## 1. INTRODUCTION

The goal of local model (LM) networks is to accurately represent a nonlinear dynamical plant over the entire operating space using a small number of local models. Since the greater part of the operating space generally consists of regions where no single model dominates, and several local models require blending, it is no surprise to discover that blending has a important role in multiple-model systems. For subsequent analysis and design it is desirable that the global dynamics of the nonlinear blended network are directly related to the underlying sub-model dynamics. Recent research (Shorten et al., 1999, Murray-Smith et al., 1999, Leith and Leithead, 1999) has highlighted the lack of this property within the conventional multiple model framework, demonstrating that, at best, the dynamics of the LM network are only weakly related to the underlying local models. Furthermore, existing multiple model systems typically employ affine models (McLoone and Irwin, 1998) which, being nonlinear, do not provide continuity with established linear theory and can also detract from the transparency of the overall network.

A novel class of velocity-based, blended multiple model system has been proposed by Leith and Leithead (1999) whereby the global dynamics are directly related to the local models employed. The underlying sub-models are velocity-based, continuous-time and linear, thus providing continuity with existing linear techniques, which is useful for analysis and controller design. Most of their work to date has concentrated on theoretical developments using simple illustrative examples.

This paper examines the key aspect of blending local models within the velocity-based multiple-model framework. A highly nonlinear dynamical system in the form of a simulated Continuous Stirred Tank Reactor (CSTR) process is used to analyse the various methods of blending the velocity-based sub-models. Results reveal that there is only one correct blending technique for velocity-based networks.

The initial concept of LM networks (Johansen and Foss, 1992, 1993, Brown et al., 1997) envisaged weighting the outputs of each local model to obtain the global network output. In the case of a heterogeneous LM network (Murray-Smith, 1994) where local models can vary from complex neural networks to simple constants within one network, weighting the local model outputs is the most obvious, and possibly only, choice for interpolation. However, this paper highlights that blending in this manner is fundamentally incorrect for the weighting of local models within a velocity-based multiple-model system. Instead, the global model dynamics should be obtained by blending the local model dynamics.

For completeness the next section outlines the salient features of the velocity-based local modelling approach. In section 3 the different ways in which the velocity-based local models can be blended are described. Section 4 briefly describes the CSTR process, used for analytical purposes in this paper. Section 5 analyses the results of the various blending methods. The paper ends with some suggestions for future work in section 6.

## 2. VELOCITY-BASED LOCAL MODELLING

For ease of reference the notation of Leith and Leithead (1999) is generally followed in this section. Consider the general nonlinear state space system:

$$\dot{x} = f(x,u) \quad (1)$$

The first-order expansion of this system about an operating point $(x_o,u_o)$ is given by:

$$\dot{\tilde{x}} = f(x_o,u_o) + \left.\frac{\partial f}{\partial x}\right|_{x_o,u_o}(\tilde{x}-x_o) + \left.\frac{\partial f}{\partial u}\right|_{x_o,u_o}(u-u_o) \quad (2)$$

Differentiating this *affine* expansion with respect to time gives the velocity-based *linear* system:

$$\ddot{\tilde{x}} = \left.\frac{\partial f}{\partial x}\right|_{x_o,u_o}\dot{\tilde{x}} + \left.\frac{\partial f}{\partial u}\right|_{x_o,u_o}\dot{u} \quad (3)$$

A similar form can be obtained by differentiating (1) with respect to time using the chain rule:

$$\ddot{x} = \frac{\partial f}{\partial x}\dot{x} + \frac{\partial f}{\partial u}\dot{u} \quad (4)$$

With the appropriate initial conditions, (1) and (4) give identical solutions and therefore there is no approximation at this stage. More importantly, however, is the fact that the velocity-based linearisation (3) is just the *frozen* form of the velocity-based system (4), at the operating point $(x_o,u_o)$. This establishes a direct relationship between the dynamics of the velocity-based form of the nonlinear system and the velocity-based linearisation. Furthermore, members of the family of velocity-based linearisations are all linear, providing continuity with established linear theory and methods.

A velocity-based, blended, multiple-model system is formed by weighting several velocity-based linearised models:

$$\begin{aligned}\ddot{\tilde{x}} &= \left(\sum_i A_i(\tilde{x}_i,u_i)\rho_i(\tilde{\psi})\right)\dot{\tilde{x}} \\ &+ \left(\sum_i B_i(\tilde{x}_i,u_i)\rho_i(\tilde{\psi})\right)\dot{u}\end{aligned} \quad (5)$$

where $A_i(\tilde{x}_i,u_i) = \left.\frac{\partial f}{\partial \tilde{x}}\right|_{\tilde{x}_i,u_i}$, $B_i(\tilde{x}_i,u_i) = \left.\frac{\partial f}{\partial u}\right|_{\tilde{x}_i,u_i}$ and $(\tilde{x}_i,u_i)$ is the linearisation or *freezing* point of the $i^{th}$ local model:

$$\ddot{\tilde{x}} = A_i(\tilde{x}_i,u_i)\dot{\tilde{x}} + B_i(\tilde{x}_i,u_i)\dot{u} \quad (6)$$

The normalised weighting function is given by $\rho_i(\tilde{\psi})$, where $\tilde{\psi}$ is the scheduling variable. The dynamics of the blended system, about the operating point $(\tilde{x}_o,u_o)$, are now considered. The velocity-based linearisation form of (5), at $(\tilde{x}_o,u_o)$, is simply obtained by *freezing* the validity function $\rho_i(\tilde{\psi})$ at the operating point. Hence, the following *linear* system is obtained:

$$\begin{aligned}\ddot{\tilde{x}} &= \left(\sum_i A_i(\tilde{x}_i,u_i)\rho_i(\tilde{\psi}_o)\right)\dot{\tilde{x}} \\ &+ \left(\sum_i B_i(\tilde{x}_i,u_i)\rho_i(\tilde{\psi}_o)\right)\dot{u}\end{aligned} \quad (7)$$

It can be seen that, with the appropriate initial conditions, the solution to (7) is initially tangential to the solution of the velocity-based multiple model system (5). The dynamics of the multiple-model system local to an arbitrary operating point are therefore the same as the dynamics of the corresponding frozen-form linear system at the same operating point. Rewriting (7) as:

$$\ddot{\tilde{x}} = \sum_i \rho_i(\tilde{\psi}_o)\left(A_i(\tilde{x}_i,u_i)\dot{\tilde{x}} + B_i(\tilde{x}_i,u_i)\dot{u}\right) \quad (8)$$

then clearly highlights this direct relationship between the frozen-form, (7), of the velocity-based blended system and the underlying local models, (6), at $(\tilde{x}_o,u_o)$. Thus at any arbitrary operating point, the global dynamics of the multiple model system are described by a straightforward weighted sum of the local model dynamics. No such direct relationship exists between the dynamics of the conventional multiple model representation and the dynamics of the first-order expansion system.

Further theoretical analysis reveals that the solution to the frozen form of the velocity-based blended network can be described by the weighted linear combination of the solutions to the local models, giving the properties desirable of a local sub-model framework. Detailed analysis of both the conventional and velocity-based nonlinear representations can be found in (Leith and Leithead, 1999, McLoone, 2000).

## 3. IMPLEMENTATION OF VELOCITY-BASED LM SYSTEM

In the velocity-based LM network the global dynamics are obtained by blending the local model dynamics. This is not the same as weighting the outputs of each local model to give the global network output, as was originally envisaged for local modelling. Indeed for a heterogeneous LM network (Murray-Smith, 1994), where local models can vary from being complex neural networks to simple constants within one network, weighting the local model outputs is the most obvious, and possibly only, choice for blending. However, if the velocity-based local models are blended in this way the results are rather different from the velocity-based structure proposed by Leith and Leithead (1999). In this regard the possible implementations of the velocity-based network are now considered.

### *3.1 Local model realisation*

Consider the following velocity-based, local state-space model:

$$\ddot{\bar{x}} = A\dot{\bar{x}} + B\dot{u}$$
$$\dot{\bar{y}} = C\dot{\bar{x}} \tag{9}$$

where $A$, $B$ and $C$ are constant matrices and/or vectors. This local model can be implemented as illustrated in Fig. 1.

When a velocity-based network contains only one local model the global output is obtained by integrating the output of that model. Since no blending of models takes place the corresponding conventional affine local model network produces the exact same output under the appropriate initial conditions. Implementation of blended multiple model networks however is not as straightforward.

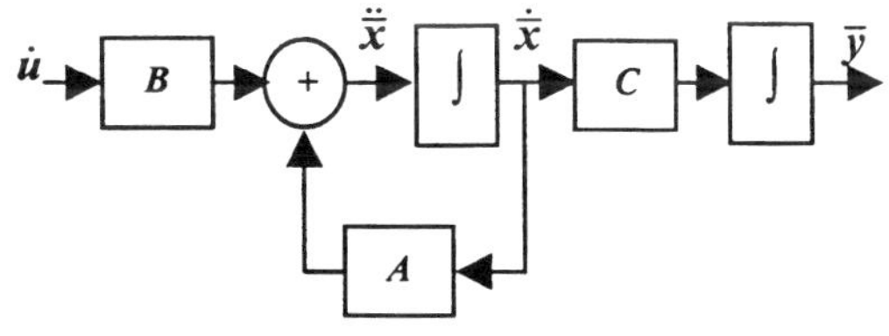


Fig. 1. Velocity-based local model implementation

### *3.2 Multiple model realisation*

For simplicity consider now a velocity-based multiple model network constructed from the blending of two local models. Three different implementations of the nonlinear network are possible since the local models can be blended in three different ways, as illustrated in Fig. 2, 3 and 4 respectively. The results in section 5 will reveal that only the third one is actually correct. In each figure, $\rho$ represents the weighting function.

The first implementation in Fig. 2 is based on weighting the output of each local model, giving the global output:

$$\tilde{y} = \sum_{i=1}^{2} \rho_i \bar{y}_i = C\sum_{i=1}^{2} \rho_i \bar{x}_i \tag{10}$$

It is worth noting that this network implementation employs local state derivative feedback. Global feedback is also possible by blending the local state derivatives, using the same weighting functions, and returning the global state derivative to each local model.

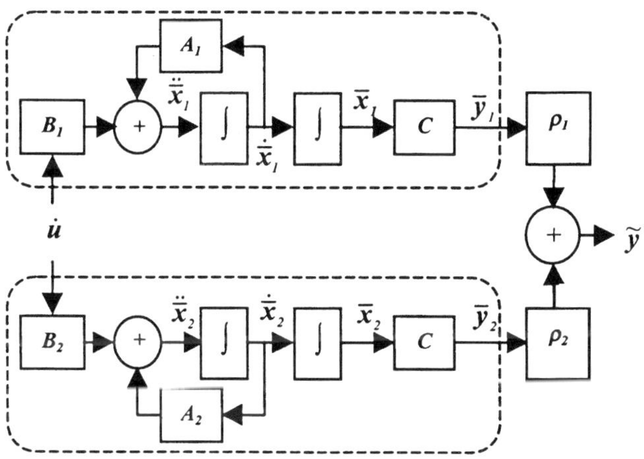


Fig. 2. Velocity-based multiple model implementation – blending $\bar{y}$ (or $\bar{x}$)

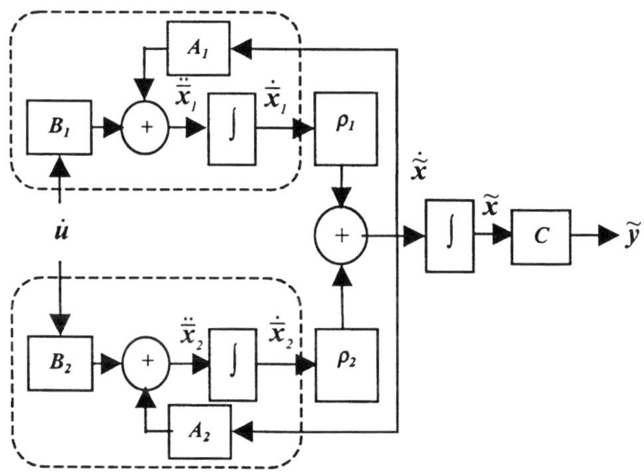


Fig. 3. Velocity-based multiple model implementation – blending $\dot{\bar{x}}$

The second implementation consists of combining the local model state derivatives, $\dot{\tilde{x}}_i$. The network output is then obtained by integrating the global state derivative:

$$\tilde{y} = C\sum_{i=1}^{2}\int \rho_i \dot{\tilde{x}}_i \quad (11)$$

Once again, global or local state derivative feedback can be used. Fig. 3 shows the network implementation with global feedback.

The third and final implementation in Fig. 4 involves weighting second derivatives $\ddot{\tilde{x}}_i$ of the local model state to give the global output:

$$\tilde{y} = C\sum_{i=1}^{2}\iint \rho_i \ddot{\tilde{x}}_i \quad (12)$$

In this case there is no difference between global and local state derivative feedback.

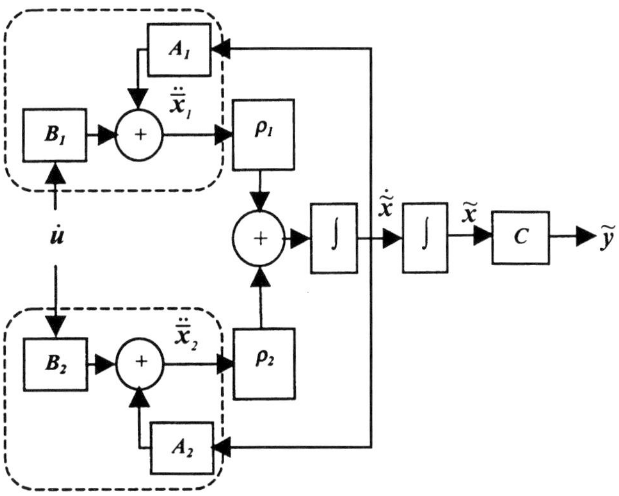


Fig. 4. Velocity-based multiple model implementation – blending $\ddot{\tilde{x}}$

The modelling capabilities of these alternative nonlinear representations are now examined using the CSTR process, which is briefly described in the next section.

## 4. THE CSTR PROCESS

The CSTR process is a single-input, single-output system where the input is the flowrate of a coolant and the output is the concentration of a product compound. The reaction that takes place to produce the compound is exothermic which raises the temperature and reduces the reaction rate. The introduction of a coolant allows manipulation of the temperature and, hence, control of the product concentration. The reaction takes place in a container of fixed volume and the product flowrate, input concentration, temperature and output flowrate are all assumed constant at their nominal values.

Table 1 Parameters for the simulated CSTR process

| | | |
|---|---|---|
| $q_f$ | product flowrate | *100* l/min |
| $C_f$ | input product conc. | *1* mol/l |
| $T_f$ | input temperature | *350* K |
| $V$ | container volume | *100* l |
| $E/R$ | activation energy | $10^4$ K |
| $T_{cf}$ | temp. of coolant | *350* K |
| $K_0$ | plant constant | $7.2 \times 10^{10}$ /min |
| $K_1$ | plant constant | $1.44 \times 10^{13}$ Kl/min/mol |
| $K_2$ | plant constant | *0.01* /l |
| $K_3$ | plant constant | *700* l/min |

The CSTR plant can be represented by the following equations:

$$\dot{T}(t) = \frac{q_f}{V}(T_f - T(t)) + K_1 C(t)\exp\left(-\frac{E}{RT(t)}\right) + K_2 q_c(t)\left(1 - \exp\left(-\frac{K_3}{q_c(t)}\right)\right)(T_{cf} - T(t))$$
$$\dot{C}(t) = \frac{q_f}{V}(C_f - C(t)) - K_0 C(t)\exp\left(-\frac{E}{RT(t)}\right) \quad (13)$$

where $q_c(t)$ is the input coolant flowrate, the internal plant state, $T(t)$, is the temperature of the solution and $C(t)$ is the output product concentration. The other parameters are their values are given in table 1.

The CSTR plant equations contain nonlinear product and exponential terms. Furthermore, the process dynamics vary considerably across the input space providing a suitably challenging and useful nonlinear system on which to examine the modelling capabilities of the various nonlinear velocity-based representations outlined in the previous section.

## 5. MODELLING RESULTS AND ANALYSIS

It is the aim of this paper to highlight the need for correct blending of the sub-models within the velocity-based LM framework. Thus, networks with two local models were employed to model the nonlinear dynamical relationship between the input coolant flowrate, $q_c$, and the output product concentration, $C$, within the operating space bounded by the input $q_c = [85\ 111]$ l/min.

A nonlinear plant representation was formed by weighting together several suitably located velocity-based linear local models. These local models were obtained by freezing the nonlinear velocity-form of equation (13) at the appropriate linearisation points, where the linearisation point for the $i^{th}$ local model is given by $(C_o^i, T_o^i, q_{co}^i)$.

The equilibrium linearisation points for the models were (*0.0620* mol/l, *448.7522* K, *90* l/min) and (*0.1298* mol/l, *432.9487* K, *110* l/min). The two models were blended together using normalised

Gaussian weighting functions. For simplicity only the output $C$ was used as a scheduling variable. The centres of the un-normalised Gaussians were manually chosen to be *0.0620* mol/l and *0.1298* mol/l, with widths of *0.02* and *0.03* respectively. While these weighting functions are not necessarily optimal, they are more than adequate for the purposes of this study.

Implementation of the velocity-based networks requires the derivative of the input $q_c$ to be available. This is difficult to achieve in practice, especially if step changes are involved. Here, a combination of a sinusoid and a constant was used to provide a suitable approximation to a step input. Reducing the sinusoid frequency results in a more accurate step change representation. With this technique the input derivative was accurately implemented and allowed the plant dynamics to be suitably excited for modelling purposes.

Four velocity-based networks were developed. In the first one the LM outputs were blended. The second and third networks both blended the LM state derivatives, but one employed local state feedback while the other used global state feedback. The final network weighted the LM second-order state derivatives.

*Simulation results: A* series of approximate step changes in $q_c$, shown in Fig. 5, were applied to each of the above nonlinear systems. The resulting product concentration in each case is displayed in Fig. 6.

These graphs show relatively poor modelling results for all the velocity-based LM networks, especially in terms of the steady-state errors. These occur because of a modelling error inherent in the velocity-based approach. Detail regarding this error can be found in (McLoone, 2000, McLoone and Irwin, 2001). This error cannot be removed by the network and will therefore accumulate over time to produce the poor steady-state results shown in Fig. 6. However, in control applications integral action can easily be used to eliminate the effect of this modelling error. The dynamic performance of the network is therefore more important.

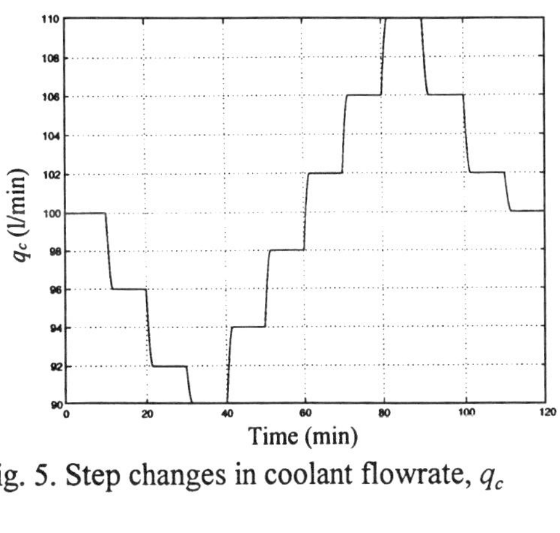


Fig. 5. Step changes in coolant flowrate, $q_c$

The results show that only the network using weighted second-order state derivatives, (12), reflects the dynamics of the CSTR model. All the other models produce inaccurate dynamics. For example, consider the operating space about the product concentration *0.095* mol/l. Here, the dynamics of the plant are slightly underdamped. The output of the network in Fig. 6(d) reflects this reasonably accurately whereas all the other networks produce significantly more underdamped responses. Note that the network using weighted LM state derivatives, (11), produces a different output when local feedback is used instead of global feedback. However neither reflect the correct dynamics of the CSTR model.

The difference between the various network representations become more apparent when equations (10)-(12) are rewritten in terms of the global second-order state derivative. Thus, equation (10) can be rearranged as follows:

$$\bar{x} = \sum_i \rho_i \bar{x}_i \Rightarrow \ddot{\bar{x}} = \frac{\partial^2\left(\sum_i \rho_i \bar{x}_i\right)}{\partial t^2}$$
$$\Rightarrow \ddot{\bar{x}} = \sum_i \rho_i \ddot{\bar{x}}_i + 2\sum_i \dot{\rho}_i \dot{\bar{x}}_i + \sum_i \ddot{\rho}_i \bar{x}_i \qquad (14)$$

Similarly equation (11) is rewritten as:

$$\ddot{\bar{x}} = \sum_i \rho_i \ddot{\bar{x}}_i + \sum_i \dot{\rho}_i \dot{\bar{x}}_i \qquad (15)$$

and equation (12) becomes:

$$\ddot{\bar{x}} = \sum_i \rho_i \ddot{\bar{x}}_i \qquad (16)$$

When the weighting function, $\rho_i$, is a constant value, $\dot{\rho}_i = 0$ and all three implementations give the same outputs assuming appropriate initial conditions are used as well as global feedback employed. However when local models are blended in LM networks the weighting functions can vary with time and thus $\dot{\rho}_i \neq 0$. In this case the above implementations will produce significantly different results.

The modelling and control of a nonlinear plant requires accurate representation of the plant dynamics. Local models linearised at particular operating points will accurately model the dynamics of the plant around those points. Intuitively, the dynamics between operating points should simply be a weighting of the local model dynamics. In order to achieve this, velocity-based local models were required. The second-order state derivative, $\ddot{x}$, reflects the dynamics of the velocity-based local model. Hence the global dynamics of the velocity-based LM network should be a blended combination of $\ddot{x}$. The LM network representation given by (12) clearly achieves this desired blending. All the other representations contain unnecessary additional terms that relate to the rate of change of the weighting functions. If the rate of change is slow then these terms can be neglected. When the rate of change is not slow these extra terms can contribute significantly to the network output thus detracting from the desired response.

## 6. CONCLUDING DISCUSSION

The choice of local models and the way in which the models are blended are very important in the local modelling process. In order to achieve accurate plant dynamics using the local modelling approach the LM network should consist of velocity-based local models whose local dynamics are blended together to give the overall global model dynamics.

The actual weighting functions also have an important part to play in the local modelling process. Here, the commonly-chosen normalised Gaussian basis functions were employed. However, alternative weighting functions are possible. Do Gaussians provide the best choice for blending? Future work involves investigating this question in more detail.

## REFERENCES


Brown, M. D., Lightbody, G. & Irwin, G. W. (1997). Local model networks for non-linear internal model control. IEE Proc.-D, 144(6), 505-514.

Johansen, T. A. & Foss, B. A. (1992). A NARMAX model representation for adaptive control based on local models. Modelling, Identification and Control 13, 25-39.

Johansen, T. A. & Foss, B. A. (1993). Constructing NARMAX models using ARMAX models. Int. J. Control 58, 1125-1153.

Leith, D. J. & Leithead, W. E. (1999). Analytic framework for blended multiple model systems using linear local models. International Journal of Control, 72, (7/8), 605-619.

McLoone, S. C. (2000). Nonlinear identification using local model networks. Ph.D. Thesis, The Queen's University of Belfast, N. Ireland.

McLoone, S. C. & Irwin, G. W. (1998). Process dynamic modelling using continuous-time local model networks. Proc. 5th IFAC Workshop on Algorithms and Architectures for real-time control, Mexico, 179-184.

McLoone, S. C. & Irwin, G. W. (2001). On nonlinear modelling using velocity-based multiple model networks. ACC01, Arlington, USA, June 25-27, 2001.

Murray-Smith, R. (1994). A local model network approach to nonlinear modelling. Ph.D. Thesis, University of Strathclyde, UK.

Murray-Smith, R., Johansen, T. A. & Shorten, R. (1999). On transient dynamics, off-equilibrium behaviour and identification in blended multiple model structures. ECC'99, Karlsruhe, Germany, F1020-7.

Shorten, R., Murray-Smith, R., Bjorgan, R. & Gollee, H. (1999). On the interpretation of local models in blended multiple model structures. International Journal of Control, 72, (7/8), 620-628.

Townsend, S. & Irwin, G. W. (1999). Local model networks for nonlinear predictive control. EUFIT'99, 7th European Congress on Intelligent Techniques and Soft Computing, Aachen, Germany, Sept. 13-16.


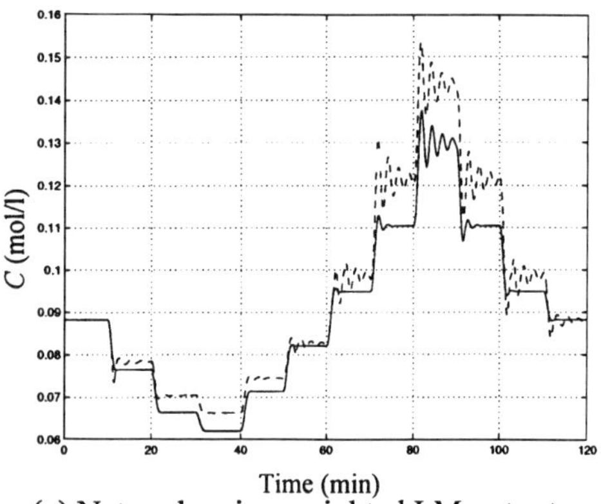

(a) Network using weighted LM outputs

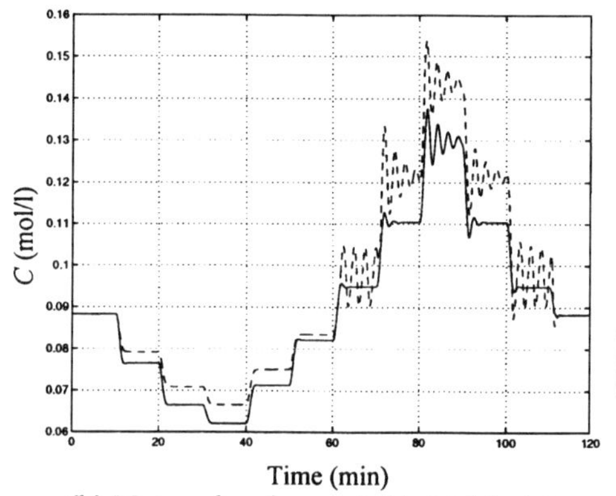

(b) Network using weighted LM state derivatives with global feedback

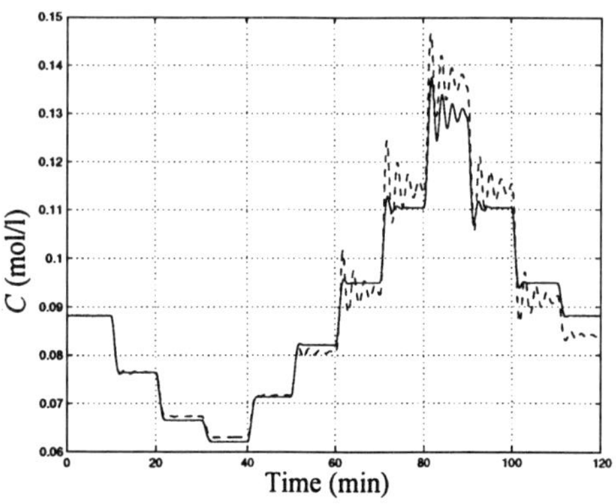

(c) Network using weighted LM state derivatives with local feedback

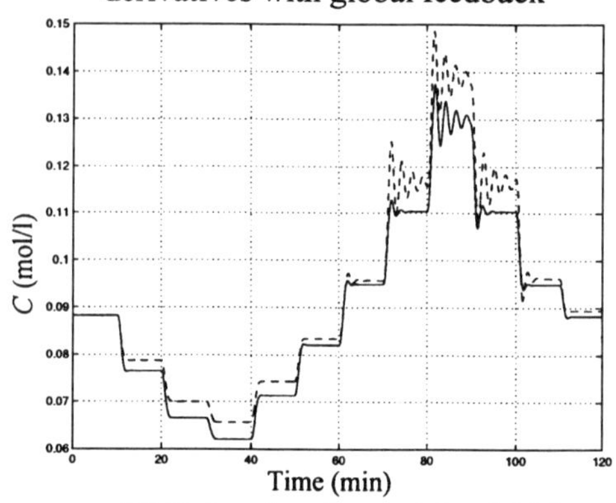

(d) Network using weighted LM second-order state derivatives

Fig. 6 Output product concentration, $C$ – in all cases, the solid line represents the actual CSTR plant output, while the dashed line represents the output of the respective network representations

IFAC
Publications
www.elsevier.com/locate/ifac

# $H_2$ OPTIMAL COMPUTER CONTROL: POLYNOMIAL TOOLBOX [1]

**Vladimír Kučera**

*Faculty of Electrical Engineering*
*Czech Technical University in Prague*
*Technická 2, 166 27 Praha 6, Czech Republic*
*FAX: + 420 2 2431 0784*
*E-mail:* kucera@fel.cvut.cz
*Suggested Track 4*

Abstract: A new algorithm for the synthesis of $H_2$–optimal computer control systems is presented. An alternative to state space methods, it is based on transfer function descriptions in terms of polynomial matrix fractions. The optimal controller is obtained by solving one polynomial matrix equation, two spectral factorizations, and three matrix fraction conversions. The algorithm can easily be implemented using Polynomial Toolbox for Matlab. *Copyright © 2001 IFAC*

Keywords: Methodology, Process optimization, Polynomial Toolbox, Transfer functions, Pole placement.

## 1. INTRODUCTION

The standard $H_2$–optimal control problem is a cornerstone of linear control theory, see (Kučera, V., 1996). It consists of stabilizing a linear system in such a way that its transfer function attains a minimum norm in the Hardy space $H_2$.

Consider the feedback configuration shown Fig. 1,

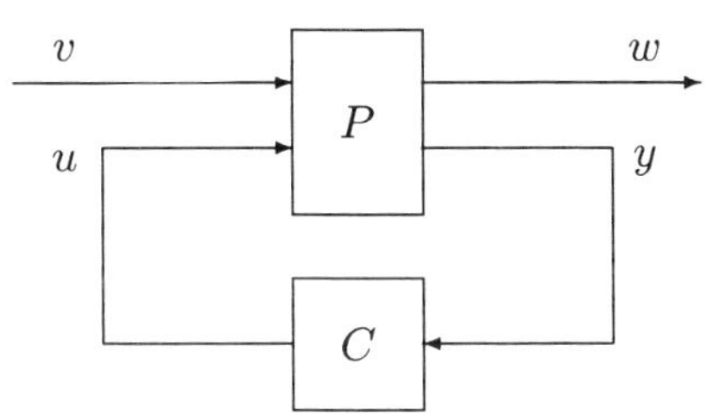


Fig. 1. Control system

where $P$ is the process and $C$ is a controller.

The process is modeled by the recurrent equations

$$\begin{aligned} x_{P\,k+1} &= Fx_{Pk} + Gu_k + G_1 v_k \\ y_k &= Hx_{Pk} + Ju_k + J_1 v_k \\ w_k &= H_2 x_{Pk} + J_2 u_k \end{aligned}$$

for $k = 0, 1, 2, ...$ where $x_P$ is the state vector of the process, $u$ is the control vector, $y$ is the measurement vector, $v$ is the vector of external signals (references and/or disturbances) and $w$ is the vector of product variables. The samples of the system variables are so labeled that $u_k$ precedes $y_k$.

The process model gives rise to the transfer function matrix

$$\begin{bmatrix} P_{11} & P_{12} \\ P_{21} & P_{22} \end{bmatrix}(z) = \begin{bmatrix} H \\ H_2 \end{bmatrix}(zI - F)^{-1}\begin{bmatrix} G & G_1 \end{bmatrix} + \\ + \begin{bmatrix} J & J_1 \\ J_2 & 0 \end{bmatrix}.$$

[1] Supported by the Ministry of Education of the Czech Republic under Research Project MSM212300013.

Denoting $C(z)$ the transfer function matrix of the controller, the control system transfer function matrix between $v$ and $w$ becomes

$$\begin{aligned} T(z) &= P_{22}(z) + P_{21}(z)\,[I - \\ &\qquad - C(z)P_{11}(z)\,]^{-1}\,C(z)\,P_{12}(z) \\ &= P_{22}(z) + P_{21}(z)C(z)\,[I - \\ &\qquad - P_{11}(z)\,C(z)\,]^{-1}\,P_{12}(z). \end{aligned}$$

Since $y_k$ is not available when $u_k$ is being calculated, the admissible controllers have a *strictly proper* rational transfer function matrix $C(z)$.

The control system is said to be stable if the states of $P$ and $C$ go to zero from all initial values when $v = 0$.

The standard $H_2$–optimal control problem is: Given a process $P$, find an admissible controller $C$ that stabilizes the control system and minimizes the $H_2$ norm of $T$, defined as

$$||T|| = \left(\frac{1}{2\pi j}\int \operatorname{tr} T'(z^{-1})T(z)\,z^{-1}dz\right)^{\frac{1}{2}}$$

where tr stands for the trace, $T'$ denotes the transpose of $T$, and the integral is a contour integral around the unit circle of the complex $z$–plane in the anti-clockwise direction.

## 2. STATE SPACE SOLUTION

The following assumptions on the plant are standard, though not necessary.

$(A.1)$ $(F, G)$ stabilizable,
$(A.2)$ $(F, G_1, H, J_1)$ strongly stabilizable and right invertible
$(A.3)$ $(F, H)$ detectable
$(A.4)$ $(F, G, H_2, J_2)$ strongly detectable and left invertible

Under these assumptions, (Kučera, V., 1991, Theorem 5.5) showed that the unique optimal controller for the $H_2$ control problem is

$$C(z) = -L(zI - F + KH + GL - JLK)^{-1}K$$

where

$$\begin{aligned} K &= (FQH' + G_1J_1')(HQH' + J_1J_1')^{-1} \\ L &= (J_2'J_2 + G'PG)^{-1}(J_2'H_2 + G'PF) \end{aligned}$$

and $P, Q$ are the unique, symmetric and non–negative definite solutions of the algebraic Riccati equations

$$\begin{aligned} P &= F'PF + H_2'H_2 - (F'PG + H_2'J_2)(J_2'J_2 + \\ &\qquad + G'PG)^{-1}(J_2'H_2 + G'PF) \\ Q &= FQF' + G_1G_1' - (FQH' + G_1J_1')(HQH' + \\ &\qquad + J_1J_1')^{-1}(J_1G_1' + HQF') \end{aligned}$$

Moreover, the matrices $F - GL$ and $F - KH$ have all eigenvalues in modulus less than 1.

The controller equations can be written in observer form as

$$\begin{aligned} x_{C\,k+1} &= Fx_{Ck} + K(y_k - Hx_{Ck}) + (G - KJ)u_k \\ u_k &= -Lx_{Ck} \end{aligned}$$

where $x_C$ is the state vector of the controller, giving rise to the control system overall equations

$$\begin{bmatrix} x_{P\,k+1} \\ x_{P\,k+1} - x_{C\,k+1} \end{bmatrix} = \begin{bmatrix} F - GL & GL \\ 0 & F - KH \end{bmatrix} \times$$

$$\times \begin{bmatrix} x_{P\,k} \\ x_{P\,k} - x_{Ck} \end{bmatrix} + \begin{bmatrix} G_1 \\ G_1 - KJ_1 \end{bmatrix} v_k$$

$$z = [H_2 - J_2L \quad J_2L] \begin{bmatrix} x_{P\,k} \\ x_{P\,k} - x_{C\,k} \end{bmatrix}.$$

This reveals the separation property of the $H_2$–optimal controller: the optimal dynamics are given by $F - GL$ and $F - KH$ where $K$ and $L$ are determined independently of each other.

## 3. POLYNOMIAL CONSTRUCTION OF $K$ AND $L$

The state feedback gain $L$ and the output injection gain $K$ can be calculated by polynomial matrix operations in place of solving the algebraic Riccati equations.

Write

$$H(zI - F)^{-1} = A_L^{-1}(z)B_{Ls}(z)$$

where $A_L, B_{Ls}$ are left coprime polynomial matrices such that $A_L$ is row reduced. Further write

$$(zI - F)^{-1}G = B_{Rs}(z)A_R^{-1}(z)$$

where $A_R, B_{Rs}$ are right coprime polynomial matrices such that $A_R$ is column reduced.

Let $S_L$ be a square polynomial matrix, whose row degrees are equal to those of $A_L$ and whose determinant is a Schur–stable polynomial (all roots in modulus less than 1), such that

$$\begin{aligned} S_L(z)S_L'(z^{-1}) &= [B_{Ls}(z)G_1 + \\ &\quad + A_L(z)J_1]\,[B_{Ls}(z^{-1})G_1 + A_L(z^{-1})J_1]' \end{aligned}$$

and let $S_R$ be a square polynomial matrix, whose column degrees are equal to those of $A_R$ and whose determinat is a Schur–stable polynomial (all roots in modulus less than 1), such that

$$\begin{aligned} S_R'(z^{-1})S_R(z) &= [H_2B_{Rs}(z^{-1}) + \\ &\quad + J_2A_R(z)]'\,[H_2B_{Rs}(z) + J_2A_R(z)]. \end{aligned}$$

The matrices $S_L$ and $S_R$ are called spectral factors and are uniquely determined up to right and left orthogonal factors, respectively.

Stronger assumptions are needed to proceed: (A.1) is replaced by

$$(A.1)'(F, G) \text{ is reachable}$$

and (A.3) is replaced by

$$(A.3)'(F, H) \text{ is observable.}$$

Under these stronger assumptions, (Kučera, V., 1981) showed that

$$K = \hat{Y}_R \hat{X}_R^{-1}$$

where $\hat{X}_R, \hat{Y}_R$ is the unique constant solution pair of the matrix polynomial equation

$$A_L(z)\hat{X}_R + B_{Ls}(z)\hat{Y}_R = S_L(z)$$

and

$$L = \hat{X}_L^{-1}\hat{Y}_L$$

where $\hat{X}_L, \hat{Y}_L$ is the unique constant solution pair of the matrix polynomial equation

$$\hat{X}_L A_R(z) + \hat{Y}_L B_{Rs}(z) = S_R(z).$$

We note that if $(A.1)'$ and/or $(A.3)'$ fails to hold, the polynomial matrix equations may have more than one constant solution pair. Hence $K$ and/or $L$ cannot be determined.

## 4. TRANSFER FUNCTION SOLUTION

Consider an $H_2$–optimal controller in standard observer form with some gains $K$ and $L$. The transfer function matrix of the controller will now be constructed directly using the transfer function matrix of the process, namely $P_{11}(z) = H(sI - F)^{-1}G + J$.

Write

$$P_{11}(z) = A_L^{-1}(z)B_L(z) = B_R(z)A_R^{-1}(z)$$

where

$$B_L(z) = B_{Ls}(z)G_2, \quad B_R(z) = H_2 B_{Rs}(z).$$

Note that $(A.1)'$ and $(A.3)'$ imply left coprimeness of $A_L, B_L$ and right coprimeness of $A_R, B_R$. Define the polynomial matrices

$$S_L(z) = A_L(z) + B_{Ls}(z)K$$
$$S_R(z) = A_R(z) + LB_{Rs}(z).$$

Following (Kučera, V., 1991, Lemmas 5.1 and 5.2) let $X_R(z), Y_R(z)$ define any polynomial matrix solution of the equation

$$A_L(z)X_R(z) + B_L(z)Y_R(z) = S_L(z)$$

and

$$W_R(z) = \Pi(S_R A_R^{-1} Y_R)(z)$$

where $\Pi(\cdot)$ denotes taking the polynomial part. Then the controller transfer matrix is provided by

$$L(zI - F + KH + GL - KJL)^{-1}K = \bar{Y}_R(z)\bar{X}_R^{-1}(z)$$

where

$$\bar{X}_R(z) = X_R(z)\bar{S}_R(z) + B_R(z)\bar{W}_R(z)$$
$$\bar{Y}_R(z) = Y_R(z)\bar{S}_R(z) - A_R(z)\bar{W}_R(z)$$

and $\bar{S}_R, \bar{W}_R$ is a right coprime polynomial matrix pair for which

$$S_R^{-1}(z)W_R(z) = \bar{W}_R(z)\bar{S}_R^{-1}(z).$$

A dual construction is possible. Let $X_L(z), Y_L(z)$ be any polynomial matrix solution of

$$X_L(z)A_R(z) + Y_L(z)B_R(z) = S_R(z)$$

and

$$W_L(z) = \Pi(Y_L A_L^{-1} S_L)(z).$$

Then the controller transfer matrix is also given by

$$L(zI - F + KH + GL - KJL)^{-1}K = \bar{X}_L^{-1}(z)\bar{Y}_L(z)$$

where

$$\bar{X}_L(z) = \bar{S}_L(z)X_L(z) + \bar{W}_L(z)B_L(z)$$
$$\bar{Y}_L(z) = \bar{S}_L(z)Y_L(z) - \bar{W}_L(z)A_L(z)$$

and $\bar{S}_L, \bar{W}_L$ is a left coprime polynomial matrix pair for which

$$W_L(z)S_L^{-1}(z) = \bar{S}_L^{-1}(z)\bar{W}_L(z).$$

Observe that

$$A_L(z)\bar{X}_R(z) + B_L(z)\bar{Y}_R(z) = S_L(z)\bar{S}_R(z)$$

and

$$\bar{X}_L(z)A_R(z) + \bar{Y}_L(z)B_R(z) = \bar{S}_L(z)S_R(z).$$

The above two equations suggest a pole placement interpretation of the state space solution for the $H_2$ control problem. In fact, the closed loop system dynamics are exhibited in terms of the roots of det $S_L$ (= roots of det $\bar{S}_L$), which are the eigenvalues of $F - KH$, and the roots of det $S_R$ (= roots of det $\bar{S}_R$), which are the eigenvalues of $F - GL$. The role of $W_R$ is to modify $S_R$ for use in the primal equation, whereas $W_L$ matches $S_L$ for use in the dual equation. In additon, the extraction of the polynomial parts $W_L$ and $W_R$ guarantees that the optimal controller has a strictly proper transfer function matrix and, consequently, is admissible.

## 5. ALGORITHMS

Two design algorithms, dual to each other, result from the interpretation of the $H_2$–optimal discrete time control as a pole placement problem in which the dynamics induced by optimization are provided by the two spectral factors $S_L$ and $S_R$. Similar results have been reported in (Kučera, V. and D. Henrion, 2000) for the $H_2$–optimal control in continuous time.

The design algorithms can be summarized as follows.

**Algorithm 1**

Data: Process $P$, either $F, G, H, J, G_1, H_2, J_1, J_2$ or $P_{11}, P_{12}, P_{21}, P_{22}$.

Assumptions: $(A.1)', (A.2), (A.3)'$, and $(A.4)$.
Result: $H_2$–optimal controller $C$.

*Step 1*: Determine left coprime polynomial matrices $A_L, B_L$ with $A_L$ row reduced, such that

$$P_{11}(z) = H(zI - F)^{-1}G + J = A_L^{-1}(z)B_L(z)$$

and put

$$\begin{aligned} N_L(z) &= A_L(z)P_{12}(z) \\ &= A_L(z)[H(zI - F)^{-1}G_1 + J_1]. \end{aligned}$$

Determine right coprime polynomial matrices $A_R, B_R$ with $A_R$ column reduced, such that

$$P_{11}(z) = H(zI - F)^{-1}G + J = B_R(s)A_R^{-1}(s)$$

and put

$$\begin{aligned} N_R(z) &= P_{21}(z)A_R(z) \\ &= [H_2(zI - F)^{-1}G + J_2]\, A_L(z). \end{aligned}$$

*Step 2*: Compute square polynomial matrices $S_L, S_R$ whose row and column degrees are respectively equal to those of $A_L, A_R$ and whose determinants are Schur–stable polynomials, such that

$$S_L(z)S_L'(z^{-1}) = N_L(z)N_L'(z^{-1})$$

and

$$S_R'(z^{-1})S_R(z) = N_R'(z^{-1})N_R(z).$$

*Step 3*: Solve the equation

$$A_L(z)X_R(z) + B_L(z)Y_R(z) = S_L(z)$$

for a polynomial matrix pair $X_R, Y_R$.
*Step 4*: Calculate $W_R$, the polynomial part of $S_R A_R^{-1} Y_R$, and determine right coprime polynomial matrices $\bar{S}_R, \bar{W}_R$ for which

$$S_R^{-1}(z)W_R(z) = \bar{W}_R(z)\bar{S}_R^{-1}(z).$$

*Step 5*: Obtain the matrices

$$\begin{aligned} \bar{X}_R(z) &= X_R(z)\bar{S}_R(z) + B_R(z)\bar{W}_R(z) \\ \bar{Y}_R(z) &= Y_R(z)\bar{S}_R(z) - A_R(z)\bar{W}_R(z). \end{aligned}$$

Then

$$C(z) = -\bar{Y}_R(z)\bar{X}_R^{-1}(z).$$

**Algorithm 2**

Data, assumptions, result: same as above.

*Step 1*: same.
*Step 2*: same.
*Step 3*: Solve the equation

$$X_L(z)A_R(z) + Y_L(z)B_R(z) = S_R(z)$$

for a polynomial matrix pair $X_L, Y_L$.
*Step 4*: Calculate $W_L$, the polynomial part of $Y_L A_L^{-1} S_L$, and determine right coprime polynomial matrices $\bar{S}_L, \bar{W}_L$ for which

$$W_L(z)S_L^{-1}(z) = \bar{S}_L^{-1}(z)\bar{W}_L(z).$$

*Step 5*: Obtain the matrices

$$\begin{aligned} \bar{X}_L(z) &= \bar{S}_L(z)X_L(z) + \bar{W}_L(z)B_L(z) \\ \bar{Y}_L(z) &= \bar{S}_L(z)Y_L(z) - \bar{W}_L(z)A_L(z). \end{aligned}$$

Then

$$C(z) = -\bar{X}_L^{-1}(z)\bar{Y}_L(z).$$

These two dual algorithms merge into one simple algorithm for processes having a *scalar* control $u$ and a *scalar* measurement $y$.

**Algorithm 3**

Data: Process $P$, same as above.
Assumptions: same as above, plus $u$ and $y$ scalar.
Result: $H_2$–optimal controller $C$.
*Step 1*: Determine coprime polynomials $A, B$ such that

$$P_{11}(z) = H(zI - F)^{-1}G + J = \frac{B(s)}{A(s)}.$$

Put

$$\begin{aligned} N_L(s) &= A(s)P_{12}(s) \\ N_R(s) &= P_{21}(s)A(s). \end{aligned}$$

*Step 2*: Compute polynomials $S_L, S_R$ whose degrees are equal to that of $A$ and whose determinants are Schur–stable polynomials, such that

$$\begin{aligned} S_L(z)S_L(z^{-1}) &= N_L(z)N_L'(z^{-1}) \\ S_R(z^{-1})S_R(z) &= N_R'(z^{-1})N_R(z). \end{aligned}$$

*Step 3*: Solve the equation

$$A(z)\bar{X}(z) + B(z)\bar{Y}(z) = S_L(z)S_R(z)$$

for a polynomial matrix pair $\bar{X}_L, \bar{Y}_L$ such that $\deg \bar{Y} < \deg A$. Then

$$C(z) = -\frac{\bar{Y}(z)}{\bar{X}(z)}.$$

Comparing with Algorithms 1 and 2 one finds that

$$\begin{aligned} \frac{\bar{Y}(z)}{A(z)} &= \frac{Y_L(z)S_L(z)}{A(z)} - W_L(s) \\ &= \frac{S_R(z)Y_R(z)}{A(z)} - W_R(s) \end{aligned}$$

is strictly proper by the definitions of $W_L$ and $W_R$. This justifies Step 3 of Algorithm 3.

## 6. POLYNOMIAL TOOLBOX

The algorithms presented rely on polynomial matrix operations only: one polynomial matrix equation, two matrix spectral factorizations, and three matrix fraction conversions to be solved. All these operations can be easily and safely implemented using the Polynomial Toolbox for Matlab, see (Polyx, Ltd., 1998).

The Polynomial Toolbox is a package for system, signals, and control analysis and design based on advanced polynomial methods. It consists of more than 200 m–files in Matlab code, which implement new original algorithms that are fast and reliable. The key features include a simple input, manipulation and display of polynomial matrices; overloaded operations and functions; solvers for numerous linear and quadratic matrix polynomial equations.

For example, Algorithm 1 can be implemented using the following m–files.

*Step 1*: `ss2lmf, ss2rmf`
*Step 2*: `spf, spcof`
*Step 3*: `axbyc`
*Step 4*: `lmf2rmf, rdiv`
*Step 5*: `*, +`

Numerous convertors enable direct cooperation with the Control System Toolbox and the Symbolic Math Toolbox. A Simulink block set for systems described by polynomial matrix fractions is also provided.

## 7. CONCLUSIONS

An alternative algorithm has been described for computing the $H_2$–optimal, discrete–time controller for a process described either by state space equations or transfer function matrices.

In contrast to the standard polynomial approach to $H_2$-optimal design, which is based on two coupled two-sided polynomial matrix equations (Kučera, V., 1980) the algorithm proposed here relies only on a single one-sided polynomial matrix equation of lower dimension. Thorough numerical experiments must however be carried out to check whether significant savings are achieved in computational time and memory requirement.

Another point that deserves further attention is an extensive numerical comparison between the polynomial approach proposed in this paper and the state-space approach (Kwakernaak, H. and R. Sivan, 1972), (Doyle, J. C., K. Glover, P. P. Khargonekar, and B. A. Francis, 1989).

## 8. REFERENCES


Doyle, J. C., K. Glover, P. P. Khargonekar, and B. A. Francis (1989). State space solutions to standard $H_2$ and $H_\infty$ control problems. IEEE Trans. Automatic Control, 34, 831–847.

Kučera, V. (1980). Stochastic multivariable control: a polynomial equation approach. IEEE Trans. Automatic Control, 25, 913-919.

Kučera, V. (1981). New results in state estimation and regulation. Automatica, 17, 745–748.

Kučera, V. (1991). Analysis and Design of Discrete Linear Control Systems. Prentice–Hall, London.

Kučera, V. (1996). A tutorial on $H_2$ control theory: The continuous time case. Polynomial Methods for Control Sytems Design (M. J. Grimble and V. Kučera, eds.) Springer, London, 1–55.

Kučera, V. and D. Henrion (2000). $H_2$ optimal control via pole placement. Preprints 3rd IFAC Symposium on Robust Control Design, Prague, Czech Republic.

Kwakernaak, H. and R. Sivan (1972). Linear Optimal Control Systems. Wiley, New York.

Polyx, Ltd. (1998). The POLYNOMIAL TOOLBOX 2.0 for use with MATLAB 5. See the web page `www.polyx.cz` or `www.polyx.com`.

IFAC Publications
www.elsevier.com/locate/ifac

# THE RESPONSE SURFACE METHODOLOGY FOR REAL-TIME DISTRIBUTED SIMULATION.

**Paul Stewart * Peter J. Fleming ****

* *Department of Automatic Control and Systems Engineering, University of Sheffield. Mappin St. Sheffield U.K. e-mail:p.stewart@shef.ac.uk tel: +44(0)114 2225670. fax: +44(0)114 2225138.*
** *Department of Automatic Control and Systems Engineering, University of Sheffield. Mappin St. Sheffield U.K.*

Abstract: Distributed real-time simulation has recently been the focus of intense development, with complex systems being represented by individual component simulations interacting as a coherent whole. The architecture is often required to run in real-time, and may be composed of physically separated simulation centres. Commercial off-the-shelf (COTS) real-time architectures exist to provide adequate bandwidth and data communication channels between the components. However if the individual models are too computationally intensive to run in real time, then the performance of the real-time architecture is compromised. In this paper, model representations are developed by the response surface methodology allowing complex systems to be included in a real-time environment. A Permanent Magnet AC (PMAC) motor drive simulation with model reference control is examined as a candidate for inclusion in a realtime simulation environment. 

Keywords: Response Surface, Real-Time, Distributed Simulation.

## 1. INTRODUCTION

Extremely complex systems can be represented by distributed simulations, leading to development of sophisticated architectures to administer the data transactions necessary for the components to operate as a unified whole. CORBA, a distributed architecture has acquired real-time extensions, while the High Level Architecture (HLA) has been developed in the United States by the Department of Defense in the first instance to provide a real-time War Gaming environment. Individual research and development centres develop models which reflect their expertise, and can subsequently join a distributed environment to construct a sophisticated representation with a great deal of detail, accuracy and model flexibility at the component level. In this manner, environments such as real-time flight simulators are constructed with levels of detail focussing hierarchically down from the airframe flight characteristics, through thermal gas turbine engine models, to component levels of fuel pumps, generators etc. Distributed components of the system are on the whole modelled in high level environments such as *Matlab / Simulink* or $Matrix_x$ / *Systembuild*. Routines may also be coded in languages such as *Fortran* or *C/C++*. The distributed architecture fulfills the task of registering the data types and protocols of these modelling environments, and providing channels for data streaming via a real-time kernel. The implementation of these architectures is beyond the scope of this document, however certain salient features must be noted before proceeding to an examination of the individual simulation components. Firstly, these architectures rely on dedicated local (LAN) or wide (WAN) area networks to provide necessary physical levels of bandwidth for the data transfer. Secondly, benchmark tests for the distributed architectures can calculate bandwidth available with any projected system. Thirdly, any individual simulation which runs slower than real-time will prevent the system from running in real-time if its operation is critical to the operation of the overall system. This last point is the most crucial, and will be considered in this paper. Many complex systems, particularly nonlinear multivariable ones require in general relatively intense computation in order to provide an accurate dynamic simulation. This problem is exacerbated by the use of high-level simulation languages to provide the model environment. There exist some potential solutions to increase the speed of the model in order to bring it up to real-time capability. Firstly, the model can be further decomposed into sub-systems to run on separate machines. This solution has a penalty in terms of both physical space, cost, and loading on the communications channels. Secondly, the item can be translated into highly optimised high level computer language code. This has the certain advantage of more rapid execution time, however the advantages of direct interaction with graphical development environments is lost. More specialised programming skills are required, and the code is not guaranteed to give the speed increase necessary. Thirdly, the model may be approximated as a lower order system. This solution is certainly a candidate solution, but requires a methodology to produce an accurate workable solution for highly non-linear systems. Finally, experimental or simulation data may be used to construct a minimal dynamic model of the system, in effect a technique analogous to image compression, with the aim of constructing a dynamic representation accurate to known statistical bounds, with real-time capability. This final approach will now be developed, as it has the advantages of a generic approach, without the disadvantages of increased hardware costs or increased system bandwidth requirement, while retaining the advantages of graphical development environments. The compression tool will be based around the response surface methodology [1], and its effectiveness in application will be examined against both experimental and simulation data for highly nonlinear electric vehicle traction drive model [2] [3]. Aside from its original application in chemical and process control, RSM has been used to realise low-cost computational solutions using Multi Objective Genetic Algorithms (MOGA) [4] [5]. The RSM in this case is used to provide objective function approximations rather than evaluating these functions directly from the gas turbine engine during an optimisation with MOGA. In this context, response surfaces are used to model the engine performance indices and is applied to the multivariable fuzzy controller design for the Rolls- Royce Spey engine. It was found that the use of the RSM greatly reduced the computational load experienced in the controller design. The work presented in this paper is the first use of the RSM in the context of system representation for dynamic distributed systems. It is found to be an extremely useful and tractable tool to enable relatively slow processes to join real-time distributed simulations.

## 2. RESPONSE SURFACE METHODOLOGY

The response surface methodology is a technique designed to optimise process control by the application of designed experiments in order to characterise a system [1]. The relationship between the response variable of interest $(y)$, and the predictor variables $(\xi_1, \xi_2, ..., \xi_k)$ may be known exactly allowing a description of the system of the form

$$y = g(\xi_1, \xi_2, ..., \xi_k) + \epsilon \qquad (1)$$

where $\epsilon$ represents the model error, and includes measurement error, and other variability such as background noise. The error will be assumed to have a normal distributeion with zero mean and variance $\sigma^2$ In general, the experimenter approximates the system function $g$ with an empirical model of the form

$$y = f(\xi_1, \xi_2, ..., \xi_k) + \epsilon \qquad (2)$$

where $f$ is a first or second order polynomial. This is the empirical or response surface model. The variables are known as *natural variables* since they are expressed in physical units of measurement. In the response surface methodology (RSM), the natural variables are transformed into *coded variables* $x_1, x_2, ...x_k$ which are dimensionless, zero mean, and the same standard deviation. The response function now becomes

$$\eta = f(x_1, x_2, ..., x_k) \qquad (3)$$

The successful application of RSM relies on the identification of a suitable approximation for $f$. this will generally be a first order model of the form

$$\eta = \beta_0 + \beta_1 x_1 + \beta_2 x_2 + ... + \beta_k x_k \qquad (4)$$

or a second order model of the form

$$\eta = \beta_0 + \sum_{j=1}^{k} \beta_j x_j + \sum_{j=1}^{k} \beta_{jj} x_j^2 + \sum_{i<j}\sum \beta_{ij} x_i x_j \qquad (5)$$

It may be neccessary to employ an approximating function greater than an order of two, based on the standard Taylor series expansion. The response surface methodology is intimately connected to *regression analysis*. For example when considering the first order model, the $\beta$ terms comprise the unknown parameter set which can be estimated by collecting experimental system data. This data can either be sourced from physical experiments, or from previously designed dynamic computer models. The parameter set can be estimated by regression analysis based upon the experimental data. The method of least squares is typically used to estimate the regression coefficients. With $n > k$ on the response variable available, giving $y_1, y_2, ..., y_n$, each observed response will have an observation on each regressor variable, with $x_{ij}$ denoting the $ith$ observation of variable $x_j$. Assuming that the error term $\epsilon$ has $E(\epsilon) = 0$ and $Var(\epsilon) = \sigma^2$ and the $(\epsilon_i)$ are uncorrelated random variables. The model can now be expressed in terms of the observations

$$\begin{aligned} y_i &= \beta_0 + \beta_1 x_{i1} + \beta_2 x_{i2} + ... + \beta_k x_{ik} + \epsilon_i \\ &= \beta_0 + \sum_{j=1}^{k} \beta_j x_{ij} + \epsilon_i, \\ i &= 1, 2, ..., n \end{aligned} \tag{6}$$

The $\beta$ coefficients in equation (6) are chosen such that the sum of the squares of the errors $(\epsilon_i)$ are minimised via the least squares function

$$\begin{aligned} L &= \sum_{i=1}^{n} \epsilon_i^2 \\ &= \sum_{i=1}^{n} \left( y_i - \beta_0 - \sum_{j=1}^{n} \beta_j x_{ij} \right)^2 \end{aligned} \tag{7}$$

The model can be more usefully expressed in matrix form as

$$\mathbf{y} = \mathbf{X}\beta + \epsilon \tag{8}$$

where

$$\mathbf{y} = \begin{bmatrix} y_1 \\ y_2 \\ \cdot \\ \cdot \\ \cdot \\ y_n \end{bmatrix}, \quad \mathbf{X} = \begin{bmatrix} 1 & x_{11} & x_{12} & \dots & x_{1k} \\ 1 & x_{21} & x_{22} & \dots & x_{2k} \\ \vdots & \vdots & \vdots & \vdots & \vdots \\ 1 & x_{n1} & x_{n2} & \dots & x_{nk} \end{bmatrix},$$

$$\beta = \begin{bmatrix} \beta_0 \\ \beta_1 \\ \cdot \\ \cdot \\ \cdot \\ \beta_n \end{bmatrix}, \quad \epsilon = \begin{bmatrix} \epsilon_1 \\ \epsilon_2 \\ \cdot \\ \cdot \\ \cdot \\ \epsilon_n \end{bmatrix}. \tag{9}$$

It is now necessary to find a vector of least squares estimators $\mathbf{b}$ which minimises the expression

$$L = \sum_{i=1}^{n} \epsilon_i^2 = \epsilon' \epsilon = (\mathbf{y} - \mathbf{X}\beta)' (\mathbf{y} - \mathbf{X}\beta) \tag{10}$$

and yields the least squares estimator of $\beta$ which is

$$\mathbf{b} = \left(\mathbf{X}'\mathbf{X}\right)^{-1} \mathbf{X}'\mathbf{y} \tag{11}$$

and finally, the fitted regression model is

$$\hat{\mathbf{y}} = \mathbf{X}\mathbf{b}, \quad \mathbf{e} = \mathbf{y} - \hat{\mathbf{y}} \tag{12}$$

where $\mathbf{e}$ is the vector of residual errors of the model.

## 3. PERMANENT MAGNET AC MOTOR AND CONTROLLER

A brief description will be presented in this section to give an overview of the complexities of a permanent magnet AC (PMAC) motor and controller, and the kind of computational overhead demanded by (for example) a *Simulink* model. The controller in question is a model reference type, which allows the torque speed envelope of the motor to be greatly increased [2,6]. The general configuration of this class of motor is three phase AC, and a non-linear transform (*Park Transform*) is usually applied to the phase currents and voltages to allow for easier analysis and control system design [7], to give the following description of the motor

$$V_q = r\, i_q + \omega L\, i_d + \omega\, \lambda + L\left(\frac{di_q}{dt}\right) \tag{13}$$

$$V_d = r\, i_d - \omega L\, i_q + L\left(\frac{di_d}{dt}\right) \tag{14}$$

where $V_d$, $V_q$ are the d and q axis voltages, $i_d$, $i_q$ are the d and q axis currents, $r$ is the phase resistance, $\omega$ is the rotor velocity, $L$ the phase inductance and $\lambda$ the back EMF constant in the reference frame as volts/radian/second. The q-axis inductance is equivalent to the armature inductance, and the d-axis inductance is equivalent to the field inductance in a field wound DC machine. In the case of the surface mount PMAC motor, these quantities are equal and are denoted by $L$. The $d$ and $q$ variables are obtained from the three phase quantities via the following definition of the Park transform

$$\begin{vmatrix} V_q \\ V_d \end{vmatrix} = \frac{2}{3} \begin{vmatrix} cos(\theta) & cos(\theta - \frac{2\pi}{3}) & cos(\theta + \frac{2\pi}{3}) \\ sin(\theta) & sin(\theta - \frac{2\pi}{3}) & sin(\theta + \frac{2\pi}{3}) \end{vmatrix} \begin{vmatrix} V_a \\ V_b \\ V_c \end{vmatrix} \tag{15}$$

where $V_{a,b,c}$ are the three phase elements. This transformation also applies to current and flux linkage quantities, and the three phase quantities may be obtained from the $d$ and $q$ axis variables by application of an inversion of the Park matrix in equation (15). High performance phase current regulation is the key to high-performance motion control with the sinusoidal PMAC motor. Fast responding current regulation combined with self synchronisation via the built-in shaft position encoder make it possible to orientate the current phasor $I$ anywhere within the $d$-$q$ reference frame subject to supply current and voltage constraints. The speed of the torque response is limited only by the source voltage and stator inductance values. The baseline approach to torque control is to map the torque command $T_e^*$ into demands for the $d$ and $q$ axis currents. Torque production is purely a linear function of $q$ axis current. These current commands are then transformed into instantaneous sinusoidal currents for the individual stator phases, using the rotor angle feedback and the basic inverse vector rotation equations. PI current regulators for each of the three phases then excite the phase windings resulting in desired current amplitudes. Examination of the motor equations (13,14) reveals cross coupling between the equations which requires correction by a feedback linearising controller to allow accurate control of the current and voltage vectors [8]. The flux weakening controller is of the model reference type, and supplies current commands based upon rotor velocity and torque demand, and is based upon the standard circle diagram representation in figure 1 [9], [10] and shows the important control principles. As speed and frequency increase, the current limit locus remains fixed, however the radius of the voltage limit locus decreases. The PWM control saturates when its duty cycle reaches maximum, and the available sinewave voltage from the inverter equals the phase voltage. This operating point is known as 'base speed' and occurs on the circle diagram at the intersection of the $q$ axis, current limit circle, and voltage limit circle. If the rotor velocity increases further, the radius of the voltage-limit circle decreases further, and maximum current is defined by a current vector terminating in the intersection of the two circles. As rotor velocity increases, the voltage-limit circle drags the current phasor further and further ahead of the $q$ axis, decreasing the torque producing current, and increasing the demagnetising negative $d$ axis current. This acceleration can increase until the point where the current vector lies entirely in the demagnetising direction, and no further torque production is possible. The motor and controller models are coded into Simulink models, together with function blocks to describe real system objects such as the

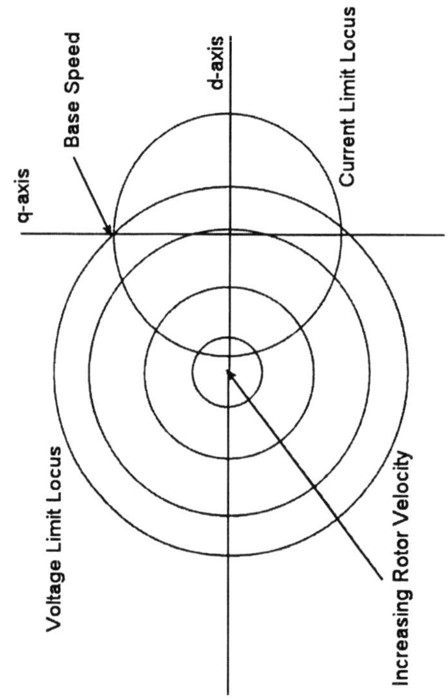


Fig. 1. PMAC motor circle diagram showing constraint locii.

three phase PWM inverter, feedback linearising controller, rotor position encoder, Park transform blocks, and blocks to describe the voltage drop due to the current dynamics [6]. It is desired that the motor and controller models participate with a gearbox/driveshaft/tyre automotive model, and an automotive vehicle body model in a distributed electric vehicle simulation. However when the Simulink model of the model reference control system was run on a Pentium 3 800MHz computer with 512Mb of system memory, it was found not to be capable of real-time operation, running outside wall clock time by a factor of five at the required sampling rate of 1kHz. It was thus decided to implement a minimal description of the controller by RSM in order to achieve real time status for the motor/controller simulation.

## 4. RESPONSE SURFACE OF THE MODEL REFERENCE CONTROLLER

The model reference controller can be converted into a response surface by viewing it as a "black box" input-output model. In this respect, the input variables are rotor velocity and torque demand, while the output variable is the angle of current vector advance. Experimental data from a dynamometer rig was utilised which examines the input-output mapping. The design of experiments to gather neccessary and sufficient information to construct an accurate response surface is ourside the scope of this paper, however the subject warrants research to design representations which are based upon the minimum amount of stored data. In this case an industrial characterisation study was conducted on a dynamometer rig. The method is to map the surface which describes the surface connecting torque, rotor velocity and current vector advance. The mapping is conducted for a quantization of $10^0$ current vector advance, and PWM modulation depth from 0% to 100% in increments of 10%. The

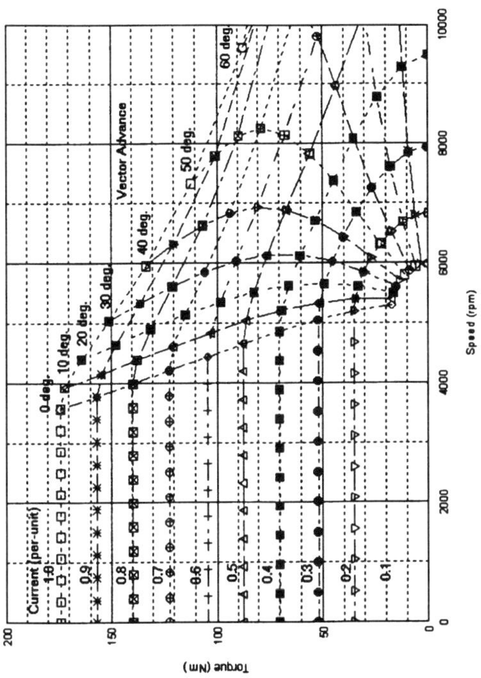


Fig. 2. Torque speed current vector advance experimental data.

PWM modulation depth is given as *per unit* current on the left hand side of figure (2). The RSM is utilised to construct a response surface which reflects the current advance profile in the flux-weakening region of the motor. The natural units $\xi_1$ (instantaneous torque in Nm) and $\xi_2$ (rotor velocity in rads/s) of the experimental dynamometer data is first transformed into the corresponding coded variables $x_1$ and $x_2$, such that

$$x_{i1} = \frac{\xi_{i1} - [max(\xi_{i1}) + min(\xi_{i1})]/2}{[max(\xi_{i1}) - min(\xi_{i1})]/2} \quad (16)$$

and

$$x_{i2} = \frac{\xi_{i2} - [max(\xi_{i2}) + min(\xi_{i2})]/2}{[max(\xi_{i2}) - min(\xi_{i2})]/2} \quad (17)$$

The second order model to be fitted to the data is

$$y = \beta_0 + \beta_1 x_1 + \beta_2 x_2 + \beta_{11} x_1^2 + \beta_{22} x_2^2 + \beta_{12} x_1 x_2 + \epsilon \quad (18)$$

. Utilising equations (10-13), we obtain the coefficient matrix

$$b = \begin{bmatrix} 38.19 \\ 25.46 \\ -7.28 \\ -11.04 \\ 6.23 \\ -5.84 \end{bmatrix} \quad (19)$$

therefore the model of the flux weakening surface is

$$\hat{y} = 38.19 + 25.46x_1 - 7.28x_2 - 11.04x_1^2 + 6.23x_2^2 - 5.85x_1x_2 \quad (20)$$

which may be slightly biased if the noise is coloured. The expression for the response surface model can now be implemented as a controller in *Simulink* and the preformance compared to the original *Simulink* model.

## 5. CONCLUSION

It is now possible to assess the usefulness of the second order model. By connecting the model reference controller and second order model in turn to the existing PMAC motor and drive model, the output can be compared to existing experimental data sets. In this case, the motor is accelerating from rest and running up without external torque loading to a rotor velocity of approximately 7500 rpm. This is a particularly arduous operating envelope, as the motor is being simulated with a coarse position feedback encoder rather than the more usual high-resolution encoder. This is responsible for the high level of ripple in the torque envelope. A visual inspection of the

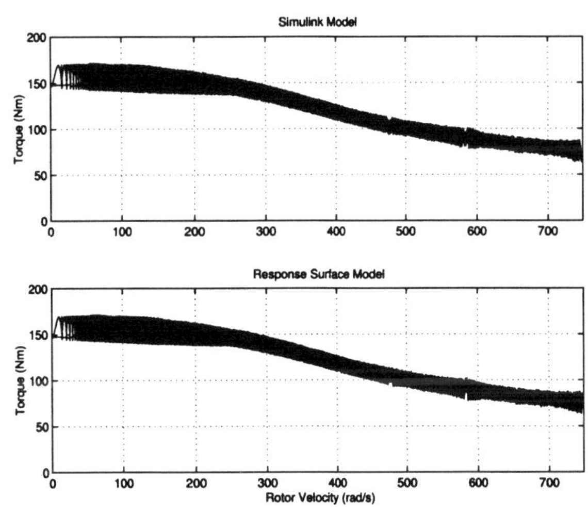


Fig. 3. Comparison of Simulink and RSM controllers in motor acceleration simulation

torque profiles produced by the two controllers

reveals little or no difference in performance of the accelerating motor. The controller command outputs were found to have a mean difference of less than $1^0$, which reflects the mean error in the initial controller design from experimental data, where the mean error of the second order fit was also $< 1^0$. The simulations were repeated for verification over a number of randomly chosen profiles over the entire torque/speed range. The error bounds were found to be $< 1^0$ in all simulations.

The second order controller obtained through the RSM was found to operate at more than ten times wall-clock time, compared to the original *simulink* controller which could not attain real-time operation. The increase in performance (> 50 times) was such that it was possible to run the motor and controller on the same machine in real-time without any modifications to the motor-drive simulation, which is in itself a valuable improvement. The final implementation was to run the RSM designed controller on a separate machine from the motor and drive to confirm its benefits for distributed simulations. The simulation ran via a 100Mb/s LAN via standard network interface cards, and bespoke data buffers and sockets programmed in both C++ and Visual Basic. No buffer under-runs were experienced, and the controller was found to successfully stream data in real-time, operating at 1kHz. A method has been presented which has been adapted from process control and optimization methodologies. The approach allows the adaptation of complex controller and systems which are too slow to join real-time simulations to be approximated in a way such that real-time simulation becomes possible. The method has been demonstrated on a PMAC flux weakening controller, and found to provide not only the performance boost in terms of run-time, but also the accuracy required by the simulation environment. The method is shown to be a useful tool in the development of complex real-time systems. More research opportunities exist for determining the correct amount of experimental samples necessary to provide a good fit for the model, and particularly a pressing need is for the examination of model fit for systems of a more complex or even multidimensional nature. Also, a study might be conducted under conditions of network loading, to verify the real time operation of the simulation.

## 6. REFERENCES


[1] Myers R.H., and Montgomery D.C., (1995) "Response surface methodology: process and product optimization using designed experiments", *John Wiley and Sons Inc. USA*, ISBN 0-471-58100-3.

[2] Stewart P and Kadirkamanathan V., (1999) "Dynamic control of permanent magnet synchronous motors in automotive drive applications", *1999 American Control Conference,* San Diego, USA. pp.1677-1681, June 2-4, 2000.

[3] Stewart P and Kadirkamanathan V., (1998) "On steady state and dynamic performance of model reference control for a permanent magnet synchronous motor". *UKACC International Conference on Control 98.* University of Wales, Swansea UK. vol.455, pp.664-669, 1-4 September, 1998.

[4] Bica B., Akat G., Chipperfield A.J. and Fleming P.J., (1998) "Multiobjective design of a fuzzy controller for a gas turbine aero-engine", *UKACC International Conference on Control '98*, Swansea, UK, no.455, vol.2, pp.901-906, September 1-4, 1998.

[5] Bica B., Ceconvciuc C., Chipperfield A.J. and Fleming P.J., (2000) "Multivariable fuzzy controller design using multiple response surfaces", *UKACC International Conference on Control 2000"*, Cambridge, UK, September 4-7, 2000.

[6] Stewart P and Kadirkamanathan V., (2001) "Dynamic model reference PI control of flux weakened permanent magnet AC motor drives" *IFAC Journal of Control Engineering Practice.* Invited Paper - in print.

[7] Pillay P., and Krishnan R., (1989) "Modelling, simulation and analysis of permanent magnet motor drives, part 1: the permanent magnet synchronous motor drive", *IEEE Transactions on Industrial Applications* vol.25, no.2, pp265-273, March/April 1989.

[8] Stewart P and Kadirkamanathan V., (2000) "Dynamic model reference control of a PMAC motor for automotive traction drives", *UKACC International Conference on Control 2000*, Cambridge UK, September 4-7, 2000.

[9] Miller T.J.E., (1993) "Brushless permanent magnet and reluctance motor drives", *University Press*, UK, 1993.

[10] Jahns T.M., (1987) "Flux weakening regime operation of an interior permanent magnet synchronous motor drive", *IEEE Transactions on Industrial Applications*, vol.1A-23, no.4, July/August 1987.

IFAC Publications
www.elsevier.com/locate/ifac

# ADAPTIVE CONCURRENCY CONTROL IN SAFETY CRITICAL REAL-TIME SYSTEMS

**Horst F. Wedde, Sabine Böhm, Wolfgang Freund**

*Informatik III*
*University of Dortmund*
*44221 Dortmund / Germany*
*horst.wedde@udo.edu*

Abstract: We present a novel transaction model for distributed safety-critical real-time applications in that transactions become increasingly critical under failures of subsequent incarnations. When a transaction-dependent criticality threshold has been reached the transaction is considered essentially critical or hard, i.e. the deadline has to be met lest the system is doomed to be in a disastrous state. The corresponding adaptive measures to provide that potentially all hard deadlines are met (criticality, sensitivity), i.e. that the system survives, are described in detail. As the major contribution of this paper we propose adaptive techniques for distributed concurrency control under safety-critical real-time requirements. In particular, we refrain from traditional features like roll-back because they are too costly. Also, tasks constituting a transaction will only be invoked whenever the needed data objects are no longer similar to their previous versions. We demonstrate how all features could be associated with available concurrency control algorithms, by modifying a recent concurrency control algorithm (*O2PL*) in a three-step procedure. We report on extensive distributed experiments. *Copyright © 2001 IFAC*

Keywords: Distributed operating systems, real-time systems, safety-critical systems, adaptive transaction management, adaptive concurrency control.

## 1. INTRODUCTION AND MOTIVATION

In safety-critical environments control operations (tasks on an elementary level), have not only to meet deadlines but most of them are critical in the sense that subsequent instances could tolerate deadline failures only to a task-specific degree. After that the next instance will be essentially critical, i.e. its failure would possibly have hazardous consequences for the system, and even for the environment. The quality of a system to handle *all* essentially critical task instances successfully has been denoted by *survivability* (Wedde and Lind 1997).Into the same vein, task deadlines in such environments cannot be guaranteed, due to the unpredictability of events. *MELODY* is a highly adaptive distributed real-time operating system for safety-critical applications in unpredictable environments. It provides for a tailored adaptive distributed *task* management. For details see (Wedde and Lind 1997), (Wedde and Xu 1992).

Different from soft real-time systems or predictable applications, typical transaction procedures like roll-back are infeasible in safety-critical systems. During the roll-back phase the resources will remain locked causing harmful effects for conflicting transactions. Thus roll-back should be avoided by all means.

With these ideas in mind we conceived adaptive techniques for distributed concurrency control.

*Previous and Related Work.* Transaction management has so far rarely been modeled in a truly distributed way taking replication into account. Among the few examples we quote (Xiong *et al.* 1998), (Gupta *et al.* 1997) discussing standard concurrency control protocols, namely 2-Phase Locking (*2PL*), and optimistic versions thereof, like the Optimistic 2-Phase Locking (*O2PL*) and Optimistic Concurrency Control (*OCC*) protocols.

Similarity has been introduced through different concepts (Kuo and Mok 1992), (Chen and Mok 1997) and (Wedde *et al.* 1999). Criticality of task instances had been introduced in (Wedde and Xu 1992), as an adaptive measure to possibly guarantee the deadlines of the most crucial task instances. Full adaptability for tasks was achieved by combining criticality with sensitivity. Since 1995 concepts related to criticality have been given increasing attention even for theoretical scheduling studies (Hamdaoui and Ramanathan 1995), (Spuri *et al.* 1995), (Koren and Shasha 1992). None of these contributions deals with hard deadlines or unpredictable task arrival.

*Organization of the paper.* In sec. 2 we will shortly recall how transactions can be composed of tasks. Section 3 is devoted to discussing how any concurrency control protocol for handling distributed replicated transactions could be modified in three steps for achieving adaptive real-time/ survivability and reliability behavior at the same time. In sec. 4 distributed experiments are described. Conclusive remarks will complete the paper.

## 2. NOVEL TRANSACTION CONCEPTS FOR SAFETY-CRITICAL APPLICATIONS

### 2.1 *Task Model*

In the *MELODY* operating system (sec. 1), tasks are read or write operations on objects that may be replicated over different sites. Read tasks access only one copy while writers update all copies of each object. Each task has only one *critical section* for its activity. *For our purpose we will assume that each task accesses one object only.*

Since it is not possible to guarantee deadlines in safety-critical environments the notion of *criticality* defines a maximum number of subsequent task instances that are tolerated to miss their deadlines. Task instances are categorized as either *non-critical, critical,* or *essentially critical.* This is expressed through a frame of criticality with thresholds $a'$ and $a''$. Each task $T_i$ is assigned an individual initial criticality value denoted by $C(T_i)$. The *relative criticality* of a task will be decreased every time an instance fails until the instance has become essentially critical. Essentially critical task instances are not allowed to fail or the whole system may be at risk. Once an instance has been successful, the criticality is reset to its initial value.

To which degree a task instance may use a private copy, or depends on the latest information, can be expressed through its *sensitivity.* Sensitivity of a task is divided into 3 categories: *essentially sensitive, sensitive,* and *robust.* As the *relative criticality* of a task decreases (becomes more critical), its *relative sensitivity* is increased and approaches the robustness threshold. Once a task instance is considered robust, it is allowed to use a private copy. The utilization of (locally available) private copies is also granted for a task instance that has become essentially critical. More technical details on criticality and sensitivity can be found in (Wedde and Lind 1997).

We employ the task-similarity concepts as specified in (Wedde *et al.* 2001*b*), sections 2.4 and 3.

### 2.2 *Transaction Model*

In this section the criticality concepts and notations for tasks are extended to transactions. We rely on the task structures defined in (Wedde *et al.* 2001*b*). Here transactions result from composing independent tasks / subtransactions and task / subtransaction sequences. Some immediate assumptions will be formulated now:

- There will be no roll-back for transactions in *MELODY* (see the motivation in sec. 1).
- A transaction instance is considered failed if it missed its deadline.
- The relative criticality of a transaction is always the minimum of the relative criticalities of its constituting tasks.
- The relative sensitivity of a transaction is always the maximum of the relative sensitivities of its constituting tasks.
- Write tasks will always write to their private memory and the results become visible to the other transactions upon commit time.

### 2.3 *Criticality and Sensitivity for Transactions*

*Task Sequences.* In order to ensure timely completion for each essentially critical task instance in a *task sequence* the preceding task has to provide a timely result at least one instance before the task under consideration becomes essentially critical. This is achieved by defining the initial criticality values for the tasks in the sequence in non-descending order. Correspondingly, the initial sensitivity values of the task sequence $TR$ would be set in non-ascending order.

*Independent Tasks.* The (initial) criticalities (sensitivities) of the tasks constituting a transaction with independent tasks can be set independently.

*Transaction Composition.* The initial criticalities of the tasks constituting a more complex transaction are set according to the definitions in the 2 previous paragraphs.

More details can be found in (Wedde *et al.* 2001*a*).

## 3. NOVEL PRINCIPLES FOR ADAPTIVE CONCURRENCY CONTROL

Three different techniques to produce distributed real-time concurrency control algorithms for manipulating transactions working in safety-critical applications will be presented. For a given concurrency control protocol *PROT* we assume a specific commit phase. The first modification of *PROT* will be a version *PROT.ADT* that heavily takes advantage of the *MELODY* operating system support and its extensions for transactions. We will explain this for the example of the *O2PL* protocol as specified in (Xiong *et al.* 1998). The second modification directly transforms *PROT* into a protocol *PROT.TSIM.* This latter protocol seeks to save, for every transaction execution, as many tasks or subtransactions as possible from being invoked. The third method transforms *PROT* into an overoptimistic version denoted by *OVO.PROT.*

### 3.1 *O2PL.ADT*

The *O2PL.ADT* protocol is the result of adapting *O2PL* to the adaptive *MELODY* environment described in sec. 2, the execution of transactions is based on the execution of its tasks, according to their criticality and sensitivity. More technical details can be found in (Wedde *at al.* 2001*a*).

### 3.2 *PROT.TSIM*

Any concurrency protocol *PROT* can be modified into *PROT.TSIM* once similarity handling as described in (Wedde *at al.* 2001*b*) is available. Upon an update of object $O$ needed by task $T$ the corresponding object and task managers determine whether $T$ needs to be executed or not. Tasks that need not be executed reduce both the system load and the resource requirements of the transaction. Due to page limitations we refer the reader to (Wedde *at al.* 2001*a*).

### 3.3 *OVO.PROT*

For *PROT* the overoptimistic version *OVO.PROT* is obtained by not setting any read lock yet aborting a transaction instead once inconsistencies have been detected. This is motivated by the expectation that the reduction of blocking times might more than outweigh such "incidental" abortions. The motivation had already been justified earlier for the *MELODY task* model (Wedde and Lind 1997).

A disadvantage of not setting any read locks is a potential data inconsistency. The values just read during the execution phase could be overwritten by another transaction before the commit phase starts. To prevent this from happening a "read-verification" phase has been introduced as a third protocol step. It will be entered immediately after the execution phase, before the commit phase is started. Here it is checked whether or not any of the read objects has been modified before the commit phase. If this is the case the transaction instance is aborted. As long as the transaction does not become essentially critical such an "incidental" abortion could be afforded. Conversely it could be expected that the previously described transaction features (criticality, sensitivity) give enough freedom for transactions to survive, despite a potentially high number of failed instances.

In the example of *OVO.2PL.ADT.TSIM* the read-verification procedure may have a somewhat relaxed effect. Since tasks need only be executed if they are dissimilar to their previous instance it is obvious that a transaction needs to be aborted during the "read-verification" phase only if any changes occurred to the objects that yield a dissimilar value. In fig.1, we have combined for *O2PL* its derivatives as described in 3.1, 3.2, 3.3. Since in all of these protocols no changes are made until the commit message is received, *there is no need for roll-back and restart.*

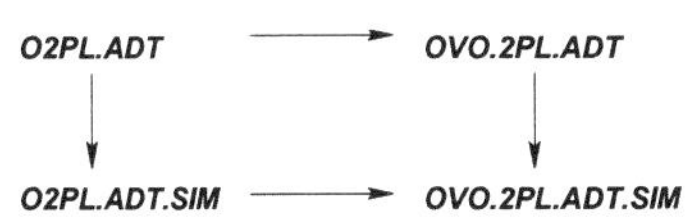


Fig. 1. Derivations of O2PL

## 4. EXPERIMENTAL EVALUATION

The major purpose of the experiments reported in this section is to study the influence of both similarity and overoptimistic transaction handling on the real-time behavior of available concurrency control algorithms. The study has an example character as we rely only on the *O2PL* protocol. Survivability being the main performance issue in safety-critical systems (as opposed to the overall deadline failure rate) we focus on adaptive transaction support. Thus we set up the experiments to run under *MELODY*, comparing the four algorithms listed in fig.1. *In (Wedde* et al.

*2001a) it has been shown that O2PL.ADT clearly outperforms O2PL.*

The *MELODY* version used here is operating under the AIX operating system on a token ring network consisting of 7 machines. By running distributed experiments instead of performing simulation the tendencies extracted are more realistic.

### 4.1 *Experimental Setup*

The distribution of transactions over the 6 sites, as well as their structure, is indicated in fig.4 (the object number is associated with each task). The experiments described here have been found representative for a very broad class of set-ups. Due to page limitations we restrict ourselves to this example discussion.

Task overall execution times (and deadlines) were engineered, and measured, between 20 msec (read) and 50 msec (write) as maximal values. There are 50 objects in the system. 20 are peripheral, 30 are derived objects. *Peripheral objects are read-only.*

For all experiments the criticality thresholds were 3 and 10, respectively while the range of the initial transaction criticality values was set to be between 6 and 8, i.e. $3-5$ consecutive task instances are allowed to miss their deadlines before becoming essentially critical. Sensitivity was initially set to 8 for all transactions, with thresholds 3 and 10.

Similarity is studied by varying both a (so far) linear similarity measure and the update frequencies of the peripheral objects. Due to page limitations and since the tendencies studied did not vary we depict our results for a similarity bound of 10% (i.e. changes of less than 10% are considered similar).

The deadline failure rate is measured through 10 runs. Since the deviations were fairly small, the average failure rates over the experimental time intervals are displayed. For survivability the first failure of an essentially critical transaction in any of the 10 runs is identified. (Note: Survivability is a worst case measure.)

The expectations are that introducing over- optimism may improve both the deadline failure rate and survivability since no read locks are set. We also expected that similarity, by introducing a flexible transaction execution policy adaptive to the environmental needs, would improve the real-time behavior since not all tasks might have to be executed. In particular, the competition for file copies would be hoped to be more moderate. In particular, if we compare *OVO.2PL.ADT* to *OVO.2PL.ADT.TSIM* the latter derivative should exhibit an improved survivability over *OVO.2PL.ADT* because the abortion rate which is due to data inconsistencies (see 3.3) may be considerably decreased.

### 4.2 *Experiment 1*

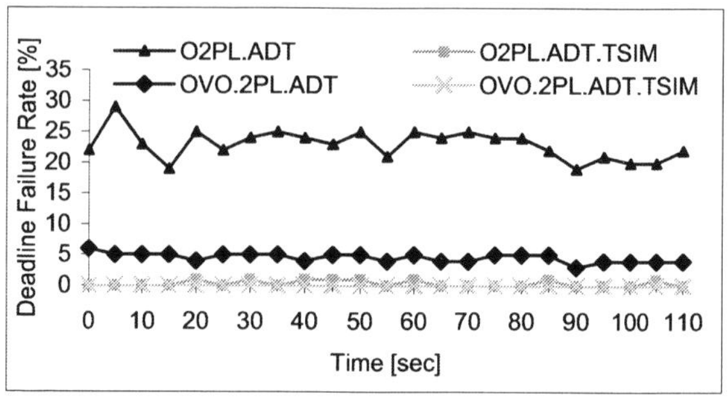


Fig. 2. Update Frequency 5 sec

In fig.2 the deadline failure rates of the 4 different versions of *O2PL.ADT* are depicted over the experimental time of 120 seconds with the update frequencies of the peripheral objects set uniformly to 5 seconds. While *O2PL.ADT* has a significant deadline failure rate of 24% in the average all other curves expose at most an average of 5%. Although the overoptimistic algorithm *OVO.2PL.ADT* clearly outperforms *O2PL.ADT* in regards to the deadline failure rate (5% vs. 24%) hardly any improvement in survivability can be seen (bold part of the curves): *O2PL.ADT* survives only 5 seconds and *OVO.2PL.ADT* dies already after 8 seconds. Utilizing the *.TSIM* versions results in a further drop of the deadline failure rate to 1% for *O2PL.ADT.TSIM* and even 0% for *OVO.2PL.ADT.TSIM* (in the average). While *O2PL.ADT.TSIM* does still not really improve significantly in survivability, it survives for 20 seconds as compared to 5 seconds for *O2PL.ADT*, *OVO.2PL.ADT.TSIM* is the only algorithm surviving throughout the whole experiment.

### 4.3 *Experiment 2*

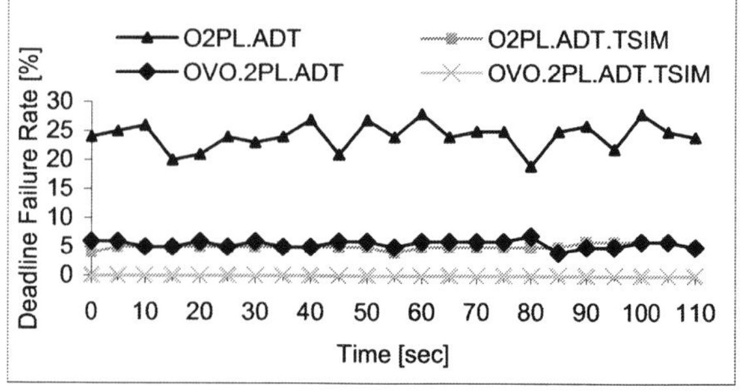


Fig. 3. Random Distribution of Update Frequencies

In this experiment (fig.3) the update frequencies of the peripheral objects have been set to different values ranging from 100 msec to 5 sec. Basically no drastic changes are visible compared to the uniform update frequencies. *O2PL.ADT*'s deadline failure rate is still high at 24% and the system dies after 3 seconds. Again, *OVO.2PL.ADT.TSIM* survives throughout the experimental time, and its deadline failure rate is still 0%. The behavior of *OVO.2PL.ADT* did not change significantly, revealing an average deadline failure rate of 6% and survivability of 5 seconds - compared to 5% deadline failure rate and a 5 second survivability in exp.1 (fig.2). The failure rate of *O2PL.ADT.TSIM* increased to 5% and the system died after 8 seconds compared to 1% and 20 seconds in exp.1.

In the other experiments, the tendencies were quite similar. The beneficial effect of similarity is particularly clear if one compares fig.2 and fig.3. In both settings *O2PL.ADT.TSIM* fares equally well or better than *OVO.2PL.ADT* while for the latter the *.TSIM* version is improved to an excellent deadline failure rate and survivability. At any rate, however, it was apparent in the experiments how much the overoptimistic measures helped to keep the deadline failure rate very low (compared to the "regular" *O2PL.ADT*). Given that the abortion rate would probably be very high without the adaptability provided by *MELODY* (in all *.ADT* versions studied here) this positive effect cannot be expected otherwise. A first indication for this conjecture comes from the observation that the overoptimistic *OVO.2PL.ADT* dies as early as *O2PL.ADT*. We will soon report separately on more detailed investigations. This is beyond the scope of this paper.

## 5. CONCLUSION AND FUTURE WORK

We defined and investigated two different techniques to transform traditional concurrency control algorithms into adaptive protocols. These build on adaptability features such as criticality, sensitivity, and similarity. For a given concurrency control protocol *PROT*, this yields versions *PROT.ADT* and *PROT.SIM*. In addition, an overoptimistic version *OVO.PROT* was defined in which read locks are abandoned while potential data inconsistencies are taken care of in a read-verification phase.

We reported on experiments comparing *O2PL. ADT* with its *.TSIM* versions and with *OVO. PROT*. Also the combined effects have been studied. The results that are reported here in an example discussion, yield an unambiguous message: For complex transaction structures only the combined adaptability features in *OVO.2PL.ADT.TSIM* give the system a good chance to survive, even under critical circumstances. A comprehensive discussion will be presented in a forthcoming paper.

We are currently working on extending our transaction model to allow a limited amount of preemption. Also included into future work is the completion of an extensive case study. This will be reported separately.

## 6. REFERENCES


Chen, D. and A.K. Mok (1997). Design of a Real-Time SQL Engine in the Distributed Environment. In: *2nd International Workshop on ARTDB-97*. Vol. 1553 of *Springer LNCS*. Springer.

Gupta, R., J. Haritsa and K. Ramamritham (1997). More Optimism about Real-Time Distributed Commit Processing. In: *Proceedings IEEE Real-Time Systems Symposium*. IEEE. IEEE. San Francisco.

Hamdaoui, M. and P. Ramanathan (1995). A Dynamic Priority Assignment Technique for Streams with (m,k)-firm Deadlines. *IEEE Transactions on Computers* **44**(12), 1143–1451.

Koren, G. and D. Shasha (1992). D-over: An Optimal On-line Scheduling Algorithm for Overloaded Real-Time Systems. In: *Proceedings 13th IEEE Real-Time Systems Symposium*. IEEE. Phoenix, Arizona, USA. pp. 290–300.

Kuo, T.-W. and A.K. Mok (1992). Application Semantics and Concurrency Control of Real-Time Data-Intensive Applications. In: *Proceedings 13th IEEE Real-Time Systems Symposium*. IEEE. Phoenix, Arizona, USA.

Spuri, M., G.Buttazzo and F.Sensini (1995). Robust Aperiodic Scheduling under Dynamic Priority Systems. In: *16th IEEE Real-Time Systems Symposium*. IEEE. Pisa, Italy. pp. 210–219.

Wedde, H.F. and J.A. Lind (1997). Building Large, Complex, Distributed Safety-Critical Operating Systems. *Real-Time Systems*.

Wedde, H.F. and M. Xu (1992). Scheduling Critical and Sensitive Tasks with Remote Requests in Safety-Critical Systems. In: *EUROMICRO Workshop on Real-Time Systems 1992 (ERTS92)*. EUROMICRO. Athens, Greece.

Wedde, H.F., S. Böhm and W. Freund (2001*a*). Adaptive Protocols For Survivability Of Transactions Operating On Replicated Objects. In: *WORDS'01*. IEEE. Rome, Italy.

Wedde, H.F., S. Böhm and W. Freund (2001*b*). Real-Time Transactions Need Their Constituting Tasks.

Xiong, M., K. Ramamritham, J. Haritsa and J.A. Stankovic (1998). Regulating Concurrent Access to Replicated Data in Distributed Real-Time Databases. In: *IEEE Workshop on Dependable and Real-Time E-Commerce (DARE)*. IEEE. Denver, CO, USA.

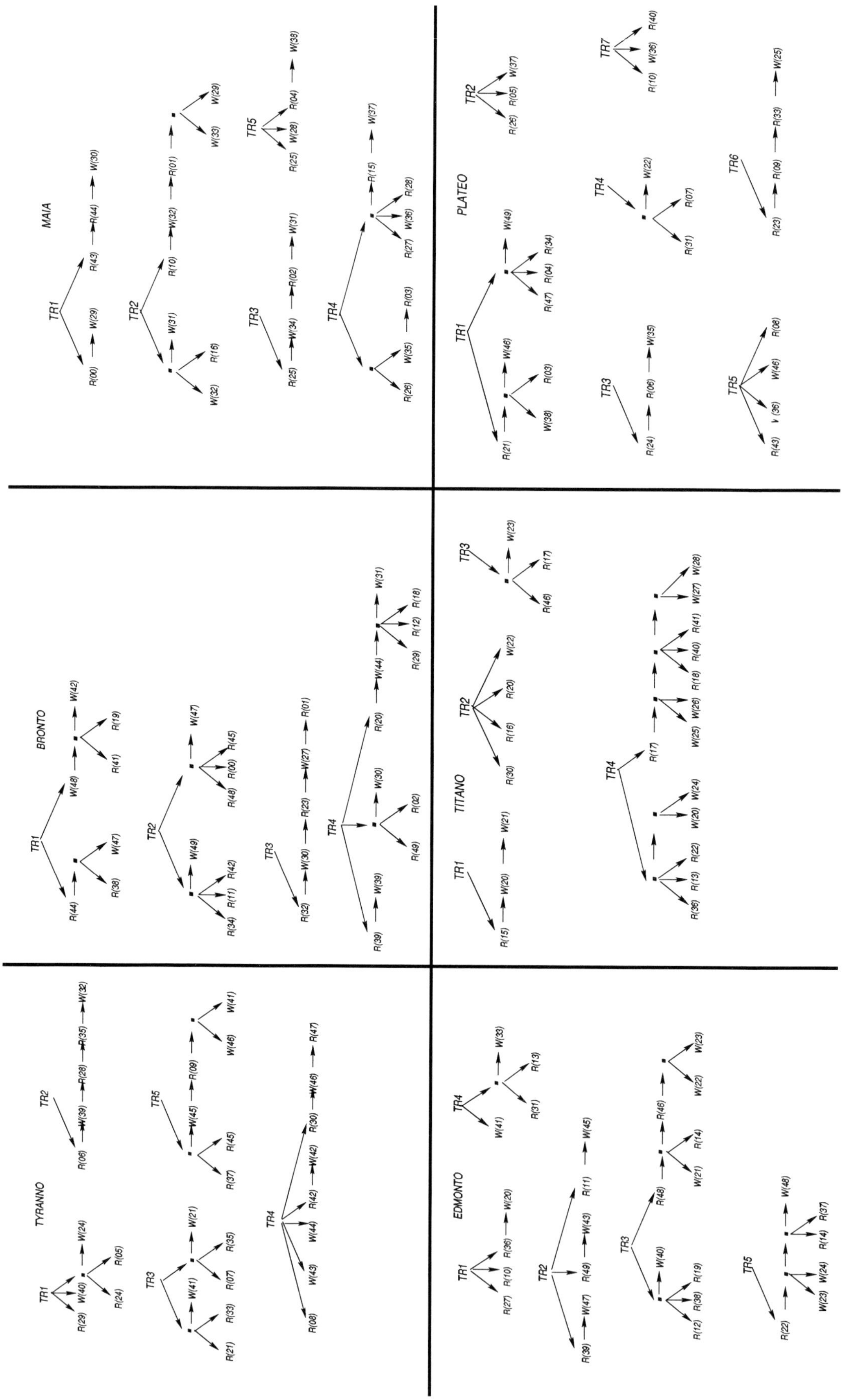


Fig. 4. Transaction Distribution

IFAC Publications
www.elsevier.com/locate/ifac

# DERIVING A FAULT TOLERANT REAL TIME PROGRAM USING PROGRAM VERIFICATION

**Grace Tsai[*] and Wolfgang A. Halang[†]**

[*] *Fairleigh Dickinson University, Computer Science, Teaneck, NJ 07666, U.S.A., tsai@alpha.fdu.edu*
[†] *FernUniversität Hagen, Faculty of Electrical Engineering, 58084 Hagen, Germany, wolfgang.halang@fernuni-hagen.de*

Abstract: The process of showing that a program satisfies some particular properties with respect to its specification is called program verification. Axiomatic semantics is a verification method that makes assertions describing properties about the states of a program. There exists a transformation from the assertions of a program's verification proof to executable assertions. The latter may be embedded in the program to make it fault tolerant. An axiomatic proof system for concurrent programs is applied to generate executable assertions in a real time distributed environment. A train set example is used as model problem. *Copyright © 2001 IFAC*

Keywords: Real time systems, temporal logic, formal methods, executable assertions, program verification, fault tolerance.

## 1. INTRODUCTION

The *application oriented fault tolerance paradigm* (McMillin and Ni, 1988) is a promising approach to provide fault tolerance in distributed environments. It features testing at the peer level, i.e., processors involved in a calculation check others involved in different parts of the calculation. An actual test comprises executable assertions embedded in the program code to ensure that, at each testable stage of a calculation, all tested processors conform to the program's specification, design, and implementation. If a difference between the expected and observed behaviour is detected, a fault has occurred and a reliable communication of this diagnostic information is provided to the system for taking appropriate actions.

Developing executable assertions for a program in a distributed environment is more complex than for sequential processing. Since state information may only be communicated by message passing, the scope of a test that may be performed by an executable assertion is limited to testing received messages about a tested processor's local state.

Earlier work on selecting executable assertions for the application oriented fault tolerance paradigm was guided by "Natural Constraints" of the problem (McMillin and Ni, 1988). Working from these natural constraints starting at the specification phase provides a global perspective on assertion development. Parallel decomposition of a problem ties in closely with the design specification of the program to solve the problem. Thus, executable assertions are closely linked to the individual components of the life cycle.

Working on the basis of this idea implies largely manual identification of appropriate assertions based on the three fundamental metrics *progress, feasibility*, and *consistency* that formulate the constraint predicate paradigm. *Progress* is required to be made at each testable step of a solution. Progress means that the state of a solution advances to the global or final solution of a problem. Each testable step of the solution advances to the goal or final solution of the problem. Each such step is a message interchange, a sub-message interchange, or a multi-message interchange depending on the type of the solution. If no progress is made, then faulty behaviour may indefinitely postpone the solution. The feasibility constraints are often immediately apparent from the problem studied. Problems in physics and engineering, such as equilibrium and Eigenvalue and, to a lesser extent, propagation problems, contain feasibility constraints as boundary conditions. It is necessary to ensure that any two processors that are correct, given the same input and executing the same executable assertion, reach the same conclusion of error or no error. However, in a system that has Byzantine faults, it is possible for a faulty processor to send inconsistent messages to different processors. The necessity to detect this establishes the purpose of the *consistency* condition.

The concept of a constraint predicate has been applied to distributed matrix multiplication (Hong and McMillin, 1991), and matrix iterative solution of simultaneous linear equations (McMillin and Ni, 1988). What is needed, however, to make the concept viable, is the application of executable assertions to real world examples to show fault tolerance. To this end, we use a proof outline from Interval Temporal Logic as a starting point, and embed executable assertions into a train set control program to generate a fault tolerant program.

Section 2 of this paper describes syntax and semantics of Interval Temporal Logic for the specification and verification of a train set control program, Section 3 deals with run-time evaluation of ITL formulae, Section 4 presents the implementation, and uses a train set example to obtain a fault tolerant real time program. Section 5 concludes the paper and discusses future work.

## 2. INTERVAL TEMPORAL LOGIC

The Interval Temporal Logic (ITL) now presented for the specification and verification of a train set control program is an extension of the Interleaving Set Temporal Logic (ISTL) (Peled and Pnueli, 1990). It adopts a partial order semantics, which considers a distributed computation as a set of partially ordered events. Hence, ITL can capture aspects of time and distribution. With the logic ISTL one cannot reason about a property within a bounded interval of time, which motivates the development of ITL. There are other temporal logics for real time systems, cp., e.g., (Moszkowski and Manna, 1983), and (Moszkowski, 1985).

The logics, (Moszkowski and Manna, 1983) and (Moszkowski, 1985) are designed for reasoning about hardware while we aim at a logic allowing to reason about a distributed real time system. In our approach, the specification of a system is to be tested to detect errors at run-time. So we need to build a mechanised support tool for the logic. For efficiency considerations, we defined ITL with a small set of syntactic forms, only, and including just the inference rules necessary for run-time evaluation. Hence, we do not adopt the logic of (Moszkowski and Manna, 1983) or (Moszkowski, 1985).

### 2.1 *Syntax*

In this subsection, we describe the syntax of the logic. We assume that a program $P$ can be identified with a *theory*, $\Sigma_P$, i.e., a collection of formulae characterising sequences of states of $P$ with arbitrary initial states.

The logic ITL is built upon a classical predicate logic $L$ as follows. The symbols of ITL are those of classical predicate logic along with operators [, ], [ ], U, E, F, X, $X^n$ and $\hat{X}$. Let $\Phi$ be the smallest set of words over the symbols of ITL such that

- If $p \in L$ then $p \in \Phi$.
- If $p, q \in \Phi$, then $p \vee q, \neg p \in \Phi$.
- If $p, q \in \Phi$, then $p\mathrm{U}q, \mathrm{E}p, \mathrm{F}p, \mathrm{X}p, \mathrm{X}^n p$, $\hat{\mathrm{X}}p \in \Phi$.
- If $p, q, r \in \Phi$, then $[p]r, [p, q]r \in \Phi$.

We call a member of $\Phi$ a *path formula*, and a formula of the language $L$ a *state formula*. A path formula can be used to characterise a sequence of states, while a state formula is interpretable only on a single state.

*Definition 2.1.* Let $\mid \sigma \mid$ denote the length (possibly infinite) of a state sequence $\sigma$, and let $\sigma(i)$ denote the $i^{th}$ state of a state sequence $\sigma$. We call $i$ a **time index** of $\sigma$.

*Definition 2.2.* A state sequence $\xi = (\xi(0), ..., \xi(\mid \xi \mid))$ refines a given state sequence, $\sigma = (\sigma(0), \cdots, \sigma(\mid \sigma \mid))$, iff $(\exists j \in [0, \mid \sigma \mid]) \ ((\sigma(j), \sigma(j+1), \cdots \sigma(j+ \mid \xi \mid)) = \xi)$. Let $R(\sigma) = \{\xi \mid \xi \leq \sigma\}$.

From the above, $R(\sigma)$ is a collection of subsequences of the state sequence $\sigma$.

### 2.2 *Semantics*

In the ITL system, formulae are quantified by fundamental operators E, F, U, X, $X^n$ , $\hat{X}$, and $X^n$, which are defined in the following semantics.

*Definition 2.3.* (Satisfaction). Let $\sigma$ be a state sequence, $i$ an integer with $0 \leq i < \mid \sigma \mid$, $f$ a state formula, and let $p, q, \phi, \psi$ be any formulae of ITL. Then, we write

| | |
|---|---|
| $(\sigma, i) \models f$ | if $\sigma(i) \models f$, |
| $(\sigma, i) \models p \vee q$ | if $(\sigma, i) \models p$ or $(\sigma, i) \models q$, |
| $(\sigma, i) \models \neg p$ | if not $(\sigma, i) \models p$, |
| $(\sigma, i) \models p\mathrm{U}q$ | if there exists $i \leq j <\mid \sigma \mid$, such that both $(\sigma, j) \models q$ and for every $k$ such that $i \leq k < j$, $(\sigma, k) \models p$, |
| $(\sigma, i) \models \mathrm{X}\phi$ | if $\mid \sigma \mid \geq i+1$ and $(\sigma, i+1) \models \phi$, |
| $(\sigma, i) \models \mathrm{X}^n\phi$ | if $(\sigma, i+n) \models \phi$, |
| $(\sigma, i) \models \hat{\mathrm{X}}\phi$ | if $(\exists n \geq i)((\sigma, i) \models \mathrm{X}^n\phi)$, |
| $(\sigma, i) \models \mathrm{F}\phi$ | if $\exists(k \geq i)(\forall j \geq k)(\sigma, j) \models \phi$, |
| $(\sigma, i) \models [p]\phi$ | if we have that if $(\sigma, i) \models p$ then $(\sigma, i) \models \phi$, |
| $(\sigma, i) \models [p, q]\phi$ | if for every $\xi \in R(\sigma)$ such that $(\xi, 0) \models p$ and $(\xi, \mid \xi \mid) \models q$, if $k_i \in \{0, 1, ..., \mid \xi \mid\})(\xi(k_i) = \sigma(i))$, then $(\xi, k_i) \models \phi$, |
| $\sigma \models \phi$ | if for all $i \in 0, ..., \mid \sigma \mid$, $(\sigma, i) \models \phi$. |

In each case, the symbol $\models$ is read "satisfies". We abbreviate $(\neg\phi \wedge \hat{\mathrm{X}}\phi)$ by $\mathrm{E}\phi$.

**Proposition 1:** Assume that $(\sigma, i) \models [p, q]\mathrm{EF}\phi$. Then, for all $\xi \in R(\sigma)$ such that $(\xi, 0) \models p$, $(\xi, \mid \xi \mid) \models q$, and there exists an index $k_i$, $\xi(k_i) = \sigma(i)$, we have $(\xi, \mid \xi \mid) \models \phi$.

**Proof:** Since $(\sigma, i) \models [p, q]\mathrm{EF}\phi$, it follows that for every $\xi \in R(\sigma)$ such that $(\xi, 0) \models p$, $(\xi, \mid \xi \mid) \models q$, and there exists $k_i \in \{0, 1, ..., \mid \xi \mid\})(\xi(k_i) = \sigma(i))$, we have $(\xi, k_i) \models \mathrm{EF}\phi$. This means that for every such subsequence $\xi$ there exists an $l > k_i$ such that $(\xi, l) \models \mathrm{F}\phi$. Equivalently, for every such subsequence $\xi$ there exists an $m \geq l$ such that $\forall(n \geq m)$. We have $(\xi, n) \models \phi$, in particular, $(\xi, \mid \xi \mid) \models \phi$.□

*Definition 2.4.* For a structure $\sigma$, an **interval** $[\phi, \psi]_\sigma$, bounded by formulae $\phi$ and $\psi$, is given by $[\phi, \psi]_\sigma = \{\xi \in R(\sigma) \mid (\xi, 0) \models \phi, (\xi, \mid \xi \mid) \models \psi\}$. The symbol $\models$ denotes the satisfaction relation of ITL.

*Definition 2.5.* If $\Sigma$ is a set of formulae of ITL, and $\sigma$ is a structure of ITL such that $\sigma \models \phi$ for all $\phi \in \Sigma$, then we say that $\sigma$ is a **model** of $\Sigma$.

*Definition 2.6.* A set of formulae of ITL is said to be **consistent** if it has a model. We call a set of consistent formulae a **theory**.

Informally, an *interval* is of the form $[p]$ or $[p, q]$ and an *interval formula* is of the form $[p]\phi$ or $[p, q]\phi$, where $p, q$ and $\phi$ are any formulae of ITL. An interval formula $[p]\phi$ ($[p, q]\phi$) is true over a state sequence $\sigma$, iff the interval $[p]$ ($[p, q]$) cannot be found or the formula $\phi$ holds on every interval $[p]$ ($[p, q]$). Thus, there are two ways to conclude that an interval formula holds. This would cause a problem in the composition of interval formulae. Thus, the following *responsiveness assertions* are proposed.

*Definition 2.7.* A *responsiveness* assertion is a path formula of the form $([p]\phi \rightarrow [p, q]\mathrm{E}F\psi)$, where $p, q, \phi$, and $\psi$ are formulae of $ITL$.

A responsiveness assertion $([p]\phi \rightarrow [p, q]\mathrm{E}F\psi)$ is true over a state sequence $\sigma$, iff the following holds: if $\phi$ holds whenever $p$ holds, then $\psi$ will occur at any $q$ following $p$. The assertion ensures bounded response of $\psi$ to $[p]\phi$ within the intervals $[p, q]$. The following Progress Rule can be applied to reason about responsiveness properties. Let $p, q, r, \phi_0, \phi_1, \phi_2$ be any formulae of ITL,

**Progress Rule**

$$\begin{array}{ll} (1) & [p]\phi_0 \rightarrow [p, q]\mathrm{E}F\phi_1 \\ (2) & [q]\phi_1 \rightarrow [q, r]\mathrm{E}F\phi_2 \\ (3) & [p, r]\mathrm{E}q \\ \hline & [p]\phi_0 \rightarrow [p, r]\mathrm{E}F\phi_2 \end{array}$$

From the above, we can conclude that if $\phi_0$ holds whenever $p$ holds, then $\phi_2$ will occur in intervals $[p, q]$, if the following premises hold:

(1) if $\phi_0$ holds whenever $p$ holds, then $\phi_1$ will occur at any $q$ following $p$,
(2) if $\phi_1$ holds whenever $q$ holds, then $\phi_2$ will occur at any $r$ following $q$,
(3) $q$ will occur within the intervals $[p, r]$,

## 3. RUN-TIME EVALUATION OF ITL FORMULAE

To check whether a program satisfies its specification denoted by ITL formula we introduce a run-time evaluation system consisting of the following components:

(1) The ITL parser and evaluation routines, which allow for the recognition and evaluation of ITL formulae.
(2) Routines to record process communications and to generate a run-time history $V_{h_i}$.
(3) The CCSP parser, which allows for user/input programs written in the language Communicating Sequential Processes (CSP) (Hoare, 1985).

Owing to page limitations, for a detailed coverage of the evaluation of ITL formulae we refer to (Tsai *et al.*, 1998).

## 4. IMPLEMENTATION

### 4.1 *Train Set Example*

Now we illustrate the proposed approach using a train set (de Lemos and Saeed and Anderson, 1992). This control problem is an example for a safety critical system involving interactions between controllers and physical processes. As depicted in Figure 1, the physical process consists of circuits, $C_p$ and $C_s$, and trains, $T_p$ and $T_s$. The circuits are divided into sections, and the crossing section ($CC$) is where the circuits intersect. Each section has a sensor, while for each train there is an actuator that can stop the train within any section. For such a safety critical system, accidents will occur if the physical specification and the logical specification do not match. We can then check whether the physical specification denoted by a run-time history $V_{h_i}$ satisfies the logical specification denoted by ITL formulae.

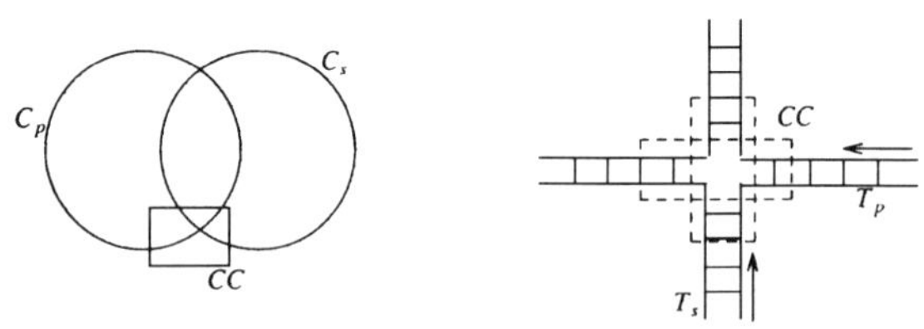


Fig. 1. Train set example

### 4.2 *Behaviour of Sensors and Actuators*

In this example, we are concerned with failures of sensors and actuators that can affect the safety of a system such as, for instance, a sensor not detecting a train or an actuator failing to stop a train. Table 1 describes the state variables to be used. The regular and failure behaviour of sensors and actuators are defined below.

Table 1. The state variables

| Variable | Comments |
|---|---|
| $On(x,c)$ | train $x$ is on circuit $c$ |
| $Ptrain(x)$ | position (section) contains the front of a train $x$ |
| $Rtrain(x)$ | set of sections reserved by a train $x$ on circuit $c$ |
| $Sens(c,i)$ | sensor of section $i$ detects a train on circuit $c$ |
| $Act(x,j)$ | is set to stop train $x$ on section $j$ |
| $Shut\_Down$ | holds to stop all trains, i.e., all actuators are set |

*Definition 4.1.* Let $T_l^x$ denote the current time when train $x$ enters section $l$, and let $T_{\bar{l}}^x$ denote the point of time immediately before $T_{l\oplus 1}^x$.

Note that addition $\oplus$ and subtraction $\ominus$ on section numbers are performed modulo the number of sections of the circuit.

*Definition 4.2.* (Sensor behaviour). Let $SS(x,c,i)$ denote the regular behaviour of a sensor, and let $FS(x,c,i)$ denote its failure behaviour. Then,

$$SS(x,c,i) = on(x,c) \wedge Sens(c, Ptrain(x));$$
$$FS(x,c,i) = on(x,c) \wedge \neg Sens(c, Ptrain(x)).$$

From the above, $SS(x,c,i)$ specifies the regular behaviour of a sensor: when a train $x$ is on circuit $c$ ($on(x,c)$), the sensor detects the train ($Sens(c, Ptrain(x))$). Correspondingly, $FS(x,c,i)$ denotes the failure behaviour of a sensor: when a train $x$ is on circuit $c$ ($on(x,c)$), the sensor fails to detect the train ($\neg Sens(c, Ptrain(x))$).

*Definition 4.3.* (Actuator behaviour). Let $SA(x,i)$ and $FA(x,i)$ denote, respectively, the regular and failure behaviour of an actuator, which is specified as follows.

$$SA(x,i) = ([T_i^x]Act(x,i)) \wedge ([T_i^x, T_{i\oplus 1}^x]Ptrain(x) = i);$$
$$FA(x,i) = ([T_i^x]Act(x,i)) \wedge ([T_i^x, T_{i\oplus 1}^x]Ptrain(x) > i).$$

Consequently, $SA(x,i)$ states that when train $x$ is on section $i$ and the actuator is set ($[T_i^x]Act(x,i)$), the train cannot enter a new section ($[T_i^x, T_{i+1}^x]Ptrain(x) = i$). Analogously, $FA(x,i)$ states that when train $x$ is on section $i$ and the actuator is set ($[T_i^x]Act(x,i)$), the train is moving beyond section $i$ ($[T_i^x, T_{i+1}^x]Ptrain(x) > i$).

### 4.3 *Safety Constraints*

The following describes the safety constraints of the system. These are the assertions to be embedded into the program.

**SC1** (Reservation constraints): For any train, the currently occupied section ($Ptrain(x)$) and the following $mcsf \oplus 1$ sections must always be reserved.

$$(\forall x \in Tr)$$
$$[T_i^x, T_{\bar{i}}^x]\{Ptrain(x), Ptrain(x) \oplus 1, \cdots$$
$$\cdots, Ptrain(x) \oplus (mcsf \oplus 1)\} \subseteq Rtrain(x)$$

**SC2** (Exclusion constraints): Mutual exclusion must be achieved for reserved sections. In other words, if $Rtrain(x)$ and $Rtrain(y)$ are the sets of sections reserved, respectively, by trains $x$ and $y$ ($x \neq y$), then $Rtrain(x) \cap Rtrain(y) = \emptyset$.

$$\forall x, y \in Tr$$
$$[T_i^x, T_{\bar{i}}^x](x \neq y \rightarrow Rtrain(x) \cap Rtrain(y) = \emptyset)$$

**SC3:** If the number of consecutive sensor failures exceeds $mcsf$, the system must be shut down.

$$\forall c \in L, \forall x, y \in Tr$$
$$[T_i^x, T_{i \oplus mcsf}^x](\neg Sens(c, i) \wedge \cdots$$
$$\cdots \wedge \neg Sens(c, i \oplus mcsf)) \rightarrow$$
$$[T_i^x, T_{i \oplus mcsf}^x] \mathrm{E}F(Shut\_Down)$$

The above formula can be derived from the Progress Rule. First, we apply the Progress Rule to derive a formula $\Psi$, which describes sensor failure on two consecutive sections as follows. To derive $\Psi = [T_i^{\overline{x}}]Ptrain(x) \rightarrow [T_i^x, T_{\overline{i \oplus 1}}^{\overline{x}}]\mathrm{E}F(on(c,x) \wedge \neg Sens(c, Ptrain(x)))$, the following premises must hold:

(1) $[T_i^{\overline{x}}]Ptrain(x) \rightarrow [T_i^x, T_i^{\overline{x}}]\mathrm{E}F(on(c,x) \wedge \wedge \neg Sens(c, Ptrain(x)))$

(2) $[T_i^{\overline{x}}]Ptrain(x) \rightarrow [T_i^x, T_{\overline{i \oplus 1}}^{\overline{x}}]\mathrm{E}F(on(c,x) \wedge \wedge \neg Sens(c, Ptrain(x)))$

(3) $[T_i^x, T_{\overline{i \oplus 1}}^{\overline{x}}]\mathrm{E}(on(c,x) \wedge \neg Sens(c, Ptrain(x)))$

Premise (1) states that the sensor of section $i$ fails within the interval $[T_i^x, T_i^{\overline{x}}]$. Likewise, Premise (2) states that the sensor of section $i \oplus 1$ fails within the interval $[T_i^x, T_{\overline{i \oplus 1}}^{\overline{x}}]$. Premise (3) states that the formula $(on(c,x) \wedge \neg Sens(c, Ptrain(x)))$ will hold in the interval $[T_i^x, T_{\overline{i \oplus 1}}^{\overline{x}}]$. If these three premises hold then, according to the Progress Rule, we can conclude $\Psi$, i.e., a sensor failure on two consecutive sections. Similarly, we can conclude a sensor failure on $(mcsf \oplus 1)$ sections by applying the Progress Rule.

**SC4:** If an actuator ever fails, the system must be shut down, i.e., if an actuator is set to stop, denoted by $([T_i^{\overline{x}}]Act(x,i))$, and the train is moving beyond section $i$, denoted by $([T_i^x, T_{i \oplus 1}^x](Ptrain(x) > i))$, then the system must be shut down.

$$(\forall x \in Tr)(\exists i \in Sc)$$
$$[T_i^{\overline{x}}]Act(x,i) \wedge [T_i^x, T_{i \oplus 1}^x](Ptrain(x) > i)$$
$$\rightarrow [T_i^x, T_{i \oplus 1}^x]\mathrm{E}F(Shut\_Down)$$

These safety constraints govern the operation of the system — if they are ever violated, then there is an error. In other words, for a safety constraint SC, if the evaluation $\Pi(V_{h_k}, SC)$ returns FALSE, then an error has occurred.

### 4.4 *The Program*

The Appendix contains a derived fault tolerant program, which shows processes train ($P_{tr}$), circuit ($P_c$), and section ($P_s$). Let $V_{h_{tr}}$, $V_{h_c}$ and $V_{h_s}$ denote respectively the histories of the processes $P_{tr}$, $P_c$ and $P_s$, and let RecvHistory denote the history received from the communicating process. The following describes these processes:

(1) Process $P_s$ informs processes $P_c$ and $P_{tr}$ upon a detection of a train entering a section. After each communication, it performs a consistency check and incorporates the received history, RecvHistory, into its history $V_{h_s}$. In the program, the following statements denote consistency check[1] and incorporation of received histories into the history of process $P_s$.
    if $\circ(V_{h_s}, RecvHistory)$ then exit;
    if $\neg(\circ(V_{h_s}, RecvHistory))$ then
    $V_{h_s} = \hbar(RecvHistory, V_{h_s})$;

(2) Process $P_{tr}$ has to obtain permission from $P_c$ to enter next sections, and from $P_c$ and $P_{DZ}$ to enter the crossing section. After the communications with $P_c$ or $P_{DZ}$, process $P_{tr}$ performs a consistency check and incorporates the received history, RecvHistory, into its history $V_{h_{tr}}$. In the program, the following statements denote consistency check and incorporation of received histories into the history of process $P_{tr}$.
    if $\circ(V_{h_{tr}}, RecvHistory)$ then exit;
    if $\neg(\circ(V_{h_{tr}}, RecvHistory))$ then
    $V_{h_{tr}} = \hbar(RecvHistory, V_{h_{tr}})$;

(3) The circuit process $P_c$ localises sensor and actuator failures. It signals minor failure if the number of consecutive sensor failures is less than or equal to a constant $mcsf$. A major failure occurs if an actuator fails to stop a train, or the number of consecutive sensor failures is greater than $mcsf$. After the communications with $P_s$ or $P_{tr}$, process $P_c$ performs a consistency check and incorporates the received history, RecvHistory, into its history $V_{h_{tr}}$. In the program, the following statements denote consistency check and incorporation of received histories into the history of process $P_c$.
    if $\circ(V_{h_c}, RecvHistory)$ then exit;
    if $\neg(\circ(V_{h_c}, RecvHistory))$ then
    $V_{h_c} = \hbar(RecvHistory, V_{h_c})$;

## 5. CONCLUSION

This paper has shown how to translate a verification proof (the safety constraints) into a fault tolerant program using the train set example. Future research will build a run-time evaluation system using the formalisms presented in this paper. This work will be extended to enable evaluation of specifications written in other suitable temporal logic languages. Also, we shall consider other formal languages such as I/O automata, process algebras or timed CSP for the run-time evaluation of fault tolerant programs.

[1] The message consistency check operator $\circ$.

## 6. REFERENCES


Hoare, C.A.R. (1985). *Communicating Sequential Processes.* Prentice Hall.

Hong, C., and B. McMillin (1991). Fault-tolerant matrix multiplication with one-iteration fault latency. *Proc. 15th IEEE COMPSAC.*

McMillin, B., and L. Ni (1988). Executable assertion development for the distributed parallel environment. *Proc. 12th IEEE COMPSAC*, pp. 284–291.

de Lemos, R. and Saeed, A. and Anderson, T. (1992). Analysis of Timeliness Requirements in Safety-Critical Systems. *Lecture Notes in Computer Science 571, Formal Techniques in Real-Time and Fault-Tolerant Systems*, pp. 171–192.

Moszkowski, B. (1985). Temporal logic for multilevel reasoning about hardware. *The Computer Journal*, pp. 10–18.

Moszkowski, B., and Z. Manna (1983). Reasoning in interval temporal logic. *Lecture Notes in Computer Science 164, Logic of Programs*, pp. 371–382.

Peled, D., and A. Pnueli (1990). Proving partial order liveness properties. *Proc. 17th Colloquium on Automata, Language and Programming*, pp. 553–571.

Tsai, G., M. Insall and B. McMillin (1998). Deriving a fault-tolerant program from verification proof. Fairleigh Dickenson University, Computer Science and Inforamtion Systems, Technical Report TR-001-98.


## APPENDIX: THE DERIVED TRAIN SET PROGRAM

```
Ps::/* Process Section (Ps) IS */
    Sn: Section_Number;
    x: Train_Number;
    Sens: Boolean; /*Signal of the sensor */
begin
    while(TRUE) {
        /* Sensor detects a train entering a section */
        [< Signal from sensor > → Sens:=TRUE;]
        [(Sens) →
            /* Set the signal Sens to FALSE after detection */
            Sens:= FALSE;
            /* Inform process Pc */
            (Pc!x, Sn);
                if ○(V_hs, RecvHistory) then exit;
                if ¬(○(V_hs, RecvHistory)) then
                V_hs = ħ(RecvHistory, V_hs);
                PI(SC1); PI(SC2); PI(SC4);
            /* Inform process PDZ */
            (PDZ, x, Sn);
                if ○(V_hs, RecvHistory) then exit;
                if ¬(○(V_hs, RecvHistory)) then
                V_hs = ħ(RecvHistory, V_hs);
                PI(SC1); PI(SC2); PI(SC4);
        ]
    } /* End while */
end.
```

```
Ptr:: /* Process Train (Ptr) IS */
    Cn: Circuit_Number;
    Sn, DZ: Section_Number;
    /* AES: Train is allowed to enter this section */
    /* AEDZ: Train is allowed to enter danger zone (DZ) */
    AES, AEDZ: Section_Number;
begin
    while(TRUE) {
        /* Obtain permission from process Pc */
        (Pc?Sn, Cn);
            if ○(V_htr, RecvHistory) then exit;
            if ¬(○(V_htr, RecvHistory)) then
            V_htr = ħ(RecvHistory, V_htr);
                PI(SC3); PI(SC4);
        [AES ≠ DZ → AES := EMPTY;
        []
        AES = DZ →
            /* Obtain permission from process PDZ */
            [(PDZ?Sn, Cn) ∧ (AEDZ = AES) →
                if ○(V_htr, RecvHistory) then exit;
                if ¬(○(V_htr, RecvHistory)) then
                V_htr = ħ(RecvHistory, V_htr);
                PI(SC3); PI(SC4);
            AES := EMPTY;
            AEDZ := EMPTY; ]
        ]
    } /* End while */
end.
```

```
Pc::/* Process Circuit (Pc) IS */
    /* NOS: Set of reserved sections */
    NOS: Set_of_Section_Numbers;
    mcsf: integer; /* max. # of consecutive sensors failures */
    NS: Reservation_Flag_Table;
    No_Failure: Boolean;
    Cn: Circuit_Number;
    Sn: Section_Number;
    /* No_of_Entries: Number of consecutive sensor failures */
    No_of_Entries: integer;
    x: Train_Number;
    Curr: integer;
begin
    while(TRUE) {
        /* Read sensor Train Tn in Section Sn from process
        section Ps */
        /* Add Sn to set of section numbers NOS */
        (Ps?x, Sn) → NOS := NOS + Sn;
            if ○(V_hc, RecvHistory) then exit;
            if ¬(○(V_hc, RecvHistory)) then
            V_hc = ħ(RecvHistory, V_hc);
                PI(SC1); PI(SC2); PI(SC3); PI(SC4);
        Localise_Sensor_Failure;
        /* Minor_failure: # of consecutive sensor failures < mcsf */
        [¬No_Failure ∧ (No_of_Entries ≤ mcsf) → Minor_Failure
        []
        ¬No_Failure ∧ (No_of_Entries > mcsf) → Major_Failure;]
        Test_Reserve_Section;
        NS := NS + Curr; /* Section curr is reserved */
        (No_Failure) ∧(Ptn!Sn, Cn) → TRUE;
            if ○(V_hc, RecvHistory) then exit;
            if ¬(○(V_hc, RecvHistory)) then
            V_hc = ħ(RecvHistory, V_hc);
                PI(SC1); PI(SC2); PI(SC3); PI(SC4);
    } /* End while */
end.
```

IFAC Publications
www.elsevier.com/locate/ifac

# A FAULT TOLERANT AND JITTER FREE DUAL RING FIELDBUS SYSTEM WITH GPS BASED TIME SYNCHRONISATION*

**T. Erdner[†], W.A. Halang[†], K.C. Chan[‡], J.K. Ng[‡]**

[†] *Faculty of Electrical Engineering, FernUniversitt,*
*58084 Hagen, Germany*
*thomas.erdner|wolfgang.halang@fernuni-hagen.de*

[‡] *Dept. of Computer Science, Hong Kong Baptist University,*
*Kowloon Tong, Hong Kong*
*jng|kcchan@comp.hkbu.edu.hk*

Abstract: A fieldbus architecture particularly designed for real time applications is described, in which the runtime performance cannot only be determined a priori, but also controlled during the entire process duration. Basically, it consists of a dual ring bus, in which two telegrams are transmitted on both channels in opposite directions, thus making delay measurements feasible. An ASIC timer module with an accuracy of $100\mu s$ is used for time synchronisation, which is fed by a GPS receiver. Only one GPS receiver is required in a system, because the timers of the bus nodes are synchronised via the field bus. By additional Hamming encoding of the telegrams, in each receiver errors can be corrected without requiring repetition of transfer. Additionally, the transfer of short data nibbles ensures high flexibility at different terminal equipment. Even in the presence of strong disturbances on the transmission channel, the signal encoding by a modified "Frequency Shift Keying" method enables the receivers to securely recognise the input signals. *Copyright © 2001 IFAC*

Keywords: Fieldbus, fault tolerance, Hamming encoding, signal coding, time synchronisation

## 1. INTRODUCTION

Generally, the processing of control algorithms is time critical, i.e., it does not render the intended results if certain timing conditions are not observed. As stressed by Lauber (Lauber, 1989), this is expressed by two fundamental requirements which the user of real time systems has, and which must be fulfilled even under extreme load or so-called worst case conditions:

- **timeliness**, and
- **predictability**.

It is inherently difficult to guarantee punctuality of a real time system, because the response times depend on the actual workload of the computer, and may differ from one external request to the next. The difficulty of accurately predicting the temporal behaviour is further exacerbated by the presence of operating system overhead and of network communication both strongly dependent on workload. Thus, the duration between the

* The work described in this paper was supported by a grant from the German Academic Exchange Service (Project No. D/0008349) and the Research Grants Council of the Hong Kong Joint Research Scheme (Project No. G_HK012/00).

arrival of external signals or sensor data and the output of corresponding actuation signals by the computer is generally unpredictable at the time a certain control loop program is written.

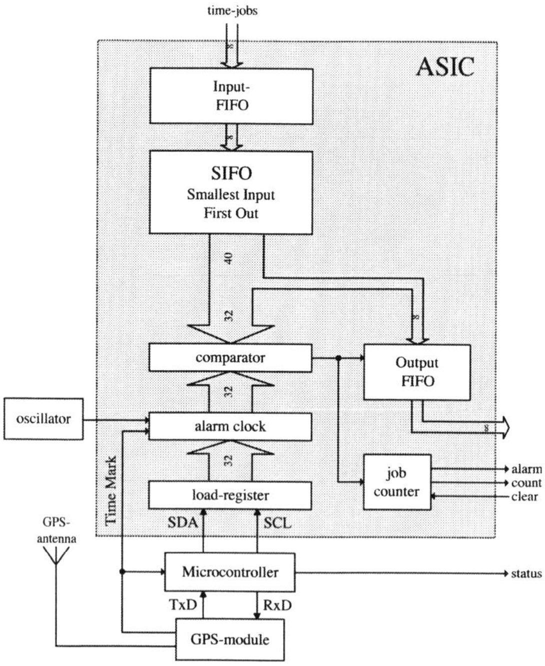


Fig. 1. High precision timer ASIC with novel SIFO memory

To achieve gains in productivity, industry increasingly requires faster automation systems, of which fieldbuses form decisive parts. The disadvantages of the presently available fieldbus systems partially result from previous demands. Since conventional bus systems were sufficient for the majority of applications, industry did not especially support "genuine" real time capabilities. Today, just fast systems are employed to meet cycle time requirements.

A further disadvantage of current systems is the kind of corrective action taken when transmission parity errors occur. In almost all systems available on the market, transmission parity errors are detected by corresponding error recognition tests, such as parity bit check, block parity check or Cyclic Redundancy Check (CRC). Fundamentally, for error handling only the error detection with the Automatic ReQuest repeat procedure (ARQ) is in use, i.e., telegrams are simply repeated. Since transmission parity errors are normally of stochastic nature, every recognised error implies a repetition and, therefore, a delay of data exchange which is not predictable.

## 2. TIME SYNCHRONISATION

Real time systems have to perform specific actions at distinct times. These instants may be given in absolute, periodic, or relative terms, but can always be transformed to absolute time values. Handling sets of such instants requires a considerable amount of system resources. This can be avoided if external devices are used for this purpose. Once instants become known, by any kind of program execution or external events, they are transferred as alarm jobs to an external high precision timer. An alarm job consists of an alarm time and a unique alarm number identifying an associated activity. The timer keeps track of all alarm jobs and can always receive new alarm jobs. Thus, a correspondingly equipped real time system can transfer alarm jobs to its timer without any delay. Since time passes in a strictly linear fashion, the timer simply needs to keep an ascendingly sorted list of alarm jobs, and to compare the alarm time of the earliest alarm job with its real time clock. When this alarm time is reached, the timer sends an alarm signal to the real time system and provides the alarm number.

In a distributed control system, each of its nodes needs an external timer as outlined above. To render expensive and with all other methods unsatisfactorily accurate synchronisation of these timers superfluous, the high precision timer developed at FernUniversitt (Halang, 1998) and implemented in form of an application specific integrated circuit (ASIC) as depicted in Fig. 1 employs the legal Universal Time Co-ordinated (UTC) as correctly broadcast via the worldwide radio navigation system operated by the U.S. Department of Defense called Global Positioning System (GPS). Thus, by design, all clocks in a distributed system run synchronously, letting the need for explicit clock synchronisation simply disappear. The costs for serial GPS modules already sank to about $ 30 (VDI, 2000).

## 3. DUAL RING BUS

As wired interconnections for stationary distributed (sub-)systems dual ring buses are foreseen, providing for fault tolerance by redundancy and for predictable data transmission delays. On such a dual ring bus every telegram is transmitted in parallel on both data channels (channel 1 and channel 2), but in opposite directions (cp. Fig. 2). The transmission delay for a telegram sent from the bus master to a slave depends on two configuration constants, only, viz., one holding for the entire bus ($T_{CCT}$), and the other one being a function of the slave's position on it ($\Delta T_{SiS}$).

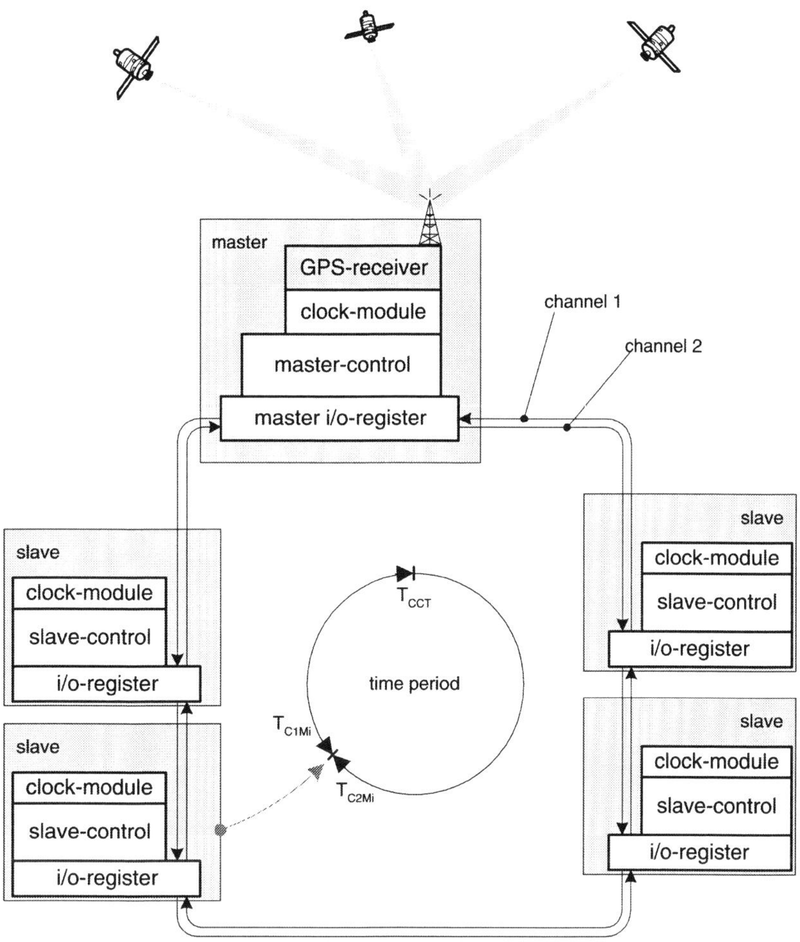


Fig. 2. Dual ring bus with GPS receiver

Definitions:

| | |
|---|---|
| $t_{ST}$ | starting time of the telegram |
| $t_{RT1}$ | receiving time of the telegram via channel 1 |
| $t_{RT2}$ | receiving time of the telegram via channel 2 |
| $T_{CCT}$ | channel circulation time of one telegram |
| $\Delta T_{SiS}$ | difference time of both telegrams in the slave module$_i$ |
| $T_{C1Mi}$ | transmission time of the telegram from master to slave$_i$ via channel 1 |
| $T_{C2Mi}$ | transmission time of the telegram from master to slave$_i$ via channel 2 |

The following equations explain the time measurement in the dual ring bus:

$$t_{RT1} = t_{ST} + T_{C1Mi} \quad (1)$$

$$t_{RT2} = t_{ST} + T_{C2Mi} \quad (2)$$

$$T_{CCT} = T_{C1Mi} + T_{C2Mi} \quad (3)$$

$$T_{CCT} = t_{RT1} + t_{RT2} - 2 \cdot t_{ST} \quad (4)$$

$$\Delta T_{SiS} = T_{C2Mi} - T_{C1Mi} \quad (5)$$

From equations (1), (2) and (5) follow:

$$\Delta T_{SiS} = t_{RT2} - t_{RT1} \quad (6)$$

add (3) and (6):

$$T_{CCT} + \Delta T_{SiS} = 2 \cdot t_{RT2} - 2 \cdot t_{ST} \quad (7)$$

$$t_{RT2} = t_{ST} + \frac{T_{CCT} + \Delta T_{SiS}}{2} \quad (8)$$

subtract (3) and (6):

$$T_{CCT} - \Delta T_{SiS} = 2 \cdot t_{RT1} - 2 \cdot t_{ST} \quad (9)$$

$$t_{RT1} = t_{ST} + \frac{T_{CCT} - \Delta T_{SiS}}{2} \quad (10)$$

Given these constants, each slave can synchronise its local clock with the clock of the master node every time it receives a telegram, provided the latter contains the information when it was sent. Temperature effects can easily be compensated for by measuring the difference between the arrival times of a telegram's versions travelling on both rings. Hence, for a distributed system interconnected by a fieldbus, only one GPS receiver is required in its master unit to provide for accurate timing synchronous with UTC.

## 4. REAL TIME ERROR CORRECTION

### 4.1 *Hamming Encoding of Data Nibbles*

Security of telegram transfers is achieved by a Hamming encoding (Swoboda, 1973). Hamming encoding allows to detect or to correct predeterminable numbers of errors in the data bits of telegrams, respectively. In the proposed design, three protection bits are provided for any four data bits, i.e., for any nibble. Thus, in any such data transfer element of seven bits, two single bit errors can be detected, or one single bit error can be corrected.

The protection bits are determined by a Hamming encoding of the data bits. In each case three data bits $x_i$ are associated by a modulo-2-sum, and the result employed as a protection bit $k_i$ ((11)...(13)). The procedure corresponds to the generation of a parity bit of a three bit data array (cp. Fig. 3). It is possible to determine up to four protection bits of one data nibble. Since three protection bits are required for simple error correction, only the fourth one is omitted. This has an effect on error recognition, but not on error correction. The fourth bit is not foreseen, because it would increase the redundancy and, therefore, the transmission time of telegrams, too.

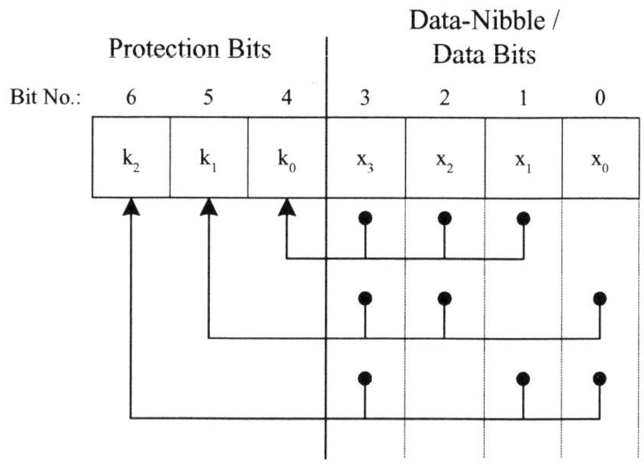


Fig. 3. Generation of the protection bits

The following equations show the calculation for the three correction bits:

$$k_0 = MOD_2(x_1, x_2, x_3) \quad (11)$$

$$k_1 = MOD_2(x_0, x_2, x_3) \quad (12)$$

$$k_2 = MOD_2(x_0, x_1, x_3) \quad (13)$$

### 4.2 *Decoding Secured Data Nibbles*

When decoding a secured data nibble, the protection bits are separated from the data nibble. From the four received data bits, three new control bits are generated according to the equations ((11)...(13)) again. These generated control bits are matched by an exclusive OR function (cp. Fig. 4) with the received protection bits and analysed according to Table 1.

In contrast to the prevailing method, viz., that a data repetition is requested when an error is recognised, the time required for a corrective action based on Hamming encoding is predictable. The price for this advantage is the higher redundancy of the telegrams and the correspondingly increased transmission time for the proper data.

## 5. SIGNAL ENCODING

Disturbances on transmission channels caused by electro-magnetic interference fields are normally impulses that overlay the traffic signal and cause amplitude errors. Correspondingly, transmitted telegrams are falsified. For this reason, a modified Frequency-Shift-Keying procedure (FSK) is employed for the transfer of signals. Here, depending on the state, not only the frequency is changed, but the DC part is changed as well. With this modification, the respective state can be obtained both via the frequency and via the DC part. Therefore, the evaluation of the frequency is reliable during low frequency troubles, since the DC part changes during these troubles. On the other hand, the evaluation of the DC part is advantageous during high frequency troubles, since these can make themselves noticeable as a change in frequency.

Additionally, through missing FSK parts one can distinguish between pure data signals and synchronisation signals (cp. Fig. 5). A corresponding evaluation logic decides on the respective state in the receiver chips and, in this way, simplifies alignment between sender and receiver.

In contrast to the prevailing method, viz., that a data repetition is requested when an error is recognised, the time required for a corrective action based on Hamming encoding is predictable. The price for this advantage is the higher redundancy of the telegrams and the correspondingly increased transmission time for the proper data.

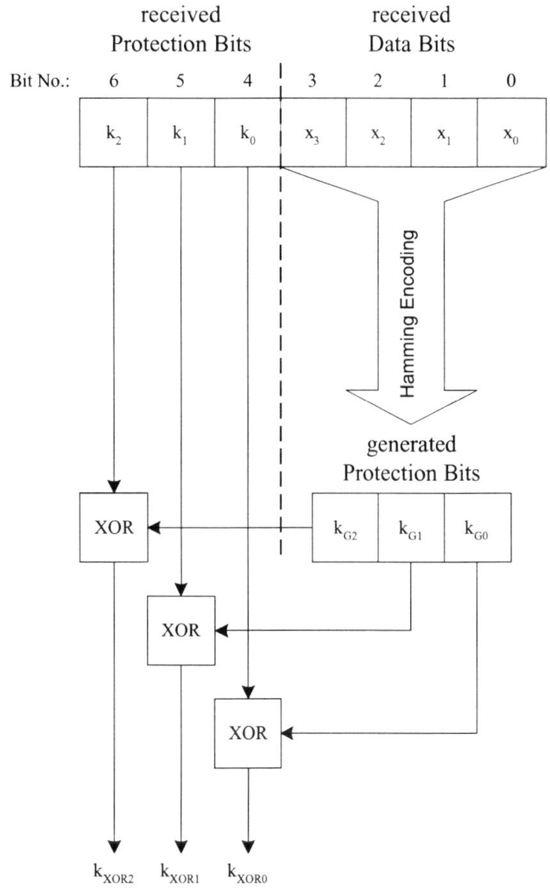


Fig. 4. Decoding of the protection bits

## 6. CONCLUSION

With a commercial and inexpensive GPS receiver and a special timer ASIC, which was developed at FernUniversitt, alarm jobs can be arranged with highly precise UTC in all units of a distributed system. By the presented time measurement procedure on the dual ring bus, all slaves can be accurately synchronised with UTC over the fieldbus.

In addition, transmission errors on the dual ring bus are correctable due to a Hamming encoding. The detection of errors is secured by employing both a Frequency-Shift-Keying procedure and an additional transfer of the same data telegram on the second channel. These mechanisms for error recognition and error correction allow for an almost secure data transfer of the telegrams, even in the presence of very strong disturbances on the transmission channels.

Therefore, this system can be used for fault tolerant, time precise and jitter free transmission in real time fieldbus applications.

## 7. FURTHER IMPLEMENTATION POSSIBILITIES

Other communication units can also be equipped with the Hamming encoding very easily. An (E)(P)ROM, in which the 7 bit Hamming code of all possible data nibbles is contained, can be implemented very easily and cheaply, and can produce the result in only one step of table look-up, which is extremely fast.

Furthermore, the modified signal encoding increases the immunity against faults, and assumes a higher information content at the same transmission rate.

## 8. REFERENCES


Halang, W.A., M. Wannemacher, R. Lichtenecker, P.A. Laplante and A.D. Stoyen (1999). High-Precision Temporal Synchronization in Multimedia Systems. In *Handbook of Multimedia Computing*, B. Furht (Ed.), pp. 545 – 556. Boca Raton: CRC Press.

Lauber, R.J. (1989). *Prozeautomatisierung I.* 2nd edition. Berlin-Heidelberg-New York-London-Paris-Tokyo: Springer-Verlag.

Swoboda, J. (1973). *Codierung zur Fehlerkorrektur und Fehlererkennung.* Mnchen: R. Oldenburg Verlag.

VDI-Nachrichten (2000). *GPS-GSM-Kombi – Mit Chip zur exakten Position.* Issue of 12 May 2000.

Table 1: Assignment of the bit errors

| Result XOR | | | Evaluation (cp. Fig. 3) | | |
|---|---|---|---|---|---|
| $K_{XOR2} = K_2 \not\equiv K_{G2}$ | $K_{XOR1} = K_1 \not\equiv K_{G1}$ | $K_{XOR0} = K_0 \not\equiv K_{G0}$ | Error Bit | Bit No. | Error |
| low | low | low | - | - | false |
| low | low | high | $k_0$ | 4 | true |
| low | high | low | $k_1$ | 5 | true |
| low | high | high | $x_2$ | 2 | true |
| high | low | low | $k_2$ | 6 | true |
| high | low | high | $x_1$ | 1 | true |
| high | high | low | $x_0$ | 0 | true |
| high | high | high | $x_3$ | 3 | true |

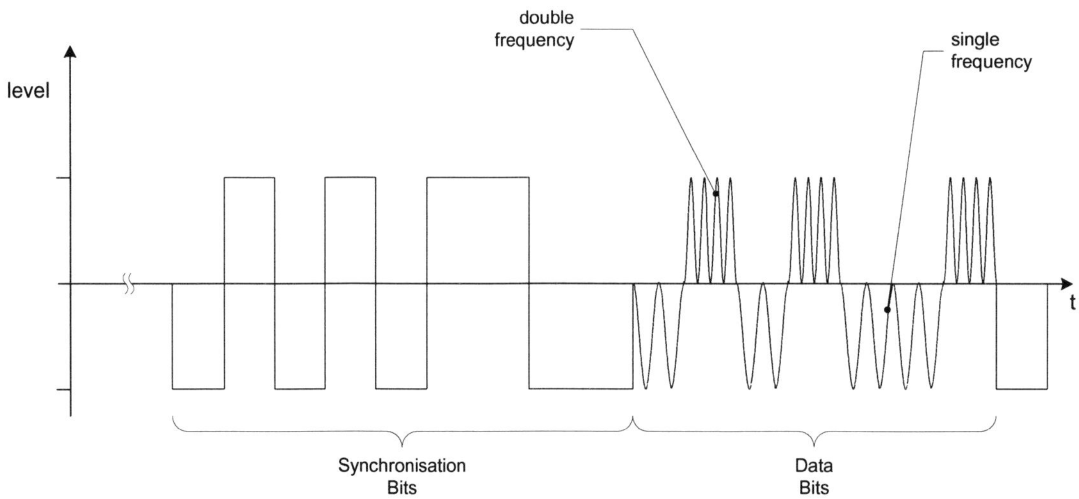


Fig. 5. Signal encoding of bits

IFAC Publications
www.elsevier.com/locate/ifac

# APPLICATION OF THE HAZAPS METHODOLOGY TO THE SAFE DESIGN OF COMPUTER-CONTROLLED PIPELESS PLANTS

F. Mushtaq* and P.W.H. Chung**

*Department of Chemical Engineering, Loughborough University,
Loughborough, Leicestershire, LE11 3TU, UK.
F.Mushtaq1@lboro.ac.uk
**Department of Computer Science, Loughborough University,
Loughborough, Leicestershire, LE11 3TU, UK.
P.W.H. Chung@lboro.ac.uk

Abstract: A way forward in multi-product manufacturing is the emerging technology of "pipeless plants". The basic idea of a pipeless plant is to move the process vessel between fixed stations for mixing, separation and other activities. The elimination of pipework for the transfer of material offers great flexibility for change, and allows a company to respond quickly to market demands and technological advances. Little research has been found on the safe design of pipeless plants, which can be split into three aspects: process, computer-control, and schedule safety. This paper is focused on the subject of computer control safety. The HAZAPS methodology, using Process Control Event Diagram (PCED) to represent control logic is described. A case study is presented to demonstrate the application of the methodology and the usefulness of the support tool. *Copyright © 2001 IFAC*

Keywords: hazard identification, computer control, pipeless plants

## 1. INTRODUCTION

In technical terms a pipeless batch plant is an arrangement of many units, such as movable reactors and other functional stations, which work co-operatively by avoiding collision or conflict (Kuroda and Ishida, 1993). To understand this it can be likened to a chemical laboratory (Niwa, 1993), where a beaker or flask could be a "mobile vessel", and the laboratory equipment such as the Bunsen burners are "processing stations". Usually all the operations required to produce a product are carried out in a single flask which is moved around to different processing stations. Therefore the basic idea of a pipeless plant is to move the process vessel between fixed stations for mixing, separation and other activities. The elimination of pipework for the transfer of material enables pipeless plants to be more flexible than conventional batch plants. Other benefits include reduced product loss, ease of cleaning, and reduced inventories by just-In-Time processing (Zanetti, 1992). Companies, mainly in Japan, already consider pipeless plants to be beneficial to their business requirements (Japan Supplement to Chemical Engineer, 1997; Kamiya and Fouchy, 1997; Shimatani and Okuda, 1992). However, the flexibility that the pipeless plant offers, which allows changes to be made quickly and easily introduces new hazards. Changes always need to be considered carefully before being implemented.

There are a small number of papers written on the subject of pipeless plants. The majority however have not considered the safety issues in any depth. These are in terms of the sharing of resources, problems with certain process operations (leading to thermal runaways), human involvement (batch processes involve numerous startups and shutdowns), movement of vessels (crossing paths on a network of tracks), and altering of set-points for control (such as level and temperature) for each batch. These issues can be examined under three main aspects of safety in the design of computer-controlled pipeless plants: process, computer-control, and schedule. Some of the issues concerned with process safety have been described elsewhere (Mushtaq and Chung, 2000a), and would pick up problems with cross-contamination, double-charging etc. in the same way as in a conventional multi-product batch plant. The movement of the vessels and in particular the relationship between scheduling and plant layout have also been examined (Mushtaq and Chung, 2000b). The focus of this paper is the computer control safety.

There are a number of hazard identification and hazard analysis methods available, which can either be qualitative or quantitative in nature (Wells, 1996).

*Qualitative methods:* Checklists, What-If Reviews, HAZOP Reviews, Preliminary Hazard Analysis, Failure Modes Analysis (FMA).
*Quantitative methods:* Event Trees, Fault Trees, Failure Modes and Effects Analysis (FMEA).

These hazard identification techniques are useful for identifying process and component failures and their effects. However, they are not appropriate for computer controlled systems as they do not consider the control logic, the operating sequence, or the control algorithm, and are not suitable for identifying potential hazards due to computer system failures or inappropriate responses from the computer system. Therefore new or modified hazard identification techniques are required for studying computer control systems. Hazard and Operability (HAZOP) Study is a method of systematically identifying every conceivable process deviation, its abnormal causes and adverse hazardous consequences in a chemical plant (Kletz, 1999). The technique has been standardised for examining process safety in continuous processes. There already exist modified versions of HAZOP for computer-related systems (Andow, 1991; Nimmo 1994) called CHAZOP or PES (Programmable Electronics Systems) HAZOP.

A framework similar to HAZOP is used that replaces existing process-related guide words and deviations with computer-related ones. Different sets of words could also be used, one set considering the hardware and logic of the system, and the other for considering human factors. During a CHAZOP the team would go through a diagrammatic representation of the system by considering all the links between different components on the diagram. Possible deviations from the design intent being investigated systematically applying the guide words to attributes such as data flow and control flow. Not all guide words are applicable, therefore inappropriate ones should be removed from the study list during the planning stage, and new ones may be added during the study at the discretion of the leader. It is also suggested that there should be a preliminary and a full CHAZOP. The preliminary to identify critical factors that influence the overall architecture and functionality of the system. The full CHAZOP to be carried out in greater detail later in the design.

There is no specific right method for carrying out a CHAZOP, and there is little experience in applying the different methodologies. However the view is that software cannot be considered alone and the framework for any study should include hardware, software and the environment in which they operate (including human interaction). The main strength of conventional HAZOP is that it facilitates systematic exploratory thinking, with the use of guide words and deviations. The problem of using the various CHAZOP schemes is that they are not only specific to a particular study, they are also influenced by their creators, and therefore the relevance of certain guide words/headings and questions is questioned. The suitability of the different schemes will only become evident through practical applications.

A general methodology has been developed based on incident analysis, i.e. using information from past incidents involving safety-critical systems (Kletz et al, 1995). The strategy behind the methodology was to derive a set of questions for a particular incident, asking why the incident occurred and what could have prevented it, and then to generalise these questions. This methodology is called HAZAPS.

## 2. HAZARD IDENTIFICATION FRAMEWORK FOR COMPUTER-CONTROLLED PROCESSES

The hazard analysis framework has been proposed as a qualitative, functional model-based approach to hazard identification for computer-controlled chemical plants at the specification stage of the control system life cycle (Chung et al, 1999). It is an extension of the HAZAPS methodology (Broomfield and Chung, 1997; Yang and Chung, 1998). Methods used include fault tree analysis, use of generic hazards list, structured design and object orientated analysis. The procedure for applying the framework is split into five stages.

### *2.1 Stage 1. Safety Requirements Identification*

The system is sub-divided and the safety-critical subsystems identified, e.g. a high temperature reactor. The safety requirements are constructed by examination of the environment and identification of potential hazards. A chemical process can be divided into smaller parts on the basis of operating units with a standard function. Safety requirements are expressed in terms of critical states that certain operating units should not reach. The control logic should prevent the system from reaching these critical states.

### *2.2 Stage 2. Process and Control Logic Representation*

The control logic to achieve a required function is specified. It is expressed at a high level so that it can be examined by a review team made of experts in different fields. The Process Control Event Diagram (PCED) has been developed for this purpose (Andow, 1991; Nimmo, 1994).

### *2.3 Stage 3. Verification of Control Logic*

The control logic has to be shown to satisfy the safety requirements of the system before being implemented. The PCED is used to determine the state transitions of the control logic. The final and all intermediate states have to be safe for the control logic to be acceptable. It this is not the case then the control logic needs to be modified and verified again.

### *2.4 Stage 4. Safety Critical Events Identification*

By the introduction of deviations from normal behaviour for each control action, e.g. no signal, the causes and consequences (response of the control logic) of each deviation can be identified. If the deviation causes the system to move into a state that violates the safety requirements, then it may be necessary to modify the control logic, or to prevent the causes of the deviation from occurring.

### *2.5 Stage 5. Application of Life Cycle Question Library*

Some safety critical events can be prevented from occurring by asking appropriate questions early in the design stage. A safety question library, based on experience from industrial incident reports, is part of the computer support tool. The library is structured so that relevant questions can be located easily.

## 3. APPLICATION OF THE FRAMEWORK

The first step of the framework is process representation. This involves the sub-division of the plant and identification of safety-critical subsystems. This fits very well with the pipeless plant concept as the plant is already divided into dedicated functional stations. In addition, for each station the pipeless plant HAZOP has already indicated the safety critical subsystems, and the safety requirements of those systems.

Unfortunately existing representations such as Piping and Instrumentation Diagrams (P&IDs) and Signed Directed Graphs (SDGs) do not represent precisely the structural and behavioural information of computer control systems. The Process Control Event Diagram (PCED) representation described in the paper complements P&IDs. It consists of seven functional levels: human, input device and display, communication 1, computer, communication 2, sensor and actuator, and process.

The PCED represents the system in terms of events, order of events, components, and control and data flow. It is made up of:

- labelled arcs: propagation of signal, or the causal action or effect
- nodes: objects, input/output variables, or computations.

The PCED model is based on the sequential behaviour of a computer, and the order of events can be read from left to right. A simple example of a PCED is shown below.

The content of the computation node is simply:

If temperature < setpoint then goto N1

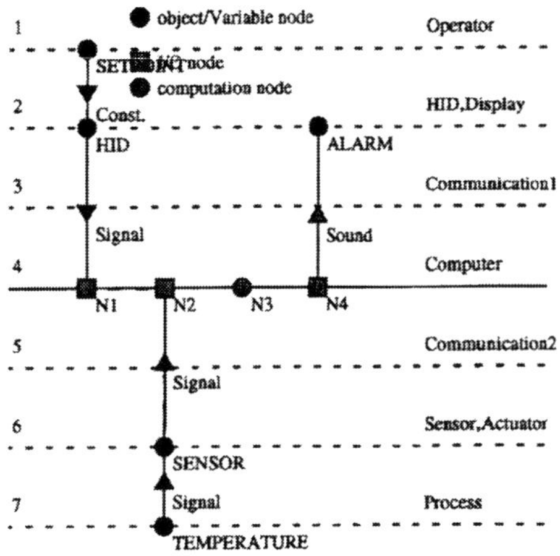


Fig. 1. Simple Example PCED

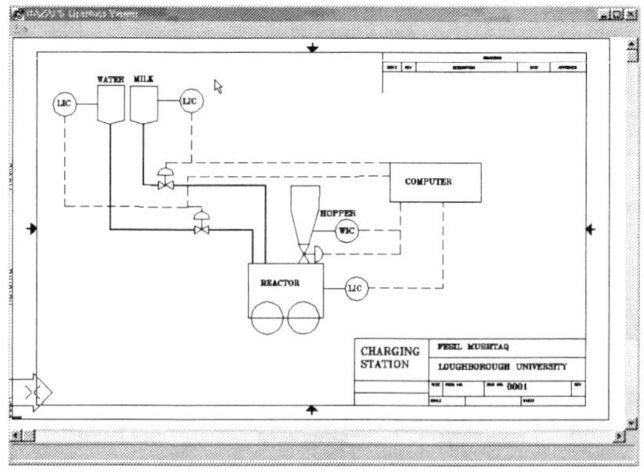

Fig. 2. Charging Station P&ID in the Graphics Viewer

Not only the control logic is expressed but also the effect of each control action on the process. This representation can be understood by people from different disciplines, and can facilitate discussion and identification of errors and hazards. This takes the form of a computer HAZOP. Possible deviations to the design intent are introduced using guide words, the same as conventional HAZOPs but with different definitions, as prompts for each action in the PCED.

This process of hazard identification is time-consuming especially for complex chemical plants. Computer-aided identification makes the process far more efficient. The support tool described implements the hazard analysis framework, records all the information generated during the analysis and produces customised reports (Chung et al, 1999).

## 4. CASE STUDY: THE CHARGING STATION IN AN INDUSTRIAL BAKERY PIPELESS PLANT

An automated charging system is to be used to feed the dry and liquid ingredients to a mobile vessel/reactor. The computer monitors the reactor level via the level control, and ensures that the correct amount of each ingredient is charged. This is achieved by the computer controlling three control valves: the milk inlet valve, the water inlet valve, and the hopper feed valve. There is an alarm to indicate a low/high level in the reactor, to signal that operator intervention is required, as well as the automatic response from the control system. The support tool allows the P&ID of the process to be viewed using the Graphic Viewer item in the TOOLS menu (Figure 2).

### *4.1 Identify Safety Requirements*

One of the requirements, Reqm-No.1, is the prevention of the overcharging of the vessel. The control logic for this requirement is input through the control logic editor (figure 3). From the control logic the PCED is generated (figure 4). The state transition for the control logic is shown in table 1.

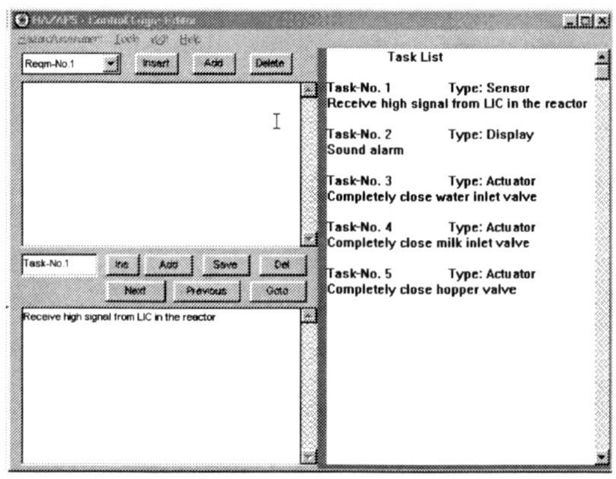


Fig. 3. Screenshot of control logic editor

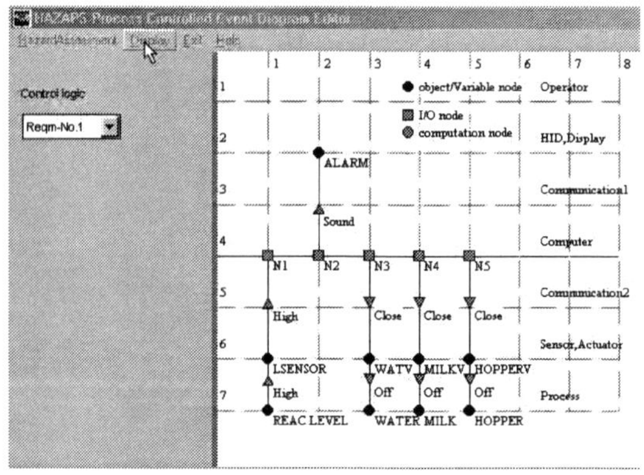


Fig. 4. Screenshot of PCED for Reqm-No.1

Table 1 State Transitions for the Control Logic

| State | Vessel Level | Alarm | Water IN (WATV) | Milk IN (MILKV) | Hopper IN (HOPPERV) |
|---|---|---|---|---|---|
| S1 | low | Silent | high | high | high |
| | | | ↓ N1 | | |
| S2 | high | Silent | high | high | high |
| | | | ↓ N2 | | |
| S3 | high | Sound | high | high | high |
| | | | ↓ N3, N4, N5 | | |
| S4 | high | Sound | closed | closed | closed |

*4.2 Safety Critical Events Verification*

Guide words are used to investigate deviations from normal behaviour for the control system. CHAZOP results are generated using the support tool to minute the review. A sample is shown in table 2 below.

Table 2 Partial CHAZOP Analysis Results

| HAZOP Item | GUIDE WORD | DEVIATION | CAUSES | CONSEQUENCES | ACTIONS |
|---|---|---|---|---|---|
| N1 | NO | No signal from liquid level sensor | Liquid level sensor is out of order | Overflow of reactor is not identified, and may not be prevented | Install duplicate sensor, require rigorous maintenance schedule |
| | MORE | Reactor liquid level outside normal limit | Liquid level sensor has not been calibrated for a long time | Batch is out of spec | Double check reactor level, with input of raw materials Recalibrate the reactor level sensor regularly |
| | LESS | Setpoint is incorrect | The operator made an error inputting the setpoint | The input of raw materials does not tally with the level sensor | In the event of the two figures not being the same operator to recheck setpoint |

*4.3 Safety Related Questions*

The next step in the analysis is the Safety Related Questions. Using the library of questions based on analysis of incident reports, safety questions are asked relevant to the identified safety critical events. The questions can be put into three categories: specification, implementation, and protection. For example the type of questions to be asked about the sensor of the level control system:

*Specification*
What state is to be monitored?
Why is this state to be monitored?
When is this state to be monitored?
How is this state to be monitored?

*Implementation*
What sensor is to be used?
How will it be installed?

*Protection*
What alarms and/or trips are associated with this task?
Why are these alarms/trips required?
What is the setting of the alarms/trips?

This is just a small sample of the type of questions to be asked about one sensor. This shows the thoroughness of the procedure. There are over 170 questions available, and therefore it would be time consuming to go through in this case study especially without the expert knowledge required. A screenshot of this section of the support tool is shown in figure 5.

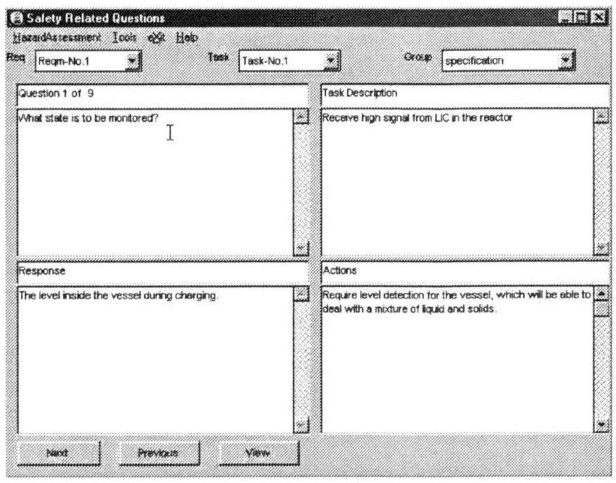


Fig. 5. Screenshot of Safety Related Questions for the level sensor in the vessel

*4.4 Improved Design for the Charging Stations*

It was decided to divide the charging station into separate stations for liquid and dry charging ingredients. It was further determined to split the liquid charging station into separate milk charging and water charging stations. By having dedicated stations charging independently many of the problems identified were dealt with, and the control is much simpler from an operational point of view. Also as the computer only has regulate the charging of one material (one control valve) at a time it is better from a safety point of view.

## 5. CONCLUSIONS

Standard hazard identification techniques do not consider the control logic, the operating sequence, or the control algorithm, and are not suitable for identifying potential hazards due to computer system failures or inappropriate responses from the computer system. CHAZOP extends the HAZOP technique to examine computer control. A qualitative, functional model-based approach for hazard identification based on the Process Control

Event Diagram (PCED) has been described. Its application and the usefulness of the support tool have been demonstrated by the use of a case study: a charging station in an industrial bakery pipeless plant. The support tool is easy to use and understand, and allows a review team to examine the control system. The framework does rely on the developer of the system to identify the top level hazards. As an improvement the starting point of the framework, the Safety Requirements Specification could be linked to the outputs of a preliminary process HAZOP analysis, especially associated with the control actions.

## REFERENCES


Andow, P. (1991). Guidance on HAZOP procedures for computer-controlled plants, *Contract Research Report*, No.26, Her Majesty's Stationary Office.

Broomfield, E.J. and P.W.H. Chung (1997). Safety assessment and the software requirements specification, *Reliability Engineering and System Safety*, 55, 295-309.

Chung, P.W.H., S.H. Yang and D.W. Edwards (1999). Hazard Identification in Batch and Continuous Computer-Controlled Plants, *Ind.Eng. Chem. Res.*, 38, 4359-4371.

Japan Supplement to Chemical Engineer (1997). Equipment Makers Showcase Technology, Internet location: http://www.che.com/ce-9705/05jp4.htm

Kamiya, T. and K. Fouhy (1997). Pipeless plants - Non-reactive processes are still the focus, Chemical Engineering, ACHEMA 97, Internet location: http://www.che.com/ *html/adlpls3.htm*

Kletz, T.A. (1999). HAZOP & HAZAN, 4th ed., *Loss Prevention*, Institution of Chemical Engineers, Rugby, England.

Kletz, T.A., P.W.H. Chung, E.J. Broomfield and C. Shen-Orr (1995). Computer Control and Human Error, *Institution of Chemical Engineers*, Rugby, England.

Kuroda, C. and M. Ishida (1993). A Proposal for Decentralized Cooperative Decision-Making in Chemical Batch Operation, *Engng Applic Artif Intell*, 6 (5): 399-407.

Mushtaq, F. and P.W.H. Chung (2000a). A systematic Hazop procedure for batch processes, and its application to pipeless plants, *Journal of Loss Prevention in the Process Industries*, 1: 41-48.

Mushtaq, F. and P.W.H. Chung (2000b). Discrete Event Simulation in the Design, Layout and Scheduling of Pipeless Batch Plants, *Hazards XV, The Process, its Safety, and the Environment – Getting it Right! Institution of Chemical Engineers,* Rugby, UK.

Nimmo, I. (1994). Extend HAZOP to computer control systems, *Chemical Engineering Progress*, October, pp 32-34.

Niwa, T. (1993). Pipeless Plants Boost Batch Processing, *Chemical Engineering*, 6/93: 102-108.

Shimatani, T. and O. Okuda (1992). 'Pipeless' Batch Chemical Plants Offer A New Approach, *Chemical Engineering*, 10/92: 181-182.

Wells, G. (1996). Hazard Identification and Risk Assessment, *Institution of Chemical Engineers*, Rugby, England.

Yang, S. and P.W.H. Chung (1998). Hazard analysis and support tool for computer controlled processes, *Journal of Loss Prevention in the Process Industries*, 11, 333-345.

Zanetti, R. (1992). The 'Pipeless' Batch Plant, *Chemical Engineering*, 6/92: 5.

IFAC Publications
www.elsevier.com/locate/ifac

# BEHAVIOURAL VALIDATION FROM A FORMAL SPECIFICATION OF SMART EQUIPMENT

**Chris Choukair, Mireille Bayart**

LAIL (CNRS UPRESA 8021), Bât. Eudil,
Université des Sciences et Technologies de Lille,
59655 Villeneuve d'Ascq Cedex - France,
Tel. : (33)3 20 43 45 65 - FAX : (33)3 20 33 71 89,
E-mail: Choukair@univ-lille1.fr, *bayart@univ-lille1.fr*

Abstract: In this paper, we propose a formal description of a component, which has been established, with objectives of system evaluation. After a short introduction to the concept of component, its external model, integrating notions of services, versions and modes, are recalled to formalise the component description. Finally, a synchronous language is used to compute the formal description, what which allows to evaluate the component. Such description can be applied to smart equipment considered alone, but also to component of global architecture. *Copyright © 2001 IFAC*

Keywords: formal specification, synchronous computation, safety evaluation

## 1- INTRODUCTION

"Distribution" and "real time" characterise current systems and new problems of design seems better ensue from it. The main objectives are the reducing of time and of costs along developments, what which consequently implies the reuse of existing equipment. On the other hand, various works are realised to analyse the safety and the robust performances of operational architecture. Considering systems like a collection of identified and interacting components which constitutes its infrastructure (Villemeur, 1988), the modelling by languages of components (Medvidovic and Taylor, 2000) could permit to introduce safety analysis but it is not always right.

Methods to analyse safety like SDM (Success Diagram Method) (Villemeur, 88) or FMEA (Blanke 96) (Failure Mode and Effects Analysis) are applied easily if architecture is defined. These methods are extremely sensible to the reconfiguration. That's why safety methods are generally employed on fixed architectures like hardware architectures. Various works on reconfiguration are realised considering only physical system, for which model is known (Gehin and Staroswiecki, 98). The interest of the presented approach will show in aim to study safety on intelligent distributed architecture before the realisation. The failures that are considered in this paper can provide of the physical part of the system but also from the hardware and software of the control system. The approach leads on the notion of external model, which has been defined for smart instrument. The external model constitutes a global view of the fundamental functionalities that the application offers to the users (Staroswiecki and Bayart, 1994). It just permits a global comprehension of the application without details. To validate the behavioural in case of failures, it's necessary to complete the model. The interest is to validate the component specification and to verify the set of configurations and their transitions in order to insure a good behaviour in case of failures during the design step.

The concept of the component is firstly presented, then its external model, used to formalise the component description, is introduced. Safety aspects of the model are described in part three. Next parts are devoted to behavioural description based on the SYNCCHARTS (André *et al.*, 96) formalism. After a short description of this formalism, the manner to derive a SYNCCHARTS representation from a part of the external model in order to complete the component description is detailed. The final section presents the integration of the behavioural aspect in regards of activities and versions of services inside a component, taking into account availability of each service. Following by the different functioning states of a component, functional and behavioural evaluation can be obtained.

## 2. THE COMPONENT

A system can be seen as an association of components. Several points of view will be considered to express the notion of component. In the paper, the following definition is adopted: A component is the smallest elementary data processing entity. The arbitrary aspect of the definition allows the user to decide on the decomposition level that he/she wants to consider according to his/her application, and the reuse capacities. This data processing could be implemented on operational architecture in software or in hardware devices. A component is characterised by the following properties:

- It owns a functional identity; each component performs a function of the application to which services are associated with the restriction that a given component renders a sole service. In case of duplication of data processing, several components provide the same service.
- It owns a geographical identity; it is not distributed over several locations
- It owns an execution identity; it executes a set of services according to specific requests.

A whole of standardisation works are realised to specify object in order to describe distributed applications, in particular with the development of industrial networks (Diedrich and Neumann, 1998) following different architecture description languages (Medvidovic and Taylor 2000). The interest of such approach is to show how to integrate safety inside the specification of the component.

## 3. THE EXTERNAL MODEL OF A COMPONENT

The external model describes the device from the point of view of the services that it is able to provide to external entities (operators, other field instruments, computers...). It is used to describe from an external point of view the component. It introduces different notions given below (Staroswiecki and Bayart, 1994; Staroswiecki and Bayart, 1996):

A **service** is defined as a procedure whose execution results in the modification of at least one datum in the device data base, or/and at least one signal on its output interface.

Services are required by the **users** who intervene on the equipment during its whole life cycle, i.e. not only during its exploitation (supervision, maintenance, technical management) but through out its life cycle, from its conception to its dismantling (initialisation, configuration,...). Two different users can be differed (Alakl and Lalanne, 1989): the **consumers** who are the most important one in terms of users needs, and the **actors** who either give information to the device or constraint it.

The description of a service consists in the description of the result, which is produced by its execution, i.e. the outputs.

In order to define the obtained values, one will have to describe the computations which are done (algorithmic or sequential procedures, qualitative or fuzzy inferences,...), its internal configuration (parameters), the variables on which they are applied (inputs) and the required resources (hardware, software). Moreover, before it can be executed, a service must verify some activation conditions. So, it is described by (figure 1):

<Service $s_i$>::=<Inputs $C_i$, Outputs $P_i$, Parameters $p_i$, Procedure $T_i$, Activation Condition $CA_i$, Resources $R_i$>

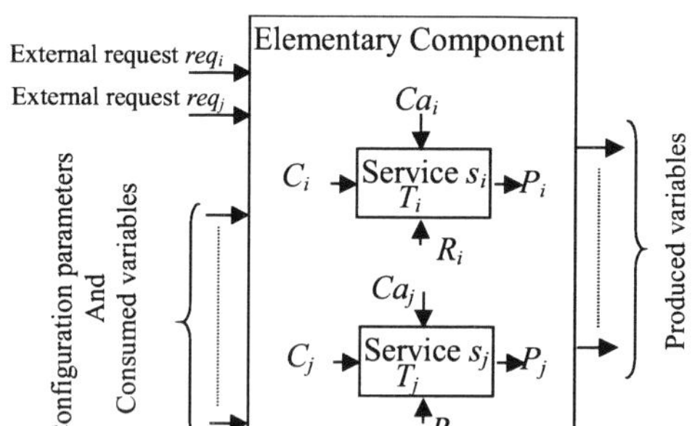

Figure 1: Services inside component

The execution of services can be either dependent (precedence, mutual exclusion,...) or independent and concurrent. Likewise, the service can have a limited duration or can end on the occurrence of simple or complex events (operator request, emergency alarm,...). Finally, the services are organised according to **user operating modes**. Few **versions** for each service have to be defined in aim to take into account different states of the component. For each version, a treatment is declared.

The availability of service is function of the resources that it uses. At precise instant the service could be nominal, degraded, or out of order. The execution of a service depends on its availability, and therefore the formal specification of any service necessitates the description of all behaviours following its availability. The version of service is associated to the global state of equipment. This global state is modified when defaults append on resources.

A **User Operating Mode** (USOM) is a coherent sub-set of services. It contains at least one service, and each service belongs at least to one USOM. Moreover, in each USOM, there exists a notion of context. The context is the subset of services of the USOM that are implicitly executed (implicit request) as long as the system remains in the given USOM. The other services of the USOM are the "*requestable*" services.

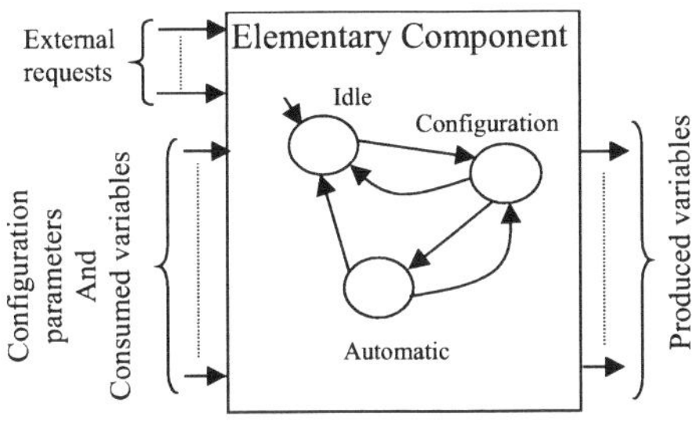


Figure 2: Services management inside component

The lack of resource in a component renders unavailable certain services of this component. According to the state of the resources of the various components, the modes of functioning of the application can be defined.

## 4. SAFETY ASPECTS OF COMPONENT MODEL

One can associates a mode of functioning of system to an image of system availability. It is a global state in regard of availability of all its external services (and so, of resources) in the current USOM chosen by user. In a formal description, faults must be detected to express system availability and to represent it with a set of modes of functioning (Choukair and Bayart, 2000).

In the nominal functioning of the application, which corresponds to a configuration, a set of external services is proposed. Each service is necessary to insure the purpose of the configuration. According to the set of hardware resources, it is possible to elaborate a management of functioning mode of a global architecture according to the state of component resources (figure 3). Each macro state represents external model for each configuration making degraded application. Each sub-state represents an USOM, which describes services that users can activate.

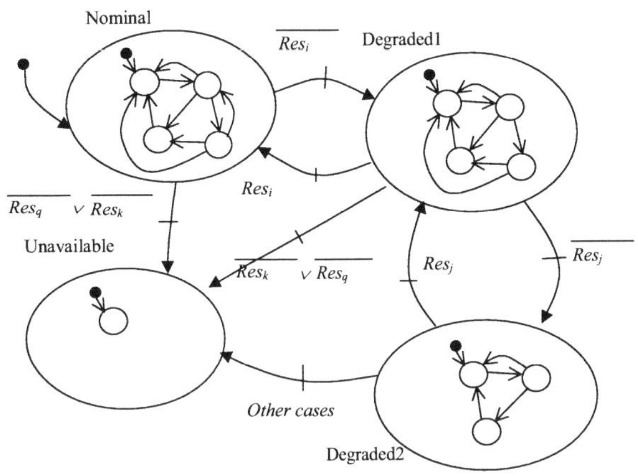


Figure 3: Management of Modes of functioning

Model changes of mode of functioning if one external service can't be executed according to unavailability of one component resource, that detection is function of capacity of implemented FDI algorithm. The study of such graph allows to analyse reconfiguration capabilities.

Nevertheless, the computation of such automaton is less complex if we are considering the versions of services like the alone expressions of modes of functioning inside the component.

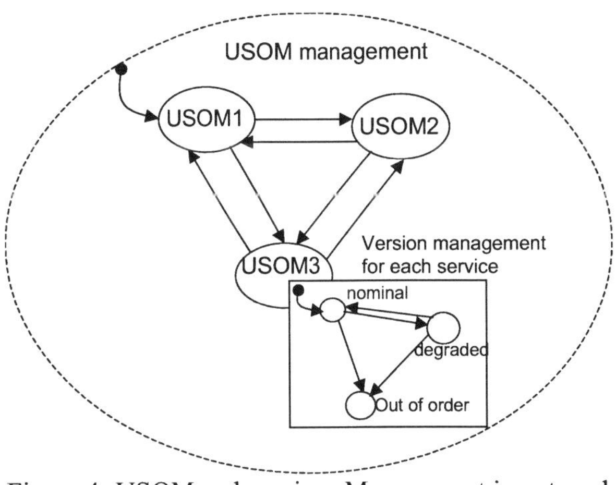


Figure 4: USOM and versions Management in external model

To specify the component, the formal language has to integrate hierarchical aspects to compute in few levels USOMs, services and versions. To compute this model, a choice must done on a language. It is practical to use an existent language, otherwise a new language has to be developed. The choice has to hold on a synchronous language well adapted to the external model.

## 5. THE SYNCCHARTS FORMALISM

"SYNCCHARTS" is an acronym for Synchronous Charts. It inherits from STATECHARTS (Harel, 1987) and ARGOS (Maraninchi, 1990). It is a graphical representation of reactive behaviours based on the synchronous paradigm. It offers enhanced pre-emption capabilities and any syncChart can be translated into an equivalent Esterel program. Recall that Esterel (Boussinot and De Simone, 1991) is a powerful imperative synchronous language dedicated to reactive system programming.

Let us have a look at the hypotheses underlying the synchronous approach (André, 1996):

- **Signals**: the system interacts with its environment through signals: input signals and output signals.
- **Global Perception**: it is assumed that all input and output signals are perceived simultaneously and that this perception is objective. So the model deals with tuples of signals.
- **Logical Time**: there is no physical time, but logical instants.
- **Zero-Delay Hypothesis**: internal operations are supposed to be executed in zero-delay. So the output signals are synchronous with the inputs that cause them.
- **Broadcasting**: all signals are instantaneously broadcast.

In SYNCCHARTS, pre-emption is a first class concept. There are two types of pre-emption: suspension and abortion. Abortion can be either weak or strong: the strong one kills the process immediately and the weak one kills the process after it executes its current reaction. SYNCCHARTS, like Esterel, deals with sequence, concurrency, pre-emption and communication in a fully deterministic way.

SYNCCHARTS is endowed with a mathematically defined semantics, fully compatible with the ESTEREL one. The main differences between SYNCCHARTS and STATECHARTS are a stricter semantics and a richer pre-emption management for the former.
Of course, SYNCCHARTS has its own graphical representation. It is a state-based description of a reactive behaviour. It supports states, hierarchy of states, concurrency and transitions of several types. The basic block is the state or **star** (Figure 5):

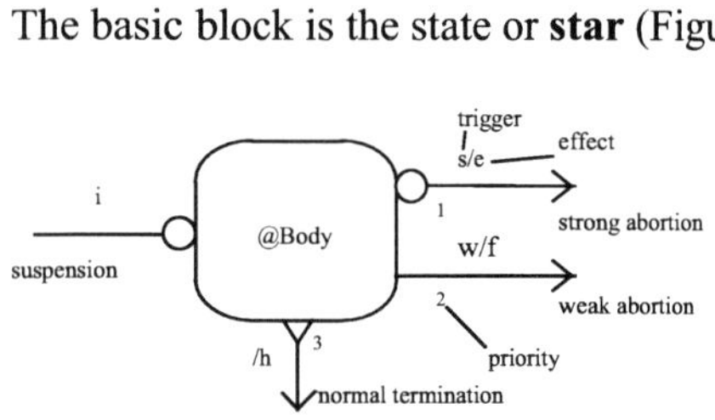


Figure 5: The star of a syncChart.

Stars are interconnected to make a **constellation**. Only one star at a time can be active in a constellation, so that a constellation can be seen as a classical state-graph. Each constellation has, at least, one initial star pointed to by an arrow. A parallel composition of constellations is a firmament or a **macro-state**. Dashed lines delimit concurrent constellations in a macro-state.

The structure of a syncChart is defined recursively: the body of a star can be a macro-state (or even an Esterel module, i.e., a textual description). This feature is convenient for supporting hierarchical descriptions.

A prototype SYNCCHARTS editor/compiler is available[1]; a commercial version also exists. The SYNCCHARTS compiler generates an Esterel program equivalent to the syncChart. Thus, SYNCCHARTS can use the rich software environment developed for Esterel (compilers, links to proof systems, interactive simulation, and efficient code generation).

## 6. THE COMPONENT FOLLOWING A SYNCHRONOUS APPROACH

In the case of a component X, this one is considered with its external model and the following syncChart is obtained (figure 6):

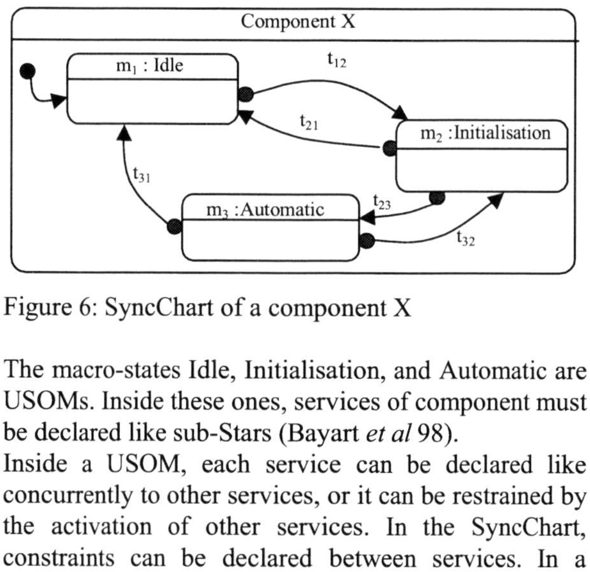


Figure 6: SyncChart of a component X

The macro-states Idle, Initialisation, and Automatic are USOMs. Inside these ones, services of component must be declared like sub-Stars (Bayart *et al* 98).
Inside a USOM, each service can be declared like concurrently to other services, or it can be restrained by the activation of other services. In the SyncChart, constraints can be declared between services. In a second time, for each service, behaviour can be specified.

### *6.1 - Behaviour per service with SYNCCHARTS*

The operational aspect of a service obliges to consider this one like an entity, which interacts with its temporal behaviour. Obviously, each service owns a typical

[1] http ://www.i3s.unice.fr/~andre

temporal model which permits to use notions like events (Vega, 1996), signals (André, 1996), and so on. The temporal attributes, that characterise a service, are events, which represent the different instants in regards of its solicitation and its execution:

- the acceptation of the request
- the taking into account of this acceptation
- the end of execution

Following these events, a temporal model of the service execution is provided. This model characterises the activity of a service after its activation. This one shows four different types of services following their periodicity and their duration or their reactivation:

- Case of non-periodic services with a limited duration or an infinite one.
- Case of periodic services with a limited number of reactivation or an infinite one.

All these cases can be interpreted with SYNCCHARTS like a specific automaton. In figure 7, the example of a non-periodic service with limited duration is shown.

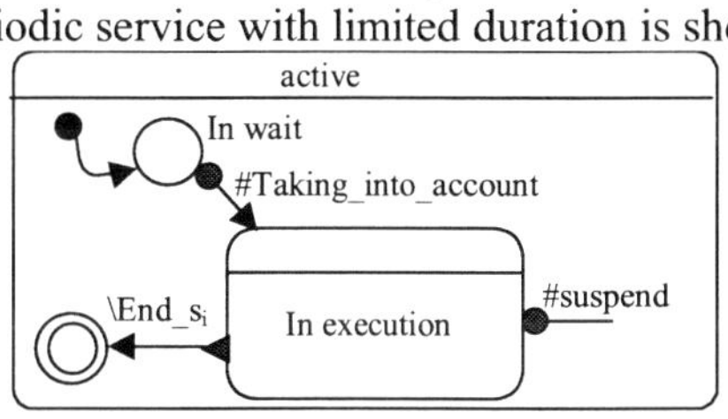


Figure 7: Example of non-periodic service with limited duration

For each type of service, a representation with SYNCCHARTS corresponding to the searched temporal behaviour can be done.

*6.2 - Activity management.*

According to the specification of our model in SYNCCHARTS, activity management of services can be integrated following their internal activity and their availability. The management of a service implies reactive mechanisms because it's a reactive entity that integrates pre-emption, suspension, concurrency of different aspects of its management, and so on. Consequently, synchronous language is used in order to specify automata as SYNCCHARTS for service management.

By using results of the European project SCEPTRE (Dorsieul and al, 1991), one can represent activity of any service with a hierarchical approach that describes different states of service activity. The activity of an external service is managed following our capacity to detect if:

- the service exists,
- it is available (if it exists),
- it is "activability" (if it is available),
- it is in waiting,
- it is in execution,
- it is suspend.

With such representation, one can elaborate a model specifying all these states and the conditions to progress inside such graph with SYNCCHARTS. One will remark that availability of a service and the execution of its activity is taken into account at different levels. A service can be either unavailable or available and non-active. But, it will be active only if it is available. According to the previous considerations, the SYNCCHARTS of figure 8 is obtained.

If the service is offered inside a current USOM, in a first level of the SYNCCHARTS (figure 8), the availability and the control of the activity of the service are specified concurrently. When the service exists it will be supposed available or degraded. It is authorised to activate it if it isn't in "Out Of Order" state. The versions of the service are taking into account inside the macro "In execution" with an adapted treatment following the current version (Nominal, Degraded1, OOO), (Choukair, 2000).

To activate the service an Activation Request may be necessary and Use Condition that characterises the service inside its USOM must be checked. The execution is effective when system takes into account the Activation Request. A suspend signal (suspend) permits to stop execution momentary. The end of the execution starts a signal (endexec), then the service returns in non-active state. Abortion is authorised to quit execution immediately.

Such description allows to validate the behavioural aspect of the component before its realisation. It can be completed in order to take into account the failures that can occur, in introducing the notion of degraded versions of services.

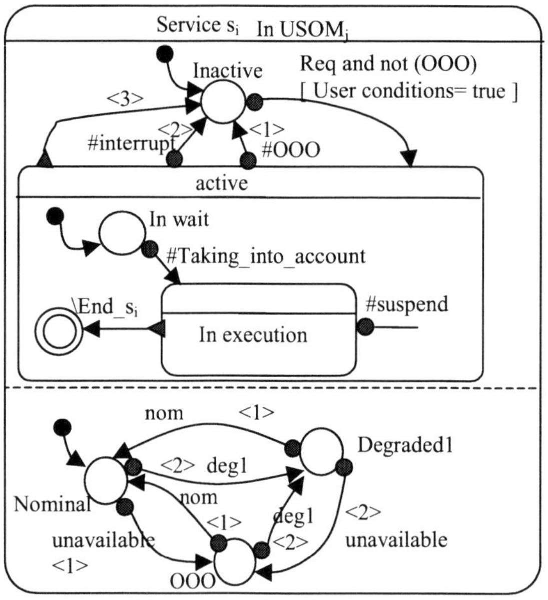


Figure 8: SYNCCHARTS model for service management.

## 7- CONCLUSION

Following the concept of component, a whole of items needs to be declared in order to formalise the component. All these different aspects of a component have to be adapted to a model. The external model allows to describe a component from the point of view of the users. It provides a functional description, which is not sufficient to validate entirely its behaviour. In that sense, a SYNCCHARTS description is proposed in order to complete the external model of component. The interest of this method rests on a hierarchical representation of user operating mode and a clear presentation of concurrent services in each mode. This language is well adapted to compute external model of component in aim to validate its functional and behavioural aspects. It is a first step to realise in design before to build architecture of components taking into account fault tolerance.

## REFERENCES


Alakl, N., J.L. Lalanne (1989), *"Architecture technologique des systèmes d'information: la méthode TACT"*, Editions d'Organisation, France.

André, C. F. Boulanger, M.A. Peraldi, J.P. Rigault and G. Vidal-Naquet (1996), Objects and synchronous programming. In: Int. Workshop on Reactive Systems Modelling, Brest, France pp; 55-62.

André, C. H. Boufaied and S. Dissoubray (1998). Synccharts : un modèle graphique synchrone pour les systèmes réactifs complexes. In: *Real-Time Systems* 1998. RTS'98, Paris, pp. 175-193.

Bayart, M., E. Lemaire, M.A. Péraldi, C. André (1998). "External Model and SyncCharts Description of an Automobile Cruise Control System", *Control Engineering Practice* 7 (10) pp. 1259-1267.

Blanke M. (1996) "Consistent Design of Dependable Control", *Control Engineering Practice*, 4 (9), pp. 1305-1312.

Boussinot, F. and De Simone, R. (1991). "The Esterel language.", *Proceeding of the IEEE*, Vol. 79 (9), pp. 1293-1304.

Choukair, C, (2000), Contribution à la conception d'architectures opérationnelles validées. PhD Thesis, Université des Sciences et Techniques de Lille1, Villeneuve d'Ascq, France.

Choukair, C., and M. Bayart (2000) Reconfigurable systems built from generic model and success diagram. In: *4th IFAC Symposium on SAFEPROCESS* Vol. 2, Budapest, pp. 768-773.

Diedrich, C. and P. Neumann (1998). "Standardisation in Automation Systems", *9th Symposium on Information Control in Manufacturing*, vol. 1, pp°51-56.

Dorseuil, A., P. Pillot (1991). *"Le Temps Réel en Milieu Industriel"*, Dunod, Paris.

Gehin, A.L., M. Staroswiecki (1998). " Analysis of the three tank benchmark reconfigurability using generic component models", In: *COSY Workshop*, Mulhouse, France.

Harel, D. (1987). "StateCharts, a Visual Formalism for Complex Systems", *Science of Computer Programming*, **8** (3) pp. 231-274.

Maraninchi, F.(1990). "ARGOS : un langage pour la conception, la description et la validation des systèmes réactifs.", PhD thesis, Université Joseph Fourier, Grenoble I, France.

Medvidovic, N. and R. N. Taylor (2000), "A Classification and Comparison Framework for Software Architecture Description Languages", *IEEE Transactions On Software Engineering*, Vol. **26** (1), pp °70-93.

Staroswiecki, M. and M. Bayart (1994). *Actionneurs intelligents*, Hermès, Paris.

Staroswiecki, M. and M. Bayart (1996). "Models and Languages for the Interoperability of Smart Instruments", *Automatica*, **Vol. 32**, pp. 853-873.

Vega, L. (1996). "Modèles de Coopération et de Communication entre Processus Temps Réel Répartis", PhD thesis, I.N.P.L., Vandoeuvre-les-Nancy, France.

Villemeur, A. (1988). *Sûreté de fonctionnement des Systèmes industriels*, Eyrolles, Paris.

IFAC Publications
www.elsevier.com/locate/ifac

# USE OF SURROGATE PULSE-RESPONSE MATRIX FOR THE UDCL

**S. H. Lam** [1]

*Department of Mechanical and Aerospace Engineering*
*D-209 Engineering Quadrangle*
*Princeton University, Princeton, NJ 08544, USA*

Abstract: The *universal dynamic control law* (UDCL) advocated by the author can achieve ODE-based control objectives without the need for *detailed* knowledge of the system itself. Thus it is very robust with respect to system uncertainties and unmodeled disturbances. The present paper shows, via an example, how to find a usable "surrogate pulse-response matrix" needed by this control law. Numerical simulations are presented. A brief discussion on design guidelines for sensors and actuators for easy UDCL implementation is also included. *Copyright © 2001 IFAC*

Keywords: Control Theory, Nonlinear Systems, Robust Control, Singular Perturbation Methods.

## 1. INTRODUCTION

Consider a $N$-dimensional system governed by the following first order ODEs:

$$\frac{d\mathbf{x}}{dt} = \mathbf{f} + \sum_{i=1}^{I} \mathbf{b}_i u^i, \tag{1}$$

where $\mathbf{x}$ is the state (column) vector, $\mathbf{f}(\mathbf{x}, t)$ and the $\mathbf{b}_i(\mathbf{x}, t)$'s are differentiable vector functions of their arguments. The second term on the right hand side represents the "forces" exerted by $I$ actuators ($I \leq N$). Physically, $\mathbf{f}$ represents the "open-loop" dynamics of the system, the $\mathbf{b}_i$'s represent the *direction* and the $u^i$'s the *amplitude* of the control forces. There are $I$ output variables $y^1, \ldots y^I$ which are related to $\mathbf{x}$ by:

$$y^i = \Psi^i(\mathbf{x}, t), \quad i = 1, \ldots, I. \tag{2}$$

where the $\Psi^i(\mathbf{x}, t)$'s are differentiable functions of $\mathbf{x}$ and $t$ but are strictly forbidden to have any dependence on the $u^i$'s. The $y^i$'s are assumed to be dimensionless and are nominally $O(1)$ entities.

The ODE system governing the $y^i(t)$'s is readily derived:

$$\frac{dy^i}{dt} = g^i + \sum_{i'=1}^{I} B^i_{i'} u^{i'} \tag{3}$$

where

$$g^i = \frac{\partial \Psi^i}{\partial t} + \mathbf{c}^i \cdot \mathbf{f}, \tag{4a}$$

$$B^i_{i'} = \mathbf{c}^i \cdot \mathbf{b}_{i'}, \tag{4b}$$

$$\mathbf{c}^i \equiv \frac{\partial \Psi^i}{\partial \mathbf{x}}. \tag{4c}$$

The matrix $B^i_{i'}(t)$ is called the *pulse-response* matrix, since it gives the response of the output variable $y^i$ to a pulsed input of $u^{i'}$. Note that $B^i_{i'}$ does not depend on $\mathbf{f}(\mathbf{x}, t)$. The control goal is to use the actuator forces to make the output

[1] Edwin Wilsey '03 Professor Emeritus.
URL: http://www.princeton.edu/~lam.
Email: lam@princeton.edu

variables honor approximately the following *ODE-based constraints*:

$$\frac{dy^i}{dt} + \phi^i(y^{i'}, t; \varpi^{i'}) = O(\epsilon), \tag{5}$$

where the $\phi^i(y^{i'}, t; \varpi^{i'})$'s are user-specified functions of the arguments (the $\varpi^i$'s are parameters to be exploited later), and $\epsilon$ is the allowable error.

The theory of the Universal Dynamic Control Law (UDCL) advocated by the author (Lam, 1997a,b, 1998, 1999a,b) is applicable to this class of problem when $N$, $\mathbf{f}(\mathbf{x}, t)$, the $\mathbf{b}_i$'s and the $\Psi^i(\mathbf{x}, t)$'s (therefore the $g^i(\mathbf{x}, t)$'s) are unknown to the controller—only $I$ and a "surrogate" of $B^i_{i'}(t)$ are assumed known—provided good real time measurements of the output variables and their time derivatives are made available to the controller blackbox.

Instead of looking for a static control law which classically gives the $u^i$'s as *algebraic* functions of the state vector and time (Ioannou and Sun, 1996; Isidori, 1995; Krstić, Kanellakopoulos and Kokotović, 1995; Marino and Tomei, 1995), the theory of UDCL provides (Lam, 1997a,b, 1998) a system of ODEs for the $u^i$'s:

$$\frac{du^i}{dt} = -\sum_{m=1}^{M} \frac{[\bar{B}^m_i]^+}{\Delta t} \{\dot{y}^m_* + \phi^m(\mathbf{y}_*, \ldots)\}, \tag{6}$$

where $M = I$ when $B^i_{i'}$ is nonsingular and $M > I$ when $B^i_{i'}$ is either singular or nearly singular, $\bar{B}^m_i$ is a "surrogate" of $B^m_i$, and $\Delta t$ is a "sufficiently small" positive user-specified time constant to be chosen by the control engineers. The matrix $[\bar{B}^m_i]^+$ is the *pseudo-inverse* (Golub and Van Loan, 1989 ) of $B^m_i$—it is identical to the conventional inverse when $M = I$. Any reasonable initial condition, such as $u^i = 0$, may be used.

The condition on the surrogate pulse-response matrix $\bar{B}^m_i$ is:

$$\sum_{m=1}^{M} [\bar{B}^m_{i'}]^+ B^m_{i'} = W^i_{i'} \tag{7}$$

where $W^i_{i'}$ is *any* $I \times I$ real matrix whose eigenvalues all have $O(1)$ positive real parts. The ideal choice for $\bar{B}^m_i$ is obviously $B^m_i$ so that $W^i_{i'}$ is simply the unit matrix. The present paper shows, via an example, how to go about finding a usable $\bar{B}^m_i$.

## 2. A SIMPLE EXAMPLE

Consider a system with $I = 2$. The control objective is to make $y^1$ and $y^2$ decay exponentially with (user-chosen) characteristic time $\tau$ toward (user-chosen) $\varpi^1$ and $\varpi^2$ approximately—provided it is possible to do so. In other words, the desired $\phi$'s are:

$$\phi^1 = \frac{y^1 - \varpi^1}{\tau}, \tag{8}$$

$$\phi^2 = \frac{y^2 - \varpi^2}{\tau}. \tag{9}$$

Measured real time data of the output variables are marked by asterisks. Thus:

$$y^i_*(t) = y^i(t) + \delta^i_0(t), \tag{10a}$$

$$\dot{y}^i_*(t) = \dot{y}^i(t) + \delta^i_1(t), \tag{10b}$$

where $\delta^i_p$ represents measurement noise—the subscript $p$ denotes the order of time derivative involved. It is assumed that all $\delta^i_p(t)$'s are "sufficiently" small, but they do not need to have zero mean. In actual implementation, it is strongly preferred that the data for $\dot{y}^i_*(t)$ be measured directly—numerical differentiation of $y^i_*(t)$ is frown upon.

The *sampling time* of the controller blackbox is $t_s$, assumed to be very small in comparison to the characteristic time scale $t_g$ of the physical system. Without loss of generality, $t_g$ is assumed to be unity.

### 2.1 *The Regular Case*

For this case, the following information is known about this matrix:

(1) When $u^1$ is pulsed, $y^1_*$ always jumps in the positive direction by $O(1)$ while the response of $y^2_*$ is not so predictable.
(2) When $u^2$ is pulsed, $y^2_*$ always jumps "the other way" in a more responsive way, while the response of $y^1_*$ is not so predictable.

The above information suggests the following nonsingular matrix as a surrogate pulse-response matrix for the system under consideration:

$$\bar{B}^i_{i'}(t) = \begin{vmatrix} 1 & 0 \\ 0 & -2 \end{vmatrix}. \tag{11}$$

The UDCL for this regular case is:

$$\frac{du^i}{dt} = -\sum_{i'=1}^{2} \frac{[\bar{B}^{i'}_i]^{-1}}{\mu t_s} \left\{\dot{y}^{i'}_* + \phi^{i'}(\mathbf{y}_*, \ldots)\right\}. \tag{12}$$

Theoretically, the value of $\mu t_s$ should be as small as possible. In practice, $\mu$ should be $O(1)$ and not smaller—in order to make sure that numerical integration of (12) using $t_s$ as integration step size will be stable.

## 2.2 *The Irregular Case*

For this case, the following information is known about this matrix:

(1) When $u^1$ is pulsed, neither $y_*^1$ nor $y_*^2$ hardly jumps at all.
(2) When $u^2$ is pulsed, $y_*^2$ continues to jump "the other way" as in the regular case, while the response of $y_*^1$ remains unpredictable.

The above information suggest that $B_{i'}^i$ is either nearly singular or precisely singular. Since $y^1$ is not responsive to the actuators, the ODE-based constraint on $y^1$ is expected to incur some *unavoidable error*. This error is formally introduced as an "extended" output variable, $y^3$:

$$y^3 \equiv \frac{dy^1}{dt} + \phi^1(\mathbf{y}, t, \varpi^i). \tag{13}$$

The best one can do is to drive $y^3$ to zero as quickly as possible. An explicit expression for $\Psi^3$ can be obtained by eliminating $dy^1/dt$ using (3). It is of interest to note that unless $B_1^1$ and $B_2^1$ are *identically* zero, the resultant $\Psi^3$ will have a weak dependence on the $u^i$'s. This weak dependence has very significant consequences, as we shall see later.

Instead of trying to achieve $y^3 = O(\epsilon)$, the UDCL tries to make $y^3$ decay to zero. A possible choice for $\phi^3$ is:

$$\phi^3 = \frac{2y^3}{\tau} \tag{14}$$

where the factor 2 is (arbitrarily) inserted to make $y^3$ decays faster than $y^1$ (and $\varpi^3$ has been set to zero).

In order for this new control objective to be achievable, the following information about the extended pulse-response matrix is assumed:

- When $u^1$ is pulsed, $y^3$ always jumps by $O(1)$ in the positive direction, while its response to a $u^2$ pulse is unpredictable.

Based on this information, the following can be used as a surrogate pulse-response matrix:

$$\bar{B}_{i'}^m = \begin{vmatrix} 0 & 0 \\ 0 & -2 \\ 1 & 0 \end{vmatrix}. \tag{15}$$

The pseudo-inverse of this $3 \times 2$ matrix is readily found to be :

$$[\bar{B}_{i'}^m]^+ = \begin{vmatrix} 0 & 0 & 1 \\ 0 & -0.5 & 0 \end{vmatrix}. \tag{16}$$

This nonsingular matrix has two finite singular-values (1 and 0.5). The UDCL for this irregular case is:

$$\frac{du^i}{dt} = -\sum_{m=2}^{3} \frac{[\bar{B}_i^m]^+}{\mu t_s} \{\dot{y}_*^m + \phi^m(\mathbf{y}_*, \ldots)\}, \tag{17}$$

where again the value of $\mu$ is to be chosen by the control engineers. However, because $\Psi^3$ may have a weak dependence on the $u^i$'s, the value of $mu$ should be a relatively larger $O(1)$ number—unless it is known that $B_{i'}^i$ is precisely singular. Note that in (17) the $m$ summation omits $m = 1$ since (8) is replaced by (13) which is the definition of $y^3$.

It is assumed that good quality real time measurements $y_*^3(t)$ and $\dot{y}_*^3(t)$ are available to the controller blackbox.

## 3. NUMERICAL SIMULATIONS

In order to program numerical simulations, the value of $N$, detailed information on $\mathbf{f}(\mathbf{x}, t)$, the $\mathbf{b}_i$'s and the $\Psi_i$'s must be known.[2] Consider a physical system with $N = 3$ (and $I = 2$). The actual $\mathbf{b}_i$'s are given by:

$$\mathbf{b}_1 = [\gamma, 0.2x_3^2, 1]^T, \tag{18a}$$
$$\mathbf{b}_2 = [0.1\gamma x_3, -2, 1]^T, \tag{18b}$$

where $\gamma$ is a parameter to be exploited later. For the sake of simplicity, the output variables are simply $y_1 = x_1 = \Psi_1$ and $y_2 = x_2 = \Psi_2$. Thus, the actual row vectors $\mathbf{c}_i$'s are given by:

$$\mathbf{c}_1 = \frac{\partial \Psi_1}{\partial \mathbf{x}} = [1, 0, 0], \tag{19a}$$
$$\mathbf{c}_2 = \frac{\partial \Psi_2}{\partial \mathbf{x}} = [0, 1, 0]. \tag{19b}$$

The actual $2 \times 2$ pulse-response matrix is then:

$$B_{i'}^i = \begin{vmatrix} \gamma & 0.1\gamma x_3 \\ 0.2x_3^2 & -2 \end{vmatrix}. \tag{20}$$

It is readily seen that when $\gamma = O(1) > 0$ and when $\gamma$ is negligibly small, this pulse-response matrix does have the properties described in the

[2] In this subsection only, superscript of column vectors will appear as a subscript to avoid confusion with exponents.

previous section for the regular and irregular case, respectively.

The following is (arbitrarily) chosen for $\mathbf{f}(\mathbf{x}, t)$:

$$f_1 = \sin(x_1 - \alpha) + \exp(-x_2^2) + x_3, \quad (21a)$$
$$f_2 = \arctan(x_1^2 + x_2 + \sin(\alpha x_3) + \alpha), \quad (21b)$$
$$f_3 = -\beta x_3 + \cosh(x_1 + x_2)\cos(x_1 + \alpha x_2), \quad (21c)$$

where $\alpha$, $\beta$ and $\gamma$ are nominally $O(1)$ parameters which may be smooth functions of time. It is important to note that $\partial f_1/\partial x_3$ is unity—a $O(1)$ constant.

The initial conditions used were chosen (arbitrarily) to be $x_1(0) = 1$, $x_2(0) = -3$, $x_3(0) = 1$. In all cases, the integration timestep used was 0.01, and it is to be interpreted as the $t_s$ of the controller blackbox. All measurements are assumed noise-free—the issue of measurement noise has been treated in a separate paper (Lam, 1999a).

### 3.1 *The Regular Case with* $\gamma = 1$

Fig. 1 shows the results for $\gamma = 1$, $\beta = 1$, and $\alpha = -2\sin(0.1\pi t)$ using $\mu = 3$, $\varpi_1 = \varpi_2 = 0$. When other $O(1)$ choices for $\alpha$ were used (*e.g.* $\alpha = 5, 1, 0, -1, -5$), the results for $x_1(t)$ and $x_2(t)$ were indistinguishable from Fig. 1. The results for $x_3(t)$ are $\alpha$-dependent and are well behaved so long as $\beta$ is positive.

When $\beta$ is negative, $x_3$ diverges—as expected—and the system eventually crashes. This (zero dynamics) instability can easily be arrested by exploiting the $\varpi^i$'s—assuming sensor measurement of $x_3(t)$ is available to the controller. For example, when $0 \geq \beta \geq -0.57$, the choice of $\varpi_1 = -0.55x_3$ and $\varpi_2 = 0$ was empirically found to stabilize $x_3$ at the expense of $x_1$ which is now $\alpha$-dependent and no longer decays toward zero. Fig. 2 shows the results of this simulation. Note that the exponential decay of $x_2$ is strictly enforced in both Fig. 1 and Fig.2.

### 3.2 *The Irregular Case with Small* $\gamma$

After introducing the new output variable $y_3$ by (13), the actual $3 \times 2$ extended $B_i^m$, which now does depend on $f_1(\mathbf{x}, t)$, is:

$$B_{i'}^m = \begin{vmatrix} O(\gamma) & O(\gamma) \\ 0.2x_3^2 & -2 \\ 1 + 0.2x_3^2\dfrac{\partial f_1}{\partial x_2} & 1 - 2\dfrac{\partial f_1}{\partial x_2} \end{vmatrix}. \quad (22)$$

Since both singular values of $B_i^m$ are now $O(1)$, the irregular problem has been "regularized" with $M = 3$. This actual matrix (22) should be compared with the surrogate matrix proposed in (15).

Simulations were performed with

$$\gamma = 0.05\cos(0.1\pi t), \quad (23)$$

a variety of the $\alpha$'s, $\beta$ of either signs, and $\varpi_1 = \varpi_2 = 0$. Since the chosen $\gamma(t)$ is merely small and not precisely zero, the original $B_{i'}^i$ is nearly singular. As a consequence, $\Psi_3$ does have a weak dependence on the $u_i$'s. This complication introduces a lower bound for $\Delta t$. In the simulations, $\mu = 3$ indeed caused the system to crash, while $\mu = 7$ was found empirically to work well in all cases. A sample simulation is presented in Fig. 3. It is seem that $x_1(t)$—which did not respond to pulsing of the $u_i$'s—has a hump before it decays toward zero, in contrast to its previous behavior as shown in Fig. 1. The new output variable $y_3(t)$ as defined by (13) always honors the ODE-based constraint as specified by (14), and is not shown. Simulations for $\beta = -1$ were also performed, and the results for $x_1(t)$ and $x_2(t)$ look essentially the same as Fig. 3.

## 4. DESIGN GUIDELINES FOR SENSORS AND ACTUATORS

Traditionally, control engineers often begin the design of the controller black box after the choices of sensors and actuators have been made by others. An obviously good idea is to invite the control engineers to play an active role in the whole design process, including the design of sensors and actuators. The following are design guidelines which can greatly facilitate the use of UDCL by the controller blackbox:

- Variables of special importance to the control tasks at hand should be selected as primary output variables and accurate real time measurements must be provided. All needed rate data should be directly measured if at all possible. All unmeasured state variables are assumed to be of secondary interest and, hopefully, are well-behaved without the need of intervention by the controller.
- The actuators should be chosen to yield a full-rank primary impulse-response matrix $B_{i'}^i$ such that a (constant or slowly varying) surrogate $\bar{B}_i^m$ can readily be identified off-line, or determined on-the-fly. Regular prob-

lems are preferred since its pulse-response matrix is independent of $\mathbf{f}$.

- If the problem is intrinsically irregular, the number of derived output variables should be as small as possible. Precisely singular $B_{i'}^{i}$ is preferred over nearly singular ones.
- The physical unit of the singularvalues of $B_i^m$ is reciprocal time. The actuators must be powerful enough such that the smallest dimensional singularvalue is $O(1/t_g)$ where $t_g$ is the characteristic time of the global system.
- The ODE-based constraints should be chosen such that the associated $u^i$'s are expected to be $O(1)$ and smooth.

It is taken for granted that the sensor measurements (including all the needed time derivatives) are reliable and have good signal-to-noise ratios, and $t_s$ is sufficiently small in comparison to $t_g$.

## 5. CONCLUDING REMARKS

The main merit of the UDCL is that no detailed knowledge of the system is needed. The Achilles' heel is that it requires not only reliable $y_*^m(t)$'s, but also $\dot{y}_*^m(t)$'s. For example, for mechanical systems measurements of accelerations are needed when the primary output variables are either positions or velocities. This is the price paid for robustness with respect to uncertainties of the system model. Ideally, the needed time derivatives of the primary output variables should be directly measured; numerical differentiation should be avoided if at all possible for obvious reasons. From the point of view of sensor technology, there is no fundamental reason to believe that time derivatives $\dot{y}^m(t)$'s are generically more difficult to directly measure than the $y^m(t)$'s themselves. Obviously, the UDCL can take advantage of any additional detailed knowledge of the system to reduce its present strong reliance on the $\dot{y}_*^m(t)$'s. This issue and the impacts of noise are discussed in more details in a separate paper (Lam, 1999a).

The most crucial information about the system needed by the UDCL is a surrogate of $B_i^m(t)$—the precise $B_i^m(\mathbf{x},t)$ itself is not necessary. Roughly speaking, the $i$-th column of the surrogate matrix $\bar{B}_i^m$ should mimic the $M$-dimensional vector representing the jump of the $y_*^m$'s in response to a unit pulse of $u^i$. The easiest problems for the UDCL are those for which a constant or nearly constant $\bar{B}_i^m$'s is adequate—as was the case in the example presented. Whenever UDCL is adopted, the task of system identification reduces to the determination of $\bar{B}_i^m(t)$ or $\bar{B}_i^m(y_*^m,t)$. For the $M = I = 1$ case, only the sign of the scalar $B_1^1(t)$ needs to be determined. If the elements of the actual $B_i^m$ has very complicated dependence on $\mathbf{x}(t)$ and $t$ and no reliable $\bar{B}_i^m(y_*^m,t)$ can be found, then the UDCL is not useful.

In essence, the UDCL ignores the unobserved variables—assuming that the zero dynamics of the problem needs no intervention. If intervention is needed, the $\varpi^i$'s are available to be exploited. The intelligent use of this degree of freedom is at the disposal of the control engineers—provided additional information and/or real time sensor measurements are available.

The highest order time derivative needed by UDCL for $y^i(t)$ is commonly known as the "relative degree" of $y^i$. In general, the theoretically achievable accuracy of UDCL degrades as the relative degree increases. Generally speaking, the UDCL is most useful for regular problems (*i.e.* all relative degrees are unity) and for irregular problems whose relative degrees are not large. It can be used also in a restricted class of optimal control problems (Lam, 1999c).

## REFERENCES


Golub, G. H. and Van Loan C. F., (1989). *Matrix Computations*, Second Edition, The Johns Hopkins University Press.

Ioannou, P. A. and Sun, J., (1996). *Robust Adaptive Control*, Prentice Hall.

Isidori, Alberto, (1995). *Nonlinear Control Systems*, 3rd Edition, Springer-Verlag.

Krstić, M., Kanellakopoulos, I. and Kokotović, P., (1995). *Nonlinear and Adaptive Control Design*, John Wiley and Sons, Inc.

Lam, S. H., (1997a). "A Robust Universal Controller," DSC-Vol. 61, Proceedings of the ASME Dynamic Systems and Control Division, Editor: Giorgio Rizzoni, Book No. H01105-1997, pp. 571-577.

Lam, S. H., (1997b). "Quasi-steady approximation and adaptive nonlinear control," Vol. 3, Computational Physics, Chemistry and Biology, Proceedings of the *15th IMACS World Congress on Scientific Computation, Modeling and Applied Mathematics*, edited by Sydow, A., Wissenschaft & Technik Verlag, Berlin, pp. 127-132.

Lam, S. H., (1998). "A New Approach to Adaptive Nonlinear Controls," *Journal of Sys-*

*tem Analysis Modeling and Simulation*, Vol. **4**, No. 2, pp. 153-167. Gordon Breach, Berlin.

Lam, S. H., (1999a). "Universal Control Law and Rate Feedbacks," presented as paper #F-2e-02-4 at the *1999 IFAC World Congress*, July 5-9, Beijing, China.

Lam, S. H., (1999b). "Control of Nonlinear Systems with Delays and Hysteresis," presented at the 1999 IEEE Hong Kong Symposium on Robotics and Control, July 2-3, Hong Kong.

Lam, S. H., (1999c). "The Scalar Optimal Control Problem," presented at the *1999 ACC*, June 2-4, San Diego.

Marino, R. and Tomei, P., (1995). *Nonlinear Control Design, Geometric, Adaptive and Robust*, Prentice Hall.

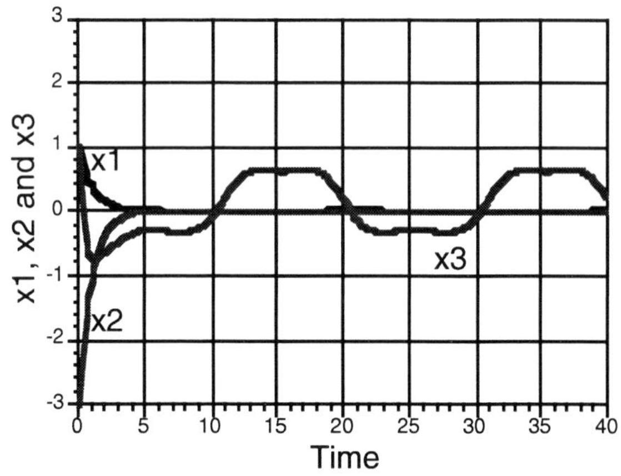


Fig. 1. The regular case ($\gamma = 1$) with $\beta = 1.0$ ($x_3$ zero-dynamics is stable without intervention); $\tau = 1$.

$\alpha = -\sin(0.1\pi t)$, $\mu = 3$ and $\varpi_1 = \varpi_2 = 0$. Measurements are noise-free.

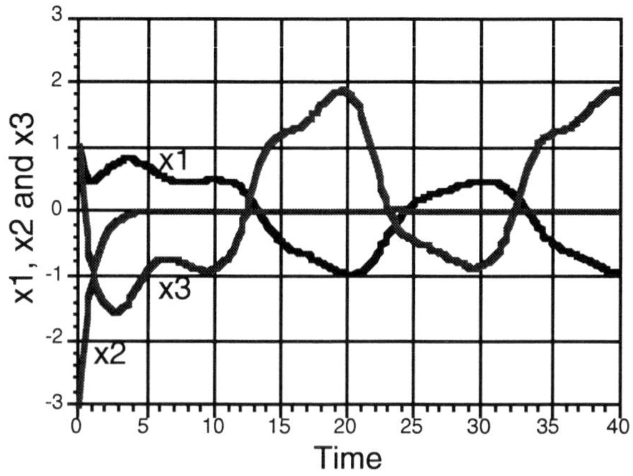


Fig. 2. The regular case ($\gamma = 1$) with $\beta = -0.57$ (unstable $x_3$ zero-dynamics stabilized by exploiting $\varpi_1$); $\tau = 1$.

$\alpha = -\sin(0.1\pi t)$, $\mu = 3$, $\varpi_1 = -0.55x_3$ and $\varpi_2 = 0$. Measurements are noise-free.

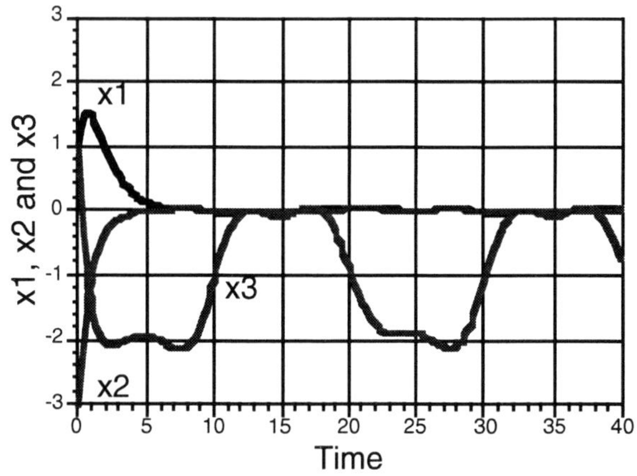


Fig. 3. The irregular case ($\gamma = 0.05\cos(0.1\pi t)$) ($N = M$, no zero-dynamics); $\tau = 1$.

$\alpha = -2\sin(0.1\pi t)$, $\beta = 1.0$, $\mu = 7$ and $\varpi_2 = \varpi_3 = 0$. Measurements are noise-free.

IFAC Publications
www.elsevier.com/locate/ifac

# GATE LEVEL IMPLEMENTATION OF HIGH SPEED CONTROLLERS AND FILTERS

**Roland Kasper * Thomas Reinemann ****

* *Roland Kasper, Prof., Dr.-Ing: Otto-von-Guericke-Universität Magdeburg, Institut für Mechatronik und Antriebstechnik, PSF 4120, D-39016 Magdeburg; Tel.: +49 (391) 67-18607, Fax 12656, e-mail: roland.kasper@mb.uni-magdeburg.de*
** *Thomas Reinemann, Dipl.-Ing: Otto-von-Guericke-Universität Magdeburg, Institut für Mechatronik und Antriebstechnik, PSF 4120, D-39016 Magdeburg;Tel.: +49 (391) 67-12952, Fax 12656; e-mail: thomas.reinemann@mb.uni-magdeburg.de*

Abstract: Today, improving a product often results in an increasing demand on its information processing capabilities. Especially in high dynamic systems these demands could be satisfied for a long time only by high speed micro controllers or signal processors. However these are very complex and expensive devices. In many applications real time operating systems come along to meet the real time requirements. These generate an additional overhead, which consumes further resources. Furthermore, it is not simple to make a clear prediction concerning the whole system performance. A more and more interesting alternative is given by FPGAs, which offer a very high speed at a price of a cheap controller. Furthermore, it is possible to make a clear prediction of the processing speed and timing. However, some obstacles are handicap the usage of ICs. The new approach presented in this paper is to transmit and process data in a bit serial manner. Therefore, a component processes only one bit during one clock cycle. This leads to very simple and fast processing elements. A library with the most common components has been built, which are needed for implementing controllers and filters, e.g. addition, subtraction, gain, multiplication, integrator, characteristic curve and so on. Presently, work runs on a simpler quantifying of signal delays.


Keywords: digital signal processing, FPGA, bit serial, computer algoritm

## 1. PROBLEMS IMPLEMENTING CONTROLLER AND FILTER ALGORITHMS

Presently, digital signal processing is dominated by micro controllers and digital signal processors. Categorically, these systems must satisfy the Shannon theorem of sampling (Föllinger, 1993). This means, the sample frequency $f_S$ has to be twice the maximum frequency occurring in the controlled system. These real time demands necessitate task execution in a limited time. Since some processor resources are only available once, they have to be shared between different tasks. Management of these resources is handled usually by a real time operating systems (Gray, 1999), (Kopetz, 1998). But high frequency signal processing increases the software overhead dramatically and consumes a too big amount of the processor's performance. As a result, a faster and often more expensive processor is needed in many cases.

Hardware like FPGAs or ASICs comes into operation for higher sampling rates. It offers a possibility to perform signal processing in a parallel manner. This means signal flows are processed simultaneously, a delay results only from the longest chain but not from the sum of all chains, as using processors.

However, some obstacles handicap the usage of hardware:

- it needs different knowledge implementing algorithms in this manner than using software
- hardware description languages (HDL) have been made for circuit designers but not for mechatronic engineers, who prefer signal flows
- missing of general function libraries. Only very simple functions (addition, subtraction) are available, therefore more complicated functions have to be designed by the applicator itself.

Simulation of hardware is more sophisticated than software. HDLs like VHDL allow the description of test environments, too. Since they are powerful languages they are more flexible than a specific processor simulator, which world ends on its pins.

## 2. BASICS OF BIT SERIAL SIGNAL PROCESSING

Today's hardware based signal processing uses bit parallel information exchange and processing, associated with costly operators with a big logic effort (Koren, 1993). A lot of logic is spent to accomplish pipelining, but this leads to complicated algorithms with big potential of errors.

These disadvantages can be avoided by a bit serial approach. Here both tasks, information processing and data exchange are performed bit serial, in difference to some algorithms shown at (Andraka, 2001). The transmission starts with the least significant bit (LSB) which follow the higher significant bits, that means time division is used.

On parallel transmission, each signal is sent via its own line, which implicitly defines its significance. A simple ripple carry adder computes the sum $s_i$ and carry $c_i$ for each pair of equal significant bits. In case of a carry in stage i it is transmitted to the next stage, where $s_{i+1}$ and $c_{i+1}$ has to be recomputed. Therefore the carry ripples through the following adders, which can change the value of their outputs and leads to a big computation time. There are faster but even more complicated algorithms to speed up operation.

### 2.1 *Adder*

The serial adder's structure is showed here as an example. Using the binary number representation, both summands are given as

$$a = \sum_{i=0}^{n-1} a_i \cdot 2^i \text{ and } b = \sum_{i=0}^{n-1} b_i \cdot 2^i$$

with the processing word width n.

The sum results from

$$s_i = (a_i + b_i + c_{i-1})\, mod\, 2 \qquad (1)$$

with the carry

$$c_i = (a_i + b_i + c_{i-1})\, div\, 2 \qquad (2)$$

considering

$$c_0 = (a_0 + b_0)\, div\, 2 \qquad (3)$$

because there is no carry in this initial situation. At the adder's input a tuple $a_i$, $b_i$ appears at each rising clock edge. At the same moment the previously sum and carry bit are stored in the D-flip-flops (fig. 1). This assures that the carry affects the sum bit's computation one clock later. The multiplexor controlling the carry FF ensures equation 3. Therefore it is switched to put out an 0 while $c_{n-1}$ is transmitted on the output of the part for the carry's computation.

Each bit of the sum can be processed in the following elements. As a result, an operand is handled in a distributed manner, which conforms a pipeline. This kind of adder consumes a very small amount of logic. It takes only two FFs and about 15 gates. Detecting an overflow or underflow needs determination of the word's boundary. Therefore a Sync-signal $j_{Adder}$ has to be chosen from a Sync-bus (see section 4). This can be achieved by calculating the signal's constant delay in bit flow.

$$j_{Adder} = (Delay + n - 1)\, mod\, n \qquad (4)$$

### 2.2 *Gain*

The serial gain presented here uses an amplification factor with a norm greater than one. The multiplicands a and b are used in binary representation again. However b is a constant amplification factor. Summarized, multiplication means the partial product $P^{(i)}$ is added up in n steps

$$P^{(i+1)} = (P^{(i)} + a_i \cdot b)\, div\, 2 \text{ with } P^{(0)} = 0 \qquad (5)$$

For this reason the i-th product bit $p_i$ results from

$$p_i = (P^{(i)} + a_i \cdot b)\, mod\, 2 \text{ with } P^{(0)} = 0 \qquad (6)$$

The operations mod and div are realized by a right shift. As a result the LSB is put out. $P^{(n)} = 0$ is required after n steps or an overflow emerges, which can be detected during operation. For $a_i = 1$ b

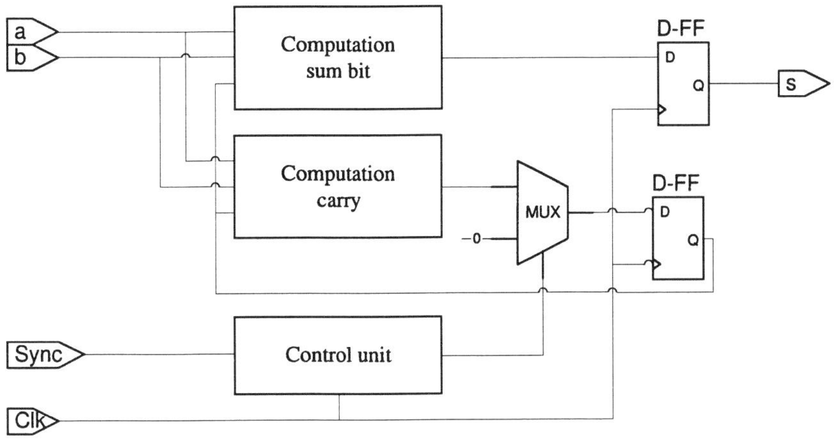


Fig. 1. Serial Adder

must be added to $P^{(i)}$. If $a_i = 0$ an 0 is added and then all bits are shifted. This corresponds to the Booth-algorithm and has been used for implementation. To implement this algorithm some advantages can be used. The term in brackets in equations 5 and 6 is implemented only once. Its multiplication can be written

$$a_i \cdot b = \sum_{j=0}^{n-1} a_i \cdot b_j \cdot 2^j \tag{7}$$

and then reduced to a conjunction of the bits originated by the binary number presentation. Figure 2 shows the gain's structure.

Condition $P^{(n)} = 0$ requires the AND gates between the adders. These prevents the LSB to be affected from the previously computed MSB. Calculation of the Sync-signal is made in the same manner like with the adder (equ. 4).

### 2.3 *Processing word width*

An essential advantage of designing algorithms in hardware emerges from the free choice of the processing word width. Whereas, this is fixed for processors, the user has here the freedom to adapt its components optimally to his needs.

For a bit serial approach, only one bit is transmitted per cycle. Therefore a higher word width consumes more time for processing and transmission an operand. As a result, the sampling frequency $f_s$, the clock frequency $f_c$ and the word width n depend on each other according to equ. 8:

$$f_s = \frac{f_c}{n} \tag{8}$$

## 3. BLOCK LIBRARY

More complex elements like differentiators, integrators etc. are needed to implement control and filter algorithms. They are composed of basic elements like adders, gains etc. and will be shown here only as an example. Furthermore other non-linear components, like characteristic curves and look-up-tables, are available that can be built from the above mentioned basic elements.

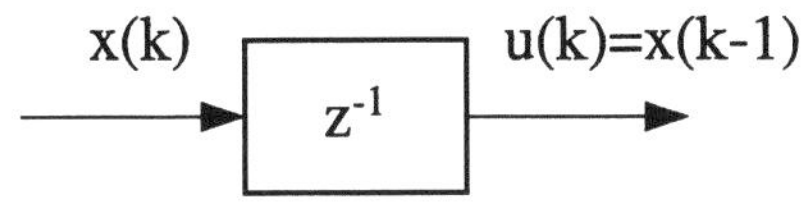


Fig. 3. Delay block

### 3.1 *Delay block*

Theory of sampling systems always needs an element to delay a value for performing operations on it during the next computation cycle. It delays the value for one sampling period $\Delta T$, which has to be a multiple of $1/f_C$. Than this element can be realized by an appropriate shift register. Further considerations in this paper are restricted to equidistant sampled discrete values:

$$x(k) \hat{=} x(k\,\Delta T) \tag{9}$$

### 3.2 *Differentiator*

The derivative of a signal $x_d$ is used widely in controllers to achieve good dynamics.

$$u_d(t) = T_V \cdot \dot{x}_d(t) = T_V \frac{dx_d(t)}{dt} \tag{10}$$

In digital realization equ. 10 means estimation of the gradient between two samples and therefore the derivative is replaced by a difference quotient. The coefficient $T_V$ and the sampling period $\Delta T$ are merged together to a single gain factor.

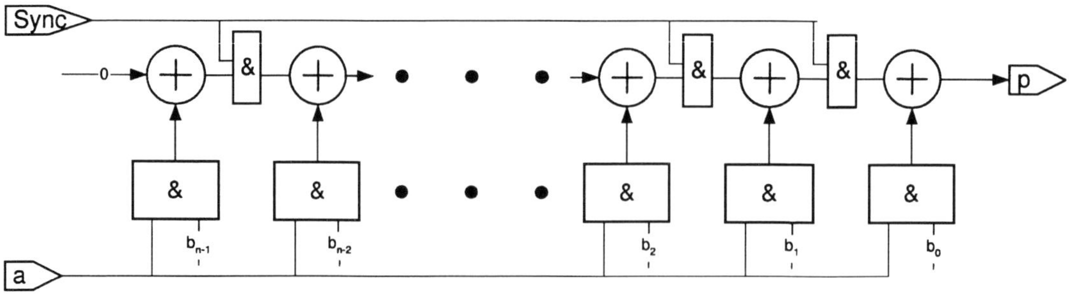


Fig. 2. Serial gain

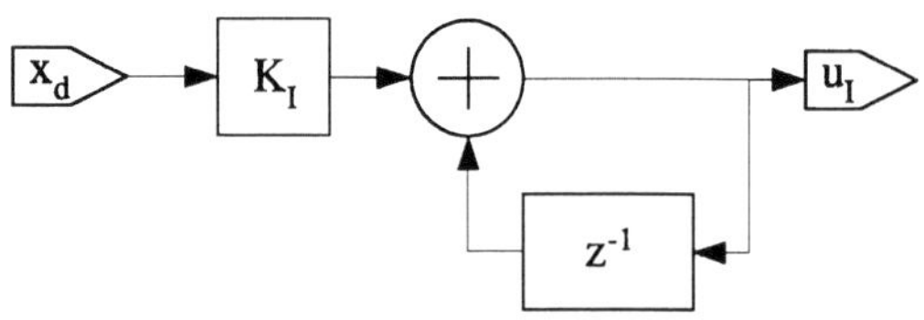


Fig. 4. Integrator

$$u_D(k) = T_V \frac{x_d(k) - x_d(k-1)}{\Delta T} \quad (11)$$
$$= K_D \left[ x_d(k) - x_d(k-1) \right] \quad (12)$$

3.3 *Integrator*

The integrator is another frequently used element. In continuous time it is defined by

$$u_I(t) = \frac{1}{T_N} \int_0^t x_d(\tau)\, d\tau \text{ or} \quad (13)$$
$$\dot{u}_I(t) = \frac{1}{T_N} x_d(t) \quad (14)$$

To obtain a realization for discrete time systems the derivative is replaced by a difference quotient again.

$$\frac{u_I(k) - u_I(k-1)}{\Delta T} = \frac{1}{T_N} x_d \Longrightarrow \quad (15)$$
$$u_I(k) = u_I(k-1) + K_I\, x_d(k) (16)$$

Equation 16 leads to a feed back of the latched adder's output, as shown in Figure 4. However, there is always a possibility for an overflow or underflow. To avoid this, an error treatment is necessary. If an error occurs then the sign changes in a wrong manner. Unfortunately, it is transmitted and therefore computed as the last bit. Thus, the whole word has to be latched to enable a substitution with a given replacement. This additional delay corresponds a dead time and worsens control quality. An error is impossible by the compromise of addends with a limited range. The usable range for the addends is $-2^{n-2} \dots 2^{n-2} - 1$. An addition of two such operands can never lead to a over or under flow. Since the addend originated by $x_d$ has a known range it has only to be scaled by an appropriate gain factor $K_I$. However the other addend is the fed back integrator part of the control input and uses the full value range. Therefore it has to be limited via a limiter. An integrator realization basing upon this approach avoids a dead time by the compromise of a halved value range. However, a big advantage of the bit serial approach is the possibility of scaling in steps of one bit, therefore the range can be simply extended.

3.4 *Filters*

Filters with a regular structure like FIR or IIR filters can be simply built, with the well known blocks like adder, gain and delay.

## 4. SYNCHRONIZATION

In this place synchronization means aligning bits to the same significance and marking any bit of an operand. Generally, hardware solutions execute parallel information processing. Therefore different signal chains can lead to different delays. That means, a certain operator can receive operands with different delays.

For bit parallel processing, the significance of a bit is determined by the transmission line. However, if transmission is bit serial, then another method is needed to determine significance. Generally, various bits have to be labeled, depending on a certain algorithm and signals with different delays have to be synchronized. As a consequence it is necessary to mark any bit of an operand. Furthermore, a result can grow to n+1 or more bits, on certain operations, e.g. an addition or multiplication. Since the frame available for transmitting operands is constant, it has to take care that the operands are fitting within this frame. Therefore it is necessary to create a possibility to mark certain bits for an error treatment.

Accordingly, there is a bus with a width of n. A signal becomes active only for one clock cycle on each line. This offers a possibility to label any bit of the word, what happens by passing the signal's delay to the component. With this information a

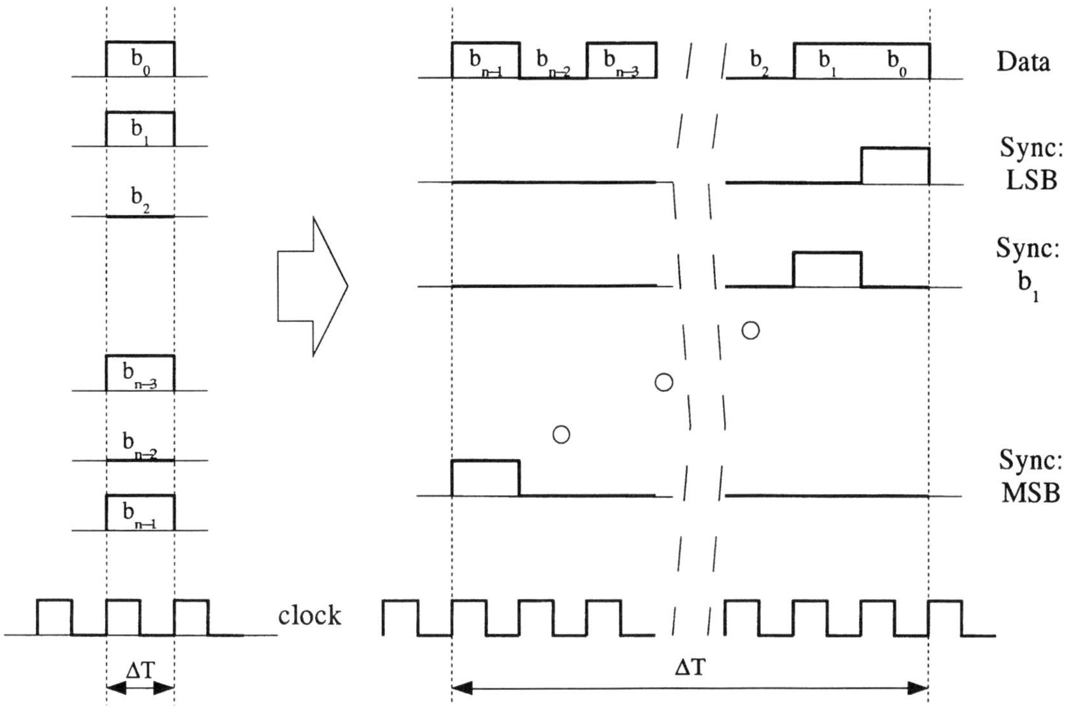


Fig. 5. Sync-Signals

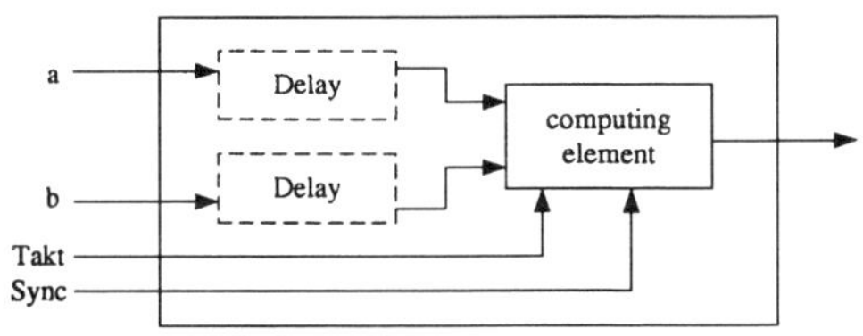


Fig. 6. Principle structure of a two inputs component

unit selects a given Sync-signal from the bus as given for example by equ. 4.

Depending on the known signal's delay it is possible to detect a difference of delays within a component, which has more than one input. The difference can be used to insert additional delay stages in the less delayed data stream (see figure 6). This method provides a simple way to synchronize data streams, if necessary. It is difficult and fault-prone to quantify delays manually especially for feed back loops. Since, an input's delay depends on its output's delay here. However it is possible to establish an equation system (equ.: 18), which contains the dependencies of the signal delays $D_{sx}$ ($D_{sui}$ manipulated variable in fig. 4), the constant delays of the operators $D_{ox}$ ($D_{oa}$ adder delay) and additional delay stages $D_{gx}$ ($D_{gxk}$, $D_{gsz}$ at adder) needed for synchronization in the operators. This is linked with a set of valid ranges for the delay stages and a sum over all of them:

$$S_{DS} = \sum_{i=0}^{n} D_{gi} \tag{17}$$

can be used to find a solution via minimization methods see (Bronstein and Semendjajew, 1991). Quantifying the delays shall be shown by an example for the integrator shown in fig. 4.

$$\begin{aligned} D_{sui} &= D_{sxk} + D_{gxk} + D_{oa} \\ D_{sui} &= D_{sz} + D_{gsz} + D_{oa} \\ D_{sxk} &= D_{sx} + D_{og} \\ D_{sz} &= D_{sui} - D_{gz} \end{aligned} \tag{18}$$

The delay block has the task to delay a sample for one clock period. Therefore the output's signal delay would be greater than the processing word width n and leads to additional delay stages at the other adder's input. This can be avoided by aligning to the prior word via subtracting $D_{gz}$. For implementation it has to be converted in an appropriate value $D_{igz}$ by equ. 19.

$$D_{igz} = n - D_{gz} \tag{19}$$

Valid ranges for the delay stages are:

$$\begin{array}{cc} D_{gxk} \geq 0 & D_{gsz} \geq 0 \\ 0 \leq D_{gz} \leq 16 & D_{suik} \geq D_{sx} \end{array} \tag{20}$$

Furthermore the delays of the processing elements are known. These are $D_{oa} = 1$ for the adder and $D_{og} = 1$ for the gain. Equation 17 has to be stated more precisely for this example in:

$$S_{DS} = D_{gxk} + D_{gsz} + D_{gz} \tag{21}$$

This linear optimisation problem can be simple solved via algebra programs. The solution is:

$D_{gxk} = D_{gsz} = 0$ and $D_{gz} = 1$. With $n = 16$ and respect to equ: 19 $D_{igz} = 15$. That means only the delay block contains additional delay stages.

As already mentioned at the introduction of the bit serial approach, each bit needs one clock cycle for transmission and processing. Therefore it is not possible to implement an algebraic loop. Since this would be require $D_{gz} = 0$ in the example, which is not a solution of the minimizing problem however.

## 5. OUTLOOK

Future work runs on implementation a plug-in for Renoir. It shall provide the automatic set up and solving of the minimization problem by a post process after generating VHDL.

## 6. REFERENCES


Andraka, Ray (2001). *Multiplikation in FPGAs.* Andraka Consulting. http://www.andraka.com/multipli.htm, last visited 20.07.2001.

Bronstein and Semendjajew (1991). *Taschenbuch der Mathematik.* Teubner Verlagsgesellschaft. Stuttgart-Leipzig.

Föllinger, Otto (1993). *Lineare Abtastsysteme.* Oldenbourg. München.

Gray, David (1999). *Introduction to the formal design of real-time systems.* Springer. London.

Kopetz, Hermann (1998). *Real-time systems: design principles for distributed applications.* Kluwer. Boston.

Koren, Israel (1993). *Computer Arithmetic Algorithms.* Prentice Hall. Englewood Cliffs.

IFAC Publications
www.elsevier.com/locate/ifac

# REAL TIME SECURE COMMUNICATIONS VIA X|X| CHAOS MODEL

H S Kwok, K S Tang, K F Man and G R Chen

*Department of Electronic Engineering*
*City University of Hong Kong*
*Email: eekman@cityu.edu.hk*

**Abstract:** A secure digital communication system based on chaotic theory is proposed to fence off the intrusion of eavesdroppers during the transmission of signal and data. With the master-slave M-synchronization technique, the transmitter and the receiver are designed using the chaotic phase shift keying. A new discrete-time modified Chua's circuit using the x|x| nonlinearity is adopted. This methodology for the design of secure communication systems can be applied to most real time communication applications, such as Internet telephone, e-commerce transactions, and digital data transfer. *Copyright © 2001 IFAC*

**Keywords:** chaos, deadbeat control.

## 1. INTRODUCTION

Today's communication in the Internet network is widespread. The communication types include email, file transfer, e-commerce and now, the emerging Internet and mobile telephone.

As the traffic of information signal and/or data being transmitted from one end to the other is prompted to attract eavesdroppers, the growing desire to secure and/or to encrypt the information has gathered force worldwide. We have witnessed the applications of chaotic systems for private or secure communication for the last ten years or so. Because of its chaotic nature, chaos as a means for secure communication is well accepted. The principles have been derived and experimental implementations have been shown.

The security of hidden signal/data in the chaotic waveforms is a subject largely governed by the dimension of chaos model. It has been reported that low order chaotic system is prompted to be attack (Pecora and Carroll, 1990). Despite the richness of the research work in the area (Hasler and Vandewalle, 1999; Kennedy and Ogorzalek, 1997), we have yet to witness a universal benchmark chaos model of the kind although the recent hyperchaos model (Grassi and Mascolo, 1999) may have shown its promise for complexity.

However, much work for chaos synchronization (Pecora and Carroll, 1990) was conducted in analog domain, while the secure digital communication is still in its infancy. Chaotic information encryption or more precisely chaotic cryptography is still a relatively new subject with methods and standards still under development.

This paper is proposed to develop a secure digital communication system that has an inherent resistance against eavesdroppers on the basis of the past work done on chaos control and chaos synchronization worldwide. A new discrete-time chaotic model on the basis of Chua's circuit was firstly proposed in (Tang and Man, 1998). In view of the fact that the complexity of chaotic waveforms is of primary interest for secure communication, the sheer limited computational power of the communication device such as mobile phone should not be overlooked. It is now a well-known fact that the more computational power is required the more drainage of battery charges is observed.

The effectiveness of the modified Chua's circuit using x|x| has been demonstrated in (Tang et. al., 2001; Tang and Man, 1998). This model presents richness in producing chaotic waveforms as required by a secure communication system.

With the master-slave M-synchronization technique (Koh and Ushio, 1997), a transmitter and a receiver for different types of signals are to be designed using the chaotic phase shift keying (CPSK) approach. This has a number of advantages over the other techniques, (i) increased flexibility in digital form not binary, to cater for wider range of information, (ii) increased rate of transmission, and (iii) deployable for sending multiple sets of data simultaneously, in effect increasing the security of the system. As a result, this modulation method can be applied to furnishing the design of wireless secure digital communication system.

The organization of the paper is as follows. In Sect 2, the discrete time model of the modified Chua's circuit using x|x| is described. The design of secure communication system based on M-synchronization is explained in Sect. 3. The testing results are given in Sect 4 and finally, conclusion remarks are given in Sect. 5.

## 2. DISCRETE-TIME CHAOTIC MODEL

Consider the following piecewise-linear modified Chua's circuit proposed in (Tang and Man, 1998),

$$\begin{aligned} \dot{x}_1 &= \alpha[x_2 - g(x_1)] \\ \dot{x}_2 &= x_1 - x_2 + x_3 \\ \dot{x}_3 &= -\beta x_2 \end{aligned} \qquad (1)$$

where $g(x_1) = ax_1 + bx_1|x_1|$, with coefficients defined in (Tang and Man, 1998).

Using a bilinear transform, the discrete-time model of Eq. (1) is obtained as

$$\begin{bmatrix} x_1(k+1) \\ x_2(k+1) \\ x_3(k+1) \end{bmatrix} = \begin{bmatrix} \phi_{11} & \phi_{12} & \phi_{13} \\ \phi_{21} & \phi_{22} & \phi_{23} \\ \phi_{31} & \phi_{32} & \phi_{33} \end{bmatrix} \begin{bmatrix} x_1(k) \\ x_2(k) \\ x_3(k) \end{bmatrix} + \begin{bmatrix} \gamma_1 \\ \gamma_2 \\ \gamma_3 \end{bmatrix} g(x_1(k)) \qquad (2)$$

or, in a more general form,

$$\mathbf{x}(k+1) = \Phi\mathbf{x}(k) + \Gamma g(x_1(k))$$

where, with sampling time T>0 and constants $\alpha, \beta, a, b$,

$$\phi_{11} = \frac{4\alpha T^2 + \varsigma_1}{\varsigma_1},\ \phi_{12} = \frac{8\alpha T}{\varsigma_1},\ \phi_{13} = \frac{4\alpha T^2}{\varsigma_1},$$

$$\phi_{21} = \frac{8\alpha T^3 + 4T\varsigma_1}{\varsigma_1\varsigma_2}, \phi_{22} = \frac{16\alpha T^2 + (8 - \varsigma_2)\varsigma_1}{\varsigma_1\varsigma_2},$$

$$\phi_{23} = \frac{8\alpha T^3 + 4T\varsigma_1}{\varsigma_1\varsigma_2}, \phi_{31} = \frac{4\alpha\beta T^4 + 2\beta T^2\varsigma_1}{\varsigma_1\varsigma_2},$$

$$\phi_{32} = \frac{8\alpha\beta T^3 + 4\beta T\varsigma_1}{\varsigma_1\varsigma_2},$$

$$\phi_{33} = \frac{4\alpha\beta T^4 + (\varsigma_2 - 2\beta T^2)\varsigma_1}{\varsigma_1\varsigma_2},$$

$$\gamma_1 = \frac{-2\alpha T\varsigma_2}{\varsigma_1}, \gamma_2 = \frac{-4\alpha T^2}{\varsigma_1},\ \gamma_3 = \frac{-2\alpha\beta T^3}{\varsigma_1},$$

$$\varsigma_1 = 8 + 4T(1 + \alpha a) + 2T^2(\beta + \alpha a + \alpha) + \alpha\beta a T^3,$$

$$\varsigma_2 = 4 + 2T + \beta T^2.$$

The nonlinear function $g(x_1(k))$ is now given by

$$g(x_1(k)) = ax_1(k) + bx_1(k)|x_1(k)| \qquad (3)$$

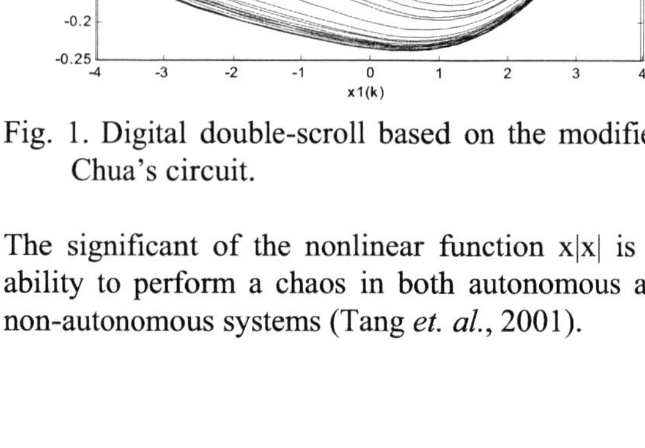


Fig. 1. Digital double-scroll based on the modified Chua's circuit.

The significant of the nonlinear function x|x| is its ability to perform a chaos in both autonomous and non-autonomous systems (Tang *et. al.*, 2001).

On the basis of Eqs. (2) and (3), a double-scroll attractor can be generated, as shown in Fig. 1.

## 3. SECURE DIGITAL COMMUNICATION

### *3.1 Chaos Synchronization*

The goal of chaos synchronization is to design a coupling between two chaotic systems, called transmitter and receiver, in such a way that their behaviours become identical after a transient time.

Consider two discrete-time autonomous chaotic systems, in the form of Eq. (2):

$$\begin{aligned} S1 &: \mathbf{w}(k+1) = \Phi_1 \mathbf{w}(k) + \Gamma_1 g_1(\mathbf{w}(k)), \\ S2 &: \mathbf{v}(k+1) = \Phi_2 \mathbf{v}(k) + \Gamma_2 g_2(\mathbf{v}(k)) \end{aligned} \tag{4}$$

where $\mathbf{w}(k), \mathbf{v}(k) \in R^3$ are the state vectors, with constant matrices.

A natural measure of the synchronization quality for systems of $\mathbf{w}(k)$ and $\mathbf{v}(k)$ is

$$\rho = \| \mathbf{w}(k) - \mathbf{v}(k) \| \tag{5}$$

where $\rho$ is the length of a vector between a point in the subspace spanned by $\mathbf{w}(k)$ and its corresponding point in the subspace spanned by $\mathbf{v}(k)$. The synchronization of these two systems implies that their sub-manifolds will converge to each other, in such a way that

$$\lim_{k \to \infty} \| \mathbf{w}(k) - \mathbf{v}(k) \| = 0 \tag{6}$$

There are two types of synchronization: mutual synchronization and master-slave synchronization. The mutual synchronization is a bi-directional coupling and the latter is unidirectional coupling. The manifolds $\mathbf{w}(k)$ and $\mathbf{v}(k)$ can be synchronized with both types using an observer approach (Koh and Ushio, 1997). To classify different synchronization states systematically, the term M-synchronization is introduced: subsystems $S_1$ and $S_2$ are said to be M-synchronization if

$$\mathbf{w}(k) - \mathbf{M}\mathbf{v}(k) = 0 \tag{7}$$

where $\mathbf{M} \in R^{3\times3}$. Clearly, $\mathbf{M} = I$ for in-phase synchronization and $\mathbf{M} = -I$ for antiphase synchronization (Zhong *et. al.*, 2001).

### 3.2 M-Chaotic Phase Shift Keying

To design a secure digital communication system, the chaotic phase shift keying (CPSK) approach (Kennedy and Ogorzalek, 1997) is adopted. Figures 2 and 3 depict the model with $T_r$ as the transmitter and $R_e$ as the receiver. The transmitter observer $T_r$ is governed by

$$\mathbf{w}(k+1) = \mathbf{M}_i^{-1}\Phi\mathbf{M}_i\mathbf{w}(k) + \mathbf{M}_i^{-1}\Gamma u(k) \tag{8}$$

where $\mathbf{w}(k) \in R^3$ is the state vector of $T_r$ at time $k$, $\Phi \in R^{3\times3}$ is a stable matrix, $\Gamma \in R^{3\times1}$, $\mathbf{M}_i \in R^{3\times3}$ denotes a nonsingular synchronization matrix, and $u(k) \in R^1$ is the nonlinear feedback function.

Each type of signals consisting of n-bits is passed into the modulator where an integer $i\ (1 \le i \le 2^n)$ is assigned, and mapping into a synchronization matrix $\mathbf{M}_i$ and a transmission vector $\eta_i \in R^{1\times3}$. The transmitting signal, $s(k) \in R^1$, is given by $s(k) = \eta_i \mathbf{w}(k)$, and the feedback function is $u(k) = g(s(k))$ where $g(\bullet)$ is defined in Eq. (3).

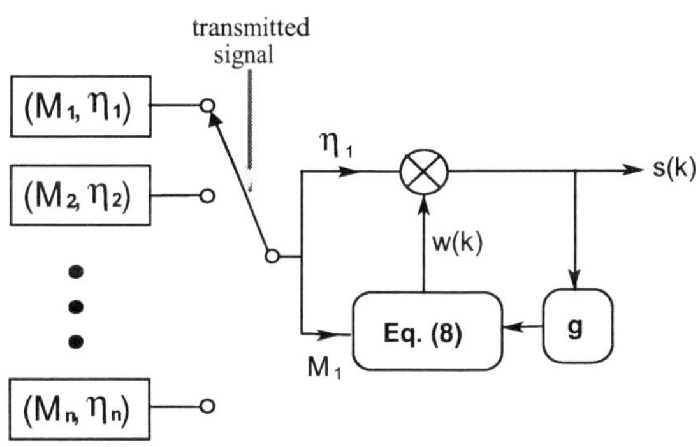


Fig. 2. Transmitter $T_r$ in M-CPSK

For the receiver, $R_e$ is described by

$$\mathbf{v}(k+1) = \left(\Phi + \mathbf{H}_i \eta_i \mathbf{M}_i^{-1}\right)\mathbf{v}(k) + \Gamma\mathbf{u}(k) - \mathbf{H}_i s(k) \tag{9}$$

with respect to Eq. (8) for some $\mathbf{H}_i \in R^{3\times1}$.

In a real time situation, Eq. (9) needs to be speedily converged in order to avoid signal mismatch. The use of deadbeat control (observer) is therefore a desirable solution so that the eigenvalues of Eq. (9) can be set to zero. It ensures that the steady state is attained within a small number of steps. $T_r$ and $R_e$ are then $\mathbf{M}_i$-synchronized, i.e. $\mathbf{M}_i\mathbf{w} = \mathbf{v}$ asymptotically.

At the demodulator, by investigating the relation between the spectra of $\mathbf{M}_i^{-1}\mathbf{v}$ and $\mathbf{w}$, it is possible to recover the transmitted signal. With respect to each value of $i$, the following error function is defined:

$$\varepsilon_i^{\,2} = \sum_{k=\tau_m}^{\tau} \left[\mathbf{w}(k) - \eta_i \mathbf{M}_i^{-1}\mathbf{v}(k)\right]^2 \tag{10}$$

where $\tau$ is the signal interval and $\tau_m$ is the minimum number of steps to reach synchronization.

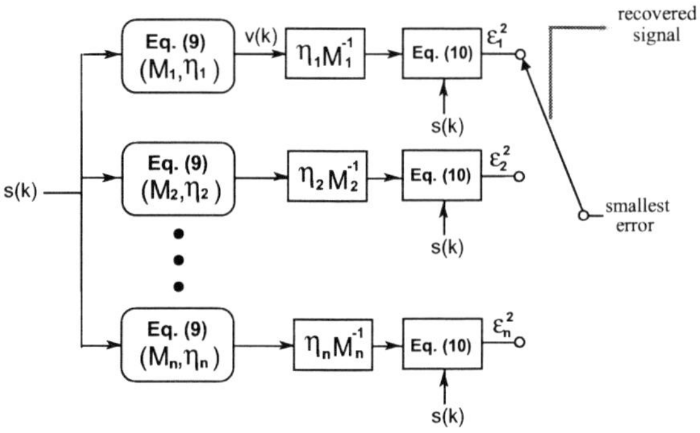


Fig. 3. Receiver $R_e$ in M-CPSK

The signal corresponding to the value of $i$ resulting the smallest $\varepsilon_i^2$ is the recovered signal. Care is needed to ensure that $\mathbf{M}_i$ and $\eta_i$ are correctly chosen, such that $\eta_i \mathbf{M}_i^{-1} \neq \eta_j \mathbf{M}_j^{-1}$ if $i \neq j$; $T_r$ and $R_e$ will then behave chaotically to make it difficult for eavesdroppers.

## 4. TESTING RESULTS

To demonstrate the proposed system, the configuration for transmitting binary signals "0" and "1" is presented.

Referring to Fig. 2: $\eta_1 = \eta_2 = [1 \quad 0 \quad 0]$; $\mathbf{M}_1 = \begin{bmatrix} 1 & 0 & 0 \\ 0 & 3 & 0 \\ 0 & 0 & 2 \end{bmatrix}$ and $\mathbf{M}_2 = \begin{bmatrix} 2 & 0 & 0 \\ 0 & -1 & 0 \\ 0 & 0 & 1 \end{bmatrix}$ are assumed. Hence, the transmitting signal $s(k)$ is equal to $w_1(k)$. If a signal of "0" is to be transmitted, four discrete data based on Eq. (8) with pair $(\mathbf{M}_1, \eta_1)$ are transmitted. $(\mathbf{M}_2, \eta_2)$ is used while a signal of "1" is transmitted. Figure 4 shows the transmitting signal with a binary sequence embedded in the chaotic carrier waveforms. These signals are not only difficult to predict, but also well disguised from eavesdroppers. The main attribute of these signals is largely derived from the nonlinear function of Eq. (3).

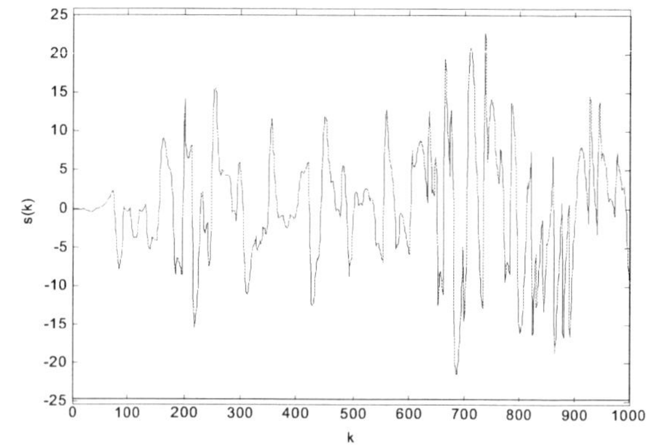


Fig. 4. Transmitting signal $s(k)$

At the receiver end, the vector $\mathbf{H}_i \in R^{n\times 1}$ in Eq. (9) is designed such that the desired eigenvalues are $(0, 0.001, -0.002)$ and $(0, 0.002, -0.001)$ for $\mathbf{H}_1$ and $\mathbf{H}_2$, respectively.

The errors between $\eta \mathbf{M}_i^{-1}\mathbf{v}$ and $\eta \mathbf{w}_1$ with $\eta = [0 \quad 1 \quad 0]$ (i.e. the spectra of $w_2(k)$ and the corresponding states in $\mathbf{M}_i^{-1}\mathbf{v}$ are compared) are depicted in Fig. 5. Due to the deadbeat control design, it only takes about 4 samples for the convergence between $\eta \mathbf{M}_i^{-1}\mathbf{v}$ and $\eta \mathbf{w}$ (i=1 or 2).

By setting $\tau_m = 3$ and $\tau = 4$ for each 4-samples window, $\varepsilon_1^{\ 2}$ and $\varepsilon_2^{\ 2}$ can be computed using Eq. (10). It can determine the recovered bit as "1" if $\varepsilon_1^{\ 2} < \varepsilon_2^{\ 2}$, and "0" otherwise. The recovered bit and the transmitted bit sequences are shown in Fig. 6, and it can be concluded that all the bits are correctly restored. It should be emphasized that the transmission rate will be increased if more pairs of $(\mathbf{M}_i, \eta_i)$ are designed.

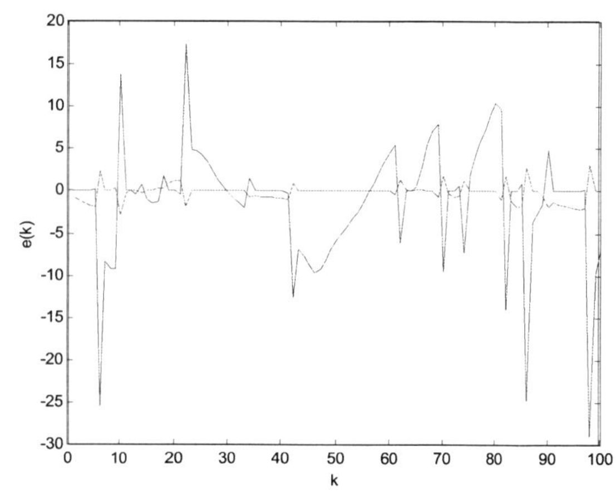


Fig. 5. Error between $\eta \mathbf{M}_i^{-1}\mathbf{v}$ and $\eta \mathbf{w}$

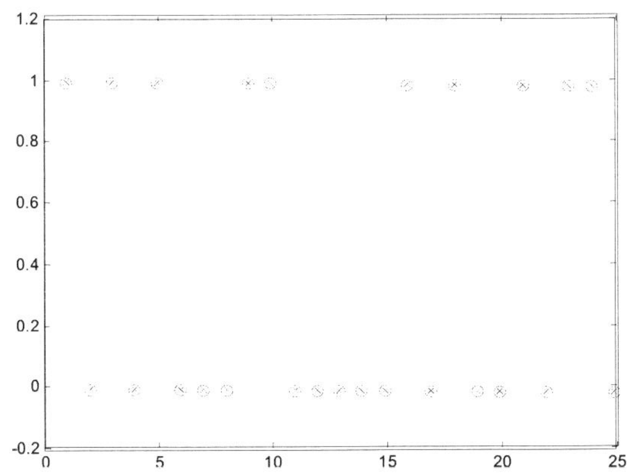


Fig. 6. Recovered bit (x) and Transmitted bit (o) sequences

## 5. SUMMARY

In this paper, the effectiveness of chaos phase shift technique together with a discrete time Chua's circuit based on x|x| nonlinearity for secure communication was demonstrated. We have shown that the transmitted signal can be recovered from the receiver end with predictable steps of computation. This finding will form a solid foundation for secure digital communication as its applications can be far and wide, and it is particular useful for real time Internet telephone and e-commerce environment.

## 6. REFERENCES


Grassi G. and Mascolo S. (1999). Synchronization of High-order Oscillators by Observer Design with Application to Hyperchaos-based Cryptography. *International Journal of Circuit Theory and Applications*, **27**, 543-553.

Hasler M. and Vandewalle J. (1999). Special Issue on Communications, Information Processing and Control Using Chaos. *International Journal of Circuit Theory and Applications*, **2**.

Hayes S., Grebogi C. and Ott E. (1993). Communication with Chaos. *Physical Review Letters*, **70**, 3031-3034.

Kennedy M.K. and Ogorzalek M.J. (1997). Special Issue on Chaos Synchronization, Control and Applications, *IEEE Transactions on Circuits and Systems Part I*, **44(10)**.

Koh C.L. and Ushio T. (1997). Digital Communication Method Based on M-synchronized Chaotic Systems. *IEEE Transactions on Circuits and Systems Part I*, **44(5)**, 383-390.

Pecora L.M. and Carroll T.L. (1990). Synchronization in Chaotic Systems. *Physical Review A*, **44(4)**, 2374-2383.

Short K.M. (1996). Unmasking a Modulated Chaotic Communication Scheme. *Int, of Bifurcation and Chaos*, **6**, 367-375.

Tang K.S., Man K.F., Zhong G.Q. and Chen G. (2001). Generating Chaos via x|x|," *IEEE Transactions on Circuits and Systems, Part I*, **48(5)**, 635-641.

Tang K.S. and Man K.F. (1998). An Alternative Chua's Circuit Implementation. *IEEE Proceedings on ISIE'98*, **2**, 441-444.

Zhong G.Q., Man K.F. and Ko K.T. (2001). Uncertainty in Chaos Synchronization. *Int. J. of Bifurcation and Chaos,* (accepted for publication).

IFAC Publications
www.elsevier.com/locate/ifac

# TRACKING VARIABLE PERIODIC SIGNALS WITH VARIABLE SAMPLING RATE – FREQUENCY DOMAIN APPROACH

**Zhenwei Cao**

*School of Advanced Technologies and Processes*
*Central Queensland University*
*Rockhampton QLD 4702 Australia*
*Email: z.cao@cqu.edu.au*

**Gerard Ledwich**

*Electrical and Electronic Systems Engineering*
*Queensland University of Technology*
*GPO Box 2434 Brisbane 4001 Australia*
*Email: g.ledwich@qut.edu.au*

**Abstract:** This paper presents two adaptive control algorithms for tracking repetitive reference signals with dynamic change in period. These two control algorithms are the adaptive feedforward tracking controller and the adaptive feedback repetitive controller. Under the assumption that the controlled plant is linear time invariant and whose steady state behaviour can be described by its frequency response, the proposed algorithms generate the controls based on the estimation of the gain and phase of the plant at selected frequencies contained in the reference signal. Hence, no structural knowledge of the plant needs to be known. It has been shown that the adaptive feedforward tracking controller provides fast transient response, and the adaptive feedback repetitive controller exhibits good steady state performance. The sequential combination of the two algorithms is studied, which demonstrates fast transient as well as good steady state performance. Experimental results are presented to illustrate the effectiveness of the proposed schemes. *Copyright © 2001 IFAC*

Keywords: time-varying, adaptive, tracking, feedforward, feedback.

## 1. INTRODUCTION

There has been a substantial amount of research done in the development of tracking control algorithms for known fundamental frequency. However, in many situations, the periodic reference changes dynamically in period. For example, the robot manipulators may have to do some repetitive jobs, which may have time-varying period due to variations in load. The design of the control algorithms, which can track a time-varying repetitive reference, is less attended.

In this paper, we present two tracking control algorithms for repetitive reference signals with dynamic change in period. Two algorithms, the adaptive feedforward tracking controller and the adaptive feedback repetitive controller, are developed based on the estimation of the gain and phase of the plant at the frequencies contained in references, which are found from the learnt input and output data by an interpolative method.

This paper is organised as follows. Section 2 describes the proposed adaptive feedforward tracking control. Section 3 presents the proposed adaptive feedback repetitive control. Section 4 explains an application to a servomotor-tracking problem. Section 5 covers the discussion. The paper concludes in Section 6.

## 2. ADAPTIVE FEEDFORWARD TRACKING CONTROL

We propose an adaptive feedforward tracking control system is shown in Figure 1, which consists of the controlled system, the frequency identifier, the system response estimator and a feedforward controller. The frequency identifier detects the reference frequency, and changes the sampling rate, so that the same number of samples per period is maintained. This means that the time-varying reference signal is time-invariant at sampling points. The system response estimator estimates the system gain and phase response at harmonics of the new reference frequency. The changes in the system response are due to sampling rate adjustment and harmonic shift of the new reference frequency. The feedforward controller generates the controls using the estimated system response at frequencies contained in the reference and the time-invariant Fourier transform of the reference. Details of the frequency identifier, the system response estimator, and the feedforward controller are discussed in the subsequent sections.

*The frequency identifier:* In some tracking applications, the reference frequency can be measured directly; therefore the identifier is not needed. However, when the time-varying reference frequency can not be measured, the frequency identifier must be utilised. The frequency identifier can be accomplished in a few ways including a recursive scheme (Tsao and Qian 1993), Phase-Lock-Loop (PLL) approach (Bodson

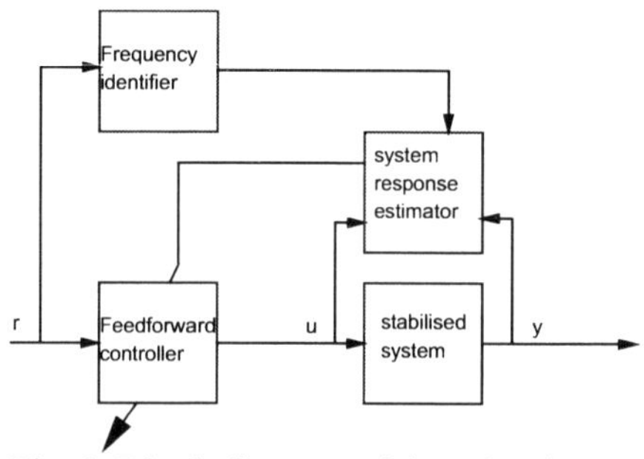


Fig. 1. Block diagram of the adaptive feedforward tracking control

et al 1997), notch filtering technique (Dotsch and Smarkman 1997), and etc. Here we use a simple digital method based on the measurement of duration between two zero-crossings in the signal. This method is implemented in TMS320C30 DSP chip.

*The system response estimator:* The objective of the system response estimator is to provide the estimation of gain and phase response at harmonics of the new reference. The design of the system response estimator should consider two factors: the sampling rate variation and the harmonics shift of the new reference. The analysis of sampled data systems (Kuo 1992) shows the following relationship:

$$F(j\omega)=\frac{\omega_s}{2\pi}H(j\omega)\bar{F}(j\omega)\,,\,\omega\in\left[-\frac{\omega_s}{2},\frac{\omega_s}{2}\right] \quad (1)$$

where $\bar{F}(j\omega)$, $F(j\omega)$ and $H(j\omega)$ are the frequency response of the continuous and discrete system and sample and hold circuit respectively, $\omega_s$ is the sampling rate. $H(j\omega)$ is usually known, for example, for the zero-order hold circuit:

$$H(j\omega)=\frac{2\pi}{\omega_s}\frac{\sin(\pi\omega/\omega_s)}{\pi\omega/\omega_s}e^{-j(\pi\omega/\omega_s)} \quad (2)$$

Now we can conclude that if we know the discrete system response $F(j\omega)$ at frequency $\omega$, we are able to calculate the continuous system response $\bar{F}(j\omega)$ at the same frequency and vice verse.

Now we consider a stabilised continuous system $\bar{F}(j\omega)$ sampled at a rate of $\omega_{s,i}$, which gives the discrete system $F(j\omega)$. The system is tracking a periodic reference with a frequency $\omega_{r,i}$, where $\omega_{r,i} = \frac{\omega_{s,i}}{N}$, $N$ is the number of samples per period and $i$ represents the time index for the reference frequency change. When the reference frequency changes to $\omega_{r,i+1}$, the new sampling rate is set to $\omega_{s,i+1} = N\omega_{r,i+1}$ and the discrete system response at the harmonics of $\omega_{r,i+1}$ under the new sampling rate $\omega_{s,i+1}$ is estimated as follows:

1. From the learnt system input and output, identify the discrete system response at harmonics of the previous reference, $\omega_{r,i}$, $F_i(j\omega), \omega = 0, \omega_{r,i}, 2\omega_{r,i}, \ldots$.
2. Using (1), convert the system response obtained from the discrete domain to the continuous domain at harmonics of $\omega_{r,i}$, $\bar{F}_i(j\omega), \omega = 0, \omega_{r,i}, 2\omega_{r,i}, \ldots$.
3. Interpolate the response at harmonics of $\omega_{r,i}$ to estimate the continuous system response at harmonics of the new reference frequency $\omega_{r,i+1}$, $\bar{F}_{i+1}(j\omega), \omega = 0, \omega_{r,i+1}, 2\omega_{r,i+1}, \ldots$.
4. Using (1), convert the estimated system response from the continuous domain back to the discrete domain, $F_{i+1}(j\omega), \omega = 0, \omega_{r,i+1}, 2\omega_{r,i+1}, \ldots$

Since the sampling rate is varied with respect to the reference frequency, if we assume perfect frequency tracking, the time-varying reference is time-invariant at sampling points. In other words, the frequency domain descriptions, which can be derived from the DFT coefficients, are time-invariant, $R(j\omega)$. Therefore the new control vector can be calculated by:

$$U_{i+1}(j\omega) = \frac{R(j\omega)}{F_{i+1}(j\omega)}, \omega = 0, \omega_{r,i+1}, 2\omega_{r,i+1}, \ldots \quad (3)$$

## 3. THE ADAPTIVE FEEDBACK REPETITIVE CONTROL

The proposed adaptive feedback repetitive control system is shown in Figure 2. The pre-compensator $F^*(z)$ is designed as the complex conjugate of $F(z)$ by Ledwich and Bolton (1991). $F^*(z)$ compensates the phase shift caused by $F(z)$, so

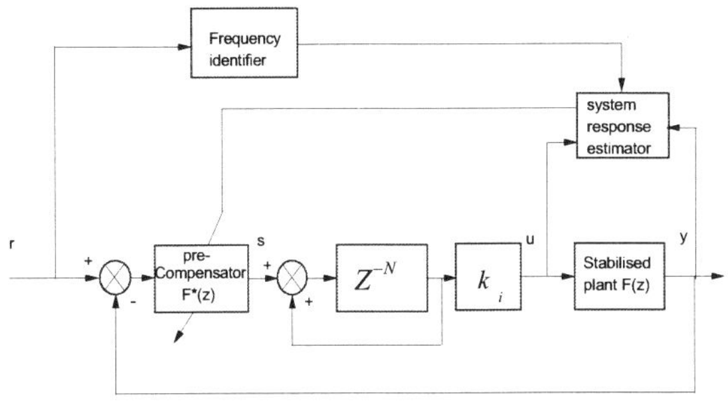


Fig. 2. Block diagram of the feedback repetitive control.

that the total phase shift in open loop forward path is ideally zero. The adaptation will converge provided the cancellation error is less than $\pm 90°$. The frequency identifier measures the reference frequency and maintains the sampling rate as multiples of the new frequency. When the sampling rate is adjusted to follow the variations in the reference, the frequency response of $F^*(z)$ needs to be changed accordingly to give the required phase cancellation for the system stability. Suppose the discrete system response at new harmonics, $F_{i+1}(j\omega), \omega = 0, \omega_{r,i+1}, 2\omega_{r,i+1}, \ldots$, is estimated using the procedures described in the Section 2. The response of $F^*_{i+1}(j\omega)$ can be calculated by simply inversing of the phase response of $F_{i+1}(j\omega)$.

## 4. APPLICATION TO A SERVO MOTOR CONTROL

The effectiveness of the adaptive repetitive control is demonstrated by applying the control schemes to a servomotor-tracking problem in simulation and experiment. The model is identified using the Feedback Digital Servo Workshop software applying the surface method. A proportional control, with a unit gain, is introduced to stabilise the system and the stabilised system has the closed loop transfer function:

$$F(s) = \frac{0.843}{0.287s^2 + s + 0.843} \quad (4)$$

*Simulation:* The simulation has been conducted using Matlab. The reference $r_1$ is a triangular waveform with a frequency $\omega_{r,1} = 0.818$ rad/s, sampled at $\omega_{s,1} = 52.36$ rad/s and the number of samples per period is $N = \frac{\omega_{s,1}}{\omega_{r,1}} = 64$. In the first

24 periods, the repetitive control is employed for the system to track the reference $r_1$ with control vector $u_1$ and the output vector $y_1$. After period 25, the reference frequency jumps to $\omega_{r,2} = \frac{\omega_{r,1}}{2} = 0.409$ rad/s and the sampling rate is changed to $\omega_{s,2} = \frac{\omega_{s,1}}{2} = 26.18$ rad/s to maintain same $N$ multiples of the reference frequency. The tracking errors are shown in Figure 3, where the

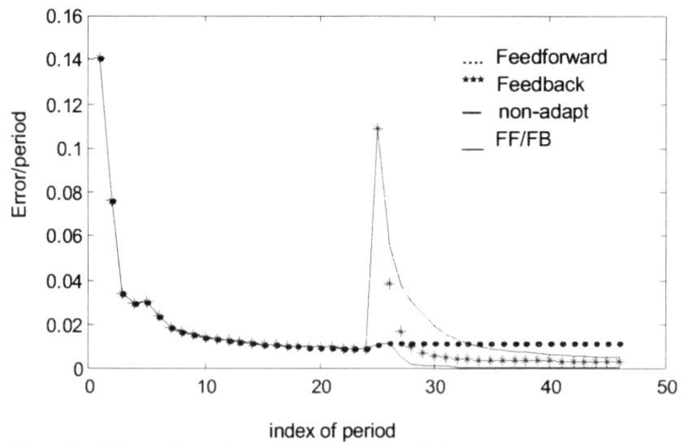


Fig. 3. Simulated results of tracking error.

adaptive feedforward tracking control has a fast convergence and the adaptive feedback repetitive control exhibits a good steady state performance. Therefore a sequential combination of the adaptive feedforward and feedback control is simulated. The adaptive feedforward tracking control is applied to the system for period 25 and 26. At period 27, when the track error can not be reduced further, the system is switched to the adaptive feedback repetitive control, where the estimated discrete system response at the new sampling rate is used in the compensator. The tracking error is added to Figure 3, which shows good features of both algorithms.

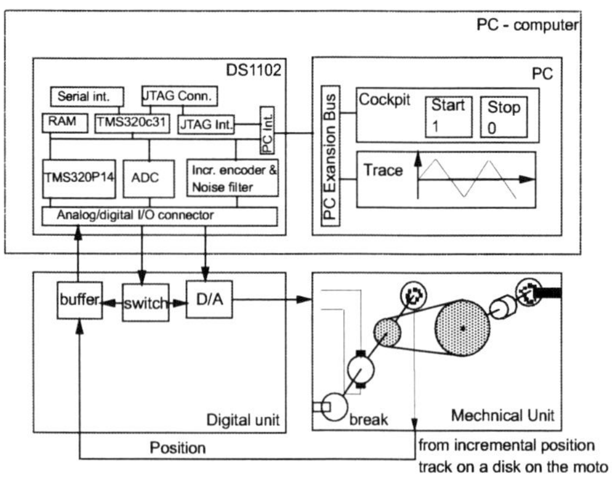


Fig. 4. Block diagram of the experimental system.

*Experiments:* The experimental system, shown in Figure 4, consists of the Feedback servomotor Model MU154C as a plant, Dspace system 1101 as a controller and a digital interface board DU154B. The interfacing board has an 8-bit analog-to-digital converter, a servo pre-amplifier and incremental position signals. The Dspace 1101 system consists of a floating point digital signal processor (TMS320C30) and an incremental decoder. All the control and adaptation algorithms are implemented in C language.

The experiment implements the combined schemes, which consists of the adaptive feedforward tracking control and adaptive feedback repetitive control. It starts with a triangular reference generated by the DS1102 control board internally, with a magnitude of $\pm 3$ rad, a frequency $\omega_{r,1} = 0.818$ rad/s, sampling rate $\omega_{s,1} = 52.36$ rad/s and $N = \frac{\omega_{s,1}}{\omega_{r,1}} = 64$. In the first 24 periods, the system is under the feedback repetitive control, which causes the system output to track the reference. The frequency identifier measures the reference period by counting samples between two zero crossing. The number of samples counted stays constant in first 24 periods, which means the reference frequency has no change. After period 25, the reference jumps to the new frequency $\omega_{r,2} = 0.409$ rad/s generated internally. The frequency identifier starts counting clock samples on the positive going zero crossing. Before the new frequency is determined, the system uses the old control input. The frequency identifier finishes counting at the negative going crossing of the reference and the number of clock samples counted is doubled, which means the halving in the reference frequency. This process makes use of the known half cycle symmetry for this reference waveform. Then the new sampling rate is set to $N$ times the identified reference frequency. The discrete system response at the new sampling rate is estimated and the controls are then generated using the adaptive feedforward control algorithm, and are applied to the plant for periods 25, 26 and 27. After period 28, the adaptive feedback repetitive control is used to achieve better steady state performance. The system output is recorded from period 23 for 100 second (x-axis scale) as shown in Figure 5, where the solid and dashed lines represent the system output and the reference respectively.

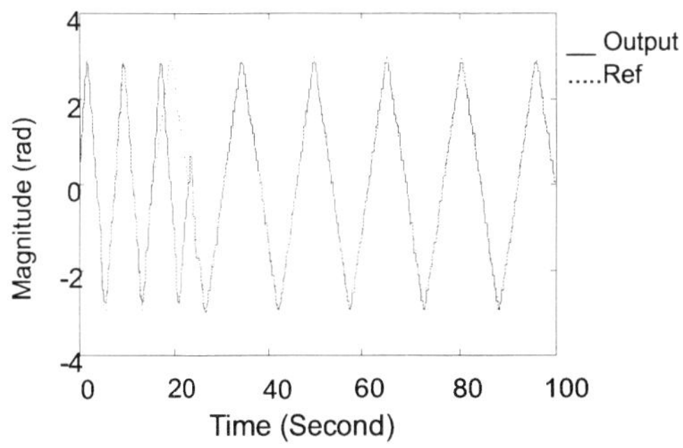


Fig. 5. Experimental tracking results of the system with combined two adaptive control schemes.

To make a comparison, the system without any adaptive schemes and using only repetitive control (Ledwich and Bolton 1991) with constant parameters is tested as well. In the first 24 periods, the system has the same repetitive control as the adaptive system. When the reference jumps to $\omega_{r,2}$ after period 25, the new frequency is identified using the same algorithm and the sampling rate is changed accordingly. However, since this system does not adopt any adaptive schemes and uses the feedback repetitive control with constant parameters in the compensator through whole process, the tracking performance is poor, shown in Figure 6, where the solid and dotted lines represent plant output and reference respectively.

The average tracking errors in each period of the signal recorded from period 23 over 100 seconds (x-axis) are shown in Figure 7. The solid and dotted lines are from the system using the combined algorithms and non-adaptive algorithms respectively. The error of period 25 is relatively large (0.3137 rad), because the old control is used until the new frequency is identified. The error of period 26 is reduced greatly to 0.0279 rad, the error of period 27 is 0.0076 rad, and the system is

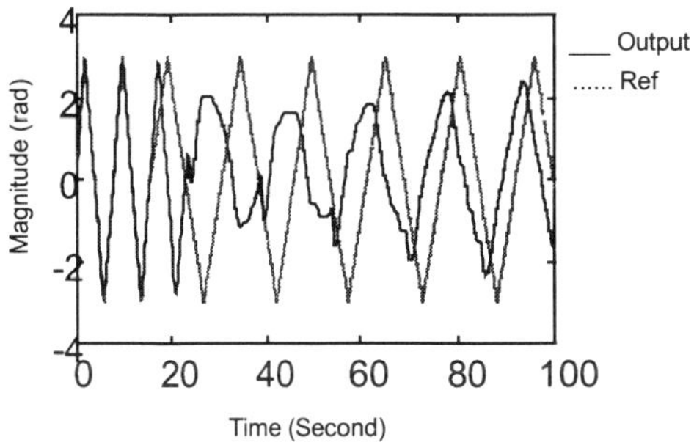


Fig. 6. Experimental tracking result of the system with non-adaptive scheme, which uses same parameters before and after the reference changes.

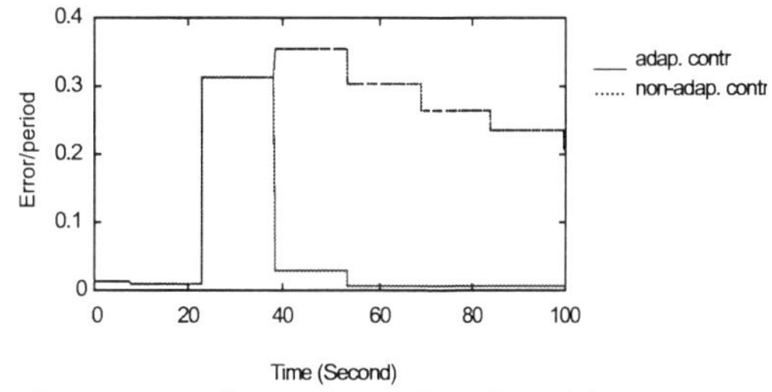


Fig. 7. Experimental results of tracking errors.

switched to the adaptive feedback repetitive control at period 28 with the estimated system response used in the compensator. The error of period 28 and 29 are 0.0076 rad and 0.0073 rad respectively, which is not significantly different from the error at period 27 (0.0076 rad). That means the system has achieved a near perfect tracking after the adaptive feedforward tracking control is applied while the estimation error is small. For comparison, the dashed line in Figure 7 shows the tracking error of the non-adaptive repetitive control, which uses constant control parameters before and after the reference change. It is obvious that the performance of the system with the adaptive schemes is superior in terms of tracking speed and convergent rate of error.

## 5. DISCUSSION

The digital method used to identify the reference frequency is used for its simplicity. However, when significant measurement noise exists, the accuracy of this method can be affected. Passive pre-filters and hysteresis may be needed to improve the measurement.

The estimation of the discrete system response is crucial for successful adaptive schemes, especially for the feedforward case. If the frequency response, learnt from only last control, is used in the interpolation as discussed in this paper, the accuracy of the interpolation will then depend on the spectral richness of the reference, the relative position of the new frequency to the old frequency and the smoothness of the system response. These factors make it difficult to give an explicit condition to assure good interpolation. However, it is often possible to have the stored frequency response of the plant at some individual points, such as, the DC gain, the system bandwidth and etc, which can be used to get accurate interpolation.

Another way to accumulate the knowledge of the frequency response of the plant is from a learning process. The learnt system discrete response generated for various reference frequencies can be stored. Then the interpolation can be performed on the stored data, which is accumulated from pre-knowledge and previous learning process. Although the interpolation based on the stored data may not be accurate initially because of less system response stored, it would be improved gradually. This initial inaccuracy can be compensated by the subsequent feedback control if necessary. The more system response stored, the more accurate the interpolation. A perfect interpolation can give the system a perfect tracking and the subsequent feedback control can be removed. Another alternative way is to use a parametric model in the estimation to fill in gaps where the signal is not rich enough.

## 6. CONCLUSION

Two adaptive tracking schemes have been developed for linear discrete systems to tracking a periodic signal with dynamic change in period. A simple zero crossing method has been used to detect the reference frequency. The sampling rate is adjusted to obtain time-invariant reference at sampling point. The system response changes caused by the sampling rate variation is then compensated by considering the effect of the sample and hold circuit. The system response at harmonics of the new reference frequency is estimated by an interpolative method. The feedforward tracking controller is designed based on the inverse of the system response and gives the system fast convergence. The adaptive feedback repetitive control is synthesised based on phase shift cancellation and provides good steady state performance. A sequential combination of the two adaptive schemes exhibits fast convergence as well as good steady state performance. Both simulations and experiments on a servomotor system have shown the effectiveness of the proposed schemes.

## REFERENCES

Bodson, M., Douglas, S. 1997, 'Adaptive algorithm for the rejection of sinusoidal disturbances with unknown frequency', Automatica, Vol. 33, No. 12, pp. 2213-2221.

Dotsch, H., Smakman, H. 1995, 'Adaptive repetitive control of a compact disc mechanism', Proceedings of the 34th Conference on Decision and Control, New Orleand, LA, December.

Feedback Instruments Ltd, 1998, 'Digital servo workshop - Advanced teaching manual'.

Kuo, B. C. 1992: "Digital control systems", HBJ Saunders, pp. 124-126.

Ledwich, G., and Bolton, A. 1991, 'Tracking periodic inputs using sampled compensators', IEE Proceedings-D, vol. 138, no. 3.

Tsao, T. C., and Qian, Y. X. 1993, 'An adaptive repetitive control scheme for tracking periodic signals with unknown period', Proceedings of the American Control Conference, San Francisco, California, pp.1736-1740 Vol 2.

IFAC Publications
www.elsevier.com/locate/ifac

# FAST COMPUTATION OF TIME-FREQUENCY DISTRIBUTIONS USING A PARALLEL DSP-BASED SYSTEM FOR SIGNAL ANALYSIS

**F. García-Nocetti F., J. Solano Gonzalez, E. Rubio Acosta, E. Moreno Hernandez**

*Departamento de Ingeniería de Sistemas Computacionales y Automatización, Instituto de Investigaciones en Matemáticas Aplicadas y en Sistemas, Universidad Nacional Autónoma de México, P.O. Box 20-726, Del. A. Obregón, 01000 México D.F., México*

Abstract: This paper describes some algorithms to estimate the spectral response of a signal based on time-frequency distributions from a Cohen's class point of view. Four distribution cases are evaluated: Wigner Ville, Choi Williams, Bessel and Born Jordan. A discrete distribution is formulated for each case and a criterion for determining the interaction between the spectral components of the signal is also given. Simplified discretised expressions for the implementation of distributions are formulated, these leading to a reduction of the computations realized when comparing to original definitions. Finally, a general parallel approach for the computation of the distributions is proposed, implemented and assessed by using a parallel DSP-based system. *Copyright © 2001 IFAC*

Keywords: Time-Frequency Distributions, Parallel DSP Architectures, Signal Analysis.

## 1. INTRODUCTION

A classic method for spectral estimation is the so-called Fourier Transform. However, its use is limited to stationary signals giving as a result a poor frequency resolution when estimating non-stationary ones.

Other types of spectral estimators, called time-frequency distributions, have been developed. Unlike conventional methods these distributions are not limited to the use of stationary signals (Cohen, 1989). Despite of this important advantage, the number of calculations involved in obtaining thc spcctral estimation increases substantially compared to the traditional methods. Therefore, it is desirable to simplify the formulation of the distributions in such a way that the computations involved can be reduced without any loss in the spectral resolution. On the other hand, there are a great variety of time-frequency distributions. It would be very useful to develop an analysis criterion such that can provide a tool for selecting the optimum time-frequency distribution according to the features of the signal under consideration. This paper deals with these issues.

## 2. TIME-FREQUENCY DISTRIBUTIONS

This section formulates the so-called Cohen's class for the time-frequency distributions and it defines some related concepts.

### 2.1. The Cohen's Class

The Cohen's class in terms of time frequency distributions (Cohen, 1989) can be formulated as follows. Let the time-frequency distribution kernel be defined as $\phi(\theta,\tau)$. This kernel will define the particular characteristics of each time-frequency distribution. Then, the Cohen's class for the time-frequency distributions with kernel $\phi(\theta,\tau)$ can be defined as:

$$TFD(t,\omega)= \frac{1}{2\pi}\int_{-\infty}^{\infty}\left[\int_{-\infty}^{\infty}\left[\int_{-\infty}^{\infty}\phi(\theta,\tau)e^{-j\theta(t-\mu)}d\theta\right]x\left(\mu+\frac{\tau}{2}\right)x^*\left(\mu-\frac{\tau}{2}\right)d\mu\right]e^{-j\omega\tau}d\tau \quad (1)$$

### 2.2. *Auto-term, Crossing Term and Crossing Term Weighting Factor*

In order to establish a comparison criterion between the different time-frequency distributions considered in this paper, it is necessary to develop a method that may determine the degree in which the various components of a signal interact when the time frequency distribution is calculated (Cardoso, *et. al.*, 1996). Consider the following signal, which is composed of a finite number of sinusoidal signals with constant amplitude $A_n$, frequency $\omega_n$ and phase $\theta_n$:

$$x(t) = \sum_{n=1}^{N} A_n e^{j(\omega_n t + \theta_n)} \tag{2}$$

The $n$ components of the signal in eq. (2) interact between them through eq. (1). The interactions of the components with themselves generate the so-called auto-terms of the distribution, which are always positive and constitute the spectral contents of the signal. On the other hand, interactions between different components generate the so-called crossing terms of the distribution, which can be positive or negative and are added to the spectral contents of the signal. Therefore it is desirable to minimize such terms.

Substituting eq. (2) in eq. (1) and grouping the auto-terms in the first summation and crossing terms in the second summation, a time-frequency distribution can be expressed as:

$$TFD(t,\omega) = 2\pi \sum_{n=1}^{N} A_n^2 \delta(\omega - \omega_n) + \sum_{\substack{n=1 \\ n \neq m}}^{N} \sum_{m=1}^{N} F_{TFD}(\omega) A_n A_m \cos((\omega_n - \omega_m)t + \theta_n - \theta_m) \tag{3}$$

where $F_{TFD}(\omega)$ is the crossing terms weighting factor, a quantitative measure for evaluating the different time-frequency distributions. The following sections describe the distributions according to their definition.

### 2.3. *The Wigner Ville Distribution*

According to its definition (Cardoso, *et. al.*, 1996; Fan, and Evans, 1994; Martin, and Flandrin, 1985), this distribution for the discrete case is given by:

$$DWVD(n,k) = 2 \sum_{\tau=-N+1}^{N-1} W(\tau) W^*(-\tau) e^{-j\frac{2\pi k\tau}{N}} x(n+\tau) x^*(n-\tau) \tag{4}$$

where $n$ represents the discrete time and $k$ the discrete frequency; both variables are normalized.

### 2.4. *The Choi Williams Distribution*

According to its definition (Cardoso, *et. al.*, 1996; Choi, and Williams, 1989), this distribution for the discrete case is given by eq. (5), where $\sigma > 0$ is a scaling factor.

$$CWD(n,k) = 2 \sum_{\tau=-N+1}^{N-1} W(\tau) W^*(-\tau) e^{-j\frac{2\pi k\tau}{N}} \sum_{\mu=-M}^{M} \sqrt{\frac{1}{4\pi\tau^2/\sigma}}\, e^{-\frac{\mu^2}{4\tau^2/\sigma}} x(\mu+n+\tau) x^*(\mu+n-\tau) \tag{5}$$

### 2.5. *The Bessel Distribution*

According to its definition (Cardoso, *et. al.*, 1996; Guo, *et. al.*, 1994), this distribution for the discrete case is given by eq. (6), where $\alpha > 0$ is a scaling factor.

$$BDB(n,k) = 2 \sum_{\tau=-N+1}^{N-1} W(\tau) W^*(-\tau) e^{-j\frac{2\pi k\tau}{N}} \sum_{\mu=-2\alpha|\tau|}^{2\alpha|\tau|} \frac{1}{\pi\alpha|\tau|} \sqrt{1 - \left(\frac{\mu}{2\alpha\tau}\right)^2}\, x(\mu+n+\tau) x^*(\mu+n-\tau) \tag{6}$$

### 2.6. *The Born Jordan Distribution*

According to its definition (Cohen, 1989), this distribution for the discrete case is given by (7), where $\alpha > 0$ is a scaling factor.

$$DBJD(n,k) = 2 \sum_{\tau=-N+1}^{N-1} W(\tau) W^*(-\tau) e^{-j\frac{2\pi k\tau}{N}} \sum_{\mu=-2\alpha|\tau|}^{2\alpha|\tau|} \frac{1}{4\alpha|\tau|} x(\mu+n+\tau) x^*(\mu+n-\tau) \tag{7}$$

## 3. EVALUATION OF THE TIME-FREQUENCY DISTRIBUTIONS BASED ON THE CROSSING TERMS WEIGHTING FACTOR

As stated previously, an ideal crossing terms weighting factor would be one that eliminates the crossing terms. But, a desirable and more realistic situation would be that the weighting factor concentrated the crossing terms due to two different frequency components of the signal around such frequencies and not around other frequencies or spread them out over a wide range of frequencies.

Substituting the signal defined by eq. (2) in eq. (1) and arranging the crossing terms in the distributions according to eq. (3), the crossing terms weighting factors of each distribution are obtained and defined by the following expressions. For the Wigner Ville distribution:

$$F_{WVD}(\omega) = 2\pi\delta\left(\omega - \frac{\omega_n + \omega_m}{2}\right) \tag{8}$$

For the Choi Williams distribution:

$$F_{CWD}(\omega) = \sqrt{\frac{\pi\sigma}{(\omega_n - \omega_m)^2}}\, e^{-\frac{\sigma}{4(\omega_n - \omega_m)^2}\left(\omega - \frac{\omega_n + \omega_m}{2}\right)^2} \tag{9}$$

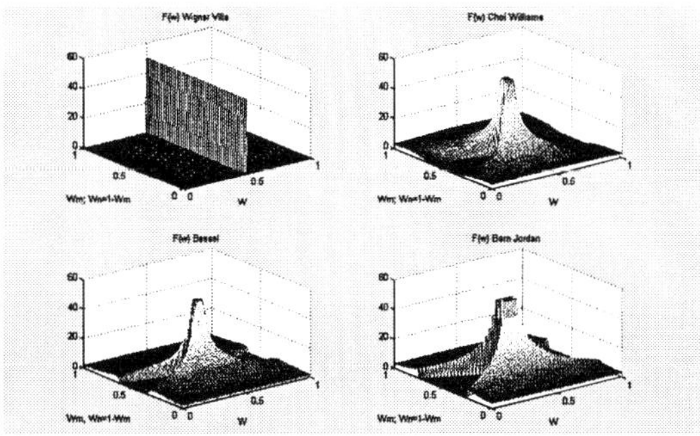

Fig. 1. Global view of the crossing terms weighting factors for the Wigner Ville, Choi Williams *($\sigma$=5)*, Bessel *($\alpha$=2)* and Born Jordan *($\alpha$=1)* distributions.

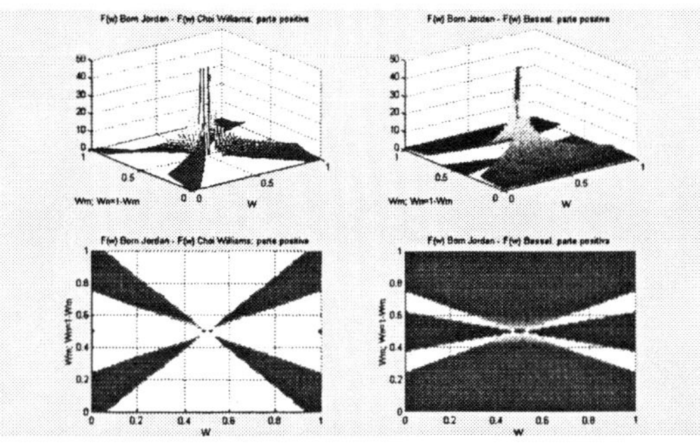

Fig. 2. Differences of the crossing terms weighting factors between Born Jordan minus Choi Williams and Born Jordan minus Bessel distribution.

For the Bessel distribution:

$$F_{BD}(\omega)=\frac{4}{\alpha|\omega_n-\omega_m|}U_0\left(\frac{\omega-\frac{\omega_n+\omega_m}{2}}{\alpha(\omega_n-\omega_m)}\right)\sqrt{1-\left(\frac{\omega-\frac{\omega_n+\omega_m}{2}}{\alpha(\omega_n-\omega_m)}\right)^2} \quad (10)$$

For the Born Jordan distribution:

$$F_{BJD}(\omega)=\frac{\pi}{\alpha|\omega_n-\omega_m|}P_{2\alpha(\omega_n-\omega_m)}\left(\omega-\frac{\omega_n+\omega_m}{2}\right) \quad (11)$$

where $P_\alpha(t)$ is a rectangular symmetrical pulse of duration $\alpha$. Figure 1 shows a global view of the crossing terms weighting factors for the Wigner Ville, Choi Williams, Bessel and Born Jordan distributions. The weighting factor is defined in terms of $\omega$ and contains the interacting frequencies $\omega_n$ and $\omega_m$ defined in signal in eq. (2). Such frequencies can be added or subtracted depending on the behavior of each distribution. Consider a normalized addition, that is $\omega_n+\omega_m=1$, then, the graphs relate the weighting factor versus $\omega_n$ and $\omega$, where $0<\omega_n<1$. Given the normalized addition, for each $\omega_n$ value, $\omega_m$ is given by $1-\omega_m$.

## 4. ANALYSIS

Figures 2 and 3 depict graphs which show the differences of the crossing terms weighting factors between Born Jordan minus Choi Williams, Born

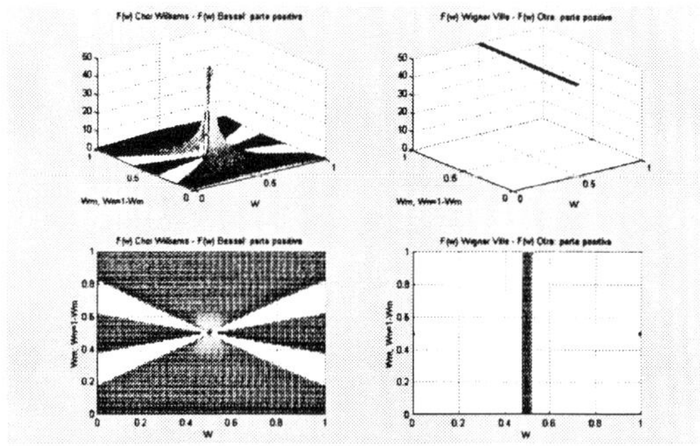

Fig. 3. Differences of the crossing terms weighting factors between Choi Williams minus Bessel and Wigner Ville minus any other distribution.

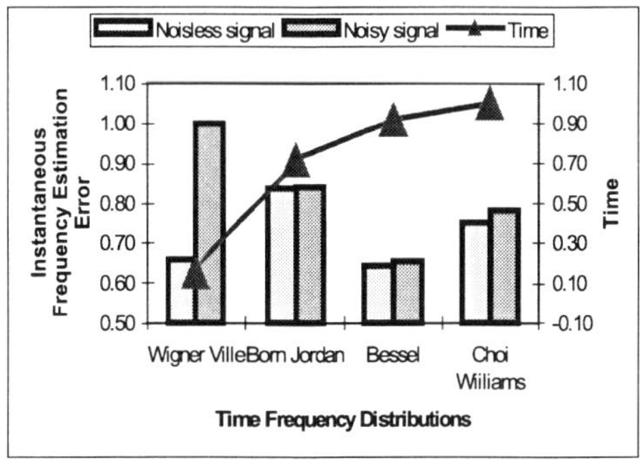


Fig. 4. Normalized instantaneous frequency estimation error and normalized execution time of noiseless signal and noisy signal using Wigner Ville, Born Jordan, Bessel and Choi Williams time-frequency distributions.

Jordan minus Bessel, Bessel minus ChoiWilliams and Wigner Ville minus any other distribution. The dark zones in the graphs correspond to points where the weighting factor of the first distribution under comparison is greater than the second one. For the purpose of this analysis scaling factors *$\sigma$=5, $\alpha$=2* and *$\alpha$=1* are considered in the Choi Williams, Bessel and Born Jordan distributions, respectively.

In general, it is observed that the weighting factor for the Bessel distribution is smaller than the Choi William's and the Born Jordan's, and that the Choi Williams's is smaller than the Born Jordan's. These results indicate that the Bessel distribution spreadsout the crossing terms better than the Choi Williams's and Born Jordan's distributions and, in consequence, estimates with more precision the spectral contents of a signal in the presence of noise. All this facts explains the reason why the Bessel distribution is less sensitive to the presence of noise (Cardoso, *et. al.*, 1996). But there is a direct relationship between the precision and the amount of calculations involved in the estimation of the spectral contents of the signal.

In the case of the Wigner Ville distribution, the crossing terms are concentrated on the average of such frequencies (due to the two frequency components of the signal). Therefore, this distribution estimates with better precision the spectral contents of noiseless signals with small bandwidth.

The figure 4 shows the normalized instantaneous frequency estimation error versus the time involved in performing the calculations, considering a simulated non-stationary Doppler signal proposed in (Cardoso, *et. al.*, 1996), that is, a band limited stochastic signal with Gaussian probability density function which models a signal sampled at the center of a normal carotid artery. It is important to point out that all the distributions are affected by a scaling factor which in turn modifies the crossing term weighting factors. As stated previously, the results have been obtained considering $\sigma=5$, $\alpha=2$ and $\alpha=1$ in the Choi Williams, Bessel and Born Jordan distributions, respectively. These optimum scaling factors have been found experimentally and they depend on the characteristics of the signal under study.

## 5. REDUCING THE COMPUTATIONAL COMPLEXITY OF THE DISCRETE DEFINITIONS

In order to evaluate the different distributions for spectral estimation, a discrete signal $x(n)$ is considered. Such a signal contains $2N-1$ elements, where $N$ is a power of $2$ and the element range is from $-N+1$ to $N-1$, therefore $x(0)$ is the central element. Based on these elements, this section presents a reduction in computational terms of the number of calculations involved in the evaluation of each of the distributions considered in this paper.

### *5.1. The Wigner Ville Distribution*

Considering eq. (4) for estimating the Wigner Ville distribution and evaluating it in $n=0$ (Boashash, and Black, 1987; Fan, and Evans, 1994), an equivalent simplified expression would be given by:

$$DWVD(0,k) = 4\,\mathrm{Re}\left[\sum_{\tau=0}^{N-1} W(\tau)W^{*}(-\tau)e^{-j\frac{2\pi k\tau}{N}}x(\tau)x^{*}(-\tau)\right] - 2|x(0)|^{2} \quad (12)$$

Assuming that $W(\tau)W^*(-\tau)$ is a single factor then, for each value of $k$ in eq. (4) evaluated in $n=0$, there are $6N-3$ complex multiplications, $2N-2$ complex additions and $1$ scalar multiplication, whereas in eq. (12) there are $3N+1$ complex multiplications, $N$ complex additions and $2$ scalar multiplications.

### *5.2. The Choi Williams Distribution*

Similarly, considering eq. (5) for estimating the Choi Williams distribution and evaluating it in $n=0$, an equivalent simplified expression would be given by:

$$DCWD(0,k) = -2|x(0)|^{2} + \quad (13)$$
$$+4\,\mathrm{Re}\left[\sum_{\tau=0}^{N-1} W(\tau)W^{*}(-\tau)e^{-j\frac{2\pi k\tau}{N}}\sum_{\mu=-N+1+|\tau|}^{N-1-|\tau|}\sqrt{\frac{1}{4\pi\tau^{2}/\sigma}}\,e^{-\frac{\mu^{2}}{4\tau^{2}/\sigma}}x(\mu+\tau)x^{*}(\mu-\tau)\right]$$

where the summation respect to $\mu$ for $\tau=0$ is $x(0)x^*(0)$.

Assuming that $W(\tau)W^*(-\tau)$ and the square root multiplied by the exponential are single factors then, for each value of $k$ in eq. (5) evaluated in $n=0$, there are $8N^2-4N$ complex multiplications, $4N^2-6N+2$ complex additions and $1$ scalar multiplication, whereas in eq. (13) there are $2N^2-2N+1$ complex multiplications, $N^2-2N$ complex additions and $2$ scalar multiplications.

### *5.3. The Bessel Distribution*

Considering eq. (6) for estimating the Bessel distribution and evaluating it in $n=0$, an equivalent simplified expression would be given by:

$$DBD(0,k) = -2|x(0)|^{2} + \quad (14)$$
$$+4\,\mathrm{Re}\left[\sum_{\tau=0}^{N-1} W(\tau)W^{*}(-\tau)e^{-j\frac{2\pi k\tau}{N}}\sum_{\mu=\max\{-2\alpha|\tau|,-N+1+|\tau|\}}^{\min\{2\alpha|\tau|,N-1-|\tau|\}}\frac{1}{\pi\alpha|\tau|}\sqrt{1-\left(\frac{\mu}{2\alpha\tau}\right)^{2}}\,x(\mu+\tau)x^{*}(\mu-\tau)\right]$$

where the summation respect to $\mu$ for $\tau=0$ is $x(0)x^*(0)$.

Assuming that $W(\tau)W^*(-\tau)$ and the square root divided by $\pi\alpha|\tau|$ are single factors then, for each value of $k$ in eq. (6) evaluated in $n=0$, there are $8\alpha N^2-8\alpha N$ complex multiplications, $4\alpha N^2-4\alpha N-2N$ complex additions and $1$ scalar multiplication, whereas in eq. (14) there are less than $4\alpha N^2-4\alpha N+1$ complex multiplications, less than $2\alpha N^2-2\alpha N-N$ and $2$ scalar multiplications.

### *5.4 The Born Jordan Distribution*

Considering eq. (7) for estimating the Born Jordan distribution and evaluating it in $n=0$, an equivalent simplified expression would be given by:

$$DBJD(0,k) = -2|x(0)|^{2} + \quad (15)$$
$$+4\,\mathrm{Re}\left[\sum_{\tau=0}^{N-1} W(\tau)W^{*}(-\tau)e^{-j\frac{2\pi k\tau}{N}}\sum_{\mu=\max\{-2\alpha|\tau|,-N+1+|\tau|\}}^{\min\{2\alpha|\tau|,N-1-|\tau|\}}\frac{1}{4\alpha|\tau|}x(\mu+\tau)x^{*}(\mu-\tau)\right]$$

where the summation respect to $\mu$ for $\tau=0$ is $x(0)x^*(0)$. The analysis is similar to Bessel's.

## 6. PARALLEL PROCESSING OF THE TIME-FREQUENCY DISTRIBUTIONS

As stated previously, the use of time-frequency distributions for spectral estimation of signals opens the possibility of analyzing non-stationary signals. However, the associated computational cost is high. Section 5 has described strategies in order to achieve a reduction in the amount of calculations involved in evaluating the original definitions for each

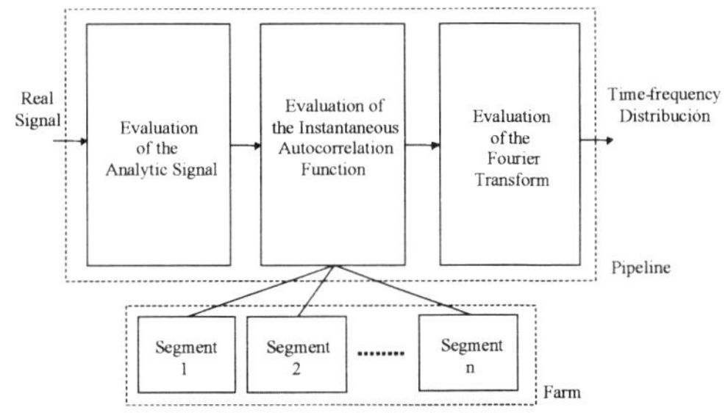


Fig. 5. Intrinsic parallel processing approach for execution of the time-frequency distributions.

distribution. In this section, the use of parallel processing techniques in order to reduce the time required to perform the evaluations is proposed. Previous work has been addressed to investigate parallel processing techniques in signal analysis (Madeira, et. al., 1999; Solano, et. al., 1999).

In this work two approaches of parallelism are presented: intrinsic parallelism and data parallelism. The first approach considers a single data-time segment being processed by a number of processors, where each processor executes a different part of the signal analysis algorithm.

In the second approach a sequence of N data-time segments is processed in parallel by using N processors, each processor executing the whole algorithm over each data-time segment. Both approaches are implemented and assessed by using a parallel DSP-based system.

### *6.1. Parallel Architecture*

A system based on the ADSP-21062 SHARC architecture has been used for implementing the time-frequency distribution algorithms. Specifically, an ASP-P15 card with 4 Analog Devices DSP is used. This development system programmed in C language and hosted in a personal computer has been used to implement the parallel approaches presented in this work.

### *6.2. Intrinsic Parallelism Approach*

In this approach, the time-frequency distribution algorithms are carried out on single data-time segment of the signal. Parallelism is achieved by means of partitioning and allocating the composing processes of the algorithms on a number of processors. For the intrinsic parallelism approach, a 3-stages pipeline is proposed. The first stage calculates the analytic signal $x_a(t)$ of the real signal. The second stage calculates the generalized time-indexed auto-correlation function $R'_x(t,\omega)$ for $t=0$ of $x_a(t)$, this being the integral respect to $\mu$ in eq. (1). Finally, the third stage calculates the Fourier transform of $R'_x(0,\omega)$, this being the time-frequency

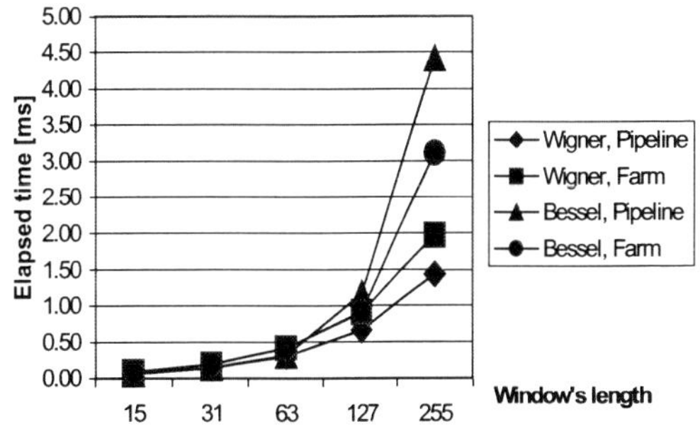


Fig. 6. Execution time vs. signal's length of Wigner Ville and Bessel distributions for pipeline and farm+pipeline parallel architectures using 4 SHARC processors.

distribution $TFD(t,\omega)$ for $t=0$ of the real signal. Figure 5 shows the pipeline structure of the process. The calculations for first and third stages are relatively simple and a Fast Fourier Transform (FFT) algorithm is used. Second stage involves a more complex processing, then this stage is being paralleled by using a parallel farm computational model in a star topology.

Here, each node calculates a set of operations of the generalized time-indexed auto-correlation function. Although the expressions for the evaluation of each of the time-frequency distributions are different, the second stage can be adapted easily adding or subtracting processors according to the needs. Figure 6 shows the time performance of the Wigner Ville and the Bessel distributions for two different parallel arquitectures (pipeline and farm+pipeline). Note that the Choi Williams and the Born Jordan distributions behave close to the Bessel distribution.

### *6.3 Data Parallelism Approach*

This approach regards the execution of a number N of data-time segments in parallel using an array of N processors. Each processor has a copy of the whole three-stage algorithm described previously and executes the analysis on a dedicated segment. As one dedicated processor is used for process each segment, the number of data-time segments to be processed depends on the available processors. A star processor array scheme has been used with N nodes, where N is the number of data-time segment to be analyzed simultaneously. Figure 7 shows the time performance for Wigner Ville and Bessel time-frequency distributions using this parallel approach.

## 7. TRUNCATING THE DISCRETE TIME FREQUENCY DISTRIBUTION.

It is possible to achieve a further reduction in the calculation time of the Choi Williams, Bessel and Born Jordan discrete distributions by truncating the summation respect μ in eq. (13), (14) and (15)

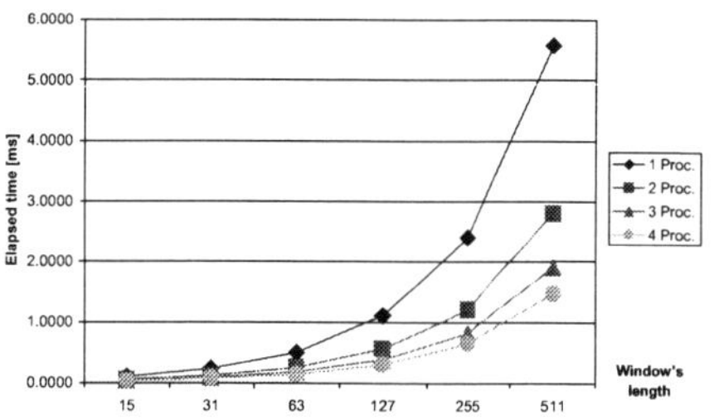


a)

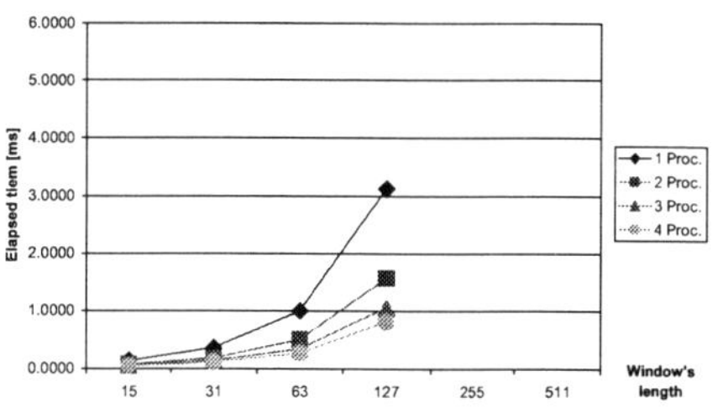


b)

Fig. 7. a) Execution time vs. Signal's length for Wigner Ville distribution using a star parallel arquitecture (1 to 4 SHARC processors). b) Similar explanation for Bessel distribution.

respectively. Although such a truncation is destructive, it can be experimentally characterized. Considering a noisless signal and a window's length of 127, the following results have been obtained. A truncation of $\mu=20$, $\mu=40$ and $\mu=20$ in eq. (13), (14) and (15) respectively, results in an increase of less than 5% in the estimation error of the pseudo instantaneous mean frequency. Similar results are valid for noisy signals with SNR up to 20dB.

## 8. CONCLUSIONS

Conventional methods for spectral estimation are limited to the analysis of stationary signals to produce a good estimate, however, these methods offer poor resolution when dealing with non-stationary signals. This paper has presented some alternative methods based on the so-called time frequency distributions for spectral analysis. Four methods based on the Cohen's class have been analyzed: Wigner Ville, Choi Williams, Bessel and Born Jordan distributions. A comparison criterion based on the crossing terms weighting factor has been proposed, this showing that the Bessel distribution spreads out the crossing terms better than the Choi Williams's and Born Jordan's distributions and, in consequence, estimates with more precision the spectral contents of a signal in the presence of noise, whereas the Wigner Ville distribution estimates with better precision the spectral contents of noiseless signals with small bandwidth. This analysis was conducted taking into account the optimum scaling factors: $\sigma=5$, $\alpha=2$ and $\alpha=1$ for the Choi Williams, Bessel and Born Jordan distributions, respectively. This work also has proposed a simplification in the complexity of the expressions utilized for calculating the time-frequency distributions giving as a result a reduction of at least half the operations involved in the original definition. Two approaches of parallelism have been presented: intrinsic parallelism and data parallelism. In the first approach parallelism has been achieved by means of partitioning and allocating the composing processes of the distribution algorithms on a number of processors. A single data-time segment has being processed by a number of processors, each processor executing a dedicated part of the signal analysis algorithm. Here, a pipeline scheme with three stages is utilized, corresponding to the second stage to deal with the more expensive computational process (evaluation of the generalized time-indexed auto-correlation function). A generalized scheme has been implemented which can be adapted easily according to the time-frequency distribution under consideration. As a result, the Wigner Ville distributions shows a better performance in a pipeline parallel arquitecture, while the Choi Williams, Bessel and Born Jordan distributions perform better in a farm+pipeline scheme. The second parallel approach, deals with several data-time segments that are processed by a number of processors. Here, the signal analysis algorithm for each data-segment has been calculated by each processor. In contrast with the first approach, multiple data-time segments are processed in parallel. This approach has shown better flexibility and scalability than the intrinsic parallelism one. An advantage of this approach is that the communications between processors are reduced only to the distribution of the input data to each one of the processors, at the beginning of the loop, and at the end of the process when each processor sends back the results to the master processor, this leading to a higher performance than the intrinsic parallelism approach. The authors acknowledge the UNAM (PAPIIT IN117999), CONACYT (27982-A), CONACYT-REDII (7350-858), for the financial support.

## REFERENCES


Boashash, B. and P. Black (1987). An Efficient Real-Time Implementation of the Wigner-Ville Distribution. *IEEE Transactions on Acoustics, Speech, and Signal Processing*. **ASSP-35**. 1611-1618.

Cardoso, J. G. Ruano and P. Fish (1996). Nonstationary Broadening Reduction in Pulsed Doppler Spectrum Measurements Using Time-Frequency Estimators. *IEEE Transactions on Biomedical Engineering*. **43**. 1176-1186.

Choi, H. and W. Williams (1989). Improved Time-Frequency Representation of Multicomponent Signals Using Exponential Kernels. *IEEE Transactions on Acoustics, Speech and Signal Processing*. **37**. 862-871.

Cohen, L. (1989). Time-Frequency Distributions -A Review. *Proceedings of the IEEE*. **77**. 941-981.

Fan, L. and D. Evans (1994). Extracting Instantaneous Mean Frequency Information from Doppler Signals Using the Wigner Distribution Function. *Ultrasound in Med. & Biol.* **20**. 429-443.

Guo, Z., L. Durand and H. Lee (1994). The Time-Frequency Distributions of Nonstationary Signals Based on a Bessel Kernel. *IEEE Transactions on Signal Processing*. **42**. 1700-1707.

Martin, W. and P. Flandrin (1985). Wigner-Ville Spectral Analysis of Nonstationary Processes. *IEEE Transactions on Acoustics, Speech, and Signal Processing*. **ASSP-33**. 1461-1470.

Madeira, M., S. Bellis, L. Beltran, J. Solano, F Garcia, W. Marnane, M. Tokhi, M. Ruano (1999). High Performance Computing for Real Time Spectral Estimation. *IFAC Journal Control Engineering Practice (Pergamon)*. **7**. 679-686.

Solano, J., F. Garcia, M. Ruano (1999). High Performance Parallel-DSP Computing in Model-based Spectral Estimation. *Microprocessors and Microsystems (Elsevier)*. **23**. 337-344.

IFAC Publications
www.elsevier.com/locate/ifac

# AN INTELLIGENT POSITION AND QUALITY DETECTION SYSTEM IN IC PRODUCTION

**Qi Qifeng Hu Yueming* Yuan Peng Wu Xinsheng**

*Department of Automatic Control Engineering*
*South China University of Technology, Guangzhou, 510640, P.R.China*
*Email: auymhu@scut.edu.cn*

**Abstract:** An intelligent quality detection and position control system in IC production is developed in this paper. The vision based fast pattern recognition and fuzzy logic techniques are well applied to satisfy the requirements of real-time industrial production. The chips' position can be detected accurately through the base-image-processing technology, which is collected by the vision sensor CCD. Implementation shows the efficiency of the system in real-time production. *Copyright © 2001 IFAC*

**Key words:** quality, detection system, positioning system, pattern recognition, fuzzy logic, hierarchies

## 1. Introduction

Chips quality detection is a necessary part of chips manufacture process. The main purpose is to automatically pick out the good chips and reject the bad ones. Due to the limitation of camera precision and binary image processing technology, most of the existing techniques cannot achieve good performance, since it may lose a lot of important messages of image, and thus cannot get high-accuracy origination and process complex image information.

The main purpose of this paper is to introduce an intelligent quality detection and position control system developed by the authors. To achieve fast pattern recognition, we propose a fuzzy logic based technique to process gray images collected by the vision sensor CCD, and it makes the fast chips allocation via gray image detection become practicable.

## 2. Principle of the experimental set

The experimental set consists of CCD camera, industrial computer, and stepping motor, as shown in Figure 1.

There is an adjustable lens on the camera. To reduce the disturbance of indoor ray, and keep the image both stable and clear, we select coaxial cold-light as

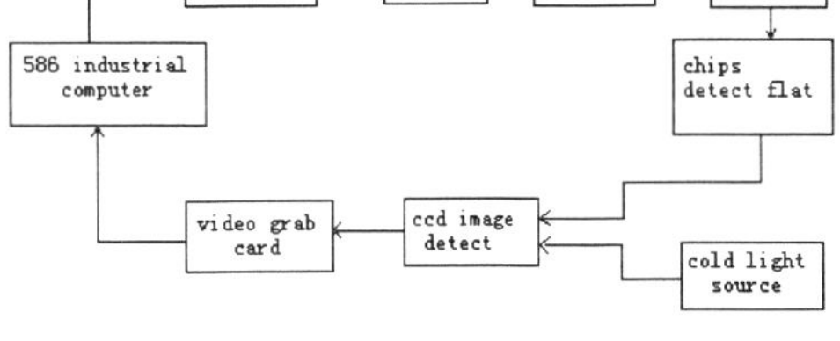


Figure 1 Schematic of the system

the light supply device, which irradiated on the chips uprightness, and reflect in the camera uprightness too. The simulated signal which caught by the CCD camera pass through the video grab card and turn into the digital signal first, then the industrial computer do image processing with fuzzy logic to calculate the position of the chip, and give the warp signal to X-Y stepping motor through computer I/O port and current amplifier, the motor will adjust the chips position to meet the camera.

## 3. Image processing

The digital image is represented by 255 grade gray color. Zero in dark and 255 in white, using 320*240 pixel to deal with the whole chips zone. In order to allocate the chip's position in high speed and accuracy, the whole process should include image pre-processing, pattern recognition, etc. The main purpose of the image pre-processing is to improve the image quality, while pattern recognition includes several methods to accelerate the searching.

### *3.1 Improvement of the image quality*

Before manipulating and extracting information from an image, it requires to obtain the best possible digital representation. Several factors can affect the quality of an image, e.g.

*(I) Random Noise.* There are two main types of random noise: Gaussian noise and Salt-and-pepper noise.

When the first type of noise appears, the exact value of any given pixel is different from each grabbed image. This type of noise adds to or subtracts from the actual pixel value. Usually we apply a low-passed spatial filter to the images to reduce the Gaussian random noise and systematic noise with small-scale variations. This technique replaces each pixel with a weighted sum of its neighborhood.

The second type of noise introduces pixels of arbitrary value (usually high-frequent value) that are generally noticeable since they are completely unrelated to the neighboring pixels. Random noise may be caused by the camera or digitizer, due to the fact that electronic devices tend to generate a certain amount of noise. If the images were transmitted, the distance between the sending and the receiving devices also magnifies the random noise because of interference. We can apply a median filter to the images so as to reduce salt-and-pepper noise. This technique replaces each pixel with the median pixel value of its neighborhood.

*(II) Systematic.* N*oise* Unlike random noise, this type of noise can be predicted, appearing as a group of pixels that should not be part of the actual image. This can be caused, for example, by the camera or digitizer or by uneven lighting. If the image was magnified, microscopic dust particles, on either the object or a camera lens, can appear to be part of the image. To solve this problem, we can first perform a morphological opening operation to remove small particles and break isthmuses between objects in the images, and then perform a morphological closing operation to remove small holes in objects.

Actually, all chip's manufacturing courses are under no-dust circumstance, so the main disturbing signal comes from image grabbed and transmission, we utilize median filter to reduce salt-and-pepper noise to improve the quality of images in this system.

### *3.2 Pattern recognition*

After the pre-processing, we start to recognize the chips. The correlation operation can be seen as a form of convolution, where the pattern-matching model is analogous to the convolution kernel. In fact, ordinary (un-normalized) correlation is exactly the same as a convolution, as given below

$$r = \sum_{i=1}^{i=N} I_i M_i$$

Unfortunately, with ordinary correlation, the resulted value increases if the image gets brighter. So in this case, we use a more complex, normalized version of the correlation function as follows

$$r = \frac{N\sum IM - (\sum I)\sum M}{\sqrt{[N\sum I^2 - (\sum I)^2][N\sum M^2 - (\sum M)^2]}}$$

With this correlation function, the value is unaffected by linear changes (constant gain and offset) in the image or model pixel values. The result reaches its maximum value of 1 when the image and model match exactly, and takes the value 0 during the model and image are uncorrelated. While the similarity is less than expected by chance, then the value is negative. To avoid the slow square-root operation, we use $r^2$ instead of $r$ in the system.

Finally, the above resulted value is converted to a percentage (where 100% represents a perfect matching), so the matching score can be evaluated by $Score = \max(r,0)^2 * 100\%$. Some of the terms in the normalized correlation function depend only on the model, hence can be evaluated once and for all when the model is defined. The only terms that need to be calculated during the search are $\sum I^2$, $\sum IM$ and $\sum I$, which require two multiplications and three additions for each model pixel.

*3.2.1 Hierarchical Search.* A reliable method of reducing the number of computations is to perform a so-called hierarchical search. Basically, a series of smaller, lower-resolution version of both the image and the model are produced, and the search begins on a much-reduced scale. This series of sub-sampled images is sometimes called a resolution pyramid. Each level of the pyramid is half the size of the previous one, and is produced by applying a low-pass filter before sub-sampling. If level 0 (the original image or model size) is 512x512, then level 1 is 256x256, level 2 is 128x128, and so on. Therefore, the higher the level in the pyramid, the lower the resolution of the image and model. The search starts at low resolution to quickly find possible matching candidates. It proceeds to higher and higher resolution to refine the positional accuracy and make sure that the matches found at low resolution were actually occurrences of the model. Because the position is already known from the previous level (to within a pixel or so), the correlation function need be evaluated only at a very small number of locations. Since each higher level in the pyramid reduces the number of computations by a factor of 16, it is usually desirable to start at as high a level as possible. However, the search algorithm must trade off the reduction in search time against the increased chance of not finding the pattern at very low

*3.2.2 Gradient Search.* Now we should handle another problem. Normally, we search a pattern inside a zone from the top to the bottom and from the left to the right. However, this way is not effective in applications, We apply another approach to judge the search track by comparing the matching score of each point with its neighbors. The steps are given as follows:

S1 Allocate the first searching place. In the process of chip's searching, each chip's interval is relatively stable, after the last chip is found, we thus can move the camera nearly up to the next chip in advance so that the searching process can be started nearly at the center of the next chip.

S2 Utilizc thc highcst level model (vaguest but smallest) to recognize the pattern in the center and the pattern around it.

S3 If the highest matching score is the central one, i.e., the chip is in this zone, then we change to lower level (clearer but slower) to do the same pattern recognition in this zone.

S4 If the central one is not the highest matching score, i.e., the chip is not in this zone, then we move the match center to its neighbor, which has the highest score.

S5 If the matching score is lower than the expectation, then back to step S2 and search again, else calculate the X-center and Y-center and out.

This method enables the searching become more intelligent and effective.

*3.2.3 Sub-Pixel Search.* The highest matching score occurs at only one pixel position, and drops off around this peak. The exact (sub-pixel) position of the model can be estimated from the matching scores found on either side of the peak. A curve is fitted to the matching scores around the peak, and the exact peak position can be calculated by the equation of the curve.

Let us suppose the exact matching point p( $x_0, y_0$ ) is allocated in the zone of p( $i_0, j_0$ ), which detected by the normal pixel matching way. Let the curve be characterized as:

$$r(x, y) = ax^2 + bxy + cy^2 + dx + ey + f$$

where a, b, c, d, e, f are coefficients to be calculated. When $r(x,y)$ meets maximum, it must satisfy the following conditions

$$\begin{cases} \dfrac{\partial r(x,y)}{\partial x} = 2ax + by + d = 0 \\ \dfrac{\partial r(x,y)}{\partial y} = 2cy + bx + e = 0 \end{cases}$$

From which we can obtain the maximum position ( $x_0, y_0$ ) as follows

$$\begin{cases} x_0 = \dfrac{2cd - be}{b^2 - 4ac} \\ y_0 = \dfrac{2ae - bd}{b^2 - 4ac} \end{cases}$$

The actual accuracy that can be obtained depends on several factors including the noise in the image and the particular pattern of the model. However, if these factors are ignored, the absolute limit on accuracy, imposed by the algorithm itself, the number of bits and precision used to hold the correlation result, is about 0.05 pixels. In a real application, accuracy better than 0.1 pixels might be achieved for low-noise images.

## 4. Experiments

To show the efficiency of the above technique, we perform real-time implementation. Figure 2 to Figure 7 give the implementation results. From the table 8 we know that the developed system can satisfy the requirements in real-time production.

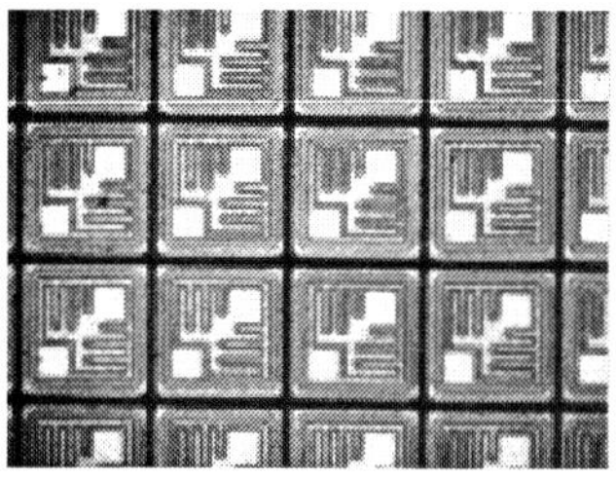

Figure 2 Search Zone

Figure 3 Sample before processing

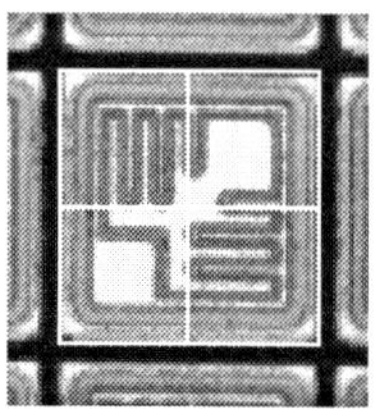

Figure 4 Sample after processing

Figure 5 Search Zone

Figure 6 Sample before processing

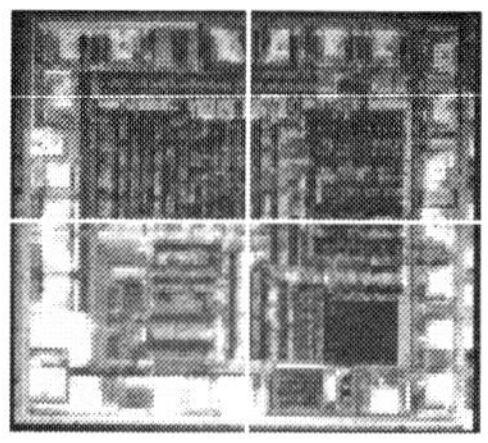

Figure 7 Sample after processing

| | Chip 1(figure 2 to figure 4) | Chip 2(figure 5 to figure 7) |
|---|---|---|
| Use time | 32 ms | 53 ms |
| Precision | 0.1 pixel | 0.1 pixel |
| Chips dimension | 62 pixel*62 pixel | 104 pixel*104 pixel |
| Search area | 320 pixel *240 pixel | 320 pixel* 240 pixel |

Table 1 Real-time experiment results

## 5. Conclusions

This paper concerns the application of the computer vision in Chips Control System, and mainly focus on the topic of how to use hierarchical search, gradient search and sub-pixel search to improve the efficiency and precision of photo-detection in chips allocated system, and to afford credible position statistics for other complex chips detection. It turns the automation with high speed in chips-detection into reality.

## REFERENCES


Hou, C.G.,.Zhao, M.T., and L.S.Ou, A Subpixel Image Matching Algorithm by Quadratic Surface Fitting, Acta Metrologica Sinica, 18(3), pp 229-229, July.1997

Huang, X.D., and Y.L.Yu, Application of computer vision in the Weld Seam Tracking Control, Control Theory and Applications, 18(1), pp 26-30, Feb.2001

Matrox Electronic System ltd., Matrox image library user Guide, Manual, No 10513-MN-0501

Zhen, N.N, Computer vision and pattern recognize, National Defence Industry Press, Beijing, 1998

IFAC Publications
www.elsevier.com/locate/ifac

# Reinforcement Adaptive Fuzzy Control for a Class of Nonlinear Uncertain Systems

*Young H. Kim, **Frank L. Lewis , and ***Chiman Kwan

* Weapons Systems Studies Center
Korea Institute for Defense Analyses
Cheong Ryang P.O. Box 250
Seoul, 130-650
Republic of Korea

** ARRI
The University of Texas
7300 Jack Newell Blvd. S.
Fort Worth, TX 76118
USA

***IAI, Inc.
7519 Standish place
Suite 200
Rockville, MD 20855
USA

**Abstracts**

The paper describes the application of reinforcement learning techniques to feedback control for a class of nonlinear systems using adaptive fuzzy system. The adaptive fuzzy system approximates the nonlinear characteristics of the plant on-line to achieve the desired tracking accuracy without any preliminary off-line learning or training phase. The reinforcement signal from a critic is used for finding proper fuzzy rules to approximate the nonlinear dynamics in the plant. The approximation of the nonlinear dynamics in the plant is performed on-line without knowing any knowledge of the nonlinear terms in the plant, whence comes the term Reinforcement Adaptive Learning (RAL). The fuzzy system based on the RAL algorithm is referred to as Reinforcement Adaptive Fuzzy (RAF) system. The RAL algorithm is derived from Lyapunov stability analysis, so that both tracking stability and error convergence can be guaranteed in the closed-loop system. *Copyright © 2001 IFAC*

Key Words : Closed-loop control, Feedback control, Intelligent control, Nonlinear dynamic system, Fuzzy system

## I. Introduction

The linear conventional controller, such as proportional derivative and proportional integral, is widely used in industrial motion control systems due to simple control structure, ease of design, inexpensive cost and robustness in the face of plant nonlinear terms. Nevertheless, the conventional control algorithm might be difficult to deal with systems with highly nonlinear characteristics. To date, many sophisticated algorithms have been used to help the linear conventional controller under the highly nonlinear and uncertain environment [9]. One of the sophisticated algorithms is to employ artificial neural networks or fuzzy systems to deal with the nonlinear terms and uncertainties in the plant [2, 3, 6, 8].

Fuzzy control has recently found extensive applications for a wide variety of industrial systems and has attracted the attention of many control researchers due to its model-free approach [6]. An important adaptive fuzzy control system has been developed to incorporate with the human expert information systematically [10]. An adaptive fuzzy system is a fuzzy logic system equipped with a learning algorithm, in which the fuzzy logic system is constructed from a collection of fuzzy IF-THEN rules, and the learning algorithm adjusts the parameters of the fuzzy logic system according to the given numerical information.

A new adaptive fuzzy control algorithm is proposed in this paper so that the stability of adaptive fuzzy control system is guaranteed and the new design methodology of a reinforcement learning techniques is shown to be model-free and intelligent in the feedback control. The proposed reinforcement control scheme consists of a performance evaluator, a critic, and an adaptive fuzzy system. The fuzzy system is exploited to provide the nonlinear and uncertain characteristics in the plant according to the reinforcement signal from a critic. The critic evaluates the effects of the plant performance of the state-dependent nonlinear terms or unpredictable uncertainties. The reinforcement signal is used for adjusting the parameters of the fuzzy logic system, such as finding proper fuzzy rules or tuning membership functions. The learning is performed on-line without any preliminary analysis on the nonlinear characteristics appearing in the plant, whence comes the term Reinforcement Adaptive Learning (RAL). The RAL algorithm is derived from Lyapunov stability analysis, so that both system stability and error convergence can be guaranteed in the closed-loop system. Stability is the main objective of all control systems.

## II. Preliminaries

### A. Stability: Nonlinear Dynamic Systems

Given $x(t) \in \Re^n$ and a nonlinear function $h(x,t): R^n \times R \to R^n$, the differential equation

$$\dot{x}(t) = h(x,t), \quad t_0 \le t, \tag{2.1}$$

has a differential solution $x(t)$ if $h(x,t)$ is continuous in $x(t)$ and $t$.

The solution $x(t)$ is said to be *Uniformly Ultimately Bounded* (*UUB*) if there exists a compact set $D$ such that, for all $x_0 \in D$, there is a, $\delta > 0$ and a number $T(\delta, x_0)$ such that $\|x(t)\| \le \delta$ for all $t \ge t_0 + T$.

### B. A Class of Nonlinear Systems

We assume the plant can be represented as the following $n$th-order nonlinear systems of the form

$$\begin{aligned} \dot{x}_1(t) &= x_2(t) \\ \dot{x}_2(t) &= x_3(t) \\ &\vdots \\ \dot{x}_n(t) &= g(x) + d(t) + u(t) \\ y(t) &= x_1(t) \end{aligned} \tag{2.2}$$

where $g(x)$ is an unknown linear or nonlinear function, $u(t)$ is the control input, $y(t)$ is the output of the plant, and $x(t) = [x_1^T \; x_2^T \; \cdots \; x_n^T]^T$ is the state vector of the plant which is assumed to be available for measurement. And $d(t)$ is the unknown external disturbance.

The control objective can be described as follows: given a desired trajectory $x_d(t) = [x_d^T \; \dot{x}_d^T \; \ddot{x}_d^T \cdots]^T$, find a control action $u(t)$ such that the plant follows the desired trajectory with an acceptable accuracy.

Define the tracking error vector as,

$$e(t) = x_d(t) - x(t). \tag{2.3}$$

The problem is thus to design a controller $u(t)$ which ensures that the tracking error should be as small as possible. In the rest of the paper, we show how to derive $u(t)$ in terms of fuzzy intelligent mechanism to achieve the above control objective.

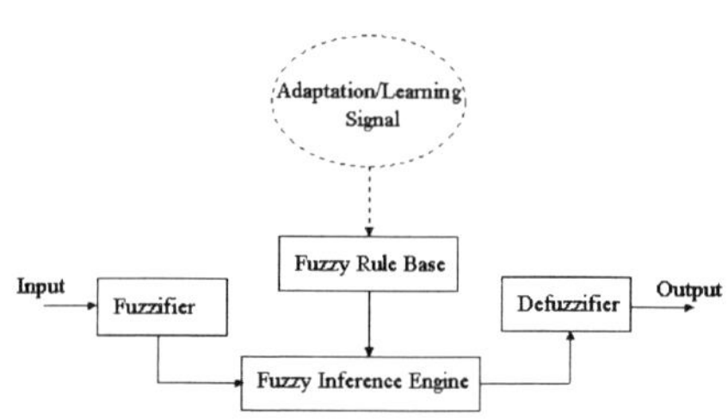


Fig. 1. Building Blocks of Adaptive Fuzzy System

### C. Fuzzy Systems as Function Approximator

The basic configuration of an adaptive fuzzy system is shown in Fig. 1. An adaptive fuzzy system can be described as a typical fuzzy logic system equipped with training or learning elements. The operations of each components can be found in [10]. Without loss of generality, we consider multi-input-single-output fuzzy system. A fuzzy system with the singleton fuzzifier, Gaussian membership functions, the product inference rule, the center average defuzzifier would be used in this paper. The output of such a fuzzy system with $M$ inference rules is in the following form [10]

$$g(x;P) = \frac{\sum_{j=1}^{M} w_j \left[ \prod_{i=1}^{n} \exp\left( -\left( \frac{x_i - c_{ji}}{s_{ji}} \right)^2 \right) \right]}{\sum_{j=1}^{M} \left[ \prod_{i=1}^{n} \exp\left( -\left( \frac{x_i - c_{ji}}{s_{ji}} \right)^2 \right) \right]} = \frac{\sum_{j=1}^{M} w_j \varphi_j(\sigma(x;p_j))}{\sum_{j=1}^{M} \varphi_j(\sigma(x;p_j))} \tag{2.4}$$

where an $w_j$ is the corresponding crisp output of $\varphi(\sigma(\boldsymbol{x};\boldsymbol{p}_j))$ with unit membership function value. The values $\boldsymbol{x}$ and $g(\cdot)$, are the input and output to the fuzzy system, respectively. The vector $\boldsymbol{p}_j$ is the set of all adjustable parameters for the Gaussian membership functions in the fuzzy system. The basis vector $\varphi(\sigma(\boldsymbol{x};\boldsymbol{p}_j))$ is defined as fuzzy basis functions [10]. Since the fuzzy system (2.4) uses only the fuzzy IF-THEN rules for the purpose of approximating the nonlinear functions, we define the fuzzy system as the un-normalized form

$$g(\boldsymbol{x};\boldsymbol{P}) = \sum_{j=1}^{M} w_j \left[ \prod_{i=1}^{n} \exp\left( -\left( \frac{x_i - c_{ji}}{s_{ji}} \right)^2 \right) \right] \quad (2.5)$$

$$= \sum_{j=1}^{M} w_j \varphi_j(\sigma(\boldsymbol{x};\boldsymbol{p}_j)).$$

The expression (2.5) can be written in a compact matrix form

$$g(\boldsymbol{x},\boldsymbol{P}) = \boldsymbol{W}^T \varphi(\boldsymbol{x},\boldsymbol{P}) \quad (2.6)$$

with $\boldsymbol{W} \in \Re^{M\times N}$ and $\varphi(\boldsymbol{x},\boldsymbol{P}) \in \Re^{M\times N}$. The number $N$ is chosen to be "1' in the expression (2.6) since we consider multi-input-single-output fuzzy system. We assume that the number of fuzzy rule M and the membership function parameters $\boldsymbol{P}$ are given in this paper, therefore, the problem becomes finding the parameters $\boldsymbol{W}$ such that the system performance meets the specified requirement.

## III. RAF Controller Design

### A. Structure

Fig. 2 shows the detailed structure of the proposed RAF controller used in this paper, illustrating the overall adaptive-learning scheme. This RAF controller is a combination of an adaptive fuzzy system that uses a reinforcement signal to update the parameters in the fuzzy system, and a performance evaluator, which uses an error based on the given desired trajectory. The ideal behavior of the RAF controller is that the performance evaluator measures the system performance for the current system states, and at the same time provides information to a critic, which supplies the reinforcement signal for the adaptive fuzzy system. Then, the fuzzy system generates the counter-signal necessary to overcome the nonlinear phenomena and the disturbances with which the performance evaluator cannot deal. The performance evaluator plus the gain term constitutes a conventional linear controller. In the architecture, the critic is defined as a pseudo-sign function, which outputs the negative or positive quantity depending on the signal from the performance evaluator. The reinforcement signal provided by the critic represents the system performance degradation due to the nonlinear terms appearing in the plant and/or additional unknown bounded disturbances. Based on the approximation property of the fuzzy system [10], a new reinforcement learning algorithm is developed on the basis of the Lyapunov theory, and the tracking stability is guaranteed in closed-loop control. The proposed scheme avoids the difficulty of getting the correct value for the fuzzy system output.

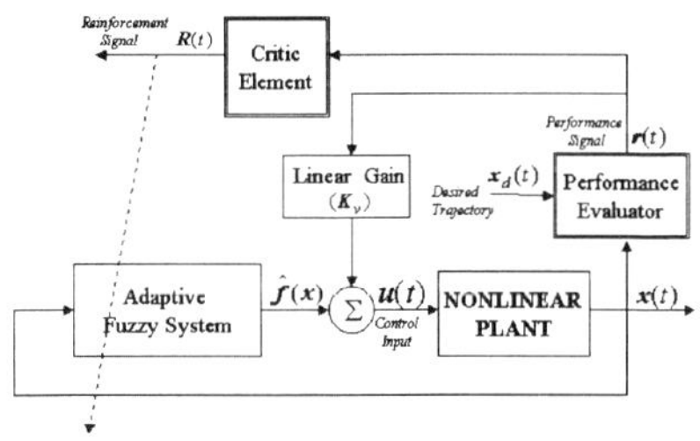


Fig. 2. Structure of the Proposed RAF Controller

### B. Stability Analysis: Lyapunov Approach

Firstly the performance evaluator is defined as output tracking error, i.e.,

$$\boldsymbol{r}(t) = [\Lambda^T \ \boldsymbol{I}]\boldsymbol{e}(t) \quad (3.1)$$

where $\Lambda = [\lambda_1 \, \lambda_2 \cdots \lambda_{n\text{-}1}]^{\mathrm{T}}$ is an appropriately chosen coefficient vector such that $s^{n-1} + \lambda_{n-1}s^{n-2} + \cdots + \lambda_1$ is Hurwitz (i.e. $\boldsymbol{e}(t) \to 0$ exponentially as $\boldsymbol{r}(t) \to 0$). The expression $\boldsymbol{I}$ is the identity vector. The performance evaluator can be viewed as the real-valued utility function of the plant performance, i.e., when $\boldsymbol{r}(t)$ is small enough, the plant performance is good.

The critic for reinforcement signal is defined in terms of the performance measure as

$$\boldsymbol{R}(t) = \boldsymbol{R}^+(t) + \boldsymbol{R}^-(t) = \sigma^+(\boldsymbol{r}) + \sigma^-(\boldsymbol{r}) \quad (3.2)$$

with $\sigma^+(\boldsymbol{r}) = \alpha^+/(1+e^{-\alpha^+\boldsymbol{r}})$ and $\sigma^-(\boldsymbol{r}) = -\alpha^-/(1+e^{\alpha^-\boldsymbol{r}})$. Its behavior characteristics are much dependent on the critic slope gain $\alpha^+$ and $\alpha^-$.

The time derivative of the performance signal can be written from (2.2) and (3.1) as

$$\dot{\boldsymbol{r}}(t) = \boldsymbol{f}(\boldsymbol{x},t) - \boldsymbol{u}(t) \quad (3.3)$$

where the nonlinear function $\boldsymbol{f}(\boldsymbol{x},t)$ is a fairly complex function of $\boldsymbol{x}(t)$ and $\boldsymbol{x}_d(t)$. It is assumed that there are no external disturbances in the plant. From the architecture of the proposed RAF controller in Fig. 2, the control input to the plant $\boldsymbol{u}(t)$ is given by

$$\boldsymbol{u}(t) = \boldsymbol{K}_v\boldsymbol{r}(t) + \hat{\boldsymbol{f}}(\boldsymbol{x},t). \quad (3.4)$$

The function $\hat{\boldsymbol{f}}(\boldsymbol{x},t)$ is provided by the adaptive fuzzy system, the fixed gain matrix is $\boldsymbol{K}_v = \boldsymbol{K}_v^T > 0$.

From (3.4), the time derivative of the performance measure signal (3.3) can be rewritten as

$$\dot{r}(t) = -K_v r(t) + \tilde{f}(x,t). \quad (3.5)$$

And the functional estimation error is defined as

$$\tilde{f}(x,t) = f(x,t) - \hat{f}(x,t). \quad (3.6)$$

According to the approximation property of fuzzy systems, the continuous nonlinear function $f(x,t)$ in (3.3) can be represented by a fuzzy system with some constant ideal values $W$ and some sufficient number of rules, i.e.,

$$f(x,t) = W^T \varphi(\sigma(x)) + \varepsilon(x) \quad \varepsilon(x) \le \varepsilon_M. \quad (3.7)$$

The functional reconstruction error $\varepsilon(x)$ is bounded by known constant $\varepsilon_M$. We assume that the ideal values $W$ are bounded by a known positive value so that

$$\|W\|_F \le W_M \quad (3.8)$$

with the value $W_M$ assumed to be known.

Let the fuzzy system functional estimate for the continuous nonlinear function $f(x,t)$ be given by

$$\hat{f}(x,t) = \hat{W}(t)^T \varphi(\sigma(x)). \quad (3.9)$$

The current value, $\hat{W}(t)$ is provided by an adaptation algorithm.

Define fuzzy system parameter errors as

$$\tilde{W}(t) = W - \hat{W}(t). \quad (3.10)$$

From (3.7) and (3.9), we have the following performance signal dynamics, which is useful for the stability proof,

$$\dot{r}(t) = -K_v r(t) + \tilde{W}(t)^T \varphi(\sigma(x)) + \varepsilon(x). \quad (3.11)$$

***Theorem 1:*** Let the control action $u(t)$ be provided by the RAF controller (3.4) and $R$ be the reinforcement signal (3.2). Let the reinforcement learning rule for tuning the fuzzy system parameters be given by

$$\dot{\hat{W}}(t) = F\varphi(\sigma(x))R(t)^T - \kappa F \hat{W}(t) \quad (3.12)$$

with $F = F^T > 0$ design matrices determining the learning rate and $\kappa > 0$ design parameters governing the speed of convergence. Then the errors $r(t)$ and $\tilde{W}(t)$ are *Uniformly Ultimately Bounded*. Moreover, the performance measure $r(t)$ can be made arbitrarily small by increasing the fixed control gains $K_v$.

***Proof:*** Consider the following positive definite Lyapunov function

$$L(t) = \sum_{i=1}^{n} \left\{ \ln\left(1 + e^{\alpha^+ r_i(t)}\right) + \ln\left(1 + e^{-\alpha^- r_i(t)}\right) \right\} + \tfrac{1}{2} tr(\tilde{W}^T F^{-1} \tilde{W}). \quad (3.13)$$

It is easy to show that this Lyapunov function candidate is positive definite, continuous, and differentiable with respect to $r(t)$ and $t$. The time derivative of Lyapunov function is given by

$$\dot{L} = \sum_{i=1}^{n} \left\{ \alpha^+ / \left(1 + e^{-\alpha^+ r_i(t)}\right) - \alpha^- / \left(1 + e^{\alpha^- r_i(t)}\right) \right\} \dot{r}_i(t) + tr(\tilde{W}^T F^{-1} \dot{\tilde{W}}). \quad (3.14)$$

Using the definition of (3.2) and applying the performance signal dynamics (3.11), we have

$$\dot{L}(t) = -R^T K_v r + R^T \tilde{W}^T \varphi(\sigma(x)) + tr(\tilde{W}^T F^{-1} \dot{\tilde{W}}) + R^T \varepsilon(x). \quad (3.15)$$

Take transpose and trace operator, then we have

$$\dot{L}(t) = -R^T K_v r + tr(\tilde{W}^T \varphi(\sigma(x)) R^T) + tr(\tilde{W}^T F^{-1} \dot{\tilde{W}}) + R^T \varepsilon(x). \quad (3.16)$$

The following property is used:

$$tr(D^T a b^T) = a^T D b \quad (3.17)$$

with $a \in \Re^n$, $b \in \Re^\ell$, and $D \in \Re^{n \times \ell}$. Inserting the adaptive-learning rule (3.12) into (3.16), we obtain

$$\dot{L}(t) = -R^T K_v r + R^T \varepsilon(x) + \kappa\, tr(\tilde{W}^T \hat{W}). \quad (3.18)$$

The expression (3.18) is bounded by

$$\dot{L}(r, \tilde{W}) \le -\alpha_m \lambda_m(K_v) \|r\| + \alpha_M \varepsilon_M + \kappa\, tr\left\{\tilde{W}^T (W - \tilde{W})\right\} \quad (3.19)$$

with $\alpha_M = \max\left(\alpha^+, \alpha^-\right)$ and $\alpha_m = \min\left(\alpha^+, \alpha^-\right)$.

Note that the relation $R(t)^T r(t) > 0$ for any $r(t) \neq 0$ has been used.

Using the following inequality

$$tr(\tilde{Z}^T \hat{Z}) = tr\left\{\tilde{Z}^T (Z - \tilde{Z})\right\} \le \|\tilde{Z}\|_F (Z_M - \|\tilde{Z}\|_F) \quad (3.20)$$

yields

$$\dot{L}(r, \tilde{W}) \le -\alpha_m \lambda_m(K_v) \|r\| + \alpha_M \varepsilon_M + \kappa \left\{ \|\tilde{W}\|_F W_M - \|\tilde{W}\|_F^2 \right\}. \quad (3.21)$$

The time derivative $\dot{L}(r, \tilde{W})$ is guaranteed to be negative if the following condition

$$\|r(t)\| \ge \frac{\kappa W_M^2 / 4 + \alpha_M \varepsilon_M}{\alpha_m \lambda_m(K_v)} \equiv B_{r(t)} \quad (3.22)$$

is satisfied . Or rewrite (3.21) such that

$$\dot{L}(r, \tilde{W}) \le -\alpha_m \lambda_m(K_v) \|r\| - (1 - \delta) \kappa \|\tilde{W}\|_F^2 \quad (3.23)$$

is guaranteed to be negative as long as the following condition

$$\|\tilde{W}(t)\|_F \ge \frac{W_M}{2\delta} + \sqrt{\frac{W_M^2}{4\delta^2} + \frac{\alpha_M \varepsilon_M}{\kappa\delta}} \equiv B_{\tilde{W}(t)} \quad (3.24)$$

holds with $0 < \delta < 1$.

The regions $B_{r(t)}$ and $B_{\tilde{W}(t)}$ are the convergence regions for the performance measure (error), and the parameter estimation errors for the adaptive fuzzy system, respectively. Therefore, $\dot{L}(\boldsymbol{r},\widetilde{\boldsymbol{W}})$ is negative outside a compact set. According to a standard Lyapunov theory extension [4], this demonstrates the *UUB* of both $\|\boldsymbol{r}(t)\|$ and $\|\widetilde{\boldsymbol{W}}(t)\|_F$. ■

It is interesting to note that $\|\boldsymbol{r}(t)\|$ can be kept arbitrarily small by increasing the gain $\boldsymbol{K}_v$ or the critic slope gain $\alpha_m$. The right-hand side of (3.22) and (3.24) can be taken as practical bounds on $\|\boldsymbol{r}(t)\|$ and $\|\widetilde{\boldsymbol{W}}(t)\|_F$, respectively. Moreover the regions given by (3.22) and (3.2m 4) represent the worst case one can have.

*Remarks:* The following remarks highlight some aspects of the proposed RAF controller.

(1) The persistent excitation of learning systems has been a challenging issue since any learning algorithms for nonlinear plant have a tendency to stagnate in a preferred region. Due to the second terms in the RAF adaptive-learning rules (3.12), the persistent excitation condition is eliminated in the proposed learning scheme. In fact, our reinforcement learning rules is in the form of $\sigma-$modification where the persistency of excitation condition is not needed [7] in adaptive control.
(2) From the stability proof, the performance evaluator $\|\boldsymbol{r}(t)\|$ which is defined as output tracking error in this paper, can be kept arbitrarily small by increasing the gain matrix $\boldsymbol{K}_v$. This demonstrates the stability definition of *Uniformly Ultimately Bounded* in a practical sense. In practical situations, the output tracking errors are not exactly equal to zero. The best we can do is to guarantee that the error converges to a neighborhood of the origin.
(3) Finally, it is emphasized that the adjustable weight parameters $\boldsymbol{W}$ in the fuzzy system may be initialized at zero, and the fixed controller, $\boldsymbol{K}_v\boldsymbol{r}(t)$ maintain the stability until the fuzzy system learns the plant. This means that there is no off-line learning or trial and error phase, which require enormous times of learning. The detailed plant dynamics are completely unknown to the fuzzy system. Here comes the terminology, **Adaptive-Learning,** in that there is no consideration on the effect of plant dynamics and the controlled plant follows the desired trajectory in one process.

## IV. Simulation Results

The simulations of Duffing forced oscillation system were performed to demonstrate the effectiveness of the proposed control scheme. The MATLAB command "ode23," which uses the second/third order Runge-Kutta method was used for solving the differential equations in the simulations.

The Duffing forced oscillation system is given by [10]

$$\begin{aligned}\dot{x}_1 &= x_2\\ \dot{x}_2 &= -0.1x_2 - x_1^3 + 12\cos t + u(t) + d(t) \qquad (4.1)\\ y &= x_1 .\end{aligned}$$

The system shows chaotic behavior if $u(t)=0$ and $d(t)=0$. The reference signal selected here is $y_d(t)=\sin t$, and the external disturbance $d(t)$ is a square wave with the amplitude $\pm 1$ and the period $2\pi$. The initial state conditions are chosen to be $x_1(0)=2$ and $x_2(0)=-2$ through the simulations. The linear gain terms are $K_v=10$ and $\lambda=5$ in the simulations.

### A. $f(\boldsymbol{x},t)$: known and $d(t)$: unknown square wave

Assuming the nonlinear function in the plant is exactly known and the unknown disturbance is existent, we select the following the controller,

$$u(t) = K_v r(t) + f(\boldsymbol{x},t) \qquad (4.2)$$

with $f(\boldsymbol{x},t) = -x_1^3 + 12\cos(t)$ and $d(t)$.

The tracking performance is shown in Fig. 3. It shows the fairly good tracking performance with the known nonlinear function and the unknown disturbance. However, it is very difficult to have the exact knowledge of the nonlinear function in the real plant. The effect of the disturbance on the performance is shown. As expected, the inexact knowledge on the plant nonlinear characteristics produces a significant steady state error. Even though the high gain $K_v$ can reduce the tracking error considerably, there still exist steady state error resulting from the nonlinear characteristics. In fact, the high gain is limited by the excitation of un-modeled high frequency component in the plant or measurement noise.

### B. RAF Controller and $d(t)$: unknown square wave

Assuming the nonlinear function and the disturbance in the plant is completely unknown, select the following the controller,

$$u(t) = K_v r(t) + \hat{w}(t)^T \varphi(\sigma(\boldsymbol{x})) \qquad (4.3)$$

with $\hat{w}(t)^T \varphi(\sigma(\boldsymbol{x}))$ provided by the fuzzy system. The adaptive-learning rule (3.12) is employed

$$\dot{\hat{w}}(t) = \boldsymbol{F}\varphi(\sigma(\boldsymbol{x}))R(t) - \kappa \boldsymbol{F}\hat{w}(t) \qquad (4.4)$$

with the parameters $\boldsymbol{F} = diag[30]$ and $\kappa = 0.01$. The initial parameter values are selected as $\hat{w}(0) = 0$. The critic is selected as

$$R(t) = R^{+}(t) + R^{-}(t) = \sigma^{+}(r) + \sigma^{-}(r) \quad (4.5)$$

with $\alpha^{+} = \alpha^{-} = 2$.

The input and the output universes of discourse are given as $x_1 \in [-2 \;\; 2]$, $x_2 \in [-2 \;\; 2]$ and $y \in [-2 \;\; 2]$. The fuzzy membership functions are chosen along each state trajectory $\boldsymbol{x} = [x_1 \;\; x_2]$ as follows:

$$G(1) = \frac{1}{1 + \exp(5 * (x_i + 1.25))} \qquad G(2) = \exp\left[-(x_i + 0.75)^2\right]$$

$$G(3) = \exp\left[-(x_i + 0.25)^2\right] \qquad G(4) = \exp\left[-x_i^2\right]$$

$$G(5) = \exp\left[-(x_i - 0.25)^2\right] \qquad G(6) = \exp\left[-(x_i - 0.75)^2\right]$$

$$G(7) = \frac{1}{1 + \exp(-5 * (x_i - 1.25))}.$$

The tracking performance is shown in Fig. 4. It shows an excellent tracking performance even with the unknown nonlinear function provided by the RAF output.

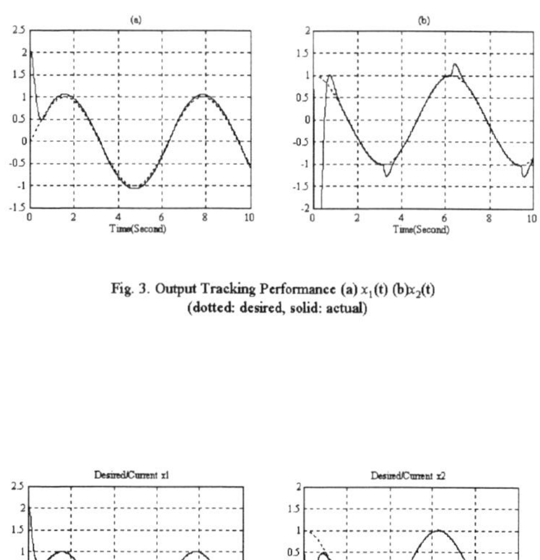


Fig. 3. Output Tracking Performance (a) $x_1(t)$ (b) $x_2(t)$
(dotted: desired, solid: actual)

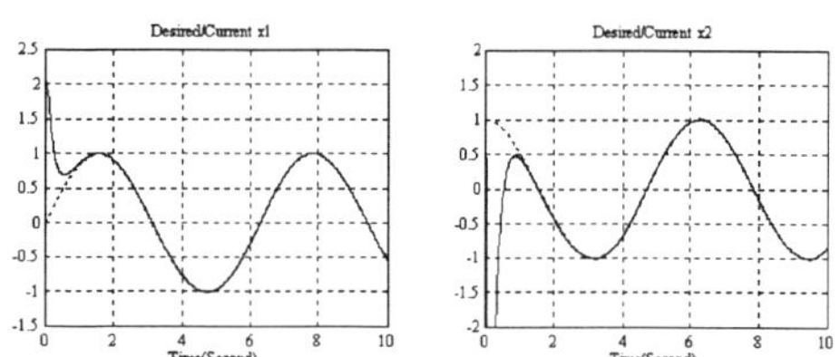


Fig. 4. Output Tracking Performance (a) $x_1(t)$ (b) $x_2(t)$
(dotted: desired, solid: actual)

## V. Conclusion

We have presented a new method of Reinforcement Adaptive Fuzzy scheme for control of nonlinear plants. The proposed control scheme is fairly simple compared with other reinforcement learning methods that require the gradient information either directly or indirectly on the critic or the plant. The reinforcement adaptive-learning rule for the fuzzy system parameter is derived, based on the Lypaunov stability. Simulation results confirm that after the fuzzy system has compensated perfectly or partially for the nonlinear characteristics of the controlled plant on-line, the output tracking of the plant follows the desired trajectory very satisfactorily even in the presence of unknown nonlinear plant dynamics and disturbances.

## References

[1] H. M. Kim and J. M. Mendel, "Fuzzy Basis Functions: Comparisons with Other Basis Functions," *IEEE Trans. Fuzzy Systems*, vol. 3, no. 2, pp. 158-168, 1995.

[2] C. C. Lee, "Fuzzy logic in control systems: fuzzy logic controller-Part I," *IEEE Trans. Syst., Man, Cybern.*, vol. 20, no. 3, pp. 404-418, 1990.

[3] C. C. Lee, "Fuzzy logic in control systems: fuzzy logic controller-Part II," *IEEE Trans. Syst., Man, Cybern.*, vol. 20, no. 3, pp. 419-435, 1990.

[4] F. L. Lewis, C. T. Abdallah, and D. M. Dawson, *Control of Robot Manipulators.* MacMillan, New York, 1993.

[5] F. L. Lewis, A. Yesildirek, and K. Liu, "Neural net robot controller with guaranteed tracking performance," *IEEE Trans. Neural Networks*, vol. 6, no. 3, pp. 703-715, 1995.

[6] J. M. Mendel, "Fuzzy logic systems for engineering: A tutorial," *Proc. IEEE*, vol. 83, no. 3, pp. 345-377, 1995.

[7] K. S. Narendra and A. M. Annaswamy, "A new adaptive law for robust adaptation without persistent excitation," *IEEE Trans. Automat. Control*, vol. 32, no.2, pp. 134-145, 1987.

[8] K. S. Narendra and K. Parthasarathy, "Identification and control of dynamical systems using neural networks," *IEEE Trans. Neural Networks*, vol. 1, no. 1, pp. 4-27, 1990.

[9] J-J. E. Slotine and Li, *Applied Nonlinear Control.* Prentice Hall, 1991.

[10] L. X. Wang, *Adaptive Fuzzy Systems and Control: Design and Stability Analysis.* Englewood Cliffs, NJ: Prentice Hall, 1994.

IFAC Publications
www.elsevier.com/locate/ifac

# NEURO-FUZZY INVERSE DYNAMICS CONTROL FOR FLEXIBLE-LINK SPACE ROBOTS

**Fuchun Sun, Lingbo Zhang, Zengqi Sun**

*Dept. of Computer Science and Technology, State Key Lab of Intelligent Technology & Systems, Tsinghua University, Beijing 100084, P.R.China*

Abstract: A neuro-fuzzy (NF)-based adaptive controller is presented in this paper for the trajectory tracking control of a flexible-link space manipulator. Based on the singular perturbation method and two time-scale decompositions, the dynamic model of a flexible space robot is reduced into a slow subsystem of an equivalent rigid-link arm and a fast subsystem of flexible mode. A dynamic NF adaptive controller based on dynamic inversion is presented for the tracking control of the equivalent rigid arm, while a fuzzy proportional-derivative (PD) controller is used to stabilize the elastic dynamics. Finally, experiment studies are carried out to show the viability and effectiveness of the proposed control approach. *Copyright © 2001 IFAC*

Keywords: Flexible-link space robots, singular perturbation, NF systems, dynamic inversion.

## 1. INTRODUCTION

Lightly weighted flexible-link manipulator as new generation of robot, shows prosperous future in the fields of modern industry, defense and space applications, etc.. There are mainly three schools of researches for flexible-link manipulators. Modal expansion is an intuitive method usually used by control engineers for controller design, which can simplify the controller design by decomposing the partial differential equation of the flexible-link manipulator in temporal and spatial domains through finite dimensional truncations. Based on these model reduction techniques, classical control methods are used to design controllers for flexible-link robots, examples of these approaches are optimal control, pole placement, robust control, $H\infty$ control, and singular perturbation control, etc.. Infinite dimensional system theory is a direct method where no model truncation and simplification is required for the controller design. Semi-group theory is used to analyze and control the dynamic behavior of such systems through evolutionary equation. Intelligent control as a model-free design approach has been used in the control of flexible-link manipulator these years. These applications include neural networks, fuzzy systems, and genetic algorithm.

This paper is concerned with the NF adaptive controller design based on dynamic inversion for a flexible-link space robot. In the proposed control structure, dynamic NF system is used to approximate the slow dynamics of a flexible-link space robot, the

This work was jointly supported by the National Science Foundation of China under Grants 60084002, the National Excellent Doctoral Dissertation Foundation of China and the Science Foundation of the College of Information science and Technology of Tsinghua University.

PD feedback control based on the tip deflection signal measured by PSD is used to suppress the tip vibration. The key properties associated with NF dynamic inversion control include: (1) assumption under which the system state should be within a compact set can be excluded; (2) dynamics nonlinearities of the space robot is not required to be known; (3) the dynamic performance of the space robot can be guaranteed, especially in the initial stage of the control process. The NF learning algorithm and the stability proof of the closed-loop system are given, the viability and effectiveness of the proposed control approach is verified by simulation research on a single-link flexible space robot.

## 2. PRELIMINARY

### 2.1 Model of a Flexible-Link Space Robot

The model of a flexible-link space robot can be written as follows

$$M(q)\ddot{q}+f(q,\dot{q})+Kq+G(q)=B(q)u(t) \qquad (1)$$

where $q=[q_r^{\mathrm{T}},y^{\mathrm{T}}]^{\mathrm{T}}\in R^n$, $q_r\in R^{n_r}$ is rigid mode vector, $y\in R^{n_f}$ the tip deflection vector, $n=n_r+n_f$; $M(q)=(M_{i,j})\,(i,j=1,2)\in R^n$ is the inertial matrix, $K$ the stiffness matrix, $f(q,\dot{q})$ the viscous terms due to Coriolis and Centripetal forces, $G(q)$ is the gravitational torques, $u$ control input, $B(q)u$ general force, $B(q)$ the input matrix with the form as follows

$$B(q)=\begin{pmatrix}I & Q\end{pmatrix}^{\mathrm{T}} \qquad (2)$$

for clamped-free boundary condition

$$Q=0 \qquad (3)$$

for pined-pined boundary condition

$$Q=\text{block-diag}\{\text{col}(\phi'_{i,1}(0),\cdots,\phi'_{i,m_i})\} \qquad (4)$$

where $\phi_{ij}(0)$ denotes the $j$th modal function of the $i$th link of a flexible-link space robot.

Based on the singular perturbation and two time-scale decomposition, the slow subsystem of the flexible space robot can be derived along the same lines as those in (Siciliano, and Book ,1998) as

$$\ddot{q}_{sr}=-M_{11}^{-1}\left(D_{fr}\dot{q}_{sr}-B_r u_s(t)\right) \qquad (5)$$

where $M_{11}\in R^{n_r\times n_r}$ is the positive definite matrix satisfying $M_m\le\|M_{11}(q_{sr})\|\le M_M$, with $M_M$ and $M_m$ constants, $q_{sr}(t),\dot{q}_{sr}(t)$ are the state vectors of the equivalent rigid arm, $u_s(t)$ is the control input for the slow subsystem, $D_{fr}$ and $B_r$ can refer to (Siciliano, and Book ,1998).

To define the fast subsystem, it is assumed that the slow variables are constant during fast transients. At the same time, we select states $\eta_1(k)=\xi-\xi_s$, $\eta_2=\varepsilon\dot{\xi}$ and a time-scale change of $\tau_s=t/\varepsilon$ results in

$$\mathrm{d}\eta_1/dt_s=\eta_2 \qquad (6)$$

$$\mathrm{d}\eta_2/dt_s=-H_{ff}\tilde{K}_{ff}\eta_1+B_f^1 u_f \qquad (7)$$

where $\varepsilon$ is a scale factor, $\xi=y(t)/\varepsilon^2\,\xi_s$, $H_{ff}$, $\tilde{K}_{ff}$, $B_f^1$ can refer to (Siciliano, and Book ,1998) for more detail, and $u_f(t)=u(t)-u_s(t)$.

The mathematical model of the equivalent rigid arm (5) can be represented in state space form as

$$\dot{x}=Ax+f(x)+B(x)u_s(t) \qquad (8)$$

where $x=(q_{sr}^{\mathrm{T}},\dot{q}_{sr}^{\mathrm{T}})^{\mathrm{T}}$ denotes the state vector of the equivalent rigid arm, and

$$A=\begin{bmatrix}0 & I\\ 0 & 0\end{bmatrix},\ B(x)=\begin{bmatrix}0\\ M_{11}^{-1}(q_{sr})B_r\end{bmatrix},\ f(x)=\begin{bmatrix}0\\ -M_{11}^{-1}(q_{sr})D_{fr}\end{bmatrix} \qquad (9)$$

For fast subsystem, (6) and (7) can also be written as

$$\dot{\eta}=A_f\eta+B_f u_f(t) \qquad (10)$$

where $\eta=(\eta_1^{\mathrm{T}},\eta_2^{\mathrm{T}})^{\mathrm{T}}$, $A_f=\begin{bmatrix}0 & I\\ -H_{ff}\tilde{K}_{ff} & 0\end{bmatrix}$,

$$B_f=\begin{bmatrix}0 & B_f^{\mathrm{T}}\end{bmatrix}^{\mathrm{T}} \qquad (11)$$

### 2.2 Dynamic NF Systems

The dynamics of a flexible-link space robot can be represented by the following dynamic T-S fuzzy model which is described by fuzzy IF-THEN rules

Rule $i$: if $z_1(t)$ is $F_{i1}$, $z_2(t)$ is $F_{i2},\cdots,z_p(t)$ is $F_{ip}$, then

$$\dot{x}=A_r x+\bar{f}_i(x)+\bar{G}_i(x)u_s(t),i=1,\cdots,m \qquad (12)$$

$$\dot{\eta}=\bar{A}_{fi}\eta+\bar{B}_{fi}u_f(t) \qquad (13)$$

where $F_{ij}(i=1,2,\cdots,m;j=1,2,\cdots,p)$ are fuzzy sets described by membership function $A_{ij}(z_j)$, $x(t)\in$

$R^{2n_r}$ the state vector, $\boldsymbol{u}_s(t) \in R^{n_r}$ and $\boldsymbol{u}_f(t) \in R^{n_f}$ the input vectors. Besides, it is assumed that $\overline{\boldsymbol{B}}_i(\boldsymbol{x}) = \overline{\boldsymbol{B}}_i$, $\overline{\boldsymbol{f}}_i(\boldsymbol{x}) = \overline{\boldsymbol{A}}_i\boldsymbol{x} + \overline{\boldsymbol{b}}_i$, where $\boldsymbol{A}_r \in R^{2n_r \times 2n_r}$ in the form of matrix $\boldsymbol{A}$ in (9), $\overline{\boldsymbol{A}}_i \in R^{2n_r \times 2n_r}$, $\overline{\boldsymbol{B}}_i \in R^{2n_r \times n_r}$, $\overline{\boldsymbol{b}}_i \in R^{2n_r}$, $\overline{\boldsymbol{A}}_{fi} \in R^{2n_f \times 2n_f}$ and $\overline{\boldsymbol{B}}_{fi} \in R^{2n_f}$ are constant matrices and vector, respectively. $m$ is the number of fuzzy rules, and $z_1(t) \sim z_p(t)$ are some measurable variables, *i.e.* the premise variables.

## 3. NF COMPOSITE CONTROL FOR A FLEXIBLE-LINK SPACE ROBOT

As discussed above, the decomposition of the flexible space robot dynamics into separate slow and fast subsystems suggests that separate control laws be designed for each subsystem and then combined into a composite control of the original system. Here, we consider a multi-rate implementation of composite control laws. If the basic sampling period $\delta$ is appropriate for the slow subsystem (8), then a faster sampling rate $T_f = \delta / v$ for the fast subsystem (10) will be considered where $V$ is a constant.

### *3.1 NF Adaptive Control Of the Slow Subsystem*

The slow dynamics of the flexible space robot is exactly the Lagrange form of the dynamics of an $n_r$ -link rigid robot arm, thus all the well-established control techniques developed for rigid robots can be applied. Here, stable NF adaptive control approach is developed for the control of the slow dynamics.

The following tracking error metric is defined for the NF adaptive controller design

$$\boldsymbol{S} = \boldsymbol{C}(\boldsymbol{x} - \boldsymbol{x}_d) \tag{14}$$

where $\boldsymbol{S} = (s_1, \cdots, s_n)^{\mathrm{T}}$, $\boldsymbol{x} = (\boldsymbol{q}_{sr}^{\mathrm{T}}, \dot{\boldsymbol{q}}_{sr}^{\mathrm{T}})^{\mathrm{T}}$, $\boldsymbol{x}_d = (\boldsymbol{q}_{sd}^{\mathrm{T}}, \dot{\boldsymbol{q}}_{sd}^{\mathrm{T}})^{\mathrm{T}}$ is the desired state, and $\boldsymbol{C} = (\boldsymbol{r}, \boldsymbol{I}) \in R^{n_r \times 2n_r}$, $\boldsymbol{r} = \boldsymbol{r}^{\mathrm{T}} > \boldsymbol{0}$.

Substituting the state equation presented by the $i$th fuzzy rule described in (12) into the derivative of (14), the error dynamics can be obtained as follows

$$\boldsymbol{G}^{-1}\dot{\boldsymbol{S}} = \sum_{i=1}^{m} \theta_i(z(t))\boldsymbol{G}_i^{-1}\boldsymbol{h} + \sum_{i=1}^{m} \theta_i(z(t))(\boldsymbol{A}_i\boldsymbol{x} + \boldsymbol{b}_i) + \boldsymbol{u}_s(t) \tag{15}$$

where $\boldsymbol{G}^{-1} = \sum_{i=1}^{m} \theta_i(z)\boldsymbol{G}_i^{-1}$, $\boldsymbol{G}_i = \boldsymbol{C}\overline{\boldsymbol{G}}_i \in R^{n_r \times n_r}$,

$\boldsymbol{A}_i = \boldsymbol{G}_i^{-1}\boldsymbol{C}\overline{\boldsymbol{A}}_i \in R^{n_r \times 2n_r}$, $\boldsymbol{b}_i = \boldsymbol{G}_i^{-1}\boldsymbol{C}\overline{\boldsymbol{b}}_i \in R^{n_r}$,

$\theta_i(z(t)) = \prod_{j=1}^{p} A_{ij}(z_j) / \sum_{i=1}^{m}\prod_{j=1}^{p} A_{ij}(z_j)$,

$z = (z_1, \cdots, z_p)$, $\boldsymbol{\theta}(z(t)) = (\theta_1(z(t)), \cdots, \theta_m(z(t)))$.

As stated in subsection 2.2, dynamic T-S fuzzy model constructed by fuzzy model identification method is only a rough approximation to the robot dynamics. For accurate trajectory tracking of the flexible-link space robot, a stable adaptive control approach using dynamic NF system will be developed, where the NF system can be adapted by learning algorithms for updating its free parameters such that the approximation performance of the NF system can be improved on-line. The following dynamic NF system with the same antecedents as that in (12) is constructed to approximate the slow dynamics shown in (8) as

Rule $i$ : if $z_1$ is $F_{i1}$, $z_2$ is $F_{i2}$, $\cdots$, $z_p$ is $F_{ip}$

then $\dot{\hat{\boldsymbol{x}}} = \overline{\boldsymbol{A}}\hat{\boldsymbol{x}} + \hat{\overline{\boldsymbol{f}}}_i(\hat{\boldsymbol{x}}) + \hat{\overline{\boldsymbol{B}}}_i(\hat{\boldsymbol{x}})(\boldsymbol{u}_s(t) - \boldsymbol{u}_p(t))$ (16)

where $F_{ij}(i = 1, \cdots, m; j = 1, \cdots, p)$ are fuzzy sets described by membership function $A_{ij}(z_j)$, $\hat{\boldsymbol{x}} = (\hat{\boldsymbol{q}}_{sr}^{\mathrm{T}}, \dot{\hat{\boldsymbol{q}}}_{sr}^{\mathrm{T}})^{\mathrm{T}} \in R^{2n_r}$ is the state of the NF system. Besides, assume that $\hat{\overline{\boldsymbol{B}}}_i(\hat{\boldsymbol{x}}) = \hat{\overline{\boldsymbol{B}}}_i$, $\hat{\overline{\boldsymbol{f}}}_i(\hat{\boldsymbol{x}}) = \hat{\overline{\boldsymbol{A}}}_i\hat{\boldsymbol{x}} + \hat{\overline{\boldsymbol{b}}}_i$, and $\hat{\overline{\boldsymbol{A}}}_i \in R^{2n_r \times 2n_r}$, $\hat{\overline{\boldsymbol{B}}}_i \in R^{2n_r \times 2n_r}$ and $\hat{\overline{\boldsymbol{b}}}_i \in R^{2n_r}$ denote the estimates of $\overline{\boldsymbol{A}}_i$, $\overline{\boldsymbol{B}}_i$ and $\overline{\boldsymbol{b}}_i$. $\boldsymbol{u}_p(t)$ is a robust control component which will be defined later. As $\hat{\boldsymbol{x}} \to \boldsymbol{x}_d$, $\boldsymbol{u}_p(t) \to 0$, then the NF system constructed is equivalent to the slow subsystem dynamics in terms of state and control input.

The tracking error metric for the dynamic NF system (19) can be defined as

$$\boldsymbol{S}_0 = \boldsymbol{C}(\hat{\boldsymbol{x}} - \boldsymbol{x}_d) \tag{17}$$

By the same derivation as that of (15), $\boldsymbol{S}_0$ dynamics in the whole space can be written as

$$\boldsymbol{G}_F^{-1}\tilde{\boldsymbol{S}}_0 = \sum_{i=1}^{m} \theta_i(z(t))\left(\hat{\boldsymbol{G}}_i^{-1}\hat{\boldsymbol{h}} + \hat{\boldsymbol{A}}_i\hat{\boldsymbol{x}} + \hat{\boldsymbol{b}}_i\right) + \boldsymbol{u}_s(t) - \boldsymbol{u}_p(t) \tag{18}$$

where $\boldsymbol{G}_F^{-1} = \sum_{i=1}^{m} \theta_i(z(t))\hat{\boldsymbol{G}}_i^{-1}$ with $\hat{\boldsymbol{G}}_i = \boldsymbol{C}\hat{\overline{\boldsymbol{B}}}_i$, $\tilde{\boldsymbol{S}}_0 = \dot{\boldsymbol{S}}_0 + r_0\boldsymbol{S}_0$, $\hat{\boldsymbol{A}}_i = \hat{\boldsymbol{G}}_i^{-1}\boldsymbol{C}\hat{\overline{\boldsymbol{A}}}_i$, $\hat{\boldsymbol{b}}_i = \hat{\boldsymbol{G}}_i^{-1}\boldsymbol{C}\hat{\overline{\boldsymbol{b}}}_i$. $r_0 > \boldsymbol{0}$ is a parameter used to assign the dynamics of $\boldsymbol{S}_0$. Obviously, matrices $\hat{\boldsymbol{A}}_i$, $\hat{\boldsymbol{G}}_i$ and $\hat{\boldsymbol{b}}_i (i = 1, \cdots, m)$ are the estimates of parameter matrices $\boldsymbol{A}_i$, $\boldsymbol{G}_i$ and vector $\boldsymbol{b}_i$, and

$$\hat{\boldsymbol{h}} = \boldsymbol{C}\overline{\boldsymbol{A}}\hat{\boldsymbol{x}} + r_0\boldsymbol{S}_0 - \boldsymbol{C}\dot{\boldsymbol{x}}_d \tag{19}$$

As defined in (Sun, 2001), dynamic inversion used in the controller design is defined as the inverse model of the dynamic NF system shown in (16) with state specified by desired dynamics. For a dynamic NF system (16), if the desired dynamics is chosen as

$$\dot{\boldsymbol{S}}_0 = -\boldsymbol{r}_0\boldsymbol{S}_0 \tag{20}$$

then dynamic inversion of the dynamic NF system (16) is obtained from (18) using definition as

$$\boldsymbol{u}_I(t) = -\sum_{i=1}^{m}\theta_i(z(t))\hat{\boldsymbol{G}}_i^{-1}\hat{\boldsymbol{h}} - \sum_{i=1}^{m}\theta_i(z(t))\left(\hat{\boldsymbol{A}}_i\hat{\boldsymbol{x}} + \hat{\boldsymbol{b}}_i\right) \tag{21}$$

The following control law is considered for trajectory tracking of the equivalent rigid arm

$$\boldsymbol{u}_s(t) = -\sum_{i=1}^{m}\theta_i(z(t))\left(\hat{\boldsymbol{G}}_i^{-1}\hat{\boldsymbol{h}} + \hat{\boldsymbol{A}}_i\hat{\boldsymbol{x}} + \hat{\boldsymbol{b}}_i\right) - \overline{\boldsymbol{K}}\tilde{\boldsymbol{x}} \tag{22}$$

where $\overline{\boldsymbol{K}} = \boldsymbol{K}_s\boldsymbol{C} \in R^{n_r \times 2n_r}$, and $\boldsymbol{K}_s = R^{n_r \times n_r}$ is assumed to be diagonal.

It is shown in (22) that the NF adaptive controller is composed of NF dynamic inversion $\boldsymbol{u}_I(t)$ and a PD control $\boldsymbol{u}_p(t)$. The dynamic inversion which acts as a feedforward controller, is used to compensate for the inverse dynamics of the equal rigid arm, while the PD control is used to enhance robustness of the system.

Subtracting (18) from (15) and using (20), we have

$$\begin{aligned}\boldsymbol{G}^{-1}\tilde{\boldsymbol{S}}_e &= \sum_{i=1}^{m}\theta_i(z(t))\left((\boldsymbol{G}_i^{-1} - \hat{\boldsymbol{G}}_i^{-1})\hat{\boldsymbol{h}} + (\boldsymbol{A}_i - \hat{\boldsymbol{A}}_i)\hat{\boldsymbol{x}} + \boldsymbol{b}_i - \hat{\boldsymbol{b}}_i\right) \\ &+ \sum_{i=1}^{m}\theta_i(z(t))(\boldsymbol{V}_i - \boldsymbol{K})\tilde{\boldsymbol{x}} \\ &= \tilde{\boldsymbol{W}}\boldsymbol{Y}(\boldsymbol{\theta}(z(t)),\hat{\boldsymbol{x}},\hat{\boldsymbol{h}}) + \sum_{i=1}^{m}\theta_i(z(t))(\boldsymbol{V}_i - \boldsymbol{K})\tilde{\boldsymbol{x}}\end{aligned} \tag{23}$$

where $\boldsymbol{S}_e = \boldsymbol{C}\tilde{\boldsymbol{x}}$, $\tilde{\boldsymbol{x}} = (\boldsymbol{q}_e^{\mathrm{T}}, \dot{\boldsymbol{q}}_e^{\mathrm{T}})^{\mathrm{T}}$, $\boldsymbol{q}_e = \boldsymbol{q}_r - \boldsymbol{q}_{sd}$

$$\boldsymbol{V}_i = \boldsymbol{G}_i^{-1}(\boldsymbol{C}\overline{\boldsymbol{A}} + \boldsymbol{r}\boldsymbol{C}) + \boldsymbol{A}_i,\ \tilde{\boldsymbol{S}}_e = \dot{\boldsymbol{S}}_e + \boldsymbol{r}\boldsymbol{S}_e$$

$$\boldsymbol{V} = \sum_{i=1}^{m}\theta_i(z(t))\boldsymbol{V}_i,\ \tilde{\boldsymbol{W}} = \boldsymbol{W} - \hat{\boldsymbol{W}} \in R^{n_r \times n_a}$$

$$\boldsymbol{W} = \left(\boldsymbol{G}_1^{-1}, \cdots, \boldsymbol{G}_m^{-1}, \boldsymbol{A}_1, \cdots, \boldsymbol{A}_m, \boldsymbol{b}_1, \cdots, \boldsymbol{b}_m\right)$$

$$\hat{\boldsymbol{W}} = \left(\hat{\boldsymbol{G}}_1^{-1}, \cdots, \hat{\boldsymbol{G}}_m^{-1}, \hat{\boldsymbol{A}}_1, \cdots, \hat{\boldsymbol{A}}_m, \hat{\boldsymbol{b}}_1, \cdots, \hat{\boldsymbol{b}}_m\right)$$

$$\begin{aligned}\boldsymbol{Y}(\boldsymbol{\theta}(z(t)),\hat{\boldsymbol{x}},\hat{\boldsymbol{h}}) &= \left(\boldsymbol{\theta}(z(t)) \otimes \hat{\boldsymbol{x}}^{\mathrm{T}}, \boldsymbol{\theta}(z(t)) \otimes \hat{\boldsymbol{h}}^{\mathrm{T}}, \boldsymbol{\theta}(z(t))\right)^{\mathrm{T}} \\ &= \left(y_1, \cdots, y_a\right)^{\mathrm{T}},\ n_a = (3n_r + 1)m\end{aligned} \tag{24}$$

The following theorem gives stable adaptive control law and learning algorithm for dynamic NF systems.

<u>**Theorem**</u> Consider the $n_r$-link equal rigid arm described by the dynamic model (9), by applying the control law (22) with the dynamic NF learning algorithm

$$\dot{\hat{\boldsymbol{W}}} = (\boldsymbol{S}_e\boldsymbol{Y}^T(\boldsymbol{\theta}(z(t)),\hat{\boldsymbol{x}},\hat{\boldsymbol{h}}) + \sigma(\boldsymbol{W}_0 - \hat{\boldsymbol{W}})) \tag{25}$$

where $\boldsymbol{\eta} = \mathrm{diag}(\eta_1, \cdots, \eta_n) > \boldsymbol{0}$ is the learning rate matrix, $\sigma > 0$ and $\boldsymbol{W}_0$ are design parameters. If the following relation

$$K_{sm} > M_G + \tfrac{3}{2}V_{\max}(1 + r_m^{-1}) \tag{26}$$

holds, where, $M_G = \frac{1}{2}\sup_{q_{sr}}\left\|\dot{\boldsymbol{M}}_{11}(\boldsymbol{q}_{sr})\right\|$, $V_{\max} = \sup_{t\geq 0}\left\|\boldsymbol{V}\right\|$, $K_{sm}$ and $r_m$ are the minimum eigenvalues of $\boldsymbol{K}_s$ and $\boldsymbol{r}$, respectively, then the tracking error metric of the robot is uniformly ultimately bounded.

Proof: See the appendix

### *3.2 Control for the Fast Subsystem*

To design controller for the fast subsystem, fuzzy system (13) is assumed to be controllable. Above all, local state feedback controller is designed for the $i$th subsystem pair $(\boldsymbol{A}_{fi}, \boldsymbol{B}_{fi})$.

Controller rule $i$ : IF $z_1$ is $F_{i1}$, ... $z_p$ is $F_{ip}$,

Then $\boldsymbol{u}_f = -\boldsymbol{K}_{fi}\boldsymbol{\eta}$, $i = 1, 2, \ldots, r$, (27)

The controller output for the fast subsystem can be written through fuzzy composition as

$$\boldsymbol{u}_f(t) = -\sum_{i=1}^{m}\theta_i(z(t))\boldsymbol{K}_{fi}\,\boldsymbol{\eta} \tag{28}$$

Substituting (28) into (13) yields

$$\dot{\boldsymbol{\eta}} = \sum_{i=1}^{m}\sum_{j=1}^{m}\theta_i\theta_j(\boldsymbol{A}_{fi} - \boldsymbol{B}_{fi}\boldsymbol{K}_{fj})\,\boldsymbol{\eta} \tag{29}$$

For asymptotic stable closed-loop system, optimal and other robust design approaches can be used to design feedback gains.

## 4. APPLICATION EXAMPLE

In this section, the above developed control approach is employed in position and angle tracking control of a single-link flexible space robot with motion platform. Parameters for space robot are as follows:

The platform with length $a$=2 m and width $b$ = 2.5 m, weights $M_0$ = 40 kg, link one with length $l_1$ = 10 m, and the bending stiffness $EI$ = 10000 kgm$^2$, weights $M_1$ = 10 kg, its end tip mass is $m_1$ = 2 kg.

The goal of control is to drive the platform of the flexible space robot to move along horizontal axis and vertical axis in inertial space, and make attitude angles of the platform and link one to follow sine signals. The desired position movement speeds of the platform are respectively:

Along horizontal axis: $\dot{x}_d = 0.1$ (m/s)

Along vertical axis: $\dot{y}_d = 0.5$ (m/s)

The desired attitude angles for the platform and link one of the flexible-link space robot to follow are chosen as

$$\theta_{0d}(t) = 0.5\sin(0.1t),\ \theta_{1d}(t) = \cos(0.2t) \qquad (30)$$

The dynamic T-S fuzzy model of the slow subsystem are derived using Gustafson Kessel fuzzy clustering approach through I/O data from the robot control process with sampling interval 0.002 sec. (Sun, et al., 2001). It has been shown in the example that the dynamic NF system with eight rules is enough to approximate the dynamic process of the equivalent rigid arm. Concrete parameters for membership functions and $\hat{G}_i^{-1}$, $\hat{A}_i, \hat{b}_i$ $(i = 1,\cdots,8)$ are omitted here for saving the paper length.

Controller parameters are chosen as

$$\boldsymbol{r} = \mathrm{diag}(75,75,75,75),\ \boldsymbol{r}_0 = \mathrm{diag}(20,20,20,20)$$

$$\boldsymbol{K} = \mathrm{diag}(50,50,200,200)$$

Paremeter vector $\boldsymbol{r}_0 = 0.9\boldsymbol{I}$ is chosen to assign the dynamics of the NF dynamic system for good dynamic performance in the initial stage of the control process.

The learning law for dynamic NF adaptive control is chosen as

$$\eta_i = 0.15/(\left\|\boldsymbol{Y}(\boldsymbol{\theta}(z(t)),\hat{\boldsymbol{x}},\hat{\boldsymbol{h}})\right\|_1 + 0.5),\ i = 1,\cdots,4$$

Figs.1 and 2 present the position and attitude angle tracking error responses of the flexible space robot using NF adaptive control during the first and last 30 seconds of operation. Figs.3 and 4 are the position and attitude angle track ing error responses of the flexible space robot using PD control. Fig.5 is the tip vibration response of the flexible space robot using NF adaptive controller, while Fig.6 using PD control. It has been shown in Fig.1 through Fig.6 that NF controller shows a better control performance than PD controller. The reason for this is two folds. One is attributed to the approximation capability of the dynamic NF system, where the inverse dynamics of the slow subsystem is compensated by NF system. The other is introduction of the PD control in the control structure for enhancing the stability and robustness of the dynamic NF control system.

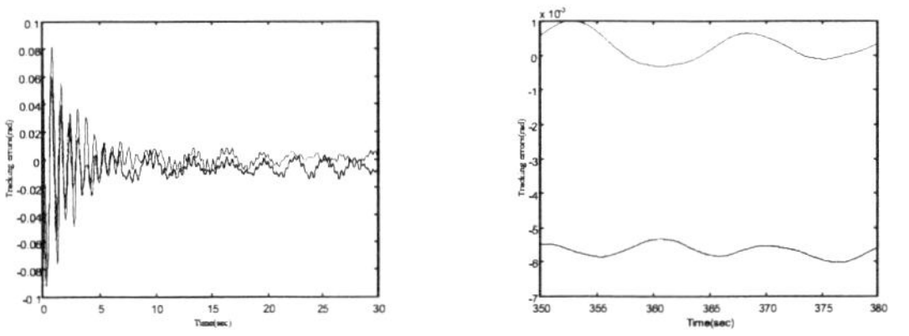

Fig. 1. Platform position tracking errors using NF control for $\tilde{x}$ (solid) and $\tilde{y}$ (dotted).

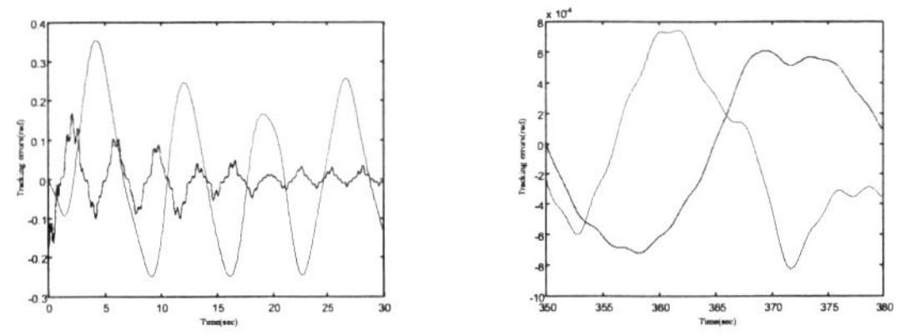

Fig. 2 Attitude angle tracking errors using NF control for $\tilde{\theta}_0$ (solid) and $\tilde{\theta}_1$ (dashed).

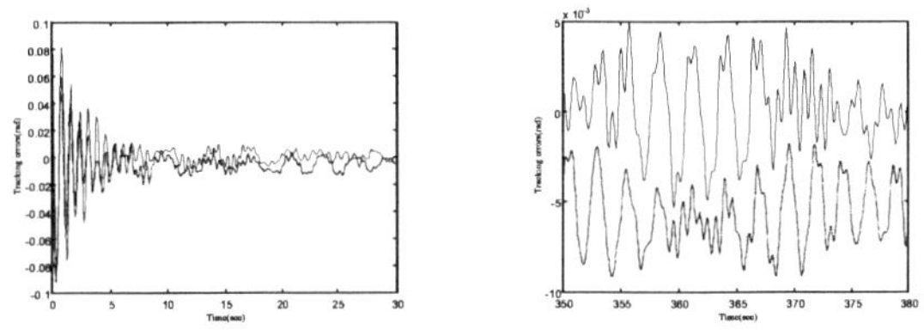

Fig. 3. Platform position tracking errors using PD control for $\tilde{x}$ (solid) and $\tilde{y}$ (dotted).

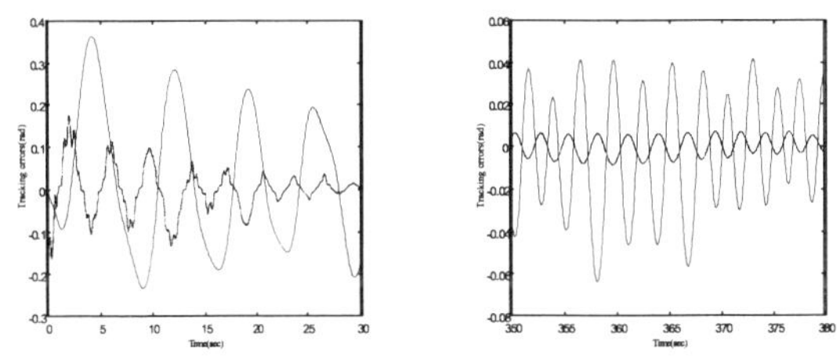

Fig. 4. Attitude angle tracking errors using NF control for $\tilde{\theta}_0$ (solid) and $\tilde{\theta}_1$ (dashed).

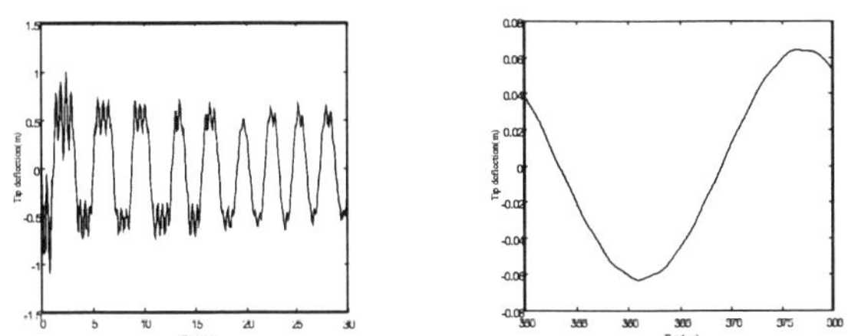

Fig. 5. Tip vibration response of the flexible link using NF control.

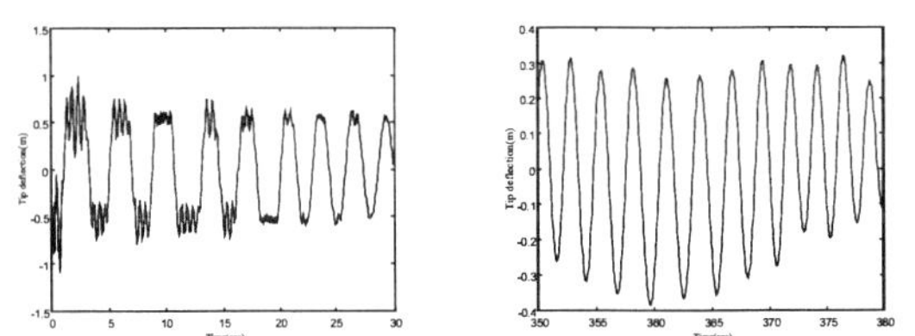

Fig. 6. Tip vibration response of the flexible link using PD control.

## 5. CONCLUSIONS

A stable NF adaptive control approach integrating the merits of dynamic neural network approach and the modeling power of the fuzzy logic system has been developed for the trajectory tracking of a flexible space robot with poorly known dynamics. Singular perturbation approach is used to decompose the dynamics of a flexible space robot into a slow subsystem for an equivalent rigid arm and a fast subsystem for flexible modes. The control law developed contains dynamic NF adaptive control for the slow subsystem and a fuzzy PD control for suppressing the tip vibration. A complete control structure and the NF learning algorithm are given, Simulation shows the superiority of the dynamic NF adaptive control approach to that using PD control.

## REFERENCES


Siciliano, B. and Book, W. J. (1998). A singular perturbation approach to control of lightweight flexible manipulators. *The International Journal of Robotics Research*,**7(4)**, 79-90.

Zhao, H. C. and Chen, D. G. (1995). Tip trajectory tracking for multilink flexible manipulators using stable inversion. *Journal of Guidance, Control, And Dynamics*, **21,** 314-320.

Sun, F. C. and Sun, Z. Q. (2001). Neuro-fuzzy adaptive control based on dynamic inversion for robotic manipulators. *Fuzzy Sets and Systems,* to appear in the recent issues of 2001.


## APPENDIX

Let the Lyapunov function is defined by

$$V=\frac{1}{2}S_e^T G^{-1} S_e+\frac{1}{2}S_0^T S_0 +q_e^T K_s r q_e+\frac{1}{2}tr(\widetilde{W}^T\eta^{-1}\widetilde{W}) \quad (A.1)$$

Differentiating $V_0$ defined in (A.1) with respect to time leads to

$$\dot{V}=\tfrac{1}{2}S_e^T\dot{G}^{-1}S_e+S_e^T G^{-1}\dot{S}_e+S_0^T\dot{S}_0+2\dot{q}_e^T K_s r q_e+tr\left(\dot{\widetilde{W}}^T\eta^{-1}\widetilde{W}\right)$$
$$=\frac{1}{2}S_e^T\dot{G}^{-1}S_e-S_e^T G^{-1}rS_e+S_e^T\widetilde{W}Y(\theta(z),\widetilde{x},\widetilde{h})+$$
$$S_e^T(V-\overline{K})\widetilde{x}-S_0^T r_0 S_0+2\dot{q}_e K_s r q_e-tr(\hat{\dot{W}}^T\eta^{-1}\widetilde{W}) \quad (A.2)$$

Since

$$S_e^T V\widetilde{x}\le\tfrac{1}{2}V_{\max}(3+r_m^{-1})\|\dot{q}_e\|^2+\tfrac{1}{2}V_{\max}(1+3r_m^{-1})\|rq_e\|^2 \quad (A.3)$$
$$-S_e^T Kx+2\dot{q}_e K_s A q_e=-\dot{q}_e^T K_s q_e-q_e^T AK_s A q_e$$
$$\le -K_{sm}(\|\dot{q}_e\|^2+\|rq_e\|^2) \quad (A.4)$$

Besides, it is easy to verify from (5) that $G^{-1}=M_{11}(q_{sr})$. For this, we have

$$\frac{1}{2}S_e^T\dot{G}^{-1}S_e\le M_G(\|\dot{q}_e\|^2+\|rq_e\|^2) \quad (A.5)$$

Substituting (A.3) - (A.5) into (A.2) leads to

$$\dot{V}\le -S_e^T M_{11}(q_{sr})rS_e-S_0^T r_0 S_0 +tr((Y(\theta,\hat{x},\hat{h})S_e^T-\dot{\hat{W}}^T\eta^{-1})\widetilde{W})$$
$$-(K_{sm}-\frac{1}{2}V_{\max}(3+r_m^{-1})-M_G)\|\dot{q}_e\|^2$$
$$-(K_{sm}-\frac{1}{2}V_{\max}(3+r_m^{-1})-M_G)\|rq_e\|^2$$
$$<-r_m M_m\|S_e\|^2-r_{0m}\|S_0\|^2-\alpha\|rq_e\|^2 -\sigma tr((W_0-W)^T\widetilde{W}) \quad (A.6)$$

Since

$$-\sigma tr((W_0-W)^T\widetilde{W})=-\frac{1}{2}\sigma\|\widetilde{W}\|^2-\frac{1}{2}\sigma\|W_0-\hat{W}\|^2 +\frac{1}{2}\sigma\|W-W_0\|^2 \quad (A.7)$$

Substituting (A.7) into (A.6), we have

$$\dot{V}<-r_m M_m\|S_e\|^2-r_{0m}\|S_0\|^2-\alpha\|rq_e\|^2 -\tfrac{1}{2}\sigma\|\widetilde{W}\|^2-\tfrac{1}{2}\sigma\|W_0-\hat{W}\|^2+\tfrac{1}{2}\sigma\|W-W_0\|^2$$
$$\le -\frac{P}{P_{\max}}V+\gamma \quad (A.8)$$

where $\alpha=K_{sm}-\frac{3}{2}V_{\max}(1-r_m^{-1})-M_G>0$ with condition given in (26), $P_{\max}=\max(M_m,1.0,2K_m r_m^{-1},\eta_m)$, $p=\min(2r_m M_m,2r_{0m},2\alpha,\sigma)$, $\gamma=\frac{1}{2}\sigma\|W-W_0\|^2$. $M_M$ and $M_m$ have been defined in subsection 2.1, $r_{0m}$ and $\eta_m$ are defined as the minimum eigenvalues of $r_0$ and $\eta$, respectively.

It is concluded from (A.8) that $\|\widetilde{x}\|$ and $\|\widetilde{W}\|$ will eventually fall into a residual set with the size $o(\gamma)$, and so will $\|\dot{q}_e\|$ and $\|q_e\|$. Moreover, $\hat{q}_{sr}$ will converge to the desired state vector $q_{sd}$ exponentially fast as required (See (20)), so $\|S\|$ will also eventually fall into a residual set with the size $o(\gamma)$. Theorem is proved.

IFAC Publications
www.elsevier.com/locate/ifac

# A FUZZY DYNAMIC UNCERTAINTY COMPENSATOR FOR INDUSTRIAL ROBOTS

**Wuwei Chen*(Visiting Scholar+) James K. Mills+ Jiaxin Chu+ Dong Sun#**

**Department of Mechanical and Automotive Engineering, Hefei University of Technology, Hefei, Anhui, P. R. China, 230069 Email: wchen@mie.utoronto.ca*

*+Department of Mechanical and Industrial Engineering, University of Toronto, 5 King's College Road, Toronto, Canada, M5S 3G8 Email: mills@mie.utoronto.ca*

*#Department of Manufacturing Engineering and Engineering Management City University of Hong Kong, Hong Kong Email: medsun@cityu.edu.hk*

Abstract: In the work reported here, a fuzzy logic controller plays the role of a compensator for a robot system, together with a computed torque control method, to improve trajectory tracking performance of an industrial robot. To improve FLC performance, the controller can be tuned automatically by adjusting the weight parameter $\alpha$ through applying a self-tuning fuzzy logic compensator (SFLC). By taking into account the full nonlinear nature of the robotic dynamics, the overall closed-loop system is shown to be bounded asymptotically stable. Experimental results demonstrate the effectiveness of the computed torque and fuzzy compensation scheme to control an industrial CRS Robotics Corporation A460 robot during a trajectory tracking task. *Copyright © 2001 IFAC*

Keywords: Fuzzy control, uncertainty, industrial robots

## 1. INTRODUCTION

Robotic manipulators with high speed and high precision trajectory tracking capabilities are indispensable in many applications. Manipulator controller design is influenced by structured (dynamic parameter uncertainty) and/or unstructured uncertainties (unmodelled dynamics), (Chen *et al.* 1993), (Okuma and Ishiguro, 1990).

Adaptive approaches have been proposed to maintain the tracking performance of robotic manipulators in the presence of structured parameters uncertainties. Although the adaptive control is effective to compensate the influence of structured parameters uncertainties, it is not clear that the adaptive methods can reduce the effect of unstructured uncertainties, (Slotine and Li, 1988).

Fuzzy logic has been widely used as a practical approach for the design and implementation of control systems, (Chen and Yin, 1997), (Malki *et al.*, 1994), (Kim *et al.*, 1998), (Llama *et al.*, 2000). Fuzzy logic controllers are suitable for many industrial processes such as linguistically controlled devices and systems.

For the case of robotic manipulators, approximate models of the robot dynamics can be derived, hence model based control is an effective approach to achieve high closed-loop behaviour. Computed torque control (CTC), using a dynamic model of the robotic manipulator, is an effective means for trajectory control (Spong and Vidyasagar, 1989). However, the tracking performance of this method in high speed operations is affected by the uncertainties mentioned above, (Okuma and Ishiguro, 1990).

This paper presents a fuzzy logic compensator (FLC) which assists the computed torque control method. The fuzzy logic compensator is used to compensate the dynamic uncertainties of the robotic manipulator. Furthermore, a control scheme for improving FLC, i.e., self-tuning fuzzy logic compensator (SFLC) is also presented. A comparison of the performance of the proposed CTC + FLC and CTC + SFLC schemes with that of the only CTC scheme is shown, clarifying the effectiveness of the proposed SFLC scheme in compensating structured and unstructured uncertainties.

## 2. DYNAMIC MODEL AND CONTROLLER DESIGN

The dynamic model for an $n$-link, serially connected, rigid-link revolute joint robot can be expressed by the following:

$$\tau = \mathbf{M}(\mathbf{q})\ddot{\mathbf{q}} + \mathbf{H}(\mathbf{q},\dot{\mathbf{q}})\dot{\mathbf{q}} + \mathbf{G}(\mathbf{q}) \tag{1}$$

where all term are defined in the relevant literature. Here, we wish to design a controller in such a way that the following control objective is achieved:

$$\lim_{t\to\infty} \boldsymbol{q}(t) = \boldsymbol{q}_d(t) \tag{2}$$

where: $\boldsymbol{q}_d(t) \in \boldsymbol{R}^n$ represents the desired joint position trajectory. In order to achieve the control we define the joint position tracking error $\boldsymbol{e} \in \boldsymbol{R}^n$ as:

$$\boldsymbol{e} = \boldsymbol{q}_d(t) - \boldsymbol{q}(t) \tag{3}$$

The computed torque controller, is given as follows:

$$\tau = \hat{\mathbf{M}}(\mathbf{q})(\ddot{\mathbf{q}}_d + \mathbf{K}_v\dot{\mathbf{e}} + \mathbf{K}_p\mathbf{e}) + \hat{\mathbf{H}}(\mathbf{q},\dot{\mathbf{q}})\dot{\mathbf{q}} + \hat{\mathbf{G}}(\mathbf{q}) \tag{4}$$

where: $\hat{\boldsymbol{M}}, \hat{\boldsymbol{H}}, \hat{\boldsymbol{G}}$ are estimates of $\boldsymbol{M}, \boldsymbol{H}, \boldsymbol{G}$ respectively, $\boldsymbol{K}_p$ and $\boldsymbol{K}_v$ are positive diagonal gain matrices. If structured and unstructured uncertainties do not exist, the system can be linearized by the application of the control given by equation (4). The closed-loop equation for the whole system is derived from equations (1) and (4) as:

$$\ddot{\mathbf{e}} + \mathbf{K}_v\dot{\mathbf{e}} + \mathbf{K}_p\mathbf{e} = 0 \tag{5}$$

In a real robotic system, structured and unstructured uncertainties exist, the dynamic equation of the closed-loop system becomes:

$$\ddot{\mathbf{e}} + \mathbf{K}_v\dot{\mathbf{e}} + \mathbf{K}_p\mathbf{e} = \mathbf{D}(\mathbf{e}) \tag{6}$$

where: $\mathbf{D}(\mathbf{e}) = \hat{\mathbf{M}}^{-1}[\Delta\mathbf{M}\ddot{\mathbf{q}} + \Delta\mathbf{H} + \Delta\mathbf{G} + \mathbf{N}]$, $\boldsymbol{N}$ is the unstructured uncertainty, $\Delta\mathbf{M} = \mathbf{M} - \hat{\mathbf{M}}$, $\Delta\mathbf{H} = \mathbf{H} - \hat{\mathbf{H}}$, $\Delta\mathbf{G} = \mathbf{G} - \hat{\mathbf{G}}$, represent errors caused by structured uncertainties.

Structured and/or unstructured uncertainties degrade the control performance of the CTC as described in Section

To overcome this shortcoming of the CTC, a fuzzy logic controller is incorporated with computed torque control. The fuzzy logic controller acts as a nonlinear compensator for the robotic manipulator, and together with a computed torque controller, completes the control scheme. The proposed control system is shown in Figure 1, with the input signal to the robotic manipulator given as follows:

$$\tau = \tau_c + \tau_f \tag{7}$$

where: $\tau$ is the robot controlling torque, $\tau_c$ is the output torque of the CTC, $\tau_f$ is the compensating torque of the FLC. To control and compensate the robotic system, a set of fuzzy logic compensators is proposed as follows:

$$\tau_{fi} = \sum_{i=1}^{n} k_{ui} u_i \quad i = 1, \ldots, n \tag{8}$$

where: $u_i$ is the output of a sub-compensator, $k_{ui}$ is a fuzzy control gain determined by the designer. Furthermore, the proposed FLC can be expressed in a fuzzy PD controller form as in [5], and can be written as:

$$u_i = k_{ei} e_i + k_{di}\dot{e}_i \tag{9}$$

where: $k_{ei}$ and $k_{di}$ are gains for magnifying $e_i$ and $\dot{e}_i$.

The proposed MIMO fuzzy compensator can be written in vector-matrix form as:

$$\boldsymbol{u} = \boldsymbol{K}_e\mathbf{e} + \boldsymbol{K}_d\dot{\mathbf{e}} \tag{10}$$

$$\tau_f = \boldsymbol{K}_u\boldsymbol{u} = \boldsymbol{K}_u(\boldsymbol{K}_e\mathbf{e} + \boldsymbol{K}_d\dot{\mathbf{e}}) \tag{11}$$

where: $\boldsymbol{K}_e$, $\boldsymbol{K}_d$ and $\boldsymbol{K}_u$ are diagonal gain matrices.

Hence, required control torque equation is given as follows:

$$\tau = \tau_c + \tau_f = \hat{\mathbf{M}}(\mathbf{q})(\ddot{\mathbf{q}}_d + \mathbf{K}_v\dot{\mathbf{e}} + \mathbf{K}_p\mathbf{e}) + \hat{\mathbf{H}}(\mathbf{q},\dot{\mathbf{q}})\dot{\mathbf{q}} + \hat{\mathbf{G}}(\mathbf{q}) + \mathbf{K}_u(\mathbf{K}_d\dot{\mathbf{e}} + \mathbf{K}_e\mathbf{e}) \tag{12}$$

With the CTC and the FLC applied to the robotic system, the error equation (6) becomes:

$$\ddot{\mathbf{e}} + \mathbf{K}_v\dot{\mathbf{e}} + \mathbf{K}_p\mathbf{e} = \Delta v \tag{13}$$

where: $\Delta v = \boldsymbol{D}(\boldsymbol{e}) - \tau_f$, represents the compensating error. With appropriate $\boldsymbol{K}_v$ and $\boldsymbol{K}_p$, $\boldsymbol{e} \to 0$, ($\boldsymbol{q} \to \boldsymbol{q}_d$) as t $\to \infty$. Thus it can be concluded that if we design the compensator so as to achieve $\Delta v \to 0$, then we obtain asymptotic tracking of the desired motion of the robot manipulators.

Since $\Delta v = 0$ implies that $\boldsymbol{D}(\boldsymbol{e}) = \tau_f$, although the structure of the uncertain function $\boldsymbol{D}(\boldsymbol{e})$ is known the exact values of the parameters of this function are not explicitly known. Assume an ideal FLC is a function whose compensating output signal $\tau_f$ exactly equals that of the uncertain function $\boldsymbol{D}(\boldsymbol{e})$, so that $\Delta v = 0$. Thus, robotic system can achieve excellent performance. Based on such a premise, the problem of designing a compensator can then be considered as a function approximation problem. As shown in Figure 1, the proposed FLC, i.e., fuzzy PD controller, employs two input signals $x(t)$ and $y(t)$, and one output signal $u(t)$, expressed as:

$$\begin{cases} x(t) = k_e e(t) \\ y(t) = k_d \dot{e}(t) \\ \tau_f(t) = k_u u(t) \end{cases} \tag{14}$$

A FLC can describe complex systems with linguistic descriptions. The information is described in terms of fuzzy sets. In contrast to conventional controllers, three steps must be carried out before the FLC can generate a desired output. These three steps are: (1) fuzzification of the input; (2) fuzzy inference based on the knowledge base; (3) defuzzification of the fuzzy control signal.

In general, for given inputs $x(t)$, $y(t)$, a basic fuzzy compensator which has two inputs and one output $u(t)$, can be described by the following rules:

***$k^{th}$ rule:*** if $x(t)$ is $A_k$ and $y(t)$ is $B_k$, then $u(t)$ is $C_k\,(k = 1, 2, \ldots, r)$ (15)

where: $A_k$ and $B_k$ are the antecedent linguistic terms, $C_k$ is the consequent linguistic terms.

Thus, $x(t)$, $y(t)$ and $u(t)$ can be defined as the following seven fuzzy subsets:

$$x(t) = \{A_k\} = \{NB, NM, NS, ZO, PS, PM, PB\}$$
$$y(t) = \{B_k\} = \{NB, NM, NS, ZO, PS, PM, PB\} \quad (16)$$
$$u(t) = \{C_k\} = \{NB, NM, NS, ZO, PS, PM, PB\}$$

where: $NB$ = Negative Big, $NM$ = Negative Medium, $NS$ = Negative Small, $ZO$ = Zero, $PS$ = Positive Small, $PM$ = Positive Medium, $PB$ = Positive Big.

Since the above control signal is in fuzzy mode, defuzzification is required to transform the fuzzy control signals into the exact control outputs. The "center of gravity method" (Chen and Yin, 1999) is applied to defuzzify the fuzzy control signal. This method can be expressed as:

$$u(t) = \sum_{k=1}^{r} h_k u_k \Big/ \sum_{k=1}^{r} h_k \quad (17)$$

$$h_k = \mu_{A_k}(x(t)) \wedge \mu_{B_k}(y(t)) \quad (18)$$

where: $h_k$ is the absolute strength of the $k^{th}$ rule fired by input variables; $\mu_{A_k}$, $\mu_{B_k}$ are the membership functions of fuzzy subsets $A_k$ and $B_k$, operator " $\wedge$ " denotes minimum.

In this work, triangular membership functions were employed for the input and output variables. The fuzzy dynamic model, is described by fuzzy IF-THEN rules which represent input-output relations of non-linear robot system. Consider an uncertain nonlinear robot system, the control signal applied to the system depends on how large the error and its derivative are. The control rules are designed and adjusted based on simulation, testing and knowledge of the characteristics and response for the robot system under investigation.

In order to improve the performance of the FLC, a weighting parameter $\alpha$ may be used to impose different weights on $x$ and $y$ to obtain the control signal [4] as follows:

$$u = < \alpha x + (1-\alpha) \times y > \quad 0 \leq \alpha \leq 1 \quad (19)$$

where: $< \cdot >$ denotes an integer value, e.g., $<1.6> = 2$. By adjusting the weight parameter $\alpha$, different FLC performance can be obtained. Through proper selection of the weight parameter $\alpha$, better fuzzy control rules can be synthesized.

By substituting equation (11) into (19), we obtain:

$$\tau_f = K_u u = K_u(\alpha \mathbf{K}_e \mathbf{e} + (1-\alpha)\mathbf{K}_d \dot{\mathbf{e}}) \quad (20)$$

Certain difficulties exist in the implementation of the FLC. For example, for a complex control plant, it is difficult to synthesize suitable control rules with manual update of the parameter $\alpha$. In addition, if changes in the plant dynamic characteristics have taken place. In order to overcome these weakness, further improvement of the control effect is required. This improvement can be achieved through introduction of a learning capability, utilizing a SFLC and automatic update of the parameter $\alpha$, based on equation (19), as shown in Figure 3.

Suppose the linguistic variable for adjusting the parameter $\alpha$ is $S_k$, i.e.,

$$\text{If} \quad A_k \text{ and } B_k \quad \text{then } S_{k(A,B)} \quad (21)$$

where:
$$x = \{A_k\} = \{NB, NM, NS, ZO, PS, PM, PB\}$$
$$y = \{B_k\} = \{NB, NM, NS, ZO, PS, PM, PB\}$$
$$S = \{S_k\} = \{C_k\} = \{NB, NM, NS, ZO, PS, PM, PB\}$$

The procedure to automatically update $\alpha$ is as follows:

( In order to reduce the on-line running time of the system, using equations (16) ~ (18), the fuzzy control rules, given in Table 1, are converted to an exact control output, to construct a look-up control table in advance.

4 As the system runs, variables $x$ and $y$ are input.

5 According to $x$ and $y$ values, utilizing the look up table, obtain the output values $S$. Thus, the parameter $\alpha$ is updated in the sampling interval according to the following:

$$\alpha_t = \alpha_{t-1} + k_c \times S \quad (22)$$

where: $k_c$ is the scaling factor.

The algorithm of the SFLC is as follows:

(1) Sample data $e$ and $\dot{e}$, calculate $x = k_e e$, $y = k_d \dot{e}$.
(2) According to $x$ and $y$ values, look up the control rule table to obtain $S$ and calculate parameter $\alpha$ utilizing (22).
(3) Fuzzify the inputs $x$ and $y$, then perform fuzzy inference and defuzzification of the fuzzy control signal, utilizing (19) to obtain the exact control output $u$.
(4) According to control output $u$, obtain torque value ($\tau_f = k_u u$), using it to compensate robot system uncertainty.

Therefore, applying this control scheme, the system control signal can be tuned automatically by adjusting the parameter $\alpha$.

## 3. STABILITY ANALYSIS: CTC + FLC SYSTEM

Here, the stability of the CTC + FLC system will be analyzed. Recall Figure 1 and the formulas of the CTC + FLC given in Section 2. The stability analysis is given as follows:

***Theorem***: If the compensating torque produced by the FLC, plus control torque produced by the CTC, is applied to the robotic system given by (1), the closed-loop control system is said to be uniformly ultimately bounded (u.u.b.).

***Proof***: Firstly, we consider the case in which the FLC can compensate structured and unstructured uncertainties partially, i.e. in (13), let $\Delta v \neq 0$,

$$\ddot{\mathbf{e}} + \mathbf{K}_v(\mathbf{e},\dot{\mathbf{e}})\dot{\mathbf{e}} + \mathbf{K}_p(\mathbf{e},\dot{\mathbf{e}})\mathbf{e} = \Delta v = D(\mathbf{e},\dot{\mathbf{e}}) - \tau_f$$

Here, we let $K_u = I$, here $I$ is a unit matrix, so the compensating torque $\tau_f$ in (11) is expressed by:

$$\tau_f = \alpha\mathbf{K}_e(\mathbf{e},\dot{\mathbf{e}})\mathbf{e} + (1-\alpha)\mathbf{K}_d(\mathbf{e},\dot{\mathbf{e}})\dot{\mathbf{e}} \quad (23)$$

Then, the closed-loop equation (13) is given as:

$$\ddot{\mathbf{e}} + (\mathbf{K}_v(\mathbf{e},\dot{\mathbf{e}}) + (1-\alpha)\mathbf{K}_d(\mathbf{e},\dot{\mathbf{e}}))\dot{\mathbf{e}} + (\mathbf{K}_p(\mathbf{e},\dot{\mathbf{e}}) + \alpha\mathbf{K}_e(\mathbf{e},\dot{\mathbf{e}}))\mathbf{e} = \mathbf{D}(\mathbf{e},\dot{\mathbf{e}}) \quad (24)$$

where: $K_p(\mathbf{e},\dot{\mathbf{e}})$, $K_v(\mathbf{e},\dot{\mathbf{e}})$, $K_e(\mathbf{e},\dot{\mathbf{e}})$, $K_d(\mathbf{e},\dot{\mathbf{e}})$ are nonlinear functions of $\mathbf{e},\dot{\mathbf{e}}$, and they are suitable diagonal matrix functions whose entries are denoted by $k_{pi}(e_i,\dot{e}_i)$, $k_{vi}(e_i,\dot{e}_i)$, $k_{ei}(e_i,\dot{e}_i)$ and $k_{di}(e_i,\dot{e}_i)$ respectively. Suppose the Lyapunov function candidate $V$, constructed in a similar way to that given in (Llama *et al.*, 2000), is as follows:

$$V(\mathbf{e},\dot{\mathbf{e}}) = \frac{1}{2}\dot{\mathbf{e}}^T\dot{\mathbf{e}} + \int_0^e \sigma^T[\mathbf{K}_p(\sigma) + \alpha\mathbf{K}_e(\sigma) + \gamma(\mathbf{K}_v(\sigma) + (1-\alpha)\mathbf{K}_d(\sigma))]d\sigma + \gamma\mathbf{e}^T\dot{\mathbf{e}}$$

(25)

Here, we use the notation $\lambda_m\{A\}$ and $\lambda_M\{A\}$, to indicate the smallest and largest eigenvalues, respectively, of a symmetric positive definite matrix $A(x)$ for all $x \in \mathbf{R}^n$. There exists a small positive number $\varepsilon$, and where $\varepsilon > 0$, such that the following nonlinear gains satisfy:

$k_{pi}(e_i,\dot{e}_i) \geq \varepsilon$, $k_{vi}(e_i,\dot{e}_i) \geq \varepsilon$, $k_{ei}(e_i,\dot{e}_i) \geq \varepsilon$, $k_{di}(e_i,\dot{e}_i) \geq \varepsilon$,

for all $e_i \in R$, $\dot{e}_i \in R$, and $i = 1,\ldots,n$.

Differentiating (25) and evaluating various terms, without showing the details, leads to proof (main parts omitted here for brevity) that $V(\mathbf{e},\dot{\mathbf{e}})$ is positive definite

## 4. EXPERIMENTAL EVALUATION

To verify the effectiveness of the proposed control law, an experiment has been conducted to test the tracking control of the robotic manipulator under the action of the proposed controller. The desired trajectory for each joint was generated with a fifth-order polynomial in time. The experimental subject for the computed torque control (CTC), computed torque control plus fuzzy logic compensating (CTC + FLC), and computed torque control plus self-tuning fuzzy logic compensating (CTC + SFLC) experiments, described in this paper, is a CRS Robotics Corporation A460 robot with a C500 controller. This robot is a six-axis vertically articulated jointed arm robotic manipulator, and uses permanent magnet DC brushed motors to activate harmonic drive gears (Figure 4). A custom designed experimental interface permits experiment set-up, data collection and display.

### *4.1 Results and Discussion*

Representative results of the various experiments are illustrated in this section (here we only show the results of joint 1 and joint 2). This provides an illustration of the relative performance of the controllers on the real-time trajectory (fifth-order polynomial) tracking performance of the robot. The real-time position tracking error is used as the major criterion for evaluating the performance of a controller. The results are shown in Figures 4 ~ 8 respectively. It can be seen from the experimental results that the CTC scheme (Figure 4) has a greater error and it is not able to suitably follow the desired position trajectory because of the influence of uncertainty.

Figures 5 ~ 6 show the position tracking errors change with different weight parameter $\alpha$ (CTC + FLC scheme). Here, we assume $\alpha$ in equation (18) is 0.6 and 0.75 respectively. In contrast, the SFLC approach based on equation (20) has shown to have smaller position tracking error (see Figure 7, CTC + SFLC scheme).

## 5. CONCLUSIONS

More and more industrial robot operations demand high-accuracy trajectory performance which may not be achieved by using the computed torque scheme. This paper presents a fuzzy logic compensating method for trajectory control of robotic manipulators. The on-line tuning of the fuzzy logic control scheme is used to compensate only the uncertainties of the robotic manipulator. A comparison of its performance with the computed torque scheme (CTC), fuzzy logic control plus computed torque scheme (FLC + CTC) and self-tuning fuzzy logic control plus computed torque (SFLC + CTC) in compensating the uncertainty effects was carried out. Stability analysis incorporating the nonlinear robot dynamics guarantees uniformly ultimately bounded stability. From the experimental results, it is verified the validity of real-time implementation of the SFLC algorithm. Therefore, the SFLC can reduce the trajectory tracking error of the robotic manipulator quite well and compensate uncertainties effectively.

## 6. REFERENCES

Chen, P.C. J.K. Mills, and K.C. Smith, "Parameter Uncertainty Compensation in Robot Trajectory Tracking: A Neural Network Approach", Proc. International IEEE System, Man and Cybernetics Conference, Le Touquet, France, 1993, pp. 17-20.

Chen, C. and Y. Yin, "Fuzzy logic control of a moving flexible manipulator", Proc. of the 1999 IEEE International Conference on Control Applications, Hawaii, USA, 1999, pp. 315-320.

Hsu, Y. G. Chen, and E. Sanchez, "A Fuzzy PD Controller for Multi-Link Robot Control: Stability Analysis", Proc. of the 1997 IEEE International Conference on Robotics and Automation, Albuquerque, New Mexico, 1997, pp. 1412-1417.

Kim, S. C. Park, and F. Harashima, "Adaptive Fuzzy Controller Design for Trajectory Tracking of a 2 D.O.F. Wheeled Mobile Robot using Genetic Algorithm", Proc. of the 1998 IEEE/RSJ Conference on Intelligent Robots and Systems, Victoria, B. C., Canada, 1998, pp. 1584-1589.

.Llama, M.A., R. Kelly, and V. .Santibanez, "Stable Computed-Torque Control of Robot Manipulators via Fuzzy Self-Tuning", IEEE Trans. on System, Man, and Cybernetics-Part B: Cybernetics, Vol. 30, No. 1, 2000, pp. 143-150.

Malki, H.A., H. Li, and G. Chen, "New Design and Stability Analysis of Fuzzy Proportional-Derivative Control Systems", IEEE Trans. on Fuzzy Sys., Vol. 2, No. 4, 1994, pp. 245-254.

Okuma, S. and A. Ishiguro, "A Neural Network Compensator for Uncertainties of Robotic Manipulators", Proc. of the 29th Conference on Decision and Control, Honolulu, Hawaii, 1990, pp. 3303-3307.

Slotine, J. and W. Li, "Adaptive Manipulator Control: A Case Study", IEEE Trans. Automatic Control, Vol. AC-33, No. 11, 1988, pp. 995-1003.

Spong, M. W. and M. Vidyasagar, *Robot Dynamics and Control*, New York, John Wiley & Sons, 1989


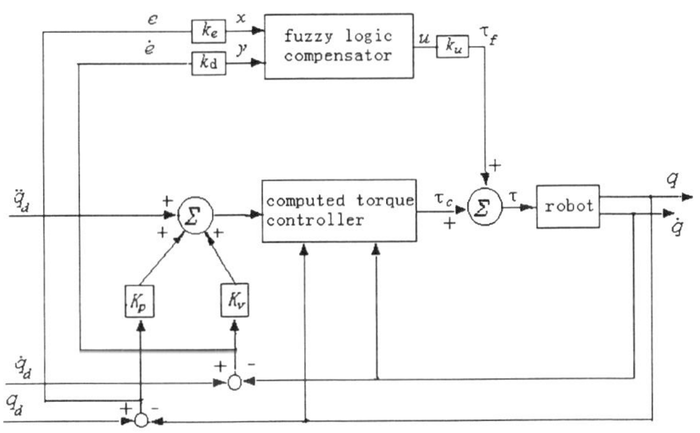


**Figure 1 Controller using the computed torque method with FLC**

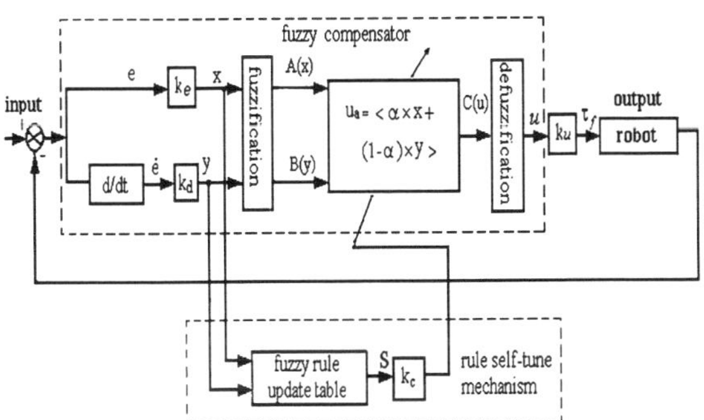


**Figure 2 Self-tuning fuzzy compensator**

**Figure 3 CRS Robotics Corporation A 460 Robot**

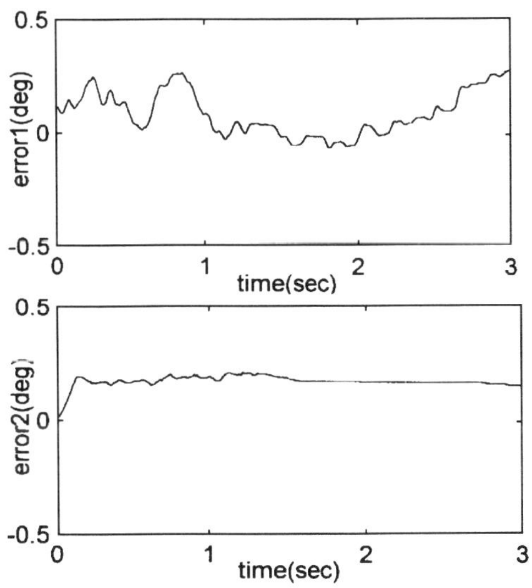


(a) joint 1 (b) joint 2

**Figure 4 Position errors of the CTC**

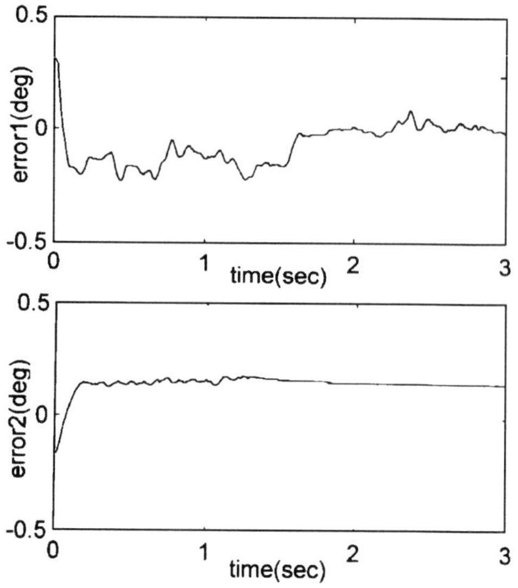


(a) joint 1 (b) joint 2

**Figure 5 Position errors of the CTC + FLC** ( $\alpha$ = 0.6)

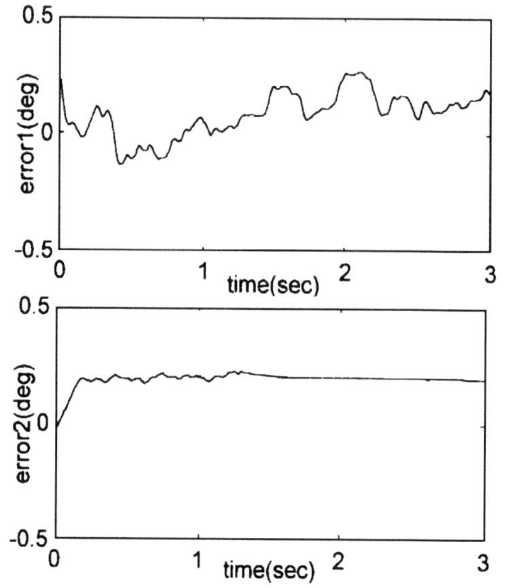


(a) joint 1 (b) joint 2

**Figure 6 Position errors of the CTC + FLC** ( $\alpha$ = 0.75)

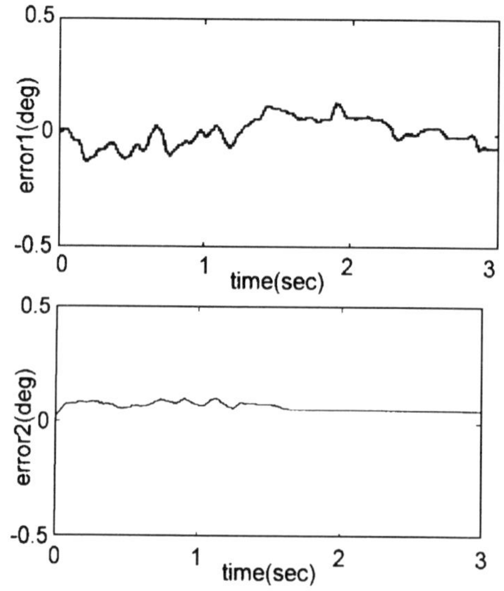


(a) joint 1 (b) joint 2

**Figure 7 Position errors of the CTC + SFLC**

IFAC Publications
www.elsevier.com/locate/ifac

# ADAPTIVE NEURAL CONTROL OF MIMO NONLINEAR SYSTEMS IN BLOCK-TRIANGULAR PURE-FEEDBACK FORM

**Shuzhi S. Ge** [1] **Cong Wang** [2]

*Department of Electrical & Computer Engineering*
*National University of Singapore*
*Singapore 117576*

Abstract: In this paper, an adaptive neural control scheme is proposed for a class of MIMO nonlinear systems in block-triangular pure-feedback form. The MIMO system is composed of interconnected subsystems. Each subsystem is in the pure-feedback form with all the nonlinearities unknown. Firstly, we design for each subsystem an adaptive neural controller by using backstepping methodology. With the help of NN approximation, there is no need to solve the implicit functions for the explicit virtual controls to cancel the unknown functions in backstepping design. Secondly, by exploiting the structure properties of the MIMO system, we conclude the stability of the whole closed-loop MIMO system by stability analysis of all the signals in the closed-loop in a nested iterative manner. Semi-global uniform ultimate boundedness of all the signals in the closed-loop of MIMO nonlinear systems is guaranteed. The outputs of the systems are proven to converge to small neighborhoods of the desired trajectories. The control performance of the closed-loop system is guaranteed by suitably choosing the design parameters. *Copyright © 2001 IFAC*

Keywords: Neural network (NN), adaptive neural control, multi-input/multi-output (MIMO) systems, pure-feedback systems, backstepping.

## 1. INTRODUCTION

Over the past decade, adaptive neural control of uncertain nonlinear systems has attracted considerable interest, and significant progress in this area has been achieved in both theoretical studies and practical applications (see, e.g., Lewis *et al.*, 1999; Ge *et al.*, 2001; and the references therein).

In nonlinear control area, the problem of controlling multi-input/multi-output (MIMO) nonlinear systems is very complicated due to the couplings among various inputs and outputs. This problem becomes more involved to be dealt with when there exist uncertain parameters and/or unknown nonlinear functions in the input coupling matrix. By using adaptive NN control design initiated in (Narendra, K. S. and Parthasarathy, 1990; Sanner and Slotine, 1992; etc.), some results have been proposed for MIMO nonlinear systems with special structural properties. Specifically, adaptive neural network controllers were proposed in (Lewis, *et al.*, 1999) for a special class of MIMO nonlinear plants, rigid body robotic systems, using several nice properties of the input decoupling matrix of robotic systems. In (Ge, *et al.*, 2001), adaptive NN control was considered for a class of MIMO nonlinear systems with a triangular structure in control inputs.

In this paper, we present an adaptive neural control scheme for a class of MIMO nonlinear systems in block-triangular pure-feedback form. The MIMO system is composed of interconnected subsystems. Each subsystem is in the pure-feedback form with all the nonlinearities unknown. With mild assumptions on the partial derivatives of the nonlinearities, and with the help of mean value

[1] Corresponding author: elegesz@nus.edu.sg.
[2] C. Wang is now with Department of Electronic Engineering, City University of Hong Kong.

theorem and implicit function theorem, we first design for each pure-feedback subsystem an full-state adaptive neural controller, where the idea of backstepping (Krstic, *et al.*, 1995) is still employed, i.e., some of the state variables are considered as "virtual controls", and intermediate control laws are designed in the constructive design procedures. With the help of NN approximation, there is no need to solve the implicit functions for the explicit virtual controls to cancel the unknown functions in backstepping design.

Secondly, we proceed to prove the stability of the whole closed-loop MIMO system. Since all the subsystems are interconnected, it is difficult to conclude directly the stability of the whole system by stability analysis of individual subsystem separately. By exploiting the structural properties of the MIMO system, we conclude the stability of all the signals in the closed-loop system in a nested iterative manner. Semi-global uniform ultimate boundedness of all the signals in the closed-loop is guaranteed. The outputs of the system are proven to converge to small neighborhoods of the desired trajectories. The control performance of the closed-loop system is guaranteed by suitably choosing the design parameters.

## 2. PROBLEM FORMULATION AND PRELIMINARIES

Consider the MIMO nonlinear system in the following block-triangular pure-feedback form

$$\begin{cases} \dot{x}_{1,i_1}=f_{1,i_1}(\bar{x}_{1,(i_1-\varrho_{j1})},\cdots,\bar{x}_{m,(i_1-\varrho_{jm})},x_{1,i_1+1}) \\ \qquad 1\le i_1\le \rho_1-2 \\ \dot{x}_{1,\rho_1-1}=f_{1,\rho_1-1}(\bar{x}_{1,\rho_1-1},\cdots,\bar{x}_{m,\rho_m-1}) \\ \qquad +g_{1,\rho_1-1}(\bar{x}_{1,\rho_1-1},\cdots,\bar{x}_{m,\rho_m-1})x_{1,\rho_1} \\ \dot{x}_{1,\rho_1}=f_{1,\rho_1}(X)+g_{1,\rho_1}(\bar{x}_{1,\rho_1-1},\cdots,\bar{x}_{m,\rho_m-1})u_1 \\ \quad \cdots \\ \dot{x}_{j,i_j}=f_{j,i_j}(\bar{x}_{1,(i_j-\varrho_{j1})},\cdots,\bar{x}_{m,(i_j-\varrho_{jm})},x_{j,i_j+1}) \\ \qquad 1\le i_j\le \rho_j-1 \\ \dot{x}_{j,\rho_j-1}=f_{j,\rho_j-1}(\bar{x}_{1,\rho_1-1},\cdots,\bar{x}_{m,\rho_m-1}) \\ \qquad +g_{j,\rho_j-1}(\bar{x}_{1,\rho_1-1},\cdots,\bar{x}_{m,\rho_m-1})x_{j,\rho_j} \\ \dot{x}_{j,\rho_j}=f_{j,\rho_j}(X,u_1,\cdots,u_{j-1}) \\ \qquad +g_{j,\rho_j}(\bar{x}_{1,\rho_1-1},\cdots,\bar{x}_{m,\rho_m-1})u_j \\ y_j=x_{j,1},\quad 1\le j\le m \end{cases} \tag{1}$$

where $x_{j,i_j}$, $i_j=1,\cdots,\rho_j$ are the states of the $j$th subsystem, $X=[\bar{x}_{1,\rho_1}^T,\cdots,\bar{x}_{m,\rho_m}^T]^T\in R^N$ ($N=\rho_1+\cdots+\rho_m$, $\bar{x}_{j,i_j}=[x_{j,1},\cdots,x_{j,i_j}]^T\in R^{i_j}$, $i_j=1,\cdots,\rho_j$ and $j=1,\cdots,m$) represents the state variables of the whole system, $u_j\in R$ and $y_j\in R$ are the system inputs and outputs, respectively; $f_{j,i_j}(\cdot)$, $g_{j,\rho_j-1}(\cdot)$ and $g_{j,\rho_j}(\cdot)$ are unknown smooth functions, $j$, $i_j$, $\rho_j$ and $m$ are positive integers, and $\varrho_{jl}\triangleq\rho_j-\rho_l$ ($j,l=1,\cdots,m$) is the order difference between the $j$th and $l$th subsystems.

*Remark 1.* Note that if $i_j-\varrho_{jl}>0$, then $\bar{x}_{l,(i_j-\varrho_{jl})}$ represents the maximum state variables of the $l$th subsystem which are embedded in the $j$th subsystem. If $i_j-\varrho_{jl}\le 0$, then it is assumed that the corresponding variable vector $\bar{x}_{l,(i_j-\varrho_{jl})}$ ($j,l=1,\cdots,m$) does not exist, and does not appear in the functions in (1).

The control objective is to design adaptive neural controllers $u_j$, $j=1,\cdots,m$ for system (1) such that (i) all the signals in the closed-loop remain semi-globally uniformly ultimately bounded, and (ii) the output $y_j$ follows the desired trajectories $y_{dj}$ generated from the following smooth, bounded reference model

$$\begin{aligned} \dot{x}_{di}&=f_{di}(x_d),\ 1\le i\le n \\ y_{dj}&=x_{dj},\ 1\le j\le m\le n \end{aligned} \tag{2}$$

where $x_d=[x_{d1},x_{d2},\cdots,x_{dn}]^T\in R^n$ are the states, $y_d=[y_{d1},\cdots,y_{dm}]^T\in R^m$ are the system outputs, $f_{di}(\cdot), i=1,2,\cdots,m$ are known smooth nonlinear functions. Assume that the states of the reference model remain bounded, i.e., $x_d\in\Omega_d$, $\forall t\ge 0$.

For the pure-feedback MIMO system (1), define

$$g_{j,i_j}(\bar{x}_{1,(i_j-\varrho_{j1})},\cdots,\bar{x}_{m,(i_j-\varrho_{jm})},x_{j,i_j+1})\triangleq \frac{\partial f_{j,i_j}(\bar{x}_{1,(i_j-\varrho_{j1})},\cdots,\bar{x}_{m,(i_j-\varrho_{jm})},x_{j,i_j+1})}{\partial x_{j,i_j+1}}$$
$$j=1,\cdots,m,\ i_j=1,\cdots,\rho_j-1 \tag{3}$$

Note that the partial derivatives $g_{j,i_j}(\cdot)$ are still unknown nonlinear functions of the system states. We have the following assumptions on the partial derivatives $g_{j,i_j}(\cdot)$ ($j=1,\cdots,m$, $i_j=1,\cdots,\rho_j$).

*Assumption 1.* The signs of $g_{j,i_j}(\cdot)$ ($j=1,\cdots,m$, $i_j=1,\cdots,\rho_j$) are known, and there exist constants $\infty>\bar{g}_{j,i_j}\ge \underline{g}_{j,i_j}>0$, $j=1,\cdots,m$, $i_j=1,\cdots,\rho_j$ such that $\bar{g}_{j,i_j}\ge|g_{j,i_j}(\cdot)|>\underline{g}_{j,i_j}$, $\forall X\in\Omega_X\subset R^N$.

The above assumption implies that smooth functions $g_{j,i_j}(\cdot)$ are strictly either positive or negative. Without losing generality, we shall assume $\bar{g}_{j,i_j}\ge g_{j,i_j}(\cdot)>\underline{g}_{j,i_j}$.

It can be easily seen the derivatives of $g_{j,i_j}(\cdot)$ ($j=1,\cdots,m$, $i_j=1,\cdots,\rho_j$) depend on states $X$. Because $f_{j,i_j}(\cdot)$, $g_{j,\rho_j-1}(\cdot)$ and $g_{j,\rho_j}(\cdot)$ ($j=1,\cdots,m$, $i_j=1,\cdots,\rho_j$) are assumed to be smooth functions, the derivatives of $g_{j,i_j}(\cdot)$ ($j=1,\cdots,m$, $i_j=1,\cdots,\rho_j$) are therefore bounded as long as the system states $X$ remain within some compact sets. Thus, we have the following assumption.

*Assumption 2.* There exist constants $g_{j,i_jd}>0$ ($j=1,\cdots,m$, $i_j=1,\cdots,\rho_j$) such that $|\dot{g}_{j,i_j}(\cdot)|\le g_{j,i_jd}$, $\forall X\in\Omega_X\subset R^N$.

The following lemma is used to establish the existence of an implicit desired function.

*Lemma 1.* Assume that $f(x,u) : R^n \times R \to R$ is continuously differentiable $\forall(x,u) \in R^n \times R$, and there exists a positive constant $d$ such that $\frac{\partial f(x,u)}{\partial u} > d > 0$, $\forall(x,u) \in R^n \times R$. Then there exists a continuous (smooth) function $u^* = u(x)$ such that $f(x,u^*) = 0$.

**Proof.** See (Ge and Wang, 2001).

## 3. ADAPTIVE NEURAL CONTROL DESIGN

In this Section, we first design for the $j$th ($j = 1, \cdots, m$) subsystem an adaptive neural controller. With the help of implicit function theorem (see, e.g., Apostol, 1968), an intermediate desired feedback control $\alpha^*_{j,i_j}$ is shown to exist which possesses some desired stabilizing properties at the recursive $i_j$th step. Then the $i_j$th-order sub-subsystem of the $j$th subsystem is stabilized with respect to a Lyapunov function $V_{j,i_j}$ by the design of a stabilizing function $\alpha_{j,i_j}$, where an RBF neural network is employed to approximate the unknown part in intermediate desired feedback control $\alpha^*_{j,i_j}$. Then, we proceed to prove the stability of the whole closed-loop MIMO system $\Sigma_{S_1}$. By exploiting the structure properties of the MIMO system, we conclude the stability of all the signals in the closed-loop in a nested iterative manner.

Step 1: Define $z_{j,1} = x_{j,1} - x_{d1}$. Its derivative is

$$\dot{z}_{j,1} = f_{j,1}(\bar{x}_{j1}, x_{j,2}) - \dot{x}_{d1} \tag{4}$$

where $\bar{x}_{j1} := [\bar{x}^T_{1,(1-\varrho_{j1})}, \cdots, \bar{x}^T_{m,(1-\varrho_{jm})}]^T$. From Assumption 1, we know that $g_{j,1} = \partial f_{j,1}/\partial x_{j,2} > \underline{g}_{j,1} > 0$ for all $X \in \Omega_X \subset R^N$. Define $\nu_{j,1}$ as

$$\nu_{j,1} = -\dot{x}_{d1} + k_{j,1} z_{j,1} \tag{5}$$

where $k_{j,1} > 0$ is a constant. It is clear that $\nu_{j,1}$ is a function of $z_{j,1}$ and $x_d$. Considering the fact that $\partial \nu_{j,1}/\partial x_{j,2} = 0$, the following inequality holds

$$\frac{\partial [f_{j,1}(\bar{x}_{j1}, x_{j,2}) + \nu_{j,1}]}{\partial x_{j,2}} > \underline{g}_{j,1} > 0$$

According to Lemma 1, by viewing $x_{j,2}$ as a virtual control input, there exists a smooth ideal control input $x_{j,2} = \alpha^*_{j,1}(\bar{x}_{j1}, \nu_{j,1})$ such that

$$f_{j,1}(\bar{x}_{j1}, \alpha^*_{j,1}) + \nu_{j,1} = 0 \tag{6}$$

Using mean value theorem (Apostol, 1963), there exists $\lambda_{j,1}$ ($0 < \lambda_{j,1} < 1$) such that

$$\begin{aligned} f_{j,1}(\bar{x}_{j1}, x_{j,2}) &= f_{j,1}(\bar{x}_{j1}, \alpha^*_{j,1}) \\ &\quad + g_{\lambda_{j,1}}(x_{j,2} - \alpha^*_{j,1}) \end{aligned} \tag{7}$$

where $g_{\lambda_{j,1}} := g_{j,1}(\bar{x}_{j1}, x_{2\lambda_{j,1}})$ with $x_{2\lambda_{j,1}} = \lambda_{j,1} x_{j,2} + (1-\lambda_{j,1})\alpha^*_{j,1}$. Note that Assumptions 1 on $g_{j,1}(\bar{x}_{j1}, x_2)$ is still valid for $g_{\lambda_{j,1}}$. Similar to Assumption 2, it is also reasonable to assume that $|\dot{g}_{\lambda_{j,1}}|$ is bounded by some constant within some compact region. For uniformity of presentation, we assume that Assumption 2 on $\dot{g}_{j,1}(\bar{x}_{j1}, x_2)$ is valid for $\dot{g}_{\lambda_{j,1}}$.

Combining (4)-(7) yields

$$\dot{z}_{j,1} = -k_{j,1} z_{j,1} + g_{\lambda_{j,1}}(x_{j,2} - \alpha^*_{j,1}) \tag{8}$$

By employing an RBF neural network $W^T_{j,1} S_{j,1}(Z_{j,1})$ to approximate $\alpha^*_{j,1}(\bar{x}_{j1}, \nu_{j,1})$, where $Z_{j,1} = [\bar{x}^T_{j1}, \dot{x}_{d1}, z_{j,1}]^T \in \Omega_{j,1}$, $\alpha^*_{j,1}$ can be expressed as

$$\alpha^*_{j,1} = {W^*_{j,1}}^T S_{j,1}(Z_{j,1}) + \epsilon_{j,1} \tag{9}$$

where $W^*_{j,1}$ denotes the ideal constant weights, and $|\epsilon_{j,1}| \le \epsilon^*_{j,1}$ is the approximation error with constant $\epsilon^*_{j,1} > 0$.

Let $\hat{W}_{j,1}$ be the estimate of $W^*_{j,1}$. Define $z_{j,2} = x_{j,2} - \alpha_{j,1}$ and let

$$\alpha_{j,1} = -c_{j,1} z_{j,1} + \hat{W}^T_{j,1} S_{j,1}(Z_{j,1}) \tag{10}$$

where $c_{j,1}$ is a positive constant to be specified later. Then, the dynamics of $z_{j,1}$ is governed by

$$\begin{aligned} \dot{z}_{j,1} &= -k_{j,1} z_{j,1} + g_{\lambda_{j,1}}(z_{j,2} + \alpha_{j,1} - \alpha^*_{j,1}) \\ &= -k_{j,1} z_{j,1} + g_{\lambda_{j,1}}[z_{j,2} - c_{j,1} z_{j,1} + \tilde{W}^T_{j,1} S_{j,1}(Z_{j,1}) - \epsilon_{j,1}] \end{aligned} \tag{11}$$

where $\tilde{W}_{j,1} = \hat{W}_{j,1} - W^*_{j,1}$. Through out this paper, define $\tilde{(\cdot)} = \hat{(\cdot)} - (\cdot)^*$.

Consider the Lyapunov function candidate

$$V_{j,1} = \frac{1}{2g_{j,1}} z^2_{j,1} + \frac{1}{2}\tilde{W}^T_{j,1}\Gamma^{-1}_{j,1}\tilde{W}_{j,1} \tag{12}$$

The derivative of $V_{j,1}$ is

$$\begin{aligned} \dot{V}_1 &= \frac{z_{j,1}\dot{z}_{j,1}}{g_{j,1}} - \frac{\dot{g}_{j,1} z^2_{j,1}}{2g^2_{j,1}} + \tilde{W}^T_{j,1}\Gamma^{-1}_{j,1}\dot{\hat{W}}_{j,1} \\ &= \frac{k_{j,1}}{g_{\lambda_{j,1}}} z^2_{j,1} + z_{j,1} z_{j,2} - c_{j,1} z^2_{j,1} - \frac{\dot{g}_{j,1}}{2g^2_{j,1}} z^2_{j,1} + z_{j,1}\epsilon_{j,1} \\ &\quad + \tilde{W}^T_{j,1} S_{j,1}(Z_{j,1}) z_{j,1} + \tilde{W}^T_{j,1}\Gamma^{-1}_{j,1}\dot{\hat{W}}_{j,1} \end{aligned} \tag{13}$$

Consider the adaptation law for $\hat{W}_{j,1}$ as

$$\dot{\hat{W}}_{j,1} = \Gamma_{W_{j,1}}[-S_{j,1}(Z_{j,1}) z_{j,1} - \sigma_{j,1}\hat{W}_{j,1}] \tag{14}$$

where $\sigma_{j,1} > 0$ and $\Gamma_{W_{j,1}} = \Gamma^T_{W_{j,1}} > 0$ are design constants.

Let $c_{j,1} = c_{j,10} + c_{j,11}$, with $c_{j,10}$ and $c_{j,11} > 0$. Then, equation (13) becomes

$$\dot{V}_{j,1} = -\frac{k_{j,1}}{g_{\lambda_{j,1}}} z^2_{j,1} + z_{j,1} z_{j,2} - (c_{j,10} + \frac{\dot{g}_{j,1}}{2g^2_{j,1}}) z^2_{j,1}$$

$$-c_{j,11}z_{j,1}^2 + z_{j,1}\epsilon_{j,1} - \sigma_{j,1}\tilde{W}_{j,1}^T\hat{W}_{j,1} \quad (15)$$

By completion of squares, we have

$$\begin{aligned}-\sigma_{j,1}\tilde{W}_{j,1}^T\hat{W}_{j,1} &= -\sigma_{j,1}\tilde{W}_{j,1}^T(\tilde{W}_{j,1} + W_{j,1}^*)\\ &\leq -\sigma_{j,1}\|\tilde{W}_{j,1}\|^2 + \sigma_{j,1}\|\tilde{W}_{j,1}\|\|W_{j,1}^*\|\\ &\leq -\frac{\sigma_{j,1}\|\tilde{W}_{j,1}\|^2}{2} + \frac{\sigma_{j,1}\|W_{j,1}^*\|^2}{2}\end{aligned}$$

$$-c_{j,11}z_{j,1}^2 + z_{j,1}\epsilon_{j,1} \leq \frac{\epsilon_{j,1}^2}{4c_{j,11}} \leq \frac{\epsilon_{j,1}^{*2}}{4c_{j,11}} \quad (16)$$

Because $-(c_{j,10} + \frac{\dot{g}_{j,1}}{2g_{j,1}^2})z_{j,1}^2 \leq -(c_{j,10} - \frac{g_{j,1}^d}{2\underline{g}_{j,1}^2})z_{j,1}^2$, by choosing $c_{j,10}$ such that $c_{j,10}^* \triangleq c_{j,10} - \frac{g_{j,1}^d}{2\underline{g}_{j,1}^2} > 0$, we have the following inequality

$$\dot{V}_{j,1} < z_{j,1}z_{j,2} - c_{j,10}^*z_{j,1}^2 - \frac{\sigma_{j,1}\|\tilde{W}_{j,1}\|^2}{2} + \frac{\sigma_{j,1}\|W_{j,1}^*\|^2}{2} + \frac{\epsilon_{j,1}^{*2}}{4c_{j,11}} \quad (17)$$

Step $i_j$ ($2 \leq i_j \leq \rho_j - 1$): Define $z_{j,i_j} = x_{j,i_j} - \alpha_{j,i_j-1}$. Its derivative is

$$\dot{z}_{j,i_j} = f_{j,i_j}(\bar{x}_{ji}, x_{j,i_j+1}) - \dot{\alpha}_{j,i_j-1} \quad (18)$$

where $\bar{x}_{ji} := [\bar{x}_{1,(i_j-\varrho_{j1})}^T, \cdots, \bar{x}_{m,(i_j-\varrho_{jm})}^T]^T$.

Define

$$\nu_{j,i_j} = -\dot{\alpha}_{j,i_j-1} + k_{j,i_j}z_{j,i_j} \quad (19)$$

where $k_{j,i_j} > 0$ is a constant. Similar to Step 1, we have $\frac{\partial[f_{j,i}(\bar{x}_{ji}, x_{j,i_j+1}) + \nu_{j,i}]}{\partial x_{j,i_j+1}} > \underline{g}_{j,i} > 0$. By viewing $x_{j,i_j+1}$ as a virtual control to stabilize the $(z_{j,1}, \cdots, z_{j,i_j})$-sub-subsystem of the $j$th subsystem, according to Lemma 1, there exists a desired feedback control $x_{j,i_j+1} = \alpha_{j,i_j}^*(\bar{x}_{j,i_j}, \nu_{j,i_j})$ such that

$$f_{j,i_j}(\bar{x}_{ji}, \alpha_{j,i_j}^*) + \nu_{j,i_j} = 0 \quad (20)$$

Using mean value theorem (Apostol, 1963), there exists $\lambda_{j,i_j}$ ($0 < \lambda_{j,i_j} < 1$) such that

$$f_{j,i_j}(\bar{x}_{ji}, x_{j,i_j+1}) = f_{j,i_j}(\bar{x}_{ji}, \alpha_{j,i_j}^*) + g_{\lambda_{j,i_j}}(x_{j,i_j+1} - \alpha_{j,i_j}^*) \quad (21)$$

where $g_{\lambda_{j,i_j}} := g_{j,i}(\bar{x}_{ji}, x_{(i+1)\lambda_{j,i_j}})$ with $x_{(i+1)\lambda_{j,i_j}} = \lambda_{j,i_j}x_{j,i_j+1} + (1 - \lambda_{j,i_j})\alpha_{j,i_j}^*$. Note that Assumptions 1-2 on $g_{j,i}(\bar{x}_{ji}, x_{j,i_j+1})$ is still valid for $g_{\lambda_{j,i_j}}$. Combining (18)-(21) yields

$$\dot{z}_{j,i_j} = -k_{j,i_j}z_{j,i_j} + g_{\lambda_{j,i_j}}(x_{j,i_j+1} - \alpha_{j,i_j}^*) \quad (22)$$

Since $\alpha_{j,i_j-1}$ is a function of $\bar{x}_{j(i_j-1)}$, $x_d$ and $\hat{W}_{j,1}, \cdots, \hat{W}_{j,i_j-1}$, $\dot{\alpha}_{j,i_j-1}$ is given by

$$\dot{\alpha}_{j,i_j-1} = \sum_{l=1}^{m}\sum_{k=1}^{i_j-1-\varrho_{jl}}\frac{\partial\alpha_{j,i_j-1}}{\partial x_{l,k}}(g_{l,k}x_{l,k+1} + f_{l,k}) + \phi_{j,i_j-1} \quad (23)$$

where $\phi_{j,i_j-1} = \frac{\partial\alpha_{j,i_j-1}}{\partial x_d}\dot{x}_d + \sum_{k=1}^{i_j-1}\frac{\partial\alpha_{j,i_j-1}}{\partial\hat{W}_{j,k}}\dot{\hat{W}}_{j,k}$ is computable.

By employing an RBF neural network $W_{j,i_j}^T S_{j,i_j}(Z_{j,i_j})$ to approximate $\alpha_{j,i_j}^*(\bar{x}_{ji}, \nu_{j,i_j})$, where $Z_{j,i_j} = [\bar{x}_{ji}^T, \frac{\partial\alpha_{j,i_j-1}}{\partial x_{j,1}}, \cdots, \frac{\partial\alpha_{j,i_j-1}}{\partial x_{j,i_j-1}}, \phi_{j,i_j-1}, z_{j,i_j}]^T \in \Omega_{j,i_j}$, $\alpha_{j,i_j}^*$ can be expressed as

$$\alpha_{j,i_j}^* = W_{j,i_j}^{*\,T}S_{j,i_j}(Z_{j,i_j}) + \epsilon_{j,i_j} \quad (24)$$

Let $\hat{W}_{j,i_j}$ be the estimate of $W_{j,i_j}^*$. Define $z_{j,i_j+1} = x_{j,i_j+1} - \alpha_{j,i_j}$ and let

$$\alpha_{j,i_j} = -z_{j,i_j-1} - c_{j,i_j}z_{j,i_j} + \hat{W}_{j,i_j}^T S_{j,i_j}(Z_{j,i_j}) \quad (25)$$

where $c_{j,i_j}$ is a positive constant to be specified later. Then, equation (45) becomes

$$\dot{z}_{j,i_j} = -k_{j,i_j}z_{j,i_j} + g_{\lambda_{j,i_j}}[z_{j,i_j+1} - z_{j,i_j-1} - c_{j,i_j}z_{j,i_j} + \tilde{W}_{j,i_j}^T S_{j,i_j}(Z_{j,i_j}) - \epsilon_{j,i_j}] \quad (26)$$

Consider the Lyapunov function candidate

$$V_{j,i_j} = V_{j,i_j-1} + \frac{1}{2g_{j,i_j}}z_{j,i_j}^2 + \frac{1}{2}\tilde{W}_{j,i_j}^T\Gamma_{j,i_j}^{-1}\tilde{W}_{j,i_j} \quad (27)$$

Consider the adaptation law for $\hat{W}_{j,i_j}$ as

$$\dot{\hat{W}}_{j,i_j} = \Gamma_{W_{j,i_j}}[-S_{j,i_j}(Z_{j,i_j})z_{j,i_j} - \sigma_{j,i_j}\hat{W}_{j,i_j}] \quad (28)$$

where $\sigma_{j,i_j} > 0$ and $\Gamma_{W_{j,i_j}} = \Gamma_{W_{j,i_j}}^T > 0$ are design constants.

Let $c_{j,i_j} = c_{j,i_j0} + c_{j,i_j1}$, where $c_{j,i_j0}$ and $c_{j,i_j1} > 0$. By using (26) and (28), and with some completion of squares and straightforward derivation similar to those employed in Step 1, the derivative of $V_{j,i_j}$ becomes

$$\dot{V}_{j,i_j} < z_{j,i_j}z_{j,i_j+1} - \sum_{k=1}^{i_j}c_{j,k0}^*z_{j,k}^2 - \sum_{k=1}^{i_j}\frac{\sigma_{j,k}\|\tilde{W}_{j,k}\|^2}{2} + \sum_{k=1}^{i_j}\frac{\sigma_{j,k}\|W_{j,k}^*\|^2}{2} + \sum_{k=1}^{i_j}\frac{\epsilon_{j,k}^{*2}}{4c_{j,k1}} \quad (29)$$

where $c_{j,i_j0}^* \triangleq c_{j,i_j0} - \frac{g_{j,i_j}^d}{2\underline{g}_{j,i_j}^2} > 0$.

Step $p$ ($p = \rho_j - 1, \rho_j$): The derivative of $z_{j,p} = x_{j,p} - \alpha_{j,p-1}$ is

$$\dot{z}_{j,\rho_j-1} = f_{j,\rho_j-1}(\bar{x}_{j(\rho_j-1)})$$

$$+g_{j,\rho_j-1}(\bar{x}_{j(\rho_j-1)})x_{j,\rho_j} - \dot{\alpha}_{j,\rho_j-2} \quad (30)$$

$$\dot{z}_{j,\rho_j} = f_{j,\rho_j}(X, u_1, \cdots, u_{j-1}) + g_{j,\rho_j}(\bar{x}_{j(\rho_j-1)})\alpha_{j,p} - \dot{\alpha}_{j,\rho_j-1} \quad (31)$$

where $\bar{x}_{j(\rho_j-1)} := [\bar{x}_{1,\rho_1-1}^T, \cdots, \bar{x}_{m,\rho_m-1}^T]^T$. To stabilize the $(z_{j,1}, \cdots, z_{j,p})$-sub-subsystem of the $j$th subsystem, there exists desired virtual feedback controls

$$\alpha_{j,p}^* = -z_{j,p-1} - c_{j,p}z_{j,p} - \frac{1}{g_{j,p}}(f_{j,p} - \dot{\alpha}_{j,p-1}) \quad (32)$$

where $c_{j,p} > 0$ is a design constant to be specified later, $\alpha_{j,p-1}$ is a function of $\bar{x}_{j(\rho_1-1)}$, $x_d$ and $\hat{W}_{j,1}, \cdots, \hat{W}_{j,p-1}$. Therefore, $\dot{\alpha}_{j,p-1}$ can be expressed as

$$\dot{\alpha}_{j,p-1} = \sum_{l=1}^{m}\sum_{k=1}^{\rho_l-1} \frac{\partial \alpha_{j,p-1}}{\partial x_{l,k}}(g_{l,k}x_{l,k+1} + f_{l,k}) + \phi_{j,p-1}$$

where $\phi_{j,p-1} = \frac{\partial \alpha_{j,p-1}}{\partial x_d}\dot{x}_d + \sum_{k=1}^{p-1} \frac{\partial \alpha_{j,p-1}}{\partial \hat{W}_{j,k}}\dot{\hat{W}}_{j,k}$ is computable.

By employing an RBF neural network $W_{j,p}^T S_{j,p}(Z_{j,p})$ to approximate $\frac{1}{g_{j,p}}(f_{j,p} - \dot{\alpha}_{j,p-1})$, where

$$Z_{j,\rho_j-1} = \left[\bar{x}_{j(\rho_j-1)}^T, (\frac{\partial \alpha_{j,\rho_j-1}}{\partial \bar{x}_{j(\rho_j-1)}})^T, \phi_{j,\rho_j-1}\right]^T \in \Omega_{j,\rho_j-1} \quad (33)$$

and

$$Z_{j,\rho_j} = \left[X^T, u_1, \cdots, u_{j-1}, (\frac{\partial \alpha_{j,\rho_j-1}}{\partial \bar{x}_{j(\rho_j-1)}})^T, \phi_{j,\rho_j-1}\right]^T \in \Omega_{j,\rho_j} \quad (34)$$

Then, $\alpha_{j,p}^*$ can be expressed

$$\alpha_{j,p}^* = -z_{j,p-1} - c_{j,p}z_{j,p} + W_{j,p}^{*T} S_{j,p}(Z_{j,p}) - \epsilon_{j,p}$$

Let

$$\alpha_{j,p} = -z_{j,p-1} - c_{j,p}z_{j,p} + \hat{W}_{j,p}^T S_{j,p}(Z_{j,p}) \quad (35)$$

Choose the practical control $u_j = \alpha_{j,\rho_j}$. Then, we have

$$\dot{z}_{j,p} = g_{j,p}[-z_{j,p-1} - c_{j,p}z_{j,p} + \tilde{W}_{j,p}^T S_{j,p}(Z_{j,p}) + \epsilon_{j,p}] \quad (36)$$

Consider the Lyapunov function candidate

$$V_{j,p} = V_{j,p-1} + \frac{1}{2g_{j,p}}z_{j,p}^2 + \frac{1}{2}\tilde{W}_{j,p}^T \Gamma_{j,p}^{-1}\tilde{W}_{j,p} \quad (37)$$

Consider the adaptation law for $\hat{W}_{j,p}$ as

$$\dot{\hat{W}}_{j,p} = \Gamma_{W_{j,p}}[S_{j,p}(Z_{j,p})z_{j,p} - \sigma_{j,p}\hat{W}_{j,p}] \quad (38)$$

where $\sigma_{j,p} > 0$ and $\Gamma_{W_{j,p}} = \Gamma_{W_{j,p}}^T > 0$ are design constants.

Let $c_{j,p} = c_{j,p0} + c_{j,p1}$, where $c_{j,p0}$ and $c_{j,p1} > 0$. By using (36) and (38), and with some completion of squares and straightforward derivation similar to those employed in Step 1, the derivative of $V_{j,p}$ becomes

$$\dot{V}_{j,p} < -\sum_{k=1}^{p} c_{j,k0}^* z_{j,k}^2 - \sum_{k=1}^{p} \frac{\sigma_{j,k}\|\tilde{W}_{j,k}\|^2}{2} + \sum_{k=1}^{p} \frac{\sigma_{j,k}\|W_{j,k}^*\|^2}{2} + \sum_{k=1}^{p} \frac{\epsilon_{j,k}^{*2}}{4c_{j,k1}} \quad (39)$$

where $c_{j,p0}^* \triangleq c_{j,p0} - \frac{g_{j,p}^d}{2\underline{g}_{j,p}^2} > 0$.

Let

$$\delta_j \triangleq \sum_{k=1}^{\rho_j} \frac{\sigma_{j,k}\|W_{j,k}^*\|^2}{2} + \sum_{k=1}^{\rho_j} \frac{\epsilon_{j,k}^{*2}}{4c_{j,k1}} \quad (40)$$

If we choose $c_{j,k0}^*$ such that $c_{j,k0}^* \geq \frac{\gamma_j}{2\underline{g}_{j,k}}$, i.e., $c_{j,k0} > \frac{\gamma_j}{2\underline{g}_{j,k}} + \frac{g_{j,k}^d}{2\underline{g}_{j,k}^2}$, $k = 1, \cdots, \rho_j$, where $\gamma_j$ is a positive constant, and choose $\sigma_{j,k}$ and $\Gamma_{j,k}$ such that $\sigma_{j,k} \geq \gamma_j \lambda_{max}\{\Gamma_{j,k}^{-1}\}$, $k = 1, \cdots, n$, then from (39) we have the following inequality

$$\dot{V}_{j,\rho_j} \leq -\sum_{k=1}^{\rho_j} \frac{\gamma_j}{2\underline{g}_{j,k}} z_{j,k}^2 - \sum_{k=1}^{\rho_j} \frac{\gamma_j \tilde{W}_{j,k}^T \Gamma_{j,k}^{-1}\tilde{W}_{j,k}}{2} + \delta_j \leq -\gamma_j V_{j,\rho_j} + \delta_j \quad (41)$$

Let $V = \sum_{j=1}^m V_{j,\rho_j}$. Its derivative is

$$\dot{V} = \sum_{j=1}^{m} \dot{V}_{j,\rho_j} < \sum_{j=1}^{m}(-\gamma_j V_{j,\rho_j} + \delta_j) < -\gamma V + \delta \quad (42)$$

where $\gamma = \min\{\gamma_1, \cdots, \gamma_m\}$ and $\delta = \sum_{j=1}^m \delta_j$ are positive constants.

*Theorem 1.* Consider the closed-loop system consisting of the plant (1), the reference model (2), the controllers (35) and the NN weight updating laws (14)(28)(38) for all the subsystems ($j = 1, \cdots, m$). Assume there exist sufficiently large compact sets $\Omega_{j,i_j}$ such that $Z_{j,i_j} \in \Omega_{j,i_j}, i_j = 1, \cdots, \rho_j, j = 1, \cdots, m$ for all $t \geq 0$. Then, for bounded initial conditions, all signals in the closed-loop system remain bounded, and the output tracking error $y_j(t) - y_{dj}(t)$ ($j = 1, \cdots, m$) converges to a small neighborhood around zero by appropriately choosing design parameters.

**Proof:** (i) From (41), we have that all $z_{j,i_j}$ and $\hat{W}_{j,i_j}$ ($j = 1, \cdots, m$, $i_j = 1, \cdots, \rho_j$) are uniformly ultimately bounded. Since $z_{j,1} = x_{j,1} - x_{dj}$ and $x_d$ are bounded, we have that the first state

$x_{j,1}$ of all the subsystems ($j = 1, \cdots, m$) remain bounded. To prove $x_{j,2} = z_{j,2} + \alpha_{j,1}$, $j = 1, \cdots, m$ remain bounded, we need to show that $\alpha_{j,1}$ is bounded. Note that because $\alpha_{j,1}$ in (10) is a function of $z_{j,1}$, $Z_{j,1}$ and $\hat{W}_{j,1}$, i.e., a function of $\bar{x}_{1,(1-\varrho_{j1})}$, $\cdots$, $\bar{x}_{m,(1-\varrho_{jm})}$, $x_d$ and $\hat{W}_{j,1}$, we may not conclude the boundedness of $\alpha_{j,1}$ immediately in case some of the states $\bar{x}_{l,(1-\varrho_{jl})}$, $l = 1, \cdots, m$, $l \neq j$ have not been proven bounded yet. We can only prove the boundedness of some $\alpha_{q,1}$ in those $q$th subsystem(s) first, where $q$ is (or $q$'s are) determined by

$$(q, i_q) = \arg \max_{(j,i_j) \in U_r} \{\rho_j - i_j\} \qquad (43)$$

where $i_q = i_j = 2, j = 1, \cdots, m$ in this step, and

$$U_r \triangleq \{(j, i_j) \mid x_{j,i_j} \text{ not proven bounded yet}\} (44)$$

By choosing the $q$th subsystem(s) in this way, we can pick only those $\alpha_{q,1}$ which are functions of bounded variables $x_d$, $\hat{W}_{q,1}$, and $x_{q,1}$. Accordingly, we conclude that $\alpha_{q,1}$ are bounded. From the boundedness of $z_{j,2}$, and $\alpha_{q,1}$, we obtain the boundedness of $x_{q,2}$. Next, we will proceed to prove the boundedness of $\alpha_{q,i_q-1}$, and subsequently $x_{q,i_q}$ by definition, where $(q, i_q)$ are determined again by (43). In fact, according to the block-triangular structure of system (1), checking condition (43) indicates which virtual control $\alpha_{q,i_q-1}$ has the minimum variable embedding from other subsystems and thus is a function of state variables which have all been proven bounded already. Accordingly, $\alpha_{q,i_q-1}$ can be proven bounded next. In this way, we can prove one by one that all $\alpha_{j,i_j-1}$ in (25) and subsequently, $x_{j,i_j}$, $j = 1, \cdots, m$, $i_j = 2, \cdots, \rho_j$ remain bounded. Thus, the states $X$ of the interconnected MIMO system (1) remain bounded.

Using (35), we conclude that $u_1$ is bounded because it is a function of bounded variables $X$, $x_d$ and $\hat{W}_{1,1}$, $\cdots$, $\hat{W}_{1,\rho_1}$. Similarly, we have that controls $u_j$, $j = 2, \cdots, m$ are also bounded. Thus, all the signals in the closed-loop system remain bounded.

Let $\varpi_j \triangleq \delta_j/\gamma_j > 0$, then (37) satisfies

$$0 \leq V_{j,\rho_j}(t) < \varpi_j + (V_{j,\rho_j}(0) - \varpi_j)\exp(-\gamma_j t) (45)$$

From (45), we have

$$\sum_{k=1}^{\rho_j} \frac{1}{2g_{j,k}} z_{j,k}^2 < \varpi_j + (V_{j,\rho_j}(0) - \varpi_j)\exp(-\gamma_j t)$$
$$< \varpi_j + V_{j,\rho_j}(0)\exp(-\gamma_j t) \qquad (46)$$

Let $\bar{g}_j = \max_{1 \leq i_j \leq \rho_j}\{\bar{g}_{j,i_j}\}$. Then, we have

$$\frac{1}{2\bar{g}_j} \sum_{k=1}^{\rho_j} z_{j,k}^2 \leq \sum_{k=1}^{\rho_j} \frac{1}{2g_{j,k}} z_{j,k}^2 < \varpi_j + V_{j,\rho_j}(0)\exp(-\gamma_j t)$$

that is,

$$\sum_{k=1}^{\rho_j} z_{j,k}^2 < 2\bar{g}_j\varpi_j + 2\bar{g}_j V_{j,\rho_j}(0)\exp(-\gamma_j t) \quad (47)$$

which implies that given $\tau_j > \sqrt{2\bar{g}_j\varpi_j}$, there exists $T$ such that for all $t \geq T$, the tracking error satisfies

$$|z_{j,1}(t)| = |x_{j,1}(t) - x_{d1}(t)| = |y_j(t) - y_{dj}(t)| < \tau_j$$

where $\tau_j$ is the size of a small residual set which depends on the NN approximation error $\epsilon_{j,i_j}$ and controller parameters $c_{j,i_j}$, $\sigma_{j,i_j}$ and $\Gamma_{j,i_j}$. It is easily seen that the increase in the control gain $c_{j,i_j}$ and NN node number $l_{j,i_j}$, and the decrease in $\sigma_{j,i_j}$ will result in a better tracking performance. ◇

## 4. CONCLUSION

An adaptive neural control scheme has been proposed for a class of uncertain MIMO nonlinear systems in block-triangular pure-feedback form. The proposed adaptive neural control scheme can be applied to a large class of uncertain MIMO nonlinear systems.

## 5. REFERENCES


Apostol, T. M. (1963). *Mathematical Analysis*, MA: Addison-Wesley, Reading.

Ge, S. S., C.C. Hang, T.H. Lee and T. Zhang (2001). *Stable Adaptive Neural Network Control*, Kluwer Academic Publishers, Norwell, USA, in print.

Ge, S. S. and C. Wang (2001). Adaptive NN Control of Nonlinear Pure-Feedback System, accepted by *Automatica*.

Krstič, M., I. Kanellakopoulos and P. Kokotovič (1995). *Nonlinear and Adaptive Control Design*, John Wiley, New York.

Lewis, F. L., S. Jagannathan and A. Yeildirek (1999). *Neural Network Control of Robot Manipulators and Nonlinear Systems*, Taylor & Francis, London.

Narendra, K. S. and Parthasarathy, K. (1990) Identification and control of dynamic systems using neural networks, *IEEE Trans. Neural Networks*, vol. 1, no. 1, pp. 4-27.

Sanner, R. M. and J. E. Slotine (1992). Gaussian networks for direct adaptive control, *IEEE Trans. Neural Networks*, vol. 3, no. 6, pp. 837-863.

IFAC Publications
www.elsevier.com/locate/ifac

# LIFT CONTROLLERS BY EVOLUTIONARY ALGORITHMS

**Agostinho Rosa, Nuno Lopes, Pedro Silva**

*Laseeb-ISR-IST Technical University of Lisbon, Portugal*

*Abstract*: High-level controllers of lifts taking into account the continuous traffic flow is still an open problem especially as real-time optimization problem. The objective of this work is to show that Evolutionary Algorithms (EA) can be used to solve successfully the practical and specific problem of evolving lift controllers. Comparative results are presented. *Copyright © 2001 IFAC*

*Keywords*: Genetic Algorithms, Adaptive Control, Simulation, Heuristic, Logical Control.

## 1. INTRODUCTION

The evolution of lift controllers is still an actual problem and goes to the demand of modern societies, where the efficient time management is of paramount importance. In fact, there are today increasing trends to reduce idle periods specially related to commuting and transport in general. The good performance of a lift system in a several story building is highly desirable and is very sensitive and also a key contribution to the objective or subjective perception of quality of life. A solution is proposed for this problem using Evolutionary Algorithms (EA), a recent method already successful applied to a multitude of optimization and search problems of large domain in the fields of engineering. The use of EAs in these kinds of problems has many advantages, where it is difficult or even impossible to find a suitable search strategy.
Section 2 presents the proposed simulation model. Section 3 describes the EA used in the simulation and also all intervening parameters. Section 4 describes a local search technical with the objective of promoting the chromosome repair procedure. Section 5 presents the test results and in Section 6 the final conclusions are presented.

## 2. SIMULATION MODEL

The application to be developed to test the EA capabilities is a real-time simulator with high degree of realism the working characteristics of a set of lifts and the responses to external requests. 4 communicating components compose the simulator:

- Floors;
- Lifts;
- Controllers;
- EA.

Figure 1 shows the relations between the four components of the simulation system, it is described below.
In the rest of the text, the expression *internal requests* is used to designate requests given by the users when entering the lift, and *external requests* refers to request for going up or down issued by users waiting in the different floors.

The floor assumes the role of external request generator, with the only responsibility of generating arriving users in different time slots. In order to perform a more realistic simulation, a Poisson distribution models the number of requests issued during a fixed time slots, where the time interval between requests follows an exponential distribution. When a floor generates a new request, it notifies the event to the controller. The controller will store this information, making it available to the EA. It is worth noting that each floor has its own generator of requests, independent of others floors. In the simulator developed, the mean time between requests of 3 minutes (for each floor).

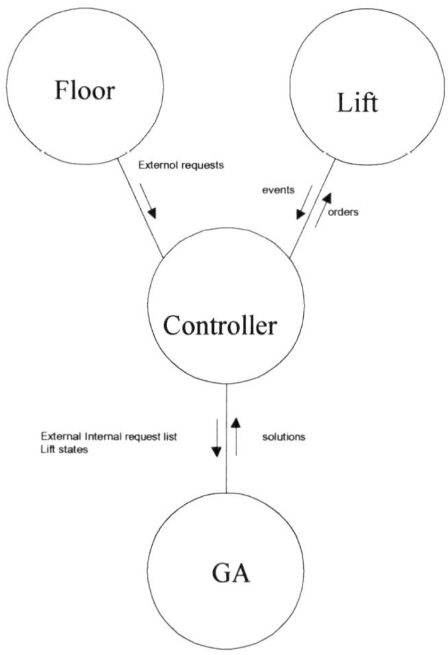


Figure 1 – Simulator Structure: components and relations.

The lift will move through different floors according to the commands of the controller. The lift will report to the controller the important events like floor arrivals and request inside the lift. Besides this reporting function the lift is passive, the actions signals are provided by the controller only. The simulator takes into account the door opening and closing times, the acceleration and damping time when approaching each floor and the stop time, the time users take to enter and leave the lift. The opening and closing times are assumed to be fixed (5 seconds) and the same for all floors. The stop time is a normal distributed random variable N(1,10) with unit standard deviation. The travel time between each floor is also fixed (10 seconds) and the same for all floors. The acceleration and damping time although depends on the number of floors in the travel between stops are assumed to be fixed and included in the travel time between two consecutive floors.

The EA component implements the genetic algorithm. The EA works as “online” and “real-time”, it starts at the beginning of the simulation and only stops when the simulation finishes. The genetic algorithm needs the information about the list of all internal and external pending requests and the current state of each lift from the controller in order to evaluate the fitness of each chromosome in the population. Based on this information the EA evolves a population of solutions for the current situation. Every 5 generations the EA will send to the controller the best solution found. The controller will use this solution to complete control the lifts. As said above the EA works permanently giving a new solution periodically and according to the demands, this means that the EA provides a dynamic control schedule adapted to changes on the demands.

As figure 1 suggests, the controller is the main and central component of the simulator, its role includes:

- Registering all external requests sent by the lifts and sends this information to the EA.
- Send to the EA the essential information about the current status of the lifts. This information arrives to the controller through a messaging system.
- Control the lifts according to the solution provided by the EA

## 3. Description of the EA

The evolutionary genetic algorithm, (Goldberg 1989) used in this problem evolves a fixed size population of individual solutions to the demands. The solution is a sequence of control orders for the lifts. The solutions have to fulfill the requests completely.

### 3.1. Coding

The chromosome coding represents the sequence of control orders. For this problem a multi-circular chromosome is chosen as shown in figure 2.

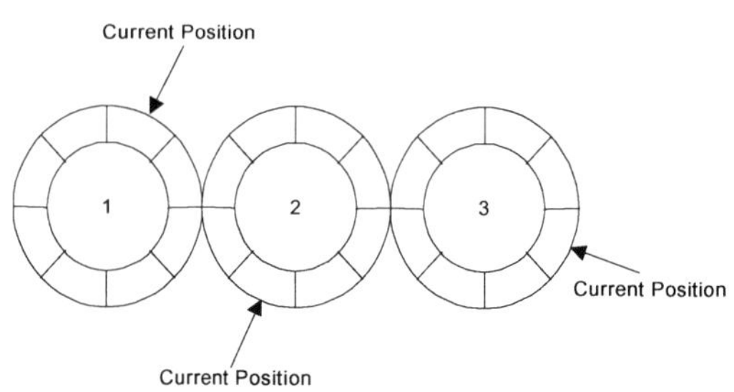


Figure 2 – Multi-Circular Chromosomes

A multi-circular chromosome is composed by several *ring components*. Each ring component is associated to one lift and represents the sequence of floor stops for the corresponding lift.
The current position is indicated by a pointer, which can be changed along the simulation. The gene set by the pointer is the next floor stop.
Chromosome dimension depends on two parameters of the problem. The number of lifts determines the number of rings components and the number of floors set the number of alleles (plus one for the idle state) of the genes and also the number of genes in each ring component.
In order to find good and complete solutions, the fitness function evaluation has to scan through a sufficient number of genes of each ring component. On the other hand it is not necessary usually to scan all the genes of the ring component in order to fulfill all requests, more important though, the list of requests is likely to change on the next stop. For completeness only each ring component contains a number of genes equal to the number of floors, allowing a complete sequence of stopping on all floors. So the total number of genes of the chromosome is given by:

$$nGenes = nElevators * nFloors$$

The idle allele is set to *–1,* which will immobilize the lift and stays idle until this allele is changed. Figure 3 shows a ring component with the sequence of floors the lift should stop.

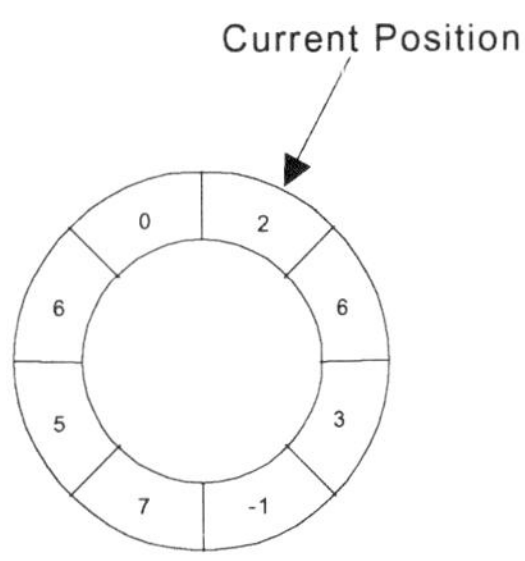


Figure 3 – Schema of a component of a chromosome. In this case, the sequence of stops for the associated lift is: 2,6,3.

At each stop the EA updates all the chromosomes in the population, by stepping forward (clockwise) all the ring component's pointers.

A new allele will be generated randomly for the old gene. Figure 4 exemplifies the updating process of a ring component when a lift effects a stop.

## *3.2. EA Parameters*

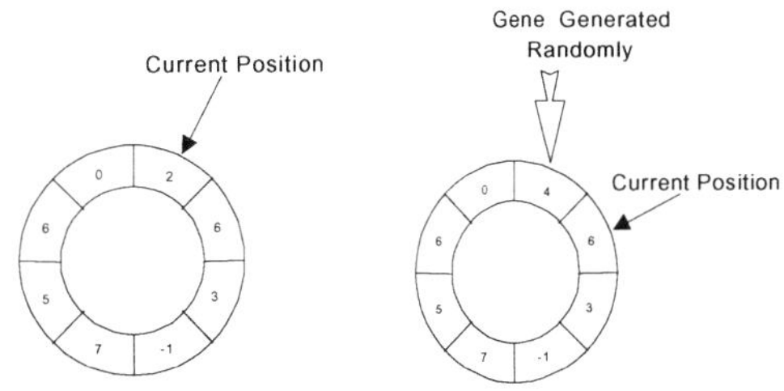


Figure 4 – Ring Component updating process when a lift stops, by stepping forward a gene

A steady-state genetic algorithm is used in this simulation, (Davis 1996). This algorithm substitutes in each generation, only a percentage of the current population a temporary population is created on each generation and added to the current population, where the worse elements will be eliminated, resulting a new population of the same size. The percentage of the substitution defined for this purpose is 25%. A population size of 250 elements was chosen empirirically to provide a good diversity in the population characteristics. The population is randomly initialized.

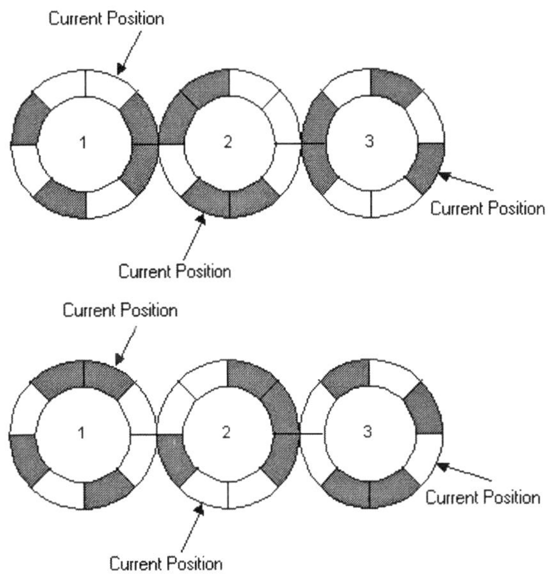


Figure 5 – Schema of uniform crossover for multi-circular chromosomes.

The selection operator is the six ways tournament, because this operator does not require fitness scaling and it is also more efficient and presents always a good performance in most problems. The mutation operator is the random bit flip.
This mutation operator introduces higher diversity when compared to for example to random bit change. The mutation probability is 5%.

The crossover operator used is the uniform crossover. Figure 5 shows the working structure of the uniform crossover operator. A random mask is generated, which corresponds to the gray genes of the first chromosome. The combination of the gray genes in the parent chromosomes results in the first child, the second child is obtained by the combination of the white genes. The crossover probability used is 95%.

### 3.3. Fitness Function

The fitness function for this problem takes into account two fundamental aspects in the chromosome evaluation: the number of internal or external requests satisfied and the total distance traveled by all lifts. The fulfillment of the internal or external requests has maximum priority. But this situation will results in a competition between the lifts, all the lifts trying to satisfy all the same requests, even a single one. In order to avoid this situation and promotes cooperation between the lifts, the inclusion of the total distance traveled in the fitness function is crucial.

The evaluation of the travel path of each lift is done through the analysis of the sequence of genes until the idle allele is found.

The fitness function expression for the chromos some is given by:

***fitness = Totaldistance +***
***(nRequests-nSatisfied Requests)*factor +***
***penalties***

The Totaldistance is equal to the some of the path of all lifts associated to the number of steps needed for the travel:

***Totaldistance=Σ Pathdistance + nsteps***

The objective of the problem is to minimize this fitness function. By minimizing the distance also minimizes the travel time.

The *factor* term is used as relative weighting factor between request satisfied and traveled distance. This value is set to

***factor = nfloors*4***

The last term of the fitness function expression is the penalties. The fitness function gives penalties in two situations. The first situation is related to the possibility that the next stop by a lift (gene 0 of the ring component) can be in the opposite direction in relation to the current travel direction. This direction inversion is not desired, since it gives the feeling of unnecessary waste of time and it is penalized by 5 units. The second situation is related to the number of *illegal genes that each chromosome holds*. An illegal gene means a gene with an allele not in the internal or external request list, leading to a useless stop of the lift; 5 units also penalize this situation.

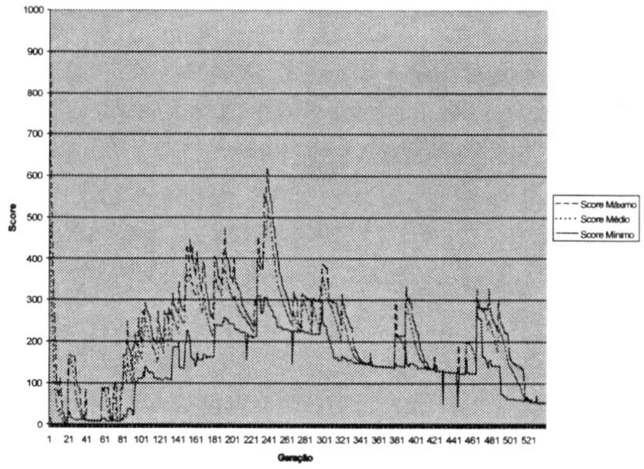


Figure 6 – Test Results with 3 lifts and 20 floors.

### 3.4. Simulations

The simulation results of one of the test are presented in figure 6, where an example of 3 lifts and 20 floors is chosen. The simulation time, equivalent to a real-time of 10 minutes is set for each simulation. The external requests were generated only during the first 3 minutes, in order to allow the observation the EA convergence process.

By analyzing the graphic one can observe the existence of several peeks in the curves. These peaks correspond to the arrival of external requests and the EA cannot provide an immediate reply. After a few generations the minimum curve goes down steadily.

Without further requests the curves has a monotonous descending trend.

## 4. LOCAL SEARCH REPAIR

For most real world specific problem it is possible to include additional search strategy taking into account the available *a priori* knowledge. In this particular case of lift control and by using the genetic algorithm described above it is possible to occur frequently in the population chromosomes with incoherent and illegal genes, reducing the efficiency of the evolution process as consequence. An example of an anomalous situation is when a chromosome contains in one of its ring component *adjacent genes* with the same allele value, which means that the lift will stop two times on the same floor. This type of situations can be minimized or even eliminated by using a local repair technical

called *gene repair*. The repair strategy consists off the implementation of a set of operators (*workers)* responsible to correct the ill conditions. The *workers* are independent of each other and are in charge of a specific type of repair and not interfering in the jog of other workers.
The workers perform their job before the fitness function evaluation. This strategy improved the quality of the population leading to a better performance of the EA.
For the lift control problem two workers is implemented, (Back 1996).

### *4.1. Worker for illegal genes*

The probability of illegal genes in the chromosome is high. A worker for repairing this problem implements the following set of rules:

- Higher priority for the first genes
- The number of genes to repair is always below or equal to the number of requests (external or internal) at each moment, starting always by the first genes.
- The illegal gene is substituted by a randomly selected allele from the request list.
- If one gene has the idle value of *–1*, there is no need to repair the remaining genes.

### *4.2. Worker for adjacent genes*

Another ill condition is the replication of adjacent genes. The solution to this situation is achieved through dispersing the adjacent genes to other positions of the ring component of the chromosome.

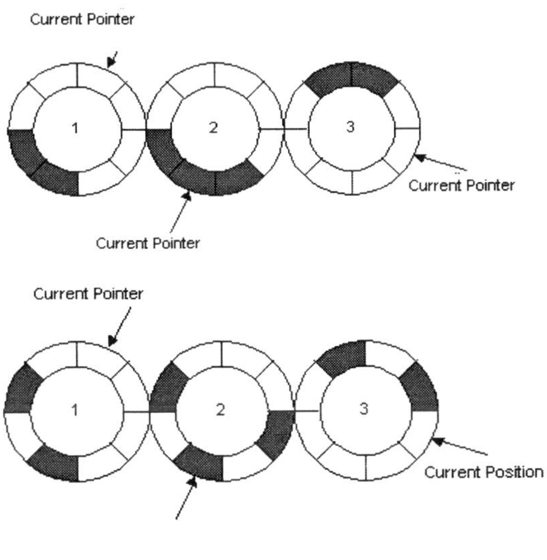


Figure 7 – Adjacent genes represented in gray in a multi-circular chromosome.

In figure 7 an example is shown where the adjacent genes are illustrated in gray.

## 5. COMPARATIVE STUDY

The objective of this section is the presentation of a comparative study of the EA performance with and without the repair process.
The tests were repeated 10 times for each situation. The results show clearly the improve in performance when the repair process is active. The test is done using a simulator with 2 lifts and 10 floors and run for a period of 10 minutes; as with previous simulations the requests arrival is off after the first 3 minutes.

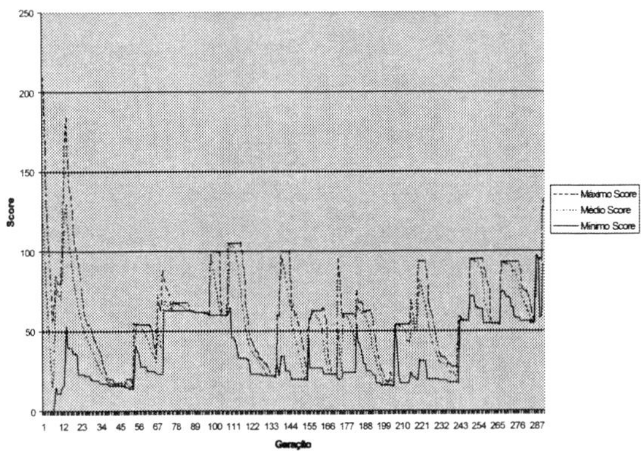


Figure 8 – Evolution of the minimum, mean and maximum *score* of the population without local search.

Figures 8 e 9 presents the obtained results for each case.
The graph of figure 8 shows the evolution of the minimum, mean and maximum *score* of the chromosomes in the situation without local search.

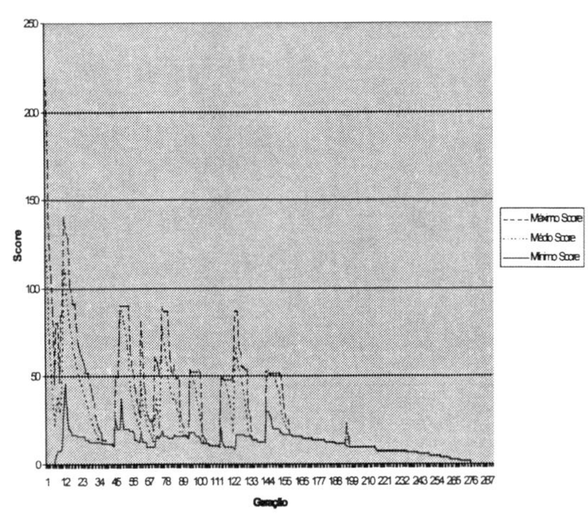


Figure 9 – Evolution of the minimum, mean and maximum *score* of the population with local search.

The graph of figure 9 shows the situation with local search.
It is clearly seen that the graph with local search is always below the one without and it converges faster to a lower minimum.

## 6. Conclusions

The real word problem under consideration in this work is the real-time adaptive control of a set of lifts in a multi story building is relatively complex dynamic problem with tight time restrictions. Therefore the search algorithms for the control solution should be fast and responsive the ever changing conditions and optimizing the performance by minimizing the mean waiting or travel time. The local search repair strategy showed added value by improving clearly the original optimization algorithms.

## References

Back T., *Evolutionary Algorithms in Theory and Practice*, Oxford University Press, 1996.

Davis L., *Handbook of Genetic Algorithms.* Van Nostrand Reinhold, 1994.

Goldberg D., *Genetic Algorithms in Search, Optimization and Machine Learning*, Addison-Wesley, 1989.

IFAC Publications
www.elsevier.com/locate/ifac

# PREDICTIVE FUNCTION CONTROL BASED ON MULTI-MODEL FOR pH PLANT

**Zhi-huan Zhang** , Shu-qing Wang***

*The National Lab. of Industrial Control Technology*
*Zhejiang University 310027, P.R.China**
*The Collage of Information Science & Engineering ,*
*Ningbo Univ. 315211, P.R.China***

Abstract: A multi-model structure for predictive function control has been presented for a pH neutralisation process. The nonlinear pH process is partitioned into several local operating region. Within each region, a local predictive function controller is constructed based on the corresponding local linear model, their outputs are combined to form a global control action by using their membership functions. i.e. the global predictive function controller (GC-PFC) consists of a set of local model predictive function controller (LM-PFC) by weighting. The GC-PFC has been applied to quality control of a pH nonlinear process. Realistic simulation shows that GC-PFC is an attractive option for nonlinear MPC.*Copyright © 2001 IFAC*

Key words: Nonlinear predictive function control, Multi-model, operating regime, the interpolation function, pH nonlinear process.

## 1. INTRODUCTION

Predictive Function Control (PFC) has been presented in the base of Model Predictive Control (MPC) and obtained widespread use in the elastic industrial robot (H.B.Kuntze *et al.,* 1986)a tracking turret(D.Cuadrado *et al.,* 1991)weapon system guidance and control (D.Cuadrado, P.Guerchet and S.Abu Ei Ata Doss, 1991) rolling mill (J.M.Compas, P. D- ecarreau, G.Lanquetin, J.L.Estival, N.Fulget, R.Martin and J.Richalet, 1994). Even though PFC was originally developed to meet the needs of fast、highly accurate track system. Recently, PFC-controller begin to the application of slowly chemical industries process, such as batch reactors, pH neutralisation process and heat exchangers. Due to enthalpic processes are well known nonlinear processes, `complete` modelling of enthalpic processes is very difficult.Therefore, we focus on the use of a relatively simple way to introduce nonlinear PFC. Nonlinear PFC is designed by using multi-model. Further, this approach is investigated through simulation of a multi-model of a typical pH process.

## 2. MULTI-MODELS

A potentially attractive method for dueling with nonlinear processes over a wide operating range is the use of a multi-models. Multi-models can briefly be described as follow: consider a process described by a state-space model in the following form

$$\dot{x} = f(x,u)$$

$$y = g(x,u)$$

Let the operating point $\phi \in \Phi$ be defined by

$$\phi = h(x,u,y)$$

and define an operation regime by a subset $\Phi_i \subseteq \Phi, \Phi_1 \cup \Phi_2 \cdots \cup \Phi_N = \Phi$ and $\Phi_i \cap \Phi_j = O$ This means that the operating set $\Phi$ is partitioned into disjunct operating regimes. To each operating regime, we allocate a local model ($M_i$). Hence, we define N local models $M_1,...,M_N$. A local model can often be simple since it is only valid in one operation regime (and possibly its neighborhood). In this work we will choose the local models as linear state-space models with a structure based on first principles insight of the process in question.

The local linear models are combined into a global nonlinear model (M) by a convex combination of weighting functions $w_1(\phi),\cdots,w_N(\phi)$

$$M = \sum_{i=1}^{N} w_i(\phi)M_i$$

$$\sum_{i=1}^{N} w_i(\phi) = 1$$

$$w_i(\phi) \in [0,1]$$

Note that model $M_i$ will be dominating model when $\phi \in \Phi_i$ This means that $w_i(\phi)$ will be large in this case. A typical choice for $w_i(\phi)$ is a normalized Gaussian function.

## 3. PREDICTIVE FUNCTION CONTROLLER BASED ON MULTI-MODEL (GC-PFC)

The global operation of a nonlinear process is divided into several local operating regions. Within each local region, a controller is now designed about each of the local models. The output of each controller is then passed through the interpolation function which effectively generates a window of validity for each of the individual controllers. The interpolated output are then summed and used to supply the control commands to the process. The resultant GC-PFC structure is as shown in Fig. 1

The global controller output is obtained as following:

$$u = \frac{\sum_{i=1}^{N} w_i u_i}{\sum_{i=1}^{N} w_i}$$

$w_i$ and $u_i$ is the membership function and output of the local controller for the ith operating region. $w_i(y)$ is given by

$$w_i(y) = \frac{M^i(y)}{\sum_{j=1}^{N} M^j(y)}$$

$$M^i(y) = \exp(-\tfrac{1}{2}(\tfrac{y-\bar{y}_i}{\gamma_i \Delta y_i})^2)$$

$$\Delta y_i = 0.5(y_{i,max} - y_{i,min})$$

$$\bar{y}_i = 0.5(y_{i,min} + y_{i,max})$$

and the i-th regime is characterized by the intervals $[y_{i,min}, y_{i,max}]$ In addition, the parameter $\gamma_i$ determines the smoothness.

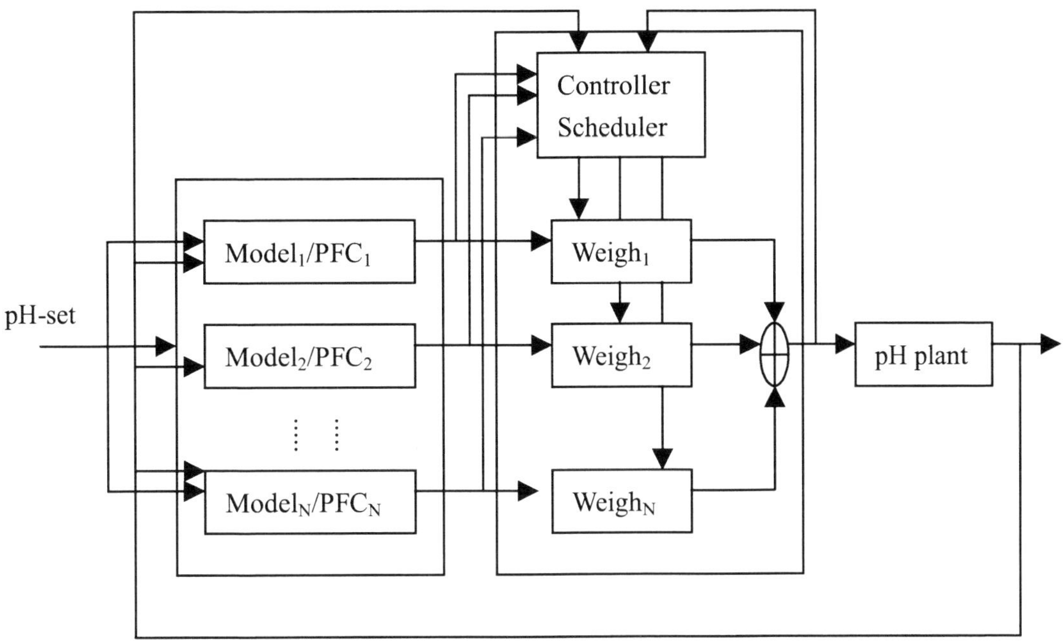


Fig. 1 Predictive Function Controller Based On Multi-Model

## 4. LOCAL PREDICTIVE FUNCTION CONTROL

Consider the following local ARMAX model

$$y_m(k) = \sum_{i=1}^{n_a} (-a_i) y_m(k-i) + \sum_{i=1}^{n_b} b_i u(k-i)$$ as

representing the process behavior. The prediction is obtained adding an autocompensation term calculated as a function of the observed differences between the model and past output.

$$\hat{y}(k+j/k) = y_m(k+j) + e(k+j)$$

The future control signal is structured as a linear combination of the basis function $B_i$, that are chosen according to the nature of the process and the reference:

$$u(k+j) = \sum_{i=1}^{n_B} \mu_i(k) B_i(j)$$

Normally these functions are polynomial type: steps, ramps or parabolas, as the majority of references can be expressed as combinations of these functions. With this strategy a complex input profile can be specified using a small number of unknown parameters. The choice of the basis functions defines the input profile and can assure a predetermined behavior (smooth signal, for instance). This can result in an advantage when controlling nonlinear systems.

The cost function to be minimized is:

$$J = \sum_{j=1}^{P} [\hat{y}(k+j) - w(k+j)]^2$$

where w(k+j) is usually a first order approach to the known reference:

$$w(k+j) = c(k+j) - \alpha^k (c(k) - y(k))$$

In order to smooth the control signal, a quadratic factor of the form $\lambda[\Delta u(k)]^2$ can be added to the cost function.

In the case of SISO processes without constraints, the control law can be obtained as follows. First, the output is decomposed into free output and forced one, and the structurization of the control signal is employed to give:

$$y_m(k+j) = y_l(k+j) + y_f(k+j)$$

$y_l(k+j)$ is the system free response, $y_f(k+j)$

is the system forced response.

Now the cost function can be written as:

$$J = \sum_{j=1}^{P} [\hat{y}(k+j) - w(k+j)]^2$$

$$= \sum_{j=1}^{P} [\mu^T y_B - d(k+j)]^2$$

Where:

$$\mu = [\mu_1(k) \cdots \mu_{n_B}(k)]^T$$

$$y_B = [y_{B_1}(j) \cdots y_{B_{n_B}}(j)]^T$$

$$d(k+j) = c(k+j) - \alpha^j (c(k) - y(k))$$

$$- y_1(k+j) - e(k+j)$$

Minimizing J with respect to the coefficients

Now the coefficients of the vector $\mu$ are computed and the control signal, taking into account the receding horizon strategy, is given by:

$$u(k) = \sum_{i=1}^{n_B} \mu_i(k) B_i(0)$$

## 5. APPLICATION TO A pH NEUTRALISATION PLANT

The neutralisation of pH represents a highly non-linear process, and offers a suitable case study for the demonstration and evaluation of GC-PFC techniques. Fig.2 shows a schematic of the plant.

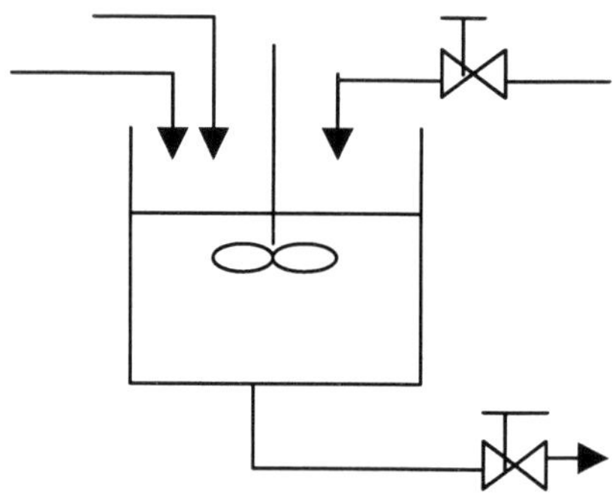

Fig.2 Schematic of pH neutralisation plant

The process consists of weak concentration acid, base and buffer streams being continuously mixed within a reaction vessel whose effluent pH is measured at a distance from the plant, introducing a time delay. The aim is control the pH value of the output stream by varying the inlet base flow rate ($q_2$). The outlet flow rate ($q_4$) is dependent upon the fluid height (h) within the vessel and the position of the outlet valve, which is set manually. Ramping the base flow rate while recording the outlet stream pH value produced an estimate of the process titration curve, as shown in Fig.3.

For this plant, the static gain between the base flow rate and the pH (the slope of the titration curve) varies considerably as these variables change. It can be seen that three regions in

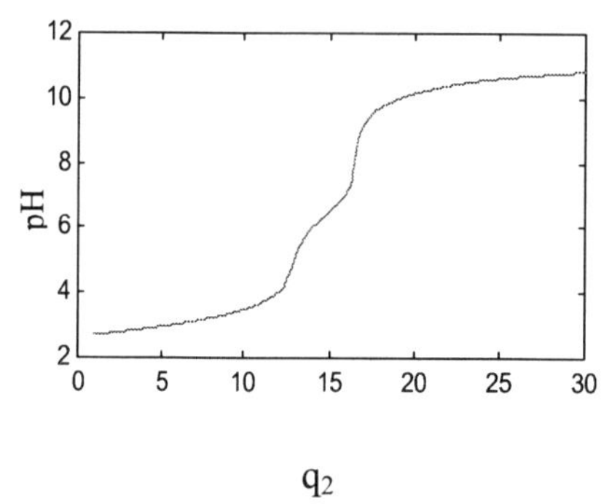


Fig.3 Titration curve for pH process

which the gain is nearly constant. Also, within each region, linear identification experiments suggested that a second-order model was sufficient to describe the dynamic process behavior. The local second-order, linear models with three were constructed to give:

$$G_1(s) = \frac{K_1 e^{-10s}}{(300S+1)(5S+1)}$$

$$G_2(s) = \frac{K_2 e^{-10s}}{(300S+1)(5S+1)}$$

$$G_3(s) = \frac{K_3 e^{-10s}}{(300S+1)(5S+1)}$$

This GC-PFC strategy was test in simulation on the pH neutralisation process. Fig.4 shows the results obtained from a series of step changes in pH from 4.0 to 10.0.

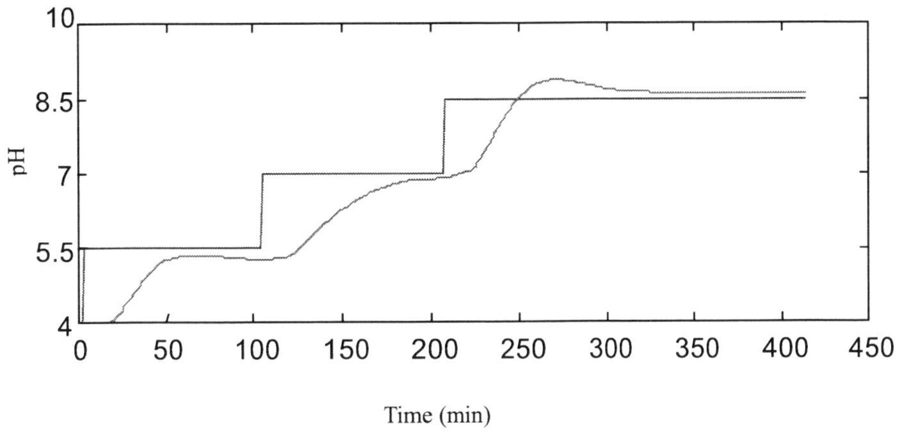


Fig. 4 Set-point tracking for local model based PFC and GC-PFC controller

In this case, the GC-PFC controller shows a better performance for large operating regions.

REFERENCES


H.B.Kuntze, A.Jacubasch, J.Richalet and C.Arber(1986). On the predictive function control of an elastic industrial robot, Proc. 25th CDC, Athens, Greece, pp:1877 - 1881.

D.Cuadrado, P.Nicodeme and A.Coic(1991). Application of global identification and predictive function control to a tracking turret. ECC91 European Control Conference, Grenoble, France. **7**, pp:2 - 5.

D.Cuadrado, P.Guerchet and S.Abu Ei Ata Doss(1991). Computer aided design of weapon system guidance and control with predictive function control technique, Proc. Conference of software for guidance and control (AGARD-CP-503). **5**.

J.M.Compas, P.Decarreau, G.Lanquetin, J.L.Estival, N.Fulget, R.Martin and J.Richalet(1994). Industrial applications of predictive function control to rolling mill, fast robot, river dam. Proc.3rd IEEE Conference on Control Application, Glasgow, UK. **3**, pp:1643-1655.

J.L.Estival and A.Huguel (1993). Predictive thickness control for aluminium cold rolling mills. Revue Generale de l'Electricite. **5**, pp:32-39.

Murray-Smith, R and Johansen, T.A. (1997). Multiple model approaches to modeling and control, Taylor and Francis.

IFAC Publications
www.elsevier.com/locate/ifac

# GLOBAL PATH PLANNING AND DYNAMIC ELUDING OBSTACLES IN SOCCER GAME

**Ping Tang*, YiMin Yang****

** The faculty of Computer, GuangDong University of Technology, GuangZhou, 510090, China. rachel@gdut.edu.cn*
***The faculty of Automation GuangDong University of Technology, GuangZhou, 510090, China.*

Abstract: The paper introduces an optimal way of path planning in soccer game. A dynamic grids algorithm and a dynamic eluding obstacles algorithm have been proposed, which include the position-prediction, and obstacle detecting and controlling backward methods. An enhanced algorithm, introduced also in the paper based on the two techniques, results in simple and effective controlling in soccer game. *Copyright © 2001 IFAC.*

Key words: robot, path planning, dynamic

## 1. INTRODUCTION

Artificial Intelligence(AI) and Robotics are two research fields with great challenges, and overlapping as well[Bourbakis,1997].The importance of the soccer game as a typical issue for multi-agent systems is that it provides a dynamic ,real-time, unknown-environment[Tambe et al.,1999]. Each soccer robot is an agent with some intelligence[Asada,1998], but does not have enough autonomy. The decision-making is accomplished entirely by the higher-rank agent[Hong,1998]. Soccer robots should be able to dynamically elude obstacles (other soccer robots) to move around smoothly in soccer field[Tambe,1998][Mansor,1992]. Therefore, path planning for soccer robots in a decision-making system is the key technique of soccer game. From the point of view of capturing information, there are two kinds of path planning: model-based path planning and sensor-based path planning. From the point of view of research approaching, there are many ways of studying path planning, including configuration space(c-space), artificial potential field, topological and grids methods[Wang,1998]. These methods have been successful in physical implementation of path planning of robotic without intelligence of avoiding collision or in a static environment. A distinct feature of robotic soccer systems is collision-allowance between soccer robots. An important aspect of optimal path planning of soccer robots is global path planning and capability of flexible, real-time responding to possible collisions. This paper proposes a optimal path planning based on dynamical grids algorithm and a dynamic eluding obstacles, which fully meets the requirements of eluding obstacles of soccer game. The proposed path planning can be implemented on a Mirosot software simulation platform, and then soccer robots will well elude obstacles in real games.

Part 2 and Part 3 of this paper discuss dynamical grids approach and dynamical eluding collision approach, respectively. Part 4 develops designing of optimal path planning. Part 5 gives the simulation results. Part 6 concludes this paper.

## 2. GLOBAL PATH PLANNING APPROACH

The grids method is commonly used in searching optimal path in eluding obstacles procedures of static path planning. Predicting obstacle positions and regarding obstacles as static in a certain period of time T, grids method searches optimal paths. This approach combining static grids method and position predicting algorithm, is known as dynamical grids approach.

The work space is divided into N×M squares[Asada et al.,1998], which each is a little bigger than the area needed by a single robot. We represent information of robots with space states, predict robot positions of n cycles, and draw lines between the predicted positions of two consecutive cycles. If the line touches a obstacles square, more than half of the square area is regarded covered, and the corresponding entities in the two dimensional array are set as 1's and the rest as 0's,called a blank squares. Optimal path searching approach is a way which seeks connected paths in the blank squares (Fig 1).

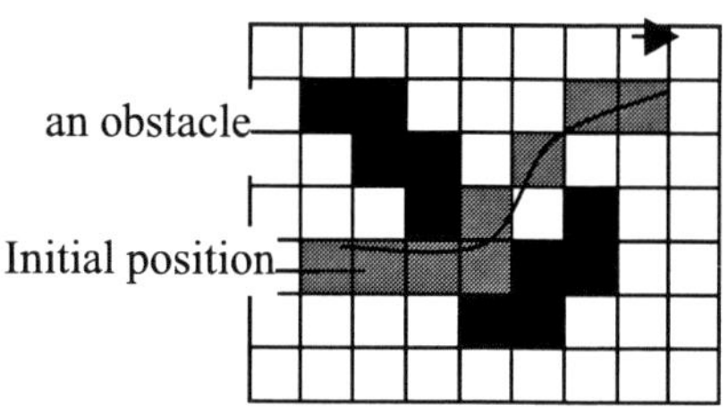


Fig 1. Diagram of the Dynamical grids approach

Though dynamical grids approach possesses advantages, such as global and good representation of decision-making , dynamical grids method searches paths in dynamical procedures by regarding the situation as static in a period of time T. T determines the effectiveness of searching . Big T results in large space of prediction, and searching may fail. Small T may not predict possible positions of robots, and the resulted paths may be invalid.

## 3. DYNAMICAL ELUDING OBSTACLES METHOD

Eluding obstacles is a key technique in path planning. Though soccer robots are not able to elude obstacles perfectly like intelligent, mobile robots, an effective method of eluding obstacles is required to ensure optimization of global path planning.

### *3.1. Inertial predicting*

In soccer robotic system, the decision-making of the major Agent is determined by the telecamera-capture information about the soccer robots on the field. The information can be regarded as static. However, soccer robots paths are dynamical. The robot positions of the next moment should be known to elude obstacles. Inertial predicting method computes v, α and β,and predicts the position of the next moment, by the present positions of robots' x and y, captured by the telecamera, and the positions of the previous moment, x' and y', based on the following inertial prediction formulae:

$$x = x' + (vt + \frac{1}{2}at^2) * \cos \beta$$

$$y = y' + (vt + \frac{1}{2}at^2) * \sin \beta$$

where x and y are the present coordinates of the robot; v is its present velocity of the robot; β is the present direction of the robot; α is the present acceleration of the robot; t is the period of time. Obviously, the less t is, the more accurate the prediction is. However, robots have no inertia if t is too little. It is important to adjust t appropriately. No matter what value t assumes, the result predicted by the inertial prediction method has some error in it. Hence, the result should be adjusted continuously. On the other hand, fuzzy reasoning method can predict the intention of the opposite soccer robots, such as fencing, defending or close-marking, and adjust the positions of our robots.

### *3.2. Collision detection and backwards control*

Inertial prediction does not guarantee to get rid of all collisions. A distinct feature of soccer robots is collision allowance. Thus, a soccer robot has to be able to quickly backward from collision- or butt-state, and has high flexibility and strong, real-time responding capability. Touch-states can only be captured by telecameras if soccer robots do not have sensors in them[Xu,1998]. While two robots touch each other in a captured image, the algorithm which determines whether the two robots really touch each other by distance and position will be very complicated because the shape of robots is cubic. Though two robots are touching one another, they may separate from each other in the next moment because they have different directions. But a telecamera cannot distinguish these circumstances. Therefore, collision detection method based on distance and position is not reliable. Our collision detection method issues an instruction to a robot, and judges in the next moment whether the robot has executed the previous instruction stored in the computer based on the field situation. If the instruction has been executed, there have been no collisions. Otherwise, check whether there are other robots around the robot. If yes, the method identifies a collision between the robot and another robot, and branches to the routine handling collision backwards control. If not , the method determines that something is wrong with the robot, and replaces it. The collision-backwards control algorithm is as follows.

(1).After a collision occurs, both sides execute collision detection routines. The robot of our side may stay at the present position (does not move) and wait for the opposite side to backward to save energy. If the opposite side does not backward, enter step (2).

(2).Evaluate the situation. Do not backward if backwards causes threat to our goal. If not, enter step(3).

(3).Determine the backwards direction. If our robot faces an obstacle, back along backward direction, If not, back along forward direction.

(4).Stop moving along the original direction, and backward along backward direction (determined in step(3)) by two wheels with the same speed.

## 4. AN ALGORITHM ON OPTIMAL PATH PLANNING

An algorithm is designed by dynamic grids approach, and possesses a good global path-planning because the computing is simple and a good representation of decision-making. But given time T, that is not appropriate, results in collisions.

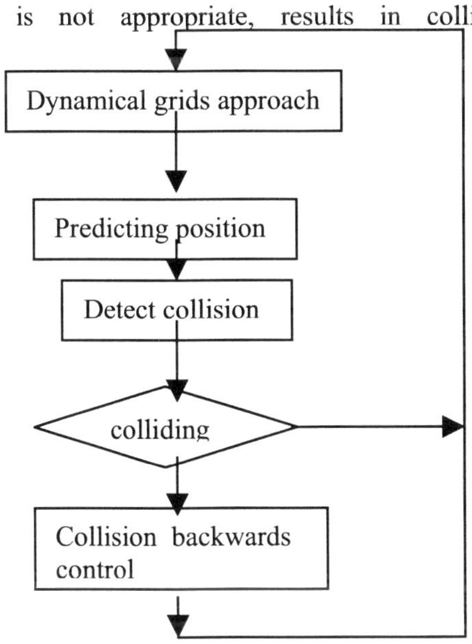


Fig.2 flow chart of the algorithm

Dynamical eluding obstacles method in part 3 can resolve the problem effectively. Combining dynamical grids approach and dynamical eluding obstacles ,we design an optimal path-planning as follows.

(1).Implement the path-planning globally by dynamical grid approach.

(2).Modifying the path-planning locally by predicting obstacle positions.

(3).Detect collision information by storing the previous instruction and checking the executing status.

(4).Implement the collision-backwards control algorithm.

(5).Return to 1.

The flow chart of the algorithm is shown as fig.2

## 5. SIMULATION

The simulation is implemented on the Mirosot software simulation platform and programmed in Visual C. In the initial state, consider soccer robots' stochastic status in the field. When the distance L between the two robots exceeds 4R (R is the diameter of robot), eluding obstacles need not to be considered (Fig 3-1). When $L \leq 4R$, global path-planning should be invoked firstly (Fig 3-2). If the robot accomplishes movement successfully, it is all right. Otherwise, the robot may collide and collision-test routine or backtracking procedure ought to run (Figures 3-3,4,5).

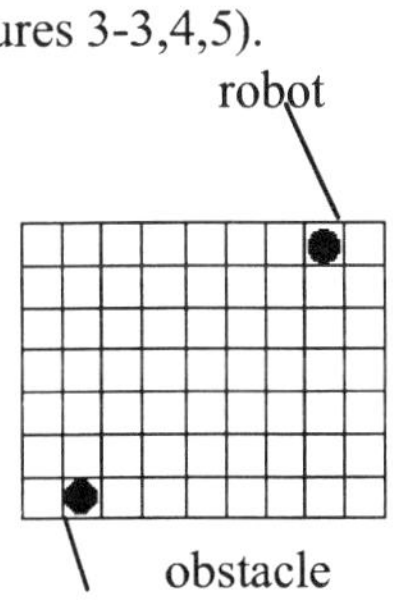


Fig.3-1 Initial state

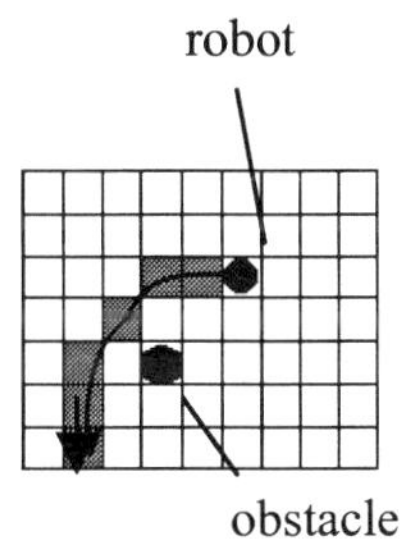


Fig.3-2 Globalplanning

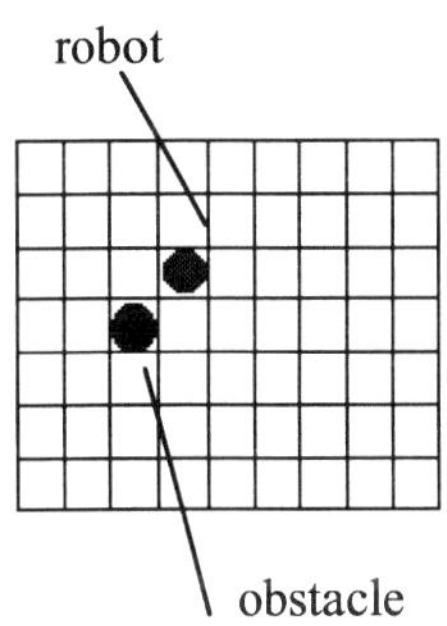


Fig.3-3collision checking

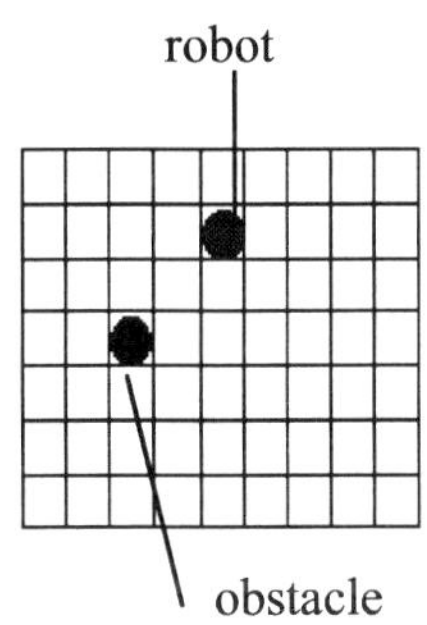


Fig.3-4 backward

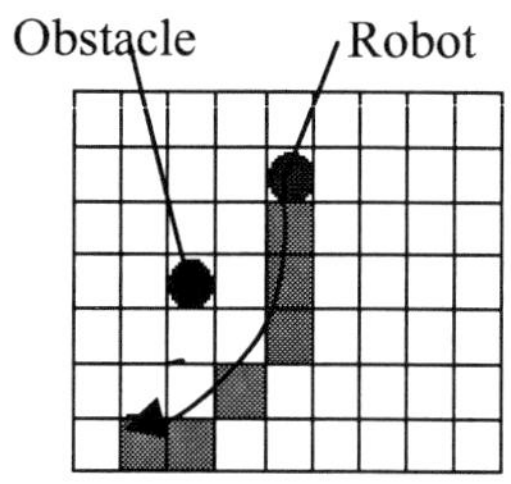


Fig.3-5 Planning again

## 6. CONCLUSION

Combining static grids and position prediction methods, the dynamical grids approach results in better tactics in global path-planning. Imitating human activities of eluding collisions and combining position predicting approach with collision detection and fuzzy control backwards techniques, dynamical eluding collision approach achieves better effects. The two methods make soccer robots move more reasonably and smoothly. We have implemented the approaches on the Mirosot software simulation platform. We took part in the First National Soccer Robots Championships in Harbin, November 24-26,1999, and received the champion in the Mirosot simulation software contest.

Aknowledgements:This research was supported by the GuangDong province foundation under grant 2KM00502G.

## 7.REFERENCES

Bingrong Hong(1998). Soccer Microrobots Game Systems based on Multiple Robots. *Technology and Application of Robots*,**5**, 7 10.

Chun Li and Zhiqiang Zheng(1999). Modular Design and Implementation of Multiple Robots Systems. *Robots* ,**5**,167 171

Nikolaos G.Bourbakis(1997). Intellligent Robotics. *Journal of Intelligent and Robotic system*,**19**,235 1997

Minoru Asada(1998). The Robocup Physical Agent Challenge: Phasel. *Applied Artificial Intelligence*,**12**,251 263.

Minoru Asada, shoichi noda and Koh Hosoola(1998). Action-Based sensor space segmentation for soccer Robot Learning. *Applied Artificial Intelligence*, **12**,149 164.

IFAC Publications
www.elsevier.com/locate/ifac

# TOWARDS SELF-ORGANISING TIME-SENSITIVE CONTROL SYSTEM'S SOFTWARE

**Leo Motus and Merik Meriste[#]**

*Tallinn Technical University, leo.motus@dcc.ttu.ee*
*[#]Tartu University, merik.meriste@ut.ee*
*Estonia*

**Abstract:** The embedded computer applications are rapidly growing in numbers, in complexity, and in dependability. This trend has been accompanied by shift of systems building paradigm from individual algorithms to interacting computing entities (e.g. agents, components, modules, etc). A number of studies focus on self-organising capabilities of such entities (agents). Unfortunately for computer control systems, many essential issues – such as, forced concurrency, time awareness of theories and software, and some others – have been almost neglected in the conventional theory. This paper provides a superficial survey of evolution of software-intensive systems towards multi-agent systems, lists subjectively some basic problems and achievements in developing time-critical systems. The authors suggest that sufficiently good theoretical and practical basis exists for building a test-bed that would assist in merging recent development trends in software engineering and artificial intelligence, and enable verification of time correctness of agent interaction. The test-bed would also assist in developing a theory for quantitative assessment of some properties of the evolving structures, and enable step-wise extension and testing of the supported functions (as the theoretical basis develops). *Copyright © 2001 IFAC*

**Key words:** multi-agent systems, time awareness, computational models, time models, test-bed.

## 1. INTRODUCTION

Everyday life in the contemporary world depends heavily on software-intensive time-, safety-, and security-critical systems. It is not easy to manage a day without using (the services) of the following devices, or systems -- vehicles, telephones, audio-video equipment, security systems, medical devices, transport system, air conditioning systems. Some other examples are less visible, or not yet functioning – for instance, distributed computer control systems, mobile multi-robot systems, automatic motorways, human and computer system collaboration in problem solving.

Many of the listed software-intensive applications may have dynamically changing goals, should preferably be able to modify dynamically their configuration – most of these changes are invoked either by the environment they are embedded in, or by the subjective will of their users (or the goal function). Some systems must have fault-tolerant and/or safety-critical design and implementation and, in addition, must be (strictly) time-sensitive so as to cope with their goals and imposed constraints.

The design and implementation methods have passed a remarkable evolution to enable the building of such applications. Procedure and algorithm centric approach in software engineering and computer science has been substituted by interaction centred paradigms (see for instance, Milner (1980), Manna and Pnueli (1991), Milner (1993)). In artificial intelligence similar shift has taken place from logic-based to agent-based models (Wegner (1997)). Interaction centred systems appear to be more powerful problem solvers than procedure and algorithm centred. It is also suggested that algorithmic models alone do not suffice to express certain expected aspects of behaviour of today's systems.

Interaction based computation involves, unlike algorithmic computation, infinite and dynamic input/output streams. Such a computation can be considered as kind of distributed control of information streams and agents' activities in a time-sensitive environment. At the same time a collection of components (agents) often shows a behaviour that is rather complex, despite of the simplicity of each single agent. This observation conforms perfectly to the software engineering empirical experience regarding the nature of system integration errors.

The integration problems, that are difficult to detect in due time, hard to explain and to resolve often hinder the successful introduction of a new system. In many cases these problems are caused by undue increase in the structural complexity of software-intensive systems caused by questionable design decisions, made by inexperienced designers. An alternative to the now prevailing method of system building seems to be the idea of building components that are capable of self-organising to optimise (or rather satisfy) the (set of) given goal function(s). This assumes the existence of well-founded concepts for structural organisation, or that thorough experience is available. Successful attempts have been made in this direction (see for example Ferber (1999), Arkin (1998)). However, it is believed that structural organisation principles and interaction methodologies are still insufficiently studied – especially for the cases of time-critical applications.

The importance of time model and time-criticality of systems has emerged because computers and software are directly interacting with components from the surrounding real world. Many tasks representing different components of the real world (and operating in different time scales) inside a computer system are to be able to co-operate under strict restrictions. Problems can be expected since one cannot assume that the static environment assumption holds in the man-made world – large share of software-intensive systems actually operate in the man-made environment. The conventional science, including the algorithm theory, assumes that axiomatic basis of a theory is fixed during the application of this theory. The validity interval of this assumption may be different in different parts of the man-made world (that extends the natural world). This is the major reason for introducing time constraints on the execution time of components, on the interaction of components, etc. (for more details see Motus and Rodd (1994)).

The phenomena of self-organisation, interaction, learning, and adaptation are studied theoretically, with tests in various applications. The obtained results, testing conditions, and implicit assumptions made by different research groups are difficult to compare. The short-term goal of this paper is to propagate the idea of developing a test-bed for studying and comparing various theories applicable to multi-agent systems – preferably on the same platform and under the same conditions. The long-term focus of this paper, and that of the suggested test-bed, is on emphasising the importance of time and on promoting theory for time-sensitive multi-agent systems. At the same time the test-bed helps to save time and resources by assessing the properties of future system before the system is implemented.

## 2. EVOLUTION OF IDEAS AND METHODS FOR SYSTEM DEVELOPMENT AND MODELLING

The evolution of software-intensive systems produces evidences that many thoughts and cautious practical steps have been taken towards applying the multi-agent paradigm to designing and implementing of those systems. The paradigm of multi-agent systems has resulted from shift of algorithm (and procedures) based computational models to interaction based computational models. This shift involves two other changes:

- a primitive component for building a system is not an algorithm, but rather a set of interacting, repeatedly activated algorithms, and
- study of infinite, and dynamic input/output sequences becomes important, since they cover non-terminating behaviour of computing processes as well as (infinite times) repeatedly activated, terminating behaviour of interacting computing processes.

In conventional computer science behaviour of interaction (as well as algorithm) based computing models is independent of time – of the time required to compute a function, and of the time at which the function is computed. Even complexity theory abstracts from elapsed time to model the number of computation steps, not to mention the influence of time-constraint interactions on the overall complexity (Motus and Rodd (1994), Wegner

(1997), Lee (1999)). There have been some dissidents who have fortunately maintained the presence of quantitative time in computational models -- for instance, Quirk and Gilbert (1977), Caspi and Halbwachs (1986), Motus (1993). More comments on the role of time models in time-sensitive systems will be further in the paper.

*The basic influencing factors.* As always, a survey is subjective and incomplete – only results that have directly influenced the authors are referred, for which the authors apologise. When forming general guidelines of thinking, the authors of this paper have been influenced by:
1. Structural synthesis of programs, that was the first attempt to automatically create a purposeful structure between the components with known properties; it started in the 70-s (see Tyugu (1988))
2. Software engineering recommendations to develop components separately from system's design, and to handle system building separately from component's design; in particular, the use of MIL languages (Module Interconnect Language), with accompanying attempts to verify some properties of the created structure
3. A number of practical developments, such as MMS (manufacturing message system) used in communication protocols (Valenzano, (1992)) to facilitate semi-automatic reconfiguration of control system when substituting minor components (e.g. intelligent sensors, or actuators), collaborating robots (Arkin, (1998)), automatic analysis and reconfiguration of computer networks, and others.

## 3. THE ESSENTIAL PROBLEM

Majority of today's theoretical and practical achievements in the agents' domain is related to individual agents in a not-time-sensitive environment. For instance, agents are used for systematic information collection in the Internet, for data mining from databases, and on the negative side of life as worms and viruses. It is natural that these applications are not time-critical, and they operate in pretty well defined environments

Extensive use of embedded software-intensive systems challenges the introduction of the multi-agent paradigm to the whole man-made world, not only to its pretty deterministic and well-defined computer network sub-domain. Characteristic to the man-made world is that a number of natural processes and man-made complexes (computers, organisations, and humans) interact in order to satisfy a designer-defined goal function. Approximations of many theories, incompletely known causal reasons, intrinsically non-deterministic behaviours, inability to guarantee satisfaction of static environment assumption during application of theories and algorithms, multitude of time counting systems and time scales are typical features of systems working in the man-made world.

Considering the said, all the formations (multi-agent systems, complexes of interacting concurrent programs, distributed computer control systems, etc) that operate in man-made world, should be, at least, aware of quantitative time and capable of pre-run-time and run-time checking for the justifiability of their operation and design. It is important to understand that correctness and efficiency of those formations depend upon the goal function and restrictions imposed upon the formation and, indirectly, on the efficiency of interactions.

In the today's world pretty good progress has been made in integrating the myriad of design methods and tools into one – the OMG (Object Management Group) initiative has lead to the UML design methodology and tools, supported with the CORBA initiative to facilitate distributed implementation of the designs. The ongoing initiatives of the OMG – real-time UML and Real-time CORBA – create hope that a designer of time-sensitive control system will soon have decent new generation design tools.

In a short, those new tools will support design and implementation of time-sensitive (time-critical) systems, provided that the designer knows how to choose a suitable time model, and the designer is given a set of required time parameters and constraints (with consistent values) for the system's components.

In the other words, in addition to advanced semi-formal design and implementation support (that is in progress already), the system developer needs formal support in selecting essential features and characteristics for the future system, when abstracting from the informal world. Development of methods and tools that would enable the system designer to choose time model, assess required parameter values, verify the consistency of parameter and constraint values, is still in a pretty messy status. There are many alternative, mostly heuristically justified approaches, and a large number of disparate methods. Such a situation hinders the progress in multi-agent systems' developing tools substantially.

The problem: *A unifying idea is needed to blend (or at least, co-ordinate) the research and development activities in communication protocols, self-learning and teaching methods, structure adaptation methods, belief processing (updating), and reasoning methods in order to facilitate purposeful self-organisation of multi-agent systems.*

The authors suggest that time awareness of computational models for agents, interactions, goal functions, multi-agent systems, etc. is one of the most feasible candidates for unifying the variety of existing methods. For instance, consumer defined time-selective communication (Quirk and Gilbert (1977), Motus and Rodd (1994)) seriously facilitates automatic structure modification. Hard real-time constraint imposed upon a deduction based reasoning method forces to apply approximate reasoning method, and eventually to adopt a reflex-based paradigm of reasoning (Motus (1991)).

The authors experience with real-time systems indicates that it is difficult to underestimate the influence of time in formal models of time-sensitive systems. Time in multi-agent systems has been considered, for instance, by Horty and Belnap (1995), and by Y.Tretyakova et alii (2001). Branching time is used for partial reasoning about joint (co-ordinated) actions of agents, in particular for verifying the obligations and conditional commitments of partners in a joint action. Essential assumption for using branching time is, that all the partners involved in a joint action follow action ladder and handshaking protocol (see also Roscoe (1998), and Hinds and Esterline (2001)).

In many practical systems this is too strong assumption – i.e. the case of competing, or not quite fair agents. Even provided that all the partners in the joint action appreciate the action ladder and handshaking, the time-sensitivity of some of the partners may invalidate the achieved decision. This leads directly to the necessity of time-selective inter-agent communication, and to more sophisticated time models (see, for instance, Motus and Rodd (1994)).

## 4. PLAN OF ACTION

In the long run the goal is to introduce an explicit and sufficiently elaborated time dimension to multi-agent systems behaviour. This should be the starting point for studying the influence of time-sensitivity, incomplete information, and limited resources upon agents, interaction structure formation, learning, negotiating, and reasoning capabilities.

Short-term activities are aimed at developing a test-bed for rapid prototyping of agents, and for studying the behavioural aspects of multi-agent systems. The planned behavioural studies are closely related to the research of system's integration errors' detection, and their avoidance. The behavioural studies start from verifying the time correctness of the agents' interactions, and gradually move deeper to less studied areas – such as, negotiating, reasoning about the negotiations, and estimating (measuring) the efficiency of the negotiated structure.

In a wider systems engineering scale the test-bed can be considered as an approximation for discovery and delivery systems, ideally required in addition to the tools for developing the delivered system (see Evans (2001)). As part of that new doctrine, the requirements are not considered disparately from the other design decisions. This conforms to the real necessity to start formal time analysis already at the requirement specification stage, since a large share of emerging errors stems from the requirements.

### *4.1 The departure point*

The research and development is based on authors' experience with interaction-based computations, and in introducing time awareness into the interacting agents and interactions. The underlying computational model stems from two branches:

- Q-model obtained by cross-fertilising the ideas of Quirk and Gilbert (1977) and Milner (1980), as elaborated in Motus and Rodd (1994) and related developments
- multi-stream interaction machines (MIM), represented here by interactive attributed automata , as published in (Wegner (1997), Meriste (1998), Meriste (1995), Wegner (1999)).

Requirements to the used time model come from Motus and Rodd (1994) and OMG requirements on time models as expected from real-time UML and real-time CORBA (OMG document (1999)). Additional discussion on comparing the two requirements sources is given by Motus and Naks (2001). It is essential that each agent can have its own sufficiently sophisticated time model that supports the three concepts of time (non-reversible increasing time, fully reversible time, a set of relative times), and the agents' interaction mechanism can support time-selective message exchange.

Preliminary experience with concurrent time-constraint software timing analysis is based on a CASE tool LIMITS that implements the ideas published in (Motus and Rodd, (1994)). LIMITS has also been tested in ESPRIT 22154 project BRIDGE. A prototype for modelling multi-stream interaction machines has been also implemented and tested (Meriste (1998)).

### *4.2 The test-bed*

The use of interaction based computational model is essential for the required analysing power of the test-bed. The authors have first hand experience with two interaction based computational models – the Q-model and attributed automata. The first prototype will enable the user to experiment with both computational models – at this stage it is too early to stick to one alternative only.

The test-bed is planned as a general-purpose framework that enables to model and assess the properties of the used algorithms and those of the overall multi-agent systems. The main focus of the test-bed is to be on studying the group behaviour of agents – assessing the structural properties of monitoring, servicing, and control of systems, where computers and real-world entities interact basically on their own (without interference from the humans) according to given strategic goal functions. Such properties may include, for instance:

- the capability of an individual agent to reason about the actions, intentions, and goals of its environment,
- the efficiency of an agent's collaboration with its environment,
- the (agent's, or) system's ability to handle dynamic changes in its environment.

As explained earlier in this paper, the properties of application systems are, in many cases, subject to imposed temporal constraints.

### *4.3 Architecture of the test-bed*

The basic architecture of the test-bed stems from the agent-based ideology – separate parts of the test-bed can be considered as agents. This also facilitates comparatively easy introduction of additional computational models (or modification of the existing models), and extension of the suite of respective analysis methods.

The test-bed is designed as a multi-agents system, comprising four clusters of agents. Agents belonging to the same cluster have, as a rule, full-scale in-cluster communication capability. It is assumed that agents from the same cluster can substantially benefit from the in-cluster co-operation. However, each agent can freely select the level of its in-cluster co-operation and the strategy (i.e. sequence of sub-goals) for achieving its final goal. There are no restrictions on shared information and depth of co-operation unless the restrictions are imposed by safety, security, or other application-specific considerations.

Each cluster has different role in the test-bed and the goal of the test-bed can be accomplished only by appropriate interaction between the clusters. One cluster forms (or models) the real-world entities that are to be monitored, serviced, and/or controlled by the delivered system. In some cases this cluster could be substituted by a physical model, or by entities of the natural environment. In other cases economical, ethical, or safety considerations might rule out the use of natural environment as part of the test-bed.

The second cluster models (or represents) a delivered system that monitors the first cluster (the environment of the computer system) – sensors, signal pre-processing, condition checking, exception handling, and respective decision-making. The third cluster models (or represents) a system that controls and/or influences the behaviour of the environment, based on the information provided by the monitoring system and as prescribed by the goal function.

The fourth cluster comprises measurement, estimation and reasoning tools, used to assess the properties of the organisation that results from the established in-cluster and inter-cluster interactions. Formation of this cluster is not a routine task since suitable tools are not readily available. The authors of this paper have some experience with methods and tools for assessing time-correctness of interactions and consistency of validity times for events and data in a system. The quality and rate of learning algorithms has been measured, mechanisms for adaptation have been studied and assessed by many authors – however on-line assessment methods are not available for a general case. The situation with measuring (or assessment) methods for efficiency of the interaction structure, and its convergence to a stable structure, is even more confusing. Several aspects may be of interest when assessing the structure of interactions – such as, fitting into given throughput limits of the communication media, fault-tolerance of the structure, security against unauthorised influence, upper limit for a response time to stimuli, probability of achieving the goal, etc.

The test-bed building process is not straightforward. Several theoretical problems are to be handled before the completion of the test-bed. The underlying interaction-based computational models still need mathematical study; some innovative ideas are discussed in (Wegner and Goldin, (1999)). The ongoing computations and incomplete specification of the environment lead to the use of co-induction. This implies substitution of conventional minimal fixed point with the maximal fixed point. Maximal fixed point correspond to the paradigm that any behaviour of the system consistent with the observations is admissible. This enables to minimise constraints on functioning of the test-bed. At the same time, it is not quite clear whether the delivered system (i.e. the application system that is being tested) can function with only minimal set of specified constraints. Another incompletely solved problem is how to introduce time-sensitivity into agents and express time-sensitivity of agent' interactions.

The test-bed building process will, hopefully, provide the actual requirements to analysing tools and properties of those tools required to assess multi-agent systems. It will also provide a basis for experimental comparison of theoretical solutions and their implementations.

## 5. CONCLUSIONS

During the last decade paradigm for designing software-intensive systems has gradually shifted from algorithm centred to interaction centred computational models. A number of interaction centred computational models has appeared that makes the multi-agent paradigm naturally usable for designing software-intensive systems.

This paper surveys the evolution of paradigms and models, and points out that time awareness is essential if a multi-agent system operates in the real world. The problem of introducing time into computational models is discussed. It is pointed out that many other aspects – such as time- selective interaction, and protocols for negotiation and reasoning -- of multi-agent systems also essentially depend upon the time-awareness of agents.

It is suggested that a test-bed for modelling and investigating the properties of multi-agent systems be built. The existing preconditions and experience for building the test-bed are discussed. The generic architecture of the test-bed is superficially described.

## REFERENCES


Arkin R.C. (1999) "Behavior-based Robotics", MIT Press, 491 pp.

Evans R.P. (2001) "Design as Discovery. Ten System Design Tenets", Proc. 5th Conference on Systemics, Cybernetics and Informatics, vol.XI, 546-551

Caspi P. and Halbwachs N. (1986) "A Functional Model for Describing and Reasoning about Time Behaviour of Computing Systems" Acta Informatica, vol.22

Ferber J. (1999) "Multi-agent Systems. An introduction to Distributed Artificial Intelligence", Addison-Wesley (the French edition in 1995)

Hinds O.P. and Esterline A.C. (2001) "Joint activity coordination and common knowledge in multi-agent/person environment", Proc. 5th Conference on Systemics, Cybernetics and Informatics, vol. III, 208-212.

Horty J. and Belnap N. (1995) "The deliberative stit: A study of Actions, Omissions, Ability, and Obligation", Journal of Philosophical Logic, vol.24, 583-644

Lee E.A. (1999) "Embedded Software – An Agenda for Research", University of California at Berkely, Memorandum M99/63

Manna Z. and Pnueli A. "The temporal logic of Reactive and Concurrent Systems", Springer Verlag, 1991

Meriste, M., Penjam J., Vene V. (1998) "Models of Attributed Automata" Informatica, vol.9, no.1, 85-105

Meriste, M., Penjam J. (1995) "Attributed Models of Computing" Proc. of the Estonian Academy of Sciences. Engineering, viol.1, no.2, 139-157

Milner R. (1980) "A Calculus of Communicating Systems", LNCS no. 92, Springer Verlag, 171pp.

Milner R. (1993) "Elements of Interaction" Communications of the ACM, vol.36, no.1, 78-89

Motus L., and Naks T. (2001) "Time models as used in Q-model and suggested for RT UML", Proc. 5th Conference on Systemics, Cybernetics and Informatics, vol. XI, 467-472.

Motus L. and Rodd M.G. (1994) "Timing analysis of real-time software", Elsevier Science Ltd, 212 pp.

Motus L. (1993) "Time concepts in Real-time Software", Control Engineering Practice, vol.1, no.1, 21-33

Motus L. (1991) "Artificial Intelligence in hard real-time – a new paradigm needed?" IFAC Workshop series, no. 9205, Artificial Intelligence in real-time control, Elsevier Science

OMG document (1999) "Enhanced View of Time. Request for Proposal", orbos/99-01-14

Quirk W. and Gilbert R. (1977) ′"The Formal Specification of the Requirements of Complex Real-time Systems" AERE, Harwell, rep. No. 8602

Roscoe A.W. (1998) "The Theory and Practice of Concurrency", Prentice-Hall

Tretyakova Y., Y.Liu, A.C. Esterline (2001) "Deontic transaction model for multiagent system", Proc. 5th World Conference on Systemics, Cybernetics and Informatics, vol. III, 224-229.

Tyugu E. "Knowledge based programming", 1988, Addison Wesley

Valenzano A., Demartini C., and Ciminiera L. (1992) "MAP and TOP communications; standards and applications"" Addison-Wesley, 541 pp

Wegner P. and Goldin D. (1999) "Interaction as a Framework for Modelling", in Chen et alii (eds) LNCS #1565

Wegner P. (1997). "Interactive Software Technology", in Handbook of Computer Science and Engineering, CRC Press, Ch. 151

IFAC Publications
www.elsevier.com/locate/ifac

# THE HARDWARE/SOFTWARE DIVIDE FOR HIGH-PERFORMANCE EMBEDDED CONTROLLERS

**Roger Goodall and Simon Jones**

*Loughborough University, Loughborough, LE11 3TU, UK*

Abstract: The paper provides a primarily qualitative overview of the hardware/software issues that relate to achieving efficient solutions for real-time embedded control. This issue has always existed, even with general-purpose processing elements, but it is likely to become much more critical for new families of programmable devices. The purpose of the paper is to identify trends relating to controllers that employ these devices. *Copyright © 2001 IFAC*

Keywords: Real-time control, co-design, embedded control,

## 1. INTRODUCTION

Ever since processor-based solutions became a viable proposition for implementing high-performance embedded digital control, there has been a tension between the electronics/software engineers who have a professional desire to produce efficient, cost-effective solutions, and the control engineers whose main preoccupation is at the overall systems level and generally have neither the will nor the skill at the electronics level and therefore want a functionally correct but easy solution to the problem of implementation.

The trend has been in favour of the transparency required by the control engineers, and the implementations are now frequently buried below layers of software compilation and general-purpose hardware schemes, a result that is some senses very effective, but at the same time undeniably inefficient from a computational viewpoint. For some applications this is not a problem, for example industrial automation and process control where sample rates are lower. However it is problematic for embedded control of electro-mechanical systems such as high-performance applications in the automotive and aerospace industries for which sample rates of 100s or 1000s of Hz are needed, and it is surprising how often one hears of limited sample frequencies due to computational limitations. While floating-point DSPs can be used, fixed-point micro-controllers seem to be most common for such applications. However it is not unusual for people to use "real" floating-point computations, perhaps not realising by how much they are restricting themselves in terms of releasing the full capability of their chosen processors. The ease of use therefore comes at a high computational cost.

However, on the horizon for control (but very much in the mainstream for electronic design) are the new generations of programmable silicon solutions such as large FPGAs and programmable ASIC devices. Effective application of these devices provides the opportunity to start again from scratch, to re-think how best to provide solutions which are both efficient <u>and</u> usable.

The paper will explore the possibilities of the trade-offs between efficiency, designability and transparency. It will identify current and future possibilities, although in many senses identifies research challenges rather than definitive solutions. It will however provide insights into the way in which controller design might (or should?) develop over the next 5 years.

## 2. BACKGROUND

A number of architectures for solving control problems have been proposed in the literature, each having advantages and disadvantages. These can be divided into three broad areas - general-purpose solutions, dedicated solutions and reconfigurable solutions.

General-purpose solutions concentrate on processors that are tailored towards the requirements of control systems, but are not dedicated to the solution of any one specific task. Lang (1984) describes a control processor based upon a logarithmic number system, which is shown experimentally to give a reduced bit width requirement when compared with linear number systems. Jaswa *et al* (1985) describe a RISC co-processor consisting of a continuous processing element performing the next-state update process and a discrete processing element for processing switch based information.

Dedicated solutions are aimed at solving one specific task. Typical of such tasks is robotics control, for which Sadayappan *et al.* (1989) describe a processor operating on a Single Instruction Multiple Data (SIMD) approach. Catthoor *et al.* (1991) detail a system of 4 different execution units operation in parallel, the application being a 6 degree-of-freedom mechanical robot. Moving away from robotics, Grout *et al.* (1995) present an ASIC for a simple Proportional-Integral (PI) controller. Analogue-to-digital and digital-to-analogue conversion is performed on-chip, and both the proportional and integral gains are set digitally.

The third main area covers a range of possible architectures, the common factor being the use of parallel computation. The approach of using general-purpose parallel processors for solving control problems (e.g. Garcia Nocetti and Fleming (1992)) exploits the parallelism inherent in control problems. Such use of standard processors is not ideal, much of the functionality remains unused and hence redundant. Thus architectures tailored towards specific control systems needs are more appropriate.

Tsunekawa and Miura (1995) describe a direct approach to the solution of state-space equations. Each processing element (PE) performs the calculations for one output or next state variable. Multiple PEs are combined to solve the complete set of equations, with each PE is a single chip. Fujioka *et al.* (1990) describe another PE, consisting of 2 multipliers, 2 adders, local memory and a reconfigurable switch circuit. A number of PEs are combined to perform multiply-accumulate operations on a different number of operands per cycle. Both of these architectures are coarse grained, with each PE offering a large amount of computation power. A medium-grained architecture is described by Chen and Rabaey (1991), with each execution unit (EXU) being capable of performing addition, subtraction, comparison and shift operations. Each chip contains 32 EXUs, divided into 4 quadrants and interconnected via a crossbar switch. A similar approach is taken by Spray and Jones (1991) using a more complex PE that is also capable of multiplication. Communication is on a local level between any of the 8 neighbouring PEs.

Previous work in application-specific processors has often failed, not because they cannot provide the raw computation needed, but because they have to perform a host of other tasks to which their architecture is unsuited. Furthermore, their lack of general-purpose programmability makes them inconvenient to deploy and they fit poorly within a system design, which is structured around the assumption of general-purpose computation being available. This is a theme that will be developed later in the paper.

## 3. FUNCTIONAL REQUIREMENTS FOR REAL-TIME CONTROLLERS

### *3.1 Typical present-day requirements.*

There is now a variety of control design methods by which appropriate control laws can be created for complex multi-variable systems, but the actual implementation of control laws is a part of the design process which most control engineers want to achieve as straightforwardly and transparently as possible. The difficulty is that there are particular numerical requirements in control system processing for which standard processor devices are not well suited, in particular arising from the high sample rates which are needed to avoid adverse effects of sample delays upon stability. These could be satisfied in μP or DSP devices using a high level language, but there is no question that this is computationally inefficient (Goodall and Goodall, 1993). Alternatively it is possible to use "hand-crafted" numerical routines, probably written in assembly language, but as mentioned control engineers generally have neither the will nor the skill to do this.

The principal requirement, which is the outcome of the process of designing a controller, is the control law itself plus its associated analogue I/O, which must be converted into a real-time processing algorithm. As explained in the previous paragraph, this must be as efficient and straightforward as possible. However there are additional requirements, often thought of as somewhat more peripheral, which are nonetheless essential for a practical control system. Probably the most important aspect is what is often called the interlocking, i.e. all the logical functions which have to be carried out to monitor logic inputs such as limit switches, to provide enable signals for power amplifiers, to switch on small indicator lights, to provide simple diagnostics, and so on. Often fault handling is needed, which may be related to simple actions such as those which form part of the interlocking, through to full implementation of fault detecting isolation and reconfiguration. Some of these requirements can be satisfied with modern software tools such as Matlab's real-time workshop. While these are generally straightforward to use they are not optimised for the various tasks which need to be carried out because they are focussed on general purpose processors and do not represent an efficient solution from the cost, performance or power dissipation point of view.

*3.2 Anticipated future requirements*

In the longer term the functionality for real-time control is likely to expand further, and it is important to identify these requirements. Fig 1 provides a vision for the full range of functions that will be needed for future embedded control applications.

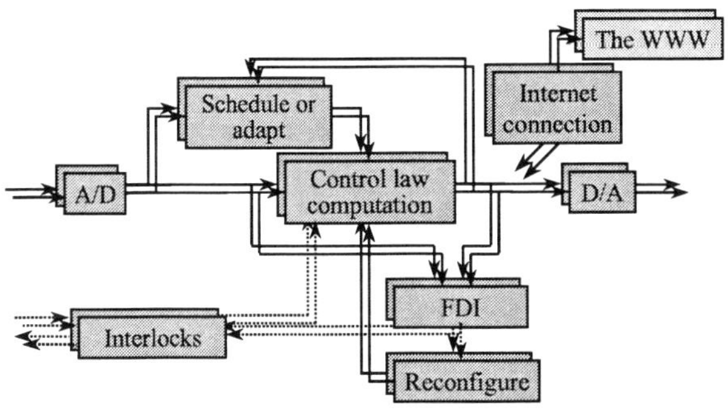


Fig. 1 Future embedded controller functionality

A primary requirement is that of interlocking which was mentioned in the previous sub-section. This is mainly a stand-alone task, largely independent of the control law processing, although it may be that some interaction with the computational function is needed, for example to limit the action of integrators until the control system is enabled. This is not of course the kind of function which a device such as a DSP is particularly efficient at handling, but it usually doesn't have the same degree of time criticality as the control law computation so it can usually be provided as a background task.

Adaptation, particularly in the form of gain scheduling, is also something that is not uncommon, and this simplest form is relatively straightforward to introduce by adjusting the coefficients in the controller. More advanced forms of adaptation depend upon the effects of the outputs, or perhaps state or parameter estimates from the system. If the adaptation process is relatively slow (ie compared with the closed loop system bandwidth), then the LTI computational framework still applies; faster adaptation requires a re-think of the way the equations are formulated, although in this case provability of stability margins etc is more difficult and so use for embedded control may be problematic.

Another function that is increasingly being incorporated is fault detection and isolation. It is shown in the figure as simply taking the controller inputs and outputs, but obviously more complex strategies are possible. It is also possible to have reconfiguration of the control law to respond to the identified fault. In the case of FDI alone the output will interact with the logical operations being carried out, perhaps to initiate a safe shutdown; if reconfiguration is included then there will be change to the coefficients (and perhaps the structure) of the control law computations, and both these links are shown on the diagram.

Those mentioned so far are principally internal functions, but external connections are also important. Many micro-controllers include a standard bi-directional serial communication peripheral, to facilitate easy connection to a computer, for example; a few include a more complex interface to provide direct connection to an external bus such as CAN or Fieldbus. However communications seem to be becoming oriented around the internet, and so it is probable that internet connectivity will become an increasingly important feature.

As control extends its influence, safety and reliability become increasingly important. Fault detection is of course an essential part of this, but almost inevitably duplication or triplication of the processing channel is needed to ensure safe and/or reliable operation in the presence of faults, and so there will probably be an interest in the future of providing a certain amount of the parallel redundancy within the processing device - Fig. 1 includes a superimposed copy of the various functions to indicate this.

## 4. SOLUTIONS USING GENERAL-PURPOSE PROCESSORS

Some comments about implementation have already been made, but this section discusses the implications in more detail, especially in relation to hardware v software issues.

*4.1 Hardware/software co-design*

The whole area of hardware/software co-design is a major subject that has seen substantial research progress over the last ten years (Ong et al 1997, Ernst 1998, Shulz et al 1998). The principles of co-design are clearly important, but this paper is more concerned with the partitioning of functionality for embedded real-time control in a fundamental sense, thereby identifying frameworks that are broadly applicable to a range of implementations. The established principles of hardware/software co-design are naturally an important issue, but it is generally recognised that different co-design approaches are used in different industry applications. A recent "round table" discussion organised by the IEEE (Anon 2000) indicated clearly that co-design is a paradigm, a way of thinking analogous with the mechatronic approach that is driving the system design in control engineering.

*4.2 Hardware v. software for general-purpose processors*

General-purpose processors such as micro-processors, micro-controllers and DSPs have a pre-determined hardware architecture with the functionality provided wholly by means of the real-time software. Of course there is a hardware choice that comes from the availability of a range of device types. Complex and expensive processors such as floating-point DSPs enable the software to be simpler, or more precisely they enable the code from less efficient software design to compute faster. A simpler device such as a fixed-point micro-processor or micro-controller may need to be programmed in Assembler language to achieve the same performance, but this requires a proper understanding of the numerical processes (not needed so much with floating-point devices) and a more substantial software development activity.

One approach is to programme a fixed-point microprocessor device in a high level language, using floating point variables so that numerical issues are not a concern, but the computational overhead is large and unsurprisingly restrictions in sample rate are found. A second approach is to use a fixed point digital signal processor (DSP) which has an architecture better targeted for computationally-intensive applications, as these offer some speed advantage over a microprocessor. Other options are to use a number of parallel processors or sophisticated floating-point DSPs. However, this doesn't result in a cost-effective solution, especially for high-volume embedded control applications.

| | | HARDWARE | | |
|---|---|---|---|---|
| | | Fixed-point µ-processor | Fixed-point DSP | Floating-point DSP |
| SOFTWARE | Assembler | 6/10 | 7/10 | 10/10 |
| | Compiled | 2/10 | 5/10 | 9/10 |
| | Code generation | 1/10 | 4/10 | 8/10 |

Table 1 Hardware and software choices

Table 1 shows the inter-relationship between hardware and software and its effect upon processing speed for typical general-purpose processors. It presents a broad, qualitative assessment using marks out of 10, and compares three different ways of programming, two of which are fairly familiar, the third less so.

Firstly there is the use of the chosen processor's Assembly language, highly efficient but more difficult to develop and not transportable to a different processor; secondly the use of a high level language, converted into object code by a compiler - relatively easy to develop and fairly transportable to another processor, but not as computationally efficient. The next two paragraphs elaborate a little upon these two options.

Computation time comparisons between low and high level programming for microprocessors were undertaken some years ago (Goodall and Goodall, 1993), and this work provided a useful indication of the difference in performance. If the equations were written in C using floating-point variables, computation took around 20 times as long compared with code written in assembly language. Using integer variables it was possible to reduce this factor so the compiled code took perhaps 3 or 4 times as long, but this approach required a good understanding of the numerical requirements and variable scaling, so it is nothing like as easy as with floating point variables.

It is important to be careful of compilers, because the user is in the hands of the persons who wrote the compiler. For example, as mentioned earlier a 32-bit floating point DSP strictly only provides 23-bit precision, and the use of "doubles" instead of "floats", i.e. 64-bit variables instead of 32, might be required. Unfortunately these higher precision variables are sometimes not supported by DSP compilers, and the author is aware of one instance where "doubles" are implemented as "floats" without any warning!

A third software option is the use of products such as Matlab's real-time workshop which generates code, or links to a DSP card directly, often straight from a Simulink description of the controller. This of course is one level higher again, and even more trust is being placed in the manufacturer of the product, but many workers have found that this provides a very easy method of implementation. However the overall efficiency may be quite low, and so although this may be very suitable for a development environment, It is probably not suitable for a high-volume, embedded control application.

### *4.3 Targetted processors*

The limitations of general-purpose processors have motivated a number of researchers to investigate special architectures, and some of these were described in Section 2. Common to much of the previous work is the absence of a real attempt to adopt a systems approach and bring together the control and electronic design requirements. Much of the hardware developed to date has tended to be adapted from a more traditional microprocessor or digital signal processor (DSP) background with relatively little analysis of the underlying requirements of control systems. The authors have carried out research into optimised architectures aimed specifically at high-performance real-time control and fully exploiting a comprehensive understanding of both control and electronics. This work has clearly shown how effective a targetted architecture can be for the control algorithm itself. It has resulted in a Control System Processor (CSP) which uses a little more than 20k gates on a modern "programmable" ASIC device but out-performs a state-of-the-art floating-point DSP by a factor of 3 or 4 in computation time (Goodall et al 2000).

### 4.4 Processor types v. functional requirements

Table 2 shows the range of requirements identified in Section 3, and provides a qualitative assessment for different processing solutions. It illustrates the fact that none of the current computing devices is ideal for all of the required functions. The micro-controller is best for the logical/interlocking operations, the targeted CSP core is best for the computationally intensive control algorithm, and the DSP falls somewhere in the middle, well supported for low quantity/development type applications, but certainly not optimised. What is really needed therefore is a device and a design approach which satisfy all the requirements in an efficient and effective manner.

| Device type | Control law computation | Adaptation etc | Logical operations |
|---|---|---|---|
| μC | X | XX | ✓✓ |
| DSP | ✓ | ✓ | X |
| CSP | ✓✓ | ✓ | XX |

Table 2 Control operations v. processor type

## 5. SOLUTIONS USING PROGRAMMABLE SILICON

New solutions have become feasible due to the imminent production of multi-million gates FPGAs (Field Programmable Gate Arrays). FPGAs are circuits that the user can program to realise an application specific function. FPGAs, which are individually programmable by downloading bit patterns or blowing fuses/setting switches, enable small volumes and inexperienced users to develop systems using desktop tools. Until recently, FPGAs had only a few hundred gates per chip, limiting the use of these devices to supporting simple logic functions. However, the rapid improvement in silicon technology now permits several million gates per chip to be realised, permitting microprocessors and other complex designs to be realised.

### 5.1 Architecture possibilities

An important factor with such designs is that now even the hardware on the chip will normally be produced by software in the form of VHDL, synthesised (i.e. compiled) into the necessary code for the programmable device. This would be complemented by application-specific code, downloaded into memory cells onto the FPGA and executed by the programmed chip in real-time.

A wide spectrum of possibilities exist for dividing the functional responsibility between the devices programming code and the real-time code, but this can be simplified by identifying three broad categories:-

(a) <u>*Single targetted processor*</u> This would be specially designed for control, probably an extrapolation from a μP/μC or a DSP to give it greater versatility. The functionality for a particular application would be totally provided by the real-time code which would run sequentially in the processor in the classical manner. The limitations identified in Table 3 still apply in some respects, although the CSP entry refers to a processor solely designed to implement the control law calculations using a very minimal instruction set, whereas this could be extended to handle some of the other control requirements.

(b) <u>*Fully-customised hardware*</u> In this case the VDHL code would fully programme the device exactly to meet a particular requirement. Its functionality would be completely pre-defined and there would be no real-time code of any kind.

(c) <u>*Multi-function hardware*</u> There is a clear inefficiency in using a single processor to carry out a variety of different functions, and an option that falls between the previous two can be considered. This would have separate hardware elements, semi-customised to meet the range of control requirements: almost certainly a CSP core to handle the computationally-intensive real-time equations; possibly a state machine to carry out the logical operations associated with interlocking; perhaps a micro-controller core by which the general "housekeeping" activities such as simple fault-checking, input/output etc. can be executed; if required an "IP core" to handle specialist functions such as internet connectivity. All would be defined in a hardware description language of some kind, but at least two of the elements would also have software which would be executed in real-time, for example in the CSP and the micro-controller.

### 5.2 Software design and verification

In general there would be a reduction in the amount of real-time software needed, compensated in part by the existence of the hardware description code, and overall there is likely to be a significant reduction in the code that needs to be written because of the targetted and efficient programmed elements on the device offering RISC-like solutions. Of course an important change would be that the software engineer would need to know how about both types of code. Alternatively a higher level definition of how the functionality is partitioned is necessary so that electronic system designer and the real-time software designer can work separately. A very efficient solution would be to have an appropriate set of pre-designed, carefully verified hardware elements for control making up a standard device, programmed for a particular application by means of a relatively small amount of purpose-designed, highly efficient, real-time software.

Software verification raises some interesting issues. In principle the much reduced real-time code would substantially ease the process of verification, but with general-purpose processors the design would normally already have been fully verified, whereas with the programmable device solution the VHDL and synthesis process would also have be to proven. On balance it seems however that implementation of controllers using this kind of approach should result in a significant reduction in verification costs, but at this stage this is conjecture.

## 6. CONCLUSIONS

The trends seem clear: the large gains in efficiency arising from the use of programmable devices such as FPGAs means that almost inevitably they will become increasingly employed for control, particularly for demanding applications where the computational capability of general-purpose devices is stretched or when more complex control algorithms would offer superior performance but the algorithms currently require excessive computation.

The research challenges are large: a method by which functionality can be most efficiently partitioned, design tools which will enable control engineers to convert their controllers into an effective solution, and validation processes that can support application for high-integrity control requirements.

At present control specialists have very little awareness of some of the possibilities outlined in this paper, but it is hoped that it will help to expose the key issues and persuade people in universities and industry to resolve the research challenges.

## REFERENCES


Anon, (2000), Hardware-software Codesign, *IEEE Design and Test of Computers*, Jan/Mar, pp 92-99

Catthoor, F., F. Franssen, K. Cools, C. Hendriks, F. Demeester, J. De Schutter and H. De Man (1991). An Application-Specific Microcoded Architecture for a Robot Control Application. *In: VLSI Signal Processing IV*, (H. S. Moscovitz and K. Yao, Eds.), pp. 452-461. IEEE press.

Chen, D. C. and J. Rabaey (1991). PADDI: Programmable Arithmetic Devices for Digital Signal Processing. *In: VLSI Signal Processing IV*, (H. S. Moscovitz and K. Yao, Eds.), pp. 240-249. IEEE press.

Ernst, R (1998). Codesign of Embedded Systems: Status and Trends, *IEEE Design and Test of Computers*, April-June, pp 45-54

Fujioka, Y., M. Kameyama and N. Tomabechi (1996). Reconfigurable parallel VLSI processor for dynamic control of intelligent robots. *IEE Proceedings - Computers and Digital Techniques*, **143**, pp. 23-29.

Garcia Nocetti, D. F. and P. J. Fleming (1992). Parallel Processing in Digital Control. *Springer-Verlag, London*

Goodall, R. M. and M. G. Goodall, (1993), Computational procedures for recursive digital filters, *Microprocessors and Microsystems*, **17**, (10), pp 587-596.

Goodall, R. M., S. R. Jones, R. Cumplido-Parra, F. Mitchell and S. Bateman, (2000), A control system processor architecture for complex LTI controllers, *6th IFAC Workshop on Algorithms and Architectures for Real-Time Control,* Palma De Mallorca, Spain, pp 167-172.

Grout, I. A., S. E. Burge and A. P. Dorey (1995). Design and Testing of a PI Controller *ASIC. Microprocessors and Microsystems*, **19**, pp. 15-22.

Jaswa, V. C., C. E. Thomas and J. Pedicone (1985). CPAC: Concurrent Processor Architecture for Control. *IEEE Transactions on Computers*, **C-34**, pp. 163-169

Ong, SA, Jozwiak L, Tiensyrja K (1997). Interactive Codesign for Real-time embedded control systems, *Proc ISIE'97*, Portugal, pp170-175, IEEE Cat 97TH8280

Sadayappan, P., Y-L. C. Ling, K. W. Olson and D. E. Orin (1989). A Restructurable VLSI Robotics Vector Processor Architecture for Real-Time Control. *IEEE Transactions on Robotics and Automation*, 5, pp. 583-599.

Shulz, S, Rozenblit J W, Mrva, M, Buchenrieder K (1998), Model-based Codesign, *IEEE Computer*, Aug, pp 60-67

Spray, A.J.C. and Jones, S.R., 'PACE: A Regular Array for Implementing Regularly and Irregularly Structured Algorithms', *Proceedings IEE Pt G,* **138**, October 1991, pp 613-619

Tsunekawa, Y. and M. Miura (1995). High Performance VLSI Architecture Suitable for Control Systems for State-Space Digital Filters using Distributed Arithmetic. *Electronics and Communications in Japan*, **5**, pp. 12-21.

IFAC Publications
www.elsevier.com/locate/ifac

# MONITORING A PROGRAM'S PERFORMANCE USING OBJECT ORIENTED CONCEPTS

**Yong-Soo Kim, Keum-Suk Lee**

*Dept. of Health & Medical Informatics Gachon Medical School Inchon, Korea*
*1198 Kuwol-dong, Namdong-gu Inchon, 405-220, Korea*
*yskim@gachon.ac.kr*
*Dept. of Computer Engineering Dongguk University Seoul, Korea*
*3-26 Phildong, Joong-gu, Seoul, 100-715, Korea*
*kslee@dgu.edu*

Abstract: The physical attributes of a computer, such as the processor speed, the size and access time of memory, and the I/O bandwidth, are fixed when the computer is delivered to the user. But the performance of a multi-process system, where multiple processes share various resources, can be enhanced by monitoring and balancing the processes' usage of the system resources. This paper applies the object oriented concept to the performance management and suggests a standard way of monitoring the system's performance. A message-exchange protocol between the user processes and performance manager is designed and a prototype system is implemented. Through the standardization, the user processes and the performance manager can be developed independently and the system's performance can also be collectively managed. *Copyright © 2001 IFAC*

Keywords: performance monitoring, object oriented, performance manager, resource manager, managing object, managed object

## 1. INTRODUCTION

The importance of performance management is often ignored during the design and the implementation phase of a system. The developers think the system can be tuned later when severe performance problems are reported (Phillip, 1988). If the systems performance can be monitored and controlled through a standard protocol and performance metrics, the components which comprise the system, the performance manager, the resource managers, and the user processes, can be developed independently but the performance in mind from the start.

This paper assumes that 1) there are resources, resource managers and users of the resources in a computer system, 2) each resource is managed by a corresponding resource manager, 3) the resource managers and the user processes actively involve in performance management, and 4) there is a performance manager that monitors and controls the performance of the resources and the user processes. The elements that take part in performance management are defined as objects, i.e. a resource manager, a user process, and a performance manager are objects that have performance data as their attributes. The performance metrics of the resources, and the methods and attributes of the objects that are defined in the "Design of an Online Performance Management System using Object Oriented Concepts" (Young-Soo and Keum-Suk, 1996a), and "Object Oriented Online Performance Management System" (Young-Soo and Keum-Suk, 1996b) were assumed in the implementation of the monitoring system for user processes.

## 2. COMPONENTS OF THE PERFORMANCE MANAGEMENT SYSTEM

The targets of performance measurement are the system resources, the user processes and the relationship between them. The resources are assigned to user processes by the resource managers. The basic resources of the system are physical resources, such as processors, memory and I/O devices. The logical resources, such as IPC(Interprocess Communication), buffers, queues, file systems and virtual memory, are derived from the physical ones. A resource manager and a user process are defined as "managed objects" and have attributes of performance metrics which are only manipulated by their methods. The performance metrics of the managed object are transferred through a protocol to the performance manager, which is defined as a "managing object". The managing object analyzes the performance data, direct the resource managers, and provides the data to the system administrators through GUI. The managed object creates its "class and objects" of its performance metrics in the managing object through a protocol and thus actively involves in performance management.

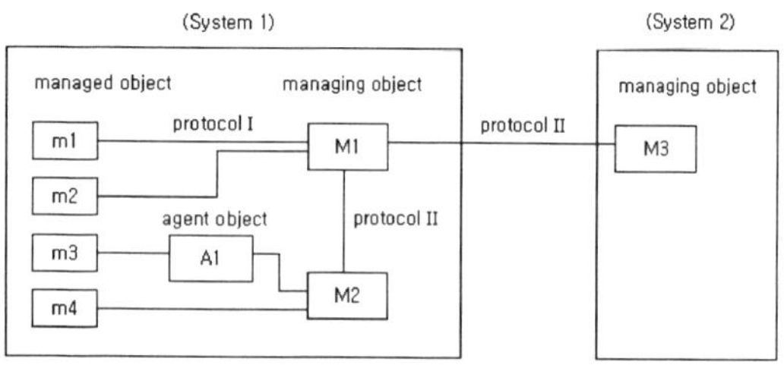


Fig. 1. Components of the system

Figure 1 shows the relationship among the objects. The protocol I defines message-exchange rules between the managed object and the managing object. There are a few merits for the performance management system designed like in Figure 1. First, because the system's performance is understood by the interaction among the user processes and the resources, it's ideal to have a protocol among them, i.e. the managed objects. But it's difficult to know the interaction among the managed objects during the design phase of a system. Even though the interaction is known later, it's not easy to modify the resource managers because they are the parts of an OS. Hence the modification of the managing object is much easier if the needs arise. Also the managed objects may converse with each other indirectly via the managing object. Second, the managing object has all the performance data of a system and thus be expanded to an expert system for performance management. Third, by defining a protocol among managing objects, the performance management for distributed systems may be possible.

## 3. DATABASE MODEL OF THE PERFORMANCE MANAGER

### *3.1 Database Model*

The performance manager receives prototypes of the performance metrics from the managed objects and builds a database by instantiating the corresponding metric classes. An object identifier is used as a key in searching the corresponding object via hashing table like in Figure 2. The synonym expression is abridged in the figure. The pointers between classes and objects are intrinsic in object oriented system and the pointers from hashing table are used for building the database. A class and an object comprise of three fields, the first field is a superclass identifier, the second field is its own identifier and the third is performance data.

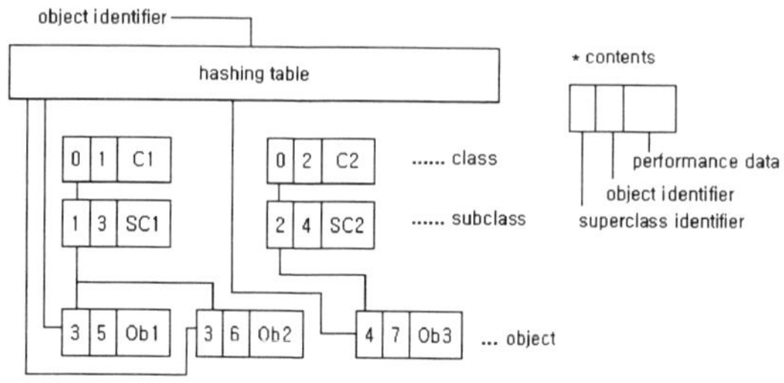


Fig. 2. Database model of the performance manager

### *3.2 Performance Data in the Database*

Each object may have more than one type of performance data. It also has current and threshold value of performance data (Young-Soo and Keum-Suk, 1996a; Young-Soo and Keum-Suk, 1996b). When the performance manager comes to select the worst performing object, it has to compare the different types performance data. So we introduce the threshold ratio as in equation(1) which is a ratio that an object's performance exceeds the threshold.

$$\text{Threshold ratio} = \frac{\text{performance data}}{\text{threshold of performance data}} \qquad (1)$$

The Class data, i. e. the third part of C1, C2, SC1, SC2 in the Figure 2 contains the object and data identifier which has the largest threshold ratio among inherited objects as in the Table 1. It also contains the number of objects and data that exceeded the threshold. This kind of design enables the performance manager to understand the overall system's performance without monitoring all objects.

Table 1 Performance data of the class

| Name | Type | Description |
|---|---|---|
| object identifier | Integer | object id that has the greatest threshold ratio |
| data identifier | Integer | data id that has the greatest threshold ratio |
| threshold value | Float | the greatest threshold ratio |
| number of objects | Integer | number of objects that exceeded threshold ratio |
| number of data | Integer | number of data that exceeded threshold ratio |

Figure 3 shows how the performance manager sets the class performance data in the database. The state of managed objects are reflected in the database of the managing object. Two tree structures on the upper part of Figure 3 show the hierarchy of managed objects. Each leaf node is an object that has an object identifier assigned by the managing object. The lower part of Figure 3 is another representation of Figure 2, which is the database governed by the performance manager. The pointers from the leaf nodes to the root are abridged in the Figure 3 as done in Figure 2.

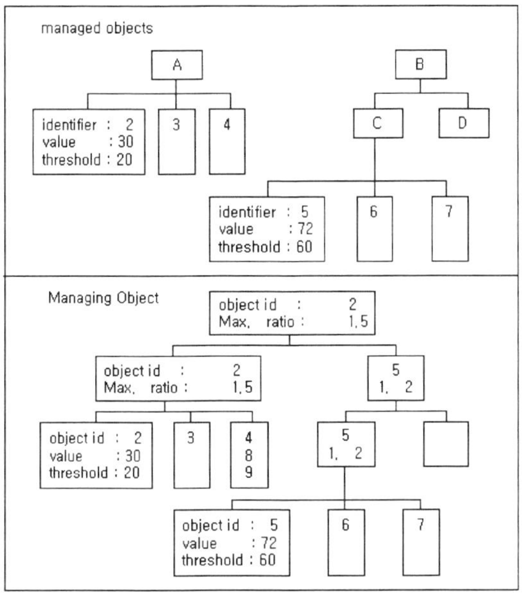


Fig. 3. An Example of the class performance database

Let's assume that the length of the message queue of the managed object "5" exceeded the threshold and sent an event message to the managing object. The managing object finds the object id "5" in its own database, sets the performance data, calculates the threshold ratio, compares the calculated threshold ratio to the greatest threshold ratio of the upper object in the hierarchy and if the former is greater than the latter, the greatest threshold ratio of the upper object is set with the former value and the comparison continues with the next higher object. This process continues up to the root node and the other fields in Table 1 are also set. Whenever the former value is not greater than the latter, the operation stops. Now, the utilization of the managed object "2" exceeds the threshold and sent an event message to the managing object. The managing object does the same processing for "2" as for "5". In Figure 3, as the threshold ratio of "2" is greater than that of "5" in the root node, the former value is set in the root node. This structure enables us to identify various levels of performance bottlenecks by checking the root or intermediate nodes. It also makes possible to reduce the monitoring overhead because the managing object do not have to visit the each leaf node to understand the overall performance of the system.

## 4. SERVICES AND PROTOCOL

### *4.1 Definition of Services*

The half-duplex flip-flop is used for conversation between the objects, which means that a requester issues next request only after it receives a reply for the previous request. The services are described in Table 2. The mo means a managed object and MO a managing object respectively in the table.

Table 2 Definition of services

| Service | Protocol | Description |
|---|---|---|
| CREATE | I | mo requests MO the creation of class/object |
| DELETE | I | MO/mo requests other the deletion of class/object |
| GET | I | MO requests the performance data from mo |
| SET | I | MO requests mo to modify the settable data |
| ACTION | I | MO sends mo the control message |
| CANCEL | I | MO requests mo the cancelation of the previous request |
| EVENT-REPORT | I | mo reports an event to MO(unsolicited message) |
| GET-NEXT | I | MO requests mo the next data |
| AUTO-REPLY | I | MO requests mo the performance data repeatedly |
| MIGRATE | II | MO migrates mo data and conversation to other MO |
| INFORM | II | data exchange among MO(like inform-request of SNMPv2) |

### *4.2 Protocol*

The protocol I is for the conversation rules between the managed object and the managing object, and the protocol II among the managing objects as shown in Figure 1. The protocol I is similar to RFC 1157 of SNMP, and CMIS(Common Management Information Service),i.e. ISO 9595/X.710 (ISO, 1991a) and CMIP(Common Management Information Protocol),i.e. ISO 9596/X.711 (ISO, 1991b). ISO/CCITT ISO 9595/X.710 defines 7 services for the user interface. This paper defines 11 services and uses the service names from ISO if the meaning is similar.

### *4.3 Format of the Protocol Data*

The PDU(Protocol Data Unit) corresponding to the protocol I for CREATE and GET message is presented in ASN.1(Abstract Syntax Notation One) (ISO, 1991c) format in the Appendix. A managed object sends M-CREATE message to the managing object to create an instance of its performance metric in the managing object. The managing object requests performance data to the managed object through M-GET message.

## 5. IMPLEMENTATION

### *5.1 Static Instrumentation*

The static instrumentation is used for easy of implementation, and a pre-compiler and a pre-linker are devised. A parameter file which describes various monitoring points is passed to the pre-compiler with the source code. The pre-compiler finds relevant monitoring locations in program source codes using the parameter file and inserts instrumentation codes there. During the pre-compiling process the instrumentation file, which contains "instrumentation sequence number", "function name or line number" and "comment on the statement" to name an important few, is produced. These function name and comment are used for the name of performance data. A stub routine which allocates a dynamic memory to store performance metrics and initiates a communication with the managing object, is added at the start of the main program. This stub routine gets the relevant information from the instrumentation file produced by the pre-compiler and the pre-linker.

Table 3 Instrumentation locations

| Specification | Description |
|---|---|
| call | all function calls |
| syscall | system function calls |
| usercall | all function calls |
| do_while | do while structure |
| do_until | do until structure |
| line number | call at the specified line, do while, do until |

Table 3 shows several monitoring points where the instrumentation can be done. The pre-linker makes a final instrumentation file using the instrumentation files produced by the pre-compiler, and the module name is included in the final file. It reads the instrumentation files sequentially as specified in the command arguments.

Figure 4 describes a process making a final instrumentation file. Assume a main program A calls subprogram B at 50th line, system interface malloc() at 65th and kill() at 80th. B calls C at 12th line and system interface read() at 30th. Here we want to monitor all function calls invoked in A and user function calls in B.

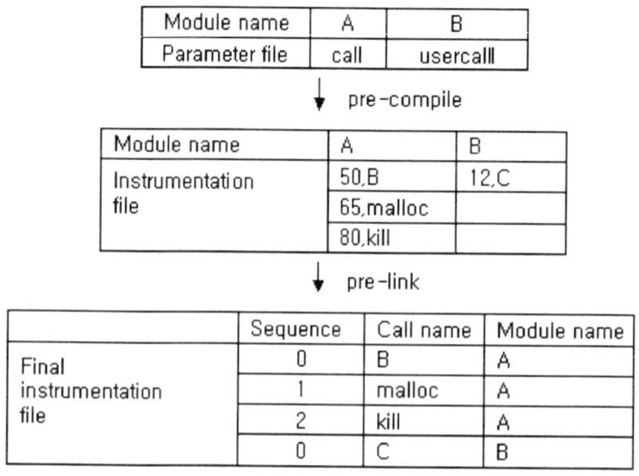


Fig. 4. Instrumentation file

### *5.2 Monitoring the User Process*

Figure 5 shows how a user process participates in the performance monitoring. The stub routine(Ostub) calculates the size of performance data from the instrumentation file and allocates it(Odata) in the shared memory①. Ostub creates a managed object(mo)②. The instrumentation code in the running process gathers performance data and stores them in Odata③. mo sends CREATE.request message to the managing object(MO), i.e. performance manager, which creates performance database as in Figure 2 ④. When the performance manager requests performance data via GET.request message, mo gets the data from Odata⑤ and transfers it to MO⑥. Events ③, ④ and ⑤ can occur at random sequence.

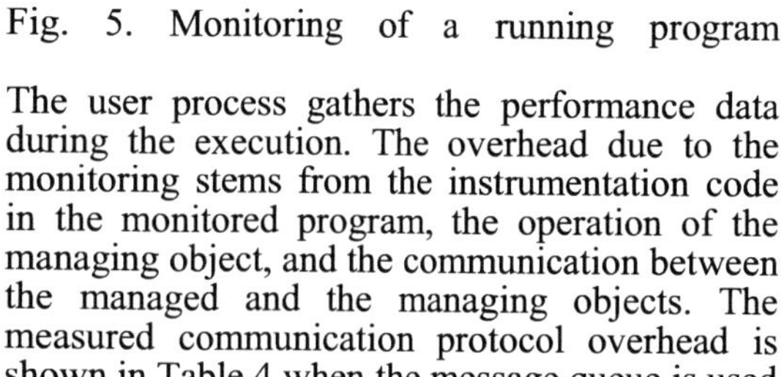


Fig. 5. Monitoring of a running program

The user process gathers the performance data during the execution. The overhead due to the monitoring stems from the instrumentation code in the monitored program, the operation of the managing object, and the communication between the managed and the managing objects. The measured communication protocol overhead is shown in Table 4 when the message queue is used under UNIX.

An average transmission time of 1,000 bytes of message is measured with the variation of the number of waiting users on different platforms. As shown in Table 4, the transmission time is proportionally increase as the number of waiting users increases.

Table 4. Average message transmission time through message queue

| H/W | OS | # of waiting processes on queue | | | |
|---|---|---|---|---|---|
| | | 0 | 4 | 9 | 19 |
| Unisys U/6000 | SVR4.2 | 0.22 ms | 1.4 ms | 2.6 ms | 4.9 ms |
| HP D350/2 | HP-UX 10.10 | 0.06 ms | 0.5 ms | 0.7 ms | 1.3 ms |
| Sun Server 1000 | Solaris 2.4 | 0.06 ms | 0.4 ms | 0.7 ms | 1.4 ms |

In Table 4, it takes about 0.22 ms of transmission time for Unisys. So it can accommodate maximum traffic of 4,545 messages/sec.

$$1{,}000 \div 0.22\text{ms} = 4{,}545/\text{sec} \quad (2)$$

Generally, the overhead due to monitoring is allowed within 2 - 5 % range in commercial online performance monitors (Candle, 1991). Here if we assume that 1 % of the processor power can be alloted to the message traffic overhead, we can get 45 message traffics per second.

$$4{,}545 \times 0.01 = 45/\text{sec} \quad (3)$$

If we assume that the arrival time of the messages has exponential distribution and the service time is constant, M/D/1 queue can be applied.

Then, Transmission density(ρ) =
Arrival rate(λ) × Service time per work(E[s])

$$45 \times 0.22 \times 0.001 = 0.01 \quad (4)$$

The probability Pn where n users exist in the queue is presented as 3 equations as in (5) (Raj, 1991).

$$P_n = \begin{cases} 1-\rho & ,n=0 \\ (1-\rho)(e^{\rho-1}) & ,n=1 \\ (1-\rho)\sum_{j=0}^{n}\frac{(-1)^{(n-j)}(j\rho)^{(n-j-1)}(j\rho+n-j)e^{j\rho}}{(n-j)!} & ,n\geq 2 \end{cases} \quad (5)$$

If we substitute the transmission density to the equation (5), the probability where no users exist in the queue is 0.99 and 0.0099 for one user. Thus the probability where two or more users exist in the queue is 0.00001 which is calculated as "1 - 0.99 - 0.0099". This value is negligible, so the performance degradation due to the increase of waiting users can be ignored.

## 6. CONCLUSION

It's possible to standardize the system performance management environment by defining the components, i.e. objects that actively take part in the performance management, and the protocol between the objects. Through the standardization, the portability and interoperability of the performance management system can also be accomplished. We showed that the user programs can actively involve in the performance management through static instrumentation. In the paper, user processes and the performance manager are defined as the managed objects and the managing object respectively and a message exchange protocol among objects is presented. If the source codes of the resource managers which are a part of an OS can be accessed, we can apply the similar instrumentation into the codes and have the resource managers take part in the performance management.

## REFERENCES


ISO(CCITT). (1991a). *ISO/IEC9595 (X.710) : Common Management Information - Service Definition.*

ISO(CCITT). (1991b). *ISO/IEC 9596(X.711): Common Management Information - Protocol Definition.*

ISO(CCITT). (1991c). *ISO/IEC 8824(X.208): Abstract Syntax Notation One.*

Candle. (1991). *Omegamon II for MVS: Volume II Advanced Reference Realtime Component.* Candle.

Phillip McKerrow. (1988). *Performance Measurement of Computer Systems.* Addison-Wesley Publishing Company.

Raj Jain, (1991). *The Art of Computer Systems Performance Analysis: Techniques for Experimental Design, Measurement, Simulation and Modeling.* John Wiley & Sons Inc.

Yong-Soo Kim, Keum-Suk Lee. (1996a). Design of an Online Performance Management System using Object Oriented Concepts. *Journal of KISS(A)*, **23**, 594-608.

Yong-Soo Kim, Keum-Suk Lee. (1996b). Object Oriented Online Performance Management System. *22nd Euromicro Conference.* 95-100.


## APPENDIX

```
Performance_PDUs ::= BEGIN
IMPORTS TimeTicks FROM RFC1155-SMI;
PDUs ::= CHOICE {
   create_request [0] IMPLICIT CreateRequest,
   create_response [1] IMPLICIT
CreateResponse,
   get_request [4] IMPLICIT GetRequest,
   get_response [5] IMPLICIT GetResponse}
```

```
CreateRequest::=SEQUENCE{
      createRequestMessage MessageType(1),
      version  VersionNumber,
      key  MessageQueueKey,
      classCount  CountType,
      classIdentifiers OBJECT IDENTIFIER,
      performanceData PerformanceData }

MessageType ::= INTEGER(0 .. 255)
VersionNumber ::= INTEGER(0 .. 32767)
MessageQueueKey ::= INTEGER
CountType ::= INTEGER(0 .. 32767)
PerformanceData ::= SEQUENCE {
dataCount CountType,
dataProtoType SEQUENCE OF DataProtoType }
DataProtoType ::= SEQUENCE {
      dataIdentifier IdentifierType,
      dataname Name,
      type DataType,
      limitValue Value,
      nomalValue Value,
      eventStatus EventStatus }
IdentifierType ::= INTEGER
Name ::= VisibleString
DataType ::= INTEGER(0 .. 255) {
      type_1(1),
      type_2(2),
      type_3(3) }
Value ::= CHOICE {
      integer INTEGER,
      real  REAL }
EventStatus::=INTEGER(0..255){
      eventReport(1),
      noEventReport(2),
      statusReport(3),
      noStatusReport(4) }

CreateResponse::=SEQUENCE{
      createResponseMessage MessageType(2),
      version  VersionNumber,
      objectIdentifer IdentifierType,
      createResult  CreateResult }

CreateResult ::= INTEGER(0 .. 255) {
      noError(0),
      versionError(1),
      formatError(2),
      overHeadError(3),
      spaceError(4) }

GetRequest::=SEQUENCE {
      getRequestMessage MessageType(17),
      objectIdentifier IdentifierType,
      requestStructure RequestStructure }
RequestStructure ::= SEQUENCE {
      dataCount CountType,
      requestDataStructure SEQUENCE OF
      RequestDataStructure }
RequestDataStructure::=SEQUENCE{
      dataIdentifier IdentifierType,
      attributeData AttributeData }
AttributeData ::= SEQUENCE {
attributeCount CountType,
attributeNames SEQUENCE OF AttributeName }
AttributeName ::= VisibleString

GetResponse ::= SEQUENCE {
      getResponseMessage MessageType(18),
      objectIdentifier IdentifierType,
      getResult GetResult,
      responseStructure ResponseStructure }
GetResult ::= INTEGER(0 .. 255) {
      noError(0),
      noSuchIdentifier(6),
      noSuchName(7),
      badValue(8),
      genError(9) }
ResponseStructure ::= SEQUENCE {
   dataCount CountType,
   ResponseDataStructure SEQUENCE OF
   ResponseDataStructure }

ResponseDataStructure ::= SEQUENCE{
      dataIdentifier IdentifierType,
      attributeStructure AttributeStructure }
AttributeStructure ::= SEQUENCE {
   attributeCount CountType,
   attributePairs SEQUENCE OF AttributePair }
AttributePair ::= SEQUENCE {
      attributeName AttributeName,
      attributeValue Value }

END
```

IFAC Publications
www.elsevier.com/locate/ifac

# DATA AND EVENT MANAGEMENT IN A MARITIME TERMINAL OF CONTAINERS[1]

**J.L. Poza, J.L. Posadas, J. Simó, A. Crespo**

*Departamento de Informática de Sistemas y Computadores (D.I.S.C.A.)*
*Universidad Politécnica de Valencia, Spain*
*{ jopolu, jposadas, jsimo }@disca.upv.es*

Abstract: Containers are currently used in the transport of products because of the point-to-point service they offer. By using specialised cranes, operations in maritime transport and land transport can be achieve quickly. To optimise operation costs, the whole management of the terminal is computerised proportionally with the complexity of the problem. This paper studies the problem from the point of view of the integration of different technologies supporting communication, data management, control, and human interface. This integration of heterogeneous intelligent components is made possible by means of a communication system that is the base of an object oriented distributed system. Using this system, "operation agents" that gain access to information in a high level and transparent way can be developed. 

Keywords: Distributed computer control systems, transportation control, information integration.

## 1. INTRODUCTION

Maritime terminal is the connection between maritime transport and transport by land. To make this process it is necessary to store containers temporarily in the yard, using specific machines, until they are sent to their destination (Fig. 1.).

The yards are normally large and, due to the environmental impact of ports and their high costs, it is necessary to optimise the utilisation of space and the resources involved in the container management. (Simarro, et al., 2001).

The whole set of computer resources and procedures is referred to as Terminal Operating System. These systems are composed of databases, and tools intended to help human operators in the planning tasks.

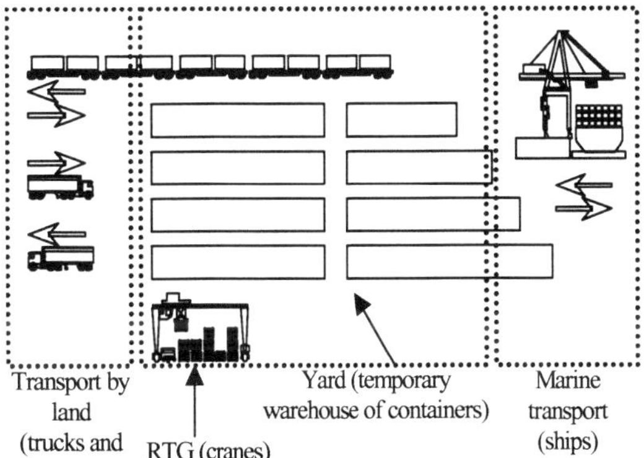


Fig. 1. Terminal of containers

[1] This work has been supported by grant 1FD97-2158-C04-03. CICYT Spanish Government.

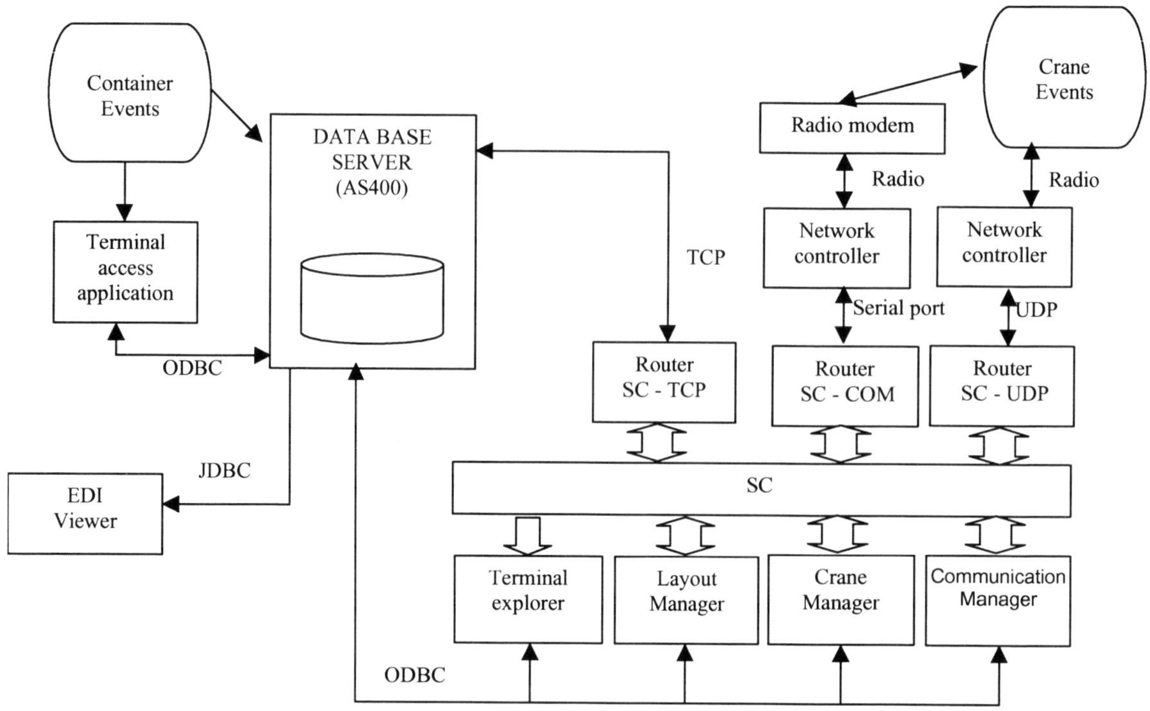


Fig. 2. Application distributed system and communication channels

The results of off-line planning are difficult to implement due the high unpredictability of the elements to control. This leads us to consider the terminal operating system as a real-time information and control system instead of a central planning system.

Commercial operative systems used in container terminals are closed and should be flexible. In many cases flexibility for operative is supported by the system, but not the flexibility for the data sources and the communication system installed on the layout.

This lack of flexibility is a serious drawback for system adaptation and migration to similar industrial systems.

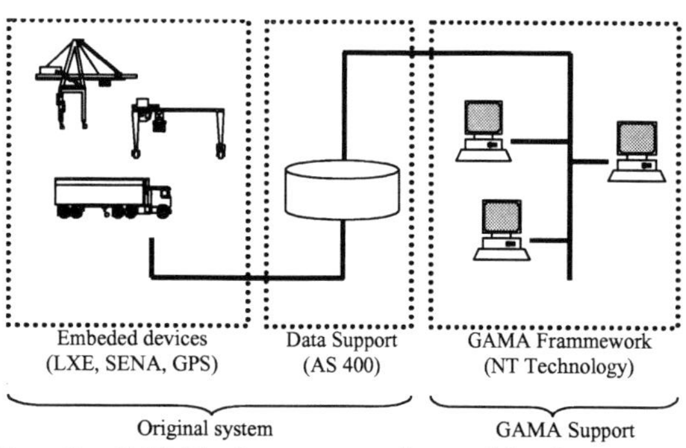


Fig. 3. GAMA interconnection with the current system.

Container terminal automatization is minimal. Distributed applications that add the support to the management of the terminal must be connected to the system as transparently as possible (Fig. 3.).

The goal is to develop an open system that distributes information through time-stamped events. This temporal control is done using the temporal firewall concept (Kopetz, 1998). The time stamp is used to compute the delays in the system and therefore to detect the bottlenecks and overload situations. The developed system is known as GAMA (Gestión Automática MArítima) (fig. 2 y 3).

## 2. GAMA SYSTEM APLICATION ARCHITECTURE

For the correct terminal management, from the point of view of quality of service and resources optimisation, four modules with very specialised functions have been proposed (Fig. 2):

- Layout Manager: determines the optimal container position in the yard based on specific distribution or location criteria.
- Crane manager: optimises machine movements in the container yard.
- Terminal explorer: high performance graphics module that shows, in real time, terminal state and machines position implied in the work with containers.
- Communication system: composed of the Communication Server (SC) and a set of routers. SC enables the connection of Win32 processes through a distributed blackboard system. Specialised routers allow the interconnection of

different platforms, outside the Win32 world, which are not compatible with SC (embedded computers and AS400 processes).

Data queries are specific for each application, thus connections are made at this level (ODBC). Nevertheless, a same event is relevant for many applications, so a distributed communication channel is necessary.

The system is organised in a distributed manner, each process can be executed in any host, belonging to the intranet, only by connecting the data bus and/or event bus (Fig. 4.).

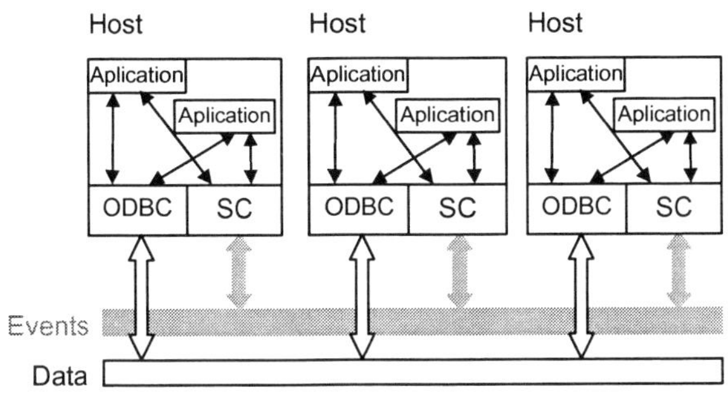


Fig. 4. Data and event bus connection.

This structure of distributed applications assumes the existence of a multilevel system of communications. This allows the communication of events produced in the containers terminal and with the data associated to the management of the containers terminal (Fig. 4.).

When a physical event is generated in the container yard, such as the arrival of a container or the movement of a crane, a logical event is produced. This logical event causes a set of messages which depends on the container yard state. These messages travel through a set of communication channels to arrive finally at one or more applications to process them. For example, the arrival of a container sends a message to the "container manager" which will have to decide the container's location in the yard. Event oriented communication is provided by the SC channel and the routers.

In many scenarios, applications have to make queries or changes in the container yard database. These data actions need a data oriented communication channel that is provided by operating system ODBC drivers.

## 3. GAMA SYSTEM COMMUNICATION STRUCTURE

Communication in distributed systems presents several problems (Hadzilacos and Toueg, 1993) (Singha and Casavant, 1991). Mainly the following:

- Synchronisation and coordination of messages and applications.
- Coherence in the format of the messages.
- Fault tolerance.
- Marking time boundary in applications.

Communication between GAMA modules is achieved in two levels defined by the communication channel used (Fig. 5.).

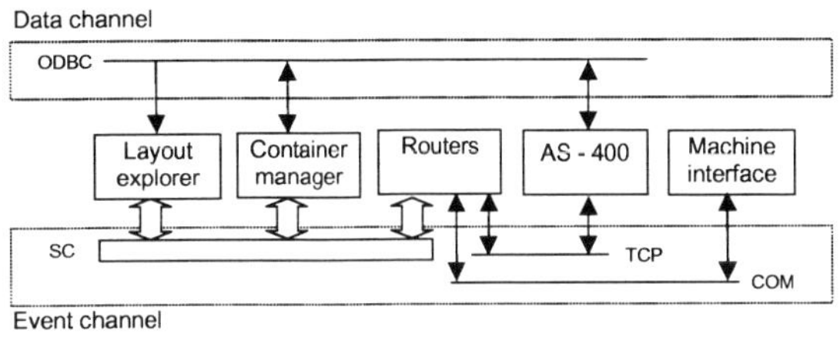


Fig. 5. Communication system diagram.

- Event level (SC and routers): offers oriented services to communication and diffusion of the events produced by the terminal. A Communication Server (SC) based in the data sharing through a distributed blackboard model (Penny, 1989) makes this diffusion. Specific routers are the necessary bridges connecting heterogeneous systems with the SC.

- Data level (ODBC): offers database access to perform queries and changes in the container yard state.

### *3.1 Event Communications: SC and Routers*

To allow the interaction between Win32 processes of the distributed system, a communication common interface called Communication Server (SC) is used. The SC system is based on the internal data representation through distributed blackboard architecture. The data structure that represents the blackboard is continually updated by the SC, showing the values of the messages sent between the processes. Through the access to the data structure of the SC, the distributed processes communicate using a common interface independently of their location.

The SC system has been developed and tested by the authors in previous works in the context of distributed systems and mobile robots (Posadas, et al., 2000).). The system has been subjected to extreme condition tests proving its reliability and high performance.

The SC system requires that each computer belonging to the system executes an SC instance (Fig. 6.). These instances are in charge of fulfilling the communication tasks, updating and control data for obtaining partial copies in different computers. As a result, processes, independently of their location, can contact all the system by making accesses local SC instance.

To guaranty the consistency between the copies of the distributed blackboard, the SC instances manage the necessary communications and the data transmission between the computers establishing permanent connections with fault detection and recovery.

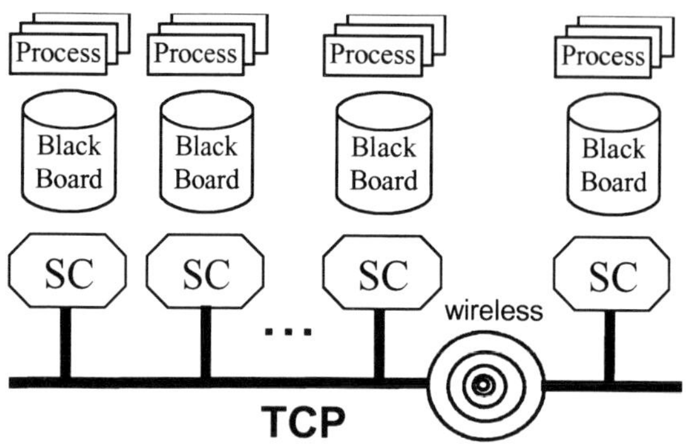


Fig. 6. SC connection system with the processes.

The system gives a bind-notification schema to associate the execution of specific code of processes with a set of events such as the updating in the system of a data in the blackboard. Thus, processes connected to the SC receive automatically the messages which they are interested in.

The configuration of the SC consists, mainly, on the specification of a object set that is used as a logic media for distributing system messages. This allows all applications access to them.

There are nodes where it is impossible to execute a SC instance (for example, a embedded PC where the operating system does not have the SC requirements, Unix systems or AS400 systems), in these cases the processes have to communicate with a remote SC through a specific gateway or router application. These nodes are called heterogeneous nodes.

In the normal operation of GAMA routers have the function to open, maintain and close the communication channels through sockets (TCP, UDP) and serial port (COM) between the heterogeneous nodes and the Communication Server (SC).

Messages transmitted between the Database Server and the Communication Server (SC) can be synchronous (with answer) or asynchronous (without answer). The management of the answers in the case of the synchronous messages is the responsibility of routers. Routers, therefore, hold a set of objects and threads keeping track of the life cycle of each communication transaction.

*3.2 Data Communication: ODBC*

Distributed applications have to make queries into the Database Server through ODBC channel.

ODBC communications use TCP through specific drivers for AS400 from Database Server and through a specific DLL (Dynamic Linked Libraries) from GAMA modules. These specific DLL's export a set of tools and data formats to help applications accessing terminal data in a uniform way independently of the database technology and definitions in the background. The programming tasks in the adaptation of GAMA system to a new database system are restricted to minimal changes in these DLL's.

## 4. DEVELOPED SYSTEM

The management of cranes and containers is an unpredictable problem due to the continuous variation of the container yard state and several factors such as weather changes and transport delays.

However, container yard behaviour can be simulated off-line, and modelled and monitored on-line. The developed system can make a snapshot of the yard state to simulate different management policies.

*4.1 Off-line System*

Starting from a yard state it is possible to perform different simulations, with different container and crane management policies, to obtain the possible evolution of the system and future results (Fig. 7.). Therefore, different policies can be evaluated before deciding to use one of them.

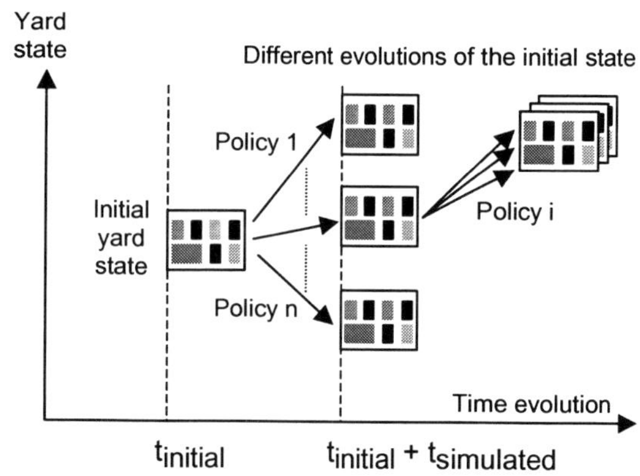


Fig. 7. Yard states evolution simulations.

*4.2 On-line System*

A first prototype of the system is already working in the terminal of the port of Valencia. Current installed applications, even being completely functional, do not require high performance in the communications. Because of this reason and other operative ones, the real plant is not the best scenario for system tests. The future introduction of the simulator on-line (Fig.8.) points to a high increase of the communication requirements and these extreme conditions must be foreseen. The system presented in this paper has been working under extreme conditions generated by the plant simulator off-line.

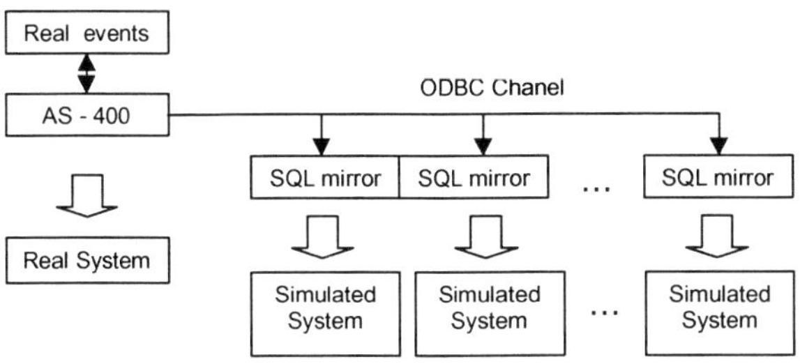


Fig. 8. Interconnection between real and simulated system

## 5. CHARACTERIZATION OF THE SYSTEM

### *5.1 Model of the System*

For each event produced, from any source, there is a route of channels and applications through the associated message has to travel. Each element of the system can be seen like an interconnected node with others based on the route that must take the message (fig.9.).

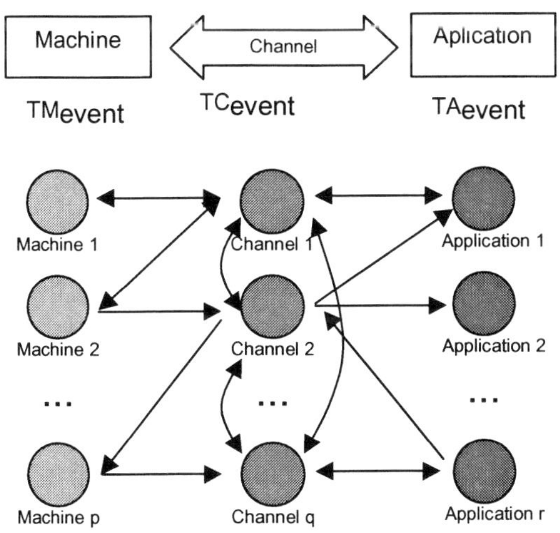


Fig. 9. Generic node graph of the system.

The time of a generic message corresponding to a specific event through the system can be characterised by the next equation (1).

$$T_{event} = TM_{event} + TC_{event} + TA_{event} \qquad (1)$$

Where $T_{event}$ is the total time of permanence of a generic event in the system: from the event is generated by a physical event source, to it is processed by one or more applications. $TM_{event}$ is the time that an event remains in a machine: from it is generated, to an embedded system inserts it in some communication channel. $TC_{event}$ is the time used by an event in crossing a generic communication channel: from the insertion of the event in the channel by the machine, to the event is extracted from the communication channel by an application. $TA_{event}$ is the time that an event remains in an application: from it enters, to it is processed.

The calculation of the time for a specific message is made directly following the route of the message by all the components of the system (fig.10.).

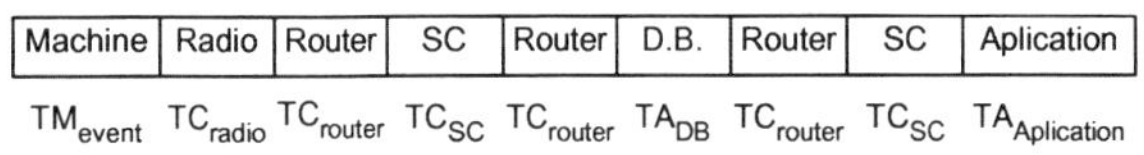


Fig. 10. Temporary sequence of a generic event through the applications.

Once the route of a message corresponding to an event has been determined, obtaining the time corresponding to the event is made using the equation that represents the specific route. Every type of message has its own equation (2).

$$T_{event} = \sum_{i=1}^{p} TM_i + \sum_{i=1}^{q} TC_i + \sum_{i=1}^{r} TA_i \qquad (2)$$

Where $T_{event}$ is the total time that an event remains in the system, taking into account differences in the time measurements depending on the transmission media used. For each event, different machines (*p*), different communications channels (*q*) and different applications (*r*) can be involved in the calculation of the total time.

Each node knows the temporary limits that must have a message related with an event. Based on this, nodes can determine the validity of the messages consulting an internal table (fig.11) in which the times have been obtained from the corresponding equation (2) applied to the specific node.

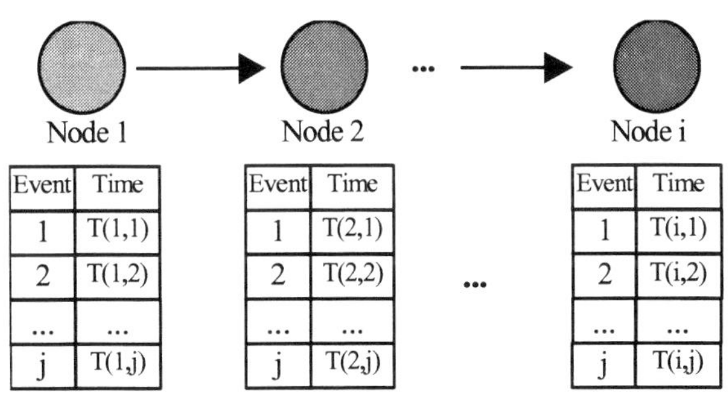


Fig. 11. Internal node table to process the "Time Firewall" value.

Content of tables can be changed depending on the system topology. Because of this reason, it is necessary to store the tables in the Database Server. When a change in the system topology occurs, as a new ODBC driver or a new network component, content in the nodes must be actualised.

Table 1. Generic time table for a node i that process j different messages.

| Type of messages | Minimum time | Maximum time |
|---|---|---|
| *Message 1* | T(I,1) min | T(i,1) max |
| *Message 2* | T(I,2) min. | T(i,2) max. |
| ... | ... | ... |
| *Message j* | T(I,j) min. | T(i,j) max. |

A well know variation of the system topology, is not a problem due to the times minimum and maximum, for a node i and a message type j, are registered in the tupla T(i,j).

*5.2 Time Control*

In order to characterise temporarily the time consumed in the route from a message through the system, the Time Firewall method is used by each application maintaining a registry of the time consumed by a message while it remains in the application.

When the message arrives, the application compares its Time Field with the corresponding internal table where temporary characteristics of the message are specified. Based on this comparison, the message will be dealt with one or another form.

When the message leaves the application, the time that has remained in the application is added to the Time Field.

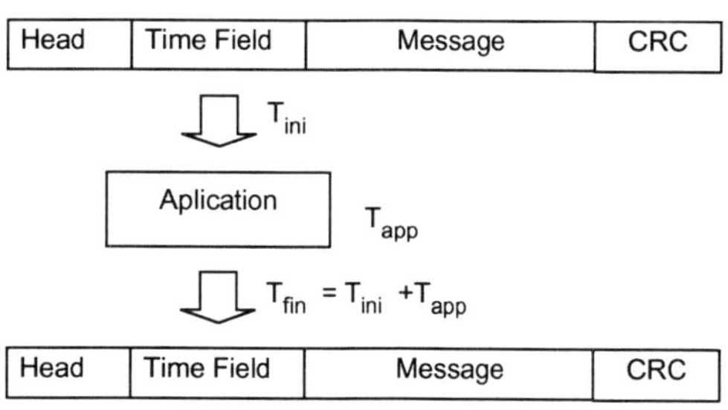


Fig. 12. Time processing in a generic message.

For every node exists a table to maintain the time intervals (Fig. 11.). With this table, the state of the communications can be temporarily characterised at any moment.

This temporary characterisation is very interesting because it allows knowing at any moment if the system is tuned and works in a synchronous form. In the case of arriving messages with a time field with very high value, it is possible to make an inspection in the route of nodes towards the origin of the bottleneck and take measures to solve it.

## 6. CONCLUSIONS

A distributed industrial on-line system with simulation support has been developed. The system can test and simulate different policies about the distribution of the containers and the management of the machines and it can demonstrate the viability in the response of the system in the real plant taking into account future communication requirements. The first prototype of the GAMA software is currently working in the port of Valencia.

The core of the GAMA software is the communication system, a programming model used to develop distributed modules in a transparent and scalable way. Routers have been developed to open the communication system to heterogeneous nodes such as currently devices installed in the plant. Time stamps, in the form of "time firewall", are proposed to look for "bottlenecks" and find the system limits under stressed simulation.

The system can be adapted easily to other similar industrial system without great relevant changes.

## REFERENCES


Simarro, R., J.L Navarro, I. Huet, E. Orellana. (2001). Simulación de una terminal marítima de contenedores. *Workshop en Metodologías de Modelado y Simulación de Sistemas*. Barcelona.

Hadzilacos, V., S. Toueg. (1993). Fault-Tolerant Broadcasts and Related Problems In: *Distributed Systems.* (S. Mullender, Ed.). ACM Press.

Singha Mukesh, T. L. Casavant (1991). Distributed computing systems. *Computer*, 24(8):12-14.

Penny, H. (1989). *Blackboard Architectures and Applications.* Edited by V. Jagannathan, Rajendra Dodhiawala, Lawrence S. Baum.

Kopetz H. (1998). The Time-Triggered Model of Computation. 0-8186-9212-X/98 *IEEE.*

Posadas, J.L., P. Pérez, J.E. Simó, G. Benet, F. Blanes (2000). Communications Structure for Sensor Fusion in Distributed Real Time Systems. *6th IFAC Workshop on Algorithms and Architectures for Real-Time Control. AARTC'2000.* Mallorca.

IFAC
Publications
www.elsevier.com/locate/ifac

# A RECURSIVE MULTI-BLOCK PLS ALGORITHM FOR MONITORING INDUSTRIAL PROCESSES

Xun Wang[1], Uwe Kruger[2*], Andrew Y.T. Leung[3], Barry Lennox[1]

[1]*Department of Mechanical Engineering, The University of Manchester, M13 9PL, U.K.*
[2]*Intelligent Systems and Control Group, Queen's University of Belfast, BT9 5AH, U.K.*
[3]*Department of Building and Construction, City University of Hong Kong, P.R. China*

Abstract: Industrial processes often present a large number of highly correlated variables that are frequently measured and consist of several operating units. For monitoring such processes multi-block partial least squares (MBPLS) has shown to be capable of dividing the variables according to operating units and can reduce the number variables that describe significant process variation with sufficient accuracy. As shown in this paper, MBPLS may run into difficulties when the process exhibits non-stationary behaviour. To overcome this deficiency, the development of recursive MBPLS is discussed including an application study that relates to a realistic simulation of a fluid catalytic cracking unit. *Copyright © 2001 IFAC*

Keywords: Multivariate Quality Control, Statistical Process Control, Large-Scale Systems, Model Reduction, Process Identification

## 1. INTRODUCTION

The task of detecting and diagnosing abnormal operating conditions in industrial processes, whether continuous or batch, is difficult. This is because such processes frequently present a large number of process variables, such as temperatures, pressures, flow-rates, compositions etc., which are regularly recorded up to several thousand times a day (Piovoso and Kosanovich, 1992). This very large amount of data is difficult to analyse and interpret simply by observation. Additionally, industrial processes are often comprised of numerous operating units which often show complex interaction.

To address the issue of detecting and diagnosing abnormal operating conditions, Multivariate Statistical Process Control (MSPC) approaches have been successfully employed (MacGregor, 1997). The MSPC techniques aim to reduce the number of variables which are required to describe significant variation of the process. One such MSPC method is MBPLS, which is an extension of the partial least squares algorithm, pioneered by Wold in the mid 1960s.

The development of MBPLS dates back to H. and S. Wold (Wold and Joreskog, 1980; Wold, *et al.*, 1987). Since then, several refinements of MBPLS have been discussed (Wangen and Kowalski, 1988; Westerhuis, *et al.*, 1998). MBPLS is designed to divide a given set of variables into a number of "blocks" and then to identify relationships between these blocks. If these "blocks" are predictor "blocks", Westerhuis, *et al.*, (1998) demonstrated that the MBPLS solution for multiple blocks can be converted into the conventional PLS solution for two blocks (one representing the predictor and one representing the response block). With MBPLS, each block can be monitored on the basis of two univariate statistics, one of which represents significant variation (the $T^2$ statistic) and the other insignificant variation (the SPE statistic) of the process unit.

* Correspondence should be sent to:
uwe.kruger@ee.qub.ac.uk

Process monitoring with MBPLS, however, may run into difficulties if the process exhibits strong non-stationary process behaviour. In such cases, the provided univariate statistics give rise to false alarms. In this paper, a recursive MBPLS or RMBPLS is developed to overcome this deficiency. The paper further discusses the application of adaptive control limits for the univariate statistics to encounter strong non-stationary process variation. It is demonstrated that the combination of both approaches can considerably reduce false alarms. However, the univariate statistics are sensitive for detecting abnormal process behaviour.

The organisation of the paper is as follows. A brief summary of the partial least squares (PLS) and the recursive PLS is presented in section 2 and 3, respectively. This is followed by a discussion of the MBPLS algorithm and the development of the RMBPLS algorithm in section 4. Then, the univariate statistics for the blocks are discussed in section 5. A summary of the monitoring approach is given in section 6 prior to an application that relates to the simulation of a fluid catalytic cracking unit. Finally, section 8 summarises and concludes this paper.

## 2. PLS ALGORITHMS

PLS decomposes a given predictor matrix, $\mathbf{X}\in\mathbb{R}^{K\times M}$, and a given response matrix, $\mathbf{Y}\in\mathbb{R}^{K\times N}$, as follows:

$$\mathbf{X}=\sum_{i=1}^{n}\mathbf{t}_k\mathbf{p}_k^T+\mathbf{F}_n=\mathbf{T}_n\mathbf{P}_n^T+\mathbf{F}_n;\qquad \mathbf{Y}=\sum_{i=1}^{n}\hat{\mathbf{u}}_k\mathbf{q}_k^T+\mathbf{E}_n=\hat{\mathbf{U}}_n\mathbf{Q}_n^T+\mathbf{E}_n \tag{1}$$

where:

- $\mathbf{t}_k$ is referred to as the $k^{th}$ t-score vector of the predictor matrix,
- $\hat{\mathbf{u}}_k$ is referred to as the $k^{th}$ predicted u-score vector of the response matrix,
- $\mathbf{p}_k$ and $\mathbf{q}_k$ are defined as the loading vectors of the predictor and the response matrix, respectively,
- $n$ represents the number of latent variables (LVs) retained,
- $\mathbf{F}_n$ and $\mathbf{E}_n$ are the residual matrices of the predictor and the response matrices, respectively,
- $\mathbf{T}_n$ and $\mathbf{P}_n$ are the score and loading matrix to reconstruct the predictor matrix and
- $\hat{\mathbf{U}}_n$ and $\mathbf{Q}_n$ are the predicted score and loading matrix to predict the response matrix.

Note that the predictor and the response matrices are constructed from process data that describe normal variation of the process.
For determining the $k^{th}$ score and loading vectors, an iterative algorithm is required. To calculate subsequent score and loading vectors, the contribution of each pair of score and loading vectors is subtracted from the predictor and the response matrix, which is referred to as the deflation procedure:

$$\mathbf{X}_k=\mathbf{X}_{k-1}-\mathbf{t}_{k-1}\mathbf{p}_{k-1}^T\qquad \mathbf{Y}_k=\mathbf{Y}_{k-1}-\hat{\mathbf{u}}_{k-1}\mathbf{q}_{k-1}^T. \tag{2}$$

The $k^{th}$ pair of score vectors, the t-score and the u-score vector, can be calculated as the product between the deflated predictor and response matrices, $\mathbf{X}_k$ and $\mathbf{Y}_k$, and the weight vectors:

$$\mathbf{t}_k=\mathbf{X}_k\mathbf{w}_k\qquad \mathbf{u}_k=\mathbf{Y}_k\mathbf{q}_k, \tag{3}$$

where:

- $\mathbf{w}_k$ is the $k^{th}$ weight vector of predictor matrix and
- $\mathbf{q}_k$ is the $k^{th}$ weight vector of the response matrix.

For each pair of score vectors, the score model is defined as follows:

$$\mathbf{u}_k=\hat{\mathbf{u}}_k+\boldsymbol{\varepsilon}_k=\mathbf{t}_k\cdot b_k+\boldsymbol{\varepsilon}_k, \tag{4}$$

where:

- $b_k$ is the regression coefficient between the $k^{th}$ pair of score vectors and
- $\boldsymbol{\varepsilon}_k$ represents the mismatch between the u-score and the predicted u-score vector.

## 3. RECURSIVE PLS ALGORITHMS

Numerous publications, for instance (Helland *et al.*, 1991; Qin, 1998), discuss the recursive calculation of the PLS factors. Given the loading matrices retaining $M$ LVs, $\mathbf{P}_M$ and $\mathbf{Q}_M$, the diagonal matrix of regression coefficients, $\mathbf{B}_M$, and a new sample point of the process $\{\mathbf{x}_1,\mathbf{y}_1\}$, Qin (1998) demonstrated that performing the PLS algorithms on $\begin{bmatrix}\mathbf{P}_M^T\\ \mathbf{x}_1\end{bmatrix},\begin{bmatrix}\mathbf{B}_M\mathbf{Q}_M^T\\ \mathbf{y}_1\end{bmatrix}$ gives the same PLS factors as performing the PLS algorithm on $\begin{bmatrix}\mathbf{X}\\ \mathbf{x}_1\end{bmatrix},\begin{bmatrix}\mathbf{Y}\\ \mathbf{y}_1\end{bmatrix}$. It should be noted that the recursive PLS or RPLS algorithm using the former approach is computationally more efficient. This is particularly true if the number of data points is considerably larger than the number of predictor and response variables.

## 4. CONVENTIONAL AND RECURSIVE PLS ALGORITHMS

MBPLS is designed to identify the relationships between a number of "blocks" (Wangen and Kowalski, 1988; Westerhuis, *et al.*, 1998). The application of MBPLS in this paper considers the division of the predictor variables into "blocks" which represent the operating units of the process. It has been shown that: *the resultant t-score vectors of the "blocks" can be combined to produce equivalent t-score vectors and u-score vectors to the conventional PLS algorithm (Westerhuis, et al., 1998). The combined scores are referred to as super t-scores.* Thus, it is pertinent to identify the PLS factors by applying conventional PLS and then perform the division into "blocks" (Westerhuis, *et al.*, 1998).

Given the above discussion, the replacement of the identification procedure of the MBPLS with the conventional PLS one also implies that the recursively updating RPLS identification procedure can be utilised to achieve a recursively updating MBPLS algorithm. This implies that the recursive identification of the PLS factors is carried out prior to the division of the predictor and response variables into blocks and lead to a recursive identification for the MBPLS factors of the "blocks". This is particularly useful if the process is time-varying or describes strong non-stationary variation.

The MBPLS algorithm used in this paper was introduced by Westerhuis, *et al.* (1998). However, utilising the RPLS algorithm implies that the super t-scores must have unit length. Hence, a modification of the MBPLS algorithm is required as discussed in Appendix I, which gives an example of the modified MBPLS algorithm for two predictor blocks and one response block. The Recursive MBPLS or RMBPLS algorithm is presented in Table 1.

Table 1 Description of the Basic Steps of the Introduced Recursive Multi-Block PLS Algorithm

1. Scale the data matrices $\{\mathbf{X}, \mathbf{Y}\}$.
2. Calculate a PLS model using the algorithm presented by Qin (1998).
3. Divide the updated PLS model into the corresponding blocks.
4. When a new sample point $\{\mathbf{x}_1, \mathbf{y}_1\}$ is available, scale it the same way as step 1. Let $\mathbf{X} = \begin{bmatrix} \mathbf{P}^T \\ \mathbf{x}_1 \end{bmatrix}$, $\mathbf{Y} = \begin{bmatrix} \mathbf{BQ}^T \\ \mathbf{y}_1 \end{bmatrix}$ and return to step 2.

In the case of non-stationary process behaviour, the mean value of the process variables changes over time. However, Qin (1998) proposed that the predictor matrix can be augmented by an offset term to account for changes in the mean values. It can be shown that adding an offset term for each "predictor" block for MBPLS gives equivalent results to the inclusion of one offset term for the "combined" predictor block for PLS.

## 5. MONITORING STATISTICS

To apply RMBPLS for condition monitoring, two univariate statistics can be constructed for each of the blocks, the $T^2$ and the SPE statistic (MacGregor, 1997). Each statistic gives rise to a monitoring chart including control limits to distinguish between normal and abnormal process behaviour as discussed by MacGregor (1997). The $T^2$ statistic for the $a^{th}$ predictor block at time instance $i$ is defined as follows:

$$ {}_aT_i^2 = {}_a\boldsymbol{\tau}_i^T \cdot \boldsymbol{\Lambda}_a^{-1} \cdot {}_a\boldsymbol{\tau}_i , \tag{5} $$

where ${}_aT_i^2$ is the corresponding $T^2$ statistic, ${}_a\boldsymbol{\tau}_i$ is a vector of t-score values at the $i^{th}$ instance in time and $\boldsymbol{\Lambda}_a$ is the covariance matrix of the retained t-score variables. The SPE statistic for the $a^{th}$ predictor block at time instance $i$ is:

$$ {}_aSPE_i = \left\| {}_a\xi_i - {}_a\hat{\xi}_i \right\|^2 , \tag{6} $$

where ${}_aSPE_i$ is the corresponding SPE statistic, ${}_a\xi_i$ is a vector of measured values of the block variables, ${}_a\hat{\xi}_i$ is a vector of reconstructed values of the block variables when $n$ LVs are retained and $\|\circ\|^2$ represents the squared length of a vector. The SPE statistic for the response block at time instance $i$ is as follows.

$$ SPE_i = \left\| \boldsymbol{\psi}_i - \hat{\boldsymbol{\psi}}_i \right\|^2 , \tag{7} $$

where $SPE_i$ is the corresponding SPE statistic, $\boldsymbol{\psi}_i$ is a vector of measured values of the block variables and $\hat{\boldsymbol{\psi}}_i$ is a vector of predicted values of the block variables using the RMBPLS model.

The control limits for each statistic can be obtained as discussed by Box (1954). To address the problem of non-stationary process behaviour, the control limits are determined adaptively as proposed by Li, *et al.* (2000) for example. For each statistic, the control limits are selected to include 95% and 99% of the $T^2$ and SPE values corresponding to normal process behaviour.

## 6. SUMMARY OF THE INTRODUCED RECURSIVE MULTI-BLOCK PLS ALGORITHM

The approach discussed in this paper is based on the development of the RMBPLS algorithm and the use of adaptive control limits. The RMBPLS algorithm can adapt time variant process behaviour and therefore, false alarms are avoided that relate to normal changes in the process behaviour. In addition, the inclusion of an offset term in the predictor variables and the use of adaptive control limits can account for non-stationary process behaviour which occurs as normal process variation. In contrast, abnormal process behaviour is diagnosed if the control limits change severely or if at least one of the univariate statistics show significant violations of the control limits. The next section presents an application study to demonstrate the performance of this approach in the presence of non-stationary process behaviour.

## 7. APPLICATION STUDY

A fluid catalytic cracking unit or FCCU is an important economic unit in refining operations. It typically receives several different heavy feedstocks from other refinery units and cracks these streams to produce lighter, more valuable components that are eventually blended into gasoline and other products. The principal feeds of this particular Model IV unit (McFarlane, *et al.,* 1993) is gas oil but heavier diesel and wash oil streams also contribute to the total feed stream. Fresh feed is preheated in a heat exchanger and furnace and then passed to the riser, where it is mixed with hot, regenerated catalyst from the regenerator. Slurry from the main fractionator bottoms is also recycled to the riser. The hot catalyst provides the heat necessary for the endothermic cracking reactions. The gaseous cracked products are passed to the main fractionator for separation. Wet gas off the top of the main fractionator is elevated to the pressure of the lights end plant by the wet gas compressor. Further separation of light components occurs in this light ends separation section.

As a result of the cracking process, a carbonaceous material, coke, is deposited on the surface of the catalyst, which depletes its catalytic property. Spent catalyst is recycled to the regenerator where it is mixed with air in a fluidised bed. Oxygen reacts with the deposited coke to produce carbon monoxide and carbon dioxide. Air is pumped to the regenerator with a high-capacity combustion air blower and a smaller lift air blower. In addition to contributing to the combustion process, air from the lift air blower assists with catalyst circulation. Complete details of the mechanistic simulation model for this particular model IV FCCU can be found in the paper by McFarlane, *et al.* (1993).

The selected predictor variables for the FCCU case study are the Wash Oil, the Diesel and the Slurry Flowrate as well as the Preheater Outlet Temperature from the feed section of the unit. Additionally, various pressures and flow-rates of the combustion and lift air blower are included as predictor variables. To simulate non-stationary process behaviour, various different AutoRegressive Integrated Moving Average (ARIMA) (Box, 1971) signals were superimposed on the variables of the feed section.

The response set included Excess Oxygen in the Flue Gas, Concentration of Carbon Monoxide in the Flue Gas, Riser Temperature, Regenerator Bed Temperature, Regenerator Standpipe Level, Reactor Pressure as well as a further eight variables measured from the FCCU system (McFarlane, *et al.,* 1993).

To apply the monitoring approach, introduced in section 4 and 5, a data set was obtained which includes a loss of the air blower capacity. The data set contains 1000 sample points and was recorded at a 1 minute sample interval. Furthermore, the loss of air blower capacity is injected after 400 minutes. For the FCCU process, it is pertinent to divide the predictor variables into two blocks, one representing the feed section of the process and one representing the air blower.

The first 200 sample points were used to identify an initial PLS model. Applying cross validation (Wold, 1978) suggests to retain 4 LVs. On the basis of this initial model, the RMBPLS algorithm was used to update the model as discussed in section 4. A comparison between the results obtained from conventional MBPLS algorithm and the RMBPLS algorithm is given in Figure 1 to Figure 4.

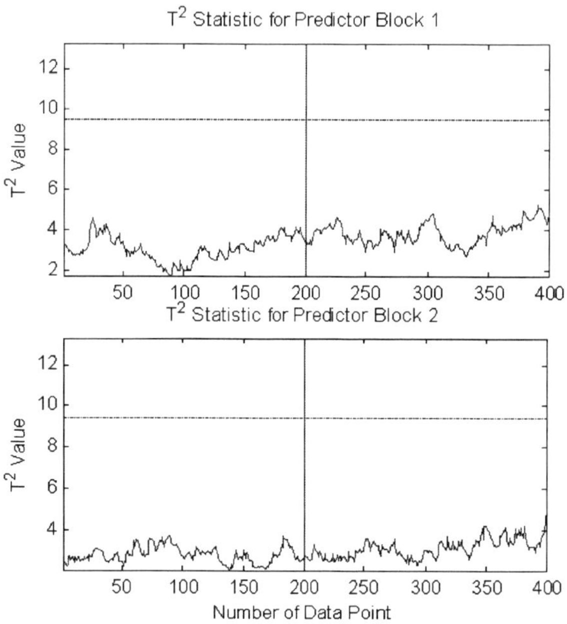


Fig. 1: $T^2$ statistic for block 1 (upper chart) and block 2 (lower chart) on first 400 sampling points

Figure 1 and 2 show that the $T^2$ and the SPE statistics remain below the control limits for the first 200 sample points and that strong violation of the SPE statistic occurs for the last 200 sample points. This is expected since the FCCU process is simulated to present strong non-stationary process behaviour. In contrast, Figure 3 and 4 show that applying RMBPLS and adapting the control limits can account for the presence of strong non-stationary process behaviour.

Although the application of RMBPLS and the adaptation of the control limits circumvents false alarms in this example study the approach is sensitive to the injected process fault. The loss of air blower capacity was felt immediately after the injection of this fault. The $T^2$ statistic of the second predictor block (corresponding to the combustion and lift air blower) shows a strong response to this anomalous process behaviour both, in terms of the actual $T^2$ value and the considerable rise of the control limit. Therefore, the introduced fault was detected and the multi-block approach correctly identified the contributing "block" to this event. However, the RMBPLS algorithm eventually accommodates the anomalous behaviour and the fault condition diminishes after about 100 minutes.

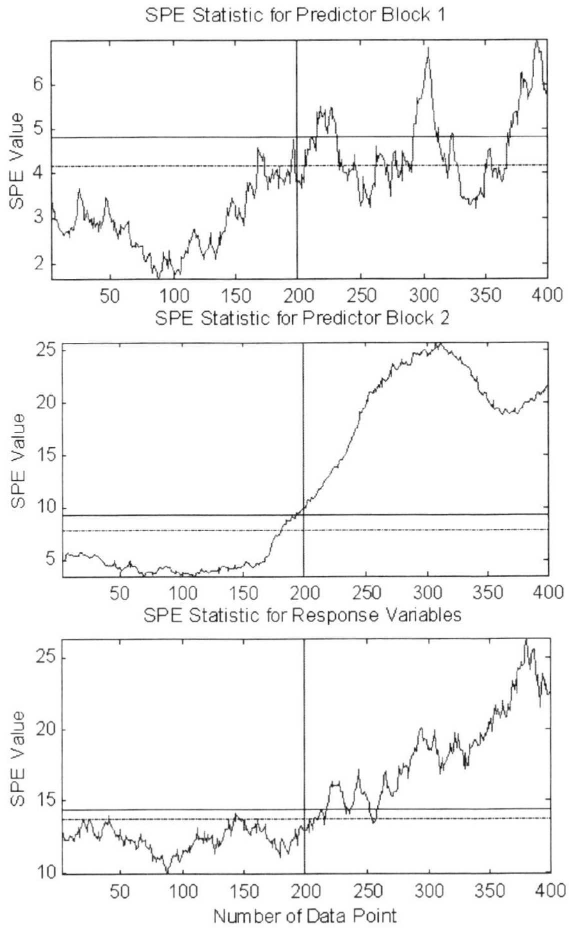


Fig. 2: SPE statistic for block 1 (upper chart), block 2 (middle chart) and response block (lower chart) on first 400 sampling points reference data set

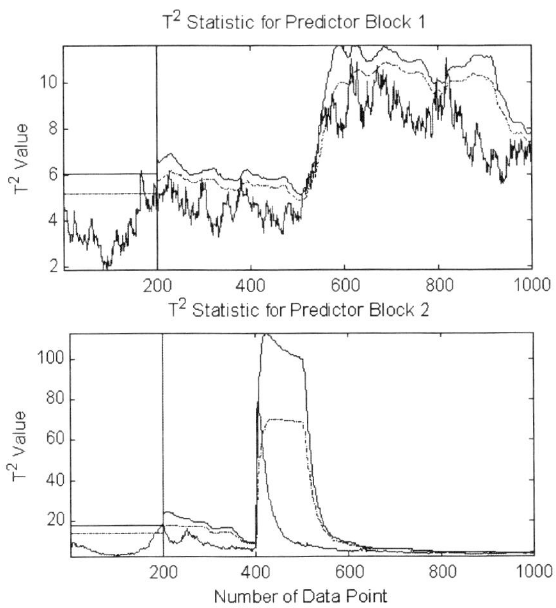


Fig. 3: $T^2$ statistic for block 1 (upper chart) and block 2 (lower chart) on complete data set (fault is injected at the 400[th] sampling point)

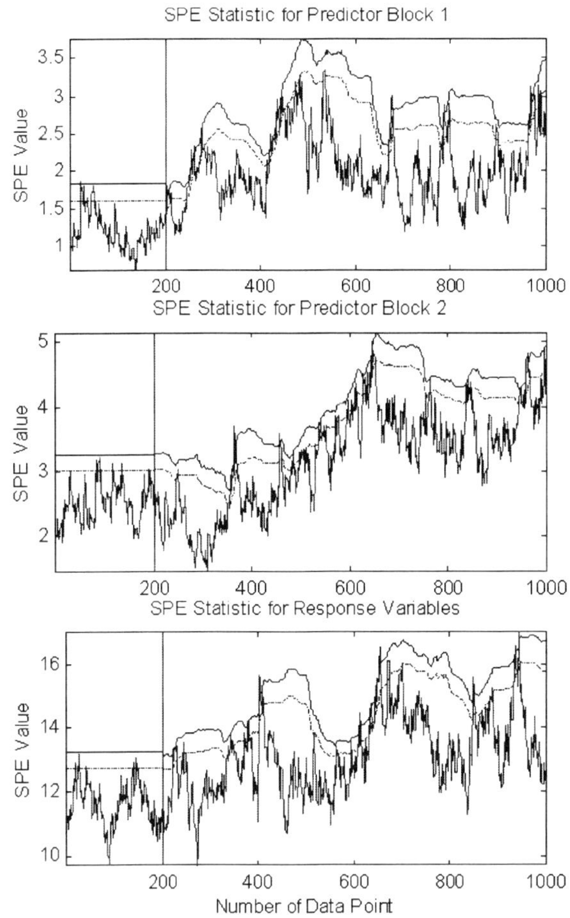


Fig. 4: SPE statistic for block 1 (upper chart), block 2 (middle chart) and response block (lower chart) on complete data set (fault is injected after the 400[th] sampling point)

## 8. CONCLUSION

The contribution of this paper is the development of a recursive multi-block partial least squares algorithm (RMBPLS). This algorithm can monitor industrial processes which exhibits strong non-stationary process behaviour without raising false alarms. It is further discussed that in the presence of non-stationary process behaviour, the control limits for the univariate statistics, the $T^2$ and the SPE statistic, should also be updated adaptively.

To illustrate this monitoring approach, an application study is considered that relates to a realistic simulation of a fluid catalytic cracking unit. It is shown that applying MBPLS without adapting the control limits leads to strong violations of the control limits although the process behaves normally. In contrast, by applying RMBPLS and adapting the control limits the violation of the control limits is circumvented. It is further shown that an injected fault could be detected and that RMBPLS identifies the correct block which corresponds to the abnormal event.

Future research will concentrate on the correct identification of the root cause of abnormal behaviour. Furthermore, attention will be drawn on how to select the number of PLS factors recursively when the process exhibits time-varying behaviour.

## REFERENCES


Box, G. E. P. (1954). Theorems on Quadratic Forms Applied in the Study of Analysis of Variance Problems, I. Effect of Inequality of Variance in the One-Way Classification. *Ann. Math. Stat.*, **25**, 2, 290-302.

Box, G. E. P., and G. M. Jenkins (1971). *Time Series Analysis*, Holden Day, San Francisco.

deJong, S. (1993). SIMPLS, and Alternative Approach to Partial Least Squares Regression. *Chemometrics and Intelligent Laboratory Systems*, **18**, 251-263.

Helland, K., E. E. Berntsen, O. S. Borgen, and H. Martens (1991). Recursive Algorithm for Partial Least Squares Regression. *Chemometrics and Intelligent Laboratory Systems*, **14**, 129-137.

Kosanovich, K. A., and M. J. Piovoso (1991). Process Data Analysis Using Multivariate Statistical Methods. Proc. of the American Control Conference, Chicago IL, June 26-28.

Li, W., H. H. Yue, S. Valle-Cervantes, and S. J. Qin (2000). Recursive PCA for Adaptive Process Monitoring. *Journal of Process Control*, 10 (5), 471-486.

MacGregor, J. F. (1997). Using On-Line Process Data to Improve Quality: Challenges for Statisticians. *International Statistical Review*, **65**, 3, 309-323.

McFarlane, R. C., R. C. Reineman, J. F. Bartee, and C. Georgakis (1993). Dynamic Simulator for a Model IV Fluid Catalytic Cracking Unit. *Computers Chem. Engng.*, **17**, 3, 275-300.

Piovoso, M. J., and K. A Kosanovich (1992). Process Data Chemometrics. *IEEE Transaction on Instrumentation and Measurement,* **41**, 2, 262-268.

Qin, S. J. (1998). Recursive PLS Algorithms for Adaptive Data Modelling. *Computers Chem. Engng.*, **22**, 4/5, 503-514.

Wangen, L. E., and B. R. Kowalski (1988). Multiblock Partial Least Squares Algorithm for Investigating Complex Chemical Systems. *J. Chemomet.*, **3**, 3-20.

Westerhuis, J. A., T. Kourti, and J. F. MacGregor (1998). Analysis of Multiblock and Hierarchical PCA and PLS Models. *J. Chemomet.*, **12**, 301-321.

Wold, H., and K. G. Joreskog (1980). *Systems under Indirect Observation*, North Holland, Amsterdam.

Wold, S. (1978). Cross Validation Estimation of the Number of Components in Factor and Principal Component Models. *Technometrics*, **20**, 397-405.

Wold, S., S. Hellberg, T. Lundstedt, and M. J. Sjostrom (1987). PLS Meeting, Frankfurt.


## APPENDIX

The MBPLS algorithm by Westerhuis, *et al.* (1998) is presented below. Note that the box represents the required modification to achieve super t-scores that have unit length.

for i = 1 to $n$

Take $\mathbf{u}$ as some column of $\mathbf{Y}$

Loop until convergence of $\mathbf{t}_T$

Determine block weight vector for $\mathbf{X}_1$ block

$\mathbf{w}_1 = \mathbf{X}_1{}^T \cdot \mathbf{u} / \mathbf{u}^T \cdot \mathbf{u}$

$\mathbf{w}_1 = \mathbf{w}_1 / \|\mathbf{w}_1\|$

Calculate t-score vector for $\mathbf{X}_1$ block

$\mathbf{t}_1 = \mathbf{X}_1 \cdot \mathbf{w}_1$

Determine block weight vector for $\mathbf{X}_2$ block

$\mathbf{w}_2 = \mathbf{X}_2{}^T \cdot \mathbf{u} / \mathbf{u}^T \cdot \mathbf{u}$

$\mathbf{w}_2 = \mathbf{w}_2 / \|\mathbf{w}_2\|$

Calculate t-score vector for $\mathbf{X}_2$ block

$\mathbf{t}_2 = \mathbf{X}_2 \cdot \mathbf{w}_2$

Combine block t-score vectors to super block $\mathbf{T}$

$\mathbf{T} = [\mathbf{t}_1 | \mathbf{t}_2]$

Determine block weight vector

$\mathbf{w}_T = \mathbf{T}^T \cdot \mathbf{u} / \mathbf{u}^T \cdot \mathbf{u}$

Determine super t-score vector

$\mathbf{t}_T = \mathbf{T} \cdot \mathbf{w}_T$

| |
|---|
| Adjust length of super block weight vector<br>$\mathbf{w}_T = \mathbf{w}_T / \|\mathbf{t}_T\|$<br>Set super t-score vector to unit length<br>$\mathbf{t}_T = \mathbf{t}_T / \|\mathbf{t}_T\|$ |

Calculate loading vector for response block

$\mathbf{q} = \mathbf{Y}^T \cdot \mathbf{t}_T / \|\mathbf{Y}^T \cdot \mathbf{t}_T\|$

Calculate new score vector of response block

$\mathbf{u} = \mathbf{Y} \cdot \mathbf{q}$

end

Determine loading vector for $\mathbf{X}_1$ block

$\mathbf{p}_1 = \mathbf{X}_1{}^T \cdot \mathbf{t}_T$

Determine loading vector for $\mathbf{X}_2$ block

$\mathbf{p}_2 = \mathbf{X}_2{}^T \cdot \mathbf{t}_T$

Calculate regression coefficient for score model

$b = \mathbf{u}^T \cdot \mathbf{t}_T$

Deflate $\mathbf{X}_1$ block using super t-score vector

$\mathbf{X}_1 = \mathbf{X}_1 - \mathbf{t}_T \cdot \mathbf{p}_1{}^T$

Deflate $\mathbf{X}_2$ block using super t-score vector

$\mathbf{X}_2 = \mathbf{X}_2 - \mathbf{t}_T \cdot \mathbf{p}_2{}^T$

Deflate response block

$\mathbf{Y} = \mathbf{Y} - b \cdot \mathbf{t}_T \cdot \mathbf{q}^T$

end

IFAC Publications
www.elsevier.com/locate/ifac

# DESIGN AND SIMULATION OF HETEROGENEOUS CONTROL SYSTEMS USING PTOLEMY II

**Johan Eker, Chamberlain Fong, Jörn W. Janneck, Jie Liu**

*Department of Electrical Engineering and Computer Sciences*
*University of California at Berkeley*
*Berkeley, CA 94720-1770, US*
*{johane,chf,janneck,liuj}@eecs.berkeley.edu*

Abstract: Complex control systems are heterogeneous from both an implementation and a modeling perspective. Design and simulation environments for such systems need to integrate different component interaction styles, like differential equations, discrete events, state machines, dataflow networks, and real-time scheduling. This paper motivates the use of Ptolemy II software environment for modeling and simulation of heterogeneous control systems. Ptolemy II advocates a component-based design methodology, and hierarchically integrates multiple models of computation, which can be used to capture different design perspectives. A Furuta pendulum control system is used as a motivating example. After designing a three-mode hybrid controller under idealized assumptions, implementation effects, like real-time scheduling and network protocols, are taken into consideration to achieve a more realistic simulation. The 3D animation package in Ptolemy II helps designers to visualize the control results. In this process of refining the design, components modeled in early phases can be reused. *Copyright © 2001 IFAC*

Keywords: Simulators, real-time computers, embedded systems

## 1. INTRODUCTION

The heterogeneity of modern embedded control systems puts high demands on design and simulation tools. A typical hybrid control system consists of a set of subcontrollers and some switching logic. The controllers themselves are conveniently described using discrete equations, the switching logic may be expressed using state machines, the controller is implemented as a task under a real-time operating system, the controlled plant is modeled as differential equations, and so on.

The above subsystems differ in the way their components interact, both in terms of the way they exchange information, and in terms of the flow of control between them—synchronous or asynchronous, buffered or unbuffered, sequential or parallel, etc. Ideally, a designer should be able to model each of these in their natural computational *domain*, based on a corresponding *model of computation*.

This requirement makes the Ptolemy II environment an interesting candidate for creating such models of heterogeneous control systems. Ptolemy II provides a wide (and extensible) range of computational models which are based on a common notion of "domain". Sub-models in different domains may be hierarchically combined, which makes the simulation model much clearer and more understandable, and facilitates the modeling of complex systems without the need for an equally complex simulation configuration.

Typically, a controller design project proceeds in several distinct phases: modeling, design, simulation and implementation. Clearly, if possible it is preferable to use the same tool through all of the above phases and reuse the components designed earlier. The extensive component libraries and hierarchical compositionality of heterogeneous models make Ptolemy II suitable for refining designs through these phases. The 3D graphical animation support allows the designer to visualize the simulation, and the retargetable code generation facility supports the implementation phase by producing e.g. Java or C code for a particular embedded control system.

This paper shows the application of these concepts and the Ptolemy II environment to the design of a heterogeneous controller by first introducing a small example problem (Section 2) to motivate the Ptolemy approach, and then presenting the pertinent modeling concepts and domains provided by Ptolemy II (Section 3). In Section 4 these concepts are then applied to the example.

## 2. MOTIVATING EXAMPLE

The inverted pendulum is a classic control problem basically for two reasons: it is nonlinear and unstable. A picture of the inverted pendulum is shown in Fig. 1. The pendulum consists of two moving parts, the arm that moves in the horizontal plane and the pendulum that moves in the vertical plane.

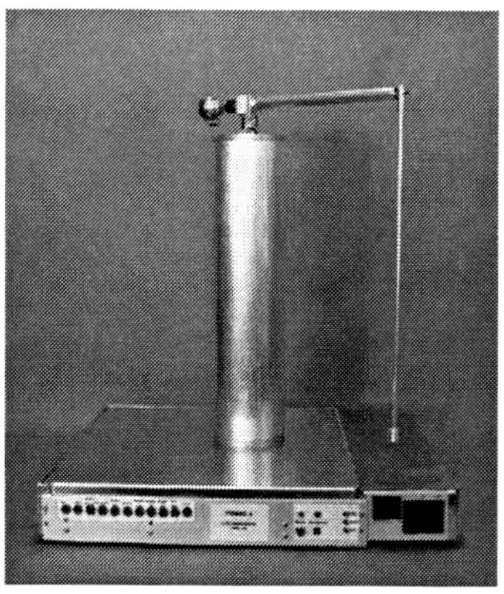

Fig. 1. The rotating inverted pendulum.

A hybrid controller can be designed to swing-up and stabilize the pendulum. The controller consists of three modal subcontrollers: a *swing-up* controller, a *catch* controller, and a *stabilizing* controller. Initially, the pendulum starts in the downward position and the swing-up mode is used to bring it to the top position. Once it is sufficiently close to the top equilibrium, the controller enters the catch mode. The task of the catch mode is to reduce the speed of the pendulum and the arm before the third mode, stabilize, is entered. The catch and stabilize modes are linear state feedback controllers. The swing-up controller uses a nonlinear energy based algorithm for bringing the pendulum to upward position. The controller switching logic can be described using a finite state machine, where each state corresponds to a control mode, and each of the subcontrollers can be described as a dataflow computation, as shown in Fig. 2.

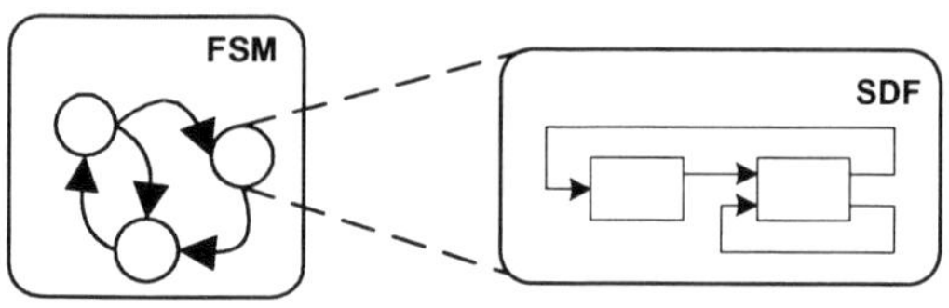


Fig. 2. The hybrid pendulum controller consisting of three subcontrollers.

The above system is straightforward to model and simulate in many modern design tools. One example of it is shown in Fig. 3, where the controller in Fig. 2 is integrated with the plant dynamics, which is modeled as ordinary differential equations (ODEs). However, this

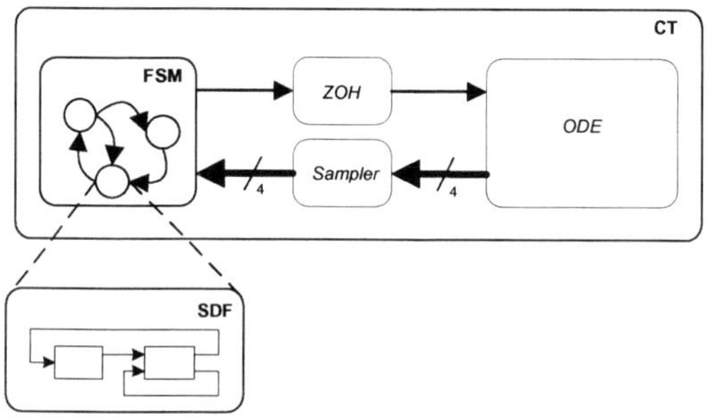


Fig. 3. A basic controller model.

model does not capture many issues related to an actual implementation. Here we make the usual assumptions that the execution time is negligible and that we have no computation and communication jitter. Of course, this is not the case in the real-world. When the controller is running on a real computer and on top of a real-time operating system (RTOS), it will compete with other tasks for resources, e.g. the CPU and I/O. This will give rise to input-output delays and variations in the sampling period. Furthermore, the actuators and the sensors are usually not directly connected to the controller, but instead some network is used for transferring data. The network is a common resource possibly shared by many other control loops, and again the loops compete for the network bandwidth. We would like to capture the above properties so that we can predict the real behavior of the embedded system, and evaluate scheduling mechanisms and communication protocols in terms of applications performance.

In this process of refining a design, designers need to gradually add design considerations to the existing model and migrate the control system from algorithms to implementation. Different design perspectives usually imply heterogeneous component interaction styles. It is desirable that a design environment can support multiple component interaction styles and the components designed in earlier phases can be reused under new interaction styles, so that the verified properties can be preserved as much as possible. We argue that integrating different models of computation will help decompose design perspectives and achieve elegant and reusable models.

## 3. PTOLEMY II COMPONENT ARCHITECTURE AND MODELS OF COMPUTATION

The Ptolemy II modeling and design environment addresses the heterogeneity and design migration issues using a component-based design methodology, disciplined component interactions, and component reuse. In Ptolemy II, we acknowledge the existence of well-defined models of computation, which guide the communication style and the execution of a composition of components. Each model of computation asserts certain properties that could be desirable for certain aspects of a system. These heterogeneous models of computation can be composed hierarchically to preserve understandability, manage complexity, and encourage component reuse. This section presents the Ptolemy II component structure and the models of computation that are useful for simulation and visualization of control system designs.

### 3.1 *Model structure*

In Ptolemy II, a model is a hierarchical aggregation of components, which are called *actors*. Actors encapsulate an atomic execution and provide communication interfaces (called *ports*) to other actors. An actor can be *atomic*, meaning that it is at the bottom of the hierarchy. An actor can be *composite*, meaning that it contains other actors. A composite actor can be contained by another composite actor, so hierarchies can be arbitrarily nested.

A port of an actor can be an input, output, or both. Communication channels among actors are established by connecting ports. A port for a composite actor can have connections both to the inside and to the outside, and thus mediates the communication from inside actors to outside actors.

A model of computation associated with a composite actor is implemented as a *domain*. A domain defines the communication semantics and the execution order among actors. The communication mechanism is implemented using *receivers*. To ensure a clear model of computation at a level of hierarchy, all receivers within one composite actor are the same. Receivers could represent FIFO queues, mailboxes, proxies for a global queue, or rendezvous points. Actors, when resident in a domain, adopt domain-specific receivers. By separating computation and communication in this way, many actors can be reused in different domains.

Actors, both atomic and composite, are executable. The execution order of the actors contained in a composite actor is controlled by a *director*. Since receivers and directors must work together, the director is also responsible for creating receivers. When a composite actor is executed, the director associated with the composite actor executes the actors of the contained model. The directors are carefully designed so that they both respect the model of computation they implement and provide a polymorphic execution interface to the outside domain. This compositionality of execution is the key to managing heterogeneity hierarchically.

### 3.2 *Domains*

A wide variety of domains have been implemented in Ptolemy II. We discuss here a subset of them, which are useful for designing control systems.

3.2.1. *Continuous time* The continuous time (CT) domain (Liu, 1998) models ordinary differential equations (ODEs), extended to allow the handling of discrete events. Special actors, which represent *integrators*, are connected in feedback loops in order to represent the ODEs. Event generators, e.g. periodic samplers, triggered samplers, and zero crossing detectors, and waveform generators, like a zero order hold, are implemented to convert between continuous-time signals and discrete events.

The execution of a CT model involves the computation of a numerical solution to the ODEs at a discrete set of time instance. In order to support the detection of discrete events and interact with discrete models of computation, the time progression and the execution order of a CT model are carefully controlled (Liu and Lee, 2000).

3.2.2. *Discrete event* In the discrete event (DE) domain of Ptolemy II, actors communicate through events placed on a (continuous) time line. Each event has a value and a time stamp. Actors process events in chronological order. The output events produced by an actor are required to be no earlier in time than the input events that were consumed. In other words, DE models are *causal.*

Discrete event models, having the continuous notion of time and the discrete notion of events, are suitable for modeling hardware and software timing properties, communication networks, and queuing systems. A good reference on discrete event systems is (Cassandras, 1993).

3.2.3. *Dataflow models* By "dataflow" we refer to a class of models that interact using asynchronous message passing (Lee and Parks, 1995). In these models, components are processes which communicate by sending messages through FIFO queues. The sender of a message need not wait for the receiver to be ready to receive the message. Most dataflow models only specify the data dependency among components, imposing only a partial ordering on the execution of components. These properties match them well with signal processing and control algorithms, and allow highly efficient execution.

Synchronous dataflow (SDF) (Lee and Messerschmitt, 1987) is a particularly restricted special case of dataflow models with extremely useful properties. In SDF, whenever a component executes, it consumes a fixed amount of input data and produces a fixed amount of outputs. For a consistent SDF model, a schedule can be computed, such that the components do not have to test for sufficient data before execution. So, for algorithms with a fixed structure, SDF is very efficient and predictable.

3.2.4. *Finite State Machines* Finite state machine (FSM) models, see (Hopcroft and Ullman, 1979), are used at two levels, within a component or across components. Within a component, state machines specify precise sequencing of component states and transitions among them. When used across components, state machines can be used to coordinate other models. The typical use is to specify operation modes and sequences among components. These components can be built using other domains (Girault et al., 1999). Hierarchically combining state machines with concurrent models makes state machines concurrent and helps to prevent the explosion of the number of states in complex systems.

3.2.5. *RTOS* The real-time operating system (RTOS) domain in Ptolemy II allows designers to explore priority-based scheduling policies and their effects on real-time software. In this domain, components are software tasks with priorities. The director of this domain implements a prioritized event dispatching mechanism and invokes tasks according to their feasibility and priority. Both preemptive and nonpreemptive scheduling, as well as static and dynamic priority assignment, can be captured.

3.2.6. *3D visualization/graphics* The graphics (GR) domain (described in more detail in (Fong, 2001)) provides a component-based infrastructure for displaying three-dimensional graphics in Ptolemy II. The GR domain follows loop-free synchronous semantics. The basic idea behind the GR domain is to arrange geometry and transform actors in a directed acyclic graph to represent the location and orientation of objects in a world scene. This topology of connected GR actors forms what is commonly known in computer graphics literature as a *scene graph.* Fig. 4 shows an example scene graph topology in Ptolemy II. The geometry actors are source actors that produce basic 3D shapes such as spheres, cylinders, and boxes. The transform actors perform linear transformations on these shapes in order to place them properly in the world scene. Hierarchically combining graphics subsystems with other domains in Ptolemy II is useful for producing animated simulations. The availability of 3D graphics complements Ptolemy's heterogeneous modeling and simulation capabilities, facilitating visualization and understanding of dynamic model behavior.

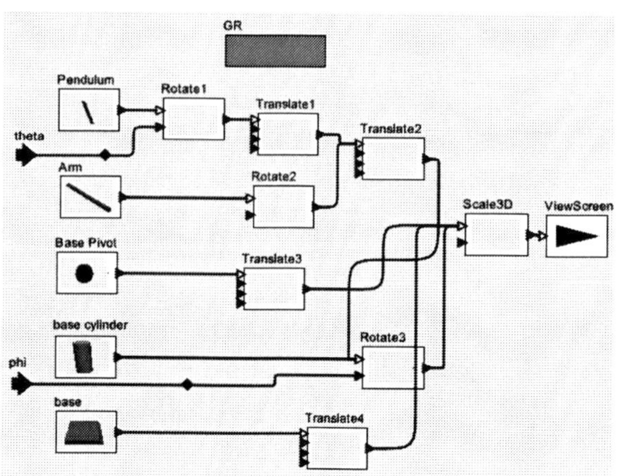


Fig. 4. GR model of the pendulum.

## 4. REFINING THE MODEL

In this section we will show how a much detailed design of the pendulum control system can be constructed from a basic model using hierarchical composition of domains. The controller is modeled as running on top of a RTOS model, while the sensor and the actuator are distributed and communicate with the controller over a network.

### 4.1 *Basic model*

The initial model described in section 2 is constructed using three domains in Ptolemy II. The pendulum is modeled in the *continuous time* domain using differential equations. The switching logic of the hybrid controller is described as a *finite state machine*, whose states refine to subcontrollers. These subcontrollers are in turn modeled in the *synchronous dataflow* domain. Here, we assume that the controller takes no time to execute.

### 4.2 *Model refinement*

This model does not reflect the effects introduced in a concrete implementation, such as controller latency and communication delay. A more accurate model would include a model of the real-time operating system and the network. This is done in two steps in Ptolemy II. First, to consider the real-time issues, we embed the controller designed in the basic model (i.e. the composite actor that contains the finite state machine and the subcontrollers) into an RTOS domain to capture the effects of the interaction between the different tasks running concurrently on the system.

The composite actor for the modal controllers is treated as a task in the RTOS domain. Some other tasks are added to capture the dynamics of concurrent tasks running on the same real-time kernel. Notice that the composite actor we built in the basic model only specifies the computational part of the controller. To actually reflect the implementation, another task, which models the I/O part of the controller, is added, similarly to the task model used in (Eker and Cervin, 1999). This I/O task may compete for resources with other I/O operations running on the system.

The next step is to include a model of the network communication. This is done by using a discrete event domain at the top level, and introducing a network actor, which models the behavior of a given network protocol. We connect both the pendulum model and the controller to the network actor so that the event going through the network will be delayed according to the load of the network. This extended model is shown in Fig. 5. The top-level domain has been changed to DE and it now contains five actors: the pendulum, the controller, the network, an actor that represents other network components, and a visualization actor. Much of the pendulum composite actor reusing the previous pendulum model, which is implemented in the CT domain. The controller is implemented using the RTOS domain described above. The network actor simulates the communication between the nodes in the control system. Arbitrarily many nodes may be connected to the network. The basic functionality of the network actor is to simulate the network behavior, i.e. model multi-node network access, packet dropping, and message latency. The animation composite actor is modeled in the GR domain, and explained further in the next section.

Fig. 6. A 3D model of the Furuta pendulum.

### 4.3 *Model animation*

To visualize the Furuta pendulum system, we model its core 3D shape in the GR domain. A lofted shape is used to represent the base that supports the moving parts. Cylinders of various cross-sections are used to represent the rotational base, the arm, and the pendulum tip. After translating and rotating these shapes into the right place, we connect them into a scene graph actor. The GR scene graph diagram is shown in Fig. 4. This scene graph specifies that the motions of the rotational base are inherited by the arm and pendulum, and that the motions of the arm are inherited by the pendulum. The whole GR model is a composite actor contained by the top-level DE model. The GR model uses two inputs from the sampler -

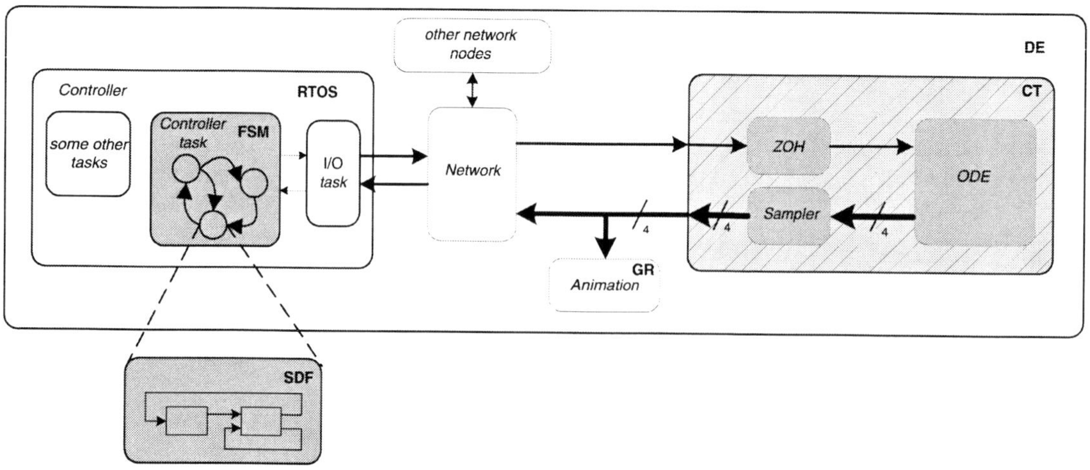


Fig. 5. A refined controller model, which also models execution times and communication latencies and allows us to study how they influence the closed loop performance.

the horizontal angle for the arm and the vertical angle for the pendulum. Since the GR model only represents object geometry, containment, and motion, the velocities are not used as inputs for the animation. It is possible that angular velocity values are needed to produce motion blur of the animation, but we did not implement it. Fig. 6 shows an image from our animated simulation of the pendulum.

## 5. CONCLUSION

We presented an approach to create models of heterogeneous control systems by following a discipline of hierarchical composition of various domains. A domain implements the rules of communication and control flow between model components in that domain. Domain interactions are well-defined, since they are based on a common framework notion used to express these models of computation. As a consequence, model components are reusable in almost arbitrary contexts.

We have demonstrated these concepts by taking a small control model and embedding its components in environments that better represent real-world influences on the actual system. We also showed how to use the same concepts to incorporate three-dimensional animation and interaction capabilities into the model, which significantly aid in understanding controller behavior and locating problems.

Further research is needed in identifying, defining, and implementing a more complete collection of domains pertinent to control system modeling. As domains form the semantical basis of the modeling framework, it is essential that they be orthogonal and that they interoperate in meaningful ways with a minimum of emergent behavior.

## References


Christos G. Cassandras. *Discrete Event Systems: Modeling and Performance Analysis.* Richard D. Irwin Publ., 1993.

Johan Eker and Anton Cervin. A Matlab toolbox for real-time and control systems co-design. In *Proceedings of the 6th International Conference on Real-Time Computing Systems and Applications*, pages 320–327, Hong Kong, P.R. China, December 1999.

Chamberlain Fong. Discrete-Time Dataflow Models for Visual Simulation in Ptolemy II. Memo M01/9, UCB/ERL, University of California at Berkeley, 2001.

Alain Girault, Bilung Lee, and Edward A. Lee. Hierarchical finite state machines with multiple concurrency models. *IEEE Transactions on Computer-aided Design of Integrated Circuits and Systems*, 18(6):742–760, June 1999.

John E. Hopcroft and Jeffrey D. Ullman. *Introduction to Automata Theory, Languages, and Computation.* Addison-Wesley, 1979.

Edward A. Lee and David G. Messerschmitt. Synchronous data flow. *Proceedings of the IEEE*, 75(9):1235–1245, September 1987.

Edward A. Lee and Thomas M. Parks. Dataflow process networks. *Proceedings of the IEEE*, 83 (5):773–801, May 1995.

Jie Liu. Continuous time and mixed-signal simulation in Ptolemy II. Memo M98/74, UCB/ERL, EECS UC Berkeley, CA 94720, July 1998.

Jie Liu and Edward A. Lee. Component-based hierarchical modeling of systems with continuous and discrete dynamics. In *2000 IEEE International Symposium on Computer-Aided Control System Design, Anchorage, Alaska, USA*, pages 95–100, September 2000.

IFAC Publications
www.elsevier.com/locate/ifac

# REAL-TIME TRANSACTIONS NEED THEIR CONSTITUTING TASKS

**Horst F. Wedde, Sabine Böhm, Wolfgang Freund**

*Informatik III*
*University of Dortmund*
*44221 Dortmund / Germany*
*horst.wedde@udo.edu*

Abstract: Transaction handling in real-time applications has been studied since quite a number of years, pursuing a variety of research topics. Considering concurrency control, in particular for replicated objects, the algorithms suggested have been moderate modifications of algorithms originally defined for non-real-time database applications. This resulted, for some of them (*O2PL* and *OCC*), in a considerable performance improvement (total deadline failure rate) compared to the original versions. As their specific time-sensitive parameter these real-time versions were designed to work on a *transaction* deadline. While transaction deadlines are to be chosen in accordance with the deadlines of the tasks constituting the transactions no attention has been paid to the latter ones. Taking the perspective of object or task/ transaction *similarity* we experimentally demonstrate that it is highly advantageous (if not indispensable) for the real-time performance to fully utilize the task level and structure of transactions, in particular for safety-critical applications. *Copyright © 2001 IFAC*

Keywords: Distributed operating systems, real-time systems, safety-critical systems, adaptive transaction management, adaptive concurrency control.

## 1. INTRODUCTION

Transaction management in distributed real-time systems for safety-critical applications is gaining more and more attention by researchers and practitioners alike. Lately also replicated real-time database management has been studied (Gupta *et al.* 1997), (Xiong *et al.* 1998). The algorithms proposed here are moderate modifications (*O2PL, OCC*) of standard concurrency protocols. Whether real-time issues are at stake or not (like (Anderson *et al.* 1998)), no particular attention has been paid to the structure of the tasks constituting the transactions. However, this may be disadvantageous whenever deadlines are hard, i.e. they have to be met lest system survivability is jeopardized. In such worst case task priorities may be used to determine that tasks that not so relevant would not be invoked, i.e. the values computed by the latest previous incarnation would be used if necessary. For a task sequence within a transaction this may save execution time.

In this paper we will present an experimental study for demonstrating how advantageous it may be to fully utilize the task level and structure of transactions. We focus on the aspect of object and task similarities (Kuo and Mok 1992), (Ho *et al.* 1997), (Mok *et al.* 1998). The idea is that as long as object values, or results of task executions, do not change in a relevant way they are considered similar to their previous values. Tasks using similar object values, or yielding similar results, have not to be invoked. We will use this concept as it was operationalized in (Wedde *et al.* 1999).

*Organization of the paper.* In section 2 a general transaction model is presented. Section 2.4 describes how similarity can be used to exploit the transaction structure. In section 3 we present some experimental results. Section 4 concludes the paper.

## 2. BUILDING TRANSACTIONS

Transaction structures get easily fairly complex as has been demonstrated in (Wedde *et al.* 2001) by analyzing the release of the *Boeing 747* landing gear. In this section we shortly describe how we build complex real-time transactions through composition of elementary structures of tasks.
Tasks are operations on one object that may be replicated over different sites. Read tasks access only one copy while writers update all copies of the item in one atomic action.

### 2.1 *Independent Tasks*

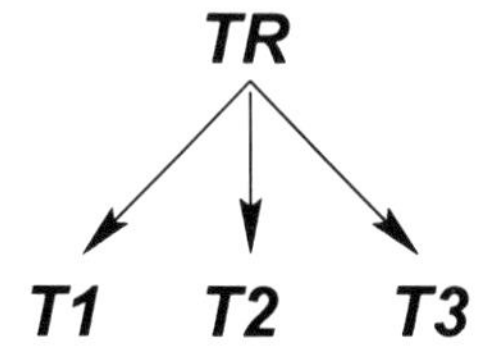


Fig. 1. Independent Tasks

The first kind of a basic transaction structure is composed of *independent tasks* (fig.1). A transaction *TR* with independent tasks is considered operating on different objects, and no task depends on the result of another one. Thus all tasks can be executed in parallel. The only restriction imposed equally on all tasks is that they have to be finished before the transaction deadline. Therefore we set the deadlines for all tasks to the transaction deadline (subtracting only the time estimated for the commit phase of the transaction).

### 2.2 *Task Sequences*

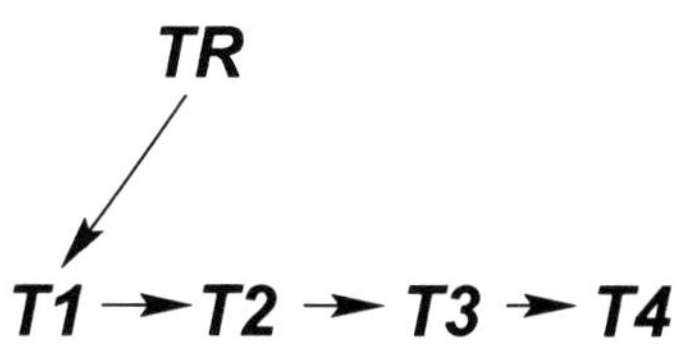


Fig. 2. Task Sequence

The second kind of elementary transaction structure has been termed *"task sequence"* (fig.2). In a transaction which is a task sequence tasks are no longer independent of each other. They have to be executed in a certain order. As long as task $T_i$ has not provided a result $T_{i+1}$ may not be started since it depends on the result of its predecessor. In order to complete the transaction by its deadline $D(TR)$ the last task in the sequence has to be finished "on time", i.e. before the transaction has to enter the commit phase (in order to meet its deadline). Taking (estimated) execution times of the tasks into account, deadlines can be computed backward (beginning at the latest task) for each of the constituting tasks. In fig.2 we find $D(T_1) < D(T_2) < D(T_3) < D(T_4) < D(TR)$.

### 2.3 *Transaction Composition*

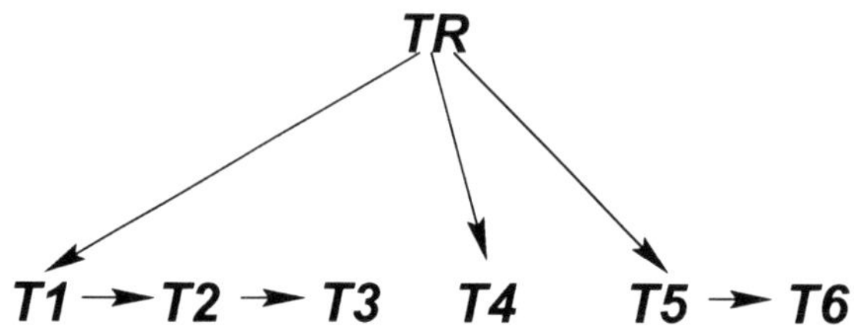


Fig. 3. Transaction Composition

Composition of elementary transactions results in a transaction which is an ensemble of independent tasks and task sequences as presented in fig.3. 2.1 and 2.2 imply that $T_3$, $T_4$ and $T_6$ will be assigned $TR$'s deadline (without commit phase time), while the deadlines of each task in both sequences could be set according to 2.2. (Sub-transactions may consist of a composition of independent subtransactions of sequences thereof.)

### 2.4 *Similarity*

Similarity is a measure of how long to consider objects to be "accurate enough" (or tasks to yield nearly accurate results). In (Kuo and Mok 1992) similarity has been defined as a binary relation on the domain of a data object. Two *views* of a transaction are similar iff *every* read task in both views uses similar values with respect to the transaction. Two task instances are similar if they are of the same type and access similar values of the same data object. This means that 2 views of a transaction are similar if *all* corresponding read events are similar. This definition implies that upon a change of one object/value yielding a dissimilar result, *all tasks of the transaction have to be executed once the transaction is invoked. Even the tasks reading only the previous value again.*

In (Wedde *et al.* 1999) object and task similarity have been defined and interrelated in a new way: In distributed safety-critical systems there are sensors at the *periphery* of the application which input a continuous stream of data into preprocessors. Here periodic tasks filter, digitize, and structure these data. All other objects are called *derived.* (They are results of task executions). To each data object $O$ originating from this process - which is a template for its instances - a similarity bound $sm(O)$ will be assigned. Every task $T_j$ will also be assigned a similarity bound $sm(T_j, O)$ which is related to the effect of $T_j$ using $O$ for an execution instance. $sm(T_j, O)$ will be sent to the site from which $O$ originates. We will assume that

$$sm(O) \leq min\{sm(T_j, O)|T_j \, needs \, O\} \quad (1)$$

If for a data object $O$ generated at a preprocessor $P_i$, or another processor, a subsequent instance is no longer similar to the previous one (as specified by $sm(O)$) - say it differs by $a(O) > sm(O)$ - then the object manager at $P_i$ will compare $a(O)$ to $sm(T_j, O)$. If $a(O) \leq sm(T_j, O)$ then $T_j$ would not be invoked. If $a(O) > sm(T_j, O)$ then a message will be sent to the site holding $T_j$ to invoke the next task instance (the task will be *marked*).

Condition 1 makes sure that task instances will be invoked whenever necessary, i.e. whenever $sm(T_j, O)$ is exceeded, since the alerting procedure at the site generating $O$ would be invoked in time. For details see (Wedde *et al.* 1999).

Once the site holding $T_j$ is marked, i.e. a new task instance should be invoked, the transaction constituted by $T_j$ is invoked instead - *yielding an aperiodic approach.* During the execution of the transaction instance only those tasks that have been *marked* will be invoked. The others are still similar to their previous incarnations. This allows not only for saving complete transaction instances if all tasks are still considered similar as is the case in (Kuo and Mok 1992), but even for skipping task or subtransaction executions resulting both in a higher flexibility, and adaptability to environmental circumstances.

## 3. EXPERIMENTAL EVALUATION

The major purpose of the experiments reported in this section is to study the influence of the new similarity concepts for transactions on the real-time behavior of available concurrency control algorithms. The study is performed in the example of the *O2PL* protocol (Xiong *et al.* 1998).

*We will denote by* **O2PL.SIM** *the algorithm which would operate like O2PL in the sense of (2.4),* ***i.e. whenever one of the tasks constituting a transaction is marked then all of its tasks will be invoked.*** *O2PL is compared to* **O2PL.TSIM**. ***Here only the marked tasks will be executed upon invoking a transaction*** *(see last paragraph of (2.4)). O2PL.SIM represents a transaction-based method, O2PL.TSIM utilizes task-based similarity handling.*

The experiments are performed under the AIX operating system on a token ring network consisting of 7 machines. One of them is serving as the interface to the environment. (It "simulates" the environment.) An object manager process creates and updates the objects, and it informs the corresponding tasks about the changes whenever its similarity bound on that object is exceeded. The tasks/ transactions will then be invoked. By running distributed experiments instead of performing simulation the tendencies extracted are more realistic, in particular regarding the impact of communication.

### 3.1 *Experimental Setup*

The distribution of transactions over the 6 sites, as well as their structure, is indicated in fig.8. The object number is associated with each task. Task overall execution times (and deadlines) were engineered, and measured, between 20 msec (read) and 50 msec (write) as maximal values.

There are 50 objects in the system. 20 are peripheral, i.e. they have been created by the object manager, 30 are derived objects. The latter are obviously managed by the transactions. *Peripheral objects are read-only.*

For all experiments the similarity bounds were set to

- Zero (*O2PL.SIM*) – i.e. tasks have to be invoked upon every update of the corresponding object.
- 5% to 10% (*O2PL.SIM05-10*) – i.e. a random distribution of similarity bounds varying between 5% and 10%.
- 0% to 10% (*O2PL.SIM00-10*) – i.e. a random distribution of similarity bounds varying between 0% and 10%.

*Note: The new variations are denoted by O2PL.TSIM00, O2PL.TSIM05-10 and O2PL.TSIM00-10, respectively.*

The deadline failure rate is measured through 10 runs. Since the deviations were fairly small, the average failure rates over the experimental time intervals are displayed.

By considering similarity on the task level we expect that saving the execution of some task instances within a transaction incarnation results in a lower deadline failure rate than executing all constituting tasks. Thus the *O2PL.TSIM* versions should perform better than the *O2PL.SIM* variations. The more the similarity bounds are relaxed the more tasks are likely to be skipped. Thus *O2PL.TSIM05-10* can be expected to show the lowest deadline failure rate since in the *O2PL.TSIM00-10* and *O2PL.TSIM00* variations (some) tasks are assigned tighter bounds resulting in a more frequent invocation. The same argument holds for the *.SIM* versions.

In all experiments presented the update frequencies of the peripheral objects were randomly distributed between 100 msec and 5 seconds (see tab.1).

| Object | Update Frequency (*msec*) |
|---|---|
| 00 | 200 |
| 01 | 200 |
| 02 | 500 |
| 03 | 500 |
| 04 | 100 |
| 05 | 500 |
| 06 | 200 |
| 07 | 200 |
| 08 | 200 |
| 09 | 100 |
| 10 | 200 |
| 11 | 200 |
| 12 | 500 |
| 13 | 5000 |
| 14 | 500 |
| 15 | 500 |
| 16 | 200 |
| 17 | 5000 |
| 18 | 5000 |
| 19 | 500 |

Table 1. Update Frequency Distribution

In experiments 1 – 3 we assume that upon every object update the respective value is changed. While in experiment 4 the probability of an update resulting in a new value is 50%.

To provide for substantiated results we varied the similarity bounds over a wide range of values. Furthermore, several transaction structures and distributions have been investigated. Throughout all experiments the tendencies observed were quite similar. Therefore we depict only one representative set of examples.

### 3.2 *Experiment 1*

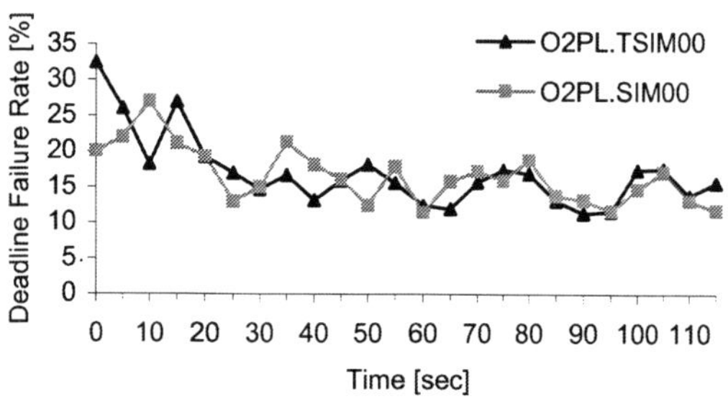


Fig. 4. SIM00 vs. TSIM00.

In the first experiment (fig.4) the *O2PL.SIM00* protocol is compared to *O2PL.TSIM00*. The 2 curves are almost identical displaying a deadline failure rate of 16% in the average.

### 3.3 *Experiment 2*

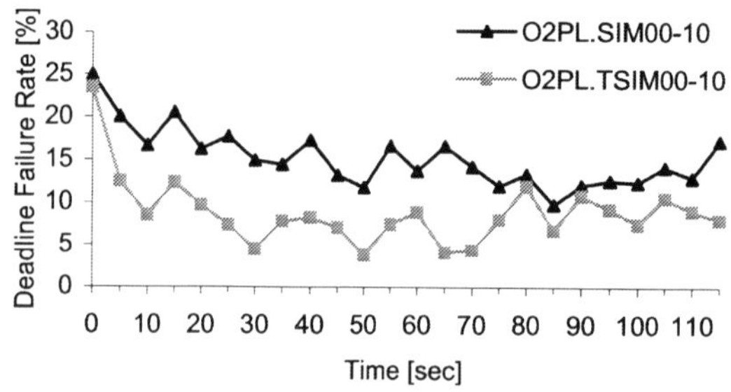


Fig. 5. SIM00-10 vs. TSIM00-10.

For the experiment depicted in fig.5 we set the similarity thresholds randomly between 0% and 10%. While *O2PL.SIM00-10* displays a deadline failure rate of 15%, *O2PL.TSIM00-10*'s failure rate is only half as high at 8%.

### 3.4 *Experiment 3*

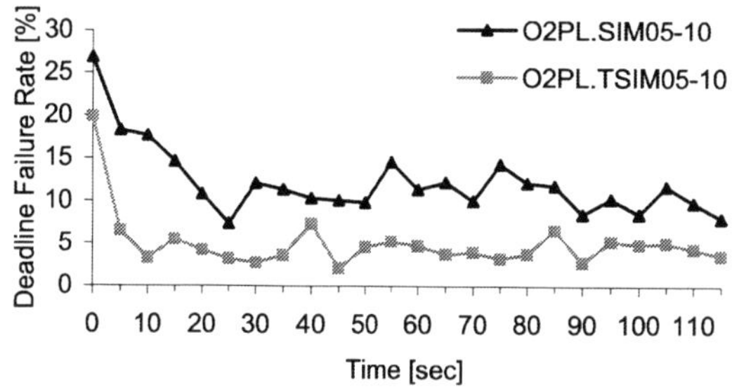


Fig. 6. O2PL.SIM05-10 vs. O2PL.TSIM05-10.

The third experiment (fig.6) shows the results when distributing similarity thresholds between 5% and 10%. While the *O2PL.SIM05-10* version has a failure rate of still 12%, *O2PL.TSIM05-10*'s failure rate drops to 5%.

### 3.5 *Experiment 4*

The last experiment (fig.7) corresponds to experiment 1. Only in this case we assume that the chances for a change in the object value are only 50%. Again, the curves show that the *.TSIM* version outperforms *.SIM*. While *O2PL.SIM00* shows the same deadline failure rate as in experiment 1 (16%) *O2PL.TSIM00*'s failure rate drops to 5%.

In experiment 1 the *.TSIM* version does not improve performance at all. The setup of this experiment is under a *high load* since tasks (and thus transactions) will be invoked very often -

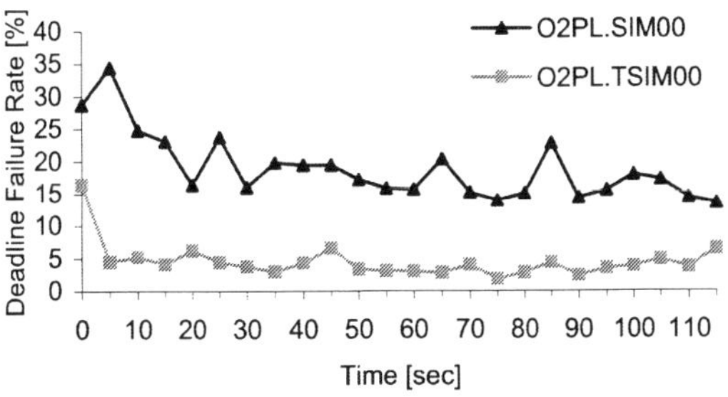


Fig. 7. Update Probability 50%.

every time the corresponding objects are updated. In this setup it is highly likely that almost every task needs to be executed within any transaction instance. Thus the additional overhead of the *.TSIM* version is not outweighed by the beneficial effects of dropping a task execution every now and then. Once the chances are increased that more tasks may be skipped either by relaxing similarity thresholds (experiments 2 and 3) or reducing the probability of object changes (experiment 4), the *.SIM* variation does not improve significantly (16% deadline failure rate in experiments 1, 3 and 4 and 12% in experiment 3). *.TSIM* reduces its deadline failure rate from 16% in experiment 1 to 8% in experiment 2 and even to 5% in experiments 3 and 4.

## 4. CONCLUSION

Basic transaction structures for task sequences and independent tasks have been defined. In this way we built a general transaction model for repli cated real-time databases which extends considerably beyond the concepts of cohorts in (Xiong *et al.* 1988). After introducing similarity of objects as much as of tasks we demonstrated in a series of distributed experiments that our task-based approach to transaction management is considerably superior to a transaction-based concept such as in (Mok *et al.* 1998). The reason is that when a transaction is invoked in our model only those of its tasks have to be executed that have been *marked* regarding their exceeded similarity threshold, in contrast to the transaction-based approach where all tasks will be invoked.

If we interpret similarity as a measure for flexibly and adaptively reacting to the environment then our task-based approach opens new avenues for safety-critical applications. (Here the environmental events occur typically in an unpredictable way.) This will be further substantiated as a general principle in a separate paper (submitted to this conference), by introducing and testing different adaptability features.

## 5. REFERENCES


Anderson, T., Y. Breitbart, H.F. Korth and A. Wool (1998). Replication, Consistency, and Practicality: Are These Mutually Exclusive. In: *SIGMOD 98*. SIGMOD. ACM. ACM. Seattle, WA, USA. pp. 484–495.

Gupta, R., J. Haritsa and K. Ramamritham (1997). More Optimism about Real-Time Distributed Commit Processing. In: *Proceedings IEEE Real-Time Systems Symposium*. IEEE. IEEE. San Francisco.

Ho, S.-J., T.-W. Kuo and A.K. Mok (1997). Similarity-Based Load Adjustments for Real-Time Data-Intensive Applications. In: *Real-Time Systems Symposium (RTSS'97)*. IEEE. San Francisco, CA, USA. pp. 144–153.

Kuo, T.-W. and A.K. Mok (1992). Application Semantics and Concurrency Control of Real-Time Data-Intensive Applications. In: *Proceedings 13th IEEE Real-Time Systems Symposium*. IEEE. Phoenix, Arizona, USA.

Mok, A.K., D. Chen and G. Liu (1998). Semantics and Resource Management of Real-Time Database Systems.

Wedde, H.F., K.-W. Lin, A.K. Mok and K. Ramamritham (1999). A Completely Integrated Approach to Developing, Implementing, Evaluating Distributed Active Database Management and its OS Support. In: *Joint IFAC WRTP'99 and ARTDB-99 Workshop*. Dagstuhl Castle, Germany.

Wedde, H.F., S. Böhm and W. Freund (2001). Adaptive Protocols For Survivability Of Transactions Operating On Replicated Objects. In: *WORDS'01*. IEEE. Rome, Italy.

Xiong, M., K. Ramamritham, J. Haritsa and J.A. Stankovic (1998). Regulating Concurrent Access to Replicated Data in Distributed Real-Time Databases. In: *IEEE Workshop on Dependable and Real-Time E-Commerce (DARE)*. IEEE. Denver, CO, USA.

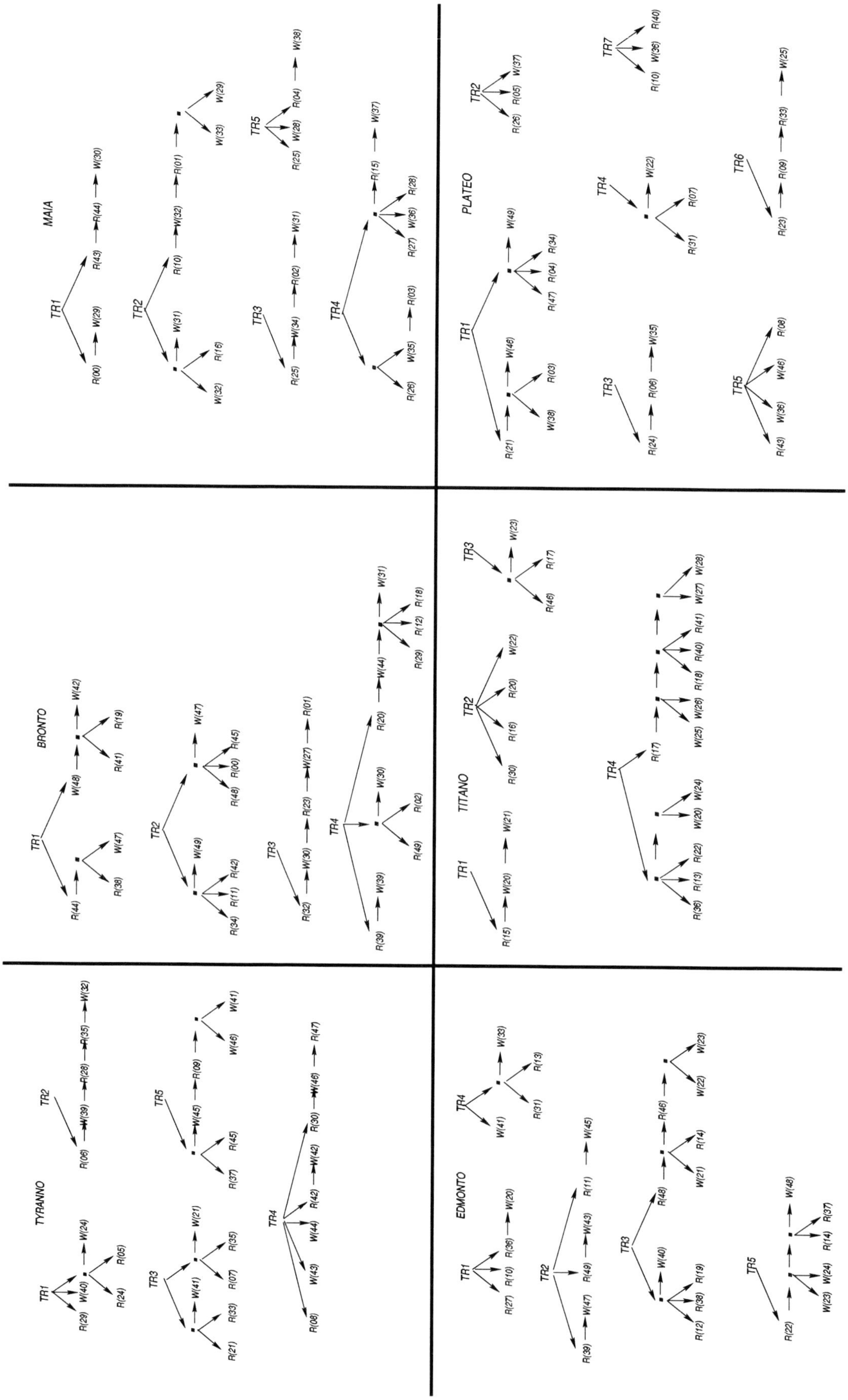


Fig. 8. Transaction Distribution

IFAC Publications
www.elsevier.com/locate/ifac

# A HEURISTIC ALLOCATION METHOD FOR MULTIPROCESSOR SYSTEMS

**Shuhua Wang* and Wolfgang A. Halang****

**ICM6, ROHDE & SCHWARZ GmbH, Postfach 80 14 69, D-81614 München, Germany*
*Shuhua.Wang@RSD.rohde-schwarz.com*
***FernUniversität Hagen, Faculty of Electrical Engineering, D-58084 Hagen, Germany*
*Wolfgang.Halang@fernuni-hagen.de*

Abstract: The allocation problem for multiprocessor systems, in essence to determine which task will be executed on which processor, is NP-hard. To cope with the complexity, two heuristic schemes to allocate communicating tasks on a multiprocessor system are proposed. The purpose of heuristic approaches is to find acceptable solutions by some simple strategies in order to avoid spending too much time in optimizing the objective functions. The new heuristic methods presented in this paper are based on list scheduling. To build the list of tasks, some concepts such as *critical job* and *static deadline* are introduced. In the task allocation process, tasks communicating with each other are allocated to the same processor whenever possible. With this method, deadlines are easier to meet when the communication costs for tasks residing on different processors are taken into consideration. 

Keywords: Real-time systems, allocation, timing requirements, objective functions, multiprocessor systems.

## 1. INTRODUCTION

The problem to be tackled in this paper is the automatic static allocation of a set of real-time tasks to a set of identical processors. The processors are connected through an interconnection network. The tasks have timing constraints, communication constraints and precedence constraints. All of these constraints have to be fulfilled.

In general, optimal allocation under time and precedence constraints in a multiprocessor system is NP-hard. To cope with the complexity, *non-guided-search* techniques which are not optimal algorithms, like *Steepest Descent*, *Tabu Search*, *Simulated Annealing* , *Threshold Accepting*, *Genetic Algorithms* could be used. These methods always maintain a current solution which is slightly modified in order to obtain a better one in the next iteration step.

Another group of techniques to cope with the complexity are *constructive heuristics*. These approaches try to build feasible solutions, i.e., all deadlines are met, based on some heuristics. They are often based on one or more global priority lists or stacks, where in each allocation step the task with the highest priority is selected to be placed on a certain processor. Therefore, they are also called *List Scheduling*. Two major attributes for assigning priorities to tasks are the *t-level* and *b-level*.

A list could also be maintained in two ways: A list is static if it is constructed before allocation starts and remains the same throughout the whole allocation process. A list is called dynamic if it will be changed during the allocation process. In other words, if the priority assignment is based on the static level, the resulting list is static; otherwise, it is dynamic.

A list scheduling algorithm is presented by Hu (1961) where tasks are assigned priorities according to their static b-levels. It is shown that the list scheduling strategy produces a very good performance when tasks are assigned priorities in the order of decreasing static b-levels. In that paper, the makespan is used as performance measure. However, the presented algorithm is only applicable under the assumptions of tree-like graph structures, uniform task execution times and negligible interprocessor communication costs. In (Sih, *et al.*, 1993), a dynamic list scheduling algorithm is proposed. The dynamic level is defined as the difference between the static level and the earliest time at which the task can be scheduled on a processor. Then a task is selected for allocation on the basis of its dynamic level. The choice of the objective functions in list scheduling algorithms is discussed in (Jonsson and Vasell 1997). Three schemes for task priority assignment are investigated. They are based on the highest static level (HSL), earliest start time (EST) and earliest finish time (EFT). All three task lists are static. The non-preemptive time-driven scheduling strategy is used to schedule tasks on a processor. The makespan is used as performance measure to evaluate the different algorithms. Simulation results show that EFT performs consistently better than EST when the number of processors is low.

The main concern of the above list scheduling algorithms is to minimize the makespan. But for a hard real-time system, to guarantee its deadlines is most important.

In (Santos, *et al.*, 1997), three lists (called stacks in the original paper) are used for ordered tasks, processors and communicating pairs. The task list is ordered by decreasing task utilizations. The list of processors is ordered by increasing processor utilizations, and the list of communicating pairs is ordered by decreasing communication costs. Rate monotonic scheduling is used to locally schedule tasks on a processor. The goal of this list scheduling method is to minimize the inter-processor communication by using some heuristics.

The new heuristic methods presented in this paper are constructive and based on list scheduling. To assign priorities to tasks, the concepts of *critical job* and *static deadline* are introduced. A list of tasks is constructed before allocation starts, and will not be changed in the allocation process, i.e., it is a static list. To cope with complexity, the proposed methods decompose the considered task set into task sub-sets of the same period length, i.e., jobs. In addition, tasks communicating with each other are allocated to the same processor whenever possible. In this way, not only the communication network load is lightened, but also deadlines are easier to meet.

The body of this paper is organized as follows: The next section will give a brief introduction to the basic concepts used in the new methods. Then an outline of new heuristic allocation algorithms is presented in Section 3. Section 4 gives evaluation results of the new methods. Finally, Section 5 summarizes the results of this paper.

## 2. FUNDAMENTAL CONCEPTS

A multiprocessor system consists of a set $P=\{p_q: 1 \le q \le m\}$ of homogeneous processors and an interconnection network. The number of processors is limited and can be derived from the user requirements. For illustration purposes, it is assumed that processors are connected to a shared broadcast bus through a token passing protocol (Wang and Färber, 2000).

A real-time application consists of a task set $\Gamma = \{\tau_i: 1 \le i \le n\}$. Each task $\tau_i \in \Gamma$ is characterized by a triple $\{C_i, D_i, T_i\}$. The worst-case execution time $C_i$ includes various overheads. The relative deadline $D_i$ is the time within which the task must complete its execution with respect to the beginning of the period. The period $T_i$ is the minimum time interval between two successive invocations of $\tau_i$. Since tasks communicate with each other, precedence constraints arise from communicating tasks. Precedence constraints between tasks in a task set $\Gamma$ are represented by a partial order $\prec$ over $\Gamma$. If task $\tau_i$ can not begin its execution before task $\tau_j$ has completed its execution, it is represented as $\tau_j \prec \tau_i$.

The task set and the intertask precedence constraints are represented by the precedence graph $G = (N, A)$. $N$ is a set of nodes representing the tasks in the set $\Gamma$. $A$ is a set of directed arcs representing the precedence constraints between the tasks in $\Gamma$, that is, if $\tau_j \prec \tau_i$, then $(\tau_j, \tau_i) \in A$. Examples of precedence graphs are shown in Fig. 1.

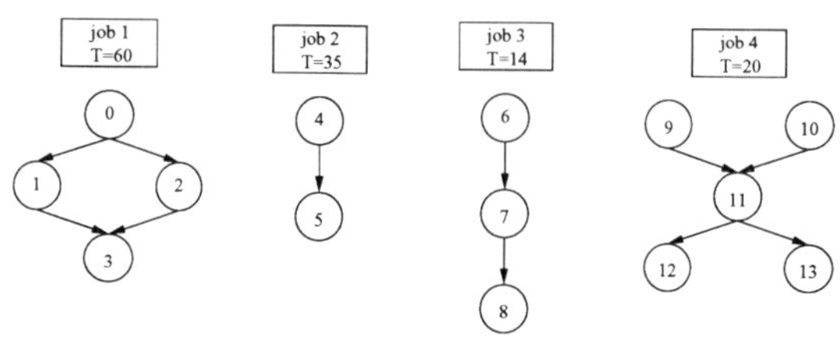


Fig. 1. Examples of precedence graphs

The basic concepts which will be used in the presented heuristic approaches are as follows:

- *Input Task*: A task which has no predecessor in the precedence graph is called an input task.
- *Output Task*: A task which has no successors in the precedence graph is called an output task.
- *Job:* A job is a collection of related tasks which have precedence constraints between them.

As an example, for the task set in Fig. 1, task *0*, task *4*, task *6*, task *9* and task *10* are input tasks. Task *3*, task *5*, task *8*, task *12* and task *13* are output tasks. Task *0*, task *1*, task *2* and task *3* compose job *1*, task *4* and task *5* compose job *2*, etc.

- *Static Deadline*: The static deadline of a task $\tau_i$ takes into account the execution times of the tasks following $\tau_i$ in the precedence graph, i.e.:

$$SD_i = \begin{cases} Min\{SD_j - C_j\}, & \forall j : (\tau_i, \tau_j) \in A \\ D_i, & otherwise \end{cases} \quad (1)$$

In other words, the static deadline $SD_i$ adds structural information to the original deadline $D_i$. If $\tau_i$ is an output task, $SD_i = D_i$. Otherwise, the static deadline $SD_i$ is equal to the original deadline $D_i$ minus the length of the longest path from $\tau_i$ to an output task. In this sense, the static deadline could also be called static b-level as in some literature. Nevertheless, the static deadline $SD_i$ reflects the execution urgency of task $\tau_i$, and will be used as a limit that must not be exceeded by a task during allocation.

- *Job Utilization*: The job utilization is the sum of the execution times of all tasks in a job, i.e.:

$$JU_k = \sum_{\tau_i \in Job(k)} C_i \quad (2)$$

- *Job Slack:* The job slack $SL_k$ is defined as:

$$SL_k = T_k - JU_k \quad (3)$$

  where $T_k$ is the period of the tasks in the job.
- *Critical Job:* The critical job is the job $j$ with the minimum job slack, i.e.:

$$SL_j = Min\{SL_k : 1 \le k \le JO\} \quad (4)$$

  where $JO$ is the number of jobs.

The job slack is the time left between finishing all tasks in the job and the period of the job.

## 3. HEURISTIC APPROACHES

In this section two heuristic approaches are introduced. The purpose of heuristic approaches is to find acceptable solutions by some simple strategy in order to avoid spending too much time in optimizing the objective functions.

The considered problem is restricted as follows: each task executes completely on one processor and tasks are structured to constitute jobs.

### *3.1 Objective Function*

The *performance* of an allocation algorithm is measured as the goodness of the outcome, often relative to the best possible outcome, if such an outcome exits. To achieve a good performance, the choice of the objective function is important. For hard real-time systems, the main concern is whether all tasks meet their deadlines. Therefore, a selected objective function should indicate the scalability of the scheduled system workload. In our case, scalability will be evaluated from three levels of a system, i.e.: task level, job level and system level.

*Definition (Task Slack)*
The slack of task $\tau_i$ is the time difference between its deadline and its worst-case response time $r_i$, i.e.:

$$TS_i = D_i - r_i \quad (5)$$

It could be calculated according to schedulability analysis methods such as ours described in (Wang and Färber, 2000).

According to this, the average task slack and maximum task slack, respectively, are defined as:

$$TS_{ave} = \sum_{i=1}^{n} \frac{TS_i}{n} \quad (6)$$

$$TS_{max} = Max\{TS_i, 1 \le i \le n\} \quad (7)$$

From the above definition, it can be seen that the task slack is the time left between finishing a task and its deadline in the worst case. Or in other words, the task slack is the maximum possible increase in response time that task $\tau_i$ can afford without sacrificing the schedulability of the system. From this point of view, it reflects the scalability of the system workload from the task level.

On the other hand, the *job slack* defined in Section 2 is a static parameter as it represents a static entity independent of the actual allocation scheme. However, the *task slack* is a dynamic parameter in the sense that it represents an entity resulting from the allocation and scheduling schemes used in a system.

*Definition (Makespan)*
The makespan of job $k$ is the maximum worst-case response time of its output tasks, i.e.:

$$F_k = Max\{r_i, \forall i : \tau_i \text{ is an output task of job } k\} \quad (8)$$

Then the average makespan and maximum makespan, respectively, are given as:

$$F_{ave} = \sum_{k=1}^{JO} \frac{F_k}{JO} \quad (9)$$

$$F_{max} = Max\{F_k, 1 \le k \le JO\} \quad (10)$$

Note: The above definition is different from the one generally used in the literature. For example, the makespan is usually defined as the longest finishing time among all tasks in a system, such as in (Tindell *et al.*, 1992). However, the set of tasks is restricted to a job in our definition. With this restriction, the makespan can be seen as the maximum reaction time of a system to an external event.

The deviation of processor utilization from the average utilization, on the other hand, reflects the distribution of workload in a system. It is defined as follows:

*Definition (Utilization Deviation)*

$$UD = \sum_{q=1}^{m} \frac{|ProU_q - ProU_{ave}|}{ProN} \quad (11)$$

where $m$ is the number of processors, $ProU_q$ is the utilization of processor $p_q$ and $ProU_{ave}$ is the average processor utilization in the system:

$$\mathrm{Pro}U_q = \sum_{\tau_i \in \mathrm{Pro}(q)} \frac{C_i}{T_i} \qquad (12)$$

$$\mathrm{Pro}U_{ave} = \sum_{q=1}^{m} \frac{\mathrm{Pro}U_q}{m} \qquad (13)$$

with *Pro(q)* being the set of tasks allocated to processor $p_q$.

*3.2 Priority Assignment*

After the objective functions are defined, the heuristic approaches can be divided into two main steps: assignment of allocation priority and task allocation. In this subsection, the problem of priority assignment is discussed. Without loss of generality, *1* is used to represent the highest priority and *n* is the lowest priority.

The rationale behind the heuristic approaches proposed here is the following:

- Allocate the most critical job first. In this way, an absolutely unschedulable system can be detected early.
- Allocate the tasks communicating with each other to the same processor if possible. In this way, precedence constraints are easier to meet, the load in the communication network is lightened, and the system is also easily shedulable.
- Allocate the remaining tasks, the input task of the next critical job or the input task of the job with the next maximum job utilization first. It will be allocated to the least loaded processor. Thus, the most critical task has a better chance to fit on a processor.

According to this, priorities can be assigned to tasks in the following manner:

- First, assign job priorities between jobs. This is done according to job slack. That is, the smaller the job slack, the higher the job priority of a job. In other words, the most critical job has the highest job priority.
- Then, within a job, tasks are assigned allocation priorities according to their static deadlines. That is, the smaller a static deadline, the higher the task priority. This is similar to the policy of deadline monotony. But the static deadline is used instead of the original deadline of a task. With this method, the sequential behavior of communicating tasks is taken into consideration.

Thus, each task in a task set is assigned a unique allocation priority.

*3.3 Task Allocation*

The proposed heuristic approaches are based on list scheduling. Tasks are placed in a list which is arranged in descending order of their priorities. Then, a task with a higher priority is examined for allocation before a task with a lower priority.

There are two possible task allocation schemes with a slight difference between them. Thc allocation procedure can be performed by the following steps (the difference between approach 1 and approach 2 is only at steps S7 and S7*):

**S1:** Arrange the jobs according to their job priorities.

**S2:** Arrange the tasks within a job according to their allocation priorities.

**S3:** Calculate the minimum number of processors needed:

$$m = \left\lceil \frac{SysU}{\mathrm{Pro}U_{def}} \right\rceil \qquad (14)$$

where $\mathrm{Pro}U_{def}$ is the maximum average utilization of the processor permitted by a user definition, and $SysU$ is the system utilization defined as:

$$SysU = \sum_{i=1}^{n} \frac{C_i}{T_i} \qquad (15)$$

**S4**: Place the tasks of the first *proN* critical jobs on the *proN* processors.

**S5**: Calculate the load of processors:

$$\mathrm{Pro}L_q = \sum_{\tau_i \in \mathrm{Pro}(q)} \frac{C_i}{T_i} \qquad (16)$$

**S6:** Calculate the job utilization of non-allocated jobs:

$$JU_k = \sum_{\tau_i \in Job(k)} C_i \qquad (17)$$

**S7**: Allocate the input task of the next critical job on the least loaded processor *prono*:

$$prono = Min\{\mathrm{Pro}L_q, 1 \le q \le m\} \qquad (18)$$

**S7***: Allocate the input task of the job with the next maximum job utilization:

$$jobno = Max\{JU_k, k \in unallocated\ job\} \qquad (19)$$

on the least loaded processor *prono*.

**S8**: Allocate communicating tasks on the same processor if possible.

**S9**: Allocate the remaining tasks according to their priorities to the least loaded processors.

**S10**: When a task is allocated to a processor, schedulability analysis is performed on that processor. If it is schedulable, it is followed by the next allocation. If it is not schedulable, it is tried to allocate the task to the second, third, ..., less loaded processor until one processor is found on which the task can be scheduled. If no allocation is possible, then the system is deemed to be unschedulable with m processors.

**S11**: Increase the number of processors in the system:

$$m = m + 1 \qquad (20)$$

and repeat the steps from **S4** to **S11** until the system is schedulable.

In the process of task allocation above, the load of processors given by equation (16) in step **S5** is different from the processor utilization given by equation (12). The load of a processor is a dynamic parameter in the sense that it varies with the process of task allocation, but the processor utilization is a static parameter, as it is calculated after the task allocation procedure is finished.

In general, the schedulability analysis used in step **S10** can be any one valid. However, the schedulability analysis algorithm presented in (Wang and Färber, 2000) is used in the above procedure. Additionally, when communicating tasks are allocated on different processors, the influence of the communication cost should be considered in the schedulability analysis. This is done according to the schedulability analysis algorithm presented in (Wang and Färber, 2000).

## 4. CASE STUDY

In this section the performance of the two heuristic approaches proposed in this paper will be evaluated using test cases.

Tindell's example (Tindell *et al.*, 1992) and Altenbernd's example (Altenbernd, 1996) are completely specified with execution times, original deadlines, periods, precedence constraints and communication requirements for all tasks involved. Therefore, they can be taken as test cases to evaluate the presented new approaches. More details about above test cases can be found in the original papers.

First, the performance of the presented heuristic allocation schemes is evaluated for the task level for different test cases.

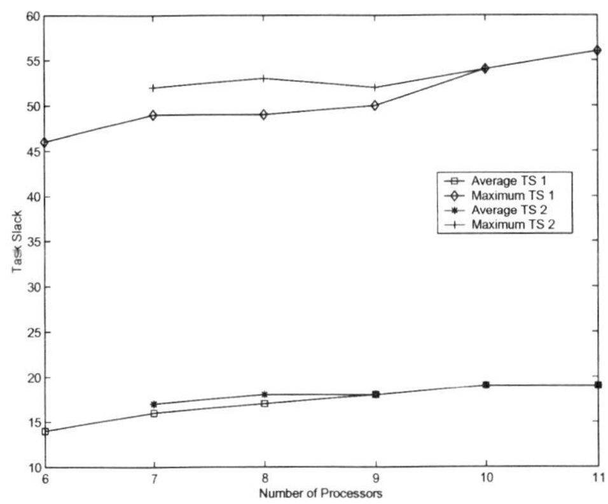


Fig. 2 Task slack of Tindell's example

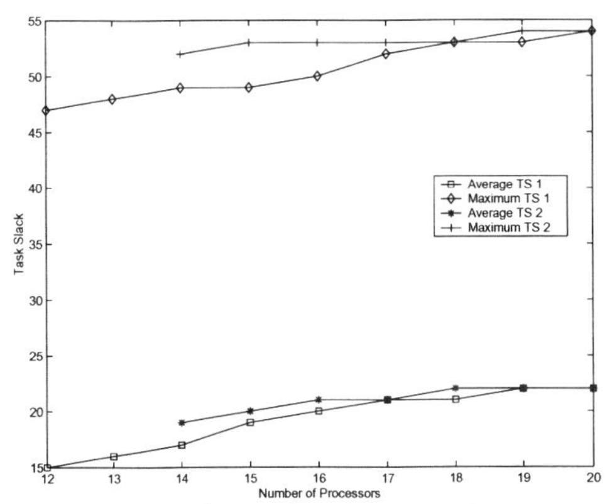


Fig. 3 Task slack of Altenbernd's example

The evaluation results for the two heuristic approaches are shown in Fig. 2 and Fig. 3. In these figures, *Average TS 1* represents the average task slack when heuristic approach 1 is used, and *Average TS 2* represents the average task slack when heuristic approach 2 is used, etc.

From these figures it can be seen that, although the average and maximum task slacks are smaller in approach 1 than in approach 2, the difference between the two approaches is not very big. More specifically, the difference is less than 10%. However, when allocation approach 1 is used, the minimum number of processors needed to keep the systems schedulable is less than for allocation approach 2.

The evaluation results at the job level for the two heuristic approaches are shown in Fig. 4 and Fig. 5. From these figures the same tendency as for the task slack can be seen. Although the average and maximum makespans are greater in approach 1 than in approach 2, the difference between the two approaches is not very significant. Furthermore, the difference will decrease when the number of processors in the system is increased.

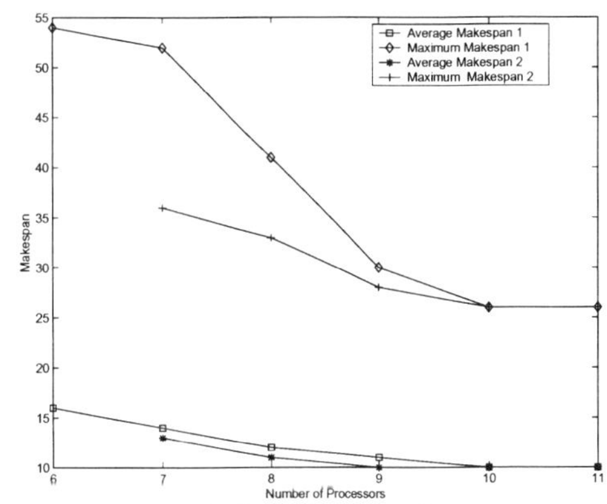


Fig. 4 Makespan of Tindell's example

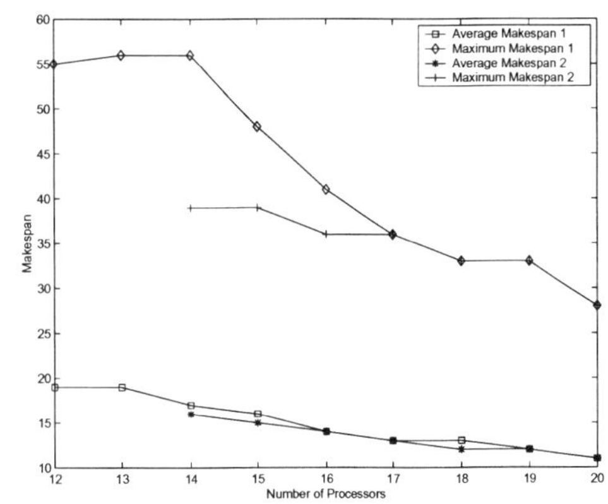


Fig. 5 Makespan of Altenbernd's example

Finally, the performance of the presented heuristic allocation schemes is evaluated at the system level for different test cases. The utilization deviation is used for this purpose. The evaluation results for the two heuristic approaches are shown in Fig. 6 and Fig. 7.

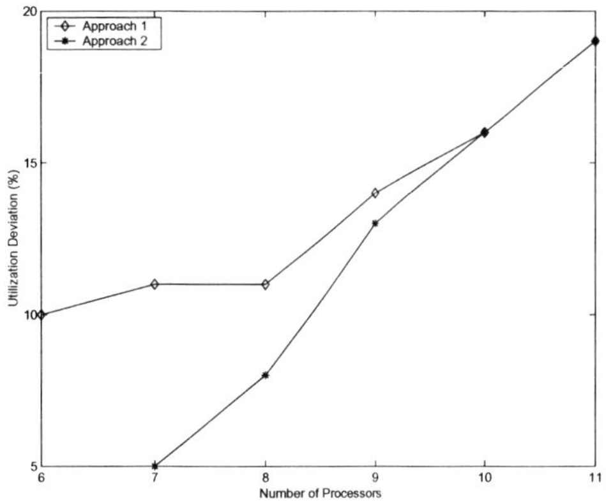

Fig. 6 Utilization deviation of Tindell's example

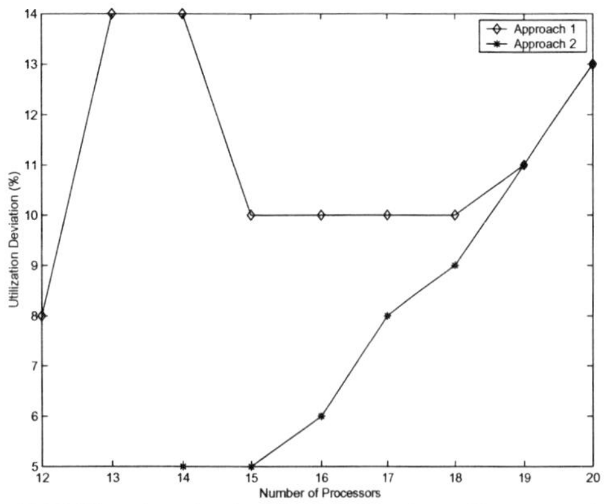

Fig. 7 Utilization deviation of Altenbernd's example

From these figures the same tendency as for the task slack and makespan can be seen. That is, the utilization deviation is greater in approach 1 than in approach 2. However, when allocation approach 1 is used, the minimum number of processors needed to keep the system schedulable is less than for allocation approach 2.

The reason for this difference lies in the fact that heuristic approach 1 can be seen as *critical job* oriented with tasks having a better chance to fit in the least loaded processor. Thus, for a schedulable system, the minimum number of processors needed is lower and the processor utilization is higher than for heuristic approach 2. On the contrary, heuristic approach 2 can be regarded as partly *maximum job utilization* oriented, and better for load balance. Therefore, its utilization deviation is less than that of heuristic approach 1.

## 5. CONCLUSION

In this paper, two heuristic approaches to allocate communicating tasks on a multiprocessor system are proposed. To cope with complexity, some concepts such as *static deadline* and *critical job* are introduced: the static deadline $SD_i$ adds structural information to the original deadline $D_i$, and the critical job is the job with the minimum job slack.

In the presented methods, first job priorities are assigned to jobs according to their job slacks, then allocation priorities are assigned to tasks within the same job according to their static deadlines. In the process of task allocation, a task with a higher priority is selected for allocation before a task with a lower priority. Additionally, tasks communicating with each other are allocated to the same processor whenever possible. In this way, the communication network load is lightened and deadlines are easier to meet. On the other hand, it is also assumed that the scheme of deadline monotony is used to locally schedule tasks on a processor. The schedulability analysis used during the task allocation process is carried out according to the algorithm presented in (Wang and Färber, 2000).

Our simulation results show that the two proposed approaches have different performance, although the difference is small. With heuristic approach 1, the maximum schedulable average processor utilization $ProU_{ave}$ is higher. On the other side, the average and maximum task slacks, the average and maximum makespans of approach 2 are smaller than those of approach 1. Therefore, the two approaches can be used for different purposes. When the main goal is to use as few processors as possible, then approach 1 is a better choice than approach 2. When the main goal is to keep task slack and makespan as small as possible, then approach 2 is more suitable than approach 1.

## 6. REFERENCES


Altenbernd, P. (1996). Timing analysis, scheduling and allocation of periodic hard real-time tasks. PhD thesis, Department of Mathematics and Computer Science, Universität Paderborn, Germany.

Hu, T. C. (1961). Parallel sequencing and assembly line problem. *Operations Research*, 9(6):841-848.

Jonsson, J. and J. Vasell (1997). On the objective function selection in list scheduling for digital signal processing. In *Proceedings of the 22th IEEE International Conference on Acoustics, Speech and Signal Processing*, pages 667-670. Munich, Germany.

Santos, J., E. Ferro, J. Orozco and R. Cayssials (1997). A heuristic approach to the multitask-multiprocessor assignment problem using the empty-slots method and rate monotonic scheduling. *Real-Time Systems*, 13(2), pages 167-199.

Sih, G. C. and E. A. Lee (1993). A compile-time scheduling heuristic for interconnection-constrained heterogeneous processor architectures. *IEEE Transactions on Parallel and Distributed Systems*, pages 625-637.

Tindell, K., A. Burns and A. J. Wellings (1992). "Allocating real-time tasks: An NP-hard problem made easy". *Real-Time Systems*, 4(2), pages 145-165.

Wang, S. and G. Färber (2000). A new algorithm to prove the schedulability of real-time systems. *Control Engineering Practice*, 8(2000), pages 689-695.

IFAC Publications
www.elsevier.com/locate/ifac

# EDROOM. AUTOMATIC C++ CODE GENERATOR FOR REAL TIME SYSTEMS MODELLED WITH ROOM

**Polo, O. R., De la Cruz, J.M., Giron-Sierra, J.M., Esteban, S.**

*Dept Arquitectura de Computadores y Automatica. Fac. Fisicas. Universidad Complutense de Madrid, 28040, Madrid, Spain.*
*Tf: (34) 91 3944389; e-mail: opolo@darc.esi.uem.es*

Abstract: The development of real-time control systems is usually part of a design cycle, involving implementation, experimental control system testing, and re-design. For a faster and better complete development, a software engineering approach, based on a modelling language, is recommended. A successful alternative is ROOM: a modelling language with a graphical syntax. In this paper EdROOM is presented: an environment developed to graphically edit ROOM models and to automatically generate, from the models, C++ real-time control code for direct experimental application. The code generated is modular and easy to debug. The paper describes the design of the environment, with the key aspects of the modelling methodology. EdROOM has been used by our research team for several applications, with satisfactory results. One of these applications concerns the control of actuators of a fast ship for vertical motion smoothing: this is described as an example of EdROOM use. 

Keywords: Real-time software engineering, real-time software development, real-time code generation

## 1. INTRODUCTION

A quick review of the evolution of programming shows how the initial problems of non-structured code (meaning long debugging efforts), led to solve them by developing and using methodological approaches of software engineering, to be able to program reliable complex applications. Following a complete discipline, which starts from models of the software application to be created, the generation of modular, well-structured code is promoted. This is the purpose, for the particular case of real-time systems, of modelling languages such STATEMATE (Harel, et al., 1990), ROOM (Selic, et al., 1994) and Real-Time UML (Powel, 1998), with a graphical syntax to express the relationships between the system components and the specification of the behaviour of all of them.

Taking ROOM to model some of the practical control applications, the experience shows that the coding task is still a hard work to do after modelling. The idea of bridging the gap between the graphical syntax and the real-time code has been developed here by creating a visual interactive environment, denoted EdROOM, for diagram-based modelling and automatic real-time control code generation.

There are other tools that generate code from graphical specification of the real-time systems. Three of the most remarkable are: Rational Rose RealTime (previously called ObjectTime), Rhapsody and STATEMATE. In comparison with them, EdROOM provides a low cost solution for developing embedded real-time system. The code generated by EdROOM is supported by the 16 bits version of RTKernel. This version has an affordable cost and offers a excellent set of characteristics for this type of systems: task-switch time of about 6μs (33 Mhz 486), pre-emptive scheduling, it is ROMable and runs on a simple PC. In addition, working with EdROOM is really an easy task; the graphical syntax is intuitive and the only programming language that the developer has to know is C++.

EdROOM provides a solution for fast development, clear graphical expression of the design ideas, homogenous code generated, easy debugging, design components reusability, encapsulation of detail levels and easy redesign.

EdROOM has been successfully used for the development of real time control systems. An interesting example of application has been the development of a real-time control system to carry on experiments with a fast ship, at towing tank scale (R.Polo, et al., 2001).

The paper has three sections. First, the design methodology is summarized. Second, the EdROOM environment is described. Finally, an example of application (the fast ship) is given.

## 2. ROOM METHODOLOGY

The main feature of real time control systems is that they are in constant interaction with their environment. The diversity of simultaneous tasks these systems perform determines their complexity. The different tasks must be scheduled with different priorities and according to events. A concurrent model must be used to define the system.

The design of this type of systems must provide the following elements: (1) the determination of the set of concurrent tasks; (2) the definition of the communication between them, including its topology and the valid protocols; (3) the specification of the behaviour of each of the tasks and the system scheduling policy. ROOM provides graphical means to define these three elements.

### *2.1. Tasks: actors and actor classes in ROOM*

In ROOM each task is assigned to an agent called actor. The actors are the basic entities that are used to define the ROOM models. ROOM uses a multilevel definition of the structure, since each actor may contain other actors within itself. Fig. 1 shows a part of the hierarchical structure of a ROOM model. The first level of the model, composed by five actors, is graphically represented by the top left scheme. The arrow that originates at one of its actors points at the scheme on the right that represents the second level of actors contained in it. The other arrow points at the scheme of third level actors.

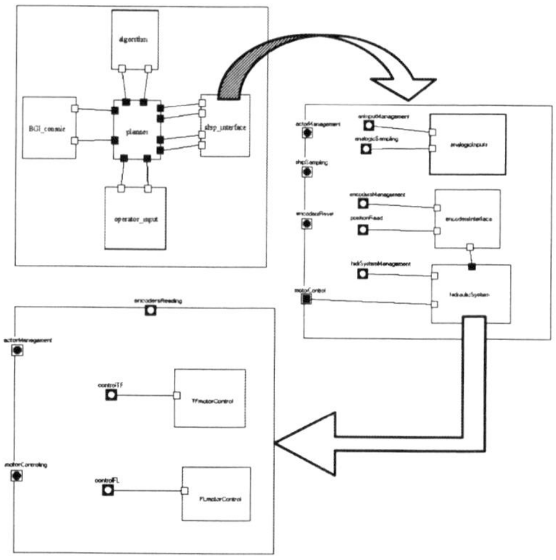

Fig. 1. Actors hierarchy in ROOM.

ROOM uses an object oriented solution to organise the relationships between the actors, which is based on the definition of **actor classes**. Each actor is an instance of an actor class; the definition of the actor class structure will determine witch actors will be components of all its instances.

### *2.2 Communication among actors*

The actors establish communication between them by means of message passing. To determine the set of messages that actors may exchange between them, the model includes some **protocol classes**. Each one defines a set of input messages and another of output messages that shall be used for communication.

An actor uses a protocol class when it adds a **port** of that class to the interface of its actor class. Fig. 2 shows an actor class with 3 ports in its external interface(port1, port2 and port4) and one more in its internal interface (port3).

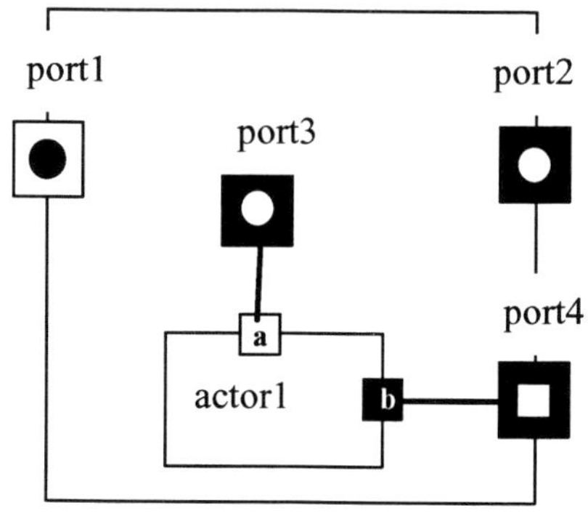


Fig. 2. Communication ports.

Communication between actors is established trough ports. There are several types of ports identified by different graphical representations. The ports with a dark border (port2, port3 and port4) use its protocol class in a conjugate way, taking the input messages as output messages and vice versa. Those with the rectangular internal border, such as port4, are relay ports and they allow to export a port from a component actor.

### *2.3. Behaviour and scheduling. ROOMCharts and priority of messages*

To define the actor behaviour, each actor class has a kind of state chart called ROOMChart. ROOMCharts are based on Harel's Statecharts formalism (Harel, 1987). They determine the transition that will be triggered by a message received at the present state. Each transition has one signal, one reception port and one guard associated to it. The transition is triggered only if there is coincidence for the signal and port of the message received and the transition, and the guard is true. The ROOMCharts allow several definition levels. This is possible because any state may contain other sub-states and define its own context. Fig. 3 shows a two level ROOMChart example where A and B are substates of top that define their own context.

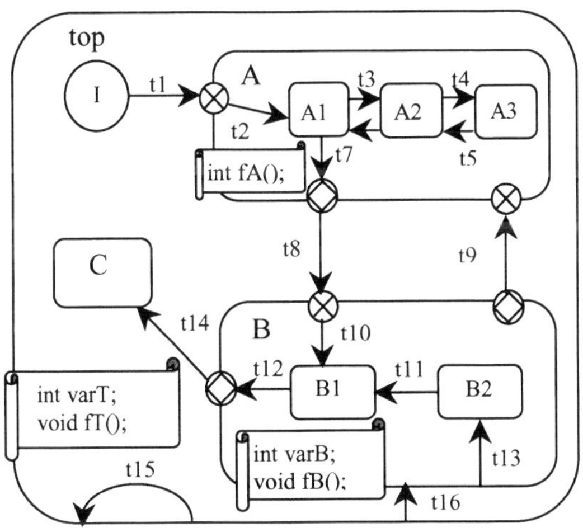


Fig. 3. ROOMChart example.

In each context, it is possible to define a set of variables and functions. These are used for the treatment of the received message when the actor is in one state of that context. State entries and exits, as well as transitions, may have associated functions that are executed when a message arrived triggers them. These functions may either belong to this context, or to any of its higher contexts.

The scheduling policy is established by means of the priority of messages sent. Each actor supports a message queue arranged by priority. In a multitasking environment, the actor that has the message with the highest priority must catch the processor time for handling it. ROOM allows synchronous and asynchronous message passing.

## 3. EDROOM: A GRAPHICAL EDITOR OF ROOM MODELS AND AUTOMATIC CODE GENERATOR

In order to take advantage of the ROOM modelling potential, the EdROOM environment has been developed. EdROOM is a CASE tool for editing ROOM models that integrates an automatic code generator

### *3.1. Model edition*

EdROOM supports the management of the actor classes, the protocol classes and the data classes by employing windows and mouse. Fig. 4 shows the edition canvas of the actor class. To add an component actor, or a port, to the actor class it is enough to drag the actor class, or the protocol class, from the lists located on the right and drop it into the canvas. The tool bar on the left is used to connect actors.

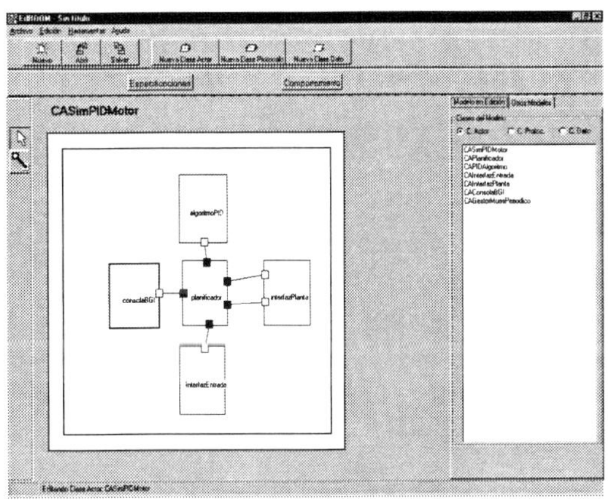


Fig. 4. Actor class edited with EdROOM.

With respect to the behaviour, EdROOM includes a supplementary window for editing the state chart of each actor class. Fig 5 shows this window. The tool bar located on the top is used to add states and transitions to the chart. By double click over one state it is possible to add its sub-states and define a new context. This is the way to specify several levels in the behaviour. Besides, each transition includes a menu for adding its trigger condition(signal, port, guard).

### *3.2. Implementation: the detail level functions and the ROOM services*

In EdROOM, the detail level functions used to handle the input messages are integrated within the design. To do that, the behaviour edition window (Fig. 5) s used to add variables and functions to the different contexts.

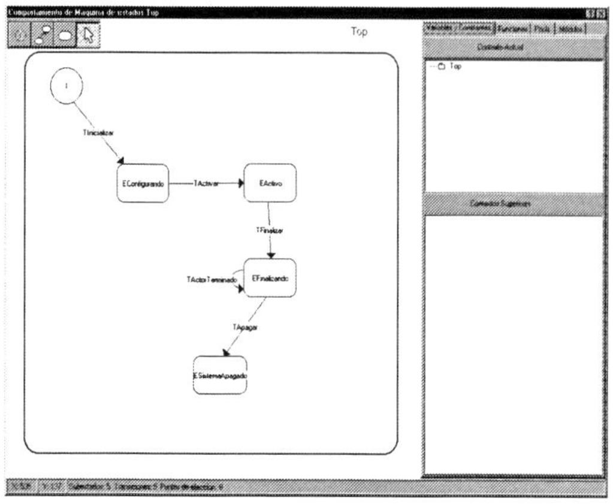

Fig. 5 ROOMChart edited with EdROOM.

The functions can be linked to the transitions and to the state entry and exit by means of menus. EdROOM includes also text edition windows to define in C++ the prototype and the body of these functions.
There is also a set of basic services that have been provided to simplify the detail level implementation. These services are: communication, scheduling, timing, and memory control. We have developed a library called mv_rtk.lib that contains them. This library is linked with the generated code and it implements all the primitives that are needed for making these services transparent to the user.

In ROOM, the scheduling is specified by assigning priorities to the messages. The library mw_rtk.lib provide the following primitive to send across a port an asynchronous message with a specific priority.

port.send(signal, priority, dataP, poolP);

The timing service is accessible using the following primitives:

**Absolute:** ROOMtimer.InformAt(time, dataP, poolP, priority);

**Relative:** ROOMtimer.Informln(interval, dataP, poolP, priority);

With respect to the memory control, EdROOM allows to define the stack space and the maximum number of messages queuing for each actor. In addition, EdROOM uses data **Pools** for emulating the behaviour of the dynamic memory avoiding its low performance.

All this services can be used in the detail level functions, that are integrated in the behaviour of the actor classes, in a transparent way by means of the invocation of the correspondent primitives.

*3.3. Code generation. Requirements and capabilities*

The generated code is a set of C++ source files that implement the structure, the communication and the behaviour of the actor classes of the model. From each actor class one C++ class with the same name is generated. The detail level functions added to the behaviour are correctly inserted in this code. To obtain the executable file, the source files are compiled and linked with the mv_rtk.lib library.

This library has been implemented to run on the low cost real time kernel RTKernel. This kernel works on Intel-based systems, and supports MS-DOS as a service. We are also developing other versions of the library, for other real-time operating systems, in order to provide portability to other computing platforms.

The generated code is clear and coherent with the design. This makes easy the debugging of the program. As an example we show the main function of a system that performs a PID control of the speed of a D.C. motor (Fig. 6).

```
#include "mv_rtk.h"
#include "PIDControlE.h"
void main(){
  Pr_Kernel kernel(ROOMVeryHighPriority);
  PIDControl   MainActor   ("top",   NULL,   10,
               ROOMNormalPriority,1024);
  MainActor.ROOMConfig();
  MainActor.ROOMStart();
  }
```

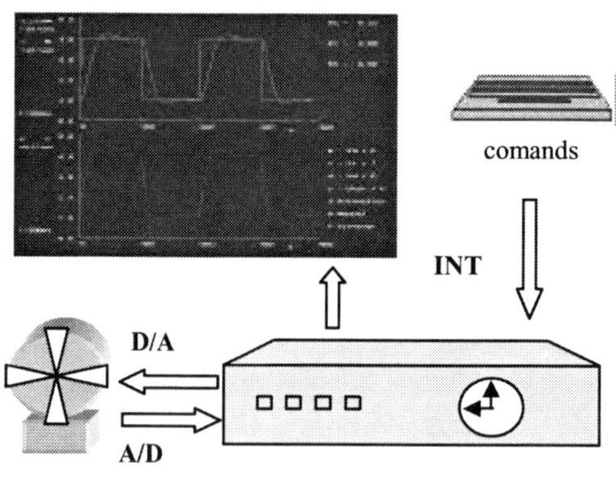


Fig. 6. Scheme of a PID control of a DC motor

The requirements to perform the development are not stringent: a Pentium 100 with 16 MB of memory, with Windows 95/98, is enough to run EdROOM. The compiler used to obtain the executables was Borland 4.52. The RTKernel licence needed to develop as many projects as you want costs only 550$. Finally, the computer needed to run the executable is a standard PC.

## 4. APPLICATION EXAMPLE: CONTROL EXPERIMENTS WITH A SHIP ON A TOWING TANK

Part of our research concerns a fast ferry. Vertical motions at high speed can have negative effects on the ship and the passengers comfort. By means of transom flaps and a T-foil near the bow, we can counteract the motions induced by waves. The flaps and the T-foil can move under control. The problem is to get a control strategy for optimal motion smoothing. A scaled down replica, 4.5 meters long has been built, for experimental studies in a towing tank institution (CEHIPAR, Madrid). The replica has flaps and T-foil. There is a step motor to move the T-foil wings and other to move the Flaps. The sensors located in the replica measure the following variables: heave, pitch, the height of the arriving wave, the drag forces ( starboard and port ) and the accelerations in several points of the replica.

It has been used an industrial PC to run the control program. The PC includes the Advantech PCL812PG data acquisition card, to perform the A/D conversions of the sensors measurements and to generate the pulses that control the step motors. Another card, the TE5312, is also connected to the PC bus to read the motor encoders value. The control system has been developed to interact with the operator commands and to display the signals on a monitor. Fig. 7 shows a schematic of the whole system. The clock inside the computer means the ROOM timing service has been used to perform the time synchronization tasks.

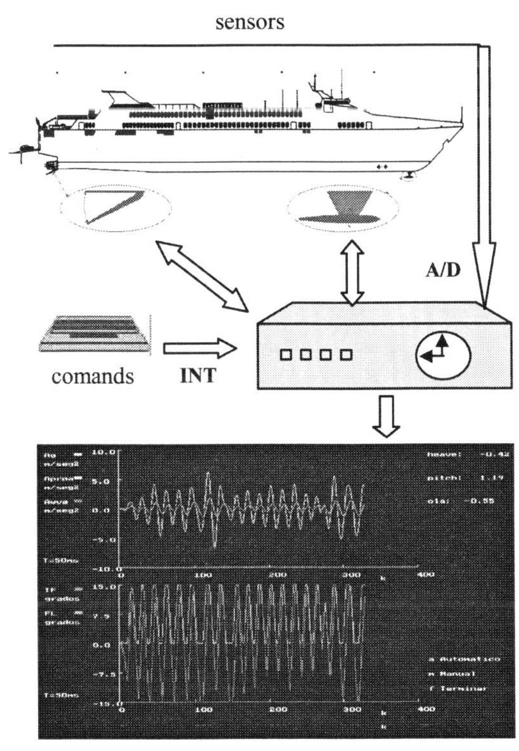


Fig. 7. Scheme of the control of the ship replica.

### *4.1. ROOM Actors structure of the system*

The model of the experimental control of the replica includes 5 inter-connected main actors (Fig 8).

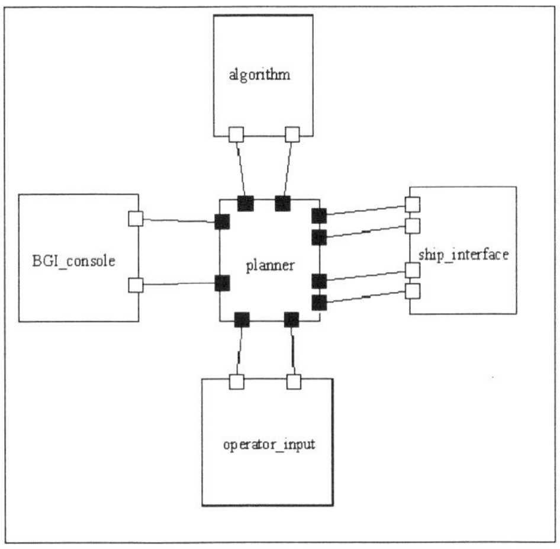


Fig. 8. The five main actors of the ROOM model of the control program.

The function of these actors are the following:

-"ship_interface" is employed to provide the interface with the step motors and the sensors.

-"BGI_console" displays the main variables in the console.

-"operator_input" is employed to handle the operator commands given with the keyboard.

-"algorithm" executes the control algorithm.

-"planner" coordinates the work of the rest of the actors in each sampling period. It manages the start-up and termination of the whole system.

The "planner" also contains other actor, called "periodic_Sampling", which manages the sampling timing. The "ship_interface" has three actor components inside too (Fig. 9): one of them performs analog measurements sampling with the PCL812PG card. Other provides access to the TE5312 card and reads the motor encoders value. The third is devoted to move the appendages in accordance with the calculated commands. This last includes also two actors more in its internal structure. One of them generates the pulses to move the T-foil and the other moves the flaps.

The hierarchical strategy promoted by EdROOM, with the definition of several levels of actors, is an important feature. A complex task can be divided into simpler ones. This has been a key point for the success of the development.

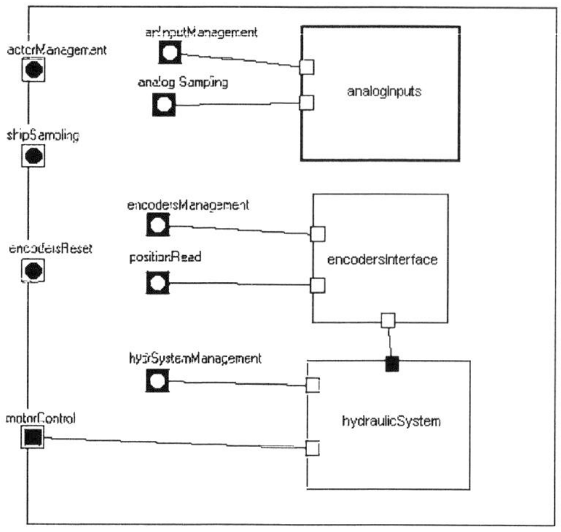


Fig. 9. Actor components of the "ship_interface" actor.

### 4.2. Experience with EdROOM

One of the advantages that we wanted to obtain was to make easy the re-design of the control program. With EdROOM its possible add actor components in a advanced stage of the development to extend the functionality of some class actor.

Before the experiments on the water, the control system has been developed and tested directly with the replica (hanging from a crane). Actually, the replica was manually balanced, to see if sensors and actuators worked correctly. Some problems (calibration, signs, noise from motors) were solved by using EdROOM on the field to modify the control program. Once the replica was on the water, new problems appeared. The control strategy (taking signals from accelerometers and applying a digital PID) has to be re-designed to include some filtering. Again, EdROOM was used for a quick re-design of the control on the field (in the towing tank, time is money).

The protocol classes and the behaviour can be added or extended without problems. Other important benefit is the possibility of checking several alternatives. This can be made by a direct substitution of an actor with another actor which has the same interface.

The strict interface definition of the ROOM actors assures the correct integration of the new actors in the system. This fact is also the key point of the reusability of the ROOM actors: we can say that actors work like "plug and play" software components. As the result of all these aspects the development of real-time systems can be stated like the evolution of a basic prototype that runs properly in each stage.

## 5. CONCLUSIONS

A visual CASE tool for real time automatic control code generation has been created. The tool, called EdROOM, runs under Ms-Windows. It follows an object oriented methodology of software engineering, called ROOM, based in the use of actors. The use of ROOM guarantees a structured and clear design of control programs and allows for the reuse of software components at design level. The executable code is generated quickly and can be used in Intel-based systems. This code runs on a low cost real time kernel called RTKernel. The library mv_rtk.lib implements the memory control, communication, timing and scheduling services, necessary to execute the ROOM model, using RTKernel primitives. This library is linked with the C++ code generated to create the executable. The implementation of this library for other RTOS paves the way to obtain platform independent applications and multi platform reusable components.

## 6. REFERENCES

Harel, D., H. Lanchover, A. Naamad, A. Pnueli, M. Politi, R. Sherman, A. Shtull-Trauring and M. Trakhtenbrot (1990). STATEMATE: A Working Environment for the Development of Complex Reactive System. *IEEE Transactions on Software Engineering*, **16**, 403-414.

Harel, D. (1987). Statecharts: A Visual Formalism for Complex Systems. *Science of Computer Programming*, **8**, 231-274.

R.Polo, O., S. Esteban, A. Grau, J.M. De la Cruz (2001). Control Code Generator used for Control Experiments in Ship Scale Model. *CAMS2001 IFAC Conference.*

Powel Douglass, B. (1998). *Real Time UML. Developing Efficient Objects for Embedded Systems.* Addison Wesley.

RTKernel 4.0 and RTKernel 4.5 Real-Time Multitasking kernel for C/C++. *User's Manual.*

Selic, Bran, G. Gulleckson and Paul T. Ward (1994). *Real-Time Object Oriented Modelling.* NewYork, John Wiley and Sons.

IFAC Publications
www.elsevier.com/locate/ifac

# USING LINUX AND ADA IN THE DEVELOPMENT OF DISTRIBUTED COMPUTER CONTROL SYSTEMS

**Alejandro Alonso** * **Roberto López** * **Juan A. de la Puente** *
**Bárbara Álvarez** ** and **Andrés Iborra** **

* *Universidad Politécnica de Madrid, Depto. de Ingeniería de Sistemas Telemáticos, ETSI Telecomunicación, E-28040, Madrid, Spain*
** *Universidad Politécnica de Cartagena, Dpto.Tecnología Electrónica, Paseo Alfonso XIII, 50, 30203-Cartagena, Spain*

Abstract: Many distributed computer control systems are non-critical in the sense that they have only soft time constraints and no stringent safety requirements. This is the case with certain teleoperated robot control systems, which can be effectively built on top of general purpose hardware and software execution platforms, instead of having to use costly and difficult-to-program real-time computers and operating systems. An approach for developing such kind of applications, based on the use of Linux and Ada on PC architectures, is presented in this paper. Distribution is based on the Distributed System Annex of Ada on top of a TCP/IP LAN. A number of guidelines on how to use them in a proper way are described with reference to an example teleoperation architecture. *Copyright © 2001 IFAC*

Keywords: Soft real-time systems, Ada, Linux, teleoperation systems, robotics.

## 1. INTRODUCTION

Many real-time control systems have critical safety requirements, which means that failures in the execution of the system may lead to unacceptable consequences, such as economic losses, environmental damages, or even danger to human lives. Therefeore, the development of these systems requires appropriate and rigorous development techniques, together with highly reliable hardware and software execution platforms. Moreover, most computer control systems have strict time requirements which must be guaranteed. This implies the use of special real-time operating systems with such special features as high kernel reliability, high resolution timers, or bounded kernel service execution time. As a result, the development of this kind of real-time control systems —usually known as *hard real-time systems*— tends to be difficult and expensive (Burns and Wellings, 2001).

On the other hand, there are other kinds of control systems which have less stringent requirements (i.e. they are *soft real-time systems*). These systems can usually tolerate isolated failures in computation or timing without catastrophic consequences.

This is the case for a large class of robot control systems which are supervised or teleoperated by humans. Examples can be found in such application areas as nuclear plant maintenance robots or ship cleaning robots. These kinds of robots usually operate at low speeds, and in the case of a failure in the computer control system, the operator has means for manually stopping the robot in a safe way.

This flexibility enables the development of these type of computer control systems to rely on general-purpose execution platforms which are easier to program and test, and more affordable too, than the specialized platforms which are required for hard real-time systems.

The rest of the paper describes a novel approach for the development of distributed soft real-time control

* Work partially supported by CICYT, projects TIC99-1043 and 1FD97-083.

systems. The approach has been used in the GOYA project (Iborra and Álvarez, 1999), which is aimed at developing a robot for automatic hull blasting and coating removal on ships. The execution platform is based on GNU/Linux and an industrial PC. The programming language is Ada (Ada, 1995). Distribution is based on the Distributed System Annex (DSA) of Ada. The particular implementations are GNAT, the GNU-NYU Ada Translator (Schonberg and Banner, 1994), and GLADE (Pautet and Tardieu, 2001). The project has been used as a test bench to explore the applicability and limits of this approach, and to compare it with other alternatives. As a result, a set of guidelines for developing applications with a sufficient degree of robustness and determinism have been obtained.

## 2. PROBLEM DEFINITION AND ANALYSIS

### 2.1 *System description*

The goal of the GOYA project is to develop an automated and environment-friendly system for hull blasting and coating removal on big ships. The system uses a new technology based on high pressure water jets mounted on a moving cleaning head. This technique is intended to replace the current paint and coating removal methods, based on open grit blasting, which is likely to be banned in the near future because of the pollution it causes.

One of the technical activities of the project is the development of a robot for maneuvering the cleaning head along the ship hull surface. The robot system is teleoperated, and is composed of the following elements:

- *Hydraulic elevation platform*, which is in charge or moving up and down the rest of the components. It can move the cleaning head on a band between 800m m and 3300 mm of height, approximately. An additional supplement with a height of 2500 mm will allow to increase the cleaning band. Its motion speed is around 0.07 m/s.
- *Positioning arm*, which moves the cleaning head in a direction perpendicular to the ship surface. Its speed is around 0.1 m/s.
- *Head positioning motor*, which moves the cleaning head in a direction parallel to the ship surface.
- *Head orientation motor*, which moves a pan&tilt head that directs the blasting with an incidence angle of $45^o$ with respect to the ship surface.

The sampling time requirements are in the order of 100-300 ms. These values are derived from the speed of the motors and the refresh rate of the teleoperation system display. Actuator deadlines are taken equal to the sampling periods. The characteristics of the system and the application are such that missing a deadline from time to time does not cause any unacceptable harm. Therefore, the system has soft real-time requirements.

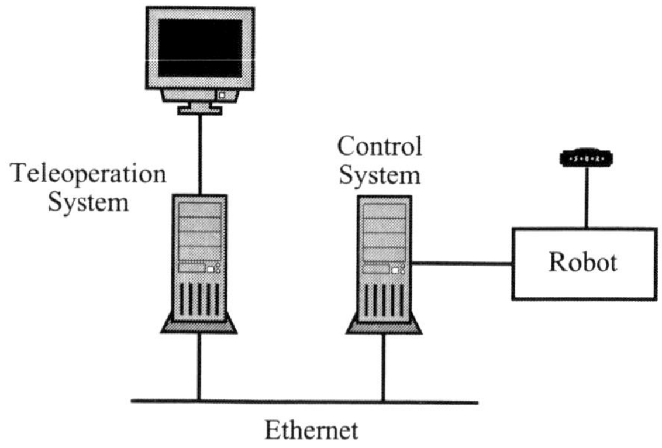

Fig. 1. System architecture

### 2.2 *Analysis of alternatives for the execution platform*

The system architecture is shown in figure 1. It is composed of two computer subsystems: the control subsystem, which drives the operation of the robot, and the teleoperation subsystem, which interacts with a human operator, who can see the state of the robot and remotely drive its operation. These subsystems are connected by a dedicated Ethernet LAN with no additional nodes. There is a safety control panel directly connected to the robot which can be used for emergency stop when required.

The control subsystem interfaces directly with the robot sensors and actuators, driving the motors and the start and end of the blasting operation, and executing the control algorithms for all of them. It also accepts commands from the teleoperation subsystem, and sends messages to it through the LAN link.

The teleoperation subsystem displays the state of the robot and accepts and interprets operator commands for the robot. The state is updated on a regular basis in response to periodic commands which are automatically sent to the control subsystem.

Two computers are used because of the different nature of the operations to be performed. Moreover, the functional requirements lead to using two different platforms: one with powerful graphical capabilities, for showing the robot state and the operational environment with enough detail, and another one with the appropriate interface hardware for interacting with the robot. The robot control subsystem runs on an industrial PC, while the teleoperation subsystem is based on a workstation with powerful graphics capabilities. The rest of this paper is focused on the design and implementation of the control subsystem, as the teleoperation subsystem has no special time and safety requirements, and can be developed with general-purpose techniques.

Three alternatives for the basic software platform were explored:

- Linux (Bovet and Cesati, 2000).
- Linux and RTLinux (Barabanov and Yodaiken, 1996).
- A dedicated real-time kernel, such as JTK (Ruiz and González-Barahona, 1999) or RTEMS(OAR, 2000), running on the bare hardware.

The advantages of the first approach are the wide range of available drivers and the possibility of using the same platform for development and execution. The availability of drivers for the most common Ethernet cards and protocols is especially interesting. The main disadvantage of this approach is the lack of a true real-time kernel, although some of the functionality for real-time threads in the POSIX standard is implemented (IEEE, 1990).

The other two alternatives rely on real-time kernels, which provide the required functionality for implementing real-time tasks. The RTLinux/Linux configuration is particularly interesting, as it enables the use of a real-time kernel while taking advantage of the Linux drivers. The main drawback of this approach is the perception that RTLinux was not mature enough for an industrial project at the start of this project, and the lack of an Ada compiler for this kernel.

The first configuration with Linux only was finally selected for the GOYA project, after carefully checking that the real-time facilities it provides are sufficient for meeting the functional and timing requirements of the system. In this way, it was not necessary to spend time learning how to use the system or developing complex drivers, and it was possible to use the large amount of information available about this operating system.

## 3. LINUX REAL-TIME BEHAVIOUR

There a number of studies on the possibility of using Linux in the development of real-time systems. One conclusion of these studies is that there are indeed some problems in using Linux as a platform for the design and development of real-time systems. The most important ones are:

- *Linux scheduler*: The basic Linux scheduler is based on the well-known time-sharing technique (Bovet and Cesati, 2000). CPU time is divided into time slices, one for each runnable process. If the process is not terminated or self-suspended when its current slice expires, a process switch occurs and the CPU is granted to another ready process. Processes are ranked according to priority, which is a process parameter which may change dynamically according to complex algorithms. This type of scheduler is clearly not appropriate for real-time systems (Burns and Wellings, 2001).

  However, Linux provides another type of scheduler, in compliance with the POSIX standard (IEEE, 1990). The alternate scheduler is identified as `SCHED_FIFO`, and it uses a priority-based preemptive method to assign the CPU to a ready process. It is possible to specify the scheduling policy to be used and a fixed real-time priority for a process. Real-time priorities are always more urgent than non-real-time priorities.
- *Timer resolution*: As documented (and measured) the resolution of Linux timers is 10 ms, which is indeed to coarse for fast real-time system, as it can lead to errors of up to 20 ms when programming timer events. However, the time requirements of the GOYA control subsystem are much longer, and it is thus feasible to use Linux timers for its implementation.
- *Dynamic memory*: This feature can increase significantly the execution time of a process, if some portion of its address space is moved onto disk. However, this can be easily overcome by using the `mlockall` system call which disables paging for the calling process.
- *Kernel preemptability*: The Linux kernel is not preemptable. This means that whenever a system call is issued, the kernel executes without preemption, no matter the priority of the runnable user processes. The effect is that real-time processes can suffer undesirable blocking.
- *Interrupt disabling*: The kernel or the device drivers may disable interrupts to prevent their being preempted by an interrupt or another kernel activity. It is clear that while interrupts are disabled, a real-time process will not run, no matter its priority.

The main consequence of this scenario is that real-time processes may suffer undesirable blocking, which causes an increase in their response times. The main problem is that it is difficult to bound the length of the worst case blocking time for a process. This undesirable effect can be reduced by removing from the kernel those modules and device drivers that are not necessary for the execution of the real-time application. In addition, it is convenient to disable all services that are not required.

In order to validate the approach, a number of experiments have been made with the target system. The results indicate that the highest blocking suffered by a real-time Linux process is tolerable, according to the time requirements of the application.

## 4. THE ADA PROGRAMMING LANGUAGE

Ada(Ada, 1995) is the implementation language selected for the GOYA project. It was considered to be the most suitable language because of the following reasons:

- *Reliability*. Ada is a highly reliable language, in which many errors can be found at compilation time, where there are less costly to fix, as opposed to C and especially C++, which very often show complex run-time errors.
- *Abstraction*. Ada is a high level language which includes many concurrency and real-time features as part of the language itself. It also includes a annex for the development of distributed programs. This makes the implementation more reliable, as the compiler is able to make a number of semantic checks. This approach is better than the use of sequential languages such as C and C++, which require handling complex operating system interfaces to provide the same features.
- *Predictability*. Ada's built-in fixed priority scheduling and intertask synchronization mechanisms enable accurate timing analysis to be performed, and predictable timing behaviour to be enforced.
- *Availability*. GNAT, the GNU/NYU Ada Translator [1] is a high quality development environment for Ada 95, which is available as free software (Schonberg and Banner, 1994). The GNAT environment includes native and cross compilers for many common platforms, as well as a graphic debugger and other useful tools.
- *Industrial penetration*. Although the use of Ada in general industry applications is much less extended than C or C++, it is the first choice language for safety-critical systems.

## 5. CONTROL SUBSYSTEM DEVELOPMENT

### 5.1 *Design overview*

The robot controller is the most relevant subsystem of the teleoperation system from a real-time point of view. The subsystem design relies on four components, one for each of the moving components of the robot, with similar internal structures. A simplified version of this structure is shown in figure 2.

Each component provides a remote interface to the teleoperation system, enabling it to perform a number of operations on the robot components. For example, the remote interface for the elevation platform is defined by the following Ada package:

```
package P_Platform.Teleoperation_Interface is
pragma Remote_Call_Interface;
  function Get_State return T_State;
  procedure Stop_Platform;
  procedure Raise_Platform;
  procedure Descend_Platform;
  procedure Move_Platform_To (A_Height : T_Position);
end P_Platform.Teleoperation_Interface;
```

Whenever a command is issued, the internal state of the robot is changed accordingly. For example, if a command is received to raise the platform, the operating state of the robot is updated and an actuation signal is sent to the elevation motor.

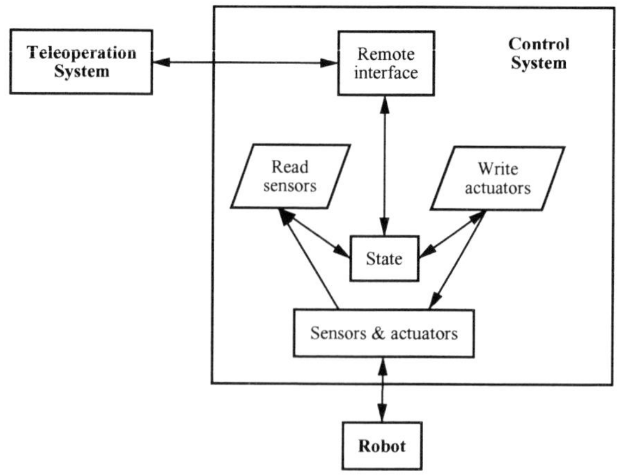


Fig. 2. Internal design of the control system components.

The active parts of a controller component are a couple of tasks. The first one is a periodic task which reads the sensors to get the current position of the platform. This information is processed in order to update the robot state and generate actuation signals, if required. The other one is a sporadic task which drives the platform actuators when required by a teleoperation command or a state change. Both tasks also check the state of the rest of robot components so that a coordinated operation is ensured.

### 5.2 *Implementation details*

The purpose of this section is to provide details on how the Ada language and the Linux operating system have been used in order to get a proper behaviour from the application.

#### 5.2.1. *Ada process configuration*

Each controller subsystem is implemented as an Ada program, which in turn is executed as a Linux process at run time. In order to have a proper behaviour, controller processes are configured as follows:

- The scheduling policy for the process is set to `SCHED_FIFO`, and the process is assigned a high priority within the allowable range. In this way, control processes have higher priorities than other user processes.
- The address space of the process is locked in memory and hence paging is disabled.

The configuration options are implemented by calling the appropriate Linux services when the process execution starts. The facilities provided in Ada for interfacing with C functions are heavily used for this purpose. Linux system calls are invoked from the main Ada procedure, as shown in the following example:

[1] http://www.gnat.org

```
procedure Configure_Ada_Process is

   function Getppid return Integer;
   pragma Import(C,getppid);

   function Mlockall (Mode : Integer) return Integer;
   pragma Import(C,mlockall);

   type Sched_Param is record
      Sched_Priority: Integer;
   end record;
   pragma Convention(C,Sched_Param);
   type Access_Sched_Param is access Sched_Param;

   function Sched_Setscheduler(Pid: Integer) return Integer;
   pragma Import(C,sched_setscheduler);

   SCHED_FIFO: constant Integer := 1;
   Parameter : Access_Sched_Param := new Sched_Param;
   Error     : Integer              := 0;

begin

   Parameter.Sched_Priority := 99;
   Error := Mlockall(2);
   Error := Sched_Setscheduler
              (Getppid, SCHED_FIFO, Parameter);

   if Error < 0 then
      Handle_Error;
   end if;

end Configure_Ada_Process
```

### 5.2.2. *Drivers in Linux*

The hardware platform includes a number of cards with D/A and A/D converters and digital I/O. There are additional cards for controlling the motors. Drivers have been developed in the usual Linux way. In order to ease their use within Ada, an interface package has been developed in Ada. This package calls the Linux drivers using the facilities for interfacing with C. It exports a number of functions for accessing the input/output devices with a higher abstraction level. This interface makes no reference to the underlying cards and eases the porting of the Ada code.

### 5.2.3. *Distribution*

The Distributed System Annex (DSA) of Ada defines several extensions to write distributed systems entirely in Ada. It is based on the use of remote procedure calls (RPC) and transparent remote invocation of object methods.

A distributed application is developed using the DS in much a similar way as it were an ordinary centralized application. The developer divides the application into several *partitions*, which are executed as different processes on the same or different computers. The packages that make up the application are assigned to different partitions with help of a configuration tool. Procedures defined in remote packages (i.e. packages in other partitions) can be called from a partition, provided they comply with some restrictions which are checked at compilation time with help of so-called *categorization pragmas*.

An example of categorization pragma can be seen in the code of the teleoperation interface that was given in section 5.1. The pragma `Remote_Call_Interface` states that the procedures declared in the package can be called remotely. The pragma `Pure` is used to categorize packages with type and constant definitions only. This implies that the package has no internal state and can be replicated in all the partitions.

GLADE (Pautet and Tardieu, 2001) is an implementation of the DSA programming model for GNAT. It provides means for configuring a distributed application and a full PCS (Partition Communication Subsystem), called GARLIC, which is a runtime library that is linked with applications to make different parts communicate together in a transparent way.

Gutiérrez and González (2000) have studied the use of the DSA in the development of real-time systems, coming to the conclusion that it is not appropriate for developing hard real-time systems. However, a proper configuration of the application and the GLADE middleware makes it possible to use it in soft real-time systems.

The GOYA controller and teleoperation subsystems communicate through remote procedure calls (RPC), according to the Ada DSA model. RPC requests are automatically received and analysed by a service task. If everything is correct, it passes the request to one anonymous task in a server pool in order to execute it.

Some of the DSA real-time problems identified by Gutiérrez and González (2000), and the way they have been coped with in GOYA are:

- *Priority type.* The type `System.Priority` represents relative priorities that are meaningful only to one processor in the system. This can be overcome by carefully assigning priorities in all computers in a consistent way. The only problem with this approach is that it is not portable.
- *Priority of the service task.* This is the task that initially handles the RPC request and allocates another task from the server pool to execute it. The priority of this task is set to `Priority'Last` in order to start the execution of the RPC as soon as possible, and also to bound the effect of priority inversion.
- *Priority of the anonymous server tasks.* The priority of tasks in the server pool is initially set to a default value. When one of these tasks is signalled to handle a RPC request, the task inherits the priority of the caller task. This priority is indeed the appropriate one for executing the request, as priorities have been set with this purpose on both sides. However, if the initial priority of the tasks in the server pool is too low, there is a potential danger of priority inversion.

In fact, this is the case with GLADE. In order to overcome this problem, the code of GARLIC has been slightly modified to set a higher initial priority to the tasks in the pool.

- *Number of tasks in the server pool.* If the number of tasks in the pool is less than the maximum number of simultaneous requests, it may happen that some of the calling tasks are blocked. This situation may compromise the system real-time behaviour if the blocked requests have a high priority. An easy way to prevent this to happen is to ensure that there are enough tasks in the pool for dealing with the maximum number of simultaneous requests. In the GOYA controller subsystem there may be up to three simultaneous requests:
  - One request to get the robot state. This kind request is periodically issued by the system.
  - One request to execute a robot command. This kind of request is sporadically issued by the operator. The system is designed so that it is not possible to issue a new command until the previous one has been received and initiated. In this case, the caller does not wait until the end of the command execution, because of its possibly long duration. The command is sent to the robot controller and the operator checks the status of the operation, looking at the robot state which is displayed on the teleoperation monitor.
  - One request to inmediately execute a robot stop command.

  Hence the system is configured to have a pool of three tasks in order to handle these requests.

### 5.3 *Testing the approach*

A number of tests have been performed in order to check the suitability of this approach. A common Linux configuration, with all unused services (including the X windows system) removed, has been used. The test results show that it is indeed possible to make real-time application tasks run with a high priority, suffering only a small amount of interference from system activities.

Additional tests have been performed in order to test the behaviour of the GLADE middleware. The tests confirmed that priority inheritance mechanism described in the documentation really works. The pool tasks execute the RPCs with the priority of the caller.

## 6. CONCLUSIONS

The experiments performed let us conclude that it is feasible to use Linux and Ada in the development of soft real-time systems with time requirements longer than 50 ms. A number of guidelines on how to configure the Ada program and GARLIC in order to get an appropriate real-time application behaviour have been presented. The use of GLADE for distribution has been found especially interesting and positive,

## 7. REFERENCES


Ada (1995). *Ada 95 Reference Manual: Language and Standard Libraries. International Standard ANSI/ISO/IEC-8652:1995*. Available from Springer-Verlag, LNCS no. 1246.

Barabanov, Michael and Victor Yodaiken (1996). Real-time linux. *Linux Journal*.

Bovet, Daniel P. and Marco Cesati (2000). *Understanding the Linux Kernel*. O'Reilly.

Burns, Alan and Andy J. Wellings (2001). *Real-Time Systems and Programming Languages*. 3 ed.. Addison-Wesley.

Gutiérrez, Javier and Michael González (2000). Towards a real-time distributed system annex in ada. In: *10th Workshop on Real-Time Ada Issues* (Michael González Harbour, Ed.).

Iborra, Andrés and Bárbra Álvarez (1999). GOYA technical specification. Technical report. UPCT.

IEEE (1990). *Portable Operating System Interface (POSIX) — Part 1: System Application Program Interface (API) [C Language] (Incorporating IEEE Stds 1003.1-1990, 1003.1b-1993, 1003.1c-1995, and 1003.1i-1995)*. ISO/IEC 9945-1:1996.

OAR (2000). RTEMS C user's guide. Technical Report Edition 1, for RTEMS 4.5.0. On-Line Applications Research Corporation.

Pautet, Laurent and Samuel Tardieu (2001). GLADE user's guide. Technical Report Version 3.14a. ACT.

Ruiz, José F. and Jesús M. González-Barahona (1999). Implementing a new low-level tasking support for the GNAT runtime system. In: *Reliable Software Technologies — Ada-Europe'99* (Michael González-Harbour and Juan A. de la Puente, Eds.). number 1622 In: *LNCS*. Springer-Verlag. pp. 298–307.

Schonberg, Edmond and Bernard Banner (1994). The GNAT project: A GNU Ada-9x compiler. In: *Tri-Ada 94 Proceedings*. ACM Press.

IFAC Publications
www.elsevier.com/locate/ifac

# USING RT-LINUX FOR DEVELOPING REAL-TIME EMBEDDED SYSTEMS (*)

**P. Mendoza, J. Vila, S. Terrasa, P. Balbastre, A. Crespo**

*Department of Computer Engineering*
*Universitat Politècnica de Valencia, Spain*
*pabmench,jvila,sterrasa,patricia,alfons@disca.upv.es*

Abstract: Linux has been effectively used as a reliable operating system in workstations and servers. However, the interest for using it in embedded systems has grown recently, due to the easiness for customizing it and the availability of the source code. This paper describes the experience of customizing RT-Linux for its use in an embedded real-time control. The main issues involved in this task are outlined and the adopted solution is described. These issues include the system configuration, the minimization effort and how to achieve real-time features through the use of RT-Linux. An application for obtaining task executions chronograms is presented as an example of application design under RT-Linux. Finally some measurements about context switching times are presented. *Copyright © 2001 IFAC*

Keywords: real-time operating systems, embedded systems

## 1. - INTRODUCTION

In the last few years Linux has proved to be a competitive operating system for a wide range of systems (workstations, web servers, general propose machines, etc.). One of the newest fields where it is becoming more appealing is embedded systems (Cass, 2001). There are several reasons why embedded systems manufacturers are adopting Linux more and more against proprietary based solutions: short time market, cost, easiness of tailoring to some particular needs, need for network connectivity, intersystem portability... However, there are also some important issues that make this market a difficult one for Linux. One of them is that most open software is released under the GNU General Public License (GPL) and so must be any software derived (even partially) from GPL software, however there are alternatives to coexist GPL software and commercial applications as is described in (Cass, 2001). Nevertheless this requirement does not bother hardware vendors, who generally do not distribute software or do not make any profit of it. A much more important issue to introduce Linux to the world of industrial and aerospace market is real-time operation. The 2.4 kernel does not provide a real-time operating system, because it is designed to optimize throughput, not to provide predictable response times. Another one is size: Linux usually requires a big amount of disk and memory, however the falling prices of memory and silicon makes embedded hardware has come to meet Linux. Finally real-time embedded systems are usually very expensive systems that work under hard physical conditions. This requires high reliability and security, and easiness to develop and test. All these features make impossible use devices such as hard disks.

Real time operations are an issue for embedded Linux vendors who want to move into industrial and aerospace market. RT-Linux (Barabanov and Yodaiken, 1996; Barabanov, 1997) is a real time operating system under Linux that runs Linux as the lowest priority task (as its idle task). It guarantees a response in a specified amount of time and also provides capabilities such as ignoring all but the highest-priority task when overloaded.

This paper describes the experience of customizing RT-Linux for its use in an embedded real-time control. The resulting platform is able to provide real-time operation through the use of RT-Linux and to work on embedded systems with a comprehensive size and using a flash memory as a disk.

The reminder of the paper is organized as follows; section 2 describes the hardware platform and

(*) This work has been supported by the Spanish Government Research Office (CICYT) under grant TIC99-1043-C03-02.

describes the Linux customization to embedded systems. Section 3 talks about RT-Linux, section 4 presents an example of how to monitor an application for RT-Linux and, finally, section 5 gives some conclusions and lines of future work.

## 2. - CUSTOMIZING LINUX FOR EMBEDDED SYSTEMS

The Linux platform presented in this paper has been originally developed for an embedded system based on an embedded industrial-PC motherboard. This section describes the customization of this platform for running a Linux operating system.

### *2.1. - Hardware platform. -*

The need for persistent storage in the industrial-PC platform for supporting the Linux operating system has been covered with ROM's (Read-Only-Memories) and Flash memories with a disk interface, also termed DOC's (Disk-On-Chip). There are several system configurations according to the following aspects: the support for the file system and for the boot device.

The file system can be targeted to a Flash memory (Fig. 1) or to a RAM disk (Fig. 2). With the file system in Flash memory, RAM memory is used to store the Linux kernel and the application processes, while persistent storage (Flash memory) is used to store the file system. With the RAM disk configuration RAM memory stores everything: the Linux kernel, the application processes, and the file system.

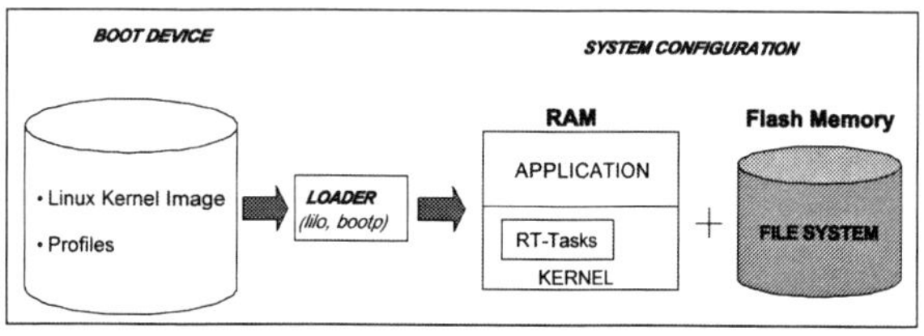


Fig. 1: File system in Flash disk

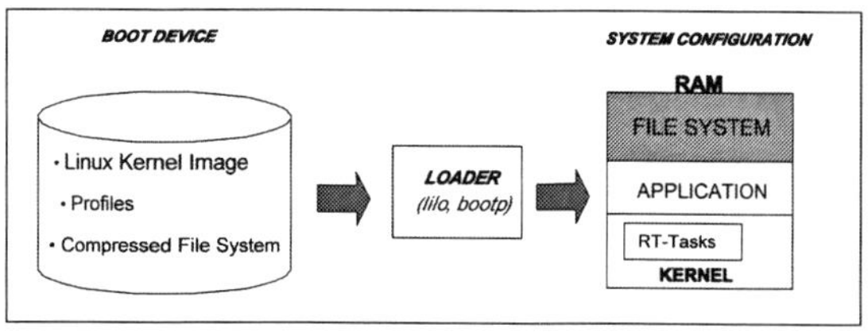


Fig 2: File system in RAM disk.

Having the file system in a persistent storage (Flash memory disk) is the most usual configuration in a Linux system and the most straightforward. The RAM disk configuration has the advantage that RAM memory is much cheaper than Flash memory and also faster, since the RAM access time is lower than the Flash access time. Furthermore, if the file system is backed in a compressed format (e.g. zip) on the Flash memory disk, the need for expensive and low capacity flash disk is reduced about a 50%, due to compression. However, the RAM disk configuration has also important drawbacks: it has to be loaded at boot time from some persistent storage, and this may take some time, especially when the file system is compressed.

A "pure" RAM disk configuration cannot be feasible when some data need to be backed on persistent storage. In this case, it is also possible to have a hybrid configuration in which the root file system stays in RAM and the Flash disk acts as boot device and also holds some files for persistent data.

According to the boot device, there are also some other possible configurations. The usual one is booting from a flash disk, but the system can also be booted from a floppy disk, from a ROM memory, or from the network. The floppy disk is only connected to the system in the very initial stage or when some serious system crash makes impossible to boot from other devices. Every boot device requires its own loader program. In our case these loaders have been all derived from the standard Linux LILO loader. In addition, booting from the network requires a special network adapter (bootp card) with a boot PROM.

When using the RAM disk file system configuration, the boot device must storage the Linux kernel and the file system, including the application files and required libraries (this is called "application profiles"). The file system is stored in a compressed format and decompressed and setup into main memory by the loader. When using the Flash file system configuration, the boot device only provides the Linux kernel.

### *2.2. - Linux customizations. -*

The main issues involved in configuring the Linux operating system to provide all the embedded system requirements can be summarized under the following items:

- Development system and programs
- Initialization and programs
- Kernel
- File system configuration and minimization

The goal of the development system is to configure the target system with the proper Linux loader, kernel and file system. This is done using a Linux development station used as a kind of cross development host. Before being configured, the target system is booted using a floppy initialization diskette that includes the network drivers and Flash disk drivers. Once the target is booted, the root file system is installed in RAM disk and a predefined directory of the host system is mounted via NFS. This directory contains the system loader, the Linux kernel and file system for the target system, and a bin directory with other Linux utilities used during the setup procedure. Once the NFS connection is established, the Flash device is formatted, mounted and initialized its contents from the directory exported via NFS.

After the usual Linux loading process the *init* process, the parent of all other Linux processes, is executed. The goal of *init* in a conventional Linux system is forking off a shell per each terminal. This is not longer needed in an embedded system, since control has to be transferred directly to the application after loading. For that reason, the standard *init* process has been replaced by a custom one that setups the environment and initializes the necessary profiles for executing the application right after. A nice feature of Linux is that this custom *init* may be even a shell script.

Since *init* is not present, the need for a an operator terminal in order to do system debugging and maintenance is assigned to a small program, called the *socket* program, that acts as a kind of minimal telnet server. This server is permanently listening on port 5000 and it forks a shell whenever a telnet connection request is detected on this port.

The customization of the Linux kernel (Beck *et al.*, 1998) for an embedded system is basically done using the standard utility *make Xconfig*. This allows including a specific driver by simply selecting it on a configuration menu. In order to reduce the kernel size is convenient to select the minimum options needed during the configuration phase. With this idea in mind, the size of a kernel for the described system is about 500Kb. It is very difficult to go below this size.

One of the most difficult parts of customizing Linux for an embedded system is to minimize the required *file system*. However this is not an easy task, since different applications may require different libraries and utilities. For example the required file system can be very different for an application written in Java, for an application written in C or for an application that makes use of networking or the X-windows system. This last case is interesting option in a Linux embedded system, since it allows using another host, running an X11 server, as a console for displaying windows.

The idea for configuring the required file system of a given application is based on the concept of "profiles". A *profile* is basically a package containing a set of libraries, application programs, configuration files, ... for allowing a given programming environment. Examples of profiles can be the networking profile, the graphic profile, the Java profile etc... Profiles are stored in compressed tar format Every profile includes a script that auto-installs itself *(/sbin/inst-pck)*. At boot time, every package is decompressed and the shell script is executed, so that package is made operative. Some profiles may require more basic profiles to be included. In particular, there is a *minimum profile* that contains the very basic file system to execute a simple shell (*ash*) and some few Unix utilities (*mount, rm, tar* ...) on the embedded system. The idea behind profiles is to combine them incrementally as shown in Fig. 3. A file system configuration tool generates a file system that contains the minimum profile, other selected profiles (as zipped tar modules) and the application files themselves, that are configured yet as another profile.

The idea of profiles allows developing the system in a modular and flexible way. The main goal is to create a file system made up of modules. Depending on the services, different packages can be installed.

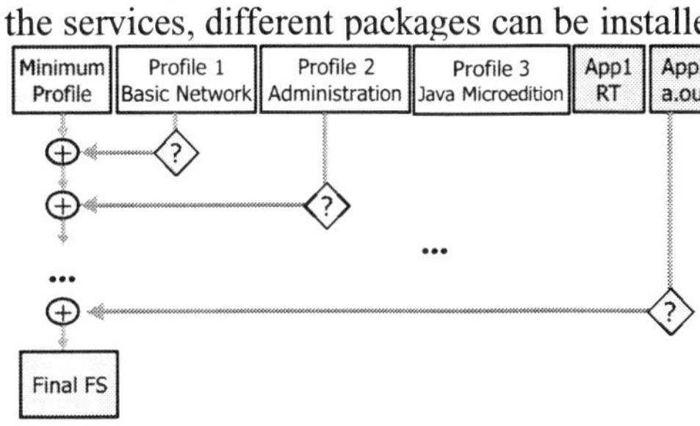


Fig 3: Combining profiles incrementally.

In the actual state of the project four profiles have been identified and implemented: the minimum profile, the network profile, the administration profile, and the Java 2 Microedition profile. The GNAT profile (Ada compiler) is now being developed. The characteristics of the existing profiles are:

- *Minimum profile*: This profile includes the basic file system and is always required. It allows tow develop application programs in C. There are two versions: with or without RT-Linux. The basic version size is about 1Mb, while the RT-Linux size is about 2.5 Mb.
- *Basic network profile*: Provides the ability to act as a network client and includes some basic TCP/IP utilities. Its size is about 130 KB.
- *Administration profile*: It includes a much more powerful UNIX shell and a large set of Unix standard commands and administration programs. Its size is 3.5 MB. It is interesting to install this profile in the development phase.
- *Java 2 Microedition profile*: J2ME Java is a minimalist Java edition that runs on small systems (like palm computers), making use of a subset of Java classes and a reduced virtual machine called "j2me". Its size is 2MB: 1700 KB for j2me and 1.3MB for classes.

## 3. - REAL TIME OPERATION: RT-LINUX

RT-Linux was developed by V. Yodaiken and M. Barabanov (1996), (Barabanov, 1997) and it is a patch to the Linux kernel that allows Linux to execute tasks with real-time constraints at kernel level. Actually the RT-Linux programming interface follows closely the POSIX standards for real-time operating systems. The last version is 3.0 and it is available for x86 and PPC architectures.

The conventional approach for allowing real-time features in Unix systems is to modify the kernel in order to make it predictable and to provide an additional set of system calls to gain access to real-time features. This set of system calls has been standardized by POSIX. Unlike this approach, RT-Linux does not modify the Linux kernel or provides additional system calls. Instead of trying to make the kernel of Linux predictable, what it does is to build a kind small microkernel or software layer, called RT-Linux, directly on the bare hardware. Linux runs on top of this layer.

RT-Linux implements the concept of RT-task and uses its own scheduler for these tasks. The default scheduler that comes with RT-Linux is a preemptive, fixed priority scheduler, but it can be easily customized. Linux, including all its kernel and user processes, is regarded by this scheduler just as another RT-task and shares the processor with other RT-tasks (Fig. 4). More precisely, Linux is the lowest priority RT-task and thus, it is executed as a background task, so it only runs when no other RT-tasks are running. RT-Linux also provides full control over interrupts, and it implements a software interrupt manager. It allows either capturing and handling interrupts or just bypassing them to Linux.

RT-Linux shares the same address space than the Linux kernel so, in theory, any Linux kernel function can be invoked from a RT-task, however this is not advisable and can lead to deadlocks.

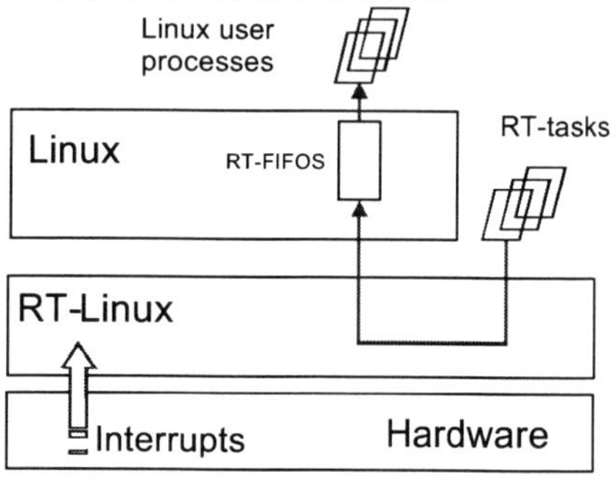

Fig. 4: RT-Linux architecture.

Developing real-time applications in RT-Linux usually requires splitting the application into two parts:

- Tasks with hard real-time constraints: they are programmed and executed as RT-tasks at kernel level.
- Tasks with soft or no real-time constraints: they are executed as "conventional" Linux user processes. Unlike RT-tasks, the graphical system and the networking facilities are available for these tasks, so those parts of an application that need these resources have to be executed as Linux processes.

Both parts of an application can communicate using a special device called RT-fifos. The execution of RT-tasks is done using the Linux facility to dynamically load new kernel modules. RT-tasks are dynamic kernel modules. Even RT-Linux is a set of kernel modules (scheduler, fifos, ...) that need to be dynamically loaded.

## 4. - MONITORING RT-LINUX APPLICATIONS

This section describes an application for RT-Linux that allows monitoring, i.e, obtaining timing chronograms (Fig. 5) of RT-tasks. It also illustrates how to develop RT-Linux applications at two levels: RT-tasks and user level processes.

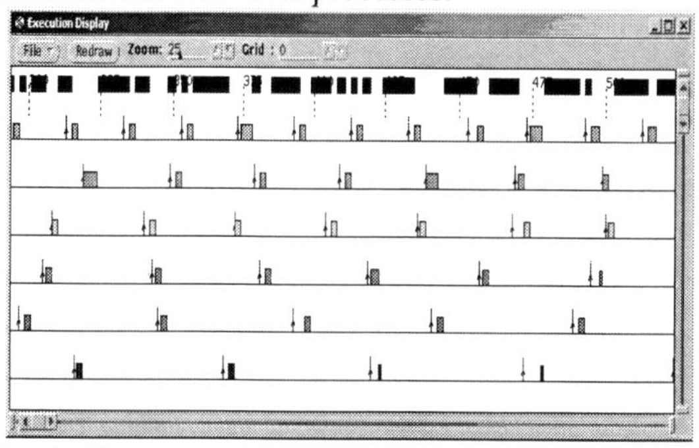
Fig. 5: Chronogram of RT-tasks.

The application is structured as shown in Fig. 6. Three main parts can be differentiated:

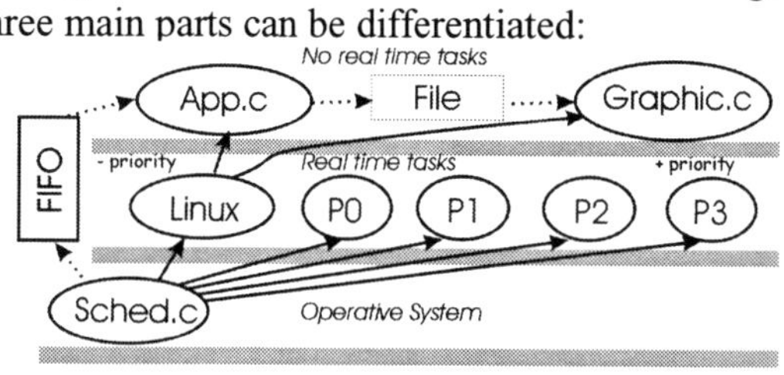

Figure 6: Monitoring application structure.

1. *Modifications to the RT-Linux scheduler.* They are located in file *"rtl_sched.c"*. The goal of these modifications is to add a RT-fifo output to the scheduler that provides information about all context switches and events that take place at the scheduler.
2. *RT-tasks.* They are implemented in file *"rt_process.c"*. This is a module for RT-Linux that creates several periodic RT-tasks with different priorities and compute times.
3. *User level programs*: It consists of two programs: *"app.c"* and *"graphic.c"*. The first program basically reads from the scheduler RT-fifo and writes the information in a file. The second program presents a graphical representation of the trace stored in the file. Note that the visualisation of a system trace cannot be real-time, but deferred, because of the small time units being represented.

Some details about the source code of the application that illustrate the use of the POSIX interface of RT-Linux and the use of RT-fifos are given next.

The first interesting point is the creation of periodic tasks, *"rt_process.c"*, (Fig. 7). As previously stated, Linux loadable modules are needed to use RT-Tasks. They are executed using command:

$ insmod rt_process.o

RT-tasks tasks have to be created inside the init_module function of the loadable module. This function creates RT-tasks using pthread_create (Lewis and Berg, 1998). Once init_module is executed, control is transferred to RT-tasks, which execute function fun. Tasks start by making themselves periodic using pthread_make_periodic_np, and then they execute a CPU burst and suspend themselves until the start of the next period with pthread_wait_np.

Creating periodic tasks is thus accomplished through the use of two functions:

- pthread_create creates a new RT-task with the semantics of POSIX threads its attributes and the function to execute every periodic activation.
- pthread_make_periodic_np is the function to make a RT-task periodic. It is not a POSIX function. Periodic tasks in POSIX are accomplished through the use of periodic signals, but they cannot be used in RT-Linux due to the fact that the clock interrupt is captured and handled by the scheduler.

```
pthread_t my_task;
pthread_attr_t attrib;
int priority, period,compute, OFFSET;

void *fun(void *arg){
  int id = (int)arg, x, dummy=1;
  hrtime_t now = gethrtime();
  x=pthread_make_periodic_np(pthread_self(),
      now + OFFSET, period);
  while(TRUE){
    pthread_wait_np();
    for (b=0; b < compute ; b++){
     for (x=0; x<100; x++)
       dummy + = 3;
    } }
  pthread_exit(0);
  return( void *)0;
}

int init_module(void){
  /*Task attributes*/
   pthread_attr_init(&(attrib));
   pthread_attr_setschedparam(&(attrib) ,
      &priority);
  pthread_create(&(my_task),
   &(my_attrib) , fun, (void *)x);

  return 0;
}

void cleanup_module(void){
  pthread_kill(my_task,-9);
}
```

Fig. 7: Source code of "rt_procces.c".

Tasks may terminate normally using the standard POSIX function pthread_exit, or abnormally using pthread_kill when the cleanup_module is invoked upon termination of a module.

The communication method between the scheduler and the Linux process can be done by means of the RT-Linux standard RT-fifos, although a shared memory mechanism has also been implemented in order to improve efficiency. This method allows the scheduler to improve context switches by three milliseconds below the accomplished with the FIFO communication.

The following subsections describe how the scheduler provides information about context switches using these two methods.

### *4.1. - Communication using RT-fifos*

The RT FIFO is a sort of Unix FIFO based communication mechanism adapted to RT-Linux. The communication is unidirectional; that is, it is like a circular buffer. Therefore, in each reading operation the data is deleted.

The communication mechanism using RT-fifos consists of the scheduler writing task information in an RT_FIFO, and the Linux process reading this data from the same RT-FIFO. The modifications of the scheduler affect functions init_module, rtl_schedule, and cleanup_module.

In the init_module function the FIFO must be initialized with function:

rtf_create(0, 4000);

The rtl_schedule implements the scheduling policy. It has to be modified to supply context switches information trough the RT-fifo

rtf_put(0,(char*)&data,sizeof(data));

With *data* being the structure to write in the FIFO. This structure contains the task identifier, and their initial and final times. The Linux process reads from the FIFO using function *rtf_put* as shown in Fig. 8.

```
struct Events . . .
int main(){
  struct Events data;
  schedfifo = open("/dev/rtf0",O_RDONLY);
  for (i=0; i<2000 ; i++){
    read(schedfifo, (char *)
      &data, sizeof(Events));
  }
}
```

Fig. 8: Source code of "app.c".

*4.2. - Communication using shared memory*

With this communication method, the scheduler writes data in some known memory addresses, and the Linux process reads from this location. Memory sharing between Linux processes and RT-Tasks is accomplished through the use of the *mbuff* functions. The main problem to implement these functions is that the *bigphysarea* patch has to be used in order to guarantee contiguous memory. This patch keeps at boot time as contiguous memory as required.

The structure of the application is similar to the FIFO example; it only consists of replacing the FIFO functions with the *mbuff* ones. As in the previous example, the *mbuff* initialization is performed in the init_module function of the file *"rtl_sched.c"* file. The synopsis is:

```
volatile char *p
p=(volatile char) m_buff_alloc("demo",1024*1024);
```

The m_buff_alloc function creates a memory area called "demo" of 1MB. It returns the initial address of the memory area. Then the scheduler, in the rtl_schedule function, can write data in the memory accessed by the pointer p:

```
P="data to write";
P++;
```

The Linux process corresponding to *"app.c"* must use m_buff_alloc too with the same parameters.

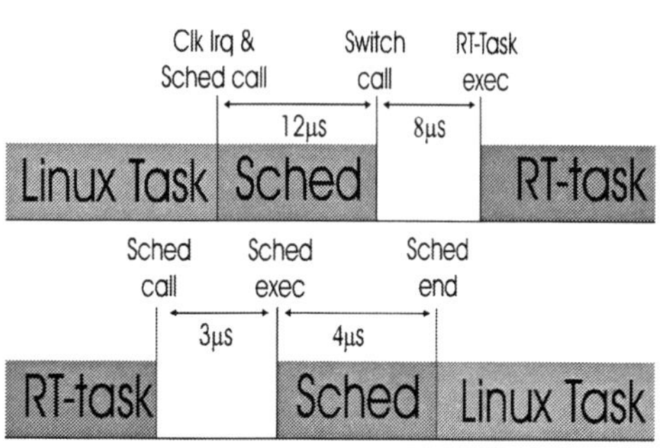


Fig. 9: Context switching times.

*4.3. - Measurements of some timing parameters*

Some measurements about context switching times in RT-Linux have been made in the execution of the previous program. It has been observed that the context switching time is different when it is caused by an interrupt or by a system call (Fig. 9).

In the case of interrupts, this time is very variable, but it can be up to 20 microseconds in the worst case. From these 20 microseconds, the scheduler uses about 12 to execute the scheduling algorithm and to provide scheduling information through the RT-fifo. The variability is imputed to the effect of the Pentium cache. Times are always higher when switching from Linux to a RT-task than when switching from a RT-task to Linux. This is because the cache invalidation effect is much more important in the first case. In the case of system calls, the context switching time is much lower: 8 μs. The scheduler requires about 4 microseconds.

To avoid the scheduling overhead due to writing in the RT-fifo, the structure of this application is being changed so the scheduler is released from doing this. Instead, all tasks make write in the RT-fifo with their id and the current time every time they wake up.

## 5. - CONCLUSIONS AND FUTURE WORK

This paper has described the experience in using Linux on embedded systems. The conclusion is that, although Linux is a general-purpose operating system, it has shown to be very flexible in being adapted to this kind of systems. The openness of the system and the availability of the entire source code make very easy to develop complex application in very short time. The major trouble has been with real-time features, but RT-Linux has also proved to be an excellent solution. Nevertheless, the "two level" programming model required by this solution requires a well-defined programming model and environment. It also requires a timing analysis method. This is part of ongoing work.

The weakest points are that the resulting system is still "not too small", since it requires some few Mbytes of memory and Flash memory and context switching times are also a little bit high. The best way to deal with the minimization problem is in, our opinion, the concept of *file system profiles* introduced by this paper. We are also working on this concept and providing tools for easing this task. Finally the kernel should have also to be further optimized for this kind of systems.

## REFERENCES


Barabanov M., Yodaiken V. (1996) *"Real-Time Linux."*, Linux Journal, March. 1996.

Barabanov M., (1997) "A linux-based Real-Time Operating System" *Phd Thesis*, New Mexico Institute of Mining and Technology.

Beck M., Böhme H., Dziadzka M., Kunitz U., Magnus R., Verworner D., (1998) *"Linux Kernel Internals". 2nd edition* Addison-Wesley

Bill Lewis, Daniel J. Berg, *"Multithread Programming with Pthreads"* Sun Microsystems

Cass, S. *"Little Linuxes"* (2001). IEEE Spectrum. On Line, March 2001

IFAC Publications
www.elsevier.com/locate/ifac

# THE CONTROL ACTION INTERVAL WITH SHARED RESOURCES

**Patricia Balbastre, Ismael Ripoll and Alfons Crespo** [1]

*Department of Computer Engineering,
Technical University of Valencia,
Camino de Vera, 14, E-46071
Valencia, Spain*

Abstract: Industrial control applications are usually developed in two phases: control design and real-time system implementation. In the control design stage a regulator is obtained and later it is translated into an algorithm in the implementation phase. Traditionally, this two phases have been developed in separate ways. Recently, some works have pointed out the necessity of the integration of the control design and its implementation. One of these works reduce the delay variance of control tasks (defined as the Control Action Interval (CAI) parameter) by means of split every task into three parts. The CAI minimization method highly reduces the delay variance and improves the control performance. However, neither of these works have considered resource sharing. The aim of this paper is to extend the CAI minimization method to deal with shared resources in fixed priority scheduled systems. 

Keywords: Real-time systems, control design, resources, timing jitter.

## 1. INTRODUCTION

Industrial control applications are usually developed in two phases: control design and real-time system implementation. In the control design stage a regulator is obtained and later it is translated into an algorithm. In the implementation phase each control algorithm (in multi-loop control systems) becomes a control task running under some timing constraints. Under the assumption of preemptive scheduling and task deadlines equal to task periods, (Liu and J.W.Layland., 1973) proved that the rate monotonic algorithm is the optimal fixed priority scheduling algorithm.

Traditionally, control design and its real-time implementation have been developed in separate ways. For example, in the control design stage it is assumed that the platform will not affect the system execution, that is, it provides an ideal behaviour. In the real-time implementation phase it is assumed that the worst case execution times (WCET) are known, the periods are fixed, deadlines are hard, etc (Isermann, 1989). However, neither of these considerations have to be necessarily true. When the systems being controlled are highly performant, the restrictions imposed by the computer platform resources need to be taken into account.

Recently, some works have pointed out the need to integrate the control design and its implementation. In (Seto *et al.*, 1989), a method to consider the period range of the tasks involved in a control design, as criteria to obtain an optimal use of the system resources, is proposed. In this integrated approach, the implementation is tuned to execute the control tasks at the maximum frequency while ensuring the system schedulability. The result of this approach is a system with a high system (CPU) utilization. In (Cervin and Ecker, 2000) on-

[1] This work was supported by the Spanish Government Research Office (CICYT) under grant TAP-99-1226-C02

line scheduling is proposed for control tasks whose execution times can vary abruptly. The proposed solution attempts to keep the CPU utilization at a high level, avoid overload, and distribute the computing resources among the tasks. The approach in (Crespo *et al.*, 1999) was to reduce the delay variance in the delivering of the control action computed by the different tasks. For that purpose, each task is partitioned into an initial or acquisition part, a mandatory part for the algorithm evaluation and a final part to deliver the action. Moreover, the average delay can be incorporated in the controller design. In this case, the target is the average degradation of control performances which is directly related to both the control effort and the delay. Thus, the scheduling policy distributes the control action interval among the different tasks in order to minimize this index.

However, neither of these published methods have considered resource sharing. Permitting shared resources is fundamental to the usefulness of scheduling techniques for the hard real-time application engineer. It is not clear how hard real-time systems could be designed and implemented without such resources (Audsley, 1991). Moreover, in control applications the control action delivery or the data acquisition use external devices that are shared between all the control tasks. These acceses must be managed in order to avoid conflicts.

In general, for hard real-time systems, establishing the schedulability of a set of tasks that require mutually exclusive access of some set of resources is an NP-hard problem (Mok, 1983). Many resource control protocols have been proposed for uniprocessor systems. The most popular techniques are based upon the priority inheritance model proposed by (Sha *et al.*, 1990). In the Priority Inheritance Protocol (PIP) when a task T is blocked, the task that has blocked T inherits its priority while using the resource. The main drawback of this protocol is that deadlock and chained blocking can occur. The Priority Ceiling Protocol (PCP) (Sha *et al.*, 1990) appears to avoid these two problems. In PCP a task T can only lock a resource if that resource, or any other locked resource, is not accessed by a task with higher priority than T. The maximum priority that a task can inherit is equal to the ceiling of the resource accessed. The ceiling of a resource is the highest priority of all tasks that could lock it. These two protocols used the Rate Monotonic (RM) algorithm (Liu and J.W.Layland., 1973). In (Baker, 1991) a general resource access protocol is proposed, the Stack Resource Policy (SRP), which can be used with both fixed and dynamic scheduling algorithms. The SRP scheduling rules state that a task is not allowed to start executing until its priority is the highest among the active tasks and its preemption level is greater than the system ceiling (Baker, 1991). The preemption level of a task $T_i$ is defined as $\Pi_i = 1/D_i$, being $D_i$ the task deadline. With the SRP a task that starts its execution will never be blocked until completion. It will only be preempted by higher priority tasks. Another interesting protocol is the Ceiling Semaphore Protocol (CSP), in which a task, when granted a resource, sets its priority to the ceiling of that resource. This is in contrast to the ordinary PCP which only raises process priorities when a process actually blocks a higher priority process.

Regarding the schedulability test of systems with shared resources, (Sha *et al.*, 1990) states that a set of n periodic tasks can be scheduled by the RM policy if

$$\forall k \quad k = 1..n \quad \sum_{i=1}^{k} \frac{C_i}{P_i} + \frac{B_k}{P_k} \leq k(2^{1/k} - 1)$$

where Bk is the blocking factor, and it is the maximum time that a task can be blocked by lower priority tasks. It depends on the resource protocol.

Another schedulability test based on the response time of tasks for fixed priority scheduling was presented in (Audsley *et al.*, 1993), adapting the formulation made in (Joseph and Pandya, 1986) to take into account the blocking factor.

$$\forall i = 1..n \quad R_i \leq D_i \text{ where}$$

$$RT_i = C_i + B_i + \sum_{j \in hp(i)} C_j \left\lceil \frac{RT_i}{P_j} \right\rceil$$

In short, the computer platform introduces nonidealities that cause a degradation of the control performance. One of these interfering factors is the variable delay introduced by the scheduler due to different task priorities. This problem has been studied and solved in (Crespo *et al.*, 1999)) for static priorities and in (Balbastre *et al.*, 2000) for the Earliest Deadline First (EDF) scheduling algorithm. The aim of this paper is to extend the method presented in (Crespo et al., 1999) to deal with shared resources using the Deadline Monotonic (DM) scheduling algorithm.

The reminder of this paper is organised as follows: in section 2 the task model used is explained as well as the assumptions made and some useful definitions. Section 3 briefly describes the minimization method explained in Crespo:99 in order to reduce the delay variance. In section 4 this method is extended to deal with shared resources. A simple example is presented in section 5 to show how the method works. Finally, the conclusions and future work are presented in section 6.

## 2. ASSUMPTIONS AND TERMINOLOGY

Each task $T_k$ is characterized by four parameters: a computation time $C_k$, a period $P_k$, a deadline relative to the release of the task $D_k$ and an offset $O_k$, relative to the release of the task as well. We consider:

- Tasks are periodic.
- $C_i \leq D_i \leq P_i$.
- Task priorities are fixed during the system execution.
- As resource control protocols can be used both PCP, SRP or CSP.

The activities involving a control loop task can be divided into the following parts:

(1) Data acquisition: data from an external sensor is acquired.
(2) Computation of the control action: it corresponds to the regulator computation and should be executed as soon as possible. At the end of this part, the basic control action is ready to be sent to the actuators.
(3) Output the control action: must be done either as soon as possible or within a fixed time interval.

The control computation can be divided into a mandatory part and one or more optional parts. The mandatory part gives a first solution to the control problem. Optional parts try to improve the solution. Nevertheless, in this paper optional parts will not be considered. Therefore, any control task $T_k$ can be partitioned into three parts: $T_{ki}$, $T_{km}$, $T_{kf}$, for initial, mandatory and final tasks, respectively. From now on, the first subscript on any parameter denotes the task identifier, and the second (if it exists) denotes the part of the task (initial, mandatory or final).

Moreover, the worst and best case response times will be useful for our purpose. So, we define:

- $wcrt_k$ is the worst case response time of $T_k$ in a system that does not use shared resources.
- $WCRT_k$ is the worst case response time of $T_k$ in a system with shared resources.
- $bcrt_k$ is the best case response time of $T_k$ in a system that does not use shared resources.
- $BCRT_k$ is the best case response time of $T_k$ in a system with shared resources.

The same definitions can be made relating to the activation of tasks (WCAT, BCAT).

In control applications it is important that the control action and the data acquisition are executed at regular instants of time. However, preemptions and blocking due to task priorities and resource locking cause a non-periodic behaviour. So, it is necessary to measure the delays of the initial and final tasks. To do this, two parameters are defined: CAI and DAI.

The CAI expresses the percentage variation of the control action of a task relative to its period. This measure gives information about the delay variation that a task will suffer in the control action delivery, and it allows comparison of the performances of a control system. The Control Action Interval (CAI) of task $T_k$ is denoted by

$$CAI_k = \frac{WCRT_k - BCRT_k}{P_k} * 100$$

In a similar way, the Data Acquisition Interval (DAI) of a task $T_k$ is denoted by

$$DAI_k = \frac{WCAT_k - BCAT_k}{P_k} * 100$$

The DAI expresses the percentage variation of the sensor acquisition relative to its period. Intuitively, if the task $T_k$ has the higher priority and it is executed under priority based scheduling, its DAI will be 0 since it will start to execute at the beginning of its period. The CAI of the same task will be also 0, and $WCRT_k = BCRT_k = WCET_k$.

## 3. MINIMIZATION METHOD

In order to reduce the jitter terms, a task decomposition criterion and its schedulability analysis is proposed in Crespo:99 for task sets executed under a priority based scheduling policy. This section briefly comments on this work. Figure 1 shows the task decomposition of a task $T_k$ in three tasks (initial $T_{ki}$, mandatory $T_{km}$ and final $T_{kf}$). The priorities assigned to these tasks are also shown in Figure 1. Priorities are grouped in three levels:

- Final parts ($T_{kf}$) are located in the highest priority block and inside the block, according to DM policy.
- Initial parts ($T_{ki}$) are assigned to the second priority block and also assigned according to DM policy.
- Mandatory ($T_{km}$) parts are placed in the lowest priority block and also ordered by DM policy.

The priority assignment of initial and mandatory parts ensures that these tasks will be executed in the right order: initial parts before mandatory ones. Since final parts have the higher priority, an offset has to be applied to the final part. In some cases, the mandatory part can complete its execution after the offset time calculated. In this case, the scheduler has to delay the start of the final part until the end of the mandatory one. Another solution is to synchronize the mandatory

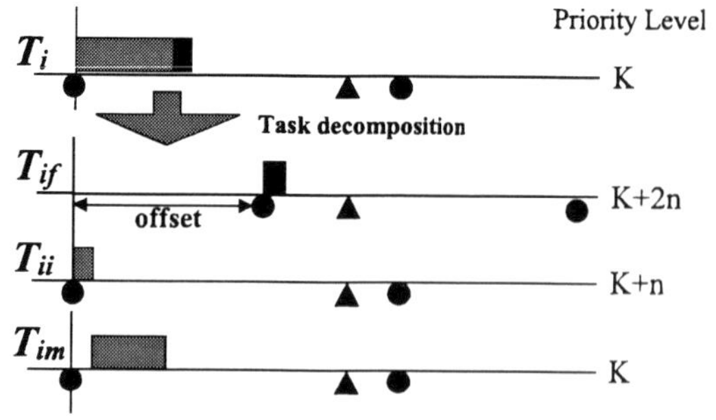


Fig. 1. Task split under DM scheduling

and the final part by means of a semaphore, so the final part has to wait until the mandatory part unlocks the semaphore.

The new task set is schedulable with the proposed scheduling algorithm if

$$\forall k \qquad wcrt_{km} \leq D_{km}$$

The final part offset follows the next expression:

$$O_{kf} = wcrt_{km} - wcrt_{kf}$$

The value of wcrt represents the worst case response time of any part (initial, mandatory or final) when the system is scheduled by a simple priority driven scheduler (a scheduler that does not delay the final part). The final part offset is adjusted to minimize the absolute delay as well as the CAI. This task formulation ensures a fixed delay of the control action and a small variable delay range for each task. The method is applied off-line, that is, once the task set is partitioned, the final part offset is calculated, as well as CAI and DAI parameters. These parameters are introduced in the regulator and finally the system is executed. This approach drastically reduces the CAI and it allows a better algorithm behaviour. The control performance obtained with this task split is illustrated in (Albertos *et al.*, 2000) and shows how it is highly improved.

## 4. MINIMIZATION METHOD WITH SHARED RESOURCES

This section contains the main contribution of the paper.

A traditional control loop task, as mentioned before, follows three main steps. Data acquisition, control action computation and control action delivery. We assume that initial and final parts will share I/O devices to communicate with the process. It is also assumed that mandatory tasks do not communicate with the process, so these tasks will not share I/O resources, but other resources, called General resources 2.

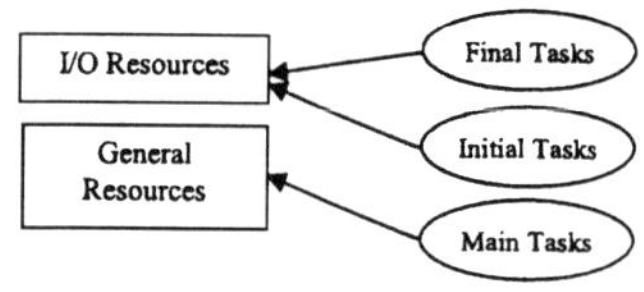


Fig. 2. Resource types

It is also assumed that final and initial parts do not share general resources, mainly for two reasons: it can be assumed that final and initial tasks only write and read data from external devices respectively. So, these tasks only communicate with the process to be controlled and it only uses I/O resources. The second reason is that a main task will not be blocked by final or initial tasks because main tasks always have lower priority than the other parts of the task set.

To deal with the problem of considering shared resources, two subproblems will be analyzed: I/O resource sharing and general resource sharing.

### 4.1 *I/O resource sharing*

These type of resources will only be used by final and initial tasks. In this situation, initial tasks could be promoted to a higher priority level whenever they lock a resource. So, final tasks will be blocked by initial tasks and by other final tasks with lower priority (longer relative deadline). Initial tasks will only be blocked by other initial tasks with lower priority.

To apply the CAI minimization method to deal with I/O resources two questions have to be considered: the schedulability test and the new offset applied to the final task (and, consequently the new CAI and DAI).

Regarding the schedulability condition, it is obvious that mandatory tasks will not suffer any blocking, since they do not share I/O resources. So, the WCRT of mandatory tasks will not change and the schedulability condition will remain as:

$$\forall k \qquad wcrt_{km} \leq D_{km}$$

As mandatory tasks do not share I/O resources, its WCRT will be the same as the wcrt in a system without resource sharing.

The final task offset is modified by the final task blocking factor. So, the offset applied to the final task is:

$$O_{kf} = wcrt_{km} - WCRT_{kf}$$

where the $WCRT_{kf}$ is given by the next expression:

$$WCRT_{kf} = C_{kf} + B_{kf} + \sum_{j \in hp(kf)} C_j \left\lceil \frac{WCRT_{kf}}{P_j} \right\rceil$$

Now, the CAI parameter is known off-line:

$$CAI_k = \frac{WCRT_{kf}}{P_k} * 100$$

The same arguments can be used to obtain the DAI parameter, using the BCAT and bcat. (Balbastre *et al.*, 2001*a*).

### 4.2 *General resource sharing*

These kinds of resources will be used by mandatory tasks. At the same time, it is supposed that final and initial tasks share I/O resources. A mandatory task that uses a General resource can suffer blocking only by other mandatory tasks with lower priority. As a consequence response times of mandatory tasks will be greater. The new worst case response time has to take into account the blocking factor:

$$WCRT_{km} = C_{km} + B_{km} + \sum_{j \in hp(km)} C_j \left\lceil \frac{WCRT_{km}}{P_j} \right\rceil$$

and the new schedulability condition is:

$$\forall k \qquad WCRT_{km} \leq D_{km}$$

Since the response times of mandatory tasks are greater, final tasks will have to wait until their mandatory tasks complete execution. So, the final task offset has to be modified, in order to maintain execution order. The new offset is:

$$O_{kf} = WCRT_{km} - WCRT_{kf}$$

It is important to note that CAI and DAI do not depend on mandatory tasks behaviour. Even if the main tasks are blocked, the variable delay of the final tasks will not change. However, an increase of the fixed delay is expected, as the final task offset is greater due to the main tasks blocking.

### 4.3 *The blocking factor*

According to SRP, the maximum blocking time $B_i$ for a task $T_i$ can be calculated as the longest critical section among those with a ceiling greater than or equal to the preemption level of $T_i$. This blocking time affects the response time of task $T_i$.

A property of SRP is that a task can not be blocked twice, that is, once it commences its execution, it can not be blocked again. If we consider a final task, it can be preempted by other final tasks with higher priorities. And it can be blocked by all the initial tasks, and final tasks with lower priorities.

Therefore, it is impossible that a final task $T_{kf}$ is blocked by an initial task $T_{ji}$ and preempted by the final task $T_{jf}$. If a final task $T_{kf}$ is blocked by an initial task $T_{ji}$, then $T_{kf}$ will execute always

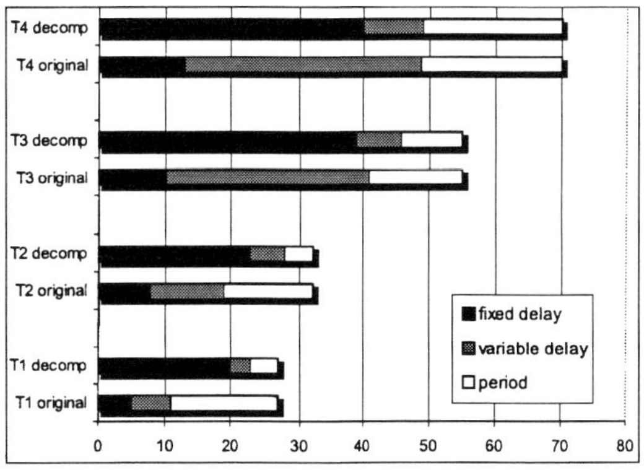


Fig. 3. Delay comparison with shared resources

before the mandatory task $T_{jm}$, and thus before the final task $T_{jf}$. The same reasoning can be applied for initial tasks $T_{ki}$.

This can be used to reduce the calculated value for wcrtk. The algorithm for a task $T_k$ is:

Let $\delta_k$ be the set of critical sections with ceiling greater than or equal to the preemption level $\pi_k$ of $T_k$.

(1) Compute blocking factor $B_k$ as $max(\delta_k)$
(2) If $B_k$ is the compute time of any initial task $T_{ji}$ then
  - If $B_k > C_{jf}$ and $C_{jf}$ contributes to $wcrt_k$ then $wcrt_k = wcrt_k - C_{jf}$
  - Else delete $C_{ji}$ from $\delta_k$. Go to 1.
(3) Else stop.

## 5. EXAMPLE

As an example, consider the following task set in the next table.

| | | | | General Resources | | | I/O Resources | | |
|---|---|---|---|---|---|---|---|---|---|
| | C | D | P | n | id | C | n | id | C |
| T1 | 5 | 27 | 27 | 1 | 1 | 2 | 1 | 2 | 3 |
| T2 | 8 | 30 | 32 | 1 | 1 | 3 | 1 | 2 | 5 |
| T3 | 10 | 45 | 50 | 1 | 1 | 3 | 1 | 3 | 6 |
| T4 | 13 | 60 | 70 | 1 | 1 | 5 | 2 | 2 | 2 |
| | | | | | | | | 3 | 5 |

All initial and final tasks share the same I/O resource 1, while mandatory task 4 use two General resources (with identifiers 2 and 3). Task 3 uses only General resource 3, while the other mandatory tasks use only General resource 2.

Figure 3 shows a comparison between delays and periods in the original task set, and the decomposed task set (after applying the minimization method). Figure 4 shows the same parameters in a task set without resource sharing.

As can be seen in the figures, task 4 has the same CAI and fixed delay in both cases, since it is the lower priority task, and it will never be blocked. Therefore, the same behaviour is expected with and without shared resources. The rest of the

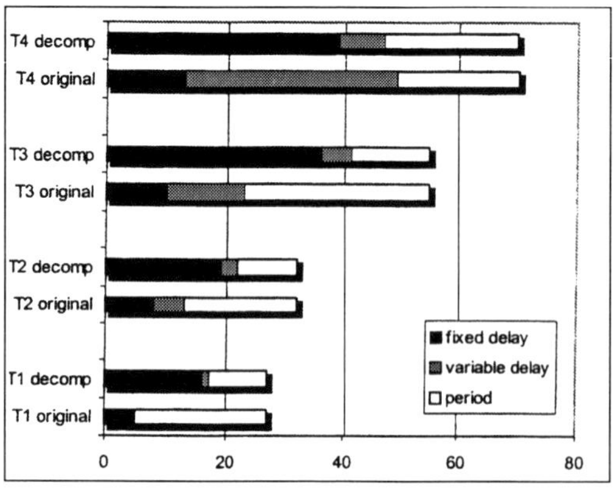


Fig. 4. Delay comparison without shared resources

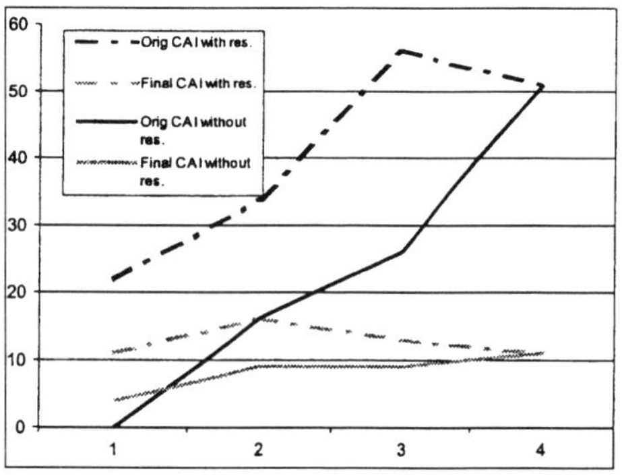


Fig. 5. CAI comparison between shared and non-shared resources

tasks present greater variable delays in the shared resource system. This increase is due to the blocking factor of final tasks. Moreover, the fixed delay is also greater, in this case due to the blocking factor of mandatory tasks.

Figure 5 shows values of variable delays relative to the period (CAI). The figure shows a comparison between shared and non-shared resources, in the original task set and in the decomposited task set.

## 6. CONCLUSIONS AND FUTURE WORK

This paper have dealt with the problem of resource sharing in real-time industrial control applications. In recent years the integration of control design and real-time scheduling has been pointed out. However neither of these works have considered resource sharing. This paper extends an existing method that reduces variable delays (due to the scheduler interference) in control tasks to deal with shared resources. The method reduces variable delays in comparison with the non-partitioned system. However, this reduction is expected to be lower than in the non-shared resource system, since the blocking between tasks increases the response time.

A future goal is to extend this method using dynamic priority scheduling algorithms such as EDF, as well as consider optional parts to improve the control action computation. Another interesting issue is the effect of overloads in this kind of partitioned systems.

## 7. REFERENCES


Albertos, P., A. Crespo, I. Ripoll, M. Vallés and P. Balbastre (2000). Rt control scheduling to reduce control performance degrading. *Proceedings of the 39th IEEE Conference on Decision and Control.*

Audsley, N. (1991). Resource control for hard real-time systems: A review. *Technical Report YCS 159 University of York - Dep of Computer Science.*

Audsley, N., A. Burns, K. Tindell, M. Richardson and A. Wellings (1993). Applying new schedulability theory to static priority preemptive scheduling. *Software Engineering Journal* **8(5)**, 284–292.

Baker, T. P. (1991). Stack-based scheduling of realtime processes. *The Journal of Real-Time Systems* **3**, 67–99.

Balbastre, P., I. Ripoll and A. Crespo (2000). Control task delay reduction under static and dynamic scheduling policies. *7th International Conference on Real-Time Computer Systems and Applications RTCSA'00.*

Balbastre, P., I. Ripoll and A. Crespo (2001*a*). Reducing delays with shared resources. *Technical Report. Dept. of Computer Engineering. Technical University of Valencia. Spain.*

Cervin, A. and J. Ecker (2000). Feedback scheduling of control tasks. *Proceedings of the 39th IEEE Conference on Decision and Control.*

Crespo, A., I. Ripoll and P. Albertos (1999). Reducing delays in rt control the control action interval. *IFAC World Congress Beijing.*

Isermann, R. (1989). Digital control systems. *Springer Verlag.*

Joseph, M. and P. Pandya (1986). Finding response times in a real-time system. *The Computer Journal* **29**, 390–395.

Liu, C.L. and J.W.Layland. (1973). Scheduling algorithms for multiprogramming in a hard real-time environment. *JACM* **20**, 46–61.

Mok, A. (1983). Fundamental design problems of distributed systems for the hard real-time environment. *MIT/LCS/TR-297, Laboratory of Computer Science, MIT.*

Seto, D., J.P. Lehoczky and L. Sha (1989). Task period selection and schedulability in real-time systems. *IEEE Real-Time Systems Symposium.*

Sha, L., R. Rajkumar and J.P. Lehoczky (1990). Priority inheritance protocols: An approach to real-time synchronization. *IEEE Trans. on Computers.*

IFAC Publications
www.elsevier.com/locate/ifac

# RECONFIGURABLE DISTRIBUTED CONTROL CONSIDERING COMMUNICATION TIME DELAY EFFECTS

**Benítez-Pérez H.*, and García-Nocetti F.****

*(*),(**) Departamento de Ingeniería de Sistemas Computacionales y Automatización, IIMAS, UNAM, Apdo. Postal 20-726. Del. A.Obregón, México D.F., 01000, México.*
*Fax: ++52 5616 01 76, Tel: (*) ++52 5622 36 39, (**) ++52 5622 35 69*
*Email: (*)* **hector@uxdea4.iimas.unam.mx** *(contact author)*
*(**) fabian@uxdea4.iimas.unam.mx*

**Abstract**: Increasingly, there is a move towards in-built intelligence for sensors and actuators to integrate "smart" peripheral elements as part of a distributed control system. The utilisation of local intelligence to provide local fault detection and fault tolerance opens the incorporation of new variables into the structure of the control law. How to integrate this information into the structure of the control law is presented in here. The main goal is to integrate a decision making procedure between peripheral elements and different control strategies. The reconfiguration of the control law based upon local health measures is decided by this procedure. *Copyright © 2001 IFAC*

**Keywords:** Distributed Systems, Process Supervision, Distributed Control, Static Scheduling.

## 1. Introduction

The emergence of smart sensor and actuator technology removes the need for centralised control with feedback loops to dumb peripheral actuators replacing it with a databus connection (Benítez-Pérez et al., 1998).

Control systems over distributed systems consider several design parameters such as partitioning allocation and communication policies. These are part of the design and deployment decision. Different aspects over the system are proposed. Firstly, the integration of smart peripheral elements as autonomous elements capable of self diagnose in order to reconfigure the control strategy. Second, the reconfiguration of the control law based upon a safety degradation measure.

Several strategies for reconfigurable control have been studied for different research groups. For instance Dussud et al., (1998) present a control law reconfiguration based upon a fuzzy decision maker working as fuzzy supervisor.

The strategy followed in here uses the information generated by smart elements in a decision making procedure. Peripheral elements generate a measure named confidence value, which indicates how the element has been degraded due to the presence of a local fault.

The decision making procedure chooses which control law needs to be used based upon the fault scenario. There are three control strategies, the first one is based upon a predictive control law, the second is based upon a PID controller and the third one is a safety value, which sends the plant to a stable operating point.

Due to the use of a distributed system in order to communicate different entities, it is necessary to propose a scheduler taking into account the behaviour of the system under fault free and fault scenarios.

A ball and beam plant, using two arrays of sensors and two actuators, has been selected as case study. This paper has been divided in seven sections. First section is the current introduction. Second section presents an overview of smart element technology. A description of the proposed scheduler is provided in third section. Fourth section describes briefly the plant model. The control design and the decision making strategy adopted is presented in the fifth section. Sixth section, shows practical results of case study. Finally, seventh section presents some concluding remarks.

## 2. "Smart" Elements Design

Smart elements are defined as peripheral elements who have the capabilities of self-diagnose faults, self-compensates disturbances and digital communications (Masten, 1997). These elements are mainly sensors and actuators. The typical configuration followed in this paper is presented in Fig. 1 (Benítez-Pérez, 1999).

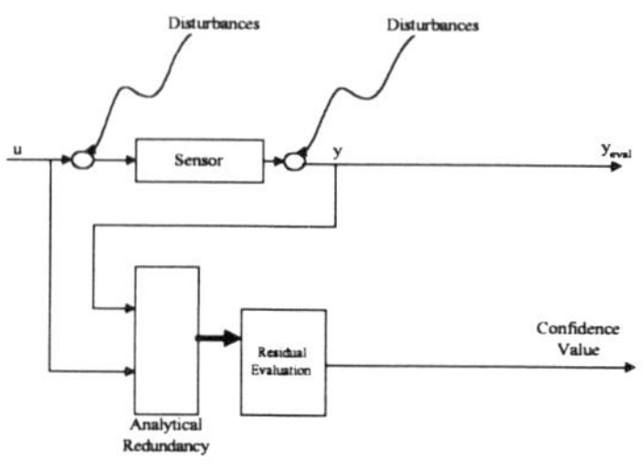


**Fig. 1 Smart Sensor Strategy**

In Fig. 1 there are two blocks to be considered: the analytical redundancy block and the residual evaluation block. First block is based upon two parameter estimation procedures using two different sampling periods. The first being faster than the second one. First procedure performs the parameters every $t_1$ seconds, while, the second one estimates the parameters every $t_2$ seconds. First procedure catches the current performance of the element, meanwhile second one catches the stable performance of the element. The difference between them gives the residual vector. This vector has a non-zero behaviour when a fault appear. PE1 estimates five parameters as well as PE2. Both vectors do not have an exact physical meaning related to the dynamics of the sensor . This is not the aim pursued in this paper. Nevertheless, the relation between residual vector and fault presence is directly proportional.

The evaluation block is performed using a fuzzy logic procedure named as residual evaluation block (Fig. 1). This block fully evaluates the residual vector based upon Mamdani fuzzy rule strategy. This system has been tuned using the experience of the designer. The result of this evaluation procedure is a confidence value which has a range from 0 (catastrophic behaviour) to 1 (fault free scenario).

The goal of this evaluation procedure is the identification of faults based upon the behaviour of whole residual vector. This idea uses the directional residuals concept explained in more detail by Gertler (1998).

The fuzzy system determines how big is the degradation in order to modify the current value of Confidence Value (CV). If the deviation of one of the residuals is considerable with respect to the current parameter this is declared as erroneous condition. Otherwise, the residual is categorised as "good".

## 3. Scheduler Design

Scheduler strategy is integrated by four different types of elements, sensors, decision making module, controller and actuators. This scheduler presents sporadic communications due to the presence of fault scenarios. There are several types of time delays to be considered (Table 1). These are related to communication, processing and capture.

Time values shown in Table 1 are from the simulation procedure of the implemented example.

There are two possible scenarios, first scenario establishes a fault free situation. Second scenario considers the communication between nodes when a fault scenario appears (Fig. 2).

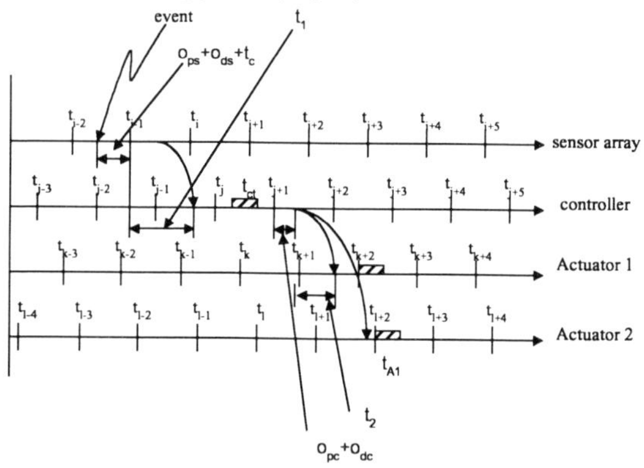


**Fig. 2 Scheduler for Fault Free Scenario**

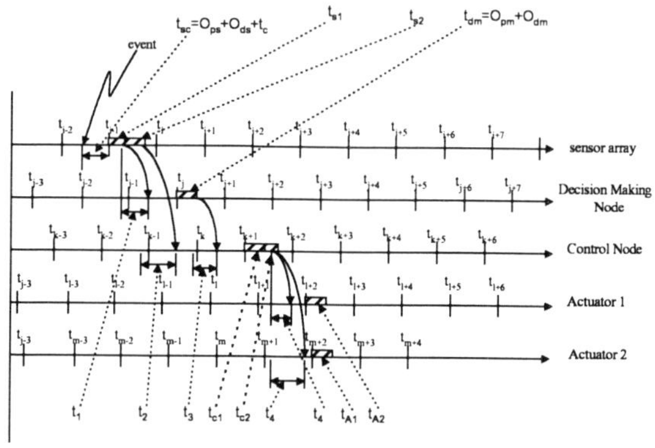


**Fig. 3 Scheduler for Fault Scenario**

For fault free scenario the time related to the sensing node is $t_1+o_{ps}+o_{ds}+t_c=6.67$ milliseconds. Time related to control node is $o_{pc}+o_{dc}+t_2=2.8$ milliseconds. When a fault scenario appears a new element is considered. Its name is decision making node. In order to incorporate this node to the system scheduler works as follows. The selected sensor transmits its value to control node. Due to its CV is less than a threshold, it transmits this value to the Decision Making node. This node processes the information and transmits its result to control node. Finally, control node decides which law is used in order to send a result to the actuators. Time delays are considered as follows, sensor uses $t_{sc}$=3 milliseconds (Fig. 3). The communication between sensor and control nodes is $t_{s2}$= 2 milliseconds and the communication between decision making and control node is $t_{s1}$= 2 milliseconds. Decision Making node spends $t_{dm}$= 1 millisecond in processing its information. Having recruit all necessary information control node process it during $t_{c1}$=2.0 milliseconds. Afterwards, this result is sent to actuators nodes. The systems spends in this action $t_{c2}+t_4$=1.57 milliseconds. Finally the actuator node spends $t_{A1}$=2.0 milliseconds in order to modify the plant. The total time spent for each scenario is 11.5 and 17.73 milliseconds, respectively. This calculation is based upon summation of every time concerned. For fault free scenario the sum is as follows $t_1+o_{ps}+o_{ds}+t_c + o_{pc}+o_{dc}+ t_2+ t_{A1}$ having as a result 11.5 milliseconds. For fault scenario the sum is next $t_{dm}+ t_{s2}+t_{s1}+ t_2+t_3+t_{c2}+t_4+ t_{c1}+t_{A1}$. This sum gives as a result 17.73 milliseconds. As the reader may realize this difference performs an overhead due to reconfiguration strategy. This is overtaken by the use of PID controller as mentioned further on.

For the case of this simulation, CANbus standard is used to establish time delays as shown in Table 1.a and Table 1.b.

Clock synchronisation is time stamping over each communication process. Time delays related to sensing, controlling and processing information are based upon the response of the dynamics tested in a PIII computer. The implementation of this scheduler is achieved using State-Flow toolbox from Matlab 5.3 (MATLAB, 1998).

## 4. Plant Design

The strategy followed in this paper is based upon Fig. 4. There is a reconfiguration approach based upon a confidence measure array generated by a group of "smart" sensors.

Each "smart" sensor produces its own confidence value. This array is evaluated by a decision making module in order to choose one of the available control laws.

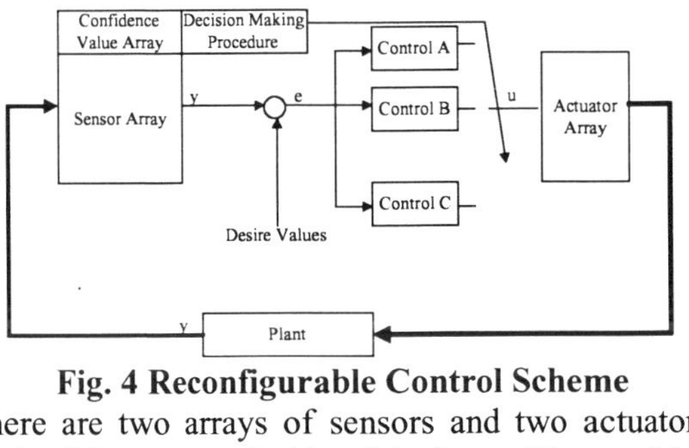


**Fig. 4 Reconfigurable Control Scheme**

There are two arrays of sensors and two actuators. Each of them on each side of the beam. The model of the plant uses one sensor who is reporting the actual position of the ball and one actuator who is moving the beam.

The plant dynamics are shown next:

$$A(z^{-1})y(t) = z^{-d}B(z^{-1})u(t-1) + c(z^{-1})e(t) \tag{4.1}$$
$$B = 0 + 0.0013z^{-1} + 0.0016z^{-2}$$
$$C = 1 + 1.5977z^{-1} + 0.8258z^{-2}$$
$$A = 1 + 2.018z^{-1} + 1.032z^{-2}$$

Where $y$ is the output, $u$ is the input, $z$ is the space variable, $A$, $B$ and $C$ are the representative polynomials of the system. $d$ is a time delay variable that for this particular case is considered to be zero. This representation must be modified to that one expressed next. In this case, $C$ term is no considered as part of the error.

$$\frac{A}{C}y(t) = \frac{B}{C}u(t-1) + e(t) \tag{4.2}$$
$$or$$
$$y^f = \frac{y(t)}{1 + 1.5977z^{-1} + 0.82558z^{-2}}$$
$$u^f = \frac{u(t)}{1 - 1.5977z^{-1} + 0.82558z^{-2}}$$

A modification under equation 4.2 must be made to this stage in order to determine a suitable representation for predictive control.

$$\tilde{A} = \Delta A = \left(1 - z^{-1}\right)A = 1 - 1.032z^{-1} + 1.032z^{-2} - 1.032z^{-3} \tag{4.3}$$

Sensors and actuators models are considered lineal by the inherent self calibration within these elements. Due to this assumption, the plant model depends exclusively the dynamical behaviour of the ball and beam. Time delays produced by communication scheduler are considered as part of the dynamic behaviour. Control laws integrate this behaviour in both cases (fault and fault free scenario).

The only considered fault scenario is the degradation of any of these peripheral elements per time. Just one sensor is considered to be faulty. The condition of the fault scenario is noise over one peripheral element. There are three main possible scenarios, fault free, non-catastrophic and catastrophic. The first scenario presents the obvious behaviour of the plant. In this case, the used control law is a predictive control strategy.

The second scenario degrades system up to a established threshold in order to reconfigure the control law. In this case, the control law is based upon a PID controller in order to stabilise the plant. Last scenario presents a catastrophic situation. The control law used is a constant value. This is due to safety conditions of the beam without considers the ball. In fact, the ball is expected to be lost but the beam is stable.

## 5. Reconfigurable Control Integration

Having defined the "smart" element strategy, the scheduler behaviour and the dynamics of the plant, next step is the integration of both into the dynamics of the control law. As mentioned in section 1, different strategies may be followed. For instance, Nilsson (1998) uses stochastic control in order to integrate random time delays due to disturbances within the system.

Alternative techniques, such as, predictive control (Camacho et al., 1998) are feasible due to incorporation of time delays to the control structure. The use of "smart" elements information into the control law is based upon the use of arrays of confidence values from peripheral elements. The approach followed is the reconfiguration of the control law through the selection of different control strategies. These are integrated to a decision making algorithm in order to switch from one control law to another (Fig. 2). In case of catastrophic conditions this switch tends to move the system to a safe mode. Just one sensor is considered to be faulty.

Remember that in this case, the measure of one sensor is given to the controller. This is to determine the actual position of the ball.

First controller considers CV at 100% of its value where the input is 100% trustable. Second control strategy establishes the input at its 50% trust. This is due to the presence of a fault. Last control law is based upon a safe value in order to maintain the system within a safe condition. For last two cases second scheduler is used (Fig. 3). The structure of the predictive control law is based upon eqn. 5.1.

$$y = Gu + f \tag{5.1}$$

Where $y$ is the current output, $f$ is the variance matrix $u$ is the current input and $G$ is a recursive polynomial based upon eqn. 5.2.

$$G_{j+1} = E_{j+1}B = \left(E_j + F_j z^{-J}\right)B \tag{5.2}$$

where $E_{j+1}$ and $F_j$ are elements of a diophantine eqn. 5.3.

$$1 = E_j\left(z^{-1}\right)\tilde{A}\left(z^{-1}\right) + z^{-1}F_j\left(z^{-1}\right) \tag{5.3}$$

Where $A$ represents the dynamics of the system. Having resolved eqn. 5.3, the control law is expressed as:

$$u = 0.001u(t-1) + 0.02y(t-1) - 0.001y(t-2) - 0.05y(t-3) \tag{5.4}$$

Second structure differs from first option in the control strategy. This second approach is based upon a PID controller tuned from the conditions of a local fault.

The appearance of a fault scenario decreases the certainty of the current position. Then, the controller is defined with this uncertainty. Constant values are tuned considering the lost of current position during a time interval. These values are k1=0.1, k2=-0.12 and k3=0.0. Where $u = k_1 e + k_2 \int e dt + k_3 \frac{de}{dt}$ is the control law, u is the control output and e is the system error. Final implementation is based upon the use of a safe constant who considers the ball is on the middle of the beam. The constant value is k=0 which makes the current position to the beam turn to be horizontal. This structure set the system to a constant value.

The decision making procedure is based upon two thresholds in order to define the boundaries of these three control laws. Table 2 shows this classification. The selection of control law a degradable condition based upon one faulty element. If every CV element are between 0.79 and 1.0 then first control law is selected. If one of the CV measurements is between 0.49 to 0.78 the control law is selected. Finally, if one of the CV measurements drops its value to below 0.48 last control law is chosen.

| | Good | Bad | Safe | Values |
|---|---|---|---|---|
| 0.79<CV<1 | ✓ | | | |
| 0.49<CV<0.78 | | ✓ | | |
| 0.0<CV<0.48 | | | ✓ | |
| Thresholds | | | | |

**Table 2 Threshold Decision Making Procedure**

## 6. Evaluated Example

The results presented in this paper have been performed in a simulation implemented in MATLAB ver 5.3.

Having defined the four major elements (peripheral elements, scheduling algorithm, plant and controller) the evaluation is performed next. Three scenarios are considered. First scenario is a fault free scenario where the response of the plant is ideal. Second scenario is a fault scenario based upon the loss of 50 percent of confidence in the "smart" sensor output. Finally a catastrophic condition is presented where CV is zero, then the control no longer uses the current sensor output. In order to be stable, the controller is switched to a safe value, which is constant. The current input of the plant is a pulse train whose indicates the current position of the ball. The output of the plant represents the current force applied to the beam. A fault is applied to one of the sensors, the fault is noise modifying the output of the sensor who is measuring the current position of the ball. Due to this condition, the presence of the fault has an effect into the system by eliminating one sensor within the configuration.

Fig. 5 presents four windows, firstly the current input related to the movement of the beam, second is the current control output, third is confidence value response from current sensor element and last window shows the response of decision making procedure.

Fig. 5 shows the response of the sensor, the plant and the controller during a fault free scenario. Fig. 6 shows the response of the system when CV presents a fault condition at 2000 seconds. This CV degrades the system 50% of the current value. In here, the control law is switched to a classical PID controller where the output of the faulty sensor is still considered. This fault is active during 500 seconds.

The response of the plant during a catastrophic scenario is presented in Fig. 7. In here, the faulty sensor is switched off. This fault appears at 2500 seconds as a degradation of the system after the appearance of first fault. This last fault is active during 500 seconds.

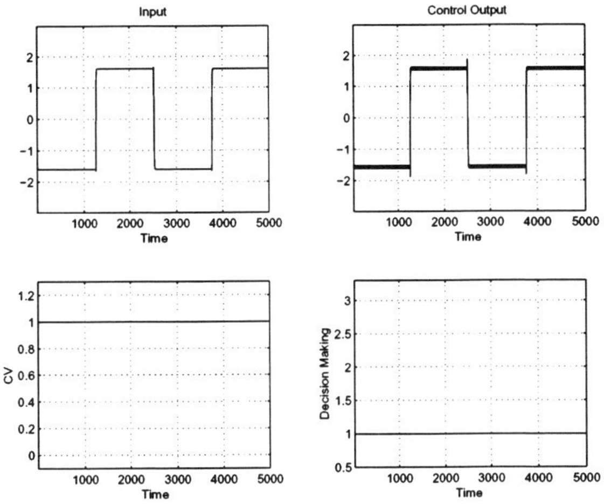


**Fig. 5 Fault Free Scenario**

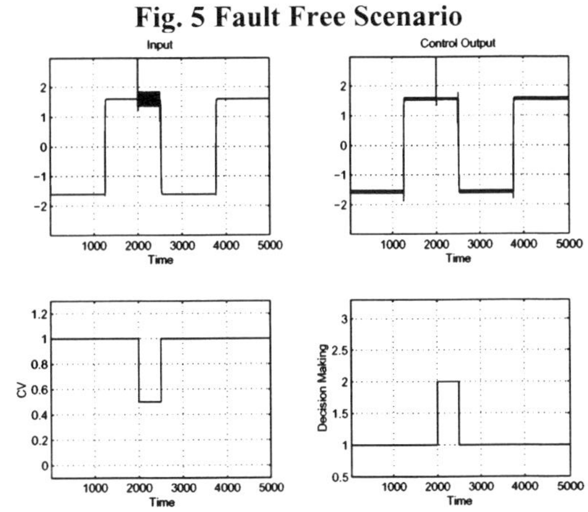


**Fig. 6 Fault Scenario Starting at 2000 Seconds**

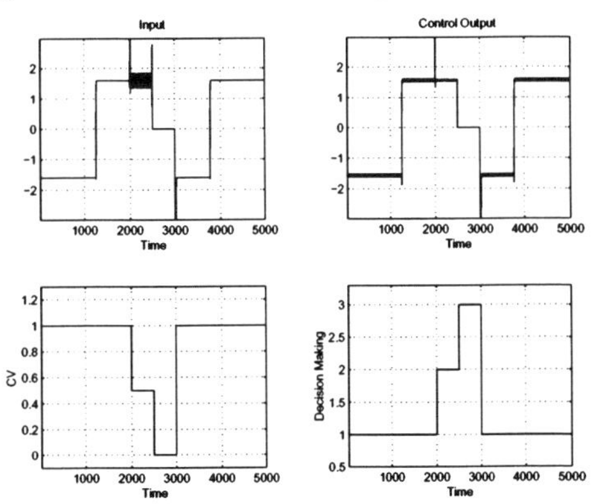


**Fig. 7 Catastrophic Scenario at 2500 Seconds**

For the case of second and third scenario there is an error which presents non-zero conditions due to the fault. This residual existence is propagated to the control scheme. At this point, the decision making procedure decides which control is going to be used. This condition can present undesirable glitches during transitions. The problem is eliminated through the use of similar controllers. This means that the control laws are designed with equivalent performance responses.

The presence of communication delays has an effect on the response of the controllers. For instance Fig. 8 shows the fault free scenario as part of the system. Solid line represents the response of the controller using the communication delays. Dashed line shows the response of the controller without considering time delays.

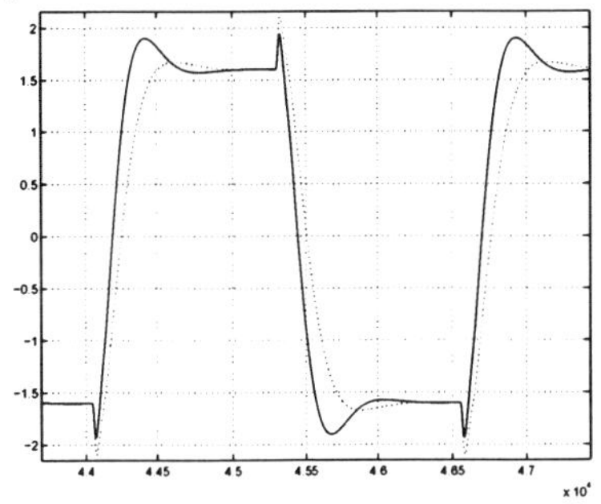

**Fig. 8 Control Result from Fault Free Scenario**

Alternatively, Fig. 9 Shows the fault scenario that degrades the system upto 50% of correct response. In this case communication delays are considered. Solid line shows the controller response taking into account the communication delays shown in Fig. 3. Dashed line shows the response of the controller without considering these time delays. The reader may appreciate the degradation suffered due to the non consideration of time delays even by the use of control reconfiguration strategy.

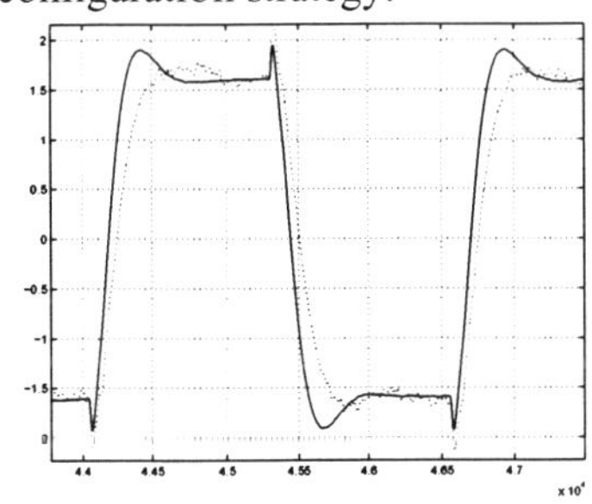

**Fig. 9 Control Result from Fault Scenario**

## 7. Conclusions

This research work has shown the implementation of an on-line control reconfiguration approach. This is based upon a local fault diagnosis procedure integrated to peripheral elements which reports the condition of each element during a fault scenario. The result of each "smart" element is used by a decision making scheme in order to choose a suitable control law. Three different control strategies are proposed. First control law uses the whole number of sensors and actuators, this is using a predictive control law. Second control considers the degradation of one sensor due to the presence of a local fault. For this case a PID controller is proposed. Finally a safe control status is performed in order to maintain the plant in a constant operating point, This drives the plant to a safe status during catastrophic conditions.

The integration of local Fault Diagnosis with a supervisor presents the possibility of on-line control reconfiguration. The strategy of having different control strategies shows how the system can degraded its performance in a safety mode. There is a potential risk of glitches when the decision making module switches from one condition to another. This is avoided by using the three control laws at the same time but just one of them closing to the feedback loop.

This research has presented a strategy of control reconfiguration based upon local fault diagnosis using smart peripheral elements. Further work is necessary from the decision making module in order to produce a smooth transition from one controller to other. Further research is proposed in order to integrate local confidence values into the control law as additional parameters.

**Acknowledgements**
The authors would like to thank to DISCA-IIMAS-UNAM and CONACYT (Num. I35561-A and CONACYT-UNAM Sub-RedII/002/2000) for their finacial support.

## 8. References

Almeida L., Pasadas R. and Fonseca J. A. (1999). Using a Planning Scheduler to Improve the Flexibility of Real-Time Fieldbus Networks, *Control Engineering Practice*, vol 7, pp. 101-108, Pergamon.

Benítez-Pérez, H., Thompson, H. A. and Fleming, P. J. (1998). Simulation of Distributed Fault Tolerant Heterogeneous Architectures for Real-Time Control, *5th IFAC Workshop on Algorithms and Architectures for Real-Time Control AARTC'98*, pp. 89-94, Cancún, México.

Benítez-Pérez, H. (1999). Smart Distributed Systems, PhD. Thesis AC&SE Dept. *University of Sheffield*, UK.

Camacho E. B. and Bordons A. (1998) *Model Predictive Control.* Springer-Verlag, USA.

Dussud M., Galichot S. and Foulloy L. (1998). Application of Fuzzy Logic Control for Continuous Casting Mold Level Control,*IEEE Transactions on Control Systems Technology*, vol.6, no.2, pp.246-56.

Gertler J. (1998). *Fault Detection and Diagnosis in Engineering Systems*, Marcel Dekker Inc.

Marcos M., Portillo J. and Bass J. M. (2000) MATLAB-Based Real-Time Framework for Distributed Control Systems, *6TH IFAC Workshop on Algorithms and Architectures For Real-Time Control, AARTC'2000*, pp. 197-202, Palma de Mallorca, Spain.

Masten, M. K. (1997). Electronics: The intelligence in Intelligent Control, *IFAC Symposium on Intelligent Components and Instrument for Control Applications*, Annecy, France, pp. 1-11.

Mathworks (1998). *System Identification Toolbox User's Guide*, MATLAB.

Nilsson, J. (1998). Real-Time Control with Delays"; PhD. Thesis, Department of Automatic Control, *Lund Institute of Technology*, Sweden.

| Var | Name | Time Consume (micro seconds) |
|---|---|---|
| c | Communication | 450 |
| b | Blocking | 50 |
| i | Interference | 0 |
| $t_c$ | Capture sensor information | 100 |
| $O_{ps}$ | Overhead time from pre-processing sensor information | 3000 |
| $O_{ds}$ | Overhead time from post-processing sensor information | 3000 |
| $t_1$ | Communication time from sensor node to control node | 575 |
| $O_{pc}$ | Overhead of Pre-processing Information from control node | 1000 |
| $O_{dc}$ | Overhead of Process Information from control node | 1000 |
| $t_{ct}$ | Control Process Time | 250 |
| $t_2$ | Communication time from control node to actuator node | 575 |
| $t_{A1}$ | Processing time from actuator 1 and 2 | 2000 |

**Table 1.a Time variable from Fault Free Scenario**

| Var | Name | Time Consume (micro seconds) |
|---|---|---|
| $O_{ps}$ | Overhead time from pre-processing sensor information | 1000 |
| $O_{ds}$ | Overhead time from post-processing sensor information | 1000 |
| $t_c$ | Capture sensor information | 1000 |
| c | Communication | 450 |
| b | Blocking | 50 |
| i | Interference | 0 |
| $t_1$ | Communication time from sensor node to decision making module | 575 |
| $t_2$ | Communication time from sensor node to control node | 575 |
| $t_{s1}$ | Processing time before sending information | 2000 |
| $t_{s2}$ | Processing time before sending information | 2000 |
| $O_{dm}$ | Overhead of Pre-processing Information | 3000 |
| $O_{pm}$ | Overhead of Process Information | 3000 |
| $t_{dm}$ | Processing time before sending information from Decision Making to Controller | 1000 |
| $t_{c1}=t_{c2}$ | Processing time from control node | 1000 |
| $t_3$ | Communication time from Decision making node to control node | 575 |
| $t_4$ | Communication time from control node to actuator node | 575 |
| $t_{A1}$ | Processing time from actuator 1 and 2 | 2000 |

**Table 1.b Time variables from Fault Scenario**

IFAC Publications
www.elsevier.com/locate/ifac

# STABILITY OF ON-LINE COMPENSATED REAL-TIME SCHEDULED CONTROL TASKS

**Pau Martí*, Ricard Villà*, Josep M. Fuertes*, and Gerhard Fohler****

**Dept. of Automatic Control*
*Technical University of Catalonia*
*Pau Gargallo 5, 08028 Barcelona SPAIN*
*Ph.: 34-3-4011679, Fax: 34-3-4017045, Email: {pmarti, villa, pepf}@esaii.upc.es*

*** Dept. of Computer Engineering*
*Mälardalen University*
*PO Box 883, SE-721 23 Västerås, SWEDEN*
*Email: gerhard.fohler@mdh.se*

Abstract: Real-time scheduling methods introduce various types of jitter in task instance execution. For real-time computer-controlled systems, the introduced sampling jitter and sampling-actuation delays may degrade the system performance and even lead to instability in the system. The degradation of the system performance can be compensated on-line by updating the controller parameters at each controller task instance execution, in what we call the compensation approach. In this paper we present stability analysis for controller tasks that perform the compensation approach. *Copyright © 2001 IFAC*

Keywords: Real-time computer systems, Real-time tasks, Scheduling algorithms, Discrete-time systems, Stability analysis

## 1. INTRODUCTION

Traditionally, real-time computer-controlled systems have been implemented using cyclic executives. A cyclic executive runs a sequence of non-preemptive tasks, invoking each task in a fixed order throughout the execution history. For an evaluation of cyclic executive models for controller tasks see Baker and Shaw [1988]. Although the advantages of using cyclic executives for computer-controlled systems have been proven, some real limitations have also been identified. Locke [1992] discusses such difficulties arising when trying to fit complex systems into the cyclic executive model.

A solution to overcome this problem is the use of more general real-time scheduling methods for computer control. Since the early work on real-time scheduling presented by Liu and Layland (1973), real-time tasks can be scheduled using a wide variety of general purpose scheduling algorithms. One common feature of almost all these algorithms is that they introduce different forms of jitter in task instance execution. Martí *et al.* (2001a) have shown that these jitters may result in a degradation of the system control performance, and even lead to instability in the system. However, in Cervin (2000) and Årzen *et al.* (2000) it is suggested and in Martí *et al.* (2001b) further developed that this degradation can be compensated by adjusting the controller parameters according to these jitters at each controller task instance execution, in the so-called compensation approach.

In this paper we present stability analysis of the on-line compensation approach. In order to be general, all types of jitter introduced by real-time scheduling methods are treated and the stability analysis is finally performed.

This paper is structured as follows: in section 2, we revise the compensation approach under study. In section 3, the problem formalization is given. In section 4, we characterize which kind of jitter patterns real-time scheduling algorithms produce. In section 5, the stability analysis is presented, and finally, section 6 outlines some conclusions and future work.

## 2. COMPENSATION APPROACH

### *2.1. Overview*

The compensation approach compensates for the degradation that sampling jitter and sampling-actuation delays may introduce in the control system performance. The basic idea is that at each controller task instance execution, the controller task parameters are adjusted according to both the actual sampling interval and the actual sampling-actuation delay. For an extensive revision of the compensation approach as well as for a jitter characterization, see Martí *et al.* (2001b).

### *2.2. Problem description*

The problem can be stated as follows. Let's suppose that all instances of a controller task that is running in isolation on a CPU are executing an appropriate control law (designed for a time-invariant system) according to a constant sampling period h (without taking into account any time delay). If the same controller task is scheduled with other tasks by a real-time scheduling algorithm on a single CPU, the actual sampling interval ($h_k$) and the actual sampling-actuation delay ($\tau_k$) may vary at each controller task k-instance execution. Consequently, in the later case, the previous control law may not be useful anymore.

However, under the compensation approach and with the adequate control law, the controller parameters will be adjusted at run-time (on-line), at each controller task instance execution. Although the control system is no longer time-invariant, the time variability that appears under the compensation approach can be analysed by studying all types of jitter that are introduced by real-time scheduling methods. Simulations have shown that the compensation approach seems to work very well. However, a proper analysis is needed in order to be able to assess when, from a stability viewpoint, this approach can be applied.

### *2.3. Example*

An inverted pendulum mounted on a motor driven car will be used. A sketch of this system can be found in Martí *et al.* (2001b). The goal of the control is to maintain the desired vertically oriented position at all times. A linear model of the plant is:

$$\begin{bmatrix} \dot\theta \\ \dot\omega \\ \dot x \\ \dot v \end{bmatrix} = \begin{bmatrix} 0 & 1 & 0 & 0 \\ \dfrac{(M+m)\cdot g}{M\cdot l} & 0 & 0 & 0 \\ 0 & 0 & 0 & 1 \\ -\dfrac{m\cdot g}{M} & 0 & 0 & 0 \end{bmatrix} \cdot \begin{bmatrix} \theta \\ \omega \\ x \\ v \end{bmatrix} + \begin{bmatrix} 0 \\ -\dfrac{1}{M\cdot l} \\ 0 \\ \dfrac{1}{M} \end{bmatrix} u(t)$$

$$y(t) = \begin{bmatrix} \theta \\ x \end{bmatrix} = \begin{bmatrix} 1 & 0 & 0 & 0 \\ 0 & 0 & 1 & 0 \end{bmatrix} \cdot \begin{bmatrix} \theta \\ \omega \\ x \\ v \end{bmatrix}$$

In all that follows, for the sake of simplicity, we will focus on the output angle ($\theta$). It can be easily justified that the open-loop system is unstable.

With the appropriate sampling period h, and using a pole placement observer design approach that meets the specified closed-loop requirements, a state feedback control law for the time-invariant system is obtained. All the following simulations have been done using the simulator presented by Eker and Cervin (1999). The step transient response of a controller task implementing the state feedback control law executing in isolation on a single CPU can be partially seen in Figure 1.

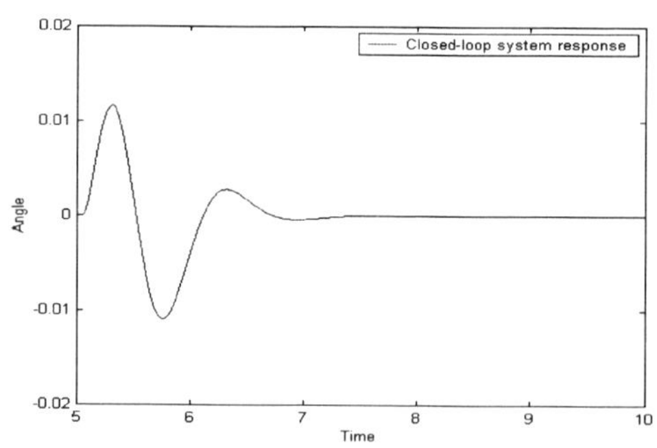


Fig. 1. Closed-loop system response

If we schedule the same controller task with other tasks by Rate Monotonic Scheduling (Liu and Layland, 1973) on a single CPU, the system becomes unstable (Figure 2) due to the jitters introduced by the scheduling method.

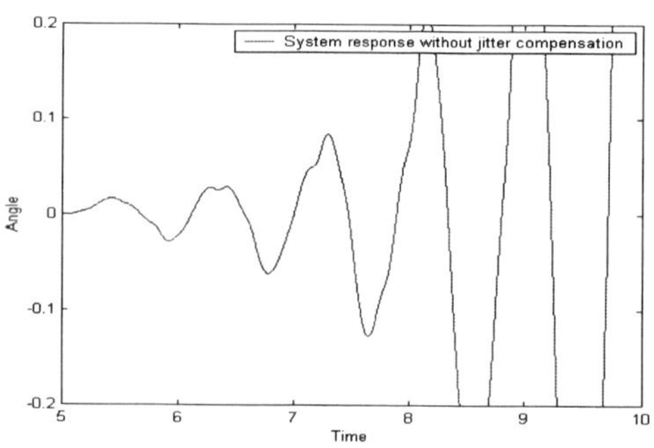


Fig. 2. Degradation of the closed-loop system response due to jitters introduced by RMS

However, if the same controller task in the same situation as before implements the compensation approach, the closed-loop system response meets again the specified requirements. It can be seen that the compensated task (Figure 3) gives almost as good response as the task that is executing in isolation on a CPU (Figure 1).

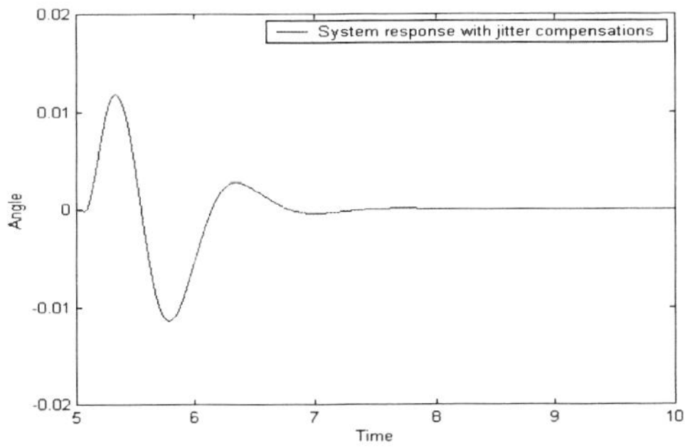


Fig 3. Closed-loop system response of a task implementing the compensation approach

## 3. PROBLEM FORMALIZATION

In order to formalize the control problem for the posterior stability analysis, we do a simplification that we remove later on. In a first approximation, we suppose that the scheduling algorithm only introduces sampling jitter. Once this case is properly formulated in what we call *irregular sampling discrete time system model*, we analyse the case with sampling-actuation delays (which can also be introduced by the scheduling algorithm) in what we call *discrete time system model with varying time-delays*. Finally, both models are combined in what we call *irregular sampling discrete time system model with varying time-delays*.

### 3.1. Irregular sampling discrete time system model

In the state space form a continuous linear time-invariant process is modelled by equations (1) and (2).

$$\frac{dx(t)}{dt} = Ax(t) + Bu(t) \quad (1)$$

$$y(t) = Cx(t) + Du(t) \quad (2)$$

In (1) and (2) $A$ is the system matrix, $x(t)$ is the state vector, $B$ is the input matrix, $u(t)$ is the control input, $y(t)$ is the system output, $C$ is the output matrix and $D$ is the direct matrix, all matrices of suitable dimension. Equation (2) is the output equation. For periodic sampling with constant period $h$, the discrete-time system can be described by (3) and (4), where $\Phi$ and $\Gamma$ are obtained from (1) and (2) as detailed in (5) and (6).

$$x(kh+h) = \Phi(h)x(kh) + \Gamma(h)u(kh) \quad (3)$$

$$y(kh) = Cx(kh) + Du(kh) \quad (4)$$

$$\Phi(h) = e^{Ah} \quad (5)$$

$$\Gamma(h) = \int_0^h e^{As} ds B \quad (6)$$

To meet the closed-loop requirements, the system specified by (3) and (4) is controlled using state feedback, where matrix $L$ (gain matrix) can be obtained by a design method such as pole placement or optimization approach. Equation (7) gives the state feedback control law.

$$u(kh) = -L(h)x(kh) \quad (7)$$

Equation (3) can be rewritten in terms of (7) as in (8).

$$x(kh+h) = (\Phi - \Gamma L)x(kh) \quad (8)$$

At the end, the closed-loop time-invariant system is characterized (at least) by equations (3), (4), and (7). To study the system stability, we have to examine the closed-loop matrix (9), where $\Phi$, $L$, and $\Gamma$ are constant matrices in terms of a constant sampling period h.

$$\Phi_{cl} = \Phi - \Gamma L \quad (9)$$

The system is stable iff the spectral radius of the closed-loop matrix $\Phi_{cl}$ is less than one.

$$\text{Stable} \Leftrightarrow \rho(\Phi_{cl}) < 1$$

The spectral radius $\rho$ of a matrix A is defined as follows:

$$\rho(A) = \max\{|\lambda| \mid \lambda \text{ is an eigenvalue of } A\}$$

However, if the on-line compensation approach is used and at each controller task instance execution the controller parameters are adjusted according to the actual sampling interval, the discrete-time system is no longer time-invariant. Therefore, the stability of the system will not only depend on a fixed closed-loop matrix $\Phi_{cl}$, but also on the sampling jitter.

If we denote the actual sampling interval of each task instance execution by $h_k$, systems described by equations (1), (2) and (7) but with irregular sampling can be modeled by equations (10), (11) and (12).

$$x(\bar{h}_{k+1}) = \Phi(h_k)x(\bar{h}_k) + \Gamma(h_k)u(\bar{h}_k) \quad (10)$$

$$y(\bar{h}_k) = Cx(\bar{h}_k) + Du(\bar{h}_k) \quad (11)$$

$$u(\bar{h}_k) = -L(h_k)x(\bar{h}_k) \quad (12)$$

where $\bar{h}_k = \sum_0^k h_k$ and matrices $\Phi(h_k)$, $\Gamma(h_k)$ and $L(h_k)$ are obtained from (5) and (6), using the same control design approach, all with $h=h_k$ at each controller task instance execution. In this approach, the stability of the system controlled by a task $\tau_i$ will depend on a product-sequence of matrices $\Phi_{clk}$, each one depending on the task instance sampling interval $h_k$.

### 3.2. Discrete time system model with varying time-delays

In a state space form a continuous linear time-invariant process with a constant time-delay $\tau$ is modelled by (13) and (14).

$$\frac{dx(t)}{dt} = Ax(t) + Bu(t-\tau) \quad (13)$$

$$y(t) = Cx(t) + Du(t) \quad (14)$$

In (13) $\tau$ is assumed to be less than or equal to the sampling period h (for further details, see Åström and Wittenmark, 1997 and Wittenmark *et al.*, 1995). For periodic sampling with constant period $h$, the discrete-time system can be described by (15) and (16).

$$x(kh+h) = \Phi(h)x(kh) + \Gamma_0(h,\tau)u(kh) + \Gamma_1(h,\tau)u(kh-h) \tag{15}$$

$$y(kh) = Cx(kh) + Du(kh) \tag{16}$$

Matrices $\Phi$, $\Gamma_0$ and $\Gamma_1$ are obtained from (13) and (14) as detailed in equations (17), (18) and (19).

$$\Phi(h) = e^{Ah} \tag{17}$$

$$\Gamma_0(h,\tau) = \int_0^{h-\tau} e^{As}ds B \tag{18}$$

$$\Gamma_1(h,\tau) = e^{A(h-\tau)} \int_0^{\tau} e^{As}ds B \tag{19}$$

As in the previous section, to meet the closed-loop systems requirements, the system specified by (15) and (16) is controlled using state feedback (20) where the gain matrix $L$ can be obtained using the same methods.

$$u(kh) = -L(h,\tau)x(kh) \tag{20}$$

Equation (15) can be rewritten in terms of (20) as in (21).

$$\begin{bmatrix} x(kh+h) \\ x(kh) \end{bmatrix} = \begin{bmatrix} \Phi(h) - L(h,\tau)\cdot\Gamma_0(h,\tau) & -L(h,\tau)\cdot\Gamma_1(h,\tau) \\ I & 0 \end{bmatrix} \cdot \begin{bmatrix} x(kh) \\ x(kh-h) \end{bmatrix} \tag{21}$$

At the end, the closed-loop time-invariant system is characterized by equations (15), (16), and (20). In this case, to study the system stability, we have to examine the closed-loop matrix (22), where $\Phi$, L, $\Gamma_0$ and $\Gamma_1$ are constant matrices in terms of a constant sampling period h and a constant time-delay $\tau$.

$$\Phi_{cl} = \begin{bmatrix} \Phi(h) - L(h,\tau)\cdot\Gamma_0(h,\tau) & -L(h,\tau)\cdot\Gamma_1(h,\tau) \\ I & 0 \end{bmatrix} \tag{22}$$

As before, the system is stable iff the spectral radius of the closed-loop matrix $\Phi_{cl}$ is less than one.

$$\text{Stable} \Leftrightarrow \rho(\Phi_{cl}) < 1$$

As before, if the on-line compensation approach is used and at each controller task instance execution the controller parameters are now adjusted according to the actual sampling-actuation delay, the discrete-time system is no longer time-invariant. Therefore, the stability of the system will not only depend on a fixed closed-loop matrix $\Phi_{cl}$, but also on the sampling-actuation delay.

If we denote the actual sampling-actuation delay of each task instance execution by $\tau_k$, systems described by equations (13) and (14) with discrete constant sampling and varying time-delays can be modeled by equations (23), (24) and (25).

$$x(kh+h) = \Phi(h)x(kh) + \Gamma_0(h,\tau_k)u(kh) + \Gamma_1(h,\tau_k)u(kh-h) \tag{23}$$

$$y(kh) = Cx(kh) + Du(kh) \tag{24}$$

$$u(kh) = -L(h,\tau_k)x(kh) \tag{25}$$

Matrices $\Phi(h)$, $\Gamma_0(h,\tau_k)$, $\Gamma_1(h,\tau_k)$ and $L(h,\tau_k)$ are obtained from (17), (18) and (19), using the same control design approach as in (20), all with $\tau=\tau_k$ at each controller task instance execution. In this approach, the stability of the system controlled by a task will depend on a product-sequence of matrices $\Phi_{clk}$, each one depending on the task instance sampling-actuation delay $\tau_k$ and the constant sampling period h.

### 3.3. *Irregular sampling discrete time system model with varying time-delays*

Combining appropriately the two models described in the two previous sections, that is, compensating both sampling jitter and sampling-actuation delays, systems described by equations (13) and (14) with irregular discrete sampling and varying time-delays can be modeled by equations (26), (27), (28) and (29).

$$x(\bar{h}_{k+1}) = \Phi(h_k)x(\bar{h}_k) + \Gamma_0(h_k,\tau_k)u(\bar{h}_k) + \Gamma_1(h_k,\tau_k)u(\bar{h}_{k-1}) \tag{26}$$

$$y(\bar{h}_k) = Cx(\bar{h}_k) + Du(\bar{h}_k) \tag{27}$$

$$u(\bar{h}_k) = -L(h_k,\tau_k)x(\bar{h}_k) \tag{28}$$

$$\bar{h}_k = \sum_0^k h_k \tag{29}$$

Matrices $\Phi(h_k)$, $\Gamma_0(h_k,\tau_k)$, $\Gamma_0(h_k,\tau_k)$ and $L(h_k,\tau_k)$ are obtained from (17), (18) and (19), using the same control design approach as in (20), all with $h=h_k$ and $\tau=\tau_k$ at each controller task instance execution. In this case, the stability of the system controlled by a task will depend on a product-sequence of matrices $\Phi_{clk}$ (as in (22)), but each one depending on the task instance sampling interval $h_k$ and sampling-actuation delay $\tau_k$.

### 3.4. *Stability condition*

In the three previous subsections, we concluded that the stability of the system depends on a product-sequence of specific closed-loop matrices $\Phi_{clk}$. Therefore the system stability depends on the expression described by (30).

$$\left\{ \prod_{k=0}^{\infty} \Phi_{clk} \right\} \tag{30}$$

As a consequence, an overview of what kind of product-sequences will apply, due to real-time scheduling methods, will be given. Afterwards, according to these types of product-sequences of matrices, we will present a stability analysis for the last case we formalized in subsection 3.3, which is the generalization of all the other cases.

## 4. JITTER PATTERNS

Real-time scheduling methods can be divided into two major categories according to the time when jitters are generated and analyzable:

- *Offline*: the whole schedule is constructed before run time (e.g., see Fohler, 1994 or Ramamritham, 1990): over some hyper-period (usually LCM of all periods of all scheduled tasks), it is known at which time each task instance will execute. Therefore, all the jitters are known or could be specified beforehand, and follow a periodic pattern over the hyper-period. Depending on how the offline schedule is constructed, two kinds of jitter periodic pattern apply: a known sampling interval constant (h) with constant sampling-actuation delays ($\tau$) through all the execution life of a controller task or a finite sequence of known sampling intervals ($h_1, h_2, ..., h_k, ..., h_n$) with a finite sequence of known sampling-actuation delays ($\tau_1, \tau_2, ..., \tau_k, ..., \tau_n$) that repeats periodically through all the execution life of a controller task if the runtime execution preserves the off-line scheduled controller instances starting times.

- *On-line*: task instances are dispatched at run time according to the scheduling algorithm that is being used. Therefore, there is not a priori knowledge of the schedule that will apply at run time nor whether the runtime dispatcher will preserve the execution pattern of controller task instances that have been previously scheduled offline. In these cases, the only available knowledge is that during run time, an infinite sequence of variable but bounded sampling intervals will apply with an infinite sequence of variable but bounded sampling-actuation delays through all the execution life of a controller task. If the system is schedulable, the only available knowledge is that each sampling interval $h_k$ and each sampling-actuation delay $\tau_k$ for all instances of a given controller task will be bounded:

$$\forall k, \quad \min(C^k_i) \le h_k \le \max(st^{k+1}_i - st^k_i)$$
$$0 < \tau_k \le \max(h_k) - \min(C^k_i)$$

where $C^k_i$ is the computation time and $st^k_i$ is the starting time of a k-instance of a i-task.

Therefore, we distinguish three types of jitter patterns:

Case 1: Constant jitters.
Case 2: Known repeating sequence of known jitters.
Case 3: Unknown infinite sequence of bounded jitters.

## 5. STABILITY ANALYSIS

Given the system specified by equation (3), (4), and (7), the system stability depends on the closed-loop matrix in (9). Taking into account that the state feedback law will be computer implemented and a real-time scheduling algorithm will schedule the corresponding controller task, the following two statements will hold:

1. The sampling period h belongs to an interval $h \in [h_{min}, h_{max}]$ with $h_{min} > C$ and $h_{max} < T_s$, where $T_s$ is the Shannon's sampling theorem limit and C shall include the minimum computation time of a controller task instance.

$$h_{min} < h < h_{max}$$

Similarly, the sampling-actuation delay $\tau$ belongs to an interval $\tau \in [\tau_{min}, \tau_{max}]$ with $\tau_{min} > 0$ and $\tau_{max} < h_{max}$.

$$\tau_{min} < \tau < \tau_{max}$$

2. The sampling period and the sampling-actuation delay are integers multiple of the clock ticksize.

$$h = m * ticksize, \ \tau = n * ticksize, \quad n, m \in N$$

Therefore, both intervals for the sampling period and for the sampling-actuation delay have a finite number of values, which we can characterize by the following sets:

$$H = \{h \mid h = m * ticksize, m \in N \text{ and } h_{min} < h < h_{max}\}$$
$$T = \{\tau \mid \tau = n * ticksize, n \in N \text{ and } \tau_{min} < \tau < \tau_{max}\}$$

The stability analysis for the jitters characterized in the previous section is depicted in the following. If jitters are constant, the following Case 1 applies.

Case 1: there is one and only one closed-loop matrix, $\Phi_{cl}$. In this case, the stability test is the following, where $\rho$ is the spectral radius:

$$\text{Stable} \quad \Leftrightarrow \quad \rho(\Phi_{cl}) < 1$$

If jitters are not constant, we can distinguish two possibilities (note that the process is no longer time-invariant and equations (26), (27), (28) and (29) can describe the new system), depicted in Case 2 and Case 3.

Case 2: A variable sampling period ($h_k \in H$) and a variable sampling-actuation delay ($\tau_k \in T$) that repeats periodically:

$$h_1, h_2, ..., h_k, ..., h_n, h_1, h_2, ..., h_k, ..., h_n, ...$$
$$\tau_1, \tau_2, ..., \tau_k, ..., \tau_n, \tau_1, \tau_2, ..., \tau_k, ..., \tau_n, ...$$

In this case, the closed-loop system will be characterized by a known finite set of matrices that will follow a known periodic pattern. Therefore, a known repeating sequence of known matrices will apply. In this case, the stability test can be performed by checking the stability of the product of the repeating sequence of matrices (Dogruel and Özgüner, 1995):

$$\text{Stable} \Leftrightarrow \rho(\Phi_{cl1} \cdot \Phi_{cl2} \cdot ... \cdot \Phi_{cln}) < 1$$

Case 3: A variable sampling period ($h_k \in H$) and a variable sampling-actuation delay ($\tau_k \in T$), but with unknown execution pattern:

$$h_1, h_2, \ldots, h_k, h_{k+1}, \ldots$$
$$\tau_1, \tau_2, \ldots, \tau_k, \tau_{k+1}, \ldots$$

In this case, the closed-loop system will be characterized by a product of an infinite number of matrices taken randomly from a finite set of matrices. In this case, the stability test applied above cannot be used. However, another test can be applied using known results of linear algebra (see for example Strang, 1980):

$$\forall\ \Phi_{clk} : \mathrm{eig}(\Phi^T_{clk} \cdot \Phi_{clk} - I) < 0 \ \Rightarrow \ \text{Stable}$$

If each matrix $\Phi_{clk}$ satisfies that all eigenvalues of $\Phi^T_{clk} \cdot \Phi_{clk} - I$ are less than zero (each $\Phi^T_{clk} \cdot \Phi_{clk} - I$ is negative definite), each matrix $\Phi_{clk}$ will guarantee immediate decay. In this case, any product of an infinite number of such matrices $\Phi_{clk}$ will guarantee stability in the system.

The presented stability analysis has covered all possible types of jitter that real-time scheduling algorithms may introduce. Therefore, the on-line compensation approach will be applicable when the closed-loop system matrix or matrices fall into one of the three cases.

## 6. CONCLUSIONS

A review of the stability of on-line compensated real-time scheduled control tasks has been done showing that three cases can be distinguished. Two of them were already known. The third case has been categorized and a sufficient stability test presented. Using this analysis, given a system to be controlled and a scheduling algorithm, it will be easy to decide if the system will behave correctly from a stability viewpoint when using the compensation approach.

Even though extensive simulations of the on-line compensation approach show good performance, more study besides the presented stability analysis is needed in order to achieve a deeper characterization of the controlled system under this approach. Therefore, on the one hand, further work will focus on study how the compensation approach behaves in terms of system responsiveness. On the other hand, we will focus on what are the implications of using the on-line compensation approach from a schedulability viewpoint.

## ACKNOWLEDGEMENTS

The presented research has received support from Spanish CICYT project ref. DPI2000-1760-C03-01.

## REFERENCES

Årzen, K.-E., A. Cervin, J. Eker and L. Sha. (2000). An Introduction to Control and Scheduling Co-Design. In *39th IEEE Conference on Decision and Control, Sydney*, Australia, December

Åström, K. J and B. Wittenmark (1997). *Computer-Controlled Systems. Third edition*. Prentice Hall.

Baker, T.P. and A. Shaw (1988). The Cyclic Executive Model and Ada. In *Proceedings of IEEE Real-Time Systems Symposium*, pages 120-129, Madrid, Spain.

Cervin, A. (2000). Towards the Integration of Control and Real-Time Scheduling Design. Licentiate Thesis. ISRN LUTFD2/TFRT—3226—SE, Dept. of Automatic Control, Lund Institute of Technology, Lund, Sweden, May

Dogruel, M and U. Özgüner (1995). Stability of a Set of Matrices: A Control Theoretic Approach. In *Proc. of the 34th Conference on Decision and Control*, New Orleans, USA, September

Eker J. and A. Cervin (1999). A Matlab Toolbox for Real-Time and Control Systems Co-Design. *Proc. 6th Int. Conf. on Real-Time Computing Systems and Applications*, Hong Kong, China, December

Fohler, G. (1994). Flexibility in Statically Scheduled Real-Time Systems. Phd Thesis. Technisch-Naturwissenschaftliche Fakultaet, Technische Universitaet Wien, Austria, April

Liu, C. and J. Layland (1973). Scheduling Algorithms for Multiprogramming in a Hard Real-Time Environment. *J.ACM*, 20, 46-61.

Locke, C.D. (1992). Software Architecture for Hard Real-Time Applications: Cyclic Executives vs. Fixed Priority Executives. In *J. of Real-Time Systems*, 4, 37-53, Kluwer Academic Publishers.

Martí P., R. Villà, J.M. Fuertes and G. Fohler (2001a). On Real-Time Control Tasks Schedulability. In *IFAC European Control Conference,* September, Porto, Portugal

Martí P, G. Fohler, K. Ramamritham and J.M. Fuertes, (2001b). Jitter Compensation in Real-Time Control Systems. *Accepted for publication to 22th IEEE Real-Time Systems Symposium*, December, London, UK.

Ramamritham, K. (1990). Allocation and Scheduling of Complex Periodic Tasks. *In 10th Int. Conf. on Distributed Computing Systems*, pages 108--115.

Strang, G. (1980) *Linear Algebra and its Applications. Second Edition.* MIT. Academic Press, Inc. USA

Wittenmark B., J. Nilsson and M. Törngren (1995). Timing Problems in Real-time Control Systems. In *Proc. American Control Conference*, USA

IFAC Publications
www.elsevier.com/locate/ifac

# QUASI TIMED-PETRI-NETS AND ITS APPLICATION IN FIELD COMMUNICATION SYSTEM

**Minrui Fei, Lixiong Li, Yunchao Qiu and Xu Li**

*School of Mechatronical Engineering & Automation, Shanghai University*
*Shanghai, 200072, P. R. China*
*Email: mrfei@srcap.stc.sh.cn*

Abstract: Owing to lack of uniform communication protocol, field instruments have less capability in interconnection. With a view to solving this problem, a Multi-protocol Field Communication System (MPFCS) is implemented by using Multi-protocol Field Communication Controller (MPFCC) which is developed by our project team. In this paper, as a protocol description tool of MPFCS, Quasi Timed-Petri-Nets is presented, and it has been successfully applied in describing different communication protocols in same model. Furthermore, performance evaluation of MPFCS is achieved via Quasi Timed-Petri-Nets. Finally, this approach has proven to be highly rational and effective through the real-time communication experiments. *Copyright © 2001 IFAC*

Keywords: Timed Petri-nets, Communication protocols, Performance evaluation, Real-time communication, Computer simulation

## 1. INTRODUCTION

With the development of modern industry, production plants in growing number are tending to install more intelligent instruments to meet the increasing demands for field measurement and control and long-distant supervision, and many instruments manufacturers have established their single-purpose communication protocols for their own products. As a result, it is very difficult to interconnect instruments of different manufacturers. With a view to solving this problem, our project team has presented the concept of field communication open policy (Fei and Liu, 1997), in the meantime, a Multi-protocol Field Communication Controller (MPFCC) was designed and corresponding Multi-protocol Field Communication System (MPFCS) was established (Liu, 1999). In MPFCS, varied instruments with different single-purpose communication protocols can work together in the entire system. Obviously, the most important technique in MPFCS is to have a general approach of protocol description.

This paper aims to present such an approach to make exact description for varied instruments, and different protocols can be packed and downloaded into MPFCC to achieve communication. In addition, it also can be used to make performance evaluation for MPFCS by establishing its entire model. Using the evaluation results, reasonable configuration mode can be chosen and optimal poll cycle can be selected when MPFCS has been running, and furthermore the design can be optimized when MPFCS is under construction. In a word, this approach of protocol description and performance evaluation will help to decrease cost and improve efficiency for field communication. The framework of this approach is given in Fig. 1.

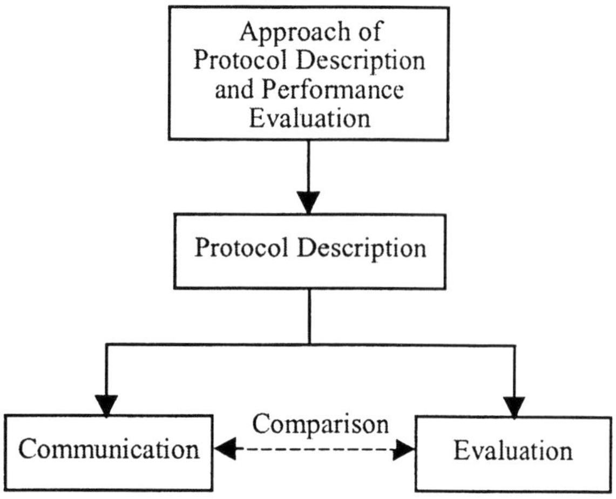


Fig. 1. Approach of protocol description and performance evaluation

## 2. MULTI-PROTOCOL FIELD COMMUNICATION SYSTEM (MPFCS)

Multi-protocol Field Communication System (MPFCS) whose hardware structure is given in Fig. 2 consists of a supervisory computer, some intelligent instruments and MPFCC. MPFCC, which is composed of main board, protocol boards, and interface boards, connects intelligent instruments in workshop with computer in control center, and as heart of MPFCS it can contain 192 communication nodes at most, and support many standard interfaces, such as RS-232C, RS-422A and RS-485.

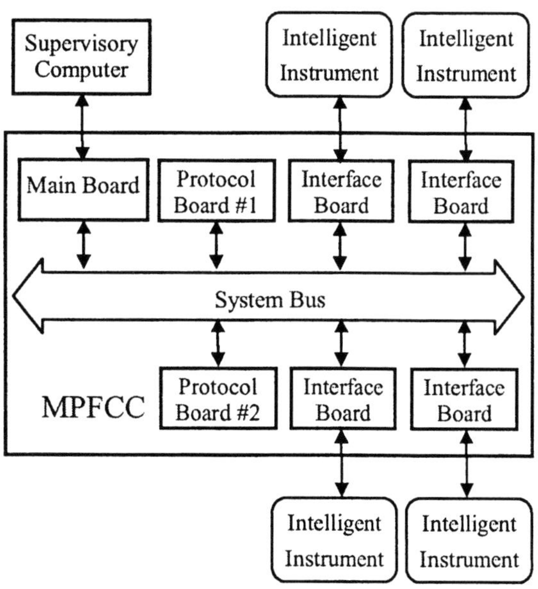


Fig. 2. Structure of MPFCS

## 3. PRINCIPLES OF QUASI TIMED-PETRI-NETS

First defined by Carl Adam Petri(1966), Petri-nets is a formal and graphical modeling tool with intuitionistic description mode, firm mathematic foundation and distinct analysis technique. These features make it sufficiently appropriate for describing numerous phenomena in communication process, such as conflict, competition and concurrence, and moreover, it can be used in performance evaluation. As a directed graphics, Petri-nets contain four main elements which are given in Table 1 as follow.

Table 1 Elements in Petri-nets

| | Graphical Representation | Function |
|---|---|---|
| Place | ○ | local state of system |
| Transition | \| ▮ ▯ | describing when to change system state |
| Directed arc | → | linking place with transition |
| Token | • | number of resource or condition |

Basic Petri-nets can only describe logic structure of system because of its lack in concept of time, so it is not appropriate for modeling field communication which is a real-time system. In order to tackle this problem, timed Petri-nets (Rami, et al., 1985), which is approximately similar to the structure of communication protocols, appeared by introducing time variables into basic Petri-nets to make up its shortage in describing time. As a result, timed Petri-nets, extended timed Petri-nets (Lee and Kim, 1987) and stochastic Petri-nets (Lin, 2000) are widely applied in modeling communication protocols.

Although timed Petri-nets can describe communication protocols as well as communication system in theory, it is very difficult to realize because timed Petri-nets still has many differences with communication system in terms of the particularity of communication protocols. Consequently, in this paper, a new Petri-nets—quasi timed-Petri-nets is proposed by extending and improving timed Petri-nets, and the improvement methods are give as follows.

1) In timed Petri-nets, number of tokens in a place represents the state of place element, but it is not able to describe the concept of data frame format which is extremely important in communication protocols. As

a result, in quasi timed-Petri-nets, place element with characters is defined to provide data frame format properties for each place element.

2) In timed Petri-nets, firing time, holding time, enabling time and delay time are defined, but there are not so many concepts of time in communication protocols. Consequently, in quasi timed-Petri-nets, concept of time is combined with transition, and the define of time is simplified, only delay time is applied. At the same time, responding time and repeating time is defined in view of the requirement of communication protocols.

3) In timed Petri-nets, token primarily represents condition to fire corresponding transition, in quasi timed-Petri-nets, besides condition, it can fire the properties in place element. In other words, if a place contains token, the token in the place element can not only fire the corresponding transition but also fire the properties of another place element. In communication protocol, the properties are data frame formats.

4) In timed Petri-nets, each place element can contain many tokens. In quasi timed-Petri-nets, we assume there is only one token in each place element because token is merely used to fire the corresponding transition and properties in another place element, and it will transfer to different place element according to corresponding transition.

Based on above-mentioned improvements, the quasi timed-Petri-nets is presented, the differences between quasi timed-Petri-nets and timed Petri-nets are given in Table 2, and the relations between quasi timed-Petri-nets and communication protocols are given in Table 3.

Table 2 Timed Petri-nets in comparison with quasi timed-Petri-nets

| | Timed Petri-nets | Quasi Timed-Petri-Nets |
|---|---|---|
| Place | without character | with character |
| Transition | event | event |
| Token | condition, many tokens | condition, only one token in a place |
| Time | firing time, holding time, enabling time | delay time |

Table 3 Relations between quasi timed-Petri-nets and communication protocols

| Quasi Timed-Petri-Nets | Communication Protocols |
|---|---|
| Place | state or data frame format |
| Transition | event |
| Token | condition<br>firing format of data frame |
| Time | transferring time,<br>responding time,<br>repeating time |

Through quasi timed-Petri-nets defined above, different communication protocols can be conveniently configured by elements in Table 3. A quasi timed-Petri-nets model of intelligent instrument CP20 is depicted in Fig. 3.

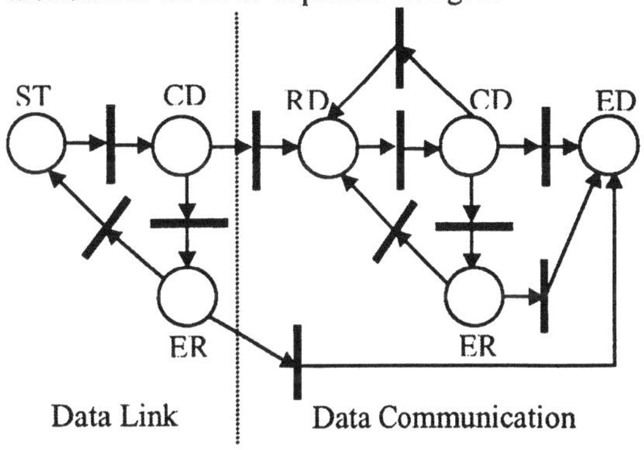


Fig. 3. Quasi timed-Petri-nets model of intelligent instrument CP20

where
ST (state): establishing the data link between MPFCC and intelligent instruments;
RD (reading data): collecting data periodically;
CD (command and data): intelligent instruments responding to the communication process;
OT (overtime responding): responding time;
ER (error): returning error message;
ED (end): ending communication.

## 4. MODELING OF MPFCS BASED ON QUASI TIMED-PETRI-NETS

Relying on quasi timed-Petri-nets and features of MPFCS, model of protocol can be successfully established as well as entire model of MPFCS which is shown in Fig. 4, and the explanations of labels in Fig. 4 are listed below the figure. This model is not only suitable for analysis in theory, it also can realize the real-time communication by describing the different communication protocols with same method.

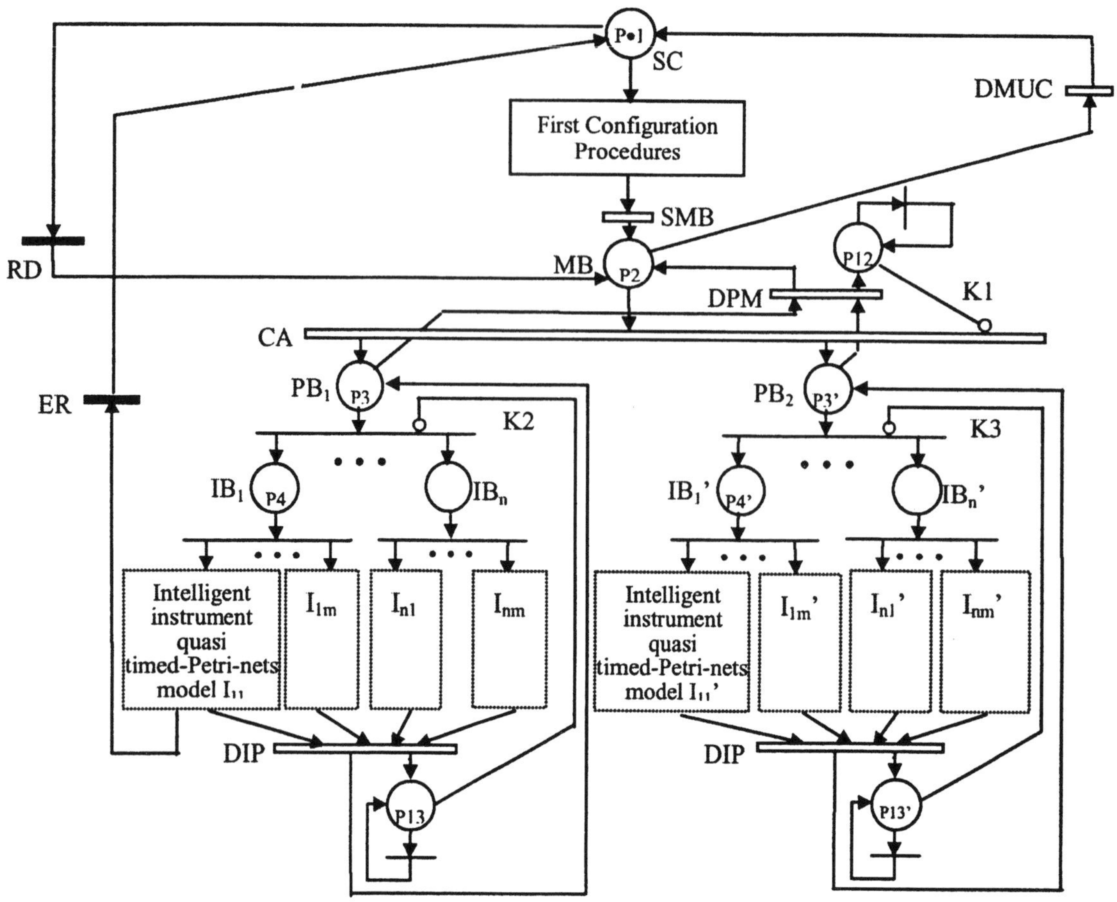


Fig. 4. Quasi timed-Petri-nets model of MPFCS

where

SC: supervisory computer;

MB: main board;

CA: assigning configuration information of each node;

PB: protocol board;

RD: reading data;

ER: error;

DIP: receiving data from node;

DPM: receiving data from protocol board;

DMUC: receiving data from main board;

$IB_1$-$IB_n$: interface board linking to protocol board;

$I_{11}$-$I_{nm}$: field intelligent instrument model created by quasi timed-Petri-nets, such as model of CP20 in Fig. 3.

## 5. APPROACH OF PROTOCOL DESCRIPTION AND ITS IMPLEMENTATION IN REAL-TIME COMMUNICATION

From the foregoing model of MPFCS created by quasi timed-Petri-nets, a new approach of protocol description is proposed. By this approach, different communication protocols of various intelligent instruments can be described and converted into uniform format according to this model, so MPFCC can make coding and decoding in spite of the protocol differences. MPFCS that contain two different intelligent instruments, CP20 and WP-C90, is given in Fig. 5 by means of this protocol description approach.

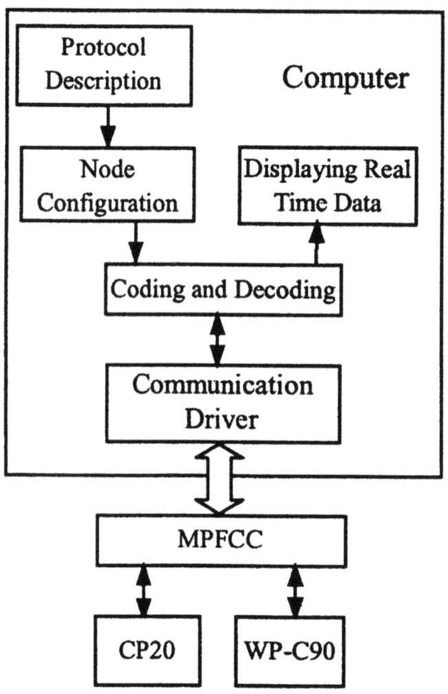


Fig. 5. Protocol description approach in real-time communication

## 6. APPROACH OF PERFORMANCE EVALUATION AND ITS SIMULATION EXPERIMENTS

It has been proven that a reachability graph of MPFCS's quasi timed-Petri-nets model, which is shown in Fig. 4, is homogeneous to a continuous time Markov chain (CTMC) (Zuberek, 1989), so analysis methods of CTMC can be introduced into the performance evaluation of MPFCS.

According to Markov theory, linear simultaneous equations are list as follow

$$\begin{cases} X \cdot R = 0 \\ \sum_i x_i = 1, 1 \le i \le n \end{cases} \quad (1)$$

where $X=(x_1,x_2, \ldots ,x_n)$ is steady probability vector and R is transition rate matrix. The solve of equation (1) is steady probability of each token.

$$P[Mi]= x_i, i=1,2,\ldots,n. \quad (2)$$

suppose λ is the firing rate and then throughput can be given by

$$S = \lambda \cdot P[M_1] \quad (3)$$

and mean delay time can be given by

$$T = \tau_{M1} / P[M_1] - \tau_{M1} \quad (4)$$

where $\tau_{M1} = 1/\lambda$

Based on formula (3) and (4) which are deduced from quasi timed-Petri-nets model of MPFCS in Fig. 4, a performance evaluation software is designed and developed. By means of this software, performance evaluation simulation is made. The simulation object is MPFCS with two different protocols (CP20 and WP-C90), and baud rate is 2400bps. The simulation results are given in Fig.6, where the relation between throughput and poll cycle is given in Fig. 6(a) and the relation between mean delay time and number of nodes is given in Fig.6 (b).

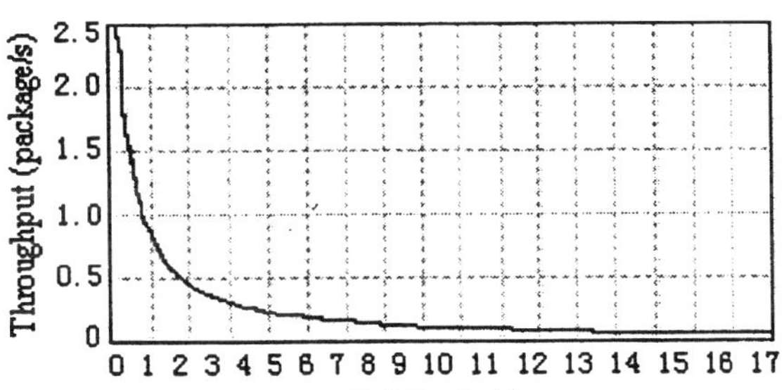


(a) Relation between throughput and poll cycle

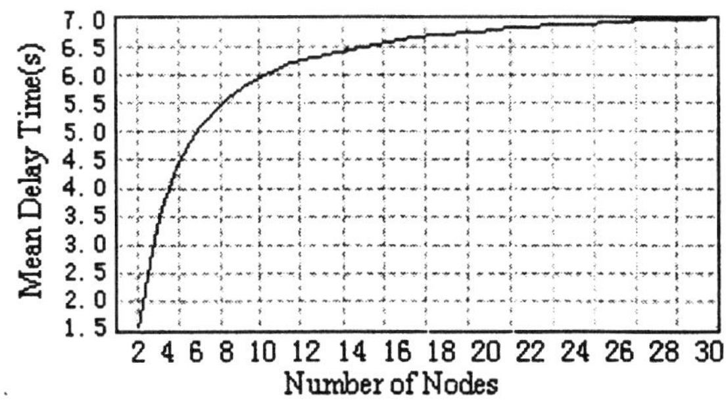


(b) Relation between mean delay time and number of nodes

Fig. 6. Results of performance evaluation simulation

From the result, it is clear that,

1) The increase in poll cycle contributes to the decrease in throughput if number of nodes and baud rate are constant.

2) The increase in nodes contributes to the increase in mean delay time if baud rate is constant.

## 7. CONCLUSIONS

In this paper, a new quasi timed-Petri-nets is constructed because of limitation of timed Petri-nets in describing communication protocols and system. Furthermore a protocol description and performance evaluation approach for MPFCS is presented according to quasi timed-Petri-nets model. This approach has been successfully applied in the real-time communication, and the performance evaluation simulation results are similar to real-time data. All these indicate that the protocol description and performance evaluation approach based on quasi timed-Petri-nets is rational and effective.

## ACKNOWLEDGEMENTS

This work was supported by Science & Technology Commission of Shanghai Municipality.

## REFERENCES

Fei, Minrui and Wen Liu (1997). Investigation and Implementation of Field Communication Open Policy. In: Proc. of *14th IFAC DCCS*, Seoul.

Lee, Dongho and Chongsang Kim (1987). Extended Timed Petri Nets Models for Performance Analysis of Communication Protocol, *IEEE-TENCON*.

Lin, Chuang (2000). *Stochastic Petri nets and its Application in Performance Evaluation*, Tsinghua University, Peking.

Liu, Wen (1999). Implementation of Multi-protocol Field Communication Controller Based on Communication Open Policy, In: *Thesis for*

*Master Degree*, Shanghai University
Petri, C.A. (1966). Communication with Automata, Technical Report RADC-TR-65--377, **1, Suppl. 1**, New York.
Rami, Raouk, et al. (1985). Performance Analysis Using Timed Petri Nets, *Protocol Specification, Testing, and Verification*.
Zuberek W M. (1989). Performance Evaluation Using Unbound Timed Petri Nets. In: *Proc. of the Third International Workshop on Petri nets and Performance Models*, 180~186, Kyoto, Japan.

IFAC
Publications
www.elsevier.com/locate/ifac

# ANN–BASED TOOLS LEAD TO IMPROVED FEEDFORWARD COAGULATION CONTROL POLICY

**Cox C.S., Adgar, A., Fletcher, I., Böhme, T.J. & Jones, C.A.***

Control Systems Centre, University of Sunderland,
School of Computing, Engineering & Technology, Edinburgh Building,
Chester Road, Sunderland, SR1 3SD, UK

* Northumbrian Water Ltd., Leat House, Pattinson Rd,
District 15, Washington, NE38 8LB, UK.

Abstract: Attempts to improve the performance of water treatment works through the application of improved control and measurement have had variable success. The most quoted reason for this is that the individual dynamic operations defining the treatment cycle are complex, highly non–linear and poorly understood. These problems are compounded by the use of faulty or badly maintained sensors. The efficient and robust operation of any industrial system is critically dependent on the quality of the measurements made. Also, the structure of the control policy and choice of the individual controller parameters are important decisions to the economic operation.
Because of their ability to capture non–linear information very efficiently, artificial neural networks (ANNs) have found great popularity amongst the 'control community' and other disciplines. This paper discusses a recent application of ANNs at surface water treatment works. The study is used to describe how the introduction of ANNs has resulted in more reliable system measurement and consequently improved coagulation control. *Copyright © 2001 IFAC*

Keywords: Water Treatment, Coagulation Control, Artificial Neural Networks, Process Control, Sensor Failure Detection.

## 1. INTRODUCTION

The majority of the drinking water consumed in the UK is treated at surface water treatment works where raw water is abstracted from rivers and reservoirs. The type of treatment it then undergoes depends on the source and the quality of its water. In general, the poorer the qualities of the raw water the more expensive it is to treat.

Prior to privatisation the majority of water treatment processes were under manual control. Since privatisation, the water industry has been seeking ways, especially via the increased use of automatic control (Bevan, 1999), to produce high quality water at reduced cost whilst at the same time 'down–sizing' its work force. The water treatment process consists of a complex group of interconnected physical and chemical systems. It is not immediately obvious how each one relates with its neighbour, however, it is well known that a problem with one process, if not addressed, will quickly result in a much larger problem in one or more of the subsequent stages.

The addition of chemicals to the water is arguably the most critical process within a surface water treatment works. Water quality legislation dictates maximum concentrations for chemical elements in drinking water and so the control of the amount of chemicals added to the water and the process of monitoring its quality are very important. The key unit operation is chemical coagulation. Successful control of this clarification process is invariably followed by problem-free operation of the secondary and tertiary treatment processes.

The key to efficient coagulation control is the addition of just sufficient coagulant chemical to the process. Too much coagulant will ensure treatment targets are achieved but at great cost, both in chemical consumed and extra sludge waste produced. Too little coagulant will result in poor treatment performances and problems in subsequent processes such as filtration.

At many works the coagulation dosage is determined by a 'jar test.' This involves simulation of the coagulation process in laboratory vessels under a range of treatment conditions in an attempt to discover the optimum treatment conditions, but primarily the coagulant dose level. Automatic strategies for using this information are desirable to both reduce costs and to improve water quality and consistency.

This paper presents one example of developing an automatic strategy where the core of the implementation has been achieved using some form of artificial neural network.

## 2. OVERVIEW OF THE CLARIFICATION PROCESS

The purification of water for domestic consumption involves several stages of treatment of the raw water to remove suspended solids, colour and bacteria before entering the distribution network. The individual treatment processes include clarification, disinfection, pH adjustment, filtration and taste and odour removal. Clarification can be roughly divided into a two-stage process, comprising of coagulation and flocculation (see Fig 1). In the coagulation stage, a coagulant chemical (a salt of a highly charged metal ion) is added to the water in a mixing vessel, often with mechanical agitation to ensure uniform distribution of the chemical. The highly charged metal ion destabilises the negatively charged impurities in the water. The species combine together to form larger particles called flocs. Flocculation involves the combination, by collision, of small particles, under natural turbulence, into larger particles. A flocculation aid (usually a natural starch or a polymer solution) may also be added to assist floc formation. The final stage of the clarification process is the separation of the large flocs formed by coagulation and flocculation from the water, usually by settlement.

## 3. COAGULATION CONTROL

Numerous factors combined to ensure that manual control was the method used for many water treatment plants in the past. The processes were even less well understood than today, control hardware and instrumentation was much less advanced and more expensive and finally there was not the stringent economic and regulatory pressures on the industry which now exist. There has been a lot of recent research into the study of feed-forward algorithms for coagulation control. In these schemes the amount of chemical coagulant required is assumed to be directly related to raw water parameters (Nahm et al., 1996).

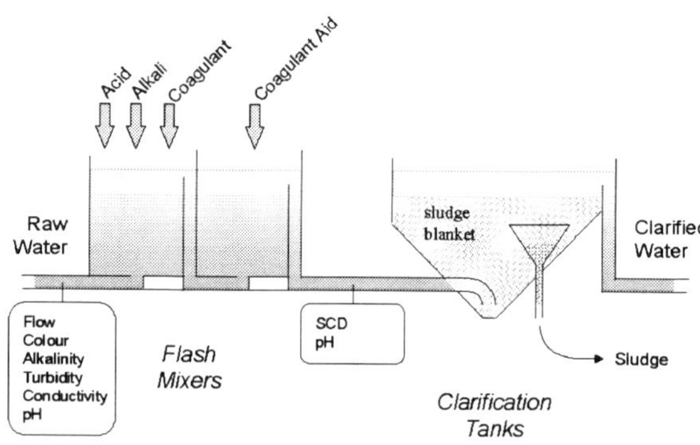


Fig 1. Schematic of the coagulation and flocculation processes in potable water treatment.

When investigating strategies to improve coagulation process control, it is desirable to have a reliable measuring method with which the effectiveness of any such strategy developed can be assessed, ideally under service conditions. The following metrics are suggested to be useful based upon the authors' understanding of the problem:

i) Reduction in chemical usage
ii) Improvement in treated water quality, and
iii) Improvement in consistency of treated water quality.

The direct comparison of control strategies, however, is always difficult due to the continually changing raw water quality.

Regression modelling has been used for determining optimum chemical dosing rates on a water treatment plant. Ellis *et al.* (1991) proposed a method based on multiple regression analysis in which coagulant dose level is predicted from correlations with physical parameters describing the influent conditions of the plant. Early work by Cox & Graham (1994) also focused upon regression methods for coagulation modelling.

The above studies were important since they provided the motivation for the search for alternative techniques that has resulted in the development of many expert system based methodologies. Neural network techniques have been applied to improve upon the linear modelling and control strategies previously relied upon. Models of the clarification process have been built to capture the relationships between certain raw water parameters and chemical coagulant dose required for treatment.

The approach taken in this research was to examine a data set of nearly 500 data samples, representing over one years' worth of 'jar test' results from a single treatment works. The data consisted of six variables that together classified the raw water quality, namely: colour, pH, turbidity, conductivity, alkalinity and temperature. The data were pre–processed using techniques to improve the statistical quality of the data, such as allowing for seasonal variations in water quality, by having different models for different seasons and removing examples of poor control performance from training data sets intended for human operator mimicking.

Trials on this 'jar test' data have been performed to investigate the benefits achieved by the neural network approaches to the modelling problem. The trials were performed using: a) Multi–layer Perceptron Network (MLP), b) Radial Basis Function Network (RBF), and c) Regression Model (based upon the approach of Cox & Graham (1994)).

The fits to process data for the three model types are all acceptable. The neural network approaches improve upon the regression model, as would be expected, this is shown in Fig 2. The neural approaches show similar performances (e.g. both having approximately 150 samples in the central |error|<2.5 band), but with considerably less spread observed in prediction error than the regression model.

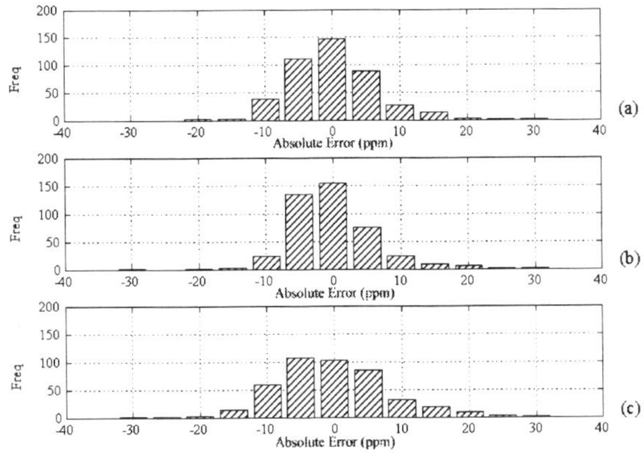


Fig 2. Comparison of model performance: a) MLP network, b) RBF network, c) regression model.

By basing treatment decisions on statistical models of previous successful treatment operation, plant efficiency will be improved, thus resulting in fewer plant upsets and more consistent water quality. Improved performance of the coagulation process can benefit other processes in the treatment scheme. Improvements can not only be seen in settled water quality (and often in filtered water quality), but also in areas such as: sludge production (reduced); sludge disposal cost (reduced); pH adjustment expense (reduced); and finally filter run length (increased).

## 4. COPING WITH SENSOR FAILURE SCENARIOS

Previous work by Adgar & Cox (1997) has suggested possible strategies to detect and identify sensors that have failed or are beginning to fail. These range from simple univariate statistical models through more complex multivariate statistical models and finally to advanced techniques such as the auto–associative neural network to be described here.

The auto–associative neural network (AANN), has been shown to be capable of sensor malfunction identification and also signal reconstruction (Dong & McAvoy, 1994; Guo & Musgrave, 1995; Hines *et al.*, 1996; Lin *et al.*, 1991). Alternative techniques such as observers based upon models of the process are notoriously difficult to implement since even after extensive research many of the unit operations in water treatment processes (especially clarification and filtration) are not well understood (Adgar & Cox, 1997).

The approach taken here is for the network to be trained from past data alone, since this information is extensively available. To ensure redundancy in the data, a comprehensive set of raw water variables were monitored including conductivity, pH, temperature, dissolved oxygen, turbidity, TOC, UV absorption and colour.

For $n$ redundant sensory measurements in $m$–dimensional space, where $n$ can be viewed as $n$–dependent equations and $m$ the inherent degree of freedom of the system, it is possible to reconstruct one or more lost sensor(s) if the relationships between the sensors is known. Usually, the $n$ redundant relationships can be described as mathematical equations with $p$ variables (where $p<n$). This inequality automatically introduces analytical redundancy which make it possible to compare (dissimilar) measurements to detect signal failure and permit reconstruction (Chow & Willsky, 1984). The reduction from $n$ variables to $p$ is called dimensionality reduction where $p$ represents the feature extraction.

In structural terms an AANN is a five–layer perceptron feed–forward network (Kramer, 1991) which can be viewed as two independent three–layer neural networks connected in series. The first network mixes and compresses the n redundant measurements into a smaller number of characteristic variables which should ideally represents the essential characteristics of the process. The second network works in the opposite way and uses the compressed information to regenerate the original n redundant measurements (as shown in Fig 3). The AANN consists of 8 inputs and outputs, 5 feature (bottleneck) hidden layer neurons, and 50 neurons for each of the other two hidden layers. The minimum number of the nodes in the feature layer, which represents the degree of freedom of the system, was determined using Principal Component Analysis

The goal of the AANN training is to generate good sensor estimates for all sensors even though some of the sensor measurements may have been corrupted. The network training also includes the following data pre–processing steps, which can be described as:

1. filter the sensor data for the entire year,
2. split the filtered sensor data into monthly terms to ensure reliable network performance,
3. randomise the sensor data for each month,
4. select one month to be trained,
5. select samples for training, test and final validation of the network,
6. colour each sensor variable of the training set with added noise terms. The noise terms are drawn from a random number generator based on Gaussian distribution with zero mean and standard deviation $1.5\varepsilon_i$, where $\varepsilon_i$ is the valid range for the i[th] sensor reading $S_i$ (Guo & Nurre, 1991).
7. Finally, 50% of the training samples of each sensor variable will be coloured to represent corrupted sensor values.

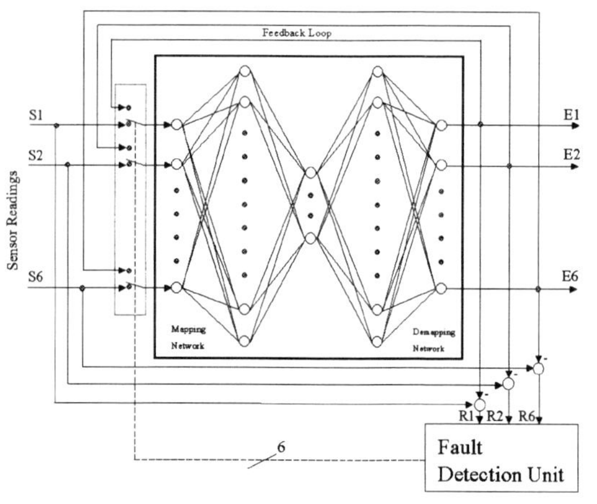


Fig 3: Auto–associative artificial neural network.

The detection of sensor failures is based on residual generation, produced by comparing the neural network outputs with their corresponding inputs. If the difference between a sensor measurement and its estimate exceeds a specified residual threshold (whilst the other residuals stay relatively low) then that sensor is declared as faulty. Once a faulty sensor measurement is detected, it will be disconnected from the input layer of network to isolate the false information. The external isolating procedure assumes that we can replace the identified faulty sensor input with a reasonably close value to the original one. An efficient technique for this is to feed the last 'good' output value back to the failed sensor input. In addition, if some of the sensor measurements are used in the control loop, e.g. pH, the controller will automatically switch to the estimated value for control purposes.

In the first example, the conductivity sensor is simulated to be degraded with a step of $-15\mu S/cm$ at T = 243.5 hours. Such a trend may be caused by the formation of a bubble between the plates in the instrument. The conductivity sensor is declared faulty by the AANN at sample 488, that is one sample (30 min) later. Fig 4 and Fig 5 show the reconstruction of the faulty conductivity sensor and the corresponding fault detection using the total residual of all sensor measurements. To identify which sensor is to be isolated and replaced with the feedback value, each individual residual must be checked.

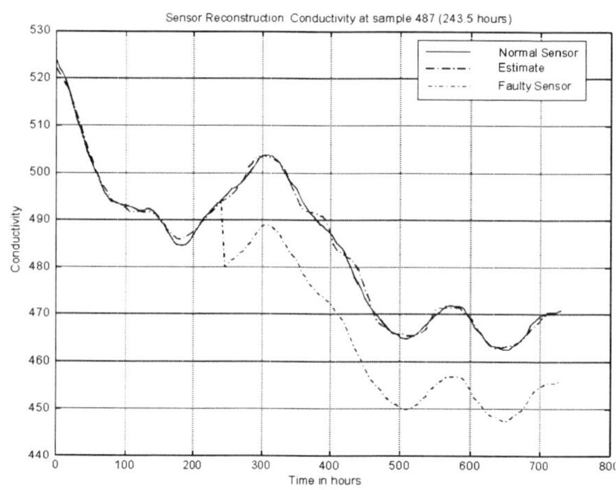


Fig 4: Reconstruction of conductivity signal.

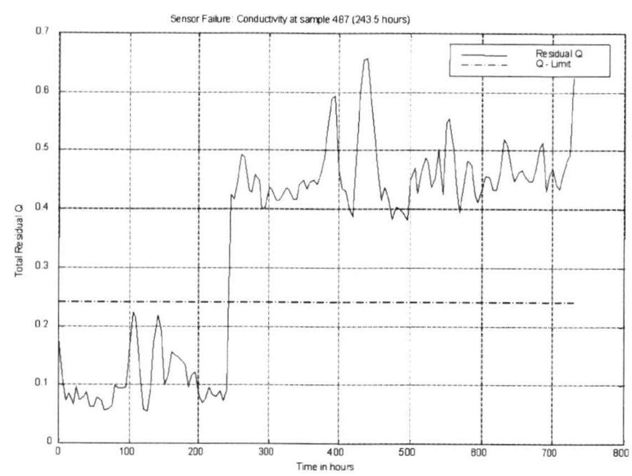


Fig 5: Total residual signal.

In the second example, the pH sensor is simulated to drift with a slow, rising ramp of 0.00004/hour. A fault like this may be observed when there is algal growth on the pH electrodes. This fault also occurs at T = 243.5 hours whilst the pH sensor is declared faulty a further 113.5 hours later (sample 714). Fig 6 and Fig 7 show the reconstruction of the faulty pH sensor and the corresponding fault detection.

## 5. CONCLUSIONS

The introduction of increasingly stringent regulations at water treatment works emphasises the need for a high degree of reliability in the operation of the individual unit operations. The response of the industry has been the establishment of a range of quality or conformance standards since their satisfaction offers potential benefits in terms of production consistency, reduced operational costs and improved safety. However, these improvements can only be realised if the process control and plant management policies are well structured and designed. This paper has identified two situations where an ANN has been used to help provide a solution that has contributed to an improvement in the overall plant performance.

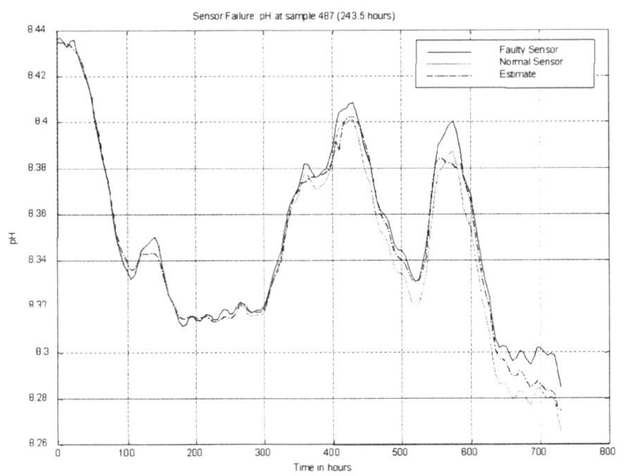


Fig 6: Reconstruction of pH signal.

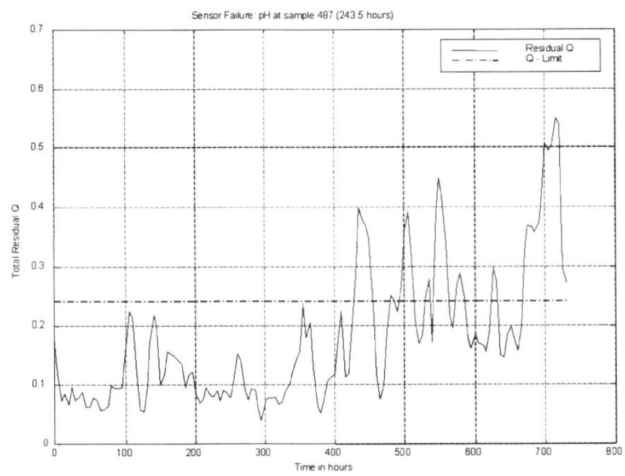


Fig 7: Total residual signal.

A new methodology that identifies sensor failure and aids signal reconstruction using an autoassociative neural network has been described. The results suggest that the validation technique is able to adequately identify sensor failure whilst also providing reasonable estimates of "corrupted measurements" for monitoring and control purposes. Some of these techniques have been proven on a small–scale pilot plant sited at one of the larger works. These new methods are presently being investigated at five different sites owned by Northumbrian Lyonnaise.

## 6. ACKNOWLEDGEMENTS


The authors would like to thank Northumbrian–Lyonnaise and the University of Sunderland for technical and financial support of this work.


## 7. REFERENCES


Adgar, A, Cox, CS (1997) Process Fault Detection and Reconstruction using Statistical and Advanced Techniques: Some Preliminary Findings. Int. Conf. Systems Engineering, Coventry, UK, 9–11th Sept. 1997, Vol. 12, No. 2, pp12–15.

Bevan, D. (1999) Monitoring, Control and reporting strategies for surface water treatment plants M.Phil. 1999, University of Sunderland.

Chow, EY, Willsky, AS. (1984) Analytical Redundancy and the Design of Robust Failure Detection System, IEEE Trans. on Automatic Control, AC–29, No. 7, pp. 603–614, 1984.

Cox, CS, Graham, J. (1994) Steps towards automatic clarification control. IEE Colloq. Advances in Control in the Process Industries: an Exercise in Technology Transfer, Computing and Control Division, London, UK, 30th Mar 1994, No. 081, pp 6/1–6/4.

Dong, D, McAvoy, T. (1994) Sensor data analysis using autoassociative neural nets. World Congress on Neural Networks, San Diego, CA, Vol. 1, pp. 161–166.

Ellis, GW *et al.* (1991) Chemical dosing of small water utilities using regression analysis. Journal of Environmental Engineering, Vol. 117, No. 3, pp 308–319.

Guo, TH, Musgrave, J. (1995) Neural network based sensor validation for reusable rocket engines., Proc. American Control Conference, Seattle, WA, June, 1995.

Guo, TH. & Nurre, J. (1991) Sensor Failure Detection and Recovery by Neural Networks, IEEE, pp. 221–226.

Hines, JW *et al.* (1996) Plant wide sensor calibration monitoring. Proc. IEEE Int. Symp. Intelligent Control, Dearborn, MI, September, 1996.

Kramer, MA. (1991) Non–linear Principal Component Analysis using Autoassociative Neural Networks, AIChE Journal, Vol. 37, No. 2, 1991.

Lin, CS *et al.* (1991) Neural networks for sensor failure detection and data recovery, Proc. Artificial Neural Networks in Engineering Conference, St. Louis, MO, November, 1991.

Nahm, ES *et al.* (1996) Development of an optimum control software package for coagulant dosing process in water purification system. Proc. SICE Annual Conference, Tottori, Jpn, Jul 24–26 1996, pp 1157–1161.

IFAC Publications
www.elsevier.com/locate/ifac

# DISTRIBUTED WIRELESS CONTROL USING BLUETOOTH

**Johan Eker**

*Department of Electrical Engineering and Computer Science*
*University of California at Berkeley*
*337 Cory Hall, Berkeley, CA 94720-1770, US*
*johane@eecs.berkeley.edu*

**Anton Cervin**

*Department of Automatic Control*
*Lund Institute of Technology*
*Box 118, SE 221 00 Lund, Sweden*
*anton@control.lth.se*

**Andreas Hörjel**

*ConnectBlue*
*Stora Varvsgatan 11 N:1*
*SE 211 19 Malmö, Sweden*
*andreas.horjel@connectblue.se*

*Abstract:* Bluetooth is currently emerging as one of the most promising personal wireless network technologies. The automation industry is also showing interest in using Bluetooth for more industrial applications. However, problems with unreliable links and time delays must be explicitly addressed in any wireless control application. This paper discusses two specific problems that may occur when using Bluetooth in a control loop: long random delays and bit errors. Simple solutions to these problems are presented. The ideas have been tried out, both in simulations and in experiments, on a rotating inverted pendulum controlled over a Bluetooth network. *Copyright © 2001 IFAC*

*Keywords:* Bluetooth, distributed control, delay compensation, bit-error filter

## 1. INTRODUCTION

In a distributed control system the sensors, the actuators and the controller are physically displaced and communicate over a network. The main advantage of this setup is flexibility. It allows us to use the same control computer for several control loops and it relieves us from the potential problem of placing the controller close to the controlled process. It also allows us to build our control systems in a hierarchical fashion. The cost for using networks has decreased radically over the last years which has made distributed solutions much more appealing. However, a downside to using a distributed approach is that delays will be introduced in the control loops due to latencies in the network communication. The delays will lead to decreased phase margins in the control loops and potentially unstable systems.

In this paper a wireless Bluetooth radio network is used. Going wireless has some obvious advantages; without cables we have a much greater freedom to physically distribute the nodes. It is possible to place actuators and sensors without worrying about the location of the control node. For example, it allows us the place the sensor and actuator on a mobile object and still run the controller from a stand-still platform. However, going wireless will also result in less dependable systems. The likelihood of a wireless link containing bit errors is several orders of magnitude larger than for a wired link. This means that the possibility of a message being corrupt or delayed due to retransmissions is much larger.

Two particular problems are studied in this paper. In the first case, we assume that we have a reliable connection, i.e. the underlying network protocol guarantees that there will be no corrupted messages. The cost for this is that messages may be retransmitted several times,

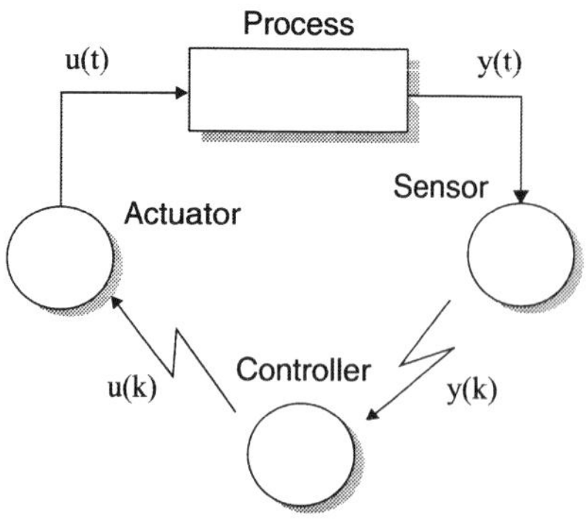


**Fig. 1** The distributed system configuration used in the paper. The communication between the sensor, the actuator, and the controller is over a wireless link.

causing lengthy delays. The implications for the controller design is that we must explicitly deal with the varying time delays. In the second case, we assume an unreliable connection. We will receive possibly scrambled messages, but we will have a short and nearly constant delay. The problem in this case is to detect and handle the bit errors. We use a model-based observer approach for error detection and correction.

The two problem cases have been simulated in MATLAB/SIMULINK and then implemented and run over a wireless Bluetooth link controlling a rotating inverted pendulum. To control the experiments, random delays and bit errors were artificially introduced by fault injection in the Bluetooth protocol stack. More details about the Bluetooth-pendulum implementation are available in (Hörjel, 2001).

## 2. THE SETUP

The distributed control configuration is shown in Figure 1. The feedback loop consists of three parts: the sensor, the controller, and the actuator. The sensor node is time-driven while the controller and actuator nodes are event-driven. The sensor samples the process periodically and sends the measurement values to the controller. Upon receipt, the controller calculates a new control signal and sends it to actuator node which outputs the value. In our setup, the sensor node and the actuator node are located in the same hardware unit, called the Remote I/O.

### 2.1 *The Network*

Bluetooth is currently emerging as one of the most promising personal wireless network technologies. It is primarily designed as a cable replacement between for example a cell phone and a PDA, but there is a growing interest to also use it in industrial applications. Up to eight Bluetooth units may be connected in a so called Piconet. The communication protocol is time divided, with a slot-length of 625 $\mu$s. One unit acts as the master, and the slaves are only allowed to transmit if they were addressed by the master in the previous slot. The radio uses the Industrial-Scientific-Medical (ISM) frequency band, which ranges in Europe and the USA from 2.400 to 2.4835 GHz. The maximal data rate is 1 Mb/s, but the useful data rate is much lower. Interference is avoided by using a frequency-hop, spread spectrum technology, with 1600 hops per second. This gives relatively short packet length and good interference protection. For more information on Bluetooth see (Bluetooth Consortium, 2001).

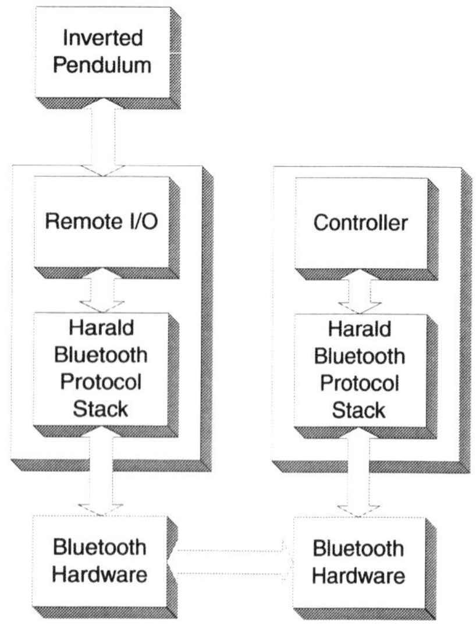


**Fig. 2** The implementation structure.

### 2.2 *Bluetooth in the Loop*

The implementation structure is shown in Figure 2. In our setup the controller is the master and the Remote I/O is a slave. The theoretical minimum round-trip delay in Bluetooth is 2·625 $\mu$s. However, the length of the data packets, the Bluetooth protocol stack, and the communication between the Bluetooth hardware and the computer hardware bring the minimum round-trip delay up to 17 ms in our implementation. Still, this is fast enough to control the inverted pendulum, which is a process with fast and unstable dynamics.

## 3. THE PENDULUM CONTROLLER

The rotating inverted pendulum used in the experiments is shown in Figure 3. The objective is to control the pendulum angle $\theta$ and the arm angle $\phi$ to zero by applying the torque $u$ to the rotating arm. The non-linear equations of

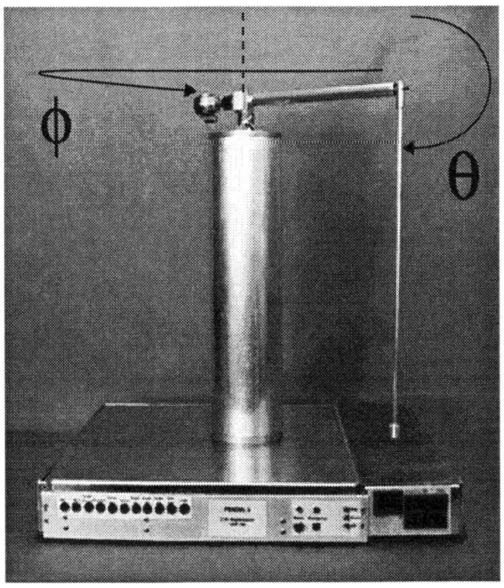


**Fig. 3** The rotating inverted pendulum used in the experiments. The objective is to stabilize the pendulum in the upright position $\theta = 0$.

motion can be written

$$\alpha\ddot{\theta} - \alpha\dot{\phi}^2 \sin\theta\cos\theta + \gamma\ddot{\phi}\cos\theta - \varepsilon\sin\theta = 0$$

$$\gamma\ddot{\theta}\cos\theta - \gamma\dot{\theta}^2\sin\theta + 2\alpha\dot{\theta}\dot{\phi}\sin\theta\cos\theta + (\beta + \alpha\sin^2\theta)\ddot{\phi} = u$$

Introducing the state vector

$$x = \begin{pmatrix} \theta & \dot{\theta} & \phi & \dot{\phi} \end{pmatrix}$$

and linearizing around the upright equilibrium a state-space description is given by

$$\frac{dx}{dt} = \begin{pmatrix} 0 & 1 & 0 & 0 \\ a_{21} & 0 & 0 & 0 \\ 0 & 0 & 0 & 1 \\ a_{41} & 0 & 0 & 0 \end{pmatrix} x + \begin{pmatrix} 0 \\ b_2 \\ 0 \\ b_4 \end{pmatrix} u$$

The full state vector is directly measurable on the process. The sampling interval for digital control was chosen as $h = 60$ ms. Discrete state-feedback controllers were designed based on an linear-quadratic (LQ) formulation where the control should minimize the cost function

$$J = \lim_{T\to\infty} \frac{1}{T} \mathrm{E}\left\{ \int_0^T \left( x^T Q_1 x + u^T Q_2 u \right) dt \right\}$$

The cost $J$ can also be measured and used to compare the performance of different control schemes for the pendulum.

## 4. COPING WITH VARYING DELAYS

Delay in the control loop always has an adverse effect on control performance. For example, a constant delay of one sample decreases the phase margin of the system by 11 to 34 degrees, assuming that the sampling interval has been chosen according to the rule of thumb $0.2 \leq \omega h \leq 0.6$ (Åström and Wittenmark, 1997). Here, $h$ is the sampling interval and $\omega$ is the cross-over frequency of the closed-loop system.

The case of constant delay is easy to cope with in the control design. The sampled-data description of the plant including the control delay will simply be of higher order, and design can be made using standard techniques. The control law has the form

$$u(k) = -L \begin{pmatrix} x(k) \\ u(k-1) \end{pmatrix}.$$

where $L$ is a constant feedback gain vector, $x(k)$ is the state vector and $u(k-1)$ is the old control signal.

The case of randomly varying delays is more difficult. A simple but non-optimal solution is to do static delay compensation for the mean delay.

The optimal LQ-controller for random sensor-to-controller and controller-to-actuator delays was derived in (Nilsson *et al.*, 1998). The controller performs dynamic delay compensation by adjusting the feedback gain according to the actual delays. The optimal control law has the form

$$u(k) = -L(\tau_k^{sc}) \begin{pmatrix} x(k) \\ u(k-1) \end{pmatrix} \qquad (1)$$

where the feedback gain vector $L$ is now a function of the sensor-to-controller delay $\tau_k^{sc}$ in the current sample.

In Figure 4, the cost $J$ as a function of the amount of stochastic delay has been computed for different control schemes. When the delay becomes very long, the dynamically compensating controller is still stable ($J < \infty$), while the other control schemes become unstable.

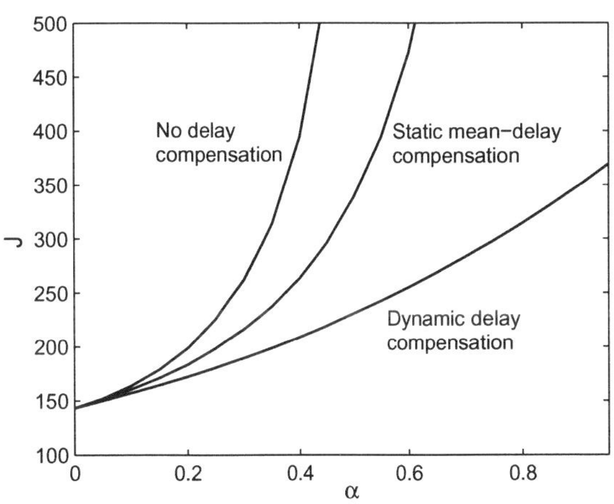


**Fig. 4** Values of the cost function $J$ for the pendulum controller under different schemes. In this plot, the control delay is uniformly distributed on $[0, \alpha h]$.

### 4.1 *Dynamic Delay Compensation with Intelligent I/O*

Dynamic delay compensation works better the shorter the controller-to-actuator delay is compared to the sensor-to-controller delay. If the controller is located *at* the actuator, the delay in the current sample can be known exactly and the correct gain can be applied. Modifying the setup is of course not a real solution, but using an I/O with only very limited computational power, this ideal behavior can be emulated closely. The idea is illustrated on the pendulum controller.

The optimal feedback gain depends on the distribution of the round-trip delay. For the case of heavy disturbances and frequent retransmissions, we assume the exponentially decaying delay distribution shown in Figure 5.

The optimal gain vector is computed using the formulas in (Nilsson *et al.*, 1998), and the result in shown in Figure 6. We notice that the optimal gain can be closely approximated by a linear function of the delay $\tau$. Approximating Eq. 1, we write

$$u(k) \approx -L_0 \begin{pmatrix} x(k) \\ u(k-1) \end{pmatrix} - \frac{dL}{d\tau} \begin{pmatrix} x(k) \\ u(k-1) \end{pmatrix} \tau$$

In the controller node, $x(k)$ and $u(k-1)$ are known, but $\tau$ is still unknown. However, the controller can precompute

$$\hat{u}(k) = -L_0 \begin{pmatrix} x(k) \\ u(k-1) \end{pmatrix}$$

and

$$\lambda(k) = -\frac{dL}{d\tau} \begin{pmatrix} x(k) \\ u(k-1) \end{pmatrix}$$

These two scalars are then sent to the Remote I/O. If the I/O keeps track of the round-trip delay $\tau$, it can do the simple adjustment

$$u(k) = \hat{u}(k) + \lambda(k)\tau$$

before applying the control signal to the process.

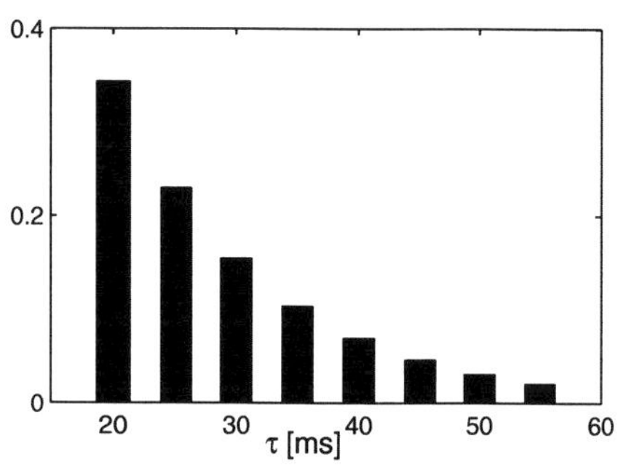


**Fig. 5** The assumed delay distribution.

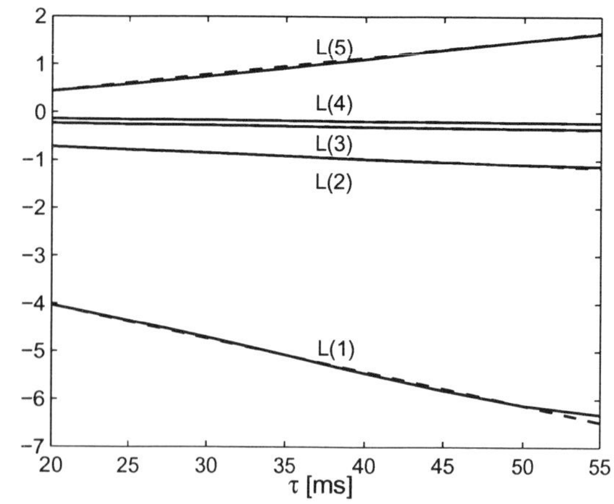


**Fig. 6** Optimal gain vector (full lines) and linear approximation (dashed lines).

### 4.2 *Simulation*

The suggested delay compensation scheme was simulated against the nonlinear pendulum model. The pendulum was disturbed by white process noise and the round-trip delay in the control loop varied between 20 and 55 ms according to the distribution shown in Figure 5. The simulation result is shown in Figure 7. This was compared to a controller which was designed for a constant delay of 20 ms (the nominal round-trip delay). The result of this simulation is shown in Figure 8. For this particular control problem, the dynamic delay compensation scheme was able to reduce the cost $J$ by about 30 %.

### 4.3 *Experiment*

The dynamic delay compensation was also tried on the distributed Bluetooth-pendulum setup. The random delays were not due to actual

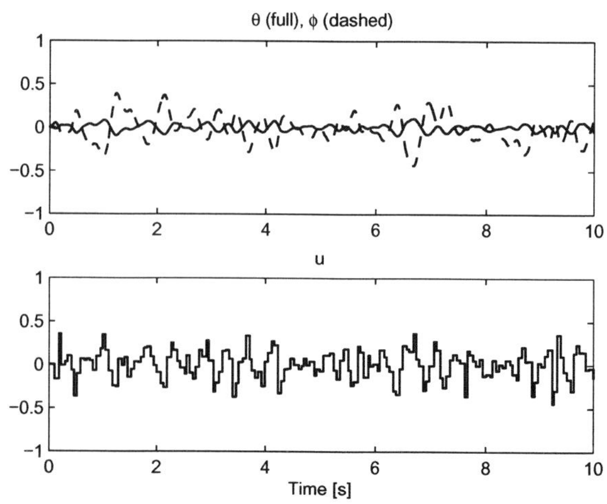


**Fig. 7** Simulation of the pendulum controller with random communication delays and dynamic delay compensation.

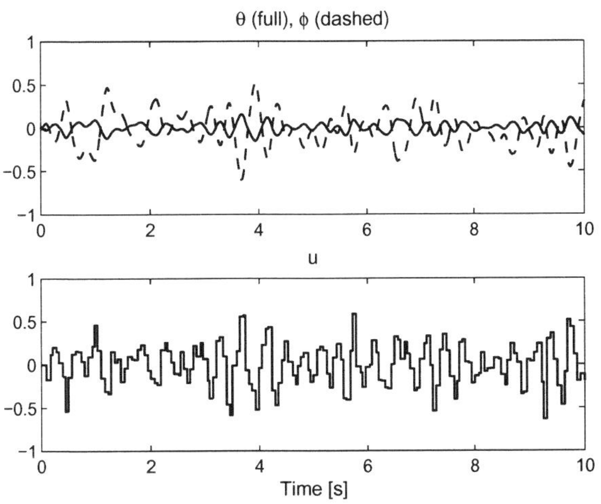


**Fig. 8** Simulation of the pendulum controller with random communication delays and static delay compensation.

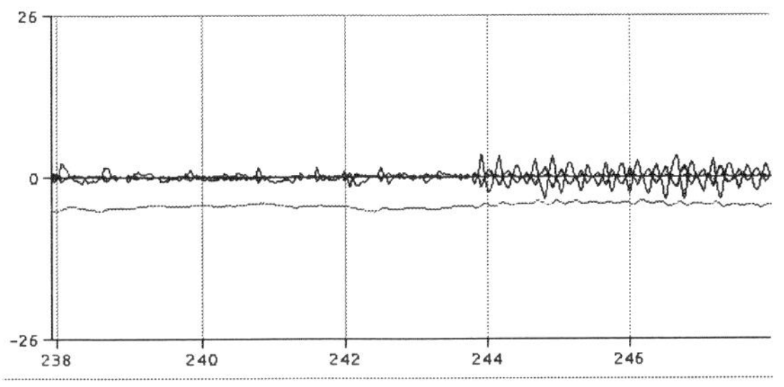

**Fig. 9** Experiment on the Bluetooth-pendulum setup with random communication delays. At time 244, the dynamic delay compensation is turned off, and the pendulum starts to oscillate. A large part of the increase in the cost is due to unmodeled friction.

disturbances but were injected in the control loop on purpose. The result of an experiment is shown in Figure 9. Due to unmodeled friction, the pendulum easily started to oscillate when the delay compensation was turned off.

## 5. COPING WITH FAULTY MESSAGES

Another problem that might occur is bit errors in the Bluetooth packets due to disturbances. These errors are currently handled by the Bluetooth baseband layer. When an error is encountered in the baseband layer, the packet is wasted and a retransmission is requested. This retransmission may need several tries before the packet can be delivered error free. The retransmission scheme can result in large delays in the loop if the disturbance is heavy. The delays will lead to decreased control performance, as explained in the previous section. An alternative solution is to reconstruct the information from the faulty packages, and hence avoid the long random delays.

Our solution to the bit-error problem is to run an observer (a pendulum model) in the

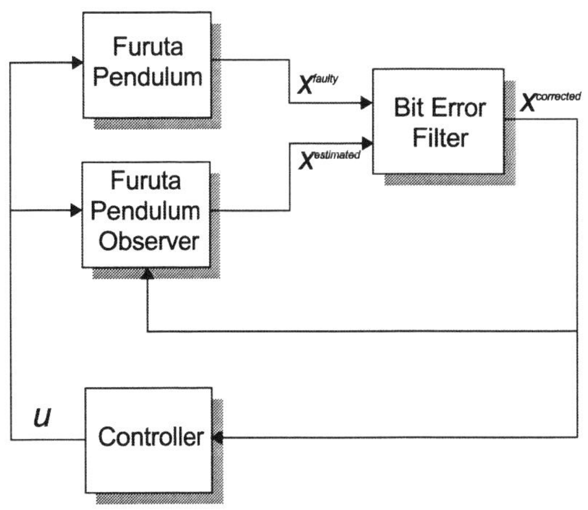


**Fig. 10** The block diagram for the bit-error filter. The bit-error corrector uses estimates from the observer to detect and correct bit errors from the process.

controller node, in parallel with the real process. When samples are received from the sensor node, the potentially disturbed measurement value is compared bitwise to the output from the pendulum model. If there is an error in one of the most significant bits, there will be a sudden jump in the measurement compared to the observer output. The idea is illustrated in Figure 10.

### 5.1 *Filter design*

The bit error filter is designed as a standard state-observer that runs concurrent with the process and checks the validity of the Bluetooth packets when they arrive at the controller node. The packets contain the 4 sampled states of the system, represented as 4 times 12 bits. The bit error corrector locates a faulty bit by first computing the difference $\Delta$ between the sampled state and the estimated state. The difference indicates if and which bit that is incorrect. Assuming that there is a single bit error, the bit position is related to a bit value $2^{n_{bit}}$ as

$$\Delta = |x^{faulty} - x^{estim}| \approx 2^{n_{bit}} \tag{2}$$

The potential faulty bit is thus computed as

$$n_{bit} = \text{round}\left(\frac{\log(\Delta)}{\log(2)}\right) \tag{3}$$

The bit is then flipped if $n_{bit}$ is above some threshold value. Tuning of the threshold values is critical to the performance of the filter. There will always be a difference between the sampled state and the observer state, so we should only attempt to correct the sudden large errors which indicate faulty bits and which are not just modeling uncertainty and measurement noise. In our experiments, the threshold values had to be set at different levels for the different process

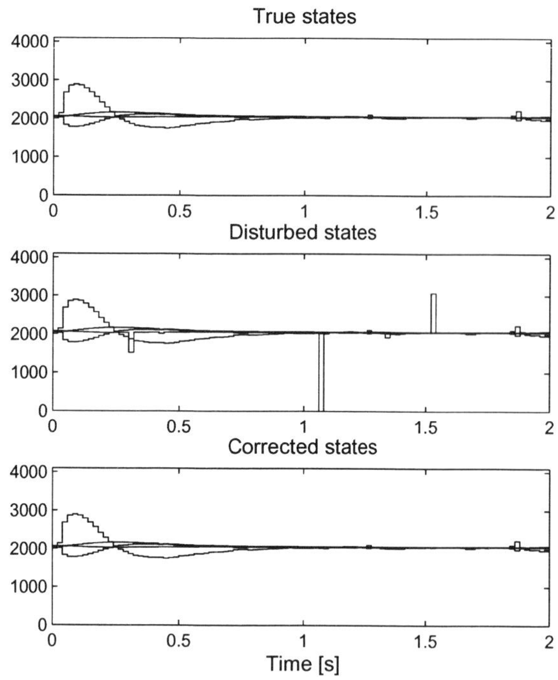


**Fig. 11** Simulation of the pendulum controller with random bit errors in the $\theta$ measurements. The large bit errors are detected and corrected, and the estimated states are very close to the true states.

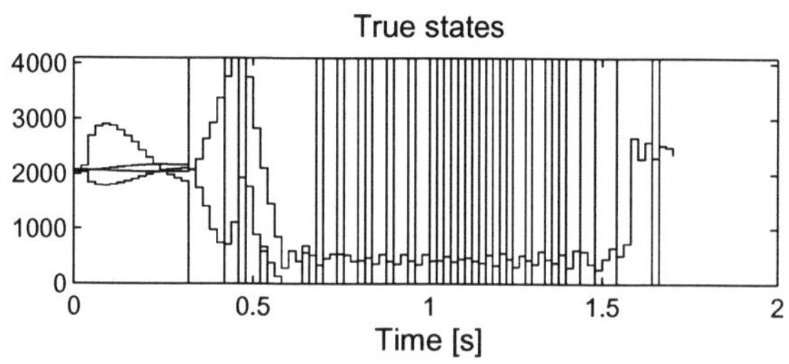


**Fig. 12** Simulation of the pendulum controller with random bit errors but without the bit-error correction. The performance deteriorates completely.

states, some measurements being noisier than the others.

### 5.2 *Simulation*

The bit-error filter was first tried out in simulations. In one particular simulation, a single bit error was injected into 5 % of the packets containing the $\theta$ measurement. The filter was then tuned to remove errors in the 8 most significant of the 12 bits. The simulation of the bit-error correction is shown in Figure 11. Without the filter, the control deteriorated completely, as shown in the simulation in Figure 12.

### 5.3 *Experiment*

The bit-error corrector was also tried on the Bluetooth-pendulum setup. Now, the round-trip delay in the loop was constant and equal to 20 ms. A bit error was injected in 5 % of the $\theta$ measurement values. The result of an experiment with and without error correction is showed in Figure 13. When the bit-error filter was turned off, the pendulum soon fell down.

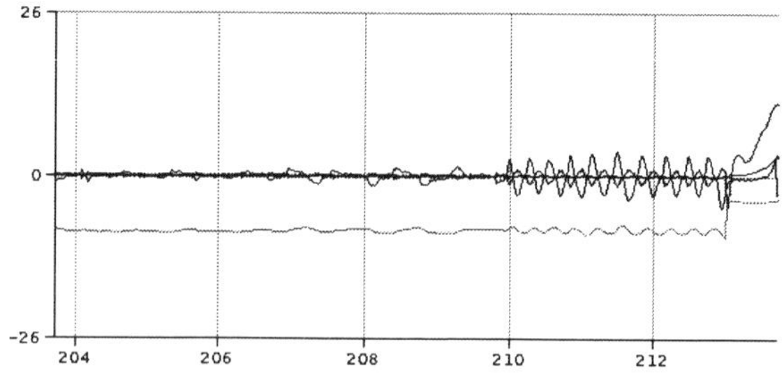


**Fig. 13** Experiment on the Bluetooth-pendulum setup with random bit errors in the $\theta$ measurements. At time 210, the bit-error correction is turned off, and the pendulum soon falls down due to the large erroneous control signals.

## 6. CONCLUSION

A distributed wireless control system has been described, where Bluetooth is used to communicate between the sensor node, the controller node, and the actuator node. Two specific problems when using Bluetooth have been identified and simple solutions have been presented. An intelligent I/O can be used to compensate for random delays that might occur due to retransmissions. An observer-based bit-error filter might be used to deal with faulty packages. The ideas have been tried out, both in simulations and in real experiments, on a rotating inverted pendulum.

## 7. REFERENCES

Åström, K. J. and B. Wittenmark (1997): *Computer-Controlled Systems*, third edition. Prentice Hall.

Bluetooth Consortium (2001): "The Bluetooth specification 1.1." http://www.bluetooth.com.

Hörjel, A. (2001): "Bluetooth in control." Master Thesis TFRT-5659 ISRN LUTFD2/TFRT--5659--SE. Department of Automatic Control, Lund Institute of Technology, Lund, Sweden.

Nilsson, J., B. Bernhardsson, and B. Wittenmark (1998): "Stochastic analysis and control of real-time systems with random time delays." *Automatica*, **34:1**, pp. 57–64.

IFAC Publications
www.elsevier.com/locate/ifac

# UPLINK POWER CONTROL IN BLUETOOTH TRANSCEIVER

**Soo Young Shin, Hong Seong Park†, Wook Hyun Kwon**

*ERC-ACI, School of Electrical Engineering & Computer Science,*
*Seoul National University, Seoul, Korea*
*† Department of Control & Instrumentation Engineering, Kangwon National University,*
*Chuncheon, Kangwon-Do, Korea*
*Email:wdragon@cisl.snu.ac.kr*

**Abstract:** In this paper, a new method that controls the uplink transmit power in Bluetooth transceiver is suggested. Two closed loop power control (CLPC) algorithms are proposed. One is the fixed step power control (FSPC). The other is the dynamic step power control (DSPC). The proposed power control algorithms are simulated under indoor radio propagation model. Simulation results show that power consumption of Bluetooth devices is reduced using the proposed power control algorithms. *Copyright © 2001 IFAC*

**Keywords:** Bluetooth; uplink; closed loop power control (CLPC); fixed step power control (FSPC); dynamic step power control (DSPC); indoor propagation; power consumption

## 1 INTRODUCTION

Recently, a new universal radio interface, Bluetooth, has been developed enabling electronic devices to communicate wirelessly via short-range ad-hoc radio connections. The Bluetooth technology eliminates the need of wires, cables and the corresponding connectors between cordless or mobile phones, head-sets, personal digital assistants (PDAs), computers, notebooks, printers, projectors, etc. Bluetooth provides wireless interconnections among the electronic devices and peripherals used by individual. This network concept is called as personal area network, PAN [1, 2, 3].

For applications of PAN, the communication range of commercial Bluetooth is set to 10 $m$ sufficient for the cable replacement [4]. Many advantages such as wirelessness, mobility, and inter-operability expand the application range of Bluetooth. For many applications in various fields, 10 $m$ is not always a sufficient range. To meet the requirement of longer communication range, the maximum range can be extended to 100 $m$.

However, to extend communication range, transmitter power should be increased. Increasing transmitter power causes more power consumption. Because Bluetooth devices are almost portable, battery-driven equipment, low power consumption must be considered because of limited battery capacity. Especially, for the devices that use battery power for other purposes such as cellular phones and PDAs, low power consumption is critical interest. The transmit power level control is one of the solutions for low power implementation [4, 5, 6].

Using power control in the Bluetooth transceiver,

power consumption can be reduced while the transmitter range of Bluetooth is extended. In spite of the importance of power control, it has not been studied yet. In the current Bluetooth specification, the power control is optional and not defined yet [7].

In this paper, the power control algorithms in Bluetooth transceiver are proposed. The control is focused on the uplink which means transmission link from mobile station to base station. The proposed power control algorithms are simulated under the indoor radio propagation model. By the simulation, the proposed algorithms show the effective reduction of power consumption in Bluetooth devices.

This paper is organized as follows. In Section 2, the Bluetooth radio interface is introduced. Section 3 shows some characteristics related to power control in Bluetooth. In Section 4, power control algorithms are proposed and the simulation model is also described. Simulation results are shown in Section 5. Finally, the conclusions are presented in Section 6.

## 2 Bluetooth Radio Interface Overview

The user scenarios envisioned for first generation Bluetooth products are typically focused on traveling business people; portable devices that contain Bluetooth radio transceivers would enable to eliminate cables and connectors at home. These user scenarios share the following requirements:

- Global operation
- Peer connectivity : i.e. there is no wired infrastructure to provide call setup and networking functions ; connections are made based on an ad-hoc fashion
- Simultaneous voice and data, for example, for multimedia applications
- The radio transceiver must be small and operate at low power - i.e. the radio must fit into small, portable devices, such as mobile phones, headsets and personal digital assistants (PDA)

To fulfill the requirements, Bluetooth operates at 2.4 GHz industrial scientific medical (ISM) band because this band is globally available and requires no license. The radio modem employs frequency hopping with 1600 hops/s to reduce the impact of jamming from other systems such as wireless LAN, micro oven and to comply with emission rule. The band is divided into 79 channels, each 1 MHz wide.

The Bluetooth radio channel is assumed to be a burst-error channel without time dispersion [6]. Such a burst-error channel is rather good in between error bursts and very bad during bursts. Thus, a combination of fast automatic retransmission query (ARQ) scheme and simple coding is employed, rather than extensive coding, to increase the link robustness. Furthermore, the frequency hopping used at radio interface also improves the jamming resistance.

Table 1: Power Classes of Bluetooth Transmitter

| Device Power Class | Maximum Output Power | Nominal Output Power | Minimum Output Power | Expected Range |
|---|---|---|---|---|
| Class1 | 100mW/ 20dBm | N/A | 1mW/ 0dBm | 100m |
| Class2 | 2.5mW/ 4dBm | 1mW/ 0dBm | 0.25mW/ -6dBm | 10m |
| Class3 | 1mW/ 0dBm | N/A | N/A | 10cm |

Bluetooth communication protocol assumes that one unit is acting as master, the call-initiating, and all other participating units are slaves. A master together with slaves (up to seven slaves are supported) form an ad-hoc piconet. In the point-to-point case, a master and a slave sends and receives alternately to provide full duplex. In the point-to-multipoint situation, a master sends every other slot and all slaves share the same time slot. Then, a master arbitrates between slaves by means of polling and is, thus, responsible for distributing the channel capacity among the slaves. Connections can be opened and closed dynamically. Connections are performed by a direct call from master to slave. Units may also respond to inquiry calls such that one unit can search for other Bluetooth units within the communication range.

## 3 Power Control Related Characteristics in Bluetooth

In this Section, some characteristics related to the power control are investigated. The power control in Bluetooth specification is also studied.

### 3.1 Transmitter Characteristics

The modulation scheme chosen for Bluetooth is Gaussian frequency shift keying (FSK) with a BT product of 0.5 and a modulation index of 0.32. Three power classes are available for Bluetooth. They have the following characteristics as shown in Table 1.

Power control is required for power class 1 equipment to limit the transmitted power over 0 $dBm$. Power control capability under 0 $dBm$ is optional and could be used for optimizing the power consumption and overall interference level [7].

### 3.2 Receiver Characteristics

The reference sensitivity level equals to -70 $dBm$ [7]. Actual sensitivity level is defined as the input level for which a raw bit error rate (BER) of 0.1 % is met. The requirement for Bluetooth receiver is an actual sensitivity level of -70 $dBm$ or better. The receiver must achieve the -70 $dBm$ sensitivity level with any class Bluetooth transmitter compliant to the transmitter specification. The receiver sensitivity of the commercial 20 $dBm$ transceiver is over -80 $dBm$ [8, 9].

### 3.3 Power Control in Bluetooth

A transceivers that wishes to take part in a power controlled link must be able to measure its own receiver signal strength and determine if the transmitter on the other side of the link should increase or decrease its output power level. A receiver signal strength indicator (RSSI) makes this possible. To implement a link with power control the transmitting module must support transmit power control and the receiving module must have a RSSI and a defined golden receiver range.

These are optional components implemented at the radio level. The capability of each device is determined during link establishment. Changes in the local device's transmit power, up and down a preset power table to higher and lower transmit powers, are controlled by the remote device. The remote device has automatic gain control (AGC) upper and lower limits stored which define golden receiver power band. The change is triggered by the remote link control (LC) layer noting that the RSSI has moved outside golden receiver power band. Then LC sends a change request to link management (LM) layer. The link management protocol (LMP) of the LM layer sends the appropriate change request message with 1 byte such as LMP_incr_power_req or LMP_decr_power_req to the baseband, which transmits it to the local device. The local LM layer changes the transmit power level up or down.

The power steps are defined as a monotonic sequence, with a maximum step size of 8 $dB$ and a minimum step size of 2 $dB$. If it is already at an upper or lower limit, a response is sent to the remote device indicating this using LMP_max_power or LMP_min_power messages. For the simulation, the power consumption of the power control procedure is assumed to be same that of the normal communication operation. The host controller interface (HCI) RSSI commands, HCI_Read_RSSI, are used for information only. The host is not involved in altering the transmit power of Bluetooth device.

## 4 Simulation Models and Power Control Algorithms

Before describing the simulation model, some assumptions used are set forth.

**Assumption 1** *There is one piconet consisting of one master and one slave.*

This assumption means that channel interference by other piconet is not considered.

**Assumption 2** *The master always sends maximum transmitter power.*

Generally, master of the piconet is powered by fixed power supply such as LAN access point (LAP), PCMCIA Bluetooth PC card. Hence, this assumption is reasonable. Then, power control is focused on the uplink, which means the link that slave sends signal to master.

**Assumption 3** *During transmission, the slave sends data with equal transmitter power.*

Actually, Bluetooth unit doesn't send continuously during Tx slot. However, rest of the Tx slot, it should perform Tx/Rx mode switching to receive data and consume power. Hence, transmitter power is assumed to be equal during one power control period.

**Assumption 4** *Both the master and the slave have the same data to be transmitted.*

### 4.1 Simulation Models

For the power control, transmitted, received power and the relation between them are required. Because Bluetooth radio communication aims indoor environment, the indoor radio propagation is considered in this paper. There were some researches for the indoor propagation model of Bluetooth transceiver [10, 11, 12, 13]. In this paper, the following indoor propagation model described in Equation (1) is adopted.

$$g(d) = \begin{cases} 20log_{10}(\frac{4\pi d}{\lambda}) + \mathbf{N}(0,\sigma) \\ ,d < d_0 \\ 20log_{10}(\frac{4\pi d_0}{\lambda}) + 10nlog_{10}(\frac{d}{d_0}) + \mathbf{N}(0,\sigma) \\ ,d \geq d_0 \end{cases} \tag{1}$$

The reference distance,$d_0$, means the line-of-sight (LOS). The first term of Equation (1) represents path loss. The second term, $\mathbf{N}(0,\sigma)$, represents shadowing factor that is a log-normally distributed random variable.

The simple transceiver model between the master and the slave is shown like Figure 1.

In Figure 1, the power control model in Bluetooth is discrete time system with sampling time $T_p$. In this model, closed loop power control (CLPC) is adopted. In CLPC, the base station (BS), that is the master of

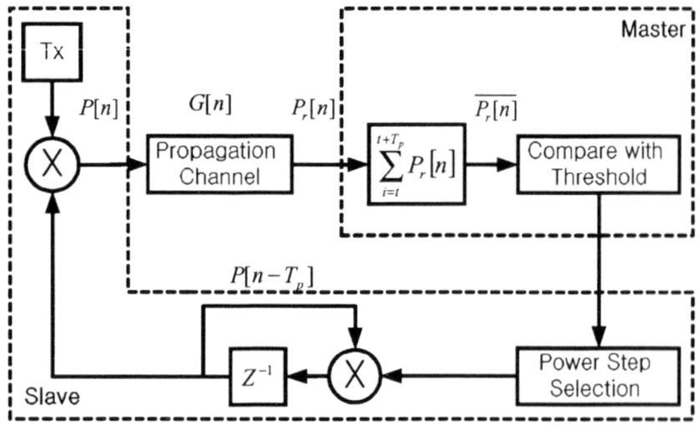


Figure 1: Uplink Closed Loop Power Control Model in Bluetooth

piconet, measures the received signal power over every period $T_p$, and compares it with the target threshold. It then transmits a power control command to the slave for transmitter power increase or decrease. $P[n]$ is the transmitter power of mobile slave and $\overline{P_r[n]}$ is the mean received power over one power control period $T_p$. The propagation channel, $G[n]$ is the indoor propagation model as described in the Equation (1). $R_0$ is the reference value compared to the received signal. Since the received power during power control period $T_p$, the time average of received signal can be obtained directly. Therefore, the system equations can be expressed as Equation (2).

$$P_r[n] = \overline{P_r[n]} = \frac{1}{T_p} \sum_{i=n-T_p}^{n} P_r[i] \tag{2}$$
$$P[n] = P[n-T_p] + \Delta[n-T_p]$$
$$\Delta[n] = R_0 - \overline{P_r[n]}$$

### 4.2 Power Control Schemes

As mentioned in previous Section, the power control in Bluetooth is optional. Hence, algorithms and parameters are not defined yet. In this Section, associated parameters and algorithms are proposed and considered.

For the power control, two important parameters should be defined. One is the power control period, $T_p$ and the other is the power control step size, $\Delta$. The power control period lies on trade-off situation between the channel utilization and the convergence time. If the period is sufficiently short, the transmitter power converges to the target level effectively, but channel utilization can be lowered and network resources may be wasted. If the period is longer, the convergence time will take more time. That makes power consumption be increased and bit error rate may be also increased. The choice of the power control period is one of the issues in the power control. The power control step size $\Delta$ is closely related to the power control algorithm. The step size can be fixed or dynamically changed according to the power control algorithms.

In Bluetooth specification, the maximum and minimum value of control step sizes are defined, 8 $dB$ and 2 $dB$, respectively. Hence, two power control algorithms are proposed. One is the fixed step size power control with step size of 2 $dB$. The other is the dynamic step size power control where the step size is determined by the difference between the actual received power strength and reference power strength. The dynamic step range obeys the maximum and minimum step size of the Bluetooth specification. The dynamic step size is determined by following Equation (3).

$$\Delta[n] = \begin{cases} 0 & , \quad 0 \le |R_0 - \overline{P_r[n]}| \le 1 \\ 2 & , \quad 1 < |R_0 - \overline{P_r[n]}| \le 3 \\ 4 & , \quad 3 < |R_0 - \overline{P_r[n]}| \le 5 \\ 6 & , \quad 5 < |R_0 - \overline{P_r[n]}| \le 7 \\ 8 & , \quad 7 < |R_0 - \overline{P_r[n]}| \end{cases} \tag{3}$$

## 5 Simulation Results

In this Section, the simulation results of power control in Bluetooth are presented. Simulation scenario is as follows. The slave moves from the master at the speed of 2.0 $m/s$. When the distance between master and slave is 100 $m$, the slave waits for 5 sec. Then, it comes back to the position of the master. The simulation parameters are as follows:

- Line-of-sight(LOS) : $d_0 = 5$m
- Wavelength : $\lambda = 0.1224Hz$
- Exponent : $n = 3.3$
- Standard deviation : $\sigma = 6$dB
- Power control period : $T_p = 1.25sec$
- Sensitivity : -84 $dBm$
- Reference power level : -75 $dBm$

Simulation result of the fixed step power control is shown in Figure 2 and 3. Figure 2 shows transmitted power of the slave, $P_{tx}$, and received of the master, $P_{rx}$. The transmitted power of slave tracks the reference power level, which is higher than the sensitivity. Initial transmitter power is assumed to be 1 $mW$.

Figure 3 shows the power command and instantaneous power in Bluetooth transceiver using the fixed step control. The fixed step power control is the bang-bang control. The 0 $dB$ power control command region means that the received power is at golden receiver range.

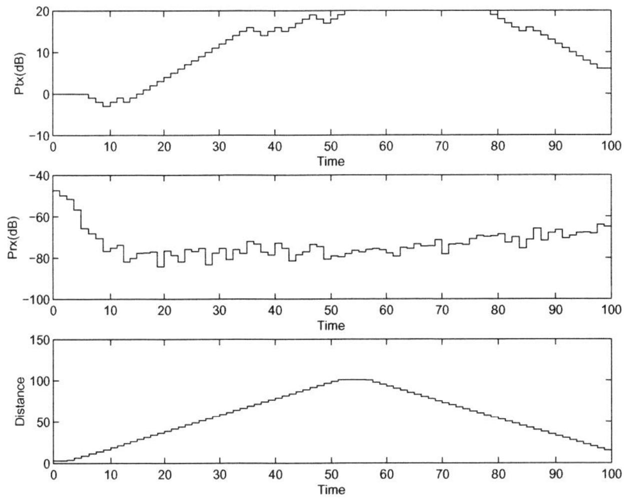


Figure 2: Tx/Rx Power of the Fixed Step Power Control

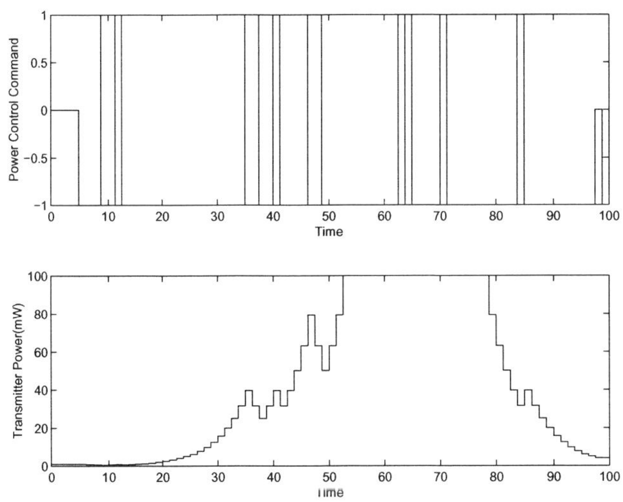


Figure 3: Power Control Command and Instantaneous Transmitter Power using Fixed Step Power Control

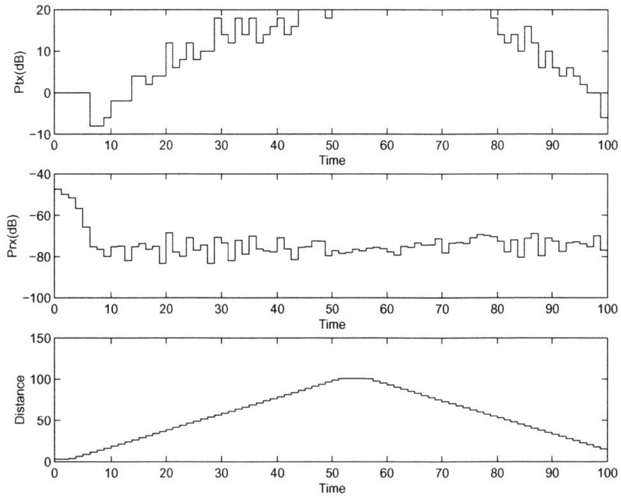


Figure 4: Tx/Rx Power of the Dynamic Step Power Control

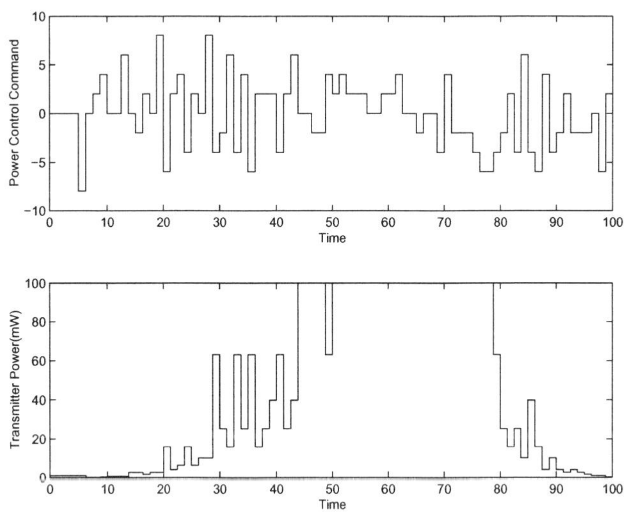


Figure 5: Power Control Command and Instantaneous Transmitter Power of Dynamic Step Power Control

Figure 4 shows the transmitted and received power of multi-step power control. Figure 5 shows the power control command and instantaneous power in Bluetooth transceiver using the dynamic power control step. The 0 $dB$ power control command region means that the received power is at golden receiver range.

Both in Figure 3 and 5, the instantaneous power of Bluetooth transceiver are equal to or below 100 $mW$ while bit error rate (BER) is satisfied. Hence, these results shows that power consumption are reduced using both FSPC and DSPC algorithms. Assume that both master and slave have the same data to send. This assumption means slave, uplink transceiver, uses half of time for data transmission. With this assumption, the power consumption of FSPC and DSPC are 2119 $mW$ and 2212 $mW$ respectively. These power consumption are about 40% those of Bluetooth transceivers without power control. This result can be translated that the mean time between two consecutive battery recharges will be extended to about 2.5 times.

As shown in previous results, the DSPC consumed slightly more power than the FSPC. However, power consumption difference is so small as to be negligible. Because the power level change is more abrupt in DSPC, power level fluctuation is more severe in the DSPC. That may cause more power consumption in the DSPC. However, abrupt power change makes effective adjustment transmitter power if mobile moves frequently and fast. This is shown in Figure 3 and 5. In a view of implementation, the FSPC is simpler than the DSPC.

## 6 CONCLUSION

In this paper, power control is introduced to the uplink Bluetooth transceiver. The two power control algorithms are suggested, the fixed step power control and the dynamic step power control. Simulation results are shown for the suggested algorithms under the indoor radio propagation model. Using the suggested power control, power consumption for the uplink communication, which means transmission from slave to master is reduced to about 40 % that of the commercial 100 $mW$ transceiver. This can be translated that the mean time between two consecutive battery recharges will be extended to about 2.5 times.

These power control algorithms are implemented at the LMP level. However, because the proposed power control is optional in the current Bluetooth specification, additional hardware and software modules are required for the implementation.

The power control command consists of 8 bits. If these 8 bits are fully used, more precise power control can be suggested using the optimal control or predictive control theories.
[1]

## REFERENCES


[1] J. Haartsen, "Bluetooth-the unicersal radio interface for ad hoc, wireless connectivity," *Ericsson Review*, vol. 1, pp. 110–117, 1998.

[2] J. Haartsen, "The bluetooth radio systems," *IEEE Personal Communications*, vol. 1, pp. 28–36, 2000.

[3] J. Haartsen, "Bluetooth-a new radio interface providing ubiquitous connectivity," *IEEE Vehicle Technology Conference*, vol. 1, pp. 107–111, 2000.

[4] J. Haartsen and Mattisson S, "Bluetooth-a new low-power radio interface providing short-range conncectivity," *Proceedings of the IEEE*, vol. 88, pp. 1651–1661, 2000.

[5] M. Kalia S. Garg and R. Shorey, "Mac scheduling policies for power optimization in bluetooth: a master driven tdd wireless system," *IEEE Vehicular Technology Conference*, vol. 1, pp. 196–200, 2000.

[6] S. Mattisson, "Low-power considerations in the design of bluetooth," *International Symposium on Low Power Electronics and Design*, vol. 1, pp. 151–154, 2000.

[7] Bluetooth SIG, "Bluetooth core specification v1.1," .

[8] http://www.cambridgesiliconradio.com, "web document, bluecore01," .

[9] http://www.digianswer.com, "web document," .

[10] T.S. Rapparport, "Wireless communication : Principle and practice," *Prentice Hall*, 1996.

[11] B. Sklar, "Rayleigh fading channels in mobile digital communication systems," *IEEE Communication Magazine*, vol. 1, pp. 136–146, 1997.

[12] S. Zurbes J.C. Haartsen, "Bluetooth voice and data performance in 802.11 ds wlan environment," *Ericsson SIG Publication*, pp. 1–13, 1999.

[13] J. Ziren, "Reliability of ieee 802.11 hi rate dsss wlans in a high density bluetooth environment," *Bluetooth'99*, pp. 1–12, 1999.


[1] *This paper is supported by Brain Korea 21*

IFAC Publications
www.elsevier.com/locate/ifac

# AN APPROACH FOR POWER PLANT CCS MODELING AND OPTIMIZATION[1]

**Li Pingkang[1] Xia Ming[2] Liu Tuo[1]**

[1] *Northern Jiaotong University, 100044 Beijing, China*
[2] *China Electric Power Research Institute, 100080, Beijing, China*

**Abstract:** An on-line approach using Matlab for power plant Coordinate Control System (CCS) modeling and optimization is provided. With the development of modern Distributed Control System (DCS) communication, and the Windows NT as Human-Machine Interface, a lot of data collected from the DCS could be utilized. Many useful third-party optimum software such Matlab could be integrated into DCS environment. Some application and considerations for power plant CCS modeling and optimization are reported. *Copyright © 2001 IFAC*

**Keywords:** fossil-fired generating unit, coordinated control system, distributed control system, modeling and optimization

## 1. INTRODUCTION

Nowadays the electricity sectors in China have been going from a monopoly setting to an open-market situation. This will most likely turn the Automatic Generation Control (AGC) and power plant load-following capability into a commercial commodity. Thus one of the most important objectives in the power plants is to produce as cost efficiently as possible. Economic objectives direct the operation of the plant and are eventually translated into optimal settings for the control systems. For the last decades control-system manufacturers of power plant applications have carried out extensive developments. Advanced distributed control systems (DCS) are widely applied into variety plant utilities. But unfortunately, many of these DCS equipped units are still turned by trial and error method, and the controllers of the DCS are still sloppy tuned (not tightly tuned). One of the main reasons for this phenomenon is that the early DCS application is not open to the customer, and is not exchangeable. The early DCS still use the traditional electronic instrument idea to deal with the control system. The complexity of the power plants makes the refinement of the controller parameters time-consuming and expensive. With the development of modern computer communication, and the Windows NT as Human-Machine Interface (HMI), a lot of data collected from the DCS could be integrated and further analyzed. The data from DCS could provide more useful information for the system. This data is available to be used in gaining efficiency. The modeling and optimization in this paper is aimed to process the information acquired from DCS to improve the plant performances and quality specifications.

[1] The work is supported by North China Electric Power Group

Matlab is used as a tool for identification and optimization of the power plant. This is possible because now many kinds of DCS use Windows NT as HMI. The experiments are accomplished by reading a large data set from the DCS, then identifying the ARX model, calculating the optimal parameters of the controllers based on the plant DCS controller structure and parameters, and back revising the DCS control system parameters. Furthermore, Matlab could also be used to model and analyze various failure and disturbance scenarios. When a disturbance occurs in the DCS, data could be analyzed in detail to determine the cause of the disturbance. With Matlab, and the System Identification Toolbox, and the Control System Toolbox, the engineers have reduced the analysis time greatly.

The power plant process is described shortly in next section. Then DCS identification for plant coordinate control system (CCS) is shown and an on-line data processing is presented. Identification and optimization results for some 300MW units of the power plant using Matlab is reported in section 4. The last section 5 is some conclusion.

## 2. THE POWER PLANT PROCESS MODELING

The overall modeling and optimization of the power plant unit involves many variables. The plant model in flow chart is sketched in Fig.1. The mass and the energy flows, including chemical process, combustion and heat transfer, mechanical and electrical process, constitutes the basic for the operation of the control system. The main variables for the control are fuel power, heat power in the furnace, steam pressure and electric power, etc. The main disturbances in the combustion process of power plants are mostly caused by stochastic variations in feed rate and the heat value of the fuel. Sometimes the load change in the plant power still becomes the disturbance (As in Automatic Generation Control -AGC- mode). The detail dynamic interaction between power plants and power system is discussed in [1]. In the routine situation, the objective of the control system is to adjust electric power while minimizing unwanted steam pressure variations. For emergency response, larger pressure variations would be permissible. Good control of pressure implies that other variables, including steam temperatures and drum water level, are all well controlled. The plant optimization is to keep these other loops independent so as to follow the AGC load instruction and to minimize the scope for common mode failure [2-5].

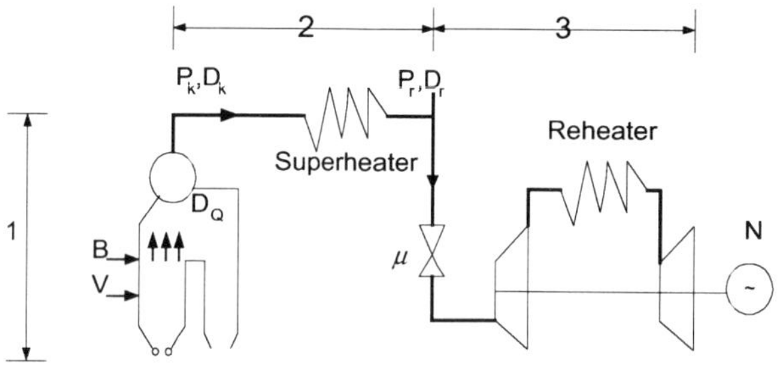

Fig.1 Power plant unit energy flow chart

1—ªBoiler firing:» 2—ªheat transfer:» 3—ªturbine power generation

The DCS control modes are also factors to effect the system performance. For better steady state control and fast load changing in both electric power and steam pressure, various hybrid schemes are developed which known as unit coordinated mode (UCM). For example, a standard Turbine Follow (TFB) control may be feed a filtered version of the frequency signal to the pressure set point to make it emulate Boiler Follow (BFT) response to frequency changes only. The realization of the UCM on a DCS environment is often called as Coordinate Control System (CCS).

The DCS structure of the Matlab Simulink realization of the power plant CCS is viewed in Fig.2. There the standard SAMA notations are developed. The circle stands for input signal from/to other sheet or transmitter, The rectangular for algorithm, and diamond for valve or actuator. Matlab Simulink does not only provide the methods how to construct control panels of 'high fidelity', but offer a new way of building complicated software systems by applying reusable building blocks. The method is well fitted to the block-diagram view of system structure, which is natural to most engineers. It does not use conventional programming techniques, instead offers interactive graphical tool for constructing hierarchical block-diagrams.

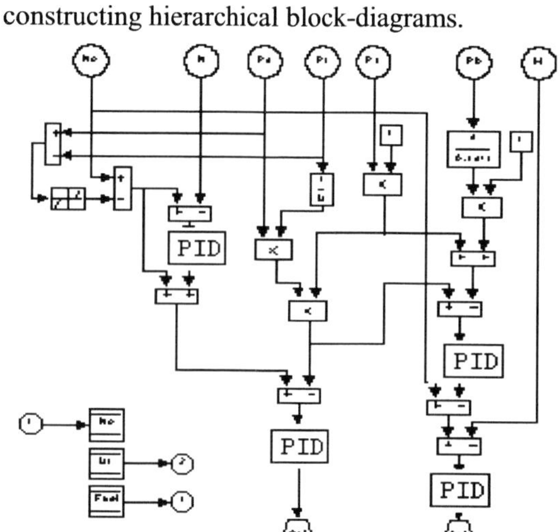

Fig. 2 Co-ordinate control of the power plant with DCS

## 3. PLANT DATA COLLECTION AND MODEL IDENTIFICATION

The Modeling and identification experiments were conducted on some 300MW units equipped with MAX 1000+plus and Simens Teleperm XP DCS. All the experiments were well organized, with the normal operation is still kept on. Open loop and closed loop tests were carried out. DCS data log configuration with 8 data per record is on-line carried out on DCS engineer workstation. More records could be configured for more information. Fig.3 is an example of the experiment data log, where the data records in upper two rows stand for First superheater temperature system; and the last two rows for Main pressure( c), Main pressure setpoint, Total fuel flow, Demand power command, First stage pressure in turbine, Boiler master control(a), Throttle press(b), Real power(d). The CCS modeling and optimization is concentrated on data marked as (a)-(d) in the Fig.

The basic step in this model identification is follows: first the firing rate was stepped up and down, while other control variables were tried to keep constant. After about one hour, that is after 1000 data points, when the response of the firing rate (changing) went back to flat, the valve position stepped up and then quickly down as shown in Fig.3 (a) and (b). The data were collected on sample rate 3 seconds. The pressure and power were recorded as in Fig.3. (c) and (d) Instead the use of the delay equations of the first or second order, which describe the plant with or without self equalizing well in many paper, an ARX model is developed here and used in the controller parameter optimization. The identified model outputs are also compared in Fig 4 (c) and (d). The pressure Pt(MPa) and power N(MW) were selected as system outputs, and firing rate Fuel and valve position Ut(MPa) were selected as inputs. A sample of identified ARX model is:

| ARX Model | | |
|---|---|---|
| | PT-y1 | NE-y2 |
| A(i,3) | -0.9583 | -0.8775 |
| A(i,4) | 0.0340 | 0.1286 |
| B(i,3) | -0.0046 | -0.0130 |
| B(i,4) | 0.0085 | 0.0241 |
| B(i,5) | -0.0283 | 0.0549 |
| B(i,6) | 0.0161 | -0.0242 |

The correspondence relation is:

y1(k)=-A(1,3)*y1(k-1)-A(1,4)*y2(k-1)+B(1,3)*x(1,k-1)+B(1,4)*x(2,k-1)+B(1,5)*x(1,k-2)+B(1,6)*x(2,k-2);

y2(k)=-A(2,3)*y1(k-1)-A(2,4)*y2(k-1)+B(2,3)*x(1,k-1)+B(2,4)*x(2,k-1)+B(2,5)*x(1,k-2)+B(2,6)*x(2,k-2);

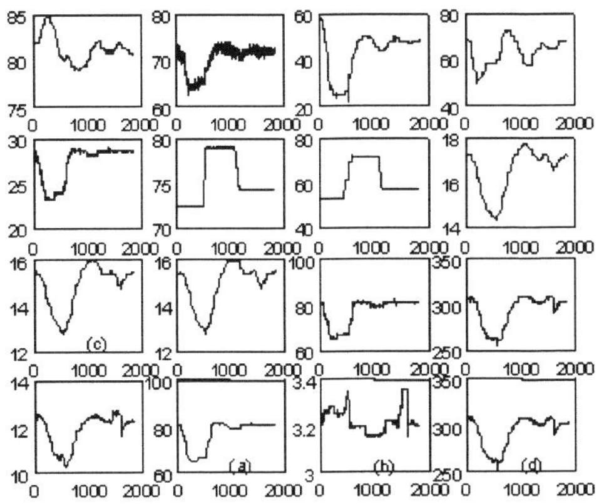


Fig.3 300MW Unit Modeling Data

From the compared plots, it is easy to known that the model is quite accurate.

After the data processing, following considerations are worth to be mentioned .

1) Both output variables have a much larger settling time than the sampling rate. This might indicate that the sampling rate is too short. But for the large DCS data set acquisition, small sample rate is still recommended. Enlarge the sample rate is easier than otherwise, and Matlab resample command could be used to fit the different data processing requirements

2) The process needs to be strongly excited. For some processes, it is very difficult to excite the process adequately because of safety reasons. In a well designed experiment, if the model structure is of finite impulse response type then the sum of the model coefficients describe the step response function of the CV depending upon the particular manipulated variable (MV). As a rule of thumb, to obtain data that describes the process "sufficiently" excited, there are at least 60 step changes required and they have to occur randomly otherwise higher frequencies may be over-presented and the process dynamics (usually lower frequencies) are difficult to extract.

3) The ARX model that identified using Matlab ARX command is of absolute type. For control system parameter tuning application, it is necessary and easy

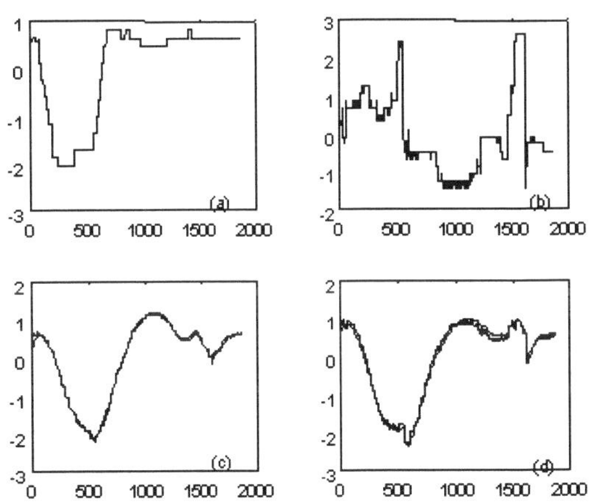


Fig.4 300MW CCS identification

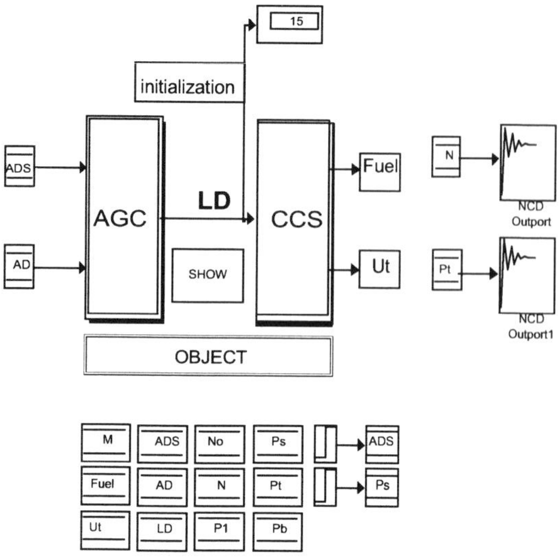


**Fig 5. Simulation and Optimization of CCS**

to transfer the ARX model to state space or other kinds of model.

## 4. SIMULATION AND APPLICATION

Traditionally, the tuning of the CCS loop parameter utilizes the SISO (single input, single output) loop techniques and thus is often tedious, since the control loops interact with one another. Here Matlab Nonlinear Control Design (NCD) toolbox is used to do the optimization. Fig.5 is the CCS controller parameter optimization structure. The plant model is created as a subsystem named as object, and the CCS subsystem has the structure as in Fig.2, and where Pt, N, Fuel and Ut are selected as the system outputs and inputs. The optimization is carried on PID controller parameters and forward gain. The main optimization steps are as follows:

1) Initializing the NCD by double click the initialization box in Fig.5, assigning the controller parameters and forward gain for adjusting;
2) Simulating the plant, checking the output plot to make sure that the process and simulation are coincident;
3) Run NCD to improve the optimization of the variables;
4) Comparing the results with what the real process possess. If the optimization is not reasonable and acceptable, then go back to check the ARX modeling and do from the first step again.

Matlab NCD need initial value for the optimum parameters. The DCS original PID parameters could be programmed in the initial file [7]. Note that if initial parameters are not suitable, the NCD could not convergence to the global optimal values. Part of the optimization results is shown in Fig.6. The upper left is the original curves, and the lower left part is the result after the optimization. The optimization process with the NCD is shown on the right part of Fig.6. Where outport_1 (upper part) is the tracking control, while the outport_2 is for setpoint disturbance rejection. This approach has applied to the plant system successfully.

## 5. CONCLUSION

Dynamic interactions between power plants and power systems are getting more attention. The complexity of the power plant makes the refinement of the controller parameters time-consuming and expensive. Optimization is needed and available. An on-line approach using Matlab for power plant Coordinate Control System (CCS) modeling and optimization is presented. With the development of modern Distributed Control System (DCS) communication, and the Windows NT as Human-Machine Interface, a lot of data collected from the DCS could be utilized. Many useful third-party optimum software such Matlab could be integrated into DCS environment. The modeling and optimization is not as difficult as before. Many algorithms are available and easy to be used in engineering practice. Some application and considerations for power plant CCS modeling and optimization are reported.

## REFERENCE

1. E.Welfonder, Dynamic interactions between power plants and power systems, Control Engineering Practice, 7(1999), p.27-40
2. N.Jalecli and L.S.VanSlyck et.al., Understanding Automatic Generation Control, IEEE Trans.Power System, Vol.7, No.3, 1992, p.1106-1122
3. J.Waddington And G.C.Maples, The control of large coal- and oil-fired generating units, Power Eng. J., 1987, January, p25
4. N.G.Anisimova and E.K.Krug, Computer-aided Process Control System, Automation and Control, Vol.60, No.7, 1999
5. Y.Wang, Adaptability of Coordinated Control of Fossil-fired Generating Unit to AGC, ELECTRIC POWER, (in Chinese) No.6, 1999
6 Pingkang Li and Ming Xia, The Auto-tuning of a Complex Control System in Power Plant□1996 International Conference on Power System Technology, Aug.1996, Beijing
7. Li Pingkang, Xiaming and Sun Jiwei, Matlab Realization for Coal-fired Power plant Control System Optimization, Journal of North China Electric Power, 2001.4 (in Chinese)

## BIOGRAPHIES

Pingkang Li was born in Shichuan, China on

December 16,1955. He received his Master and Bachelor degree in 1984.7 and 1982.1 respectively. Now he is a professor in Northern Jiaotong University, Beijing, China. His research includes the automation of the industrial process, intelligent control and devices. He was a visiting professor to Purdue University between 1992.9 and 1994.11. He is a senior supervisor for North China Electric Power Institute. Email:bepclpk@public.bta.net.cn

Xia Ming was born in Jiangshu, China on 1957. He received his Master degree in 1985.7. Now he is a senior Engineer in China Electric Power Research Institute, Beijing, China. His research area is mainly in process control engineering.

Liu Tuo was born in Beijing, China on Oct.1956. She is an associate professor in Northern Jiaotong University, Beijing, China. Hers research area is mainly in electronics engineering.

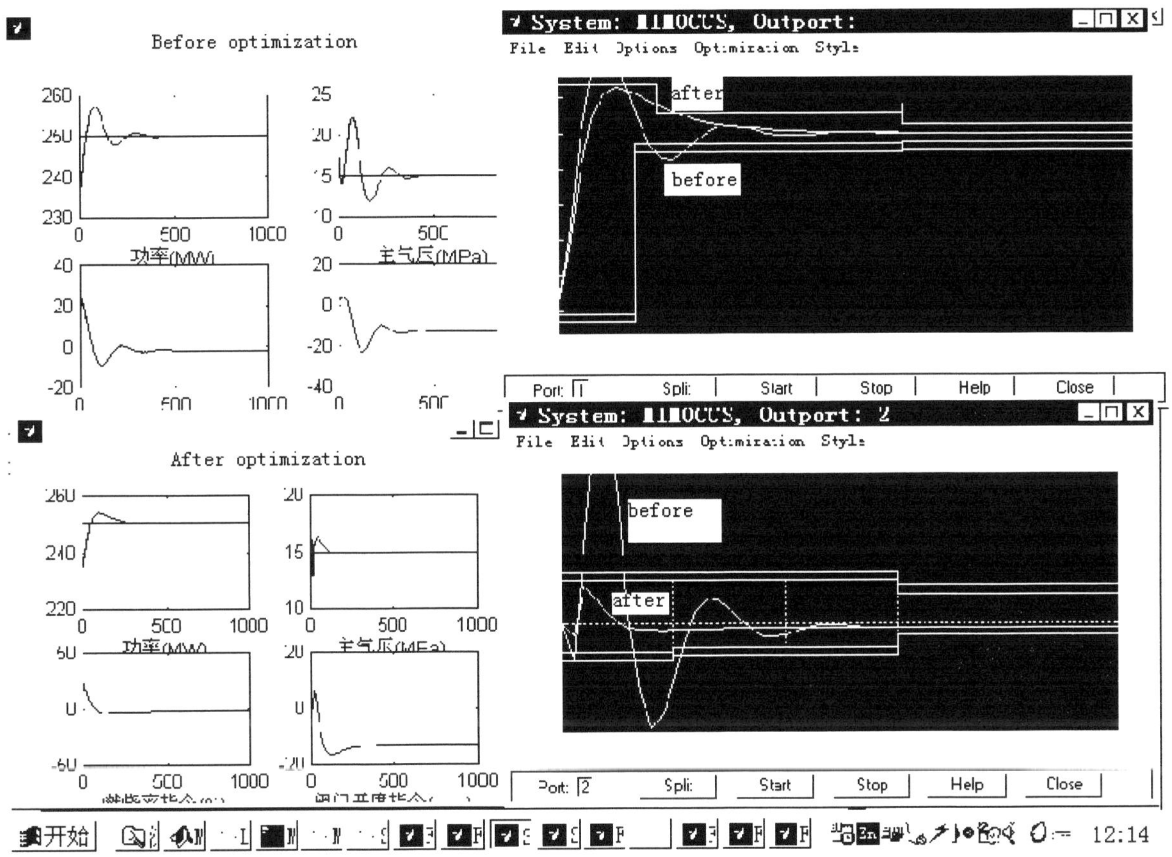


Fig. 6 Power Plant CCS Optimization Results

IFAC
Publications
www.elsevier.com/locate/ifac

# A NEW PROCESSOR ARCHITECTURE FOR REAL-TIME SYSTEMS

**Masato Uchiyama*1 Tsutomu Itou*2 Junichi Sato*3**
**Nobuyuki Yamasaki*4 Yuichiro Anzai*5**

Anzai and Yamasaki Lab.
Faculty of Science and Technology
Keio University
3-14-1, Hiyoshi, Kouhoku-ku,
Yokohama Kanagawa, 223-8522, JAPAN

Abstract: It is difficult to realize a fine-grain real-time system because of overheads caused by restrictions of traditional processor architectures. This paper describes a processor architecture designed for reducing the overheads by applying a simultaneous multithreading (SMT) architecture to real-time systems. The processor has eight hardware contexts in the processor and it can hold 32 contexts in a high-speed on-chip memory for context backup. The processor provides fast context switching capability up to 39 threads. The processor has a mechanism which selects instructions in an issue stage of the SMT architecture, based on priority. Because the mechanism executes several threads in parallel, the processor can improve the total throughput and can provide flexible thread scheduling. *Copyright © 2001 IFAC*

Keywords: Real-time, Processor, Parallel processing, Priority

## 1. INTRODUCTION

Nowadays, many real-time systems are used widely in various fields. However, there are many problems caused by scheduling overheads, frequent context switching, etc. in current real-time systems. In order to realize a fine-grain real-time system, it is necessary to reduce the overhead of context switching and to guarantee to execute a thread with high priority by improving processor architecture.

Hardware multithreading mechanisms reduce the overhead of context switching. High-speed on-chip memory for context backup provides the fast context switching. It is considered that the total throughput can be improved by the adoption of a Simultaneous Multithreading (SMT) architecture (Tullsen *et al.*, 1995) that fuses a superscalar architecture with a multithreading architecture. This can execute eight threads in parallel. In order to guarantee the execution of a thread with high priority, priority-based arbitration is introduced to deal with various access conflicts.

1. utiyama@ny.ics.keio.ac.jp
2. itomu@ny.ics.keio.ac.jp
3. satojun@ny.ics.keio.ac.jp
4. yamasaki@ics.keio.ac.jp
5. anzai@ics.keio.ac.jp

## 2. BACK GROUND

### 2.1 *Related Work*

A processor architecture for real-time systems (Uchiyama *et al.*, 2000) was proposed in our laboratory. The processor adopted Block

Multithreading (Gunther *et al.*, 1997) to support real-time operations. The processor has four pipelines and shared functional units (FUs). Eight threads are held in the processor. These threads can be assigned to each of the four pipelines. However its performance is not good when there are few threads, and it needs a special cache system. It demands four read ports of a cache tag table and an instruction cache for itself.

### 2.2 *Multithreading*

On the traditional single-thread architecture (e.g. Superscalar), processor utilization is reduced by a dependency of instructions and long latency instructions. Since a multithreaded processor can switch contexts very fast, it can execute a thread while other threads wait for some operations. Thus processor utilization and total throughput can be improved.

When an interrupt occurs in a real-time system, the processor can reduce the overhead of context switching, and can reduce the response time. Of course some context switching occurs in the case of normal thread scheduling. The scheduling time is reduced by the mechanism. The previous study has adopted these mechanisms for real-time systems.

### 2.3 *SMT Processor*

SMT (Tullsen *et al.*, 1995) is a technique that fuses a superscalar architecture with a fine-grain multithreading. Various threads can issue their instructions at each clock cycle in an SMT processor. Several threads can be executed in parallel so that, while a long latency operation is waited, instructions of other threads can be executed. So the SMT processor can achieve high total throughput of all threads. Flexible systems can be constructed by executing various threads including a scheduler, an interrupt handler, real-time threads, normal threads, etc. in parallel. The SMT processor can achieve the same average performance as a superscalar processor, even if there is only one thread.

### 2.4 *Priority*

Issue instructions are selected from the instructions of several threads at each clock cycle in an SMT processor. The threads of real-time systems have priority and are scheduled according to priority. It is considered that if priorities are introduced to the issue instruction selection, the selection mechanism will become useful for real-time systems.

When there is a shared resource (like an SMT processor), it causes a situation where several threads request the resource at the same time. In this situation, an execution time of the thread that is selected earlier is short and the execution time of the thread that is selected later is long. This case also needs arbitration based on priorities.

### 2.5 *Cost*

Recently, the number of transistors on a chip is increasing fast and it is expected to continue in future. It is preferable to design a processor architecture to make good use of extra transistors to gain performance, than to cut down the total number of transistors. The multithreading is typical of an architecture that gain performance by using many transistors.

## 3. APPROACH

The processor proposed in this paper is based on the SMT architecture for real-time systems. The processor has several advantages. First, it provides fast context switching capability, when an interrupt occurs or when thread scheduling is executed. Second, it executes several threads including a scheduler, real-time threads, in parallel. Third, it can improve total throughput.

In the processor, if a conflict that requires shared resources occurs, then an arbitration of requests is based on priority of a real-time system. When it selects issue instructions, the selection mechanism introduces a priority of a real-time system with processor statements. These two mechanisms can help the real-time capability.

The processor is also designed for multimedia extension. The processor also integrates Responsive Link (Yamasaki and Matsui, 1997) that is a real-time network standard in the future, then it will be possible to use it for various real-time systems.

## 4. DESIGN

### 4.1 *Overall*

The instruction set architecture is compatible with MIPS in the processor, because of its simplicity and its applicability to embedded systems. Some additional instructions are added for thread control and synchronization.

A DDR-SDRAM (Double Data Rate SDRAM) interface with 128-bit width is designed for the main memory.

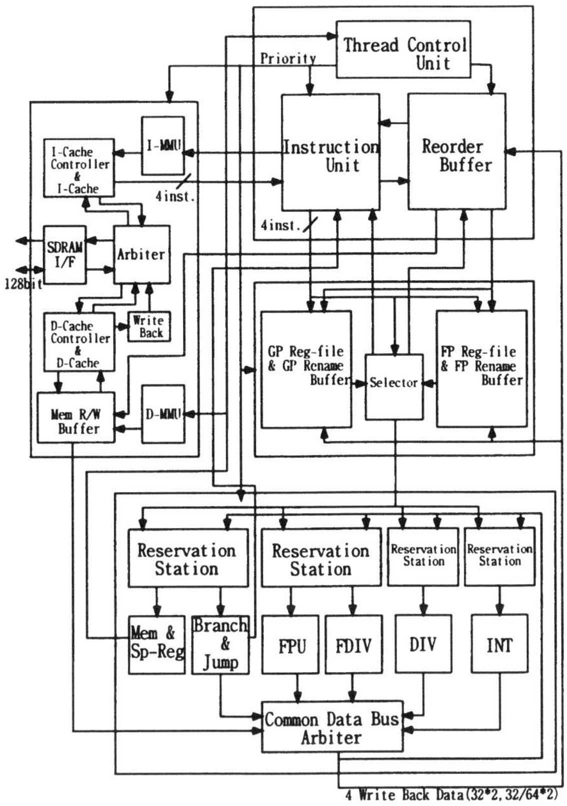


Fig.1 Overview of Processor

| | |
|---|---|
| IF1 | 1. Instruction Fetch Stage 1 |
| IF2 | 2. Instruction Fetch Stage 2 |
| IA | 3.Instruction Decode and Analysis Stage |
| IS | 4.Instruction Select Stage |
| REG | 5.Register Rename Stage |
| EXE | 6.Execution Stage |
| WB | 7.Write Back Stage |
| COM | 8.Commit Stage |

Fig.2 Pipeline

Figure.1 shows an overview of the processor.

The fetch bus is 128 bit wide, so that it can fetch four instructions per clock cycle. Four fetched instructions are decoded in parallel, then held in ana instruction buffer. Instructions in the instruction buffer are selected and issued to FUs through renaming buffer, and entered reorder buffer on a parallel of renaming. Instructions are divided broadly into two categories: integer instructions and floating-point instructions. They are executed at each Integer Unit (IU) and Floating-Point Unit (FPU). The IU contains branch units, load/store units that are shared by a FPU, a synchronization unit, and a reservation station of a thread control unit. There are as many reorder buffers as threads. Each reorder buffer commits 2 instructions per clock cycle in order.

The pipeline is divided into eight stages as shown in Figure.2. The operations on each stage are as follows:

IF1: (Instruction Fetch Stage 1) A fetch request is sent. The virtual address is converted to the physical address in the MMU.

IF2: (Instruction Fetch Stage 2) Four instructions are fetched from the instruction cache. If a cache miss occurs, only the thread can't send a fetch request until the instructions are received.

IA: (Instruction Decode and Analysis Stage) Four fetched instructions are decoded. If they contain a branch or a jump instruction, it predicates branch address and decides a next fetch address. Then it decides whether the instruction is valid or invalid. Decoded instructions are put in the instruction buffer.

IS: (Instruction Select Stage) Four instructions are selected from the instructions that are ready for issue in the instruction buffer. Selected instructions are issued to renaming buffer.

REG: (Register Rename Stage) The destination registers of issued instructions are renamed to GP (32 bit width) or FP (64 bit width) renaming buffer. The renamed number of destination registers are used in the following pipeline stages. The source data of issued instructions are obtained from register file. On a parallel of renaming, issued instructions enter reorder buffer. The entry number of reorder buffer is used for ordering of instructions in part of FUs. Renamed instructions are put in the reservation station of each FU.

EXE: (Execute Stage) The instruction is executed. This stage is not necessarily a single clock cycle. (e.g. a multiplier instruction) The branch instruction is also executed in this stage. Whether the branch is taken or not taken is decided.

WB: (Write Back Stage) Four execution results (e.g. ALU result data, load data, branch signal) are selected. Result data is written back to renaming buffer and finish states are sent to reorder buffer through common data bus.

COM: (Commit Stage) This stage is divided from WB stage for coherence control. Four instructions are selected and they are committed. Execution results are written back to each register set from renaming buffer and store data is written to the data cache or the main memory. An exception handling and a recovering of a branch misprediction start as soon as their corresponding instructions are selected to commit in this stage.

The number of clock cycles executed by the IS stage, the REG stage, the EXE stage, and the COM stage are not always constant. The number of clock cycles from the IF1 stage to the COM stage is not constant.

4.2 *Hardware context*

A multithreading architecture provides fast context switching capability with several hardware contexts. The mechanism of a hardware context switching is described next.

A hardware context consists of a general register set and special registers (e.g. Program Counter), and thread status. The processor holds eight hardware contexts and all instructions that these contexts can issue. It is then possible to realize flexible scheduling. It can switch contexts very fast by hardware, but there are only 8 sets of hardware contexts. If the total number of threads is over eight, the cost of context switching for threads in main memory is high, because an access to main memory is caused. Instead it saves the contexts to on-chip memory that it can access faster than the external main memory with very wide bandwidth (over 500-bits). The processor can save or restore contexts very fast. The backup-memory holds 32 contexts, so it can provide fast context switching capability up to 39 contexts. To distinguish from threads in on-chip memory, a thread in the hardware context is named an active thread.

Hardware contexts are managed with a thread table (Figure.3).

The STATE field indicates a thread state. If the STATE is RUN, the thread is executed. If the STATE is READY, the thread can be executed but it is not executed. If the STATE is TWAIT, the thread is instructed to stop fetch and issue of instructions and it waits for finishing the execution of issued instructions. If the KEEP bit is enabled, the thread can't be saved to backup memory, then the thread is kept in hardware context. It is used to prevent a context switch from stopping the execution of a high priority thread. The PBASE field holds base priority, and the PRIOR field holds current priority. These two fields and the TIMER field help to thread scheduling by software.

There are also some instructions to manage threads.

| 27 | 26 17 | 16 15 | 14 | 13 12 | 11 10 | 9 0 |
|---|---|---|---|---|---|---|
| ENABLE | I D | STATE | KEEP | PBASE | PRIOR | TIMER |

Fig.3 Thread table entry

There are instructions to read the thread table, to set the thread states including the KEEP bit, to set the PBASE field and PRIOR field, to set the PC of start point, and to save or restore contexts. The policy of save/restore contexts fundamentally depends on the scheduling policy of the operating system. The response time to an interrupt is important in real-time systems because applications often require to output the result as soon as input from the outside world. A hardware interrupt handler is provided to reduce the response time. As soon as the interrupt handler detects interrupts with a special number set by software, the interrupt handler signals to the thread control unit to restore the context of an exception handler.

The thread table is placed in the thread control unit (Figure.1). A bus that transports eight set of priorities is wired to almost all units in the processor from the thread control unit. An instruction does not hold priority but hold only its thread number. When the instruction needs priority, it refers to the bus wired to the thread control unit to get its priority with its thread number. In this way, the instruction is not controlled based on the current priority of the IA stage, but is controlled with the current priority of its thread.

Thread control instructions are executed in the thread control unit, but its reservation station is placed in IU. The instruction is issued to the reservation station similar to other reservation stations in IU. The reservation station is implemented as a FIFO queue and out-of-order execution is prohibited.

4.3 *Access control based of priorities*

The processor controls instructions according to the priority of their thread. The arbitration of accesses to a shared resource is based on priority and guaranteeing an execution of the thread with a high priority.

4.3.1 *Instruction fetch and issue* The instruction fetch and issue mechanism is implemented in instruction unit (Figure.1). The mechanism consists of two selection parts (Figure.4). In the first selection part, a thread that sends a fetch request to the instruction cache is selected at each clock cycle. It is implemented in fetch thread select unit. In the second selection part, four instructions that are issued to FUs are selected at each clock cycle. It is implemented in issue instruction select unit.

The thread that sends a request is selected from all of the threads that are ready for a fetch. The selection is basically based on priority. However, when all threads have the same priority, it is considered that the performance can be improved with another selection mechanism that takes account of whether the fetch address of each thread is a result of the branch prediction or is unrelated to branch prediction. A sequence of four instructions is fetched at a time with 128-bit width. The fetch address of the next clock cycle can't be decided until it is known whether the fetched four instructions include branch instruction or not and the branch address has been predicted. Consequently, the interval from the previous fetch request of a thread to the next fetch request of the same thread is four clock cycles. If there are few threads, the rate of instruction fetch is not so good. To improve the rate, the fetch mechanism introduces another prediction. The prediction acts likes a prefetch. The mechanism is as follows:

(1). The thread sends a fetch request as soon as it is signaled that the previously request hits the instruction cache or that the previous requested data will have been received at the next clock cycle. The fetch address is generated by incrementing the PC.

(2). When the previously requested instructions are received and analyzed, it is decided whether the predicted instructions are necessary or unnecessary. If the instructions are unnecessary, it sends a cancel signal to the instruction cache.

(3). If the cancel signal is not in time and the predicted instructions are received, they are discarded.

A thread can fetch four instructions per two clock cycles with the prediction mechanism.

The number of issue slots per clock cycle is four. The selection policies of instructions which are issued to FUs have various possible patterns. The selection policies can be classified according to how these four slots are assigned to threads. If all four slots are assigned to only a thread and it is switched to an other thread each cycle, the policy is similar to TDM (Time Division Multiplexing) scheduling. If each slot is assigned to a fixed thread, the number of slots that are assigned a thread is decided based on the priority of the thread. Furthermore, it is considered several policies, a policy that a slot is assigned to several threads and they are switched with the TDM scheduling, a policy that a slot is assigned to a fixed thread and the thread that no slots is assigned can issue instructions when an issue slot becomes empty, and so on.

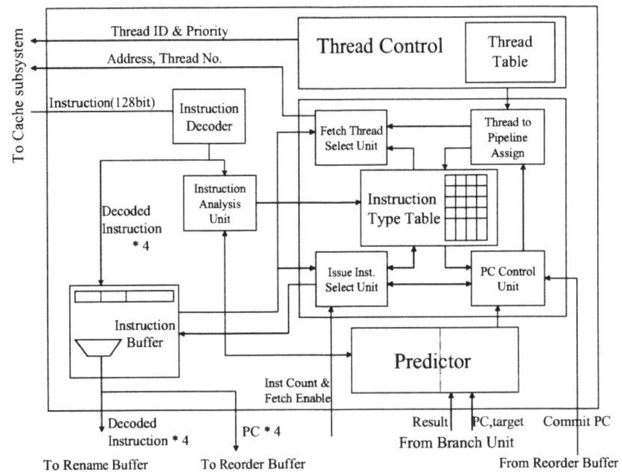


Fig.4 Instruction fetch and issue mechanism

It is confirmed that the selection policy that takes account of the number of instruction in the reorder buffer is effective (Tullsen *et al.*, 1996)

On the other hand, there is a restriction that any selection mechanism must finish within a clock cycle time. A good performance improvement is expected with a complex selection policy, but its complexity leads to increasing the clock cycle time. It is necessary to make comparisons on real programs in order to select effective selection policies.

4.3.2 *Execution of instructions in EXE stage* Instructions in each reservation station are sorted according to their priorities. Then the instruction with the highest priority which is ready to be executed is executed. The instructions are sorted every clock cycle. The mechanism prevents the execution of a thread with a high priority from being disturbed by a thread with a low priority.

4.3.3 *Access to memory* The cache subsystem consists of an instruction cache and a data cache. The main memory is a shared resource so there will be conflicts of requests from several threads. So it is necessary to prevent the execution of a thread with a high priority from being disturbed by the execution of a thread with a low priority. As in the reservation station, the prevention is implemented by selecting threads based on priority.

Each thread has an individual virtual address space in the processor. The virtual address is converted to the physical address in the MMU. The MMU has not only addresses but also thread numbers as TLB tags. In the instruction cache, the fetch request is held after an address conversion. Then a conflict of memory accesses is arbitrated. In the data cache, operations are held in the write buffer or the read buffer at load/store unit one of the FUs. If the operation is a read, it looks up in the write buffer whether its read address and its thread number are matched or not matched. If the address and the thread number are matched, the data is read from the write buffer. Otherwise the operation is held in the read buffer and it sends a read request to the data cache after an address conversion in the MMU. If the operation is a write, the operation is held in the write buffer at once. Operations in the write buffer wait for a signal that is generated in the reorder buffer at the time that instruction is ready to commit. Signaled operations in the write buffer are arbitrated, and the selected operation accesses main memory after an address conversion in the MMU. When cache misses are caused at the instruction cache and at the data cache at the same clock cycle, it is necessary to arbitrate requests. The arbitration is based on priority. If a request from the instruction cache and a request from the data cache have the same priority, the request from the instruction cache is selected first.

The caches have four-way banks and the caches are physical address cache. The throughput of a thread with high priority is improved by introducing a policy of cache-bank replacement based on priority.

### 4.4 *Register Set*

The processor has eight register sets. A register set consists of 32 GP registers that is 32-bit wide for integer operations, 32 FP registers that is 64-bit wide for floating-point operations, and special registers. These register sets are saved to or restored from the memory for the context backup through a bus that is 512-bit wide, and then it can finish both save and restore operations only for four times of the memory accesses.

### 4.5 *Implementation*

The processor has been specified in Verilog-HDL and simulated. It is implemented on a FPGA board to be evaluated with real programs and to be verified a performance. Some software for the processor, including a real-time OS has been developed on this board. Finally, it will be implemented in a chip. The processor is intended as a CPU core for a Responsive Processor (Yamasaki and Matsui 1997) that is a System-On-a-Chip for parallel distributed real-time control systems with Responsive Link: a real-time communication protocol.

## 5. SUMMARY

A processor architecture which introduces several mechanisms for executing real-time systems is described in this paper.

The processor adopts SMT architecture. The processor can switch contexts very fast by holding several threads in the processor as hardware contexts. Thus the processor can reduce the overhead of thread scheduling. The mechanism of instruction fetch and issue can issue instructions flexibly from several threads. The mechanism can improve the total throughput and provide flexible thread scheduling. There are conflicts of accesses to shared resources in various parts of the processor. Arbitrations that are based on priority can prevent the execution of a thread with high priority being disturbed by the execution of a thread with a low priority.

The processor can improve the performance of real-time systems, and it can support new scheduling policies.

## REFFERENCES

Gunther, Bernaed K. (1997). Multithreading with distributed functional units. In: *IEEE Transaction on computers.* Vol.46 No.4 pp.399-411.

Tullsen, Dean M., Susan J. Eggers and Henry M. Levy (1995). Simultaneous multithreading: Maximizing on-chip parallelism. In: *Proceedings of the 22nd annual International Symposium on Computer Architecture.* pp.392-403

Tullsen, Dean M., Susan J. Eggers, Joel S. Emer and Henry M. Levy (1996). Exploiting choice: Instruction fetch and issue on an implementable simultaneous multithreading processor. In: *Proceedings of the 23rd annual International Symposium on Computer Architecture.* pp.392-403

Uchiyama, Masato, Tsutomu Itou, Nobuyuki Yamasaki and Yuichiro Anzai (2000). Design and implementation of multithreaded processor for real-time operation. In: *Technical Report of IEICE.* Vol.99 No.725. pp.29-36

Yamasaki, Nobuyuki and Toshihiro Matsui (1997). In: *Proceedings of IEEE/RSJ International Conference on Intelligent Robots and Systems.* Vol.2. Grenoble, France. pp.793-798

IFAC Publications
www.elsevier.com/locate/ifac

# AN ARCHITECTURAL-FRAMEWORK FOR A TESTING ENVIRONMENT FOR GEOGRAPHICALLY DISTRIBUTED TEAMS[1]

**Juan Garbajosa[±], Hector Garcia[±], Maria Alandes[±], Maria-Angeles Mahillo[¥], Abraham García[†], Belen Magro[¥]**

*±Technical University of Madrid. E. U. Informatica. Dpt. OEI. Madrid. Spain*
*email: [jgs, hgarcia]@eui.upm.es*
*¥Technical University of Madrid. E. U. Informatica. Dpt. LPSI. Madrid. Spain*
*email: amahillo@eui.upm.es*
*†Technical University of Madrid. E. U. Informatica. Dpt. IA. Madrid. Spain*
*email: abraham@eui.upm.es*
*¥Invensys. Madrid. Spain. email: belen.magro@invensys.com*

Abstract: Testing, particularly system testing, is an essential process for critical systems. Most of the applications that contain control components fall under this category.

For these applications the definition of system tests relies necessarily on the experience and knowledge of the test engineer. For this reason it is of significant importance to provide him with assistance. With this objective of providing assistance, a model-based environment, *TOPEN*, has been developed considering a number of specific application-domains. TOPEN also provides support for geographically distributed teams.

This paper describes the architectural framework that is underlying the TOPEN environment. *Copyright © 2001 IFAC*

Keywords: Assisted Test Procedures distributed development, system testing, test and operations environment, networking.

## 1. INTRODUCTION

Testing is an essential process in software and systems engineering. Testing is especially important for critical systems, and most of the applications that contain control components can be considered under this category. Testing of *user/system* requirements is the focal point in our research work. User requirements are the main issue for processes referred as acceptance testing, system testing, and validation. It may be necessary to notice that terminology and definitions concerning testing may vary depending on the sources but for the purpose of this paper it does not seem necessary or useful to go further into detail concerning testing concepts. During the rest of this paper the term *system testing* will be used with the meaning of the process dealing with the objective of testing user/system requirements. The test batteries will be called *tests procedures*, and the system to test, *unit under test* (*UUT*).

With the objective of providing assistance in the definition and execution of system test procedures, a model-based environment has been developed and named *TOPEN*. TOPEN stands for

[1] This research work was performed within the ARCO project. The Inter-ministerial Commission for Science and Technology of the Government of Spain, CICYT, partially supports the Project ARCO Ref. TIC98-0782.

*Test and Operation Environment.* TOPEN has been devised considering a number of specific applications domains, achieving a highly application-domain independence.

### *1.1. Software engineering context*

Vardanega (1999) classifies user/system requirements of critical applications into two levels. The first level is the, traditional, functional level. It includes those features that the customer understands that the system shall include. The second level comprises requirements derived from the functional set, such as the real time set. A full testing of a system requires validating not only the real time set, but also the functional set, since real time requirements are derived from the functional set, and not the other way round. It is possible to produce a traceability graph that relates second-level requirements to first-level requirements. In this form we can know, for a given functional requirement, which real time requirements should be correctly implemented, and conversely.

The environment presented within this paper, TOPEN, has been devised to assist in the definition and execution of test procedures corresponding to the first level of requirements. The view of the system that the test engineer shall get will be similar to that of the operator once the system becomes operational.

### *1.2. Technical Context*

Our main concern is related to high reliability systems characterised by features such as a complex dynamic behaviour or interaction with humans. A networked system is a good example in which a high number of elements, some of them implementing critical functions, interoperate; interaction with an operator may also exist. Producing tests for such systems becomes a difficult task for the test engineer.

The research and development of tools and environments to provide assistance to the test engineer has been addressed in the past few years, but still specific tools are required. As examples we have some results published as (ESA, 1996), (Garbajosa and Wolff, 1997) and (Tufarolo, *et al.*, 1999).

Operation of complex systems is often characterised by sequences of operations which interact to reach the goal of the system. Tests can be only fully automatically generated under a large set of constraints as explained by Dalal, *et al.*, (1999), Cunning and Rozenblit (1999) and Sutcliffe (1998). The definition of test procedures relies on the knowledge and experience of the test engineer, who must be also an expert in programming techniques. A way to improve the test engineer work is providing him with assistance by means of tool support.

Another problem that nowadays raises up when developing software, is the fact that the development team might be distributed across different organisations, and distributed geographically. Internet has helped to design simpler infrastructures. However, specific environments have specific requirements and particular solutions must be provided. Therefore it seemed quite necessary to support distributed testing process. According to Blackburn, *et al.*, (2000), already highlighted by Haag, *et al.*, (1997) the main flaw in concurrent software process tools is lack of management of the process. As long as the main objective of our work was not to provide facilities to manage the process the approach followed was to implement a platform to support the process in a distributed fashion and to guarantee that the process could be managed by third parties. As it is explained below this is achieved ensuring that external tools can be easily integrated. Among other functions these tools can perform the process and configuration management when and as required.

### *1.3. Business Context*

Tools and environments for system testing have a high cost because of the complexity of the functions and all the interfaces to the UUT. Therefore there exist a strong interest in defining a framework for these tools and environments. This framework must be independent, as much as possible, from the application domain and it shall be the reuse baseline. It should remarked that in the satellite-checking field, testing tools and environments are more and more often reused for operations. This issue has been considered in our approach so that the investment performed can be paid off.

### *1.4. Framework Definition Process*

The definition of an application framework is always a complex issue. The process considered previous works such as (Garbajosa, *et al.*, 1997) and (Valera, 1999). Both of them corresponds to the satellite-checking ground segment. A new application domain, network management, was introduced during the development of the TOPEN prototype described in (Garbajosa, *et al.*, 2000a, 2000c). This prototype was used to produce the first version of the framework. As a result, it was possible to compare different application-domain models and various implementation approaches.

### 1.5. TOPEN background

Within this article we present an architectural framework for the *TOPEN*. Former works (Garbajosa, *et al.,* 2000a, 2000b) have described the environment objectives, strategy, and architecture; features and deployment of the implemented model–based approach can be found in (Garbajosa, *et al.*, 2000b and 2000c). The way concurrent-testing process and other tele-testing features are supported has been presented in (Garbajosa, *et al.*, 2001).

### 1.6. Article structure

Section 1 of this article is *Introduction*, this one. In section 2, *TOPEN: Functions, Architecture and operation,* the philosophy underlying the environment and the architecture of TOPEN are described to facilitate the understanding of section 3. In section 3, *A Framework for the Environment,* the basic architectural patterns are introduced providing details of the framework. Section 4 presents some *Conclusions*.

## 2. TOPEN: FUNCTIONS, ARCHITECTURE AND OPERATION

Within this section a general background is provided in order to facilitate the understanding of the next section. In section 1.5 a number of references that provide further information on the subject are mentioned.

### 2.1. Command and model Concept

To test a system requires executing a number of commands and measuring the results. According to the terminology in the space field, where originally TOPEN comes from, these concepts are called telecommand and telemetry, as presented in figure 1. In terms of object orientation this is represented as method and result of the method execution.

The second issue concerns the use of a model describing the UUT. This model has not been used for automatic generation as in (Dalal, *et al.,* 1999) and (Benjamin, *et al.*, 1999), but for providing assistance to the test engineer in the test definition and execution tasks as described in (Garbajosa, 2000b).

### 2.2. Architecture

The architecture has been implemented in terms of components. There are three main components:

- Man-Machine Interface (MMI) offers a graphic representation of the system. Test procedures are defined using this MMI.
- Code Generator (CG) generates the code that will be sent to the UUT.
- Mission Information Base (MIB) stores the UUT model and test management information.

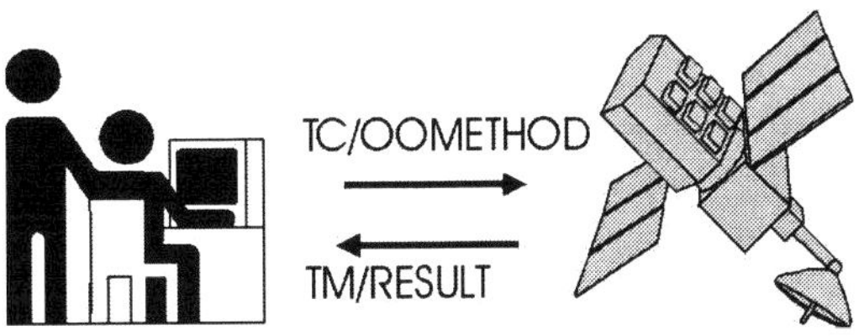


Figure 1. Telemetry and telecommand

It should be considered that the TOPEN prototype has been built for a specific application domain that is **network management**. A network management system will be composed of an agent that performs the management, and a number of network elements as presented in figure 2. Managers are the different test engineers or operators that in our system are represented by windows in the Man machine interface.

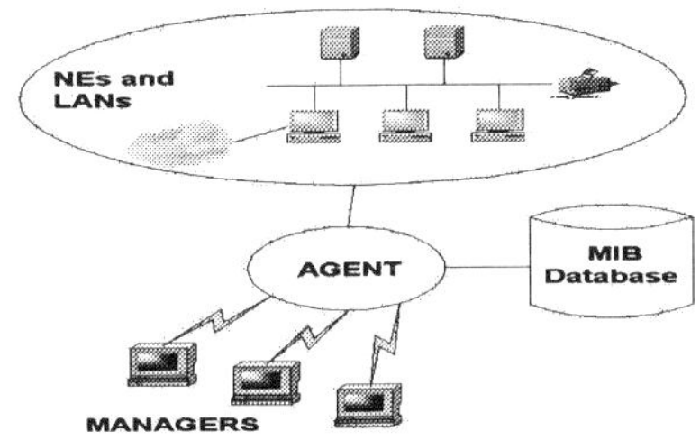


Figure 2. Network management system

Architecture modules interact exchanging a flow of *commands, replies, notifications and alarms*. This is a general scheme, application (*domain*) *independent*. In figure 3 we can see the data model for the *reply* command. As the application domain for the TOPEN prototype is network management the exchanged information is structured in *protocol data units* (PDUs) from the Common Management Information Protocol interchanged by Common Management Information Service Element CMISE peer entities. PDUs are specified using ASN.1 and the external macros OPERATION and ERROR from Remote Operations Protocol, defined by ITU (1998) and ISO/IEC (1989).

A reply corresponds to the execution of a command that might have attempted to *create* a network element, *get result*, *get list* of network elements, etc. Each box has the associated attributes. Every change on UUT item's state is notified to the TOPEN, as well as those replies to executed commands, following reverse order of those steps defined above.

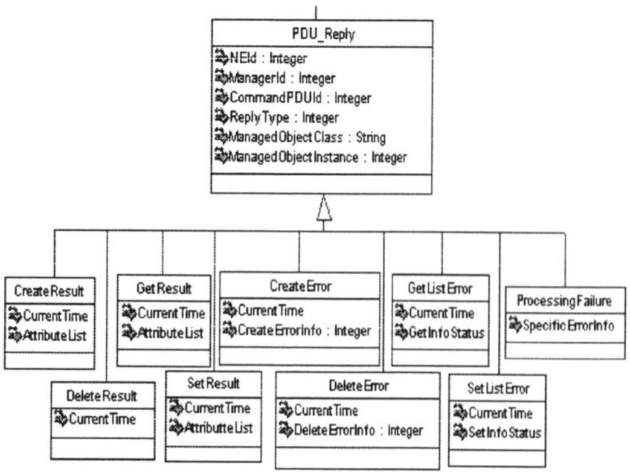


Figure 3. Example of reply PDU

*2.3. Operational Support to concurrent definition of tests procedures*

MMI is the TOPEN's interface to the user, offering two different interfaces: Manager Panel Interface (MPI) and Test Procedure Panel Interface (TPPI) as drawn in figure 5. The Manager Panel Interface (MPI) is a purely graphical interface, user-friendly. MPI is made of a number of windows. Each window represents the view of a manager, in terms of the network management system description in figure 2. Each window is like a subschema. The schema would be the whole system. This is the base to implement *concurrent testing* in terms of several testers that might be in different locations. TPPI, as opposed to MPI windows, provides a global view. This view includes all the partial views coming form each MPI window. In figure 4 TOPEN's graphical interface is shown.

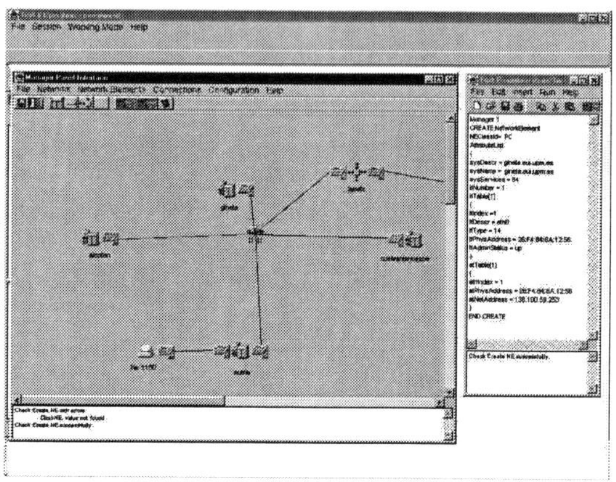

Figure 4. TOPEN's Man Machine Interface

The test engineer will find a number of objects, such a network element (NE), and a number of associated operations, such as *create* a network element. This follows the philosophy of having an object and then an associated number of methods. As shown on the right of figure 4, the Test Procedure Panel Interface (TPPI) is a command line interface for advanced users which offers the possibility of defining Test Procedures (TP) using a Test Procedure Language (TPL). This test procedure language can be one such as Atlas. But in our case to use an in-house developed TPL was preferred. For TPPI a syntax-based editor is provided.

The MMI has two additional modules. These are the Model checker (MOCER), to assure that test procedures are built consistently with the system model and MCI that manage the MPI window system from a logical point of view.

*2.4. Component distribution and support to tele-testing*

As described above, the TOPEN component architecture has been designed to work in a distributed way. The communication technology that has been used is the CORBA environment for the communication among the MMI, MIB and CG. The fact of using CORBA allows us to provide the architecture components with a large degree of independence with respect to a specific implementation.

Two possibilities are supported in the client: using CORBA/HTTP and HTTP *standalone*. The first choice is more solid but requires a larger infrastructure in the client. The second is lighter but also easier to have. Figure 5 shows a graphical representation of the tele-testing architecture.

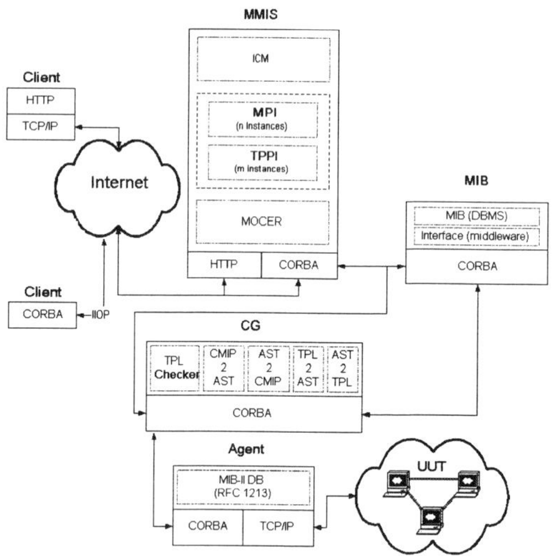


Figure 5. Teletesting Architecture

## 3. A FRAMEWORK FOR THE ENVIRONMENT

The first goal when designing the TOPEN architecture was to create a general architecture for testing complex systems. The patterns used have been extracted from (Gamma *et al.* 1997). The framework has been defined and structured following a similar fashion as described in (Schmidt, 1999).

Within this section two basic patterns are presented. The first concerns the general architecture of TOPEN. The Code generator plays a role of adapter for MIB and MMI. The Code Generator acts as an adapter between the MMI and the MIB. See figure 6. The MMI needs to use the MIB but their interfaces are different, due to the different representations. Thus, the CG component will adapt all the requests coming from the MMI. The Code Generator acts as a mediator between the different parts of the TOPEN translating the information exchanged among these parts. Depending on the representations/languages used in the architecture, the Code Generator will use one kind of translator or another. The interactions among the different parts of the translator will remain the same. The Adapter Pattern might be also used when the MMI and the Agent exchange PDUs, and all this traffic has to be stored in the MIB.

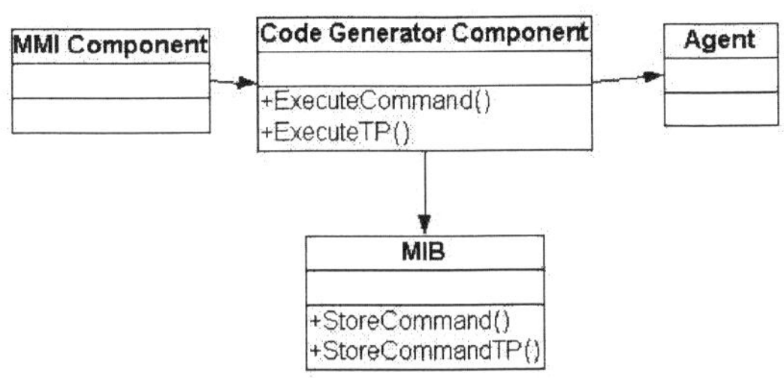


Figure 6. TOPEN Basic architectural framework pattern

Patterns for the Code Generator and MMI architecture have also been developed. The MMI is comprised of three subcomponents: the MPPI, the TPPI and ICM. The relationship among them and with the UUT can be specified using the Chain of Responsibility Pattern. This pattern expresses the behaviour that the ICM has when it receives a request from the UUT (through the CG). The ICM will decide whether the request goes either to a Manager, a set of Managers or to the TPPI. In this way the UUT does not need to know a priori who is the receiver of the request.

The CG component is mainly composed of two subcomponents, each of them specified as a Bridge Pattern. The Bridge Pattern is especially useful when we want to separate the implementation from the interface.

Another issue of importance, from a framework point of view, is the information exchanged between the components. TOPEN has a set of commands that are executed in the system that is operated and tested. The managers do not need to know how the system performs each action, but they need to know the set of commands that can be applied to the system. The Command Pattern encapsulates a request as an object, thereby letting parameterise clients with different requests, queue or log requests, and support undoable operations. It does not need to know anything about the operation being requested or the receiver of the request. It decouples the object that invokes the operation from the one having the knowledge to perform it. This pattern describes the behaviour of executing a command from a manager in our framework. The manager does not need to know what the command does exactly and who does it. Figure 7 presents a command pattern. Figure 8 is an instantiation of the pattern for a NMS application domain.

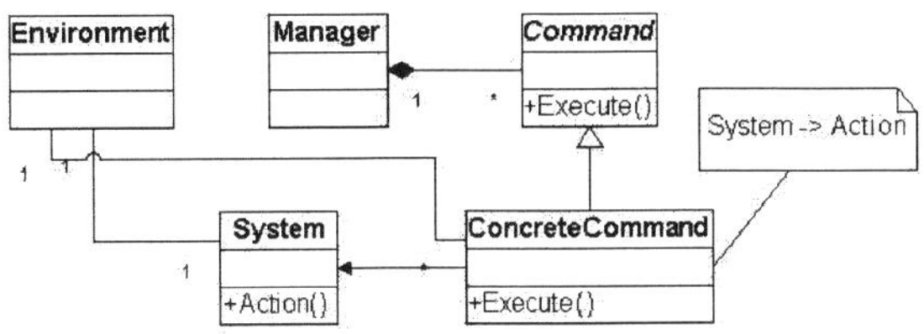


Figure 7. Command Pattern

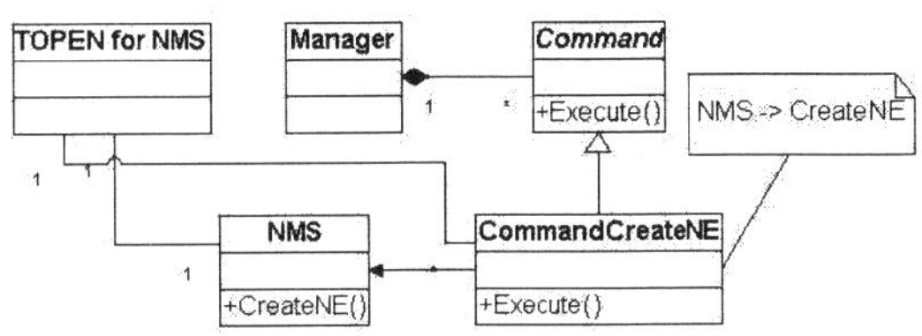


Figure 8. Command pattern for a NMS.

## 4. CONCLUSIONS

Within this paper we have presented the basic guidelines of the design of the TOPEN environment and its architectural framework pattern. TOPEN provides assistance based on a system model of the UUT. The framework supports several testers working concurrently and low-cost remote connections.

The TOPEN architecture framework has proven to be flexible enough to be able to support different application domains. Actually the system test definition process is somehow independent from the application domain, though the knowledge within the tests is not. The specification of the model requires a great part of the effort to implement a TOPEN for a new application.

The work remaining has to focus on how to deploy the application framework to produce lower cost TOPEN implementations.

ACKNOWLEDGEMENTS

Authors are indebted to the reviewers for their helpful comments. Authors are also indebted to Bryan Melton and Serge Valera, of ESA/ESTEC TOS-GCE for a number of interesting discussions that became fundamental to obtain some of the results. Authors also appreciate the support from Francisco Sanchis, Francisco Arizmendi, Mario Piattini, Esperanza Marcos, Carlos Cuvillo. Authors are also are grateful to those students who have taken part in the implementation of the environment, especially Angelica Gonzalez, Esther Alonso, Antonio García y Ahmad Mareie.

REFERENCES

Benjamin, M., D. Geist, A. Hartman, Y. Wolfsthal, G. Mas, R. Smeets (1999). A Study in Coverage-Driven Test Generation. In: *Proc. ACM DACS'99.* ACM Press.

Blackburn, J., G. Scudder, and L. N. Van Wassenhove (2000). Concurrent Software Development. In: *Comm ACM. Dec 2000.* ACM Press.

Cunning, S.J. and J.W. Rozenblit (1999). Test Scenario Generation from a Structured Requirements Specification. In: *Proc. IEEE ECBS'99.* IEEE Press.

Dalal, S. R., A. Jain, J.M. Karunanithi, J.M. Leaton, C.M. Lott, G.C. Patton, and B.N. Horowitz (1999). Model-Based Testing in Practice. In: *Proc. IEEE Int. Conf. on Software Engineering. ICSE '99.* ACM Press.

ESA (1996) *Proc. Fourth International Symposium on Space Mission Operations And Ground Data Systems.* European Space Agency SP-394. ESA.

Tufarolo, J., J. Nielsen, S. Symington, R. Weatherly, A. Wilson and T.C. Hyon (1999). Automated Distributed System testing: designing and RTI Verification system. In: *Proc. of the 1999 Winter Simulation Conf.* ACM Press.

Gamma *et al.* (1997) *Design Patterns – Elements of Reusable Object-Oriented Software.* Addison-Wesley Professional Computer series.

Garbajosa J., and M. Wolff. Automatic Generation of Satellite Test Procedures. In: *Proc. International Symposium Computing Tools, Systems Engineering and Competitiveness.* Association Aeronautique et Astronautique de France (AAAF), Paris, March 1997.

Garbajosa, J., Piattini, M., Mahillo, M.A., Alandes, M. (2000a). Assisting the Definition and Execution of Test Suites for Complex Systems. In: *Proc. IEEE Int. Conference on Engineeering of Computer Based systems. ECBS'2000.* IEEE Press.

Garbajosa, J., H. Garcia, M. Alandes, M.A. Mahillo, M. Piattini (2000b). An Approach to Use System Models to Assist in the Definition of Test Procedures. In *Real-Time Programming 2000.* Ed. A. Crespo and J. Vila. Elsevier.

Garbajosa, J., M. Alandes, F. Arizmendi, C. Cuvillo, H. García, F. Sanchis, M.A. Mahillo, M. Piattini and E. Marcos (2000c). Model-Based Environments for Acceptance Testing of Complex-Systems. *In Proc. International Conference of Software and System Engineering.* CNAM, Paris, France.

Garbajosa, J., J. Gil, H. García, M. Alandes, A. Gutierrez (2001). Migrating a Corba-based Tele-testing Architecture to Internet. In: *1st IFAC conference on Telelematics Applications in Automation and Robotics.* Weingarten University, Weingarten, Germany.

Haag, Z., R. Foley and J. Newman (1997). Geographically Distributed Software Engineering: An Initial Evaluation. In: *Proceedings of the 23rd EUROMICRO Conference '97 New Frontiers of Information Technology.* Euromicro.

ISO/IEC (1989). *ISO/IEC 9072-1:1989 Information processing systems -- Text communication -- Remote Operations -- Part 1: Model, notation and service definition.* ISO.

ITU (1988). *ITU - X.219 (11/88). Remote Operations: Model, notation and service definition.* ITU-T recommendation. ITU.

Schmidt, D.C. and J.C. Hu, (2000). Developing Flexible and High Performance Web Servers with Frameworks and Patterns. In: *Proc. International Conference on Software Engineering ICSE'00.* ACM Press.

Sutcliffe, A. G., Maiden, N. A.M., Minocha, S., and D. Manuel (1998). Supporting Scenario-Based Requirements Engineering.IEEE Transactions on Software Engineering. **Vol. 24**, No. 12. pp. 1072-1088.

Valera, S. (1999). EGSE and Mission Control System. In: *The PROBA EGSE/SCOS2 Experience.* ESTEC. European Space Agency.

Vardanega, T., (1999). Development of On-Board Embedded Real-TimeSystems: An Engineering. Approach. ESA Technical Report, STR-260. ESA.

IFAC Publications
www.elsevier.com/locate/ifac

# A SWINGUP CONTROL OF AN INVERTED PENDULUM WITH CART POSITION CONTROL

**Yusuke Otani*1, Takuya Kurokami, Akira Inoue*2 and Yoichi Hirashima*3**

*Department of Systems Engineering, Okayama University, 3-1-1 Tsushima-naka, Okayama 700-8530, JAPAN*

**1 yusukeo@suri.sys.okayama-u.ac.jp, *2 inoue@suri.sys.okayama-u.ac.jp*
**3 hira@suri.sys.okayama-u.ac.jp*

Abstract: Mechanical systems such as robots or manipulators have characteristics of nonlinearity and underactuatedness. These characteristics make the systems difficult to be controlled and to overcome the difficulties are strongly required by control engineers in industry. Inverted pendulum also has the two characteristics. Especially in a swing-up of an inverted pendulum, the effect of the nonlinearity and underactuatedness appears strongly. This paper gives a control scheme to achieve both of the swing-up control of a pendulum and the position control of the cart carrying the pendulum. The control law is derived by using a Lyapunov function and making the derivative of the Lyapunov function with respect to time be negative. An experimental result is given to show the effectiveness of the scheme.


Key words: Inverted pendulum, Swing-up control, Lyapunov function, Cart position control, Energy control

## 1. INTRODUCTION

Mechanical systems such as robots or manipulators have characteristics of nonlinearity and underactuatedness. These characteristics make the systems difficult to be contorolled and contorl schemes to overcome the difficulties are strongly required by control engineers in industry.

An inverted pendulum driven by a moving cart has a simple structure, but has one control input, which is a force generated by a motor driving the cart, controls both of the angle of the pendulum and the position of the cart. This means that the system is underactuated. Also the equation of the motion of the pendulum includes nonlinear functions, $\cos\theta$ and $\sin\theta$ of the angle $\theta$ of inclination of the pendulum. In the swing-up control, where the pendulum moves from the downward stable equilibrium position (bottom position) to the inverted unstable equilibrium position (upright position), the pendulum swings in a wide angle range, and the nonlinearity in the equation has an essential effect in the movement of the pendulum. Many authors have considered a swing-up control of the pendulum and position control as a clue for obtaining controllers for robots and manipulators having nonlinearity and underactuatedness. And they have studied and proposed many swing-up schemes.

Among these schemes, a swing up control law by energy control, which was first proposed by Wiklund *et al.*(1993), is simple, effective and robust to uncertainties and disturbances.

The method swings and accelerates the pendulum in a manner that increases its total mechanical energy. The total energy of the pendulum consists of kinetic energy measured by the angular velocity of the pendulum and the potential energy given by the height of the center of mass of the pendulum, which is measured by the angle of the pendulum. The potential energy is maximum at the upright position and a swing-up controller is designed to cause a movement to increase the energy. Then it is expected that by the movement the energy reaches to the maximum which corresponds to the upright position.

However, swing-up by energy control has two problems. First, if the pendulum receives more energy than the potential energy at the upright position, then the pendulum passes the position and does not stay at the unstable position. It is necessary to accelerate the pendulum to give it exactly the correct energy corresponding to the upright position. Astrom and Furuta (2000) gave switching strategies to solve this problem. Henmi *et al.*(2000) also discussed this problem.

The second problem is that the energy method controls only the angle of the pendulum and does not control the position of the cart carrying the pendulum or the position of the pivot. To control the position as well as the swing up, Chung and Hauser (1995) added a PD position controller to the swing-up controller. Inoue *et al.*(1999) used a P control law to control the cart position in swing-up control of a double inverted pendulum.

To solve the second problem, Wei *et al.*(1995) and Lin *et al.*(1996) proposed control laws to restrict the displacement of the cart. For the first and second problems, Lozano and Fantoni (1998) used a Lyapunov function to derive a control law controlling the cart position and swinging up the pendulum. The Lyapunov function is given by the sum of the squares of mechanical energy, cart position and cart velocity. However, simulation results given in their paper show that their control law causes an oscillation and does not give a smooth movement.

To remove the oscillation and to get a smooth controlled trajectory, this paper defines a new Lyapunov function and proposes a new control law for the swing up and the position control derived by the Lyapunov function. The Lyapunov function uses the absolute value of the mechanical energy

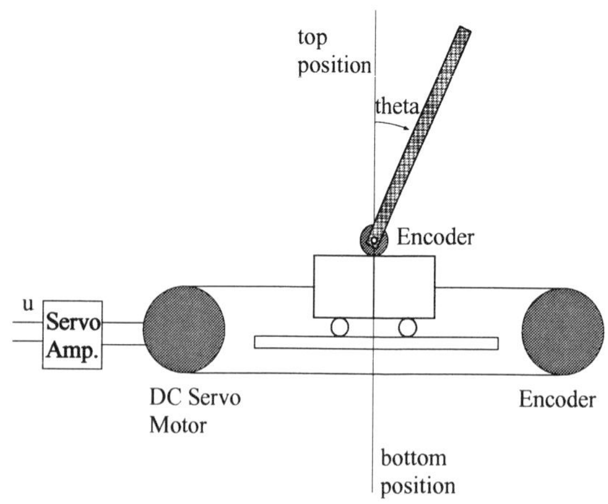


Figure 1: Controlled system

instead of the square of the energy. Using the proposed control law, an experimental result shows smooth movements of the pendulum and the cart.

The next section explains the controlled system and its mathematical model. In the third section, a control law for the swing up and the cart position control is proposed. Finally, in Section 4, a simulation result and an experimental result are presented.

## 2. CONTROLLED SYSTEM AND MATHMATICAL MODEL

The controlled system is shown in Fig.1, where the pendulum is swung by a cart and the cart is driven by a DC servo motor. The control input of this system is velocity command input $v[volt]$ to a servo amplifier driving the DC motor. The outputs are the angle $\theta[rad]$ of the pendulum and the position $x[m]$ of the cart measered by optical encoders.

The control objective is to swing up the pendulum from the downward stable equilibrium (bottom) position to the inverted unstable (upright) position, balance the pendulum at the upright position and control the cart position to the starting position.

In this paper, the equation for the cart position $x[m]$ is approximated as

$$\dot{x} = av \tag{1}$$

where the coefficient $a$ is the ratio of the attained velocity$[m/sec]$ to the velocity command input$[volt]$ and its value is obtained by preliminary experiments. This approximation is justified by the fact that in the experiment a sufficiently powerful DC servo motor is used and the cart velocity $\dot{x}$ follows the velocity command input $v$ instantenously. To simplify the design procedure the new variable $u = \dot{v}$ is used as the input variable. Hence, the mathematical model for the cart is

$$\ddot{x} = au \tag{2}$$

At the experiment, first the control input $u(t)$ is calculated and the input $u$ is integrated, then the integrated value $v(t)$ is inputted to the amplifier as the velocity command input.

The mathematical model for the pendulum is obtained by using the Lagrange equation as

$$\ddot{\theta} = \frac{mgl}{I}\sin\theta - \frac{ml}{I}\ddot{x}\cos\theta \tag{3}$$

where the frictional force is neglected and the following notations are used:

- $x$: the displacement of the cart$[m]$
- $\theta$: the angle from the upright position of the pendulum$[rad]$
- $l$: the length from the pivot to the center of mass of the pendulum$[m]$
- $m$: the mass of the pendulum$[kg]$
- $g$: the acceleration of gravity$[m/s^2]$
- $I$: the moment of inertia of the pendulum at the pivot$[kg \cdot m^2]$

## 3.SWING-UP CONTROL LAW

To derive the control law to swing up the inverted pendulum and control the cart displacement to the initial position, a new Lyapunov function is defined.

The Lyapunov function consists of two parts: First is the absolute value of the total mechanical energy of the pendulum;

$$E_p = \frac{1}{2}I\dot{\theta}^2 + mgl(\cos\theta - 1) \tag{4}$$

The absolute value of energy has the maximum value $|E_p| = 2mgl$ at downward stable equilibrium $(\theta = \pi)$ and the minimum value of $|E_p| = 0$ at upright position $(\theta = 0, \dot{\theta} = 0)$. The second part is the sum of the squares of the cart displacement and the cart velocity:

$$E_c = rml(x^2 + \dot{x}^2) \tag{5}$$

where $r > 0$ is a weighting factor. The Lyapunov function in this paper is

$$V = \frac{1}{2}(|E_p| + E_c)^2 \tag{6}$$

The function is positive except at the upright unstable equilibrium position $(\theta = 0, \dot{\theta} = 0, x = 0, \dot{x} = 0)$.

A control law is derived in order to make the derivative of $V$ with respect to time be negative definite. The derivative of the absolute value $|E_p|$ is

$$\frac{d}{dt}|E_p| = \begin{cases} \dot{E}_p & \text{when } E_p > 0 \\ -\dot{E}_p & \text{when } E_p < 0 \\ \text{not defined when } E_p = 0 \end{cases}$$
$$= \begin{cases} \text{sgn}(E_p)\dot{E}_p & \text{when } E_p \neq 0 \\ \text{not defined when } E_p = 0 \end{cases} \tag{7}$$

where $\text{sgn}(E_p)$ shows the sign of the value $E_p$. In actual physical systems, the case when the total energy $E_p$ is equal exactly to zero would happen at probability 0, that is, when the energy varies from $E_p > 0$ to $E_p < 0$, the energy takes the value $E_p = 0$ on the infinitesimally small time interval. Hence the case $E_p = 0$ can be neglected and

$$\frac{d}{dt}|E_p| = \text{sgn}(E_p)\dot{E}_p \tag{8}$$

Using this formula and equations(2) and (3),

$$\begin{aligned} \dot{V} &= (|E_p| + E_c)(\text{sgn}(E_p)\dot{E}_p + \dot{E}_c) \\ &= (|E_p| + E_c)[\text{sgn}(E_p)(I\dot{\theta}\ddot{\theta} - mgl\sin\theta\dot{\theta}) \\ &\quad +2rml(x\dot{x} + \dot{x}\ddot{x})] \\ &= -ml(|E_p| + E_c)[\text{sgn}(E_p)\dot{\theta}\cos\theta\ddot{x} \\ &\quad -2r(x\dot{x} + \dot{x}\ddot{x})] \\ &= -ml(|E_p| + E_c)\{[\text{sgn}(E_p)\dot{\theta}\cos\theta - 2r\dot{x}]\ddot{x} \\ &\quad -2rx\dot{x}\} \end{aligned} \tag{9}$$

To make this derivative negative definite, the control input is defined as

$$u = \frac{u_a|\text{sgn}(E_p)\dot{\theta}\cos\theta - 2r\dot{x}| + 2rx\dot{x}}{\text{sgn}(E_p)\dot{\theta}\cos\theta - 2r\dot{x}} \tag{10}$$

where $u_a$ is a weighting factor. In the experiment to achieve the control objective with control input in the restricted values $(-5 \leq v \leq 5)$ and to obtain fast convergence to the upright position of the pendulum and the starting position of the cart with a smooth controlled trajectory, the values of parameters $r$ and $u_a$ are selected simulation and preliminary experiments.
Except in the case

$$\text{sgn}(E_p)\dot{\theta}\cos\theta - 2r\dot{x} = 0, \tag{11}$$

the derivative of $V$ is

$$\dot{V} = -ml(|E_p| + E_c)u_a|\text{sgn}(E_p)\dot{\theta}\cos\theta - 2r\dot{x}| < 0 \tag{12}$$

Hence, the value of $V(t)$ decreases along the trajectory of the motion of the pendulum and the cart and the state $(\theta, \dot{\theta}, x, \dot{x})$ approaches to the upright position and the initial cart position.

The case where equation(11) holds occurs at the infinitesimally small time interval. Hence the case of (11) also can be neglected. However, giving consideration to the computer implementation, the control law(10) must be avoided from dividing by a small number which causes a large value of input $u$. Hence, in the computer program to calculate the input $u$, a small number $\epsilon > 0$ is selected and the control law (10) is modified to

$$u = \begin{cases} \frac{u_a|\mathrm{sgn}(E_p)\dot{\theta}\cos\theta - 2r\dot{x}| + 2rx\dot{x}}{\mathrm{sgn}(E_p)\dot{\theta}\cos\theta - 2r\dot{x}} \\ \text{when } |\mathrm{sgn}(E_p)\dot{\theta}\cos\theta - 2r\dot{x}| > \epsilon \\ \\ \frac{u_a|\mathrm{sgn}(E_p)\dot{\theta}\cos\theta - 2r\dot{x}| + 2rx\dot{x}}{\epsilon\,\mathrm{sgn}(\mathrm{sgn}(E_p)\dot{\theta}\cos\theta - 2r\dot{x})} \\ \text{when } |\mathrm{sgn}(E_p)\dot{\theta}\cos\theta - 2r\dot{x}| \le \epsilon \end{cases} \quad (13)$$

**Remark** If the weight $r$ in the Lyapunov function(6) is selected as $r = 0$, then the control law becomes

$$u = u_a \mathrm{sgn}(E_p)\mathrm{sgn}(\dot{\theta}\cos\theta) \quad (14)$$

which is the same control law as the one obtained by Wiklund and others(Wiklund, *et al.*, 1993). Hence it can be said that the control law obtained in this paper is an extension from the one based on the energy control(Wiklund, *et al.*,1993).

After the pendulum is swung up close to the upright position, that is,

$$|\theta| \le \theta_a,$$

the control law is switched to linear feedback in order to stabilize the pendulum to the upright position and the cart to the initial position. The linear feedback law is obtained by pole placement method.

## 4. SIMULATION AND EXPERIMENTAL RESULTS

In order to show the performance of the control law given in this paper, a simulation run and an experiment are conducted. In the simulation the values of the coefficients of the controlled pendulum and the values of the parameters in the control law are chosen to be the same values as in the experiment and are given in Table 1, where both tunable parameters $r$ and $u_a$ are chosen by the trial and error.

Simulation results are shown in Fig.2 and Fig.3. Fig.2 shows the cart position, the angle of the pendulum and the velocity command input $v = \int u dt$ to the amplifier. The angle starts at the downward position $\theta = -\pi$ and it swings up at $t = 4$, to $\theta = 0$. The figure shows that, during the swing up control, the cart moves only $0.08[m]$ from the initial position and it can be said that the cart position is also controlled well to keep the initial position $x = 0$. Fig.3 shows the values of the Lyapunov function $V(t)$, the total mechanical energy

Table1: Parameters of the system and control law

| controlled Inverted Pendulum | Control law | Switching angle | Poles of Stabilizing law | Sampling Time |
|---|---|---|---|---|
| $l=0.15$ | $r=2.31$ | $\cos\theta_a = 0.99$ | $\begin{bmatrix} -7 \\ -6 \\ -5 \\ -4 \end{bmatrix}^T$ | $5[msec]$ |
| $m=0.15$ | $u_a=8.5$ | | | |
| $a=0.232$ | $\epsilon=0.001$ | | | |
| $I=0.0045$ | | | | |

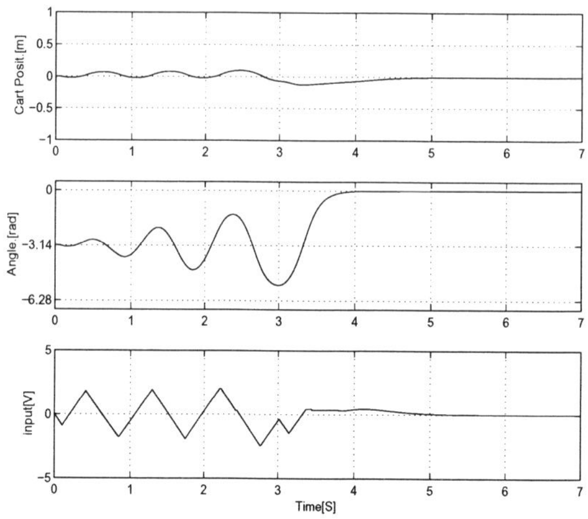


Figure 2: Simulation Results 1

$E_p$ and the sum $E_c$ of squares of the cart position and the velocity. In the figure it is shown that the Lyapunov function decreases monotonically, which means that the controlled system approaches the upright position and the initial cart position $(\theta = 0, \dot{\theta} = 0, x = 0, \dot{x} = 0)$.

The experimental system is shown in Fig.4. A personal computer is used as the controller. The computer measures the angle of the pendulum and the position of the cart as outputs of the controlled system through an optical encoder counter. The control law is calculated in the computer and fed into the controlled system as the input of the system through a digital signal buffer, D/A converter and servo amplifier as velocity command voltage.

The experimental results are given in Fig.5 and Fig.6 in the same way as Fig.2 and Fig.3. Fig.5 shows that the cart and the pendulum move smoothly and the cart position is controlled to the initial position.

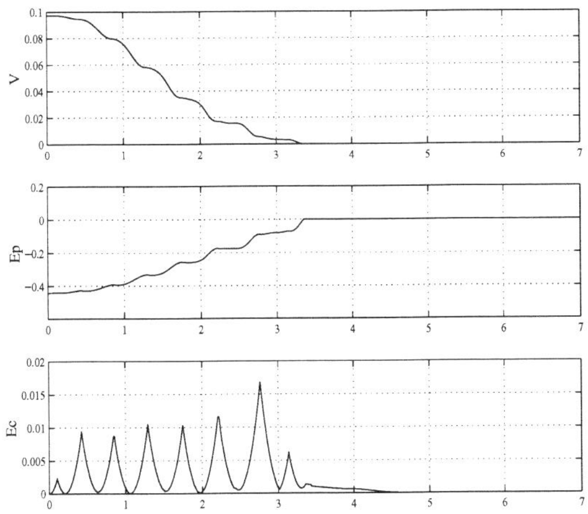


Figure 3: Simulation Results 2

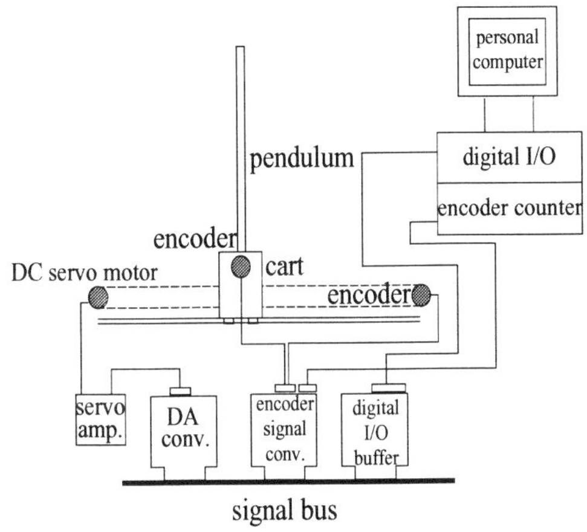


Figure 4: Experimental system

Fig.7 compares the result produced by the method in this paper with a result given by a conventional control law. The figure shows an experimental result obtained by using the swing-up control law based on the energy control proposed by Wiklund *et al.*(1993). Since the control law does not control the cart position until the pendulum reaches the upright position, the cart travels almost $1.0[m]$ from the initial position during the swing up control.

The reason why the proposed method gives better results than the energy method is as follows; The proposed control law consists of not only energy of the pendulum but also variables of cart position. That is, the control law is given as the feedback law which regulates both of the pendulum and the cart displacement, whereas the conventional method takes only the energy into account and controls the pendulum angle only.

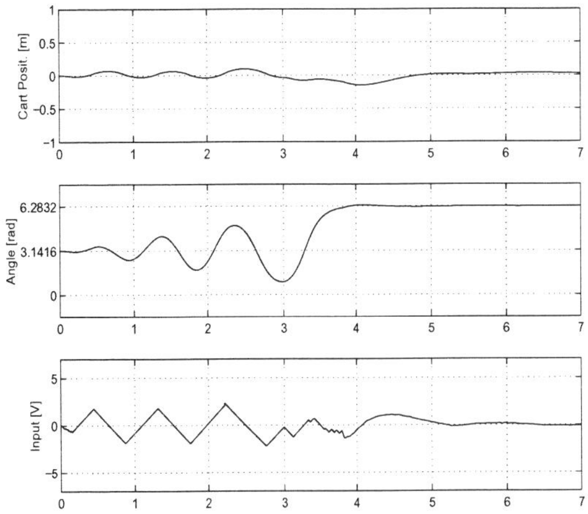


Figure 5: Experimental Results 1

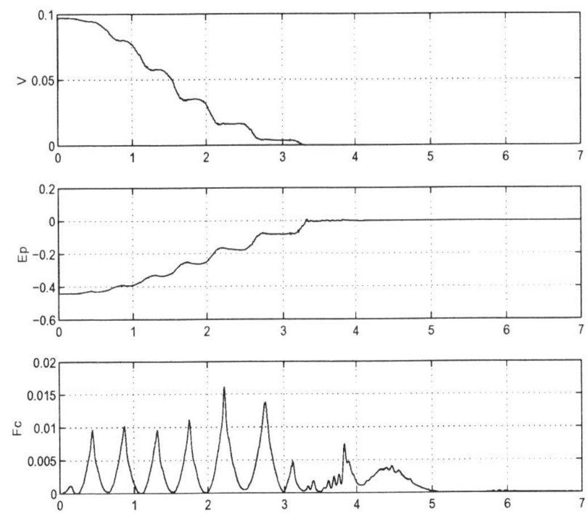


Figure 6: Experimental Results 2

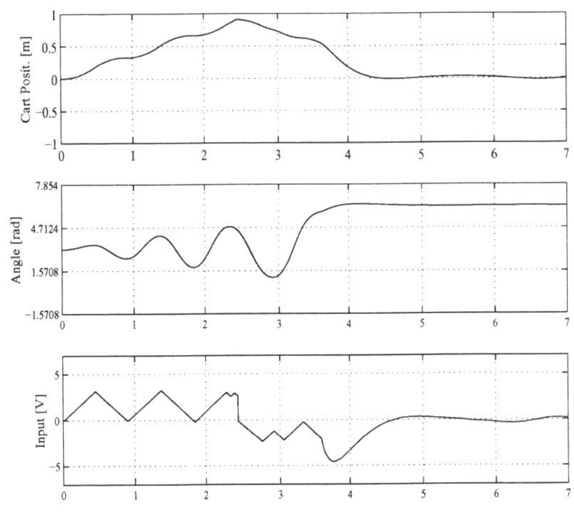


Figure 7: Conventional method

## 5. CONCLUSION

In this paper a nonlinear control to swing up a pendulum and control the cart to the initial position is proposed. To derive the control law, a new Lyapunov function is defined as the square of the sum of the absolute value of the total mechanical energy of the pendulum and the square of the cart position and the cart velocity. The function is positive definite and takes the value 0 only at the upright position of the pendulum and the initial position of the cart. The control law is obtained so as to make the derivative of the function with respect to time be negative definite, to decrease the value of the function and to make the controlled system approach the upright position of the pendulum and the initial position of the cart.

If the weight given to the terms of the cart position and the cart velocity in the Lyapunov function is made 0, then the proposed control law reduces to the law based on energy control given by Wiklund and others (1993). Hence the control law in this paper is an extension of the law based on energy control. Simulation and experimental results are given to show the effectiveness of the proposed swing up control law.

## REFERENCES


Astrom, K.J., and K. Furuta (2000). Swinging up a pendulum by energy control, *Automatica*, **36**, No.2, pp.287-295.

Chung, C.C., and J. Hauser (1995). Nonlinear Control of a Swinging Pendulum, *Automatica*, **31**, No.6, pp.851-862.

Henmi, T., X. Xin, T. Ochiiwa and M. Kaneda (2000). Swing up control for Inverted Pendulum via Hybrid Control Schemes. Preprint of *Japanese Joint Automatic Conference*, pp.13-16. (*In Japanese*)

M.Wiklund, A.Kristenson and K.J.Astrom (1993). A New Strategy for Swinging up an Inverted Pendulum, Preprint of *IFAC 12th world Congress*, Sydney, Australia, **9**, pp.151-154.

Inoue, A., T. Emoto, and Y. Hirashima (1999). Swing-up Control of Serial Double Inverted Pendulums Using Robustness of Sliding Mode Control. Proceedings of *JSME Dynamics and Design Conference*, **13**, pp.9-10. (*In Japanese*)

Lin, Z., A. Seberi, M. Gutmann and Y. A. Shamash (1996). Linear Controller for an Inverted Pendulum Having Restricted Travel: A High- and-Low Gain Approach, *Automatica*, **32**, No.6, pp.933-937.

Lozano, R., and I. Fantoni (1998). Passivity Based Control of the Inverted Pendulums. In:*Perspectives in Control* (Normand-Cyrot Ed.). pp.83-95, Springer.

Wei, Q., W. P. Dayawansa and W. S. Levine (1995). Nonlinear Contoroller for an Inverted Pendulum Having Restricted Travel. *Automatica*, **31**, No.6, pp.841-850.

IFAC Publications
www.elsevier.com/locate/ifac

# APPLICATION OF PCA FOR DETECTING AND DIAGNOSING ABNORMAL PROCESS BEHAVIOUR

**Uwe Kruger[1*] and David J. Sandoz[2]**

[1]*Intelligent Systems and Control Group, Queens University of Belfast, BT9 5AH, U.K*
[2]*Control Technology Center Ltd., Manchester Incubator Building, M13 9XX, U.K.*

Abstract: In this paper, the application of principal component analysis (PCA) for the monitoring of industrial processes is discussed. Particular focus is drawn on the detection and the diagnosis of abnormal process behaviour. Although numerous publications present examples of PCA applications, it is shown in this paper that the diagnosis of abnormal process behaviour may be incorrect and misleading. To give an examples of such misleading diagnosis, an application study relating to the realistic simulation of a fluid catalytic cracking unit is presented. *Copyright © 2001 IFAC*

Keywords: Industrial Production Systems, Large-Scale Systems, Statistical Process Control, Fault Detection, Fault Diagnosis

## 1. INTRODUCTION

The pressure of increasing competition and market fluctuation is forcing the chemical process industry to use resources ever more efficiently. However, it is necessary to guarantee a defined level of product specification whilst producing at low-running costs. Additionally, operating the process reliably and safely has increasingly gained attention because of the related risk of increased cost level and maybe even loss of production.

Key elements in achieving the above objectives are to observe and control the industrial process appropriately. Observing, or monitoring, entails evaluating how the process currently operates whilst controlling addresses the issue of keeping the product quality within the defined specification.

Difficulties arise frequently due to the typically large number of highly correlated process variables that are measured and recorded. In order to overcome this problem, several multivariate statistical methods (Kresta, *et al.*, 1991; MacGregor, *et al.*, 1991; Wise, *et al.*, 1996), such as Principal Component Analysis (Wold, *et al.*, 1987; Jackson, 1991) have been successfully developed and applied over the last decade. These techniques describe significant process variation by a reduced number of variables.

Numerous publications discuss the application of PCA for detecting and diagnosing abnormal process behaviour, e.g. (Dunia, *et al.* 1996; Martin and Morris, 1996; Vedam and Venkatasubramanian, 1999). For the diagnosis of anomalous process behaviour, however, PCA may give an incorrect and misleading indication of the root cause. This deficiency is demonstrated here by an application study concerning the simulation of a fluid catalytic cracking unit.

The organisation of this paper is as follows. A brief description of PCA is presented in section 2 prior to the discussion of monitoring plots for detecting abnormal process behaviour. Then, the plots for

* Correspondence should be sent to:
uwe.kruger@ee.qub.ac.uk

diagnosing abnormal process behaviour are described in section 4. This is followed by an application study concerning the simulation of a fluid catalytic cracking unit (FCCU) in section 5. Finally, the conclusions of this paper are presented in section 6.

## 2. PRINCIPAL COMPONENT ANALYSIS

PCA is used to describe significant variation of a process by defining a considerably reduced dimensional variable set. In fact, PCA carries out a basis transformation from the Euclidean bases into a rotated basis. For a matrix in which measured data points are stored as row vectors in subsequent order (data matrix), the rotation results in the product of two matrices. These matrices are a matrix, which contains the new co-ordinates, and a matrix that contains the new basis vectors of the rotated basis. This matrix product can be further described as the sum of rank one component matrices. These component matrices are constructed in order to maximise the contribution of each of them towards the original data matrix. The component matrices are arranged so that the first one contributes mostly to the variation of the data matrix, the second component matrix has the second largest contribution and so on. In cases where the process data are highly correlated, only the first few component matrices are required to estimate the original data matrix. The remaining component matrices capture accordingly noise and insignificant process variation. The new co-ordinates are also defined as scores or principal components (PCs), whilst the new basis vectors are referred to as the loading vectors. In fact, the loading vectors are the eigenvectors of the covariance matrix obtained by data stored in the data matrix. A detailed description of PCA is presented in (Jackson, 1991; Wise, *et al.*, 1996) for example.

## 3. DETECTION OF ABNORMAL PROCESS BEHAVIOUR

For highly correlated process variables, the application of PCA splits the data matrix into a sum of component matrices that describes significant variation and a sum of component matrices that represents insignificant variation and noise. Wise, *et al.* (1996) outlined that each matrix corresponds to a statistic, denoted as Hotelling's $T^2$ or $T^2$ statistic, and SPE statistic. Assuming that:

- the data matrix contains $m$ process variables (column vectors of the data matrix),
- the first $r$ component matrices represent significant contribution to the data matrix and
- the remaining $m$-$r$ component matrices represent insignificant contribution to the data matrix,

the $T^2$ and SPE statistics for the $k^{th}$ instance in time may be defined as:

$$\mathrm{T}_k^2 = \sum_{i=1}^{r} \frac{t_{ki}^2}{\lambda_i^2}$$
$$\mathrm{SPE}_k = \sum_{j=1}^{m} \frac{\left(z_{kj} - \hat{z}_{kj}\right)^2}{\sigma_j}, \quad (1)$$

where:

- $\mathrm{T}_k^2$ and $\mathrm{SPE}_k$ represent the $T^2$ and SPE statistic, respectively,
- $t_{ki}$ is the value of the $i^{th}$ score at time instance $k$,
- $\lambda_i$ is the standard deviation of the $i^{th}$ score variable,
- $z_{ki}$ is the value of the $j^{th}$ process variable at time instance $k$,
- $\hat{z}_{ki}$ represents the "reconstructed" value of $z_{ki}$ if only $r$ component matrices are used to reconstruct the data matrix and
- $\sigma_j$ represents the standard deviation of the mismatch between $z_{ki} - \hat{z}_{ki}$.

The values of the $T^2$ and the SPE statistic can respectively be depicted in monitoring plots with a time base. To test the hypothesis that the process behaves normally a hypothesis test has to be carried out for each statistic. The confidence limits for the hypothesis test are usually selected to include 95% and 99% of the values of the $T^2$ and SPE statistic which correspond to normal process behaviour.

## 4. DIAGNOSIS OF ABNORMAL PROCESS BEHAVIOUR

If at least one of the above statistics indicate that the process behaves abnormally, the question of what has caused this event arises. For PCA, abnormal process behaviour may be diagnosed by using contribution charts for each univariate statistic (Wise, *et al.*, 1996; Martin and Morris, 1996). These are bar charts for which each process variable represents a bar and the bar height corresponds to its value. The contribution of the process variables to the $T^2$ statistic at the $k^{th}$ instance in time can be determined as discussed by Wise and Gallagher:

$$\mathbf{t}_k = \mathbf{P}_r^T \mathbf{z}_k$$
$$\mathbf{t}'_k = \mathbf{\Lambda}_r^{-1} \mathbf{t}_k$$
$$T_k^2 = \mathbf{t}'^T_k \mathbf{t}'_k, \quad (2)$$
$$\tilde{\mathbf{z}}_k = \mathbf{P}_r \mathbf{t}'_k$$

where:

- $\mathbf{z}_k$ is a vector storing the value of each process variables at time instance $k$,
- $\mathbf{P}_r$ is a transformation matrix to project $\mathbf{z}_k$ onto the rotated co-ordinate system,
- $\mathbf{t}_k$ represent the first $r$ co-ordinates (scores) of $\mathbf{z}_k$ in the rotated co-ordinate system,
- $\mathbf{\Lambda}_r$ is the covariance matrix of the score variables,

- $\mathbf{t}'_k$ is a vector storing the scaled elements of $\mathbf{t}_k$ and
- $\widetilde{\mathbf{z}}_k$ represents a vector storing the contribution of each process variable towards the $T^2$ statistic.

An alternative approach is to provide a bar chart for the obtained value for each process variable at the $k^{th}$ instance in time as advocated by Miller, *et al.* (1998). A contribution chart for the SPE statistic can directly be obtained from equation (1). Assuming that the mismatch for each process variable represents a normally distributed random variable, the contribution chart for the SPE statistic can be augmented to include upper and lower confidence limits for including 95% and 99% of the mismatch corresponding to normal process behaviour.

## 5. APPLICATION STUDY

A fluid catalytic cracking unit or FCCU is an important economic unit in refining operations. It typically receives several different heavy feedstocks from other refinery units and cracks these streams to produce lighter, more valuable components that are eventually blended into gasoline and other products. The principal feeds of this particular Model IV unit, described by McFarlane, *et al.* (1993) is gas oil, but heavier diesel and wash oil streams also contribute to the total feed stream. Fresh feed is preheated in a heat exchanger and furnace and then passed to the riser, where it is mixed with hot, regenerated catalyst from the regenerator. Slurry from the main fractionator bottoms is also recycled to the riser. The hot catalyst provides the heat necessary for the endothermic cracking reactions. The gaseous cracked products are passed to the main fractionator for separation. Wet gas off the top of the main fractionator is elevated to the pressure of the lights end plant by the wet gas compressor. Further separation of light components occurs in this light ends separation section.

As a result of the cracking process, a carbonaceous material, coke, is deposited on the surface of the catalyst, which depletes its catalytic property. Spent catalyst is recycled to the regenerator where it is mixed with air in a fluidised bed. Oxygen reacts with the deposited coke to produce carbon monoxide and carbon dioxide. Air is pumped to the regenerator with a high-capacity combustion air blower and a smaller lift air blower. In addition to contributing to the combustion process, air from the lift air blower assists with catalyst circulation. Complete details of the mechanistic simulation model for this particular model IV FCCU can be found in (McFarlane, *et al.*, 1993).

The selected predictor variables for the FCCU case study are the Wash Oil, the Diesel and the Slurry Flowrate as well as the Preheater Outlet Temperature from the feed section of the unit. All of these variables belong to the feed section of the unit. To simulate realistic disturbance conditions, various different AutoRegressive Integrated Moving Average (ARIMA) (Box and Jenkins, 1971) signals were superimposed on the Slurry Flowrate and the Preheater Outlet Temperature. The Diesel Flowrate received only an AutoRegressive Moving Average (ARMA) (Box and Jenkins, 1971) signal and the Wash Oil Flow was held constant.

The response set included Excess Oxygen in the Flue Gas, Concentration of Carbon Monoxide in the Flue Gas, Riser Temperature, Regenerator Bed Temperature, Regenerator Standpipe Level, Reactor Pressure as well as a further six variables measured from the FCCU system (MacFarlane, *et al.*, 1993).
The FCCU simulator is augmented to include several pre-programmed faults that could be applied on command. The first fault was a step change to the coke formation factor of the feed, which simulated a plug of heavier-than-normal feed entering the unit. The second fault simulates a disruption in the flow of regenerated catalyst between the regenerator and riser, which is typically caused by partial or complete plugging of steam injectors located in this line.

For the identification of a PCA model, a reference data set containing 3127 sample points at a sample rate of 30 minutes was simulated. Cross validation (Wold, 1978) reveals that 3 PCs are required to represent significant variation of the FCCU process. A data set for each fault was recorded, containing a sequence of normal operation prior to the injection of the fault.

The results for detecting and diagnosing the first fault are presented in Figures 1 to 3. Since this fault simulates a change in composition of the feed –a plug of heavier feed–, its effect is felt immediately in the riser and subsequently in other parts of the unit that are affected by a change in riser conditions. A plug of heavier will cause a rapid increase in the amount of coke deposited on the catalyst in the riser and transported to the regenerator, having a direct effect on oxygen consumption and production of carbon monoxide. Therefore, the concentration of excess oxygen and carbon monoxide in the flue gas are most dominantly affected by this event.

Both the $T^2$ and the SPE statistic (Figure 1) correspond to this abnormal event immediately after the fault was injected. For diagnosing this event, the contribution charts of the $T^2$ statistic (Figure 2) by Wise and Gallagher (upper chart) and Miller, *et al.* (1998) (lower chart) and the SPE statistic (Figure 3) are presented. For each contribution plot, the first 11 "bars" represent the response variables and the last 4 "bars" represent the predictor variables. Note that the $9^{th}$ and the $10^{th}$ bar correspond to the concentration of excess oxygen and carbon monoxide, respectively. The contribution charts for the $T^2$ by Wise and Gallagher and the SPE statistic provide and incorrect diagnosis since, the 3rd predictor variable and the 9th and 10th response variable appear to be associated with the root cause of this event. It should be noted that there is not controller feedback and hence, the predictor variables cannot be affected by the abnormal process behaviour. Only the contribution chart by Miller, *et al.* (1998) provides a correct diagnosis of this event.

The results for detecting and diagnosing the second fault are presented in Figures 4 to 6. Any change in flow of regenerated catalyst will affect the material balance in the standpipe, and hence its level. A change in regenerated catalyst flow will also affect catalyst-to-feed ratio in the riser, resulting in a change in the amount of coke deposited on spent catalyst and subsequently the level of oxygen usage in the regenerator. Therefore, the standpipe level and the excess oxygen in the flue gas are most dominantly affected by this event.

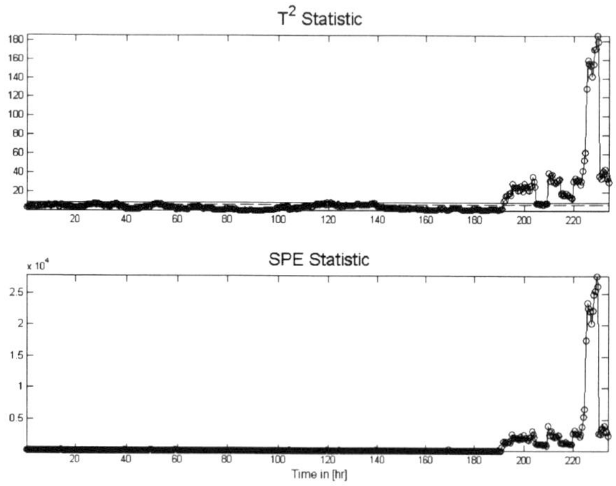


Fig. 1. $T^2$ and SPE statistic on data set representing the step change in the coke formation factor.

The disturbance in the regenerated catalyst flow is detected immediately after the fault was injected by both the $T^2$ and the SPE statistic (Figure 6). The diagnosing of this abnormal process behaviour is again presented by the contribution charts of the $T^2$ statistic (Figure 5) by Wise and Gallagher (upper chart) and Miller, *et al.* (1998) (lower plot) and the SPE statistic (Figure 6). Note that the 9th and the 11th bar correspond to the concentration of the excess oxygen and the standpipe level, respectively.

A confusing diagnosis is provided by the contribution charts of the $T^2$ statistic by Wise and Gallagher and the SPE statistic. These charts indicate that the 2nd and 3rd predictor variables as well as the 3rd 9th and 11th response variable are affected by this event. This is again incorrect because there is no direct mechanistic path back to any part of the feed system. However, a correct diagnosis is provided by plotting the values of the process variables in a bar chart as proposed by Miller, *et al.* (1998).

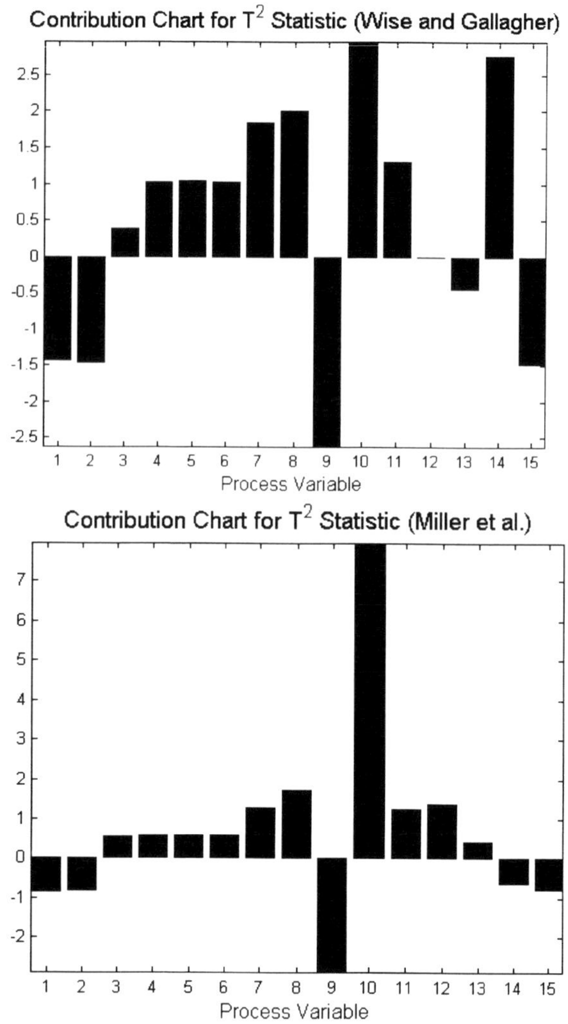


Fig. 2. Contribution chart for $T^2$ statistic by Wise and Gallagher and Miller et al. corresponding to the step change in the coke formation factor.

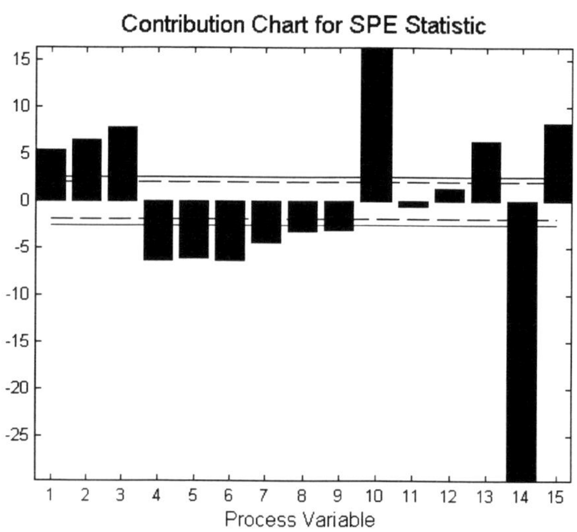


Fig. 3. Contribution chart for SPE statistic corresponding to the step change of the coke formation factor

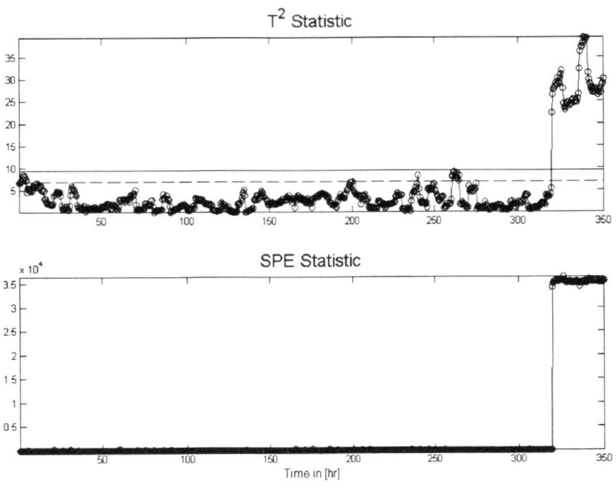


Fig 4. $T^2$ and SPE statistic on data set representing the disruption of the regenerated catalyst flow.

## 6. CONCLUSIONS

In this paper, the application of PCA for process monitoring is described and the detection and diagnosis of abnormal process behaviour briefly reviewed. The contribution of this paper is that PCA may lead to an incorrect and misleading diagnosis of abnormal process behaviour. This is particularly true for when the diagnosis is based on the mismatch between the measured values of the process variables and the reconstructed values from the PCA model.

An example of these problems is presented on the basis of a realistic simulation of a fluid catalytic cracking unit for which two fault conditions are considered. In both of the simulated fault conditions, some of the PCA diagnosis charts provide a confusing diagnosis and only the diagnosis chart advocated by Miller, *et al.* (1998) gives a correct diagnosis. However, this diagnosis chart relates to the measured values of the process variables. In cases where only the mismatch between the measured and reconstructed values of the process variables indicate anomalous process behaviour, the diagnosis chart by Miller, *et al.* (1998) is not applicable.

## REFERENCES


Box, G.E.P. and G.M. Jenkins, (1971). *Time Series Analysis*, Holden Day, San Francisco, (1971)

Chen, Q., R.J. Wynne, P.R. Goulding and D.J. Sandoz (2000). The Application of Principal Component Analysis and Kernel Density Estimation to Enhance Process Monitoring, *Control Engineering Practice*, **8**, 531-543

Dunia, R., S.J. Qin, T.F. Edgar and T.J. McAvoy (1996). Identification of Faulty Sensors Using Principal Component Analysis, *Process Syst. Eng.*, **42**, 2797-2812

Jackson, J. E. (1991), A User's Guide to Principal Components, Wiley, New York


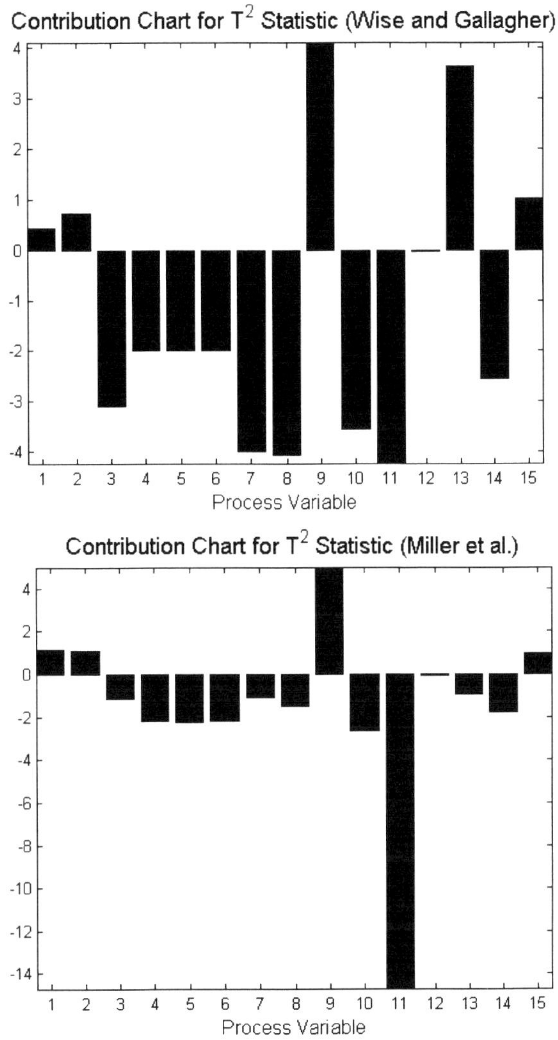


Fig. 5. Contribution chart for $T^2$ statistic by Wise and Gallagher and Miller et al. corresponding to the disruption of the regenerated catalyst flow.

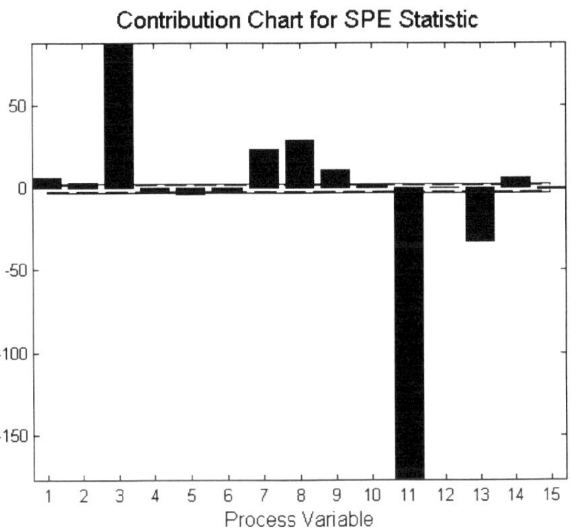


Fig. 6. Contribution chart for SPE statistic corresponding to the disruption of the regenerated catalyst flow.


Kresta, J.V., J.F. MacGregor and T.E. Marlin (1991). Multivariate Statistical Monitoring of Process

Operating Performance, *The Canadian Journal of Chemical Engineering*, **69**, 35-47

MacGregor, J.F., T.E. Marlin, J. Kresta and B. Skagerberg (1991). Multivariate Statistical Methods in Process Analysis and Control. AIChE Symp. Proc., AIChE Publ. No. P-67, New York, 79-99

Martin, E.B. and A.J. Morris (1996). An Overview of Multivariate Statistical Process Control in Continuous and Batch Process Performance Monitoring, *Transactions of the Institute of Measurement and Control*, **18**(1), 51-60

McFarlane, R. C., R. C. Reineman, J. F. Bartee, and C. Georgakis (1993). Dynamic Simulator for a Model IV Fluid Catalytic Cracking Unit. *Computers Chem. Engng.*, **17**(3), 275-300.

Miller, P., R. E. Swanson and C. E. Heckler (1998). Contribution Plots: A Missing Link in Multivariate Quality Control, *Applied Mathematics and Computer Science*, **8**(4), 775-792

Vedam, H. and V. Venkatasubramanien (1999). PCA-SDG based Process Monitoring and Fault Detection, *Control Engineering Practice*, **7**, 903-917

Wise, B.M., N.B. Gallagher and J.F. MacGregor (1996). The process Chemometric Approach to Process Monitoring and Fault Detection, *Journal of Process Control*, **6**(6), 329-348

Wise, B.M. and N.B. Gallagher. PLS-Toolbox, Version 2.0, www.eigenvector.com

Wold, S. (1978). Cross Validatory Estimation of the Number of Components in Factor and Principal Component Models, *Technometrics*, **20**, 397-406

Wold, S., K. Esbensen and P. Geladi (1987). Principal Component Analysis, *Chemometrics and Intelligent Laboratory Systems*, **2**, 37-52

IFAC Publications
www.elsevier.com/locate/ifac

# FUZZY GENETIC MODELLING OF AIR-CONDITIONING SYSTEM FOR FAULT DIAGNOSIS

**C.Y.Kung and H.N.Lam**

Department of Mechanical Engineering
The University of Hong Kong
Pokfulam Road, Hong Kong
Phone: 852- 28592896
Fax: 852-28585415
E-mail: kung_chi_yau@hotmail.com

Abstract: This paper presents a methodology based on a combination of fuzzy logic and genetic algorithms which has been employed to analyse the operating conditions of a simulated variable-air-volume (VAV) air-conditioning system and to diagnose the faults when they are detected. The approach has been tested experimentally with simulated faulty process behaviour of the system and the results are presented.


Keywords: fuzzy logic, genetic algorithms, air-conditioning systems, fault detection

## 1.INTRODUCTION

Modern buildings are now being equipped with increasingly complex facility systems, the optimal operation of which has become more difficult to monitor in order to achieve the design conditions with maximum energy efficiency. For a heating, ventilating and air-conditioning (HVAC) system, although the sensor and control signals can provide potentially valuable information about the status of the system, and building automation systems (BAS) have the ability to monitor, store and report these signals, the routine checks carried out in practice, however, can only serve to ascertain that the values of the set points and critical variables are kept within the predefined limits. This may be adequate for the superficial detection of faults, but the task of identifying the underlying causes of the faults often requires a detailed and continuous analysis of the trend data up to the point of failure. Thus, there is a need for a robust fault detection and diagnosis tool that can be used to assist the building operators in monitoring the facility systems to ensure proper operation.

In this study, a typical variable-air-volume air-conditioning system is considered. Some commonly encountered system faults which will cause deterioration of operation are used as examples to illustrate the application of the fault detection and diagnosis scheme.

For the study of fault detection and diagnosis in connection with building facility systems, a number of researchers have proposed different methods.

Philip Haves, Timothy I. Salsbury and Johnathan A. Wright (Haves et al., 1996) employed the first principle model to describe a condition-monitoring scheme which involves the estimation of the values of model parameters which are expected to change in case a fault exists.

Henk C. Peitsman and Vincent E. Bakker (Peitsman et. al., 1996) applied the black-box models for fault detection and diagnosis. It described the behavior mathematically between the inputs and outputs of a process for a VAV system simulation.

Won-Yong Lee, John M. Home, Cheol Park and George E. Kelly used the residual and recursive parameter identification methods (Lee et. al., 1996a) and artificial neural network (ANN) (Lee et. al., 1996b) together to work for fault detection and diagnosis. Autoregressive with exogenous input (ARMAX) and autoregressive with exogenous input (ARX) models with both single-input/single-output (SISO) and multi-input/single output (MISO) structures were examined.

Arthur L Dexter and Mourad Benounarets (Dexter et. al., 1996) applied a qualitative mathematical fuzzy model to implement the FDD scheme in which the uncertainties and imprecision related to describe the behaviour of the system can be included. The method used reference models describing normal and faulty operation that are generated from simulation data.

C. Y. Kung and H. N. Lam (2000) applied fuzzy clustering method for the fuzzy modelling of the air-conditioning systems and the fuzzy decision tree induction method for knowledge extraction and the building of rules for the detection processes.

In this study, a typical variable-air-volume air-conditioning system is considered. The MATLAB environment is used to develop a dynamic modelling and simulation analysis of the system under normal and faulty operations. Typical fault conditions are modelled based on fuzzy logic with membership functions and fuzzy rules determined by using an optimization technique based on genetic algorithms. Testing of the entire fault detection and diagnosis scheme is carried out within the MATLAB environment.

## 2. COMPUTER SIMULATION OF VAV SYSTEM

The reference VAV air-conditioning system used in this study for evaluating the fault detection and diagnosis methods is shown in Figure 1. The system has outdoor and return air dampers, cooling and heating coils, an air-filter section, a supply fan, a return fan, and air plenum sections. The supply air is ducted to four VAV boxes, which supply conditioned air to four different zones. Each of them can have different room set temperatures. The return air fan takes air from the conditioned space and discharges it into the mixing section and/or the outside through a motorized exhaust air damper. Both supply and return fans are fitted with variable frequency controllers for regulating the static pressure in the ducts.

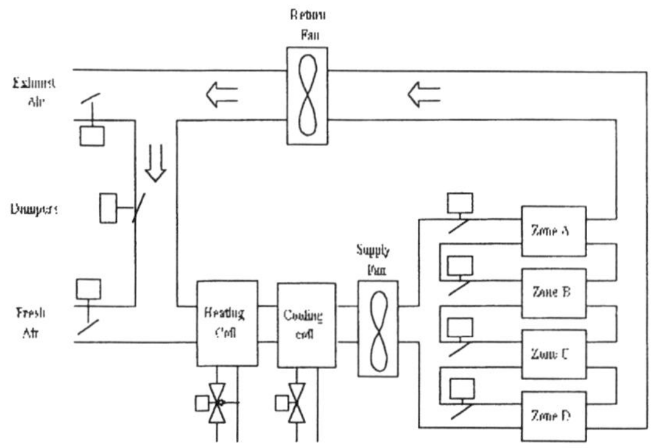


Fig.1 Block Diagram of the VAV system

## 3. FAULT MODELLING

There are many kinds of faults associated with the installation or the operation of an air-conditioning system. These are commonly found in the cooling coil and the variable-air-volume terminal boxes. Two types of degradation faults, fouling on the cooling coil and the leakage of air ducts have been selected for study.

These faults are chosen because they are not easy to detect. Once they exist, they will not cause sudden nuisances to the environment. Moreover, it can be seen from Figures 9, 10 and 11 that the behaviour of the system at faulty operation is quite similar to the correct operation for the whole operating profile. Also, once the leakage is detected, it is rather difficult to check the degree of leakage which will affect the normal operation of the system.
As fouling is caused by the contaminating substance like calcium carbonate inside the coil which results in a reduced rate of heat transfer, this can thus be modelled by changing the value of the heat transfer coefficient of the coil by a certain factor. For the air leakage problem, this can be

modelled by introducing a split within the transportation duct.

## 4. GENETIC ALGORITHMS

Genetic algorithms (GAs) are general search and optimization algorithms (Goldberg, 1989) inspired by the natural biological processes like gene crossover and mutation as shown. In this paper, genetic algorithms have been used to optimize the membership values of the fuzzy sets and the selection of fuzzy rules for the fault detection and diagnosis system. Generally speaking, genetic algorithms can be divided into two main categories, the binary genetic algorithm (BGA) and the real-coded genetic algorithm (RCGA), or known as continuous parameter genetic algorithm. The latter formulation has been employed in this study.

### 4.1 Continuous Parameters genetic algorithm

It is generally convinced that binary genetic algorithm can solve many optimization problems. However, if the parameters in the specific problem are continuous, it is inevitable that the size of the chromosomes will grow quickly. Thus, it is more logical to represent them by floating point numbers (Michalewicz et. al., 1993) (Michalewicz, 1996). In addition, since the BGA has its precision limited by the binary representation of parameters, using real numbers instead will easily allow representation to the required system precision. The RCGA also has the advantage of requiring less storage than the binary genetic algorithm because a single floating point number represents the parameters instead of n-bit integers. This is because as n increases, the storage requirement will become significant. It follows therefore that the RCGA is superior to the BGA in terms of accuracy of representation and reduced storage requirement.

### 4.2 Genetic Operations of the rule selection problems

In genetic algorithms, a rule set S is treated as an individual. In this paper, fuzzy logic is applied to evaluate the fitness of the chromosomes. Figure 5 shows the flow chart of the evaluation procedure. First, one chromosome is chosen from a new population and each bit inside the chromosome is converted to become the output of each fuzzy rule. The training data are then put into the rules for the evaluation of the fitness and the whole working profile is shown below.

The following genetic operators are employed to generate the population set. (Steps 1 to 6 are iterated until a pre-specified termination condition is achieved.)

Step 1: Population Initialization

Generate an initial population containing the Nvar strings where Nvar is the number of variables in the population. In this operation, according to the number of predetermined fuzzy sets N, each chromosome in the population is generated by assigning a number 1 to N randomly to specify the fuzzy sets each variable belongs to.

Step 2: Evaluation of rules by fuzzy logic

Step 3: Rule extraction and termination check

Once the fitness is evaluated, if a pre-specified stopping condition is satisfied, the flow stops.

Step 4: Parents Selection

In this paper, roulette wheel selection has been applied. First, the sum of the fitness values over all the individuals in the current population is calculated. They are then mapped onto the intervals [0, Sum]. The size of each individual corresponds to the fitness value of the associated individual. To select an individual, a random number is generated in the interval [0, Sum] and the individual whose segment spans the random number is selected.

Step 5: Crossover, Mutation

For each selected pair, uniform crossover strategy has been used to select the site for crossover. This strategy makes every locus a potential crossover point. A crossover mask, the same length as the chromosome structures is created at random and the value of the bits indicates which parent will supply the offspring with which bits.

In this paper, Wright's heuristic crossover (Wright, 1991) has been employed to find the product of crossover. If C1, C2 are the parents, a random number r [0,1] is chosen so that the new offspring will become $r \times (C1 - C2) + C1$ in which C1 is the parent with better fitness.

For mutation, random mutation has been applied. For two parents C1 and C2, a random number is selected from the domain [C1, C2].

Step 6: Elimination of weak members. Back to Step 2.

## 5. FAULT DETECTION AND DIAGNOSIS SCHEME

A fault detection and diagnosis scheme is used to determine the degree of belief (or degree of firing) of each of the possible rules in the models obtained from the training data. The degree of belief is calculated from a measure to which the data are members of the fuzzy sets associated with the antecedent and consequent of the rule. The value of the belief will be 0, if the associated rule fails to describe the relationship between the data and 1 if it describes the data correctly.

An overview of the detection process is shown in Figure 2.

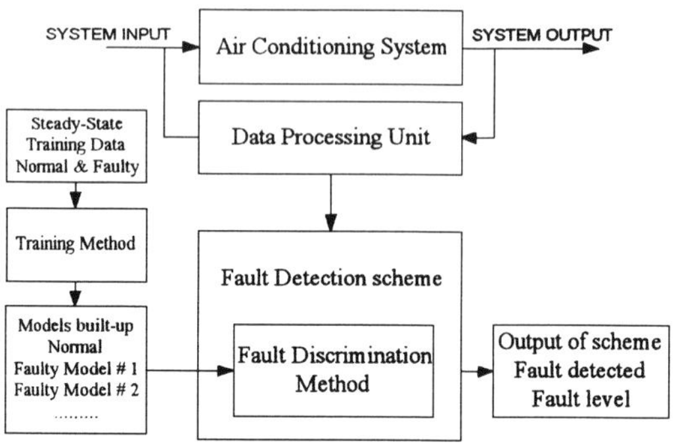


Figure 2 The fault detection and diagnosis procedure

Also, in actual practice, after the data are collected, they are generally given some treatment to make them more suitable and rational for fault detection purposes. The treatment used is the application of a moving average filter, which can remove the high frequency noises.

The fault detection scheme used in this paper is based on the method proposed by Xu and Lu [Xu et. al., 1987) for a general fuzzy model defined by the rules generated from the fuzzy genetic algorithm

If ( $X_1$ is $A_{1,s1}$ ) and ( $X_2$ is $A_{2,s2}$ ) ...
and ( $X_m$ is $A_{n,sm}$ )

THEN ( Y is $B_s$ )
where the reference sets for $X_1, X_2, \ldots, X_m$ and Y are $A_{1,1}$ to $A_{1,r1}$, $A_{2,1}$ to $A_{2,r2}$,...etc. and $A_{m,1}$ to $A_{m,rm}$ and $B_1$ to $B_r$ respectively.

If there are N sets of data,

$[\, X_1(1), X_2(1), \ldots, X_m(1), Y(1) \,]$
........
........
$[\, X_1(N), X_2(N), \ldots, X_m(N), Y(N) \,]$

then the estimates of the degree of belief are given by

$$r(s_1, s_2, \ldots, s_m) = \bigcup_{K=1}^{N} r_K(s_1, s_2, \ldots, s_m)$$

where

$$\hat{r}_K(s_1, s_2, \ldots, s_m) = MIN[Poss_{1,s1}(K), \ldots, Poss_{m,sm}(K), Poss_s(K)]$$

i = 1...m ( number of inputs )
j = 1...$r_i$ ( number of reference sets )

and

$$Poss_{ij}(K) = \text{Possibility}\,[A_{ij} \mid X_i]$$

J = 1...r (number of reference sets)

## 6. APPLICATION OF THE SCHEME

The performance of the diagnosis is assessed using a detailed dynamic simulation of a 4-zone variable-air-volume air-conditioning system for a typical office building.

By using the procedures described above, the reference models can be obtained off-line using input-output training data generated from the simulation program. For the genetic algorithm session, a crossover probability of 0.7 and a mutation probability of 0.05 have been used. Also, the maximum number of generations has been tested experimentally to be around 1000 and 1200. The three operating models, i.e. normal operating, supply air leakage and chilled water leakage, can be identified.

The models generated include a 15% and a 30% fouled cooling coil models, and a 2.5% and a 5 % leakage models. The effects of the fouled cooling coil and the leakage fault can be observed from Figure 4 and Figure 6.

Some samples of the rules obtained are shown below.

RULES for the faulty condition:

1. IF m is VL and T is L THEN t is L
2. IF m is LM and T is LM THEN t is L
3. IF m is MH and T is LM THEN t is M
4. IF m is MH and T is MH THEN t is H
5. IF m is H and T is L THEN t is H
6. IF m is H and T is H THEN t is H

RULES for the normal condition:

1. IF m is L and T is L THEN t is L
2. IF m is LM and T is LM THEN t is L
3. IF m is M and T is LM THEN t is L
4. IF m is M and T is MH THEN t is M
5. IF m is VH and T is L THEN t is H
6. IF m is VH and T is H THEN t is H

where m is the supply air flow rate, T is the outside air temperature and t is the supply air temperature and the reference sets for m, T and t are {VL, L, LM, M, MH, SH, H, VH}, {VL, LM, MH, H} and {L,M,H} respectively.

Based on the rules obtained, the training data of a specific fault are used as the test data to be compared with the normal operating condition and the other faults. The results of the diagnosis are given in the Table 1.

For testing of the fault detection and diagnosis scheme, a building/HVAC emulator can be used, which is capable of simulating the behaviour of a building and its HVAC system in real time. A computer hardware interface allows the emulator to be connected to a real BAS for process monitoring and control. The operating data of the air-conditioning system under normal and fault conditions can be automatically logged into the BAS for further processing based on the fault detection and diagnosis scheme.

## 7. CONCLUSIONS AND DISCUSSIONS

The work presented in this paper is on the development of a fault detection and diagnosis scheme for a variable-air-volume air conditioning system based on fuzzy logic and genetic algorithm. The proposed scheme can be used for diagnosing a series of faults which are commonly found in the air conditioning system. Due to constraints of time and resources, the performance of the fault detection and diagnosis scheme has only been studied using simulated data. The results showed

Table 1 Degree of belief estimated from simulated test data

| Degree of Belief | Air Leak | Fouled Coil | Normal |
|---|---|---|---|
| Normal | 42% | 20% | 100% |
| Air Leak | 100% | 60% | 72% |
| Fouled Coil | 55% | 100% | 32% |

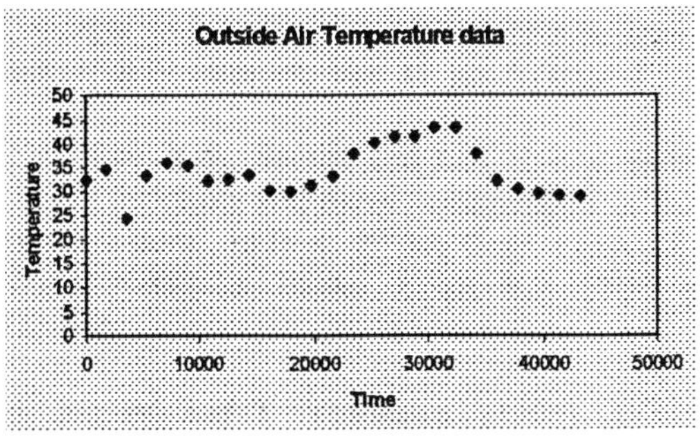


Figure 3 Outdoor Air Temperature Training data

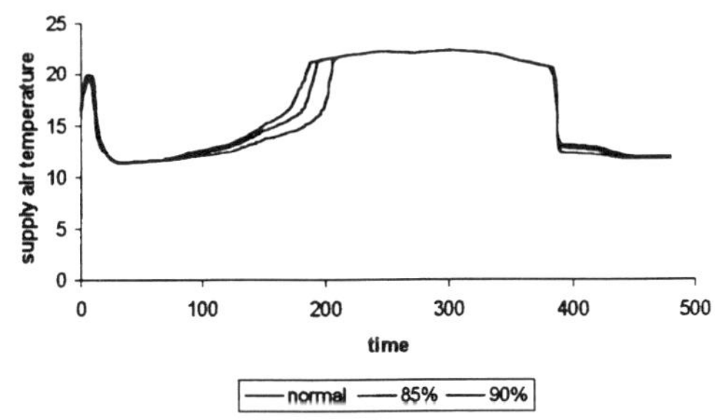


Figure 4 The variation of supply air temperature under the effect of air leakage

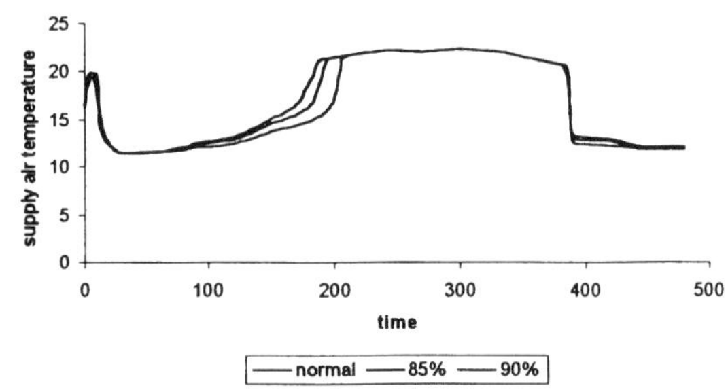


Figure 5 The variation of supply air flow rate under the effect of air leakage

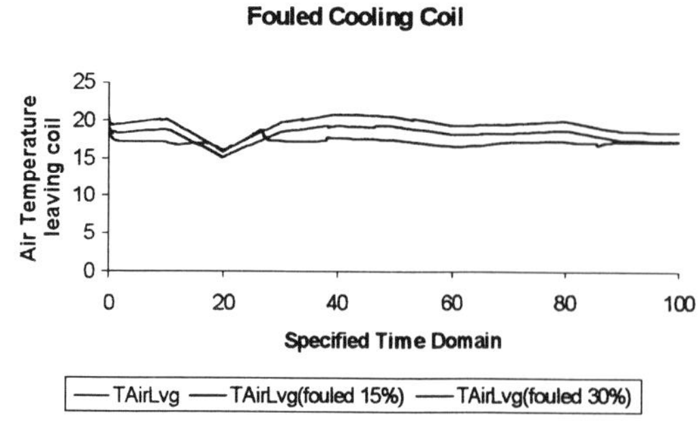


Figure 6 The variation of supply air temperature under the effect of fouled coil

that the proposed scheme is capable of identifying certain kinds of faults in the system.

Also, it is undoubtedly that the scheme can be worked more accurately if the generation of the rule models are based on real processing data and the database is covered enough on all the faulty and normal operating conditions.

It is envisaged that the successful incorporation of an automatic fault detection and diagnosis scheme into a real BAS will be a significant contribution towards reliable system operation, maintenance of comfort within the built environment, enhancement of energy efficiency and cost savings.

## 8. REFERENCES


Arthur L Dexter, Mourad Benounarets, (1996). "A Generic Approach to Identifying Faults in HVAC Plants" ASHRAE Technical Data Bulletin, volume 12, number 2

A. Wright, (1991). Genetic Algorithms for Real Parameter Optimization. Foundation of Genetic Algorithms, p.205-218

C.-W. Xu and Y.-Z. Lu, (1987). "Fuzzy model identification and self-learning for dynamic systems", IEEE Trans. Syst., Man, Cybern., vol. SMC-17, July/Aug, pp. 683-689

C. Y. Kung and H. N. Lam (2000) "Fuzzy Modelling of Air-Conditioning systems for Fault Detection", IASTED International Conference on Modelling, Identification and Control P.542-547

Cezary Z. Janikow and Zbigniew Michalewicz, (1993). An Experimental comparison of Binary and Floating Point Representations in Genetic Algorithms, Proceedings of International Forth International Conference on Genetic Algorithms, P.31-36

D. E. Goldberg. (1989). Genetic Algorithms in Search, Optimization, and Machine Learning, Addison-Wesley

Henk C. Peitsman and Vincent E.Bakker, (1996). "Application of Black-box Models to HVAC Systems for Fault Diagnosis", ASHRAE Technical Data Bulletin, volume 12, number 2

K. F. Man et al. (1996), Genetic Algorithms: Concepts and Applications IEEE Transactions on Industrial Electronics, Vol. 43, no. 5 October

K-M. Lee, K-A. Seong, and L-K Hyung, (1992). "Fuzzy matching and fuzzy comparsion in fuzzy expert system", in Proc. 2nd Int. Conf. Fuzzy Logic Neural Networks, Iizuka, Japan, July 17-22, pp. 313-316

M. Sugeno and T. Yasukawa, (1993). A fuzzy logic approach to qualitative modeling, IEEE Trans. Fuzzy Syst., vol. 1, pp. 7-31

Philip Haves, Timothy I. Salsbury and Jonathan A. Wright, (1996). "Condition Monitoring in HVAC Subsystems Using First Principles Models", ASHRAE Technical Data Bulletin, volume 12, number 2

R. R. Yager and D. P. Filev, (1994) Essentials of Fuzzy Modeling and Control, New York: Wiley.

Won-Yong Lee, John M. Home, Cheol Park and George E. Kelly, (1996a). "Fault Detection in Air-Conditioning Using Residual and Recursive Parameter Identification Methods", ASHRAE Transaction 102(1)

Won-Yong Lee, John M. Home, Cheol Park and George E. Kelly, (1996b). "Fault Diagnosis of an Air-Handling Unit Using Artificial Neural Networks", ASHRAE Transaction 102(1)

Z. Michalewicz (1996). Genetic Algorithms + Data Structures + Evolution Programs 3 rd Edition, Springer-Verlag

IFAC
Publications
www.elsevier.com/locate/ifac

# A MODEL REDUCTION METHOD FOR MONITORING LARGE-SCALE PROCESSES

**Uwe Kruger[1*], Xun Wang[2] and S. Joe Qin[3]**

[1]*Intelligent Systems and Control Group, Queens University of Belfast, BT9 5AH, U.K*
[2]*Department of Mechanical Engineering, The University of Manchester, M13 9PL, U.K.*
[3]*Department of Chemical Engineering, The University of Texas at Austin, TX 78712, U.S.A*

Abstract: In this paper, a novel identification method, termed as latent variable least squares (LVLS) is introduced. LVLS describes linear steady-state process behavior by a reduced dimensional model. This is particularly useful for large-scale processes, which often present a considerable number of highly correlated process variables. Additionally, it is shown that LVLS can be utilized for process monitoring. In an application study on a real industrial process, it is shown that LVLS performs better than partial least squares, which is a competitive method that is similar in approach.


Keywords: Model Reduction, Industrial Production Systems, Statistical Process Control, Large-Scale Systems, Fault Detection

## 1. INTRODUCTION

The importance of detecting and diagnosing abnormal process behavior in its early stages has been widely recognized in process engineering. However, the instrumentation of large-scale processes often present a considerable number of highly correlated process variables which are frequently sampled, (MacGregor, et al., 1991). This makes it difficult for even an experienced operator to detect abnormal process operation by eye. However, the recorded data has embedded within it information that can reveal the current state of process operation. In fact, the difficulty lies in extracting this information from the data, (Piovoso and Kosanovich, 1992).

The problem of high correlation has led to the use of data reduction methods, such as partial least squares (PLS) which aim to describe significant process variation by a significantly reduced set of "latent" process variables, (Wise, et al., 1996). Furthermore, PLS provides a linear process model on the basis of measured process variables, (Wold, et al., 1984). The size of these models therefore, remain high dimensional. In this paper, the development of an identification method for determining a reduced dimensional model which describes linear steady-state process behavior is introduced. This new identification method, termed as latent variables least squares or LVLS, obtains a process model from latent variables of the input or predictor variables and from latent variables of the output or response variables of the process.

It is shown in this paper, that the number of latent variables required to describe significant process behavior for highly correlated process variables, is considerably smaller than the total number of process

* Correspondence should be sent to: uwe.kruger@ee.qub.ac.uk

variables. This implies that the size of the resultant LVLS process model is greatly reduced, revealing that:

1. the LVLS model is considerably smaller than the corresponding PLS model,
2. the reduced set of latent PLS process variables may not be adequately represent a reduced dimensional process model,
3. the LVLS model presents pairs of decoupled latent variables and
4. for process monitoring, the reduced dimensional LVLS model is more manageable than a high dimensional PLS model.

First, the principles of LVLS are introduced prior to the determination of the LVLS model. This is followed by a discussion of the properties of the LVLS model and a comparison of LVLS and PLS. Application of the LVLS model to process monitoring, as one of its potential uses, is then discussed. This is followed by an application study on the basis of data provided from a real industrial process, which includes a comparison between PLS and the LVLS.

## 2. LATENT VARIABLE LEAST SQUARES

A linear steady-state model for given predictor and response matrices can be written as follows:

$$\mathbf{Y} = \mathbf{XB} + \mathbf{E}, \tag{1}$$

where $\mathbf{Y}$ and $\mathbf{X}$ are the response and the predictor matrices of dimension $K \times N$ and $K \times M$, respectively, $\mathbf{B}$ is the regression matrix and $\mathbf{E}$ represents the residuals of the response matrix.

To minimize the residual matrix $\mathbf{E}$, the regression matrix $\mathbf{B}$ is determined with respect to the following cost function:

$$J_{min}^{(\mathbf{E})} = \min_{\mathbf{B}} \left\{ \left\| \mathbf{Y} - \mathbf{XB} \right\|_2^2 \right\}, \tag{2}$$

where $J_{min}^{(\mathbf{E})}$ is the cost function to be minimized and $\| \ \|_2^2$ represents the squared value of the Frobenius norm of a matrix.

The latent variables of the predictor and the response matrix are determined as:

$$\mathbf{t} = \mathbf{Xw} \text{ and } \mathbf{u} = \mathbf{Yq}, \tag{3}$$

where $\mathbf{t}$ and $\mathbf{u}$ are the latent variables or score vectors and $\mathbf{w}$ and $\mathbf{q}$ are weight vectors of unit length for the predictor and the response matrix, respectively. The elements of the t-score and the u-score vectors are further referred to as the t-scores and the u-scores, respectively.

The corresponding cost function for determining a linear score model, $J_{min}^{(\mathbf{e})}$, is then:

$$\mathbf{u} = \mathbf{t}b + \mathbf{e}$$
$$J_{min}^{(\mathbf{e})} = \min_{b,\mathbf{q},\mathbf{w}} \left\{ \left\| \mathbf{u} - \mathbf{t}b \right\|_2^2 \right\}, \tag{4}$$

where $b$ is the regression coefficient and $\mathbf{e}$ represents the residuals of the u-score vector. Assuming the u-score and the t-score vectors are predetermined, equation (4) clearly leads to the following solution:

$$b = \frac{\mathbf{t}^T \mathbf{u}}{\mathbf{t}^T \mathbf{t}}. \tag{5}$$

Incorporating equation (3) and (5) into equation (4) leads to:

$$J_{min}^{(\mathbf{e})} = \min_{\mathbf{w},\mathbf{q}} \left\{ \left\| \mathbf{Yq} - \mathbf{Xw} \frac{\mathbf{w}^T \mathbf{X}^T \mathbf{Yq}}{\mathbf{w}^T \mathbf{X}^T \mathbf{Xw}} \right\|_2^2 \right\} \tag{6}$$

Given that the t-score vector can be used to predict the u-score vector, the q-weight vector can be made dependent upon the w-weight vector as shown in Appendix A. The derived relationship between the weight vectors is as follows:

$$\mathbf{q} = \frac{\mathbf{Y}^T \mathbf{Xw}}{\sqrt{\mathbf{w}^T \mathbf{X}^T \mathbf{YY}^T \mathbf{Xw}}} \tag{7}$$

Substituting this relationship into equation (6) yields:

$$J_{min}^{(\mathbf{e})} = \min_{\mathbf{w}} \left\{ \left\| \left[ \mathbf{Y} - \mathbf{X} \frac{\mathbf{ww}^T \mathbf{X}^T \mathbf{Y}}{\mathbf{w}^T \mathbf{X}^T \mathbf{Xw}} \right] \frac{\mathbf{Y}^T \mathbf{Xw}}{\sqrt{\mathbf{w}^T \mathbf{X}^T \mathbf{YY}^T \mathbf{Xw}}} \right\|_2^2 \right\} \tag{8}$$

## 3. SOLVING THE LVLS COST FUNCTION

Equation (8) highlights the fact that the independent variable in the expression for minimizing the cost function, $J_{min}^{(\mathbf{e})}$, is the weight vector $\mathbf{w}$. Hence, equation (4) may be rewritten as follows:

$$J_{min}^{(\mathbf{e})} = \min_{\mathbf{w}} \left\{ \left\| \mathbf{u} - \mathbf{Xw}b \right\|_2^2 \right\}, \tag{9}$$

To minimize equation (9) an iterative approach is developed containing the steps shown below. The minimum value of the cost function for the $i^{th}$ iteration step given by:

$$J_{min}^{(\mathbf{e}^{(i)})} = \min_{\mathbf{w}^{(i)}} \left\| \mathbf{u}^{(i)} - \mathbf{Xw}^{(i)} b^{(i)} \right\|_2^2. \tag{10}$$

The basis for determining minimum value of the cost function for the $(i+1)^{th}$ iteration step is the perturbation of the weight vector $\mathbf{w}^{(i)}$:

$$\begin{aligned} \mathbf{e}^{(i)}\left(\mathbf{w}^{(i)}\right) &= \mathbf{u}^{(i)} - \mathbf{Xw}^{(i)} b^{(i)} \\ \mathbf{e}^{(i)}\left(\mathbf{w}^{(i+1)}\right) &= \mathbf{u}^{(i)} - \mathbf{Xw}^{(i+1)} b^{(i)}, \\ \mathbf{w}^{(i+1)} - \mathbf{w}^{(i)} &= \Delta \mathbf{w}^{(i+1)} \end{aligned} \tag{11}$$

where $\Delta \mathbf{w}^{(i+1)}$ represents the perturbation of $\mathbf{w}^{(i)}$, $\mathbf{e}^{(i)}$

is the prediction error for the $i^{th}$ u-score vector and $\mathbf{w}^{(i+1)}$ is $(i+1)^{th}$ w-weight vector. Using a first order Taylor approximation for equation (11) yields:

$$\mathbf{e}\left(\mathbf{w}^{(i+1)}\right) = \mathbf{e}\left(\mathbf{w}^{(i)}\right) - \mathbf{X}b^{(i)}\Delta\mathbf{w}^{(i+1)}. \qquad (12)$$

Since the $(i+1)^{th}$ value of the cost function can be approximated as:

$$J^{\left(\mathbf{e}^{(i+1)}\right)} = \mathbf{e}\left(\mathbf{w}^{(i+1)}\right)^T \mathbf{e}\left(\mathbf{w}^{(i+1)}\right), \qquad (13)$$

the optimal w-weight vector, $\mathbf{w}^{(i+1)}$, for minimizing $J^{\left(\mathbf{e}^{(i+1)}\right)}$ given by:

$$\begin{aligned} \Delta\mathbf{w}^{(i+1)} &= \left[\mathbf{X}^T\mathbf{X}\right]^{\perp}\mathbf{X}\left(\frac{\mathbf{e}^{(i)}}{b^{(i)}}\right) \\ \mathbf{w}^{(i+1)} &= \frac{\mathbf{w}^{(i)} + \Delta\mathbf{w}^{(i+1)}}{\left\|\mathbf{w}^{(i)} + \Delta\mathbf{w}^{(i+1)}\right\|_2} \end{aligned}, \qquad (14)$$

where $\| \ \|_2$ represents the norm of a vector and $\perp$ denotes the generalized inverse. Applying equations (3), (5) and (7) to equation (9), gives the following expression for the minimum value of the cost function at the $(i+1)^{st}$ iteration step:

$$J_{min}^{\left(\mathbf{e}^{(i+1)}\right)} = \left\|\mathbf{u}^{(i+1)} - \mathbf{X}\mathbf{w}^{(i+1)}b^{(i+1)}\right\|_2^2. \qquad (15)$$

Equations (3), (5), (7) and (14) give a summary of the steps, required for the iteration procedure. The difference of two subsequent score or weight vectors can be used for the check of convergence. The iteration procedure continuous for all cases where the difference is larger than a given threshold but is terminated otherwise. The iteration procedure can be initialized by setting the u-score and the t-score equal to some column of the response and the predictor matrix, respectively, and the w-weight vector as follows:

$$\mathbf{w} = \begin{pmatrix} \frac{1}{\sqrt{M}} \\ \vdots \\ \frac{1}{\sqrt{M}} \end{pmatrix} \qquad (16)$$

The series of iteration steps will converge against the minimum value for the cost function in equation (9).

Geometrically, equations (7) and (14) illustrate that the iteration procedure "rotates" the weight vector, **w**, and then "adjusts" the weight vector, **q**, relative to **w** to obtain the best possible score model by minimizing equation (9).

After the weight and score vectors and the regression coefficient is computed, the contribution of the predicted u-score vector, $\tilde{\mathbf{Y}}$, is subtracted (deflated) from the response matrix. This deflation procedure has to be carried out prior to determining subsequent weight and score vectors as well as corresponding regression coefficients. The complete LVLS algorithm is presented in Appendix B.

The contribution of the t-score vector to the predictor matrix can also be calculated as follows:

$$\begin{aligned} J_{\min}^{(p)} &= \min_{\mathbf{p}}\left\{\left\|\mathbf{X} - \mathbf{t}\mathbf{p}^T\right\|_2^2\right\} \\ \mathbf{p} &= \frac{\mathbf{X}^T\mathbf{t}}{\mathbf{t}^T\mathbf{t}} = \frac{\mathbf{X}^T\mathbf{X}\mathbf{w}}{\mathbf{w}^T\mathbf{X}^T\mathbf{X}\mathbf{w}} \\ \tilde{\mathbf{X}} &= \mathbf{t}\mathbf{p}^T \end{aligned}, \qquad (17)$$

where **p** is an independent vector and $\tilde{\mathbf{X}}$ represents the contribution of the **t** vector to **X**. The p-vectors, however, are not required for the score models.

In the case where the process variables are highly correlated, only a few score models are required to adequately represent the steady-state behavior of process. The forgoing discussion outlines use of the LVLS method in reducing the dimension of the model. Because the score models are independent of each other, the LVLS method gives rise to a decoupled steady-state process model. That the latent variable PLS model cannot necessarily be used for the same purpose is elaborated upon later on in this paper.

## 4. PROPERTIES OF THE LVLS FACTORS

The scores and the weight vectors of the LVLS algorithm (LVLS factors) have similar properties to the PLS score and weight vectors (PLS factors). The properties of the LVLS factors are:

- the t-score vectors are mutually orthogonal
- the $j^{th}$ u-score vector is orthogonal to every other t-score vector apart from the $j^{th}$ one ($j$=1, ... $n$).
- the q-weight vectors are mutually orthogonal
- the $j^{th}$ w-weight vectors is orthogonal to every other p-vector apart from the $j^{th}$ one.
- with LVLS, a regression matrix between the predictor and the response matrix can be established, **Y** = **XB** + **E**, which is equivalent to regression matrix determined by the ordinary least squares (OLS) solution. This regression matrix is as follows:

$$\mathbf{B} = \mathbf{W}\tilde{\mathbf{Q}}^T, \qquad (18)$$

where the w-weight and the $\tilde{q}$ -vector –see Appendix A– are stored as column vectors in **W** and $\tilde{\mathbf{Q}}$, respectively.

## 5. COMPARISON BETWEEN LVLS AND PLS

As is the case for LVLS, PLS also determines score and weight vectors. However, the PLS score vectors are seen to have maximum covariance and do not provide the most accurate score model. Geladi et al.

(1986) and Hoskuldsson (1988) have given a good account of the principles of PLS.

Figure 1 and 2 show the LVLS and PLS score model for the first score model based on data obtained from an industrial fluidized bed reactor. It is evident from these figures that the LVLS score model does give a sufficiently accurate representation of the linear steady-state behavior of the process, whilst the PLS score model does not.

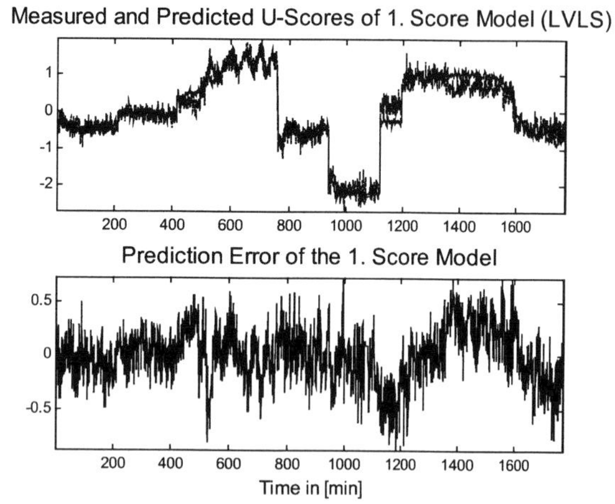


Fig. 1: Performance of first LVLS score model

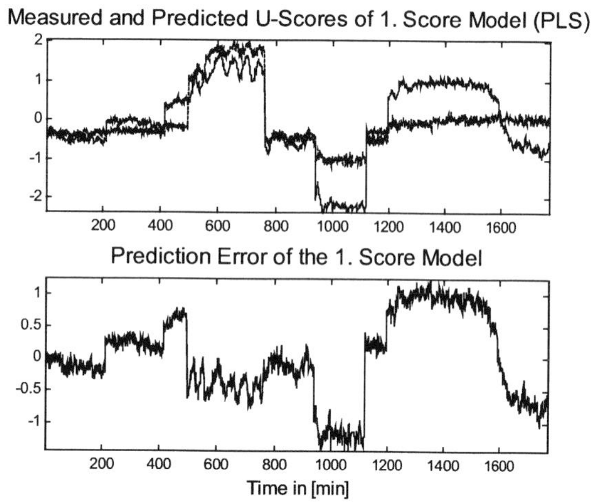


Fig. 2: Performance of first PLS score model

The poor performance of the PLS score model is a result of the fact that it determines the t-score and u-score vectors as having a maximum covariance. In contrast, the LVLS model, which provides the best possible score model.

## 6. MONITORING CHARTS FOR LVLS

MacGregor et al. (1991) highlighted the fact that in PLS, the monitoring of large-scale processes could be carried out by plotting the univariate statistics, in particular the $T^2$ and the $Q$ statistic, with a time base.

In the LVLS method, the $T^2$ statistic can be determined on the basis of the t-score vectors, whilst the $Q$ statistic can be determined with respect to the residuals of the score models.

The t-scores and u-scores as well as the residuals of the score models for the $k^{th}$ instance in time are determined as follows:

$$\boldsymbol{\tau}_k = \mathbf{W}^T \mathbf{x}_k \text{ and } \boldsymbol{\upsilon}_k = \mathbf{Q}^T \mathbf{y}_k$$
$$\boldsymbol{\varepsilon}_k = \boldsymbol{\upsilon}_k - \begin{bmatrix} b_1 & \cdots & 0 \\ \vdots & \ddots & \vdots \\ 0 & \cdots & b_n \end{bmatrix} \boldsymbol{\tau}_k , \qquad (19)$$

where $\boldsymbol{\tau}_k$ and $\boldsymbol{\upsilon}_k$ are the t-scores and the u-scores, respectively, $\mathbf{W}$ and $\mathbf{Q}$ are matrices in which the w- and q-weight vectors are stored as column vectors, $\mathbf{x}_k$ and $\mathbf{y}_k$ are the measured values of the predictor and the response variables at time instance $k$, $b_1$, ... $b_n$ are the regression coefficients of the score models, $n$ represents the total number of score models and $\boldsymbol{\varepsilon}_k$ represents the residuals of the score models.

In equation (19), the definition of the univariate statistics for LVLS is as follows:

$$T_k^2 = \sum_{j=1}^{n} \frac{\tau_{jk}^2}{\sigma_j} ,$$
$$Q_k = \sum_{j=1}^{n} \varepsilon_{jk}^2 \qquad (20)$$

where $T_k^2$ represents the $T^2$ statistic, $Q_k$ represents the $Q$ statistic at time instance $k$ and $\sigma_j$ is the variance of the $j^{th}$ t-score vectors.

The confidence limits of the $T^2$ and the $Q$ statistics are calculated on a central Chi-Squared distribution with $n$ degrees of freedom as discussed by Box (1954).

The definition of the univariate statistics shows that residuals exist for each of the $n$ score model. In contrast, the PLS model leads to a total number of $N$ residuals (one for each response variable). This highlights another benefit of LVLS. With LVLS, only $n$ "latent" response variables need to be considered instead of $N$ physical response variables.

## 7. APPLICATION STUDY

The industrial process considered here, produces two solvent chemicals and consists of several operating units. The core elements of this plant are five parallel operating fluidized bed reactors each of which produces the solvents by complex exothermic chemical reactions. These reactors are fed with two different streams of five different reactants.

The reactors consist of a large shell and a total number of 37 vertically oriented tubes in which the chemical reactions are carried out supported by fluidized catalyst. To remove the heat of the exothermic reaction, oil is circulated around the tubes.

Normal operating data as well as data containing process abnormalities was obtained for a particular reactor over a period of several weeks. The data set for capturing normal process operation (reference data) was selected with care. It was important to ensure that the reference data did not capture abnormal process behavior. The size of the reference data needed to be large enough to ensure that any normal process variation was adequately represented. Each data set represents the process in steady state operation with a sampling interval of 1 minute.

In applying LVLS to identify a reduced dimensional model, as described in sections 2 and 3, the predictor variables are the feed streams of the five reactants and an additional stream for regulating the pressure in one of the adjacent units of the reactor. The response variables consist of the individual tube temperatures.

The contribution of the score models to the predictor and the response matrices is shown in table 1, which illustrates that the first three score models are the dominant contributors. The remaining three score models can be omitted as their respective contributions are insignificant.

Table 1 Contribution of each score model to the predictor (first two columns) and response matrix (last two columns)

| % Contr. | Tot % Contr. | % Contr. | Tot % Contr. |
|---|---|---|---|
| 22.54 | 22.54 | 76.62 | 76.62 |
| 24.91 | 47.45 | 4.88 | 81.50 |
| 31.96 | 79.42 | 1.49 | 82.99 |
| 18.28 | 97.70 | 0.50 | 83.49 |
| 1.23 | 98.93 | 0.18 | 83.67 |
| 1.07 | 100.00 | 0.01 | 83.68 |

The forgoing discussion clearly demonstrates the fact that the steady-state behavior of the reactor can be adequately described by 3 decoupled linear score models. In contrast, a PLS model would lead to a 37 by 6 regression matrix, see equation 1.

The abnormal event observed in the obtained data represents a large unmeasured disturbance. This event arose from a drop in steam pressure supplied to one of the adjacent units of the reactor. The $T^2$ and the $Q$ statistic of the corresponding data set, containing 2000 data points, are presented in figure 3 and 4. The drop in steam pressure occurred about 1300 minutes into the data set and did not affect the steam flow to this unit. The predictor variables are thus not effected by this event. This can also be observed from the $T^2$ statistic in figure 3. However, the enthalpy balance within the reactor did change thus, affecting the tube temperatures. That the score models cannot describe this event is clearly evident from the $Q$ statistic in figure 4.

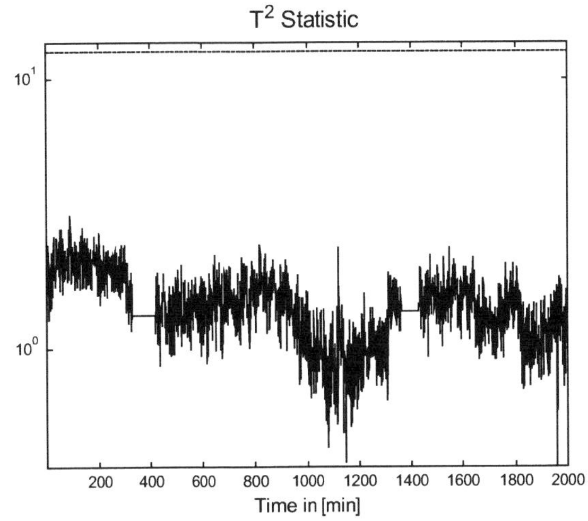


Fig. 3: $T^2$ Statistic representing to the drop in steam pressure

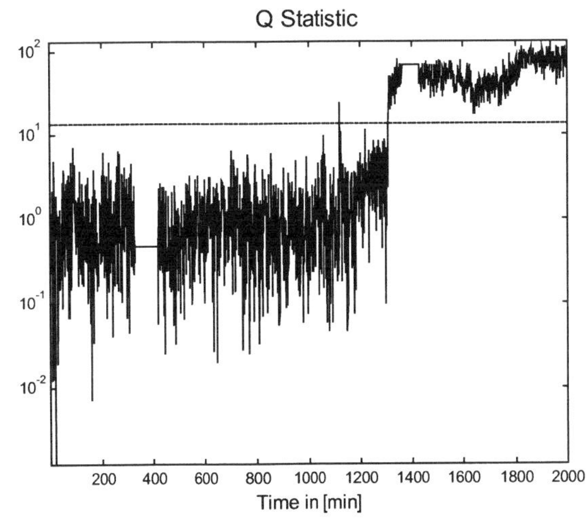


Fig. 4: $Q$ Statistic representing to the drop in feed steam pressure

## 8. CONCLUSIONS

In this paper, a novel method for identifying a linear steady-state model of large-scale processes is introduced. This method, termed as latent variable least squares or LVLS, is designed as a model reduction tool for use in cases where the process variables are highly correlated.

The monitoring of large-scale processes is discussed in this paper as a possible application of LVLS. It is found that the LVLS is similar in approach to partial least squares (PLS), an established technique for process monitoring. PLS, like LVLS, addresses the data reduction aspect by a reduced number of latent variables. However, the final PLS model describes linear steady-state behavior only on the basis of the physical predictor and response variables.

The paper highlights the following advantages of LVLS over PLS:

1. it provides a model which is considerably smaller than the corresponding PLS model,

2. the reduced dimensional PLS model may not adequately represent a reduced dimensional process model
3. it presents pairs of decoupled linear models, and
4. it is easier to manage the LVLS model relative to the PLS model with much higher dimensions

The application study on a real industrial process shows that the LVLS model is able to detect abnormal process behavior on the basis of univariate statistics derived from latent variables and residuals of the LVLS model.
Future research on LVLS focuses on the use of its decoupling effect in process control and on the geometric analysis of the corresponding score and weight vectors.

## APPENDICES

*Appendix A: Relationship between the weight vectors*

Assuming that the predicted u-score vector, $\hat{\mathbf{u}} = \mathbf{t}b$, is already calculated, the contribution of this vector to the response matrix is given by:

$$\widetilde{\mathbf{Y}} = \hat{\mathbf{u}}\widetilde{\mathbf{q}}^T = \mathbf{t}b\widetilde{\mathbf{q}}^T, \qquad \text{(A1)}$$

where $\widetilde{\mathbf{Y}}$ represents the contribution to the response matrix. By setting $\mathbf{q}$ as an independent variable, the maximum contribution of $\widetilde{\mathbf{Y}}$ to $\mathbf{Y}$ can be determined as follows:

$$J_{min} = \min_{\mathbf{q}}\left\{\left\|\mathbf{Y} - \mathbf{t}b\widetilde{\mathbf{q}}^T\right\|_2^2\right\}, \qquad \text{(A2)}$$

which leads to the solution:

$$\widetilde{\mathbf{q}} = \frac{\mathbf{Y}^T\mathbf{t}}{\mathbf{t}^T\mathbf{t}b} = \frac{\mathbf{Y}^T\mathbf{t}}{\mathbf{t}^T\hat{\mathbf{u}}}. \qquad \text{(A3)}$$

By setting the length of $\widetilde{\mathbf{q}}$ to one, equation (A3), the q-weight vector is equal to:

$$\mathbf{q} = \frac{\mathbf{Y}^T\mathbf{X}\mathbf{w}}{\sqrt{\mathbf{w}^T\mathbf{X}^T\mathbf{Y}\mathbf{Y}^T\mathbf{X}\mathbf{w}}} \qquad \text{(A4)}$$

*Appendix B: The LVLS algorithm*

The steps of the LVLS algorithm presented below include; preliminary steps to initiate the iteration process in the first three steps, an iteration procedure for minimizing equation (9) as is summarized in the fourth step and deflation of the response matrix as the fifth step.

1. select a column vector of $\mathbf{Y}$ as $\mathbf{u}$
2. Define initial w-score vector

$$\mathbf{w} = \begin{pmatrix} \frac{1}{\sqrt{M}} \\ \vdots \\ \frac{1}{\sqrt{M}} \end{pmatrix}$$

3. select a column vector of $\mathbf{X}$ as $\mathbf{t}$
4. loop until convergence of $\mathbf{t}$ or $\mathbf{u}$
   4.1. Calculating regression coefficient
   $b = \mathbf{t}^T \cdot \mathbf{u} / \mathbf{t}^T \cdot \mathbf{t}$
   4.2. Determine predicted u-score vector
   $\hat{\mathbf{u}} = b \cdot \mathbf{t}$
   4.3. Update weighting vector $\mathbf{w}$
   $$\Delta\mathbf{w} = \left[\mathbf{X}^T\mathbf{X}\right]^{-1}\mathbf{X}\left(\frac{\mathbf{e}}{b}\right)$$
   $$\mathbf{w} = \frac{\mathbf{w} + \Delta\mathbf{w}}{\left\|\mathbf{w} + \Delta\mathbf{w}\right\|_2}$$
   4.4. Compute t-score vector
   $\mathbf{t} = \mathbf{X} \cdot \mathbf{w}$
   4.5. Calculate q vectors
   $$\widetilde{\mathbf{q}} = \frac{\mathbf{Y}^T\mathbf{t}}{\mathbf{t}^T\mathbf{t}b} \text{ and } \mathbf{q} = \frac{\mathbf{q}}{\left\|\mathbf{q}\right\|_2}$$
   4.6. Compute u-score vector
   $\mathbf{u} = \mathbf{Y} \cdot \mathbf{q}$
5. Deflate of the response matrix
   $\mathbf{Y} = \mathbf{Y} - \hat{\mathbf{u}} \cdot \widetilde{\mathbf{q}}^T$

## REFERENCES


Box, G. E. P. (1954), "Some theorems on quadratic forms applied in the study of analysis of variance problems, I. Effect of inequality of variance in the one-way classification", *Ann. Math. Stat.*, **25**(2), 290-302

Geladi, P. and Kowalski, B. R. (1986), "PLS Tutorial", *Analytica Chimica Acta*, **186**, 1-17

Hoskuldsson, A. (1988), "PLS regression methods", *Journal of Chemometrics*, **2**, 1-17

MacGregor, J. F., Marlin, T. E., Kresta, J. and Skagerberg, B. (1991), "Multivariate statistical methods in process analysis and control", AIChE Symp. Proc., AIChE Publ. No. P-67, New York, 79-99

Piovoso, M. J. and Kosanovich, K. A. (1992), "Process Data Chemometrics", *IEEE Transaction on Instrumentation and Measurement*, **41**(2), 262-268

Wise, B. M., Gallagher, N. B. and MacGregor, J. F. (1996), "The process chemometric approach to process monitoring and fault detection", *Journal of Process Control*, **6**(6), pp. 329-348

Wold, S., Esbensen, K., Wold, H. and Dunn, W. J. (1984), "The collinearity problem in linear regression. The partial least squares (PLS) approach", *SIAM Journal of Scientific Statistical Computations*, **5**, 735-743

IFAC Publications
www.elsevier.com/locate/ifac

# FAULT DIAGNOSIS BASED ON QUALITATIVE BOND GRAPH AND GENETIC ALGORITHMS

**C.H. Lo, Y.K. Wong*, A.B. Rad and Y.L. Ip**

*Department of Electrical Engineering*
*The Hong Kong Polytechnic University, Hung Hom, Kowloon, Hong Kong*

Abstract: Most of the engineering systems now become more complex and sophisticated, reliability and safety turn into an important issues when operating these systems. In this paper, genetic algorithms are proposed to search for possible fault components among a system of qualitative equations. Qualitative bond graph is adopted as the modeling scheme to generate a set of qualitative equations. Results from diagnosing faults in the floating disc system will be presented and discussed. Fault diagnosis is activated by fault detection mechanism when a discrepancy between measured abnormal behavior and predicted system behavior is observed. *Copyright © 2001 IFAC*

Keywords: Artificial intelligence; Bond graphs; Fault diagnosis; Fault location; Genetic algorithms.

## 1. INTRODUCTION

Fault diagnosis is one of the significant issues in modern automatic control engineering since most of the engineering systems now become more complicated and sophisticated. The growing demand of system reliability, safety and low-cost operation even makes the fault diagnosis more essential. The process of fault diagnosis starts when a dis□repancy between observed abnormal behavior and the expected behavior or desired behavior is measured. An initial set of fault candidates is then generated, and additional measurements are suggested to help refine the initial candidate set. Many early approaches to diagnosis are of rule-based (Wang, *et al.*, 1996) in nature. They are likely to be incomplete and inflexible and required priori knowledge of the fault instead of actually inferring the fault. Better approaches to fault diagnosis is based on a model of the system's structure and behavior. This model-based fault diagnosis (Wang, *et al.*, 1996) enables more complex cause-effect reasoning and hence a more robust diagnostic system can be developed.

Quantitative and qualitative are two major approaches to model-based fault diagnosis. The quantitative approach relies on physical laws and precise mathematical models. The main problems to this method are the difficult and expensive to obtain a precise numerical model and the sensitivity of the diagnostic system to modeling error. Usually, the effect of modeling errors obscures the effect of fault which results in false alarm (Ghiaus, 1999; Wang, *et al.*, 1996). The qualitative approach may be based on physical laws and expert knowledge/rules and is more applicable when mathematical models are not available.

Bond graph modeling technique is suitable to model a system which will have different power transfer. It is a method for modeling mechanical, hydraulic, thermal and electrical systems in a unified approach. Karnopp, Margolis and Rosenberg (Karnopp, *et. al.*, 2000) give a detailed account on the bond graph method. Qualitative bond graph modeling scheme proposed by Wang and Linkens (Wang, *et al.*, 1996) is adopted in this paper. It is basically derived from classical bond graph with the integration with qualitative reasoning. After the integration, a set of qualitative equations can be generated systematically and easily, without the concern about the causality assignment and algebraic manipulation. With the origin of bond graph, the structural information about component locations and their interconnections can be described explicitly by the qualitative equations. Hence, the diagnosis mechanism can easily localize system faults with the reasoning about the relations between system faults and system structure. Finally, the effect of modeling errors can be eliminated since the qualitative model contains no numerical information.

Diagnosis accomplishes along the set of qualitative equations. The method of solving the qualitative equations is different from their numerical counterpart. For example, if $a = b + c$, if both $a$ and $b$ are know value, we cannot calculate $c$ by simply reasoning the equation as $c = a - b$. This characteristic requires testing all the possible candidates repeatedly when solving the equations. However, as the system becomes more complex, the number of qualitative equations increases and the testing process becomes time-consuming and inefficiency. In this paper, genetic algorithms (GA) are used as searching engine for possible faults is proposed in order to improve efficiency. Hypothesized faults are coded in a chromosome and through genetic operations, an initial set of fault candidates is generated. With the additional measurements, the initial set of fault candidates can then be refined. Section 2, gives a brief account on qualitative bond graph. Section 3 presents an example on fault diagnosis based on qualitative bond graph. Section 4 gives an outline of GA and its integration with the diagnosis mechanism. Application of GA and qualitative bond graph on the floating disc system will be discussed in Section 5. Section 6 concludes the paper.

## 2. QUALITATIVE BOND GRAPH

### 2.1 Bond Graph

Bond graph provides a systematic and formal way for modeling dynamic systems. A bond graph simply consists of subsystems linked together by lines representing power bonds (detailed description on bond graph can be found in (Karnopp, *et. al.*, 2000)), as in Fig. 1.

Fig. 1. A power bond. Positive power flow from element A to B.

The half arrow indicates the direction of actual power flow (positive power flow). The product of power variables, effort ($e$) and flow ($f$), results the power on the bond. Usually, flow represents either: current, flow rate or velocity; effort represents either: force, voltage or pressure. Bonds are numbered and the numbered effort and flow corresponds to that numbered bond.

The bond graph modeling language consists of nine primitive entities, three are passive elements, resistance ($R$), capacitance ($C$), and inertia ($I$); and two distribution elements, transformer ($TF$) and gyrator ($GY$); and two ideal sources, effort source ($S_e$), flow source ($S_f$); and two kinds of junctions, serial junction (*1-junction*) and parallel junction (*0-junction*). Ideal sources are1-port active elements. Passive elements are also 1-port elements: R for power dissipation, both C and I for energy storage and their constitutive laws are defined as follow:

$$e_i = R \times f_i; \quad f_i = C\frac{d}{dt}e_i; \quad e_i = I\frac{d}{dt}f_i \tag{1}$$

where $i$ represents the numbers of bonds.

The 2-port elements, $TF$ and $GY$, transfer energy between two parts of a system and their constitutive laws are defined as follow (in both case power is conserved):

$$\text{For } TF: \quad e_i = \frac{b}{a}e_j, \qquad f_i = \frac{a}{b}f_j \tag{2}$$

$$\text{For } GY: \quad e_i = \frac{b}{a}f_j, \qquad f_i = \frac{a}{b}e_j \tag{3}$$

where $i$ and $j$ represent the numbers of bonds between the 2-port element. Transformer relates the effort at one port to the other port, while gyrator relates the effort at one port to the flow at the other port.

Bond graphs consists of two types of connections: serial and parallel junctions, and classified as multi-port elements as they can connect $n$ elements in a node. For the serial junction (*1-junction*), the flow is common to all bonds while the algebraic sum of all efforts on the bonds is zero. For the parallel junction (*0-junction*), the algebraic sum of all flows on the bonds is zero while the effort is common to all bonds. For example, the equation of 1-junction with $m$ input elements and $n$ output elements is:

$$\begin{aligned} e_{in1} + e_{in2} + \cdots + e_{inm} &= e_{out1} + e_{out2} + \cdots + e_{outn} \\ f_{in1} = f_{in2} = \cdots = f_{inm} &= f_{out1} = f_{out2} = \cdots = f_{outn} \end{aligned} \tag{4}$$

And the equation of 0-junction with $m$ input elements and $n$ output elements is:

$$\begin{aligned} f_{in1} + f_{in2} + \cdots + f_{inm} &= f_{out1} + f_{out2} + \cdots + f_{outn} \\ e_{in1} = e_{in2} = \cdots = e_{inm} &= e_{out1} = e_{out2} = \cdots = e_{outn} \end{aligned} \qquad (5)$$

### 2.2 *Qualitative Bond Graph*

Qualitative bond graph uses qualitative values instead of numerical data for system parameters. The qualitative values are drawn from the set {[1], [+], [0], [-], [-1], [?]}. For measured variables, [+] and [-] denote the positive and negative value in the measurement space, respectively. [0] is the boundary between [+] and [-], while [?] denotes an uncertain value. [1] and [-1] represent large positive value and large negative value, respectively. For power variables, $e$ and $f$, [1], [-1] and [0] represent different abnormal states caused by system faults. For the system components, $R$, $C$ and $I$, [1] denotes component blocked (i.e. obstruct power delivery), while [0] denotes component leakage or short circuit (i.e. power loss) (Ghiaus, 1999; Wang *et. al.*, 1996).

The qualitative operators used to infer qualitative value among a set of qualitative equations are the same as standard operators on real numbers, {+, -, ×, /}, and they are defined in Fig. 2 (Wang *et. al.*, 1996). However, the qualitative operations are different than operation on real numbers. For example, [1] = [1] + [0], [1] = [1] + [+] and [1] = [1] + [1]. So, with the equation: [1] = [1] + [X], a set of solutions results: [X] = {[0], [+], [1]}. But, solving the equation [1] = [1] + [X], with [X] = [1] – [1], results the solution [X] = [0]. Special inferring method is employed to solve qualitative equations.

[X] + [Y]:

| [Y] \ [X] | -1 | - | 0 | + | 1 | ? |
|---|---|---|---|---|---|---|
| -1 | -1 | -1 | -1 | - | 0 | ? |
| - | -1 | - | - | ? | + | ? |
| 0 | -1 | - | 0 | + | 1 | ? |
| + | - | ? | + | + | 1 | ? |
| 1 | 0 | + | 1 | 1 | 1 | ? |
| ? | ? | ? | ? | ? | ? | ? |

[X] – [Y]:

| [Y] \ [X] | -1 | - | 0 | + | 1 | ? |
|---|---|---|---|---|---|---|
| -1 | 0 | + | 1 | 1 | 1 | ? |
| - | - | ? | + | + | 1 | ? |
| 0 | -1 | - | 0 | + | 1 | ? |
| + | -1 | - | - | ? | + | ? |
| 1 | -1 | -1 | -1 | - | 0 | ? |
| ? | ? | ? | ? | ? | ? | ? |

[X] * [Y]:

| [Y] \ [X] | -1 | - | 0 | + | 1 | ? |
|---|---|---|---|---|---|---|
| -1 | 1 | 1 | ? | -1 | -1 | ? |
| - | 1 | + | 0 | - | -1 | ? |
| 0 | ? | 0 | 0 | 0 | ? | ? |
| + | -1 | - | 0 | + | 1 | ? |
| 1 | -1 | -1 | ? | 1 | 1 | ? |
| ? | ? | ? | ? | ? | ? | ? |

[X] / [Y]:

| [Y] \ [X] | -1 | - | 0 | + | 1 | ? |
|---|---|---|---|---|---|---|
| -1 | 0 | + | 1 | 1 | 1 | ? |
| - | - | ? | + | + | 1 | ? |
| 0 | -1 | - | 0 | + | 1 | ? |
| + | -1 | - | - | ? | + | ? |
| 1 | -1 | -1 | -1 | - | 0 | ? |
| ? | ? | ? | ? | ? | ? | ? |

Fig. 2. Qualitative operators: Addition, Substraction, Multiplication and Division.

With the definition of qualitative operators, qualitative bond graph equations can be formulated in discrete form. For example, the equation of $C$ elements:

$$f_i = C\frac{d}{dt}e_i \qquad (6)$$

it becomes:

$$f_i(k \cdot \Delta t) = C \times \frac{[e_i(k \cdot \Delta t) - e_i((k-1)\Delta t)]}{\Delta t} \qquad (7)$$

In discrete system, Δt is the sampling period and it is always positive. So, it can be neglected since it will not have any effect to the qualitative behavior. The symbol $k$ denotes a certain number of time steps while $(k\text{-}1)$ denotes the past time step of $k$. Then, the qualitative representation of $C$ elements becomes:

$$f_i(k) = C \times [e_i(k) - e_i(k-1)] \qquad (8)$$

Similarly, the qualitative representation of $I$ elements is:

$$e_i(k) = I \times [f_i(k) - f_i(k-1)] \qquad (9)$$

All other qualitative bond graph equations will have the same form as their original bond graph (Section 2.1), except that the proportionality constants (b/a) of $TF$ and $GY$ equations can be omitted because they are always positive. Detailed description on qualitative bond graph can be found in (Wang *et. al.*, 1996).

## 3. FAULT DIAGNOSIS BASED ON QUALITATIVE BOND GRAPH

In this section, a very simple example: a tank (Ghiaus, 1999), is used to demonstrate the process of fault diagnosis among a set of qualitative equations. Fig. 3 shows the structure of the tank, with an input flow source $Q_{in}$, an output control valve $R$ and the capacity of the tank $C$. The measured variable is the height, $h$, of the tank (liquid level) which is related to the pressure as $P = f(h)$. In this example, volume flow rate and pressure are the flow and effort variables respectively. The qualitative bond graph equations are as follow:

$$\left.\begin{aligned} e_1 &= e_2 = e_3 \\ f_1 &= f_2 + f_3 \\ f_2 &= C \times (e_2 - e'_2) \\ e_3 &= R \times f_3 \end{aligned}\right\} \qquad (10)$$

All the power variables are considered at time $t$, except $e'_2$, which is the past time step of $t$.

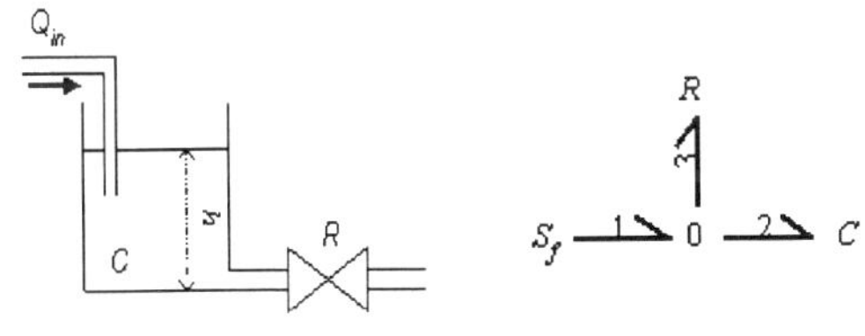


Fig. 3. A tank and its bond graph.

The measured variable is $h$ which correspond to the power variable $e_2$. Assume that the $h$ is abnormally small, and so, set the effort $e_2$ = [-1]. Assigning normal values to all system parameters before hypothesized a fault. Then a component fault is considered: the input flow rate is abnormally small, i.e. $f_1$ = [-1]. The inference is as follows:

$$\left.\begin{array}{ll} e_1 = e_2 = e_3 & e_2 = [-1] \Rightarrow e_1 = e_3 = [-1] \\ f_1 = f_2 + f_3 & \\ f_2 = C \times (e_2 - e'_2) & f_2 = [+]\,([-1] - [+]) = \\ & [+][-1] = [-1] \\ e_3 = R \times f_3 & [-1] = [+] \times f_3 \end{array}\right\}$$

Assume that $f_3$ = [-1], and then evaluate $f_1 = f_2 + f_3$ again.

$$f_1 = f_2 + f_3 \qquad \Rightarrow \qquad [-1] = [-1] + [-1]$$

Hence, all the equations are satisfied. The hypothesis of input flow is abnormally small is considered as a fault candidate. Other faults are being tested: $R$ = [0], $R$ = [1], $C$ = [1], $C$ = [0]...etc. The other fault that satisfy all the equations is $C$ = [0]. Additional measurements or direct inspection are required to refine the fault candidates. As the system becomes more complex, the number of components grows and the one-by-one testing method becomes time-consuming and inefficiency. Wang and Linkens (Wang *et. al.*, 1996) suggested to improve efficiency by decomposing the bond graph model into several segments. When a component is considered as a fault candidate, the other components within the segments are fault candidates as well. However, it will make the modeling task sometimes even difficult. A genetic algorithms based searching engine is proposed in this paper to preserve the simplicity of modeling task and improves efficiency.

## 4. GENETIC ALGORITHMS

### *4.1 Overview of GA*

Genetic algorithms (GA) were proposed by Holland in 1975 (Holland, 1975). It is a search algorithm based on the mechanism of natural selection and genetic reproduction (crossover and mutation). GA is performed over a population of chromosomes, and their components are called genes. Each chromosome represents a solution to the problem and has a finite number of genes which depends on the coding strategy. As analogy to the survival of the fittest law, the fitness of each chromosome is evaluated within a population by the fitness (objective) function. According to the fitness value, the best-fit chromosome is being selected for reproduction to generate the offspring and the worst is rejected. A higher fitness value indicates a better solution. This allows the GA to improve or optimize its solution. GA can be applied to solve non-linear, discontinuous, multi-objective optimization problems (Bolc, *et. al.*, 1992; Mitchell, 1996), in artificial intelligence (e.g. learning systems).

### *4.2 The GA based searching engine*

Fault diagnosis involves a global search through a space of possible fault candidates. GA can tackle large search spaces of complex systems without having to perform exhaustive searches (Khoo, *et. al.*, 2000). During the search, each chromosome represents a potential solution to the fault diagnosis process. All components of the bond graph model are encoded into a chromosome as shown in Fig. 4. The fault type of each component is assigned randomly or by commonsense reasoning (which can improve convergence).

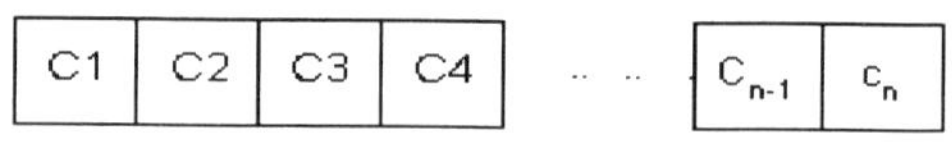


Fig. 4. System components (from 1 to $n$) are encoded in a chromosome (C: Component).

The set of qualitative equations acts as constraints in the genetic search for fault candidates. The following fitness function is used to determine the fitness value of the chromosomes:

$$\text{fitness of a chromosome} = \frac{\sum_{1}^{n} (\text{fitness of a component})}{n} \qquad (11)$$

where $n$ is the number of components in a chromosome. Component satisfies its corresponding equations will have fitness value of "1". Otherwise, its fitness value will be "0". Chromosomes with highest fitness value will have a higher probability of being selected for reproduction through randomly performed crossover or mutation. The searching process ends when a set of possible fault candidates is generated. In Linkens, *et. al.*, 1995 and Wang, *et. al.*, 1996, additional measurements are suggested in order to refine the initial fault diagnosis results automatically. These additional measurements will be added as extra constraints to the searching engine.

## 5. FAULT DIAGNOSIS OF FLOATING DISC SYSTEM

The proposed fault diagnosis algorithm is tested on the floating disc system. Fig. 5a shows the cross-section of the system and Fig. 5b shows the real system. The system consists of a power supply circuit, a controlling circuit, a d.c. motor, an analogue infra-red distance sensor and the disc. The fan output is controlled by varying the duty cycle of the input voltage to the fan. The average output power will be decreased when the off-time is longer than the on-time provided that the period of the pulse is constant, and vice versa.

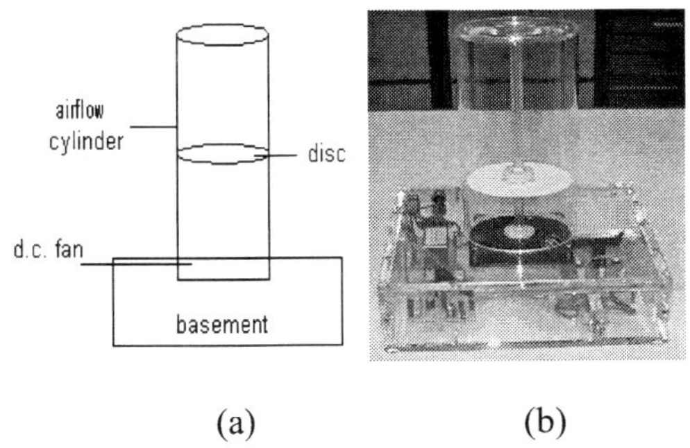


Fig. 5. (a) Cross-section of the floating disc system. (b) The real system.

The height of the disc is regulated by the speed of the fan. If the fan speed is increasing, the height of the disc will increase. If the fan speed decreases, the height of the disc will decrease. The infra-red sensor is used to measure the height of the distance in centimeter. A standard fuzzy logic controller controls the disc level. Fig. 6 shows the bond graph of the system and the qualitative equations are as follows:

$$e_1=e_2+e_3+e_4$$
$$f_1=f_2=f_3=f_4$$
$$e_2=I1\times(f_2-f'_2)$$
$$e_3=R1\times f_3$$
$$e_4=f_5$$
$$f_4=e_5$$
$$e_5=e_6+e_7+e_8$$
$$f_5=f_6=f_7=f_8$$
$$e_6=I2\times(f_6-f'_6)$$
$$e_7=R2\times f_7$$
$$e_8=e_9=e_{10}$$
$$f_8=f_9+f_{10}$$
$$f_9=C1\times(e_9-e'_9)$$
$$e_{10}=e_{11}+e_{12}$$
$$f_{10}=f_{11}=f_{12}$$
$$e_{11}=I_{disc}\times(f_{11}-f'_{11})$$
$$e_{12}=R_{cover}\times f_{12}$$

where $I1$ and $R1$ are the armature inductance and resistance respectively, $R2$ is the axial friction, $I2$ is rotor inertia, $C1$ is the capacitance of the airflow cylinder, $I_{disc}$ is the inertia of the disc and $R_{cover}$ is the flow resistance of the cylinder. The measured variable is the effort $e_9$, the height of the disc.

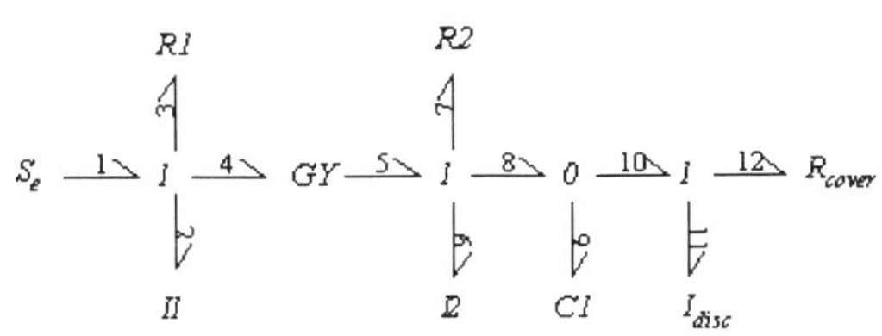


Fig. 6. Bond graph of the floating disc system.

System components are encoded into the chromosome as shown in Fig. 7. The initial population is generated randomly and the size of the population is 10. The propagation of qualitative values in the system of qualitative equations is presented in Section 3. Assume the disc is at the 12cm position, i.e. $e_9$ = [1], which is caused by a component fault. Fig. 8 shows the control response results with the set point level 8cm. The resistance of the motor is lower at 38s (i.e., $R1$ = [0]) by varying the rheostat which is connected in series with the motor in order to simulate the fault. After 30 generations, the initial fault candidates are then generated as follows:

Fault candidates: *R1, I1, I2, R2*

In the test, decreasing the resistance of the rheostat simulates a leakage or short circuit state of the armature resistance of the motor. Since the armature resistance is lowered, more voltage (effort) can be applied to the motor; hence the speed of the motor is increased. Now, more power is applied to the motor. As the speed increases, the disc level is also increased (i.e., $e_9$ = [1]) which causes the abnormal behavior of the system. Besides the actual fault ($R1$), more fault candidates are inferred. Wang and Linkens (Linkens, *et. al.*, 1995; Wang, *et. al.*, 1996) have suggested for additional measurements to refine the initial fault candidates automatically.

| | | | | | | |
|---|---|---|---|---|---|---|
| I1 | R1 | I2 | R2 | C1 | $I_{disc}$ | $R_{cover}$ |

Fig. 7. Chromosome is coded with system components.

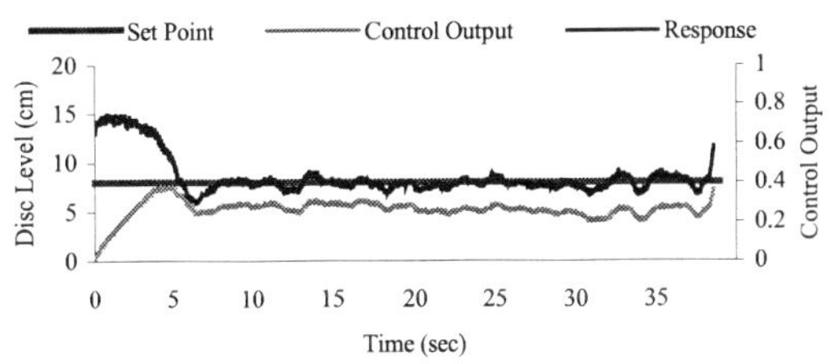


Fig. 8. The control response of the floating disc system. Fault is applied at 38sec.

## 6. CONCLUSIONS

This paper proposed the use of genetic algorithms as the search engine to find the initial fault candidates. The approach can diagnose a fault without any prior knowledge of the possible faults. Qualitative bond graph is used as the formal modeling scheme which provides a unified approach to model different energy domain subsystems together. The proposed approach is tested on a newly constructed floating disc system and the fault can be inferred correctly. Additional measurements are necessary to refine the initial fault candidates automatically.

## 7. ACKNOWLEDGEMENTS

The authors gratefully acknowledge the financial support from the Hong Kong Polytechnic University through grant G-V872.

## REFERENCES

Bolc, L. and J. Cytowski. (1992). *Search Methods for Artificial Intelligenc,* Chapter 9, pp.201-217. Academic Press, London.

Ghiaus, C (1999). Fault diagnosis of air conditioning systems based on qualitative bond graph. *Energy and Buildings*, **30**, pp.221-232.

Holland, J.H. (1975). *Adaptation in Natural and Artificial Systems*. Ann Arbor, MI: Univeristy of Michigan Press.

Karnopp, D.C., D.L. Margolis and R.C. Rosenberg (2000). *System Dynamics: Modeling and Simulation of Mechatronic Systems*. John Wiley & Sons, Inc, 3rd edition, New York.

Khoo, L.P., C.L. Ang, J. Zhang (2000). A fuzzy-based genetic approach to the diagnosis of manufacturing systems. *Engineering Applications of Artificial Intelligence*, **13**, pp.303-310.

Linkens, D.A. and H. Wang (1995). Fault diagnosis based on a qualitative bond graph model, with emphasis on fault locations. In: *IEE Colloquium on Qualitative and Quantitative Modeling Methods for Fault Diagnosis*, pp.1-6.

Mitchell, M. (1996). *An Introduction to Genetic Algorithms.* MIT Press, London.

Wang, H. and D.A. Linkens (1996). *Intelligent supervisory control: a qualitative bond graph reasoning approach.* World Scientific Publishing, Singapore.

IFAC Publications
www.elsevier.com/locate/ifac

# SATELLITE FAULT DIAGNOSIS BASED ON OPTIMAL PARITY VECTOR METHOD

**Jing Yang     Hong Yue Zhang**

*Beijing University of Aeronautics and Astronautics, Beijing 100083, China*
*Email: hoyzhang@public3.bta.net.cn*

Abstract: In this paper, a fault diagnosis approach is proposed to improve the integrity of Global Navigation System. Optimal parity vector is used to detect and diagnose satellite faults and improve the effectiveness of satellite fault diagnosis, especially for small fault. Simulation results show that correct detection rate and correct diagnosis rate of using optimal parity vector method is higher than that of using generalized likelihood test for detecting small fault. And, optimal parity vector method can also shorten time spent on detecting fault. Thereby this method improves system reliability. *Copyright © 2001 IFAC*

Keywords: Global Positioning Systems, Fault Detection, Fault Diagnosis, Navigation Systems, Satellite

Integrity of GPS has long been an important problem studied by many investigators, since it determines whether GPS can be a sole navigation system. Although control station on ground can process satellites data, and inform user when satellite is faulty, however, time to inform user always lags time that fault happens about several hours. RTCA/DO-208 (RTCA, 1991) has published the GPS Minimum Operational Performance Standards for supplemental navigation, in which time to alarm for non-precise approach, terminal and en route are $10s$, $30s$ and $30s$ respectively, therefore monitor by ground can't satisfy the need of these occasion completely. In addition, in some important situation such as approach and landing, it requires that GPS not only gives position solution with high precision, but also keeps the integrity of navigation data. Position measurement with high precision can be realized by carrier phase difference technique, but the security and usability of satellite are the pre-condition of the integrity of position solution. Therefore, integrity monitoring of GPS is the base for GPS as a sole navigation system.

Integrity monitoring methods include Receiver Autonomous Integrity Monitoring (RAIM), receiver integrity monitoring assisted by other navigation equipment and ground monitoring. Many different RAIM schemes have been suggested over past years, among which three methods have received special attention in recent years, least squares residuals method (Kozel and Cardoza, 1996), parity method (Pervan et al., 1996), approximate horizontal protection level (Hunzinger et al., 1997), etc. Brown (1992) has proved the equivalence of former two methods. Because residual is affected by different error, noise and fault source jointly, it is difficult to select proper false alarm rate and false detect rate. After satellite fault has been detected, faulty satellite should be isolated. In some cases, there are some troubles in distinguishing the effect of fault and noise on measurement residual using analytic methods, so it is difficult to diagnose specific false satellite reliably. It is necessary to seek an effectual way to isolate satellite fault. In this paper, the effect of fault and uncertain factor are taken into account, optimal parity vectors for specified faulty satellites are designed to diagnose satellite faults.

## 1. GPS MEASUREMENT MODEL

In general, for Global Position System with 24 satellites, there are always more than four satellites visible in most time and region. However using only 4 satellites can determine user's position, extra visible satellites give system redundancy information. So parity vector can be used to detect satellite fault (Chen, 1996).

According to positioning principle of GPS, measurement model can be given as

$$\mathbf{Z} = \mathbf{H}\mathbf{X}_u + \varepsilon \qquad (1)$$

where $\mathbf{X}_u \in \Re^n$ is estimate state vector, includes user's position and clock error of receiver. Here $n = 4$, is the dimension of state vector.

$\mathbf{H} = \begin{bmatrix} e_{11} & e_{12} & e_{13} & 1 \\ e_{21} & e_{22} & e_{23} & 1 \\ \cdots & \cdots & \cdots & \cdots \\ e_{m1} & e_{m2} & e_{m3} & 1 \end{bmatrix}$ is geometry matrix ($m$ is number of visible satellites), $\varepsilon$ is measurement noise vector of satellites with Gaussian distribution of zero mean. $\mathbf{Z} \in \Re^m$ is measurement vector calculated according to pseudo-range, position and clock error of all visible satellites.

## 2. GENERALIZED LIKELIHOOD TEST METHOD

Define the following parity equation from (1)

$$\mathbf{p} = \mathbf{V}'\mathbf{Z} \qquad (2)$$

where " ' " denotes transpose. $\mathbf{p}$ is parity vector residual, $\mathbf{V}$ is parity matrix need to be determined. Obviously, it is that $\mathbf{p} = \mathbf{V}'\mathbf{H}\mathbf{X}_u + \mathbf{V}'\varepsilon$. In order that $\mathbf{p}$ is independent of $\mathbf{X}_u$, and only relates to noise or potential fault, it requires that

$$\mathbf{V}'\mathbf{H} = 0 \qquad (3)$$

then

$$\mathbf{p} = \mathbf{V}'\varepsilon \qquad (4)$$

When satellites are normal, parity vector is only a function of noise. p is also a vector with Gaussian distribution of zero mean. But, if some satellite is faulty, measurement equation (1) should be

$$\mathbf{Z} = \mathbf{H}\mathbf{X}_u + \gamma_f + \varepsilon \qquad (5)$$

where $\gamma_f$ is fault vector with component corresponding to faulty satellite not being zero and others being zeros. Then residual of parity vector becomes

$$\mathbf{p} = \mathbf{V}'\gamma_f + \mathbf{V}'\varepsilon \qquad (6)$$

Here, parity vector depends not only on noise, but also on fault. Because parity vectors of normal and faulty condition are different, it is the foundation of fault detection. In general, GLT method can be used to evaluate residual of parity vector, then to detect and diagnose fault. At first, design $m-n$ uncorrelated parity vectors $\mathbf{v}_i$ ($i = 1,\dots,m-n$), then form parity matrix $\mathbf{V} = (\mathbf{v}_1,\cdots,\mathbf{v}_{m-n})$. Each column $\mathbf{v}_i$ of parity matrix is a parity vector direction in parity space. On the assumption that the length of each parity vector $\|\mathbf{v}_i\|$ is equal, the $i$th component of residual vector can be expressed as

$$p_i = \mathbf{v}_i'\gamma_f + \mathbf{v}_i'\varepsilon \text{ , } (i = 1,\dots,m-n) \qquad (7)$$

The second term on the right side of (7) is the effect of noise in residual, which is not dependent on fault. The first term is the effect of fault. Hence, for fault with high ratio of signal to noise, residual is mainly determined by the first term on the right side of (7). For fault with given direction and size, if direction of parity vector is closer to direction of fault, corresponding component of residual is greater. Here vector is closer means that inclination between two vectors is smaller. So the absolute value of residual component is greater means this parity vector components is more sensitive to specified fault, vice versa.

However, when GLT method is used to detect fault, decision function is a quadratic form of $m-n$ residual components, that is $DF_{DGLT} = \mathbf{p}'(\mathbf{V}'\mathbf{V})^{-1}\mathbf{p}$. If parity matrix meets the constrained conditions of Potter (1977) and Suman, $\mathbf{V}'\mathbf{V} = \mathbf{I}_{m-n}$, and covariance matrix of model error is identity matrix, then decision function of fault detection is $DF_{DGLT} = \mathbf{p}'\mathbf{p} = \|\mathbf{p}\|^2$. It is obviously that $DF_{DGLT}$ is $\chi^2(m-n)$ distribution. $DF_{DGLT}$ integrates $m-n$ residual components sensitive to specific fault, as a result, GLT method reduces fault detection ability. Simulation results show that fault isolation rate is lower than detection rate. That is, when fault is detected, it can't always isolate fault correctly, especially for soft fault.

## 3. OPTIMAL PARITY VECTOR METHOD

Jin and Zhang (1999) proposed an optimal parity vector method. This method is designed to detect and isolate fault of specified sensor in a redundant sensor set. In the following, a brief introduction of this method is given.

Consider a redundant measurement system $\mathbf{Z} = \mathbf{H}\mathbf{X}_u + \mathbf{G}\gamma + \varepsilon$. From (5), one can see $\mathbf{G}\gamma = \gamma_f$. Where $\gamma = (\gamma_1, \cdots, \gamma_m)'$, $\gamma_i$ $(i = 1, \ldots, m)$ is fault size of $i$ th sensor. $\mathbf{G} = (\mathbf{g}_1, \cdots, \mathbf{g}_m)$, $\mathbf{g}_i$ $(i = 1, \ldots, m)$ is the $i$ th input vector of sensor fault, then measurement equation can be rewritten as

$$\mathbf{Z} = \mathbf{H}\mathbf{X}_u + \sum_{i=1}^{m} \mathbf{g}_i \gamma_i + \varepsilon \tag{8}$$

In order to design parity vector sensitive to specified sensor, design a new cost function. Then cost function to design an optimal parity vector of the $i$ th sensor is

$$\max_{v'H=0} \frac{(\mathbf{v}'\mathbf{g}_i)^2}{[\|\mathbf{v}'\|^2 + \sum_{j \neq i} (\mathbf{v}'\mathbf{g}_j)^2]} \tag{9}$$

Where numerator is fault sensitivity of specified sensor $j$, and denominator is sensitivity to noise and fault of all other sensors. In addition, constrained condition of parity is still kept. Consider the constrained condition of parity, that is $\mathbf{v}'\mathbf{H} = 0$, parity vector $\mathbf{v}$ can be expressed as a linear combination of all columns of $\mathbf{V}$

$$\mathbf{v} = \mathbf{V}\mathbf{c} \tag{10}$$

where all columns of $\mathbf{V}$ make up a set of basis in parity space, $\mathbf{c}$ is a weighting vector composed of weighting coefficients of each column of $\mathbf{V}$, that is the coordinate vector of $\mathbf{v}$ under this basis. Substituting (10) into (9), the extremum problem can be rewritten as

$$\max_{c \neq 0} \frac{(\mathbf{u}'\mathbf{c})^2}{\mathbf{c}'\mathbf{M}_B\mathbf{c}} \tag{11}$$

where $\mathbf{M}_B = \mathbf{V}'(\mathbf{I} + \mathbf{G}\mathbf{G}' - \mathbf{g}_i\mathbf{g}_i')\mathbf{V}$ is a symmetrical matrix with dimension $m-n$, $\mathbf{u} = \mathbf{V}'\mathbf{g}_i \in \Re^{m-n}$. If $\mathbf{M}_B$ is a positive definite matrix, then

(1) When $\mathbf{c} = k \cdot \mathbf{M}_B^{-1}\mathbf{u}$, (11) achieves maximum.

(2) Optimal parity vector is $\mathbf{v} = k \cdot \mathbf{V}\mathbf{M}_B^{-1}\mathbf{u}$, where $k$ is an arbitrary real number not equal to zero.

When determining decision threshold, it needs to normalize residual, therefore, it is not essential to select a specific $k$.

Suppose $\mathbf{v}_i$ is an optimal parity vector sensitive to the $i$ th sensor fault, then corresponding residual is:

$$p_i = \mathbf{v}_i'\mathbf{y} \tag{12}$$

In order to determine detection threshold, assume that $\varepsilon$ has a normal distribution of multiple variables, that is $\varepsilon \sim N_m(0, \Xi)$. So residual obtained from (12) is a normal random variable as follows:

(1) If $\gamma = 0$, $p_i \sim N_1(0, \mathbf{v}_i'\Xi\mathbf{v}_i)$;

(2) If $\gamma \neq 0$, $p_i \sim N_1(\mathbf{v}_i'\mathbf{G}\gamma, \mathbf{v}_i'\Xi\mathbf{v}_i)$

Let normalized residual be $p_i^* \triangleq p_i \big/ \sqrt{\mathbf{v}_i'\Xi\mathbf{v}_i}$, then

(1) If $\gamma = 0$, $p_i^* \sim N_1(0,1)$;

(2) If $\gamma \neq 0$, $p_i^* \sim N_1(\mathbf{v}_i'\mathbf{G}\gamma \big/ \sqrt{\mathbf{v}_i'\Xi\mathbf{v}_i}, 1)$

The $m$ normal residuals $p_i^*$ $(i = 1, \ldots, m)$ are sensitive to their own sensor fault. Then normalized residual with max absolute value can be determined. If $|p_k^*| = \max_i |p_i^*|$, then the $k$ th normalized residual is the maximum one. Decision function of fault detection $DF_{DOPT} = (p_k^*)^2$ has a distribution of $\chi^2(1)$. For given false alarm rate $\alpha$, one can determine detection threshold of OPT method $T_{DOPT}$. If $DF_{DOPT} \leq T_{DOPT}$, there is no sensor fault. If $DF_{DOPT} > T_{DOPT}$, the $k$ th sensor is faulty, and can be isolated simultaneously.

## 4. SIMULATION RESULTS

Take Beijing as observation station, then position of GPS user is that latitude $\phi = 40°$ N, longitude $\lambda = 116°$ E, altitude $h = 0m$, observe time is 1000s. Given position and velocity of user, GPS simulator can output position and velocity of all visible satellites, and corresponding pseudo range, rate of pseudo range change. Measurement noise of receiver is $\sigma = 3.5m$. Simulation result shows that there are seven satellites visible in the observation time. Satellite numbers are 0, 3, 4, 6, 7, 11 and 15. In simulation, fault can be inserted into different satellite data, then GLT method and OPT method are used to detect satellite fault respectively. Suppose that false alarm rate $P_{FA}$ is 0.001, then decision function of fault detection $DF_{DOPT}$ using OPT method obeys $\chi^2(1)$ distribution, detection threshold $T_{DOPT}$ is 10.827. Decision function of fault detection $DF_{DGLT}$ using GLT method obeys $\chi^2(m-4)$ distribution, and detection threshold is calculated as $T_{DGLT} = \sigma^2\chi^2(3) = 199.283m^2$.

If all satellites are normal, fault decision functions for

seven satellites using OPT method is shown in Fig. 1. (In figure, solid line is fault decision function; dashed line is decision threshold.) Because fault decision function is lower than detection threshold, there is no fault reported. If at t=100s, there happens a step fault of satellite 7. Fault amplitude is $4\sigma$. In this case fault decision function is shown in Fig. 2. It is obviously that $(p_7^*)^2$ is the biggest one 100 second after. Because $DF_{DOPT} = (p_7^*)^2 > T_{DOPT}$, fault of satellite 7 can be detected and isolated.

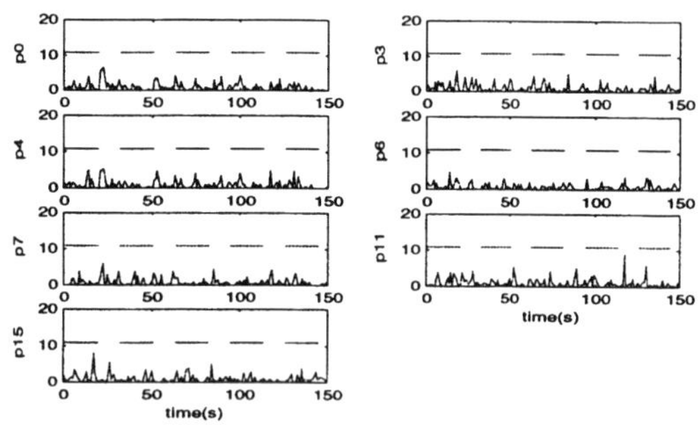


Fig. 1 Fault decision function without fault

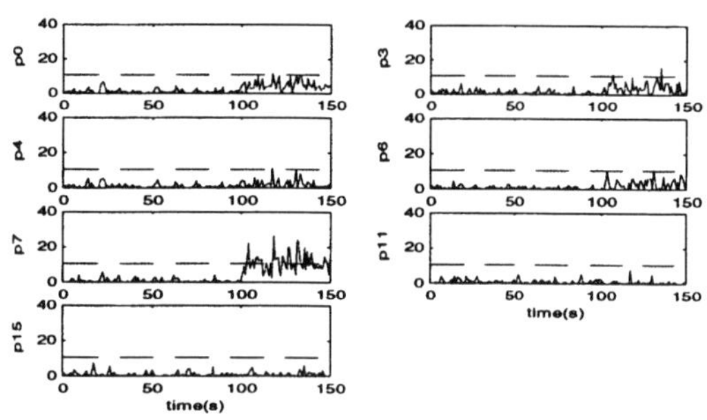


Fig. 2 Fault decision function with fault of satellite 7

In case of $\sigma = 3.5m$, $P_{FA} = 0.001$, compare the simulation results of the following two fault types:

(1) Step fault: insert step faults with different amplitude to different satellite, then simulate with Monte Carlo method 2000 times. Results are showed in Table 1 ~ Table 4, where $j$ is number of faulty satellite, $\gamma_j$ is ratio of fault and noise. Table 1 and Table 2 are the correct fault detection percent and correct fault isolation percent using GLT method respectively. Table 3 and Table 4 are the correct fault detection percent and correct fault isolation percent using OPT method respectively.

(2) Ramp fault: insert ramp faults with different slope to different satellite, then simulate with Monte Carlo method 1000 times. Compare of detect time with two methods is shown in Table 5. When ramp fault which slope is $0.25\sigma$ is inserted into satellite 7, Fig. 3 is the fault decision function of satellite 7. It is shown that detection function increase slowly after 100s because of satellite 7 fault till 116s that decision function exceed detect threshold. Then fault can be detected and isolated, and time to detect fault is 16s.

Table 1. Correct fault detection percent of GLT

| $j$ | $\gamma_j$ | | | |
|---|---|---|---|---|
| | 3 | 5 | 7 | 10 |
| 0 | 0.00 | 0.00 | 0.00 | 2.85 |
| 3 | 0.50 | 88.90 | 100.00 | 100.00 |
| 4 | 0.25 | 55.60 | 100.00 | 100.00 |
| 6 | 7.95 | 100.00 | 100.00 | 100.00 |
| 7 | 5.15 | 100.00 | 100.00 | 100.00 |
| 11 | 5.90 | 100.00 | 100.00 | 100.00 |
| 15 | 0.00 | 4.35 | 84.25 | 100.00 |

Table 2. Correct fault isolation percent of GLT

| $j$ | $\gamma_j$ | | | |
|---|---|---|---|---|
| | 3 | 5 | 7 | 10 |
| 0 | 0.00 | 0.00 | 0.00 | 1.55 |
| 3 | 0.35 | 84.85 | 98.70 | 99.90 |
| 4 | 0.15 | 47.70 | 92.50 | 97.85 |
| 6 | 5.90 | 100.00 | 100.00 | 100.00 |
| 7 | 3.70 | 99.90 | 100.00 | 100.00 |
| 11 | 3.70 | 99.65 | 99.95 | 100.00 |
| 15 | 0.00 | 2.80 | 79.35 | 98.20 |

Table 3. Correct fault detection percent of OPT

| $j$ | $\gamma_j$ | | | |
|---|---|---|---|---|
| | 3 | 5 | 7 | 10 |
| 0 | 0.00 | 0.00 | 0.001 | 16.25 |
| 3 | 3.60 | 99.90 | 100.00 | 100.00 |
| 4 | 1.10 | 96.20 | 100.00 | 100.00 |
| 6 | 34.45 | 100.00 | 100.00 | 100.00 |
| 7 | 23.85 | 100.00 | 100.00 | 100.00 |
| 11 | 26.70 | 100.00 | 100.00 | 100.00 |
| 15 | 0.00 | 23.25 | 99.60 | 100.00 |

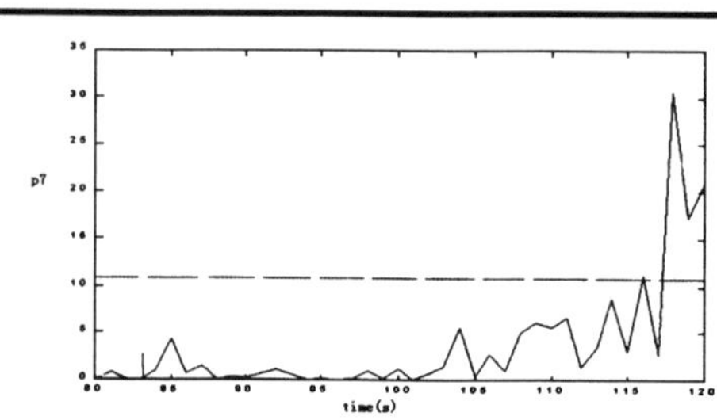

Fig. 3 Fault decision function of satellite 7 with slope fault

Table 4. Correct fault isolation percent of OPT

| $j$ | $\gamma_j$ | | | |
|---|---|---|---|---|
| | 3 | 5 | 7 | 10 |
| 0 | 0.00 | 0.00 | 0.0005 | 12.90 |
| 3 | 3.05 | 94.45 | 96.65 | 97.25 |
| 4 | 0.80 | 82.05 | 89.25 | 89.95 |
| 6 | 34.45 | 100.00 | 100.00 | 100.00 |
| 7 | 23.85 | 100.00 | 100.00 | 100.00 |
| 11 | 26.00 | 99.20 | 99.50 | 99.30 |
| 15 | 0.00 | 20.75 | 92.75 | 96.15 |

Table 5. Detecting time comparison of OPT and GLT ( Time: s )

| $j$ | Fault Slope | | | |
|---|---|---|---|---|
| | OPT | | GLT | |
| | $0.25\sigma$ | $0.50\sigma$ | $0.25\sigma$ | $0.50\sigma$ |
| 0 | 55.38 | 30.22 | 63.94 | 34.46 |
| 3 | 21.86 | 12.30 | 25.36 | 14.84 |
| 4 | 22.33 | 13.18 | 27.15 | 15.81 |
| 6 | 18.20 | 10.52 | 21.38 | 12.83 |
| 7 | 18.72 | 10.84 | 21.89 | 13.20 |
| 11 | 18.65 | 10.77 | 21.87 | 13.14 |
| 15 | 29.23 | 16.29 | 33.74 | 19.29 |

Conclusions can be drawn from Table 1~Table 5.

(1) With increasing amplitude of step fault, correct detection rate and correct diagnosis rate increase.
(2) Because direction cosine vectors generated by visible satellites and receiver are different, correct detection rate of satellite faults with bad geometric constellation is relative low for two methods.
(3) Because of noise, even if satellite fault is detected, fault can also not be isolated correctly, especially when fault is small. That is, fault isolation rate is smaller than detection rate.
(4) For small fault, the correct detection percent and the correct diagnosis percent of OPT method are much larger than that of GLT method.
(5) Simulation results of slope fault show that with increasing of fault slope, time to detect fault of both methods is shorten. For the same slope fault, detection time is shorter for OPT method than that of GLT method.

## 5. CONCLUSIONS

In this paper OPT method is used to diagnose satellite fault. $m$ optimal parity vectors are designed for $m$ visible satellites, each optimal vector is sensitive to its corresponding satellite fault. Then $m$ scalar residuals can be obtained. By comparing absolute values of $m$ scalar residuals, faulty satellite can be detected more precisely. The following conclusions can be obtained from simulation results:

(1) If fault amplitude is small, correct detection rate and correct diagnosis rate can be increased remarkably using OPT method.
(2) If fault amplitude is high, correct detection rate and correct diagnosis rate will be decreased appreciably using OPT method.
(3) If satellite has ramp fault, OPT method can detect and isolate fault in time when fault happens at early time. Thus it improves reliability of system.

## REFERENCES


RTCA (1991). Minimum operational performance standards for airborne supplemental navigation equipment using Global Position System (GPS), *RTCA/DO-208*, RTCA, July 1991.

Kozel, B. J., M. A. Cardoza (1996). Evaluation of GPS Precise Positioning Service autonomous integrity monitoring performance for military application, *Proceedings of the IEEE PLANS Symposium*, 1996, 83-91.

Pervan, B. S., D. G. Lawrence, C. E. Cohen, et al. (1996). Parity space methods for autonomous fault detection and exclusion using GPS carrier phase, *Proceedings of the IEEE PLANS Symposium*, 1996, 649-656.

Hunzinger, J. F., S. D. Morgern, J. Studenny (1997). CDGPS RAIM: Algorithm and protection radius calculation, *ION'97*, 1055-1064.

Brown, R. G. (1992). A baseline GPS RAIM scheme and a note on the equivalence of three RAIM methods, *Navigation: Journal of the Institute of Navigation*, 39, 301-316.

Chen, Y. H. (1996). Study on navigation system simulation and implementation, Master dissertation, Beijing: Beijing University of Aeronautics and Astronautics, 1996. (in Chinese)

Potter, J.E., M.C. Suman (1977). Thresholdless redundancy management with arrays of skewed instrument, AGARDOGR APH224, 15-1~15-25.

Jin, H., H. Y. Zhang (1999). Optimal parity vector sensitive to designated sensor fault, *IEEE Trans. On AES*, 35, 1122-1128.

IFAC Publications
www.elsevier.com/locate/ifac

# FUZZY INVERSE REASONING CONTROL OF A MIMO PLANT

**D.T. Pham and D. Li**

*Intelligent Systems Laboratory, Manufacturing Engineering Centre,*
*School of Engineering, Cardiff University*
*PO BOX 925, Cardiff CF24 0YF UK*

Abstract: Fuzzy inverse reasoning (FIR) control deduces control actions to yield the desired output of a plant by performing fuzzy inverse reasoning, which involves solving a fuzzy relational equation describing the dynamics of the plant to be controlled. Algorithms for solving fuzzy relational equations are available for equations containing a single decision variable. However, in practice, fuzzy relational equations may have more than one decision variable and thus cannot be handled by existing algorithms. This paper presents an approach to FIR control based upon decomposing a multiple-decision-variable equation into a number of single-decision-variable equations. An existing algorithm can then be used to solve the equations in order to produce the appropriate control actions. *Copyright © 2001 IFAC*

Key words: fuzzy logic, intelligent control, inverse dynamics control, process control, reasoning.

## 1. INTRODUCTION

Fuzzy inverse reasoning (FIR) aims to deduce a sufficient condition to achieve a specified goal. It searches for sufficient condition $A'$ for a goal $C'$ from fuzzy relations such as $A \Rightarrow C$. If $A'$ is found, the given goal is considered true. FIR can be expressed as follows (Arnould and Tano, 1995; Lee, 1990):

$$\begin{array}{l} \textit{goal}: y \textit{ is } C' \\ \textit{fuzzy relation}: \textit{If } x \textit{ is } A \textit{ Then } y \textit{ is } C \\ \hline \textit{sufficient condition}: x \textit{ is } A' \end{array} \tag{1}$$

Consider that the given goal is a desired process output and the sufficient condition to be derived is a control action. FIR can thus be used to produce a control action for a specified process output. The technique of determining control actions by using FIR is called FIR control or fuzzy backward reasoning control (Pham and Li, 1999). In FIR control, the fuzzy relation represents the dynamics of a controlled process the output of which can be expressed using a fuzzy relational equation:

$$X \circ \mathbf{R} = Y_d \tag{2}$$

where $\mathbf{R}$ is a fuzzy relation $X \Rightarrow Y$ representing the process dynamics, $X$ is a control action and $Y_d$ is a given goal. FIR is performed by solving a fuzzy relational equation such as Equation (2) in order to obtain a control action $X$ that is sufficient to yield $Y_d$.

FIR ensures that the produced control actions are sufficient for the desired process output. Algorithms exist for solving linear fuzzy relational equations that contain a single decision variable to obtain the appropriate control actions.

In practice, plants to be controlled often have multiple input variables and multiple output variables. When used to describe the dynamics of such multiple-input-multiple-output (MIMO) plants, a fuzzy relation will contain more than one input in its antecedent part. Consequently, there will be more decision variables included in the corresponding fuzzy relational equation and existing algorithms cannot be applied to the equation to determine control actions.

This paper introduces an approach to FIR control that can deal with MIMO plants. The idea is to decompose the fuzzy relational equations that describe such plants into single-decision-variable equations and employ an available algorithm to deduce control actions. The control of a three-tank plant will be used to illustrate the proposed approach. The rest of this paper is organised as follows. Section 2 describes the three-tank plant. Section 3 discusses the decomposition of MIMO processes where there are interactions between the different process variables. Section 4 presents the experimental results obtained.

## 2. THREE-TANK PLANT

The plant is illustrated in Fig. 1. The tanks all have the same cross-sectional area. They are linked by connecting pipes. Liquid levels in the tanks are regulated by manipulating the inlet flows of which there are two, one into Tank I (the left-most tank) and the other into Tank II (the right-most tank). Two pumps, $P_1$ and $P_2$, drawing liquid from a reservoir control the flow rates of the inlets. The flows in the pipes between Tank I and Tank II and the middle tank, Tank III, are manipulated using ball valves $V_1$ and $V_3$. The plant outlet flow which is regulated by ball valve $V_2$ is from a pipe connected to Tank II. Disturbances to the plant are in the form of leaks from the tanks which are controlled by ball valves $V_4$, $V_5$ and $V_6$, respectively. Pressure sensors are used to measure the levels of liquid in the tanks.

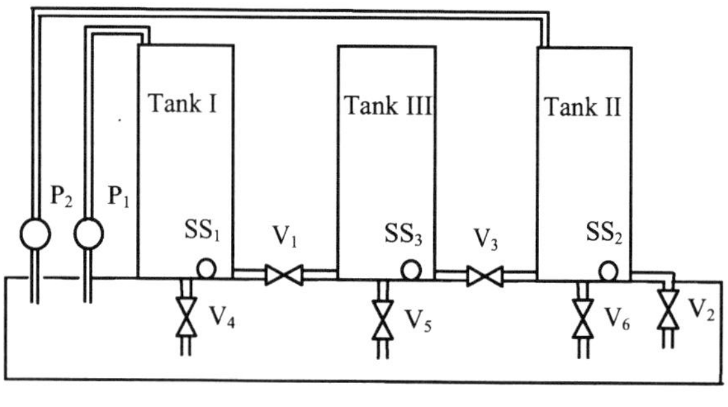


Fig. 1. A three-tank plant

When the tanks are coupled together through the connecting pipes, the effects of $q_1$ and $q_2$ will interact with one another. For example, assume that the plant is in a steady state initially. A new desired level for Tank I is set that is higher than the current level. The level $h_2$ for Tank II is to remain the same. $q_1$ must increase to bring the level $h_1$ in Tank I to its new set point. This is the *direct effect* of $q_1$ on $h_1$. However, because the tanks are coupled, $q_1$will also disturb $h_2$. A reduction in $q_2$ is required to compensate for the increase in $h_2$ due to $q_1$. The change in $q_2$, in turn, reduces $h_1$. This is called the *indirect effect* of $q_1$ on $h_1$. These interactions can be seen from the following mathematical model of the plant:

$$A\frac{dh_1}{dt} = q_1 - az_1 \cdot S_n \cdot \mathrm{sgn}(h_1 - h_3)\sqrt{2g\,|\,h_1 - h_3\,|} \quad (3a)$$

$$A\frac{dh_1}{dt} = az_1 \cdot S_n \cdot \mathrm{sgn}(h_1 - h_3)\sqrt{2g\,|\,h_1 - h_3\,|} - az_3 \cdot S_n \cdot \mathrm{sgn}(h_2 - h_3)\sqrt{2g\,|\,h_2 - h_3\,|} \quad (3b)$$

$$A\frac{dh_1}{dt} = q_2 - az_3 \cdot S_n \cdot \mathrm{sgn}(h_2 - h_3)\sqrt{2g\,|\,h_2 - h_3\,|} - az_2 S_l \sqrt{2gh_2} \quad (3c)$$

where $az_i$ represents the opening of the $i^{\text{th}}$ valve, A is the cross-sectional area of a tank, $S_n$ is the cross-sectional area of the connection pipes and $S_l$ is the cross-sectional area of the outlet pipe.

The interactions between $q_1$ and $q_2$ make it difficult to implement FIR control for the three-tank plant.

## 3. DIMENSION REDUCTION FOR FUZZY INVERSE REASONING CONTROL

The three-tank plant has 2 inputs, $q_1$ and $q_2$, and 2 outputs, $\dot{h}_1$ and $\dot{h}_2$. The fuzzy relational equation that describes the dynamics of the process has 2 decision variables and is not a linear equation. This fuzzy relational equation needs to be transformed into linear equations that can be solved using existing algorithms to obtain the appropriate control action.

The concept of a *decoupler* (Stephanopoulos, 1984) in conventional control is employed to decompose the interactions in the three-tank plant. A decoupler cancels the effect of the actions of one control loop, $u_k(t)$, on the outputs of other loops, $y_i(t)$ $(i \neq k)$ and hence decomposes interactions between $u_k(t)$ and $u_i(t)$ $(i \neq k)$. The decoupler works by providing an additional action due to $u_k(t)$ on $y_i(t)$ $(i \neq k)$ to compensate for the influence of $u_k(t)$ on $y_i(t)$ $(i \neq k)$. In the time domain, it maintains $y_i(t)$ equal to $y_i(t-1)$ when $u_k(t)$ $(i \neq k)$ changes.

Two FIR controllers have been designed based on the concept of the decoupler. One has $q_1$ as the input variable and $\dot{h}_1$ as the output variable. It also has two auxiliary variables, $h_1$ and $h_3$. The fuzzy relation employed by this controller is a relation from $q_1$, $h_1$ and $h_3$ to $\dot{h}_1$, called $\mathbf{R}_1$. The fuzzy relational equation for the controller is:

$$[Q_1 \quad H_1 \quad H_3] \circ \mathbf{R}_1 = \dot{H}_1 \tag{4}$$

where $Q_1$, $H_1$, $H_3$ and $\dot{H}_1$ are fuzzy variables corresponding to $q_1$, $h_1$, $h_3$ and $\dot{h}_1$. When a new set point $h_{1sp}$ is chosen, $\dot{h}_1(t+1) = h_{1sp} - h_1(t)$ is calculated and a goal $\dot{H}_1$ is formed. By solving Equation (4), control actions $q_1$ can be deduced. When $\dot{h}_1(t+1)$ is computed, the auxiliary variable $h_3$ remains unchanged so that the controller works by producing an effect to counteract the influence of $q_1$ on $h_3$, i.e. to cancel the indirect effects of $q_1$ on $h_1$.

The auxiliary fuzzy variables $H_1$ and $H_3$ are computed from the measured data, $h_1$ and $h_3$. Equation (4) is then rewritten as a linear equation, namely,

$$Q_1 \circ \mathbf{R}_1' \mid_{H_1, H_3} = \dot{H}_1 \tag{5}$$

$\mathbf{R}_1'$ is calculated from $\mathbf{R}_1$, $H_1$ and $H_3$. A 3-dimensional matrix is defined as $\mathbf{X} = [Q_1 \quad H_1 \quad H_3]$. Entry $x_{ijk} = q_1^i \wedge h_1^j \wedge h_3^k$ is computed from the antecedent part of some fuzzy rule "IF $Q_1$ is $A^i$ and $H_1$ is $B^j$ and $H_3$ is $C^k$, THEN $\dot{H}_1$ is $D^l$", where $q_1^i$, $h_1^j$ and $h_3^k$ are degrees of membership of fuzzy sets $A^i$, $B^j$ and $C^k$, respectively. The fuzzy relational matrix $\mathbf{R}_1$ is a 4-dimensional matrix. Element $r_{ijkl}$ of $\mathbf{R}_1$ is the truth degree of the given fuzzy rule.

$\dot{h}_1^l$ can be calculated from Equation (4) using the composition rule of inference, shown in Equation (6).

$$\begin{aligned}
\dot{h}_1^l = {} & [(x_{111} \wedge r_{111l}) \vee (x_{112} \wedge r_{112l}) \vee \cdots\cdots \vee (x_{11K} \wedge r_{11Kl})] \vee \\
& [(x_{121} \wedge r_{121l}) \vee (x_{122} \wedge r_{122l}) \vee \cdots\cdots \vee (x_{12K} \wedge r_{12Kl})] \vee \\
& \cdots\cdots\cdots \\
& [(x_{1J1} \wedge r_{1J1l}) \vee (x_{1J2} \wedge r_{1J2l}) \vee \cdots\cdots \vee (x_{1JK} \wedge r_{1JKl})] \vee \\
& \cdots\cdots\cdots \\
& \cdots\cdots\cdots \\
& [(x_{I11} \wedge r_{I11l}) \vee (x_{I12} \wedge r_{I12l}) \vee \cdots\cdots \vee (x_{I1K} \wedge r_{I1Kl})] \vee \\
& [(x_{I21} \wedge r_{I21l}) \vee (x_{I22} \wedge r_{I22l}) \vee \cdots\cdots \vee (x_{I2K} \wedge r_{I2Kl})] \vee \\
& \cdots\cdots\cdots \\
& [(x_{IJ1} \wedge r_{IJ1l}) \vee (x_{IJ2} \wedge r_{IJ2l}) \vee \cdots\cdots \vee (x_{IJK} \wedge r_{IJKl})]
\end{aligned} \tag{6}$$

From the associativity and commutativity of the t-norm and t-conorm ($\wedge$ and $\vee$) in the calculation, $\dot{h}_1^l$ can be rearranged as in Equation (7).

$$\begin{aligned}
\dot{h}_1^l = {} & \{[q_1^1 \wedge (h_1^1 \wedge h_3^1 \wedge r_{111l})] \vee \cdots\cdots \vee [q_1^1 \wedge (h_1^1 \wedge h_3^K \wedge r_{11Kl})]\} \vee \\
& \cdots\cdots\cdots \\
& \{[q_1^1 \wedge (h_1^J \wedge h_3^1 \wedge r_{1J1l})] \vee \cdots\cdots \vee [q_1^1 \wedge (h_1^J \wedge h_3^K \wedge r_{1JKl})]\} \vee \\
& \cdots\cdots\cdots \\
& \cdots\cdots\cdots \\
& \{[q_1^I \wedge (h_1^1 \wedge h_3^1 \wedge r_{I11l})] \vee \cdots\cdots \vee [q_1^I \wedge (h_2^1 \wedge h_3^K \wedge r_{I1Kl})]\} \vee \\
& \cdots\cdots\cdots \\
& \{[q_1^I \wedge (h_1^J \wedge h_3^1 \wedge r_{IJ1l})] \vee \cdots\cdots \vee [q_1^I \wedge (h_1^J \wedge h_3^K \wedge r_{IJKl})]\}
\end{aligned} \tag{7}$$

Equation (7) can be rewritten as:

$$\dot{h}_1^l = \begin{pmatrix} q_1^1 & q_1^2 & \cdots & q_1^I \end{pmatrix} \circ \begin{pmatrix} r_{1l}' \\ r_{2l}' \\ \vdots \\ r_{Il}' \end{pmatrix} \tag{8}$$

where $r_{il}' = [(h_1^1 \wedge h_3^1 \wedge r_{i11l}) \vee \cdots \vee (h_1^J \wedge h_3^K \wedge r_{iJKl})]$.

Applying the same procedure to all elements of $\dot{H}_1$, $\dot{h}_1^l$ $(l = 1, \ldots L)$, yields a 2-dimensional fuzzy relational matrix

$$\mathbf{R}_1' = \begin{pmatrix} r_{11}' & r_{12}' & \cdots & r_{1L}' \\ r_{21}' & r_{22}' & & r_{2L}' \\ \vdots & & & \vdots \\ r_{I1}' & r_{I2}' & \cdots & r_{IL}' \end{pmatrix}.$$

Thus, given a new set point $h_{1sp}$, it is possible to apply an algorithm for solving linear fuzzy relational equations to Equation (4) to produce control actions $q_1$.

The other FIR controller has a similar structure. It has $q_2$ as the input variable and $\dot{h}_2$ as the output variable. The auxiliary variables are $h_2$ and $h_3$.

## 4. EXPERIMENTAL RESULTS

The designed FIR controllers have been applied to regulate the liquid levels in the tanks under different operating conditions. First, the plant is set up as follows: the three tanks are empty and fully connected, the outlet valve opening is in the medium range, there is no leakage and the set points $h_{1sp}$ and $h_{2sp}$ are 200 (mm). With the system starting from a steady state, a leak is introduced. A disturbance is simulated by suddenly increasing the inlet flow rate for Tank II by 20% and held there for about 120 seconds when the plant is in the steady state.

The following design specifications are set for all the different operating conditions:

$$\text{Steady state error } (\varepsilon) \leq 2.50\ (\%) \qquad (9a)$$
$$\text{Rise time } (\tau) \leq 90\ (\text{sec}) \qquad (9b)$$
$$\text{Maximum overshoot } (M_p) \leq 5(\%) \qquad (9c)$$

The closed-loop system is shown in Fig. 2. The two FIR controllers, FIR 1 and FIR 2, are used to manipulate $h_1$ and $h_2$ by controlling $q_1$ and $q_2$, respectively.

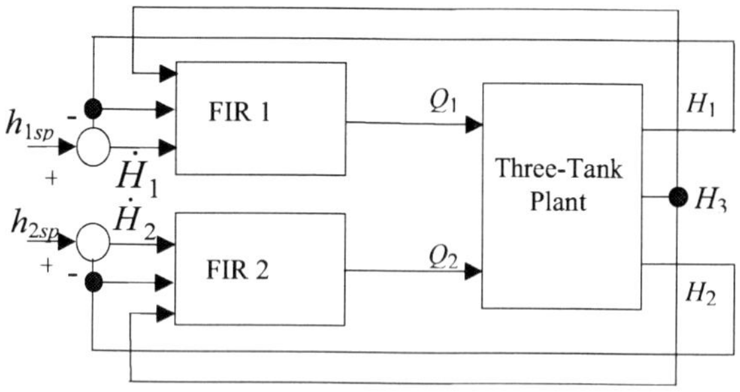


Fig. 2. Structure of the closed-loop system for the three-tank plant

The membership functions of the normalised (unitless) input and output variables are displayed in Figs. 3, 4 and 5, respectively.

Fig. 6. shows time domain plots of the liquid levels $h_1$ and $h_2$. It can be seen from these plots that the FIR controllers could successfully regulate the liquid levels and satisfy the design specifications.

Having to function under different operating conditions usually causes sensitivity problems in a closed-loop system. Although these can be solved by adjusting the controller parameters through gain scheduling (Åström, 1989), the adjusted parameters may still not be able to match the actual operating condition in cases where uncertainty exists. Because of the robustness of fuzzy logic control, a FIR controller is expected to perform well under different operating conditions. Leakage is introduced to simulate operation condition variation. Because of the leakage, the rise time became longer and the state errors were greater. This is shown in Table 1 and Table 2.

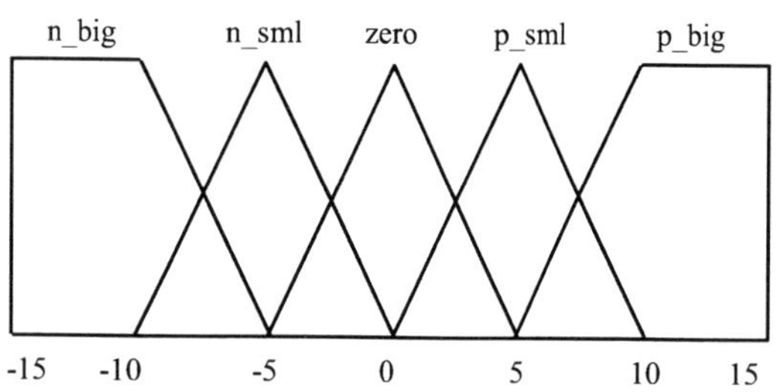


Fig. 3. Membership functions of inlet flow rate $Q_1$ *and* $Q_2$,

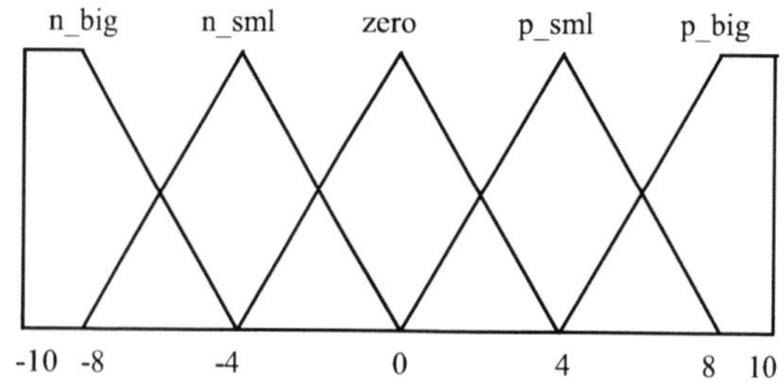


Fig. 4. Membership functions of liquid level $H_1$, $H_2$ and $H_3$,

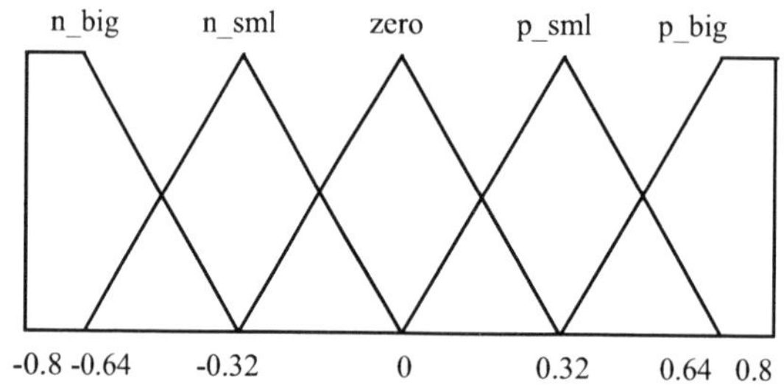


Fig. 5. Membership functions of derivatives $\dot{H}_1$ and $\dot{H}_2$

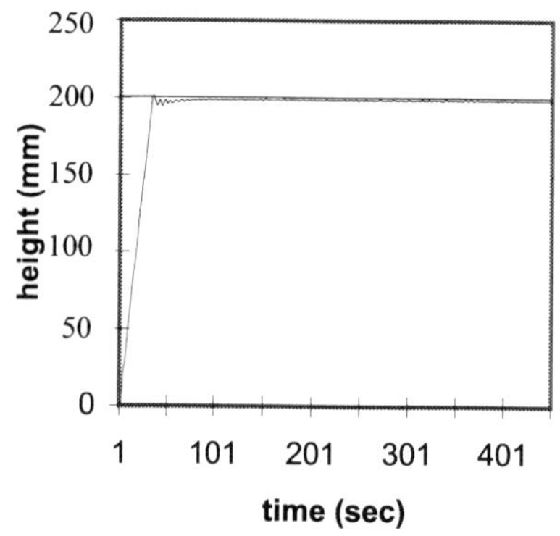


(a)

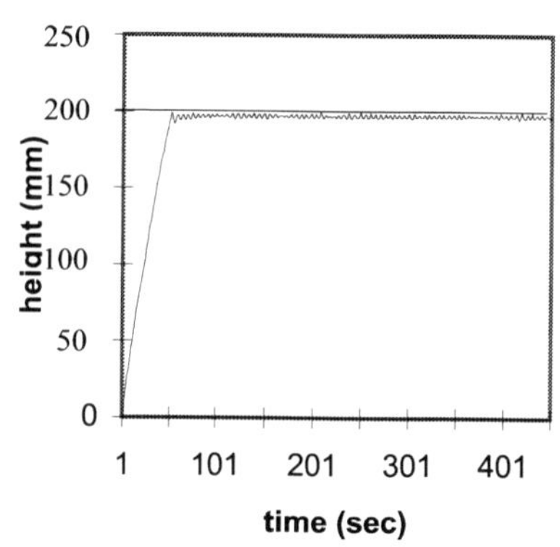


(b)

Fig. 6. Responses of the three-tank-plant with the set points raised from 0 to 200 mm
(a) level of Tank I (b) level of Tank II

Table 1. Performances of FIR controllers

| Control | Tank | Performance | | |
|---|---|---|---|---|
| | | τ (sec) | $M_p$ (%) | ε (%) |
| FIR | I | 35 | 0.0 | 0.95 |
| | II | 52 | 0.0 | 1.15 |

Table 2. Performances of FIR and PI controllers in the presence leakage

| Control | Tank | Performance | | |
|---|---|---|---|---|
| | | IAE | ε (%) | τ (sec) |
| PI | I | 482 | 1.95 | 34 |
| | II | 733 | 3.55 | 60 |
| FIR | I | 376 | 1.55 | 38 |
| | II | 716 | 2.75 | 71 |

For comparison, a PI control is designed to control the liquid levels when there is leakage. The Integral of Absolute Error (IAE) criterion, namely,

$$\mathrm{IAE} = \int_0^{100} | e(t) | \, dt \tag{10}$$

where $e(t)$ is the error between the set point $y_{sp}$ and the process output $y(t)$, is adopted to evaluate the sensitivity of the closed-loop system. The performances of the FIR and PI control schemes are shown in Table 2. It can be seen that FIR control is less affected by leakage than PI control. It is more effective than PI control at minimising the IAE. FIR control also yields smaller steady state error than PI control.

The results with a disturbance added to Tank II are depicted in Fig. 7 and Fig. 8. Fig. 7 illustrates that the controllers produce small control actions to reject the disturbance. Fig. 8 shows the effects of the disturbance on the liquid level of Tank II. It can be seen that the disturbance causes little change in the liquid level.

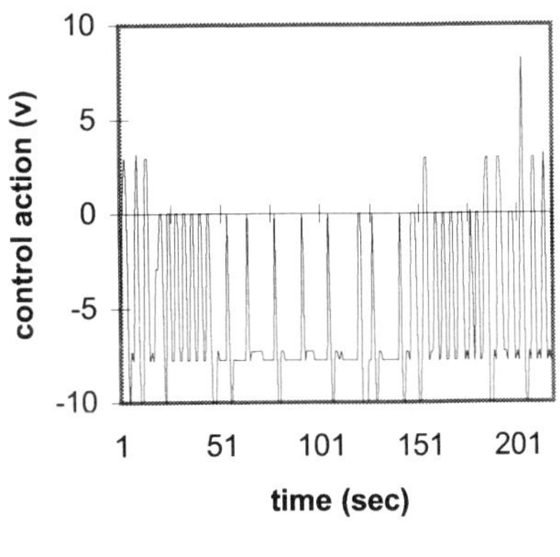


Fig. 7 Responses of the three-tank plant to a disturbance -- control actions

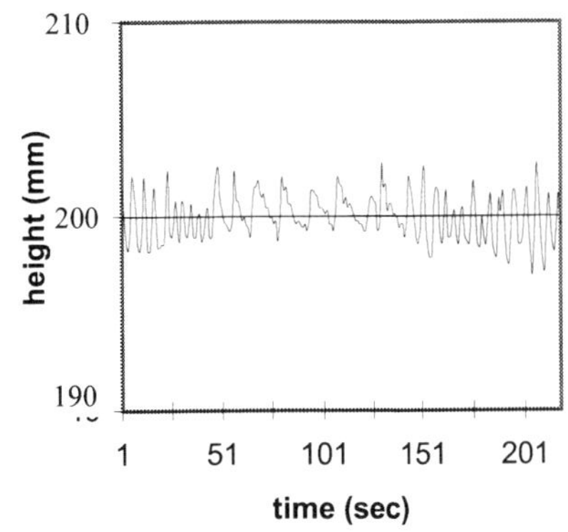


Fig. 8 Responses of the three-tank plant to a disturbance -- level of Tank II

## 5. CONCLUSION

The proposed FIR control approach derives control actions by solving a fuzzy relational equation. It involves transforming a multiple-decision-variable equation into linear equations with a single decision variable. The approach can deal with MIMO process control problems. It has been successfully demonstrated on a three-tank plant to regulate liquid levels. Experimental results have shown that FIR controllers can provide satisfactory steady state and transient responses. The controllers are robust in the sense of being able to handle operating condition changes and reject disturbances.

## REFERENCES


Arnould, T. and S. Tano (1995). Interval-valued fuzzy backward reasoning, *IEEE Trans. on Fuzzy Systems*, **Vol. 3**, **No. 4**, pp.425-437.

Åström, K. J. and B. Wittenmark, (1989) *Adaptive Control*, Addison-Wesley Publishing Company, London.

Lee, C. C. (1990). Fuzzy logic in control systems: fuzzy logic controller—part II, *IEEE Trans. on Systems, Man, and Cybernetics*, **Vol. 20**, **No. 2**, pp.419-437.

Pham, D. T. and D. Li (1999) Fuzzy backward reasoning control, *Proc. 6th UK Workshop on Fuzzy Systems*, Brunel University, Uxbridge, UK, September, 1999, pp.21-26.

Stephanopoulos, G. (1984) *Chemical Process Control*, Prentice-Hall, London.

IFAC Publications
www.elsevier.com/locate/ifac

# DESIGN OF FUZZY-LOGIC-CONTROL SYSTEM BASED ON GAUSSIAN-BASIS-FUNCTION NEURAL NETWORK

**Lixiong Li, Ziyuan Huang and Minrui Fei**

*School of Mechatronical Engineering & Automation, Shanghai University*
*Shanghai, 200072, P.R. China*
*Email: lilix@public6.sta.net.cn*

Abstract: In this paper, a fuzzy logic system is constructed by Gaussian basis function neural network. In the process of design, a learning rate self-adjusting algorithm is presented to guarantee that the basis function has a reasonable distribution, meantime to reduce the disturbance of noise and improve the control precision. Finally, a nonlinear object is simulated by this approach. The results demonstrate that: compared with normal fuzzy logic control system, this self-learning fuzzy logic control system has stronger learning ability and better control performance. *Copyright © 2001 IFAC*

Keywords: Gaussian function, Fuzzy logic, Neural networks, Process identification, Learning algorithms, Self-adapting algorithms

## 1. INTRODUCTION

With the successful application of fuzzy logic system and neural network in the intelligent control field, the approach of synthesizing the language describing ability of fuzzy logic system and the learning ability of neural network has become a research hotspot. As a kind of network with local approach ability, Gaussian basis neural network can be close to a nonlinear continuous function defined on a compact set as an arbitrary precision (Wang, et al., 1992). Especially, a fuzzy logic system with the structure of multiplication inference, Gaussian or triangular membership functions, singleton output and center of gravity defuzzification method is equivalent to a Gaussian basis neural network (abbreviated to GBNN) (Lee, et al., 1994; Wang, 2000). So it is a satisfying choice to replace a fuzzy logic controller with Gaussian basis neural network. Based on the above-mentioned idea, a self-learning fuzzy logic control system based on Gaussian basis function neural network is designed. At the same time, a self-adjusting factor algorithm for basis function parameters is given to improve learning performance.

## 2. SYSTEM STRUCTURE

Obviously, it is very difficult even impossible for us to get a precise model for a complicated controlled object, and classic control theory is helpless to control it. So control expert introduce some latest results of ideation science and neural science into control field, and developing intelligent control theory to solve the problem. Traditional method to control complicated object is to identify it by a neural network and control the identification model by another neural network (Wang and Tong, 1998). Here the unknown object is identified by a Gaussian basis function neural network (abbreviated to GBNI). And another Gaussian basis function neural network (abbreviated to GBNC) is to control the real object via fuzzy rules, in addition, as a neural network, GBNC is of self-learning ability and its error gradient back-propagation routeway is given by GBNI. As Fig. 1 shows, it is a closed-loop control system.

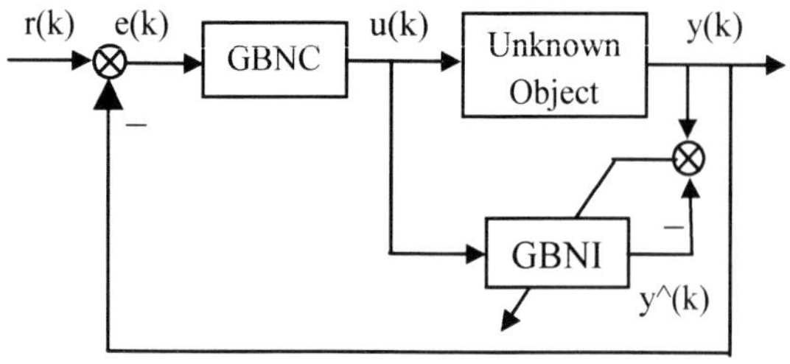


Fig.1. Structure of control system

## 3. IDENTIFICATION ALGORITHM OF GBNI

GBNI is composed by GBNN whose structure is shown in Fig. 2, its input and output relation can be described by

$$\hat{y}(k) = W_I \cdot \alpha(\bar{x}) \tag{1}$$

where $W_I = \lfloor w_1^I, w_2^I, \cdots w_r^I \rfloor$ is joint weight vector and

$\alpha(\bar{x}) = [\alpha_1(\bar{x})\cdots\alpha_r(\bar{x}))]^T, \alpha_i(\bar{x}) = e^{-\frac{\|\bar{x}-c_i\|^2}{\sigma_i^2}}$, (i = 1,2...r) is Gaussian basis function vector and

$\bar{x} = [u(k)\cdots u(k-m+1), y(k-1)\cdots y(k-n+2)]$ is input vector.

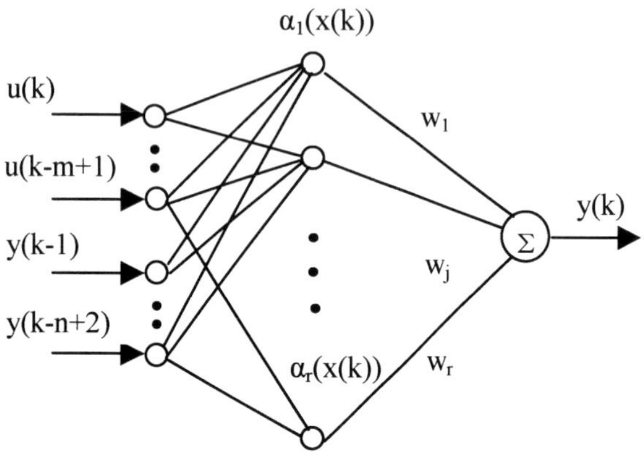


Fig.2. Network structure of GBNI

The main function of GBNN is to model the controlled object dynamically, that is to say, $\hat{y}(k) \approx y(k)$ after finite times learning. The target function for identification learning is defined as

$$E_I = \frac{1}{2}\sum (y(k) - \hat{y}(k))^2 \tag{2}$$

The gradient-descent method is used to learn.

$$w_j^I(k+1) = w_j^I(k) - \eta_w^I \frac{\partial E_I(k)}{\partial w_j^I(k)} \tag{3}$$

$$c_j^I(k+1) = c_j^I(k) - \eta_c^I \frac{\partial E_I(k)}{\partial c_j^I(k)} \tag{4}$$

$$\sigma_j^I(k+1) = \sigma_j^I(k) - \eta_\sigma^I \frac{\partial E_I(k)}{\partial \sigma_j^I(k)} \tag{5}$$

where

$$\frac{\partial E_I(k)}{\partial w_j^I(k)} = (\hat{y}(k) - y(k)) \alpha_j^I(\bar{x}(k)) \Big/ \alpha^T(\bar{x}(k))\alpha(\bar{x}(k)) \tag{6}$$

$$\frac{\partial E_I(k)}{\partial c_j^I(k)} = (\hat{y}(k) - y(k)) w_j^I(k) \alpha_j^I(\bar{x}(k)) \frac{\|\bar{x} - c_j^I\|^2}{\sigma_j^{I^3}} \tag{7}$$

$$\frac{\partial E_I(k)}{\partial \sigma_j^I(k)} = (\hat{y}(k) - y(k)) w_j^I(k) \alpha_j^I(\bar{x}(k)) \frac{\sum(\bar{x}(k) - c_j^I(k))}{\sigma^{3I}_j(k)} \tag{8}$$

## 4. GBNC AND ITS LEARNING ALGORITHM

### *4.1 Initializing GBNC by fuzzy rules*

In above system (shown in Fig. 1), GBNC replace a fuzzy logic controller both in structure and function. In order to make good use of experts' experiences based on language rules, in this paper, fuzzy rules are converted into correlative network parameters that can be accepted by GBNC such as the central value and width of Gaussian basis function and the network weight matrix. This kind of initial network value can accelerate the convergence rate of learning algorithm and avoid falling into local minimum during the training. Similar to GBNI, GBNC is also constructed by GBNN. Its network structure is shown in Fig. 3.

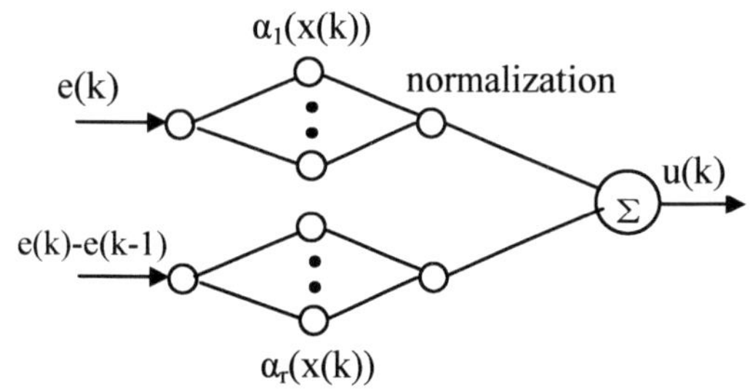


Fig.3. Network structure of GBNC

Suppose fuzzy rule as

*Rule (l): if* $x_1$ *is* $\underset{\sim}{A_1}$ *and ...* $x_M$ *is* $\underset{\sim}{A_M}$ *then y is* $\bar{y}^l$.

Then a fuzzy logic system with the structure of multiplication inference, Gaussian or triangular membership functions, singleton output and center fuzzy eliminator can be described as (Wang, 2000),

$$\bar{y} = \frac{\sum_{l=1}^{M} \bar{y}^l \prod_{i=1}^{n} \mu_{\underset{\sim}{A}}(x_i)}{\sum_{l=1}^{M} \prod_{i=1}^{n} \mu_{\underset{\sim}{A}}(x_i)} \tag{9}$$

Let $p_i(\bar{x}) = \frac{\prod_{i=1}^{n} \mu_{\underset{\sim}{A}}(x_i)}{\sum_{l=1}^{M} \prod_{i=1}^{n} \mu_{\underset{\sim}{A}}(x_i)}$ be fuzzy basis function and then formula (9) will be

$$\bar{y} = \sum_{l=1}^{M} \bar{y}^l p_i(\bar{x}) \tag{10}$$

If $\mu_{\underset{\sim}{A}}$ is Gaussian or triangular membership function, then $p_i(\bar{x})$ can be approximate to Gaussian basis function $\alpha_i(\bar{x}) = e^{-\frac{\|\bar{x}-c_i\|^2}{\sigma_i^2}}$, and $\alpha_i(\bar{x})$ can be got by making $\|p_i(\bar{x}) - \alpha_i(\bar{x})\|$ be minimized. Thus the fuzzy logic controller can be replaced by GBNC, and the network can be initialized by experts' experiences (fuzzy rules).

### *4.2 Self-adjusting learning rate factor algorithm for GBNC*

From Fig. 1, the online control target function of GBNC can be defined as

$$E_C = \frac{1}{2}(r(k) - y(k))^2 \tag{11}$$

Similar to GBNI, the learning algorithm for GBNC is gradient-descent method

$$w_j^C(k+1) = w_j^C(k) - \eta_w^C \frac{\partial E_C(k)}{\partial w_j^C(k)} \tag{12}$$

$$c_j^C(k+1) = c_j^C(k) - \eta_c^C \frac{\partial E_C(k)}{\partial c_j^C(k)} \tag{13}$$

$$\sigma_j^C(k+1) = \sigma_j^C(k) - \eta_\sigma^C \frac{\partial E_C(k)}{\partial \sigma_j^C(k)} \tag{14}$$

where

$$\frac{\partial E_C(k)}{\partial w_j^C(k)} = \frac{\partial E_C(k)}{\partial y(k)} \cdot \frac{\partial y(k)}{\partial u(k)} \cdot \frac{\partial u(k)}{\partial w_j^C(k)} \tag{15}$$

$$\frac{\partial E_C(k)}{\partial c_j^C(k)} = \frac{\partial E_C(k)}{\partial y(k)} \cdot \frac{\partial y(k)}{\partial u(k)} \cdot \frac{\partial u(k)}{\partial c_j^C(k)} \tag{16}$$

$$\frac{\partial E_C(k)}{\partial \sigma_j^C(k)} = \frac{\partial E_C(k)}{\partial y(k)} \cdot \frac{\partial y(k)}{\partial u(k)} \cdot \frac{\partial u(k)}{\partial \sigma_j^C(k)} \tag{17}$$

Because the controlled object is unknown, $\frac{\partial y(k)}{\partial u(k)}$ in formula (15)~(17) is to be got by GBNI's back propagation,

$$\frac{\partial y(k)}{\partial u(k)} \approx \frac{\partial \hat{y}(k)}{\partial u(k)} = \sum_{1}^{r} w_j^I(k) \cdot \alpha_j(x^I(k)) \cdot \left( \frac{u(t) - \sigma_j^I(k)}{c_j^I(k)} \right) \tag{18}$$

In the design of fuzzy logic system, it is sure to guarantee that fuzzy logic control rules are complete and the fuzzy sets don't interfere with each other. That is to say, the membership functions should cover the whole input space and they don't overlap with each other. Meantime, in order to reduce the disturbance of noise, low sensitivity membership functions are used when the error is fairly big, and when the error is close to zero, high sensitivity membership functions are applied to improve the control precision (Sun, 1997).

Since basis functions of GBNC are equivalent to membership functions of fuzzy logic system, the above-mentioned design principles are also available to GBNC's design. Based on this idea, self-adjusting learning rate factor algorithm is presented for basis function parameters in GBNC. In formula (13), self-adjusting learning rate factors $\eta_c^C(x)$ is introduced to guarantee that location of each Gaussian basis function is rational. In addition, in formula (14), self-adjusting learning rate factor $\eta_\sigma^C(x)$ is introduced to guarantee that the shape of each Gaussian basis function is reasonable.

The intersection point of two adjacent Gaussian basis function $\alpha_i(x)$ and $\alpha_j(x)$ is $\left( x, \exp\left(-\left(\frac{c_i - c_j}{\sigma_i \pm \sigma_j}\right)^2\right) \right)$, so the superposition degree for those can be defined as

$$d = \max_{\substack{j=1,2..r \\ j\neq i}} \left( \frac{c_i - c_j}{\sigma_i \pm \sigma_j} \right)^2 \qquad (19)$$

In order to guarantee that the adjacent Gaussian basis functions don't interfere with each other, the value of $d$ should be in reasonable range. Let $\tau$ be the ideal value for superposition degree. In the process of GBNC learning, $\eta_c^C$ should be adjusted automatically according to Table1 and Table2 to make $d$ be distributed near $\tau$.

Table1 Adjusting direction of $\eta_c^C$ ($d<\tau$)

| | $c_i$-$c_j$ >0 | $c_i$-$c_j$ <0 |
|---|---|---|
| $\partial E_C/\partial c_i$ >0 | $\eta_c^C$ ↙ | $\eta_c^C$ ↗ |
| $\partial E_C/\partial c_i$ <0 | $\eta_c^C$ ↗ | $\eta_c^C$ ↙ |

Table2 Adjusting direction of $\eta_c^C$ ($d>\tau$)

| | $c_i$-$c_j$ >0 | $c_i$-$c_j$ <0 |
|---|---|---|
| $\partial E_C/\partial c_i$ >0 | $\eta_c^C$ ↗ | $\eta_c^C$ ↙ |
| $\partial E_C/\partial c_i$ <0 | $\eta_c^C$ ↙ | $\eta_c^C$ ↗ |

To realize the adjustment, the self-adapting learning rate factors $\eta_c^C(x)\in[0,1]$ is introduced as:

$$\eta_c^C = \frac{1}{\pi} arcctg(\lambda\cdot(\tau-d_i)\cdot \mathrm{sgn}(\frac{\partial E_C}{\partial c_i^C}\cdot(c_i^C(k)-c_j^C(k)))) \qquad (20)$$

where $\lambda$ is a positive constant.

In addition, $\eta_\sigma^C$ should be adjusted by Table 3 to use varied sensitivity Gaussian basis functions in accordance with different error. And another self-adapting learning rate factor $\eta_\sigma^C(x)\in[0,1]$ is introduced as,

$$\eta_\sigma^C = \frac{1}{\pi} arcctg(\lambda\cdot \mathrm{sgn}(\frac{\partial E_C}{\partial \sigma_j^C})\cdot \|c_i\|) \qquad (21)$$

where $\lambda$ is a positive constant

Table3 Adjusting direction of $\eta_\sigma^C$

| | $\partial E_C/\partial \sigma_i$ >0 | $\partial E_C/\partial \sigma_i$ <0 |
|---|---|---|
| $\|c_i\|$ is big | $\eta_\sigma^C$ ↙ | $\eta_\sigma^C$ ↗ |
| $\|c_i\|$ is small | $\eta_\sigma^C$ ↗ | $\eta_\sigma^C$ ↙ |

## 5. SIMULATION EXAMPLES

The controlled object can be described as

*y(k)=f(y(k-1),y(k-2),u(k))=0.5y(k-1)+0.8sin(1.5y(k-1)+y(k-2))+1.2u(k)*

and the input signal is *r(k)=1*. During the simulation, the normal fuzzy logic controller, Gaussian basis neural-fuzzy controller (GBNI+GBNC) and Gaussian basis neural-fuzzy controller (GBNI+GBNC) with self-adjusting learning rate factor are used to control, the simulation results are shown in Fig. 4.

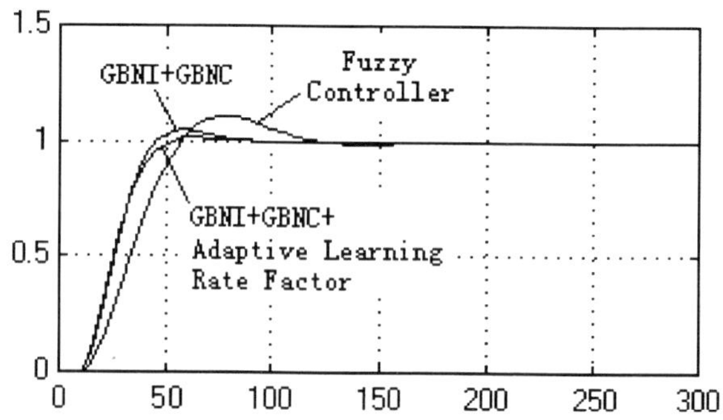


Fig.4. Simulation results

Simulation results indicate that: compared to general fuzzy logic controller, the Gaussian basis neural-fuzzy controller (GBNI+GBNC) has better control performance because of its learning ability, moreover, the learning process will be accelerated if the self-adjusting learning rate factor is adopted. Thus the self-learning algorithm and self-adjusting learning rate factor is effective.

## 6. CONCLUSIONS

The Gaussian basis neural network can identify unknown nonlinear objects, and it can replace fuzzy logic controller both in structure and function. As a result, two GBNN (GBNI and GBNC) are used to construct a self-learning fuzzy logic control system in this paper. Furthermore, a learning rate self-adjusting algorithm is proposed for Gaussian-basis function, which is proved to be effective by simulation. In conclusion, the self-learning fuzzy logic control system based on Gaussian basis neural network is a satisfying neural-fuzzy system.

## ACKNOWLEDGEMENTS

This work was supported by developing foundation

of Education Commission of Shanghai Municipality.

## REFERENCES

Lee, H. M, et al. (1994). A Fuzzy Rule-based Neural Network Model for Revising Approximate Domain Knowledge. In: *IEEE Conf. On Systems Man Cybernet.*

Sun, Z.Q. (1997), *Theory and Application of Intelligent Control.* Tsinghua University, Beijing.

Wang, L X. et al. (1992). Fuzzy Basis Functions ,Universal approximation, Orthogonal Least Squares Learning, *IEEE Trans. on Neural Networks*, No.5,

Wang, S.T. (2000). Neural-fuzzy System and its Application. Beijing University of Aeronautics & Astronautics, Beijing.

Wang, Y. N, and T. S. Tong (1998). Self-Learning Control System Using Neural Networks. In: *Proc. of Int. Conf. Modeling. Simulation. & Control.* USA.

IFAC Publications
www.elsevier.com/locate/ifac

# FUZZY SLIDING MODE LOAD FREQUENCY CONTROL OF MULTIAREA INTERCONNECTED POWER SYSTEMS

Meng Xiangping* Meng Xiangyin** Liu Dayou** Tan Wanyu*

* Department of Electrical Engineering, Changchun Institute of Technology,
Kuanping Road 4, Changchun, Jilin, P. R. China, 130012
E-mail: mxpt@public.cc.jl.cn
** Department of Computer Science, Jilin University,
Qianwei Road 10, Changchun, Jilin, P. R. China, 130012

Abstract: A fuzzy sliding mode load frequency control method for multiarea interconnected power systems is proposed in this paper. The method using sliding mode surface with integral model makes the systems can reach on switching surface from any initial state. After reaching on switching surface, systems are controlled by PI controller based on new area control error. The coefficient of switching control is adjusted by fuzzy control model automatically. When considering generation rate constraint and the governor deadband nonlinearity, the proposed control method can make systems have good dynamic performance. The design of controller in each area only considers the state information of itself area, and doesn't consider the one of the others, thus make decentrilised control be easier. Simulation results show that the proposed method is simple, effective, and can ensure that the overall system is asymptotically stable. *Copyright © 2001 IFAC*

Keywords: Load frequency control, Sliding mode control, Fuzzy control, Robustness, Decentralised control, Power system.

## 1. INTRODUCTION

Load frequency control is very important in power system operation and control for supplying sufficient and reliable electric power with good quality. Many control strategies have been proposed to achieve better performance. The main purpose of load frequency control of multiarea interconnected power system is to obtain satisfactory area frequency and tie-line power. Because power system always exists uncertainties such as parameter uncertainties, structure uncertainties and load disturbances, the robustness in the load frequency control is very important. In order to effectively control the dynamic response and consider the generation rate constraint simultaneously, the variable structure control strategies (Chan 1981, Kumar 1987, and Tripathy 1982 et al) are investigated to design the load frequency controllers, mainly due to their characteristics insensitive to internal parameter variation and external disturbance within the sliding surface. The variable structure controller proposed by Chan (1981) is insensitive to the parameter variation, a change of 20% in the nominal value of $T_t$, but under the presence of GRC, the result is unsatisfactory and, especially, when the load disturbance reaches 1%, the system may be unstable.

This paper proposes a fuzzy sliding mode load frequency control method. The sliding surface with the integral model is designed to make the system trajectories sliding on the switching surface from any initial state. The system obtains a satisfactory dynamic performance at the presence of internal parameter variations and external disturbance. Even under the existence of the generation rate constraint and the governor deadband nonlinearity, the system is still asymptotically stable. The simulation results prove that this method is effective.

## 2. DYNAMIC MATHEMATICS MODEL

Electric power systems are complex nonlinearity dynamic systems. Since a power system is only exposed to small changes in load during its normal operation, the linearized model can be used and will be sufficient to represent the power system dynamic around the operating point. The block diagram of the linearized plant model is shown in Fig. 1.

The list of symbols in Fig. 1

$\Delta f_i$ the incremental frequency deviation (Hz);

$\Delta P_{gi}$ incremental change in generator output (p. u. MW);

$\Delta Xe_i$ incremental change in generator value position (p. u. MW);

$\Delta P_{di}$ load disturbance (p. u. MW);

$T_{gi}$ governor time constant (s);

$T_{ti}$ turbine time constant (s);

$T_{pi}$ power system model time constant (s);

$K_{pi}$ power system gain;

$K_{ri}$ reheat coefficient;

$R_i$ speed regulation gain;

$ACE$ area control error ;

$ACEN$ new area control error ;

$\beta_i$ frequency bias setting ;

The system dynamic model is given by

$$\dot{X} = AX + Bu + F\Delta P_d(t) \tag{1}$$

where $X = [X_1, X_2, \cdots, X_n]^T$ , $u = [u_1, u_2, \cdots, u_n]^T$ ,

$\Delta P_d = [\Delta P_{d1}, \Delta P_{d2}, \cdots, \Delta P_{dn}]^T$ ,

$X_i$, $u_i$, $\Delta P_{di}$ are state vector, control vector and load disturbance of subsystem i, respectively. and

$$X_i = [\Delta f_i, \Delta P_{gi}, \Delta P_{ri}, \Delta P_{ti}, \Delta P_{tiei}, \int ACE_i dt, \int ACEN_i dt, ACEN_i]$$

$A$, $A_{ii}$, $A_{ij}$, $B$, $B_i$, $F$, $F_i$ see Meng (2000).

Governor deadband nonlinearity can be expressed by describing function (Tripathy 1982) and the natural period of deadband nonlinearity is about 2 second. The describing function can be described by Fourier function approximately as follow,

$$F(x, \dot{x}) = 0.8x - \frac{0.2}{\pi}\dot{x}$$

Where the governor deadband is 0.05%.

## 3. FUZZY SLIDING MODE CONTROL

In the multiarea interconnected power system, each area needs a local load frequency controller in order to maintain the area frequency and tie-line power. The local load frequency controller only cares for the state variables of corresponding area without considering about any feedback information of others. The stable problem of n-area power systems becomes the problem of overall system stability when all the local load frequency controllers work together (Chang, 1998).

### *3.1 Switching Surface Design*

The system dynamic model of area i is given as follows:

$$\dot{X}_i = A_{ii}X_i + B_i u_i + F_i \Delta P_{di} + \sum_{\substack{j=1 \\ j \neq i}}^{n} A_{ij} X_j \tag{2}$$

If eqn.2 has uncertainties, the model can be rewritten as

$$\dot{X}_i = (A_{ii} + \Delta A_{ii})X_i + (B_i + \Delta B_i)u + (F_i + \Delta F_i)\Delta P_{di} + \sum_{\substack{j=1 \\ j \neq i}}^{n} A_{ij} X_j \tag{3}$$

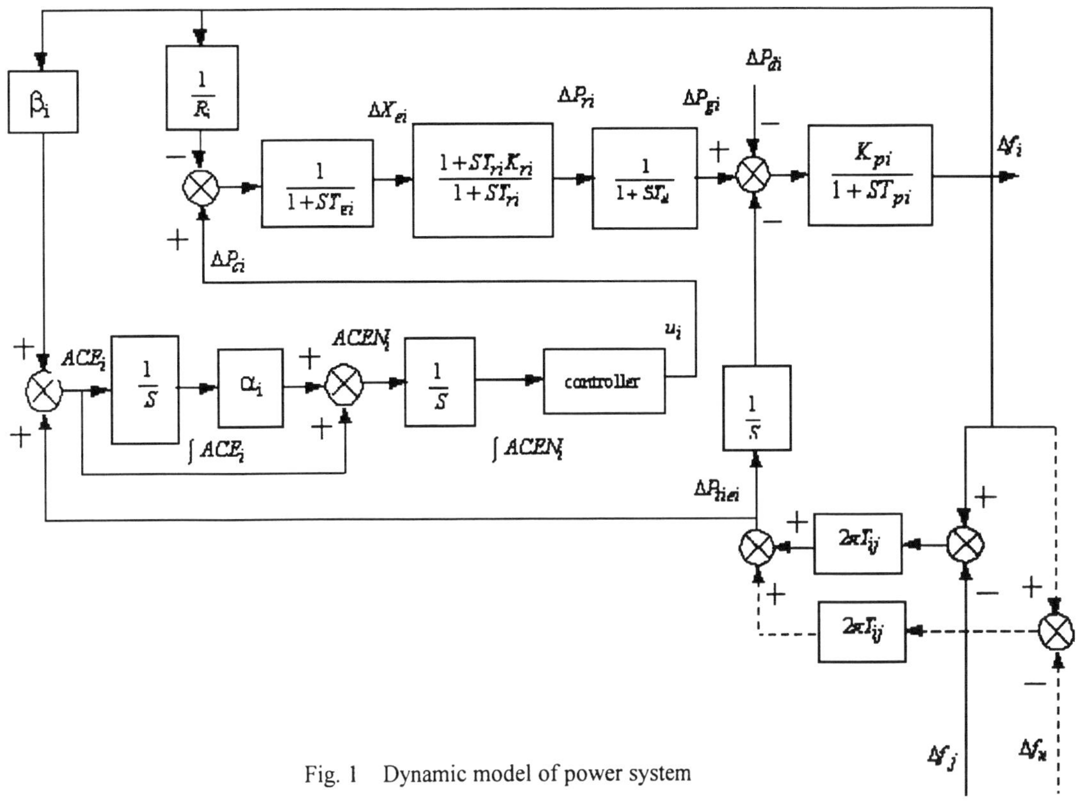


Fig. 1 Dynamic model of power system

Where $\Delta A_{ii}, \Delta B_i, \Delta F_i$ denote the uncertainties introduced by system parameter $K_{pi}, T_{pi}, T_{ti}, T_{gi}$, $R_i$ and load disturbance $\Delta P_{di}$. Reformulate eqn.2

$$\dot{X}_i = A_{ii}X_i + B_i(u_i + f_i(X_i,t)) \tag{4}$$

Where

$$f_i(X_i,t) = B_i^{+}\Delta A_{ii}X_i + B_i^{+}\Delta B_i u_i + B_i^{+}(F_i + \Delta F_i)\Delta P_{di}(t) + B_i^{+}\sum_{\substack{j=1\\ j\neq i}}^{n} A_{ij}X_j$$

and $B_i^{+} = \left(B_i^T B_i\right)^{-1} B_i^T$ is the pseudo inverse. According to eqn.4, an integral operation switching surface is designed directly from the nominal values of the system parameters $A$ and $B$.

The switching surface with integral operation of the subsystem i is designed as follows (Shyu, 1996):

$$S_i = C_i\left[X_i - \int_0^t (A_{ii} - B_iK_i)X_i(\tau)d\tau\right] = 0 \tag{5}$$

Where $C_i$ is a switching vector and $K_i$ is a state feedback gain matrix. From eqn.5, if the state trajectory of eqn.4 is trapped on the switching surface of eqn.5, i.e.

$$S_i = \dot{S}_i(t) = 0$$

the equivalent system dynamic of eqn.4 is governed by the following equation

$$\dot{X}_i = (A_{ii} - B_iK_i)X_i \tag{6}$$

The design of sliding mode control in eqn.5 implies selections of a manifold S=0 as well as the control enforcing sliding mode in S=0. The sliding mode equation does not depend on the disturbance. The desired dynamic properties may be provided by a proper choice of the vector $C_i$ and $K_i$. To simplify the system design, we can choose $C_i$=$B_i^T$.

Remark 1: Choosing $K_i = [0\,0\,0\,0\,0\,0\,0\; k_{iI}\; k_{iP}]$, then the PI control based on *ACEN* is obtained as $u_i = -k_{iP}ACEN_i - k_{iI}\int ACEN_i dt$ . Therefore, on the sliding surface, the system control switches to the PI control based on the new area control error, which not only attains desired performance of PI control, but also can adjust the incremental change in the tie-line power $\Delta P_{tie}$, incremental frequency derivation $\Delta f$, area control error *ACE*, the accumulation of time error $\varepsilon$ and inadvertent interchange *I*.

When the trajectories are on the sliding surface, the system has complete robustness to the internal parameter variation and external disturbance. The system dynamic performance can be improved. But we can find that the integral item of eqn.5 is not sure to be zero at initial position. In order to make the sliding mode condition exist throughout the whole control process, the following condition must be satisfied:

$$S_i \equiv 0 \quad \text{if } t \geq 0$$

Choose $W_i = \int_0^t (A_{ii} - B_i K_i) X_i(\tau) d\tau$,

Then $\dot{W}_i = (A_{ii} - B_i K_i) X_i$

and define $W_i(0) = X_i(0)$

Eqn.5 becomes

$$S_i = B_i^T (X_i - W_i) \tag{7}$$

From above, we can get $S_i(0) = 0$. It is obvious that system state maintains on manifold S = 0 from any initial instant.

*3.2 Design of Sliding Mode Controller*

Based on the switching surface (7), reaching condition and existing condition, the controllers are designed as following:

$$u_i = u_{eqi} + u_{si} \tag{8}$$

Where $u_{eqi}$ is the continuous control which ensure the trajectories slide to the origin on the sliding surface. $u_{si}$ is the switching control which guarantees the trajectories reaching the sliding surface and makes manifold $S_i = 0$.

Let $\dot{S}_i = 0$, then the equivalent control can be obtained

$$u_{eqi} = -K_i X_i = -k_{ip} ACEN_i - k_{iI} \int ACEN_i dt$$

When the system state reaches the sliding mode surface, the trajectories driven by the equivalent control $u_{eqi}$ are completely insensitive to internal parameter variation and external disturbance, which is equal to the PI control based on *ACEN*. The control of $u_{si}$ is to guarantee the system state sliding on the manifold $S_i = 0$ and the existence of the sliding mode.

Let $u_{si} = -h_i \operatorname{sgn}(S_i)$, where $\operatorname{sgn}(S_i)$ is the sign function, and $h_i$ is defined as $|f_i(X_i, t)| < h_i$, then

$$S_i \dot{S}_i = S_i \left[ B_i^T (\dot{X}_i - \dot{W}_i) \right]$$

$$\leq -B_i^T B_i h_i |S_i| + B_i^T B_i |S_i| |f_i(X_i, t)| < 0$$

Obviously the sliding mode arises at the initial instant, there after, the system is controlled by $u_{eqi}$, namely based on the new area control error $ACEN_i$. When there exist parameter uncertainties or external disturbance, the switching control $u_{si}$ will take charge to maintain $S = 0$, therefore, the system has desired robustness.

Remark 2: The designs of sliding surface and switching control above need only the state information of self-area, which make the system design very simple. The proof of stability of interconnected power system can be found in Meng (2000).

*3. 3 Adaptive Fuzzy switching control*

Let $h_i^*$ is a optimal value of $h_i$, and it satisfy the existence condition of the sliding mode, i.e. $|f_i(X, t)| < h_i^*$, but because of the uncertainty and complexity of unknown systems, the value of $h_i^*$ is not obtained exactly, we proposed a adaptive fuzzy control method to estimate the optimal value $h_i^*$, and can obtain the optimal control force $u_i^*$.

Suppose to exist a specialized $\hat{h}_i$, $\hat{h}_i$ satisfy sliding mode condition, and

$$\hat{h}_i - h_i^* = \varepsilon \tag{9}$$

where $\varepsilon$ is a small positive number, then $\hat{h}_i$ can be written as follow:

$$\hat{h}_i = \hat{v}^T W \tag{10}$$

where $\hat{v}$ is the optimal vector to obtain the minimal control force of sliding mode control.

Define error vector as

$$\tilde{v} = v - \hat{v} \tag{11}$$

Choose Liapunov function as

$$V_i = \frac{1}{2}(S_i^2 + \frac{1}{\gamma}\tilde{v}^T \tilde{v}) \tag{12}$$

where $\gamma$ is a positive constant, then

$$\dot{V}_i = S_i \dot{S}_i + \frac{1}{\gamma}\tilde{v}^T \dot{v} \tag{13}$$

Combine equ.2 and equ.4, obtain

$$\dot{V}_i = S_i(C_i[\dot{X}_i - (A_{ii} - B_iK_i)X_i]) + \frac{1}{\gamma}\tilde{v}^T\dot{v}$$

$$\le -C_iB_i|S_i|[h_i - |f_i|] + \frac{1}{\gamma}\tilde{v}^T\dot{v} \quad (14)$$

$$= -C_iB_i|S_i|[h_i^* + \varepsilon - |f_i|] + \frac{1}{\gamma}\tilde{v}^T[\dot{v} - \gamma C_iB_i|S_i|W]$$

Because $h_i^*$ is the optimal value with respective to sliding mode existence condition, then adaptive law is defined as:

$$\dot{v} = \gamma C_iB_i|S_i|W \quad (15)$$

then $\dot{V}_i = -C_iB_i|S_i|[h_i^* + \varepsilon - |f_i|] < 0$ (16)

The membership function center of fuzzy inference machine will acquire the minimal control force of fuzzy controller automatically from equ.15. Moreover, when parameters change or external disturbances appear, and $S_i(t) \neq 0$, then lump uncertainty $f_i(X,t)$ will change, so fuzzy inference machine and adaptive law will find a new value of $\hat{h}_i$ automatically.

## 4. SIMULATION EXPERIMENT AND RESULT ANALYSES

Consider a four-area hydrothermal interconnected power system (Kumar, 1987) with the parameters as follows:

$f = 60Hz$, $T_{p1} = T_{p2} = 20s$, $T_{g1} = T_{g2} = T_{g3} = 0.08s$,

$T_{r1} = T_{r3} = 10s$, $\alpha_i = 0.035$, $K_{ri} = 0.5$, $T_{t1} = 0.3s$, $T_{t2} = T_{t3} = 0.03s$, $R_1 = R_2 = R_3 = R_4 = 2.4$,

$K_{p1} = K_{p2} = 120Hz/p.u.$, $\beta_1 = \beta_2 = 0.425$,

$2\pi T_{12} = 2\pi T_{13} = 2\pi T_{14} = 2\pi T_{23} = 0.545$, $T_1 = 48.7$,

$T_2 = 5$, $T_3 = 0.513$, $D_3 = D_4 = 8.33\times10^{-3}$,

$H_3 = H_4 = 5s$, $K_{p3} = 1/D_3$, $K_{p4} = 1/D_4$,

$T_{p3} = 2H_3/fD_3$, $T_{p4} = 2H_4/fD_4$,

$\beta_3 = D_3 + 1/R_3$, $\beta_3 = D_3 + 1/R_3$, $\beta_4 = D_4 + 1/R_4$.

The GRC of 0.1 p. u. per minute is considered, . The governor deadband nonlinearity is also considered in the simulation. We can obtain the different results under the different system parameters using the control method given above.

From the above simulation results, we can see as follows:

(1) The generation rate constrain (*GRC*) and the deadband nonlinearity would significantly influence the dynamic response of power systems (See Fig. 2 and Fig. 3);

(2) When all parameters of the system change in the range of +20%, the system can obtain desired robustness and the fluctuation of the system response is small at the presence of GRC and the deadband nonlinearity (See Fig. 4 );

(3) The sliding mode exists from the initial position and the system is asymptotically stable. And adaptive fuzzy switching control can adjust $h_i$ value automatically (See Fig. 5);

(4) Fuzzy sliding mode load frequency control based on *ACEN* can adjust curves of the accumulation of time error $\varepsilon$ and inadvertent interchange $I$ to zero compared with the method based on *ACE* (See Fig. 6 and Fig. 7).

Moreover, the system performance is greatly influenced by the parameters of the integral gain $k_{iI}$, proportion gain $k_{iP}$ and $\alpha$.

## 5 CONCLUSION

This paper develops a fuzzy sliding mode load frequency control approach, which has the characters of simple implementation, good robustness and desired dynamic performance. When we choose $C_i = B_i^T$, the sliding surface can be designed easily. Furthermore, the system trajectories reach the switching surface from any initial state, which ensure the overall system is asymptotically stable. The design of controller in each area only considers the state information of itself area and doesn't consider the one of the others, thus make decentrilised control be easier.

## REFERENCES

A. Kumar, O.P. Malik and G.S. Hope(1987). Discrete variable structure controller for load frequency control of

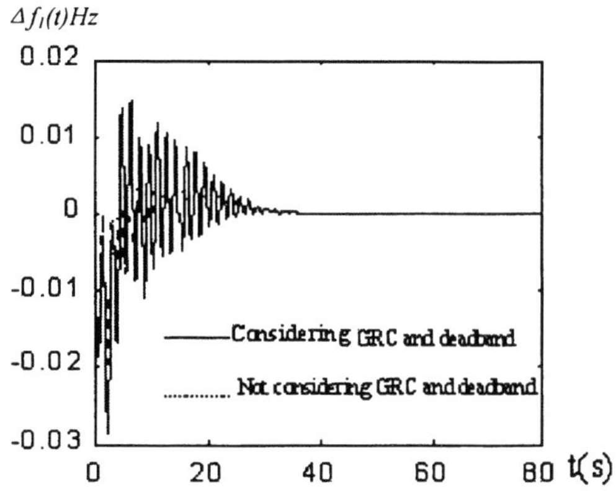


Fig .2 Frequency deviation in area 1

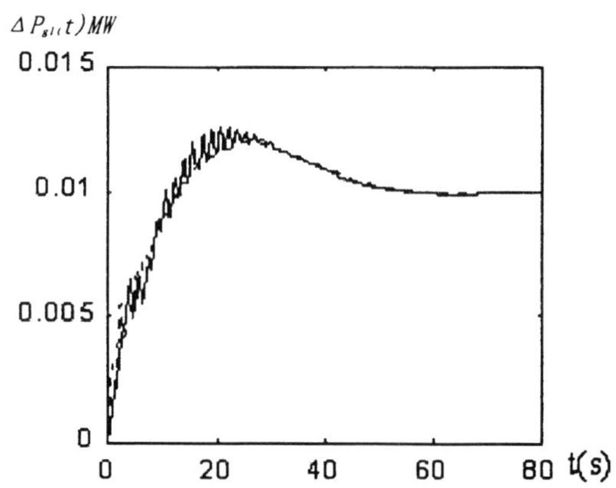


Fig. 3 Incremental change of generator power in area 1

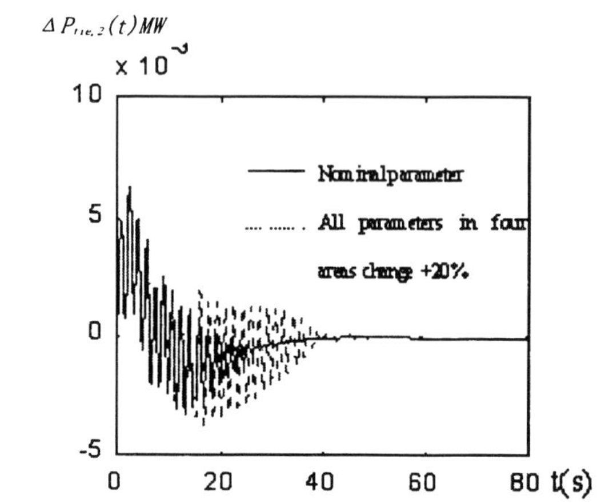


Fig. 4 Dynamic responses of tie-line power in area 2

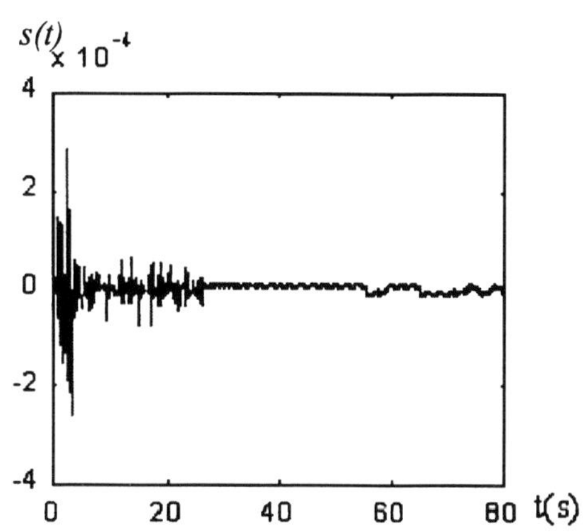


Fig. 5 Dynamic curve of and sliding mode s

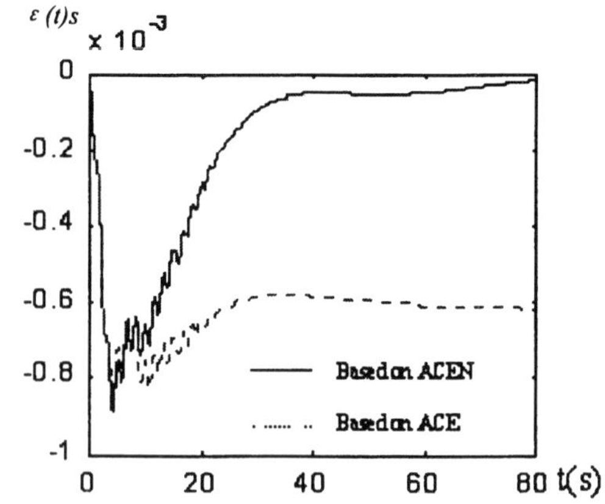


Fig.6 Dynamic curve of accumulation of time error $\varepsilon$

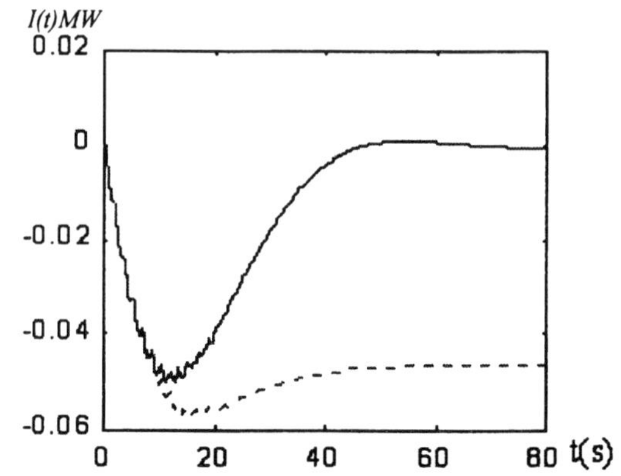


Fig. 7 Dynamic curve of inadvertent interchange $I$


multiarea interconnected power systems, *IEE Proc. C*, Vol.134, pp.116-122

W. C. Chan, S. Men, and Y. Y. Hsu(1981). Automatic generation control of interconnected power systems using variable-structure controllers. *IEE Proc. C.* Vol.128, pp. 269-29

S.C. Tripathy, G.S. Hope and O.P. Malik(1982) Optimization of load-frequency control parameters for power systems with reheat steam turbines and governor deadband nonlinearity, *IEE Proc., Pt.C*, Vol. 129, No.1, January

C. S. Chang, W. H. Fu, and F. S. Wen(1998). Load frequency control using genetic algorithm based fuzzy gain scheduling of PI controllers, *Electric Machines and Power Systems*. Vol. 26, pp. 39-52

K.K. Shyu and H.J. Shieh(1996). A new switching surface sliding mode speed controller for induction motor drive systems, *IEEE Trans. PE*, Vol. 11 , pp. 660-667, 1996

X. P. Meng, C. F. Xue, H. G. Zhang(2000), PI Sliding Mode Load Frequency Control of Multi-Area Power Systems, *Proceedings of The Chinese Society for Electrical Engineering*, Vol.21, No. 3, pp. 6-11

IFAC Publications
www.elsevier.com/locate/ifac

# SELF-TUNING FUZZY CONTROL OF MIMO SYSTEM AND ITS APPLICATION

**Liu Changliang, Niu Yuguang, Liu Jizhen, Jin Xiuzhang**

*North China Electric Power University*
*Baoding, HeBei, China*

Abstract: Against the serious coupling among system variables, long time delay of outputs response and time changing of model of the low speed pulverizer of power plant, a kind of self-tuning fuzzy control algorithm is presented. In the algorithm, the coupling among system variables has been eliminated, the proportion of the deviation and its differential can be adjusted automatically according to the control process, and accurate object model is not needed when setting the controller parameters. The fuzzy control algorithm in this paper has been used into power plant and satisfactory control result acquired. *Copyright © 2001 IFAC*

Key words: Fuzzy Control, Self-tuning control, MIMO, Power station control, Simulation

## 1. INTRODUCTION

The low speed pulverizer (*LSP*) of power plant is a typical *MIMO* system with three inputs and three outputs. It is used mainly in 125MW, 200MW and some 300MW power units in china. Due to the serious coupling among system variables, the long time delay of outputs response and model changing with time, perfect control result can't be achieve with the conventional *PID* control system, most *LSP* system of power plants in china can only run at manual mode at present. With the development of fuzzy control theory, fuzzy control has been largely used for process of power plant and other industrial process, and satisfactory results obtained (Adel Ben-Abdennor, 1996; Bolis, V., C., 1995; G. V. S. Raju, 1990).

The inputs of the LSP system are position of cold air valve, hot air valve, recycle air valve and speed of coal feeder. The outputs are inlet pressure, outlet temperature and coal mass of the pulverizer. Generally, the control variables are recycle air valve, hot air valve and coal feeder speed. The cold air is used to protect pulverizer when outlet temperature is too high and it closed normally. The coupling among process variables is very serious. For example, too much coal in the pulverizer will result in blockage. In this case, the speed of feeder should be adjusted to reduce the input coal flow. On the one hand, this will bring the increase of outlet temperature; hot air valve should be adjusted to keep the temperature. On the other hand, the regulation of hot valve will affect inlet pressure, recycle air valve should be adjusted to keep the pressure.

This paper presents an approach to realize a fuzzy based coordinated control scheme for *LSP* of power plants. A *LSP* system of *200MW* fossil power plant is used in this paper. The schematic is shown in Fig.1.

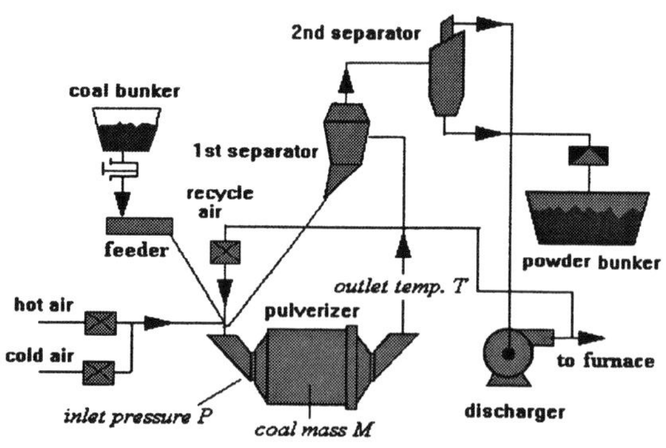


Fig.1 Schematic diagram of low speed pulverizer system of 200MW power unit

## 2. THE SELF-TUNING FUZZY CONTROLLER FOR LSP SYSTEM

Generally, the fuzzy controller is composed of many control rules that stipulate a certain action be initiated in a specific condition. The number of rules is a function of membership and system variables. If the membership of fuzzy set is *m*, the system variable number per rule is *n*, then the total number of rules is $m^n$. In the *LSP* system, there are six system variables, the errors and their rates of P, T, M. If the error and its rate is fuzzified as 9 levels, the total number of rules is $9^6$=531,441. That means the number of rules is too large to be used into practice. G. V. S. Raju and Jun Zhou introduced a kind of hierarchical fuzzy control algorithm (G. V. S. Raju and Jun Zhou, 1991; 1993). The number of control rules can be decreased largely in the controller.

In this paper, the control rules are substituted with fuzzy control formula. The configuration of self-tuning fuzzy controller of LSP system is shown in Fig.2.

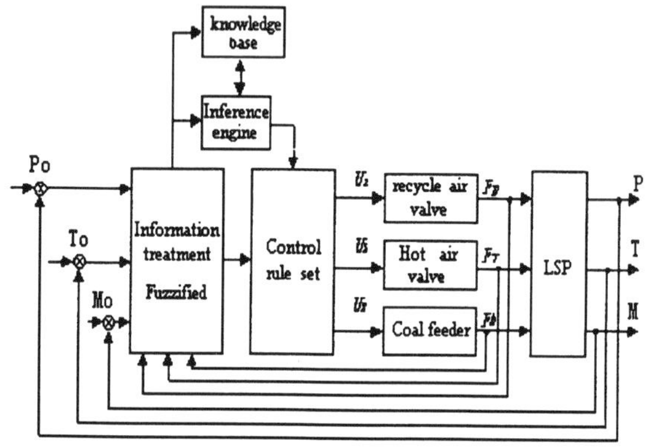


Fig.2 The configuration of self-tuning fuzzy controller

Where $P_0$, $T_0$, $M_0$ are the set points of inlet pressure, outlet temperature, coal mass; *P*, *T*, *M* are their measure values; $U_1$, $U_2$, $U_3$ are the controller outputs; $F_P$, $F_T$, $F_M$ are the feedback values, they are used into signal trace when auto/manual switched. The controller is composed of four parts: (1) the information treatment and fuzzify, it is used to acquire the position feedback and process values, calculate the errors and error rates and fuzzify them. (2) the control rules set, it provides all the control rules in various conditions. (3) knowledge base, it provides all parameters of controller and data about the LSP system at normal conditions and abnormal conditions. (4) inference engine, it matches the operation status and the control rules. The error $e_P$ and error rate $ec_P$ of inlet pressure, the error $e_T$ and error rate $ec_T$ of outlet temperature, the error $e_m$ and error rate $ec_m$ of coal mass in the pulverizer defined as following:

$$e_P(k)= P_0(k) - P(k) \quad (1)$$
$$ec_P(k)=e_P(k) - e_P(k-1) \quad (2)$$
$$e_T(k)= T_0(k) - T(k) \quad (3)$$
$$ec_T(k)= e_T(k) - e_T(k-1) \quad (4)$$
$$e_M(k)= M_0(k) - M(k) \quad (5)$$
$$ec_M(k)= e_M(k) - e_M(k-1) \quad (6)$$

Based on the actual dynamic characteristic of LSP system, the fuzzified fields of system variables can be choose as follow: $e_P \in$ [-200Pa, 200Pa], $ec_P \in$ [-20Pa/Sec, 20Pa/Sec], $e_T \in$ [-15 °C, 15°C], $ec_T \in$ [-4°C/min, 4°C/min], $e_M \in$ [-10%, 10%], $ec_M \in$ [-3%/min, 3%/min]. The system variables are fuzzified as {-4, -3, -2, -1, 0, 1, 2, 3, 4}, where -4, -3, -2, -1, 0, 1, 2, 3, 4 mean respectively large negative, medium negative, small negative, zero negative, zero, zero positive, small positive, medium positive, large positive sets. The errors and its rates after fuzzified denoted as $E_P$, $EC_P$, $E_T$, $EC_T$, $E_M$, $EC_M$. According to the habit of operators, the partnership between process variables and actual objects is: inlet pressure-recycle air valve, outlet temperature-hot air valve, coal mass in the pulverizer-coal feeder speed. Dues to the temperature of recycle air and outlet temperature is near, the regulation of recycle air valve only affects the inlet pressure. The regulation of hot air affects the output temperature and inlet pressure, its affection to coal mass can be ignored. The regulation of feeder speed affects the coal mass and outlet temperature, its affection to inlet pressure can be ignored. When some process variable deviate its set point, all the control variables should be changed at same time to ensure the stabilization of the system. For example, if the coal in pulverizer is lacking, the LSP will run at uneconomical status. If the coal in pulverizer is too much, the coal blockage will occur. The optimal coal mass changed with the kind, humidity, size of coal. The speed of feeder is adjusted to keep the optimal coal, meanwhile, the hot air value is adjusted to keep the outlet temperature and the recycle air valve is adjusted to keep the inlet pressure.

The transfer functions of the LSP system in Fig.1 can be get by the identification test. The result at 100% power load is as follows:

$$\begin{bmatrix} P \\ T \\ M \end{bmatrix} = \begin{bmatrix} \dfrac{0.4}{10s+1} & \dfrac{2}{8s+1} & 0 \\ 0 & \dfrac{0.5}{(80s+1)^3} & \dfrac{-0.65e^{-40S}}{(25s+1)^2} \\ 0 & 0 & \dfrac{0.3e^{-30S}}{(20s+1)^2} \end{bmatrix} \begin{bmatrix} x_1 \\ x_2 \\ x_3 \end{bmatrix}$$

Fig.3 The transfer function of LSP

Where x1, x2 are the positions of recycle air valve and hot air valve, x3 is the feeder speed.

Based on the analysis above, the fuzzy control formula of recycle air valve, hot air valve, coal feeder can be calculated as:

$$dU_1 = K_{11}(\alpha E_P + (1-\alpha) EC_P) - K_{12}(\beta E_T + (1-\beta) EC_T) \quad (7)$$

$$dU_2 = K_{22}(\beta E_T + (1-\beta) EC_T) + K_{23}(\gamma E_M + (1-\gamma) EC_M) \quad (8)$$

$$dU_3 = K_{33}(\gamma E_M + (1-\gamma) EC_M) \quad (9)$$

Where, $\alpha$ 、$\beta$ 、$\gamma$ are coefficients between 0 and 1 that reflect the portion of error and its rate. For fast process, the value should be large, for slow process, the value should be small. That means the fast process should be controlled mainly by error, slow process controlled mainly by error rate. $K_{11}$, $K_{12}$, $K_{22}$, $K_{23}$, $K_{33}$ are coefficients that reflect the coupling relation among process variables. A group of values of $\alpha$ 、$\beta$ 、$\gamma$ is shown in Table 1.

Table 1 Value of $\alpha$ 、$\beta$ 、$\gamma$

| E | $\alpha$ | $\beta$ | $\gamma$ |
|---|---|---|---|
| /E/=1 | 0.7 | 0.3 | 0.35 |
| /E/=2 | 0.8 | 0.4 | 0.45 |
| /E/=3 | 0.9 | 0.5 | 0.55 |

The utilization of adjustable coefficients $\alpha$ $\beta$ 、$\gamma$ in the control rules can improve the control performance effectively. When error is large, the controller will change output fast to shorten the control period. Otherwise, the controller will change output mainly according to the error rate to reduce excess regulation. The value of $\alpha$ 、$\beta$ 、$\gamma$ is related to system dynamic characteristics, they can be got by means of simulation.

The value of $K_{ij}$( $i,j$=1,2,3 )can be got according to the relationship among system variables. For example, when output temperature is high, the hot air valve should close, the inlet pressure will decrease, so the recycle valve should open. To keep inlet pressure, the effect of recycle valve should balance the effect of hot air valve. Because the gain from hot air valve to inlet pressure is 2, the gain from recycle valve to inlet pressure is 0.4, $K_{12}/K_{22}$=2/0.4=5. For the same reason, $K_{23}/K_{33}$=0.65/0.5=1.3. They only reflect the approximate static relation. Accurate values are not needed.

A group of values of α 、β 、γ is shown in Table 2.

Table 2. The Value of $K_{ij}$

| $K_{11}$ | $K_{12}$ | $K_{22}$ | $K_{23}$ | $K_{33}$ |
|---|---|---|---|---|
| 6 | 5 | 1 | 1.3 | 1 |

When outlet temperature is too high, the speed of feeder should increase to prevent coal fire. So, the formula (9) should be modified as:

$$U_3 = K_{32}[\beta E_T + (1-\beta)EC_T] + K_{33}[\gamma E_M + (1-\gamma)EC_M] \quad (9A)$$

If $|E_T|$=4 then $K_{32}=1.0$ else $K_{32}$=0。

In actual control, the increment control formula is used. The controller's output can be calculated as:

$$U_i(k+1) = U_i(k) + K_i\, dU_i \quad (i=1,2,3) \qquad (10)$$

Where, $K_i$ is the proportion factor. It is used to de-fuzzify the controller's fuzzy logic outputs. In abnormal condition, fast regulation needed. For example, if the inlet pressure is bigger than atmospheric pressure, the coal powder will leak into air, when recycle valve has closed to minimal position, the hot air valve and feeder speed should be decreased. Too high outlet temperature will result in coal powder fire, the hot air valve should be closed fast, and the feeder speed should be increased. Too much coal in the pulverizer will result in blockage, the feeder speed should be decreased fast. In these failure cases, the fuzzy controller's output should change at maximal speed. For example, the control rules when inlet pressure is abnormal are written as follows:

if $P>P_m$ and $M>M_m$ and $U_1<=U_{1i}$

then $dU_2=-4$, $dU_3=-4$ (11)

if $P>P_m$ and $M<M_m$ and $U_1<=U_{1i}$

then $dU_2=-4$, $dU_3=-2$ (12)

Where $P_m$ is the high limit of inlet pressure, $M_m$ is the high limit of coal mass, $U_{1i}$ is the low limit of recycle valve.

## 3. SIMULATION RESARCH AND APPLICATION IN POWER PLANT

To evaluate the performance of the self-tuning fuzzy control scheme, set point disturbance is simulated. The response of inlet pressure $P$ to *100Pa* set point disturbance, outlet temperature $T$ to *10 'C* set point disturbance and coal mass $M$ to *10%* set point disturbance is shown in Fig.4 (a)-(c).

When three disturbance act on LSP system at same time, the response of *P, T, M* are shown in Fig.4 (d).

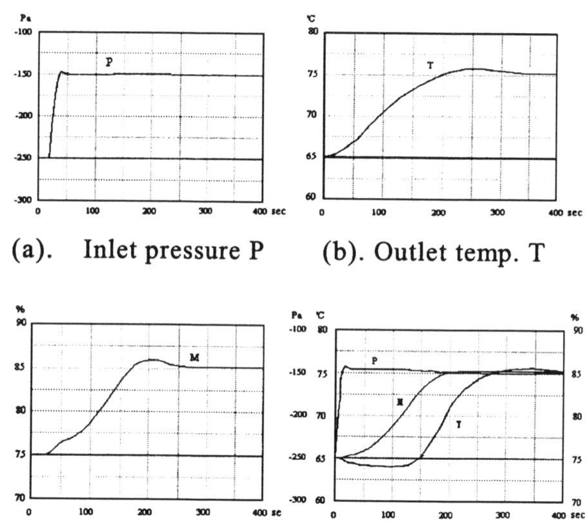


(a). Inlet pressure P (b). Outlet temp. T

(c). Coal mass M (d). P, T, M

Fig.4. The responses of P, T, M to set point disturbance

Because the model of LSP system is time changing, we need to check the robust performance of the controller. After the model changed to transfer function in Fig.5, the response of *P, T, M* to set point disturbance is shown in Fig.6.

$$\begin{bmatrix} P \\ T \\ M \end{bmatrix} = \begin{bmatrix} \frac{0.6}{15s+1} & \frac{1.5}{12s+1} & 0 \\ 0 & \frac{0.6}{(90s+1)^3} & \frac{-0.5e^{-60s}}{(25s+1)^2} \\ 0 & 0 & \frac{0.4e^{-50s}}{(15s+1)^2} \end{bmatrix} \begin{bmatrix} x_1 \\ x_2 \\ x_3 \end{bmatrix}$$

Fig.5. Another transfer function of LSP

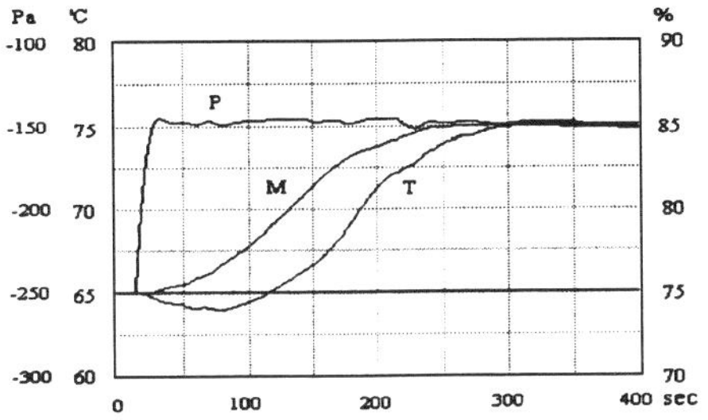


Fig.6. The responses of P, T, M to set point disturbance after parameter changed

It is shown in fig.4-6 that good control performance can be get with the self-tuning fuzzy controller and the control performance has little change when the parameters of LSP system changed. At present, the self-tuning fuzzy controller has been used into *125MW* power unit of *Taizhou* Power plant, *Zhejiang* province, *200MW* power unit of *Jiujiang* power plant, *Jiangxi* province, and many other power plants in china.

Generally, the coal mass in pulverizer is reflected by pressure difference *DP* between inlet and outlet. When coal mass changed, *DP* will change after 3~5 minutes delay, so it can't be used into real time control of *LSP* system. It can only reflect the coal mass in pulverizer at static state. So a special instrument is made to measure the coal mass. It is installed on the seat of LSP inlet, and used to measure the amplitude and frequency of vibration, then the signal is transferred to microprocessor. The vibration component corresponding to coal mass of pulverizer is separated and transferred into 4-20mA signal. It can reflect the change of coal mass in pulverizer quickly. So, the vibration signal is used to reflect the dynamic coal mass in the *LSP*, *DP* signal is used to correct the static value. The real time running curve is shown in Fig.7.

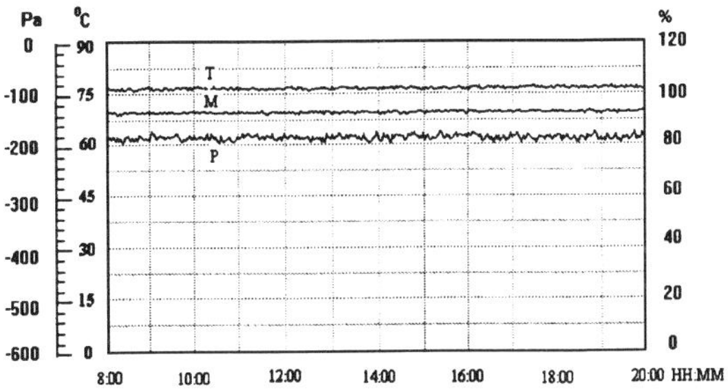


In actual control, the static errors exist because the *0* in error fuzzy set is corresponding to a region of process value. To the *LSP*, the small static error of *P, T, M* is permitted, so it doesn't matter. If one wants to eliminate the static error, the fuzzy control algorithm can be modified. One method is to increase the fuzzified grades of error, the static error will be decreased. Another method is to adopt integral algorithm (for example, *PID*) when fuzzified error is equal to zero, the error will be eliminated.

The self-tuning fuzzy controller for *LSP* can be applied to other *MIMO* system. The controller's output can be written as:

$$dU_i = \sum_{j=1}^{n} K_{ij}(a_{ij}E_j + (1-a_{ij})EC_j)$$

$$U_i(K+1) = U_i(K) + K_i dU_i$$

Where $i=1,2...n$. $K_{ij}$ is the coefficient that reflect the coupling relation between $y_j$ and $y_i$. $E_j$ is the fuzzified error of $y_j$, $EC_j$ is the fuzzified error rate of $y_j$. $\alpha_{ij}$ is the adjustable coefficient and can be calculated as:

$$\alpha_{ij} = \alpha_{0ij} + E_j / (E_j + EC_j)(1 - \alpha_{0ij})$$

$$0 \le \alpha_{0ij} \le 1$$

## 4. CONCLUSION

In this paper, a self-tuning fuzzy control algorithm for *LSP* of power plant has been studied and applied to many power plants. The simulation results and application in actual power plant indicate that the self-tuning fuzzy controller exhibits good robustness. The controller can be applied to other *MIMO* system without much difficulty. However, the values of controller's parameter still need further investigation.

REFERENCE


Adel Ben-Abdennor (1996). An autonomous Control system for boiler-turbine units, *IEEE transaction on energy conversion,* Vol.11, No.2, pp401-406.

Bolis, V. C. and Maffezzoni (1995). Synthesis of the overall boiler-turbine control system by single loop auto-tuning technique, *Control Engineering Practice*, Vol., 3, No., 6, pp761-771.

G. V. S. Raju and Jun Zhou (1993). Adaptive hierarchical Fuzzy controller, *IEEE transaction on system, man, and cybernetics,* Vol. 23, No.4, p973-980

G. V. S. Raju and Jun Zhou (1991). Hierarchical Fuzzy controller, *Int. J. Contr.,* vol.54, No.5, pp1201-1216, 1.

G. V. S. Raju and J.Zhou (1990). Fuzzy logic control for steam generator feedwater control, *proc. Amer. Contr. Conf.,* San Diego, CA., pp1491-1493.

IFAC Publications
www.elsevier.com/locate/ifac

# SENSOR FUSION BASED INTELLIGENT MOTION CONTROL FOR NONHOLONOMIC WHEELED MOBILE ROBOTS

**Gen'ichi Yasuda* and Hiroyuki Takai****

**Department of Mechanical Engineering, Nagasaki Institute of Applied Science, 536 Aba-machi, Nagasaki 851-0193, Japan*

***Department of Information Machines and Interfaces, Hiroshima City University, 3-4-1 Ozuka-higashi, Asaminami-ku, Hiroshima 731-3194, Japan*

Abstract: This paper presents a sensor fusion based intelligent motion control scheme for nonholonomic wheeled mobile robots under unstructured environments. The proposed scheme is composed of path planning and tracking control modules, considering kinematic constraints in steering control of nonholonomic vehicles. First, the path planning module generates an obstacle-free short path as a sequence of control vectors of orientation using a genetic algorithm. Then, the tracking control module produces the references for steering control using a sensor fusion network. Simulations of path planning and preliminary experiments with internal and external sensors are presented to show the effectiveness of the proposed control scheme. *Copyright © 2001 IFAC*

Keywords: Nonholonomic mobile robots, path planning, genetic algorithms, tracking control, sensor fusion.

## 1. INTRODUCTION

A most fundamental problem in mobile robot design is the navigation algorithm that needs to plan the path and execute the movement in a safe and efficient way under an unknown environment. For autonomous navigation of real mobile robots, kinematic constraints of nonholonomic vehicles and limited computing capacities of onboard computer systems should be considered. Also the goal orientation should be considered in path planning to perform complex robotic tasks efficiently at the goal position. It is well known that accurate positioning and docking of nonholonomic vehicles such as a typical AGV with two driven wheels and a caster wheel is difficult; there exists no stabilizing $C^1$ feedback control for such a vehicle (Samson and Ait-Abderrahim, 1991). Because holonomic vehicle designs are more complex and therefore more expensive, it may be disadvantageous to use them unless the application really needs the full maneuverability. When the robot environment is dynamic in multiple mobile robot systems, such as a robotic soccer playground, other robots also may be obstacles. The path planning and trajectory generation algorithm should be simple and efficient enough to be used in real time under dynamic constraints of moving obstacles by onboard computer systems (Yasuda and Takai, 1998).

In this paper, a genetic searching algorithm for the path planning with obstacle avoidance is proposed, considering kinematic constraints in steering control of nonholonomic two-wheeled vehicles (Takai, *et al.*, 2001). In path tracking control, references for steering control are produced by learning using a sensor fusion network based on multisensor fusion of internal and external sensors, and they are recalled using the internal sensor data when external sensors cannot be used. Preliminary experiments were conducted, where gyro sensors and infrared photo reflectors were used as internal and external sensors, respectively.

## 2. SENSOR-BASED PATH PLANNING USING A GENETIC ALGORITHM

The basic assumptions concerning a robot and its environment in the path planning are as follows:

(1) The position and orientation of the goal is known in advance;
(2) The environment is planar, and the robot and obstacles are represented as points;
(3) The robot can detect its position and orientation;
(4) The robot can detect obstacles in a circular detectable area using external sensors;
(5) The vehicle is nonholonomic with limited directional movements;
(6) The robot moves at a constant linear velocity.

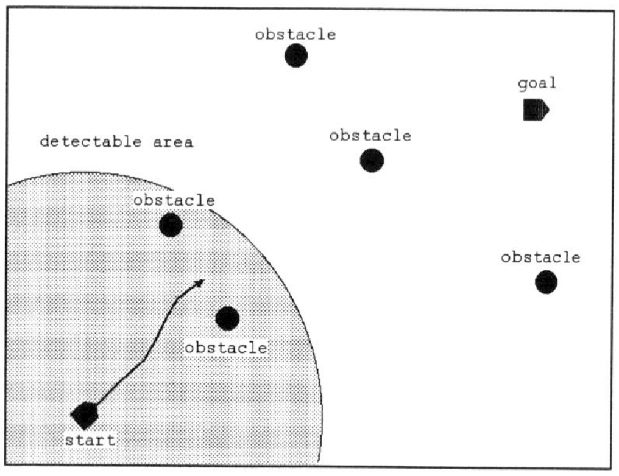


Fig. 1. Example of a robot and its environment with limited detectable area.

In the genetic algorithm, a path is represented as a set of vectors of orientation with equal length. The gene means the change of orientation of the path; the resultant path is composed of polygonal lines, where the number of via-points, or genes, is fixed. To reduce the length of code of the gene, the change of orientation is restricted to 5 values from 45 deg to -45 deg, and the orientation of the robot is restricted to 16 values as shown in Figs. 2 and 3.

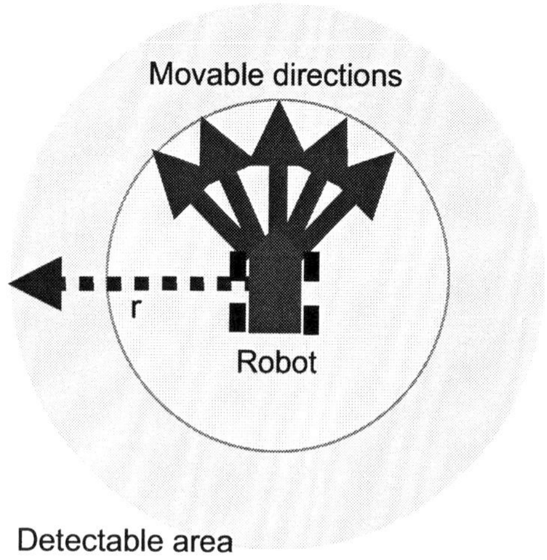


Fig. 2. Movable directions and detectable area of a robot.

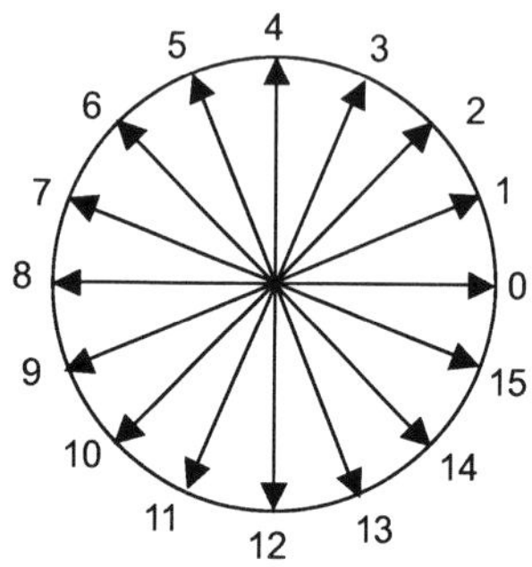


Fig. 3. Possible orientations of a robot.

The fitness value of a chromosome, or planned path is defined using the distance parameters from the path to the goal and obstacles as follows:

$$fitness = \alpha \frac{L}{\sum_{i=1}^{N} l_i + l_{end}} + \beta \frac{\sum_{i=1}^{N} d_i}{NL} + \gamma \cdot f \frac{x}{l_{end}} \tag{1}$$

where

$L$ distance between start position and goal position;

$l_i$ distance from $i$ th via-point to goal position;

$l_{end}$ distance between final point and goal position;

$d_i$ minimum distance between $i$ th via-point and obstacles;

$N$ number of genes;

$x$ projection from final position to goal orientation;

$f$ if orientation at final position coincides with

goal orientation, $+1$; otherwise $-1$;
$\alpha, \beta, \gamma$ weighting factors.

The fitness value is a weighted sum of the following three terms: 1) the ratio of distance of the planned path to that of the straight line, both to the goal position, 2) the sum of distance between via-points and the nearest obstacles, and 3) the deviation from the goal orientation. The 1st, or $\alpha$ term is used for short path consideration. The 2nd, or $\beta$ term is used for safe path consideration; if the term is large, the planned path is free from obstacles. The 3rd, or $\gamma$ term is the special term for the proposed path planning algorithm. It is used to align the robot with the goal orientation, substantially calculating the cosine of the angle between the robot's heading and the goal orientation. If the robot approaches the goal from a counter direction, the variable $f$ becomes minus and the path is rejected.

For the initial population, genes are generated according to uneven probability distribution like Table 1, because sudden large steering is seemingly undesirable for nonholonomic vehicles. Conventional basic GA operators such as one-point crossing over, point mutation, and combined roulette and elite selection regimes, were applied. The genetic algorithm parameters are listed in Table 2. The mean processing time for path planning was 0.48 sec using SunUltra60 (300 MHz UltraSparc2 CPU, with 256 Mbytes of memory).

Table 1 Probability of gene generation.

| Gene | Probability [%] |
|---|---|
| A | 10 |
| B | 25 |
| C | 30 |
| D | 25 |
| E | 10 |

Table 2 Genetic algorithm parameters.

| | |
|---|---|
| Population size | 20 |
| Generation number | 100 |
| Gene number | 8 |
| Crossover probability | 0.8 |
| Mutation probability | 0.05 |

For real-time consideration, the total generation number is fixed at a value required for the convergence of the fitness value. Path planning with obstacle detection is re-initiated before the robot reaches the final control points to perform smooth movements without stopping. After the cyclic iteration of planning and movement, it moves along the straight line to the goal when the robot is near the goal. Hence the cycle time of the sensor based path planning is defined as follows.

$$cycle\ time = T \cdot \frac{N_2}{N_1} \qquad (2)$$

where $T$ is the time required to pass all the control points of the planned path, $N_1$ is the number of passed control points, and $N_2$ is the number of all the control points of the planned path. To speed up the execution of path planning enough to be used in real time, the number of via-points of planned path and the cycle time are optimally determined with respect to the sensing range. In simulations, the sensing range is twice as large as the total distance of planned path, and the cycle time is $0.5T$. Examples of planned paths under a static environment with four obstacles are illustrated in Figs. 4 (a) ~ (d), which reveal the effects of the three terms of the fitness function.

From extensive simulations, the following relations are confirmed among the three terms of the fitness function for suitable path.

$$\alpha \text{ term} > \beta \text{ term} > \gamma \text{ term} \qquad (3)$$
$$\alpha \text{ term} > (\beta \text{ term} + \gamma \text{ term}) \qquad (4)$$

Furthermore the relative amounts of the three terms change according to different environmental conditions as summarized in Table 3. For satisfactory convergence of the algorithm, the weighting factor $\beta$ is set to a relatively large value if the distance between the start position and the goal position is long, because the first and third terms are relatively large. On the other hand, the weighting factors $\alpha$ and $\gamma$ are set to relatively large values if the distance is short. The weighting factors should be adjusted according to different environmental conditions.

Table 3 Relations between distance to goal and the relative amounts of the three terms of the fitness function.

| | In case of short distance | In case of long distance |
|---|---|---|
| $\alpha$ term | small | large |
| $\beta$ term | large | small |
| $\gamma$ term | small | large |

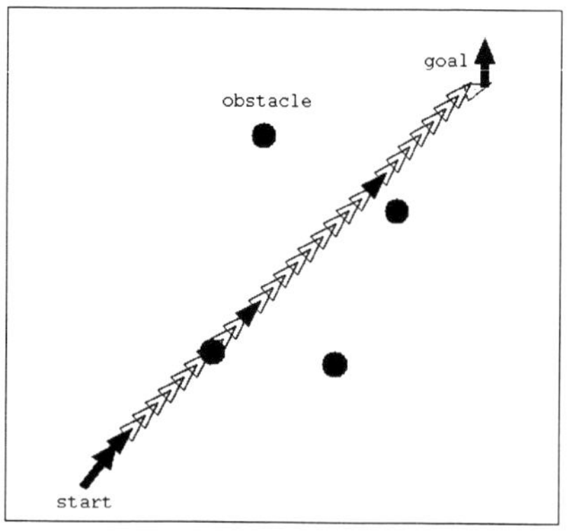


(a) $(\alpha,\beta,\gamma)$=(11,0,0)

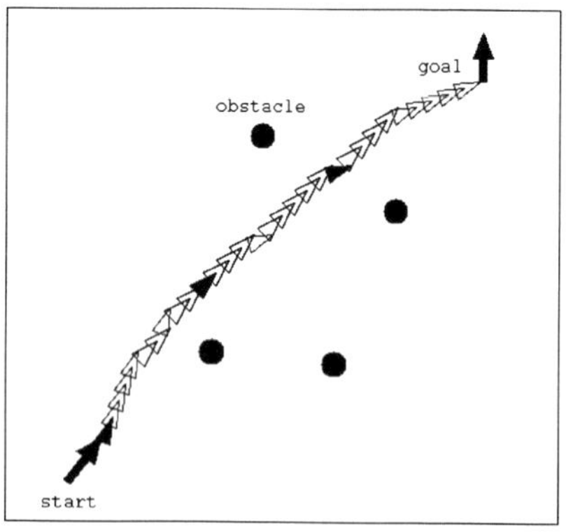


(b) $(\alpha,\beta,\gamma)$=(11,10.5,0)

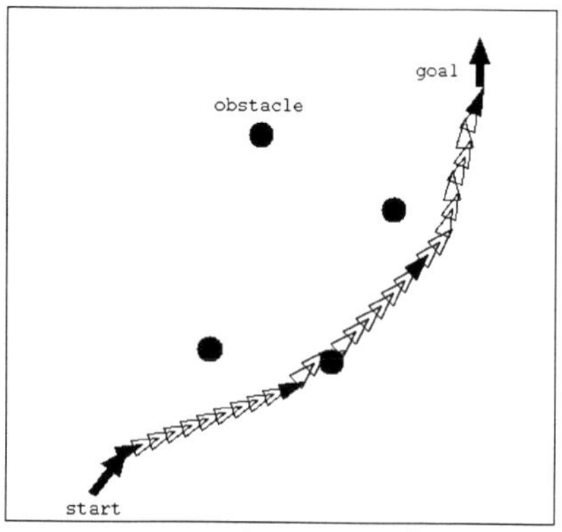


(c) $(\alpha,\beta,\gamma)$=(11,0,5)

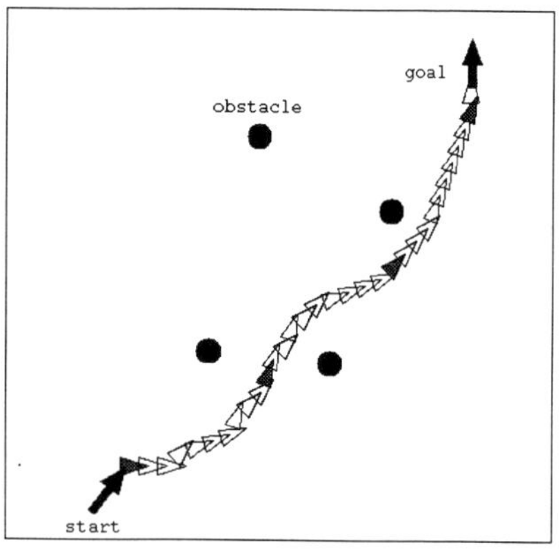


(d) $(\alpha,\beta,\gamma)$=(11,10.5,5)

Fig. 4. Effects of weighting factors on planned path.

## 3. TRACKING CONTROL SCHEME

From the path planning module, current references, or distance and orientation is given to the tracking control module. Then, it calculates the desired control variables for the motors to drive the robot along the path using a sensor fusion network at each control cycle. The overall block diagram of the proposed tracking control scheme is shown in Fig. 5.

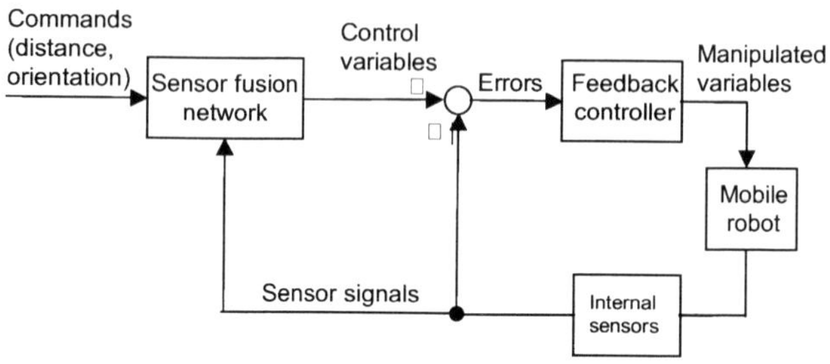


Fig. 5. Block diagram of tracking control scheme.

The sensor fusion network inputs are the commands received from the path planning module, as well as the sensor signals from the internal sensors. As internal sensors, a piezoelectric vibratory gyroscope (Tokin CG-16D) and two orthogonally mounted variable capacitance type linear accelerometers (Analog Devices ADXL202) were used, because the odometry using rotary encoders bring about large errors due to wheel slippage. The sensor fusion network outputs are the desired control variables, and their values are compared with the internal sensor signals. The errors of control variables are fed back into the conventional feedback controller to calculate manipulated variables, or PWM duty cycle of motor power applied to the DC motors of the two-wheeled mobile robot. The tracking controller does not need to calculate the position and orientation of the mobile robot, because integral of inertial sensor data induces rather large errors.

In the training phase of the sensor fusion network, multiple internal and external sensors are fused. The network receives the external reference signals from the external sensors. External sensors measure the displacement and linear and angular velocities of the mobile robot, and at the same time, internal sensory information is memorized as the reference for robot motion control such as steering control. During tracking control, the network with the embedded associative memories learns relations between the planned path, the robot movement, and the sensor outputs. Also the sensor fusion network learns relations between the distinction code, the robot movement, and the sensor outputs. The reference values are used in path tracking without external sensors when commands are given in terms of distance and orientation, or curvature of the path, from the upper level. The network can output the control variables for the robot movement corresponding to inputted distinction code.

An experimental procedure for training the sensor fusion network is designed as follows. The linear and angular velocities are measured using an infrared sensor system by moving on the floor with black-striped markers. The measured velocities are fused with the current outputs of the internal sensors, and the network is trained using the back propagation method to produce desired control variables, with relations to motor power and motor speed to drive the robot.

## 4. PRELIMINARY EXPERIMENTS

Our experimental mobile robot is an independently driven two-wheeled type. The wheel diameter is 65 mm, and the wheel separation distance is 160 mm. Fig. 6 shows the experimental set up, and Fig. 7 shows the experimental mobile robot. The preliminary experiments were conducted as follows. A line for line tracing was drawn on the floor. The black-striped markers with equal spaces of 10 mm were drawn on the right and left of the tracing line. The mobile robot detects the tracing line and black-striped markers during line tracing using dedicated OMRON EE-SY121 infrared photo reflectors. The signal timing intervals from black-striped marker reading, measured by the input capture module of the single-chip microcontroller, show the robot velocities.

Fig. 8 shows an experimental result of the characteristics of the gyro sensor. The sensor outputs a constant voltage (2.55 V) when the angular velocity is zero. The output voltage is confirmed to be reliably in proportion to the angular velocity. Table 4 shows the measurements during the straight-line and curved-line movements, where the linear velocity is measured exactly taking into account the influence of wheel slippage, using the photo reflectors. The calculated values of gyro sensor measurement, which are calculated from the measured linear velocity, the radius of curvature of the circle and the electric specifications of the sensor, are in good agreement with the measured ones, although the accuracy of the gyro sensor was $\pm 20$ %. Thus the experimental results explain relations between the robot movement and the sensor outputs.

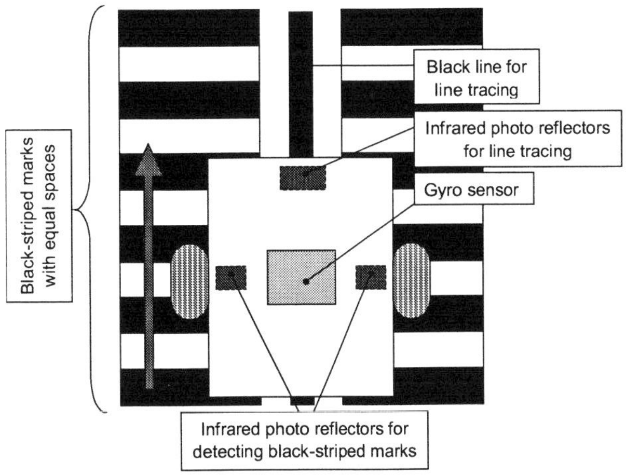


Fig. 6. Sketch of experimental set up.

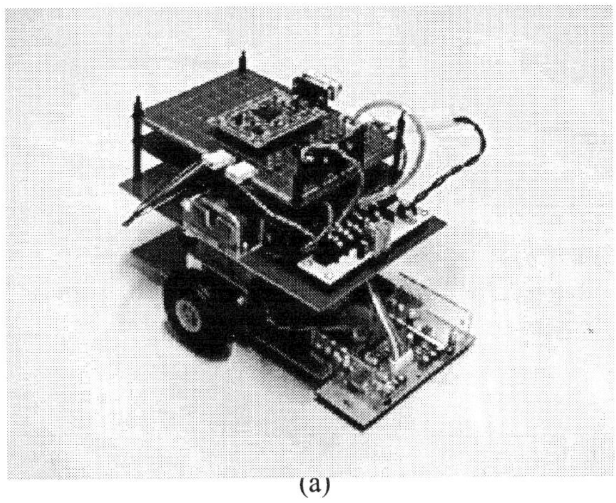

(a)

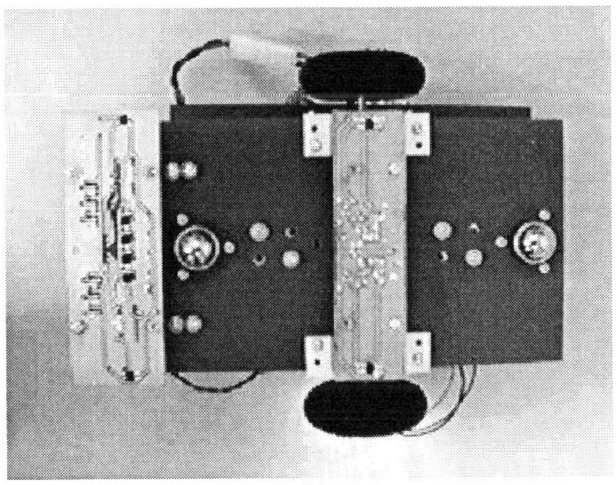

(b)

Fig. 7. Top (a) and bottom (b) views of experimental mobile robot.

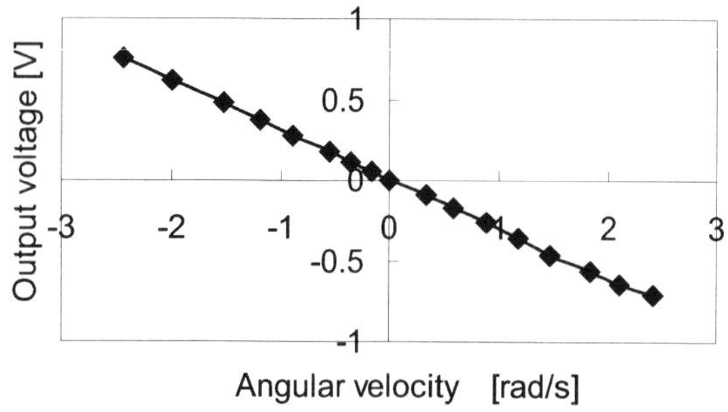


Fig. 8. Relationship between angular velocity and output voltage of the gyro sensor.

Table 4 Measurements while moving on black-striped markers.

| | Linear velocity [m/s] | Gyro sensor output (measured) [V] | Gyro sensor output (calculated) [V] |
|---|---|---|---|
| Straight line | 0.379 | 2.55 | 2.55 |
| Curved line (clockwise) | 0.376 | 2.39 | 2.36 |
| Curved line (counter-clockwise) | 0.383 | 2.73 | 2.74 |

(PWM period = 10 ms; PWM duty cycle = 50 %; circle radius = 0.5 m; space between markers = 10 mm)

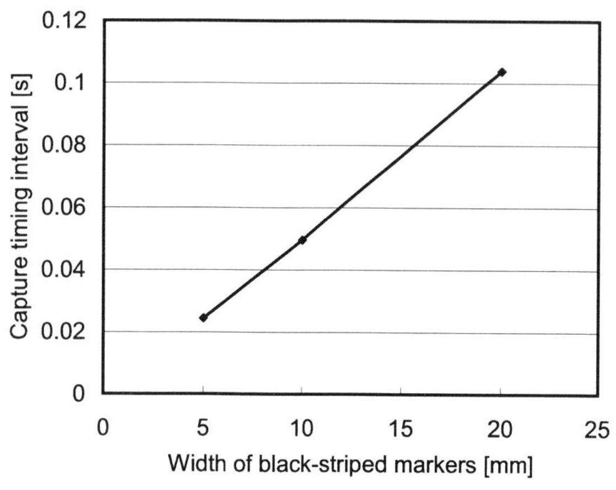


Fig. 9. Capture timing interval while moving on black-striped markers with different widths.

Fig. 9 shows that the capture timing interval measured by the input capture module is nearly in proportion to the width of black-striped marks while moving at constant speed. From the experimental results, it is confirmed that the infrared sensor system can measure widths of black-striped markers accurately. It means that the robot can read the mark-space code using external sensors, when the robot runs to keep constant speed. If the distinction code is drawn in the mark-space code, when the distinction code was read, the robot can learn relations between the robot movement, the sensor signals, and the distinction code. The experimental results show the effectiveness of the proposed measurement system using the infrared photo reflectors and the gyro sensor for intelligent motion control of mobile robots based on external abstract information.

## 5. CONCLUSIONS

The path planning algorithm with the limited directional movement and the real-time tracking control scheme were examined for autonomous navigation of intelligent mobile robots. The proposed path planning considers the goal orientation, and the genetic algorithm can get a suitable obstacle-free path in a short time without complex computation. The tracking control scheme makes the sensor fusion network learn relations between the planned path, the robot movement, and the sensor signals. Straight-line and circular movements on black-striped markers were used to show the feasibility of the learning method for the sensor fusion network, where the internal sensors are fused with the external sensors. The proposed control scheme is simple and efficient enough to be used in real time by onboard computer systems of autonomous mobile robots. The whole control scheme with associative neural networks is now being implemented on the microcontroller-based architecture.

## REFERENCES

Samson, C. and K. Ait-Abderrahim (1991). Feedback control of a nonholonomic wheeled cart in cartesian space. In: *Proceedings of the IEEE International Conference on Robotics and Automation*, 1136-1141.

Takai, H., G. Yasuda, and K. Tachibana (2001). Integrated path planning and steering control with multisensor fusion for intelligent mobile robots. In *Proceedings of the 6th International Symposium on Artificial Life and Robotics*, 301-304.

Yasuda, G. and H. Takai (1998). A microcontroller-based architecture for locally intelligent robot agents. In: *Proceedings of the 4th JSAI Sig-AI-Challenge*, 11-15.

IFAC Publications
www.elsevier.com/locate/ifac

# PATH PLANNING FOR REDUNDANT MANIPULATORS WITH LOAD DISTRIBUTION CONSIDERATION

**S. Tahmasian*, and M.J. Sadigh****

*Graduate student, Mech. Eng. Dept., Isfahan University of Technology, Isfahan, Iran
**Assistant professor, Mech. Eng. Dept., Isfahan University of Technology, Isfahan, Iran, E-mail address: mjsadigh@istt.org*

Abstract: This paper presents a novel method for path planning of redundant manipulators. The idea is to use load distribution as an objective function for resolving kinematic redundancy. The method is based on imposing a convenient relation among joint torques as a dynamic constraint equation to system. An algorithm for path planning of redundant manipulators subject to dynamic constraints, and an algorithm for finding a starting point with good load distribution is presented. These two algorithms are combined to develop the main algorithm, which can find a good starting point and generates a path in joint space such that the required load distribution objective is satisfied. The results for path planning of a 3-DOF planar manipulator using this method are presented. *Copyright © 2001 IFAC*

Keywords: Redundant manipulators, Path planning, Inverse kinematic problem, Saturation, Constraint satisfaction, Load distribution

## 1. INTRODUCTION

Redundant manipulators are increasingly employed in industry due to their good dexterity and manipulability characters. However, as a result of the kinematic redundancy, path planning for redundant manipulators is a complicated task. There are two main aspects in path planning of these systems: motion planning, which deals with existence of a feasible path, and redundancy resolution, which deals with selecting a single configuration among a set of many possible ones. Redundancy resolution can be achieved by satisfying additional requirements such as optimization of an index function related to joint speed, joint torques or consumed energy.

Numerous research in path planning of redundant manipulators, have involved instantaneous redundancy resolution at velocity level by using psudo inverse of Jacobian matrix or optimizing a position dependant scalar performance criterion (Whitney, 1969; Liegeois, 1977; Hollerbach and Suh, 1987). As opposed to this type of redundancy resolution which are purely kinematic, one can think of kinetic resolution in which a path is chosen which can satisfy some dynamic objectives such as a pre-specified relation among joint torques or optimization of a scalar index involving joint torques e.g. consumed energy and power (Saramago and Junior, 2000; Schmitt *et al.*, 1985), joint torques or a weighted norm of joint torque vector (Schmitt *et al.*, 1985; Hollerbach and Suh, 1987; Hu *et al.,* 1995).

Load distribution among joint actuators is a practically important objective. It proves helpful both in control and maintenance of a manipulator. Good load distribution can prevent saturation of a single actuator while the others are below their saturation limit. At the same time, if one actuator is frequently loaded relatively much higher than the others, the possibility of mechanical failure at that joint increases rapidly. Load distribution can be used as an objective for kinetic resolution of redundant manipulators.
This paper presents a method, which takes advantage of some linear functions among joint torques (which are referred to as dynamic constraint equations) for kinetic resolution of redundancy. These constraint equations are chosen such that satisfying them results in a good load distribution among joint torques.

The method of kinetic resolution is discussed in the next section. The third section presents the method of choosing the constraint equations. Then an algorithm for finding a good starting point with good relative load distribution is proposed. Numerical results for a typical redundant manipulator with some numerical considerations are given in last section.

## 2. KINETIC RESOLUTION

Let $\mathbf{q} \in R^n$ define the joint space coordinates and $\mathbf{X} \in R^m$ $(m < n)$ is the vector of task space coordinates of a n-DOF redundant manipulator. The path planning problem can be stated as: find a path $\mathbf{q}(t)$ in joint space which moving the manipulator along that moves the end effector from an initial configuration $(\mathbf{X_0}, \mathbf{X_0})$ to a final configuration $(\mathbf{X_f}, \mathbf{X_f})$ through the specified path $\mathbf{X}(t)$. To accomplish this, one has to solve the forward kinematic equation of the manipulator:

$$\mathbf{f}(\mathbf{q}) = \mathbf{X}(t) \tag{1}$$

for the unknown vector of $\mathbf{q}$ in several stations along the path. Instead of solving the set of non-linear equations (1), one may consider integrating the time derivative of that:

$$\mathbf{J}\dot{\mathbf{q}} = \dot{\mathbf{X}}(t) \tag{2}$$

in which $\mathbf{J}_{m\times n}$ denotes the Jacobin matrix of the system. For a redundant manipulator equations (2) represent m equations which has to be solved for n $(n > m)$ unknown $\dot{\mathbf{q}}$. One approach to resolve this kinematic redundancy is to add $r = n - m$ additional equations to equations (2).

Let us consider a set of dynamic constraint equations:

$$\mathbf{L}\boldsymbol{\tau} = \boldsymbol{\alpha} \tag{3}$$

which can enforce a specified relation among the joint torques along the path to form those r additional equations. In equation (3) $\mathbf{L}_{r\times n}$ and $\boldsymbol{\alpha}_{r\times 1}$ can be constant or functions of $\mathbf{q}$ , $\dot{\mathbf{q}}$ and $\mathbf{X}(t)$.

To solve equations (2) and (3), one should differentiate equations (2) to obtain:

$$\mathbf{J}\ddot{\mathbf{q}} = \ddot{\mathbf{X}}(t) - \dot{\mathbf{J}}\dot{\mathbf{q}} \tag{4}$$

and substitute $\boldsymbol{\tau}$ from the equation of motion of manipulator in equation (3).

The equation of motion of the manipulator can be written as:

$$\mathbf{M}\ddot{\mathbf{q}} + \mathbf{h} = \mathbf{B}\boldsymbol{\tau} \tag{5}$$

where $\mathbf{M}_{n\times n}(\mathbf{q}, t)$ is the generalized mass matrix and $\mathbf{B}_{n\times n}(\mathbf{q}, t)$ is a non-singular matrix and $\mathbf{h}_{n\times 1}(\mathbf{q}, \dot{\mathbf{q}}, t)$ denotes the nonlinear terms of equations of motion. Substituting $\boldsymbol{\tau}$ from equation (5) into equation (3), one gets:

$$\mathbf{L}\mathbf{B}^{-1}\mathbf{M}\ddot{\mathbf{q}} = \boldsymbol{\alpha} - \mathbf{L}\mathbf{B}^{-1}\mathbf{h} \tag{6}$$

Considering equations (4) and (6), we will have the following set of equations:

$$\overline{\mathbf{M}}\ddot{\mathbf{q}} = \overline{\mathbf{h}} \tag{7}$$

where $\overline{\mathbf{M}} = \begin{bmatrix} \mathbf{J} \\ \hdashline \mathbf{L}\mathbf{B}^{-1}\mathbf{M} \end{bmatrix}$ and $\overline{\mathbf{h}} = \begin{bmatrix} \ddot{\mathbf{X}}(t) - \dot{\mathbf{J}}\dot{\mathbf{q}} \\ \hdashline \boldsymbol{\alpha} - \mathbf{L}\mathbf{B}^{-1}\mathbf{h} \end{bmatrix}$.

This set of equations represents n second order non-linear differential equations, which can be solved to obtain the desired path in the joints space that satisfies both equations (1) and (3) simultaneously.

### *2.1. NUMERICAL CONSIDERATIONS*

One should note that solution of equation (7) needs an initial condition $(\mathbf{q_0}, \dot{\mathbf{q}}_\mathbf{0})$, which can be obtained using $(\mathbf{X_0}, \mathbf{X_0})$. In this situation, this set of initial conditions is not unique and one can choose among many possible joint space initial conditions such that some additional requirements such as optimization of manipulability or dexterity index are satisfied. However, once a set of initial conditions are chosen a unique path will be generated.

On the other hand, this set of equations may has no answer at some points where the matrix $\overline{\mathbf{M}}$ become singular. Besides the real singularities, at which the Jacobian matrix becomes rank deficient, there are other points for them the matrix $\overline{\mathbf{M}}$ has no inverse. These points which can be referred to as virtual (non- real) singularities can not be detected from manipulator configuration and are to be avoided by controlling the condition number of $\overline{\mathbf{M}}$ along the generated path.

## 3. CONSTRAINT EQUATIONS

Actuator loads of a manipulator with n DOF are exactly distributed if the actuators forces or torques satisfy the following equations:

$$\frac{|\tau_1|}{\tau_{max_1}} = \frac{|\tau_2|}{\tau_{max_2}} = \cdots = \frac{|\tau_n|}{\tau_{max_n}} \quad (8)$$

where $\tau_{max_i}$ $(i=1,\ldots,n)$ is the saturation limit of i-th actuator and is considered to be positive. Equation (8) can be rewritten in the following form:

$$\mathbf{L}\boldsymbol{\tau} = \boldsymbol{\alpha} \quad (9)$$

in which

$$L_{ij} = \begin{cases} \dfrac{\text{sgn}(\tau_i)}{\tau_{max_i}} \; ; \; j=i \\ -\dfrac{\text{sgn}(\tau_i)}{\tau_{max_i}} \; ; \; j=i+1 \\ 0 \; ; \; \text{otherwise} \end{cases} , \begin{cases} i=1,\ldots,n-1 \\ j=1,\ldots,n \end{cases}$$

This set of constraint equations represents $n-1$ equations, which if used in previously discussed algorithm can guarantee load distribution. However, normally r , degree of redundancy, is smaller than $n-1$. So, practically it is not possible to use equations (9) as the constraint equations for load distribution.

Instead of this, we may consider to keep the actuator loads within a bound. For instance, we may think of an acceptable bound for joint torques defined as:

$$\varepsilon(1-R_I) \le \tau_i' \le \varepsilon(1+R_I) \quad ,i=1,\ldots,n \quad (10)$$

where

$$\varepsilon = \sqrt{\sum_{i=1}^{n} \tau_i'^2 \Big/ n} \, ,$$

$$\boldsymbol{\tau}' = \begin{bmatrix} \dfrac{|\tau_1|}{\tau_{max_1}} & \dfrac{|\tau_2|}{\tau_{max_2}} & \cdots & \dfrac{|\tau_n|}{\tau_{max_n}} \end{bmatrix}^T$$

and $0 < R_I < 1$ is an arbitrary coefficient called "relaxation index". In equation (10), $\varepsilon$ is an indication of the necessary actuator effort at a specified point of the path, and $R_I$ specifies the width of the acceptable bound or the desired accuracy (Fig. 1).

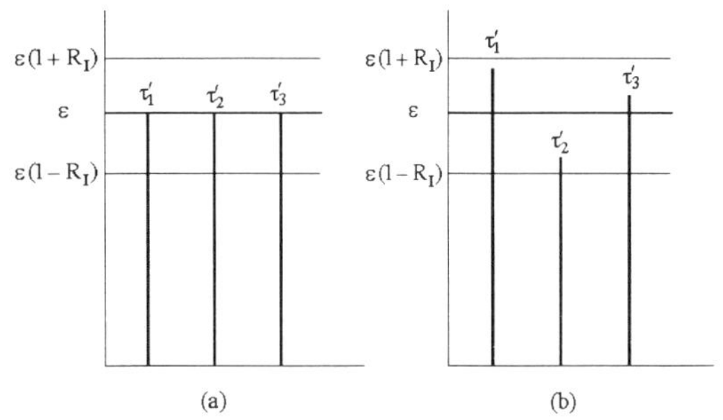


Fig. 1. a) Exact load distribution
b) Approximate load distribution

If the path starts from a starting point with fairly good load distribution ($\boldsymbol{\tau}^{*}$) which satisfies condition stated in equation (10), and the path is generated such that the set of constraint equations (3) with the following $\mathbf{L}$ and $\boldsymbol{\alpha}$ defined as:

$$L_{ij} = \begin{cases} \dfrac{\text{sgn}(\tau_j)}{\tau_{max_j}} \; ; \; i \le j \le m+i \\ 0 \; ; \; j<i \text{ or } j>m+i \end{cases} \quad (11)$$

and:

$$\boldsymbol{\alpha} = \mathbf{L}\boldsymbol{\tau}^{*} \quad (12)$$

Then there will be a neighborhood of the starting point in which the generated path satisfies the load distribution condition (equation 10). The length of this is controlled by the desired accuracy (choice of $R_I$ ).

In this approach, exact load distribution can be regarded as an approximate one with $R_I = 0$. This means that the smaller the relaxation index the better the load distribution. However, one should note that making $R_I$ smaller reduces the length of neighborhood for which the load distribution constraint (equation (10)) is satisfied.

Now the problem is how to find a starting point for which the system has acceptable load distribution and how to decrease the magnitudes of forces or torques which are out of the acceptable bounds. This problem is solved using an algorithm named "Torque Modification Algorithm" (TMA) which will be discussed in next section.

## 4. TORQUE MODIFICATION ALGORITHM

For the considered manipulator let us assume that $m'$ number of it's actuator forces or torques are exceeded the allowable limits according to inequality (10). Then we should change the magnitudes of these forces or torques such that all of them are laid inside the allowable limits. For this, we consider the $\tau_j = \max(\tau_i)$, $(i = 1, \ldots, n)$ and attempt to decrease it's magnitude, till it lays inside the acceptable bound. This can be done by multiplying $\tau_j$ by a factor $0 < k < 1$ to get:

$$\tau_j^* = k\,\tau_j \tag{13}$$

After decreasing the magnitude of $\tau_j$ as discussed above, one should determine the rest of $\tau_i$'s $(i = 1, \ldots, n, i \neq j)$ using equations (4) and (5). These equations can be written as:

$$\begin{bmatrix} \mathbf{M} & -\mathbf{B} \\ \mathbf{J} & \mathbf{O}_{m\times n} \end{bmatrix} \begin{bmatrix} \ddot{\mathbf{q}} \\ \boldsymbol{\tau} \end{bmatrix} = \begin{bmatrix} -\mathbf{h} \\ \ddot{\mathbf{X}}(t) - \dot{\mathbf{J}}\,\dot{\mathbf{q}} \end{bmatrix} \tag{14}$$

where $\mathbf{O}_{m\times n}$ denotes zero matrix. Now let us rewrite equation (14) using the new magnitude of $\tau_j$ as follow:

$$\begin{bmatrix} \mathbf{M} & -\widetilde{\mathbf{B}} \\ \mathbf{J} & \mathbf{O}_{m\times(n-1)} \end{bmatrix} \begin{bmatrix} \ddot{\mathbf{q}} \\ \widetilde{\boldsymbol{\tau}} \end{bmatrix} = \begin{bmatrix} -\widetilde{\mathbf{h}} \\ \ddot{\mathbf{X}}(t) - \dot{\mathbf{J}}\,\dot{\mathbf{q}} \end{bmatrix} \tag{15}$$

where $\widetilde{\mathbf{B}}$ is the same as $\mathbf{B}$ unless that it's j-th column is removed, and $\widetilde{\boldsymbol{\tau}}$ and $\widetilde{\mathbf{h}}$ are defined as:

$$\widetilde{\boldsymbol{\tau}} = \begin{bmatrix} \tau_1 & \cdots & \tau_{j-1} & \tau_{j+1} & \cdots & \tau_n \end{bmatrix}^T \tag{16}$$

and:

$$\widetilde{\mathbf{h}} = \mathbf{h} - \mathbf{B}^j \tau_j^* \tag{17}$$

where $\mathbf{B}^j$ denotes the j-th column of $\mathbf{B}$.

Using equation (15) one can determine the rest of joint torques as:

$$\begin{bmatrix} \ddot{\mathbf{q}} \\ \widetilde{\boldsymbol{\tau}} \end{bmatrix} = \begin{bmatrix} \mathbf{M} & -\widetilde{\mathbf{B}} \\ \mathbf{J} & \mathbf{O}_{m\times(n-1)} \end{bmatrix}^+ \begin{bmatrix} -\widetilde{\mathbf{h}} \\ \ddot{\mathbf{X}}(t) - \dot{\mathbf{J}}\,\dot{\mathbf{q}} \end{bmatrix} \tag{18}$$

in which $[\ ]^+$ denotes the psudo inverse of $[\ ]$.

After determining $\boldsymbol{\tau}$ using equations (13) and (18), the new acceptable bound for forces or torques, are calculated according to new $\boldsymbol{\tau}$, using equation (10). It is then investigated whether $\|\boldsymbol{\tau}\|_\infty$ is laid inside this limit or not. If it is laid out of this bound one should repeat the processes till $\|\boldsymbol{\tau}\|_\infty$ is laid inside the acceptable bound.

## 5. NUMERICAL RESULTS

In this section we first present the main algorithm for path planning, then numerical results are presented for a 3-DOF planar manipulator and finally numerical results are given for investigation of the effect of $R_I$.

### *5.1 MAIN PATH PLANNING ALGORITHM*

The complete algorithm for path planning can be stated in following steps:

1- Find a set of initial value $\mathbf{q_0}$, $\dot{\mathbf{q}}_0$ and $\ddot{\mathbf{q}}_0$ using equations (1), (2) and (4) corresponding to $\mathbf{X_0}$, $\dot{\mathbf{X}}_0$ and $\ddot{\mathbf{X}}_0$.
2- Calculate $\boldsymbol{\tau_0}$ using inverse dynamics (equation (5)).
3- Modify the vector $\boldsymbol{\tau_0}$ to find $\boldsymbol{\tau}^*$ using TMA discussed in section 4.
4- Calculate $\mathbf{L}$ and $\boldsymbol{\alpha}$ using equations (11) and (12).
5- Generate the path using equation (7), till either the whole path is generated or one of the joint forces or torques lies out of the acceptable boundary.
6- Repeat steps 3, 4 and 5 till the whole path is generated.

### *5.2 PATH PLANNING FOR A 3-DOF PLANAR MANIPULATOR*

The algorithm is employed to generate the path in joint space for a 3-DOF planar manipulator (Fig. 2) with following characteristics:

$m_1 = 15\ (kg)$, $m_2 = 12\ (kg)$, $m_3 = 9\ (kg)$

$l_1 = 0.5\ (m)$, $l_2 = 0.4\ (m)$, $l_3 = 0.3\ (m)$

$\boldsymbol{\tau}_{\mathbf{max}} = [200 \quad 50 \quad 15]^T\ (N.m)$

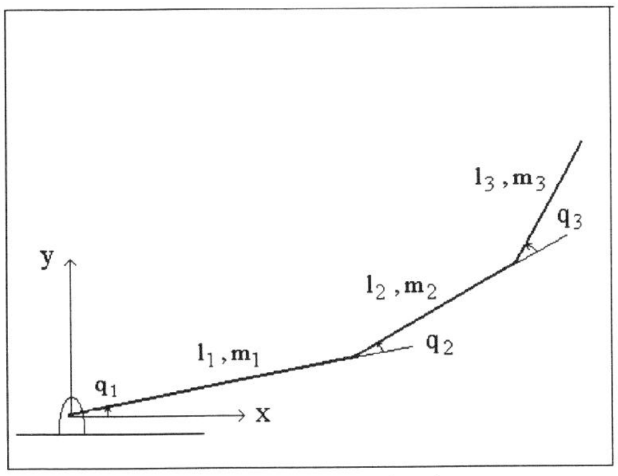


Fig. 2. A three DOF planar manipulator

The end-effector of the manipulator is supposed to move along the x-axis from point $\mathbf{X} = [1 \quad 0]'$ to point $\mathbf{X} = [0.3 \quad 0]'$. The whole motion takes place in horizontal plane. The manipulator is in rest both at $\mathbf{X_0}$ and $\mathbf{X_f}$ . $\mathbf{q_0}$ and $\dot{\mathbf{q}}_0$ are chosen as $[18 \quad -66.905 \quad 78.233]'$ (deg) and $\mathbf{0}$ respectively. The generated path, in joint space, $\mathbf{q}(t)$, and in task space, $\mathbf{X}(t)$, are shown in figures (3) and (4) respectively. The joint torques, necessary for this motion, are presented in figure (5). As can be seen from the results the generalized path has satisfied both tracking of task trajectory and load distribution. The amount of joint torques, their mean value $\varepsilon$ and the deviation factor $\beta$ , which is defined as:

$$\beta = \left( \sum_{i=1}^{n} \sum_{j=i+1}^{n} (\tau_i' - \tau_j')^2 \right)^{\frac{1}{2}} \tag{19}$$

are presented in table (1). The figures given in the first column indicates the time when torque modification is done (at least one of the torques are laid out of allowable limit). As the results show, the torque modification algorithm not only reduces $\beta$, but improves $\varepsilon$ as well.

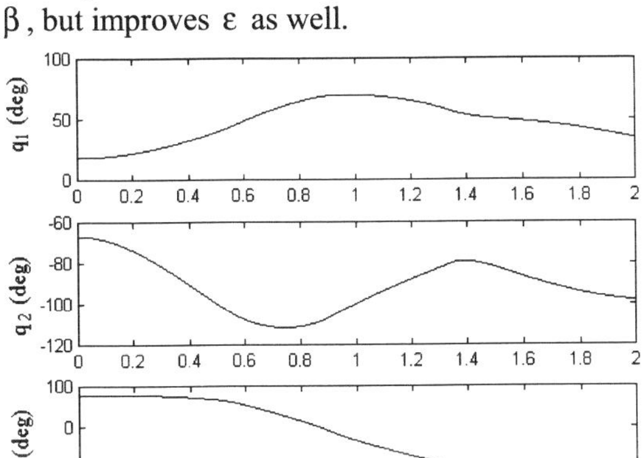


Fig. 3. Manipulator path in joint space

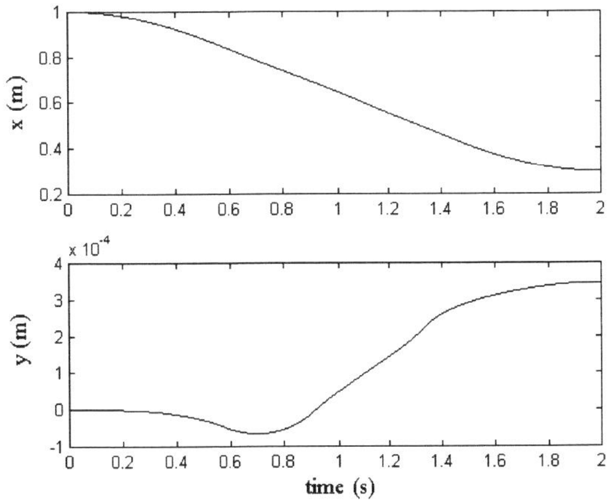


Fig. 4. Trajectory of the path in task space

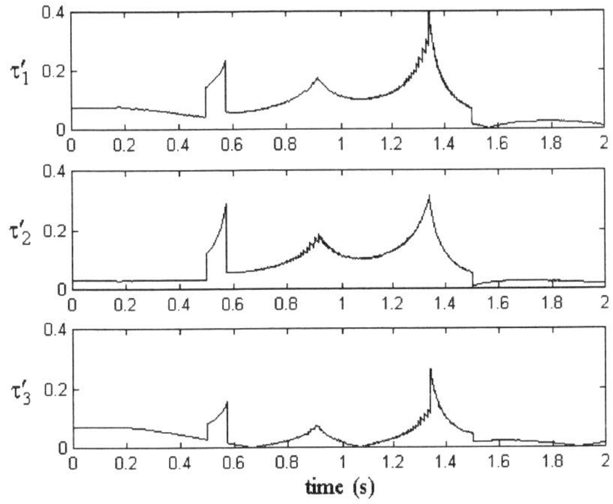


Fig. 5. Required joint torques for the generated path

Table 1. The magnitudes of torques mean value and deviation factor before and after using torque modification algorithm

| | Before TMA | | After TMA | |
|---|---|---|---|---|
| Time | $\varepsilon$ | $\beta$ | $\varepsilon$ | $\beta$ |
| 0 | 0.3276 | 0.3930 | 0.0743 | 0.0901 |
| 0.2236 | 0.0601 | 0.0612 | 0.0581 | 0.0547 |
| 0.2508 | 0.0586 | 0.0583 | 0.0566 | 0.0517 |
| 0.2720 | 0.0570 | 0.0550 | 0.0552 | 0.0488 |
| 0.2904 | 0.0555 | 0.0521 | 0.0536 | 0.0458 |

### *5.3 EFFECT OF* $R_I$

As discussed earlier, $R_I$ plays an important role both on $\beta$ and the length of the generated path before torque modification becomes necessary. To show this effect, several $\boldsymbol{\tau}_0$ for starting point of the example given in section 5.2 are calculated with different values of $R_I$. Figures (6) and (7) show the deviation factor $\beta$ and the number of steps for convergence. On the other hand, effect of $R_I$ on the length of the neighborhood in which the joint torques arc within the acceptable bound is investigated by starting from similar point with different values of $R_I$. Figure (8) shows the time elapse of this neighborhood ($t_i$) for different values of $R_I$. As can be seen from the results, increase of $R_I$ makes the torques modification algorithm faster and increases the time elapse of path generation before next torque modification is necessary, which means it makes the whole algorithm faster. However, increase of $R_I$ increases $\beta$, which means reduction of the accuracy of load distribution.

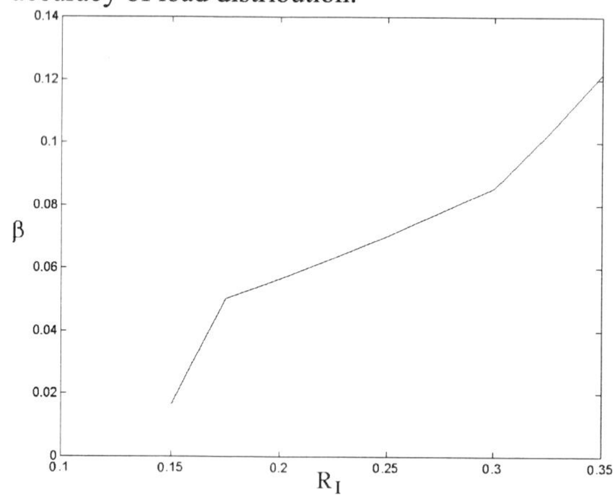


Fig. 6. Effect of $R_I$ on load distribution deviation factor

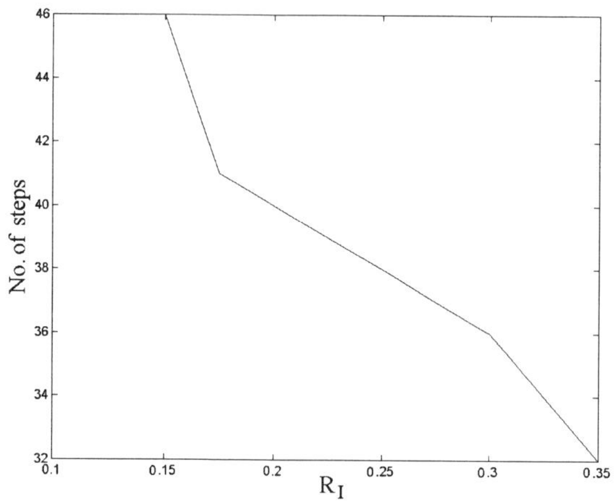


Fig. 7. Effect of $R_I$ on calculation effort

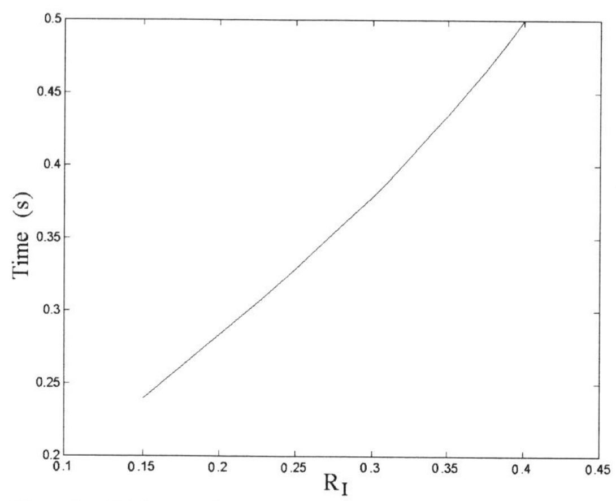


Fig. 8. Effect of $R_I$ on the length of generated path before torque modification

## 6. CONCLUSIONS

This paper proposes an algorithm for path planning of redundant manipulators with load distribution as the objective function for kinematic resolution. To this end, two major algorithms were developed for finding a good starting point and kinetic resolution of redundant manipulators subject to dynamic constraints. As most of the main algorithm is based on integration of a set of n second order ODE, the calculation effort for path planning is much less than similar algorithms, which are based on optimization of some performance index. The versatility of the algorithm and some major numerical considerations were presented through numerical results for a 3-DOF planar manipulator.

## 7. REFERENCES

Ahuactzin, J.M. and K.K. Gupta (1999). The Kinematic Roadmap: A Motion Planning Based Global Approach for Inverse Kinematics of Redundant Robots. *IEEE Transactions on Robotics and Automation*, vol. **15**, No. 4, pp. 653-669.

Hollerbach, J.M. and K.C. Suh (1987). Redundancy Resolution of Manipulators through Torque Optimization. *IEEE Journal of Robotics and Automation*, vol. **RA-3**, No. 4, pp. 308-316.

Hu, B., C.L. Teo and H.P. Lee (1995). Local Optimization of Weighted Joint Torques for Redundant Robot Manipulators. *IEEE Transactions on Robotics and Automation,* vol. **11**, No. 3, pp. 422-425.

Liegeois, A. (1977). Automatic supervisory control of the configuration and behavior of multibody mechanisms. *IEEE Trans. Syst., Man, Cybern.*, vol. **SMC-7**, pp.868-871.

Saramago, S.F.P. and V.S. Junior (2000). Optimal Trajectory Planning of Robot Manipulators in the Presence of Moving Obstacles. *Mechanism and Machine Theory*, vol. **35**, pp. 1079-1094.

Schmitt, D., A.H. Soni, V. Srinivasan and G. Naganathan (1985). Optimal Motion rogramming of Robot Manipulators. *Journal of Mechanisms, Transmissions, and Automation in Design*, vol. **107**, pp. 239-244.

Whitney, D.E. (1969). Resolved motion rate control of manipulators and human prostheses. *IEEE Trans. Man-Machine Syst.*, vol. **MMS-10**, pp. 47-53.

IFAC Publications
www.elsevier.com/locate/ifac

# IMPLEMENTATION OF A NEURO-FUZZY SENSOR BASED ROBOT NAVIGATOR

**W.Y. Wong, D.W. Yuen and T.P. Leung**

*Department of Mechanical Engineering*
*The Hong Kong Polytechnic University*
*Hung Hom, Kowloon, Hong Kong.*
*anny.wong@polyu.edu.hk, mmdwyuen@inet.polyu.edu.hk,*
*vptleung@polyu.edu.hk*

Abstract: This paper presents a fuzzy based neural network navigation method, which utilizes fuzzy logic and neural network learning for navigation of mobile robot in uncertain environments. There are five fuzzy logic controllers represent the five basic motion behaviors. These five fuzzy logic controllers generate training data for five neural networks called behavior controller. In order to select an appropriate behavior controller at the present instant, an additional neural network, called behavior selector, concludes the current situation and selecting different behavior controller to perform different action. The effectiveness of the proposed method is verified by a series of MATLAB program simulations. *Copyright © 2001 IFAC*

Keywords: Fuzzy systems, Mobile robots, Robot navigation, Neural networks, Sensor fusion

## 1. INTRODUCTION

A mobile robot that navigates in an unstructured environment must be able to fuse information from multiple sensors to carry out a number of tasks. The sensors may consist of ultrasonic, infrared sensors, laser rangefinder and even CCD camera. The task may be one or more of the followings: building global or local maps, estimating its existing location, avoiding obstacles and planning an optimum path to its goal position. In model-based navigation, an *a priori* map is assumed and therefore offline or online path planning is possible. In sensor-based navigation, only the location of the robot and the goal position are known at any time and the robot relies entirely on its multiple sensors to navigate in the unknown environment. This paper presents a neuro-fuzzy controller designed for sensor-based navigation.

In our investigation, a mobile robot with 24 ultrasonic sensors as shown in fig.1 is used. The robot is also equipped with a digital compass that measures the heading angle of the mobile robot. The 24 sensors are grouped into 8 suites and the reason for doing so will be explained in a later section.

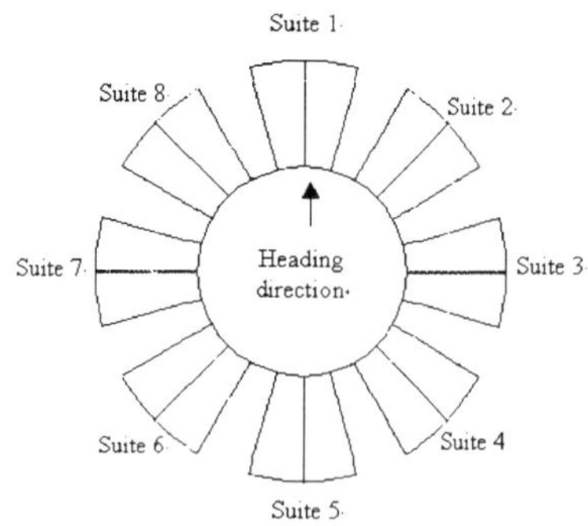


Fig. 1 Sensor arrangement of mobile robot

Many approaches had been put forward by researchers in sensor-based navigation. For example, the vector field histogram method pioneered by Borenstein and Koren (1991) is a major approach. In recent years, owing to the advancement in soft computing technologies that showed promises in handling noisy and incomplete information, mimicking human reasoning and learning abilities, fuzzy logic and neural networks were increasingly being employed in sensor-based navigation.

The overall control function is decomposed into a number of simple units, called behaviors (Brooks, 1986). Arrue, *et al.* (1997) generated robot motion control actions using different fuzzy reactive navigation behaviors and then combined them with virtual perception memory. Hee and Hyung (1995) utilized fuzzy logic and reinforcement learning for robot navigation in uncertain environment. The robot exhibited only the avoidance behavior and the goal-seeking behavior that are independently designed at design stage. At running stage, a behavior selector with a bistable switching function chose a behavior at each action step so that the mobile robot can go towards the goal position without colliding with obstacles. One rule base is responsible for one situation and therefore the fuzzy rule controller can be easily built.

Our design is an improvement on Hee and Hyung's method (1995). Using only two behaviors to handle robot navigation makes each rule base large and complicated. Therefore Hee and Hyung relied on reinforcement learning method to build the fuzzy rule bases. The resulting rule bases were no longer easy for human to interpret. Later addition of other behavior such as avoiding repeated paths would also require much retraining. In our design, we decompose the overall control function into behaviors that are as simple as possible. Therefore, even if new behaviors are required to be added to the robot, the rule bases for the basic behavior controller need not be retrained. In addition, the rule base for each behavior is simple enough to be designed intuitively. Tuning of the membership function by neural networks or other means will not make rule interpretation difficult.

A normal indoor environment consists of walls, corridor (narrow paths), isolated obstacles and dead–end etc. We formulate five different basic behaviors. First, corridor following is used to simulate the mobile robot navigating in a corridor. Second, the left wall following and right wall following are used to simulate the robot moving along a wall. Third, at dead-ends, the robot will need to perform a U-turn which can be decomposed into two right angle turns. Therefore, one behavior is right angle turning. When facing an obstacle, the robot will either turn left(left wall following) or turn right(right wall following) or U-turn(right-angle turning) so there is no explicit obstacle avoidance behavior. The goal seeking and obstacle avoidance behavior will just direct the robot move towards the goal position.

## 2. FUZZY CONTROLLERS

### *2.1 Fuzzifier*

A multi-input-multi-output system (MIMO) is used for the fuzzy rule base. A MIMO system with a inputs and b outputs can be decomposed into b a-inputs-single-output subsystems, and each of b subsystems has a set of subrule base. For our five basic behaviors, we have five fuzzy controllers. The inputs to each controller are the sensor readings and the output of each controller is the heading velocity and the incremental steering angle of the robot.

### *2.2 Fuzzy member function*

Although our robot has 24 sensors, in order to reduce the dimension of input space, sensors suite is introduced. There are 8 suites, each suite has three sensors and takes their minimum value as the representative range. All sensor suites readings (crisp) are converted into fuzzy singletons N, M, and F within the specified universe of discourse through a fuzzifier. The universe of discourse is set to be 8 units and trapezoid and triangular shape membership functions are chosen. All sensor suites are in this form of membership function for every behavior. Table 1 shown the linguistic term of the range value. The membership function of the heading velocity (Hvelocity) factor and the incremental steering angle (Sangle) for different behaviors are also determined.

Table 1 Linguistic descriptions of range values

| Fuzzy singleton | Linguistic term | Value in unit distance |
|---|---|---|
| N | Near | (0, 0, 2, 4) |
| M | Medium | (2, 4, 6) |
| F | Far | (4, 6, 8, 8) |

### 2.3 *Fuzzy rules controller*

After fuzzifying all sensor readings, all these readings can be fed into the fuzzy rules controller. The controller contains control rules, which are related to the input sensor readings. The fuzzy logic controller integrates all rules and concludes a command action to the robot to perform different behaviors such as wall following, goal seeking. For example, IF $d_n$ (n denote sensor suit number) is Far, and $\theta$ is ZZ, THEN Sangle is ZZ and Hvelocity is F. Table 2 shown one of the fuzzy rules for the left wall following behavior.

The five basic behaviors are corridor following, right-angle tuning, left wall following, right wall following, and goal seeking and obstacle avoidance. Table 3 shown the number of sensor suites, control variables and number of fuzzy rules used by the five behavior controller.

Table 2 Fuzzy rules for left wall following

| Rule no. | R7 | R8 | Sangle | Hvelocity |
|---|---|---|---|---|
| 1 | N | N | PB | L |
| 2 | N | M | NS | H |
| 3 | N | F | NB | L |
| 4 | M | N | PS | H |
| 5 | M | M | ZZ | F |
| 6 | M | F | NS | H |
| 7 | F | N | PB | L |
| 8 | F | M | PS | H |
| 9 | F | F | NM | H |

Table 3 Overview of the five behavior controllers

| | Sensor suites | Control variables | Fuzzy rules |
|---|---|---|---|
| Corridor following | 7,8,2,3 | Steering Velocity | 81 |
| Right angle turning | 1,3,5,7 | Steering | 46 |
| Left wall following | 7,8 | Steering Velocity | 9 |
| Right wall following | 2,3 | Steering Velocity | 9 |
| Goal seeking and obstacle avoidance | 7,8,1,2,3 | Steering Velocity | 89 |

Table 4 Comparison between fuzzy logic and neural network controllers

| | Their need, | Time consumed in, | Difficulty in, |
|---|---|---|---|
| Fuzzy logic controller | Rule base | Tuning of rules | Tuning of rules |
| Neural network controller | Training pattern | Training progress | Obtain training pattern |

## 3. NEURAL NETWORK CONTROLLER

Comparing fuzzy controllers with neural network controllers, a fuzzy controller has the advantage that the rule base is easy to establish. On the other hand, a neural network controller requires less memory to implement and also executes faster (Table 4). However, it is difficult to prepare data for its training. In our work, we tried to get the best of both worlds. We operate each tuned fuzzy rule controller under a test environment and collect all inputs and outputs the robot experienced in its navigation process. The sensor range data will form the input pattern for the neural network controller and the outputs of the fuzzy controller form the target value. The neural network controller is tested in untrained environment to verify its performance. In our study, five feedforward neural network controllers were trained by backpropagation algorithm to replace the five fuzzy controllers.

## 4. BEHAVIOR SELECTOR

Based on the potential field theory, Hee and Hyung (1995) calculated the changes of attractive and repulsive potentials and devised an expression for behavior selection. Some researchers suggested the construction of a fuzzy rule base to perform behavior blending (Dorigo and Colombetti, 1998). We take an alternative approach - a neural network is trained to select the appropriate behavior when the robot is placed in a certain environment.

The resulting system contains two main parts, one is Behavior Selector, and another is Behavior Controller. The function of behavior selector is to conclude the environmental situation at each time interval and select appropriate behavior controller.

The behavior selector is a neural network that has 24 inputs and 5 neurons in its output layer for selecting appropriate behavior controller. The network is a four-layer log-sigmoid / log-sigmoid / log-sigmoid / log-sigmoid network. The first hidden layer has 30 neurons, second has 30 neurons and last one has 20 neurons. Fig. 2 shown the neural network architecture of the behavior selector.

To operate correctly the behavior selector should respond with a 1 in the position of the behavior controller being presented to the network. All other values in the output vector should be 0.

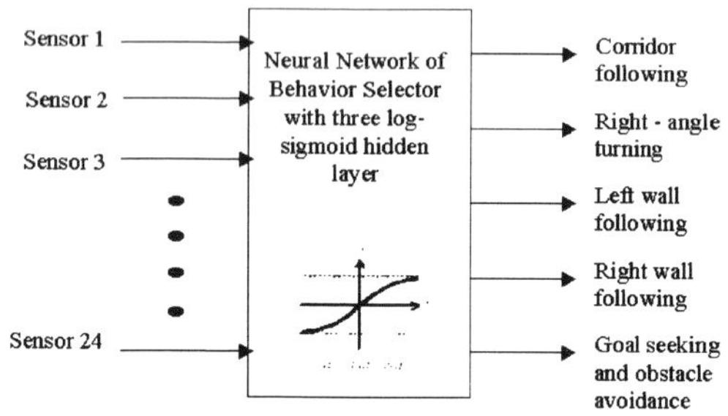


Fig. 2 Neural network architecture of the behavior selector

(a) Corridor following

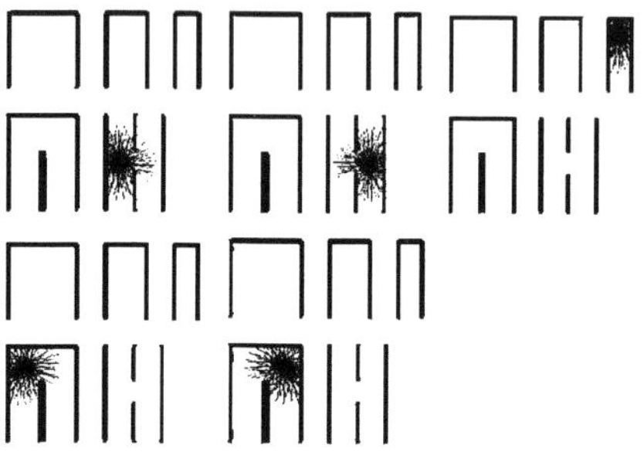

(b) Right angle turning

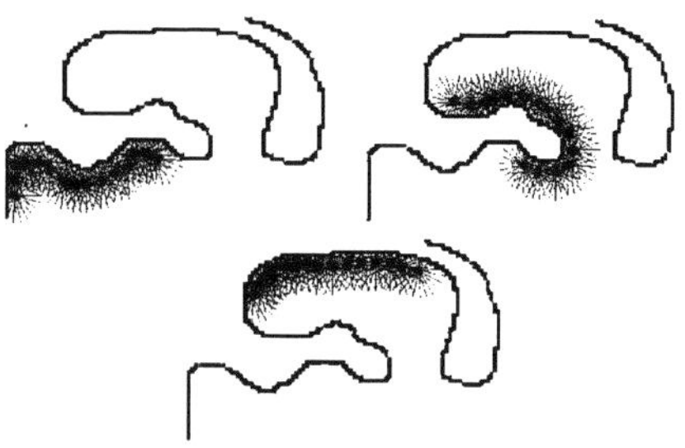

(c) Left wall following

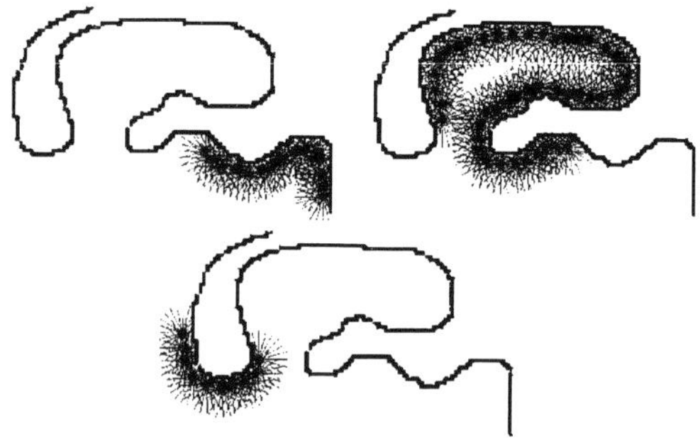

(d) Right wall following

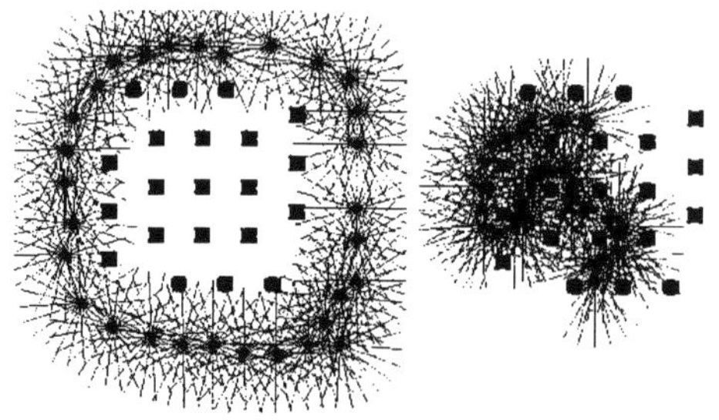

(e) Goal seeking and obstacle avoidance

Fig. 3(a) – (e) Data points for training behavior selector

### *4.1 Training of Behavior selector*

We design different environments and place the robot at locations where we believe that a certain behavior should be selected. For example, in fig. 3(a), the robot should always select the corridor following behavior. The sensor data are collected and used to train the behavior selector.

## 5. SIMULATIONS

In our simulation, it is assumed that the boundary of all obstacles has been expanded to include the radius of the robot so the robot becomes a point robot occupying only one pixel (one unit area). The size of environment is restricted. A circle indicates position of the mobile robot, the center point represents the robot body. The maximum speed of the mobile robot was assumed to be 3 pixels/sec (it is adjustable). The measurement range of the sensor was assumed to be 8 pixels radius. In order to acquire information about environment, a sensor simulator, which outputs the range data of ultrasonic sensors was used for the sensor simulator, it was assumed that ultrasonic sensor can ideally detect the nearest obstacle within their beams. Among the range data incoming from sensors, meaningful range, d is defined as $d = \min\{d_{max}, \max(d, d_{min})\}$. Here $d_{max}$ and $d_{min}$, denote the maximum and minimum ranges of ultrasonic sensor, respectively. Beside the range sensors, a 'digital compass' is installed on the robot. It can produce the angle difference (θ) between the goal position, θ with respect to the global framework. Also, the motion of mobile robot can be realized by the control of its heading velocity (v) and an incremental steering angle, (Δψ).

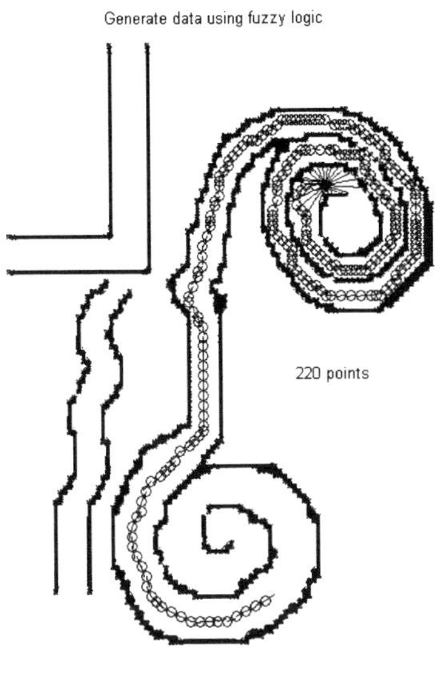


Fig. 4 Fuzzy controller in corridor following

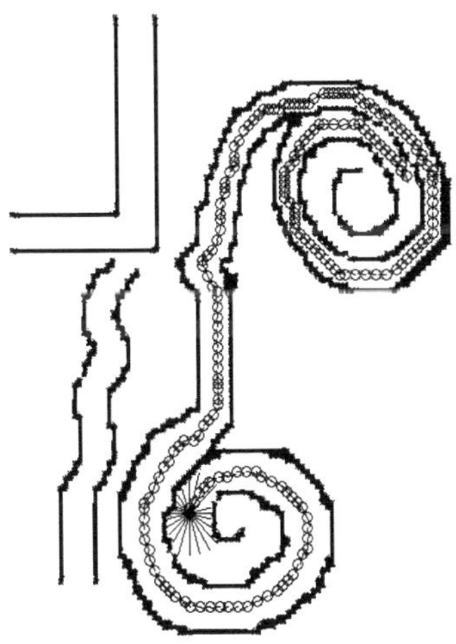

Fig. 5 Trained neural network controller in corridor following

## 6. RESULTS

### *6.1 Fuzzy logic controllers*

Fig.4 shown the performance of the fuzzy logic controller in a corridor. Satisfactory result can be obtained and the mobile robot keeps walking along the center of the corridor most of the time. The other four behavior controllers are also designed in similar way.

### *6.2 Neural network controller*

The fuzzy logic controller is used to generate training data for its corresponding neural network controller.

Fig. 5 showed the trained neural network controller in corridor following. It is similar to the fuzzy controller. The trained neural network is also tested in an unknown environment (fig. 6) and found that satisfactory result can be obtained. Fig. 7 and fig. 8 showed the performance of the left wall following and right wall following neural controllers in an unknown environment.

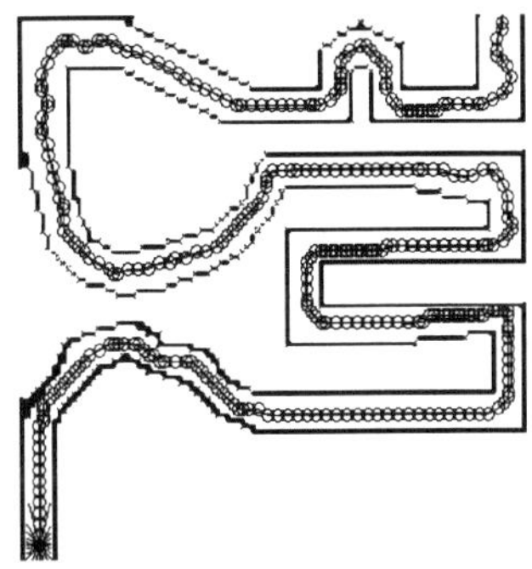

Fig. 6 Corridor following in an unknown environment for neural network controller

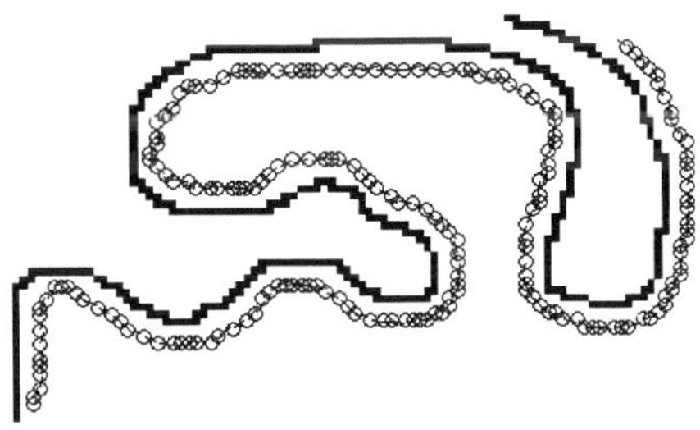

Fig. 7 Left wall following in an unknown environment for neural network controller

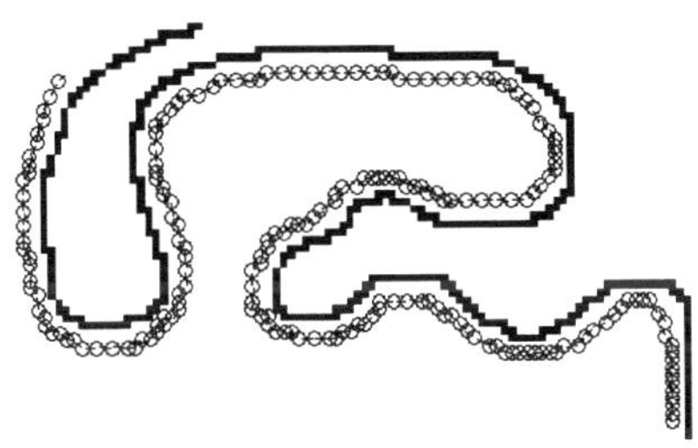

Fig. 8 Right wall following in an unknown environment for neural network controller

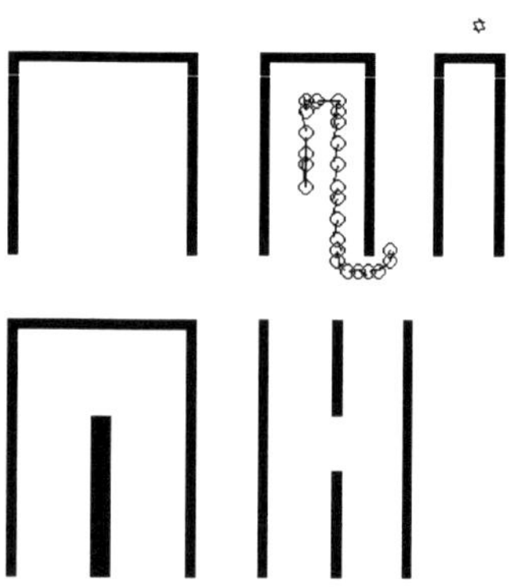

Fig. 9 Escape from dead end environment

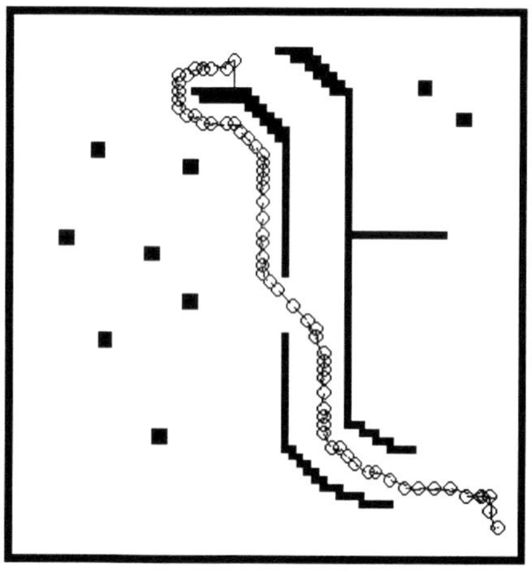

Fig. 10 Behavior selector navigate in an unknown environment

Fig. 9 and 10 showed the performance of the behavior selector in different environment. In fig. 9, the behavior selector enabled the robot to perform right angle turning and escape from the dead end.

### *6.3 Comparison of Fuzzy and Neural Network Controllers*

Table 5 showed that the neural network controllers execute much faster than the fuzzy controllers.

Table 5 Comparison between fuzzy logic and neural network

| | Left wall following (30 steps) | Right wall following (30 steps) | Corridor following (30 steps) |
|---|---|---|---|
| Fuzzy logic controller | 18.65 s | 19.45 s | 60.98 s |
| Neural network controller | 16.44 s | 16.79 s | 20.02 s |

## 7. CONCLUSIONS

This paper proposes a fuzzy based neural network navigation method, which utilizes fuzzy logic and neural network learning for navigation of mobile robot in uncertain environments. Five behaviors controller and one behavior selector have been created. Fuzzy logic controller is used to create the result and the neural network is trained by these results. We found that satisfactory results can be obtained for the five behavior controllers in unknown environment.

## REFERENCES


Arrue, B.C. *et al.* (1997). Fuzzy Behaviors Combination to Control a Non-Holonomic Mobile Robot Using Virtual Perception Memory. In: *IEEE International Conference on Fuzzy System.*

Borenstein, J. and Y. Koren (1991). The Vector Field Histogram - Fast Obstacle-Avoidance for Mobile Robots. In: *IEEE Journal of Robotics and Automation,* Vol. 7, No. 3, June, pp. 278-288.

Brooks, R.A. (1986). A Robust Layered Control System for a Mobile Robot. In: *IEEE J. Robotics and Automation*, vol 2, n. 1, pp 14 –23.

Dorigo, M. and M. Colombetti (1998). *Robot Shaping: an Experiment in Behavior Engineering*, The MIT Press, Cambridge, MA.

Hee, R.B. and Hyung, S.C. (1995). A Sensor-Based Navigation For a Mobile Robot Using Fuzzy Logic And Reinforcement Learning. *IEEE Systems, Man and Cybernetics*, vol. 25, Issue 3, pp464 – 477.

Nagata, S., Sekiguchi, M., and Asakawa, K (1990). Mobile Robot Control by a Structured Hierarchical Neural Network, *IEEE Control Systems Magazine.*

Saffioti, A., Ruspini, E.H., and Konolige K. (1997). Using Fuzzy Logic for Mobile Robot Control. In: International Handbook of Fuzzy Sets and Possibility Theory(Dubois D., Prade, H. and Zimmermann H.J. Eds.), Kluwer Academic.

IFAC Publications
www.elsevier.com/locate/ifac

# A SELF-ORGANIZING NEUROFUZZY SYSTEM AND ITS APPLICATION TO QUALITY PREDICTION FOR HYDROCRACKING PROCESSING

*Li Jia, Erguo Li and Jinshou Yu*

*Research Institute of Automation,*
*East China University ofScience and Technology,*
*Shanghai,200237, People's Republic of China*

Abstract: In this paper a self-organizing neurofuzzy system (SONFS) is proposed. It is a feed-forward multiplayer network and learns in two main phases. In the first, the system utilizes improved nearest-neighborhood clustering scheme to determine the number of if-then rules and the parameters of fuzzy antecedent membership functions. Then supervised learning is utilized to adjust parameters to minimize the network output error. On the basis of the two-phase learning, the on-line learning algorithm is also proposed. Lastly, the SONFS is used in quality prediction of hydrocracking processing. Simulation results show that SONFS can provide better generalization results and simple model structure. *Copyright © 2001 IFAC*

Keywords: Fuzzy logic inference, Neural networks, Neurofuzzy system, Improved nearest-neighborhood clustering, Quality prediction

## 1. INTRODUCTION

Recently, it is becoming increasingly important to consider how to design optimal fuzzy rules from training data, in order to construct a reasonable and suitable fuzzy system model for identifying the corresponding practical systems (Yan and Masaharu, 2000). Developed methods based on trial-and-error, clustering techniques, mathematical programming and optimization theory have been suggested to extract fuzzy rules. However, trail-and-error is not suitable or even feasible when there is no linguistic knowledge or expert knowledge available. And slow convergence and local solution, very common in algorithm based on optimization theory, often impose practical constraints in using the corresponding fuzzy design methods (Mauricio and Fernando, 1999).

Bringing the learning abilities of neural networks to automate and realize the design of fuzzy systems has been recognized as a powerful approach (Wang and Mendel, 1992; Werbos, 1992). This integration brings the low-level learning and computation power of neural networks into fuzzy logic systems, and brings the high-level, human-like thinking and reasoning of fuzzy logic systems into neural networks (Lin and Lee, 1991). Such an integration of neural networks and fuzzy systems into a functional system provides a new approach for designing fuzzy systems. Some neurofuzzy systems are capable to learn and to provide if-then fuzzy rules in linguistic or explicit form. However, most of the current neurofuzzy approaches only consider parametric identification or learning and the structure learning problems dealing with the partition of input-output space, the number of membership function and the number of fuzzy if-then rules are not concerned completely. It means the number of partitions of input space and the number of fuzzy rules must be determined before parameter's learning and the structure of the network can't be changed dynamically. This creates some difficulties in practical application because the training data only cover most parts of the whole input space and new partition need be added continuously on line.

To avoid these problems, Wang and Rong (1999) proposed a self-organizing neural-network-based fuzzy system with two-phase hybrid learning algorithm. It utilizes the nearest-neighborhood clustering scheme for both structure learning and initial parameters setting and a gradient descent method for fine tuning the parameters of the network. However, the designers think, the width of every membership function is identical. Simulation results show that the width of membership function should be chosen according to itself clustering space. Moreover, in their work, the consequent of fuzzy rule are determined by the average of desired output of all points belonging to same clustering space. Simulation results show that the method is good and simple for systems of which the difference of the maximum output and the minimum output is small only.

In this paper a class of self-organizing neurofuzzy system (SONFS) is proposed. The system is a feed-forward multiplayer network and learns in two main phases. In the first, the system utilizes improved nearest neighbor- hood clustering scheme to determine the number of if-then rules and the parameters of fuzzy antecedent membership functions. It partitions the input space according to the distribution of the training data, which greatly reduces the number of rules compared to the conventional grid partitioned neurofuzzy systems. And it uses the average, the maximum and minimum point of desired output of all points belonging to same clustering space to determine the consequent of fuzzy rule not only the average point so it is better than the scheme proposed by Wang and Rong (1999). Compared to the nearest- neighborhood clustering scheme used by Werbos (1992), it chooses the width of antecedent membership function according to respective distribution of clustering space. On the basis of the first phase, supervised learning is utilized to adjust parameters to minimize the network output error. Moreover, for time varying processes the system can add new rule nodes by on-line learning.

This paper is organized as follows. In section 2, the structure of self-organizing neurofuzzy system is introduced. Learning algorithm is described in section 3. Section 4 discuss the application of self-organizing neurofuzzy system in Quality prediction. Section 5 concludes the features of the presented algorithm.

## 2. STRUCTURE OF SELF-ORGANIZING NEUROFUZZY SYSTEM

In this section, we will describe the structure and functions of the proposed self-organizing neurofuzzy system model. Before proceeding, some important terms must be defined. For a multiple input and single output system (MISO), a typical format for a fuzzy inference system consists of a collection of fuzzy IF-THEN rules in the following form:

$$R_j : IF \quad x_1 \quad is \quad A_1^j \quad AND \quad x_2 \quad is \quad A_2^j \quad AND \quad x_M \quad is \quad A_M^j\,,$$
$$THEN \quad y = B^j\,, \quad j = 1,2,\cdots,N$$

where $x_i$ is the ith input variable, $A_i^j$ is a fuzzy set corresponding to the *i*th antecedent of the *j*th rule (*j*=1,2,...,N and *i*=1,2,...,M), N is number of IF-THEN rules and M is the number of input variable, y is the output variable and $B^j$ is fuzzy sets of the *j*th rule.

The self-organizing neurofuzzy system is shown as Fig.1. It consists of five layers. The first layer is input layer. Each node in this layer is called an input linguistic node and corresponds to one input variable.

The node only transmits input values to the next layer directly. Nodes in second layer are called input term nodes, each of which correspond to one linguistic label of an input variable. Each node in this layer calculates the membership value specifying the degree to which an input value belongs to a fuzzy set. A Gaussian membership function is used in this layer, which is described

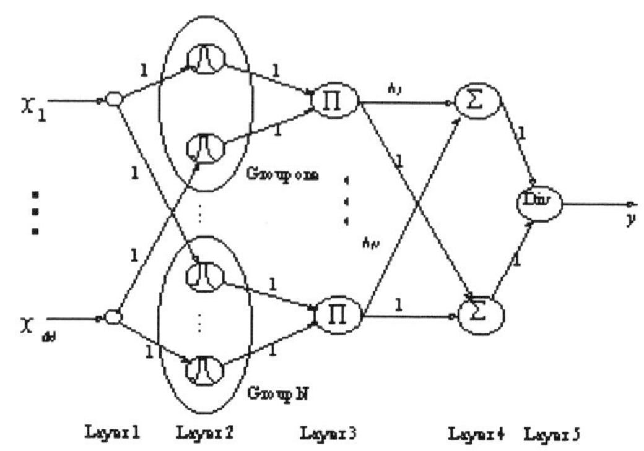

Fig.1 the structure of self-organizing neurofuzzy system

as:

$$\mu_{A_i^j}(x_i)=\exp\left(\frac{(x_i-c_{ji})^2}{\sigma_{ji}^2}\right) \qquad (1)$$

where $c_{ji}$ and $\sigma_{ji}$ respectively are the center and width of the membership function.

The third layer consists of N neurons, which compute the fired strength of a rule. Multiplicative inference is used, so the output of this layer is:

$$y_j=\prod \mu_{A_i^j}=\exp\left(\sum_{i=1}^{M}\frac{(x_i-c_{ji})^2}{\sigma_{ji}^2}\right), \qquad (2)$$
$$i=1,2,\cdots,M \quad j=1,2,\cdots,N$$

There are two neurons in fourth layer. One of them connects with all neurons of the third layer through the weight $h_j$ representing the consequence of the $j$th rule.and another one connects with all neurons of the third layer through unity weights. The last layer has a single neuron to compute $y$. It is connected with two neurons of the fourth layer through unity weights. The integral function and activation function of the node can be expressed as:

$$y=\frac{\sum_{j=1}^{N} h_j y_j^{(3)}}{\sum_{j=1}^{N} y_j^{(3)}} \qquad (3)$$

## 3. LEARNING ALGORITHM

In this section, the two-phase hybrid learning schemes for the proposed self organizing neurofuzzy system will be discussed. This learning scheme combines improved nearest-neighborhood clustering method and gradient descent algorithm to partition the input/output spaces, find proper fuzzy logic rules and adjust membership functions. In the first phase, the improved nearest- neighborhood clustering scheme is used to partition the input/output space by clustering training data. In other words, this learning phase is to locate initial membership functions in the input/output spaces. In the second phase, gradient descent is utilized to adjust parameters to minimize the network output error. There are no rule nodes in the initial structure of the self-organizing neurofuzzy system. They are all created during the learning process.

Utilizing above two-phase hybrid learning scheme, a suitable fuzzy model is obtained. But for most industrial process, the plant is always varying with time so that the on-line learning scheme should be performed in order to satisfy the need of time varying systems.

### *3.1 Nearest-neighborhood clustering method*

The improved nearest-neighborhood clustering method can be stated as: given a set of input training data, $x_i(t)$, $I$=1,2,...,M and output training data, $y(t)$, $t$=1,2,...P, try to find proper fuzzy partitions and membership functions in the input/output spaces. The learning algorithm can be summarized as follows:

1) Initialization: Define the cluster radius according to the distribution of input data. Choose the training data pairs $\{x_1(1),x_2(1),\cdots,x_M(1),y(1)\}$ as the first cluster and set cluster center $c_{1i}=x_i(1)$, the number of input data of the first cluster $N_1=1$, the output of the first cluster $Y_1(i)=y(1)$, $i=1,2,\cdots,N_1$ and the number of clusters at this time, $N=1$.

2) Consider the $k$th input/output data pairs, $\{x_1(k),x_2(k),\cdots,x_M(k),y(k)\}$ and computer the distance of the $k$th input data point $x_i(k),(i=1,2,\cdots,M)$ to cluster center $c_{ji},(j=1,2,\cdots,N,i=1,2,\cdots,M)$ and find the smallest one. This is equivalent to finding a cluster (category) that $x_i(k),(i=1,2,\cdots,M)$ could belong to. The chosen category is indexed by $J$, where

$$D_J=\|x(k)-c_J\|=\min_{i\le j\le N}\|x(k)-c_j\| \qquad (4)$$

If $D_J>r$, the $k$th input data doesn't belong to all the existing cluster and new cluster need to be established with its center located at $x_i(k),(i=1,2,\cdots,M)$, setting $N$=$N$+1, $c_{Ni}=x_i(k)$, $N_N=1$, $Y_N(i)=y(k), i=1,2,\cdots,N_N$ and keeping other information unmodified.

If $D_J\le r$, the $k$th input data belongs to the $J$th cluster and the $k$th input/output data should be used to modify the $J$th cluster, setting $N_J=N_J+1$, $Y_J(N_J)=y(k)$ and keeping other information unmodified.

3) Let $k=k+1$ then go to step 2) until all training data pairs are clustered to corresponding category. After finishing the cluster learning, we can get the number of IF-THEN rules N. The width of membership functions can be calculated as:

$$\sigma_{ji}=\min_{\substack{k=1,2,\cdots,N\\ k\neq j}}\frac{|c_{ji}-c_{ki}|}{\rho} \qquad (5)$$

where $\rho$ is overlap parameter, usually $1\le\rho\le 2$ (Lin and Lee, 1991).

In order to obtain the center of consequence of $j$th rule, we computer the maximum, minimum and average output of the $j$th cluster, which can be determined respectively by:

$$MAX\_Y_j = \max_{1\le i\le N_j}(Y_j(i)) \quad MIN\_Y_j = \min_{1\le i\le N_j}(Y_j(i))$$

$$AVE\_Y_j = \frac{\sum_{i=1}^{N_j} Y_j(i)}{N_j}$$

And the corresponding input of them are labeled by: $MAX\_U_j, MIN\_U_j$ and $AVE\_U_j$. Then the center of consequence of each rules can be got by following formula:

$$h_j = (MAX\_Y_j - AVE\_Y_j)\exp\left(-\xi_{j1}\|MAX\_U_j - x\|^2\right) + (MIN\_Y_j - AVE\_Y_j)\exp\left(-\xi_{j2}\|MIN\_U_j - x\|^2\right) + AVE\_Y_j \quad (6)$$

where $\xi_{j1}$ and $\xi_{j2}$ are adjusted parameters and will be discussed in next section.

4) End

### 3.2 *Supervised learning for rule extraction*

In supervised learning phase input and output data are presented to adjust parameters and obtain more better fuzzy model. It's goal is to minimize the error function:

$$E = \frac{1}{2}[y(t) - d(t)]^2 \quad (7)$$

where $y(t)$ is the current output and $d(t)$ is the desired output. For a training data pair, starting at the input nodes, a forward pass is used to compute the activity levels of all the nodes, a backward pass is used to compute $\partial E / \partial y$ for all parameters.

That is, in order to train $c_{ji}$, we use

$$c_{ji}(k+1) = c_{ji}(k) - \alpha\frac{\partial E}{\partial c_{ji}} + \beta(c_{ji}(k) - c_{ji}(k-1)) \quad (8)$$

where $\alpha$ is learning factor which effect the speed of learning and $\beta$ is inertia factor. From (1)-(7):

$$\frac{\partial E}{\partial c_{ji}} = 2(y-d)(h_j - y)(x_i - c_{ji})H_j / \sigma_{ji}^2 \quad (9)$$

where

$$H_j = \exp\left(-\frac{\sum_{i=1}^{M}(x_i - c_{ji})^2}{\sigma_{ji}^2}\right) / \sum_{j=1}^{N}\exp\left(-\frac{\sum_{i=1}^{M}(x_i - c_{ji})^2}{\sigma_{ji}^2}\right) \quad (10)$$

Similarly, the update law of $\sigma_{ji}$, $\xi_{j1}$ and $\xi_{j2}$ are

$$\sigma_{ji}(k+1) = \sigma_{ji}(k) - \alpha\frac{\partial E}{\partial \sigma_{ji}} + \beta(\sigma_{ji}(k) - \sigma_{ji}(k-1)) \quad (11)$$

$$\xi_{ji}(k+1) = \xi_{ji}(k) - \alpha\frac{\partial E}{\partial \xi_{ji}} + \beta(\xi_{ji}(k) - \xi_{ji}(k-1)),\ i = 1,2 \quad (12)$$

where

$$\frac{\partial E}{\partial \sigma_{ji}} = 2(y-d)(h_j - y)\sum_{i=1}^{M}(x_i - c_{ji})^2 H_j / \sigma_{ji}^3 \quad (13)$$

$$\frac{\partial E}{\partial \xi_{j1}} = (y-d)H_j\|MAX\_U_j - x\|^2(MAX\_Y_j - AVE\_Y_j)\exp\left(-\xi_{j1}\|MAX\_U_j - x\|^2\right) \quad (14)$$

$$\frac{\partial E}{\partial \xi_{j2}} = (y-d)H_j\|MIN\_U_j - x\|^2(MIN\_Y_j - AVE\_Y_j)\exp\left(-\xi_{j2}\|MIN\_U_j - x\|^2\right) \quad (15)$$

### 3.3 *On-line learning algorithm*

After the initial membership functions ad fuzzy logic rules have been found in the nearest-neighborhood clustering and supervised learning phase, the whole network structure is established. But for most industrial process, the plant is always varying with time so that the so called on-line learning scheme should be performed in order to satisfy the need of time varying systems. In this section, we will discuss the on-line learning algorithm.

The on-line learning scheme is same as the off-line learning and combines two learning phase existing in off-line. It is summarized as:

*Step one*: Compute the error between the output of self-organizing neurofuzzy system and the output of practical process, $e=|y-d|$.
If $e \le E_{\text{limit}}$, where $E_{\text{limit}}$ is the limitation of error, this means the fuzzy model can meet the practical need and need not be modified.
If $e > E_{limit}$, this means the fuzzy model can not satisfy the practical need so go to step two.

*Step two*: Compute the nearest clustering by (15).
If $D_j > r$, this means the current input doesn't belong to all the existing clusters. Add a new cluster according to the method introduced in section 3.1.
If $D_J \le r$, this means the current input belongs to one of the existing clusters and no new cluster need. Modify the parameters according to the formulates listed in section 3.2.

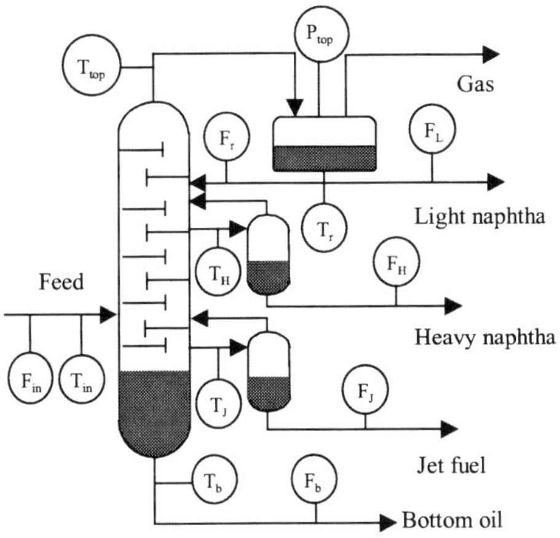


Fig.2. The schematic representation of the hydrocracking fractionator

## 4. QUALITY PREDICTION BASED ON SELF-ORGANIZING NEUROFUZZY NETWORK

### *4.1 Description of industrial background*

Hydrocracking is one of the most important processes in the petroleum industry. It upgrades heavy oil value by making high quality products, such as gasoline or kerosene. The purpose of the main fractionator of a hydrocracking process is to split a feed produced from the former process into three product streams of different molecular weight, that is, light naphtha, heavy naphtha and jet fuel. Their endpoints are the key indicators to value the product quality. Figure 2 shows a schematic representation of the main hydrocracking fractionator. According to the analysis of technological mechanisms, the endpoints are related with the 13 variables shown as Fig. 2, which can be measured and recorded on-line.

### *4.2 Quality prediction model based on self-organizing neurofuzzy system*

In this section, the jet fuel endpoint will be studied. Obviously, selecting all thirteen variables as the input of self-organizing neurofuzzy system is totally unnecessary. By statistical regression analyzing step by step, jet fuel endpoint number is mainly affected by the following six measurable variables: the reflux temperature $x_1$, the feed quality $x_2$, the reflux quality $x_3$, the light naphtha flow rate $x_4$, the heavy naphtha flow rate $x_5$ and the jet fuel flow rate $x_6$. So its model structure is represented as:

$$y = f(x_1, x_2, x_3, x_4, x_5, x_6)$$

where y is the jet fuel endpoint.

The proposed self-organizing neurofuzzy system is used to establish the process. There are 223 sets of sample data of thirteen operating variables in different operating states. 173 pairs of them are used as off-line training data sets and another 50 pairs are used as on-line testing data sets, which verify the fuzzy inference power of the self-organizing neurofuzzy system. In the training phase, all training data are scaled in the intervals [-1, +1], setting neighborhood radius $r$=0.125, overlap parameter $\rho = 2$, learning factors $\alpha = 0.2$ and $\beta = 0.01$ and the initial value of $\xi_1$ and $\xi_2$ are chose in the intervals [0,15]. At the same time, the mean square error is used to evaluate the performance of fuzzy model, it is be defined as:

$$MSE = \frac{1}{P}\sum_{k=1}^{P}[y(k) - d(k)]^2$$

where $P$ represents the number of training data.

After the off-line learning, 173 sets training data are clustered into 41 categories, that is the number of IF-THEN rules of self-organizing nurofuzzy system is 41 which is less than conventional grid partitioned neurofuzzy system's. And MSE is $5.07\times10^{-3}$ after the nearest-neighborhood clustering learning and it is $3.900\times10^{-3}$ after 50 epoch supervised learning. In order to verify the generalization of the presented fuzzy model, another 50 sets is used to test it. In on-line learning phase based on off-line learning another 2 rules are added. The estimated values are shown in Fig.3.

From above example, the proposed self-organizing neurofuzzy system possesses simple structure, better generalization ability and can be applied to quality prediction successfully.

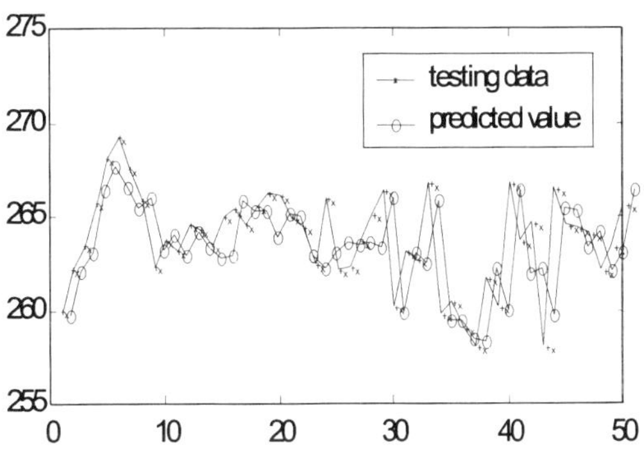


Fig.3 The comparison between the output of fuzzy predictive model and the testing data

## 5. CONCLUSION

The self-organizing neurofuzzy system and adaptive scheme was introduced in this paper as integrated method to design fuzzy system. The main features of the proposed approach are: 1) The users need not provide the initial fuzzy partitions, membership functions and fuzzy logic rules. They are all created during the learning process; 2) It partitions the input space according to the distribution of the training data, which reduces the number of rules greatly compared to the conventional grid partitioned neuro-fuzzy systems; 3) It uses the average, the maximum point and minimum point of desired output of all points belonging to same clustering space to determine the consequent of fuzzy rule not only the average point; 4) Supervised learning algorithm is utilized to adjust parameters to minimize the network output error according to the need of modeling precision; 5)on-line learning scheme is also performed so it can satisfy the need of time varying systems; 6) It is used in quality prediction of hydro-cracking processing. Simulation results show that SONFS can provide better generalization results and simple model structure.

## REFERENCES

Lin C.T. and Lee C.S.G. (1991). Neural networks based fuzzy logic control and decision system. *IEEE Trans. Compute*, **40**, 1320-1336.

Mauricio M.F and Fernando F.G (1999). Design of fuzzy systems using neurofuzzy networks. *IEEE Trans. Neural Networks*, **10**, 815-827.

Wang L.X. and Mendel J.M. (1992). Generating fuzzy rules by learning from examples. *IEEE Trans. On systems, Man, Cybernet*, **22**, 1414-1427

Wang Y.W and Rong G.R (1999). A self-organizing neural network based fuzzy system. *Fuzzy set and systems*, **103**, 1-11.

Werbos P.J. (1992). Neural control and fuzzy logic: connections and designs. *Int. J. Approx. Reasoning*, **6**, 185-219.

Yan Y.S and Masaharu M.M (2000). A new approach of neuro-fuzzy learning algorithm for tuning fuzzy rules. *Fuzzy sets and systems*, **112**, 9-116.

IFAC Publications
www.elsevier.com/locate/ifac

# SELF-ORGANIZING RADIAL-BASE-FUNCTION-NETWORK BASED INTERNAL MODEL CONTROL SYSTEM

*Li Jia, Erguo Li and Jinshou Yu*

*Research Institute of Automation,*
*East China University ofScience and Technology,*
*Shanghai,200237, People's Republic of China*

Abstract: A novel algorithm on internal model control based on Gaussian radial base function neural network (GRBFNN) is presented. The main result of the paper is the consideration of the adaptive selection of the shape parameter and variable center vector. The performance of this internal model control system and its ability to cope with disturbance are evaluated. And a temperature real-time control system is also operated. The control results show that the proposed algorithm can be introduced to time delay, non-linear and time varying systems. *Copyright © 2001 IFAC*

Keyword: Neural network, Gaussian radial base function, Nearest-neighborhood clustering algorithm, Internal model control, Adaptive control

## 1. INTRODUCTION

The internal model control (IMC) approach was developed by Monari in 1982 (Garda and Morarl, 1982) . It has two important advantages: it explicitly takes into account model uncertainty and it allows design engineers to trade off control system performance against robustness. Recently a novel technology, directly using artificial neural networks for internal model control, becomes popular. The characteristics of neural network show that they could be effective in internal model control problem. In particular, the ability of neural network to represent arbitrary nonlinear mappings attracts more and more researchers' attention.

The artificial neural network based internal model control system consists of a neural network based internal model and a neural network based inverse model of the controlled process. Compared with conventional internal model control the artificial neural network based internal model control has one obvious advantage that an exact mathematical model is not required.

Several scholars have proposed algorithms for artificial neural network internal model control. But most of them employ back-propagation method and its enhanced algorithms to train neural network to learn the system's input/output relationship and the corresponding inverse relationship (Liao and Ning 1998). However, it is known those methods have drawback on convergence and local minimum, which hinder its application in industrial process control. In Hunt's paper (1991), the Gaussion radial base function was used as active function in network hidden units, which overcame the drawbacks of back-propagation method and simplified the

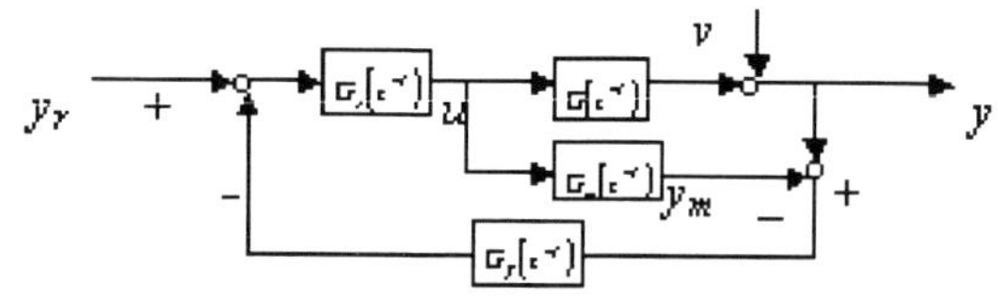

Fig.1 the structure of internal model control system

problem. Naturally Gaussion radial base function based neural network produces a partition of the input space and has the best approximation property. But in that paper, adaptive selections of the shape parameter and variable center vector(s, c), which affect the speed of convergence, weren't considered exactly. In this work, considering this problem, a two-phase hybrid learning algorithm which utilizes nearest-neighborhood clustering scheme for structure learning and initial parameters setting and adaptive gradient descent method for better tuning the parameters of (s, c) is proposed. In addition, a feedback filter is added to obtain better robustness and disturbance.

## 2 GAUSSIAN RADIAL BASE FUNCTION NETWORK BASED INTERNAL MODEL CONTROL SYSTEM

### *2.1 Internal model control*

The internal model control is now well known and is presented in Fig.1.Here the discrete transfer function $G(z^{-1})$, $G_m(z^{-1})$, $G_c(z^{-1})$ represent the process to be controlled , the model of the process and the internal model controller respectively. The function $G_f(z^{-1})$ denotes a feedback filter to be discussed later in this paper, $y_r$, $y$, $u$, $y_m$ and $v$ refer, respectively , to the set-point , to the output of process, to the control value, to the output of model and to the disturbance.

The IMC has been shown to have a number of desirable properties and a detailed analysis has been given by Morari and Zafiriou. Here, we don't discuss these properties.

### *2.2 Gaussian potential function network*

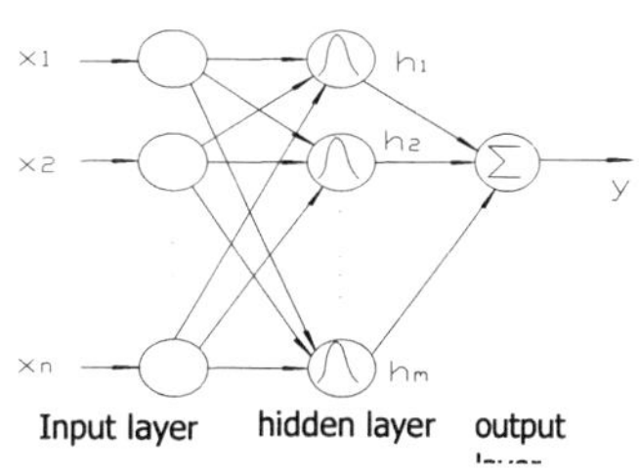


Fig.2 Gaussian potential function networks

In the present work Gaussian radial base function neural network (GRBFNN) is used to construct internal model and internal controller. Gaussian radial base function neural network is a very useful feed-forward network and has been widely used in the area of functional approximation and adaptive signal filter. In theory, Gaussian potential function network can represent arbitrary nonlinear relations with any accurate degree as other multi-layer feed-forward network. However, the difference between them is the property of base function: the multi-layer feed-forward network normally uses sigmoid function as its base unit and easily traps into local minimum, but Gaussian radial base function produces a partition of the input space and so can obtain higher speed of convergence in learning.

Structurally, GRBFNN can be considered as a three-layer feed-forward network, as shown in Fig.2. The networks consist of many inputs and a single output. The number of input unit $x_i$ is equal to dimension of the network input vector. Each linear input unit is given an activation value corresponding to the relevant value of the input vector. The linear output unit is fully connected to the hidden units. The network output $y$ is a weighted sum of the activation levels of the $m$ hidden units. The activation level of a hidden unit in a Gaussian network depends only on the distance between the input vector and the center of the Gaussian function of that unit. The center of the hidden unit is denoted as $C_j$ . More precisely, the activation function of the hidden unit $i$ is defined as:

$$h_j(k) = exp\left( -\sum_{i=1}^{N} \frac{(x_i(k) - c_j(k))^2}{2s_j^2(k)} \right),$$

$$(1 \leq j \leq M,\ s_j(k) > 0) \qquad (1)$$

Here the $s_j$ represent the band width of the hidden unit function (Light, 1992) .

### *2.3 Internal model control system*

The control system of adaptive internal model control based on GRBFNN is as shown in Fig.3. Here the operators GPFNM and GPFNC represent the model of the process and the internal model controller respectively. The error signal used to adjust the network weights and shape parameters is the difference between the process output and network internal model output. Thus the network is forced towards copying the process dynamics. For internal controller modeling the error signal used to adjust the network is defined as the difference between the network input and the process output. This tends to force the transfer function between these two signals to unity.

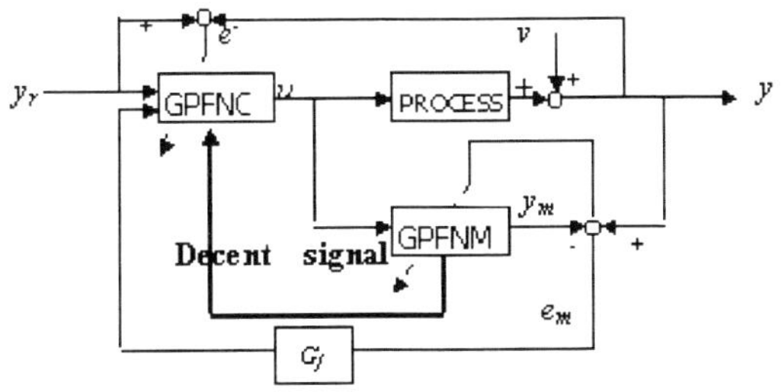


Fig.3 Internal model control system based on Gaussian potential network

In this work, a feedback filter is added to assure that the system keeps stable and possesses better dynamic response when the process model is mismatching. Its typical form is shown as follows:

$$G_f = \frac{1-\alpha_f}{1-\alpha_f z^{-1}} \qquad (0 \le \alpha_f < 1)$$

Simulation results show that filter constant $\alpha_f$ can be used as an adjustable parameter: when it is smaller, the process possesses faster response but robustness is worse; inversely, when it is bigger, the process possesses better robustness but response becomes slower. So how to select it must be based on real controlled process, compromising robustness with dynamics. Additionally, the feedback filter can be given high order structure to obtain much better control result.

## 3. THE CONSTRUCTION OF GPFNM AND GPFNC

### *3.1 GPFNM*

Consider the controlled process governed by the following nonlinear differential equation:

$$y(k) = f(y(k-1),\cdots,y(k-n),u(k-1),\cdots,u(k-m)) + v(k)$$

Here f() is nonlinear function, y(k), u(k) and v(k), respectively represent the output, the input and the noise of the process; the operators n and m is the delay order of the output and input respectively. The GPFN is used for the construction of process model and the learning laws used for training the network are considered as follows:

As shown in Fig.2, the signal of input is defined as

$$x_i^m = \begin{cases} y(k-i) & (1 \le i \le n) \\ u(k-i+n) & (n+1 \le i \le n+m) \end{cases} \tag{2}$$

The signal of hidden units is defined as

$$h_j^m(k) = exp\left(-\sum_{i=1}^{N} \frac{(x_i^m(k) - c_j^m(k))^2}{2s_j^{m^2}(k)}\right) \tag{3}$$

The signal of output units is defined as:

$$y_m(k) = \sum_{j=0}^{M} w_j^m(k) \cdot h_j^m(k) \tag{4}$$

Two-phase hybrid learning algorithm which utilizes nearest-neighborhood clustering scheme for structure learning and initial parameters setting and adaptive gradient descent method for better tuning the parameters of (s, c) are used in our work. The algorithm is stated as following:

Firstly, the nearest-neighborhood clustering scheme (Wang and Rong, 1999) is used to determine N which is the number of the hidden units and its initial parameters, namely (s,c). Here only the input training data u(k), k=1,2,…,P, where P is the number of training data pairs, are needed, the method can be summarized as:

1) Initialization: Define the cluster radius according to the distribution of input data. Choose the training data $\{x_1(1), x_2(1), \cdots, x_M(1)\}$ as the first cluster and set cluster center $c_{1i} = x_i(1)$, the number of input data of the first cluster $N_1 = 1$ and the number of clusters at this time, $N = 1$.

2) Consider the *k*th input data, $\{x_1(k), x_2(k), \cdots, x_M(k)\}$ and computer the distance of the *k*th input data point $x_i(k), (i = 1,2,\cdots,M)$ to cluster center $c_{ji}, (j = 1,2,\cdots,N, i = 1,2,\cdots,M)$ and find the smallest one. This is equivalent to finding a cluster (category) that $x_i(k), (i = 1,2,\cdots,M)$ could belong to. The chosen category is indexed by $J$, where

$$D_J = \|x(k) - c_J\| = \min_{i \le j \le N} \|x(k) - c_j\| \tag{5}$$

If $D_J > r$, the *k*th input data doesn't belong to all the existing cluster and new cluster need to be established with its center located at $x_i(k), (i = 1,2,\cdots,M)$, setting $N=N+1$, $c_{Ni} = x_i(k)$, $N_N = 1$ and keeping other information unmodified.

If $D_J \le r$, the *k*th input data belongs to the *J*th cluster and the *k*th input data should be used to modify the *J*th cluster, setting $N_J=N_J+1$, and keeping other information unmodified.

3) Let $k=k+1$ then go to 2) until all training data pairs are clustered to corresponding category. After finishing the cluster learning, we can get the number of hidden units N. The width of activation functions can be calculated as:

$$s_j = \frac{r}{\rho} \tag{6}$$

where $\rho$ is overlap parameter, usually $1 \le \rho \le 2$.

After finishing the first step, we can get the number of hidden units and the width of activation function which construct the initial neural network. Then adaptive gradient descent method is used for better parameters, which is described as follows:

In order to determine the weights, let

$$J_m = \frac{1}{2}[y(k) - y_m(k)]^2 \quad (7)$$

be the measure of error.

Implement a gradient descent to W in J, we should make the weight changes according to :

$$\frac{\partial J_m}{\partial w_i^m} = -[y(k) - y_m(k)] \cdot \frac{\partial y_m(k)}{\partial w_i^m} = e_m(k) \cdot h_i^m(k) \quad (8)$$

$$w_i^m(k) = w_i^m(k-1) + \eta_1^m \cdot \left(\frac{\partial J_m}{w_i^m}\right) + \alpha_1^m[w_i^m(k-1) - w_i^m(k-2)] \quad (9)$$

Hunt (1991) didn't clearly consider adaptive selecting (s,c). However (s,c) are very important parameters in GPFN and affect the convergence speed. According to this a method for adaptive selecting (s,c) based on gradient descent is presented in this paper. This tends to make the center locate in the proper area of input space and so higher convergence speed could be obtained. Implementing a gradient descent to c and s in E respectively

$$\begin{aligned}\frac{\partial J_m}{\partial c_i^m} &= -[y(k) - y_m(k)] \cdot \frac{\partial y_m(k)}{\partial h_i^m} \cdot \frac{\partial h_i^m}{\partial c_i^m} \\ &= -e_m(k) \cdot w_i^m(k) \cdot h_i^m(k) \cdot \frac{\Sigma(x^m(k) - c_i^m)}{s_i^{m2}}\end{aligned} \quad (10)$$

$$\begin{aligned}\frac{\partial J_m}{\partial s_i^m} &= -[y(k) - y_m(k)] \cdot \frac{\partial y_m(k)}{\partial h_i^m} \cdot \frac{\partial h_i^m}{\partial s_i^m} \\ &= -e_m(k) \cdot w_i^m(k) \cdot h_i^m(k) \cdot \frac{\Sigma(x^m(k) - c_i^m)^2}{s_i^{m3}}\end{aligned} \quad (11)$$

We could get the changes of s and c according to

$$c_i^m(k) = c_i^m(k-1) + \eta_2^m \cdot \left(\frac{\partial J_m}{c_i^m}\right) + \alpha_2^m[c_i^m(k-1) - c_i^m(k-2)] \quad (12)$$

$$s_i^m(k) = s_i^m(k-1) + \eta_3^m \cdot \left(\frac{\partial J_m}{s_i^m}\right) + \alpha_3^m[s_i^m(k-1) - s_i^m(k-2)] \quad (13)$$

Here learning factor $\eta^m_q$ is a very important parameter largely affecting the convergence speed and the trend of osciallion. Generally, it is selected in the range from 0 to 0.1, namely, $0<\eta^m_q<0.1$. Moreover, it should be selected much bigger in the precondition of guaranteeing stability because when it is bigger the change of weights and shape parameters is faster and so the speed of convergence is faster. In addition, varying learning factor can be introduced: $\eta^m_q$ is bigger in beginning and becomes more and more small along with the development of learning. Practically, $\eta^m_q$ can be selected according to the following adaptive method.

$$\begin{cases} \eta^m{}_I(k) = \lambda_1 \eta^m{}_I(k-1), & \dfrac{\partial J_m}{\partial w_i^m(k)} \cdot \dfrac{\partial J_m}{\partial w_i^m(k-1)} > 0 \\ \eta^m{}_I(k) = \lambda_2 \eta^m{}_I(k-1), & \dfrac{\partial J_m}{\partial w_i^m(k)} \cdot \dfrac{\partial J_m}{\partial w_i^m(k-1)} < 0 \end{cases}$$

The selection of $\eta^m_2$, $\eta^m_3$ are same as $\eta^m_q$. Generally, $\lambda_1$ is bigger than 1 and $\lambda_2$ is smaller than 1 (Shu, 1996).

### *3.2 GPFNC*

In internal model control system, the internal model controller is the inverse of the model of the process to be controlled. Consider the inverse process governed by:

$$u(k) = f^{-1}(y(k+1), y(k), \cdots, y(k-n), u(k-1), \cdots, u(k-m))$$

In control system the process model inverse rather than the inverse of the real process is used. The inverse of the function of f depends on the future process output value y(k+1). In order to obtain a reliable approximation, we replace this value with the set-point value. Finally, to approximate the inverse of the process we define the controller network output vectors as:

$$u(k) = f^{-1}(y_r(k+1), y_m(k), \cdots, y_m(k-n), u(k-1), \cdots, u(k-m))$$

In order to enhance robustness and resistance, A feedback value from feedback filter $e_m \cdot G_f = (y(k) - y_m(k)) \cdot (1-\alpha_f)/1-\alpha_f z^{-1}$ should be added into the network input. We define the network input vector as:

$$x_i^c(k) = \begin{cases} y_m(k-i) & (0 \le i \le n) \\ u(k-i+n) & (n+1 \le i \le n+m) \\ y_r(k+1) & (i = m+n+1) \\ \alpha_f \cdot x_{m+n+2}^c(k-1) + (1-\alpha_f)(y(k) - y_m(k)) & (i = m+n+2) \end{cases} \quad (14)$$

The signal of hidden units is defined as

$$h_j^c(k) = exp\left(-\sum_{i=1}^{N} \frac{(x_i^c(k) - c_j^c(k))^2}{2 s_j^{c2}(k)}\right) \quad (15)$$

The signal of output units is defined as:

$$u(k) = \sum_{j=0}^{M} w_j^c(k) \cdot h_j^c(k) \quad (16)$$

The number of hidden units and initial parameters (s,c) of GPFNC can be established as GPFNM's, here we don't discuss it repeatedly.

Let

$$J_c = \frac{1}{2}[y_r(k) - y(k)]^2 \quad (17)$$

be the measure of error.

Implement gradient descent respectively, we get:

$$\frac{\partial J_c}{\partial w_i^c} = -[y_r(k) - y(k)] \cdot \frac{\partial y(k)}{\partial u(k)} \cdot \frac{\partial u(k)}{\partial w_i^c} = e_m(k) \cdot h_i^c(k) \cdot \frac{\partial y(k)}{\partial u(k)} \quad (18)$$

$$w_i^c(k) = w_i^c(k-1) + \eta_i^c \cdot \left(\frac{\partial J_m}{w_i^c}\right) + \alpha_i^c[w_i^c(k-1) - w_i^c(k-2)] \quad (19)$$

$$\begin{aligned}\frac{\partial J_c}{\partial c_i^c} &= -[y_r(k) - y(k)] \cdot \frac{\partial y(k)}{\partial u(k)} \cdot \frac{\partial u(k)}{\partial h_i^c} \cdot \frac{\partial h_i^c}{\partial c_i^c} \\ &= -e_m(k) \cdot w_i^c(k) \cdot h_i^c(k) \cdot \frac{\Sigma(x^c(k) - c_i^c)}{s_i^{c2}} \cdot \frac{\partial y(k)}{\partial u(k)}\end{aligned} \quad (20)$$

$$c_i^c(k)=c_i^c(k-1)+\eta_2^c\cdot\left(\frac{\partial J_m}{c_i^c}\right)+\alpha_2^c\left[c_i^c(k-1)-c_i^c(k-2)\right] \quad (21)$$

$$\frac{\partial J_c}{\partial s_i^c}=-\left[y_r(k)-y(k)\right]\cdot\frac{\partial y(k)}{\partial u(k)}\cdot\frac{\partial u(k)}{\partial h_i^c}\cdot\frac{\partial h_i^c}{\partial s_i^c}$$
$$=-e_m(k)\cdot w_i^c(k)\cdot h_i^c(k)\cdot\frac{\Sigma\left(x^c(k)-c_i^c\right)^2}{{s_i^c}^3}\cdot\frac{\partial y(k)}{\partial u(k)} \quad (22)$$

$$s_i^c(k)=s_i^c(k-1)+\eta_3^c\cdot\left(\frac{\partial J_m}{s_i^c}\right)+\alpha_3^c\left[s_i^c(k-1)-s_i^c(k-2)\right] \quad (23)$$

Here, $\partial y(k)/\partial u(k)$ is required to be estimated. During learning the derivates can be calculated explicitly by the process model output $y_m$ substituting the process output y, namely $\partial y(k)/\partial u(k)=\partial y_m(k)/\partial u(k)$.

From(2), (3), (4)we obtain

$$\frac{\partial y_m}{\partial u}=\sum_{j=1}^{M}w_j(k)\cdot h_j(k)\cdot\frac{u(k)-c_j(k)}{s_j^2(k)} \quad (24)$$

## 4. SIMULATION RESULT

(1) The performance of the proposed internal model control system was evaluated through several simulations. Let us consider two processes:

Non-linear process:

$y(k)=(0.875(y(k-1)+u(k-1))/(1+y(k-1)y(k-1))$

Time-delay process:

$$G(s)=\frac{2e^{-4s}}{1+4s}$$

For nonlinear process the set-point signal is selected as sine-signal. GPFNM consists of two input units, four hidden units learned by nearest-neighborhood clustering algorithm and one output unit (for the sake of simplicity the construction is written as 2-4-1) and GPFNC consists of six input units, eight hidden units learned by nearest-neighborhood clustering algorithm and one output unit (6-8-1). The sine response of the proposed control system is shown in Fig.4. The identification of internal model is shown in Fig.5. From the above simulation results one may conclude that the algorithm of adaptive internal model control based on Gaussian radial basel network possess better control result and can be introduced to nonlinear systems.

For time delay process the set-point signal is selected as square-wave signal. The construction of GFNM and GPFNC is respectively chosen as 6-6-1 and 8-8-1. To evaluate the disturbance rejection performance of the proposed algorithm, white noise id injected into the process. The mean value of white noise is zero and its variance is

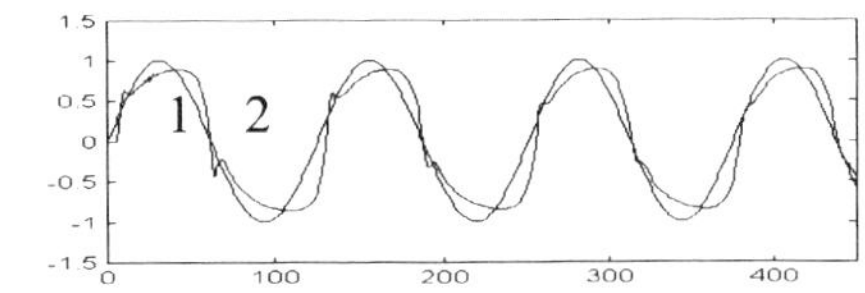


Fig.4 Sine response(curve 1-sine signal $y_r$, 2-the output of process controlled y )

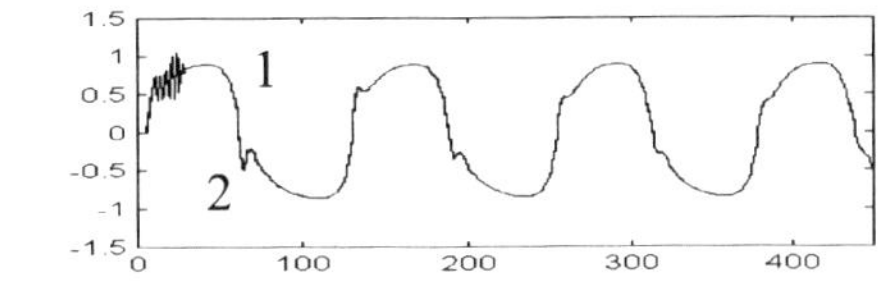


Fig.5 The identification of internal model control curve (curve 1-the output of controlled process y, curve 2-the output of internal model $y_m$)

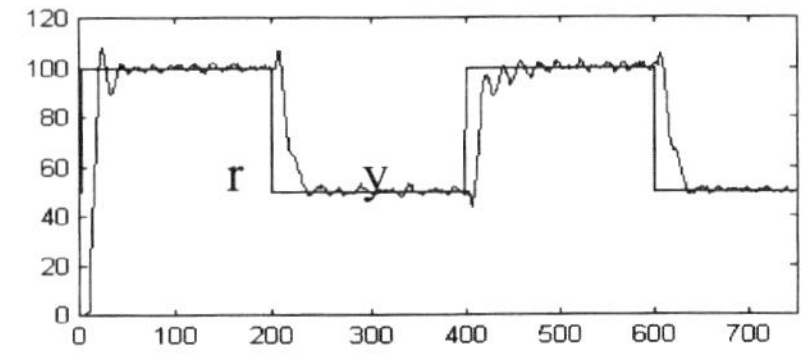


Fig.6 The square response of the algorithm proposed in this paper (curve r: set-point value; curve y: Square response)

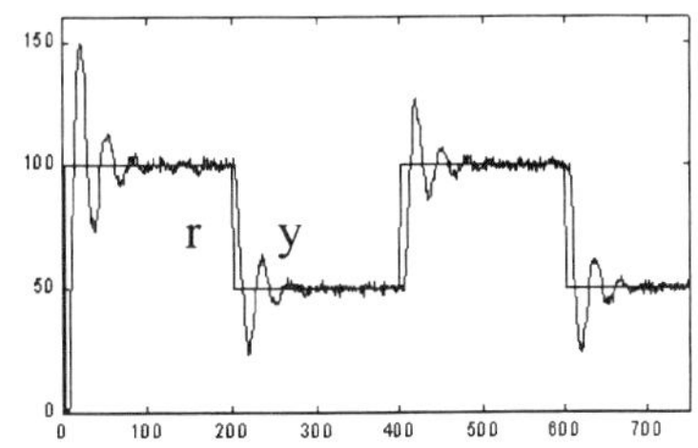


Fig.7 The square response of PID (curve r: set-point; curve y: Square response)

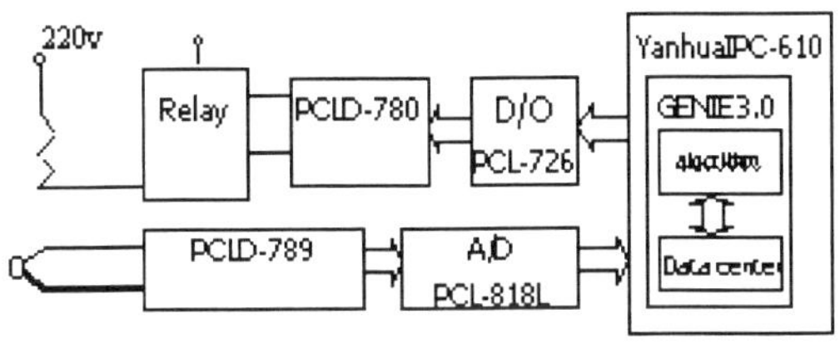


Fig.8 Temperature control system

0.01. The square-wave response of the proposed algorithm and common PID algorithm are shown in Fig.6 and Fig.7.

The overshoot algorithm is found to be too big and the control result is worse. These simulations demonstrated that the algorithm of adaptive internal model control based on Gaussian radial base possesses better response speediness and disturbance rejection than PID algorithm and can be used to time delay systems.

## 5. EXPERIMENT RESULTS

The proposed algorithm of adaptive internal model control based on Gaussian radial base network is implanted to control a laboratory electrical heating furnace. This system is designed in Fig.8. The temperature of laboratory electrical heating furnace is sampled by thermocouple and then magnified by PCLD-789D before being sent to analogue to analogue-to-digital (A/D) converter. At the same time the digital signal is sent to

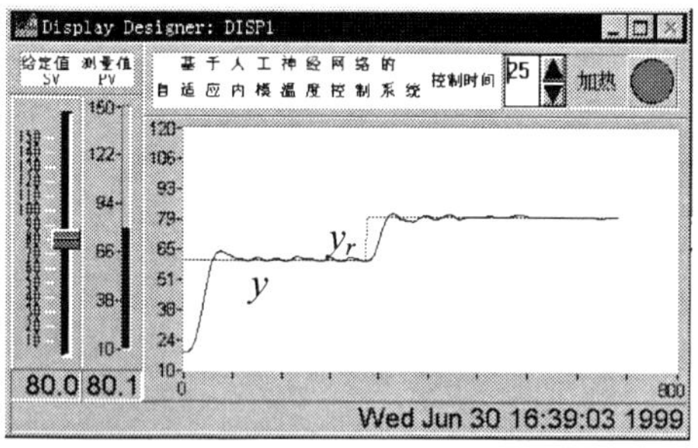


Fig.9 variable temperature control (curve $y_r$:set-point value; curve y; the output of the furnace)

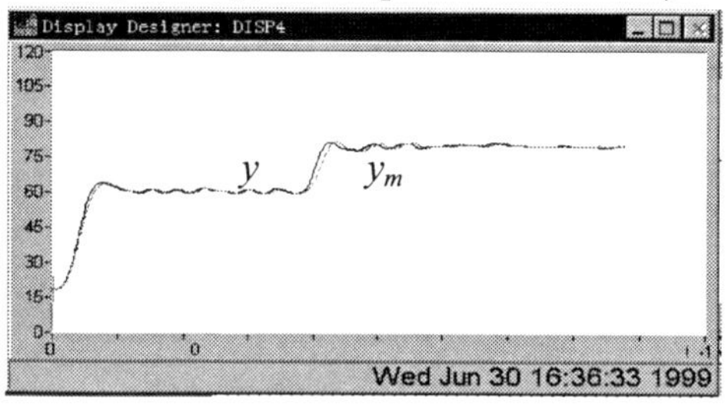


Fig.10 The identification of internal model (curve y; the output of the furnace, curve $y_m$: the output of the internal model)

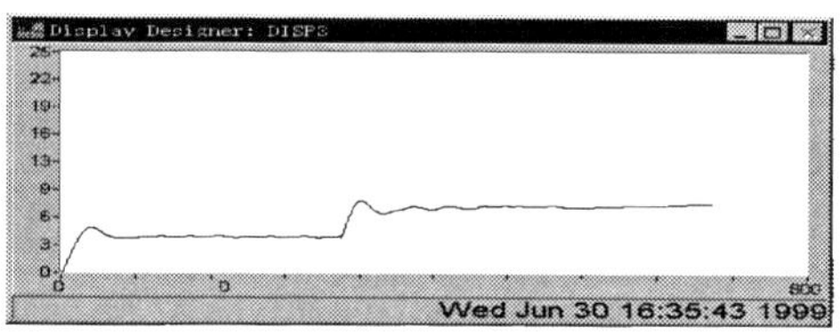


Fig.11 the variation of the control value

computer which completes the control algorithm and gets new control value. Then the new control value is used to control the on-off of the relay after being sent digital-to-analogue ( D/A) converter. The aim of the experiment is to test the performance and robustness of the proposed algorithm.In the real-time control experiment, by nearest neighborhood clustering algorithm the construction of GPFNM and GPFNC is respectively chosen as 8-8-1 and 11-12-1. The real control result are shown in Fig.9, Fig.10 and Fig.11.

## 6. CONCLUSION

The algorithm of adaptive internal model control based on Gaussian potential networks presented in this paper possessess better performances both in speediness and robustness. Moreover the performance of it can be adjusted through the feedback filter. The simulation and experiment results demonstrate that it can be introduced to time delay systems, non-linear systems and time varying systems.

## REFERENCE

Garda C.E. and Morarl, M. (1982). Internal model control. 1. A unifying review and some new results. *Ind. Eng. Chem. Pro. Des. Dev*, **21**, 308-323

Hunt K.J. (1991). Neural networks for nonlinear internal model control. *IEE. Pro-D*, **138**, 431-438

Liao M.W. and Ning X. (1998). Research on neural network internal model control. *Electrical transmission automation*, **20**, 24-28

Light W.A. (1992). *Some aspect of radial basis function approximation in approximation theory: spline function and applications.* NATO ASI Series,Boston, MA:Kluwer Academic Publishers.

Shu D.Q. (1996) *Presetting control systems and application.* Beijing,China: Machinery Industry Press

Wang Y. and Rong G. (1999). A self-organizing neural network based fuzzy system. *Fuzzy set and systems*, **103**, 1-11.

IFAC Publications
www.elsevier.com/locate/ifac

# A fuzzy controller to overcome EA accommodation

**Q. M. Zhu, X. W. Sun, and A. G. Pipe**

*Faculty of Computing, Engineering and Mathematical Sciences,*
*University of the West of England,*
*Frenchay Campus, Bristol BS16 1QY, UK*
E-mail: quan.zhu@uwe.ac.uk

Abstract: Electroacupuncture (EA) apparatus can output a sequence of pulsed electrical current to stimulate the nerves and muscles of patients. During an EA therapy, nerve systems can progressively develop adaptation to the electric stimulus and lead to a gradual fade away of the EA sensation, which is called Electroacupuncture Accommodation (EAA). The phenomenon occurs quite often in EA. Since the accommodation reduces the therapeutic effect dramatically, an extra dosage of electricity is normally supplemented to recover the therapeutic progress. In this study a fuzzy controller implemented in a PC-based EA instrument has been designed to generate the optimum electric output to overcome EAA. *Copyright © 2001 IFAC*

Keywords: Electroacupuncture and accommodation, fuzzy control.

## 1. INTRODUCTION

Electroacupuncture (EA) apparatus is used to output the required electric stimulation to patients via needles. A commonly used stimulating waveform is the biphasic pulse, shown in Figure 1. According to the patients' disorders the electric parameters such as amplitude, pulse width and frequency are selectable with the apparatus. The electrical pulses acting on nerves and muscles can elicit excitatory transmission and muscular contraction, thereby improve nutrition condition of nerves and muscles, and accelerate recovery of the nervous and muscular functions (Low and Reed, 1998). However, after a certain time (varies between patients and therapeutic conditions) with a fixed electric stimulation pattern, nerve systems can adapt themselves to the conditions resulting in a gradual fading away of the effect (Han, *et al.*, 1981). This effect is termed electroacupuncture accommodation, and needs to be overcome to retain the healing effect.

In EA, according to the nature of the disorder, pathography, and the patient's tolerance, stimulation levels are decided between strong, medium and weak. Stronger stimulus is applicable for paralysis and chronic disorders, medium stimulus is used for the most general cases; whilst weak stimulus is for spastic paraplegia and the acupoints around the eye (Chi, 1991). Accommodation can occur in every situation where EA therapy is used. To overcome stimulation adaptations, therapists often increase the electric dosage to different degrees. Basically the deeper the accommodation symptom, the more

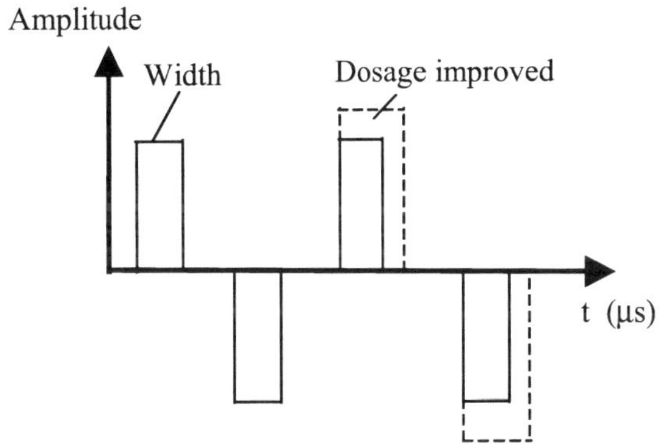


Fig. 1 Stimulating waveform used in EA

dosage should be applied. The product of the pulse amplitude and width of the stimulating pulse can reflect the quantity of the applied dosage (Trubatch and Harreveld, 1972). Therapists can increase the values of the two parameters to strengthen the stimulation as shown in Figure 1. However the two parameters play different roles in stimulating nerves and muscles; the former provides a more direct and critical action than the latter. Therefore the pulse amplitude is used for a strong adjustment and the pulse width for a fine adjustment. An optimum combination of the two parameters is important in overcoming accommodation.

Medicine has already proved to be an appropriate area for applications of fuzzy logic principles (Hudson and Cohen, 1994; Passino and Yurkovich, 1998; Zimmermann, 1996). Expert knowledge and patient feedback information in medicine are sometimes vague and many observations are very subjective. In EA therapy, imprecision, subjectivity and uncertainty are present in the whole process. The sensations of stimulus in EA can only be described verbally and depend very much on the subjective estimation of patients. Further the relationship between disorders and electric applications is generally far from crisp and unique. Fuzzy logic is a very pertinent means to provide reasoning methods for approximate inference in overcoming electroacupuncture accommodation. Fuzzy logic allows to encapsulate the knowledge and skills of the best human expert, yielding as close to the optimal automated system that is consistently and relatively available for use by others. The approach presented in this study incorporates expert knowledge of overcoming EAA into an EAAO system. The linguistic description of the accommodation from a treated patient together with the current stimulating level are fed to the EAAO system, processed through the fuzzy controller, and then the EAAO system outputs an optimum combination of the pulse amplitude and the pulse width.

The rest of the presentation is organised as follows. In section 2 electrical stimulation and the accommodation problem are described. In section 3 the design of the EAAO system based on fuzzy logic control is described and an case illustration is presented. The final section draws conclusions.

## 2. EFFECTS OF NERVES AND MUSCLES UNDER EA STIMULATION

In EA treatment, pulsed current is applied to the patient's body between the two acupuncture needles. The two needles perform as two electrodes, one is the anode, and the other is the cathode. Electricity applied in EA therapy leads to nerves and muscles actions. The applied current with sufficient range in magnitude variation can stimulate a motor nerve and so produce contraction of the ignited muscle (Forster and Palastanga, 1985). When a nerve fibre is in its resting state there is a potential difference of some 70 millivolts between the interior and exterior of the fibre. The inside of the fibre is negative with respect to the outside. If an EA current of sufficient amplitude is passed through tissue containing a fibre, the potential difference set-up across the fibre may be sufficient to cause depolarisation of the fibre membrane, stimulating the nerve (Low and Reed, 1998). To trigger the nerve impulse, a certain minimal amount of electrical charge is needed. This idea is illustrated with the strength-width curves shown in Figure 2 (Forster and Palastanga, 1985; Ward, 1986). The minimum current needed to fire a nerve impulse is small with longer stimuli but becomes greater with shorter pulse lengths. This means that long pulse width needs small current and shorter pulse width needs larger currents to provoke a nerve impulse. The current can convey nerve impulses to stimulate the muscle fibres. If the stimulating sequence is repeated, it will ignite a series of rhythmical muscle contractions and relaxations. The muscle contractions can be observed in the jumping of the needles, especially the cathode needle, and simultaneously the patient can feel the tingling sensation or even pain. Since stimulation is limited by a patient's tolerance, the electric dosage of every stimulating pulse must be kept below the tolerance of patients.

Normally precautions are taken in EA therapy to prevent accident from high amplitude electric stimulation. It is well understood that higher electric stimulation will affect more nerves leading to more intense tingling and will eventually spread to more motor nerves causing increased muscle contraction. In EA current density is in the range of 5mA to 60mA and pulse width from 0.01 to 0.4ms. Although both parameters have a strengthening effect in current stimulation, the pulse amplitude acts more directly than the width. A very small increase in stimulation amplitude produces relatively large increases in the driving force to muscle contraction. In the other words, increasing applied current causes more motor units to be stimulated strongly, thus giving rise to more widespread muscle contractions. Further

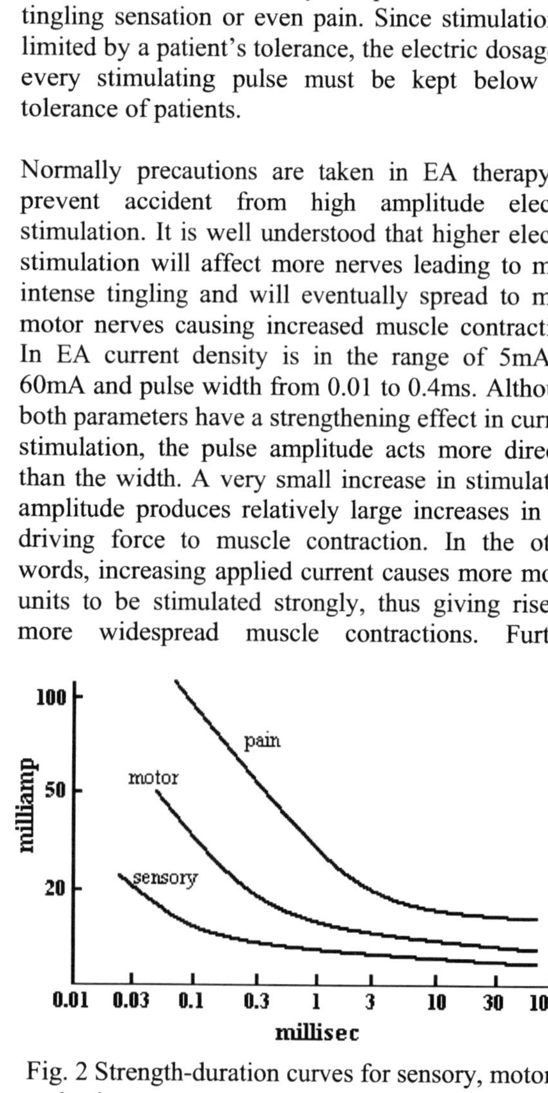


Fig. 2 Strength-duration curves for sensory, motor and pain responses

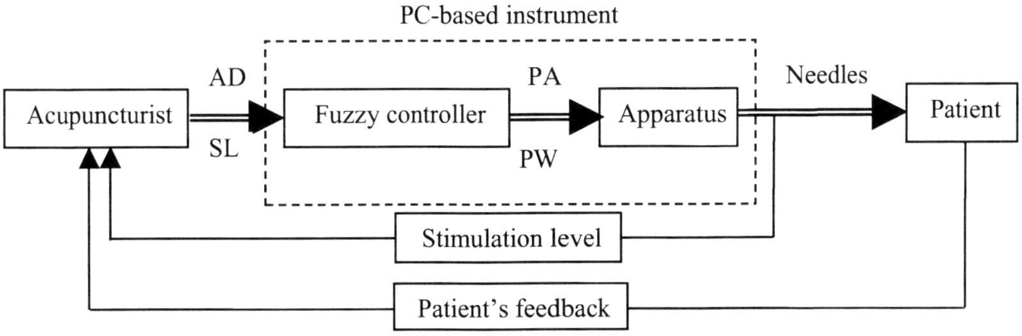


Fig. 3 Fuzzy based EA operation loop

increasing current above a certain level will cause unbearable pain, therefore caution must be exercised (Snyder-Mackler and Robinson, 1989).

The EAA is characterised by the gradual attenuation of the stimulating sensation of patients. The nerve fibre is able to adjust itself so that the threshold potential may rise in response to the applied electric charge. This may lead that the stimulating pulses do not have enough energy to provoke a nerve impulse any more, but may do so at a higher electric intensity. This illustrates the need to match the electric parameters with the patient's therapy situation.

The traditional EAAO is operated according to the acupuncturist's experience. When a patient states the accommodation and its degree, the acupuncturist uses the applied Stimulating Level (SL) and the Accommodation Degree (AD) as major reference information to determine additional electric dosage to overcome it. Afterwards, a new electrical stimulating pulse is set up, manually by the acupuncturist, with specifications on Pulse Amplitude (PA) and Pulse Width (PW) and then applied to the patient.

## 3. A FUZZY APPROACH TO OVERCOME EAA

EA therapy operation is based on a combination of the patient's description of sensation and the therapist's subjective judgement. Accordingly a fuzzy approach is suitable to enhance the consistencies and reliability of the measure to overcome EAA. A corresponding scheme of fuzzy based EA operation loop is proposed in Figure 3. The stimulating level (SL) and the accommodation degree (AD) are two major factors to determine the electric dosage applied to overcome the EAA and are chosen as the inputs of the fuzzy controller as shown in Figure 3. The pulse amplitude (PA) and the pulse width (PW) in Figure 3, which represent the supplementary electric dosage, are the outputs of the controller. The SL is based on qualitative assessment according to the stimulating level applied to the patient; strong stimulus will has a higher value than medium or weak stimulus. The AD value depends on the accommodation degree. Patients are asked to describe the severity of their accommodation feelings and, then the acupuncturist rates the AD level according to patients' descriptions. The final compensate output by the apparatus via acupuncture

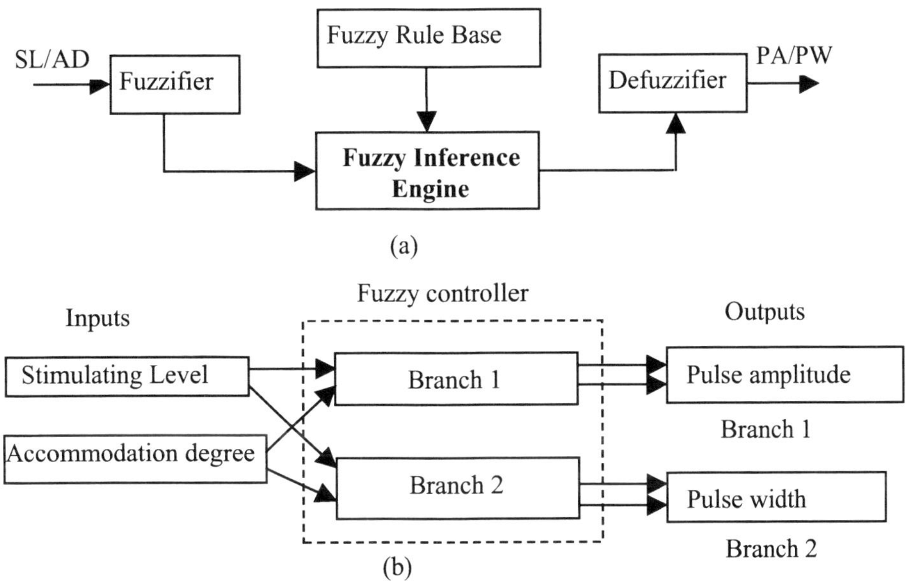


Fig. 4 (a) Structure of the fuzzy controller (b) Basic configuration of the two branches

needles represent the optimum increased dosage of the stimulating electricity.

*3.1 Fuzzy control system design*

Fuzzy sets and fuzzy logic applied to control or decision-making problems form a field of knowledge called fuzzy logic control or fuzzy decision making. It is very effective when high precision is not required (Bojadziev and Bojadziev, 1995). Those linguistic rules from expert's experience, intuitive and general manner can be directly translated to a rule base and incorporated into a fuzzy logic decision-making inference engine. Thus the final crisp decisions can be designed to be optimal.

The fuzzy controller block in Figure 3 is expanded in Figure 4. As illustrated fuzzification is the first step for translating real variables, measured or estimated, into linguistic variables. At the second step a fuzzy inference engine is started that derives a conclusion from a pre-designed rule base representing the decision strategy. The fuzzy inference engine consists of two components, which are aggregation and composition. The most commonly used method max-min has been selected as the inference operation rule. At the defuzzification stage the linguistic output variables generated from the fuzzy inference engine are translated into crisp quantities as final decision values. In regard to defuzzification methods, the Centre-Of-Gravity (COG) is one of the most commonly used methods, which usually leads to continuous output with a small change to the inputs, therefore it is selected in this study.

The relationship between the input pair Simulation Level (SL) and Accommodation Degree (AD), and the output pair Pulse Amplitude (PA) and Pulse Width (PW) can be well described by a set of fuzzy membership functions. The SL input is based on qualitative assessment according to the stimulating level applied to the patient; strong stimulus will have a higher value than medium or weak stimulus. The AD input value depends on the accommodation degree. Knowledge is accumulated to establish the rules for designing the fuzzy inference engine. The rules, derived through expert knowledge and constructed to represent a human expert "in-the-loop", are based on following considerations

(1) Light accommodation needs small amount of supplementary electric dosage.
(2) Serious accommodation needs big amount of supplementary electric dosage.
(3) Strong stimulating treatment tends to use pulse amplitude to overcome the relating accommodation.
(4) Weak stimulating treatment tends to use the pulse width to perform a soft adjustment, and.
(5) Medium stimulating level needs both parameters to reach an optimum combination.

The input and output fuzzy regions are divided up by the following fuzzy sets

Stimulating Level: SL={0, 5, 10}={S, M, B},
Accommodation Degree: AD={0, 2.5, 5, 7.5, 10}={VS, S, M, B, VB},
Pulse Amplitude: PA= {0, 0.5, 1, 1.5, 2}={VS, S, M, B, VB},
Pulse Width: PW={0, 12.5, 25, 37.5, 50}={VS, S, M, B, VB}.

VS, S, M, B, VB are fuzzy set names and defined as VS = very small, S = small, M = medium, B = big, VB = very big. The corresponding membership functions are defined as $\mu_{VS}$, $\mu_S$, $\mu_M$, $\mu_B$ and $\mu_{VB}$

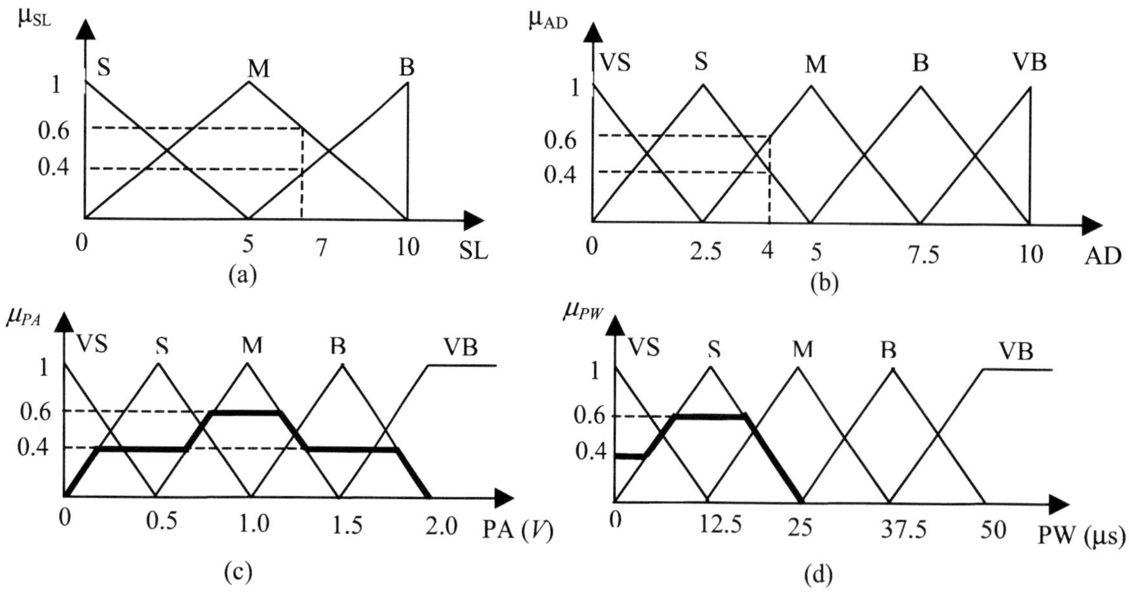


Fig. 5 Linguistic variables and terms of inputs and outputs: (a) Stimulating level (b) Accommodation degree (c) Pulse amplitude (d) Pulse width

respectively and the numerical values in the brackets are the centres of the functions.

The triangular fuzzy membership function, which is widely used in fuzzy environments for the sake of computational efficiency and ease of data acquisition, is selected in the rule base organisation. Figure 5 shows the divisions of the input and output spaces into fuzzy regions and the membership functions. The linguistic variables shown in Figure 5 have been quantified by the membership functions. The input universes of SL and AD are the same, i.e. [0,10] and the output universes of PA and PW are [0, 2] and [0, 50] respectively.

*3.2 A simulated example*

Consider an EA operation having been run under a therapy condition with a series of pulses of 10.16 voltage and 200μs (equivalently SL = 7), the patient is experiencing accommodation after a while, 5 minutes say. With the judgement of the acupuncturist, the accommodation degree is classified as AD = 4. The acupuncturist therefore inputs SL = 7 and AD = 4 to the PC based EAAO system. The following computations are automatically manipulated within the EAAO system.

There are four active roles for branch 1, which are

1 IF the *Stimulating level* is big AND the *Accommodation degree* is small THEN the *Pulse amplitude* is medium;
2 IF the *Stimulating level* is big AND the *Accommodation degree* is medium THEN the *Pulse amplitude* is big;
3 IF the *Stimulating level* is medium AND the *Accommodation degree* is small THEN the *Pulse amplitude* is small;
4 IF the *Stimulating level* is medium AND the *Accommodation degree* is medium THEN the *Pulse amplitude* is medium.

Similarly four rules that link SL and AD to PW are

1 IF the *Stimulating level* is big AND the *Accommodation degree* is small THEN the *Pulse width* is very small;
2 IF the *Stimulating level* is big AND the *Accommodation degree* is medium THEN the *Pulse width* is very small;
3 IF the *Stimulating level* is medium AND the *Accommodation degree* is small THEN the *Pulse width* is small;
4 IF the *Stimulating level* is medium AND the *Accommodation degree* is medium THEN the *Pulse width* is small.

The values of the corresponding membership functions are

$$\mu_{SL}(B) = 0.4 \qquad \mu_{SL}(M) = 0.6$$
$$\mu_{AD}(M) = 0.6 \qquad \mu_{AD}(S) = 0.4 \tag{1}$$

The dotted lines in Figure 5 show the mappings from SL and AD descriptions to the values of their membership functions.

The aggregation of the IF part is

$$\begin{aligned} \alpha_1 &= \mu_{SL}(B) \wedge \mu_{AD}(S) = \min(0.4, 0.4) = 0.4 \\ \alpha_2 &= \mu_{SL}(B) \wedge \mu_{AD}(M) = \min(0.4, 0.6) = 0.4 \\ \alpha_3 &= \mu_{SL}(M) \wedge \mu_{AD}(S) = \min(0.6, 0.4) = 0.4 \\ \alpha_4 &= \mu_{SL}(M) \wedge \mu_{AD}(M) = \min(0.6, 0.6) = 0.6 \end{aligned} \tag{2}$$

The compositions of the aggregated IF part and THEN part in PA is

$$\begin{aligned} \mu_{PA}^{comp}(1) &= \max[\min(\alpha_1, \mu_{PA}(\mathrm{M})), \min(\alpha_4, \mu_{PA}(\mathrm{M}))] \\ &= \max[\min(\alpha_4, \mu_{PA}(\mathrm{M}))] \\ \mu_{PA}^{comp}(2) &= \max[\min(\alpha_2, \mu_{PA}(\mathrm{B}))] = \min(\alpha_2, \mu_{PA}(\mathrm{B})) \\ \mu_{PA}^{comp}(3) &= \max[\min(\alpha_3, \mu_{PA}(\mathrm{S}))] = \min(\alpha_3, \mu_{PA}(\mathrm{S})) \end{aligned} \tag{3}$$

The compositions of the aggregated IF part and THEN part in PW is

$$\begin{aligned} \mu_{PW}^{comp}(1) &= \max[\min(\alpha_1, \mu_{PW}(\mathrm{VS})), \min(\alpha_2, \mu_{PA}(\mathrm{VS}))] \\ &= \max[\min(\alpha_1, \mu_{PA}(\mathrm{M}))] \\ \mu_{PW}^{comp}(2) &= \max[\min(\alpha_3, \mu_{PW}(\mathrm{S})), \min(\alpha_4, \mu_{PA}(\mathrm{S}))] \\ &= \max[\min(\alpha_4, \mu_{PA}(\mathrm{M}))] \end{aligned} \tag{4}$$

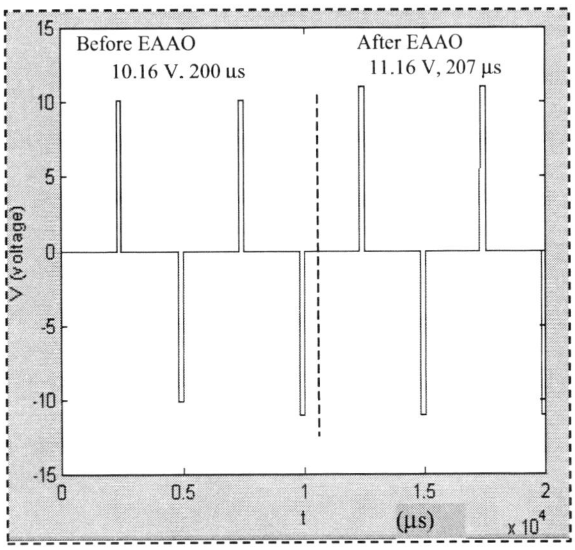


Fig. 6 Simulation of the EAAO system

The obtained composite functions are indicated with bold lines in Figure 5.

By using COG in defuzzification, the crisp parameters for the pulse amplitude and width are PA = 1.0 V and PW = 7.09 μs respectively Therefore the new stimulating pulse will be increased with amplitude of 1.0 V and width of 7.09 from last stimulating pulse. The above analysis has been simulated through a Matlab program to produce a time response sequence shown in Figure 6. The new output is a series of stimulating pulses with amplitude 10.16+1 = 11.16 V and width 200+7.09 = 208 μs and indicates that after the update, the stimulating pulse amplitude and width have increased by 9.84% and 5.5% respectively.

With the EAAO system specifications and formulations, the completed relationships of the dual inputs to each output are shown graphically in Figure 7.

## 4 CONCLUSIONS

In this study, a fuzzy EAAO system has been presented, which could act as an in-the-line expert who can judge, and chose the desired electric application to overcome the accommodation. The medical knowledge in the system is stored in the form of fuzzy logical relationship between symptom and the electric parameter applications. The system can improve the accuracy of the electricity application and the efficiency of overcoming the EAA, thus improve the healing. However, at the moment, the feedback of the accommodation degree is only presented by patient's description. There are still some difficulties in sensing patient's accommodation degree for a full automatic control.

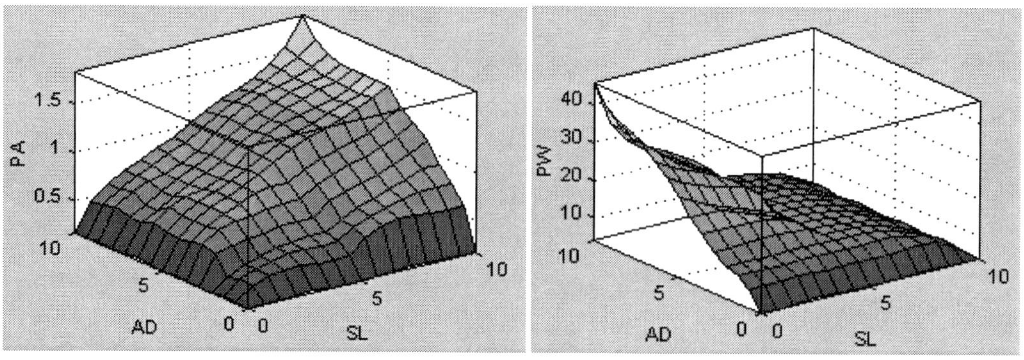


(a) (b)

Fig. 7 I/O characteristic field of the dual inputs and outputs (a) pulse amplitude (b) pulse width

## REFERENCES


Bojadziev, G. and Bojadziev, M. (1995). *Fuzzy sets, fuzzy logic, applications*. World Scientific Publishing Co Pte Ltd, USA.

Chi, X. (1991). *Manual for electroacupuncture therapy*. Science and Technology Press, China.

Forster, A. and Palastanga, N. (1985). *Clayton's electrotherapy theory and practice*, Ninth edition. Bailliere Tindall, UK.

Han, J. S., Li, S., and Tang, J. (1981). Tolerance to electroacupuncture and its cross tolerance to morphine. *Neuropharmacology*, 20, 593-596.

Hudson, D. and Cohen, M. (1994). Fuzzy Logic in medical expert systems. *IEEE Engineering in Medicine and Biology*, 13 (5), 693-698.

Low, J and Reed, A. (1998). *Electrotherapy explained: principles and practice*. Butterworth-Heinemann, UK.

Passino, K. and Yurkovich, S. (1998). *Fuzzy control*. Addison Wesley Longman Inc, USA.

Snyder-Mackler, L. and Robinson, A. (1989). *Clinical electrophysiology*. Williams and Wilkins, USA.

Trubatch, J. and Van Harreveld, A. (1972). Spread of iontophoretically injected ions in a tissue. *Journal of theoretical biology*, 36, 355-366.

Ward, A. (1986). *Electricity fields and waves in therapy*. Science Press, Australia.

Zimmermann, H. (1996). *Fuzzy set theory and its applications*, third edition. Kluwer Academic publishers, USA.

IFAC Publications
www.elsevier.com/locate/ifac

# NOVEL DESIGN METHOD FOR THE BALL MILL PULVERIZING SYSTEM BASED ON FUZZY REASONING AND AUTO-TUNING PID CONTROL

**[*]Xiao-Feng Li  Ju-Hua Zhang  Deng-Ying Zhu  Cui Zhang**

[*]Instrument and Power Plant Control Department. Guangdong Power Test and Research Institute. 73 Meihua Road, Guangzhou 510600. China.
Email: leexfeng@cmmail. com

Abstract: The pulverizing process of a ball mill is rather complex. It is difficult to obtain the optimal control by the conventional control methods, so a new fuzzy self-optimizing control method is presented. The method divides the controller into two parts: the fuzzy adaptive PID control and the supervisory fuzzy control. The self-optimizing inference sub-system of supervisory fuzzy control is proposed for solving imprecise constrains optimization problems, and fuzzy self-optimizing interval and fuzzy self-optimizing step are applied to speed the rate of self-optimizing and improve the accuracy of self-optimizing. The fuzzy adaptive PID controller combines fuzzy self-adjusting PID control with fuzzy rough-tuning mechanism. The operation result shows that it is quite suitable for the control of mill and a large amount of electric energy is saved. *Copyright © 2001 IFAC*

Keywords: Fuzzy Self-Optimizing, Fuzzy Rough Tuning, Mill Level Control, Specifications on Phase and Gain Margin (SPGM), Ball Mill Pulverizes System.

## 1. INTRODUCTION

In most processes there are some extra degrees of freedom that can be used for optimization purposes. The development of optimization techniques requires a reasonably accurate description of the process. The optimal operation point can be difficult to maintain if measure imprecision and model uncertainty are present. Self-optimizing control is an approach to solve this problem by turning the optimization into a set point problem. The key idea is to find a measurable variable with assured value at optimal operation. If this variable can be found, a feedback control loop will be closed to keep the variable at the set point, and to keep indirectly the process an optimal operation.

[1]This work has been supported by the Guangdong Power Group Co., contract JA6098009.

Unfortunately, some variables are difficult to quantitatively measure for many industrial processes including a ball mill for pulverizing system. The ball mill is of strong nonlinearity, uncertainty, time-variance and serious interconnection (i. e. some of the variables may not be linearly independent) etc. Therefore, it is impossible to establish a precise model for the ball mill. The interactions between operating conditions and the quantity of coal in the mill are complicated and have not been quantitatively studied in a systematic way. Thus in such a multi-variable system, the controlled variable (quantity of coal) is difficult to measure directly, but it can be acquired through the imprecision relationships between quantity of coal and other easily measured parameters. However, years of industrial practice and research have led to the knowledge for this relationship. This knowledge is represented in the qualitative descriptions and the semi-empirical correlation. In these circumstances, the conventional techniques do not contribute as much as expected to an acceptable (economic) performance of the optimizing control for the ball mill process systems.

Fuzzy relational model puts the processing of thinking, analyzing and inference into a linguistic analyzing mathematical module, which enables us to convert the natural language into algorithms for computer to accept and operate.

In this paper, we present an industrial application case, in which a fuzzy relational modeling and self-optimizing approach are applied to a ball mill with available knowledge sources, to optimize the process operation.

## 2. MEASUREMENT OF MILL LOADING AND OPERATION CHARACTERISTICS OF THE BALL MILL

The ball mill, as compared to other types of pulverizing facilities, has the advantages of high reliability, long service life, and low maintenance and is widely used in China. Its shortcoming is of high-energy consumption for pulverization. The consumption is high at no-load running. With the increase of pulverization output increase, the consumption increases a little. So the unit consumption reduces as pulverization output increases, and it is desirable to keep the mill working at high output level with high economical advantage.

Until recently it was the general understanding that reliable measurement of the mill loading was impossible. Conventional methods such as pressure difference between inlet and outlet, power of driving motor, temperature at outlet, etc. all involve significant disadvantages concerning practical handling and information value.

In fig. 1, the relation between the power input to the driving motor(1), vibration magnitude signal of the front end of the axial(2), difference in pressure between that at the inlet and that at outlet(3) and the pulverization output(4) with respect to the quantity of coal in the mill are shown, where the horizontal axis is the quantity of the mill. It can be seen that, in operation, the raw coal feeder continuously sends coal into the mill. The quantity of coal in the mill increases and the driving motor power also increases. When pulverization output (rate of pulverized coal output) continuously increases, the maximum of the pulverization output is on the right of the motor power.

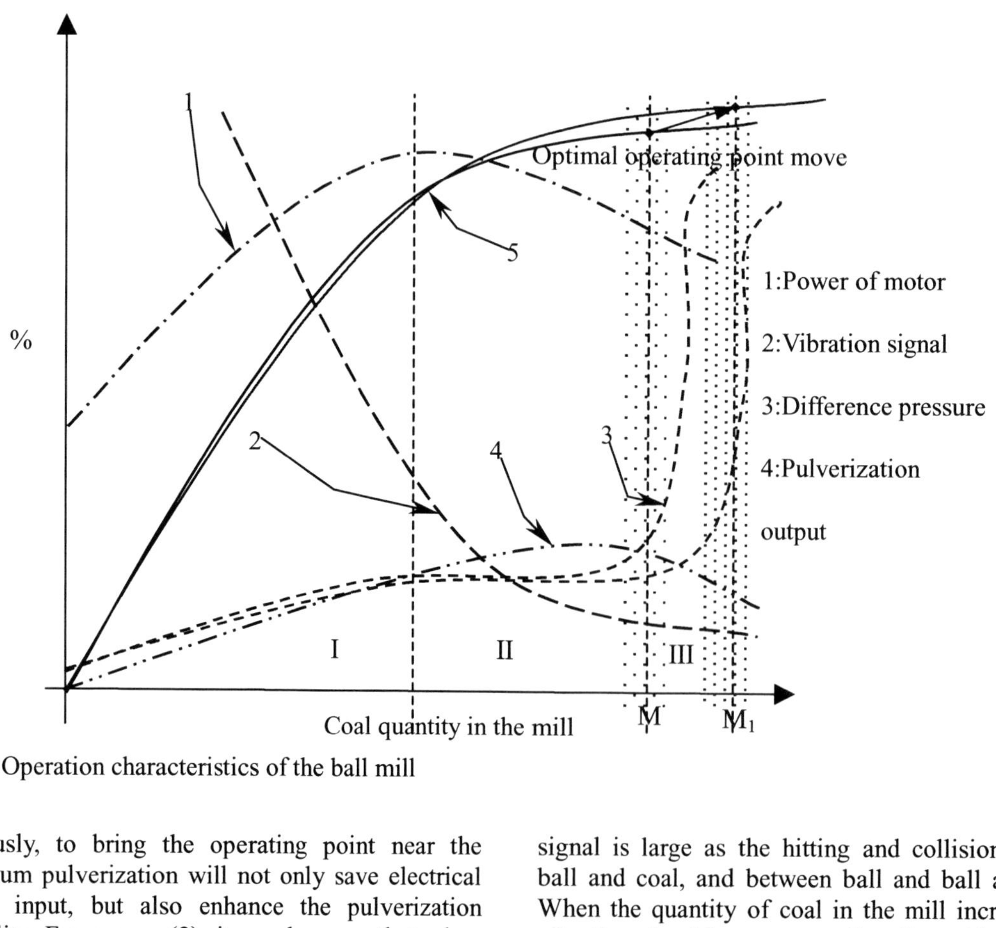

Fig. 1. Operation characteristics of the ball mill

Obviously, to bring the operating point near the maximum pulverization will not only save electrical energy input, but also enhance the pulverization capability. From curve(2), it can be seen that when the quantity of coal in the mill is small, vibration signal is large as the hitting and collision between ball and coal, and between ball and ball are heavy. When the quantity of coal in the mill increases, the vibration signal becomes smaller. Curve(4) show that when coal quantity increases, the pulverization output increases , and when the pulverization output passes the maximum, the pulverization output drop and the quantity of in-feed raw coal exceeds the pulverization capability, the quantity of coal in the mill increases rapidly, and the mill will be choked. Curve(3) show that before the pulverization output reaches the maximum, the pressure difference changes slowly and steadily. When the pulverization output passes the maximum, and the in-feed raw coal further increases, the pressure difference increase slowly at first and then rapidly till the mill is blocked. From the field tests it has been found that the equipment based on piezoelectric vibration pickup was preferable for measuring ball mill material level, which only give the imprecision correlation.

In fig. 1, the operation can be divided into three regions: I, II and III. In region I, the unit energy consumption of the mill is high with low pulverization. In region III, the mill is vulnerable to block. In region II, the operation is most desirable. Thus, to achieve optimal results requires an operating point close to the region III. If blockage occurs the process must be stopped and the mill manually cleaned, thus operators run the process far away from this optimal point. This results in high operational energy costs.

We nominally operate at the optimum but the optimal operating point has moved due to some unknown variations of the coal quality. We want to compute the optimal move in the available manipulative variable in order to follow the real optimum. With model uncertainty and imprecisely measurable variables it may be difficult to tell which direction the free variable should be move in order to bring the process closer to the real optimum. In this case, a fuzzy self-optimizing control replaced the fixed setpoints control [Li 2001].

## 3. FUZZY SELF-OPTIMIZING ON THE MILL

The objective here is to maximize the quantity of coal in the mill while maintaining the manipulated operation variables within a desired range. The optimizing control problem can be formulated as:

$$\max_{ML} E(ML) \tag{1}$$

Subject to the inequality constraints:

$$X_{PL}<X_{P}<X_{PH},$$
$$X_{TL}<X_{T}<X_{TH},$$
$$X_{\Delta P}\leq X_{\Delta PH},$$

The optimization objectives are not always clearly defined in the pulverizing process of the ball mill because the quantity of coal in the mill is difficult to measure directly. From the section 2, we know that the mapping exists between the quantity of coal in the mill and the measurable variables. They only give the imprecision correlation. But We can fuse them together and make a credible quantity of coal.

In the new fuzzy self-optimizing approach, a self-optimizing step is represented as the variant step length and self-optimizing interval is represented as the variant advance interval. It is expected that the specific self-optimizing step length (OS) and self-optimizing interval (OT) depend on the concentrations of ΔP, ml-OP, but the exact forms of the dependence are unknown:

$$OS=F_S(\Delta P, \text{ml-OP}) \tag{2}$$
$$OT=F_T(\Delta P, \text{ml-OP}) \tag{3}$$

Fuzzy sets and membership functions of the process variables are used to characterize degree to approach the steady operating constraint value. For example, equation 4 is a membership function representing a fuzzy constraint "the pressure difference ΔP less than or approximately equal to the operating constraint value of pressure difference $\Delta P_L$", with a tolerance τ:

$$\mu_{\Delta P}(SP)=1/\{1+\exp[(\Delta P-\Delta P_L)/\tau]\} \tag{4}$$

Table 1 Rule set for adjusting the self-optimizing step length (OS). Z stands for zero

| OS | | $\Delta P<\Delta P_{limit}$ | | |
|---|---|---|---|---|
| | | S | M | B |
| ml-OP | SS | S | M | B |
| | S | SS | S | M |
| | M | Z | S | S |
| | B | Z | SS | S |
| | BB | Z | Z | SS |

Table 2 Rule set for adjusting the self-optimizing interval (OT). I stand for infinite

| OT | | $\Delta P<\Delta P_{limit}$ | | |
|---|---|---|---|---|
| | | S | M | B |
| ml-OP | SS | B | S | S |
| | S | B | M | S |
| | M | I | M | M |
| | B | I | B | M |
| | BB | I | I | B |

Thus, the fuzzy linguistic models are used to approximate the unknown functions $F_S$ and $F_T$. The model consists of if-then rules. In the fuzzy set, two kinds of rules are used: "optimizing step kind" and "optimizing interval kind". The optimizing step kind rule is with the following form:

If (ml-OP is $X_i$ and $\Delta P-\Delta P_L$ is $Y_i$) Then (increment of the optimizing step is $\Delta SP_i$).

The optimizing interval kind rule is with the following form:

If (ml-OP is $X_j$ and $\Delta P$-$\Delta P_L$ is $Y_j$) Then (optimizing interval is $T_j$).

Thus a set of rules, as show in table 1 and table 2, may be used to adjust the step length (OS) and the interval (OT), respectively. The rule base of the model contains rules for all possible combinations of the antecedent linguistic terms (SS, S, M, B, etc.), and the membership functions have been extracted from the process measurements by fuzzy set.

In this approach, the step length and the interval of self-optimizing are variants based on the distance between the variable of the operating point and the steady operating constraint value, which are represented as descending and ascending functions, respectively. The setpoints at each step are the variables to be adjusted by the fuzzy self-optimizing system and become the new values.

When the variable of the operating point moves from the inside to the boundary of a constraining region, the step length function decreased gradually close to zero and the interval function increased gradually close to infinite, the fuzzy self-optimizing procedure forces the system to slow down the ascending rate of the optimizing set-point. This means that it changes from the fully allowable to completely forbidden state. When the variable of the operating point close to constraint value, the step length closed to zero and the interval closed to infinite, the self-optimizing schedule halted.

The constraint value of the variable is with a little margin away from the steady operating boundary. This gives the system some degree of tolerance to cope with stochastic fluctuation of process variables.

## 4. FUZZY ROUGH-TUNING MECHANISM

Fuzzy self-adjusting PID controller schemes have enjoyed considerable success both in academia and industry since their development [He 1993, Li 2000]. They have been widely applied to systems subject to stochastic disturbance. For the global optimizing control, it turned out that it is very hard to achieve acceptable control performance by the fuzzy self-adjusting PID systems. They cannot effectively deal with the strong nonlinearities of global optimizing control. Due to the strong nonlinearities of the optimizing control, it is well motivated to study control schemes in which optimal linear controllers are applied in combination with various control scheduling methods. Thus, a controller scheduling procedure based on the interpolation of the PID control law rather than the PID parameter of the control law is more successful for the optimizing control. This procedure bears a close resemblance to an approach in the modeling context, in which a nonlinear model is formed by interpolation between local linear models.

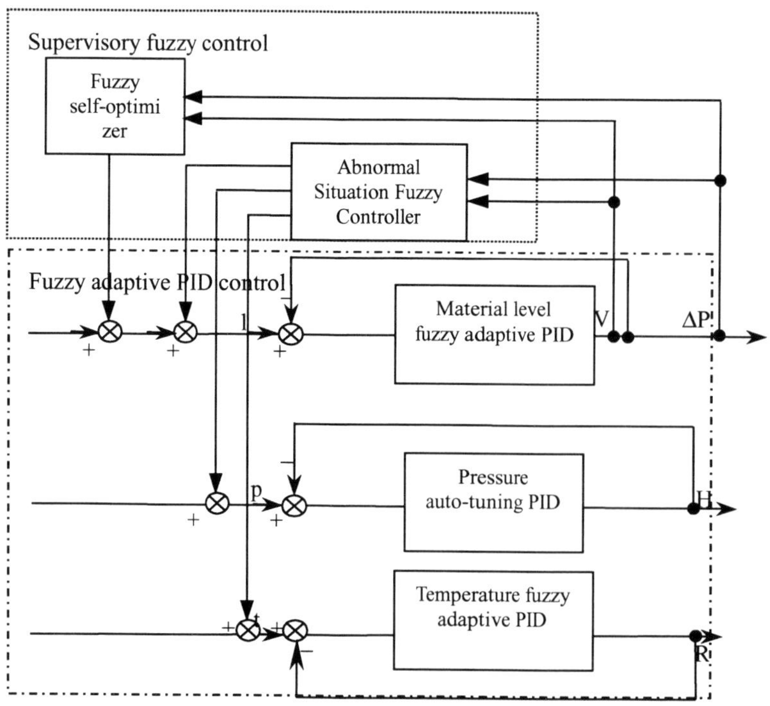


Fig. 2. Control scheme for discussion of propose algorithm

In this section, a new Fuzzy self-adaptive PID controller based on fuzzy rough-tuning approach will be presented that can effectively deal with changes in operating condition occurrence. To cope with the nonlinearities, a set of PID control laws can be designed by using the SPGM tuning formulas for various stationary points of the self-optimizing course and the global control is implemented by smoothly aggregating the laws with their relevant contribution with respect to the various operating conditions. The aggregation of control structure yields a nonlinear control system for the optimizing pulverizes operation. The essence of fuzzy rough-tuning mechanism is that at every time instance, the mechanism evaluates the trend of the operating conditions to detect the possible deviation from a prescribed operating condition. The aggregation of control structure yields a nonlinear control system for the optimizing pulverizes operation. If a deviation is found, the appropriate control law according to the local operating condition of the deviation will be generated instantaneously to adapt it.

It is well known that the optimizing set-point of the material level (ML-SP) has mostly influence on quantity of coal in the mill. The relationship between optimizing set-point and control law will change when the operating condition changes, and may result in a new set of PID parameters. Therefore, we are able to build the fuzzy rule in the following form.

IF set-point is $SO\text{-}SP_n$ THEN XP is $XP_n$ and $T_I$ is $T_{in}$ and $T_D$ is $T_{dn}$.

Where $XP_n$ , $T_{in}$ and $T_{dn}$ are the linguistic values of PID parameters; $SO\text{-}SP_n$ is the linguistic value of set-point.

Table 3 Fuzzy rough-tuning rules

| ML-SP | Material Level Loop | | |
|---|---|---|---|
| | XP | $T_I$ | $T_D$ |
| **S** | B | S | S |
| **B** | S | B | B |

The particular examination that was used in Table 3 is given in Table 4. The universes of discourse of $SO\text{-}SP_n$ is {82%, 97%}; the universes of discourse of $XP_n$ is {86.014, 89.139}, the universes of discourse of $T_{Dn}$ is {50.37, 142.31}, and the universes of discourse of $T_{In}$ is {115.75, 297.86}.

## 5. APPLICATION ON BALL MILL PULVERIZER SYSTEM

Fig. 3 is the real time curves in manual operation. It can be seen that the curves jump up and down demonstrating the improper action of the manual control with unsteady ball mill operation.

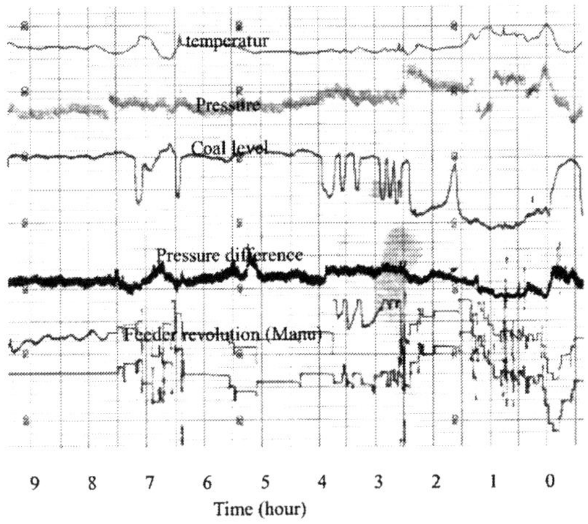


Fig. 3. The real-time curve of the parameters during operation.

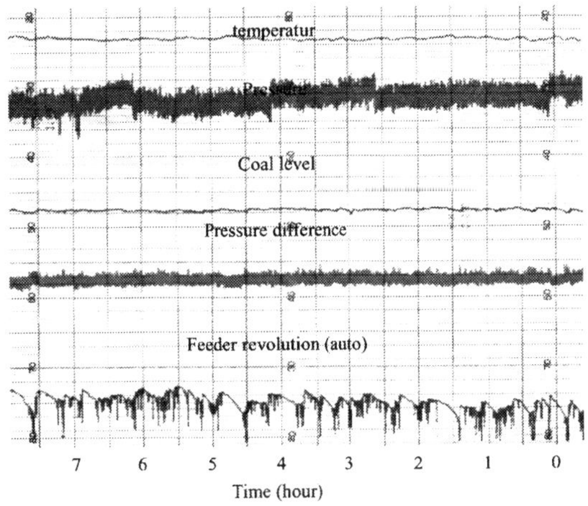


Fig. 4. The real-time curve of the parameters during SPGM+SFT.

Table 4 Tuning value of material level control loop under difference coal quantity

| | $A_m$=2.5;$\Phi_m$=30° | | | $A_m$=2.7;$\Phi_m$=45° | | | $A_m$=3; $\Phi_m$=60° | | |
|---|---|---|---|---|---|---|---|---|---|
| SO-SP | XP | $T_I$ | $T_D$ | XP | $T_I$ | $T_D$ | XP | $T_I$ | $T_D$ |
| **97%** | 76.833 | 234.57 | 202.57 | 89.139 | 297.86 | 142.31 | 94.899 | 359.03 | 94.757 |
| **82%** | 71.403 | 91.152 | 69.361 | 86.014 | 115.75 | 50.37 | 93.558 | 139.52 | 33.644 |

Results of the SFT PID controllers are shown in Fig. 4. The SFT PID controllers can keep very small deviations of Temperature, Pressure, Coal level and Pressure difference respectively.

On Aug. 18, 2000, the optimization results were tested in the mill operation under FSO+FST+FRT control, which is shown in Fig.5. Fig. 6 shows the variation of coal level under FSO+FST control.

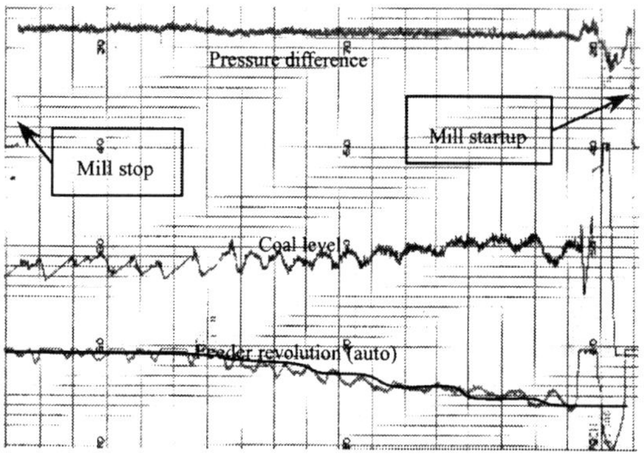


Fig.5. The real-time curve of the parameters during FSO+FST+FRT.

Fig. 6. shows the variation of coal level under FSO+FST control. It can be clearly seen that the stability of the coal level is widely improved by the FSO+FST+FRT control. In the same way, the coal level deviation from the setpoints of fuzzy self-optimizing is tightly regulated and optimizing time is shortened.

The FSO+FST+FRT control system proves itself to be capable of promoting solvent quality, saving energy and providing more reliable function. The results were encouraged, increasing in production rate, saving in energy within a reasonable operating condition. In the trial production period from 2000 Jan to 2000 Nov, the fuzzy self-optimizing control system managed to get unit energy consumption for pulverizing reduced by 14.87%. Compared to the manual operation, the FSO+FST+FRT control system saves about RMB 560, 000 in energy and produce costs per year in the No. 1 boiler of Meixin Power Plant.

## 6. CONCLUSIONS

The ball mill control system has been running in the Meixin Power Plant since January 4 in 2000. The running results show that it has possessed characteristics of reasonable design, operating convenience, quick response and good stability.

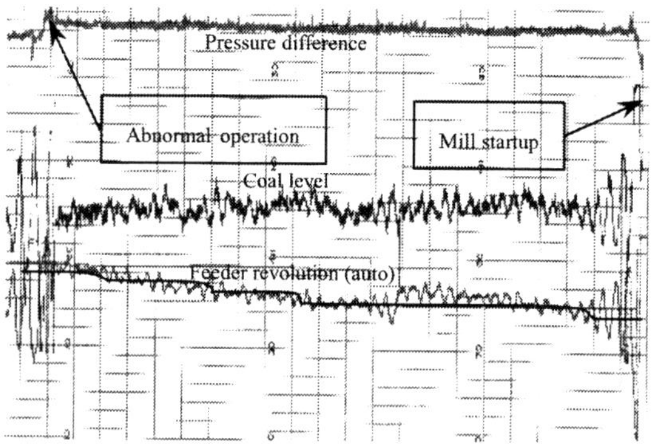


Fig.6. The real-time curve of the parameters during FSO+FST.

An optimization strategy for the operation of a complex pulverizing system is realized by Fuzzy self-optimizing control and fuzzy adaptive PID control system. This new fuzzy control system has been successfully commissioned, and remarkable improvement of the system behavior is reached. It is an ideal control system for the ball mill pulverizing system.

## REFERENCES


He S. Z., S. H Tan, C. C Hang. and P. Z.Wang (1993). Control of Dynamical Processes Using A on-line Rule-Adaptive Fuzzy Control System, *Fuzzy Sets and System,* **54,** 11-22.

Li X. F. (2000). Application of The Fuzzy Adaptive PID to A Power Plant. *IFAC Symposium on Power Plant and Power Systems Control 2000,* Brussels, Belgium. April 26-29.

Li X. F. (2000) The Realization of Fuzzy Self-Tuning PID Control Based on DCS, *Industrial Instrumentation & Automation (Chinese)* **155,** 31-33.

Li X. F. (2001). Fuzzy Self-Optimizing Control System and Its Application. *Control and Instruments in Chemical Industrial (Chinese),* **179,** 27-30.

Narendra K., S. (1995) Adoption and Learning Using Multiple Models, Switching and Tuning. *IEEE trans. on control system,* 37-51.

Skogestad S., etc. (1999) Plantwide Control: The Search for The Self-Optimizing Control Structure. *14th IFAC World Congress,* Beijing, July 1999, 325-330.

IFAC Publications
www.elsevier.com/locate/ifac

# A SELF-TUNING PID CONTROLLER BASED ON EXPERT SYSTEM

**Xing Jianchun, Wang Ping, Wang Lin**

Research Institute of Control and Measurement of Nanjing Engineering Institute
210007, Nanjing , Jiangsu P.R.China , xjc@ieee.org

**Abstract:** This paper introduces a PID controller with self-tuning function. It may provide optimal PID control parameters. The self-tuning algorithm based on the expert system and pattern recognition has quick convergence speed through the application of the new characteristic variables and the distinctive pre-tuning formulae. The uniformed self-tuning formula is also presented, which makes it easy to implement and maintain the expert system based on the single chip micro-processor. *Copyright © 2001 IFAC*

**Keywords:** PID control, error criteria, control algorithms, expert system

## 1. INTRODUCTION

With the development of electrical technology, computer technology and control theory, the conventional controller has been replaced successfully with the digital controller in the recent years. But the Proportional Integral Derivative(PID) algorithm is still widely used in the process industries(Zhuang and Atherton, 1993). This is due to several reasons: many process engineers are much more familiar with PID controller than with sophisticated modern control concepts like state-space control; they have much experience in application and tuning of the conventional analogue PID controllers; the application of complex control algorithms up to now requires special training of the user; with the great majority of processes, PID control shows good performance and is known to be nearly as robust as state-space control regarding process parameter changes. However, the tuning of digital PID controller is still a difficult and time-consuming chore and the presented self-tuning methods can't ensure that the parameters obtained from self-tuning are optimal. For these reasons, the application of the digital PID controller in the process industries has been greatly limited. The implementation of self-tuning of the digital PID controller is a emergency project.

In the last few years, some methods of tuning the parameters of PID controller have been developed and several commercial self-tuning controllers are available. However there are some practical problems in these self-tuning algorithms, for example, the pole-placement algorithm requires the accurate plant model that it is always difficult to be obtained; the algorithm of numerical optimisation or identification method is very complex and the disturbance must be introduced in the tuning procedure.

In this paper, we will introduce a self-tuning method base on expert system. In this method, a pre-tuning formula is given to improve the convergence speed of the expert self-tuning algorithm. The rule collection and the uniform self-tuning formula of the expert self-tuning are introduced. The self-tuning procedure is close-loop, that is to say, the tuning procedure will cycle until the optimal parameters are obtained. In some present self-tuning algorithm, the self-tuning procedure is open-loop, the control effect of the self-tuning parameters is judged by the control engineer but not by the controller and the refining of the PID parameter is also executed by the control engineer.

## 2. CONTROLLER STRUCTURE

The structure of the self-tuning controller is shown in the Fig. 1. The expert self-tuning controller consists of the following parts: pattern recognition, dynamic information database, static information database, inference mechanism, PID control algorithms, etc. The pattern recognition function recognizes the model of the error signal between the set value and the process value, the result is written into the dynamic database. The static information database stores the expert rules. The dynamic information database store the error model, the self-tuning procedure information, tuning target, etc. The inference mechanism calculate the tuning value(△Kp、△Ti、△Td) of the PID control parameters

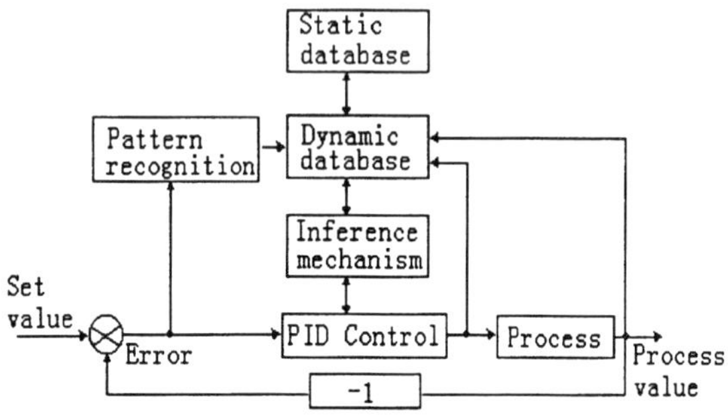


Fig. 1 The structure frame of the self-tuning controller

according to the information in the dynamic and static information database.

## 3. INTEGRAL PERFORMANCE CRITERIA AND CONTROL TARGET

The conventional performance index, such as rise time, overshoot, damping, etc., always only reflects the performance of the transient procedure in some aspect, also these performance indexes conflict each other. However, the error integral performance index is a comprehensive index, it may reflect the general performance of the transient procedure. But there are many error integrating styles, minimising the different error integral criteria, you may obtain different PID control parameters and different transient procedures.

The common description of the error integral performance criteria is:

$$J_n(\theta) = \int_0^{\infty} \{t^n e(\theta,t)\}^2 dt \qquad (1)$$

here, θ is the parameters collection which influence the error integral J, e( θ ,t) is the error signal between the set value and the process value, it is the function of θ and t, n=0,1,2,3... .

The choice of n=1 in the above integral criteria, which is known as the ISTE criterion, gives satisfactory result.

Tab.1 Relation between the error integral criteria and the characters of the transient procedure.

| Control target | Error integral criteria | The characters of the Transient procedure |
|---|---|---|
| 0 | $J_3(\theta) = \int_0^{\infty} t^6 e^2(\theta,t)dt$ | Almost no overshoot, The rising speed is very slow. |
| 1 | $J_2(\theta) = \int_0^{\infty} t^4 e^2(\theta,t)dt$ | The rising speed is slow, The overshoot is About 5% |
| 2 | $J_1(\theta) = \int_0^{\infty} t^2 e^2(\theta,t)dt$ | The rising speed is fast, The overshoot is Moderate, about 10% |
| 3 | $J_0(\theta) = \int_0^{\infty} e^2(\theta,t)dt$ | The rising speed is Very fast, the overshoot Is serious, about 20% |

A large amount of simulation experiments has been made, for the first-order plus dead time (FOPDT) plant model which is widely used to represent the practical plant procedure, using the Equ.1, in which the n is chosen as 0,1,2,3.

The transfer function of a FOPDT model is

$$G_p(s) = \frac{k \cdot e^{-\tau s}}{1 + T \cdot s} \qquad (2)$$

The simulation experiment result denotes that: if the error integral criteria is determined, for various T, k, $\tau$ in the Equ.(2), using the control parameters obtained by minimising the error integral criteria, the transient procedure is approximate. That is to say, the performance of the transient procedure is only related with the error integral criteria when the optimal PID control parameters are used. Therefor, we can obtain the required transient procedure by choosing the proper error integral criteria. The Tab.1 shows the relation between the error integral criteria and the performance of the transient procedure.

## 4. PRE-TUNING FORMULAE

The converging speed of the self-tuning algorithm is an important index to evaluate a self-tuning PID controller. Because the iteration number of the self-tuning is bigger, the disturbance introduced by the self-tuning algorithm is serious. The presented self-tuning algorithm always provided two ways to

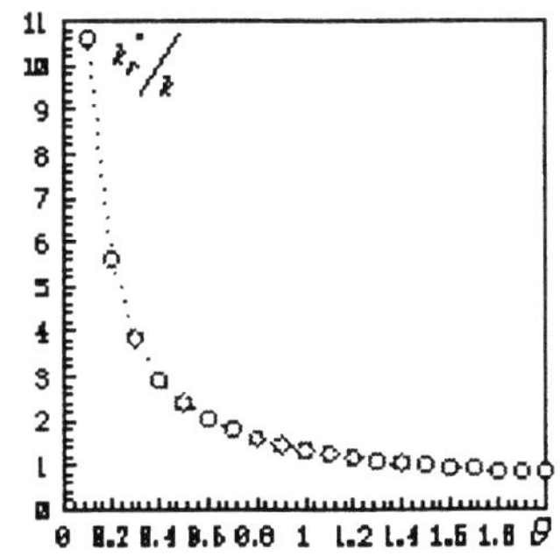

Fig. 2.a

Relation between $k_p^*/k$ and $\theta$

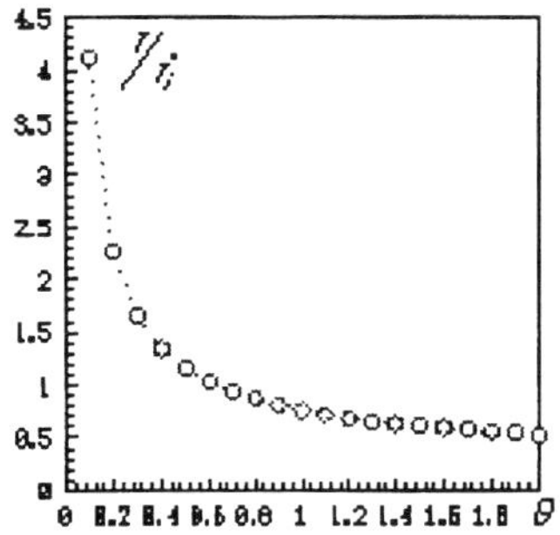

Fig. 2b

Relation between $T/T_i^*$ and $\theta$

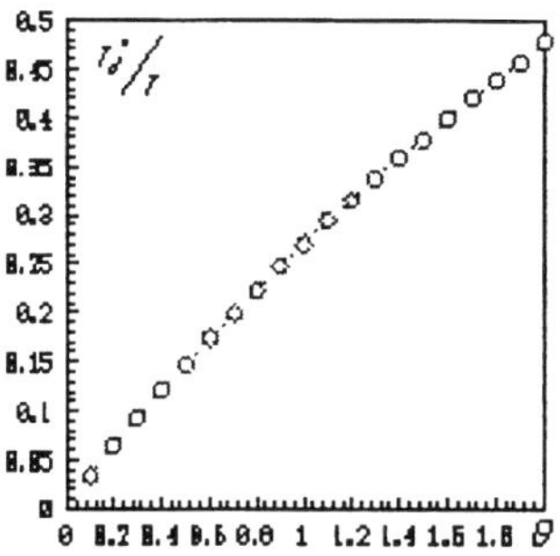

Fig. 2c

Relation between $T_d^*/T$ and $\theta$

Tab.2 The coefficients of the pre-tuning formulae

| Control target | a2 | a1 | a0 | b2 | b1 | b0 | c2 | c1 | c0 |
|---|---|---|---|---|---|---|---|---|---|
| 0 | -.210 | 1.036 | .030 | .027 | .413 | 1.329 | -.031 | .246 | .014 |
| 1 | -.239 | 1.254 | .028 | .042 | .243 | 1.191 | -.032 | .274 | .002 |
| 2 | -.272 | 1.439 | .019 | .042 | .193 | 1.154 | -.022 | .236 | .003 |
| 3 | -.293 | 1.558 | .022 | .043 | .168 | 1.126 | -.016 | .202 | .005 |

initialise the PID parameters: entering the initial parameters by the user or obtaining the initial parameters according to the presented simply self-tuning formulae. For this reason, the initial parameters may be far away form the convergence parameters, and the number of self-tuning is so bigger that it can not meet the need of the practical procedure. In our research, we have derived a group of formulae by parameter optimisation to provide the initial parameters. These formulae may provide the better initialising parameters.

In the simulation experiments, it is found that: there are some explicit relation between the optimal PID parameters $k_p^*$、$t_i^*$、$t_d^*$ obtained by minimising the error integral criteria and the FOPDT plant model parameter K ,T and $\tau$ .These relations are showed in figures 2a,2b,2c.

According to these curves, a group of pre-tuning formulae are derived as Equ.(3). In the formulae $\theta = \tau / T$ and the coefficients are shown in the Tab.2.

$$\begin{cases} K_p = K/(a_2\theta^2 + a_1\theta + a_0) \\ T_i = T \cdot (b_2\theta^2 + b_1\theta + b_0) \\ T_d = T \cdot (c_2\theta^2 + c_1\theta + c_0) \end{cases} \quad (3)$$

To verify the pre-tuning formula, a simulation experiment is made.

The plant model:

$$G_p(s) = 1/(1 + 48 \cdot s) \cdot e^{-20s}$$

Experiment condition :

control period : SP=0.5s
simulation step : SS=0.1
simulation time : St=200s
control target : OS=3

The results show that the optimal PID parameters obtained by minimising the criteria are

$$K_p^* = 2.314 \quad T_i^* = 72.615 \quad T_d^* = 5.479$$

The initial parameters according to the pre-tuning formulae are

$$K_p = 2.172 \quad T_i = 69.843 \quad T_d = 5.394$$

That is to say, the initial parameters $K_p$ , $T_i$ , $T_d$ obtained by the pre-tuning formula are near to the optimal PID parameters $K_p^*$ , $T_i^*$ , $T_d^*$ .The self-tuning algorithm can converge rapidly if these initial parameters are sued.

## 5. CHARACTERISTIC VARIABLES OF THE TRANSIENT PROCEDURE

The present expert system based PID self-tuning algorithms always select the overshoot, damping, oscillation period as the characteristic variables. In our research, we presented a group of characteristic variables to speed up the convergence of the self-tuning algorithm. The definition of these four characteristic variables is as Fig3.

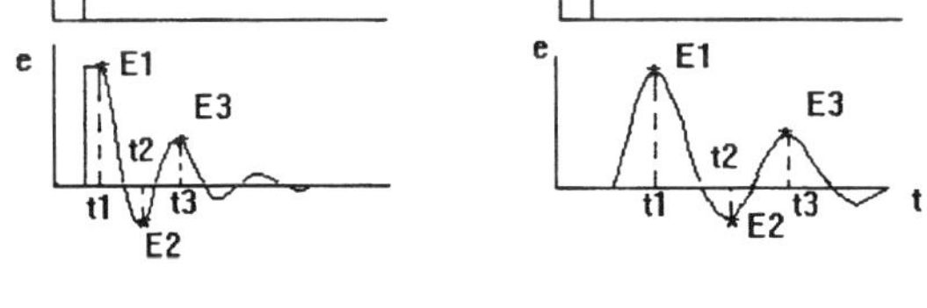

Overshoot: $OV = -\frac{E_2}{E_1}$

Damping: $DA = \frac{E_3 - E_2}{E_1 - E_2}$

Peak Rate : $PR = -\frac{E_3}{E_2}$

Time Rate: $TR = \frac{t_3 - t_2}{t_2 - t_1}$

## 6.THE UNIFORMED SELF-TUNING FORMULAE AND PRODUCTIVE RULES

If the characteristic variables of the transient procedure are directly used to verify the PID control parameters, it is difficult to express the difference between the practical response and the desired response when the different control target is utilised. Further, to speed up the convergence of self-tuning, when this error is big, the tuning should be big; and to ensure the stability of the procedure, when this error is small, the tuning should be small. So, the following definitions are introduced.

set $SROV = \left| \frac{ov - ov^*}{ov^*} \right|$

here, the OV is the practical overshoot, $OV^*$ is the desired overshoot.

if $SROV \le 0.1 * \gamma$ then
$Dov = 0$
if $0.1 * \gamma < SROV \le 0.2 * \gamma$ then
$Dov = 1 * sign(ov - ov^*)$
if $0.2 * \gamma < SROV \le 0.3 * \gamma$ then
$Dov = 2 * sign(ov - ov^*)$
if $SROV \ge 0.3 * \gamma$ then
$Dov = 3 * sign(ov - ov^*)$

here, $\gamma$ is the coefficient related with control target. For the control target 0、1、2、3, the $\gamma$ orderly is 2.0、2.0、1.2、1. $sign(\cdot)$ is the symbol function.

$$sign\ (x) = \begin{cases} -1 & x < 0 \\ 0 & x = 0 \\ 1 & x > 0 \end{cases}$$

Similarly, the damp has the following definition:

set $SRDA = \left| \frac{da - da^*}{da^*} \right|$

here, da is the practical damp, $da^*$ is the desired damp.

if $SRDA \le 0.1 * \gamma$ then
$Dda = 0$
if $0.1 * \gamma < SRDA \le 0.2 * \gamma$ then
$Dda = 1 * sign(da - da^*)$
if $0.2 * \gamma < SRDA \le 0.3 * \gamma$ then
$Dda = 2 * sign(da - da^*)$
if $SRDA \ge 0.3 * \gamma$ then
$Dda = 3 * sign(da - da^*)$

According to the above definitions, the uniformed formula is driven as following

$$\begin{cases} \Delta k_p = \alpha_{kp} \cdot \gamma \cdot (|Dov| + |Dda|) \cdot kp(n-1) \\ \Delta t_i = \alpha_{ti} \cdot \gamma \cdot (|Dov| + |Dda|) \cdot ti(n-1) \\ \Delta t_d = \alpha_{td} \cdot \gamma \cdot (|Dov| + |Dda|) \cdot td(n-1) \end{cases} \quad (4)$$

here, $\alpha_{kp}$, $\alpha_{ti}$, $\alpha_{td}$ are the coefficients determined by the expert rules according to the number and the direction of the tuning. While $kp(n-1)$, $ti(n-1)$, $td(n-1)$ are the $kp$, $ti$, $td$ after the previous tuning.

Then, after tuning, control parameters should be:

$$\begin{cases} k_p(n) = k_p(n-1) + \Delta k_p \\ t_i(n) = t_i(n-1) + \Delta t_i \\ t_d(n) = t_d(n-1) + \Delta t_d \end{cases} \quad (5)$$

For different control target and different expert rules, the application of the uniformed formula of self-tuning algorithm will make it easy to implementing and maintaining the expert rules.

The self-tuning controller has adapted the productive rules like "if...then..." to express the expert's experience and knowledge. Several rules have been given as the example of the productive rules.

Rules 1: if(DOV==0 and PR<0.2) then
{ $\alpha_{kp}$=0; $\alpha_{ti}$=0; $\alpha_{td}$=0.07; }

Rules 2: if(DOV==3 and DDA=3 and tdr >2.8) then
{ $\alpha_{kp}$=0; $\alpha_{ti}$=0; $\alpha_{td}$=-0.0252; }

Rules 3: if(DOV>=0 and PR<PrDn) then
{ $\alpha_{kp}$=0; $\alpha_{ti}$=0.0265; $\alpha_{td}$=0; }

Rules 4 : if(OV>=HOV and DA>HDA and

PR>=PrDn) then

{ $\alpha_{kp}$=-0.018; $\alpha_{ti}$=0; $\alpha_{td}$=0;}

...

The controller calculates the $\Delta k_p$、$\Delta t_i$、$\Delta t_d$ according to the coefficients given by the productive rules, then the tuning of the control parameters $k_p(n)$ 、 $t_i(n)$ 、 $t_d(n)$ may be obtained.

## 7. TESTING THE SELF-TUNING ALGORITHM

To test the self-tuning, an experiment of the temperature control of an electric oven has been made. The FOPDT plant model of the electric oven is

$$G_p(s) = e^{-214.3s} / 1 + 717.6S$$

Test condition:

control target OS=1
control period SP=4s

Because there is a long dead time, the PI control algorithm has been used in the controller.

The initialization parameters according to the pre-tuning are

Kp=1.210 Ti=717.5

After five times self-tuning, the controller parameters converged to Kp=1.068 Ti=873.4, and the desired response was obtained. The self-tuning procedure is shown in Tab.3. The step response are shown in Fig. 4.

The experiment proves that the converging speed of this self-tuning algorithm is fast, and the control effect is satisfactory.

Tab.3 Self-tuning procedure of the electric oven control

| n | Kp | Ti | OV(%) | DA | TR | PR | DOV | DDA |
|---|---|---|---|---|---|---|---|---|
| 1 | 1.210 | 717.5 | 10.1 | 0.114 | 1.297 | 0.103 | 3 | 3 |
| 2 | 1.160 | 777.2 | 8.70 | 0.097 | 1.114 | 0.201 | 3 | 3 |
| 3 | 1.112 | 831.2 | 7.0 | 0.078 | 1.975 | 0.208 | 2 | 2 |
| 4 | 1.081 | 873.4 | 5.0 | 0.066 | 0.830 | 0,385 | 0 | 1 |
| 5 | 1.068 | 873.4 | 5.0 | 0.064 | 0.057 | 0.343 | 0 | 0 |

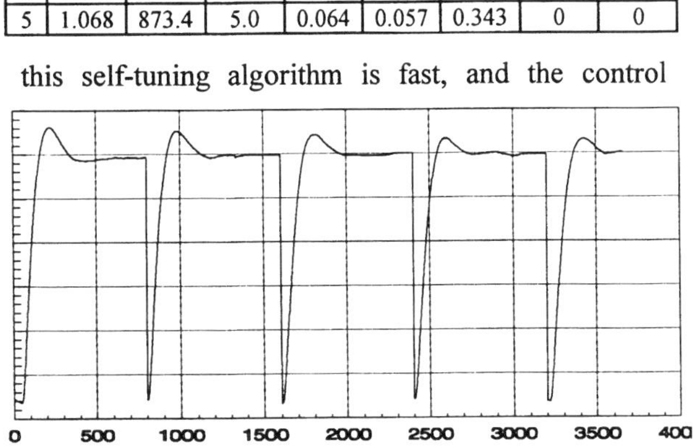


Fig. 4 Step response of the experiment control process

## 8. CONCLUSION

A self-tuning controller based on the pattern recognition and the expert system has been presented. The self-tuning algorithm structure is reasonable, and it has broken through the conventional pattern recognition method, so that it has quick converging speed and satisfactory performance of the transient procedure. This algorithm has been implemented in a digital controller(Xing, et al., 1997), and control performance is good enough for a variety of industry plants(Xing, et al., 1999).

The defect of this method is that the disturbance must be introduced into the control procedure, even though the margin of the disturbance may be set by the control engineer. The method of solving this problems is to wait for the natural disturbance during the tuning procedure, that means the self-tuning time is longer and the performance of the control system may be not satisfactory before the optimal control parameters were obtained.

## REFERENCES


Hang, C.C. and K.K.Sin(1991). A Comparative Performance Study of PID Auto-tuners. *IEEE Trans. Control System*, AUGUST pp41-47

EDGAR H. BRISTOL(1977). Pattern Recognition: an Alternative to Parameter Identification in Adaptive Control. *Automatica*, Vol.13, pp197-202.

Zhuang, M and D.P. Atherton(1993). Automatic tuning of optimum PID controller. *IEE PROCEEDINGS-D*, Vol.140, No.3, pp216-224

KRAUS, T.W. and T.J.MYRON "Self-Tuning PID Controller Uses Pattern Recognition Approach" CONTROL ENGINEER, JUNE 1984, pp.106-111

Xing Jianchun, Wang Lin, and Wang Ping(1997). Intelligent Digital Controller. *Proceedings of the IEEE First International Conference on Intelligent Processing Systems*, pp775-778, Beijing.

Xing Jianchun, Fang Zhonghua, Zhong Weiyang, Wang Ping and Wang Lin(1999). Application of intelligent controllers on the power plant boiler control, *Proceedings of the IFAC 14th World Conference*, Vol.O , pp345-349, Beijing

IFAC Publications
www.elsevier.com/locate/ifac

# FUZZY LOGIC IN SIMULATION OF LOCKHEED L1011 AIRCRAFT THREE-RATE MODEL

**Igor Astrov and Ennu Rüstern**

*Department of Computer Control, Tallinn Technical University,
Ehitajate tee 5, 19086 Tallinn, Estonia, email: igor.astrov@dcc.ttu.ee*

Abstract: This paper describes a three-rate continuous-time stochastic control system as state-space (SS) type decentralized models of multi-input/multi-output (MIMO) stochastic subsystems with optimal linear-quadratic-Gaussian (LQG) regulators and fuzzy logic (FL) control structure. An illustrative example – FL control of three-rate decentralized stochastic model of the Lockheed L1011 aircraft at cruise flight conditions – was carried out using the proposed three-rate SS decentralization technique. This example demonstrates that the research technique results in simplified low-order MIMO autonomous subsystems with low-order LQG regulators running at different rates and shows the quality of the proposed technique. The simulation results were obtained using the software package Simulink.


Keywords: Simulation, multirate, aircraft control, models, fuzzy logic.

## 1. INTRODUCTION

The nature of the multirate dynamics of the process makes it very attractive for applications, since the multirate phenomena are complex and difficult to model using more conventional methods. The multirate systems have received a great deal of attention for many years and the research activity has been motivated by the increasingly demanding industrial tasks in the late 1950s.

Recently, the topic of multirate linear systems has been developed by various researchers.

Tangirala *et al.* (1999) discussed the application of lifting techniques with respect to asymptotic setpoint tracking. A continuous-time approach for multirate controller design is proposed in this work. It is indicated that the continuous-time design framework provides a fair benchmark for comparing the performance of single-rate and multirate systems.

Gu and Tomizuka (1999) described a multi-rate feedback/feedforward system applicable to the design and analysis of tracking control system under measurement constraints. In this system, the feedback controller is updated at the slow rate, while the feedforward controller is updated at the fast rate with a desired sequence, which is obtained by interpolating the slow sampled sequence of the desired trajectory.

Sandell *et al.* (1978) claimed that with respect to designing decentralized controllers for many physical large-scale systems a good combination of engineering judgement and analysis can be used to define in a reasonable way a special structure for the dynamic system.

In Davies and Clarke (1995), a parallel genetic algorithm (GA) was applied within a practical flight control system framework (the sixth order linearized model of the Lockheed L1011 Tristar aircraft at cruise flight conditions) and produced an output feedback controller which satisfied the design requirements such that an acceptable eigenstructure assignment is performed to achieve desired aircraft handling qualities by projecting the desired eigenstructure into an output feedback constrained linear quadratic regulator (LQR) subspace using a cost function minimisation approach. This GA has been successfully implemented using a small transputer network.

In Astrov *et al.* (2000), a decentralization technique is presented to solve the problem of partitioning a centralized system into a decentralized hierarchical structure suitable for integrated flight/propulsion control implementation for a modern fighter aircraft in the longitudinal short approach to landing task. The block diagram of a comparison of functioning of the given initial centralized system with all subsystems and block diagram of functioning of all subsystems with the reference input vector and optimal LQRs are offered. This example demonstrates that the research technique results in simplified low-order subsystems with a specified structure that match the performance characteristics of a high-order centralized system.

In the last years FL control was successfully applied to various engineering problems.

From an intuitive point of view it is clear that small, medium and large labels of fuzzy sets can be ideally suited to control of three-rate subsystems with various dynamics. Thus, the study of the control of three-rate systems directly leads to the study of FL control of decentralized systems.

This paper expands the basic ideas presented in (Astrov *et al.*, 2000) and describes a three-rate continuous-time stochastic control system as SS type models of MIMO stochastic subsystems with FL control structure and the block diagram of FL control for decentralized autonomous subsystems with optimal LQG regulators.

The goal of this paper is to show the applicability of the proposed three-rate FL control technique to the lateral axis stochastic model of the Lockheed L1011 Tristar aircraft at cruise flight conditions.

The contribution of the paper is twofold: to develop new decentralization schemes appropriate for real-time multirate FL control applications, and to present the results of three-rate FL control for any chosen SS type of MIMO stochastic model in simulation form.

## 2. STATE EQUATIONS FOR THREE-RATE STOCHASTIC SYSTEMS

Consider the stochastic continuous-time control system described by the state and output equations

$$\dot{\boldsymbol{x}}(\tau) = \boldsymbol{A}\boldsymbol{x}(\tau) + \boldsymbol{B}\boldsymbol{u}(\tau) + \boldsymbol{v}(\tau) \quad (1)$$

$$\boldsymbol{y}(\tau) = \boldsymbol{C}\boldsymbol{x}(\tau) + \boldsymbol{w}(\tau) \quad (2)$$

where $\boldsymbol{x}(\tau) \in R^n, \boldsymbol{u}(\tau) \in R^m, \boldsymbol{y}(\tau) \in R^P, \boldsymbol{v}(\tau) \in R^n, \boldsymbol{w}(\tau) \in R^P$ are the state, control input, output, noise of excitation of states and noise of measurements vectors, respectively.

The linear transformation $\boldsymbol{q}(\tau) = \boldsymbol{T}\boldsymbol{x}(\tau)$, where $\boldsymbol{T}$ is a nonsingular $n \times n$ matrix, transforms the system equations (1)-(2) to

$$\dot{\boldsymbol{q}}(\tau) = \boldsymbol{J}_c\boldsymbol{q}(\tau) + \tilde{\boldsymbol{B}}\boldsymbol{u}(\tau) + \boldsymbol{T}\boldsymbol{v}(\tau) \quad (3)$$

$$\boldsymbol{y}(\tau) = \tilde{\boldsymbol{C}}\boldsymbol{q}(\tau) + \boldsymbol{w}(\tau) \quad (4)$$

where

$$\boldsymbol{J}_c = \boldsymbol{T}\boldsymbol{A}\boldsymbol{T}^{-1}, \tilde{\boldsymbol{B}} = \boldsymbol{T}\boldsymbol{B}, \tilde{\boldsymbol{C}} = \boldsymbol{C}\boldsymbol{T}^{-1}$$

The equations (3)-(4) may be written in terms of submatrices as

$$\dot{z}_1(\tau) = \boldsymbol{A}_1 z_1(\tau) + \boldsymbol{B}_1\boldsymbol{u}(\tau) + \boldsymbol{T}_1\boldsymbol{v}(\tau) \quad (5)$$

$$\dot{z}_2(\tau) = \boldsymbol{A}_2 z_2(\tau) + \boldsymbol{B}_2\boldsymbol{u}(\tau) + \boldsymbol{T}_2\boldsymbol{v}(\tau) \quad (6)$$

$$\dot{z}_3(\tau) = \boldsymbol{A}_3 z_3(\tau) + \boldsymbol{B}_3\boldsymbol{u}(\tau) + \boldsymbol{T}_3\boldsymbol{v}(\tau) \quad (7)$$

$$\boldsymbol{y}(\tau) = \boldsymbol{C}_1 z_1(\tau) + \boldsymbol{C}_2 z_2(\tau) + \boldsymbol{C}_3 z_3(\tau) + \boldsymbol{w}(\tau) \quad (8)$$

where
small eigenvalues of a matrix $\boldsymbol{A}_1$: $|\lambda(\boldsymbol{A}_1)| < \gamma_{s_{\max}}$, intermediate eigenvalues of a matrix $\boldsymbol{A}_2$: $\gamma_{i_{\min}} < |\lambda(\boldsymbol{A}_2)| < \gamma_{i_{\max}}$, large eigenvalues of a matrix $\boldsymbol{A}_3$: $|\lambda(\boldsymbol{A}_3)| > \gamma_{f_{\min}}$,

$$\boldsymbol{J}_c = \begin{bmatrix} \boldsymbol{A}_1 & 0 & 0 \\ 0 & \boldsymbol{A}_2 & 0 \\ 0 & 0 & \boldsymbol{A}_3 \end{bmatrix}, \tilde{\boldsymbol{B}} = \begin{bmatrix} \boldsymbol{B}_1 \\ \boldsymbol{B}_2 \\ \boldsymbol{B}_3 \end{bmatrix}, \tilde{\boldsymbol{C}} = [\boldsymbol{C}_1 \quad \boldsymbol{C}_2 \quad \boldsymbol{C}_3],$$

$$\boldsymbol{T} = \begin{bmatrix} \boldsymbol{T}_1 \\ \boldsymbol{T}_2 \\ \boldsymbol{T}_3 \end{bmatrix}, \boldsymbol{q}(\tau) = \begin{bmatrix} z_1(\tau) \\ z_2(\tau) \\ z_3(\tau) \end{bmatrix}$$

*Definition 1.* A function with a large derivative which is quickly decreasing is said to be the "fast" function; a function with a small derivative which is slowly decreasing is said to be the "slow" function. A function is said to be the “intermediate” function if the derivative of this function is intermediate between small and large values.

Consider the first time interval $0 \langle \tau_f \le \tau_{fi}$. Applying a method of the Euler to (5), (6) and assuming that $z_1(0) = 0$ and $z_2(0) = 0$, yields

$$z_1(\tau_f) \approx \tau_f \boldsymbol{B}_1 \boldsymbol{u}(\tau_f) + \tau_f \boldsymbol{T}_1 \boldsymbol{v}(\tau_f),$$
$$z_2(\tau_f) \approx \tau_f \boldsymbol{B}_2 \boldsymbol{u}(\tau_f) + \tau_f \boldsymbol{T}_2 \boldsymbol{v}(\tau_f) \qquad (9)$$

It follows from (7)-(9) that the state equations for a "fast" subsystem can be represented by

$$\dot{z}_f(\tau_f) = \boldsymbol{A}_f z_f(\tau_f) + \boldsymbol{B}_f \boldsymbol{u}_f(\tau_f) + \boldsymbol{T}_f \boldsymbol{v}_f(\tau_f) \qquad (10)$$
$$\boldsymbol{y}_f(\tau_f) = \boldsymbol{C}_f z_f(\tau_f) + \boldsymbol{D}_f \boldsymbol{u}_f(\tau_f) + \boldsymbol{w}_f(\tau_f) \qquad (11)$$

where

$$\boldsymbol{A}_f = \boldsymbol{A}_3, \boldsymbol{B}_f = \boldsymbol{B}_3, \boldsymbol{T}_f = \boldsymbol{T}_3, \boldsymbol{C}_f = \boldsymbol{C}_3,$$
$$\boldsymbol{D}_f = \tau_f(\boldsymbol{C}_1 \boldsymbol{B}_1 + \boldsymbol{C}_2 \boldsymbol{B}_2), \boldsymbol{V}_f = \tau_f(\boldsymbol{C}_1 \boldsymbol{T}_1 + \boldsymbol{C}_2 \boldsymbol{T}_2),$$
$$z_f(\tau_f) = z_3(\tau_f), \boldsymbol{u}_f(\tau_f) = \boldsymbol{u}(\tau_f), \boldsymbol{v}_f(\tau_f) = \boldsymbol{v}(\tau_f),$$
$$\boldsymbol{y}_f(\tau_f) = \boldsymbol{y}(\tau_f), \boldsymbol{w}_f(\tau_f) = \boldsymbol{w}(\tau_f) + \boldsymbol{V}_f \boldsymbol{v}(\tau_f)$$

Consider the second time interval $\tau_{fi} \langle \tau_i \le \tau_{is}$. According to the Definition 1, a variable $z_3$ can be considered as a "fast " function, achieving on this interval steady of meaning. Inserting $\dot{z}_3(\tau_i) \approx 0$ in (7) gives

$$z_3(\tau_i) \approx -\boldsymbol{A}_3^{-1} \boldsymbol{B}_3 \boldsymbol{u}(\tau_i) - \boldsymbol{A}_3^{-1} \boldsymbol{T}_3 \boldsymbol{v}(\tau_i) \qquad (12)$$

Using the implicit formula of the Euler for (5) and assuming that $z_1(0) = 0$ leads to

$$z_1(\tau_i) \approx \tau_i [\boldsymbol{I} - \tau_i \boldsymbol{A}_1]^{-1} \boldsymbol{B}_1 \boldsymbol{u}(\tau_i) + \tau_i [\boldsymbol{I} - \tau_i \boldsymbol{A}_1]^{-1} \boldsymbol{T}_1 \boldsymbol{v}(\tau_i) \qquad (13)$$

Taking into account the equations (6), (8), (12) and (13), the "intermediate" subsystem can be written as follows

$$\dot{z}_i(\tau_i) = \boldsymbol{A}_i z_i(\tau_i) + \boldsymbol{B}_i \boldsymbol{u}_i(\tau_i) + \boldsymbol{T}_i \boldsymbol{v}_i(\tau_i) \qquad (14)$$
$$\boldsymbol{y}_i(\tau_i) = \boldsymbol{C}_i z_i(\tau_i) + \boldsymbol{D}_i \boldsymbol{u}_i(\tau_i) + \boldsymbol{w}_i(\tau_i) \qquad (15)$$

where

$$\boldsymbol{A}_i = \boldsymbol{A}_2, \boldsymbol{B}_i = \boldsymbol{B}_2, \boldsymbol{T}_i = \boldsymbol{T}_2, \boldsymbol{C}_i = \boldsymbol{C}_2,$$
$$\boldsymbol{D}_i = \tau_i \boldsymbol{C}_1 [\boldsymbol{I} - \tau_i \mathrm{A}_1]^{-1} \boldsymbol{B}_1 - \boldsymbol{C}_3 \boldsymbol{A}_3^{-1} \boldsymbol{B}_3, \boldsymbol{V}_i = \tau_i \boldsymbol{C}_1 [\boldsymbol{I} - \tau_i \boldsymbol{A}_1]^{-1} \boldsymbol{T}_1 - \boldsymbol{C}_3 \boldsymbol{A}_3^{-1} \boldsymbol{T}_3,$$
$$z_i(\tau_i) = z_2(\tau_i), \boldsymbol{u}_i(\tau_i) = \boldsymbol{u}(\tau_i), \boldsymbol{v}_i(\tau_i) = \boldsymbol{v}(\tau_i),$$
$$\boldsymbol{y}_i(\tau_i) = \boldsymbol{y}(\tau_i), \boldsymbol{w}_i(\tau_i) = \boldsymbol{w}(\tau_i) + \boldsymbol{V}_i \boldsymbol{v}(\tau_i)$$

Consider the last time interval $\tau_s \rangle \tau_{is}$. According to the Definition 1 it is possible to consider variables $z_2$ and $z_3$ as the "intermediate" functions achieving on this interval steady of meanings. Then, it leads that these variables in (6)-(7) are expressed by

$$z_2(\tau_s) \approx -\boldsymbol{A}_2^{-1} \boldsymbol{B}_2 \boldsymbol{u}(\tau_s) - \boldsymbol{A}_2^{-1} \boldsymbol{T}_2 \boldsymbol{v}(\tau_s),$$
$$z_3(\tau_s) \approx -\boldsymbol{A}_3^{-1} \boldsymbol{B}_3 \boldsymbol{u}(\tau_s) - \boldsymbol{A}_3^{-1} \boldsymbol{T}_3 \boldsymbol{v}(\tau_s) \qquad (16)$$

It follows from (5), (8) and (16) that the "slow" subsystem can be written as

$$\dot{z}_s(\tau_s) = \boldsymbol{A}_s z_s(\tau_s) + \boldsymbol{B}_s \boldsymbol{u}_s(\tau_s) + \boldsymbol{T}_s \boldsymbol{v}_s(\tau_s) \qquad (17)$$
$$\boldsymbol{y}_s(\tau_s) = \boldsymbol{C}_s z_s(\tau_s) + \boldsymbol{D}_s \boldsymbol{u}_s(\tau_s) + \boldsymbol{w}_s(\tau_s) \qquad (18)$$

where

$$\boldsymbol{A}_s = \boldsymbol{A}_1, \boldsymbol{B}_s = \boldsymbol{B}_1, \boldsymbol{T}_s = \boldsymbol{T}_1, \boldsymbol{C}_s = \boldsymbol{C}_1,$$
$$\boldsymbol{D}_s = -\boldsymbol{C}_2 \boldsymbol{A}_2^{-1} \boldsymbol{B}_2 - \boldsymbol{C}_3 \boldsymbol{A}_3^{-1} \boldsymbol{B}_3, \boldsymbol{V}_s = -\boldsymbol{C}_2 \boldsymbol{A}_2^{-1} \boldsymbol{T}_2 - \boldsymbol{C}_3 \boldsymbol{A}_3^{-1} \boldsymbol{T}_3,$$
$$z_s(\tau_s) = z_1(\tau_s), \boldsymbol{u}_s(\tau_s) = \boldsymbol{u}(\tau_s), \boldsymbol{v}_s(\tau_s) = \boldsymbol{v}(\tau_s),$$
$$\boldsymbol{y}_s(\tau_s) = \boldsymbol{y}(\tau_s), \boldsymbol{w}_s(\tau_s) = \boldsymbol{w}(\tau_s) + \boldsymbol{V}_s \boldsymbol{v}(\tau_s)$$

## 3. EXAMPLE

In this Section, the proposed three-rate research technique is tested through various simulations.

The lateral axis model of the Lockheed L1011 Tristar aircraft at cruise flight conditions is given in (Vaccaro, 1995). The model includes actuator dynamics and a washout filter on the yaw rate. The state variables are defined as follows: $x_1$ = rudder deflection (radians); $x_2$ = aileron deflection (radians); $x_3$ = bank angle (radians); $x_4$ = yaw rate (radn/sec); $x_5$ = roll rate (radn/sec); $x_6$ = sideslip angle (radians); $x_7$ = washout filter state. The system inputs are: $u_1$ = rudder command (radians); $u_2$ = aileron command (radians). The system outputs are: $y_1$ = washed-out yaw rate (radn/sec); $y_2$ = roll rate (radn/sec); $y_3$ = sideslip angle (radians); $y_4$ = bank angle (radians).

The matrices $(\boldsymbol{A,B,C})$ of the state-space model (1)-(2) are

$$\boldsymbol{A} = \begin{bmatrix} -20 & 0 & 0 & 0 & 0 & 0 & 0 \\ 0 & -25 & 0 & 0 & 0 & 0 & 0 \\ 0 & 0 & 0 & 0 & 1 & 0 & 0 \\ -0.744 & -0.032 & 0 & -0.154 & -0.0042 & 1.54 & 0 \\ 0.337 & -1.12 & 0 & 0.249 & -1 & -5.2 & 0 \\ 0.02 & 0 & 0.0386 & -0.996 & -0.000295 & -0.117 & 0 \\ 0 & 0 & 0 & 0 & 0 & 0 & -0.5 \end{bmatrix},$$

$$\boldsymbol{B} = \begin{bmatrix} 20 & 0 \\ 0 & 25 \\ 0 & 0 \\ 0 & 0 \\ 0 & 0 \\ 0 & 0 \\ 0 & 0 \end{bmatrix}, \boldsymbol{C} = \begin{bmatrix} 0 & 0 & 0 & 1 & 0 & 0 & -1 \\ 0 & 0 & 0 & 0 & 1 & 0 & 0 \\ 0 & 0 & 0 & 0 & 0 & 1 & 0 \\ 0 & 0 & 1 & 0 & 0 & 0 & 0 \end{bmatrix}$$

The "fast"-subsystem (10)-(11) matrices are

$$A_f = \begin{bmatrix} -20 & 0 \\ 0 & -25 \end{bmatrix}, B_f = \begin{bmatrix} 20.0172 & 0 \\ 0 & 25.0273 \end{bmatrix}, C_f = \begin{bmatrix} 0.0374 & 0.0013 \\ -0.018 & 0.0466 \\ 0.0009 & 0.0001 \\ 0.0009 & -0.0019 \end{bmatrix},$$

$$D_f = \begin{bmatrix} -0.1871 & -0.0081 \\ 0.09 & -0.2917 \\ -0.0043 & -0.0003 \\ -0.0045 & 0.0117 \end{bmatrix}, T_f = \begin{bmatrix} 1.0009 & 0 & 0 & 0 & 0 & 0 & 0 \\ 0 & 1.0011 & 0 & 0 & 0 & 0 & 0 \end{bmatrix},$$

$$V_f = \begin{bmatrix} -0.0094 & -0.0003 & 0 & 0.25 & 0 & 0 & -0.25 \\ 0.0045 & -0.0117 & 0 & 0 & 0.25 & 0 & 0 \\ -0.0002 & 0 & 0 & 0 & 0 & 0.25 & 0 \\ -0.0002 & 0.0005 & 0.25 & 0 & 0 & 0 & 0 \end{bmatrix}$$

The "intermediate"-subsystem dynamic (14)-(15) is described by

$$A_i = \begin{bmatrix} -0.5 & 0 \\ 0 & -1.0855 \end{bmatrix}, B_i = \begin{bmatrix} 0 & 0 \\ -1.5294 & -1.5879 \end{bmatrix},$$

$$C_i = \begin{bmatrix} -1 & -0.0154 \\ 0 & 0.7353 \\ 0 & 0.0113 \\ 0 & -0.6774 \end{bmatrix}, D_i = \begin{bmatrix} -0.2896 & -0.0344 \\ 0.017 & 0.0165 \\ 0.2572 & 0.0102 \\ -1.0182 & -1.0611 \end{bmatrix},$$

$$T_i = \begin{bmatrix} 0 & 0 & 0 & 0 & 0 & 0 & 1 \\ -0.0765 & -0.0635 & -0.0978 & 2.5987 & 1.2819 & 2.7508 & 0 \end{bmatrix},$$

$$V_i = \begin{bmatrix} -0.0145 & -0.0014 & 0.0196 & 0.4328 & 0.0187 & 0.5475 & 0 \\ 0.0009 & 0.0007 & -0.0024 & 0.0112 & 0.0278 & -1.9246 & 0 \\ 0.0129 & 0.0004 & 0.0157 & -0.3511 & 0.0013 & 0.3775 & 0 \\ -0.0509 & -0.0424 & 0.9314 & 1.7717 & 0.8962 & -0.0611 & 0 \end{bmatrix}$$

The "slow" subsystem (17)-(18) is specified by the following state-space matrices

$$A_s = \begin{bmatrix} -0.0092 & 0 & 0 \\ 0 & -0.0882 & 1.2695 \\ 0 & -1.2695 & -0.0882 \end{bmatrix}, B_s = \begin{bmatrix} -1.9312 & -1.0936 \\ 0.2379 & 0.0972 \\ 2.5776 & 0.0518 \end{bmatrix},$$

$$C_s = \begin{bmatrix} 0.0383 & -0.0095 & -0.2699 \\ -0.0092 & -0.4168 & 0.6075 \\ 0.0036 & 0.2209 & -0.0177 \\ 0.9992 & 0.4989 & 0.2937 \end{bmatrix}, D_s = \begin{bmatrix} 0.0592 & 0.0239 \\ -1.0541 & -1.029 \\ -0.0151 & -0.0165 \\ 0.9554 & 0.9891 \end{bmatrix},$$

$$T_s = \begin{bmatrix} -0.0966 & -0.0437 & 0.8949 & 2.9922 & 0.8906 & -0.2125 & 0 \\ 0.0119 & 0.0039 & 0.0011 & -0.4553 & -0.0758 & 4.3611 & 0 \\ 0.1289 & 0.0021 & 0.1327 & -3.4128 & 0.0558 & -0.3409 & 0 \end{bmatrix},$$

$$V_s = \begin{bmatrix} 0.003 & 0.001 & 0.0014 & -0.037 & -0.0182 & -0.0391 & -2 \\ -0.0527 & -0.0412 & -0.0663 & 1.7605 & 0.8684 & 1.8635 & 0 \\ -0.0008 & -0.0007 & -0.001 & 0.0272 & 0.0134 & 0.0288 & 0 \\ 0.0478 & 0.0396 & 0.0611 & -1.6219 & -0.8001 & -1.7168 & 0 \end{bmatrix}$$

Simulation results for the case of the offered block schemes (see Fig. 1) are given in Figs. 3-14.

A control structure (see Fig. 2), which consists of the FL controllers and switches, is designed to control

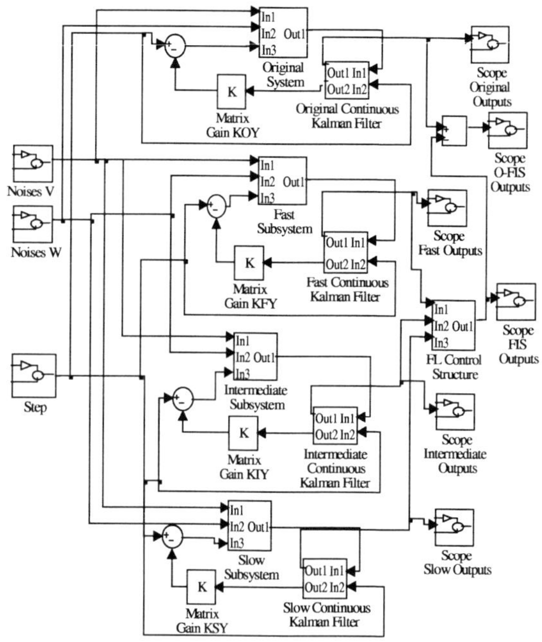


Fig. 1. The block schemes for the initial centralized control system and for the three-rate decentralized control subsystems.

the appropriate outputs of three-rate optimal control subsystems to achieve the three-rate optimal control goal. Each of these FL controllers uses Mamdani-style fuzzy inference with the twelve rules base.

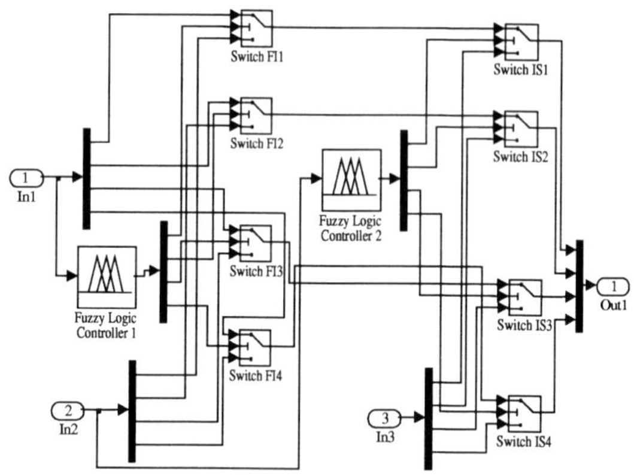


Fig. 2. The block scheme of FL control structure.

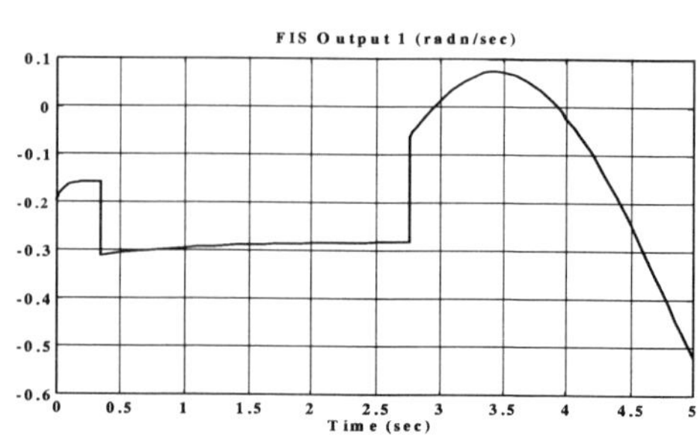


Fig. 3. Step response to the output $y_1$.

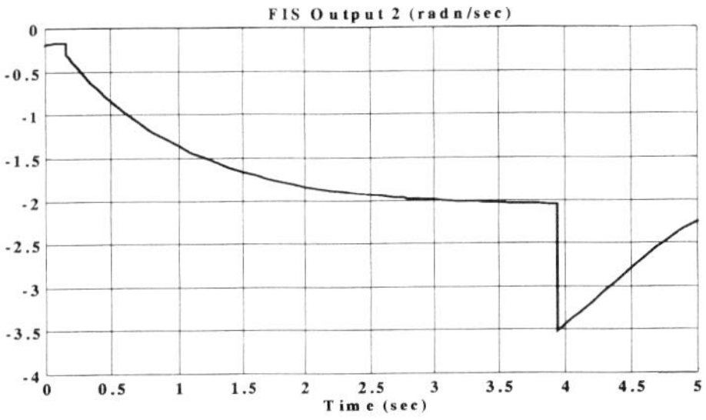


Fig. 4. Step response to the output $y_2$.

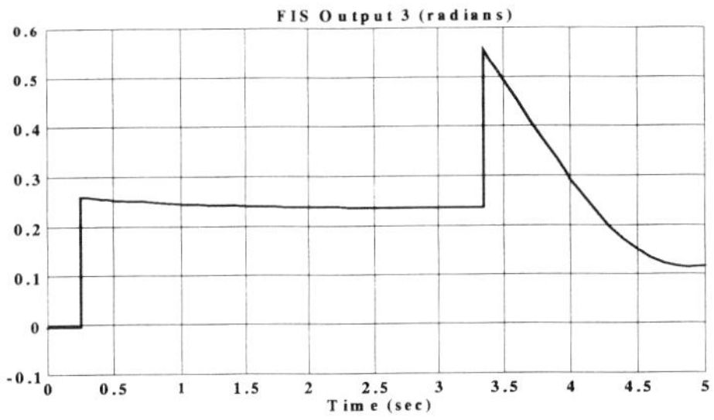


Fig. 5. Step response to the output $y_3$.

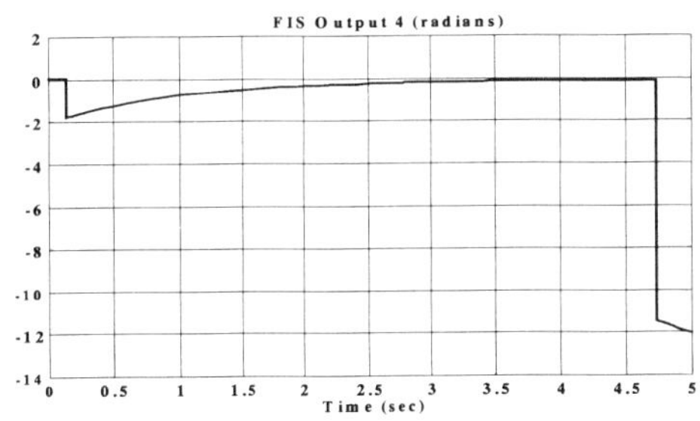


Fig. 6. Step response to the output $y_4$.

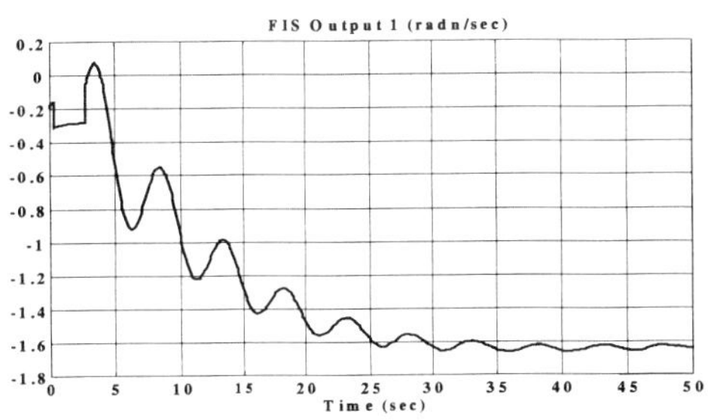


Fig. 7. Step response to the output $y_1$.

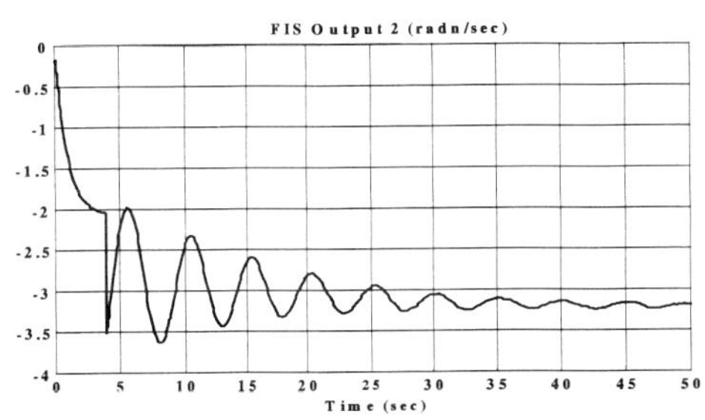


Fig. 8. Step response to the output $y_2$.

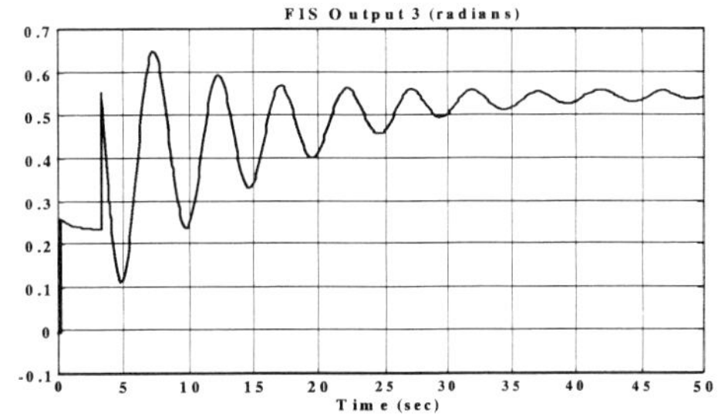


Fig. 9. Step response to the output $y_3$.

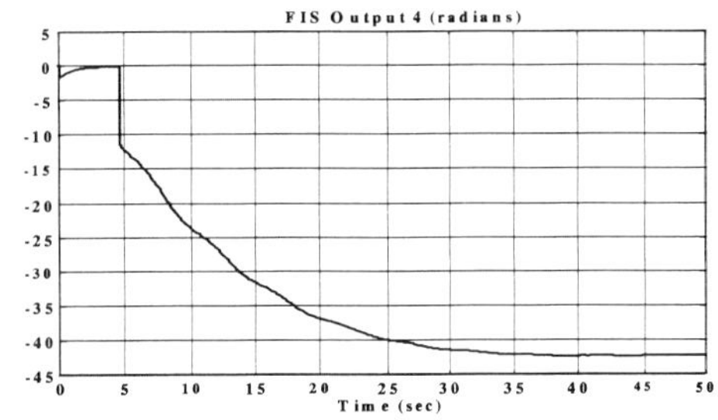


Fig. 10. Step response to the output $y_4$.

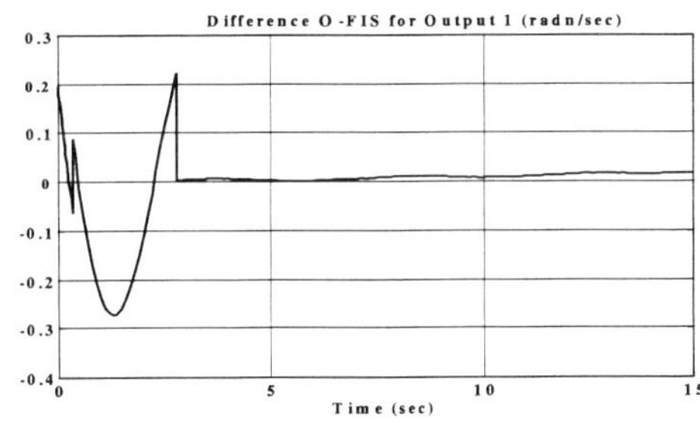


Fig. 11. Step response difference for the output $y_1$.

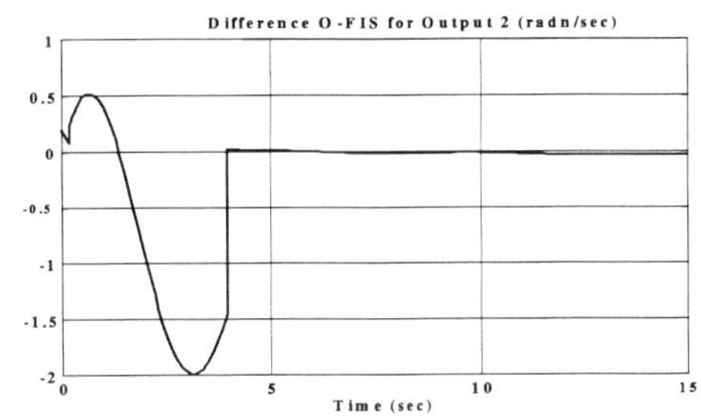


Fig. 12. Step response difference for the output $y_2$.

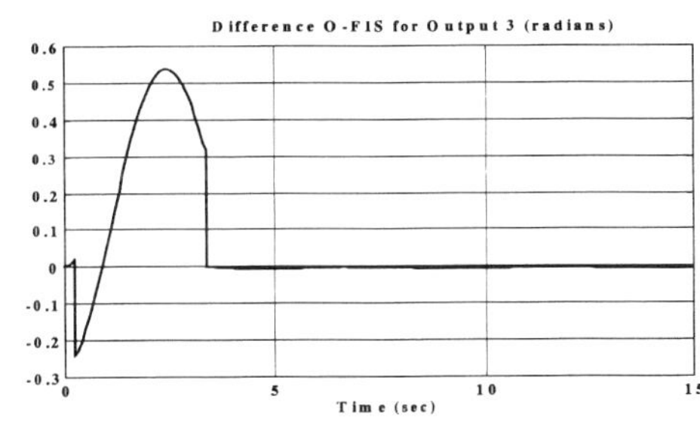


Fig. 13. Step response difference for the output $y_3$.

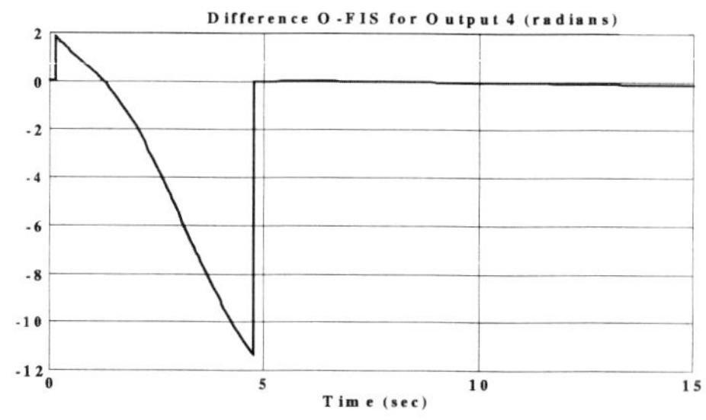


Fig. 14. Step response difference for the output $y_4$.

It can be seen that it is possible to accept an actuation time of "original", "fast", "intermediate" and "slow" regulators as 40, 0.4, 5 and 40 seconds, respectively. It is clear that the autonomous subsystems, which are controlled by local regulators, give the best parameters of quality for process of regulation than the initial system, which is controlled by one global regulator.

Comparison of the "original" control system with received control subsystems in case of identical weighting matrices shows that the step responses of the "original" system and "slow" subsystem lie at close quarters during action of the "slow" subsystem, and the step responses of the "original" system and "fast" subsystem can sometimes practically coincide during action of the "fast" subsystem, but step responses of the "original" system and "intermediate" subsystem are sharply various during action of the "intermediate" subsystem.

Note that good results were obtained using only simple FL controllers in the sense of a minimum rule base, with only three membership functions, with "max-min" inference and "centroid" defuzzification.

These results support the theoretical predictions well and demonstrate that this research technique would work for real-time systems.

## 4. CONCLUSIONS

In this paper a three-rate SS decentralization technique is proposed. The obtained subsystems not only have reduced dimensions of SS matrices, but also low-order LQG regulators with various speeds of actuation (fast, intermediate and long response times) and are completely shared in time. Further analysis of the decentralized control subsystems can be produced separately with the help of modern computer-aided control analysis software.

In example of this paper, the centralized stochastic control system of seventh order (the SS type of MIMO stochastic model of the Lockheed L1011 Tristar aircraft at cruise flight conditions) is partitioned on the "fast" subsystem of second order, on the "intermediate" subsystem of second order and on the "slow" subsystem of third order. This example shows the computational procedure and the applicability of this research technique for real-time systems.

The block diagrams of functioning of all decentralized subsystems with the optimal LQG regulators and FL control structure, and of initial system with the optimal LQG regulator are offered. The continuous-time process simulations were carried out in a MATLAB/Simulink environment.

Although many of the details inevitably relate with this particular system, there is sufficient generality for this research technique to be applied to other MIMO stochastic control systems.

A significant advantage of this approach is the relative simplicity that may be easily used for more complex systems than the model of a Lockheed L1011 Tristar aircraft, and may be improved considerably with the use of more powerful computers to be ideally suited to real-time control applications.

Potential applications are evaluated by control exploration and by computer simulations, using SS type of MIMO stochastic models with multirate phenomena.

## REFERENCES

Astrov, I., J. Em, A. Pedai and E. Rüstern (2000). Decentralized partitioning and simulation of three-rate stochastic continuous-time models. In: *Proc. 4th Portuguese Conference on Automatic Control*. Guimarães, Portugal. pp. 226-231.

Davies, R. and T. Clarke (1995). Parallel implementation of a genetic algorithm. *Control Eng. Practice*, **3**, 11-19.

Gu, Y. and M. Tomizuka (1999). High performance tracking control system under measurement constraints by multi-rate control. In: *Proc. 14th World Congress of IFAC*. Beijing, P.R. China. pp. 67-71.

Sandell, N.R., P. Varaiya, M. Athans and M.G. Safonov (1978). Survey of decentralized control methods for large scale systems. *IEEE Trans. Automatic Control*, **23**, 108-128.

Tangirala, A.K., D. Li, R. Patwardhan, S.L. Shah and T. Chen (1999). Issues in multirate process control. In: *Proc. American Control Conference*. San Diego, California. pp. 2771-2775.

Vaccaro, R.J. (1995). *Digital control: a state-space approach*. McGraw-Hill, New York.

IFAC Publications
www.elsevier.com/locate/ifac

# REAL TIME ENVIRONMENTAL MONITORING CENTRE

**Ileana Hamburg, Gabriel Vladut and Stefan Balanica**

*Institut Arbeit und Technik, Wissenschaftszentrum Nordrhein-Westfalen*
*Gelsenkirchen, Germany*
*Tel: +49 2 09 17 07 265 / Email: hamburg@iatge.de*

*IPA CIFATT, Institut Automation Engineering*
*Craiova, Romania,*
*Tel: +40-51-418882 /Email: ipa@rdslink.ro*

*"Gh. Asachi" University of Iasi, Computer Science Department*
*Iasi, Romania*
*Email: dbalan@cs.tuiasi.ro*

**Abstract**: In this paper some results of European projects about the data acquisition systems, modelling of environment pollution by developing environmental regional monitoring centres which comply with regulatory requirements and protect public health are presented.
The heart of each centre is a real time system for monitoring and detection of noxis, for prognosis and assisted decisions based on information about factors of industrial risks and meteorological conditions. The system will use fuzzy methods and neural networks for decisions and data mining for data analysis.
The development of the centres is done within two EU-cooperation projects: the INCOT and the MONITOR and they will be used by the Agencies of Civil Protection and Environment Protection in Romania in the regions Oltenia, Iasi, Arges. 

**Keywords**: environment pollution, public health, real time monitoring centre, fuzzy methods, neural networks, data mining

## 1. INTRODUCTION

The increasing importance of environmental legislation combined with public health concern is putting pressure on all plants and organizations from Central and East European countries, where these aspects have been neglected till now - to consider environmental pollution and to try to avoid or minimize health risks in the polluted areas or affected work places.

One method is to develop and use real-time systems for monitoring and prognosis in technological risk conditions also with a complete analysis about the impacts, particularly in populated areas.

The systems should support (correlated with the monitoring units, with intervention and supervision offices) the finding of optimal solutions when the chemical alarm occurs; the corresponding personnel can get information about specific measures they have taken at their work places.

In the last part of this paper such a system is described which will be developed within the EU-cooperations MONITOR and TEMPUS-INCOT. Such systems will be integrated in monitoring centres and will be used in very polluted regions from Romania and coordinated by regional Agencies of Civil Protection and Environment Protection. Every monitoring centre has local modules which are very close to the technical installation that has to be monitorized, a central module at the regional Agency of Civil Protection and Environment Protection and a monitoring unit which will be situated within a regional informatic centre (see Figure 2).

In the following chapter we present some methods which are used for the system functionality.

## 2. USING FUZZY METHODS AND NEURAL NETWORKS FOR DECISION MAKING

Fuzzy sets are generally defined by a membership function taking values in the closed unit interval. A wide range of results based on extensions and generalizations of Zadeh's original work have been developed (Zadeh, 1994).
Fuzzy sets can be used for a range of applications, including decision making. Most information required in decision making, as well as many other applications, are known with different degrees of certainty, as well as being imprecise. Therefore, a measure of the degree of certainty or reliability with which the information is known, is likely to improve the quality of information representation. Augmented fuzzy sets can be used to represent information and the degree of certainty with which it is known by replacing the standard fuzzy set membership function with an ordered pair of numbers, both of which are between 0 and 1. Both elements of the ordered pair are membership functions, with the second element giving a measure of the certainty in the assessment of the value of the first element.

Augmented fuzzy sets can be used to present information in an easily accessible and comprehensible form which gives a good indication of the reliability of the information, e.g. about environmental or other impacts, and facilitates decision making (Hersh et al., 1998).

A neural network is a computational model, consisting of many simple processing elements operating in parallel. Their function is determined by network structure, connection strengths, and the processing performed at computing elements or nodes. Each processor is connected to a number of its neighbours and the number of interconnections between processors is generally greater than the number of processors. This gives the neural network considerable robustness and fault tolerance, e.g. damage to a few neurons or interconnections generally has relatively little effect on the performance of the system.

Neural networks can be used to analyze large quantities of data to establish patterns and characteristics, a wide range of applications, including sustainable design, business, image processing, consumer electronics, system modelling, industrial control, quality control and process planning. Whether or not the use or artificial neural networks in a particular application is appropriate depends on a number of factors, including the information processing capacity required, cost implications and system requirements in terms of speed of operations, reliability, ease of maintenance and ease and modification. Another important factor is whether expertise in the use of neural networks is available in the organisation or appropriate training in their use can be obtained, to avoid the problems similar to the results of the use of sophisticated techniques by personnel without appropriate skills.

Learning (training) allows a neural network to modify its behaviour in response to the environment. The learning ability of a neural network is determined by its structure or architecture and the algorithmic method chosen for training.

The method used in the INCOT project is the backpropagation algorithm, in which the optimal connection weights which minimize the global error can be obtained from a gradient optimization method.

## 3. ANALYSIS OF DATA BY USING DATA MINING

The field of data mining is relatively new and therefore, new approaches are still being developed. Data mining can be used to predict future behaviour from examples of related behaviour in the past, as in the case of pattern design solutions. Therefore, data is collected over a period of time and good answers are stored to act as patterns to form the basis of future decisions, for instance in the areas of product design and development and customer relations, amongst other applications.

New technologies, like the Web, facilitate the use of data mining techniques, due to their well-developed capacities to develop interactivity, dynamicity and portability of information.

However, it should be recognized that all predictive techniques have their limitations. Therefore

solutions should be validated to ensure that they are appropriate. In addition appropriate training and experience is required to make the best use of such techniques (Hamburg et al., 2000).
A data mining module, consisting of two components for the extraction and analysis of data and data prediction respectively, is being implemented as part of the INCOT project; it will be linked to a neural network module and a fuzzy module, based on the techniques discussed in section 2.
Data, from different sources and in different formats, are extracted and after being represented in an appropriate format stored to form a data warehouse which is a part of the data mining module. When transactional distributed data bases, such as Oracle, mySQL and msQL are used, local support is required, whereas local operational data bases, such as FoxPro, Access, Excel and dBase, use a transfer level.

## 4. THE REAL-TIME MONITORING CENTRE

The main role of such a centre is the monitoring of the pollution units, the areas that might be affected by the pollution factors for keeping them at a tolerance acceptance level and to minimize their effect. Regarding the type of the pollution agent and its aggressive impact on human factor, complex systems have to be developed to eliminate their effects.

A special case of pollution in Craiova, Oltenia (which has to be solved urgently) is a chemical plant located near-by a power-plant unit and so the noxious effect is cumulated (Vladut et al., 1995).
In order to find a solution for monitoring, results of the following existing systems have to be considered:

- a local dispatcher that determines the chemical focus (in the unit and power plant)
- chemical alarm dispatcher that puts in use the chemical alarm plant if an accident occurs that affects not only the unit but also the near-by areas.

Regional aspects like the following are also important:

- to locate the place where the focal is situated,
- to determine the quantity of noxis,
- to determinate and to present the meteorological parameters,
- to determine the cloud shape and its evolution using a mathematical algorithm,
- to inform in a real time the decision factors in the unit, power plant and in the region.

In the following we present some main steps of the project MONITOR which has as main objective the development of monitoring centres (first in the region Oltenia):

- to characterize the industrial object through out the protection terms and maximum level of tolerance accepted for noxious emission that has impact over air, water, soil and human beings,
- to analyze the position of the populated regions around the industrial objective and geographic conditions,
- to define the events that can be considered as a real danger to the industrial objective: earthquakes, floods but also the effects on the population by the industrial objective,
- to extend and update an existing statistic data base with the main events that took place some years ago and their impact,
- to establish a graphic chart about the noxious emission,
- to develop a system for processing data in order to reach the proposed goal and to complete the existing equipment network with necessary ones.

In the following we give a short description of the main modules of the monitoring centre (Figure 2). The development for the region Oltenia will start in May 2001 after some ongoing analyses in Romania as well as in Germany (collecting European experience) are finished.

The local modules (could be one or more) are situated within the plant where the noxious process is expected. Their role is to monitor the technological parameters, framing through the limits, avoiding an accident (like the break of a conduct followed by noxious emission), or alarming in case of it, locating the exact place and the quantity of noxious emission. For the risk study, pressures, debits, temperatures are observed in the established areas.

A primary analysis is done by using linear methods and also neural networks. This module has a local data base in order to find suitable rules and make fast decisions. The local modules communicate locally with corresponding applications but also with the central module.
Procedures are necessary in order to correlate local data with central data.

The data received from the local points are structured and centralized by the DB Manager of the central module situated at the Agency of Civil Protection and Environment Protection.
By using data mining procedures (with neural networks and fuzzy methods) deep analyses should be done once a month. This module assures also the communication with the monitoring point.
The possible effects of pollution, by accidents (particularly chemical ones or within power plants)

and the impacts towards populated areas are being emphasized by a program packet that presents the evolution of the toxic cloud and the dispersion analysis through a map. On the screen the noxious and gas emission is being presented also with the speed and wind direction and the stability class. The required data that allow a visualization of the medium concentration are being provided by packet programs that modulates the pollution emission processes for the gas, liquid or biphasic stages, evaporation, dispersion, heat radiation, explosions and fires. The medium concentrations for a long period of time and for an existing receiver are important as a distinct field; every checkpoint of the module network (concentric towards the pollution source) has two plane co-ordinates and an attached concentration class (Figure 1).

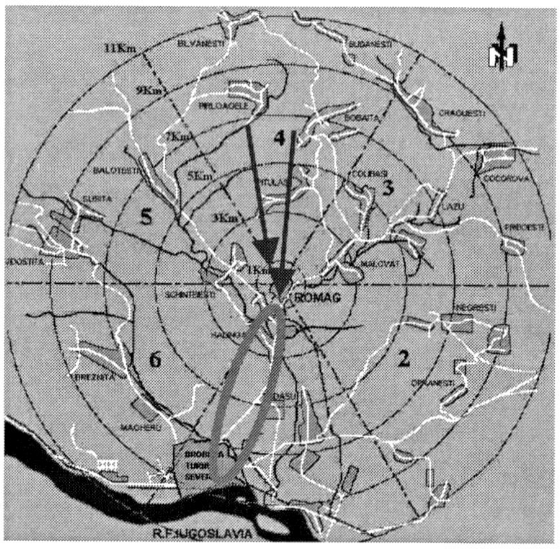


Fig. 1: Simulation for toxic cloud evolution

Time studying evolution of the medium concentrations, in a geographical context facilitates:

- geographical results from pollution cloud dimensions,
- information about impacts referring medium, population by processing a data base that contains information about the attributes of the affected objects (e.g. the number of inhabitants that are affected by lethal concentrations).

Potential beneficiaries of the centre in Oltenia are highly pollution industrial agents: Romag Drobeta, Arpechim Pitesti areas, Petromidia, RAFO Onesti, Chimcomplex, Oltchim Ramnicu Valcea, pollution cities, APM, IPM etc.

## 5. CONCLUSIONS

Civil protection and environment protection are important aspects that have been neglected in Central and East European countries till now. Fast, wide and not very expensive solutions are needed for improving them.
Eu-cooperations like the projects INCOT and MONITOR could help to develop such solutions by using experience of other countries.

## REFERENCES

Hersh, M. A., Hamburg, I. and Hamburg, P. (1998). The application of fuzzy methods to environmentally compatible production in manufacturing industry. In: *Analele Universitatii din Craiova*, **Vol. 25**, pp. 1-7, Craiova.

Hamburg, I. and Hersh, M.A. (2000). The use of soft methods and data mining to support engineering design. *Lucrarile Simpozionului Universitaria Ropet* 2000, pp. 267-270.

Vladut, G. and Ionica, M. (1995). Analytic sensors and systems for environment pollution control, *ICI* Bucharest.

Zadeh, L. (1994). Fuzzy Logic, Neural Networks, and Soft Computing. In: *Communications of the ACM*, **Vol. 37**, pp. 77-82.

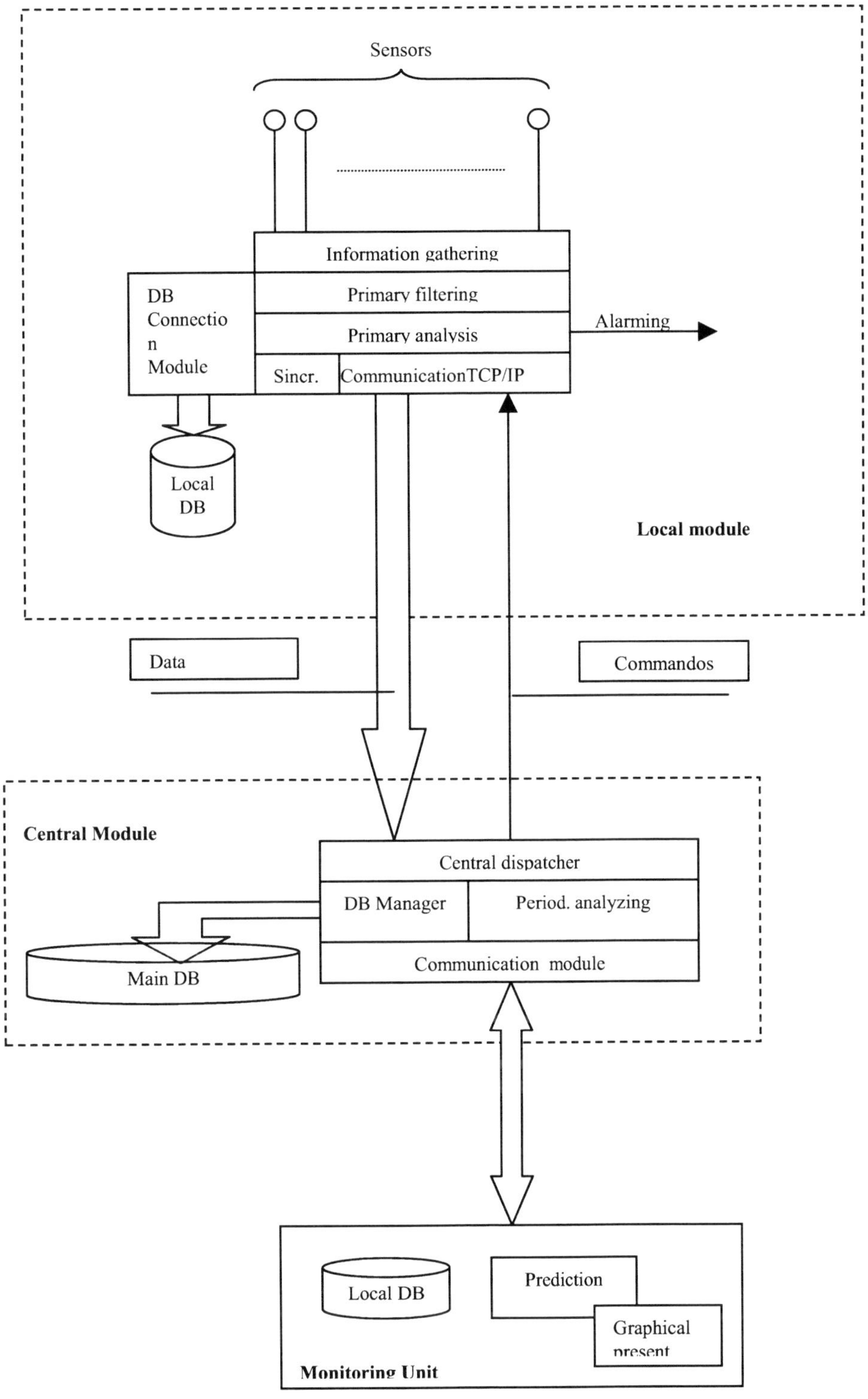


Fig. 2: The main modules of the monitoring centre

IFAC Publications
www.elsevier.com/locate/ifac

# HOW TO SUPPORT REAL-TIME TRAFFIC OF URGENT APERIODIC MESSAGE USING FF [1]

**Zhi Wang[1*] Yue Zhou[2] Yeqiong Song[3] Jiming Chen[1] Youxian Sun[1]**

*1 National Laboratory of Industrial Control Technology, Zhejiang University, Hangzhou, 310027, China*
*2 Shenyang Institute of Automation, Chinese Academy of Sciences, Shenyang, 110016, China*
*3 LORIA–ENSEM, 2, av. de la Forêt de Haye, 54516 Vandoeuvre-lès-Nancy, France*
[*] Author of correspondence: *E-mail wangzhi@iipc.zju.edu.cn or* zhiwang_iipc@yahoo.com

Abstract: This paper deeply investigates how to support real-time traffic of urgent aperiodic messages using Foundation Fieldbus (FF). This paper shows that, despite FF providing such mechanisms as PT priority and PT rotation period, etc., FF can't effectively guarantee real-time traffic of urgent aperiodic message that is shown in its worst-case response time. The main contribution of this paper is to find the deficiency of current mechanism and proposes a new one. Further, this paper gives methodologies for setting SCT under the new mechanism. *Copyright © 2001 IFAC*

Keywords: Foundation Fieldbus, real-time, aperiodic traffic, response time, token circulation period.

## 1. INTRODUCTION

The key to success in using a distributed real-time system for applications, such as industrial process control, automation manufacturing, and spaces vehicle system etc, is the timely execution of computation tasks that usually reside on different nodes and communicate with one another to accomplish a common goal. End to end deadline guarantees of these hard real-time tasks are possible only if a communication network support the timely delivery of inter-task messages (Kopetz, 1997; Buttazzo, 1997; Fieldbus foundation, 1996).

Fieldbus, as infrastructure of communication to support real-time traffic among field devices in factory floor, is generally characterized by the obligation to respect stringent temporal constraints, which must be met in order to guarantee correctness and safe. Consequently, a significant issue is devoted to taking account of temporal property of fieldbus to guarantee timing requirement of field devices. Therefore, scholars, such as Burns, Shin, Song and Tovar, pay lots of attentions on evaluating and analyzing temporal property of many popular fieldbuses, such as CAN、PROFIBUS、P-NET and WorldFIP (Song, 1996; Simonot and Song 1996; Tindell and Burns, 1997; Zuberi, et al., 2000; Zhao and Stankovic, 1990; Kim and Jeong, 1998; Shin and Chou, 1996; Tovar and Vasques, 1999; Wang, 2000; Wang and Sun, et al., 2001).

FF is one of the most popular fieldbus standards. FF not only supports real-time traffic of its application through explicitly distinguishes traffics into periodic and aperiodic, but also provides complete control function through specifying user layer protocol, including Function Block (FB) and Device Description (DD) (Fieldbus foundation, 1996). However, FF lacks temporal property of traffics, particularly response time of aperiodic messages, which is the main focus of this paper. This paper addresses worst-case response time of urgent aperiodic messages and sets PT rotation period.

## 2. FOUNDATION FIELDBUS

### *2.1 Message and token*

For guaranteeing temporal constraints of its applications, FF explicitly distinguishes periodic and aperiodic messages and applies different strategies. to support these messages. FF applies centralized media access mechanism, and implements it using Link Active Schedule (LAS). LAS manages schedule of all messages in FF through giving token, which standards for access to FF. Any fieldbus device can only

[1] This work is supported by PRA SI 01-04 and NSFC-60084001.

transmit its waiting messages after receiving token. For effectively managing periodic and aperiodic messages, FF provides two types of tokens.

The first one is scheduled token, refers to compel data (CD), which only support periodic messages. According to its stored predefined time, LAS broadcasts CD. CD includes a specific address, through which the node corresponding the address will transmit a periodic message after receiving CD.

The second token is unscheduled token, refers to pass token (PT), which gives aperiodic messages a opportunity. Within left communication time between transfers of periodic messages, LAS sends PT to fieldbus devices according to address of these fieldbus devices.

*2.2 Link active schedule implementation*

LAS includes temporal properties of all periodic messages within a segment FF. According to these properties and a proper schedule algorithm, LAS generates a network schedule. In FF, the network schedule is stored in a schedule table (ST), which is made up of a set of basic schedule table, known as microcycle. The LAS gives access right to periodic messages through sending CD according to its schedule stored in each microcycle within the ST, then gives access right to aperiodic message through sending PT if time left in this microcycle is enough. The portion of a microcycle reserved for periodic message is denoted as periodic window, whereas the time left after the periodic window is denoted as aperiodic window. The transfers for two types of messages depend on schedules stored in their respective windows. Once all microcycles have been performed in the ST, LAS repeats the same network schedule from the beginning.

Two important parameters are associated with the ST, microcycle and macrocycle. The former imposes the maximum rate at which LAS performs a set of scans, and the latter is minimum duration during which the sequence of microcycles is repeated. Usually, the microcycle and the macrocycle are respectively set equal to the highest common factor (HCF) and the lowest common multiple (LCM) of the required scan periodicities.

Table.1 Example set of periodic messages

| Variable Identifier | Periodicity (ms) |
|---|---|
| A | 1 |
| B | 2 |
| C | 3 |
| D | 4 |
| E | 4 |
| F | 6 |

Assume that within a FF there are 6 periodic messages with arrival rate in Table.1 to be transferred. According rule of HCF and LCM, the microcycle and macrocycle are set to 1ms and 12 ms respectively. A feasible schedule meeting real-time requirements of these periodic messages is illustrated as Fig.1, where we consider $C_p$=0.2ms for communication time of each periodic message.

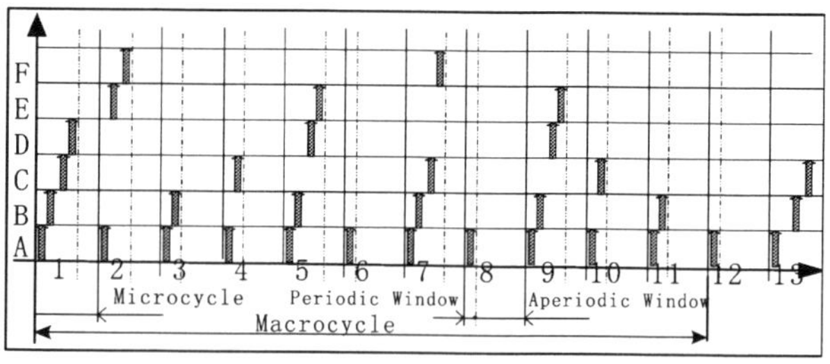


Fig.1 A feasible schedule for periodic messages in table.1

*2.3 Procedure of aperiodic message transfer*

In order to meet requirement of aperiodic messages with different levels of criticality in temporal aspect, FF provides PT priority and PT circulation period, the latter is the duration that PT consecutively twice reaches a particular node with least address.

PT priority is differentiated Urgent, Normal and Available. Correspondingly, aperiodic messages are also classified with three types. For an aperiodic message, only when its priority is not less than current PT priority and its transfer time is less than maximum token hold time (MTHT), set in PT frame, is the aperiodic message allowed to transmit. PT returns LAS when no aperiodic message with proper priority exists or MTHT expires.

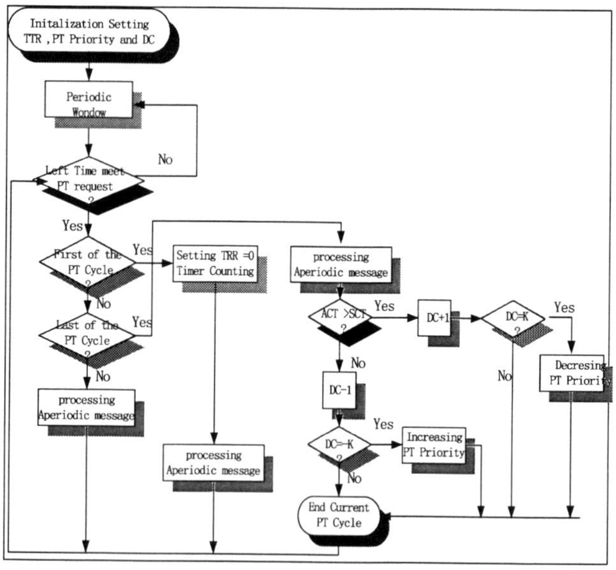


Fig.2 Schedule of LAS for aperiodic messages

PT circulation period includes setting PT circulation period (SCT) and actual PT circulation period (ACT), where the former is set online or offline by an operator and the latter is measured online. To enhance real-time response of critical aperiodic message and to adapt load in FF, LAS online changes PT priority according to the difference of between SCT and ACT. ACT being larger than SCT, LAS increases PT priority unless PT priority is urgent. Otherwise, LAS decreases PT priority unless PT priority is available.

Thus, FF as far as possible meets timing requirement of all aperiodic messages through aperiodic message priority, PT priority, SCT and ACT. Besides, in order to decrease the fluctuation of PT priority, FF provides difference counter (DC) by which PT priority is adjusted automatically. In this paper, DC is set 1. LAS schedule for aperiodic message is shown in Fig.2.

## 3. MODEL OF APERIODIC MESSAGE COMMUNICATION

### *3.1 Network and message model*

Let N be the number of nodes in a distributed system connected by FF and one FF connection per node. Assume the addresses of nodes ranges from 1 to N.

Assume there are *np* periodic messages ($Mp^i$, $i = 1,\cdots,np$) in this system. Note that FF utilizes centralized schedule scheme according to temporal characteristic of periodic messages, so $np$ is irrespective of N and *i* is irrespective of node address.

$$Mp^i = (Cp^i, Tp^i, Dp^i) \quad (1)$$

Where, $Tp^i$ and $Dp^i$ denote periodicity and deadline of message $Mp^i$ respectively, $Cp^i$ corresponds transmitting time of message $Mp^i$.

The nature of an aperiodic message is in its irregular arrival interval, and in no arrival period existing. However, for deducing its worst-case response time, we assume there is a minimum arrival interval for it, and assign an arrival period equaling to its minimum arrival interval.

Considers a node *k* (*k*=1,...,*N*), in which $na_k^u$ urgent aperiodic messages exist. The $i^{th}$ urgent aperiodic message in node *k* is denoted as $Mu_k^i\ (i = 1,\cdots,na_k^u)$

$$Mu_k^i = \left(Ca_k^{ui}, Ta_k^{ui}, Da_k^{ui}\right) \quad (2)$$

Where, $Ca_k^{ui}$, $Ta_k^{ui}$ and $Da_k^{ui}$ represent communication time, arrival period and deadline of $i^{th}$ aperiodic messages within node *k* respectively.

As previous described, there are normal aperiodic messages and available aperiodic messages in a node. For simplifying analysis, we group them together and only regard as normal aperiodic messages.

Let the number of normal aperiodic messages in node *k* is $na_k^n$. We don't intend to guarantee deadline for normal aperiodic messages, a normal aperiodic message is merely characterized as follows,

$$Mn_k^i = \left(Ca_k^{ni}, Ta_k^{ui}\right)\ \left(i = 1,\cdots,na_k^u\right) \quad (3)$$

Where, $Ca_k^{ui}$ and $Ta_k^{ui}$ represent communication time, arrival period of $i^{th}$ normal aperiodic messages within node *k* respectively.

For a message transfer, whether periodic message or aperiodic message only needs two consecutive phases, getting access to FF and transmitting message. Completing a message transfer hence needs at least two frames, CD (PT) frame and data frame, which is shown in Fig.3.

Therefore, the communication time of a message can be calculated as follows,

$$C_p = bps^{-1} \cdot \left(len(CD\_DLPDU) + len(DATA\_DLPDU)\right) + 2 \times t_r \quad (4)$$

$$C_a = bps^{-1} \cdot \left(len(PT\_DLPDU) + len(DATA\_DLPDU)\right) + 2 \times t_r \quad (5)$$

Where, $C_p$ and $C_a$ denote communication time of any periodic message and any aperiodic message respectively. *len*() denotes number of *bits* in a frame, CD_ DLPDU, PT_DLPDU and Data_DLPDU denotes number of *bits* in CD, PT and a data frame respectively, $t_r$ denotes the minimum duration between transmission of two frames.

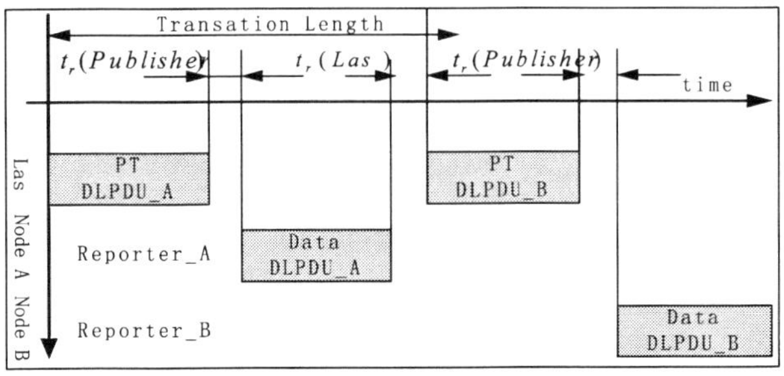


Fig.3 Elementary transaction for messages

### *3.2 PT visiting model*

For describing behavior of a sequence by which PT visits a node, we introduce a pair of subscripts, visit(c, i). Within visit(c, i), *c* and *i* respectively indicate the order of PT cycle and the order of the node being visited. PT cycle refers to a procedure that PT visits nodes from the least address to the largest address. The notion in the rest of the paper will often use is the natural ordering of PT visiting. visit(c, i) is followed by visit(c, i+1) if $(i \neq N)$, otherwise by visit(c+1, 1). visit(c, i − 1) is the previous visiting before visit(c, i) if $(i \neq 1)$, otherwise visit(c−1, N) is.

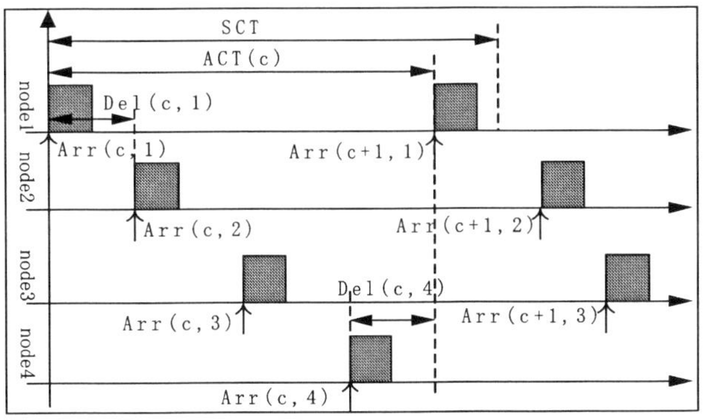


Fig.4 Relationship between the above parameters

Let Arr(c, i) as the instant when PT arrives to node i for its $c^{th}$ visit. That is also the instant of beginning visit(c, i). Let Del(c, i) as the elapsed time from the beginning of visit(c, i) to the beginning of visit(c, i+1). Then Del(c, i) owns the following property, $\mathrm{Del}(c,i) = Arr(c,i+1) - Arr(c,i)$ if $(i \neq N)$; otherwise, $\mathrm{Del}(c,N) = Arr(c+1,1) - Arr(c,N)$.

From the above descriptions, it is apparent that SCT is the setting time elapsed from the beginning of $Arr(c,1)$ to $Arr(c+1,1)$, and ACT is the measured value of $Arr(c,i+1)-Arr(c,i)$. The relationship between the above parameters is illustrated in Fig.4.

*3.3 Queue delay for an urgent aperiodic message*

When an urgent aperiodic message, say message M, arrives a node, say node X, message M cannot be transferred until PT returns node X. Within node X, aperiodic messages with different priorities are transferred according to their respective priority, but aperiodic messages with same priority are transferred according to FCFS rule. Therefore, queue delay of message M consists of waiting PT arrival to node X, transmitting other urgent aperiodic messages arrive earlier in node X, and transmitting itself. Message M will suffer worst-case queue delay if it arrives node X while PT just leaves node X. Further, during waiting PT arrival to node X, other node may send urgent and normal aperiodic messages or only urgent aperiodic messages, that depends on PT priority.

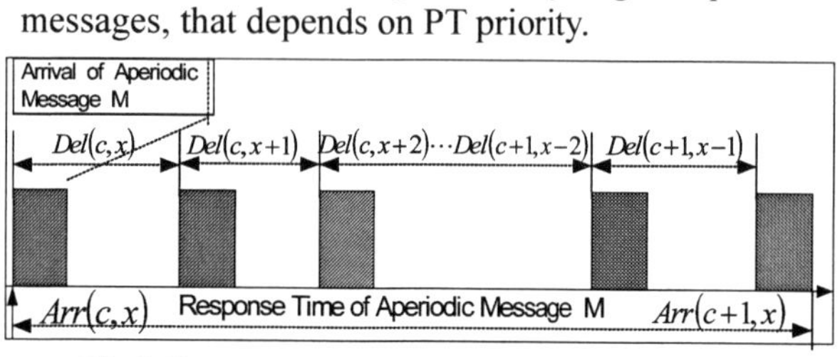


Fig 5. Response procedure of an aperiodic message

Let $Ra(c,x)$ denote response time of urgent aperiodic messages in node X during $visit(c,x)$.

$$Ra(c,x)=\sum_{i=x+1}^{N} Del(c,j)+\sum_{i=1}^{x-1} Del(c+1,j)+Del(c,x) \qquad (6)$$

Note that Eq(6) seemly does not conform to the response procedure of urgent aperiodic message illustrated in Fig.5, because we consider the worst-case scenario with all urgent aperiodic message in node X arrive simultaneously just PT leaving this node. However, in calculation of $Del(c,x)$, we conclude all urgent aperiodic message in node X, so Eq(6) is just for convenience.

4. RESPONSE TIME OF APERIODIC MESSAGE

Because of LAS changing PT priority online, so we must analyze response time of an urgent aperiodic message under different conditions. Besides, LAS sends aperiodic message only within aperiodic window, whose communication capability varies in different instants, so we analyze aperiodic window first.

4.1 *Communication capability of aperiodic window*

Further, for calculating communication capability of aperiodic window, we must consider arrival pattern of all periodic message by which a network schedule in ST is generated. A periodic window's content of network schedule for periodic message determines communication capability of its corresponding aperiodic window.

Let *MicP* denote time length of a microcycle, $N_{Mic}$ denote number of microcycles within ST. If ST is constructed by rule of HCF and LCM, $N_{Mic}$ equals to LCM/HCF. Let $W_a^i$ denote time length of $i^{th}$ aperiodic window, which equals to time length of a microcycle subtracting transfer time of periodic messages scheduled in $i^{th}$ microcycle.

$$W_a^i = MicP - \sum_{j=1}^{np}(table[j,i^*]) \times Cp_j) \qquad (7)$$

Within Eq(7), $table[j, i^*]$ denotes whether periodic message $j$ is scheduled in $i^{*th}$ microcycle, $i^*=[(i-1)\ Mod\ N_{Mic}]+1$.

Let $MNum(c,x)$ denote number of microcycles for transmitting urgent aperiodic messages in node $x$ from the beginning of $visit(c,x)$ to the beginning of $visit(c+1,x)$.

$$MNum(c,x)=\min(\psi)\cap\sum_{i=maw}^{\psi} W_a^i \geq \sum_{k=1}^{N}\sum_{j=1}^{na_1^u} Ca_k^{uj} \qquad (8)$$

Note that Eq(8) only explains how to calculate $MNum(c,x)$, particularly load of aperiodic messages in the right part of the inequation in Eq(8). This part will be adapted according to actual permitted load of urgent aperiodic messages or normal aperiodic messages.

Another worthwhile to note that *maw* in Eq(8) implicates *maw* being the instant that aperiodic window owns the less communication capability. That is,

$$maw=\max\left(MNum(c,x) \middle| MNum(c,x)=\min(\psi)\cap\sum_{i=maw}^{\psi} W_a^i \geq \sum_{k=1}^{N}\sum_{j=1}^{na_k^u} Ca_k^{uj}\right) \qquad (9)$$

In the following section, *maw* denotes the worst-case instant unless explicitly stated.

*4.2 Response time of urgent aperiodic message*

*4.2.1 PT priority keeping urgent*

Assume a scenario that urgent aperiodic messages in node X arrive just PT finishes its $c^{th}$ visit to node X and leave, and PT priority keeps urgent when PT arrive node 1 for its $(c+1)^{th}$ visits.

In this case, LAS only sends urgent aperiodic messages from $visit(c,x)$ to $visit(c+1,x)$. Let $\overline{M}Num(c,x)$ denote number of microcycles that urgent aperiodic messages in node $x$ have to wait, let $\overline{R}a(c,x)$ denote response time of these aperiodic messages.

$$\overline{M}Num(c,x)=\min(\psi)\cap\sum_{i=maw}^{\psi} W_a^i \geq \sum_{k=x+1}^{N}\sum_{j=1}^{na_k^u} Ca_k^{uj}+\sum_{k=1}^{x}\sum_{j=1}^{na_k^u} Ca_k^{uj} \qquad (10)$$

$$\overline{R}a(c,x)=(\overline{M}Num(c,x)-1)\times MicP + \sum_{j=1}^{np} table\left[j,(maw+MNum(c,x))^*\right]\times Cp_j+\sum_{j=1}^{na_x^u} Ca_x^{uj} \qquad (11)$$

*4.2.2 PT priority keeping normal*

Assume a scenario that urgent aperiodic messages in node X arrive just PT finishes its $c^{th}$ visit to node X, and PT priority keeps normal when PT arrives node 1 for its $(c+1)^{th}$ visit.

In this case, LAS not only sends urgent aperiodic messages but also normal aperiodic messages from $visit(c,x)$ to $visit(c+1,x)$. Let $\hat{M}Num(c,x)$ denote number of microcycles urgent aperiodic messages in node $x$ have to wait, let $\hat{R}a(c,x)$ denote response time of these aperiodic messages.

$$\hat{M}Num(c,x) = \min(\psi) \cap \sum_{i=maw}^{\psi} W_a^i \geq \sum_{k=x+1}^{N}\left(\sum_{j=1}^{na_k^u} Ca_k^{uj} + \sum_{j=1}^{na_k^n} Ca_k^{nj}\right) + \sum_{k=1}^{x}\left(\sum_{j=1}^{na_k^u} Ca_k^{uj} + \sum_{j=1}^{na_k^n} Ca_k^{nj}\right) \quad (12)$$

$$\hat{R}a(c,x) = (\hat{M}Num(c,x) - 1) \times MicP + \sum_{j=1}^{np} table\left[j, (maw + \hat{M}Num(c,x))^*\right] \times Cp_j + \sum_{j=1}^{na_x^u} Ca_x^{uj} \quad (13)$$

*4.2.3 PT priority changing from normal to urgent*

Assume a scenario that urgent aperiodic messages in node X arrive just after PT finishes its $c^{th}$ visit to node X under PT priority being normal, and PT priority changes to urgent when PT arrives node 1 for its $(c+1)^{th}$ visit.

In this case, LAS not only sends urgent aperiodic messages but also normal aperiodic messages from $visit(c,x)$ to $visit(c,N)$. However, LAS only sends urgent aperiodic messages from $visit(c+1,x)$ to $visit(c+1,x)$. Let $\tilde{M}Num(c,x)$ denote number of microcycles that urgent aperiodic messages in node $x$ have to wait, let $\tilde{R}a(c,x)$ denote response time of these aperiodic messages.

$$\tilde{M}Num(c,x) = \min(\psi) \cap \sum_{i=maw}^{\psi} W_a^i \geq \sum_{k=x+1}^{N}\left(\sum_{j=1}^{na_k^u} Ca_k^{uj} + \sum_{j=1}^{na_k^n} Ca_k^{nj}\right) + \sum_{k=1}^{x}\sum_{j=1}^{na_k^u} Ca_k^{uj} \quad (14)$$

$$\tilde{R}a(c,x) = (\tilde{M}Num(c,x) - 1) \times MicP + \sum_{j=1}^{np} table\left[j, (maw + \tilde{M}Num(c,x))^*\right] \times Cp_j + \sum_{j=1}^{na_x^u} Ca_x^{uj} \quad (15)$$

It is obvious that urgent aperiodic messages in node 1 will suffer worst-case response time.

$$\tilde{M}Num(c,1) = \min(\psi) \cap \sum_{i=maw}^{\psi} W_a^i \geq \sum_{k=2}^{N}\left(\sum_{j=1}^{na_x^u} Ca_k^{uj} + \sum_{j=1}^{na_x^n} Ca_k^{nj}\right) + \sum_{j=1}^{na_1^u} Ca_1^{uj} \quad (16)$$

$$\tilde{R}a(c,1) = (\tilde{M}Num(c,1) - 1) \times MicP + \sum_{j=1}^{np} table\left[j, (maw + \tilde{M}Num(c,1))^*\right] \times Cp_j + \sum_{j=1}^{na_1^u} Ca_1^{uj} \quad (17)$$

*4.2.4 PT priority changing from urgent to normal*

Assume a scenario that urgent aperiodic messages in node X arrive just after PT finishes its $c^{th}$ visit to node X under PT priority being urgent, and PT priority changes to normal when PT arrives node 1 for its $(c+1)^{th}$ visit.

In this case, LAS only sends urgent aperiodic messages from $visit(c,x)$ to $visit(c,N)$. However, LAS not only sends urgent aperiodic messages but also normal aperiodic messages from $visit(c+1,x)$ to $visit(c+1,x)$. Let $\tilde{M}Num(c,x)$ denote number of microcycles urgent aperiodic messages in node $x$ have to wait.

$$\tilde{M}Num(c,x) = \min(\psi) \cap \sum_{i=m}^{\psi} W_a^i \geq \sum_{k=x+1}^{N}\sum_{j=1}^{na_k^u} Ca_k^{uj} + \sum_{k=1}^{x}\left(\sum_{j=1}^{na_k^u} Ca_k^{uj} + \sum_{j=1}^{na_k^n} Ca_k^{nj}\right) \quad (18)$$

$$\tilde{R}a(c,x) = (\tilde{M}Num(c,x) - 1) \times MicP + \sum_{j=1}^{np} table\left[j, (m + \tilde{M}Num(c,x))^*\right] \times Cp_j + \sum_{j=1}^{na_k^u} Ca_x^{uj} \quad (19)$$

*4.3 The improvement of act counting mechanism*

Eq(10)~(19) indicates that worst-case response time of urgent aperiodic messages in different condition badly varies, and that can't be avoided by PT priority and PT rotation period. PT priority changes to urgent only when the current PT cycle ends and the current ACT being less than SCT. The worst-case scenario in section 4.2.2 actually should not occur if proper SCT can be designed, but that is difficult if it is possible. Further, even scenario in section 4.2.2 never occurs, scenario in section 4.2.3 frequently occurs and response time illustrated in Eq(16)~(17) is larger even it is not the worst. Scenario in section 4.2.2 can't be avoided by setting less SCT unless setting SCT as small as 0, but the latter will cause normal aperiodic message lost opportunity of transmission. Setting SCT as 0 will maintain PT priority at urgent.

Although we assume minimum arrival period for aperiodic message, it is just for convenience of analysis. In fact, the following case may occur: none aperiodic message arrives in a PT cycle, that causes PT maintaining or changes priority to normal. Then in next PT cycle, all aperiodic messages arrive with maximum rates, consequently urgent aperiodic messages suffer larger response time.

Therefore it is necessary to propose a new ACT counting mechanism to enhance real-time capability of FF for urgent aperiodic messages. It is obvious that the new mechanism should timely respond real-time requirement of urgent aperiodic messages and load in FF. This paper introduces a new one that ACT equals the elapsed time between PT twice consecutively arriving a particular node. Thus, ACT is assigned to all nodes instead of node 1 only.

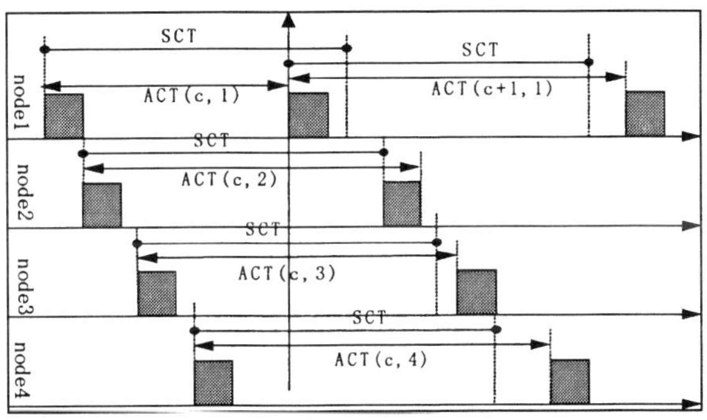


Fig.6 Relationship between the above parameters

Let $ACT(c,x)$ as actual PT rotation period, which is

illustrated in Fig.6. It is obvious that new mechanism will maintain $ACT(c,x)$ upper bound, and can change its upper bound through setting proper SCT.

Assume during $Arr(c+1,1)$ and $Arr(c+2,1)$ , $ACT(c,x)$ being larger than SCT occurs, then PT priority changes to urgent. Further, assume PT priority maintains urgent at least until $Arr(c+2,1)$ . In this scenario, the load of normal aperiodic messages PT priority prevented is determined by $ACT(c,x)$.

Let $\vec{R}a(c,x)$ denote response time of urgent aperiodic messages in node X. Assume PT priority changes to Urgent at node y, let $Lap(x,y)$ denote urgent aperiodic transmitting from $visit(c,y)$ to $visit(c+1,x)$ if x is larger than x; otherwise from $visit(c+1,y)$ to $visit(c+1,x)$.

$$Lap(x,y)=\begin{cases}\sum_{k=x}^{y}\sum_{j=1}^{na_k^u} Ca_k^{uj} & (y<X)\\ \sum_{k=x}^{N}\sum_{j=1}^{na_k^u} Ca_k^{uj}+\sum_{k=1}^{y}\sum_{j=1}^{na_k^u} Ca_k^{uj} & (y>X)\end{cases} \quad (20)$$

$$\overline{M}Num(x,y)=\min(\psi)\cap\sum_{i=maw}^{np} Wa_x^i \geq Lap(x,y) \quad (21)$$

For $\vec{R}a(c,1)$, there are different positions when ACT being larger than SCT occurs, among which $ACT(c,2)$ is the most effective position preventing normal aperiodic messages.

$$\vec{R}a(c,1)\leq SCT+(\vec{M}Num(2,N)-1)\times MicP+ \quad (22)$$
$$\sum_{j=1}^{np} table\left[j,(maw+\vec{M}Num(,N))^{*}\right]\times Cp_j+\sum_{j=1}^{na_1^u} Ca_1^{uj}$$

Note that we analyze $\vec{R}a(c,1)$ is just for convenience of example. In fact, $\vec{R}a(c,x)$ has the following result. For convenience, let $delay(x)$ denote transmitting delay of urgent aperiodic message in node from ((x-1) mod N) to ((x +N-1) mod N).

$$delay(x)=(\vec{M}Num(x+1,(x+N-1) \bmod N)-1)\times MicP+\sum_{j=1}^{na_x^u} Ca_x^{uj} \quad (23)$$
$$+\sum_{j=1}^{np} table\left[j,(maw+\vec{M}Num(x+1,(x+N-1) \bmod N))^{*}\right]\times Cp_j$$

Therefore,

$$\vec{R}a(c,x)\leq SCT+delay(x) \quad (24)$$

SCT should guarantee real-time requirement of any urgent aperiodic message, SCT must meet the following condition,

$$\min(Da_k^j)\geq SCT+\max(delay(x)) \quad (25)$$

We can easily determine the upper bound of SCT from the above analysis.

$$SCT\leq\min(Da_k^j)-\max(delay(x)) \quad (26)$$

## 5. CONCLUSIONS

Supporting real-time traffic of critical aperiodic messages using FF is a complicated issue since it concerns for SB, ACT, SCT and PT priority, etc. This paper first deduces an integrated message transmission model, which integrates periodic messages and aperiodic messages together. Then formula for response times of urgent aperiodic messages is given in this paper. Through the formula, deficiency of ACT counting mechanism in meeting real-time traffic of urgent aperiodic messages is found. Finally, this paper proposes a new mechanism and gives a methodology for setting SCT. The ongoing work is to optimize aperiodic messages priorities under some criterion, such as response time, loss rate, utilization, etc.

## REFERENCES


Buttazzo, G.C. (1997) Hard real-time computing systems predictable scheduling algorithm and application, *Kluwer academic publishers.*

Fieldbus Foundation. (1996). Foundation Specification System Architecture, *Fieldbus Foundation, Austin, Texas.*

Kim, Y. and Jeong, S. (1998) Pre-run-time scheduling method for distributed real-time systems in a FIP environment , *Control Engineer Practice*, **6**(1) pp.103-109.

Kopetz, H. Real time system design principles for distributed embedded application, *Kluwer Academic Publishers.*

Shin, K. G. and Chou, C. (1996) Design and evaluation of real-time communication for fieldbus based manufacturing system, *IEEE Transactions on Robotics and Automation,* **12**(3).

Simonot, F. and Song, Y.Q. (1997) Real-Time Communication Using TDMA-based Multi-Access Protocol, Computer Communication, **20**(6), pp435-448

Song, Y.Q. (1996) Performance analysis of periodic and aperiodic real-time message transmission in FIP networks, *Proc. of 9th international conf. on Parallel and Distributed Computing Systems*, pp499-506,

Tindell, K. and Burns, A. (1997) Analysis of hard real-time communication, *Real-Time Systems,* **9**(2), pp.147-173.

Tovar, E. and Vasques, F. (1999) Supporting real-time distributed computer control system with multi-hop p-net network. *Control Engineering Practice,* **7**(11), pp.1015-1025.

Wang, Z. (2000) Modeling and analysis of fieldbus-based distributed real-time system. The dissertation for the degree of doctor, *Chinese Academy of Sciences.*

Wang, Z., Sun, Y.X., Wang, T.R. (2001) Response Time of Urgent Aperiodic Message in Foundation Fieldbus, *International Symposium on Distributed Computing and Applications to Business, Engineering and Science*, pp.153-159.

Zhao, W. and Stankovic, J. (1990) A window protocol for transmission of time constrained messages, *IEEE Transactions on Computers,* **39**(9), pp.1186-1203.

Zuberi, K.M. and Shin, K.G. (1997) Design and implementation of efficient message scheduling for CAN, *IEEE Trans. on Computers,* **49**(2), pp.182-188.

IFAC Publications
www.elsevier.com/locate/ifac

# STATIC TASK ALLOCATION IN REAL-TIME DISTRIBUTED COMPUTER CONTROL SYSTEMS

**J.C.B. Leite, J.R.A. Torreão, O.G. Loques and A.M. Barroso**

*Instituto de Computação, Universidade Federal Fluminense*
*Rua Passo da Pátria, 156 - CEP 24210-240 - Niterói - Brazil*
*{julius, jrat, loques, melon}@ic.uff.br*

Abstract: The task allocation problem consists in defining a space and time sharing of system resources that can meet all logical and operational constraints of the tasks being considered. For real-time tasks with predictable behavior (periodic), static allocation provides a reliable way to meet temporal scheduling requirements, together with other operational constraints, such as task replication for fault-tolerance. Unfortunately, finding an optimal feasible allocation is known to be NP-hard, and thus heuristic approaches have been proposed. In this paper, the use of a new heuristic to generate static allocations for critical periodic tasks in real-time distributed control systems is described. *Copyright © 2001 IFAC*

Keywords: distributed computer control systems, real-time systems, scheduling algorithms, heuristic search, real-time operating systems.

## 1. INTRODUCTION

Reliable real-time distributed systems have to meet logical, temporal and operational constraints. The achievement of this goal depends on the particular allocation of the application tasks to the distributed processing nodes. This allocation has to be performed in order to meet temporal requirements, together with operational constraints, such as task replication for fault-tolerance and local availability of specific hardware resources. Task allocation can be performed dynamically or statically. In the dynamic case, the tasks-to-nodes mapping may undergo changes during system operation. Examples of algorithms for dynamic task allocation are given in (Bestavros, 1996; Shin and Hou, 1994; Stankovic, *et al.*, 1985). For periodic or sporadic tasks, with predictable behavior, static allocation provides a simple and reliable way to meet strict operational constraints. Unfortunately, there are $m^n$ ways of assigning $n$ tasks to $m$ processors. Finding an optimal feasible allocation is known to be NP-hard and thus, heuristics have been used. Approaches for finding good allocations in polynomial time with the use of heuristics can be found in (DiNatale and Stankovic, 1995; Santos, *et al.*, 1997; Tindell, *et al.*, 1992). In this paper, the use of one such heuristic for static allocation of critical periodic tasks in a distributed real-time computer control system is described. This type of approach could be a valuable tool for practical real-time system design, given its speed in finding a near-optimal solution. In order to evaluate the advantages of the proposed heuristic, called microcanonical optimization ($\mu O$), a study is performed comparing the quality of the achieved allocations using $\mu O$ to the ones obtained through simulated annealing (Aarts and Korst, 1989), Tabu search (Reeves, 1993) and Tsallis's annealing (Penna, 1995).

## 2. PRELIMINARY CONCEPTS

Task allocation in distributed systems is a combinatorial optimization problem. The set of all possible assignments of tasks to processors defines the solution space $U$. An allocation $\Omega \in U$ is known as a solution of the problem. In order to represent the quality of each allocation, a cost function $E(\Omega)$: $U \rightarrow \Re$ is defined that quantifies how well this allocation satisfies the optimization goals. The greater the value returned by $E(\Omega)$, the farther is the solution from the desired optimum.

In order for a heuristic to be able to explore the solution space of a problem, it is necessary to define the possible candidates for evaluation in the next step. The set of such solutions is named the neighborhood set of $\Omega$, and is noted by $V(\Omega)$. An allocation $\Omega^* \in V(\Omega)$ is known as a neighbor of $\Omega$. In this paper, two kinds of moves are considered through which $\Omega$ can be transformed in a neighbor $\Omega^*$:

i. Simple move: a task assigned to node $n_i$ is moved to node $n_j$, where $i \neq j$;
ii. Double move: a task assigned to node $n_i$ changes position with a task assigned to node $n_j$, where $i \neq j$.

## 3. HEURISTICS EVALUATED

A brief description of the evaluated heuristics is given in the sequel. Further details could be found in the references indicated in the text.

### 3.1 Simulated Annealing

The simulated annealing (SA) heuristic has its origin in statistical mechanics. The term annealing refers to a metallurgical process by which a metal is heated up to high temperatures (and thus, high energy states), and then slowly cooled down, bringing it to a low energy state, to yield a material that is free of stress.

The simulated annealing heuristic emulates the physical annealing process to achieve minimum cost solutions, with the cost function playing the role of the physical energy (Aarts and Korst, 1989). Applied to task allocation, SA starts by considering an arbitrary allocation as the current solution ($\Omega_{curr}$). At each step, the algorithm chooses an arbitrary neighbor solution $\Omega^* \in V(\Omega_{curr})$ and makes $\Omega_{curr} = \Omega^*$, if one of the following conditions holds:

i. $E(\Omega^*) < E(\Omega_{curr})$

ii. $P_N$ is greater than a random number generated in the interval [0,1], where

$$P_N = e^{-\frac{E(\Omega^*) - E(\Omega curr)}{T}} = e^{-\frac{\Delta E}{T}} \quad (1)$$

If none of the two conditions holds for the chosen $\Omega^*$, then $\Omega_{curr}$ is retained, and another neighbor solution is considered. $P_N$ is a measure derived from the Boltzmann distribution, and $T$ is a control parameter that assumes a high value at the beginning of the algorithm. Note that, the higher $T$, the easier is the second condition to be satisfied. $T$ is kept constant for a number of iterations, and the process is repeated until a stopping condition (corresponding to thermal equilibrium) is signaled. T is then gradually reduced and the algorithm proceeds as before, stopping when a minimum $T$ value is reached. The rate of cooling and the initial and final values of the control parameter comprise the annealing schedule. Problems of convergence are discussed in (Aarts and Korst, 1989; DiNatale and Stankovic, 1995). It should be pointed out that approximate, non-optimal solutions, can be obtained in polynomial time, and are sufficient for most applications, but the adequate choice of the annealing schedule is fundamental for the quality of the results.

### 3.2 Tsallis's Annealing (TA)

The slow convergence towards a solution close to the global optimum is a well-known shortcoming of SA. In order to speed up the process, variants, such as the fast simulated annealing (Szu and Hartley, 1986) have been proposed. In (Penna, 1995), a new annealing strategy has been introduced based on the generalized statistical mechanics proposed by C. Tsallis (Stariolo, 1994). The algorithm is basically the same as described for the SA, but with the Boltzmann measure (1) now replaced by

$$P_N = \left[1 - (1-q)\frac{\Delta E}{kT}\right]^{\frac{1}{(1-q)}} \quad (2)$$

where the parameter $q$ is referred to as the Tsallis constant, and $k$ is another constant. For negative values of $q$, it has been claimed that good solutions to the traveling salesman problem were obtained in fewer steps than with both the classical and the fast annealing algorithms (Penna, 1995).

### 3.3 Tabu Search

The well known Tabu search heuristic avoids being trapped at a local minimum by keeping track of solutions already visited and prohibiting the return to these "Tabu" solutions for a given interdiction period (Reeves, 1993). There are many variants of this heuristic, and hence it is convenient to point out the main features of the implementation employed in this work. The implementation starts by considering an arbitrary task allocation as the current solution ($\Omega_{curr}$). At each step, the algorithm picks the neighbor solution $\Omega^* \in V(\Omega_{curr})$ with the lowest energy value and makes $\Omega_{curr} = \Omega^*$ if one of the following conditions holds:

i. $\Omega^*$ is not a Tabu solution;
ii. $\Omega^*$ is a Tabu solution, but $E(\Omega^*)$ is the lowest energy found by the algorithm up to the present moment.

The Tabu search algorithm finishes after a predefined number of consecutive iterations without improvement of the best solution so far encountered.

### *3.4 Microcanonical Optimization*

The microcanonical optimization heuristic consists of two phases, which are alternately applied (Linhares and Torreão, 1998; Torreão and Roe, 1995). The first one, called *initialization*, performs a simple local search, accepting solutions, which reduce the cost function. This means that, starting from an arbitrary task allocation, $\Omega_{curr}$, the algorithm chooses an arbitrary neighbor solution, $\Omega^* \in V(\Omega_{curr})$, and makes $\Omega_{curr} = \Omega^*$ if and only if $E(\Omega^*) < E(\Omega_{curr})$. The initialization phase stops when the algorithm is close to a local minimum. In order to proceed with the search of possible better solutions, $\mu O$ then starts its second procedure, named *sampling*. During this phase, the algorithm accepts solutions whose cost is within a small band of values around the cost attained in the initialization. Explicitly, given the current solution, $\Omega_{curr}$, the algorithm picks an arbitrary neighbor solution, $\Omega^* \in V(\Omega_{curr})$, and makes $\Omega_{curr} = \Omega^*$ if and only if $E(\Omega^*) \in [E_S - \alpha, E_S + \beta]$, where $E_S$ is the final solution cost of the previous initialization, and $\alpha$ and $\beta$ are small values determined in the sequel. The sampling phase finishes after a predefined number of iterations, and a new initialization is executed. The algorithm thus proceeds, alternating the two phases, until no further improvement of the cost function is possible. In what follows, more details on the two phases of $\mu O$ are given.

<u>Initialization Phase:</u> To attain a low energy configuration, a random search is started whereby, from an initial allocation, each move that leads to a lower energy is accepted. If the proposed move entails an energy increase, the allocation is rejected, but the change in energy that would result is stored in a vector variable. This vector will thus keep a rough map of the topology of the cost function in those regions of the solution space explored at the initialization. This map will be useful for the definition of the parameters of the following sampling phase. The initialization stops when a given number of consecutive moves are tried without a decrease in energy. The last task in this phase is to check if the energy of the final solution is smaller than the minimum energy obtained in previous initializations. If not, a variable that stores the number of consecutive iterations without improvement is incremented. When this variable reaches a predefined maximum value, the whole algorithm is made to stop.

<u>Sampling Phase:</u> At the start of this phase, the energy band to which the solutions will be restricted must be defined. The limiting values of such band, $\alpha$ and $\beta$, are obtained from the maximum and the initial values, $E_{DMAX}$ and $E_{DI}$, of a variable called the "demon" (Creutz, 1983), whose function is to generate small and controlled disturbances on the current solution, allowing it to escape from a local minimum but without straying too much from there. Thus, when a move is proposed which leads to a configuration of higher energy, it will only be accepted if the demon can supply the required energy increase. Similarly, a move that leads to a decrease in energy will only be accepted if the demon can accommodate the released energy. The result is that, at a given sampling phase, the accepted solutions will lie within the band of values $[E_S - E_{DMAX} + E_{DI}, E_S + E_{DI}]$ (see Figure 1), where $E_S >> E_{DMAX} > E_{DI}$ is the final energy obtained in the previous initialization, and $E_{DI}$ and $E_{DMAX}$ are two low entries in a ordered vector compiled during that phase. The idea is to choose, for the demon parameters, values that are of the order of the uphill moves proposed in the initialization. The sampling procedure stops when a given number of iterations is reached. Details and a pseudo-code for the two procedures can be seen in (Barroso, 1998).

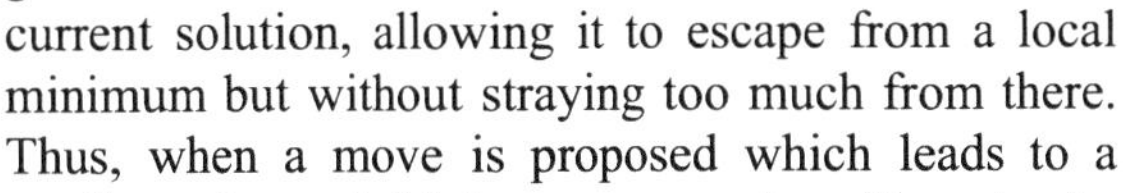


Fig. 1. Allowed energy states for the system during the sampling. $E_S$ denotes the cost of the solution attained in the previous initialization.

## 4. THE DISTRIBUTED SYSTEM MODEL

In order to compare the quality of the solutions achieved by the previously described heuristics, the distributed system model proposed in (Tindell, *et al.*, 1992), where forty three tasks have to be allocated to eight nodes (processors), was adopted as a benchmark. Its communication system uses a bus topology and a token-passing access control protocol (of course, other topologies and access control rules could be used). Each computing node has well-known processing power and memory capacities; some of them may have specialized hardware, *e.g.*, for I/O. A set of independent periodic tasks with hard real-time requirements can be allocated and potentially executed in each node. To meet fault tolerance constraints, five critical tasks are replicated; clearly, replicas of a task have to be allocated to different nodes. For each task, the period, the worst-case execution time, the memory needs, the size and destination of messages, and any special hardware requirements are defined. In addition, a limited number of messages can be generated at each task activation (the length of these messages is limited and known). It is also assumed that the intra-processor message transmission time is negligible. For inter-processor task-to-task communication, the time limit for message queueing depends on the system load and on the bus speed capacity. This time can be obtained by calculating the Token Rotation Time (TRT) associated to each particular task-to-processor mapping. The token rotation time grows with the number of messages sent through the bus. All messages are transmitted by a task only at the end of its execution and must be delivered at the destination before the beginning of its next period. A task deadline (D) is hence computed considering the following cases (P is the task period):

i. The task has only messages to be delivered locally (only intra-processor communication involved):
$$D=P \quad (3)$$
ii. The task has messages to be delivered remotely (inter-processor communication involved):
$$D=P-TRT \quad (4)$$

The execution requirements of a task mapped to a given processor has to be guaranteed in spite of the processing load imposed by the other tasks allocated to the same processor. Thus, for a given system mapping, a local schedulability test has to be performed for each processor. Considering the Deadline Monotonic algorithm as the local scheduling policy (other policies could be assumed), this test is easily accomplished (Audsley, 1990). Some allocation constraints other than task deadlines are also imposed to the solutions. The best allocation is the one that satisfies all constraints and cannot be improved.

Each allocation constraint has an associated energy (cost) corresponding to a penalty for its non-accomplishment. The total energy of the solution could be expressed as a linear combination of these partial energies (Tindell, 1992), defined as follows:

$$E_{total}(\Omega) = \sum_{i=1}^{n} K_i.E_i \quad (5)$$

where,

i. $E_{total}(\Omega)$: Energy of the solution $\Omega \in U$;
ii. $E_i$: Energy contribution related to allocation constraint $i$;
iii. $K_i$: Weighting factor related to allocation constraint $i$.

In this model, the following constraints were considered:

1. Some tasks have to be allocated to specific processor types, *e.g.*, due to specialized hardware for I/O;
2. Task replication can be used to implement a fault-tolerance technique. Clearly, replicas of a task have to be allocated to different nodes;
3. The total memory requirement for a specific task set at a given node cannot exceed its capacity;
4. The task set allocated to a given node has to have a feasible execution schedule;
5. The token rotation time grows with the number of messages sent through the bus. Although this is not mandatory, a small TRT is desirable.

The choice of the coefficients $K_i$ depends on the relevance of the associated factor to the overall optimization process. For example, the minimization of the *TRT* is desirable, but does not necessarily leads to a feasible allocation. Thus, its weight can be tuned by a proper choice of $K_i$.

## 5. COMPARATIVE EVALUATION

The model proposed in (Tindell, *et al.*, 1992) is representative of a class of distributed computer control real-time systems. Its use provides a common basis for comparisons and allows reproducible results to be obtained. It should also be pointed out that the implementations of the several heuristics done in this paper tried to get the best results in any case, that is to say, they were not biased.

### *5.1 Methodology*

For the set of tasks and processors described, several runs of simulations were performed on a Sun Ultra-1 workstation. The measures considered for the performance evaluation of the heuristics were the following:

i. $T_H$: Average execution time (in seconds);
ii. $P_F$: Percentage of results which satisfy all constraints;
iii. $E_F$: Average energy of the results which satisfy all constraints;
iv. $T_{TRT}$: Average TRT of the results which satisfy all constraints (in ms);
v. $P_{MIN}$: Percentage of the results that correspond to the global optimum.

All measured values have a small maximum error (indicated in Table 1) with a confidence level of 95%. The weighting factors took the values $K_1$ = 100,000, $K_2$ = 15,000, $K_3$ = 100, $K_4$ = 10,000, e $K_5$ = 10. These values were chosen to reflect the importance of each constraint in the total energy of the solution evaluated. Each $E_i$ was obtained as in (Tindell, *et al.*, 1992). The stopping criteria and the parameters of the compared heuristics are described in (Barroso, *et al.*, 1998).

### 5.2 Discussion

Table 1 shows the best results obtained for each heuristic. It can be seen that the microcanonical optimization outperforms the other heuristics in all of the evaluation criteria considered. Indeed, the $\mu O$ approach finds, with greater probability (see $P_F$ in Table 1) and in lesser time (see $T_H$, also in Table 1), solutions that:

i. Satisfy all imposed constraints;
ii. Provide for better use of the bus (see $T_{TRT}$);
iii. Correspond to the best reachable allocation (see $E_F$).

Table 1 Performance figures for the heuristics

| Figures | SA | µO | TA | Tabu Search |
|---|---|---|---|---|
| $T_H$ | 54±1 | 19±1 | 27±1 | 34±1 |
| $P_F$ | 77±1 | 86±1 | 61±2 | 51±2 |
| $E_F$ | 8024±27 | 7880±14 | 7944±27 | 8466±52 |
| $T_{TRT}$ | 8,92±0,03 | 8,76±0,02 | 8,83±0,03 | 9,41±0,06 |
| $P_{MIN}$ | 34±3 | 49±3 | 49±4 | 7±2 |

Finally, Figures 2 to 5 show sample curves for the time evolution of the execution of the various heuristics implemented. These samples are a good indication of the general trend; that is, convergence is reached faster through the use of the $\mu O$ algorithm. The use of a local search in the first phase certainly speeds-up the algorithm. Clearly, there is not a heuristic that will outperform all the others in every problem. However, accordingly to the results obtained, the $\mu O$ heuristic can be a very useful tool in the design of real-time distributed systems. It is our intent to add other refinements to the model proposed by Tindell, *et al.* (1992), and to produce results that also minimize jitters and that take other task models into account.

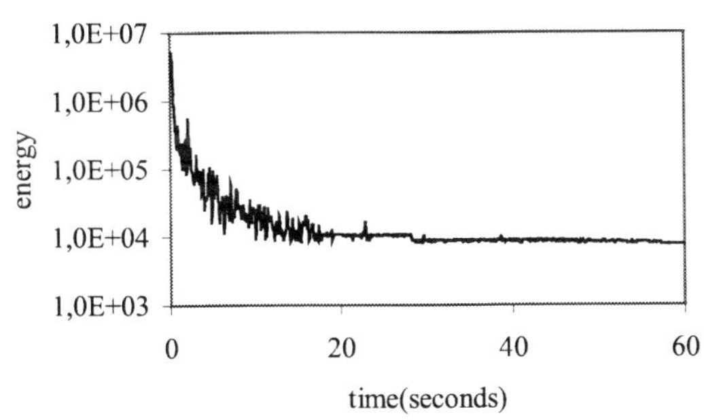


Fig. 2. Sample time-evolution curve for SA.

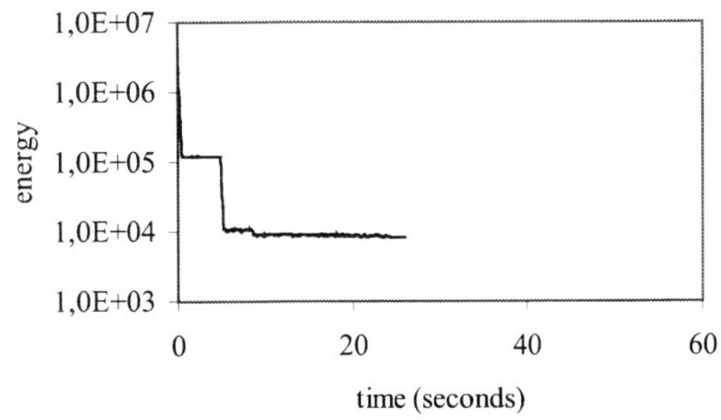


Fig. 3. Sample time-evolution curve for TA.

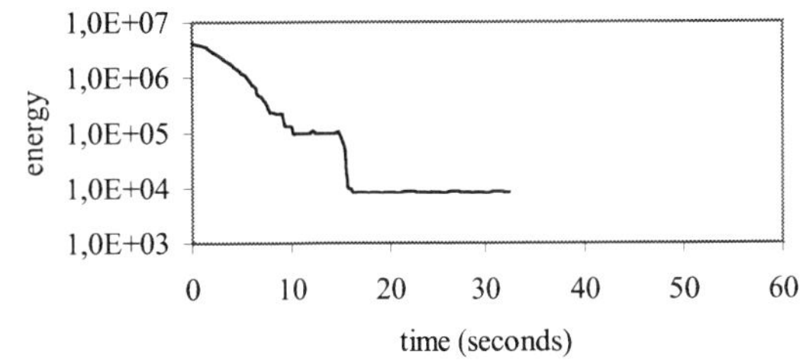


Fig. 4. Sample time-evolution curve for Tabu search.

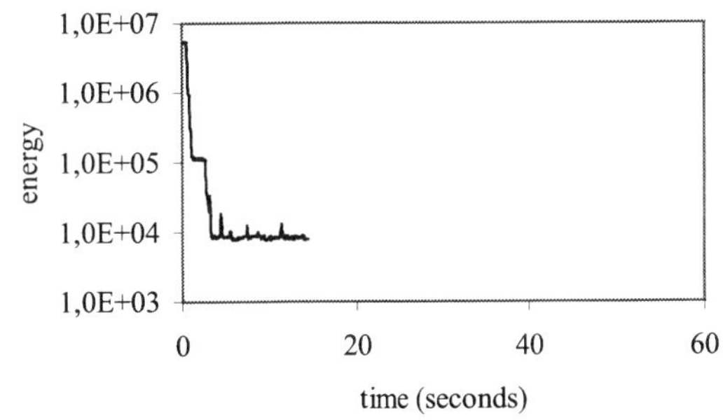


Fig. 5. Sample time-evolution curve for $\mu O$.

## 6. CONCLUSIONS

The use of heuristics for task allocation in real-time computer control systems still is an open area of research. To our knowledge, the work by Tindell, *et al.* (1992), which used simulated annealing as described in this paper, was the first to explore the subject. Recently, Di Natale and Stankovic (1995) have used simulated annealing to minimize jitter for periodic tasks, subjected to arbitrary deadlines, precedence, and exclusion constraints, in distributed systems. In (Santos, *et al.*, 1997) a heuristic approach was applied to the problem of assigning a set of preemptible resource-sharing and blockable real-time tasks to a set of heterogeneous processors communicating through an interprocessor network. Here, the use of a new heuristic to define a static allocation of critical periodic tasks in a distributed real-time system was reported. In order to evaluate the benefits of the proposed heuristic, termed microcanonical optimization, a study was performed comparing the quality of the allocations achieved by the use of $\mu O$ to those obtained through Simulated Annealing, Tabu search and Tsallis's annealing. It could be seen from the experiments that microcanonical optimization was able to outperform the others heuristics in finding better allocations in less time. To this date, this heuristic has been successfully experimented in three different problems: in the Traveling Salesperson Problem (Linhares and Torreão, 1998), in linear gate assignment in VLSI circuits (Linhares, *et al.*, 2000) and in the task allocation context here described. It is hoped that these promising results encourage the application of the $\mu O$ heuristics to other problems in the real-time systems design area.

## ACKNOWLEDGMENTS

This work has been partially supported by the brazilian research funding agencies CNPq, Finep and Faperj.

## REFERENCES

Aarts, E. and J. Korst (1989). *Simulated annealing and Boltzmann machines*. John Wiley and Sons, USA.

Audsley, N. (1990). Deadline monotonic scheduling. *Technical Report YCS146/90*, Department of Computer Science, University of York, England.

Barroso, A., J.R.A. Torreão, J. Leite and O. Loques (1998). A heuristic approach to task allocation in real-time distributed systems. *Technical Report* RT-07/98 (www.ic.uff.br/reltec.htm), Universidade Federal Fluminense, Niterói, RJ, Brazil.

Bestavros, A. (1996). Load profiling in distributed real-time systems. *Technical Report* 96-017, Department of Computer Science, Boston University, USA.

Creutz, M. (1983). Microcanonical Monte Carlo simulation. *Physics Review Letters*, **50**, 1411.

DiNatale, M. and J.A. Stankovic (1995). Applicability of simulated annealing methods to real-time scheduling and jitter control. *16th IEEE Real-Time Systems Symposium*, Pisa, Italia.

Linhares, A. and J.R.A. Torreão (1998). Microcanonical Optimization Applied to the Traveling Salesman Problem. *Intl. J. Modern Phys. C: Physics and Computers*, **9**(1), 133-146.

Linhares, A., H.H. Yanasse and J.R.A. Torreão (2000). Linear gate assignment: a fast statistical mechanics approach. *IEEE Trans. on Computer-Aided Design of Integrated Circuits and Systems*, **18**(12), 1750-1758.

Penna, T.J.P. (1995). Traveling salesman problem and Tsallis statistics. *Physical Review E*, **51**(1), R1-R3.

Reeves, C.R. (1993). *Modern heuristic techniques for combinatorial problems*. John Wiley and Sons, USA.

Santos, J., E. Ferro, J. Orozco and R. Cayssials (1997). A heuristic approach to the multitask-multiprocessor assignment problem using the empty-slots method and rate monotonic scheduling. *Real-Time Systems*, **13**, 167-199.

Shin, K. and C. Hou (1994). Design and evaluation of effective load sharing in distributed real-time systems. *IEEE Transactions on Parallel and Distributed Systems*, **5**, 704-719.

Stankovic, J., K. Ramamrithan and S. Cheng, (1985). Evaluation of a flexible task scheduling algorithm for distributed hard real-time systems. *IEEE Transactions on Computers*, **C-34**, 1130-1143.

Stariolo, D.A. and C. Tsallis (1994). In: *Annual Review of Computational Physics II* (D. Stauffer (Ed)), World Scientific, Singapore.

Szu, H. and R. Hartley (1986). Fast simulated annealing. *Physics Letters*, **A 122,** 157-162.

Tindell, K.W., A. Burns and A.J. Wellings (1992). Allocating hard real-time tasks: an NP-hard problem made easy. *Real-Time Systems*, **4**, 145-165.

Torreão, J.R.A. and E. Roe (1995). Microcanonical optimization applied to visual processing. *Physics Letters, A* **205**, 377-382.

IFAC Publications
www.elsevier.com/locate/ifac

# A NEW METHOD TO BUILD THE FF SCHEDULE TIME LIST

**Yue Zhou[1] Zhi Wang[2] Hai-bin Yu[1] Tian-ran Wang[1]**

*1 No.3 Laboratory, Shenyang Institute of Automation, Chinese Academy of Sciences*
*No.114 Nanta Street, Shenyang 110016 P. R. China*
*2 National Laboratory of Industrial Control Technology, Zhejiang University,*
*No.20 Yugu Road Hangzhou, 310027, China*
*E-mail: zhouyue@ms.sia.ac.cn*

Abstract: In a single segment Foundation Fieldbus system, a link active scheduler, according to a pre-defined schedule time list, controls all remote periodic message transfers between different field devices on bus. The remote periodic message not only has time constraints but also often has precedence constrains. In this paper a new method to build the schedule time list is proposed, in which both time constraints and precedence constrains are considered and described by modifying release time and deadline. The method proposed is based on slack time monotonic and local search. An example of application shows the process of building FF schedule time list. *Copyright © 2001 IFAC*

Keywords: precedence constraints, time constraints, slack time monotonic, local search.

## 1. INTRODUCTION

Foundation Fieldbus (FF) is the standard developed by the Fieldbus Foundation. It is an all digital, serial, two-way communication system which interconnects field equipment such as sensors, actuators and controllers. It is a Local Area Network (LAN) for instruments used in both process and manufacturing automation with built-in capability to distribute the control application across the network (Fieldbus Foundation, 1996).

FF system set the Function Blocks (FBs) down in field devices to perform control application, thereby, which really realizes distributed process control. At the same time, it brings forward the real-time communication requirements for some message transfers, i.e., these messages are sternly demanded to successfully transmit to their destinations within a precise and bounded interval to guarantee FF control system run normally. For example, those remote periodic messages used to refresh data of FBs input/output buffers need to be real-time transmitted on bus.

FF protocol is based on a centralized communication media access control mechanism, where a link active scheduler (LAS), according to a schedule time list pre-defined, controls all remote periodic message transfers between different field devices on bus (Zhou, et al., 2001). Hence, it is crucial how to build a schedule time list considering all remote periodic messages transfers in order to meet the real-time communication requirements, i.e., their time constraints.

[1] This work is supported by NSFC (60084001).

The majority of the methods to build FF schedule time list have assumed that all periodic messages are independent, and are based on critical instant (Tovar, 1999; Wang, 2000). Critical instant is the time that all periodic messages simultaneously send request to access the bus. It can occur for independent periodic messages and is the worst-case. However, for the periodic message transfers, FF systems also have to consider precedence constraints caused by the execution sequence of FBs. This means that all periodic messages cannot send simultaneously request to access the bus. Hence, based on critical instant, the methods to build FF schedule time list become pessimistic and unreasonable.

In this paper a new method to build the schedule time list is proposed, in which both time constraints and precedence constrains are considered and described by modifying release time and deadline. The method is based on slack time monotonic and local search. An example of application shows the process of building FF schedule time list.

## 2. REMOTE PERIODIC MESSAGE MODEL

It is known that FF system is composed of interrelated and independent FBs to perform control functions desired. Moreover all field devices have a common sense of time, they can be cyclically scheduled to execute the FBs in a determined sequence and publish their messages, which may be used by FBs located in other field devices, at a determined time. And these FBs will be executed only after the required inputs are scheduled to be available (Fayad, 2000).

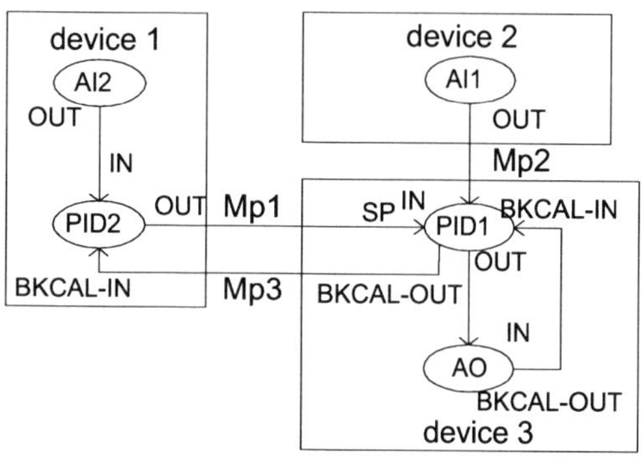


Fig.1 A cascade loop field devices configuration

The engineer only chooses FBs type, place and links by configuration pre-run-time to perform some control application. Fig.1.shows a cascade loop field devices configuration. In this configuration, we have 5 FBs (AI1, AI2, PID1, PID2, AO), 3 external links (Mp1, Mp2, Mp3) and 3 internal links in 3 field devices. Each external link corresponds with a remote periodic message, which has time constraint and is periodically transmitted on bus. Each internal link corresponds with a local periodic message, which is deal with locally and not transmitted on bus, so it will not be considered in this paper. To facilitate the description in following sections, remote periodic message is simply called as message.

From Fig.1, PID2 is executed after AI2 finishes, then, message Mp1 sends request to access the bus, i.e., message Mp1 is released; after PID2 and AI1 finish, and message Mp1and Mp2 arrive device3, PID1 will be executed. As before, this means that all messages cannot be simultaneously released. Hence the precedence constraints of messages have to be considered in the process of building the schedule time list.

To solve precedence constraints, an algorithm was presented (Chetto, et al., 1990). The base idea of their approach is to transform a set J of dependent tasks into a set J* of independent tasks by an adequate modification of timing parameters. The transformation algorithm ensures that J is schedulable and the precedence constraints are obeyed if and only if J* is schedulable.

Therefore, according to the requirement of control system, the control scan time, i.e., period of each control loop, is set. Then, considering their execution sequence, execution time and external links, FBs' start times are set and referred to system management, which will not be introduced in this paper. After FBs' start times were set, release time, at which message send requirement to access the bus, and deadline for each message should be modified to meet time constraints and precedence constraints.

Without loss of generality, a single segment of FF has n devices, mp FBs, and np messages. Moreover, the start time and execution time of each FB are known. In this case, $Rp_i$ and $Dp_i$ of $Mp_i$ are modified as,

$$Rp_i = S_{FBj} + C_{FBj}, \forall i : FB_j \rightarrow Mp_i \quad i = 1,\ldots,np, and \quad j = 1,\ldots,mp \tag{1}$$

$$Dp_i = \min(S_{FBj}), \forall j : Mp_i \rightarrow FB_j \quad i = 1,\ldots,np, and \quad j = 1,\ldots,mp \tag{2}$$

Where $FB_j \rightarrow Mp_i$ denotes that $Mp_i$ is released by $FB_j$, $S_{FBj}$ and $C_{FBj}$ denote the start time and execution time of $FB_j$, $Mp_i \rightarrow FB_j$ denotes $Mp_i$ is received by $FB_j$, of course, a message can be received by several FBs simultaneously. Equation (1) and (2) mean, for all messages, the release time $Rp_i$ of message $Mp_i$ cannot exceed the end time of the FB producing $Mp_i$, at the same time, the deadline

$Dp_i$ of $Mp_i$ cannot lag behind the earliest start time of the FBs receiving $Mp_i$.

Therefore, considering time constraints and precedence constraints, each message can be described as,

$$Mp_i = (Rp_i, Cp_i, Tp_i, Dp_i),\quad i = 1,\cdots,np \qquad (3)$$

Where $Rp_i$, $Cp_i$, $Tp_i$ and $Dp_i$ denote the release time, maximum transfer time, period (which is equal to control scan time of the control loop in which $Mp_i$ lies) and deadline of message $Mp_i$.

From above, the message in same control loop may be look as independent by modifying their release time and deadline. And the messages between different control loops are also independent. Thus, all messages will be look as independent in following sections. The FF schedule time list will be built based on independent messages, but not on critical instant.

## 3. FF SCHEDULE TIME LIST BUILDING

### *3.1 Priority assignment*

In a single processor system, adequate task scheduling algorithms are applied to meet the real-time application constraints, i.e., to guarantee to meet task's deadline. Usually, the tasks are assigned priorities with respect to their features. The priority-based approach is an important scheduling to meet real-time requirement of task (Buttazzo, 1997).

Single processor task scheduling becomes analogous to message scheduling over a single segment FF system. In a single processor computing system, a set of tasks shares a common resource-processor; similarly, in a single segment FF control system, a set of message transfers shares a common resource-bus. Hence, the priority-based approach is adopted for message scheduling in this paper.

*Definition 1:* Slack time, is the maximum time a message can be delayed on its buffer from its release to complete within its deadline. For message $Mp_i$, its slack time $T_{Si}$ is,

$$T_{Si} = Dp_i - Rp_i - Cp_i,\ i = 1,\ldots,np \qquad (4)$$

*Arithmetic:* Slake Time Monotonic (STM), assigns priorities to messages according to their slack time. Messages with shorter slack time will have higher priorities.

This means, more shorter slack time is, more shorter waiting time allowed on buffer is, more stringent message is, and then more higher priority is. After their release times and deadlines of are modified, all messages are assigned priorities according to STM arithmetic in this paper. Thus, during building FF schedule time list, the messages with higher priority are assigned the bus resource prior to those with lower priority during building FF schedule time list.

### *3.2 Local search*

In FF system the messages scheduling is non-preemptive. That is, a message, once started, is transmitted on bus until completion. As previously stated, all messages are assigned time resource, i.e., scheduled, according to their priorities. It may occur that one and more messages cannot meet their deadlines in some time segment because more time resources have been assigned to higher priority messages, which is call as local dispute. In this paper, a simple and effective method is proposed to solve the local dispute.

*Method:* Local search, appropriately modify the scheduling sequences of the messages causing local dispute. The modifying rule is exchanging the scheduling sequences arranged between lower priority messages causing the local dispute as much as possible until a schedulable sequence is searched.

From above, the method of local search is a kind of search of the scheduling sequences of the partial periodic messages in local time segment. The method will be applied in the middle of building FF schedule time list.

### *3.3 Schedule time list building*

To simplify the description, the two concepts are introduced as follows:

*Definition 2*: Macrocycle. is the lease common multiple of period of all messages.

As before, LAS will cyclically schedule all messages according to a pre-defined schedule time list built. The cyclic interval is the macrocycle, which regulates the size of the list and adequately reflects the messages scheduling.

Thus, each messages $Mp_i$ will be scheduled $n_i = Macrocycle / Tp_i$ times in a macrocycle, and the j[th] release time $Rp_{ij}$ and the corresponding deadline $Dp_{ij}$ are described as,

$$\begin{cases} Rp_{ij} = Rp_i + (j-1)Tp_i, \\ Dp_{ij} = Dp_i + (j-1)Tp_i, \end{cases} j = 1,\ldots,n_i \qquad (5)$$

*Definition 3*: Block time, represents the delay of a message transfer with respect to its release time.

The block time will be zero if and only if the message is released while the bus is free, and the same time, the message transfer finish time cannot exceed the other higher priority messages' start time if it start to transmit at its release time. Otherwise, the message will suffer from block by other higher messages directly and/or indirectly. This means the message has to wait because the bus has been occupied by other higher messages and/or the interval not occupied by other higher messages is not enough to perform the message transfer though the bus is free. So it is possible that some time segments are free though some messages are eager to be transmitted on bus.

In this paper the method to build FF schedule time list is based on STM and local search, which assigns priorities to messages and solves the local dispute respectively. All messages compose the set of messages. The process of building the list is as follows,

(1) All messages are ordered by their decreasing priorities. From the highest to the lowest message, repeatedly do following steps.
(2) Beginning from the referenced zero within a macrocycle.
(3) The message may be scheduled at its release time if the bus is free at the moment and the message transfer cannot affect the scheduling of the higher priorities, i.e., its transfer finish time cannot exceed other higher priority messages' start times which have been determined. Or else,
(4) Wait until the instant when the enough bus resource is available for the message transfer. At the same time, the message can meet its deadline if it start to transmit at the instant, the message is scheduled to execute at the instant, i.e., after block time undergone with respect to its release time. Or else,
(5) Do the local search until the schedulable sequence of these messages causing local dispute is found. If this is not possible, the set of messages is not scheduled.

Finally, if all messages are schedulable, that is, they all can meet their deadline, their scheduling times, i.e., the start instants when messages are transferred, compose a schedule time list which is referred to the LAS.

Note that, if one and more messages are not scheduled by using the method, it is necessary to reset the start time of FBs, to change the place of FBs, or/and to turn the single segment into multi-segment FF system.

## 4. AN EXAMPLE OF APPLICATION

Considering a single segment FF system shown in Fig.2, 3 independent control loops consist of 11 FBs in 7 field devices. 6 messages, corresponding with 6 external links, need real-time communication.

As before, considering execution sequence, execution time and external links, FBs start times are set. After FBs start times were set, release time and deadline for each message should be modified, and the message model considering precedence constraints will be obtain. Table 1 shows the time parameters of 6 messages.

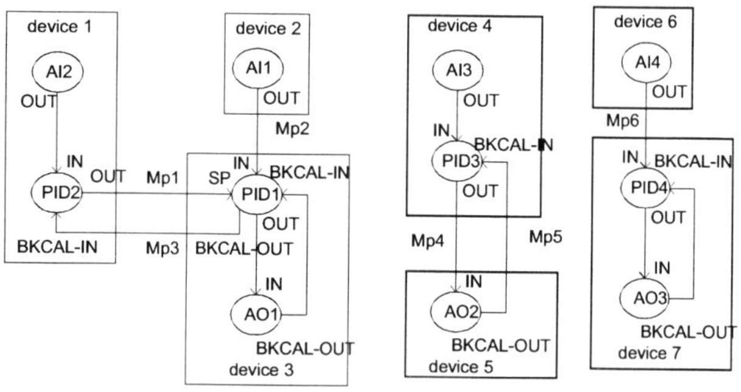


Fig.2 A single segment FF system configuration

Table 1 Time parameters of messages

| Message | $Rp_i$ /ms | $Cp_i$ /ms | $Dp_i$ /ms | $Tp_i$ /ms | $T_{Si}$ /ms |
|---|---|---|---|---|---|
| Mp1 | 100 | 20 | 152 | 300 | 32 |
| Mp2 | 35 | 20 | 152 | 300 | 97 |
| Mp3 | 240 | 20 | 300 | 300 | 40 |
| Mp4 | 105 | 15 | 150 | 200 | 30 |
| Mp5 | 170 | 15 | 200 | 200 | 15 |
| Mp6 | 20 | 15 | 35 | 100 | 0 |

Thus, 6-5-4-1-3-2 is the sequence ordered by decreasing priorities according to STM arithmetic. Then, applying the scheduling method proposed in this paper, the schedule time list is built and the scheduling process within a macrocycle is shown in Fig.3.

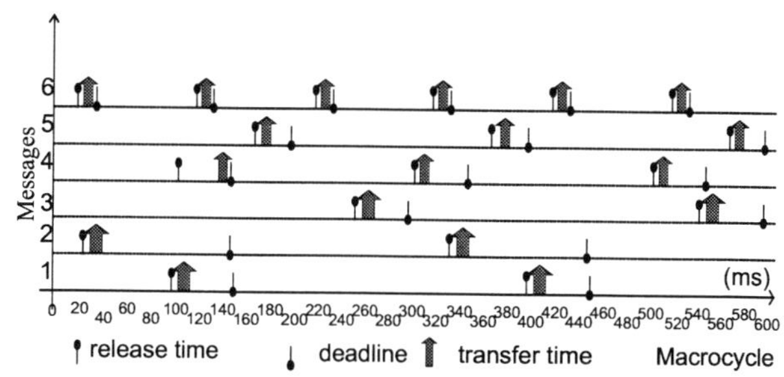


Fig.3 Scheduling process of messages

Note that, though message 4 is higher than message 1 at priority level and is arranged more early than message 1, message 1 is transmitted before message 4 at 100 ms because the local dispute occurs. That is, if message 4 and 6 are scheduled at 105ms and 120 ms respectively, the message 1 are scheduled at 135ms, which makes message 1 not to meet its deadline. Thus, from 100ms to 150ms, the scheduling sequence of message 4 and message 1 against their priority order is the result of local search.

## 5. CONCLUSIONS

In this paper, a new method to build FF schedule time list for remote periodic messages is proposed. Both time constraints and precedence constrains are considered and described by modifying release time and deadline. The new approach to build FF schedule time list, based on slack time monotonic arithmetic, which is used to assign priorities for remote periodic messages, and local search method, which is used to solve the local dispute, is presented to meet time constraints and precedence constraints. Finally, the building process of the list is showed by an example of application, which will be useful to engineers and researchers.

## REFERENCES


Buttazzo,G.C. (1997) *Hard real-time computing systems predictable scheduling algorithm and application*. Kluwer academic publishers.

Chrtto,H., Silly,M. and Bouchentouf,T. (1990). Dynamic scheduling of real-time tasks under precedence constraints, *Real-Time Systems,* **2**(3), pp. 181-194.

Fayad,C.A. (2000). Process control performance on a foundation fieldbus system. *ISA EXPO/2000*, 21-24 August.

Fieldbus Foundation. (1996). *Technical overview*, pp.1. Fieldbus Foundation, Austin, Texas.

Tovar,E. (1999). S*upporting real-time communications with standard factory-floor network.* pp. 97-101. The dissertation for the degree of Doctor. Proto.

Wang,Z. (2000). *Modeling and analysis of fieldbus-based distributed real-time system*. pp. 52-53. The dissertation for the degree of Doctor. *Chinese Academy of Sciences*.

Zhou,Y., Wang, Z., Yuan, M.Z. and Yu, H.B. (2001). The communication media access control mode of fieldbuses. *Industrial instrumentation & automation*, **159**(3), pp.61-64.

IFAC Publications
www.elsevier.com/locate/ifac

# A CORBA-BASED ARCHITECTURE FOR STRATEGIC PROCESS CONTROL

**Ricardo Sanz Miguel Segarra** [**]
**Angel de Antonio** [**] **Idoia Alarcón** [***]

*Universidad Politécnica de Madrid*

*José Gutierrez Abascal 2*
*28006 Madrid*
*Spain*

*ricardo.sanz@etsii.upm.es*

[**] *SCIlabs Ingenieros S.L.*
[***] *Instituto de Ingeniería del Conocimiento*

Abstract: Strategic decision making in complex continuous process plants has been a matter for humans since ever. Automated decision support systems are emerging to help human operators in making reliable, fast and economically advantageous decisions. Construction of these systems are, however, extremely complex due to the needs of integrating heterogeneous information sources or knowledge sources. This paper presents such an application in the cement sector built atop the ICa integrated control architecture. *Copyright © 2001 IFAC*

Keywords: Integration, distributed control, supervision, intelligence, real-time object oriented systems.

## 1. INTRODUCTION

Strategic decision making in complex continuous process plants (chemical, oil, cement, *etc.*) has been a topic restricted for humans since ever (Sheridan *et al.*, 1983). The reasons for this restriction —lack of automation indeed— are grounded in the unpredictability derived from plant complexity (Åström *et al.*, 2000).

Strategic control issues are those related with the top level management of the plant. They are oriented to reach global objectives that, in many cases, are not suitable to be integrated in an automated planner due to their heterogeneity and abstractness. Examples are *safety*, *production*, *stability* or *maintainability* In these plants, the responsibility for this type of decisions is always of a human that decides what to do in any problematic situation.

Automation, however, is desirable —in a general sense— in any type of plant and for any type of task if this automation does not mean the sacrifice of any one of those top level objectives. Partial automation is achieved for some of the objectives but no total solution is available in general because these plants are mostly unique: there are not two equal refineries or cement plants (at least after some time of operation).

The flexibility offered by present day information technology helps bridge the gap between an heterogeneous collection of information sources without sacrificing dependability or performance.

Automated decision support systems are emerging to help human operators in making reliable, fast, and economically advantageous decisions (Petrov and Stoyen, 2000).

In this paper we are not going to propose "the ultimate utility function" for this type of application, but to show an example of how a suitable integration technology can lead to a specifically tailored decision support system that can provide an integrated plant view for strategic decision making.

Not surprisingly, the construction process for these tailored systems is extremely complex due to the needs of integrating heterogeneous information sources (new and legacy systems) into a single whole application. Extreme complexity is reached when the system is designed even for integration with future (not yet existing nor specified) systems as is the common need in complex plants.

The **PIKMAC** system described in this paper was developed to support human-centered operation of a cement plant during periods when this operator is the only person in the plant (*i.e.*the only person capable of making strategic decisions in real-time). This application exploits the integrational capability of CORBA middleware (OMG, 2000) to gather heterogeneous information that is fused into simple quality, economy and maintenance views. This application is deployed atop the **ICa** integrated control architecture (Sanz *et al.*, 1999*a*) that has been specifically built for the control systems domain.

This system was developed inside the DIXIT project funded by the European Commission. The members of the DIXIT consortium were IIC (es), daCapo (se), Repsol (es), Lafarge Ciments (fr), Rambøll(dk), Eusristic Systémes (fr) and UPM (es).

## 2. CEMENT PLANT OPERATION

Cement production is a strange business from our point of view. They burn rocks and sell the final ashes making a lot of money with it. Final chemical composition is critical for the quality of the product (it gets hard by chemical reaction with water) and the main cost is not raw materials (they are usually rocks from a nearby mountain) but energy and salaries.

Energy is obtained from the combustion of different products: coal, fuel, oil, waste, *etc.*(during one of the demonstrations of this application they were burning peach). Each type of fuel has its own qualities (energy per kilogram, cost per kilogram) and side effects (specially in kiln controllability).

Cement plants suffer —as any other industrial plant— the global business objective of reduction of human personnel in all types of tasks. But this must be done with a minimum sacrifice in the rest of the strategic objectives.

Some of the the key factors of success for cement industry are to gain capability to quickly react to customer needs; be able to employ different fuels of poor quality (heavy fuel, recycled oils, waste, *etc.*) without altering the quality of the final product; and reach the capability to compete at a larger geographic scale by permanently streamlining the production costs, and optimizing the value created by the company (DIXIT Consortium, 1998).

## 3. OPERATIONAL OBJECTIVES FOR THE CONTES PLANT

To advance in the achievement of these capabilities Lafarge Ciments managers decided to develop a new generation of IT applications, taking advantage of existing process automation and supervision applications already installed, to be used by all types of plant personnel.

The ideas behind this technology programme were:

- To extract from the huge amount of data continuously stored in CIM.21 process database and other plant databases, the synthetic information relevant for decision making at any moment;
- To derive through an explicit (mathematical) model or an implicit (neural net) models high level value added information consistent with the overall target objectives of the plant (cost, quantity, quality);
- To provide decision support to control room operator to take better decisions in case of failure, in a way that incorporate commercial data, economic factors and human resources constraints to the pure technical data usually taken into account until now;
- To gather in a single user interface all the information, heterogeneous in nature, to facilitate the global control of the production;
- To offer an innovative presentation paradigm and exploration tool making easier production global performance comparison for different moments (for example now *vs* one month ago) or different process configurations (fuel types, clinker quality, etc);
- To facilitate the real time dissemination of production performance information according to formats that can be shared and understood by all the plant department people (management, maintenance, commercial, production, quality, *etc.*).

The plant selected for the development of the **PIKMAC** tools is placed in Contes, France. This plant had another challenge for **PIKMAC** because it was operated by only one person during night and week-end shifts (this means that he was the only person in plant during that time).

## 4. THE INTEGRATED CONTROL ARCHITECTURE

The **PIKMAC** application was deployed using a CORBA middleware specially suited for control purposes. The name for this middleware is **ICa** Broker and it is the cornerstone of the Integrated Control Architecture (**ICa**).

**ICa** is an ongoing, long-term project at the Universidad Politécnica de Madrid, with a basic objective: simplify the construction, deployment and maintenance of software intensive, distributed controllers.

The Integrated Control Architecture follows the specifications developed by OMG (www.omg.org) for distributed object systems, being UPM one of its members. On top of these specifications, **ICa** adds control design patterns (Sanz *et al.*, 1999*b*) and class libraries to support these patterns.

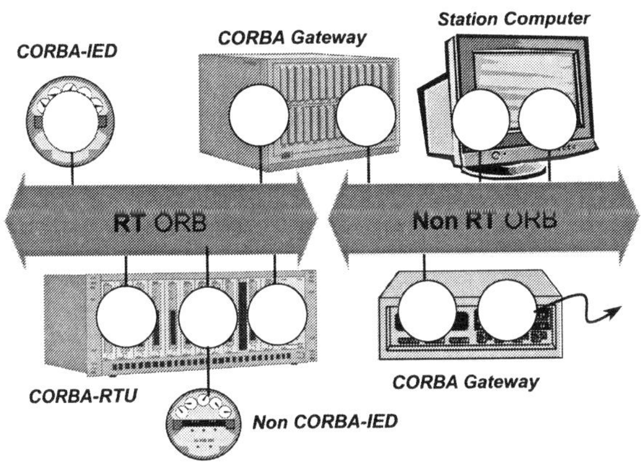


Fig. 1. A CORBA-based domain architecture provides the required functionality to deal with the special requirements of the distributed-embedded domain.

The election of CORBA as a basis for **ICa** is grounded in its extensibility and the capabilities it offers for real-time and embedded systems (OMG, 1998; OMG, 1999*b*). It is not easy to find these capabilities in other enterprise integration architectures like Microsoft DNA, Java EJB or XML/SOAP.

Figure 1 shows how CORBA subsystems running atop heterogeneous protocols (for example TCP/IP and IEC 60870) can be seamless interoperated.

## 5. THE PIKMAC DECISION SUPPORT SYSTEM

**PIKMAC** stands for *Process Information and Knowledge Modeling for Advanced Control.* This demonstration application was demonstrated in the Lafarge Ciments plant of Contes (France).

Its purpose is to keep operators informed to perform a better strategic control of the process in terms of *maintenance*, *quality* and *cost*. **PIKMAC** is based on the fact that a lot of process information is continuously acquired (sensor measurements, control system variables, operator commands, automatic test laboratory, *etc.*) but remains under-used in most cases.

This information concerns all the parts of the cement production process —raw material mill, kiln, cooler and clinker mill— and covers a wide range of process behaviour characteristics.

While several applications could be designed in order to efficiently support the plant operators and process engineers only three integrated applications were demonstrated in **PIKMAC**:

- *Production Performance Synthetic Indicator* (PPSI): provides real time estimations of production performance (quantity and cost) using a Global Production Control concept.
- *Quality Deviation Early Detector* (QDED): estimates continuously key quality parameters making possible the early detection of non-optimal situations.
- *Alarm Management Operator Assistant* (AMOA): help the operator —in particular during night and weekends— to deal with alarm situations to optimize calls to maintenance people.

### 5.1 Global Application Structure

The global structure of the application is very simple: it is a collection of active and passive agents running over the **ICa** broker. These agents provide different types of functions and their interaction capabilities are expressed by means of CORBA Interface Definition Language (IDL) (OMG, 2000).

The system is depicted in Figure 2 showing the collection of agents that composed the application.

The domain architecture for **PIKMAC** was designed by Lafarge Ciments personnel to follow their own ideas on this new set of IT tools for plant management. The people in charge of the DIXIT architecture did only map that conceptual architecture to a specific CORBA implementation based on **ICa** agents.

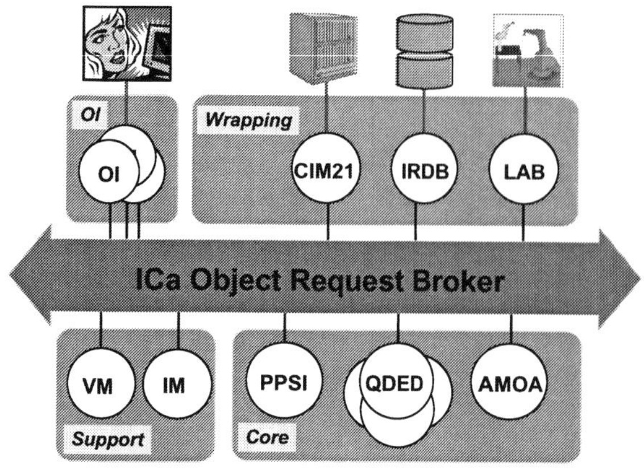


Fig. 2. **PIKMAC** is built as a collection of active CORBA objects that provide specific pieces of functionality: core systems, legacy application wrapping, operator interface and system support.

**PIKMAC** agents can be grouped in four categories:

- *Core systems*: they provide the basic functionality of the **PIKMAC** demonstrator. They are QDED, PPSI and AMOA.
- *Data sources wrappers*: they wrap external data sources to be exploited in a CORBA environment. They are CIM.21, LAB and IRDB.
- *Operator interface*: they are the user interface for the system. There is only one type of agent (OI) but it can be replicated in any number of hosts.
- *System Support*: they provide hidden functionality for the rest of the agents. They are the **ICa** Monitor and the VarManager.

### *5.2 Data Sources Wrappers*

The three main data sources for **PIKMAC** are: i) the real-time process database of the CIM.21 control system, ii) the incident report database (IRDB) and iii) the automated laboratory (Lab). All they are legacy systems that can be accessed using their specific APIs. They appear in **PIKMAC** as conventional CORBA objects resulting from wrapping part of the APIs.

### *5.3 PPSI*

The PPSI agent implements the core functionality for PPSI service. It perform calculations of process throughput and cost per processed unit. These calculations are done online in a continuous manner and uses sampled process data gathered from the plant.

The PPSI implements a sophisticated cost calculation model which takes into account the long processing time for the raw material in cement production.

### *5.4 AMOA*

AMOA is built entirely on G2, an expert systems development tool from Gensym (www.gensym.com). It uses DIXIT's G2-ORB Bridge to connect to the plant data sources and other applications like the operator interface through the **ICa** ORB.

The main component in AMOA is the *process reasoning module*. It gathers and analyze the real data coming from the plant, generating reports regarding present and possible future failures. In case of a problem situation, AMOA will generate reports informing the user about the real root causes of the problem, based on the process analysis it does. AMOA will also guide the user in the task of deciding whether a maintenance team is to be called or not, and which is the maintenance team that must be called if it is the case.

The interaction with AMOA can be done using the generic **PIKMAC** user interface (that provides a simple synthetic view) or the more specialized AMOA native G2 interface.

### *5.5 QDED*

The QDED Agent society predicts some critical quality properties of cement (free lime percentage, SO3 ratio, C3S ratio and C3A ratio) as the cement is produced in the factory.

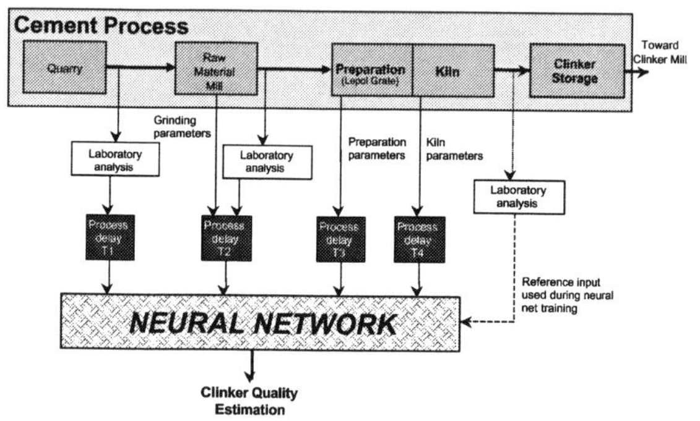


Fig. 3. The QDED neural network uses inputs from all the cement process with proper delays to estimate present quality.

QDED agents use neural networks trained on historical data to make online predictions (see Figure 3). The core issue here is to provide reliable estimates of these parameters, avoiding the delay and the cost derived from slow and expensive automatic laboratory analysis.

A single network is trained for each of these tasks, and the final version of QDED thus require all four agents running simultaneously to provide estimates for all four parameters.

### 5.6 *Operator interface*

The **PIKMAC** system is deployed over an heterogeneous collection of computing equipment. In the demonstration application most parts are running on Alpha/UNIX and Intel/NT platforms.

The user interface runs on NT computers providing a synthetic view of the plant state from the three perspectives: cost, quality and maintenance. Figure 4 shows the main user interface for the quality section. It provides numeric and visual information about the status of quality elements for the cement provided by a remote QDED agent.

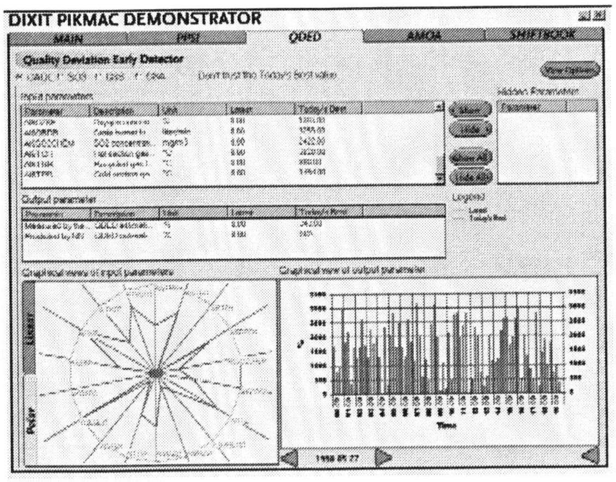


Fig. 4. **PIKMAC** has a user-friendly human-machine interface that runs on Windows NT platforms. This figure shows the part of the OI that contains quality information from QDED.

There is no fixed number of operator interfaces than can be run in an installation, thanks to the brokerage mechanisms provided by the CORBA middleware.

This agent is built using native Microsoft technology elements (*i.e.* COM components, OLE Automation and Active-X controls) that are connected to the CORBA world by means of COM-CORBA interoperability mechanisms.

### 5.7 *System Support*

There are two agents that provide support for the rest of the system. **ICa** Monitor continuously monitors the state of the systems controlling the particular status of any agent.

**PIKMAC** VarManager is a real-time database with added features for DIXIT applications: it gathers data from data sources upon schedule, can process this information and calculate derived data and also supports subscription services for any data it handles.

## 6. CONCLUSIONS

While this is an "applications´´ paper (nothing extremely new is shown here) it demonstrates how integration architectures and technologies help develop future plant-wide integrated control systems in an easy and modular way. All these systems were developed by five development groups in four countries and put to work together in a matter of hours (only for demonstration purposes).

This **PIKMAC** application can only be considered a demonstration of the technology and not a full fledged application. More work is necessary to make it a dependable decision support system. Extensions to support expert systems justification (Guida and Zanella, 1997) or fault tolerance (OMG, 1999*a*; Butler and Finelli, 1993) are obviously necessary to exploit it in a real context.

CORBA technology is here to stay and offers a clear opportunity for control system developers for leveraging previous developments in an easy way. But some contribution to OMG is needed from the controls community. While CORBA is been widely used in real-time settings (see Figure 5) not many industrial applications are described that pose critical requirements for the ORB infrastructure.

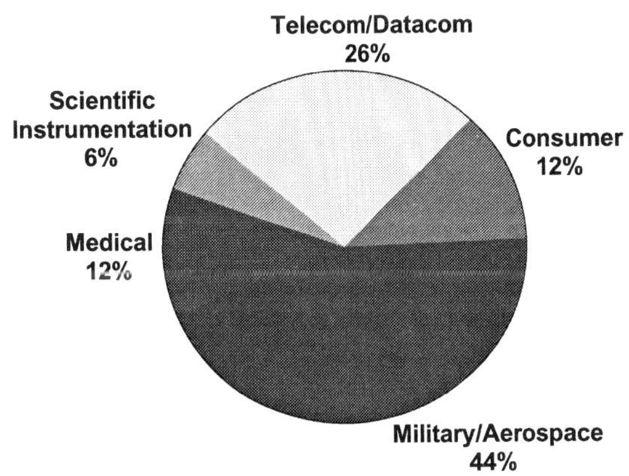


Fig. 5. Current use of ORB middleware in embedded and real-time applications (from (Czerny, 2000)).

Our contribution is necessary for this technology and the result we can obtain from it are extremely high for the deployment of new control technologies in large, complex and distributed plants.

## 7. ACKNOWLEDGEMENTS

The development of the **ICa** framework was partially funded by the Comission of the European Union and the Comisión Interministerial de Ciencia y Tecnología through ESPRIT project DIXIT.

## REFERENCES


Åström, Karl, Isidori, Alberto, Albertos, Pedro, Blanke, Mogens, Schaufelberger, Walter and Sanz, Ricardo, Eds.) (2000). *Control of Complex Systems*. Springer. Berlin.

Butler, Ricky W. and George B. Finelli (1993). The infeasibility of quantifying the reliability of life-critical real-time software. *IEEE Transactions on Software Engineering* **19**(1), 3–12.

Czerny, Paul (2000). Current uses of middleware in embedded/real-time applications. In: *1st OMG Workshop on Real-time and Embedded Distributed Object Computing*. Falls Church, USA.

de Antonio, Angel, Ricardo Sanz, Fernando Matia and Miguel Segarra (2000). Tackling control systems complexity. In: *Mobile Robotics in Healthcare* (Nikos Katevas, Ed.). IOS Press.

DIXIT Consortium (1998). PIKMAC user manual.

Guida, Giovanni and Marina Zanella (1997). Bridging the gap between users and complex decision support systems: the role of justification. In: *Proceedings of Third IEEE International Conference on Engineering of Complex Computer Systems, ICECCS'97*. pp. 229–238.

OMG (1998). Minimum CORBA specification. Technical Report orbos/98-08-04. Object Management Group. Falls Church, USA.

OMG (1999*a*). Fault tolerant CORBA joint revised submission. Technical Report orbos/99-10-05. Object Management Group. Falls Church, USA.

OMG (1999*b*). Real-time CORBA adopted specification. Technical Report orbos/99-06-02. Object Management Group. Falls Church, USA.

OMG (2000). Common object request broker architecture and specification. release 2.4. Technical report. Object Management Group. Falls Church, USA.

Petrov, P.V. and A.D. Stoyen (2000). An intelligent-agent based decision support system for a complex command and control application. In: *Proceedings of Sixth IEEE International Conference on Engineering of Complex Computer Systems, ICECCS'2000*. pp. 94–104.

Sanz, Ricardo, Miguel J. Segarra, Angel de Antonio and José A. Clavijo (1999*a*). ICa: Middleware for intelligent process control. In: *IEEE International Symposium on Intelligent Control, ISIC'1999*. Cambridge, USA.

Sanz, Ricardo, Miguel J. Segarra, Angel de Antonio, Fernando Matía, Agustín Jiménez and Ramón Galán (1999*b*). Patterns in intelligent control systems. In: *Proceedings of IFAC 14th World Congress*. Beijing, China.

Sheridan, T.B., T. Vámos and S. Aida (1983). Adapting automation to man, culture and society. *Automatica* **19**(6), 605–612.

IFAC Publications
www.elsevier.com/locate/ifac

# PRACTICE OF REAL-TIME SOFTWARE ARCHITECTURES: COLLIDER, SATELLITES AND TANKS COMBINED

**Jan van Katwijk * Jean-Jacques Schwarz ** Janusz Zalewski *****

* *Faculty of Information Technology and Systems*
*Delft University of Technology*
*Mekelweg 4, 2628 CD Delft, The Netherlands*
J.vanKatwijk@twi.tudelft.nl
** *Laboratoire L3i, Institut National des Sciences Appliquees*
*F-69621 Villeurbanne, France*
jjs@iuta.univ-lyon1.fr
*** *School of Electrical Engineering and Computer Science*
*University of Central Florida*
*Orlando, FL 32816-2450, USA*
jza@ece.engr.ucf.edu

Abstract: Three real-time mission-critical applications are discussed from the following domains: scientific (high-energy physics experiments), commercial (satellite control), and military (embedded simulations). Problems facing designers of real-time software architectures are grouped into two categories: architectural patterns and design tools. Consistent solutions used by the authors and applied in these applications are presented. *Copyright © 2001 IFAC*

Keywords: Real-time systems, real-time computing, software architecture, design patterns, safety related systems, software tools.

## 1. INTRODUCTION

The objective of this paper is to address the technological challenges and prove that respective solutions exist in engineering software architectures for large, complex, real-time mission-critical systems in three sectors: scientific (Gaspar et al., 2001), commercial (Vardanega, 2001), and military (Al-Daraiseh et al., 2000). The three applications discussed are all mission-critical in a sense that the success or failure of the entire enterprise is linked strongly to a computer system itself. The complexity of respective requirements is so high that none of the traditional approaches to software design are adequate to fully represent the needs of the developers. In the following, we define first the problems via their specific characteristics and technological challenges (Section 2). Then we describe our approach to developing respective software architectures (Sections 3) and outline the architectural solutions themselves (Sections 4–6).

## 2. OVERVIEW OF REQUIREMENTS

### 2.1 *HEP Experiments*

The online monitoring and control of high-energy physics (HEP) experiments, conducted in big accelerators called particle colliders, deals with a variety of aspects in macro (such as material and energy flow) and micro scale (such as a particle beam control). It involves hundreds of requirements of which a few important are summarized below (Gaspar et al., 2001):

- Functionality. This can be split into 3 categories: (1) The number of quantities that must be constantly measured and controlled is at the order of several hundred thousands. (2) The primary function, read-out of a significant event data (approx. 100 KB) multiple times per second, is organized through a dedicated bus that connects ca. 3000 electronics modules and 100 embedded processors. (3) Data reduction has to be done with the rate at which data are read from the detector to about 5 events per second, which means about 500 KB/s.

- Timeliness. Data handling imposes response times of the order of hundreds of milliseconds. Deadlines are firm and if missed result in losing physics data.
- Safety. In control, the response time can be of the order of several seconds or even minutes but it represents a hard constraint since the safety of people and equipment are involved.
- Reliability. A fault in the system should not prevent the monitoring of important quantities. A fault in one part of the detector should not prevent the rest of the system from functioning.

For the implementation to meet these requirements, a global development strategy must exist that provides: (1) A method to decompose the system into smaller, manageable, units. (2) Distributed model that allows transparent change and communications among all units. (3) Modeling tools to help describe and implement the model and the behavior of each unit.

### 2.2 *Commercial Satellites*

A new paradigm for the development of mission-critical real-time software embedded on-board the control platforms of satellite systems takes into account the integration of all the control functions allocated to the platform of the space segment, the so-called *spacecraft bus*, to a single processing node. In this view, the increasingly crucial role of the payload, as the value added to the spacecraft, and its impact on the architecture of the on-board system has to be evaluated (Vardanega, 2001).

The current trend to increase the payload component embedded in the on-board system stems from the commercial objective to use clusters of satellites for the implementation of a wide range of global multimedia services. For satellite systems in this configuration, virtually all of the customer-oriented services are allocated to the payload elements. Consequently, as the platform is to operate as an independent carrier of the value embedded in the payload, drastic reduction of the ground control functions is anticipated, with the subsequent transfer of most control functions to the spacecraft bus.

In this situation, the control functions allocated to the spacecraft are distributed across several processing nodes, so that the spacecraft bus has to be regarded as a distributed system. In such case, one processing node in the platform integrates all the traditional on-board control functions (such as attitude and orbit control and data handling) and co-ordinates with other nodes that control the activity of the payload elements embedded in the spacecraft. The platform retains the role of traffic manager for all the communication between the spacecraft and the ground centre.

In a view of this requirement, the challenge is to design a software process to accomplish three key objectives:

- Increase by a factor of 3-5 the software productivity, in source lines delivered per time unit.
- Increase the autonomous operation capability of the software product with reduction of service outage.
- Master the functional complexity and real-time criticality of the system.

A respective engineering approach will serve well the application needs, if it meets the above-listed process requirements.

### 2.3 *Embedded Simulation*

Current simulation devices used in military training reach the level of sophisticated simulators with motion platforms, and realistic audio and video stimuli. Still, trainers are designed for individuals. The first step toward collective training is to link these individual stations together to function as a total crew trainer. The next step is to link the crew trainers together to allow training at the unit level, up to the targeted force structure.

Yet, even in this environment it is still desirable to include other forces not physically present. This leads to the requirement to mix live (physically present) and virtual (not physically present) elements in the total training scenario (Al-Daraiseh et al., 2000).

To alleviate the lack of availability of trainers, the trend is to embed training capabilities within the weapons system. One required characteristic of embedded trainers is to be autonomous, requiring no additional resources to complete its function. The issue which needs a solution is the operation of this autonomous trainer and how it interacts with other trainers/weapons as the computer generated elements are replaced by live trainees.

An example of an individual trainer is the embedded system for a tank. In addition to the usual vehicle subsystem that handles the operation of a vehicle, it has two simulation related subsystems:

- The trainer subsystem that provides the interface between the trainees and the system during the training mode, as well as the interface between the vehicle and the simulated forces.
- The computer generated force subsystem that provides all the virtual players for training, mission rehearsal and course of action analysis.

The primary challenges from a research viewpoint are extending the capability from the autonomous trainer to collective uses. The actual embedding of the training functions into the vehicle is primarily dependent on the evolution of the following capabilities: image generation, miniaturization of components, enhancement of flat panel displays, and wireless communications.

Current computer generated force systems generate state information that is broadcast to all the other participants. As long as the communications bandwidth is sufficient, with minimal latency, entities can be added by just adding additional sources to the network. However, with additional goals to increase the number of entities involved in a given training exercise, to have entities participate from worldwide locations, and to have participants from actual vehicles while they are moving, it becomes clear that some drastic steps have to be taken to enhance architectural capabilities to achieve these goals.

## 3. AUTHORS' APPROACH

There are several common characteristics of the applications surveyed above. Two of them are especially challenging to designers: distribution and complexity (van Katwijk and Zalewski, 2001). Below we discuss new paradigms for solutions in both areas: (1) real-time architectural patterns and, (2) automatic support for a software process via real-time design tools.

### 3.1 *Architectural Patterns*

It is not clear what are the principles of real-time software architectures for complex distributed systems and what are the good and bad practices. The state of the art is such that almost every block diagram with interconnected boxes and a couple of arrows pointing from one box to another is claimed to be a description of a software architecture. The challenge is to determine what constitutes an architecture of a distributed real-time software (Zalewski, 2001).

In an ideal architectural framework, multiple software units can be created to access various sources and destinations of data and to exchange information among themselves. Any single unit can perform individual functions and communicate with every other unit. All this should operate in a dynamic environment, where the organization of components and their interconnections may change during execution. Adding or deleting new components should have no impact or minimal impact on the operation, in a sense that no degradation of functionality should occur due to such dynamic changes. The architecture should be expandable during the run of an experiment or operation of a process.

One view of a compliant architecture, which comes from control engineering, is presented in Fig. 1. Logically it forms a tree, with a root at the experiment hardware sensor/actuator layer. Only this root cannot change, everything else is flexible and includes all units necessary for fully dynamic operation, including modules for data collection and control, computations, user interface, databases, and communications.

Each unit is autonomous and can run idependently as well as be integrated into a larger entity. Communication links can operate individually or be lumped, as in Fig. 1, into a middleware layer, with program units communicating partially or exclusively via this layer.

### 3.2 *Real-Time Design Tools*

Developing software architectures for modern real-time applications, such as those described above is too complex to be done manually by individual designers. Therefore automatic tools are needed to assist in the development process. To be fully useful in development of software architectures, the automatic tools have to provide support in the following four dimensions (Fig. 2):

- internal, related to all aspects of expressing real-time models via the specific notation and respective transformations
- horizontal, related to means of communication with other models and other tools (for example, via TCP/IP protocol suite)
- vertical, related to the next and previous phases of the development process, with respect to support for (1) code generation and prototyping, and (2) design verification against the requirements
- diagonal, related to the use of architectural models in different projects and different processes.

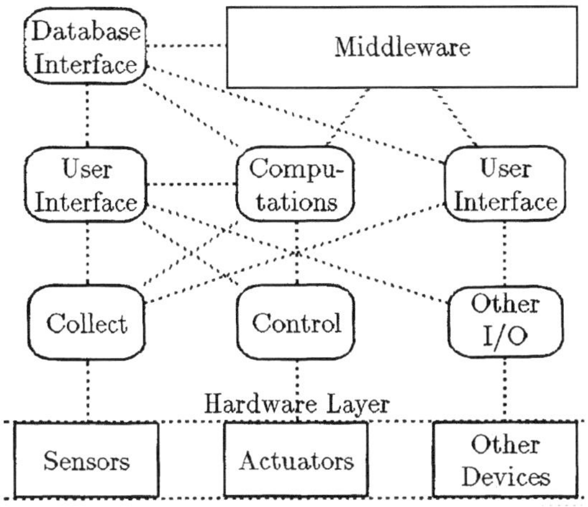


Fig. 1. Architecture of a Distributed Real-Time System

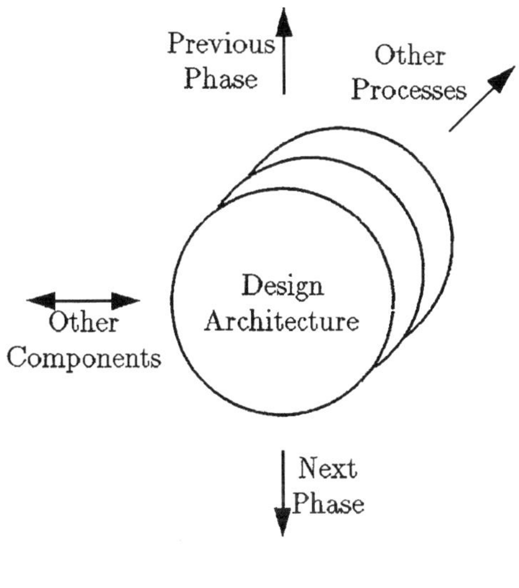


Fig. 2. Issues for Ideal Real-Time Design Toolsets

Although software tools exist that partially support these objectives, thare is no single comprehensive toolset that would support the whole spectrum as depicted in Fig. 2.

## 4. THE DELPHI EXPERIMENT

DELPHI (DEtector with Lepton, Photon and Hadron Identification) is one of the four experiments built for the 27 km long LEP (Large Electron-Positron) collider at CERN, the European Organization for Particle Physics (Gaspar et al., 2001). The electron-positron collisions take place inside the vacuum pipe in the center of DELPHI and the products of the annihilations fly radially outwards. The products are tracked by several layers of detectors and read out via some 200,000 electronic channels. A typical event requires about 1 million bits of information and may occur million times per second.

The Online System of DELPHI is responsible for the control and monitoring of all parts of the experiment, from safety concerns to real-time data read-out. It involves more than 500 processes distributed over around 40 workstations. Another 150 embedded processors are used for data-acquisition and trigger readout, slow-control and safety monitoring, and overall control.

### 4.1 *Subsystems and Software Architecture*

The Data Acquisition System (DAS) is responsible for gathering the data produced by the sub-detectors and storing it for later use. DAS is designed hierarchically where at each level the data read-out from the electronic channels is digitized, compressed, formatted and passed to the next level (Fig. 3). At the leaves (Front End Electronics) there are normally many processors working in parallel at a very high rate reading the data as fast as possible out of the detector. The data coming out of one collision inside the detector is called an event (Event Builder). Central DAS is in charge of data filtering, storage identification and overall data acquisition control.

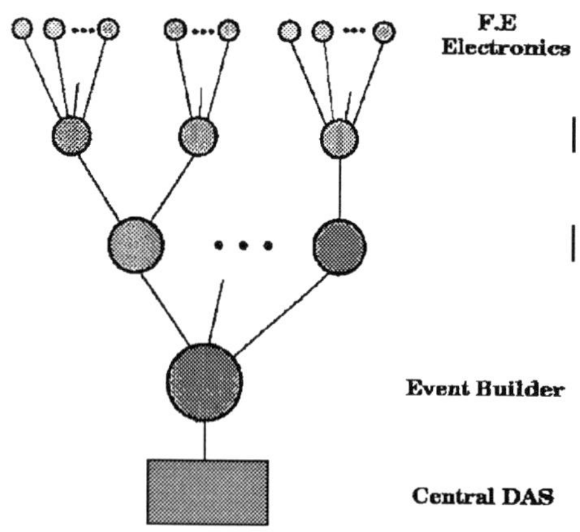


Fig. 3. HEP Data Acquisition System

The role of the Trigger System is to decide very fast after a collision has happened if it is an interesting one or not. In such complicated detectors it is very important to always make sure the detector is in a good state. This type of monitoring is partially done by the Slow Control System and the Data Acquisition System.

The software architecture is composed of building blocks organized in a tree-like structure (Fig. 4) consistent with the one presented in Fig. 1. All these blocks are able to communicate with each other across processes and machines. At the leaves of the tree there are blocks which handle real devices and where users can insert their code. Other blocks contain the model of the system and its dynamic behavior (control blocks). New blocks can be inserted at any level, from any of the three domains: communication, control and user interfacing.

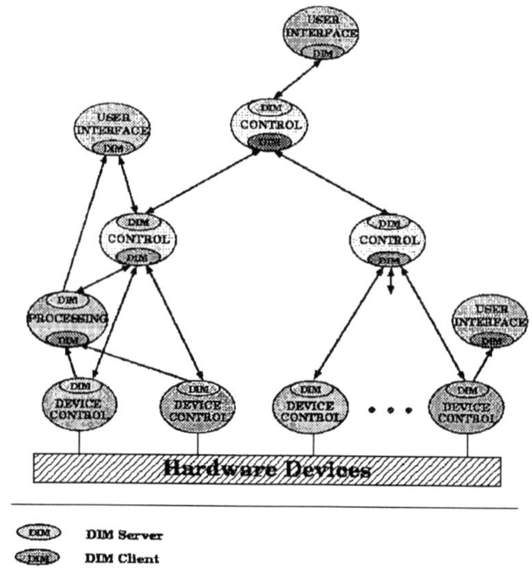


Fig. 4. Control and Monitoring Framework

In industrial environments like this, it is essential to integrate in the communication framework some specific services. They must deal with commands to be sent to the devices in several "running" modes (Fig. 5):

- individual action: a single order is sent;
- polling: orders are sent simultaneously to several devices periodically;
- diary: orders are sent accordingly to a timetable;
- context: one or more orders are sent according to information received from another device;
- alarm: a mode, when the device takes the initiative.

### 4.2 *Software Development*

Due to unavailability of coherent sets of commercial tools suitable for dealing with such a complex architecture, the generic tool framework was proposed capable of handling the control and monitoring of all aspects of the physics experiment.

First, a uniform communication model must exist to be usable by all blocks for all types of data exchanges. Each building block should at any time publish the information it produces to the communication system so that it becomes available to the other blocks. DIM

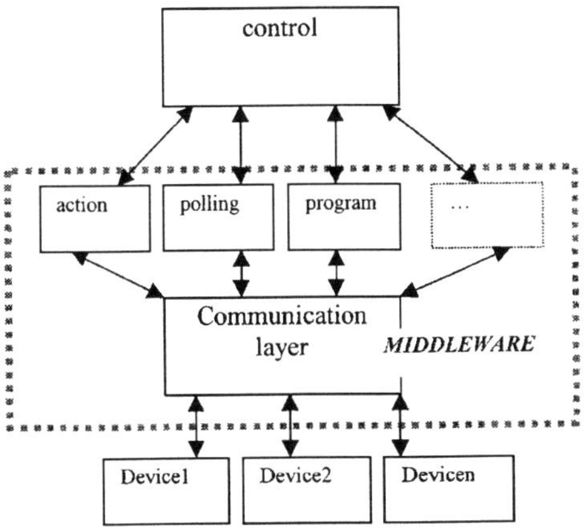


Fig. 5. Communication Infrastructure

(Distributed Information Management System), is the middleware developed for this communication layer.

The full control system is built as a hierarchy of simpler blocks where the behavior of the different parts is defined. The smallest entity representing a hardware piece or a simple software task is called an object. This concept and its implementation called SMI++ (State Management Interface) allows the user to specify the objects and their dynamic behavior via finite state machines.

The third aspect of the architecture, user interfacing, becomes simpler since all the necessary information is accessible via the same communication system, i.e. DIM. The Delphi User Interface (DUI) is designed as a hierarchy of building blocks, where at higher level only the states of the main parts are controllable and more details can be obtained via pull-down menus.

Tools for the three main areas, communication (with middleware DIM), control (with SMI++) and user interfacing (DUI) were built as inherent parts of the generic framework (Fig. 4) that allow the design and implementation of an online system according to the proposed architecture. Due to the limited space, in this paper we only mention the DIM (Gaspar et al., 2001).

The main purpose of the DIM is to make data available where and when necessary, both for control and for monitoring, independently of where it is produced. In order to accomplish its mission DIM was designed to provide efficient communication mechanism and transparency. No matter where a process runs, it should be able to communicate with any other process, using a mechanism that is independent of where the processes are located.

To each of the running modes, listed in previous section, are associated software modules the designer can use to develop his application in a uniform manner. The modules can be implemented in different manners depending on the software technology used (functional, OO). Such a paradigm, with some adaptation at the communication level is at the basis of the design of DIM.

## 5. SATELLITES AND SOFTWARE DESIGN

Satellite systems typically perform their operation under the supervision of a control center. The ground center exercises its authority on the on-board system by issuing telecommands (TC), and the receipt of status and verification data through response telemetries (TM). The set of communication-related services is collectively termed Data Handling Control subsystem.

Complementary to the DHC functionality are the on-board functions of Attitude determination and Orbit Control (AOC). An orbiting satellite system acquires and maintains its assigned orbital position and trajectory by the utilization of sensors and actuators.

Traditionally, AOC and DHC are allocated to physically disjoint subsystems that coexist, as blackboxes, within the same platform. However, the rising of more powerful processor technology fosters the transition to single computer platforms. Systems based on this concept are termed Integrated Control and Data Systems (ICDS).

### 5.1 *Software Architecture*

The typical organization of on-board processing activities is depicted in Fig. 6. The diagram shows three elements: acquisition, processing and commanding.

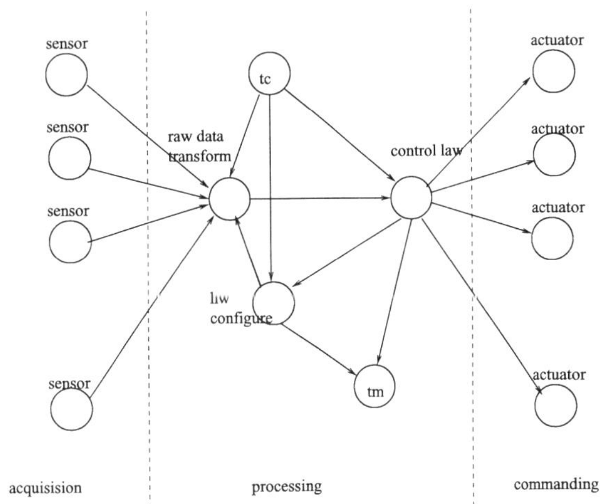


Fig. 6. On-board Processing Activities

The generic architecture of a distributed real-time system can be instantiated for this typical problem domain:

- The root of the tree is formed by the hardware layer, facilitating both data acquisition and commanding;
- The *collect* function from the diagram in Fig. 1 is instantiated as *raw data transform* in Fig. 6;
- The *control* function from Fig. 1 is instantiated as the *control law* function from Fig. 6.
- The *computations* function in Fig. 1 takes part of the functionality of *hw configure* and *control law*.

Instantiating the user interface as shown in Fig. 1 for the functionality in Fig. 6 is done through its *TM* and *TC* functions. It shows that the middleware component, depicted in Fig. 1 has its role in the satellite architecture to provide some API for the telemetry between ground station and satellite.

Summarizing, the generic architecture from Fig. 1 can directly be instantiated to the architecture, depicted in Fig. 7, the latter being *the* architecture for on board processing activities in satellites.

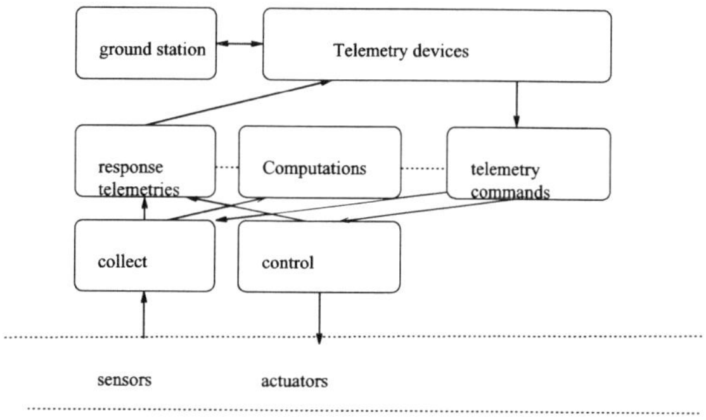


Fig. 7. Instantiated Architecture

### 5.2 *Real-Time Design Tools*

Vardanega (2001) proposed to achieve the rather ambitious goals of increasing productivity, increasing the autonomous operation capability, and mastering the functional complexity by introducing two key enhancements to the software process typically applied in the domain of embedded satellite software:

- The notion of *Computational Model*, as means to inject the concern of *real-time design* into the design definition phase as seen by the PSS-05, the classical V-shaped development model, especially designed for the induistry in this domain.
- The concept of *unified design framework*, as the basis for a *feedback-based iterative design process* which allows requirements analysis, design definition and real-time verification of the V-shaped development model to progress incrementally within a single, unfractured development phase.

The development strategy that was proposed is comprised of the following sequential steps:

- Developing a model for the system, taking into account the computational model. As notation HRT-Hood is used while taking the computational model into account, and the concerns of real-time design are injected into the design definition.
- *Static real-time analysis*, in which the anticipated real-time behavior of the program is verified against the real-time commitments of the system. For the notation used in the computational model, a variety of verification tools were developed.
- *Feedback to design*, in which the design process is consolidated and guided to completion by interpretation of the results from the previous step. The underlying assumption is that the verification tools provide more information than a simple *yes* or *no*.

A validation for that approach was found through the following mapping from the existing ICDS development toolset onto the generic model presented in Section 3.2:

Internal support. Aspects of expressing real-time models via specific notations and transformations are supported via HRT-Hood. Its support for hierarchical decomposition, maintaining semantic invariants is low, however.
Horizontal support. The satellite system may be considered an element in a distributed system, that is formed by the computer system in the satellite and the systems comprised in the ground station. Current tools only provide partial support for development with on-line connections between the components.
Vertical support. Support for the vertical dimension is fairly strong. Helped by the low level of the design notation, code skeletons can be generated easily. Since the (annotated) specifications essentially describe the complete tasking structure with the communication and synchronization framework, the design itself is an excellent candidate for a first order verification of temporal constraints.
Diagonal support. Support for reuse of architectural designs is very desirable but not very strong, since most projects are developed from scratch.

## 6. SOFTWARE FOR EMBEDDED SIMULATION

The Distributed Embedded Simulation system is extremely complicated in its entirety, as ultimately it may involve many thousands of objects, artificial and live, cooperating on a common goal. In this section, we concentrate on a small fraction of it, consisting of a training system developed for combat vehicles. Through an interface between the combat vehicle, such as a tank, and the training system, virtual environments can be generated for the purpose of embedded training, mission rehearsal, battlefield visualization, command coordination and simulation based acquisition.

### 6.1 *Basic Architecture*

The architecture of an individual simulator is consistent with the generic template in Fig. 1 and involves the following four building blocks (Al-Daraiseh et al., 2000):

- SMC, Simulation Management and Control
- CGF, Computer Generated Forces
- IG, Image Generation System, and
- IT, Intelligent Tutor and After Action Reporting.

Several different types of messages are being exchanged among modules and consist of periodic messages sent at frequencies of 1–30 Hz, event-driven messages and initialization messages. Communication between these modules uses traditional TCP/IP protocol suite. Fig. 8 shows the top-level architecture expressed in ROOM (Real-time Object-Oriented Modling technique, the basic notation for ObjecTime tools).

The real issue during development is the modeling and prototyping of each of the four components and their mutual interaction, as well as communication with external

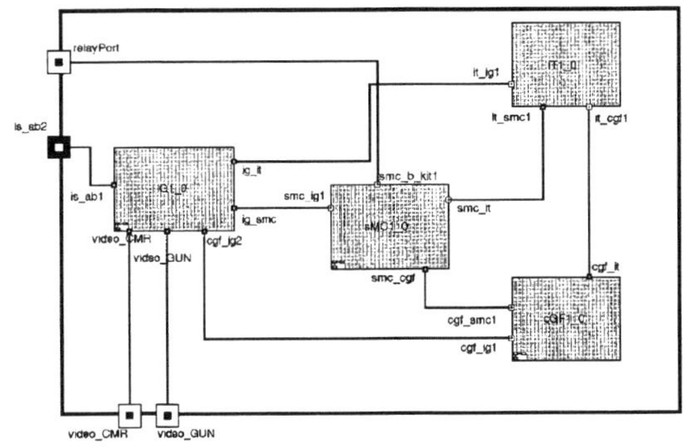

Fig. 8. High-Level Diagram of the Architecture

entities. Rather than developing an in-house solution, a set of commercial tools has been investigated to provide an array of capabilities listed in Fig. 2.

### 6.2 *Tools Use*

Several software tools operating already in the commercial marketplace were studied and used, with respect to the project-specific criteria (Fig. 8):

- SES/workbench for performance modeling and efficiency in the internal dimension
- G2 expert system for interoperability and connectivity in the horizontal dimension
- ObjecTime Developer for testability and traceability in the vertical dimensions
- Rhapsody, a UML-based design tool for reusability and portability in the diagonal dimension.

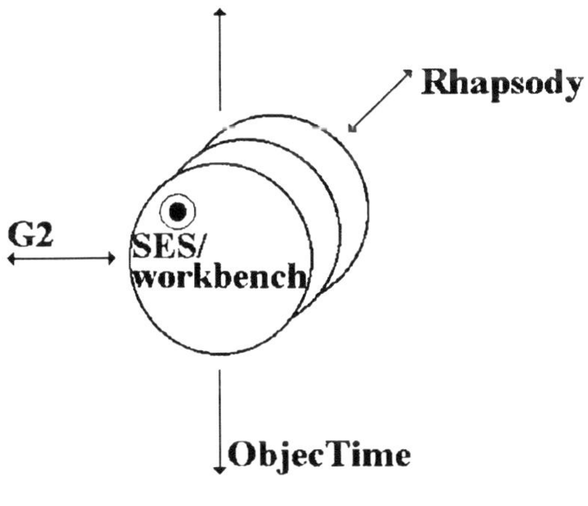


Fig. 9. Mapping the Tools on the Adopted Model

Since it is not possible to provide details of all experiments in this paper, we concentrate only on the design and prototyping. Fig. 10 illustrates the behavioral aspects for different states of the simulation and the different events permitted in each state, expressed in ROOM.

In the Initialization state, settings are chosen and the values of associated variables are transmitted to the different modules. After initialization the system enters a Running mode and the transmission of operational messages starts (event-driven or periodic). Periodic messages are implemented with a timer. Eevent-driven messages are injected through the simulation GUI. The system can also go in a Paused state, when only periodic messages are able to go around. A resume message puts the system back in the Running mode. From either the Paused or the Running mode the system can be commanded to exit.

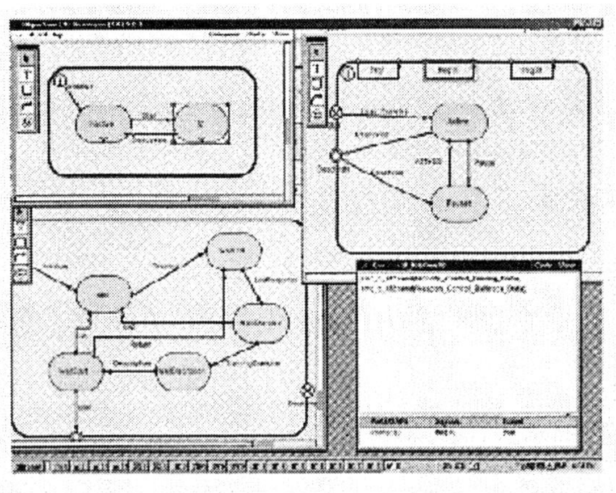

Fig. 10. Statecharts Describing Module Behavior

## 7. CONCLUSION

This paper presented the authors' approach to solving architectural problems in large real-time, mission critical software systems. Two major issues were addressed: real-time architectural patterns and real-time design tool support.

Application of these concepts to three important and demanding real-time systems: experiment control in high-energy physics, commercial satellites, and embedded simulation, were discussed. Sufficiently expressive real-time design patterns and corresponding features in tool support were particularly helpful in developing software architectures for these applications.

## REFERENCES

Al-Daraiseh A., Al Mazid A., Croeze T., Zalewski J., Dolezal M., Green G. and Bahr H. (2000), *High-Level Tool Support for Integration Architecture in Distributed Embedded Simulation*, Proc. 2nd Conf. on Simulation Methods and Applications, Society for Computer Simulation, San Diego, Calif., pp. 33-36

Gaspar C., Franek B., Schwarz J-J. (2001), *Architecture of a Distributed Real-Time System to Control Large High-Energy Physics Experiments*, pp. 147–163, in (van Katwijk and Zalewski, 2001)

van Katwijk J., Zalewski J., Eds. (2001), *Parallel and Distributed Real-Time Systems*, Nova Science Publishers, Huntington, NY

Vardanega T. (2001), *On the Distribution of Control Functions in New-Generation On-board Embedded Real-Time Systems*, pp. 127–145, in (van Katwijk and Zalewski, 2001)

Zalewski J. (2001), *Real-Time Software Architectures and Design Patterns: Fundamental Concepts and Their Consequences*, Annual Reviews in Control, Vol. 25, No. 1, pp. 133-146

IFAC Publications
www.elsevier.com/locate/ifac

# PROGRAMMING LANGUAGES FOR USE IN SAFETY RELATED APPLICATIONS

**Janusz Zalewski* and Wolfgang A. Halang†**

* *University of Central Florida, Dept. of Electrical and Computer Engineering, Orlando, FL 32816, U.S.A., jza@ece.engr.ucf.edu*
† *FernUniversität Hagen, Faculty of Electrical Engineering, 58084 Hagen, Germany, wolfgang.halang@fernuni-hagen.de*

Abstract: Programmable electronic systems are being used in almost all application sectors to perform non-safety and increasingly to perform safety functions as well. Although software based solutions are usually superior to hardwired ones for reasons of efficiency and flexibility, there is a certain reluctance of the certification authorities when it comes to licensing computer based systems which are classified as safety critical. Despite many attempts to overcome problems of software safety (IEC 61508, IEC 880, VDE 0801, IDS 00-55, RTCA/DO-178), up to now neither precise guidelines supporting the software development process are available, nor are there serious efforts being made to develop programming languages dedicated to the implementation of safety critical functions. To improve this unsatisfactory situation, i.e., to meet both economic and safety requirements, it is necessary to design appropriate language concepts with consequent regard to safety aspects. Accordingly, four subsets of a real time language suitable for the implementation of safety related systems are proposed, whose definitions fulfil the respective requirements of the four Safety Integrity Levels.


Keywords: Safety related systems, safety requirements, safety integrity levels, real time programming languages, PEARL, verification.

## 1. INTRODUCTION

Computer based solutions are permanently gaining significance in almost all application fields, not least because of the rapid progress in microelectronics enabling cheap and flexible implementations. Although it is quite natural that industry tends to benefit from these economic advantages, there is, however, a type of system designated as safety related for which this trend does not prevail. The reason for this is to be found in the growing awareness of safety hazards, and as a consequence of the demand for highly dependable systems. The latter holds true especially if malfunctions of the concerned systems may result in disastrous accidents, as it is characteristic for automation systems.

Software based solutions are still considered to be less trustworthy than conventional hardwired components, which is justified by the longer tradition of hardware and, therefore, many years of experience in the development of strategies to cope with corresponding failures. Actually, evaluating software with respect to dependability or safety aspects is a task which has been neglected for a long time. However, owing to various efforts during recent years, there exists a number of methods and guidelines belonging to different application fields and, among others, having the principal intention to define requirements for programming languages for use in safety related systems (IEC, 1998), (DIN, 1995), (MOD, 1991), (IEC, 1986), (RTCA, 1992). The guidelines agree on the following general requirements:

- programming languages should be fully and unambiguously defined or shall possess a formally defined syntax, respectively,
- be user and problem oriented,
- be of a high level with preference to safety oriented standardised subsets,
- enable structured programming, restricting branches and loops and avoiding jumps and computed branches,
- should support simple and uniform addressing techniques, and
- strong typing, i.e., enforce explicit declaration, initialisation and type conversions,
- but not comprise dynamic variables or other dynamic objects,
- language structures should be modular, interfaces between modules should be simple, uniform and fully defined, and
- procedures should be simply structured, having a minimum number of parameters, only; communication should exclusively be permitted via parameters; recursive procedure calls should be not allowed.

Unfortunately, these guidelines are restricted to the description of shortcomings or desirable features with respect to existing languages. They are often rather imprecise, thus not offering substantial support during system development. Actually, no new language has been developed on the basis of these standards, nor have the mentioned guidelines so far motivated the re-design of existing languages for use in safety related applications. Consequently, even though the process of verification is inevitable for any code segment implementing a safety related function, no progress based on the cited standards is evident or is to be expected.

Indeed, because verification is the main prerequisite to enable the certification of larger software based solutions, only serious improvements aiming to support the process of program verification will be a step in the right direction. The importance of this becomes even more evident when we consider the fact that for practicability reasons the workload involved in verifying safety related software has to be restricted to a reasonable limit. The present state of the art implies, for example, that formal methods only apply for such verification processes when the code sequences are quite short.

Summing up, up to now approaches aimed at improvements in the field of software safety obviously have not yielded results which enhance the confidence in software based solutions. The certification authorities are, therefore, still reluctant to license exclusively software based safety related systems and, consequently, system designers often prefer to employ conventional hardwired solutions despite their manifest disadvantages with respect to the mentioned economic aspects.

In an attempt to remedy this unsatisfactory situation, in this paper we introduce several programming language concepts which we think stand a better chance of enhancing trustworthiness and dependability of software based safety related systems. To this end, we first discuss some inherent characteristics of software, focusing particularly on the differences with respect to the nature of software and hardware failures. We then present a basic principle which is decisive in the development of new language concepts for safety related programming. The following section defines four increasingly restricted subsets of the real time programming language PEARL 90 (DIN, 1998) recommended for use in safety related applications, and classifies these subsets according to their suitability with reference to the four discrete levels of increasing safety integrity requirements (SIL 1,...,SIL 4) introduced in the Draft International Standard IEC 61508-1 (IEC, 1998).

## 2. CAUSES AND NATURE OF SOFTWARE FAILURES

Although always implemented with the latest technology, the architecture of digital computers has remained essentially unchanged since the times of Konrad Zuse and John von Neumann in the 1940s. The prevailing architectural model matches the technical possibilities of that time. In other words, while the technology with which hardware is built and the capabilities of hardware have tremendously advanced within the last 60 years, all high level programming languages developed so far, and presently in use, reflect to different extents the way programs are executed by Von Neumann machines, allowing maximum flexibility at the price of facilitating errors and being unsafe. As a consequence of this and due to the inflationary increase in the demand for functionality, the complexity of software is exploding exponentially. This, in turn, leads to error-prone software development and results in a lack of software dependability.

In society, there is a growing concern for safety and environmental hazards, and an increasing demand for dependable technical systems which prevent loss of human lives and environmental disasters. To enable a flexible adaptation of system functions to new needs and to enhance the productivity of system development processes, computer based systems are increasingly being applied for both control and automation functions under real time constraints. These systems have the special property of hardware and software being closely coupled to form complex mixed-technology systems such as manufacturing, process or traffic control systems.

When analysing malfunctions of computer based safety related systems, it can be observed that often the reasons are not to be found in random hardware failures, i.e., failures occurring at random times or resulting from a variety of degradation mechanisms in the hardware. Rather, they result from special system states during operation, which had not been considered in the specification or implementation phases of the software. Naturally, programs are immaterial and, consequently, do not degrade in a similar way to hardware components. Software failures, which must be classified as systematic, are due to errors, mistakes or omissions in any development life cycle phase. Systematic failures include (IEC, 1998):

- failures due to errors in the safety requirements specification, and
- failures due to errors in the design, implementation, etc. of software.

Thus, provision against these types of failures should already have been taken into account in the early phases of system development.

Software has to be valid and correct. The latter means that software has to fulfil a given problem specification, i.e., it must exclude systematic failures based on program design, coding, or the use of software tools. Correctness of software cannot be achieved by testing, reviews, audits, inspections, walkthroughs or other heuristic methods, because these inherently lack the rigour necessary to be able to detect all errors contained in a program. And, since by definition correct software meets its specification, establishing software correctness alone is not sufficient. When a program does not behave as expected, often its very specification is incorrect, incomplete or inconsistent. Thus, it and the resulting software cannot be valid. Consequently, to produce proper software based systems it is essential to invest high validation efforts already in the specification phase. Unfortunately, at this stage there are no usability checklists or measures for assessing whether a specification is valid.

## 3. FOSTERING SAFETY BY SIMPLICITY

Although sometimes supported by computerised tools, the proof of a program's correctness depends for its success on human cognizance. In order to minimise the cognitive effort necessary to produce it, and to maximise its trustworthiness, it is essential to employ user oriented programming concepts. By their very nature, programming languages are human-computer interfaces *par excellance*. This property is seldom recognised, possibly due to the programmers' familiarity with the Von Neumann architecture and their acquired comfort with machine oriented interfaces. Object oriented and visual higher level programming languages have been developed in parallel with, and reflecting the need for, graphical user interfaces for a wider population of end users. However, in order to support the process of system verification as far as possible, we need to go a step further and develop a new brand of "subject oriented" rather than "object oriented" programming languages to ensure that programming and non-programming system developers and certification licensors can be given a common ground at the interface.

The assertion that a program is apparently free of software faults is the major prerequisite for granting a safety licence by authorised certification institutions. Since everybody grasps a simple thing instantly and as intended, to enable verifiability of software, we have chosen *simplicity* as an appropriate, fundamental and particularly human oriented design concept. The ease with which simple systems are understood and their clear and distinct behaviour when executed is a pre-condition to achieve social consensus which constitutes the essence of verification. The fact that everybody understands an item under consideration immediately as intended increases the trustworthiness of the result. Thus, simplicity is a fundamental feature to increase the confidence in computer based systems.

If varied pragmatic interpretation is possible or required due to vagueness or ambiguity of expression, an interpreted item is not simple. This leads to confusion as, unless specifically prevented, people seldom interpret the same thing in the same way. Conversely, we can conclude that an item of consideration is simple, if its meaning can be taken literally as given and requires no extra cognitive processing effort to arrive at. Applied to software, a program may be regarded as simple if it can be verified in a mechanised way. Such a mechanised software verification is a complete test, because it would not require intellectual understanding and can be carried out by a computer.

Irrespective of the fact that in reality strict causality is often questionable, people usually think in terms of actions or events and their results, which they tend to perceive as causes and effects. Therefore, cause/effect tables or decision tables appear to be highly appropriate as a form of software representation oriented at human cognition. This is further substantiated by the property that tables are visual means. If such diagrams of causes and effects are simple and follow accepted ergonomic standards, everybody will develop the same mental model of the operations of the formulated software. This cannot be guaranteed if text or words are used for programming since people tend to use their idiosyncratic experience to interpret them.

Hence, visualisation is *a conditio sine qua non* of interface design supporting shared understanding across a varied user population. As they have finite size, cause/effect tables also lend themselves to a complete test. Because of their more accurate fit to human cognition and because the results of complete tests are absolutely sure, programs thus verified may be permitted to be employed in automation systems which have to meet the requirements of Safety Integrity Level 4.

## 4. SAFETY INTEGRITY LEVELS AND INHERENTLY SAFE LANGUAGES

Guided by the principle of simplicity, but relaxing the requirements gradually, we now define a sequence of programming language subsets for implementing software that has to fulfil the demands of Safety Integrity Level 4 through 1.

### 4.1 *PEARL 90*

The textual language PEARL 90 (DIN, 1998) is one of the very few genuine high level real time languages. Owing to its clear concepts and its modular, block oriented structure, PEARL 90 is especially suitable for industrial process control applications and is also increasingly used in academia for teaching purposes. Its algorithmic part has approximately the same facilities as Pascal. Additionally, PEARL 90 provides the data types **clock** an **duration** and the following statements for task scheduling and control:

- **activate** for task activation,
- **terminate** for task termination,
- **suspend** for temporary blocking of a task,
- **continue** for releasing a suspended task,
- **resume** for the combination of suspend and continue, and
- **prevent** for annihilation of all scheduled future activations of a task.

The above statements are coupled to time related or event driven scheduling conditions. The synchronisation mechanism and the protection of shared data in the language are based upon a concept of generalised semaphores. Exception handling is founded on internal signals in an unstructured manner. PEARL 90 defines a generic input/output concept in the form of so-called **DATION**s (**DAT**a sta**TION**s), which serve the specification of device topology, data access, control information etc.

### 4.2 *Enhancing Safety*

In the following sections, we offer four nested subsets of PEARL aimed at enhancing the safety of real time control software, and to be viewed as potential candidates for consideration in this domain. These subsets are associated with the safety integrity levels as follows:

| SIL | PEARL Subset |
|---|---|
| SIL 1 | *HI-PEARL* |
| SIL 2 | *Safe PEARL* |
| SIL 3 | *Verifiable PEARL* |
| SIL 4 | *Table PEARL* |

Although the subsets are traditional procedural languages, they are simple and subject oriented, because oriented at the absolutely necessary they comprise the simplest and most well understood language constructs, only. The four PEARL 90 subsets are becoming progressively restrictive towards higher safety integrity levels. The assignment of the subsets to the different safety integrity levels results from the simplicity and clarity of the verification methods available for the programming paradigms on which the single languages are based. As outlined above, the simplicity of the software verification process directly correlates to the trustworthiness of computerised systems. Basically, the language subsets differ in some "unsafe" language features that are gradually prohibited. The main advantage of this approach is that one does not have to learn a different language for each safety level and the compiler can help to verify if the programs respect certain safety requirements. The use of language subsets to implement safety critical systems also allows to mix code with different levels of safety integrity, which enables a seamless connection between safety critical and uncritical parts of a system.

### 4.3 *Table PEARL*

The subset Table PEARL defined below consists of only one executable statement allowing the formulation of the rules which constitute cause/effect tables. As was already pointed out in Section 3, software represented in this form can be verified with the highest certainty. Therefore, it is licensable for applications having to meet Safety Integrity Level 4. For the formal definition of this and the other subsets we use an extension of the classical Backus Naur form (EBNF), brackets [ ] to denote optional syntactical expressions, as well as the symbols * and + to denote repetition of the correspondingly marked expressions at least zero times or once, respectively.

| Table PEARL | | |
|---|---|---|
| ⟨*module*⟩ | ::= | `MODULE` [⟨*var-decl*⟩;]* [⟨*rule*⟩;]* `MODEND`; |
| ⟨*var-decl*⟩ | ::= | `DCL` ⟨*name*⟩ : ⟨*type*⟩; |
| ⟨*rule*⟩ | ::= | `IF` ⟨*bool-expression*⟩ `THEN` ⟨*variable*⟩ := ⟨*expression*⟩ `FIN` |

### 4.4 *Verifiable PEARL*

The derivative Verifiable PEARL defined in the following table comprises just those language constructs which are necessary to program in a textual form the interconnection patterns of Function Block Diagrams, as defined in IEC 61131-3 (IEC, 1992); in other words, parameter passing and procedure invocation. Corresponding programs are rather easy to verify with rigorous methods of high trustworthiness (Halang et al., 1998). Hence, they can be certified for Safety Integrity Level 3.

<table>
<tr><th colspan="3">Verifiable PEARL</th></tr>
<tr><td>⟨module⟩</td><td>::=</td><td>MODULE [(⟨name⟩)]<br>[⟨problem⟩] MODEND;</td></tr>
<tr><td>⟨problem⟩</td><td>::=</td><td>PROBLEM [⟨proc-spec⟩]*<br>[⟨var-decl⟩]* ⟨proc-stmt⟩+;</td></tr>
<tr><td>⟨proc-spec⟩</td><td>::=</td><td>SPC ⟨name⟩ : PROC ⟨lst-of-par⟩<br>⟨return-type⟩; [GLOBAL];</td></tr>
<tr><td>⟨var-decl⟩</td><td>::=</td><td>DCL ⟨name⟩ : ⟨type⟩;</td></tr>
<tr><td>⟨proc-stmt⟩</td><td>::=</td><td>⟨proc-name⟩ ⟨lst-of-par⟩;</td></tr>
<tr><td>⟨lst-of-par⟩</td><td>::=</td><td>( ⟨par⟩ [,⟨par⟩]* )</td></tr>
<tr><td>⟨par⟩</td><td>::=</td><td>⟨const⟩ | ⟨var-name⟩</td></tr>
</table>

### 4.5 *Safe PEARL*

Only in very rare cases all features of high level programming languages are necessary to formulate the functionalities required in control engineering. Therefore, in the following table we define an inherently safe language subset restricted to those executable statements which are really necessary and, thus, indispensible, viz., procedure calls, assignments, conditional selections, and loops with bounded numbers of iterations. The correctness of programs formulated with this subset can be proven with adequate tool support by, for instance, higher order logic, which is a typified variant of Church's higher predicate logic. Therefore, this subset qualifies for usage under the conditions of safety Integrity Level 2. The attribute READ within the declaration of an interface variable ensures that other modules are only permitted to read the declared variable but not to write it. Variables, procedures and types are just names within this language subset. Expressions contain the usual Boolean and arithmetical operations.

### 4.6 *HI-PEARL*

The derived subset called H(igh)I(ntegrity)-PEARL preserves the obvious advantages of the original language, but is much more suitable for implementing safety related software, especially with respect to schedulability, since it is verifiable with respect to its functionality and timeliness. As a consequence, HI-PEARL constitutes a programming language feasible for implementing systems classified as safety related on SIL 1.

<table>
<tr><th colspan="3">Safe PEARL</th></tr>
<tr><td>⟨module⟩</td><td>::=</td><td>MODULE [(⟨name⟩)]<br>[⟨mod-extension⟩];<br>[⟨itrfc-spec⟩]*<br>[⟨system-part⟩]<br>⟨problem-part⟩<br>MODEND;</td></tr>
<tr><td>⟨mod-extension⟩</td><td>::=</td><td>EXTEND (⟨name⟩ [,⟨name⟩]+)</td></tr>
<tr><td>⟨itrfc-spec⟩</td><td>::=</td><td>INTERFACE (⟨name⟩)<br>⟨itrfc-decl⟩* ;</td></tr>
<tr><td>⟨itrfc-decl⟩</td><td>::=</td><td>⟨itrfc-var-decl⟩<br>| ⟨itrfc-proc-decl⟩</td></tr>
<tr><td>⟨itrfc-var-decl⟩</td><td>::=</td><td>SPC ⟨name⟩ : ⟨type⟩ READ;</td></tr>
<tr><td>⟨itrfc-proc-decl⟩</td><td>::=</td><td>SPC ⟨name⟩ : PROC<br>⟨lst-of-par⟩ ⟨return-type⟩;</td></tr>
<tr><td>⟨system-part⟩</td><td>::=</td><td>SYSTEM;<br>⟨name⟩ : ⟨name⟩;</td></tr>
<tr><td>⟨problem-part⟩</td><td>::=</td><td>PROBLEM;<br>[⟨declarations⟩]+</td></tr>
<tr><td>⟨declarations⟩</td><td>::=</td><td>⟨var-decl⟩<br>| ⟨proc-decl⟩</td></tr>
<tr><td>⟨var-decl⟩</td><td>::=</td><td>DCL ⟨name⟩ : ⟨type⟩;</td></tr>
<tr><td>⟨proc-decl⟩</td><td>::=</td><td>⟨name⟩ : PROC<br>⟨lst-of-par⟩ ⟨return-type⟩;<br>⟨body⟩<br>END;</td></tr>
<tr><td>⟨lst-of-par⟩</td><td>::=</td><td>( ⟨par-decl⟩ [, ⟨par-decl⟩]* )</td></tr>
<tr><td>⟨par-decl⟩</td><td>::=</td><td>⟨name⟩ : ⟨type⟩</td></tr>
<tr><td>⟨body⟩</td><td>::=</td><td>[⟨var-decl⟩]+<br>[⟨statement-seq⟩]</td></tr>
<tr><td>⟨statement-seq⟩</td><td>::=</td><td>[⟨statement⟩;]*</td></tr>
<tr><td>⟨statement⟩</td><td>::=</td><td>(⟨statement-seq⟩)<br>| ⟨assignment⟩<br>| ⟨conditional⟩<br>| ⟨for-loop⟩</td></tr>
<tr><td>⟨assignment⟩</td><td>::=</td><td>⟨variable⟩ := ⟨expression⟩</td></tr>
<tr><td>⟨conditional⟩</td><td>::=</td><td>IF ⟨bool-expression⟩<br>THEN ⟨statement-seq⟩<br>[ELSE ⟨statement-seq⟩] FIN</td></tr>
<tr><td>⟨for-loop⟩</td><td>::=</td><td>FOR ⟨variable⟩<br>[FROM ⟨constant⟩]<br>[BY ⟨constant⟩]<br>TO ⟨constant⟩<br>REPEAT ⟨statement-seq⟩ END</td></tr>
</table>

For formulating critical regions, the structured **LOCK** mechanism, with a time-out clause, is introduced and the concept of semaphores is removed. The time-out clause ensures that the waiting time before entering a critical region is limited. In case that the lock cannot be carried through before this time limit is exceeded, an alternative action takes place. The mentioned time-out clause and the lock's execution time limit enable the determination of an upper time execution limit for each lock statement at compile time. To ensure that no synchronisation or I/O operation takes arbitrarily long — in case multiple processes claim a shared resource — shared objects may only be referenced within the framework of a **LOCK** statement. To prevent the formation of arbitrarily long cascades of exceptions and interrupts, with each exception specification a constant value is associated, bounding the number of occurrences and the time for handling the exception within one single activation of the process provoking it. The minimum time period between two interrupt signals on one line equivalent to the maximum frequency of the interrupt signal has to be stated in the interrupt's specification. To disable iterations that prohibit an analysis of schedulability of programs and, hence, lead to undeterministic behaviour **WHILE** and **REPEAT** loops are removed. Instead, iteration statements have to explicitly specify their execution times and the maximum number of repetitions. The **GOTO** statement which is unsuitable for structured programming as well as directly or indirectly recursive procedure calls are banned.

## 5. CONCLUSION

Despite the existence of standards like IEC 61508-1 or DIN V VDE 0801, software based systems are still considered to be less trustworthy than conventional hardwired components, which is justified in part by the longer tradition of hardware engineering and many years of experience in the development of strategies for coping with corresponding failures. This situation is due to

- the standards being unclear about the characteristics of environments for safety critical programming, and
- no programming language (and related programming environment) based on these guidelines being available.

To overcome this situation, a dedicated subset of PEARL 90 has been proposed for each of the 4 Safety Integrity Levels defined in IEC 61508-1, following the principle of simplicity: the more *safety critical* a system is, the more *simple* the related control software needs to be. Accordingly, on the highest Safety Integrity Level (SIL 4) the safety/integrity of a "software" system can be verified just by looking at a *Cause Effect Table*, i.e., without employing formal proofs. Control software for SIL 3 systems can be graphically constructed, based on already proven function blocks. This will simplify the process of software development as compared to textual languages and still keep the task of safety proofs relatively easy. Programs written in the subset for SIL 2 have the main advantage of being formally verifiable. Finally, the subset HI-PEARL for the lowest Safety Integrity Level (SIL 1) can be easily analysed for schedulability, which is assured by eliminating some "unsafe" language constructs such as GOTO, WHILE and REPEAT. Although they are not as simple as the subset for SIL 2, HI-PEARL programs are verifiable as well.

However, much more important than the technical details mentioned above is the realisation of the important rôle the human component plays in the process of engineering safe computerised systems, which can and will lead to future research. In this paper we have drawn attention to the fact that programming languages are best perceived as human-computer interfaces which are subject to ergonomic principles that need to be followed in design of end user interfaces. We suggested, therefore, that higher level programming languages ought to be "subject" (i.e., human user) rather than "object" (i.e., system) oriented. We have identified simplicity as a fundamental design principle for programming languages, fostering human cognitive performance and, thus, safety.

## 6. REFERENCES


Dijkstra, E.W. (1989). The next forty years. Personal note EWD 1051.

DIN 66 253-2 (1998). *Programmiersprache PEARL 90.* Beuth Verlag, Berlin.

Halang, W.A. A.H. Frigeri, R. Lichtenecker, U. Steinmann and K. Wendland (1998). *Methodenlehre sicherheitsgerichteter Echtzeitprogrammierung.* Schriftenreihe der Bundesanstalt für Arbeitsschutz und Arbeitsmedizin – Forschung – Fb 813. Verlag für neue Wissenschaft, Bremerhaven.

Interim Defence Standard 00-55 (1991). *The Procurement of Safety Critical Software in Defence Equipment.* British Ministry of Defence.

IEC 880 (1986). *Software for Computers in Safety Systems of Nuclear Power Stations.* International Electrotechnical Commission, Geneva.

IEC 61131-3 (1992). *Programmable Controllers, Part 3: Programming Languages.* International Electrotechnical Commission, Geneva.

Draft International Standard IEC 61508-1 (1998). *Functional Safety of Electrical/Electronic/ Programmable Electronic Systems: Generic Aspects — Part 1: General Requirements.* International Electrotechnical Commission, Geneva.

Krebs, H., and U. Haspel (1984). Ein Verfahren zur Software-Verifikation. *Regelungstechnische Praxis rtp* **26**, 73–78.

RTCA/DO-178B (1992). *Software Considerations in Airborne Systems and Equipment Certification.* (www.rtca.org)

DIN V VDE 0801 (1995). *Grundsätze für Rechner in Systemen mit Sicherheitsaufgaben.* Beuth Verlag, Berlin.

IFAC Publications
www.elsevier.com/locate/ifac

# ENHANCING INTEGRITY OF EMBEDDED CONTROL SYSTEMS

**Matjaž Colnarič**

*University of Maribor, Slovenia*
*Faculty of Electrical Engineering and Computer Science*
*colnaric@uni-mb.si*

Abstract: This paper gives an overview of a holistic project dealing with the consistent design of embedded control systems with special emphasis on their integrity aspects. It shows how existing methods can be adapted and reasonably employed, whenever possible, without having to resort to new innovations. Firstly, the hardware issues are dealt with and extensively elaborated, particularly the peripheral interfaces. Secondly, the operating system is briefly addressed, and finally, programming issues are considered including descriptions of the specific programming language, exception handling, and how to deal with temporal overload. *Copyright © 2001 IFAC*

Keywords: Real time, embedded, robust, safety critical systems, peripheral interfaces, exception handling, overload, fault tolerance.

## 1. INTRODUCTION

Computer control systems are increasingly being used in safety critical applications where the integrity of the systems depends substantially on them. Any failures could have severe consequences: they may result in massive material losses or endanger human safety.

Dependability, being an inherent requirement in the design of control systems, is usually insufficiently supported. Techniques intended to improve dependability are predominantly based on testing, and the quality achieved depends mainly on the experience and the intuition of the designers.

This approach, however, is inadequate. It is necessary to consider the safety measures for such systems already in the early design phases, which means that common commercial off-the-shelf control computers are usually unsuitable for safety critical applications.

In the late 1980ies, the International Electrotechnical Commission (IEC) started the standardisation of safety issues in computer control (IEC, 1992). They identified four safety integrity levels, SIL 4 being the most critical. This paper, however, is concerned with applications falling into the least demanding first level which already allows employment of computer control systems based on generic microprocessors.

It is desirable that these systems should formally be proven correct or even safety licensed. Owing to their complexity, however, this is very difficult if not impossible to achieve.

Instead, in this paper we present certain contributions to several areas of control systems design, with the aim of improving both functional and temporal correctness. The implementation of long standing, but often neglected viable solutions will be shown, rather than devising new methods and techniques.

As verification of functional correctness is more established than that of temporal correctness, although equally as important, special emphasis will be given to the latter.

## 2. HARDWARE DESIGN

Hardware architectures and their implementations provide the lowest level of control system design which must be inherently as safe as possible. It is not enough that the hardware platform is tested after it is built; support for integrity must be considered an important guideline in its design.

In our laboratory several hardware prototypes have been designed and built; all are based on asymmetrical multiprocessor system architecture.

Briefly, it consists of two types of processors: task processors, on which application tasks are being processed, and dedicated OS kernel processor, allowing for parallel execution of OS routines (Halang, 1986; Colnarič *et al.*, 1994; Colnarič *et al.*, 1998). Similar implementation can be found in some other projects (Stankovic and Ramamritham, 1991; Cooling, 1993).

Events from the environment are normally introduced to the system by interrupts. Their handling obviously delays execution of the tasks, if they are run on the same processor, and thus prevents temporal determinism and run-time estimation. In our case, events are handled in parallel with the task execution; the latter is only preempted if it is absolutely necessary in order to provide for all tasks to meet their deadlines, and under consistent supervision of the operating system.

Furthermore, parallel execution of system routines on a separate processor guarantees better system performance since the task processor is almost exclusively dedicated to task program execution. For the same reason this implementation also allows more complex OS service.

Communication between the task and the kernel processor is done via a few locations of a dual-port RAM accessible from both sides. This is feasible since the system is not supposed to be geographically distributed, and the number of task processors is limited (for the time being we are considering three task processors). Task and kernel processors are synchronised using polling and, if necessary, hardware signals.

The earliest deadline first (EDF) scheduling policy was adopted, being optimal for the proposed architecture. Conceptually, it already introduces run-time schedulability analysis which upon the occurrence of a task detects whether it will eventually cause overload, and thus initiates handling before this occurs. We refer to (Colnarič *et al.*, 1994) for details of the modified EDF scheduling and schedulability analysis used in our model.

Scheduling and run-time schedulability analysis are done on the kernel processor, thus introducing no delay. It was possible, for this reason, to add some additional functions which handle the overload situations without increasing the run time overhead.

The latest attempt was to employ symmetrical task processors (Colnarič and Verber, 2000) what is achieved by de-coupling the peripheral interfaces from the task processors. In this architecture, application tasks are no longer physically mapped onto only one processor. Apart from better utilisation of the system, this has another advantage with respect to integrity issues: the task processors can be implemented diversely.

In this case, to improve control system robustness, redundant diversely designed critical task instances (however, only of those that calculate certain values and do not directly interfere with peripherals — if any) can be run simultaneously on diverse task processors. A simple voting mechanism in the form of another task (which must be designed for robustness) is comparing the results and, in the case of satisfying given conditions, selects and propagates the best effort. If the results are incompatible an adequate action such as silent fail or exception is taken. If within a certain time limit no results are available that fully comply with the requirements, the performance may be degraded gracefully.

### *2.1 Peripheral Interfaces*

Adding intelligence to peripheral interfaces is another possibility to improve the integrity of control systems in the hardware domain.

Instead of the passive memory mapped peripheral interfaces common in generic process control computers, the implementation of units autonomous by adding certain processing facilities capable of performing desired functions is proposed, which are distributed over local area networks. Technological advances enable their implementation in form of application specific integrated circuits (ASICs). Such hardware-based automata are much easier to verify in a well established way. In certain cases, performance could also be improved.

Employing built-in local intelligence, fault tolerance can be significantly enhanced by, for example, self-diagnostics, self-check at initialisation time, and integrity checking during operation. In case of an unsuccessful initialisation the supervisory system can scan for alternative or redundant sub-units. If irregular behaviour is detected during operation, the system can be reconfigured dynamically. If necessary, performance could be degraded gracefully, preserving essential integrity and safety. In the case of a non-recoverable error, or a failure of a task processor, the fail-safe property can be guaranteed by controlled and smooth driving of the controlled process into a safe state.

To cope with a transient system overload when specified deadlines cannot be met by task processors due to an exceptional irregular situation, system performance can also be gracefully degraded using peripheral interfaces. If a peripheral unit periodically sending data to the environment, for example, does not receive data on time, it can either re-use the last valid data item if this is adequate, or even predict the new one by extrapolation from preceding values. If the problems persist, the faults should not be tolerated anymore; instead, a fail-safe shut-down should be performed.

Similarly, an intelligent peripheral unit can deal with possible transient irregular output or input values. Since it is directly connected to the environment, it is in the best position to resolve such conflicts. If, for example, an output value is calculated which exceeds the physical possibilities of, e.g., analog outputs, the peripheral interface may decide whether it is reasonable to clip it to the maximal value and/or to report a warning or even an error.

Further, these processing facilities added to peripheral interfaces can consistently support the exact timing of input or output operations by requesting certain actions to be committed at a certain instant, or timestamping the input data. This is very important for the stability of control systems and, therefore, should be more consistently supported by high level programming languages and operating systems used in process control (see Section 4). This feature must be supported in turn by an adequate time synchronisation mechanism. A more detailed description of this implementation can be obtained from (Colnarič and Verber, 2000).

## 3. OPERATING SYSTEM

In asynchronous real time control systems concurrent processing is typically based on the task concept. To support tasks, the operating system kernel and a corresponding co-processor, on which the kernel is running, need to provide some basic tasking functions like initialisation, activation, termination, synchronisation, suspension and resumption. It is based on the feasible earliest deadline first algorithm. Its schedulability can be analysed (Kligerman and Stoyenko, 1986) a priori, provided reliable and realistic worst case estimates of the task execution times are available beforehand.

This processor assignment scheme also handles overload situations as early as possible by run-time schedulability analysis. It employs a look-ahead algorithm which uses information provided by the earliest deadline first dispatcher and extracted from the schedules of task (re-) activations.

Considering the limited sizes and complexities of operating system kernel functions, as implemented in our prototype, we felt encouraged to try to prove them correct using formal methods. The concept of this operating system has been thoroughly elaborated and formally verified (Tol, 1995) using the mechanism of Hoare-style logic (Dijkstra and Feijen, 1984).

## 4. PROGRAMMING

The next subsequent layer in the control application design is programming of tasks. For this purpose, languages and tools should be used, which to the highest extent prevent the programmer to use the programming means that could potentially endanger temporal predictability and system's integrity.

A language MiniPEARL has been devised with the aim of providing for consistent programming of temporally predictable applications. It is a subset of the standardised language PEARL (DIN, 1981). Recently, it has been extended to include basic elements of object orientation (ObjectPEARL, (Verber and Colnarič, 2000).)

In the following subsections, some general features of MiniPEARL with respect to safety issues will be enumerated, then exception handling will be proposed, and the possibility of dealing with temporal overload in ObjectPEARL will be explained.

### *4.1 MiniPEARL*

MiniPEARL is a strictly typed and structured programming language. Some features were added to improve tasking operations and support for real time specific programming. Further, constructs which may result in temporally unpredictable behaviour were renounced or modified. For instance, there are no GOTOs that could result in unstructured code, and no dynamic features like pointers and recursion that could cause memory management problems.

In the sequel, some characteristics of MiniPEARL that add to temporal predictability of program executions will be enumerated.

• *Each loop block is tightly bounded.* In the REPEAT statement, lower and upper counts of loop iterations must be present and defined with compile-time constant expressions, so that the longest execution time of a loop can be estimated. In addition, the WHILE part is renounced and replaced by an explicit IF-EXIT statement within a loop.

• *Each synchronisation construct is temporally bounded.* In classical programming languages synchronisation measures (for example critical regions or semaphores) may introduce arbitrarily long delays. In MiniPEARL each such command is temporally guarded and an explicit action is defined for the case that a time-out occurs.

• *Explicitly asserted execution time.* In some cases, because of the nature of a program, estimation may yield too pessimistic execution times. To resolve this problem, additional information about the program flow must be given by the programmer (Puschner and Koza, 1989; Park, 1993). These methods, however, require complex knowledge of the program structure and are unfeasible in all situations. To overcome this problem, the execution time of a code part can explicitly be asserted by the system developer. In such a case, the estimated results are overridden by the asserted ones by the analyser. However, to guarantee that the actual execution time will not be longer than the declared one, the block must be guarded by time-out control, and time-out action must be present.

• *Updating the residual task execution time.* The scheduling algorithm in the OS kernel processor relies on the residual execution time of each task. This time is computed as maximum execution time of the task minus cumulative running time. However, as the program advances, it is possible to estimate the residual execution time more accurately. The new UPDATE statement can be used explicitly, and whenever necessary, to assert a new estimation of residual execution time.

### *4.2 Handling of Exceptions*

When a system behaves beyond its normal specifications it is called to be in an exceptional state. Such a state indicates an abnormal situation and possible danger to its integrity. It is thus important that it is handled consistently and the system brought into a safe operating state, or smoothly stopped.

Exception handling is also one of the crucial problems to be solved if temporally predictable behaviour is required. When an exception occurs in a program, the latter is inevitably delayed by the execution of associated routines (handlers) causing a problem with respect to the a priori determined task execution time. Therefore, exceptions should be prevented by all means, whenever and wherever possible (Black, 1983).

A closer examination of the possible sources of exceptions in hard real time systems leads to the following classification into three categories: preventable, avoidable and catastrophic exceptions.

The best way to handle exceptions is to prevent them from occurring at all — as far as this is possible. Some exceptions are preventable by imposing certain restrictions on the use of potentially dangerous features. Normally, these restrictions are essential to provide for temporal predictability in program execution like renouncing dynamic structures, pointers or virtual memory.

There are situations in which irregular and unwanted events cannot be prevented from occurring. Often programmers tend to service wanted events directly in their programs, and unwanted ones as exceptions: it is, however, irrelevant whether an event is wanted or not. If such events can be anticipated during the design phase, they should be included into the specifications to be handled adequately. Thus, dealing with them becomes part of the application software instead of relying on some universal system exception handling.

Unfortunately, in industrial embedded systems situations often occur which can be neither prevented nor anticipated. This is the case when programs do not follow their specifications due to hardware failures, residual software errors, or wrong specifications. It is of utmost importance that such catastrophic situations are resolved in the most appropriate way. The need for consistent solutions to the exception handling problem is exacerbated by the fact that exceptions often result from critical system states, which is when computer control is needed most.

Instead of the usual exception handlers, our proposed approach is close to automatic backward or forward recovery using recovery blocks (Cristian, 1989).

The principle of backward recovery is to return to a previous consistent system state after an inconsistency is detected by consistency tests called post-conditions.

The forward error recovery technique tries to obtain a consistent state from partly inconsistent data. Which data are usable can be determined by consistency tests, error assumptions, or with the help of independent external sources.

The proposed solution is as follows: a block (explicit block, task, procedure, loop, or any other block structure) consists of several alternative statement sequences. Each alternative can have its own pre- and/or post-conditions, represented by Boolean expressions. When the program flow enters a surrounding block, the state variables are stacked which are modifiable by the alternatives which might fail. Then, the first alternative statement sequence is executed, whose pre-condition (if it exists) is fulfilled. Finally, its post-condition is checked, and if this is also fulfilled, execution of the block is successfully terminated. If the post-condition is not fulfilled, the next alternative is checked for its pre-condition and eventually executed. If necessary, the values of the state variables recorded at the beginning of the block are first re-stored. For the details and syntax refer to (Colnarič *et al.*, 1995).

A severe problem arises if there are any actions triggered such as commencing a peripheral process which causes an irreversible change in the initial process state inside an alternative that has failed. In this case, backward recovery is generally impossible.

The backward recovery alternatives should contain diversely designed and coded programs to cope with specification errors and to eliminate possible implementation problems or residual software errors. When problems are expected, alternative design solutions or redundant hardware resources may be employed. A further possibility is to gradually assert less restrictive pre- and/or post-conditions and, thus, to degrade performance gracefully in exceptional situations.

The execution of a block is unsuccessful if there is no alternative with pre- and post-conditions fulfilled. If the block is nested inside an alternative on the next higher level, this alternative fails as well (the failure is propagated), and control is given to the next one on this level, thus providing a chance to resolve the problem in a different way. At the highest level, however, the last alternative must not have any pre- or post-conditions. It must solve the problem by applying some safe action like employing robust fault tolerance measures or performing smooth shut-down. Very basic electrical or even mechanical back-up systems can be engaged in the case of extreme safety critical applications as an ultimate alternative.

### *4.3 Handling of Overloads*

Employing the earliest deadline first strategy, on arrival of each new request a new schedule is calculated from the ready tasks' parameters, so that all their deadlines are met. However, provision should be made for dynamic overload, when the system is incapable of serving all given tasks, which should actually not happen in properly designed applications.

When schedulability is analysed at run-time, an overload situation may be detected during the re-scheduling of ready tasks upon occurrence of a new one. Each scheduled task is verified as to whether the sum of its execution time and the ones of all tasks scheduled to run before it is shorter than its response time. If for any task this does not hold an overload situation has occurred.

Theoretically, it should always be possible to prevent overload situation by proper a-priori schedulability analysis (Kligerman and Stoyenko, 1986) which should be carried out at design time. Unfortunately, in real life this is sometimes untrue for several reasons:

- A controlled system may issue service requests invoking tasks more frequently than specified.
- Due to hardware or software failures in a control system it may not behave according to the specifications which may cause the tasks to be improperly executed and their actual execution times delayed.
- Owing to task complexity, there will always be errors in run-time estimation or schedulability analysis.
- A priori schedulability analysis could yield pessimistic results and is, thus, impractical and unpopular with application designers (the implied pessimism calls for more expensive hardware platforms).

Some overload situations can be prevented. Frequency of service requests, for instance, can be monitored by a simple hardware watchdog timer: set to a minimum specified time and triggered upon occurrence of a request, it always must have run out before the next request arrives. In the case of an override, the request can be delayed if possible, or ignored, or an alert is reported so that any violation of the specifications can be detected before the system is jammed.

Once an overload situation is detected, it can only be handled with the same resources if the complexity of task processing is reduced. For this reason, the application programmer must have already considered the minimum acceptable performance of the tasks in the early stages of the design, i.e., how the full system functionality can gracefully be degraded in order to survive the overload situation. It is not the intention of this paper to go into the details of methods for graceful degradation, but to present some ideas of how it can be implemented.

We refer to the overload situation as the 'critical mode'. In this mode, there are two groups of tasks that must be handled: (1) tasks that have already started execution (the actually running task and all those which have already been executing but were pre-empted), and (2) tasks that have not been scheduled to run, yet.

To handle the latter, tasks may have two bodies: a regular one which fully implements the task's requirements, and an alternative one which performs the most essential functions, only, which may possibly even be empty. The operating system is supplied with task execution times for both normal and critical mode.

There may be tasks that were pre-empted during their previous execution on a processor in the normal mode; they have already entered their regular body. If a situation becomes critical later, they cannot re-start and enter the alternative body, but must reduce their complexity if possible, and terminate as soon as possible.

To allow for this, it is proposed to implement procedures (in the form of class methods) in two ways again: the normal and the critical alternative. This can be realised using virtual methods in a similar way as polymorphism: according to the mode reported by the OS kernel from the Virtual Method Table (VMT) the adequate method address is retrieved and executed.

By using this syntax the designer is encouraged to consider the design of alternative task bodies and methods from the early stages of design on. For details of implementation and syntax it is referred to (Colnarič and Verber, 2001). Furthermore, through automated execution time analysis the designer is always aware of the execution time savings achieved by a certain implementation and can, thus, decide whether further performance degradation is necessary.

## 5. CONCLUSIONS

This paper does not introduce many novel solutions to improve the integrity of embedded control systems classified for the lowest safety integrity level as defined by IEC. Its main contribution is to demonstrate how already established principles can improve safety.

Apart from providing solutions to resolve problems that occur during execution, when designing a safety critical application it is also important to try and prevent the potential causes of errors, non-determinism and non-predictability at their very sources.

In the current state of safety critical system design the ultimate guideline should still be simplicity: only this way reasonable verifyability can be assured. It is very important that any measure introduced does not essentially add to the complexity of the design.

## REFERENCES


Black, Andrew P. (1983). Exception handling: The case against. Technical Report TR 82-01-02. Department Of Computer Science, University of Washington.

Colnarič, Matjaž and Domen Verber (2000). Design and programming of peripheral interfaces for embedded real-time control systems.. In: *Proceedings of WRTP 2000*. Palma de Mallorca, Spain. pp. 29–34.

Colnarič, Matjaž and Domen Verber (2001). Dealing with tasking overload in object oriented real-time applications design. In: *Sixth International Workshop on Object-oriented Real-time Dependable Systems WORDS2001*. Rome, Italy.

Colnarič, Matjaž, Domen Verber and Wolfgang A. Halang (1995). Supporting high integrity and behavioural predictability of hard real-time systems. *Informatica, Special Issue on Parallel and Distributed Real-Time Systems* **19**(1), 59–69.

Colnarič, Matjaž, Domen Verber, Roman Gumzej and Wolfgang A. Halang (1998). Implementation of embedded hard real–time systems. *International Journal of Real–Time Systems* **14**(3), 293–310.

Colnarič, Matjaž, Wolfgang A. Halang and Ronald M. Tol (1994). Hardware supported hard real-time operating system kernel. *Microprocessors and microsystems* **18**(10), 579–591.

Cooling, Jim (1993). Task scheduler for hard real-time embedded systems. In: *Proceedings of Int'l Workshop on Systems Engineering for Real-Time Applications*. IEE. Cirencester, London. pp. 196–201.

Cristian, Flaviu (1989). Exception handling. In: *Dependability of Resilient Computers* (T. Anderson, Ed.). Blackwell Scientific Publications. Oxford.

Dijkstra, E. W. and W. H. J. Feijen (1984). *Een methode van programmeren.*. Academic Service.

DIN (1981). *DIN 66 253: Programming Language PEARL, Part 1: Basic PEARL*. Berlin.

Halang, Wolfgang A. (1986). On methods for direct memory access without cycle stealing. *Microprocessing and Microprogramming* **17**, 277–283.

IEC (1992). *Functional Safety of electrical/electronic programmable electronic systems: Generic aspects. Part 1: General requirements.*

Kligerman, Eugene and Alexander Stoyenko (1986). Real-time Euclid: A language for reliable real-time systems. *IEEE Transactions on Software Engineering* **12**(9), 941–949.

Park, Chang Yun (1993). Predicting program execution times by analyzing static and dynamic program paths. *Real-Time Systems* **5**(1), 31–62.

Puschner, Peter and Christian Koza (1989). Calculating the maximum execution time of real-time programs. *Real-Time Systems* **1**(2), 159–176.

Stankovic, Jack and Krithi Ramamritham (1991). The spring kernel: A new paradigm for real-time systems. *IEEE Software* **8**(3), 62–72.

Tol, Ronald M. (1995). Formal Design of a Real-Time Operating System Kernel. PhD thesis. Rijksuniversiteit Groningen. Netherlands.

Verber, Domen and Matjaž Colnarič (2000). Object oriented extension to development of hard real-time systems. In: *Proceedings of SCI 2000*. Orlando, Florida. pp. 429–434.

IFAC Publications
www.elsevier.com/locate/ifac

# A PROGRAMMABLE LOGIC CONTROLLER FOR SAFETY RELATED APPLICATIONS

**Marek Śnieżek**

*Rzeszów University of Technology, Department of Electrical Engineering, 35-959 Rzeszów, Poland, sniezekm@prz-rzeszow.pl*

Abstract: To architecturally support the programming of safety related control applications in the graphical language Function Block Diagram and the verification of such software meeting the requirements of Safety Integrity Level SIL 3, a dedicated, low complexity execution platform is presented. Its hardware is fault detecting to immediately initiate emergency shut-downs in case of malfunctions. By design, there is no semantic gap between the programming and machine execution levels, enabling the safety licensing of application software by extremely simple, but rigorous methods, viz., diverse back translation and inspection. Operating in a strictly periodic fashion, the platform exhibits fully predictable real time behaviour. *Copyright © 2001 IFAC*

Keywords: Safety related control, Safety Integrity Level 3, Function Block Diagrams, software verification, programmable logic controller.

## 1. INTRODUCTION

In this paper a workable programmable electronic system for use in industrial automation is presented, which can be licensed for the higher Safety Integrity Level 3 in full, i.e., not only its hardware, but also its software, which is more important as software dependability still needs to catch up with the one already achieved by hardware. The leading idea followed throughout its design, and the originality of the approach, was to select programming and verification methods of utmost simplicity and, hence, highest trustworthiness, and to custom-tailor an execution platform for them. Thus, the semantic gap between software requirements and hardware capabilities is closed, relinquishing the need for not safety licensable compilers and operating systems.

## 2. FUNCTION BLOCK DIAGRAMS

A programming paradigm allowing for software to be easy to grasp and to verify with respect to both source and object code is available in form of the graphical language Function Block Diagram (FBD) defined within the standard IEC 61131-3 (IEC, 1992). With its long tradition in control engineering, graphical programming in form of function block diagrams as depicted in Figure 1 is already well established in automation technology.

The function block diagram language consists of four different structural elements, only:

(1) instances of functions and function blocks, i.e., rectangular symbols,
(2) dataflow lines, i.e., connection lines,
(3) names, i.e., identifiers, and
(4) (external) connection points.

Functions and function blocks are highly application dependent and re-usable elementary units of

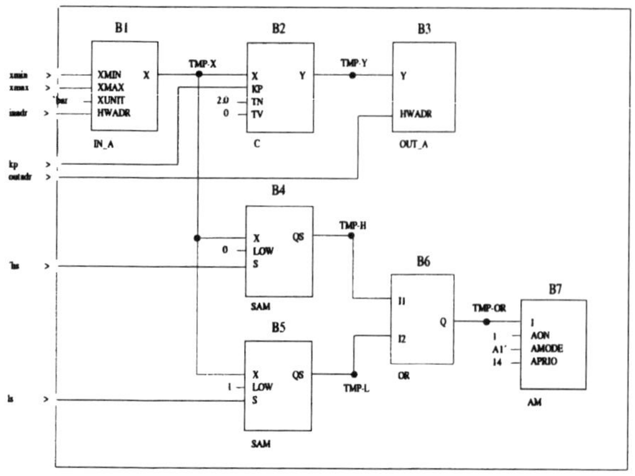


Fig. 1. Function block diagram of a program for pressure regulation and supervision

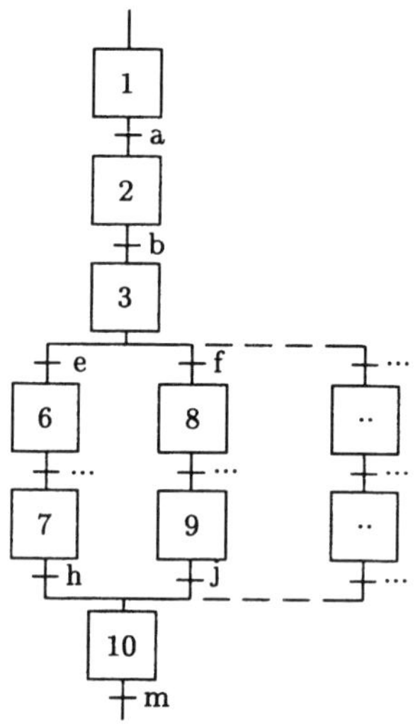


Fig. 2. A sequential function chart

application programming on a higher abstraction level. In principle, they are subroutines having inputs and outputs of arbitrary data types, and are able to perform arbitrary processing functions. Functions do not have any internal states. After being executed they yield exactly one data element as a result, which may be multivalued. Multiple named instances, i.e., copies, can be created of function blocks. Each instance possesses an associated designator and a data structure, which contains its output and internal variables as well as possibly its input variables. All values of the output variables and the internal variables in such a data structure persist from one execution of a function block instance to the next. Therefore, invocation of a function block with the same arguments does not necessarily yield the same output values. This is necessary to be able to express feedback and internal storage behaviour. Only the input and output values are accessible outside a function block instance, i.e., a function block's internal variables are hidden from the outside and are, thus, strictly protected. By the connection lines within a function block diagram a data flow is represented. Graphical, system independent program development in form of function block diagrams is very easy, and takes place in two steps:

(1) *only once* building a library of functions and function blocks, and
(2) *application specific* interconnection of functions and function block instances.

Many automation programs including safety related ones have the form of sequence controls composed of steps and transitions. While in a step, an associated program, called action, represented as a function block diagram is being executed. For safety related applications, linear sequences of steps and alternative branches of such sequences as shown in Figure 2 are permitted, only. Parallel branches in sequential function charts must either be implemented by hardware parallelism or already resolved by the application programmer in form of explicit serialisation. Also, for clarity as well as for easy comprehension and verification only non-stored actions may be used. All other types of actions as defined in IEC 61131-3 can be expressed in terms of non-stored ones and reformulated sequential control logic.

## 3. AN EXECUTION PLATFORM

When designing an execution platform as closely matching and supporting software represented in form of function block diagrams and sequential function charts, it was not the objective to save hardware costs, but to facilitate the understandability of object programs and their execution process. This led to the architecture depicted in Figure 3 with, conceptually, two different processors: a control flow processor (master) and a function block processor (slave). These two processors are implemented by separate physical units. Thus, a clear and physical separation of concerns is achieved: execution of function blocks in the slave processor, and all other tasks, i.e., execution control, sequential function chart processing, and function block invocation, assigned to the master. This concept implies that application code is restricted to the control flow processor, on which project specific safety licensing can concentrate. To enable the detection of faults in the hardware, a dual-channel configuration is chosen as displayed in Figure 3, which also supports diversity in form of different master processors and different slave processors. All processing is simultaneously performed on two processors each and all data communicated are subject to comparison.

The function block processors perform all data manipulations and take care of the communication with the environment. The master and slave

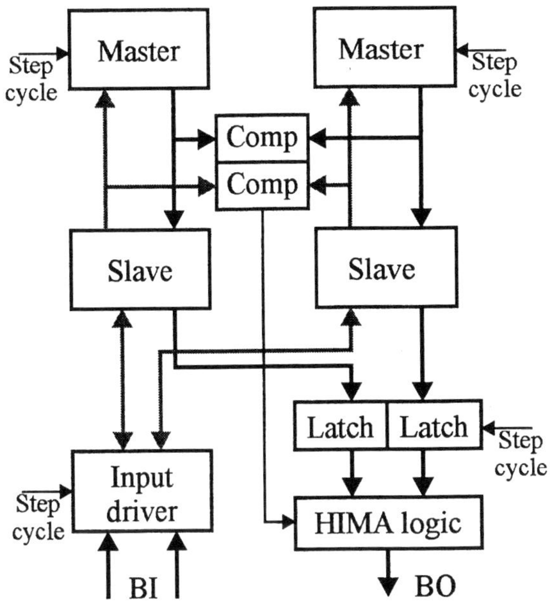


Fig. 3. Block diagram of fault detecting master-slave PES

processors communicate with each other through FIFO-queues. Clearly, the masters' and slaves' programs, though co-ordinated via communication, can be separated. This separation enables to transfer data access and data protection issues from software to hardware, thus increasing the controller's dependability. The master and slave processors execute programs in co-ordination with each other as follows. The master processors request the slaves to execute a function block by sending the latter's identification and the corresponding parameters and, if need be, also the block's internal state values via one of the FIFO-queues to the slave processors. Here the object program implementing the function block is performed and the generated results and new internal states are sent to the master processors through the other FIFO-queue. The elaboration of the function block ends with fetching these data from the output FIFO-queue and storing them in the masters' RAM memories.

It is stressed that the results and internal states are stored in the masters' memories. The slaves' memories, if needed at all, are only used temporarily while elaborating function blocks. Hence, the slaves may be viewed as memoryless function co-processors or dedicated calculators. A number of fail safe comparators checking the outputs from the master processors before they reach the slaves and vice versa completes a fault detecting two-channel configuration. Any inequality detected by the comparators generates an error signal (see below) which stops the entire system and sets the outputs to safe states. These states are provided by fail safe hardware (HIMA logic).

To prevent any modification by malfunctions, there is no program RAM, but all programs are provided in read only memories (ROMs). The code of the function blocks resides in mask programmed ROMs, which must be produced under supervision of and released by the licensing authorities, after the latter have established the correctness of the modules and their translation into object code. On the other hand, the sequences of module invocations together with the corresponding parameter passing, representing application programs at the architectural level, can be written into (E)PROMs by the user. This part of the software is subject to project specific verification again to be performed by the licensing authorities, which finally still need to install and seal the (E)PROMs in the target systems.

Besides program memory, the masters' address spaces also comprise RAM memory and FIFO-input/output registers, command registers, two step related registers each, viz., step identifier and step initial address, and transition condition registers. Furthermore, there are program counters and single bit step-clock-occurred registers, which are not programmer accessible. Additionally, in the masters' address spaces other units are memory mapped to create and receive control signals for the access of ROM, RAM, and FIFO-queues. To fulfill their purpose, the master processors need just two instructions, viz., MOVE and STEP. The MOVE instruction has two operands, which directly point to locations in address space. Thus, the memories and the above mentioned registers can be read and written. A read from a FIFO-input register implies that the processor has to wait when the input FIFO-queue register is empty. In case of writing into an output FIFO queue register, the processor also has to wait when the register is full. Execution of a MOVE implies program counter incrementation.

The programs executed by the master processors consist of sequences of steps. Behind the program segment of each step a STEP instruction, with a next-step-address as operand, is inserted, which checks whether the segment was executed within a step cycle frame or not. The step cycle is a periodic signal generated by the system clock and establishing the basic time reference for operation as programmable logic controller. The length of the cycle is selected in a way as to accommodate during its duration the execution of the most time consuming step occurring in an application (class). If the execution of a segment does not terminate within a step cycle, an error signal is generated, which indicates an overload situation or a run time error. Then, program execution is stopped immediately, and suitable error handling is carried through by external fail safe hardware. Normally, however, segment execution terminates before the instant of the next step cycle signal. Then, the processors wait until the end of the present cycle period. When the clock signal finally occurs, the step-clock-occurred registers are set. According to the contents of the transition condition registers it is decided, whether the step segment is executed once more, or whether the execution of the logically subsequent step is commenced, i.e., whether the program counters are re-loaded from the step-initial-address registers, or if another segment's initial program address is loaded from the STEP instruction's operand called next-step-address. Since only one step is active at any given time, and since program branching is only possible in this restricted form within the framework of executing STEP instructions, this mechanism very effectively prevents erroneous access to code of other (inactive) steps as well as to program locations other than the beginnings of step segments.

The design objective for providing FIFOs is to implement easily synchronisable and understandable communication links, which decouple the master and slave processors with respect to their execution speeds. The FIFO-queues consist of a fall-through memory and two single bit status registers each, viz., FULL and EMPTY, which indicate the filling states of the FIFOs. The status registers are not user accessible. They are set and reset by the FIFO control hardware and, if set, they cause a MOVE to a FIFO's input port or from an output port, respectively, to wait until space in the FIFO becomes available or data arrive. The comparison for equality of the outputs from the two master processors and of the inputs from the two slave processors, respectively, is carried out by the two fast comparators placed into the FIFO-queues. Since the responsibility for detecting errors in the system rests on these comparators, they need to meet high dependability requirements and are, therefore, implemented in fail safe technology as described in (Schuck, 1987). A comparator is connected to two FIFOs' outputs. The first data elements from each input queue are latched and subsequently compared with each other. If both latches do not hold the same value, then an error signal is generated, which stops the operation of the entire system. Otherwise, the value is transferred into both output FIFOs.

Communication with external technical processes takes place through fault detecting input/output driver units attached to the slave processors. Output data words generated by the two slaves are first checked for equality in a fail safe comparator and, subsequently, they are latched in an output port. If output data are not identical, an error signal is generated leading to a system stop. To achieve full determinism of execution time behaviour, the basic cycle was introduced as maximum step execution period. Although it exactly determines a priori the cyclic execution of the single steps, the processing instants of the various operations within a cycle, however, may still vary and, thus, remain undetermined. Since a precisely predictable timing behaviour is only important for input and output operations, temporal predictability is achieved as follows. Input data are read by the drivers at the beginning of each cycle and stored en bloc in two independent RAM buffers assigned to the respective slaves. The cycle start is signaled by the step-clock-occurred register. Only after that, the data are made available for further processing, thus providing predictability in timing. Output data generated by the slaves are latched in registers before the end of every cycle. When the step-clock-occurred register is set, the data are first checked for equality in fail safe comparators and, subsequently, they are transferred to output ports to become effective to the environment. However, if output bytes are not identical, an error signal is generated leading to a system stop.

The FIFO-queue and output comparators mentioned above are the components of a global comparator unit, which also receives operation monitoring signals from processor watch-dog timers and some correctness signals from other units. Based on all these, a global correctness signal, in other words, a negated global error, is generated and fed back to all units of the programmable electronic system. Naturally, each one can operate if this signal indicates "no error". Otherwise, the system stops and the outputs are set to safe states. The global error signal is also output, allowing to trigger some external hardware as well.

## 4. SOFTWARE SAFETY LICENSING

Matching their development process, the verification of programs constructed in the form of function block diagrams is carried out in two steps:

(1) Before being released, first all functions and function blocks contained in a library are verified employing appropriate, usually formal methods. Such a rather expensive safety licensing needs to be carried through only once for a certain application area after a suitable set of function blocks has been identified. The licensing costs justified by the safety requirements can, therefore, be spread over many implementations, leading to relatively low costs for each single automation project. In general, rather few library elements are sufficient to formulate all programs in a particular area of automation, e.g., 67 for chemical process engineering (VDI/VDE, 1995). As the details of the function blocks' imple-

mentation on the slave processors are part of the architecture, they remain invisible from the application programming point of view.

(2) Then, for any given application program, only the correct implementation of the corresponding interconnection pattern of invoked functions and function block instances (i.e., a certain dataflow) needs to be verified.

Application software is safety licensed by subjecting the object code loaded into the master processors to diverse back translation, a verification method developed in the course of the Halden experimental nuclear power plant project (Krebs and Haspel, 1984). This technique consists of reading machine programs out of computer memory and giving them to a number of teams working without any mutual contact. All by hand, these teams disassemble and decompile the code, from which they finally try to re-gain the specification. A safety licence is granted to a software if its original specification agrees with the inversely obtained re-specifications. Of course, in general this method is extremely cumbersome, time consuming, and expensive. This is due to the semantic gap between a specification formulated in terms of user functions on one hand and the usual machine instructions carrying them out on the other. Applying the programming paradigm of function blocks, however, a specification is directly mapped onto sequences of procedure invocations and parameter passing. It takes only minimum effort to verify a master program by interpreting such code, which just implements a particular module interconnection pattern, and by re-drawing the corresponding graphical program specification. Diverse back translation is especially well suited for the verification of the correct implementation of graphically specified programs on the architecture introduced above for the following reasons:

- The method is essentially informal, easily comprehensible, and immediately applicable without any training. Thus, it is extremely well suited to be used on the application programming level by people with most heterogeneous educational backgrounds. Its ease of understanding and use inherently fosters error free application.
- The effects of high complexity utility and compiler-like programs, whose correctness cannot be established, are verified, too.
- Since graphical programming based on application oriented function blocks has the quality of specification level problem description, and because by design there is no semantic gap in the architecture of the execution platform between the levels interfacing to humans and to the machine, diverse back translation leads back in one easy step from machine code to problem specification.
- For this architecture, the effort required to utilise diverse back translation for the safety licensing of application programs is by several orders of magnitude smaller than for the von Neumann architecture, once a certain set of function blocks has been proven correct.

The application of back translation is now to be illustrated by working out a relatively simple, but realistic example. The program representation levels function block diagram and object code for the master processor are shown in full detail. It will become evident that it is straightforward and very easy to draw a function block diagram from a given object program establishing the feasibility of back translation as a software verification method.

Figure 1 shows a typical industrial automation program in graphical form. It performs supervision and regulation of a pressure. The program is expressed in terms of standard function blocks as defined in the guideline (VDI/VDE, 1995). An analogue measuring value, the controlled variable, is acquired by a function block of type IN_A from the input channel with address IN-ADR, and scaled within the range from XMIN to XMAX to a physical quantity with unit XUNIT. The controlled variable is fed into a function block of type C performing proportional-integral-differential (PID) regulation subject to the control parameters KP, TN, and TV. The resulting regulating variable is converted to an analogue value by a type OUT_A output function block, and switched onto the channel addressed by OUT-ADR. In addition, the controlled variable is also supervised, with the help of two instances of the SAM limit switch standard function block type, to be within the limits given by the parameters LS and HS. If the controlled variable is outside of this range, one of the QS outputs of the two SAM instances becomes logically true and, hence, the output of the type OR function block as well. This, in turn, causes the type AM alarm and message storing function block to create a timed alarm record. The inputs of the standard function blocks comprised by the program which are neither fed by externally visible inputs of the program itself nor internally by outputs of other standard function blocks are given constant values.

The object code of this example program for the master processor is listed in Table 1. It shows a (readable) assembly language version in which, for denotational simplification, MOVE instructions to the (memory mapped) FIFO-input and FIFO-output registers are denoted by GET and PUT, respectively. Of the different function block types instantiated in the example, C, SAM, and AM have internal state variables, viz., C has 3 and the other two types have 1 each. This object code illustrates that all function block instance

Table 1. Object code representation of the example program

| | |
|---|---|
| GET ROM-loc-ID-IN_A | GET ROM-loc-0 |
| GET RAM-loc-XMIN | GET RAM-loc-HS |
| GET RAM-loc-XMAX | GET RAM-loc-B4-isv |
| GET ROM-loc-BAR | PUT RAM-loc-TMP-H |
| GET RAM-loc-INADR | PUT RAM-loc-B4-isv |
| PUT RAM-loc-TMP-X | |
| | GET ROM-loc-ID-SAM |
| GET ROM-loc-ID-C | GET RAM-loc-TMP-X |
| GET RAM-loc-TMP-X | GET ROM-loc |
| GET RAM-loc-KP | GET RAM-loc-LS |
| GET ROM-loc-2.0 | GET RAM-loc-B5-isv |
| GET ROM-loc-0.0 | PUT RAM-loc-TMP-L |
| GET RAM-loc-B2-isv1 | PUT RAM-loc-B5-isv |
| GET RAM-loc-B2-isv2 | |
| GET RAM-loc-B2-isv3 | GET ROM-loc-ID-OR |
| PUT RAM-loc-TMP-Y | GET RAM-loc-TMP-H |
| PUT RAM-loc-B2-isv1 | GET RAM-loc-TMP-L |
| PUT RAM-loc-B2-isv2 | PUT RAM-loc-TMP-OR |
| PUT RAM-loc-B2-isv3 | |
| | GET ROM-loc-ID-AM |
| GET ROM-loc-ID-OUT_A | GET RAM-loc-TMP-OR |
| GET RAM-loc-TMP-Y | GET ROM-loc-1 |
| GET RAM-loc-OUTADR | GET ROM-loc-A1 |
| | GET ROM-loc-14 |
| GET ROM-loc-ID-SAM | GET RAM-loc-B7-isv |
| GET RAM-loc-TMP-X | PUT RAM-loc-B7-isv |

invocations occurring in a program are directly mapped onto procedure calls. Each of them commences with a GET instruction, which transfers the identification (e.g., ID-C) of the corresponding block out of an appropriate ROM location to the slave's input FIFO. Then, the input parameters are supplied by reading appropriate ROM (for constants) or RAM (for program parameters and intermediate values) cells. Finally, if there are any, the values of the procedure's internal state variables are read from appropriate RAM locations. There is a set of correspondingly labelled (e.g., RAM-loc-B2-isv*i*) locations for each instance of a function block with internal states. When the slave processor has received all these data, it executes the procedure and returns, if there are any, values of output parameters and/or internal state variables, which are then stored into corresponding RAM locations. A connection between an output of one function block and an input of another one is implemented by a PUT and a GET instruction: the former storing the output value in a RAM location for a temporary value (e.g., TMP-X), and the latter loading it from there. In other words, each connection in a function block diagram gives rise to exactly one transfer from the slave's output FIFO to a RAM cell, and to one or more transfers from there to the slave's input FIFO. The implementation details of the various procedures are part of the architecture's firmware and, thus, remain invisible.

According to the above described structure of the masters' object programs, the process of back translation — disassemble and decompile object code — turns out to be very easy. To perform back translation, first the STEP instructions are searched, which clearly separate the different (sequential) steps contained in a program from each other. The code between two STEP instructions corresponds to one function block diagram. Then, the first GET instruction is interpreted. It identifies a function block instance to be drawn into the function block diagram to be set up. By comparing the subsequent GETs with the function block's description contained in the library used, correct parameter passing can be easily verified. Moreover, for each such GET which corresponds to a proper parameter (and not to an internal state variable) a link is drawn into the diagram. There are two kinds of links. The first one are connections from program inputs or constants to inputs of function blocks, or from function block outputs to program outputs. The second kind are, so to speak, half connections, namely, from function block outputs to named connection points in the diagram, or from such points to function block inputs. When the diagram is completely drawn, the names of these points can be removed. With respect to the internal state variables, it needs to be verified that the corresponding locations in the master processors' RAM are correctly initialised, and that the new values resulting from a function block execution are written to exactly the same locations from where the internal states were read in the course of the block's invocation. The process of function block identification, parameter passing verification, as well as drawing of the block's symbol and of the corresponding connections is repeated until a STEP instruction is reached, which terminates the step and, thus, the corresponding function block diagram.

## REFERENCES


IEC 61131-3 (1992). *Programmable Controllers, Part 3: Programming Languages.* International Electrotechnical Commission, Geneva.

Draft International Standard IEC 61508-1 (1998). *Functional Safety of Electrical/Electronic/ Programmable Electronic Systems: Generic Aspects — Part 1: General Requirements.* International Electrotechnical Commission, Geneva.

Krebs, H., and U. Haspel (1984). Ein Verfahren zur Software-Verifikation. *Regelungstechnische Praxis rtp* **26**, 73–78.

Schuck, H. (1987). *Analoger Fensterkomparator in Fail-safe-Technik.* Dissertation, Technische Universität Braunschweig.

VDI/VDE 3696 (1995). *Herstellerneutrale Konfigurierung von Prozeßleitsystemen.* Beuth Verlag, Berlin.

IFAC Publications
www.elsevier.com/locate/ifac

# A FAST AND FAIL SAFE COMPARATOR OF BINARY DATA WITH TERNARY RESULT AND STATUS OUTPUT

**Josef von Stackelberg**

*FernUniversität, Faculty of Electrical and Computer Engineering, 58084 Hagen, Germany, josef@stackelberg.com*

Abstract: In safety related computer control, redundancy is used to cope with errors of any kind. To detect errors, redundantly produced results must be compared imposing the responsibility for system safety on the comparators. Since comparison reduces the degree of redundancy, comparators are best implemented in form of hardware devices with fail safe behaviour. Currently available fail safe comparator circuits, however, constitute bottlenecks in safety related computing systems due to their low processing speeds. Moreover, in case of malfunctions, such comparators do not distinguish between comparison errors and comparator errors. To solve these problems, a novel fail safe comparator of two binary inputs is presented, which does not only indicate a result, but also its status. Built in a modified CMOS technology, it can match the execution speed of digital computers. In contrast to all earlier designs of fail safe comparators, by employing ternary logic it provides three different output values, allowing to distinguish between the three indications "inputs equal and comparator working properly", "inputs unequal and comparator working properly" as well as "comparator malfunctioning". The comparator consists of 37 modules. Its circuitry structure ensures triple error protection at any time. 

Keywords: Fail safe comparator, high speed comparator, ternary logic, result and status indication, safety related computing.

## 1. INTRODUCTION

As a result of the progress in computer technology, but also with respect to sensors and actuators, it became possible to elegantly master even complex control and automation functions by software based solutions. The advantage of programmable digital equipment consists in the fact that there are always defined and reproducible machine states, and that desired changes in control sequences can be achieved without considerable manual effort. An important field in control engineering, however, that, with the exception of some individual solutions, has remained almost untouched by this development are safety related systems. In particular, when human lives, expensive equipment or the environment is to be protected, i.e., when a high degree of safety is required, semiconductor based computer technology cannot offer solutions matching the plentitude of applications found elsewhere. Primarily, this is due to the fact that the classical control components switches, relays, thermoswitches etc. have high probabilities of assuming a certain "natural" switching state in case of failure, namely the state of disconnection. This allows to design controllers performing safety related tasks in such a way that in case of component failures the entire controllers assume states being safe for persons and the processes controlled ("fail safe principle"). Semiconductor devices, the other hand, which are practically used everywhere these days, assume unpredictable states such as short-circuits or complete disruptions in cases of defect. Therefore, it is impossible to associate an unambiguous error state with a semiconductor component. One way to solve this problem is to use redundant controller structures.

In technical applications, this means that control components are replicated to fulfill the same task. Furthermore, switching to a state other than a safe one is subject to monitoring with majority or unanimity voting. Controlled devices or processes are brought to safe states in case of disagreement. Depending upon the number of components working concurrently, one speaks of single or multiple error protection. For instance, when three components are used and the failure of two is safely and correctly tolerated, double error protection is given. As the design of redundant controllers necessarily leads to parallel structures and parallel task processing, the generated results need to be monitored by comparators embedded at defined points between the modules working concurrently. Hence, the responsibility for system safety rests on the comparators. Since comparison reduces the degree of redundancy, comparators are best implemented in form of hardware devices with fail safe behaviour.

Currently available fail safe logic families, of which comparator circuits can be built, constitute, however, bottlenecks in safety related computing systems due to their low processing speeds. The short outline of the only two presently marketed families of fail safe logic in the next section will show that they only allow for comparators working in the kHz range, whereas computing equipment is 3 to 5 orders of magnitude faster. Moreover, at their outputs, they only provide two values signifying either

(1) inputs equal *and* comparator properly working, or
(2) inputs unequal *or* comparator malfunctioning,

i.e., it is not differentiated between comparison errors and comparator errors. If these alternatives were distinguished they could be handled in more specific ways.

To solve these two problems of insufficient speed and indiscriminate output information, a novel fail safe comparator of two binary inputs was developed, which does not only indicate a result, but also its functional status. Built in a modified CMOS technology, it works in the MHz range and can, thus, match the execution speed of digital computers. In contrast to all earlier designs of fail safe comparators, by employing ternary logic it provides three different output values, allowing to distinguish between the three indications

a inputs equal *and* comparator working properly,
b inputs unequal *and* comparator working properly, and
c comparator malfunctioning,

These signal states may be sensed and further processed by conventional semiconductor circuits without any problems.

## 2. INDUSTRIAL FAIL SAFE MODULES

By module we mean a unit capable of realising certain, mainly simple, logic functions (And, Or, Not). Fail safe modules are developed in special technologies which assure predictable behaviour of the outputs in case of any failure. Normally, such behaviour is equivalent to fall into the switched-off state. Galvanic separation of inputs and outputs is a common feature. There exist a few families of fail safe modules applying different principles of operation. Here we shall mention two of them, Planar and MagLog, which are well established in the market.

### 2.1 *HIMA Planar Logic*

This logic (HIMA, 1991; HIMA, 1992) is built by Paul Hildebrandt GmbH & Co. KG of Brühl, Germany. Its operation principle is to convert slowly varying input signals into alternating ones (modulation). The alternating signals are processed internally according to a module's function. The result, still an alternating signal, is then converted back into a slowly varying signal presented at the output (demodulation). Besides gates, this fail safe logic family includes a hold unit, a delay, input and output amplifiers. The modules are built of discrete elements, for which the behaviour can be more easily predicted than for integrated circuits. The settling time after input changes, however, is quite long, viz., 3 to 10 ms.

*Conjunction* If the input contacts of this circuit are closed (high state), transistors transfer the alternating signal from a generator to a transformer and further to the output. A rectifier and a capacitor provide constant voltage at the output. If at least one of the input contacts is open (low state), transfer of the signal is not possible, and there is no alternating voltage coming to the transformer. So, the voltage at its secondary side becomes zero. The capacitor discharges after a while through a resistor and the input resistance of the next module. The output voltage is zero. The same situation arises if any of the transistors fails. The type of failure does not matter (shorting or break). So, the module reacts in the same way both to zero input and to failures. Low level at the output denotes the fail safe state.

The capacitors connecting successive stages are protected from shortings by resistors. Due to a special technology, shortings of resistors are eliminated. Shorting of the output to supply voltage is not possible due to the particular construction of the transformer and the lay-out of the circuit board. It should be clear that the use of alternating signals and the transformer assure fail safe behaviour. Any other standard element would not guarantee zero output in case of failure.

*Disjunction* All elements of the circuit implementing the Or function are passive. At each input three diodes are used in series. As long as not more than two of the diodes fail, the circuit remains operational. A resistor protects the diodes from shorting.

### 2.2 *GTI MagLog 24*

This *Mag*(netic) *Log*(ic) (GTI, 1993) is built by GTI Industrial Automation, Apeldoorn, The Netherlands. In it a toroidal core (such as in core memories) with a few windings coiled around plays an essential rôle. An additional simple circuit involves a transistor and a pulse modulator consisting of a winding on another core, a coil, and a diode. The cores (permalloy) have large hysteresis providing two stable states. A core with appropriate windings is capable of performing simple logic functions (And, Or, Not). It also separates the circuits Galvanically.

*Operational Principle* Logic high state is defined as a sequence of short pulses of 1 kHz frequency. Lack of pulses denotes low state. The pulses appearing at the input windings of the MagLog And gate are phase shifted. Since the windings are coiled in different directions, the pulses continuously re-magnetise the core, so positive and negative pulses appear at the output winding. Naturally, pulses only appear at the output if pulses are present at both inputs. Positive output pulses trigger a transistor which transfers them to the input of another module (second core). If one or both of the input sequences is missing, no output sequence appears. The Or function is obtained by coiling the two input windings in the same direction on a core, and connecting another winding of the opposite direction to the pulse generator. The output is then provided by a forth winding.

## 3. FAST FAIL SAFE COMPARATOR WITH TERNARY OUTPUT

Since the two families of industrial fail safe modules outlined above offer far too slow switching speeds as required by a feasible co-operation in environments characterised by integrated semiconductor circuits, a novel fail safe comparator (von Stackelberg and Halang, 1998) of two binary digits as depicted in Fig. 1 was designed, which combines high speed in the MHz range with a ternary output to indicate both comparison result and functional status. It consists of altogether 37 individual modules arranged in six stages with respect to functional operation, and four levels deep to implement fail safe behaviour. The individual modules are, to a certain extent, basic circuits of binary and ternary digital logic and, on the other hand, circuits especially designed for the purpose considered here. The comparator is operated with two voltages, viz., +1.5 V and +5 V relative to reference potential. As a digital circuit, there are only these three states determined by the operating voltages and zero potential — intermediate states do not occur. As customary in digital electronics, these voltage levels are associated with logical signal states. In this special case these are the ternary values:

**minus** for the most negative voltage level, i.e., the reference potential 0 V,
**zero** for the voltage level lying between the two extremes, i.e., for +1.5 V, and
**plus** for the most positive voltage level, i.e., for +5 V.

As binary values are used:

**LOW** for the more negative one of the two voltage levels applied to the circuit, which specifically may be 0 V or +1.5 V, and
**HIGH** for the more positive one, which specifically may be +5 V.

In the sequel, we shall use the terms mentioned above instead of the potentials at the inputs and outputs of the individual circuits and the comparator.

In the first stage of the comparator, there are four antivalence gates as shown in Fig. 2, with each of their inputs wired together. This means that the signals at the inputs E1 and E2 of the comparator are evaluated four times. Since the input signals come from binary circuits whose "LOW" and "HIGH" are represented by the voltages 0 V and +5 V, the antivalence gates are also applied to these voltages. The result of such a gate is "LOW" if both inputs are the same, and "HIGH" if both inputs are different.

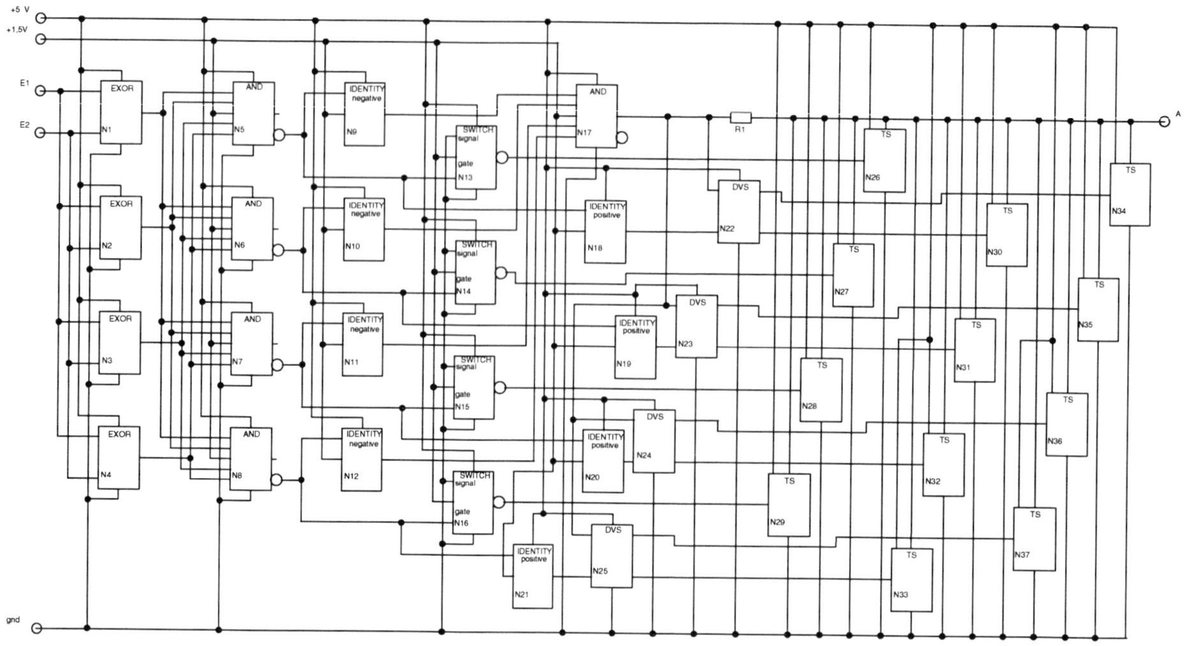


Fig. 1. Structure of the fail safe comparator

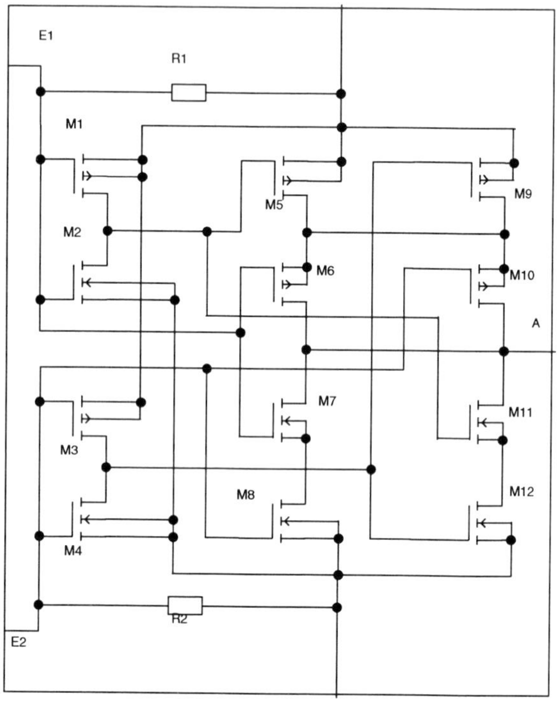


Fig. 2. Binary antivalence gate

The resulting states "LOW" and "HIGH" are fed into four parallel ternary conjunction gates (Fig. 3) of the second stage. There, the "LOW" signal corresponds to "minus" and the "HIGH" signal corresponds to "plus". The result of ternary conjunction is "minus", if all of its inputs are "minus", and "plus" if all of its inputs are "plus". In all other cases, the outputs supply the signal "zero". Since ternary negations are attached at the conjunction gates' outputs of the second stage, the results of the third, fourth and fifth stages assume the following values: "plus" is yielded if both inputs of the comparator recognise the same signal, and different signals at the comparator's inputs result in "minus". Again, the four conjunction gates of the second stage are connected in parallel to have the result of the operation available four times.

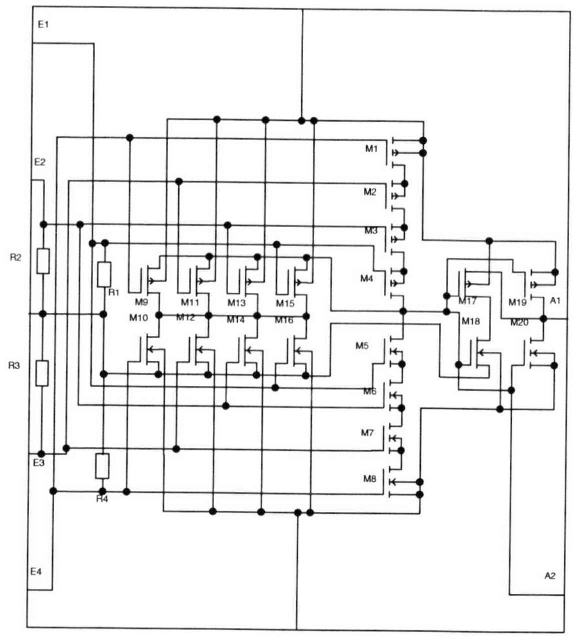


Fig. 3. Ternary conjunction gate

In the third stage of the comparator four identity gates as depicted in Fig. 4 are arranged in parallel, too. The function of an identity element is just to convey the digital signal at its input to its output. However, since the comparator's identity gates are attached to the +1.5 V potential with their negative supply voltage connections, and ternary signals are present at their inputs, there is a shift in signals: "plus" at the inputs becomes "plus" at the outputs, and both "zero" and "minus" at the

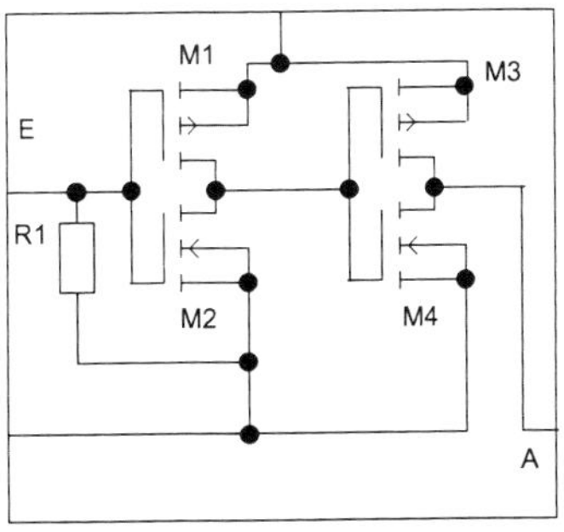


Fig. 4. Ternary identity gate with shift of signal

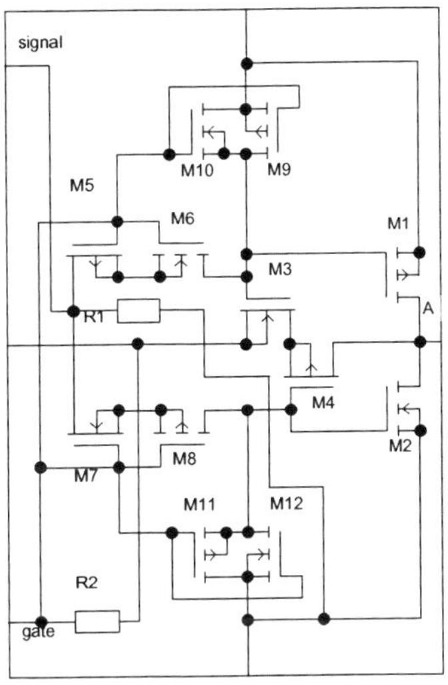


Fig. 5. Switch function

inputs become "zero" at the outputs. The reason for this is to generate the ternary signals demanded at the output of the comparator from the binary signals at its input. Depending upon the comparator's input signals, the following states result at the outputs of the identity elements: if both comparator inputs recognise the same signal, "plus" is present at the outputs of the identity elements of the third stage; if both comparator inputs recognise different signals, "zero" is generated. The "zero" signal from the second stage's conjunction gates indicating a disturbance in the first stage of the comparator is evaluated in the fourth stage and is irrelevant on this signal path.

Switch functions (Fig. 5) are employed in the fourth stage of the comparator. A switch is a basic ternary circuit switching the signal present at its "signal" input to its output in case it detects a "zero" signal at its "gate" input. If the input at "gate" is not equal to "zero", "zero" is presented at the output. Since the switch gates used in the comparator have a ternarily negating output, a "minus" signal at their "signal" inputs is transformed to "plus" at the outputs when the "gate" inputs detect "zero". Therefore, a "zero" generated at the output of a conjunction gate in the second stage leads to a "plus" at the output of the corresponding switch gate in the fourth stage.

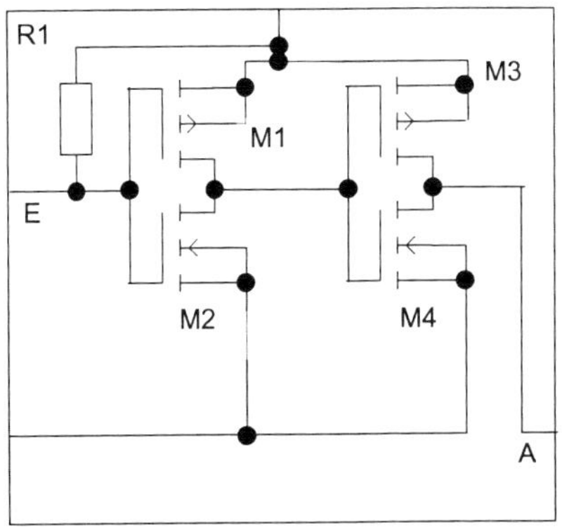


Fig. 6. Ternary identity gate without shift of signal

In the fifth stage of the comparator there are three different circuits, a conjunction gate, four identity gates (Fig. 6) and differential voltage switches (Fig. 7) each. The conjunction gate's task is to gather the signals produced in the third stage for the comparison states, and to direct them to the comparator's output, which means that "plus" signal is presented at this output for the state "equality detected by comparison and comparator in proper order" only when all four identity gates of the third stage indicate "plus". The identity gates of the fifth stage fulfill the same task as the identities in the third stage, and with each identity gate of the fifth stage the output signal of a conjunction gate in stage two is associated as well. The reason for this is to appropriately prepare the signals generated by the conjunction gates for the differential voltage switches connected behind the identity gates, as the differential voltage switches set, upon detecting different digital signals at their inputs, their outputs to the higher one of the two input signals, respectively. Both outputs are set to "minus" for equal input signals. The differential voltage switches close a feedback loop between the comparator output and the conjunction gates of the second stage to check the output conjunction gate for proper functioning.

There is a resistor in the sixth stage of the comparator located between the output of the collective conjunction gate of stage five and the output of the comparator to provide correct measuring conditions for the differential voltage switches even when a threshold switch is in conducting state. The function of the threshold switches (Fig. 8) is to put the output of the comparator to "minus" reference potential if the input signal exceeds a given value relative to the reference potential. The inputs of the threshold switches are driven by the four outputs of the switches in the fourth stage and by the four times two outputs of the differential voltage switches of the fifth stage.

Fail safe behaviour of the comparator is achieved as a result of combining the following measures.

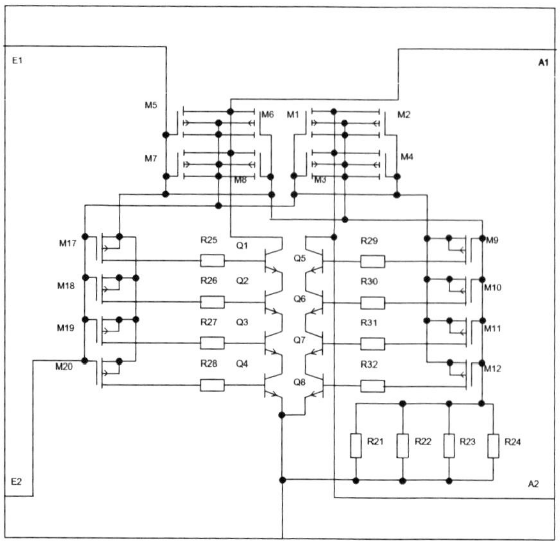


Fig. 7. Differential voltage switch

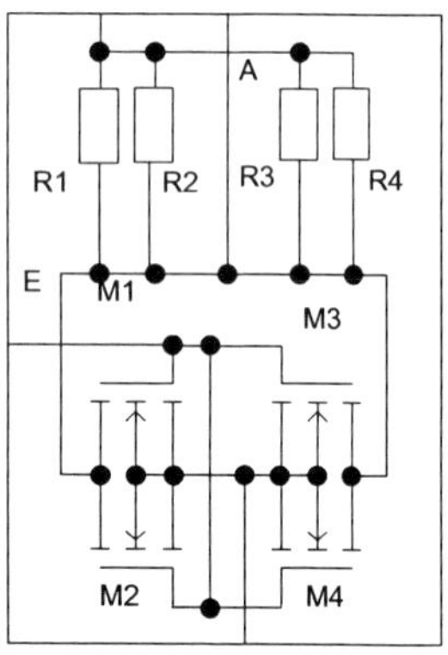


Fig. 8. Threshold switch

(1) Each function is provided four times, viz., in four levels in stages one through six.

(2) With their signal comparison between the second and the fifth stage, the differential voltage switches constitute a form of functional monitoring which provides a plausibility check between the comparator and the result of the initial Boolean operation on its inputs.

(3) The modules with the differential voltage switches and the threshold switches are identically replicated four times, since their function is not monitored further. Hence, each device in each module is present four times. Devices whose low resistance state leads to the safe state of the comparator output signal are connected in parallel, while devices whose high resistance state represents the safe state for the comparator output signal are connected in series.

(4) By means of a resistor each input of each module is placed to that potential which constitutes the safe state for the comparator output if the corresponding controlling signal fails. This case could occur, for instance, when a previous output assumes high resistance state. These measures ensure that the comparator always provides either the correct comparison result, or that it correctly signals the comparator to be out of order, which holds under the condition that not all four equal modules in one stage show the same error. This applies analogously with regard to the devices of the differential voltage switch and threshold switch modules, too.

## 4. SIMULATION RESULTS

A simulation model of the comparator was created with the tool MicroSIM PSPICE. Since presently it is not possible to simulate a ternary circuit in digital simulation mode — ternary logic is still not very well known — all simulations were run in analogue mode as transient analyses. As input signals to the comparator two pulses with the parameters 1 ns rising time, 1 ns falling time, 9 ns pulse width, and 20 ns period were defined. The second pulse arrives 4 ns after the first one. The simulation results show that the signals at the comparator's inputs have to be stable for at least 2 ns to obtain a correct output signal. The temporal behaviour of the comparator is only to a lesser degree determined by the propagation delays being less than 0.5 ns, but more by the charging processes of the gate capacities and by uncontrolled oscillations upon signal changes, mainly caused by the longitudinal transistors typical for ternary circuits. The comparator has been simulated in different cases, such as varying the level stability of the power supplies and the input signals as well as the limits of the output current.

## 5. REFERENCES

GTI Industrial Automation (1993). *An Introduction to MagLog 24 Inherently Fail-Safe Logic Technology. Eliminating the Unexpected.* Apeldoorn.

Paul Hildebrandt GmbH & Co. KG (1991). *Main Catalogue — The HIMA-Planar System.* Brochure HK 91.11. Brühl.

Paul Hildebrandt GmbH & Co. KG (1992). *Fail-Safe Electronic Controls — The HIMA-Planar System.* Brochure TI 92.08. Brühl.

Stackelberg, J. von, and W.A. Halang (1998). Ausfallsicherheitsgerichteter Binärstellenvergleicher mit Ergebnis- und Zustandsanzeige. German patent registration no. 198 57 396.0.

IFAC Publications
www.elsevier.com/locate/ifac

# FPGA-BASED LOGIC CONTROLLERS FOR SAFETY CRITICAL SYSTEMS

**Marek Wegrzyn**

*University of Zielona Gora, Institute of Computer Engineering and Electronics*
*ul. Podgorna 50, 65-246 Zielona Gora, POLAND*
*tel. (+48 68) 3282 484; fax (+48 68) 3244 733, M.Wegrzyn@iie.pz.zgora.pl*

*Abstract*: In the paper a synthesis method of logic controllers for safety critical systems into field programmable logic is presented. The specification of the system in the form of symbolic, *if-then* or *if-then-else* conditional decision rules is used for directly mapping to netlist format, or transformed into a HDL format that is accepted by standard FPGA simulators and synthesis tools. In addition, models of Logic Controllers are verified using the well-developed Petri net theory, and then they are translated through automated processes into selected FPGA specification format. 

Keywords: Process Control, Programmable logic controllers, Petri-nets, Rule-based systems, Digital systems, Logic arrays.

## 1. INTRODUCTION

The main aim of this paper is a presentation of two methods of synthesis of Logic Controller (LC) programs into Field Programmable Logic (FPL) for safety critical systems.

The behavioural specification of Logic Controller programs (Halang and Jung, 1994), related with subset of IEC 1131-3 standard, is modelled as Petri net. The textual, rule-based description of Petri net, called Petri Net Specification Format – v.2 (PNSF2), is used as an entry format for dedicated CAD tool. In the another approach, controllers are described by state tables, and then it is modelled in HDL. Finally, in both methods, LCs are implemented into FPL as a dedicated Reprogrammable Logic Controllers.

In the first of presented methodologies, a rule-based behavioural specification of Petri net model is directly transformed into hardware library (on the netlist level) by means of Xilinx implementation system (Wegrzyn, *et al.*, 1998; Wegrzyn and Adamski, 1999). The *if-then* decision rules (non-procedural conditional statements) are mapped into Xilinx XNF format after some simple additional symbolic transformations. From the conceptual point of view, a set of event-oriented (Petri net transition-oriented) *if-then* conditionals (Adamski and Wegrzyn, 2000) is easily replaced by the equivalent local state-oriented (Petri net place-oriented) *if-then-else* decision rules. In such a way, the Petri net image in FPGA structure is straightforward and simple. Standard design tools perform the final multi-level, combinational optimization, placement and routing.

A control unit, represented by state table (or decision table), can be also behaviourally specified by means of using hardware description languages (HDLs). Its specification, modelling, validation and FPL synthesis can be directed towards modern, well-accepted VHDL or Verilog-based synthesis tools. The second proposed method of structured implementation of state table is based on self-evident mapping of decision rules into HDL. The *if-then-else* conditionals may serve as both flexible and formal intermediate forms.

## 2. BACKGROUND AND MOTIVATION

### *2.1. Field Programmable Gate Arrays*

Field Programmable Logic is electronic circuit, which can be configured by the user to perform particular combinational or registered logic functions. The design process is greatly simplified by FPL compilers by some new CAD tools, which are recently available, like Xilinx Foundation Express (supporting VHDL and Verilog). The effective simulation allows the Logic Controller to be debugged before the device is programmed. If a design change is needed, it is a simple matter to re-edit the original specification and then re-program or exchange the old device. The most flexible and high-density FPL devices are Field Programmable Gate Arrays (FPGAs). Fig. 1 depicts a conceptual diagram of a typical FPGA.

FPGAs were introduced in 1985 by the Xilinx company, and since then, many different FPGAs have been developed by a number of companies: Actel, Altera, Plessey, AMD, QuickLogic, Algotronix, Concurrent Logic, Crosspoint Solutions, and others. As shown in the figure (Fig. 1), it consists of a two-dimensional array of logic blocks that can be connected by general interconnection resources. The interconnection comprises segments of wire. The segments may be of various lengths. In the interconnection there are programmable switches that enable one to connect the logic blocks to the wire segments, or one wire segment to another. Logic circuits are implemented in the FPGA by partitioning the logic into individual logic blocks, and then interconnecting the blocks as required via the switches (Xilinx, 2000).

To facilitate the implementation of a wide variety of circuits, it is important that an FPGA is as versatile as possible. This means that design of the logic blocks, coupled with that of the interconnection resources, should facilitate the implementation of a large number of digital logic circuits. There are many ways of designing an FPGA, involving trade-offs in the complexity and flexibility of both logic blocks and interconnection resources. Logic-block

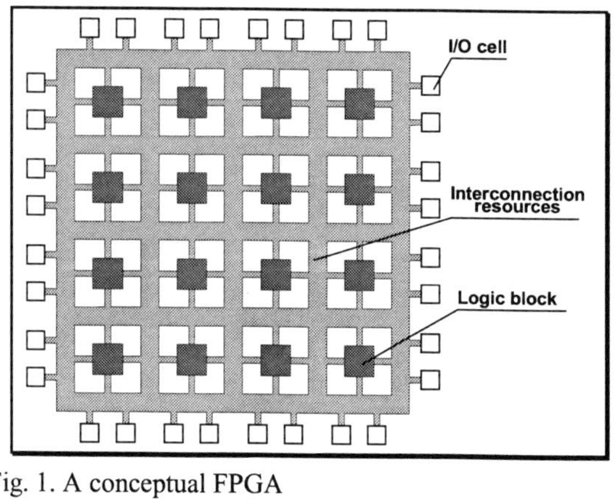


Fig. 1. A conceptual FPGA

architectures can be designed in many different ways: as simple gates (e.g., 2-input NAND gates), or as more complex structures, such as multiplexers or look-up tables. In some FPGAs, a logic block corresponds to an entire PAL-like structure. Fig. 2 shows a Xilinx FPGA with more details. The variety and complexity of currently produced FPGAs make them suitable for use in a wide set of applications. Two significant advantages of FPGAs are as follows: lower prototype costs and shorter production time.

As user-programmable application-specific integrated circuits, they provide a valuable compromise, which combines the benefits of standard microcontrollers with many of the benefits of other semi-custom logics. The design process is greatly simplified, due to the powerful FPGA compilers, as well as some new modern and effective CAD tools, which have recently become available. The effective simulation allows a logic controller to be simulated and debugged before the device is programmed. If design changes are needed, it is a simple matter to re-edit the original specification and then re-program the former device. It could be expected that after some time industrially oriented complex PLD and FPGA would be introduced. Advanced PLDs or FPGAs, togethei with modern CAD tools, may easily replace some of the old fashioned programmable controllers (Mandado, *et al.*, 1996). The Xilinx design software is called Xilinx Foundation Series, and it supports design entry, design simulation, and design implementation. Design entry occurs by schematic capture or HDL specification.

However, devices address the issues that are essential to the safety critical systems market (e.g. aerospace and defence):

- Commercial manufacturing strengths result in more efficient process flows.
- Reliability of supply. Controlled mask sets and processes insure the same quality devices, every time, without variation, which remain in production for an extended time.

The Xilinx QPRO family of ceramic and plastic QML products (Qualified Manufacturers Listing), certified to MIL-PRF-38585 and complemented by ISO-9000 certification, provides system designers

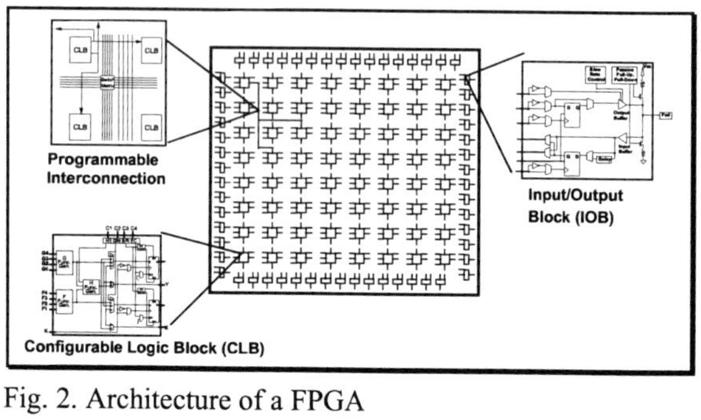


Fig. 2. Architecture of a FPGA

with advanced programmable logic solutions for next generation designs. The QPRO family also includes select products that are radiation hardened for use in satellite and other space applications. Designers can confidently design with Xilinx for High-Reliability systems with the knowledge they are getting unsurpassed quality and reliability, and long-term commitment to the safety critical applications market.

The QPRO family provides a wide variety of devices. The broad range of devices is available in various speed and package options. Both military temperature and full QML/SMD versions are available as standard off-the-shelf products. The Virtex members of the QPRO family offer FPGAs with high-performance up to 200 MHz, and densities greater than 1,000,000 system gates, and even larger devices planned for the future. Dramatic increases in silicon efficiency result from optimizing the new architecture for place-and-route efficiency and exploiting an aggressive 5-layer-metal 0.22 μm CMOS process. These advances make QPRO Virtex FPGAs powerful and flexible alternatives to mask-programmed gate arrays.

### *2.2. Safety Critical Control Applications*

Programmable electronic systems for safety critical control and regulation applications are rather a new scientific and engineering research field that stands at the beginning of its treatment.

The draft of international standard, IEC 61508 (IEC, 1998), defines four Safety Integrity Levels (SIL) for the categorization of safety related applications according to their respective risk potentials.

Control unit of digital circuits can be described in several forms. However, for safety critical controls only very-well verified methods can be applied (Kershaw, 19987; Cullyer and Pygott, 1987), e.g. interpreted Petri nets, cause/effect tables, decision tables, and HDL models.

Control oriented, 1-bounded Petri net is considered as a formal model for the set of decision rules. It is used also as a distributed, local state based model for the generated HDL descriptions. In the considered interpreted Petri net model, a particular combination of Boolean external inputs, and eventually some internal conditions, have to be true, for the transition to be fired.

Conditional logic is treated as an intermediate behavioural level language for the description of concurrent digital systems. Implicitly, it gives the *procedural* specification of distributed, concurrent digital system, implemented in FPL as a single State Machine with Data Path. The global states (Petri net markings) of Concurrent State Machine (direct hardware implementation of Petri net) are formed from local states (Petri net places). The places are encoded in such a way, that the global state code is implicitly obtained by superposition of the local state codes. A rule-based specification, considered in the paper, is composed from discrete local state symbols (places $P$), input signal symbols ($X$) and output signal symbols ($Y$) of the controller. The name of transition ($t$) is treated only as a rule label.

The textual description of developed implementation in Field Programmable Logic is not procedural, as Petri net evidently is, but rather *declarative*. The designed control part of a digital system is treated as a reasoning system implemented in the regular structure (for example - array) that is built from logic cells. The decision rule specification, as a textual equivalent of Petri net, is easily represented by using HDLs, and then mapped into hardware. The manual or automatic translation is straightforward.

The simplest and most effective technique for state encoding of state machines for FPGAs implementation is one-hot state assignment. On the other hand, as a Petri net place encoding is to use one-to-one mapping of places onto flip-flops (Wegrzyn, *et al.*, 1998).

## 3. FIELD PROGRAMMABLE LOGIC BASED DESIGN

### *3.1. General architecture of the controller*

Safety critical applications require very rigorous design process and its realization. Each step should be formally verified. For this purpose two models are used: Petri net model and HDL model. Implementation of those compatible blocks (Fig. 3) is based on such models. For better testing of correctness each output is in complementary form, i.e. affirmation $Y$ and negation $\overline{Y}$. Comparator compares outputs from two control units; the same values on both $Y$ and $\overline{Y}$ outputs indicate error.

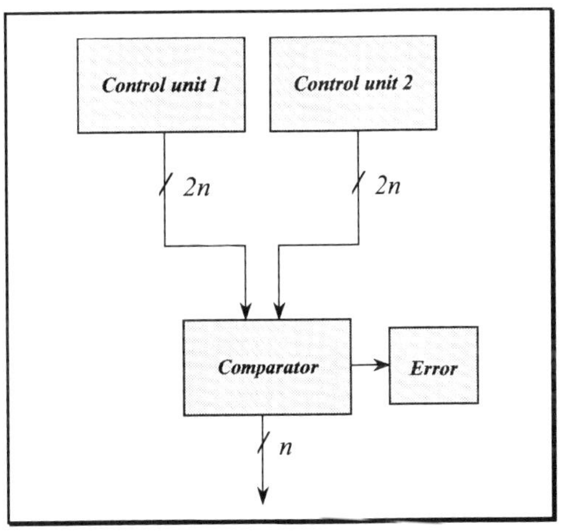

Fig. 3. General architecture of the controller

The final outputs are provided from the comparator. Higher reliability could be met by using separated chips for each control unit.

### 3.2. Example

Fig. 4 depicts the controlled part of a designed simple reactive system. The controlled system consists of two carriages ***1*** and ***2*** that go on the left (***L***) and right (***R***) sides. The carriages start on signals *M1* and *M2*, respectively. When both carriages go concurrently, they can reach points *D* and *E*. Because, that carriage ***1*** has higher priority, it goes to point *B*, and then goes back to *A*, where it waits for next pressing of bottom *M1*. In this time carriage ***2*** waits in point *E* until carriage ***1*** reaches point *D* in return way. Then the carriage ***2*** goes to *B* and back to *C*, where it waits for next pressing of bottom *M2*. At the time, on the common way only one carriage can be located.

### 3.3. Controller design

In this section, the specification of industrial controller that is related with interpreted Petri net (Fig. 5), and state or decision table (Table 2), is developed.

There is one fundamental difference between a state diagram and a state table, a difference that makes state diagram design simpler but also more error prone (Wakerly, 1994):

- A state table is an exhaustive listing of the states for each state/input combination. No ambiguity is possible.
- A state diagram contains a set of arcs labelled with transition expressions. Even when there are many inputs, only one transition expression is required per arc. However, when a state diagram is constructed, there is no guarantee that the transition expressions written on the arcs leaving a particular state cover all the input combinations exactly one.

This difference is very important for safety critical systems.

In the considered Field Programmable Logic implementation, transitions between local states occur simultaneously. It is realized by means of using the same clock for all logic blocks (synchronous state machine). To implement concurrent state machine, a global state register can be used to reflect all current, concurrently related, local states. As an alternative option it is possible to distribute the local state register into several single logic blocks. In such a case, each Petri net place may be implemented separately as two local states: active - if place is marked, passive - if place is empty.

The direct mapping of Petri net into FPL is based on correspondence between a transition and a simple combinational circuit and correspondence between a place and a clearly defined subset of state register. In dealing with concurrency the designer is confronted with some problems that will not arise in the logic synthesis of sequential systems. One of them is the concurrent local state encoding.

The one-hot-like method produces fast designs with a simple combinational part, especially for implementations in FPGA (Wegrzyn, *et al.*, 1998).

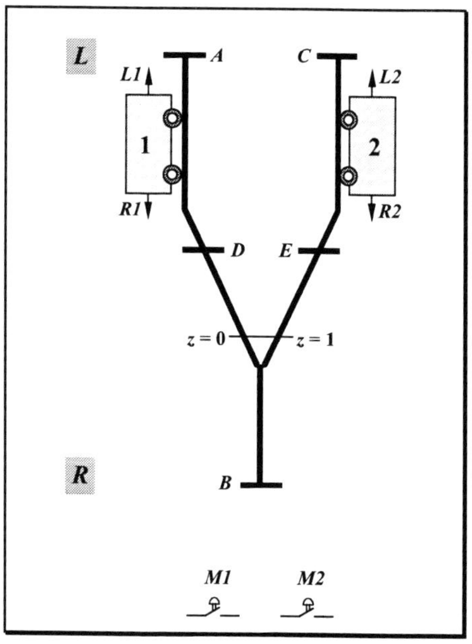


Fig. 4. The controlled part of reactive system

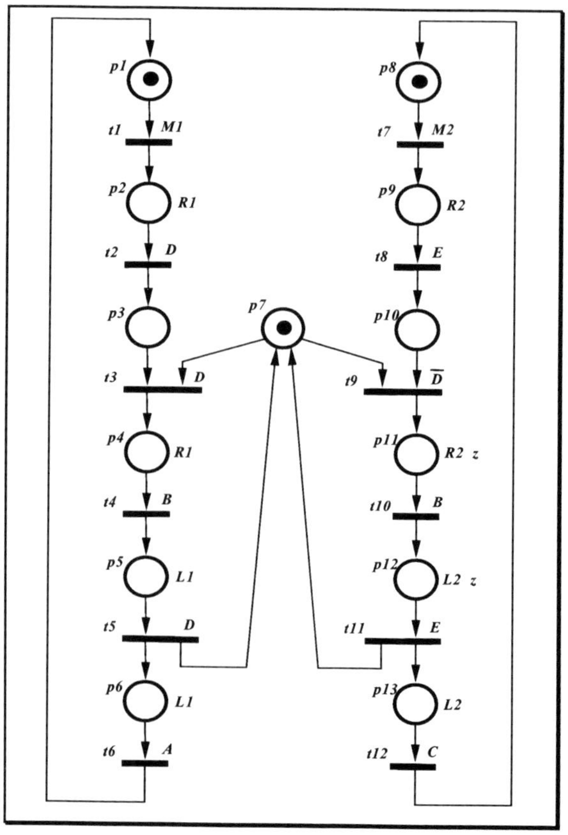


Fig. 5. Petri net model

It is not possible to assume that all flip-flops attached to places, are set to 0 since several places can be marked simultaneously. Some efficient state assignment techniques are referred to or presented in (Adamski, 1999). They are suited to FPGAs that are register-reached. The encoding procedure assigns to each place or macroplace a Boolean term. This is combined from all Boolean terms, which are assigned to higher-level macroplaces it belongs to, and a particular 'private' part. The state variables could be eventually shared among subnets but all concurrent subnets should not have orthogonal codes (Adamski, 1999).

The use of concurrent state machine techniques in FPL design has several advantages. They create near-optimal controller implementations (in terms of performance), tailored to the exact requirements. Practical designs often fit in a single FPGA package since there are no redundant elements. Any particular function, which is not included in the library, can be effectively implemented. Controller outputs respond to inputs with high speed, in predictable and repeatable way, without glitches. The behavioural specification, which is mapped in the programmable logic can be formally verified or validated by means of advanced CAD tools.

The standard CAD tools have been combined with dedicated design environment PeNCAD (Wegrzyn, *et al.*, 1998). As an intermediate format the rule-based textual specification, called PNSF2, is used (Fig. 6). The design experiments have been developed by means of using HDL simulators and synthesis tools and Xilinx implementation tools, and tested on Xilinx FPGA demoboard with XC4005E device.

Application Specific Logic Controllers are dedicated devices embedded within an application and implemented with programmable logic (Mandado, *et al.*, 1996; Wegrzyn, *et al.*, 1998). To define the behaviour of reactive system, it is necessary to specify the set of allowed sequences of conditions, actions and events. The importance of this approach lies on its portability to any VHDL simulator, for example Aldec Active-CAD. It makes possible to validate of the indented behaviour of the designed microsystem in an user-friendly environment. For example, VHDL description of the logic controller is validated using "testbenches", which models roughly the controlled subsystem behaviour.

The Reprogrammable Logic Controller is considered as an abstract reasoning system (rule based system) implemented in reconfigurable hardware (FPGA or CPLD). The mapping between inputs, outputs and local internal states of the system is described in a formal manner by means of logic rules (represented as sequents) with some temporal operators, especially with operator *next* (Adamski, 1999). The correctness preserving synthesis, based on Gentzen

```
.clock CLK
.inputs M1 M2 A B C D E
.comb_outputs L1 L2 R1 R2 z
.part CarriageControl
.places p1 p2 p3 p4 p5 p6 p7
.places p8 p9 p10 p11 p12 p13
.transitions t1 t2 t3 t4 t5 t6
.transitions t7 t8 t9 t10 t11 t12
.net
 t1: p1 * M1 |- p2;
 t2: p2 * D |- p3;
 t3: p3 * p7 * D |- p4;
 t4: p4 * B |- p5;
 t5: p5 * D |- p6 * p7;
 t6: p6 * A |- P1;
 t7: p8 * M2 |- p9;
 t8: p9 * E |- p10;
 t9: p10 * p7 * !D |- p11;
 t10: p11 * B |- p12;
 t11: p12 * E |- p13 * p7;
 t12: p13 * C |- p8;
.MooreOutputs
 p2 |- R1;
 p4 |- R1;
 p5 |- L1;
 p6 |- L1;
 p9 |- R2;
 p11 |- R2 * z;
 p12 |- L2 * z;
 p13 |- L2;
.marking p1 p7 p8
.e
```

Fig. 6. A part of PNSF2 specification

calculus, is treated as a formal transformation of the initial set of compound rules (*Specification)* into another set of compound rules *(Implementation).* On the other hand, the decision table can be prepared (Table 2). The table described all transitions between local states of the controller.

Table 1. Decision table for control unit

| *Transition* | *Current local states* | *Conditions* | *Next local states* |
|---|---|---|---|
| t1 | p1 | M1 | p2 |
| t2 | p2 | D | p3 |
| t3 | p3, p7 | D | p4 |
| t4 | p4 | B | p5 |
| t5 | p5 | D | p6, p7 |
| t6 | p6 | A | p1 |
| t7 | p8 | M2 | p9 |
| t8 | p9 | E | p10 |
| t9 | p7, p10 | !D | p11 |
| t10 | p11 | B | p12 |
| t11 | p12 | E | p7, p13 |
| t12 | p13 | C | p8 |

Each row of the table is implemented as the *if-then* conditionals:

```
t3: if ((p3 * p7) and (D)) then p4;
```

Outputs of the controller are shown in Table 3.

Table 2. Outputs table

| *Local state* | *Outputs* | *Local state* | *Outputs* |
|---|---|---|---|
| p2 | R1 | p12 | L2, z |
| p4 | R1 | p13 | L2 |
| p5 | L1 | p9 | R2 |
| p6 | L1 | p11 | R2, z |

Example implementation of the output **R1** is based on the rule:

```
if (p2 and p4 )then R1;
```

Such approach is more suitable for safety critical systems because it is simpler and easy to verify on either RTL or gate levels. In addition, there is a simple transformation for implementation of a system described by a cause/effect table. For such specification, a ROM-based approach is performed and easy realized by using look-up table (LUT) based logic block FPGAs (Xilinx, 2000).

Directly mapping into netlist format, that is based on Petri net model or decision table, characterizes additional advantage. Setting of the controller into the stable and safe state (e.g. on an accident) should be realized by resetting of all flip-flops. That action is easy implemented in such approach in FPGA devices.

### *3.4. Functions and functional blocks implementation*

By using FPGAs there are other advantages. In FPGAs there are possibilities for realization either control unit or data path (Mandado, *et.al.*, 1996; Halang and Adamski, 1997). Examples of standard functions and functional blocks are showed in Table 5. Easy-to-read models are suitable for documentation and simulation purpose. However, for synthesis purpose a more detailed model should be prepared in order to make possible a verification of the realization.

Table 3. Examples of functions and functional blocks

| *Function* | *Description* |
|---|---|
| *NOT* | Inverter |
| *AND* | 4-input AND |
| *MUX8* | 8- input multiplexer |
| *SHL* | Left shift |
| *ROL* | Left shift with rotation |
| *CTUD* | Counter up/down |

## 4. CONCLUSIONS

This paper has been presenting a approach for modelling and design of discrete logic controllers, which is based on Petri nets, and decision table. The proposed methodology combines the well-known theory of Petri net and its formal verification, with mapping of the net into reprogrammable logic devices. The Petri net is directly mapped into FPL without explicit enumeration of all possible global states. The specification is given in terms of local state changes.

The proposed method of Petri net-based realization of Logic Controllers considers explicit hierarchy included in netlist formats, or hierarchical constructs of VHDL or Verilog. The syntactic and semantic compatibility of Petri net descriptions and decision rules with structured netlist or HDLs statements are as close as possible.

## REFERENCES


Adamski, M. (1999). Application Specific Logic Controllers for Safety Critical Systems. *Proceedings of the 1999 IFAC Triennial World Congress*, Beijing, China, **Vol.Q**, pp.519-524.

Adamski, M. and Wegrzyn, M. (2000). Interpreted Petri Net Approach for Design of Dedicated Reactive Systems, *Proceedings of the 6th IFAC International Workshop on Algorithms and Architectures for Real-Time Control, AARTC'2000*, Palma de Mallorca, Spain, 15-17.04.2000, pp.179-184.

Cullyer, W.J. and Pygott, C.H. (1987). Application of formal methods to the VIPER microprocessor. *IEE Proceedings*, Vol.134, Pt.E, No.3, May 1987, pp.133-141.

Halang, W.A. and Jung, S.-K. (1994), A Programmable Logic Controller for Safety Critical Systems. *High Integrity Systems*, 1, 2, pp.179-193.

Halang W.A. and Adamski, M. (1997). A Programmable Electronic System for Safety Related Control Applications, *Proceedings of the International Conference on Safety and Reliability, ESREL'97*, Lisbon, Portugal, European Safety and Reliability Association (ESRA), 17-20.07.1997, pp.349-356.

IEC (1998). *Draft of International Standard IEC 61508-1: Functional Safety of Electrical/Electronic/Programmable Electronic Systems. Generic Aspects – Part 1: General Requirement.* International Electrotechnical Commission. Geneva.

Kershaw, J. (1987). *The VIPER Microprocessor.* Report 87014. Royal Signals and Radar Establishment. Malvern, Worc., London, HMSO.

Mandado, E., Marcos, J. and Perez, S.A. (1996). *Programmable Logic Devices and Logic Controllers.* Prentice Hall, London.

Wakerly, J.F. (1994). *Digital Design Principles and Practices.* Prentice-Hall, Inc., Englewood Cliffs, New Jersey.

Wegrzyn, M. and Adamski, M. (1999). Hierarchical Approach for Design of Application Specific Logic Controller. *Proceedings of the IEEE International Symposium on Industrial Electronics ISIE'99*, Bled, Slovenia, 12-16.07.1999, Vol.3, pp.1389-1394.

Wegrzyn, M., Adamski, M. and Monteiro, J.L. (1998). The Application of Reconfigurable Logic to Controller Design. *Control Engineering Practice*, Pergamon, Vol.6, 1998, pp.879-887.

Xilinx (2001), The Programmable Logic Data Book, *http://www.xilinx.com/databook.*

IFAC Publications
www.elsevier.com/locate/ifac

# OPTIMAL CONSTRUCTION OF LOCAL MODEL NETWORKS USING GENETIC ALGORITHM

**S. K. Sharma, G. W. Irwin and S. McLoone**

*Intelligent Control & Systems Research Group*
*School of Electrical & Electronic Engineering*
*The Queen's University of Belfast*
*Belfast BT9 5AH*
*UK*

Abstract: Local Model (LM) networks represent a complex nonlinear dynamical system by a weighted sum of locally valid, simpler sub-models defined over small regimes of the operating space. Constructing such networks requires the determination of the appropriate regimes and the local model parameters. This paper describes a new soft computing approach to simultaneously determine the local model parameters and the appropriate operating regimes.

Structured genetic coding is used to obtain the optimal number of ARX local models and their parameters. Centres and covariances of Gaussian interpolation functions are found through genetic evolution, with fuzzy logic used to dynamically evolve the genetic algorithm. Several modifications to the classical genetic algorithm are proposed to determine the locally valid sub-models. The parallel model prediction error is minimized to produce good generalization from the identified LM network. The modelling potential of the resultant LM network is investigated on practical data from an experimental pH neutralisation process. The same data was also used in a recent comparative study of a hybrid training algorithm and tree type construction (McGinnity and Irwin, 1999) for LM networks. It is concluded that the LM network from the new GA based construction formed a significantly better nonlinear dynamical model of the pH process. *Copyright © 2001 IFAC*

Keywords: Genetic algorithm, local model networks, pH neutralisation process, Gaussian function, optimization.

## 1. INTRODUCTION

Local model (LM) networks (Johansen and Foss, 1993; Johansen and Foss, 1995) have generated considerable interest in the research literature for nonlinear system identification. Here the system operating space is decomposed into overlapping regimes, assumed to cover the full range of operation. A simple (typically linear) local sub-model is then placed at the centre of each operating regime and the required global model is formed by blending these local models. To do this each local model is weighted by a validity function, such as a normalised Gaussian function. The global nonlinear model at any operating point is then formed from the sum of these sub-models weighted by their respective validity functions with nearby sub-models contributing most and distant ones making no contribution.

The identified nonlinear models produced by such operating regime decompositions have the significant advantage of transparency compared to black-box models like neural networks. Thus, linear modelling and control results can be exploited for nonlinear systems, as can *a priori* knowledge, which is important for practical applications.

Determining the appropriate number of local regimes and their placement within the operating space is a key issue. However, in some applications the number of local regimes can be chosen from *a priori* knowledge with positioning included in an optimization process (McLoone, *et al.,* 1998; Jordon and Jacobs, 1994) for training. Alternative construction algorithms have also been suggested which take a top-down approach by iteratively increasing the number of local models, based on searching over possible refinements, until a satisfactory modelling accuracy is reached (Johansen and Foss, 1995; Nelles, 1997; Friedman, 1991).

This paper presents a new soft computing approach which simultaneously finds the optimal number of local models, their parameters and the parameters of the validity or interpolation functions. A network of local linear ARX models and Gaussian interpolation functions is constructed from practical data from an experimental pH neutralisation process. This same application was the subject of a comparative study of optimisation based training and tree type construction (McGinnity and Irwin, 1999) for LM networks, providing a valuable comparison for the new GA approach. It is concluded that the LM network produced by the new GA based construction formed a significantly better nonlinear dynamical model of the pH process.

## 2. LM NETWORKS

The general discrete LM network representation of a single-output nonlinear dynamical plant, based on operating space decomposition into $N$ regimes, is given by the weighted sum:

$$\hat{y} = \sum_{i=1}^{N} \rho_i(\psi) f_i(\boldsymbol{x}) \tag{1}$$

where $\hat{y}$ is the LM network output, $\boldsymbol{x}$ is a vector of past plant inputs and outputs and $\rho_i(\psi)$ is the relative validity of the *i*th local model, $f_i(\boldsymbol{x})$. Here

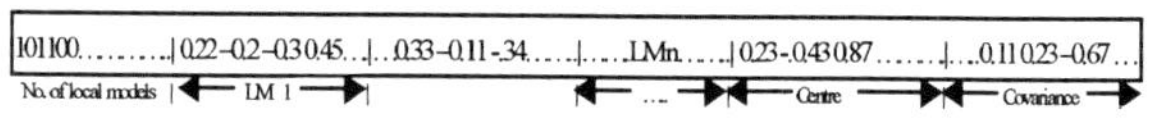


Fig. 1. Genetic representation of LM network.

$\psi$ is the vector of scheduling variables, typically a subset of $\boldsymbol{x}$.

In this paper the interpolation functions, $\rho_i(\psi)$ are normalized Gaussian functions such that:

$$\rho_i(\psi) = \frac{\exp(-\|\psi - \boldsymbol{c}_i\|^2 / 2\sigma_i^2)}{\sum_{i=1}^{N} \exp(-\|\psi - \boldsymbol{c}_i\|^2 / 2\sigma_i^2)} \tag{2}$$

while the local linear models are of ARX form with input order $m$ and output order $n$ given by:

$$\begin{aligned} f_i(\boldsymbol{x}) &= a_{0i} + a_{1i}\hat{y}(k-1) + \ldots\ldots + a_{ni}\hat{y}(k-n) \\ &\quad + b_{1i}u(k-1) + \ldots\ldots + b_{mi}u(k-m) \\ &= \theta_i^T \boldsymbol{x} + \alpha_i \end{aligned} \tag{3}$$

where $u(k)$, $y(k)$ are the plant input and output at the kth sampling instant. In order to avoid overfitting on noisy data, the parallel model error is used for training.

For eqns (2), (3) the unknown parameters for each regime are then the interpolation function centre, $c_i$, covariance $\sigma_i$ and the local model parameters $\theta_i$ and $\alpha_i$.

The identification of local operating regimes for an unknown plant is difficult since *a priori* knowledge of the plant is required. It is therefore very difficult to determine the normalised Gaussian activation regions. In this paper, a new approach is introduced for finding such operating regimes based on genetic algorithms.

## 3. GENETIC CONSTRUCTION

### 3.1 *Structured Genetic Coding*

The structured genetic coding of a chromosome has two segments. The higher segment is coded in binary and represents the number of local models where a '1' denotes that a particular local model is selected. The lower segment is coded in decimal and represents the parameters of ARX type local models along with the parameters of the interpolation functions (Figure 1).

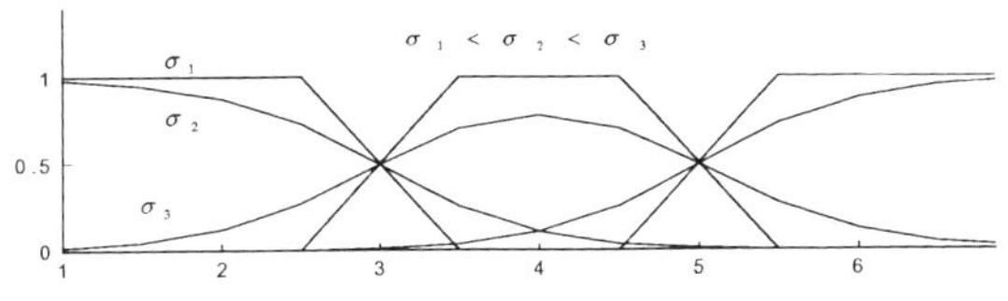


Fig. 2. Effect of varying covariance on interpolation function.

This type of encoding simultaneously describes the optimal number of local models (from a given maximum number), the parameters of these local models and the parameters of the interpolation functions. Rather than selecting the parameters randomly, when a particular local models is reselected, the weight values previously found are again used in order to give better convergence.

### *3.2 Model Construction*

To illustrate the role of the centre $(c_i)$ and covariance $(\sigma_i)$ of each interpolation function, consider the case of three local models where the total operating space is divided into three distinct regions. With centres kept at the middle of each operating region, the effect of the varying $(\sigma_i)$ is shown in figure 2. It can be seen that the operating regions are more distinct for smaller σ values. This consideration has been used in the GA to select centres and covariances so that each local model is distinct at some point in the operating space.

In applications the scheduling variables are scaled to the same range to allow uniform application of the GA operators, to provide better resolution and to speed up the algorithm.

### *3.3 GAs operators*

In the pH application a steady-state GA (Michalewicz, 1994) is applied to a population of 20 chromosomes. For each successive generation the two best chromosomes in the population are retained, producing offspring to replace the two worst chromosomes. Crossover and mutation, with the following modifications, are applied to the whole population.

Uniform crossover (Wasserman, 1993), where bits are interchanged on the basis of a generated mask, is applied to the binary coded part. In the decimal coded part, arithmetic crossover is applied between selected chromosomes, as detailed in reference (Michalewicz, 1994).

Half of the remaining population is subjected to biased mutation, where a random number is added to the previous values. For the binary coded part of the chromosome, '1' is replaced by '0' and vice-versa for a selected bit. A random number between ± 1 is added to the previous parametric values for the local models. A random substitution of $\pm(1+r)/2N$ is added to the centres and $\pm(1+r)/4N$ is added to the covariances. Here $N$ is the number of local models selected for a particular chromosome and r is a random number between 0 and 1. As $N$ increases, the centre and covariance values are confined to smaller ranges, an arrangement which provides fine-tuning.

The other half of the remaining population is subjected to unbiased mutation, where selected values are randomly replaced. For the binary coded part, '1' or '0' is randomly placed at selected bits. In case of the local model parameters, selected values are replaced by a random number between $\pm 1$.

Selected Gaussian centres are reinitialised to a random value within their operating ranges. The Gaussian covariances are likewise randomly set within $1/4N$ to $1/2N$ of the operating range. With this arrangement the GA is forced to search for interpolation functions with distinct centres and covariances chosen so that each model validity function retains a value of unity somewhere in the operating regime.

Fuzzy logic was used to dynamically select the chromosomes (Sharma and Tokhi, 2000) with a correctly classified pattern and a normalised mean-square-error (J) (Eqn. 4) as antecedents and a fitness value as the consequent. The J value was modified by multiplying by the number of local models selected in a chromosome so that the one with a smaller J also had less local models selected.

Once an optimised number of local models had been selected, after running the algorithm for 1000 generations say, a new group of chromosomes was formed using the parameter values for the optimised number of local models, along with the centres and covariances. Fine-tuning of these local model parameters and associated and Gaussian interpolation functions was done through genetic evaluation by running the algorithm for 1000 more generations. This speeded up the learning process as it was only necessary to deal with the subset of models selected from the 10 specified initially.

## 4. APPLICATION TO A pH NEUTRALISATION PROCESS

The pH neutralisation process is a highly nonlinear one in which acid, buffer and base streams are mixed in a tank and the effluent pH is measured. It offers a suitable case study for the demonstration and evaluation of LM networks techniques. Here practical data from an experimental pilot plant at the

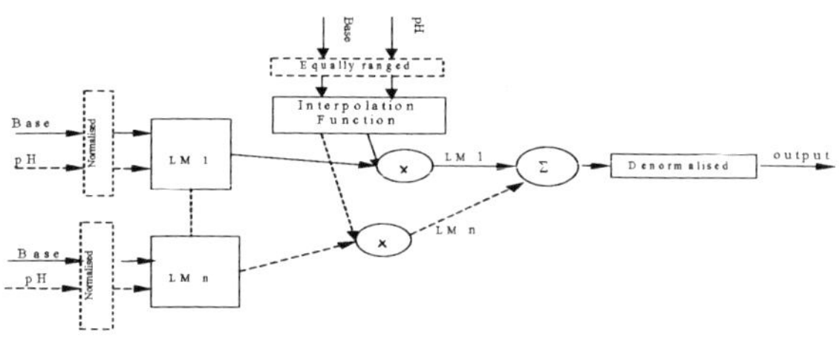


Fig. 3. LM network for pH neutralisation process.

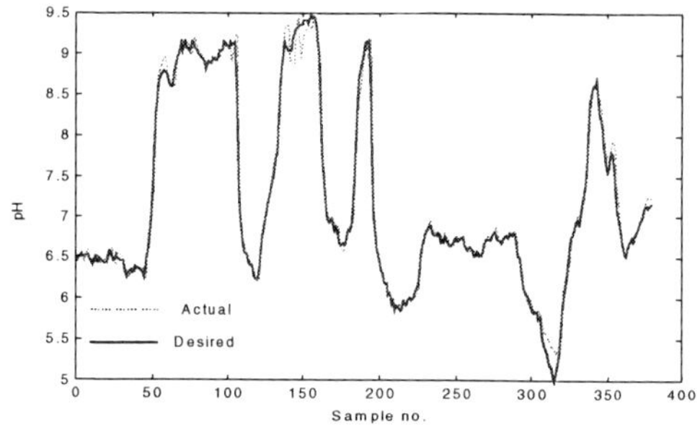


Fig. 4. Variation in pH from validation data set and predicted pH values from the LM network (GA construction).

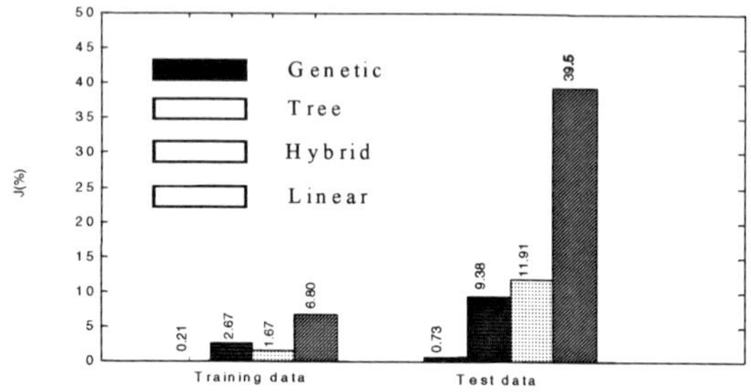


Fig. 5. Model quality measures for the LM network (GA construction) and the LM networks and single linear model from reference (McGinnity and Irwin, 1999).

University of California, Santa Barbara (Lightbody, *et al.,* 1997) was used. Acid and buffer flow rates were kept constant at their nominal values, with the base flowrate manipulated to excite the plant dynamics across the operating range. Thus, separate training and test sets had been generated using random amplitude steps in base flow rate with a superimposed prbs signal.

Figure 3 shows the LM network for the pH process. This process was the subject of a recent study of a hybrid training algorithm and tree type construction (McGinnity and Irwin, 1999) for LM networks. For quantitative comparison, the mean sum-squared error normalised by the data set variance is used.

$$J(\%) = \frac{((y-\hat{y})^T (y-\hat{y}))/n}{((y-\bar{y})^T (y-\bar{y}))/n} \times 100\% \qquad (4)$$

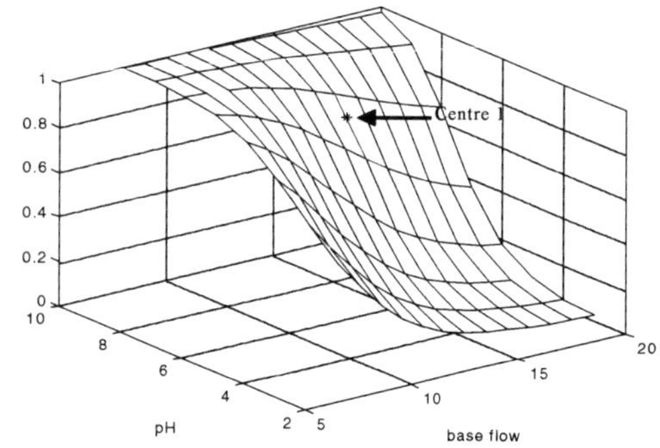


Fig. 6. Validation function $\rho_1$ for local model 1.

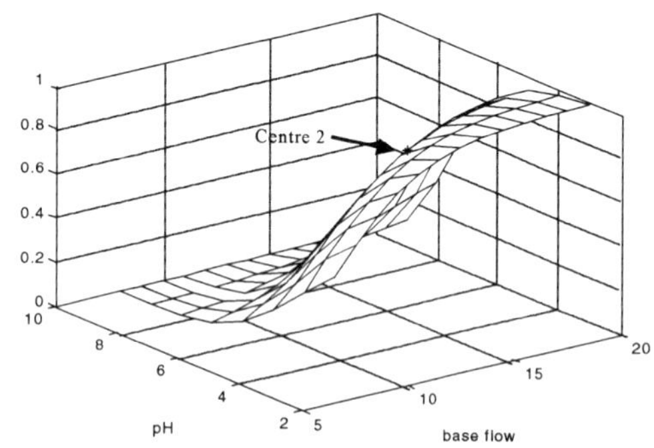


Fig. 7. Validation function $\rho_2$ for local model 2.

Here $y$ is the vector of $n$ output data values with mean $\bar{y}$, and $\hat{y}$ is the vector of predicted outputs. A low $J$ value signifies good modelling performance.

## 5. RESULTS

For quantitative comparison, the same data sets as in reference (McGinnity and Irwin, 1999) were utilised both for training and testing purposes. Figure 4 shows the pH variation in the validation data set along with the excellent multiple-step-ahead predictions from the LM network resulting from GA construction.

The normalised parallel modelling errors (J%) for both the training and validation data sets produced by GA construction appear in figure 5. For completeness, the corresponding results from (McGinnity and Irwin, 1999) are also included. These include a single linear model and two LM networks of five models, one trained using a hybrid optimisation approach with initial centres and widths selected using k-means clustering and hybrid nearest-neighbour techniques (McLoone, *et al.,* 1998), the other employing regime decomposition as proposed by Johansen and Foss (1995). In all cases second-order ARX linear models were employed. Figure 5 illustrates the significantly better modelling performance produced by the LM network formed by GA construction.

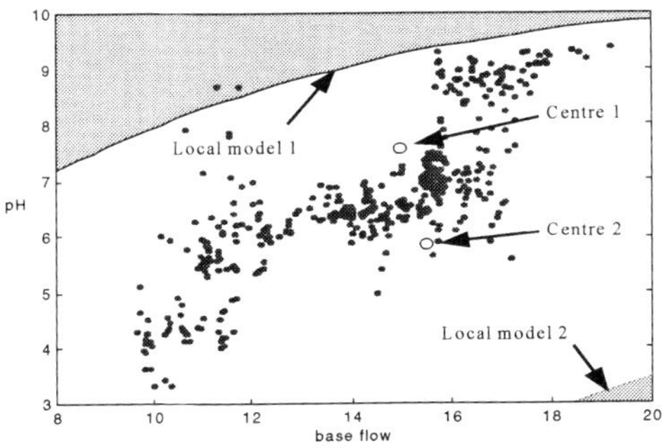


Fig. 8. Plan view of validation functions and training data.

Further, in the earlier work (McGinnity and Irwin, 1999) no consideration was given to the optimisation of the number of local models. Instead *a priori* knowledge of the steady-state titration curve was exploited to decide on the local models and initial values for the operating regions. As a result of the improved flexibility in the selection of centres and covariances with GA construction only two local models were now needed to represent the nonlinear pH neutralisation process.

Figures 6 and 7 show the variations in the two individual model validation functions over the pH, base operating plane. Figure 8 contains a plan view of these two model validity functions, along with the distribution of the training data. It is seen that the GA optimally positions the centres in regions where more data points are clustered. It reveals the areas where a single local model is valid (shaded) and also the region (unshaded) where the two local models are blended together to model the dynamics of the nonlinear pH process.

The two local models concerned are:

$$f1 = 0.2666 + 0.3173\,\hat{y}(k-1) - 0.0334\,\hat{y}(k-2) + 0.443\,u(k-1) + 0.0657\,u(k-2)$$

$$f2 = -0.7354 + 0.9247\hat{y}(k-1) - 0.4914\hat{y}(k-2) + 0.3405u(k-1) - 0.1974u(k-2)$$

The centres and covariances for interpolation function $\rho_1$ are (7.5839, 14.9463) and (1.3559,1.4040), while those for $\rho_2$ are (5.8498, 15.4736) and (1.1740,1.3589) respectively.

## 6. CONCLUSIONS

A novel approach using soft computing methods for structural and parameter identification of LM networks comprising ARX models and Gaussian interpolation functions is presented. Comparison with previously LM construction approaches applied to nonlinear dynamical modelling of a pH neutralisation process from experimental data shows that the new technique significantly reduces the multi-step-ahead or parallel dynamic modelling error.

Further, instead of requiring *a priori* knowledge to decide on the local models and initial values for the operating regions, the improved flexibility in the selection of centres and covariances with GA construction meant that only two local models were now needed to represent the nonlinear pH neutralisation process.

Future work includes the study of the more complex application of turbogenerator modelling *(Irwin, et al.,* 1998*)* where both the gain and dynamics vary across the operating space.

## REFERENCES


Friedman, J. H. (1991). Multivariate adaptive regression splines (with discussion). *Annals of Statistics*, Vol. **19**, No. **1**, pp. 1-141.

Irwin, G. W., Brown, M. and Lightbody, G. (1998). Nonlinear identification of turbogenerator by local model networks. In: *Proc IEE International Control Conference,* Publication No. **455**, pp. 1462-1467.

Johansen, T. A. and Foss, B. A. (1993). Constructing NARMAX models using ARMAX models. *International Journal of Control*, Vol. **58,** No.**5**, pp. 1125-1153.

Johansen, T. A. and Foss, B. A. (1995). Identification of nonlinear system structure and parameter using regime decomposition. *Automatica,* **31**, pp. 321-326.

Jordon, M. and Jacobs, R. A. (1994). Hierarchical mixtures of experts and the EM algorithm. *Neural Computation*, Vol. **6**, pp. 181-214.

Lightbody, G., O'Reilly, P., Irwin, G.W., Kelly, K. and McCormick, J. (1997). Neural modelling of chemical plant using MLP and B-spline networks. *Control Eng. Practice*, Vol. **5**, No. **11**, pp. 1501-1515.

McGinnity, S. and Irwin, G. W. (1999). Comparison of two approaches for multiple-model identification of a pH neutralisation process. In: *Proc. European Control Conference*, Karlsruhe, Germany, paper F 267, session BP15, 31 Aug.-3 Sep.

McLoone, S., Brown, M. D., Irwin, G. W. and Lightbody, G. (1998). A hybrid linear/nonlinear training algorithm for feedforward neural networks. *IEEE Trans. on Neural Networks*, Vol. **9**, No. **4**, pp. 669-684.

Michalewicz, Z. (1994). *Genetic Algorithms+Data Structures=Evolution Programs.* Springer-Verlag, New York.

Nelles, O. (1997). Orthonormal basis functions for nonlinear system identification with local linear model trees (LOLIMOT). In: *Proc.11$^{th}$ IFAC Symposium on System Identification*, Vol. **2**, pp. 667-672.

Sharma, S. K. and Tokhi, M. O. (2000). Genetic evolution: A dynamic fuzzy approach. In: *Proc. IEEE International Conference on Fuzzy Systems*, Texas, USA, Vol. **2**, pp. 748-752, 7-10 May.

Wasserman, P. D. (1993). *Advanced methods in neural computing.* Van Nostrand Reinhold, New York, pp 93-94.

IFAC Publications
www.elsevier.com/locate/ifac

# A NEW ESTIMATION METHOD FOR RADIAL BASIS FUNCTION-BASED MODELS

**H. Peng *, T. Ozaki **, and Y. Toyoda*****

* *College of Information Engineering, Central South University, Changsha, 410083 P. R. China. He is currently a visiting researcher at the Institute of Statistical Mathematics, 4-6-7 Minami Azabu, Minato-ku, Tokyo 106-8569, Japan. E-mail: peng@ism.ac.jp*

** *The Institute of Statistical Mathematics, 4-6-7 Minami Azabu, Minato-ku, Tokyo 106-8569, Japan. E-mail: ozaki@ism.ac.jp*

*** *Bailey Japan Co. Ltd., 511 Baraki, Nirayama-cho, Tagata-gun, Shizuoka 410-2193, Japan. E-mail: toyoda@bailey.co.jp*

Abstract: A structured nonlinear parameter optimization method (SNPOM) used to RBF network and RBF-ARX model parameter estimation is presented. This is an off-line nonlinear model parameter optimization method, which is developed based on the Levenberg-Marquardt method (LMM) for nonlinear parameter optimization and the Least Squares method (LSM) for linear parameter estimation. The SNPOM can accelerate the computational convergence of parameter optimization search process of the RBF-based models comparing with other algorithms. Some applications illustrate the usefulness. *Copyright © 2001 IFAC*

Keywords: nonlinear system; identification; parameter estimation; radial basis functions; RBF models; RBF-ARX models; state-dependent models.

## 1. INTRODUCTION

Many control methods with resorting to on-line (model or controller) parameter estimation have been proposed for nonlinear system control problems. However, for the processes with more-quickly-varying nonlinear dynamics, such as thermal power system (Prasad *et al.* 1998), the reliability, fast parameter convergence and computationally higher cost of on-line estimation are still difficult-to-solve problems. This paper is focused on the nonlinear system globally modeling and optimization for real-time control without resorting to on-line parameter estimation.

In the model based real-time control strategies for nonlinear systems, the radial basis function (RBF) network offers a considerable alternative as the internal model of the controller, because of its satisfactory nonlinear dynamics approximation precision, simple topological structure and fast learning. However, in actual applications, to obtain satisfactory approximation precision the RBF network model requires usually many RBF centers so that it may result in the parameter estimation difficulties. The state-dependent autoregressive model with functional coefficients can globally describe complex nonlinear dynamics (Priestley 1980). Using a set of RBF networks to approximate the coefficients of the state-dependent autoregressive model yields the RBF-AR or RBF-ARX model (Vesin 1993, Ozaki *et al.* 1997), which has the advantages of state-dependent ARX model in nonlinear dynamics description and RBF network in function approximation. In general, the RBF-ARX model has no need of too many RBF centers comparing with a single RBF network model.

The RBF or RBF-ARX model parameter estimation includes selection of the RBF centers and the scaling factors, and estimation of all the weights of RBF networks. There are mainly three types of methods to estimate the RBF-based models. The first is to optimize all the parameters of the model regardless of parameter features by using some nonlinear parameter optimization algorithms such as the Levenberg-Marquardt Method (LMM) (Marquardt 1963), which are generally based on an exhaustive search in the solution space and therefore require extensive computation. The second is to first select the basis function centers by setting them equal to a subset of the input vectors randomly or by some algorithms (see *e.g.* Ozaki *et al.* 1997), and then estimate the linear weights by the standard least squares method. Obviously, that procedure could not give the optimal parameters. The third as will be presented in this paper is some automatic estimation method that can optimize all the model parameters simultaneously and it also can accelerate the computational convergence of optimization search process comparing with the first method. Golub and Pereyra (1973), and Ruano *et al.* (1992) proposed an algorithm similar to the third method above, which optimizes a modified objective function including only the nonlinear parameters of the estimated model. It resulted in a greater complexity in the computation of the objective function derivatives.

## 2. RBF-BASED MODELS

### 2.1 *The state-dependent ARX model*

We consider the nonlinear systems that can be described by a nonlinear ARX model below

$$\begin{aligned} &y(t) = f(\mathbf{X}(t-1)) + e(t) \qquad (1) \\ &\mathbf{X}(t-1) = [y(t-1),\cdots,y(t-n_y),\mathbf{u}^{\mathrm{T}}(t-1),\cdots,\mathbf{u}^{\mathrm{T}}(t-n_{\mathbf{u}})]^{\mathrm{T}} \end{aligned}$$

where, $y(t) \in \Re$ is the output, $\mathbf{u}(t) \in \Re^r$ is the input vector which includes the manipulated variable and disturbance signals, for time series $\mathbf{u}(t) = \mathbf{0}$, and $e(t)$ is the white noise. Various types of particular functions have been applied to approximate the unknown nonlinear map $f(\bullet)$ that maximized the likelihood of the model, in which the state-dependent ARX model below based on the idea of state-dependent AR model (Priestly 1980) may be considered as a general description:

$$\begin{aligned} y(t) = \phi_0(\mathbf{X}(t-1)) &+ \sum_{i=1}^{p_y} \phi_{y,i}(\mathbf{X}(t-1))y(t-i) \\ &+ \sum_{i=1}^{p_\mathbf{u}} \Phi_{\mathbf{u},i}(\mathbf{X}(t-1))\mathbf{u}(t-i) + e(t) \end{aligned} \qquad (2)$$

where $\mathbf{X}(t-1)$ is regarded as the state vector at time $t$, in many cases maybe only the output series or the input series or an other signal series are contained in which. In the model (2), $\phi_0(\mathbf{X}(t-1))$ , $\{\phi_{y,i}(\mathbf{X}(t-1)),\ i=1,2,\cdots,p_y\}$ and $\{\Phi_{\mathbf{u},i}(\mathbf{X}(t-1)),\ i=1,2,\cdots,p_\mathbf{u}\}$ are called state-dependent coefficients of the model. The basic idea of the state-dependent ARX model (2) is to achieve the local linearization of the general NARX model (1) by introducing a locally linear ARX model with state-dependent coefficients. The state-dependent model (2) provides a useful framework for general nonlinear system modeling, however, the problem is how to specify the functional form of its coefficients.

### 2.2 *The RBF-ARX model*

In this paper, a set of RBF (radial basis function) networks are used to approximate the state-dependent coefficients of the model (2), because RBF network could approximate any functions by using members of a family of basis functions, moreover, the locality of the basis functions makes the RBF networks much more suitable for learning local variations. The Gaussian RBF networks are selected as the approximation of coefficients of the model (2), and the model derived is called the RBF-ARX model as follows

$$\begin{cases} y(t) = \phi_0(\mathbf{X}(t-1)) + \sum_{i=1}^{p_y} \phi_{y,i}(\mathbf{X}(t-1))y(t-i) \\ \qquad + \sum_{i=1}^{p_\mathbf{u}} \Phi_{\mathbf{u},i}(\mathbf{X}(t-1))\mathbf{u}(t-i) + e(t) \\ \phi_0(\mathbf{X}(t-1)) = c_0^0 + \sum_{k=1}^{m} c_k^0 \exp\{-\lambda_k \|\mathbf{X}(t-1) - \mathbf{Z}_k\|_2^2\} \\ \phi_{y,i}(\mathbf{X}(t-1)) = c_{i,0}^y + \sum_{k=1}^{m} c_{i,k}^y \exp\{-\lambda_k \|\mathbf{X}(t-1) - \mathbf{Z}_k\|_2^2\} \\ \Phi_{\mathbf{u},i}(\mathbf{X}(t-1)) = \mathbf{c}_{i,0}^\mathbf{u} + \sum_{k=1}^{m} \mathbf{c}_{i,k}^\mathbf{u} \exp\{-\lambda_k \|\mathbf{X}(t-1) - \mathbf{Z}_k\|_2^2\} \\ \mathbf{Z}_k = (z_{k,1}, z_{k,2}, \cdots, z_{k,n_\mathbf{X}})^\mathbf{T} \end{cases} \qquad (3)$$

where, $p_y$, $p_\mathbf{u}$, $m$, and $n_\mathbf{X} = \dim\{\mathbf{X}(t-1)\}$ are the orders, $\mathbf{Z}_k (k=1,2,\cdots,m)$ are the centers of RBF networks, $\lambda_k (k=1,2,\cdots,m)$ are the scaling parameters, $c_k^0 (k=0,1,2,\cdots,m)$ , $c_{i,k}^y (i=1,2,\cdots,p_y\ ;\ k=0,1,2,\cdots,m)$ , and $\mathbf{c}_{i,k}^\mathbf{u} (i=1,2,\cdots,p_\mathbf{u}\ ;\ k=0,1,2,\cdots,m)$ are the scalar or vector coefficients, and $\|\bullet\|_2$ denotes the vector 2-norm. In model (3), $\mathbf{X}(t-1)$ the signal to make the model coefficients vary with system working point also with time may be:

- output signal;
- input signal;
- external signal;

- computed signal based on the measured ones;
- composition of the above signals.

Besides, using a RBF neural network to approximate equation (1), fields

$$y(t)=\theta_0+\sum_{k=1}^{p}\theta_i\exp\{-\lambda_k\left\|\mathbf{X}(t-1)-\mathbf{Z}_k\right\|_2^2\}+e(t) \quad (4)$$

The RBF-ARX model (3) with the Gaussian RBF network style coefficients has a basic structure similar to linear ARX model. In fact, model (3) contains a component like equation (4); therefore, the RBF-ARX model (3) may be regarded as a more general nonlinear model than RBF neural network. At any working point, a locally linearized ARX model can be easily obtained after fixing the state vector $\mathbf{X}(t-1)$ in (3). It is useful that can allow us to use some linear model-based control method to build a controller, but that merit is not available for RBF network or other nonlinear models such as Hammerstein model etc. It is also conveniently implemented in real-time control cases, because on-line parameter estimation is avoided.

In the application of RBF neural networks, a large number of centers are usually necessary to get a satisfactory fitting to a nonlinear system; it may result in a curse of dimensionality, which is one of major potential difficulties. However, in the RBF-ARX model, the complexity of the model is partially dispersed into the AR part, so the centers in the model need not to be too many. If using the proposed optimization algorithm, the computational burden of parameter estimation of the RBF-ARX model will be largely reduced comparing with that of RBF networks.

## 3. PARAMETER OPTIMIZATION

The identification of RBF or RBF-ARX model includes the order selection and all parameters estimation. Using general optimization algorithm to estimate entire parameters will involve a large amount of computation because the number of the parameters estimated is usually very much. In this paper, based on the Levenberg-Marquardt Method (LMM) (Marquardt 1963) and the Least Squares Method (LSM), an unconstrained structured nonlinear parameter optimization method (SNPOM) is presented for the parameter estimation of the RBF-based model. The main point of the SNPOM is that divide the parameter search space into two subspaces (*i.e.* linear parameter subspace and nonlinear parameter subspace); the search centers around the optimization in the nonlinear subspace; and at each iteration in the optimization process, the search in the nonlinear (or linear) subspace is executed based on the result searched in linear (or nonlinear) subspace. The search in nonlinear subspace uses a method similar to the LMM, and the search in linear subspace uses the LSM. This algorithm is implemented as follows.

### 3.1 *Parameter classification*

For the RBF-ARX model (3), the linear part of the parameters is

$$\begin{aligned}\boldsymbol{\theta}_{\mathrm{L}} &\triangleq \{c_i^0, c_{j,i}^y, \mathbf{c}_{k,i}^{\mathbf{u}} \mid i=0,1,\cdots,m;\\ &\quad j=1,2,\cdots,p_y;\ k=1,2,\cdots,p_{\mathbf{u}}\} \quad (5)\\ \boldsymbol{\theta}_{\mathrm{L}} &\in \Re^{(m+1)\times(p_y+p_{\mathbf{u}}+1)}\end{aligned}$$

and the nonlinear part of the parameters is

$$\boldsymbol{\theta}_{\mathrm{N}} \triangleq (\lambda_1,\lambda_2,\cdots\lambda_m,\mathbf{Z}_1^{\mathrm{T}},\mathbf{Z}_2^{\mathrm{T}},\cdots,\mathbf{Z}_m^{\mathrm{T}})^{\mathrm{T}},\ \boldsymbol{\theta}_{\mathrm{N}}\in\Re^{m+m\times n_{\mathbf{X}}} \quad (6)$$

For the RBF network (4), the linear part of the parameters is

$$\boldsymbol{\theta}_{\mathrm{L}} \triangleq (\theta_0,\theta_1,\cdots,\theta_p)^{\mathrm{T}},\ \boldsymbol{\theta}_{\mathrm{L}}\in\Re^{p+1} \quad (7)$$

and the nonlinear part of the parameters is

$$\boldsymbol{\theta}_{\mathrm{N}} \triangleq (\lambda_1,\lambda_2,\cdots\lambda_p,\mathbf{Z}_1^{\mathrm{T}},\mathbf{Z}_2^{\mathrm{T}},\cdots,\mathbf{Z}_p^{\mathrm{T}})^{\mathrm{T}},\ \boldsymbol{\theta}_{\mathrm{N}}\in\Re^{p+p\times n_{\mathbf{X}}} \quad (8)$$

In general case, rewrite the model (3) or (4) estimated as

$$y(t)=f(\boldsymbol{\theta}_{\mathrm{L}},\boldsymbol{\theta}_{\mathrm{N}},\mathbf{X}(t-1))+e(t) \quad (9)$$

or $$y(t)=\boldsymbol{\Phi}(\boldsymbol{\theta}_{\mathrm{N}},\mathbf{X}(t-1))^{\mathrm{T}}\boldsymbol{\theta}_{\mathrm{L}}+e(t) \quad (10)$$

where, $\boldsymbol{\theta}_{\mathrm{N}}$ is a vector including all nonlinear parameters, $\boldsymbol{\theta}_{\mathrm{L}}$ is a vector including all linear parameters, and equation (10) is the regression form of (9), which is linear with regard to $\boldsymbol{\theta}_{\mathrm{L}}$.

### 3.2 *Initialization*

Choose the order. The choice approach of suitable order is presented in Section 3.4. For RBF-ARX model (3), the orders are $p_y$, $p_{\mathbf{u}}$, $m$, and $n_{\mathbf{X}}$. For RBF network (4), the orders are $p$ and $n_{\mathbf{X}}$. Select initial value $\boldsymbol{\theta}_{\mathrm{N}}^0$ via a prior knowledge. Fixing $\boldsymbol{\theta}_{\mathrm{N}}^0$ and estimating $\boldsymbol{\theta}_{\mathrm{L}}^0$ by the LSM, yields:

$$\begin{cases}\boldsymbol{\theta}_{\mathrm{L}}^0=\left[\sum_{t=\tau}^{M}\bar{\boldsymbol{\Phi}}_{t,0}\bar{\boldsymbol{\Phi}}_{t,0}^{\mathrm{T}}\right]^{-1}\sum_{t=\tau}^{M}\bar{\boldsymbol{\Phi}}_{t,0}\bar{y}(t)\\ \bar{\boldsymbol{\Phi}}_{t,0}=\boldsymbol{\Phi}(\boldsymbol{\theta}_{\mathrm{N}}^0,\bar{\mathbf{X}}(t-1))\end{cases} \quad (11)$$

here $\{\bar{y}(i),\bar{\mathbf{X}}(i-1)\mid i=\tau+1,\tau+2,\cdots,M\}$ is the measured data-set, $\tau$ is the biggest time lag of the variables in $\mathbf{X}(t-1)$, and $M$ is the length of the data.

### 3.3 *Optimization*

Take the objective function as

$$V(\boldsymbol{\theta}_N,\boldsymbol{\theta}_L) \triangleq \frac{1}{2}\|\mathbf{F}(\boldsymbol{\theta}_N,\boldsymbol{\theta}_L)\|_2^2 \quad (12)$$

$$\mathbf{F}(\boldsymbol{\theta}_N,\boldsymbol{\theta}_L) \triangleq \begin{bmatrix} f(\boldsymbol{\theta}_L,\boldsymbol{\theta}_N,\overline{\mathbf{X}}(\tau))-\overline{y}(\tau+1) \\ f(\boldsymbol{\theta}_L,\boldsymbol{\theta}_N,\overline{\mathbf{X}}(\tau+1))-\overline{y}(\tau+2) \\ \vdots \\ f(\boldsymbol{\theta}_L,\boldsymbol{\theta}_N,\overline{\mathbf{X}}(M-1))-\overline{y}(M) \end{bmatrix} \quad (13)$$

The optimization problem is to compute

$$(\hat{\boldsymbol{\theta}}_N,\hat{\boldsymbol{\theta}}_L) = \arg\min_{\boldsymbol{\theta}_N,\boldsymbol{\theta}_L} V(\boldsymbol{\theta}_N,\boldsymbol{\theta}_L).$$

Let $k(=0,1,2,\cdots,k_{\max})$ be iteration step. Denote the Jacobian matrix of $\mathbf{F}(\boldsymbol{\theta}_N{}^k,\boldsymbol{\theta}_L{}^k)$ with respect to $\boldsymbol{\theta}_N{}^k$ as

$$\mathbf{J}(\boldsymbol{\theta}_N{}^k) = (\partial\mathbf{F}(\boldsymbol{\theta}_N{}^k,\boldsymbol{\theta}_L{}^k)/\partial\boldsymbol{\theta}_N{}^k)^{\mathrm{T}}$$

The innovation strategy of the nonlinear parameters $\boldsymbol{\theta}_N{}^k$ is

$$\boldsymbol{\theta}_N{}^{k+1} = \boldsymbol{\theta}_N{}^k + \beta_k\mathbf{d}_k \quad (14)$$

where $\mathbf{d}_k$ is the search direction, $\beta_k$ is a scalar step length parameter that is the distance to the minimum. In order to increase robustness of the search process, based on the LMM, the $\mathbf{d}_k$ in (14) is obtained from a solution of the linear set of equations

$$[\mathbf{J}(\boldsymbol{\theta}_N{}^k)^{\mathrm{T}}\mathbf{J}(\boldsymbol{\theta}_N{}^k)+\gamma_k\mathbf{I}]\mathbf{d}_k = -\mathbf{J}(\boldsymbol{\theta}_N{}^k)^{\mathrm{T}}\mathbf{F}(\boldsymbol{\theta}_N{}^k,\boldsymbol{\theta}_L{}^k) \quad (15)$$

where the scalar $\gamma_k$ controls both the magnitude and the direction of $\mathbf{d}_k$. When $\gamma_k$ tends to zero, the $\mathbf{d}_k$ will tend towards the Gauss-Newton direction. As $\gamma_k$ tends to infinity, $\mathbf{d}_k$ tends towards a steepest descent direction. The size of $\gamma_k$ is determined at each iteration by using a method similar to that of the function '*lsqnonlin*' in Matlab Optimization Toolbox.

Following the update of $\gamma_k$, a solution of (15) is used to obtain a search direction $\mathbf{d}_k$. A step length of unity $\beta_k$ in (14) is then taken in the direction $\mathbf{d}_k$, which is obtained by a line search procedure similar to the mixed quadratic and cubic polynomial interpolation and extrapolation method (Coleman *et al.* 1999).

The optimization calculation centers around the search of $\boldsymbol{\theta}_N{}^{k+1}$ at each iteration, and in any case after computed $\boldsymbol{\theta}_N{}^{k+1}$ from (14), the linear parameters $\boldsymbol{\theta}_L{}^{k+1}$ is updated immediately by the LSM as follows

$$\begin{cases} \boldsymbol{\theta}_L{}^{k+1} = \left[\sum_{t=\tau}^{M}\overline{\boldsymbol{\Phi}}_{t,k+1}\overline{\boldsymbol{\Phi}}_{t,k+1}^{\mathrm{T}}\right]^{-1}\sum_{t=\tau}^{M}\overline{\boldsymbol{\Phi}}_{t,k+1}\overline{y}(t) \\ \overline{\boldsymbol{\Phi}}_{t,k+1} = \boldsymbol{\Phi}(\boldsymbol{\theta}_N{}^{k+1},\overline{\mathbf{X}}(t-1)) \end{cases} \quad (16)$$

The line search procedure of determining the step length $\beta_k$ in (14) ensures that

$$V(\boldsymbol{\theta}_N{}^{k+1},\boldsymbol{\theta}_L{}^{k+1}) < V(\boldsymbol{\theta}_N{}^k,\boldsymbol{\theta}_L{}^k)$$

at each iteration in the parameters $\boldsymbol{\theta}_N{}^{k+1}$ and $\boldsymbol{\theta}_L{}^{k+1}$ updated by (14) and (16), therefore, the $\boldsymbol{\theta}_N{}^{k+1}$ and $\boldsymbol{\theta}_L{}^{k+1}$ are the preferable parameters that make the objective function (12) descend at $k+1$th iteration.

*Remark 1:* In the SNPOM, the optimization of the linear parameters $\boldsymbol{\theta}_L$ can be obtained easily by the formula (16), which are used to adjust the search direction and step length to ensure that the objective function descend in all parameters not only in the nonlinear part $\boldsymbol{\theta}_N$ at each iteration. In the computing efficiency, the SNPOM is much better than general methods of optimizing all parameters regardless of parameter type, especially in the case that the number of linear parameters is more than that of nonlinear parameters in a model.

*Remark 2:* About the choice of initial values of the nonlinear parameters in the RBF-ARX model (3) or RBF network (4), a procedure is that choose a subset of the input vector $\mathbf{X}(t-1)$ evenly or randomly as initial centers $\mathbf{Z}_k$, and the initial scaling parameters $\lambda_k\,(k=1,2,\cdots,m)$ can be selected as

$$\lambda_k = -\log\varepsilon_k / \max_{t-1}\{\|\mathbf{X}(t-1)-\mathbf{Z}_k\|_2^2\}, \varepsilon_k \in [0.1 \sim 0.0001]$$

so that the linear parameters approach bounded and stable when the signal $\mathbf{X}(t-1)$ moves the centers far away.

### 3.4 *Determining order*

The order of model (9) to be identified may be determined by the AIC. Repeat the above SNPOM for model (9) in different orders to look for the final model by comparing their AIC values and model dynamics. For the RBF-ARX model (3) and RBF network (4), the AIC is defined as follows

$$AIC = M\log\hat{\sigma}_e^2 + 2(s+1), \text{ as } M \gg p$$

where $\hat{\sigma}_e^2$ is the model residual variance under the chosen orders, $p$ is the largest one in the orders of regression parts, and $s$ is the number of all parameters estimated.

## 4. APPLICATIONS

### 4.1 *Modeling of a complex nonlinear time series*

Use the Mackey-Glass equation below to generate a well-known chaotic benchmark time series in order to compare performance of the RBF-AR(*p,m,n*) model(17) (here, $v_j(t) = 0$) and RBF(*p,n*) network (4) as well as the different optimization methods (the selected equation parameters are $a = 0.2$, $b = 0.1$, $c = 10$ and $\tau = 20$):

$$\dot{x}(t) = \frac{ax(t-\tau)}{1+x^c(t-\tau)} - bx(t)$$

The original Mackey-Glass series is shown in Fig.1, in which the front 500 data are used to train the models, and the back 500 points are used to test the models. Fig.2 shows the convergent process of prediction error variance of the RBF-ARX(5,3,2) and RBF(5,5) model during parameter search iteration using the SNPOM, LMM, GNM (Gauss-Newton method) and TRM (Trust Region method) respectively. It is clear that the SNPOM has the fastest convergence, and the model estimation precision using SNPOM is also better than that of other methods. For test data and by SNPOM the AIC and parameter number of RBF-AR(5,3,2)/RBF(5,5) are -7880/-7836 and 30/31, so the RBF-AR(5,3,2) model exhibits better modeling precision and faster parameter convergence than that of RBF(5,5) network.

### 4.2 *Modeling of the Nitrogen Oxide (NOx) decomposition process in thermal power plants*

The purpose of NOx decomposition process control in thermal power plants is to reduce the NOx concentration in flue gas from boiler of the plants in order to protect environments. The process has the nonlinear dynamics dependent on load demand of the plants, and the nonlinear characteristics are mainly due to variation in gain with load. The general NOx signal sensor has larger time delay (about 120 seconds, for 20 seconds sampling period the discrete NOx signal time-delay is 6 steps). For overcoming the time-delay influence to controller, use a RBF-ARX(*p,q,m,n*) model to predict 6-step-ahead NOx output of the process as follows

$$y(t) = \phi_0(\mathbf{X}(t-1)) + \sum_{i=1}^{p} \phi_{y,i}(\mathbf{X}(t-1))y(t-i) + \sum_{j=1}^{6}\sum_{i=0}^{q-1} \phi_{v,i}^{j}(\mathbf{X}(t-1))v_j(t-d-i) + e(t) \quad (17)$$

where $y$ is the NOx signal to be predicted, $v_j$ are the input or disturbances of the process respectively;

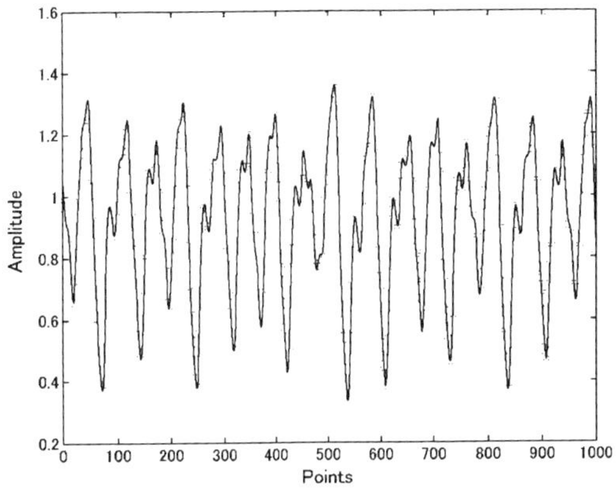


Fig. 1. A chaotic series generated from the Mackey-Glass equation

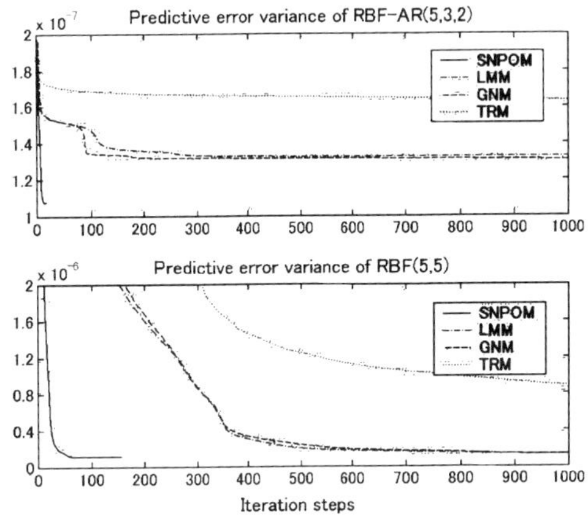


Fig. 2. Comparison of various optimization methods.

$\mathbf{X}(t-1) = [x(t-1), x(t-2), \cdots, x(t-n)]^{\mathrm{T}}$ is the load demand series, $p, q, m$ and $n$ are the orders, $d$ is the time-delay of the process, and their RBF-coefficients are similar to that in model (3).

The measured data showed in Fig.3 is used to identification of the model (17). The working point of the process is dependent on load demand $x(t)$. The parameter search process using the presented SNPOM to estimate model (17) is given in Fig.6, which shows fast convergence through only a few iterations. Note that the RBF-ARX(6,6,1,2) model identified has 89 parameters estimated, in which there are only 3 nonlinear parameters, so the search iterations carried out mainly for nonlinear parameters are very few. Fig.4-5 show the 6-step-ahead prediction of the NOx signal based on the recursively estimated linear ARX model (the result is provided by Japan CHUBU Electric Power Co.) and the RBF-ARX(6,6,1,2) model (17) estimated by SNPOM respectively. The multi-step prediction precision of the RBF-ARX model is much better than that of the recursively estimated linear ARX model as shown in Fig.4-5. Due to nonlinearity and fast change of working point of the process, the multi-step-ahead prediction based on the recursively-estimated linear ARX model gave larger predictive errors especially during large load variations as seen from Fig. 4.

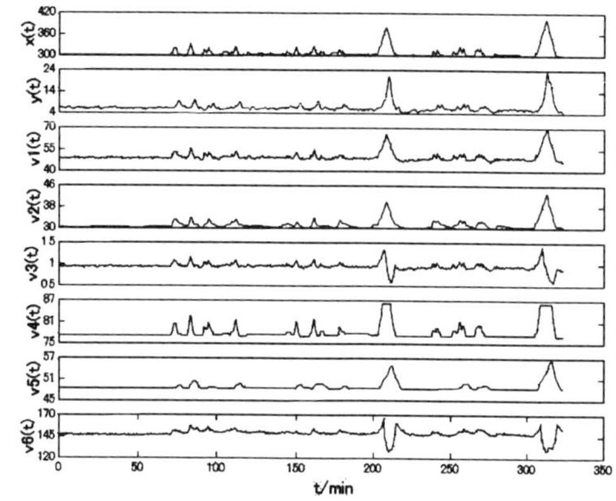


Fig. 3. The observed data of the process ($d = 6$)

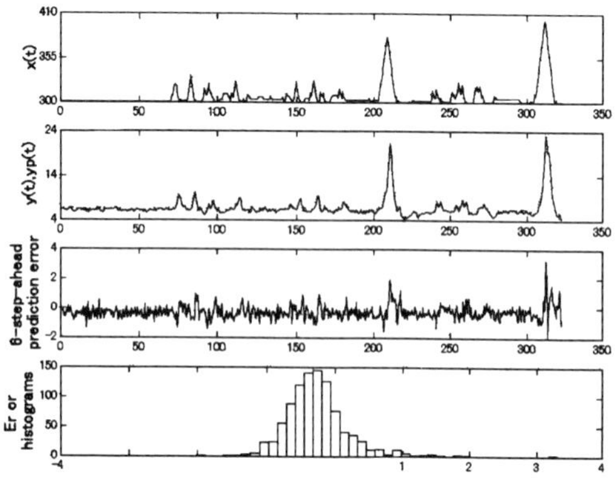


Fig. 4. The 6-step-ahead prediction using linear ARX model estimated by recursive least squares method; y(t): actual NOx signal (solid), yp(t): 6-step-ahead-predicted NOx signal (dotted).

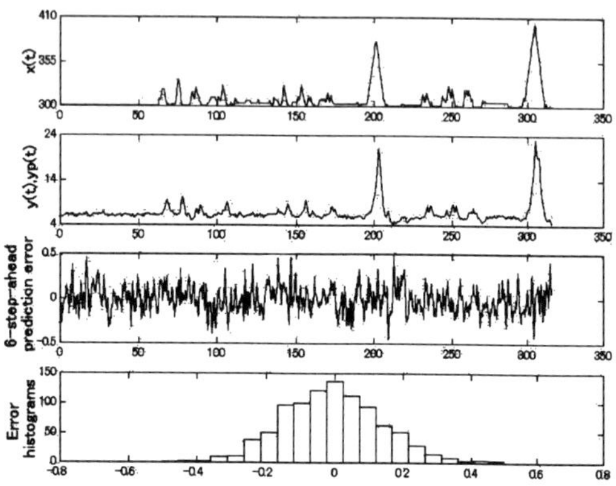


Fig. 5. The 6-step-ahead prediction using RBF-ARX(6,6,1,2) model (17) estimated by SNPOM; y(t): actual NOx signal (solid), yp(t): 6-step-ahead-predicted NOx signal (dotted).

## 5. CONCLUSIONS

The structured nonlinear parameter optimization method (SNPOM) presented in this paper was an off-line analytic iteration optimization method. It gave fast convergence and satisfactory modeling accuracy comparing with other classical methods. It may be used to the parameter optimization of the RBF-based models and other similar nonlinear models whose parameters estimated can be divided linear parameter set and nonlinear parameter set, especially in the cases that the number of linear parameters is more than that of nonlinear parameters. The RBF-ARX model estimated off-line by the SNPOM is a type of global models, which may effectively describe some smooth nonlinear dynamics. It synthesized the merits of RBF net and autoregressive model, so it could use less parameters comparing with general neural networks.

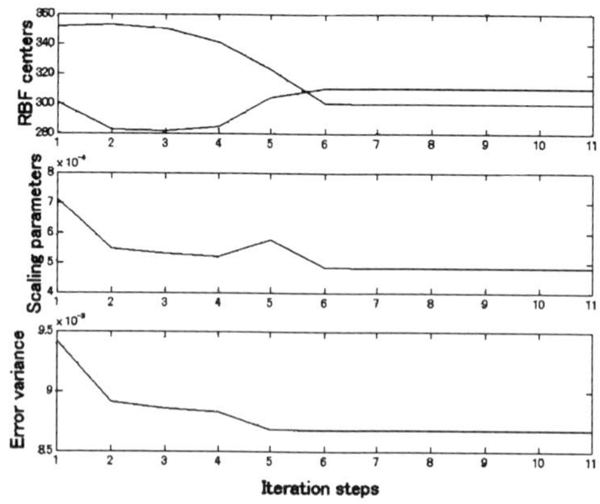


Fig. 6. Parameter search process using SNPOM for RBF-ARX(6,6,1,2) model (17)

## 6. REFERENCES

Coleman, T., Branch, M. A., and Grace, A., 1999, *Optimization toolbox user's guide.* The MathWorks, Inc.

Golub, G. H., and Pereyra, V., 1973, The differentiation of pseudo-inverses and nonlinear least squares problems whose variables separate. *SIAM J. Numerical Analysis,* **10**, 413-432.

Marquardt, D., 1963, An algorithm for least-squares estimation of nonlinear parameters. *SIAM J. Applied Math.* **11**, 431-441.

Ozaki, T., Sosa, P. V., and Haggan-Ozaki, V., 1997, Reconstructing nonlinear dynamics from time series: with application to epilepsy data analysis. *Technical Reports 1997/02* (University of Manchester/UMIST, Manchester Center for Statistical Science).

Prasad, G., Swidenbank, E., and Hogg, B.W., 1998, A local model networks based multivariable long-range predictive control strategy for thermal power plants. *Automatica,* **34**, 1185-1204.

Priestley, M. B., 1980, State dependent models: a general approach to nonlinear time series analysis. *Journal of Time Series Analysis,* **1**, 57-71.

Ruano, A. E., Fleming, P. J., and Jones. D. I., 1992, A connectionist approach to PID autotuning. *IEEE Proceedings, Part D.,* **139**, 279-285.

Vesin, J., 1993, An amplitude-dependent autoregressive signal model based on a radial basis function expansion. *Proceedings of International Conference of ASSP,* 3, Minnesota, 129-132.

IFAC Publications
www.elsevier.com/locate/ifac

# COMPLETELY SUPERVISED TRAINING ALGORITHMS FOR B-SPLINE NEURAL NETWORKS AND NEURO-FUZZY SYSTEMS

**A. E. Ruano[1,2], C. Cabrita[1], J. V. Oliveira[1], L. T. Kóczy[3]**

*1. ADEEC, FCT, Universidade do Algarve*
*Campus de Gambelas, 8000 Faro, Portugal*
*2. Institute of System & Robotics, Portugal*
*3. Department Of Telematics, Technical University of Budapest,*
*Sztoczek U.2., Budapest, H-1111, Hungary*

Abstract: Complete supervised training algorithms for B-spline neural networks and fuzzy rule-based systems are discussed. By introducing the relationships between B-spline neural networks and certain types of fuzzy models, training algorithms developed initially for neural networks can be adapted to fuzzy systems. The standard training criterion is reformulated, by separating its parameters into linear and nonlinear ones. By employing this reformulation with the Levenberg-Marquardt algorithm, a new training method, offering a fast rate of convergence, and a robust performance compared with the standard Error-Back Propagation algorithm. 

Keywords: Training Algorithms, Neural Networks, B-Splines, Fuzzy Systems; Linear and Nonlinear Separability

## 1 INTRODUCTION

[1]B-spline neural networks offer definite advantages over more commonly known neural networks, such as multilayer perceptrons or RBF networks. B-spline networks store the information locally, which means that learning in one part of the input space minimally affects the rest of it. Thus, they are suitable for on-line adaptive modelling and control. Their lattice-based structure makes them transparent, which means that, in contrast to other networks, it is possible to understand the knowledge stored in these networks. As this is an advantage that is also pointed out to fuzzy rule-based models over conventional neural networks, it is not surprising that, at a high level, the basic information principles of B-spline networks and fuzzy systems are the same. This parallel between these two different model types allows learning algorithms derived for B-spline networks to be employed for fuzzy models, and this property is exploited in this paper.
The normal design process for B-spline networks involves two different phases: the determination of the best topology, which can be seen as a system identification problem, and the determination of its parameters, a parameter estimation phase. This paper addresses only the latter issue. That is, the goal is to determine which are the optimal values of the spline knots and the linear parameters, according to a certain training criterion, and given a specified number of sub-modules, spline orders, and number of knots. In fuzzy rule-based terms, this can be translated in determining the support of the antecedent and consequent linguistic terms.
If sufficient prior knowledge exists about the problem, it can be used to determine the network knots. If this is not the case, heuristics are usually employed for their determination. The linear parameters are usually iteratively computed using LMS or NLMS rules (Brown and Harris, 1994), or directly obtained as the solution of a linear least squares problem. This means that the usual approaches are hybrid schemes. In terms of fuzzy models, different approaches are usually employed; see, e.g. (Eklund , Zhou, 1999) for a comparison. Because these models are differentiable with respect to all their parameters, complete supervised training algorithms for off-line training can be applied. Due to the structure of these three-layer networks, we can separate the nonlinear and linear parameters. This enables the use of an alternative training criterion, only dependant on the former, that are the spline knots for B-spline networks, and rule antecedents for fuzzy models.

[1] The authors wish to acknowledge the support of ICCTI, Praxis (Portugal) and OMFB (Hungary) for this work.

## 2 B-SPLINE NEURAL NETWORKS

B-spline neural networks belong to the class of networks termed *lattice-based associative memories networks* (AMN). They are composed of three layers: a *normalized input space* layer, a *basis functions* layer and a *linear weight* layer.

### 2.1 *Normalized Input Layer*

The normalized input layer is usually a grid on which the basis functions are defined. The grid is defined by vectors of *knots* on each input axis.
The interior knots, for the $i^{th}$ axis, are $\lambda_{i,j}, j = 1,\cdots,r_i$ :

$$x_i^{\min} < \lambda_{i,1} \le \lambda_{i,2} \le \cdots \le \lambda_{i,r_i} < x_i^{\max}, \qquad (1)$$

where $x_i^{\min} / x_i^{\max}$ are the min/max of the $i^{th}$ input, resp.
At the extremes of the axes, a set of $k_i$ *exterior knots* is given as:

$$\lambda_{i,-(k_i-1)} \le \cdots \le \lambda_{i,0} = x_i^{\min} \qquad (2)$$

$$x_i^{\max} = \lambda_{i,r_i+1} \leq \cdots \leq \lambda_{i,r_i+k_i} \quad (3)$$

These exterior knots are needed to generate the basis functions that are close to the boundaries. The $j^{th}$ interval of the $i^{th}$ input is defined as:

$$I_{i,j} = \begin{cases} [\lambda_{i,j-1} \quad \lambda_{i,j}[ \ for \ j = 1,\cdots,r_i \\ [\lambda_{i,j-1} \quad \lambda_{i,j}] \ if \ j = r_i + 1 \end{cases} \quad (4)$$

This way, within the range of the $i^{th}$ input, there are $r_i$+1 intervals, which means that there are $p' = \prod_{i=1}^{n}(r_i + 1)$ $n$-dimensional cells in the grid.

### 2.2 The basis functions layer

The output of the hidden layer is determined by a set of $p$ basis functions defined on the $n$-dimensional grid. The shape, size and distribution of the basis functions are characteristics of the particular AMN employed, and the support of each basis function is bounded. In B-splines neural networks, the *order* of the spline implicitly sets the size of the basis functions support and its shape. The univariate B-spline basis function of order $k$ has a support, which is $k$ intervals wide. Hence, each input is assigned to $k$ basis functions.

The $j^{th}$ univariate basis function of order $k$ is denoted $N_k^j(x)$, and it is defined by the following relationships:

$$N_k^j(x) = \left(\frac{x - \lambda_{j-k}}{\lambda_{j-1} - \lambda_{j-k}}\right) N_{k-1}^{j-1}(x) + \left(\frac{\lambda_j - x}{\lambda_j - \lambda_{j-k+1}}\right) N_{k-1}^j(x) \quad (5)$$

$$N_1^j(x) = \begin{cases} 1 \ if \ x \in I_j \\ 0 \ otherwise \end{cases} \quad (6)$$

Multivariate basis functions are formed by taking the *tensor product* of the univariate basis functions:

$$N_{\mathbf{k}}^j(\mathbf{x}) = \prod_{i=1}^{n} N_{\mathbf{k},i}^j(\mathbf{x}_i) \quad (7)$$

The number of basis functions of order $k_i$ defined on an axis with $r_i$ interior knots is $r_i$+$k_i$. Thus, the total number of basis functions for a multivariate B-spline is $p = \prod_{i=1}^{n}(r_i + k_i)$. This number depends exponentially on the size of the input, therefore B-splines are only applicable for problems where the input dimension is small.

### 2.3 The weight layer

The output of an AMN is a linear combination of the outputs of the basis functions. The linear coefficients are the adjustable weights, and as the mapping is linear, finding the weights is just a linear optimization problem. The output is therefore:

$$y = \sum_{i=1}^{p} \mathbf{a}_i \mathbf{w}_i = \mathbf{a}^T \mathbf{w}, \quad (8)$$

where $\mathbf{a}_i = N_{\mathbf{k}}^i(x), \ i = 1,\ldots,p$. As only $p'' = \prod_{i=1}^{n} \mathbf{k}_i$ are active at any one time, (8) can be reduced to:

$$y = \sum_{i=1}^{p''} \mathbf{a}_{act(i)}(\mathbf{x}) \mathbf{w}_{act(i)}, \quad (9)$$

$\mathbf{a}_{act(i)}(\mathbf{x})$ denoting the $i^{\text{th}}$ active basis function for input $\mathbf{x}$.

### 2.4 B-spline neural networks design

The architecture described above is the standard B-spline architecture. In order to lower its high computational complexity, it is beneficial to employ, instead of one single multivariate B-spline model covering all the $n$-dimensional input space, a linear combination of sub-models, each one with lower input dimension.

Let us denote the $i^{\text{th}}$ sub-model as $S_i(\mathbf{x}_i)$. Here, $\mathbf{x}_{\underline{i}}$ denotes the components of $\mathbf{x}$ that belong to the set of input variables ( $\underline{i}$ ) which compose the $i^{\text{th}}$ sub-model. If $\#(\underline{i}) = 1$, we have a univariate sub-model; if $\#(\underline{i}) > 1$, a multivariate sub-model is obtained. The output of such a network is:

$$y(\mathbf{x}) = \sum_{u=1}^{n_u} S_u(\mathbf{x}_{\underline{u}}) \quad (10)$$

The sets of input variables satisfy:

$$\#\left(\bigcup_{\underline{u}=1}^{n_u} \underline{u}\right) \leq n \quad (11)$$

If the weight vector and the basis function vector are partitioned accordingly, $\mathbf{w} = \bigcup_{u=1}^{n_u} \mathbf{w}_{\underline{u}}$ and $\mathbf{a} = \bigcup_{u=1}^{n_u} \mathbf{a}_{\underline{u}}$, then (9) can be transformed in:

$$y = \sum_{u=1}^{n_u} \sum_{j=1}^{p''_j} \mathbf{a}_{\underline{j}} \mathbf{w}_{\underline{j}} \quad (12)$$

Therefore, the design of a network of this type involves:

- The determination of the number of its sub-modules;
- For each sub-model, the set of its inputs;
- The order of the splines for each input;
- The number of the interior knots for each input;
- The location of the interior knots for each input;
- The values of the linear output weights.

Examples of these constructive algorithms are the *ASMOD* (*Adaptive Spline Modelling of Observed Data*) algorithm (Weyer and Kavli, 1995) and the *MARS (Multivariate Adaptive Regression Splines*) algorithm (Friedman,1991). This paper addresses only the two last points in the design phase, the location of the interior knots and the determination of the linear weights.

## 3 FUZZY SYSTEMS AND THEIR RELATIONSHIP TO B-SPLINES NEURAL NETWORKS

In a very concise way, fuzzy models can be considered as those models that operate on fuzzy sets (representing linguistic labels) by eliciting interrelationships between these information granules. Between other features, fuzzy models offer a methodology for conveniently representing and efficient processing of qualitative and linguistic (commonsense) knowledge. Actually, this has been recognized has one of their principal merits when compared with other classes of models.

### 3.1 Mamdani Model

Fuzzy rule-based models are the most popular and widely used type of fuzzy models. Essentially, this model includes a set of fuzzy (or linguistic) rules, such as the following Single-Input Single-Output one:

$$R^{(i)} ::= \text{if } (x \text{ is } X^{(i)}) \text{ then } u \text{ is } U^{(i)}, \quad (13)$$

where $i$=*1,..r*, $r$ being the number of rules, $X^{(i)}$, and $U^{(i)}$ are linguistic terms represented by fuzzy sets. The pair $x$ is the input of the rule, and $u$ its output. The rule itself can be viewed as an implication. Fuzzy models exhibiting such type of rules are referred to as Mamdani models, cf. (Driankov, 1993).

In this study the following assumptions are taken:

- The input value $x$ is a crisp numeric quantity;
- The linguistic terms of the rule antecedents and consequents are modelled by B-splines of order 2, i.e., triangular membership functions;
- For implementing logic connectives such as the conjunction and implication, the *t-norm* used is the product;
- The membership function of the output fuzzy set $U$, is

computed pointwisely by taking the sum over all the rule's output fuzzy sets, $U^{(i)}$; $i = 1, 2, \ldots r$. That is, for all $u$ in the output space:

$$\mu_U(u) = \sum_{i=1}^{r} \mu_{U^{(i)}}(u) \quad (14)$$

- The linguistic-to-numeric conversion (the so-called "defuzzification" process) is obtained by the centre of gravity method.

In this formulation it is implicitly assumed that there exists only one rule for each input linguistic term. However, this can be generalized by attaching to each rule a confidence value, $c_{ij} \in [0,1]$, where a value of 0 means that no credibility is given to a rule, where a value of 1 means that total credibility is given to the rule.

Under these assumptions (by lack of space, the justification is not presented here), by defining the centre of the $j^{th}$ output set as:

$$u_j^c = \sum_{k=1}^{N} u_k \mu_{U^{(j)}}(u_k) \Big/ \alpha, \quad (15)$$

the output of a single input Mamdani FM can be recast in the form of a linear regression model, *i.e.*, $u^* = \varphi^T \theta$, where the vector $\varphi$ contains observable data and $\theta$ includes unknown model parameters can be recast as:

$$\varphi^T = \left[\mu_{X^{(1)}}(x) \ \ldots \ \mu_{X^{(i)}}(x) \ \ldots \ \mu_{X^{(n)}}(x)\right] \quad (16)$$

and

$$\theta^T = \left[w_1 \ \ldots \ w_i \ \ldots \ w_n\right], \quad (17)$$

where:

$$w_i = \sum_{j=1}^{m} c_{ij} u_j^c \quad (18)$$

The generalisation for the multi-input, single output case is straightforward. Assuming $k$ inputs, and $n_i$ linguistic terms for the $i^{th}$ input, eqs. (16) and (17) can be expressed as:

$$\varphi^T = \left[\mu_{X^{(1)}}(\mathbf{x}) \ \ldots \ \mu_{X^{(i)}}(\mathbf{x}) \ \ldots \ \mu_{X^{\left(\prod_{i=1}^{k} n_i\right)}}(\mathbf{x})\right] \quad (19)$$

$$\theta^T = \left[w_1 \ \ldots \ w_i \ \ldots \ w_{\prod_{i=1}^{k} n_i}\right], \quad (20)$$

where:

$$\mu_{X^{(i)}}(\mathbf{x}) = \prod_{l=1}^{k} \mu_{X^{(i)}}(\mathbf{x}_l) \quad (21)$$

Therefore, under the assumptions previously introduced, a Mamdani fuzzy model is equivalent to a B-spline network with one multivariate sub-model, where splines of order 2 are used for every dimension. The network basis functions are equivalent to the linguistic terms of the rule antecedents, and the network output linear parameters are related with the linguistic terms of the rule consequents and their confidence values in the way expressed by (18). This way, the training algorithms described below are also applicable to fuzzy models that satisfy those assumptions.

### 3.2 *TSK Model*

The Takagi-Kang-Sugeno (or simply Sugeno) model is a fuzzy model where the consequents of the rules are real-valued functions; polynomials being the most common, and widely used. For the SISO case, a typical Sugeno model has a set of rules, such as the following:

$$R^{(i)} ::= \text{if}(x \text{ is } X^{(i)}) \text{ then } u^{(i)} = w_0^{(i)} + w_1^{(i)}x; \quad i = 1,\ldots,r \quad (22)$$

$r$, being the number of rules.

Given a crisp input datum $x$, the output of this model is:

$$u^* = \sum_{i=1}^{r} \mu_{X^{(i)}}(x) u^{(i)} \Big/ \sum_{i=1}^{r} \mu_{X^{(i)}}(x) \quad (23)$$

When the linguistic terms of the rule-antecedents $X(i)$ are modelled by B-splines, for all $x$ in the input space, $\sum_{i=1}^{r} \mu_{X^{(i)}}(x) = 1$. Therefore, (23) can be written as $u^* = \varphi^T \theta$ with

$$\varphi^T = \left[\mu_{X^{(1)}}(x) \ \cdots \ \mu_{X^{(r)}}(x) \ \mu_{X^{(1)}}(x)x \ \cdots \ \mu_{X^{(r)}}(x)x\right] \quad (24)$$

and

$$\boldsymbol{\theta}^T = \left[w_0^{(1)} \ \cdots \ w_0^{(r)} \ w_1^{(1)} \ \cdots \ w_1^{(r)}\right] \quad (25)$$

As before, the generalisation for the multi-input, single output case is straightforward. Therefore, the training algorithms proposed for B-spline neural networks can also be adapted to TSK models.

## 4 TRAINING ALGORITHMS

Bearing in mind the relationships between B-spline neural networks and Mamdani and TSK fuzzy models, the following discussion will be carried out in terms of the neural networks terminology. The knot placement determines the support of the antecedent linguistic terms, and the neural linear parameters are related with the consequent part and the confidence values as expressed by (18), for the Mamdani case, and are directly obtained in (25), for the TSK model.

The goal here is to determine the knot placement and the linear output vector. For that purpose, a minimization criterion will be employed, that is related only with the quality of the fitting, as the network complexity is constant. The training criterion that will be employed is the usual *Sum-of-the-Square of the Errors* (SSE):

$$\Omega = \frac{\|\mathbf{t} - \mathbf{y}\|^2}{2} = \frac{\|\mathbf{e}[k]\|^2}{2}, \quad (26)$$

where $\mathbf{t}$ stands for the target vector, $\mathbf{y}$ for the output vector, $\mathbf{e}$ for the error vector, and $\| \ \|$ denotes the 2-norm. We shall assume that there are $m$ patterns in the training set.

The most used method to minimize (26) is the Error-Back-Propagation (BP) (Werbos, 1974) algorithm, which is a steepest descent algorithm. Denoting as $\boldsymbol{\lambda}$ the vector of all the interior knots, and as $\mathbf{z}$ the vector of the design parameters:

$$\mathbf{z} = \left[\mathbf{w} \ \ \boldsymbol{\lambda}\right]^T, \quad (27)$$

each iteration of the BP algorithm is:

$$\mathbf{z}[k] = \mathbf{z}[k-1] - \eta \mathbf{g}[k], \quad (28)$$

where $\mathbf{g}$ stands for the gradient vector of (26):

$$\mathbf{g}[k] = \frac{\partial \Omega(\mathbf{z}[k])}{\partial \mathbf{z}^T[k]} \quad (29)$$

The BP algorithm is a first-order method as it only uses derivatives of the first order. If no line-search is used, then it has no guarantee of convergence and the convergence rate obtained is usually very slow. If a second-order method is to be employed, the best algorithm (Gill et. al., 1981) to use is the Levenberg-Marquardt (LM) algorithm (Marquardt, 1963), which explicitly exploits the underlying structure (sum-of-squares) of the underlying optimization problem.

Denoting as $\mathbf{J}$ the Jacobean matrix:

$$\mathbf{J}[k] = \frac{\partial y(\mathbf{z}[k])}{\partial \mathbf{z}^T[k]}, \quad (30)$$

the LM update, $\mathbf{s}[k] = \mathbf{z}[k] - \mathbf{z}[k-1]$, is given as the solution of:

$$\left(\mathbf{J}^T[k]\mathbf{J}[k] + \alpha \mathbf{I}\right)\mathbf{s}[k] = -\mathbf{g}[k] = -\mathbf{J}^T[k]\mathbf{e}[k] \quad (31)$$

In (31), $\alpha$ is a regularization parameter, which controls both the search direction and the magnitude of the update. The good results presented by the LM method (compared with other second-order methods such as the quasi-Newton and conjugate gradient methods) are due to its explicit exploitation of the underlying characteristics of the optimization problem, a sum-of-square of errors.

Using a similar point of view, the specific topology of the

neural network can be also explicitly exploited in the training algorithm. In Section 2.3 it has been pointed out that the network output is just a linear combination of the outputs of the basis function layer. Let us denote the basis function output matrix as **A**. Then, the network output vector is just:

$$\mathbf{y} = \mathbf{A}(\boldsymbol{\lambda})\mathbf{w} \tag{32}$$

For any combination of the interior knots, the optimal value of the linear parameters is:

$$\hat{\mathbf{w}}(\boldsymbol{\lambda}) = \mathbf{A}(\boldsymbol{\lambda})^{+}\mathbf{t} \tag{33}$$

Replacing (33) in (26), a new training criterion is obtained, which is independent of the linear output vector, **w**:

$$\Psi(\boldsymbol{\lambda}) = \frac{\left\|\mathbf{t} - \mathbf{A}(\boldsymbol{\lambda})\mathbf{A}(\boldsymbol{\lambda})^{+}\mathbf{t}\right\|^{2}}{2} = \frac{\left\|\left(\mathbf{P}_{\mathbf{A}_{\perp}}\right)\mathbf{t}\right\|^{2}}{2}, \tag{34}$$

where $\mathbf{P}_{\mathbf{A}_{\perp}}$ is the orthogonal projection matrix onto the complementary space spanned by the columns of **A**.

This reformulated criterion was first proposed in the field of neural networks in (Ruano et. al., 1991) for MLPs, and in (Ferreira and Ruano, 2000) for RBFs networks. In fact, it can be applied to any neural network with a layered structure, where the output layer is linear, and where the activation functions in the hidden layer(s) are differentiable.

To minimize (34), its gradient vector (or its Jacobean matrix) is needed, if the BP (or the LM method) is used. The gradient vector of $\Psi$, $\mathbf{g}_{\Psi}$, is given by:

$$\begin{bmatrix} \mathbf{0} \\ \mathbf{g}_{\Psi} \end{bmatrix} = \mathbf{g}_{\Omega}\big|_{\mathbf{w}=\hat{\mathbf{w}}} = -\begin{bmatrix} \mathbf{A}(\boldsymbol{\lambda}) \\ \mathbf{J}_{\Psi}^{T} \end{bmatrix}\mathbf{e}_{\Psi}, \tag{35}$$

where $\mathbf{J}_{\Psi}$ denotes the partition of the Jacobean matrix related with the knot vector $\boldsymbol{\lambda}$, provided that the value of the linear parameters are their optimal ones:

$$\mathbf{J}(\boldsymbol{\lambda})\big|_{\mathbf{w}=\hat{\mathbf{w}}} = \left[\mathbf{A}(\boldsymbol{\lambda})\big|_{\mathbf{w}=\hat{\mathbf{w}}} \quad \mathbf{J}_{\Psi}\right] \tag{36}$$

and $\mathbf{e}_{\Psi} = \mathbf{P}_{\mathbf{A}(\boldsymbol{\lambda})_{\perp}}\mathbf{t}$. It can be proved that $\mathbf{J}_{\Psi}$ is a suitable Jacobean matrix for $\Psi$ (Ruano et al., 1991).

Therefore, to minimize $\Psi$, we need to obtain derivatives that are computed in exactly the same manner as for $\Omega$, but primarily the linear output weights are replaced with their optimal values using (33). It should be noted that, if the BP algorithm is used as training algorithm, the use of the new criterion (34) is more costly, in terms of computational complexity, than the standard criterion (26), due to the computation of the optimal linear output weights (33). If the LM method is used, however, each iteration is actually cheaper.

For B-spline networks and fuzzy systems, the standard BP and LM algorithms need to be modified, so that the order relation (1) between the knots for each input dimension is maintained. In order to maintain the same search direction, the update vector ( $s[k]$ for the LM method and $\eta\mathbf{g}[k]$ for the BP algorithm), when violation of (1) is detected ( $\boldsymbol{\lambda}_{i+1}[k] < \boldsymbol{\lambda}_{i}[k]$ ), is reduced of a factor *g* so that the position of the ($i$+1)$^{\text{th}}$ knot is located half-way between the previous distance of the two corresponding knots:

$$g = \frac{\boldsymbol{\lambda}_{i+1}[k-1] - \boldsymbol{\lambda}_{i+1}[k-1]}{2\left(\mathbf{s}_{i+1}[k] - \mathbf{s}_{i+1}[k]\right)} \tag{37}$$

Therefore, when there is a violation of the order relation, the learning rate for the BP algorithm is updated at that iteration, and a primitive type of line-search algorithm is incorporated in the LM algorithm.

### 4.1 Jacobean computation

Irrespectively of the method used and of the training criterion employed we need to compute the Jacobean matrix with respect to the knot vector. This will be done below, in a pattern by pattern basis. As a B-spline network can be composed of several sub-models, and as the outputs of the basis functions are just dependent of the knot vectors associated with its own sub-model, the total Jacobean matrix can be expressed as:

$$\mathbf{J}_{\boldsymbol{\lambda}} = \left[\mathbf{J}_{\boldsymbol{\lambda}_{1}} \quad \cdots \quad \mathbf{J}_{\boldsymbol{\lambda}_{n}}\right], \tag{38}$$

where $\mathbf{J}_{\boldsymbol{\lambda}_{i}}$ is the partition of the Jacobean matrix corresponding to the knot vector associated with the $i^{\text{th}}$ sub-model:

$$\mathbf{J}_{\boldsymbol{\lambda}_{i}} = \frac{\partial \mathbf{y}_{i}}{\partial \boldsymbol{\lambda}_{i}^{T}} = \frac{\partial \sum_{j=1}^{p_{i}^{n}} \mathbf{A}_{j}\mathbf{w}_{j}}{\partial \boldsymbol{\lambda}_{i}^{T}} \tag{39}$$

The derivatives of univariate basis functions will be derived first, followed by the derivatives of a sub-model.

#### 4.1.1 *The derivatives of univariate basis functions*

The goal is to compute the derivatives of the basis functions with respect to the interior knots $\boldsymbol{\lambda}$. First of all, the following conditions apply:

$$\frac{\partial N_{1}^{j}}{\partial \lambda_{i}} = 0, \ \forall i \tag{40}$$

$$\frac{\partial N_{k}^{j}}{\partial \lambda_{i}} = 0, \ i < j-k, \ i \le 0, \ i > j, \ i > r_{i} \tag{41}$$

When $k$=2, $\lambda_{j-1} = \lambda_{j-k+1}$. Therefore, the 3 non-null derivatives of a B-spline of order 2 are given by:

a) $$\frac{\partial N_{2}^{j}}{\partial \lambda_{j-2}} = \frac{x - \lambda_{j-1}}{\left(\lambda_{j-1} - \lambda_{j-2}\right)^{2}} N_{1}^{j-1}(x)$$

Referring to (5), notice that the derivative of the B-splines are 0, as the splines are of order 1, and the derivative of the coefficient of the 2$^{\text{nd}}$ term in (5) is also 0 as it does not depend on $\lambda_{j-k}$. Notice also that $\frac{\partial N_{2}^{j}}{\partial \lambda_{j-2}}$ is 0, if $x \in I_{j}$, and $\frac{x - \lambda_{j-1}}{\left(\lambda_{j-1} - \lambda_{j-2}\right)^{2}}$ if $x \in I_{j-1}$.

b) $$\frac{\partial N_{2}^{j}}{\partial \lambda_{j-1}} = \frac{\lambda_{j-2} - x}{\left(\lambda_{j-1} - \lambda_{j-2}\right)^{2}} N_{1}^{j-1}(x) + \frac{\lambda_{j} - x}{\left(\lambda_{j} - \lambda_{j-1}\right)^{2}} N_{1}^{j}(x)$$

This derivative is $\frac{\lambda_{j-2} - x}{\left(\lambda_{j-1} - \lambda_{j-2}\right)^{2}}$ if $x \in I_{j-1}$, and $\frac{\lambda_{j} - x}{\left(\lambda_{j-1} - \lambda_{j-2}\right)^{2}}$ if $x \in I_{j}$.

c) $$\frac{\partial N_{2}^{j}}{\partial \lambda_{j}} = \frac{x - \lambda_{j-1}}{\left(\lambda_{j} - \lambda_{j-1}\right)^{2}} N_{1}^{j}(x)$$

This derivative is 0, if $x \in I_{j-1}$, and $\frac{x - \lambda_{j-1}}{\left(\lambda_{j} - \lambda_{j-1}\right)^{2}}$ if $x \in I_{j}$.

When $k \ge 3$, $k$+1 derivatives must be computed. These are:

a) $$\frac{\partial N_{k}^{j}}{\partial \lambda_{j-k}} = \frac{x - \lambda_{j-1}}{\left(\lambda_{j-1} - \lambda_{j-k}\right)^{2}} N_{k-1}^{j-1}(x) + \frac{x - \lambda_{j-k}}{\lambda_{j-1} - \lambda_{j-k}} \frac{\partial N_{k-1}^{j-1}}{\partial \lambda_{j-k}}$$

Notice that the derivative of the last term in (5) is 0, since its coefficient does not depend on $\lambda_{j-k}$, and $\frac{\partial N_{k-1}^{j}}{\partial \lambda_{j-k}} = 0$, because of (41).

b) $$\frac{\partial N_{k}^{j}}{\partial \lambda_{j-k+1}} = \frac{x - \lambda_{j-k}}{\left(\lambda_{j-1} - \lambda_{j-k}\right)} \frac{\partial N_{k-1}^{j-1}(x)}{\partial \lambda_{j-k+1}} + \frac{\lambda_{j} - x}{\left(\lambda_{j} - \lambda_{j-k+1}\right)^{2}} N_{k-1}^{j}(x) + \frac{\lambda_{j} - x}{\lambda_{j} - \lambda_{j-k+1}} \frac{\partial N_{k-1}^{j}}{\partial \lambda_{j-k+1}}$$

c) If k>3,

$$\frac{\partial N_k^j}{\partial \lambda_i}=\frac{x-\lambda_{j-k}}{\left(\lambda_{j-1}-\lambda_{j-k}\right)}\frac{\partial N_{k-1}^{j-1}(x)}{\partial \lambda_i}+ +\frac{\lambda_j-x}{\lambda_j-\lambda_{j-k+1}}\frac{\partial N_{k-1}^j}{\partial \lambda_i},\ j-k+1<i<j-1$$

d)

$$\frac{\partial N_k^j}{\partial \lambda_{j-1}}=\frac{\lambda_{j-k}-x}{\left(\lambda_{j-1}-\lambda_{j-k}\right)^2}N_{k-1}^{j-1}(x)+\frac{x-\lambda_{j-k}}{\lambda_{j-1}-\lambda_{j-k}}\frac{\partial N_{k-1}^{j-1}(x)}{\partial \lambda_{j-1}}+ +\frac{\lambda_j-x}{\lambda_j-\lambda_{j-k+1}}\frac{\partial N_{k-1}^j(x)}{\partial \lambda_{j-1}}$$

e)

$$\frac{\partial N_k^j}{\partial \lambda_j}=\frac{x-\lambda_{j-k+k}}{\left(\lambda_j-\lambda_{j-k+1}\right)^2}\frac{N_{k-1}^j(x)}{\partial \lambda_j}+\frac{\lambda_j-x}{\lambda_j-\lambda_{j-k+1}}\frac{\partial N_{k-1}^j}{\partial \lambda_j}(x)$$

Notice that the first term in (5) is 0 since the coefficient does not depend on $\lambda_j$, and $\frac{\partial N_{k-1}^{j-1}(x)}{\partial \lambda_j}=0$ because of (41).

### 4.1.2 *The derivatives of a sub-model*

For a univariate sub-model, the Jacobean matrix, for pattern $p$, is just:

$$\mathbf{J}_{p,\underline{i}}=\frac{\partial N_{k_i}^{j_i}\left(\mathbf{x}_{p,\underline{i}}\right)}{\partial \lambda_{\underline{i}}^{T}}\mathbf{w}_{\underline{i}},\ j_i\in\left[1\ \ r_i+k_i\right],\ \lambda_{\underline{i}}\in\left[1\ \ r_i\right]\ (42)$$

For a multivariate sub-model, if we denote the output matrix of the basis functions related with variable $m$ (out of the variables $\mathbf{x}_i$) as $\mathbf{A}_{.,\mathbf{i}_m}$,

$$\mathbf{A}_{.,\mathbf{i}_m}=N_{k_{\mathbf{i}_m}}^{j_{\mathbf{i}_m}}\left(\mathbf{x}_{.,\mathbf{i}_m}\right),\ j_{\mathbf{i}_m}\in\left[1\ \ r_{\mathbf{i}_m}+k_{\mathbf{i}_m}\right] \quad (43)$$

and the output matrix of the multivariate basis functions related with $\mathbf{x}_i$ as $\mathbf{A}_{.,\mathbf{i}}$, this last matrix can be obtained as a tensor product of the constituting univariate basis functions:

$$\mathbf{A}_{.,\mathbf{i}}=\mathbf{A}_{.,\mathbf{i}_1}\otimes\cdots\mathbf{A}_{.,\mathbf{i}_m}\otimes\cdots\otimes\mathbf{A}_{.,\mathbf{i}_{n_i}} \quad (44)$$

In this case, the Jacobean matrix is given by (45), as $\frac{\partial A_{.,\mathbf{i}_m}}{\partial \lambda_{\mathbf{i}_k}}=0,\ \ m\neq k$.

### 4.1.3 *The derivatives of a TSK model*

The two alternative training criteria, $\Omega(\boldsymbol{\lambda},\mathbf{w})$ (26) $\Psi(\boldsymbol{\lambda})$ (34) are directly applied to the TSK model, considering that $\mathbf{w}$ is given by (20) and $\mathbf{A}(\boldsymbol{\lambda})=\boldsymbol{\mu}(\boldsymbol{\lambda})$, given by (21).

$$\mathbf{J}_{.,i}=\begin{bmatrix}\frac{\partial A_{.,\mathbf{i}_1}}{\partial \lambda_{\mathbf{i}_1}}\otimes A_{.,\mathbf{i}_1}\otimes\cdots\otimes A_{.,\mathbf{i}_{n_i}}\\ \cdots\\ A_{.,\mathbf{i}_1}\otimes\cdots\otimes\frac{\partial A_{.,\mathbf{i}_m}}{\partial \lambda_{\mathbf{i}_m}}\otimes\cdots\otimes A_{.,\mathbf{i}_{n_i}}\\ \cdots\\ A_{.,\mathbf{i}_1}\otimes\cdots\otimes A_{.,\mathbf{i}_m}\otimes\cdots\otimes\frac{\partial A_{.,\mathbf{i}_{n_i}}}{\partial \lambda_{\mathbf{i}_{n_i}}}\end{bmatrix}^T\mathbf{w}_i \quad (45)$$

With respect to the gradient vector and the Jacobean matrix of the two criteria, the only point that needs to be redefined is the computation of $\mathbf{J}_{\Psi}$, for the TSK model. Considering (24), the Jacobean partition of the TSK model ($\mathbf{J}_{\Psi_{TSK}}$), corresponding to $\mathbf{J}_{\Psi}$, for B-splines/Mamdani models ($\mathbf{J}_{\Psi_{BS}}$), can be given as:

$$\mathbf{J}_{\Psi_{TSK}}=\left[\mathbf{J}_{\Psi_{BS}}\quad \mathbf{J}_{\Psi_{BS}}\times\mathbf{X}_1\quad\ldots\quad\mathbf{J}_{\Psi_{BS}}\times\mathbf{X}_k\right], \quad (46)$$

where the operation $\mathbf{J}_{\Psi_{BS}}\times\mathbf{X}_i$ has the meaning that every column of $\mathbf{J}_{\Psi_{BS}}$ is multiplied, in an element-by element basis, by the column associated with the $i^{th}$ input.

## 5 RESULTS

The case presented is an academical problem, with two interior knots. This was chosen so that the training results can be presented graphically. Different training algorithms, starting from different initial points, are employed. For all the examples, a termination criterion, ensuring convergence to a local minimum wass used. When (26) is employed, the initial values of the linear weights are set to zero.
This example illustrates an inverse kinematic transformation between Cartesian coordinates and one of the angles of a two-links manipulator.

### 5.1 *Mamdani / B-splines*

A B-spline composed of one bivariate sub-model is used here. For both dimensions only one interior knot per dimension is considered. Three different initial positions are considered:

$$\lambda_{1,1}=\lambda_{1,2}=0.2,\ \ \lambda_{1,1}=\lambda_{1,2}=0.9,\ \ \lambda_{1,1}=0.5;\lambda_{1,2}=0.3\,.$$

The minimum of the performance surface is approximately located at $\lambda_{i,1}=\left[0.7\ \ 0.58\right]\vee\left[0.58\ \ 0.7\right]$, and the corresponding MSE is 0.0146. The optimal values of the linear parameters are :

$$\hat{\mathbf{w}}=\left[3.2\ \ 2.1\ \ 0.38\ \ 1.8\ \ 1.2\ \ -0.71\ \ 0.25\ \ -0.66\ \ -83\right]$$

The four following figures illustrate the performance of the LM method, when applying the new training criterion (fig. 1) and the standard training criterion (fig. 2), and the BP method, minimizing the new (fig. 3) and the standard criterion (fig. 4). The results obtained are summarized in Table 1.
Analyzing Table 1 and figs. 1 to 4, it can be concluded that no algorithm, for any run, converges to any of the global minima, although most of them converge to a local minimum not very far from the global one. In the 3rd run, only the LM method, minimizing the standard criterion, converges to this specific local minimum. Comparing both LM methods, the one minimizing the new criterion achieves a faster rate of convergence. The BP method, minimizing the new criterion, converges to the same minima that the LM method, minimizing the same criterion, converges to, although with a much slower rate of convergence. The standard training method, the BP algorithm minimizing the standard criterion, has a very poor performance. No run converges, and the 2nd run is completely oscillatory, in terms of the knot evolution. It should be highlighted, however, that this behaviour is by no means evident if only the MSE evolution was inspected (see fig. 5), as it is common practice.

Table 1. Summary of the results: B-splines

| Method | I point | F Point | I MSE | F MSE | Iter |
|---|---|---|---|---|---|
| LM NC | 0.5, 0.3 | 0.55, 0.55 | 0.025 | 0.016 | 7 |
| LM NC | 0.2, 0.2 | 0.55, 0.55 | 0.045 | 0.016 | 9 |
| LM NC | 0.9, 0.9 | 0.82, 0.82 | 0.038 | 0.027 | 4 |
| LM SC | 0.5, 0.3 | 0.55, 0.55 | 4.077 | 0.016 | 11 |
| LM SC | 0.2, 0.2 | 0.55, 0.55 | 4.077 | 0.016 | 12 |
| LM SC | 0.9, 0.9 | 0.82, 0.82 | 4.077 | 0.027 | 8 |
| BP NC .01 | 0.5, 0.3 | 0.55, 0.55 | 0.025 | 0.016 | 17 |
| BP NC .01 | 0.2, 0.2 | 0.55, 0.55 | 0.045 | 0.016 | 21 |
| BP NC .01 | 0.9, 0.9 | 0.82, 0.82 | 0.038 | 0.027 | 5 |
| BP SC .005 | 0.5, 0.3 | 0.82, 0.57 | 4.077 | 0.029 | 100 |
| BP SC .005 | 0.2, 0.2 | 0.11, 0.08 | 4.077 | 0.096 | 100 |
| BP SC .005 | 0.9, 0.9 | 0.84, 0.84 | 4.077 | 0.033 | 100 |

### 5.2 *TSK Model*

The minimum of the performance surface is approximately located at $\lambda_{i,1}=\left[0.7\ \ 0.7\right]$, and the corresponding MSE is 9 $10^{-4}$. As there are 25 values in the weight vector, these are not presented here. TSK presents, as usually, better modelling capabilities than B-splines/Mamdani model.
Analysing Table 2 we can observe that only 1 run, out of 12, converges to the global minimum. This is achieved in the 2nd

run of the LM method, minimizing the standard criterion. Comparing the two criteria, for both methods, it is evident that the new criterion exhibits a faster rate of convergence to local minima. On the other hand, comparing the two methods, for both criteria, it is also clear that the LM method outperforms significantly the BP technique. It should be noted, however, that the first run of BP algorithm minimizing the new criterion, although it did not achieve convergence within the pre-specified maximum number of iterations (100), it would eventually converge to the global minimum if this restriction was removed. The correspondent run, for the LM method minimizing the new criterion converged to a local minimum.

Table 2. Summary of the results: TSK (The MSE unit is $10^{-3}$)

| Method | I point | F Point | I MSE | F MSE | Iter |
|---|---|---|---|---|---|
| LM NC | 0.5, 0.3 | 0.75, 0.65 | 4.7 | 1.5 | 13 |
| LM NC | 0.2, 0.2 | 0.76, 0.76 | 8.4 | 1.5 | 22 |
| LM NC | 0.9, 0.9 | 0.9, 0.9 | 3.8 | 3.8 | 2 |
| LM SC | 0.5, 0.3 | 0.79, 0.79 | 4077 | 3.1 | 20 |
| LM SC | 0.2, 0.2 | 0.7, 0.7 | 4077 | 0.9 | 50 |
| LM SC | 0.9, 0.9 | 0.84, 0.84 | 4077 | 2 | 24 |
| BP NC .01 | 0.5, 0.3 | 0.7, 0.68 | 4.7 | 0.9 | 100 |
| BP NC .01 | 0.2, 0.2 | 0.64, 0.58 | 8.4 | 1.5 | 100 |
| BP NC .01 | 0.9, 0.9 | 0.9, 0.9 | 3.8 | 3.8 | 3 |
| BP SC .005 | 0.5, 0.3 | 0.75, 0.44 | 4077 | 21 | 100 |
| BP SC .005 | 0.2, 0.2 | 0.04, 0.03 | 4077 | 166 | 100 |
| BP SC .005 | 0.9, 0.9 | 0.84, 0.84 | 4077 | 26 | 100 |

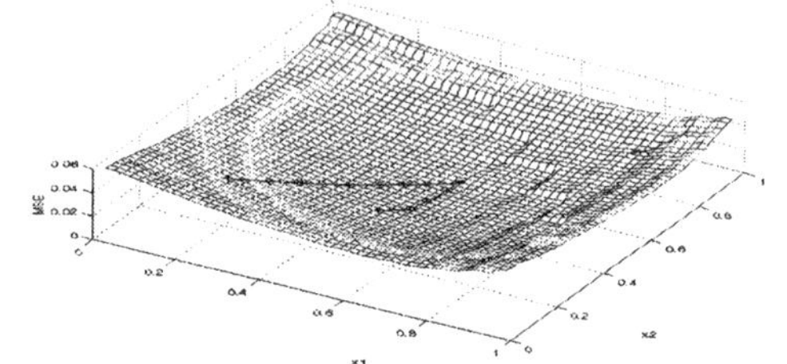

Fig. 1 – Performance of the LM method, new criterion

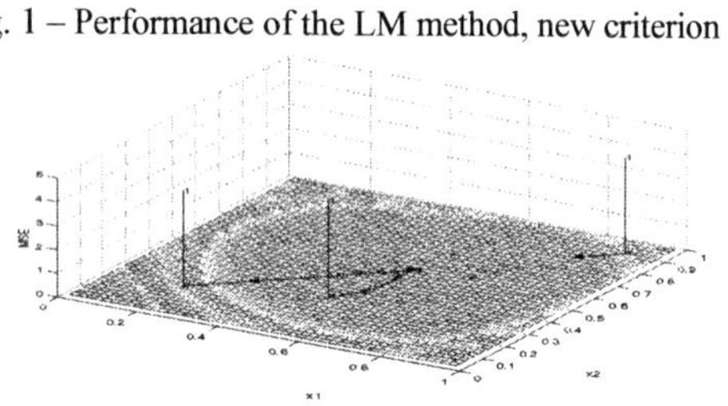

Fig. 2 – Performance of the LM method, standard criterion

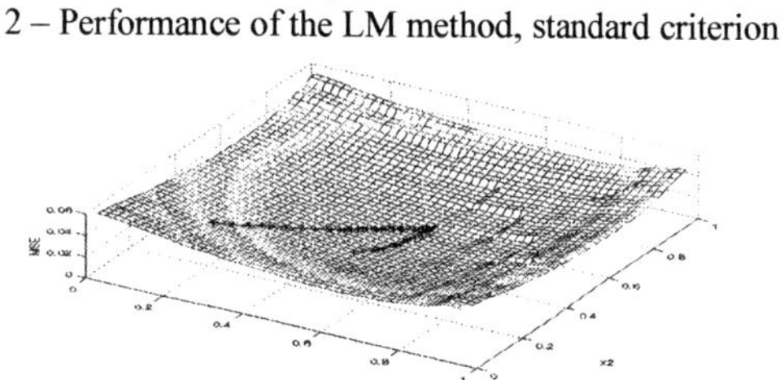

Fig. 3 – Performance of the BP method, new criterion

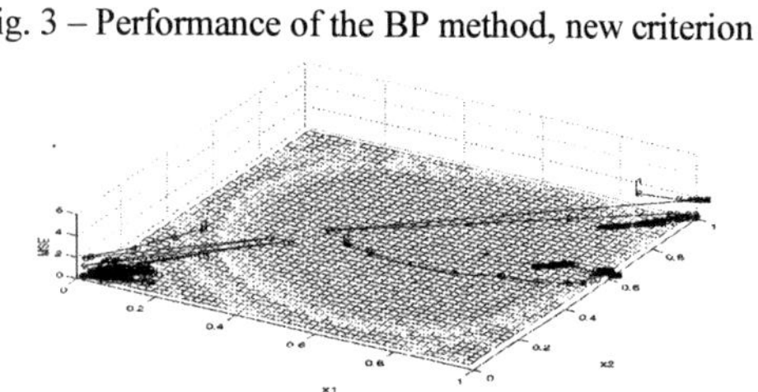

Fig. 4 – Performance of the BP method, standard criterion

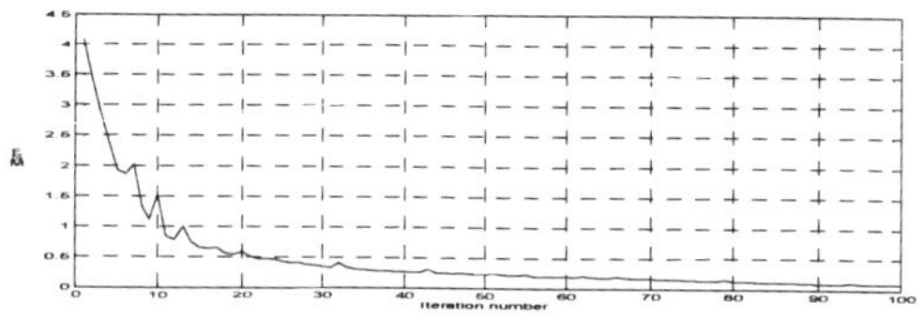

Fig. 5 – MSE evolution, 2nd run (BP, standard criterion)

## 6 CONCLUSIONS

In this paper it has been shown that the performance of the BP algorithm, minimizing the standard sum of the square of the errors criterion is very poor, and that the Levenberg-Marquardt algorithm, minimizing a reformulated criterion only dependent on the nonlinear model parameters, is a reliable method and achieves a very fast rate of convergence. Making use of the relationships between fuzzy models and B-spline neural networks, this algorithm, could also be applied to Mamdani and TSK fuzzy models. The simple example employed showed also that the training performance surface has several local minima. This clearly indicates the need of performing different training sessions, with different initial conditions, for the same problem.

This paper dealt with only two (actually the last two) of a series of steps needed in the design process of B-spline and fuzzy models. The higher steps in the design process are needed to alleviate the computational complexity associated this kind of models. Meta-heuristics, such as Genetic Programming algorithms can, in principle, offer a valid alternative to existing techniques. Due to their stochastic operation, their combination with the algorithms presented here, has the potential of overcoming one drawback pointed out, which is the existence of several local minima in the training performance surface.

## REFERENCES


Brown, M., C. Harris (1994). *Neurofuzzy Adaptive Modelling and Control*, Prentice-Hall.

Driankov D., H. Hellendoorn, M. Reinfrank (1993). *An Introduction to Fuzzy Control*, Springer-verlag, New York.

Eklund, P., J. Zhou (1999). Comparison of learning strategies for adaptation of fuzzy controller parameters. *Fuzzy Sets and Systems*, Vol. 106. 3, 321-333.

Gill, P.E., W. Murray, M.H. Wright (1981). *Practical optimization*, Academic Press, London.

Ferreira, P.M., A. E. Ruano (2000). *Exploiting the Separability of Linear and Nonlinear Parameters in Radial Basis Function Networks*, Symposium on Adaptive Systems for Signal Processing, Control and Communications, Lake Louise, Canada, 2000, 321-326

Friedman, J. H. (1991). Adaptive Spline Networks In *Advances in Neural Information Processing Systems 3*", (Lippmann, R. P., Moody, J.E., Touretzky D.S., Morgan Kaufman eds.), 675-683

Marquardt, D. (1963). An Algorithm for Least-Squares Estimation of Nonlinear Parameters. *SIAM J. Appl. Math.*, 11, 431-441

Ruano, A.E., Fleming, P.J., Jones, D.I. (1991). A New Formulation of the Learning Problem for a Neural Network Controller, *Proc. 30th IEEE CDC*, Bristol, UK, 1, 865-866

Werbos, P. (1974). *Beyond Regression: New Tools for Prediction and Analysis in the Behavioral Sciences*, PhD. Dissertation, Appl. Math., Harvard University, USA.

Weyer E., T. Kavli, (1995). The ASMOD Algorithm. Some New Theoretical and Experimental Results. SINTEF Report STF31 A95024, Oslo.

IFAC
Publications
www.elsevier.com/locate/ifac

# NEW LEARNING ALGORITHM OF NEURO-FUZZY-NETWORK [1]

**Zhen-lei Wang* Jian-hui Wang* Shu-sheng Gu***

(**School of Information Science and Engineering, Northeastern University, Box 131, Shenyang, 110004,China*)

Abstract: Training of neuro-fuzzy-networks by conventional error backpropagation methods introduces considerable computational complexities due to the need for gradient evaluations. In this paper, the concepts coming from the theory of stochastic learning automaton are used. This method eliminates the need for computation of gradients and hence affords a very simple implementation, particularly for implementation on low-end platforms such as personal computers. And the neuro-fuzzy-network training by a learning automaton approach is applied to a nonlinear multivariable system—the three-tank-system. The simulation result is given. *Copyright © 2001 IFAC*

Keywords: neuro-fuzzy-network, stochastic learning automaton, nonlinear multivariable system, and error backpropagation

## 1. INTRODUCTION

Neuro-fuzzy-network is used in diverse areas. The universal approximation of the neuro-fuzzy-network was proved by Wang (1995). And the applications in the control and identification of complex systems are posed in many papers (Horikawa, et al, 1992; Ishibuchi, et al, 1993; Sundareshan and Condarcure, 1998; Zhang and Morris, 1999). Despite the wide uses of the neuro-fuzzy-network, a major problem in a successful deployment of the network is the complexity in training. Earlier procedures attempt to extend the backpropagation approach that has been proved very popular for static feedforward networks. Later the GAs is used to train the neuro-fuzzy-network (Wang and Gu, 2000). The computation of the required updating equations for achieving desired dynamical behavior can however be very tedious. The importance of bypassing the computation of gradients is clearly evident. A new learning scheme based on the theory of stochastic learning automaton is posed in the paper.

Research on learning automaton dates back to the early work of Tsetin (1962) and has been developed since then by a number of others in various contexts (Lakshmivarahan, 1981; Narendra and Thathachar, 1989). Popularly investigated versions of these ideas, known as reinforcement learning methods, have been used in studies of animal learning. Reinforcement learning is based on the intuitive idea that if an action

[1]Supported by Liaoning Province Science and Technology Foundation (No. 002011)

on a system results in an improvement in the state of the system, then the probability of repeating that action is strengthened.

In comparison with more popularly used neuro-fuzzy-network training procedures, reinforcement learning is bounded between supervised learning methods and unsupervised learning methods (such as competitive learning algorithm which don't require a knowledge of external signals for the evolution of the learning process). In order to guide the training in the proper direction, reinforcement learning methods use a scalar value or an indicator function that indicates whether the chosen inputs (or actions) have steered the network in the direction of accomplishing the learning task.

An outline of the paper is as follows. In Section 2, we shall show the architecture of the neuro-fuzzy-network. In Section 3, the training procedure is given in detail. In Section 4, we design controller for a nonlinear multivariable system using the neuro-fuzzy-network. Section 5 is the conclusion.

## 2. ARCHITECTURE OF THE NEURO-FUZZY-NETWORK

The neuro-fuzzy-network used in this paper is the modification of the network proposed by Zhang and Abraham (1998). Figure 1 is the general architecture of the network. Layer1 is called input layer. $\{x_1 \cdots x_n\}$ is the input of the network. Layer2 is dealing layer, and it converts the input $\vec{x} = \{x_1 \cdots x_n\}^T$ to vector $\vec{z} = \{z_1 \cdots z_m\}^T$ according to $\vec{z} = \boldsymbol{W}\vec{x}$, where $\boldsymbol{W} \in R^{m \times n}$. Layer2 is used to get the prompt input to Layer3. Layer3 is the fuzzification layer. The operation is:

$$\mu_{A_i^k}(x_i) = \exp\left[-(\frac{z_i - a_i^k}{\sigma_i^k})^2\right] \quad (1)$$

where $a_i^k$ is the center of the fuzzy membership function and $\sigma_i^k$ is the width value of the fuzzy membership function. The fuzzy membership function of output is defined as

$$\mu_{B_j^k}(y_j) = \exp\left[-\left(\frac{y_j - b_j^k}{\delta_j^k}\right)^2\right] \quad (2)$$

In (1) and (2), $A_i^k$ and $B_j^k$ are fuzzy sets in $\boldsymbol{U}_i \subset \boldsymbol{R}$ and $\boldsymbol{V}_j \subset \boldsymbol{R}$, respectively, and $x_i \in \boldsymbol{U}_i$ and $y_j \in \boldsymbol{V}_j$ are linguistic variables for $i = 1,2,\ldots,m$ and $j = 1,2,\ldots,p$. Layer4 is the pessimistic-optimistic-operation layer. The operations are

$$u^k = \prod_{i=1}^{m} \mu_{A_i^k}(z_i) \quad (3)$$

and

$$v^k = \left[\prod_{i=1}^{m} \mu_{A_i^k}(z_i)\right]^{1/n} \quad (4)$$

where layer input vector is $\vec{z} = (z_1, z_2, \cdots, z_m)$. Layer5 is the compensatory operation layer. The operation is

$$C(\beta_1, \beta_2) = \beta_1^{1-\eta}\beta_2^{\eta} \quad (5),$$

where $\beta_i$ is the output of the layer4 ($\beta_i$ *equals* $u^k$ *or* $v^k$) and $\eta \in [0,\ 1]$. Layer6 is the defuzzification layer. The operation in this layer is

$$f_j(\vec{z}) = \frac{\sum_{k=1}^{N} b_j^k \delta_j^k [\prod_{i=1}^{m} \mu_{A_i^k}(z_i)]^{1-\eta+\eta/m}}{\sum_{k=1}^{N} \delta_j^k [\prod_{i=1}^{m} \mu_{A_i^k}(z_i)]^{1-\eta+\eta/m}} \quad (6),$$

where $N$ is the number of the rules, $j = 1,2,\ldots,p$, $i = 1,2,\ldots,m$.

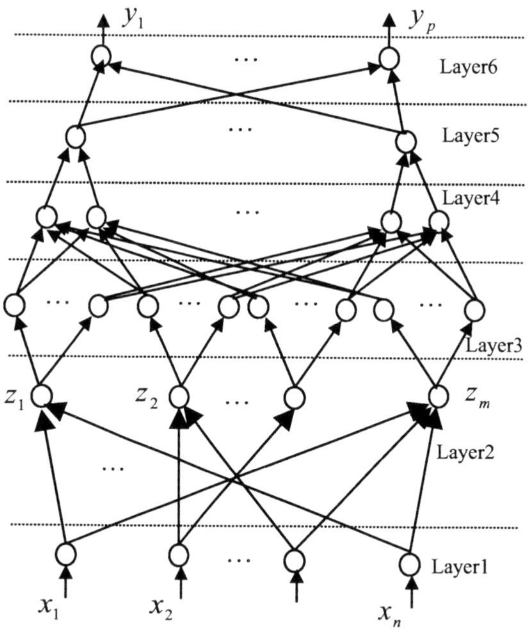


Fig. 1 Architecture of the modified neural-fuzzy-network

## 3. TRAINING PROCEDURE

### *3.1 Basic Updating Rules for Learning Automaton*

A learning automaton interacts adaptively with the environment it operates in and updates its actions at each stage based on the response of the environment to these actions. Hence an automaton can be defined by the triple $(\boldsymbol{\alpha}, \boldsymbol{\beta}, T)$. And $\alpha$ denotes the set of actions $\boldsymbol{\alpha} = \{a_1, a_2, \dots, a_r\}$ available to the automaton at any stage, $\boldsymbol{\beta} = \{\beta_1, \beta_2, \dots \beta_m\}$ is the set of observed responses from the environment, which are used by the automaton as inputs and $T$ is an updating algorithm which the automaton uses for selecting a particular action from the set $\boldsymbol{\alpha}$ at any stage.

For a stochastic learning automaton, the updating algorithm specifies a rule for adjusting the probability $p_i(n)$ of choosing a particular action $a_i$ at stage $n$. A functional relation of the form may generally describe such a rule:

$$p_i(n+1) = F(p_i(n), a(n), \beta(n)) \qquad (7)$$

The learning procedure at each stage hence consists of two sequential steps. In the first step the automaton chooses a specific action $a(n) = a_i$ from the finite set of actions available. And in the second step, the probabilities of choosing the actions are updated depending on the response of the environment to the action in the first step, which influences the choice of future actions.

An alternative way of specifying the updating algorithm is to define a state vector for the automaton and consider the transition of the state due to a certain action, which enables one to state the updating rule in terms of state transition probabilities. This approach has been quite popular in the development of learning automaton theory (Varshavskii and Vorontsova, 1963). For our application to neuro-fuzzy-network training, however, the action probability updating approach, with the updating algorithms specified in the form of (7), provides a simple and convenient framework.

For execution of training, the feedback signal from the environment, which triggers the updating of the action probabilities by the automation, can be given by specifying an appropriate "error" function. The environmental response set $\beta(n)$ at any stage $n$ can then be selected as the binary set $\beta(n) = \{0,1\}$, with $\beta = 0$ indicating that the action selected is not considered satisfactory by the environment and $\beta = 1$ indicating that the action selected is considered satisfactory. For a stochastic automaton with $r$ available actions, the updating rules can then be specified in a general form as follows.

For the selected action at the $n th$ stage $a(n) = a_j$, if $\beta = 1$ then

$$p_j(n+1) = p_j(n) + \sum_{\substack{i=1 \\ i \neq j}}^{r} \gamma_i(p(n))$$

$$p_i(n+1) = p_i(n) - \gamma_i(p(n)) \qquad for\ i \neq j \quad (8)$$

Whereas if $\beta = 0$, then

$$p_j(n+1) = p_j(n) - \delta_j(p(n))$$

$$p_i(n+1) = p_i(n) + \frac{1}{r-1}\delta_i(p(n)), \qquad i \neq j \quad (9)$$

Where

$$\sum_{i=1}^{r} p_i(n) = \sum_{i=1}^{r} p_i(n+1) = 1$$

The functions $\gamma_i(\bullet), \delta_i(\bullet)$ are appropriately selected continuous-valued nonnegative functions.

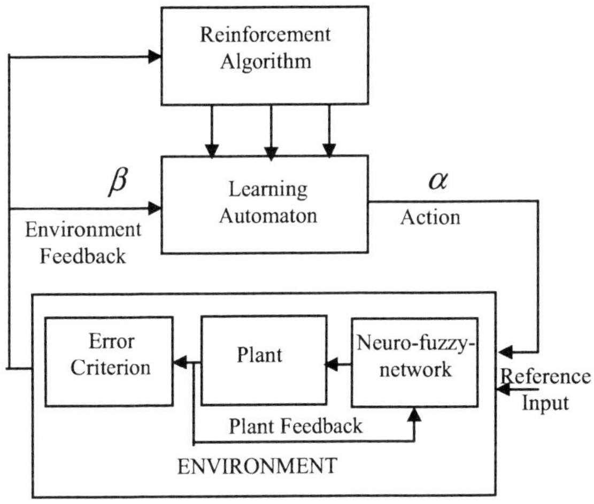


Fig. 2 Learning configuration for neuro-fuzzy-network controller

*3.2 Neuro-fuzzy-network Training*

The network training is a problem of nonlinear parameter optimization. And the parameters are changed in three directions—increment, decrement or keeping of the value. The parameters that need to be trained in neuro-fuzzy-network are $\{a_i^k, b_j^k, \sigma_i^k, \delta_j^k, w_{ij}, \eta\}$. The automaton actions are defined as either an increment or a decrement to any of the network parameters $\{a_i^k, b_j^k, \sigma_i^k, \delta_j^k, w_{ij}, \eta\}$. If the number of the parameters is $N$, this corresponds to a set of $2N$ single parameter updating actions.

The environment for this learning configuration comprises of the neuro-fuzzy-network itself with an appropriately specified error functional $E$ defined over the time interval $[t_0, t_f]$ by

$$E = \sum_{i \in K} \int_{t_0}^{t_f} f(y_i(t), y_i^d(t))dt \qquad (10)$$

Where $K$ denotes the set of designated output nodes of the network and $y_i^d$ denote the desired output signals. In this paper, we adopt $f(y_i, y_i^d) = |y_i - y_i^d|$ to define the error functional $E$.

The (7) and (8) is special in the training neuro-fuzzy-network. Suppose $E(n)$ and $E(n+1)$ are the errors in present and in next step after one of actions (without losing general, e.g. $i$ *th* action) is used, if $\beta = 0$, then

$$\delta = 1/(1+\exp(-\tau_1(E(n+1)-E(n))/E(n))))-0.47,$$

$$p_j(n+1) = p_j(n) + \frac{\delta}{r-1} p_i(n) \ (j \neq i)$$

$$p_i(n+1) = (1-\delta)p_i(n).$$

If $\beta = 1$, then

$$\gamma = 1/(1+\exp(-\tau_2(E(n)-E(n+1))/E(n))))-0.47,$$

$$p_j(n+1) = (1-\gamma)p_j(n) \ (j \neq i)$$

$$p_i(n+1) = p_i(n) + \gamma(1-p_i(n))$$

where $\tau_1 > 1, \tau_2 > 1$.

A probability of selection is initially assigned to each action. At first, a uniform distribution is used at the beginning for the action probabilities. As learning progresses, the probability associated with each action is changed. The more successful a particular action is at reducing the error, the more likely its selection will be in the future stages.

## 4. APPLICATION IN THE COMPLEX SYSTEM CONTROL

The control and modeling of nonlinear multivariable systems play a more and more important part in the scope of the advancing automation of technical processes. Due to the ever increasing requirements of process control (e.g. response time, precision, transfer behavior) nonlinear controller designs are necessary (Zhang and Abraham, 1998). But the traditional control designs need the precise mathematics model of the nonlinear system. This restricts the application of the nonlinear controller in industry. We adopt the modified neuro-fuzzy-network as the controller of nonlinear multivariable system—the three-tank-system. It does not need the precise mathematics model. The learning configuration for neuro-fuzzy-network controller is show in Figure 2.

### *4.1 Equation of the nonlinear system*

The equations of the three-tank-system can be written as:

$$d\boldsymbol{H}/dt = \boldsymbol{A}(\boldsymbol{H},t) + \boldsymbol{BQ}$$

$$\boldsymbol{Y} = \boldsymbol{CH}$$

Where

$$\boldsymbol{Q} = (Q_1, Q_2)^T,$$

$$\boldsymbol{A}(\boldsymbol{H}) = (-Q_{13}-Q_{10}, Q_{32}-Q_{20}, Q_{13}-Q_{32})^T / S,$$

$$\boldsymbol{B} = \frac{1}{S}\begin{bmatrix} 1 & 0 \\ 0 & 1 \\ 0 & 0 \end{bmatrix}, \quad \boldsymbol{C} = \begin{bmatrix} 1 & 0 & 0 \\ 0 & 1 & 0 \end{bmatrix},$$

$h_i$: liquid levels of the tanks (Unit: cm), $Q_{10} = \alpha_0 S_n (2gh_1)^{1/2}$, $Q_{13} = \alpha_1 S_n \operatorname{sgn}(h_1 - h_3)(2g|h_1 - h_3|)^{1/2}$, $Q_{32} = \alpha_3 S_n \operatorname{sgn}(h_3 - h_2)(2g|h_3 - h_2|)^{1/2}$, $Q_{20} = \alpha_2 S_n (2gh_2)^{1/2}$, $Q_1, Q_2$: supplying flow rates [cm$^3$/sec], $S$=150: Section of cylinder (cm$^2$), $S_n$=0.5:section of leak opening (cm$^2$), $g$=9.8m/s$^2$ is earth acceleration, $\operatorname{sgn}(x)$ is the sign of the argument $x$; $\alpha_i$ is the out-flow coefficient (correcting factor, dimensionless, real values ranging from 0 to 1).

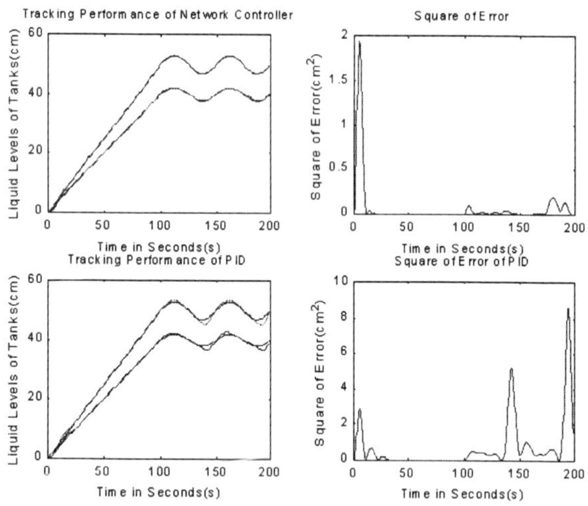


Fig. 3 Tracking Performance of Neuro-fuzzy-network Controller and PID

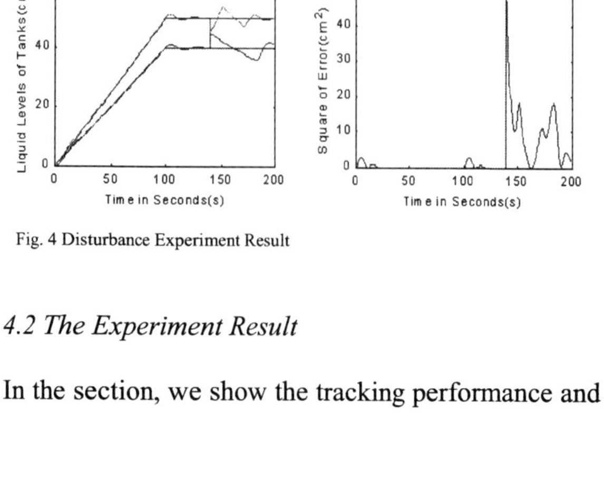


Fig. 4 Disturbance Experiment Result

### *4.2 The Experiment Result*

In the section, we show the tracking performance and

robust performance of the neuro-fuzzy-network control system. And compare the control effect with the traditional PID. The initiative states and the outflow coefficients of the tanks are $h_1 = h_2 = h_3 = 0$, $a_0 = a_1 = a_2 = a_3 = 0.3$, respectively. $T = 0.5s$ is the sample time.

And the reference signals are given in (11) and (12):

$$r_1(k+1) = r_1(k) + 2.5 \quad (k < 201)$$
$$r_1(k) = r_1(201) + 3 * \sin(2\pi(k-1)/100) \quad (200 < k < 401) \tag{11}$$
$$r_2(k+1) = r_2(k) + 2 \quad (k < 201)$$
$$r_2(k) = r_2(201) + 2 * \sin(2\pi(k-1)/100) \quad (200 < k < 401) \tag{12}$$

The tracking performances delivered by neuro-fuzzy-network controller and PID are shown in Figure 3.

In disturbance experiment, keep the value of the reference signals, change the liquid levels of tank1 and tank2 manually (Liquid in tank1 leaks from the leak clique and add some water to tank2.), which will cause disturbances to the system. The experiment result is shown in Figure 4.

## 5. CONCLUSION

In the paper, the feasibility of neuro-fuzzy-network with a learning automaton approach is demonstrated. The learning automaton approach enables the neuro-fuzzy-network to gain experience in the operating environment and be trained based on the experience. The principal advantage of this learning algorithm is that it requires no complex computations (such as gradient evaluations). So the algorithm affords a simple implementation.

The efficiency of the training approach was demonstrated in the design of controllers for nonlinear multivariable plant. The simple design and better control performances show its ability in the complex dynamical plant control.

## REFERENCES


Horikawa, S., et al (1992). On fuzzy modeling using fuzzy neural networks with the backpropagation algorithm. *IEEE Trans.On Neural Networks*, **Vol. 3**, pp. 801-806.

Ishibuchi, H., et al (1993), Learning of fuzzy neural networks from fuzzy inputs and fuzzy targets. *Proc. 5th IFSA World Congr.* , **vol. 1**, pp. 147-150.

Lakshmivarahan, S. (1981). *Learning Algorithms: Theory and Applications*, Springer-Verlag, New York.

Narendra, K. S., M. A. L. Thathachar (1989). *Learning Automata: An Introduction,* Englewood Cliffs, NJ: Prentice-Hall.

Rescorla, R. A., A. R. Wagner (1972). A theory of Pavlovian conditioning: Variations in the effectiveness of reinforcement and nonreinforcement. *Classical Conditioning II: Current Research and Theory*, A. H. Black and W. R. Prokasy, Eds. New York: Appleton-Century-Crofts.

Rumelhart, D. E., et al (1986). Learning internal representations by error propagation. *Parallel Distributed Processing: Explorations in the Microstructure of Cognition*, D. E. Rumelhart and J. L. McClelland, Eds. Cambridge, MA: MIT Press, pp.45-76.

Sundareshan, M.K., T. A. Condarcure (1998). Recurrent neural network training by a learning automation approach for trajectory learning and control system design. *IEEE Trans. On Neural Network*, **vol. 9**, pp. 354-368.

Tsetlin, M. (1962). On the behavior of finite automata in random media. *Automatic Remote Control*, **vol.22**, pp.1210-1219.

Varshavskii, V. I., I. P. Vorontsova (1963). On the behavior of stochastic automata with variable structure. *Automatic Remote Control*, **vol. 24**, pp.327-333.

Wang, L.X. (1995). *Adaptive Fuzzy Logic System and Control—Designs and Stability Analysis*, pp.242-246, State Defence Industry Press, Beijing .

Wang, Z. L., S. S. Gu (2000). FNN identifier based on real-valued genetic algorithms. *Journal of Northeastern University (Nature Science)*, **vol. 21**, pp. 354-356.

Zhang, J., A. J. Morris (1999). Recurrent Neuro-Fuzzy Networks for Nonlinear Process Modeling. *IEEE Trans. On Neural Network*, **vol. 10**, pp. 313-326.

Zhang Y. Q., K. Abraham (1998). Compensatory neurofuzzy systems with Fast learning algorithms. *IEEE Trans. On Neural Network*, **vol. 9**, pp. 83-105.

# AUTHOR INDEX